# INTELLIGENT ENGINEERING SYSTEMS THROUGH ARTIFICIAL NEURAL NETWORKS

## SMART ENGINEERING SYSTEMS: NEURAL NETWORKS, FUZZY LOGIC AND EVOLUTIONARY PROGRAMMING

## VOLUME 6

## ASME PRESS SERIES ON INTELLIGENT ENGINEERING SYSTEMS THROUGH ARTIFICIAL NEURAL NETWORKS

Editor:
C. H. Dagli, University of Missouri–Rolla, Rolla, Missouri, U.S.A.

Intelligent Engineering Systems Through Artificial Neural Networks, Volume 1, edited by Cihan H. Dagli, Soundar R. T. Kumara, and Yung C. Shin, 1991

Intelligent Engineering Systems Through Artificial Neural Networks, Volume 2, edited by Cihan H. Dagli, Laura I. Burke, and Yung C. Shin, 1992

Intelligent Engineering Systems Through Artificial Neural Networks, Volume 3, edited by Cihan H. Dagli, Laura I. Burke, Benito Fernández, and Joydeep Ghosh, 1993

Intelligent Engineering Systems Through Artificial Neural Networks, Volume 4, edited by Cihan H. Dagli, Benito Fernández, Joydeep Ghosh, and R. T. Soundar Kumara, 1994

Intelligent Engineering Systems Through Artificial Neural Networks, Volume 5, edited by Cihan H. Dagli, Metin Akay, C. L. Philip Chen, Benito Fernández, and Joydeep Ghosh, 1995

**Intelligent Engineering Systems Through Artificial Neural Networks, Smart Engineering Systems: Neural Networks, Fuzzy Logic and Evolutionary Programming**
**Volume 6, edited by Cihan H. Dagli, Metin Akay, C. L. Philip Chen, Benito Fernández, and Joydeep Ghosh, 1996**

# INTELLIGENT ENGINEERING SYSTEMS THROUGH ARTIFICIAL NEURAL NETWORKS

## VOLUME 6

## SMART ENGINEERING SYSTEMS: NEURAL NETWORKS, FUZZY LOGIC AND EVOLUTIONARY PROGRAMMING

**Proceedings of the Artifical Neural Networks in Engineering (ANNIE '96) conference, held November 10–13, 1996, in St. Louis, Missouri, U.S.A.**

**EDITORS**

CIHAN H. DAGLI
University of Missouri–Rolla

METIN AKAY
Rutgers University

C. L. PHILIP CHEN
Wright State University

BENITO R. FERNÁNDEZ
JOYDEEP GHOSH
University of Texas–Austin

NEW YORK ASME PRESS 1996

ISBN 0-7918-0051-2
Library of Congress Cataloging-in Publication Number 92-30949

# CONTENTS

## Section 2 : Learning Algorithms and Training

# PART V : ADAPTIVE CONTROL 531

# PREFACE

As a follow up to the previous five volumes of Intelligent Engineering Systems Through Artificial Neural Networks, edited by Dagli, C.H. et. al., this volume contains the edited versions of the technical presentations of ANNIE '96, the sixth international gathering of researchers interested in engineering applications of neural networks, fuzzy logic and evolutionary programming, held on November 10-13, 1996 in St. Louis, Missouri, U.S.A. Attracting researchers from twenty countries, the conference covered the theory of artificial neural networks, fuzzy logic and evolutionary programming and their applications in the engineering domain. The papers included in this volume provide a forum for researchers in the field to exchange ideas on smart engineering systems design or stated in two other ways; soft computing or computational intelligence.

An extended version of each paper selected for inclusion was reviewed by two referees, then revised, edited, and condensed to the format seen here. The papers in this edited book are grouped into six categories:

1) Neural Networks
2) Fuzzy Logic
3) Evolutionary Programming
4) Pattern Recognition
5) Adaptive Control
6) Smart Engineering System Design

Seven plenary sessions scheduled for ANNIE '96. Dr. Bob Narendra talked about "Neural Networks in Control: Theory and Practice" on the morning of the first day. Dr. Erol Gelenbe's topic at noon was titled "A Unified Framework for Neural Networks, Genetic Algorithms, and Their Engineering Applications". In the second day's plenary session, "Evolutionary Computation: The New Synthesis" was introduced by Dr. David Fogel. During the luncheon Dr. Malvin C. Teich talked about "Fractal Behavior of the Electrocardiogram: Distinguishing Heart-Failure and Normal Patients Using Wavelet Analysis". Dr. Rahmat Shoureshi from the Colorado School of Mines was the banquet speaker. He discussed the topic "Intelligent Control in Practice". "Why Function Approximation Works so Well in Neural-Net Computing and How to Make it Work Even Better: A Critique and a Perspective" was the title of Dr. Yo-Han Pao's plenary talk on the third day. Dr. Tadashi Iokibe talked on "Fusion of Fuzzy and Chaos, and its Application in Industrial Field: Time Series Prediction" at noon on the same day.

The editors would like to thank once again the keynote and luncheon speakers, and the authors for their contributions. We would also like to recognize the organizing committee members of ANNIE '96 for their excellent support in promoting the conference internationally. Further, we wish to express our gratitude to all the referees for their excellent and timely review efforts which made this edited book possible within a short period of time.

Space will not permit for the recognition of all contributors, but special mention must me made of the co-sponsors of the conference for their assistance, the session chairs who contributed to the quality of the discussions in the presentations; Dr. Yildirim Omurtag, chairman of the Engineering Management Department, for his continuous assurance and support; and the collaboration of Cynthia Clark, Barbara Signorelli, and the remaining staff of ASME Press during the printing of this book.

We would especially like to express our most sincere thanks to Refia Dagli for her timely volunteer contributions in the preparation of the camera ready manuscript of the book for her tireless help and enthusiasm during this most difficult phase of the conference. She made the publication of this volume possible and her dedicated efforts help us to continue the ANNIE tradition and created the ANNIE'96. We would like to mention also our appreciation to the senior staff of the Engineering Management Department for their timely and continuous support.

Lastly, but most importantly, we would like to thank all of our families for their patience and sustenance during the many long hours that it took to create ANNIE '96 and, subsequently, this book.

Cihan H. Dagli
*University of Missouri - Rolla*

Metin Akay
*Rutgers University*

C.L. Philip Chen
*Wright State University*

Benito Fernández
*University of Texas at Austin*

Joydeep Ghosh
*University of Texas at Austin*

# PART I : NEURAL NETWORKS

# Section 1 : Architectures

# FLEXIBLE MODULAR ARCHITECTURE FOR CHANGING ENVIRONMENTS

**Viswanath Ramamurti and Joydeep Ghosh** [1]
Department of Electrical and Computer Engineering
The University of Texas at Austin, Austin, TX 78712-1084.
E-mail: {viswa,ghosh}@pine.ece.utexas.edu

**ABSTRACT**

This paper introduces a neural network capable of dynamically adapting its architecture to model a task that can change with time. This network has its roots in the mixture of experts framework. Modules or Experts are grown or pruned depending on the (changing) complexity of the modeling problem. On-line updates of all parameters make it suitable for time varying environments. Simulation results are presented in support of the proposed architecture.

## 1 INTRODUCTION

There are many situations where a learning system capable of adapting to changing environments is desirable. For example, one may want to model a plant process which undergoes seasonal variations in its input-output behavior. In the neural network community, the problem of realizing a time varying mapping is commonly known as the sequential learning problem. Neural networks such as the MLPs are known to exhibit "Catastrophic Interference (CI)", i.e., they tend to forget previously learned mappings rather quickly when exposed to new mappings. CI is exhibited by MLPs mainly because such networks make global parameter updates when they learn. Several researchers have tried to alleviate CI by reduced coupling or unlearning in MLP networks or by using networks of cells with localized responses. The Resource Allocation Network (RAN) [Platt (1991)], which is a constructive version of the RBF network, is one such network. Unfortunately, it lacks a pruning mechanism which can lead to unbounded growth in the network. Also, the RAN network, just like the RBF network or the MLP network, outputs the expected value of the target for any given input. This may not be suitable for problems involving one to many mappings [Bishop (1995)]. This would mean that in a changing task scenario, if the output space changes for the same input space, the RAN would not be able to remember the previously learned mapping.

In this paper we present a modular architecture based on the mixture of experts framework that is capable of both growing and pruning simultaneously. In this process, the architecture maintains a suitable model complexity at all times. Also, the architecture is inherently capable of modeling one to many mappings.

The paper starts by briefly describing the mixture of experts network [Jacobs et al. (1991)], [Xu et al. (1995)]. This is followed by a short description of our previous work in determining the appropriate model complexity for solving static problems [Ramamurti and Ghosh (1996b)]. We then introduce an on-line algorithm to train the network and present two techniques to model tasks which vary with time.

## 2 MIXTURE OF EXPERTS NETWORK

The mixture of experts network is a modular architecture which works on the principle of divide and conquer. The model employs probabilistic methods to "softly" divide the input space into overlapping regions on which "experts" act. Each of the M experts looks at the common input vector $x$ to form an output $\hat{y_j}(x), j = 1...M$. The gating network also looks at the same input vector $x$ and produces outputs $g_j(x) \geq 0$, $\sum_j g_j(x) = 1$, which weight the expert

---

[1]This research was supported in part by ARO contracts DAAH04-94-G-0417 and 04-95-10494, ATP grant 442 and NSF grant ECS 9307632.

network outputs to form the overall output $\hat{y} = \sum_j g_j \hat{y}_j$. Each expert is typically a single layer network with a possible output non-linearity. For reasons explained in [Ramamurti and Ghosh (1996b)], we choose the localized model of [Xu et al. (1995)] as the gating network.

The localized gating model proposed in [Xu et al. (1995)] is of the form $g_j(x,\nu) = \frac{\alpha_j P(x|\nu_j)}{\sum_i \alpha_i P(x|\nu_i)}$, where $P(x|\nu_j) = (2\pi)^{-n/2}|\Sigma_j|^{-1/2} exp\{-0.5(x-m_j)^T \Sigma_j^{-1}(x-m_j)\}$. Thus the $j$th expert's influence is localized to a region around $m_j$, with width determined by $\Sigma_j$.

The expert networks are single layer networks with identity output activation function for regression problems. The parameter vector for expert $j$ is denoted by $\theta_j$ ($\hat{y}_j = \theta_j^T x$) .

The EM formulation for estimating the gating network and expert network parameters $\alpha_j, \nu_j, \theta_j$ from the set of training patterns $t$ is given by:

E step:

$h_j^{(k)}(y^{(t)}|x^{(t)}) = \frac{g_j^{(k)}(x^t,\nu)exp\{-0.5(y-\hat{y}_j)^T(y-\hat{y}_j)\}}{\sum_i g_i^{(k)}(x^t,\nu)exp\{-0.5(y-\hat{y}_i)^T(y-\hat{y}_i)\}}$.

M step (gating network):

$\alpha_j^{(k+1)} = \frac{1}{N}\sum_t h_j^{(k)}(y^{(t)} \mid x^{(t)})$

$m_j^{(k+1)} = \frac{1}{\sum_t h_j^{(k)}(y^{(t)}|x^{(t)})}\sum_t h_j^{(k)}(y^{(t)} \mid x^{(t)})x^{(t)}$.

$\Sigma_j^{(k+1)} = \frac{1}{\sum_t h_j^{(k)}(y^{(t)}|x^{(t)})}\sum_t h_j^{(k)}(y^{(t)} \mid x^{(t)})[x^{(t)} - m_j^k][x^{(t)} - m_j^k]^T$.

M step (expert network):

$\theta_j^{k+1} = arg \overset{min}{\theta_j} \sum_t h_j^{(k)}(y^{(t)}|x^{(t)})\|y - \hat{y}_j\|^2$.

This model has been employed for solving a number of regression [Ramamurti and Ghosh (1996b)] and classification [Ramamurti and Ghosh (1996a)] problems.

## 3 MODEL SELECTION IN MIXTURE OF EXPERTS

In general, the optimal network complexity in a neural network is not known *a priori*. This has lead researchers to develop algorithms that dynamically adapt network structure by adding or removing network weights. In this section we briefly present techniques to grow and prune experts while training a mixture of experts network on a given static modeling task. Please see [Ramamurti and Ghosh (1996b)] for details.

### 3.1 The Growing Mixture of Experts Network

Let us assume that at some instant of time there are $m$ experts in the network. The mixture of $m$ experts network is trained using the EM algorithm described in the previous section until the MSE on the validation set fails to decrease. The network parameters obtained are saved. We define a weighted MSE $E_j$, for each expert $j$ ($j$= 1 .. m), on the validation set samples $p$. For every expert, $E_j = \frac{\sum_p h_j^p(y_p - \hat{y}_p)^2}{\sum_p h_j^p}$, where the $h_j^p$s are obtained by evaluating the E-step expression for the validation set samples $p$. $E_j$ gives a measure of how well a given expert performed on the validation set. If $E_j$ is large, it probably means that the localized model is not able to approximate well the target function for the weighted samples associated with expert $j$. In this case, addition of an expert to the localized space spanned by expert $j$ would reduce the output error due to the added flexibility.

Therefore, we add an expert $m$+1 which tries to reduce the error due to expert $c$, where $c = \arg\max_j E_j$ . The key idea is to perform "weighted 2-means" on the training samples $t$ (weights = $h_c^t$). The two means obtained are the new means associated with the two experts, $m$+1 and $c$.

### 3.2 Pruning and Growing Mixture of Experts

From section 2, it is observed that the prior $\alpha_j$ is computed in the M-step as $\alpha_j^{(k+1)} = \frac{1}{N}\sum_t h_j^{(k)}(y^{(t)} \mid x^{(t)})$ i.e., $\alpha_j$ is proportional to the sum of $h_j^t$s over all patterns 't' in the training set, $h_j^t$ being the posterior probability of selecting expert 'j' given input $x^{(t)}$ and its corresponding output $y^{(t)}$. We therefore see that in every iteration of the EM-algorithm, $\alpha_j$ directly gives us a measure of how important expert 'j' is in relation to the other experts. Hence, when it is desired to prune an expert, the obvious candidate is that expert which has the least value of $\alpha_j$.

The mixture of experts network with a certain initial number of experts $m$, is trained until convergence. The expert corresponding to lowest $\alpha_i$ is pruned. If there is no significant change in network performance, pruning is continued. Network growth can be performed at any time using the procedure outlined in the previous subsection. There is no strict order in which pruning and growing should be performed. Results using growing as well as the simultaneous pruning and growing techniques given in [Ramamurti and Ghosh (1996b)] clearly show the superiority of these techniques over static models.

## 4 AN ON-LINE ALGORITHM TO TRAIN THE MIXTURE OF EXPERTS NETWORK

The training algorithm described in section 2 to train the mixture of experts network with the localized gating network is a batch algorithm. In order to employ a network for sequential learning, one needs an on-line algorithm. The EM iterations to train the gating network parameters bears a close resemblance to the procedure of training the parameters of a mixture of Gaussian densities to model an unknown probability density [Duda and Hart (1973)], which has attracted several attempts at on-line versions. We take the approach of [Traven (1991)] to derive the on-line algorithm to perform structural adaptation in the mixture of experts network.

All equations in the EM update rule and the batch mode growing technique are of the general form $\gamma_t = \frac{\sum_{k=1}^{t} h_k \gamma(x_k)}{\sum_{k=1}^{t} h_k}$.

For this form, it can be shown that $\gamma_{t+1} = \gamma_t + \lambda_{t+1}[\gamma(x_{t+1}) - \gamma_t]$, where, $\lambda_{t+1} = \frac{h_{t+1}}{\sum_{k=1}^{t+1} h_k}$.

Considering only the L most recent values of $x_k$, the denominator of $\lambda_{t+1}$ can be approximated by the recursion $D_{t+1} = (1 - 1/L)D_t + h_{t+1}$.

The gating network parameter update equations now become

$\alpha_{j,t+1} = \alpha_{j,t} + (1/L)[h_j(y^{(t)} \mid x^{(t)}) - \alpha_{j,t}]$.

$m_{j,t+1} = m_{j,t} + \lambda_{t+1}[x^{(t)} - m_{j,t}]$.

$\Sigma_{j,t+1} = \Sigma_{j,t} + \lambda_{t+1}[(x^{(t)} - m_{j,t})(x^{(t)} - m_{j,t})^T - \Sigma_{j,t}]$.

As the expert networks are linear, their parameters can be updated using either of the well known LMS or RLS algorithms [Haykin (1991)]. The RLS algorithm converges faster compared to the LMS algorithm. Comparison of the on-line algorithm with the batch algorithm on two synthetic data sets described in [Ramamurti and Ghosh (1996b)] is shown in table 1.

**Table 1. Results using static networks.**

| | Number of Experts | Average MSE Batch | Average MSE On-line |
|---|---|---|---|
| Gabor data-set | 20 | 0.0503 | 0.0310 |
| Multivariate data-set | 25 | 0.0086 | 0.0065 |

## 5 ON-LINE GROWING AND PRUNING

The algorithm developed in section 3 for structural adaptation is again a batch algorithm. The growing procedure described in section 3 with the associated "weighted 2-means" algorithm turns out to be efficient, yielding a modest number of experts for a given static problem. In the static case, the input space spanned by each expert remains essentially similar undergoing only minor refinement from epoch to epoch during training. While growing a new expert, the idea was to provide additional modeling flexibility for samples associated with the space spanned by the expert to be split, by softly dividing the space into two and having separate experts for the two subspaces.

A similar approach when extended to an on-line setting would work well if the incoming training data all come from a fixed distribution or from a distribution where changes are slow with respect to time. In this section we first present an on-line extension to the batch-mode growing technique.

The on-line error estimate for experts is given by $E_{j,t+1} = E_{j,t} + \lambda_{t+1}[(y^t - \hat{y^t})^2 - E_{j,t}]$. In a scenario of changing environments, if the error corresponding to a particular expert increases beyond a pre-fixed threshold, the need for an additional expert is detected. In order to grow an additional expert, we follow the same procedure as in batch mode to create the new parameters with the exception of the gating network means. To create the gating network means, we employ an on-line version of the weighted 2-means algorithm as follows.

- Initialize the mean vector corresponding to the new expert to be equal to the mean vector corresponding to the expert to be split.
- Add a small random perturbation to the two means.
- For a window length T of input patterns, make parameter updates to all the experts except the expert being split and the new expert.
- For a window length T of input patterns, if the posterior $h_j$ corresponding to one of the two new experts is the highest among the posteriors of all the experts for an input pattern, make parameter updates for this expert also.

The window length T is chosen in such a way as to separate the two means. After this window length of T samples, the two experts become normal experts. Pruning is performed by monitoring the parameter $\alpha_j$. When $\alpha_j$ becomes very small, the corresponding expert is pruned.

The above algorithm was applied to a problem where the modeling task changes slowly with time. For simulation purposes, an artificial data set was first constructed. Uniformly spaced samples from the Gabor function $G(x,y) = \frac{1}{2\pi(0.5)^2} \cdot \exp(-\frac{x^2+y^2}{2(0.5)^2}) \cdot \cos(2\pi(x+y))$ were first obtained in the range $-1.5 \leq x \leq 1.5$, $-1.5 \leq y \leq 1.5$ as a 49 by 49 grid. The data set was divided into three overlapping subsets each consisting of 1225 samples as shown in figure 1. Random samples from the first subset are presented to the network first for a certain length of time followed by random samples from the second subset and finally by random samples from the third subset. The following restrictions were imposed during training.

- The number of experts in the network should be always less than 20.
- The threshold for pruning an expert, $\alpha_j = 0.0001$.
- The error threshold $E_j = 0.03$ for growing an expert. Also an expert is grown only if the corresponding $\alpha_j > 0.01$.

Initial number of experts in the network was taken to be 10. Staying within the above constraints, for the first data set the network converges to an MSE of 0.0089 with 15 experts. When the network is subsequently exposed to the second data set, the network prunes away 2 experts and adds on 3 experts to give an MSE value of 0.0152. When the network is next exposed to the third data set, the network prunes away five experts and gives an MSE of 0.0088 with 11 experts.

## 5.1 Modeling Changing Environments

In a setting where either the input space, the output space or both could change with time rather quickly, a more easily adaptable scheme is required. An alternative way to grow a mixture of experts based on the novelty of the training patterns is described here. This approach of growing a neural network is not new. A similar approach based on input novelty has been used before in [Platt (1991)], [Roberts and Tarassenko (1994)],[Schaal and Atkeson (1996)]. The present approach differs from those cited by incorporating output novelty within

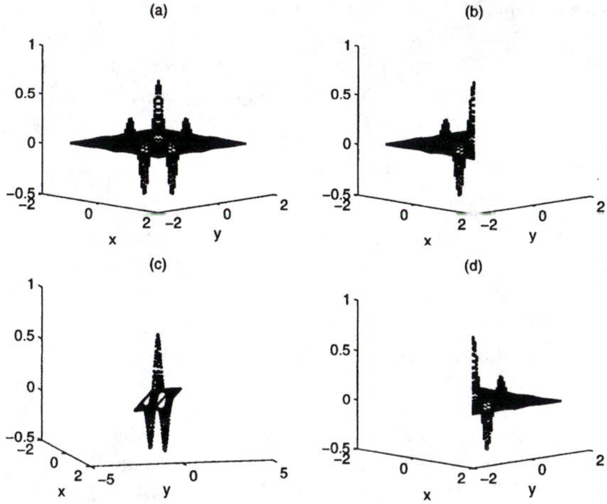


**Figure 1. (a) Global data set. (b),(c),(d) overlapping subsets of the global data set.**

the framework of novelty detection. Also, a significant advantage of the present network over the other architectures is its inherent ability to model inverse problems.

Let the network at any instant of time have $m$ experts. A new expert $m+1$ is added to the network if for the current training pattern $x_t$, $min[(x_t - m_j)^T \Sigma_j^{-1}(x_t - m_j)] < \delta$, where the $m_j$s and $\Sigma_j$s are the means and covariances corresponding to the current set of experts. The mean of the new expert is initialized to the current input pattern. $\alpha_{m+1}$ is set equal $\alpha_c/2$ where $c$ corresponds to that expert whose mean is closest to the input pattern. $\alpha_c$ is also set to $\alpha_c/2$ so that the $\alpha_i$s sum to one. The new expert is given random initial weights. The covariance matrix is initialized to a fixed constant diagonal matrix based on prior knowledge about the data. Once a new expert has been added to the network, the network parameter updates are continued using the RLS algorithm. This growing algorithm is capable of adapting to changes in the input space by creating new experts for input samples not seen earlier by the network. However, in a modeling task with changing environments, there could be occasions when the output space changes and the input space does not change. When this happens, "Catastrophic Interference" takes place in networks which adapt only to changes in the input space. In order to take into account changes in the output space, while training the network, the gating network is made to see the combined input-output space. The expert networks continue to look at the input space alone. This way distinct experts are created for changes in the output space also. While testing the network for unseen patterns, the target outputs are ofcourse not available. In this case we treat the gating network inputs as input vectors with certain missing elements [Ahmad and Tresp (1993)]. The easiest way to deal with a missing feature element is to choose the expected value of the missing feature. This expected value will be different for the different experts as each input vector is weighted differently for different experts. However, no extra computation is required as the expected values of the missing elements are already computed in the form of means of the corresponding elements in the gating network.

The application of the above growing procedure on a real life data set is now described. The data set employed is adapted from the "building" data set from the PROBEN1 set of benchmark suite [Prechelt (1994)]. The problem is to predict the hourly consumption of electrical energy, hot water, and cold water, based on the date, day of the week, time of day, outside temperature, outside air humidity, solar radiation and wind speed. The data set is spread over six months from September to February. On observing the data, it is seen that the hot water requirement has a strong correlation with the outside temperature. Upon ignoring the temperature parameter, the hot water requirement also varies according to the months. Therefore without the temperature parameter, for similar values across the rest of the inputs, the hot water requirement on average is different for different months. In order to test the growing technique in a problem involving both the changes in input and output space, the data set was divided into three segments each containing data corresponding to two consecutive months. The outside temperature was assumed to be not measured and the network was made only to predict the hot water requirement. Two experiments were conducted. In the first experiment the three segments of the data set were fed to the network sequentially, twice. In the second experiment only the first and the third segment of the data set were fed sequentially and once again twice. The second experiment was carried out to measure network performance when the output space changes abruptly.

The three data segments contained 1462, 1464 and 1282 samples respectively. For the first experiment, at the end of feeding all the three data segments once, the resulting network had grown to 33 experts starting from one expert at the beginning of the experiment. When the data segments were fed for the second time, with the addition of just one more expert, the network gave an average MSE of 0.0037. For the second experiment, when the first and the third data segments were fed to the network for the first time, the network had grown to 27 experts. When the same data segments were fed for the second time, with the addition of no more experts the network gave an average MSE of 0.0028.

## REFERENCES

Ahmad, S., Tresp, V., 1993, "Some Solutions to the Missing Feature Problem in Vision," In Hanson, S. J., Cowan, J. D., Giles, C. L., editors, *Advances in Neural Information Processing Systems 5.* Morgan Kaufmann.

Bishop, C. M., 1995, *Neural Networks for Pattern Recognition* Clarendon Press - Oxford.

Duda, R. O., Hart, P. E., 1973, *Pattern Classification and Scene Analysis* Wiley, New York.

Haykin, S., 1991, *Adaptive Filter Theory* Englewood Cliffs, NJ: Prentice-Hall.

Jacobs, R. A., Jordan, M. I., Nowlan, S. J., Hinton, G. E., 1991, "Adaptive mixtures of local experts," *Neural Computation*, Vol. 3, pp. 79–87.

Platt, J. C., 1991, "A resource allocation network for function interpolation," *Neural Computation*, Vol. 3, pp. 213–225.

Prechelt, L., 1994, "PROBEN1 — A Set of Benchmarks and Benchmarking Rules for Neural Network Training Algorithms," Fakultät für Informatik, Universität Karlsruhe, D-76128 Karlsruhe, Germany Anonymous FTP: /pub/papers/techreports/1994/1994-21.ps.Z on ftp.ira.uka.de.

Ramamurti, V., Ghosh, J., 1996a, "Advances in using Hierarchical Mixture of Experts for Signal Classification," Proc. of ICASSP 96.

Ramamurti, V., Ghosh, J., 1996b, "Structural Adaptation in Mixture of Experts," Proc. of ICPR 96, to appear.

Roberts, S., Tarassenko, L., 1994, "A Probabilistic Resource Allocating Network for Novelty Detection," *Neural Computation*, Vol. 6, pp. 271–284.

Schaal, S., Atkeson, C. G., 1996, "From isolation to cooperation: An alternative view of a system of experts," In Touretzky, D., Mozer, M., Hasselmo, M., editors, *Advances in Neural Information Processing Systems 8.* The MIT Press.

Traven, H. G. C., 1991, "A Neural Network Approach to Statistical Pattern Classification by "Semiparametric" Estimation of Probability Density Functions," *IEEE Transactions on Neural Networks*, Vol. 2.

Xu, L., Jordan, M. I., Hinton, G. E., 1995, "An Alternative Model for Mixture of Experts," In Tesauro, G., Touretzky, D. S., Leen, T. K., editors, *Advances in Neural Information Processing Systems 7.* The MIT Press.

# Vibrating Potential Field for Large Scale Traveling Salesman Problem

**Hiroshi YOKOI and Yukinori KAKAZU**
Complex Sysem Engineering, Hokkaido University
N13 W8 Kita-ku, Sapporo, 060 Japan.

**Takafumi MIZUNO and Masatoshi TAKITA**
National Institute of Bioscience and Human Technology
1-1 Higashi, Tsukuba, 305 Japan.

***ABSTRACT:***
Large scale combinatorial optimization is practical requirement. However, combinatorial explosion has disturbed finding solution for such large scale problem as well known as NP hard. The large scale problem has been caused by multi-body's problem. For such problem, this paper proposes transition technique from multi-body's problem into two-body's problem. The proposed technique is so called as vibrating potential field(VPF). The VPF is computation frame work for interaction of multi-processors, and it consists of field techniques and conservation techniques of processors' states. This paper shows details of VPF and difficulties of using such field technique. The fundamental function of VPF is shown by multi-bodies attractive and repulsive interactions through computation. The computation also shows a possibility of VPF for JSP with parallel tasks problem and TSP of 10000 over city's problem.

## INTRODUCTION

Until now, we are tossed about by the number. Just now it is time to be free from restriction of number. In the information processing methodology, growth of number to be processed lets grow difficulty of finging solution of problem. Such growth of number has been most essential problem, known as combinatorial explosion, and it has been lay on a combinatorial optimization for a long time. For this problem, three types approaches, such as, pallarelism (connectionist approach, pattern recognition), specialization approaches and hard ware speed up approaches have been applied. However, almost these strategy tries to short cut the sequence of processing or quick computation. Therefore, those mechanizm still could not be free from combinatorial explosion (expornential growth of number to be processed). In this paper, we propose new information processing to be free from the problem.
The cause of problem is that the traditional method needs count one by one relation among all number of data. If the method can be arbitrator and process (or count) all number of data at once, the combinatorial explosion can be solved. Then, we introduced the concept of field for such problem, and design theoretical basis for large scale combinatorial optimization. Appling concept of field to the combinatorial optimization, one methodology named as vibrating potential field (VPF) is proposed.
The proposed VPF offers following advantages (V1) to (V4):

(V1) Transition from multi-bodies problem to two-bodies problem.
(V2) Integration of information by shared memory (field information)
(V3) Generation of global information using information integration
(V4) Observation of global information using local information

From such advantages described above, unsupervised parallel distributed control is achieved. Using VPF and life's kinetics, the computer simulations show the behaviors of the proposed model that conserves its own energy level, and tunes its configurations or location of its parts. To show the powerful searching characteristics of VPF, the proposed model is applied to the Job Shop Scheduling Problem(JSP) and Euclidean traveling salesman problem(TSP). Chapter 2 shows the over view of combinatorial optimization as objective domain, and requierment for field method. Chapter 3 describes mathematical representation of VPF. Chapter 4 shows the potential of VPF for the JSP and the Euclidean TSP. This paper shows the VPF achieves smart computation independant for number to be processed.

## COMBINATORIAL OPTIMIZING PROBLEM

Combinatorial optimization is one of the most fundamental problem for dicision making on industry, (task planning, marketting and management, etc.). The traditional operations research have been focused on combinatorial optimization problem using LP, DP, BB, etc. The problem is defined as finding optimul sequence r, sutisfying MinMax{O(r)+C(r)} in the set of sequence R. In this definition, r is element of R. r is a sequence of task. O(r) is cost function. C(r) is constraints on each tasks. By an simple methodology or strictly analysis, the combinatorial explosion became large disterbance for problem solving. Therefore, how to design a smart methodology without large computation cost becomes our subject. For this requierment, some approximate approaches are discussed, such as Heuristics approach specialized for each concrete problem, NN, Evolutional Computation, Fuzzy logic, and Simulated Annealing. However those approaches have the problem of expornential growth of computational number.

TSP is one famous ventimark test for combinatorial optimization. For TSP, those similar approaches are applied. However even those approximate approach also have the problem of expornential growth of computational number. In other words, the problem is derived from sequencial process of one by one.Therefore, desired methodology must performs parallel processing with bundling all data to be process without sequencial process.

For astronomy computation, powerfull computation is requierd to calculate large number of interactive dynamics of particles.Those are simulation technique for computation of collisions of $10^6$ order particles. For this requirement, mean field method and tree method [Suginohara 91] is introduced. The problem is finding dencity distribution of particles for the initial condition of particles conformation. The procedure consists 3 ways. 1) Definition of interaction or dynamics among particles. 2) Setting equations of motion for each particles, and pursuing movement numerically. 3) Repeating 1) and 2). The only difficulty of this problem is computation cost for one cicle takes the Nth power N (N is number of particels). For this problem, tree method is introduced, that is one of approximation. The tree method calculate interaction force for one particle by bundled forces of particles at a distant place without calculating interaction among all particles. As shown above, how to bundle objective processes is key point for large scale computation.

## VIBRATING POTENTIAL FIELD

For the requierment of bundling approach, one field method is introduced, that is vibrating potential field(VPF). We have been interested in information processing that are inspired from biology. One approach was reported as bionic model [Yokoi 92], that is the first paper on modeling life and appled to TSP. The expanded approach of the bionic model was amoeba model [Yokoi 95], that was forcused on the capability of multi celluar amoeba, and also applied to TSP to investigate searching capability.

The common methodology of these two papers are using field technique. Those result sugesseted that the adaptability and flexibility of life is compressed in the information transition through field, especialy wavy information. Therefore, this paper closes up the VPF as a field technique and shows capability to solve large scale computation.

### A field of information propagation

Interaction among many units is called multi-bodies problem. To translate the multi-bodies problem into a simple problem, the whole field **H**, called VPF, is formed by piling up individual fields generated by units. Using VPF, such multi-bodies problem is achieved through interaction between one unit and VPF as two bodies interaction problem. Two bodies interaction offers the special feature that a unit observes the status of other units without knowledge of relationship among them.

The VPF mathematically described as equation (1) using information (**h,w,E,hl**) on individual field for each unit that propagated from unit and unit coordinate axis ($\psi,\chi,\tau,\varphi$) that prepares individual space for each information. In the equation (1), the information is defined only as a propagating wavy signal, and described using potential function on space and time intervals.

$$\mathbf{H}(\boldsymbol{r}, \phi, t) = \sum_{i}^{\text{Unit number}} \left\{ \mathbf{h}_i(\boldsymbol{r}, t)\,\psi(\phi) + \mathbf{w}_i(\boldsymbol{r}, t)\chi(\phi) + \sum_{j} \mathbf{hl}_{ij}(\boldsymbol{r}, t)\varphi_i(\phi)\varphi_j(\phi) \right\} + \mathbf{E}(\boldsymbol{r}, t)\tau(\phi) \quad \text{------ (1)}$$

Where, $H(\boldsymbol{r}, \phi, t)$ is the vibrating potential field (whole field). $\phi$ is total coordinate axis. i is unit number. $\mathbf{E}(r,t)$ is energy function of environment. $\mathbf{h}_i(r,t)$, $\mathbf{w}_i(r,t)$, and $\mathbf{hl}_{ij}(r,t)$ are information propagated form unit i (potential function). $\boldsymbol{r}$ is position vector. $\boldsymbol{v}$ is velocity vector. The information on each field $\mathbf{h}_i(r,t)$, $\mathbf{w}_i(r,t)$, $\mathbf{hl}_{ij}(r,t)$, $\mathbf{E}(r,t)$ is spreading type information of unit that dominates interaction among units. From $H(\boldsymbol{r}, \phi, t)$, a unit interacts with other units on $\psi(\phi)$, $\chi(\phi)$, $\tau(\phi)$, $\varphi(\phi)$ according to the amplitude of each individual field. Such spreading type information is mathematically described using partial differential equation $u=u(r,t)$ as equation (2).

$$\alpha\frac{d^2u}{dt^2} + \beta\frac{du}{dt} + \gamma\nabla^2 u + \delta\nabla u + \varepsilon u = 0, \qquad \nabla^2 = \frac{\partial^2}{\partial r_x^2} + \frac{\partial^2}{\partial r_y^2} + \frac{\partial^2}{\partial r_z^2} \quad \text{------ (2)}$$

Unit communication is mediated by the multiple VPF **H** in equation(1). However, since the description of **H** is only a summation of unit information and the objects that we deal with have many properties, the propagating information would be mixted up and lead to confusion. Therefore, to avoid this, we prepared the unit coordinate axes to be at a right angle to each other. Each unit coordinate axis is described as a wave equation. Boundary conditions should be different for each property to avoid confusion.

$$\frac{\kappa(a)}{2}\frac{d^2\zeta(\phi)}{d\phi^2} + \mathbf{E}_{env}(\boldsymbol{r}_i, t)\,\zeta(\phi) = 0, \quad (\zeta(\phi) = \psi(\phi), \chi(\phi), \varphi(\phi), \tau(\phi)) \quad \text{------ (3)}$$

The advantages of using wave equation are that the VPF can maintain some different kinds of interactions using vertical axes, and each unit can get the field information by easy operation form principle of superposition.

$$\zeta(\phi) = 0, \; (|\phi| = a, b, c, d) \quad \text{------ (4)}$$

Where, $\psi(\phi)$, $\chi(\phi)$, $\varphi(\phi)$, $\tau(\phi)$ are unit coordinate axis. $\mathbf{E}_{env}$ is interaction energy that consists of out flowing energy form units. $\kappa(a)$ is parameter which determines the spatial distance among unit coordinate axis. a, b, c, d are initial coefficients.

**An information processing unit.**

To attain cooperation in the movement of units, unit action is defined according to the field information and unit's stored energy, and is realized as a result of transition of such external stimulus on unit internal parameters. Therefore, necessary values for decision of unit action $\mathbf{p}(t)$ consist of field information $\mathbf{H}(r,\phi,t)$, outflowing energy $\mathbf{E}_{out}=\mathbf{E}_{out}(r,t)$ from unit to environment, unit parameter $\mathbf{K}(t)$. Then Lagrangean derives equations of motion $\mathbf{p}(t)$ regarding $\mathbf{H}(r,\phi,t)$, $\mathbf{E}_{out}$, $\mathbf{K}(t)$ as a potential function by using a convolution as equation (5).

$$\dot{\boldsymbol{p}}(t) = \boldsymbol{M} \cdot \nabla \oint \mathbf{H}(\boldsymbol{r}, \phi, t)\big(\zeta(\phi) + \varphi(\phi)\big)\, d\phi \cdot \mathbf{E}_{out} \cdot \mathbf{K}(t) \quad \text{------ (5)}$$

$\boldsymbol{M}$ is the inertia mass. Setting 10000 units, and attractive-repulsive potential function on equation (1), equation (2) and equation (5), computer simulation of interaction among 10000 units can be shown in Figure 1.The computational cost depends on size of field to calculate equation (2), however don't depends on number of units. ( If we use simple calculation, the computational cost growth in proportion to $n^n$. n is number of units.)

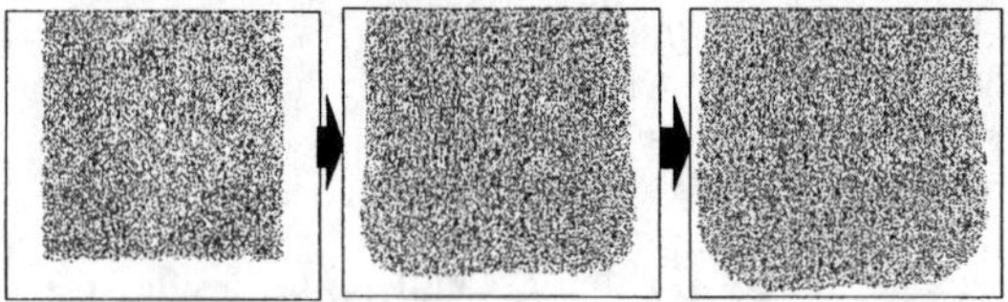

Figure 1. Computer simulation of interaction among 10000 units

**Unit Information (Potential Functions)**

The unit information in this paper deals with amplitude of propagating information in a field. Two types of information are applied, one is the density type that simply diffuses with decline, the other is the wavy type that propagates all around with gradual decline, and suffers resonance, reflection and refraction depending on a boundary condition of the field. The unit outputs such information, and influences surrounding units according to the distance among them. Although such propagating input won't actuate creatures directly, all creatures use it and make decisions according to it as well as the information from environment.

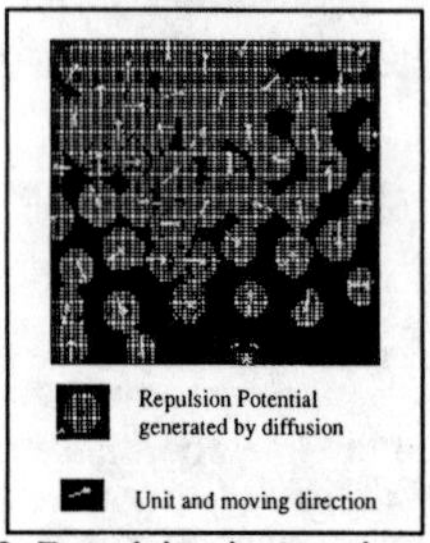


Figure 2. Repulsive interaction among units

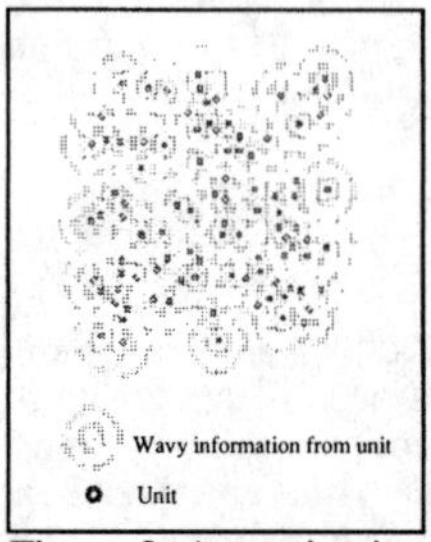


Figure 3. Attractive interaction among units

Setting $\alpha,\beta,\gamma,\delta,\varepsilon$ on the partial differential equation (2), density type information is defined as $\mathbf{h}(\mathbf{r},t)$ and $\mathbf{E}(\mathbf{r},t)$ as shown in equation (6) and equation (7), wavy type is define as $\mathbf{w}(\mathbf{r},t)$ as shown in equation (8), elastic potential among units is defined $hl(\mathbf{r},t)$ as shown in (9). A propagating information is automatically estimated along to the equation (2), each potential function must be set to an initial value of information (amplitude). Using those two types potential functions, interaction among unit is shown in Figure 2 and Figure 3. Figure 2 shows repulsive interaction through density type potential function. Figure 3 indicates attractive interaction through wavy type potential function. Those two interactions can be realized through both potential funcitons.

The density type of fields can be achieved by setting parameters on the information field u at the equation (2).

$$h(\boldsymbol{r},\ t) = u_{h,1}(\boldsymbol{r},\ t) - u_{h,2}(\boldsymbol{r},\ t) \qquad \text{------ (6)}$$

$$E(\boldsymbol{r},\ t) = E_{env}(\boldsymbol{r},\ t) \qquad \text{------ (7)}$$

| *Parameter on density type of fields:* | *Boundary condition on density type of fields:* |
|---|---|
| $u = u_{h,1}(\boldsymbol{r},\ t),\ \alpha = 0,\ \beta = 1,\ \gamma = -1,\ \delta = -1,\ \varepsilon = 0$ | $u_{h,1}(\boldsymbol{r}_i,\ t) = \oint H(\boldsymbol{r}_i,\ \phi, t)\,\chi(\phi)\,d\phi$ |
| $u = u_{h,2}(\boldsymbol{r},\ t),\ \alpha = 0,\ \beta = 1,\ \gamma = -1,\ \delta = -2,\ \varepsilon = 0$ | $u_{h,2}(\boldsymbol{r}_i,\ t) = 2\oint H(\boldsymbol{r}_i,\ \phi, t)\,\chi(\phi)\,d\phi$ |
| $u = E_{env}(\boldsymbol{r},\ t),\ \alpha = 0,\ \beta = 1,\ \gamma = -1,\ \delta = -1,\ \varepsilon = 0$ | $u_{h,1}(r_i,t)$, $u_{h,2}(r_i,t)$: amplitude of information at unit position $r_i$. |

The wavy type of field is

$$w(\boldsymbol{r}, \boldsymbol{n}) = u_w(\boldsymbol{r},\ t) \qquad \text{----- (8)}$$

| *Parameters on wavy type of field:* | *Boundary condition on wavy type of field:* |
|---|---|
| $u = u_w(\boldsymbol{r},\ t),\ \alpha = 1,\ \beta = 1,\ \gamma = -1,\ \delta = -1,\ \varepsilon = 0$ | $u_w(\boldsymbol{r}_i,\ t) = \exp(j\omega_0 t)$ |

The elastic type potential function is defined as equation (9)

$$hl_{ij}(\boldsymbol{r}, t) = l_{ij}(t)\,(\boldsymbol{r} - \boldsymbol{r}_i) \qquad \text{----- (9)}$$

The initial values are defined through the objective function that is introduced in next section.

**Objective function**

Adaptive behavior of units using local information requires an objective function for decentralized control. An ideal objective function of a decentralized system leads the units to an orderly configuration depending on environment. In this section, the objective function for each unit is set by using the concept of the mechanics of living creatures from the field of statistical mechanics.

Unit motion **p**(t) is depending on parameter **K**(t). According to the mechanics of living creatures, each unit conserves order by minimizing the differential calculated gain of entropy ds. ds is derived by input-output relation of environment energy. Using state values on each unit i are inflowing heat $\mathbf{E}_{in}=\mathbf{E}_{in}(r_i,t)$, stored energy $\mathbf{Q}=\mathbf{Q}_i(t)$, work $\mathbf{W}=\mathbf{W}_i(t)$, $\mathbf{E}_{out}=\mathbf{E}_{out}(r_i,t)$, spatial heat environment $\mathbf{E}_{env}=\mathbf{E}_{env}(r,t)$. The inflowing heat $\mathbf{E}_{in}$ is input energy from environment that is transmitted through wave propagation and threshold TH as shown in equation (10)

$$\mathbf{E}_{in} = \oint H(\boldsymbol{r},\phi,t)\,\chi(\phi)\,d\phi - TH \qquad \text{----- (10)}$$

Since such state values are connected with each state values, numerical computation is applied to obtain them. Now, the differential calculated gain of entropy ds is define as equation (11) using state values $\mathbf{E}_{in}$, $\mathbf{E}_{out}$,and emvironmental tempareture $T(\mathbf{E}_{env})$.

$$ds = \frac{d\mathbf{E}_{out} - d\mathbf{E}_{in}}{T(\mathbf{E}_{env})} \qquad \text{----- (11)}$$

The parameter **K**(t) is derived as equation (12) for minimize ds.

$$\frac{\partial \mathbf{K}(t)}{\partial t} = -\frac{\partial\, ds}{\partial \mathbf{K}(t)} \qquad \text{----- (12)}$$

# APPROACH TO TRAVELING SALESMAN PROBLEM

As shown in JSP, setting potential function to be adoped to the problem makes VPF problem solver. This section shows powerfull capability to the TSP. The methodology for TSP has been reported at the [Yokoi 95], where amoeba searching behavior model (ASBM) is proposed and applied to small size TSP. Expanding ASBM and applying to large size problem, this paper investigates a size effect delived from TSP.

**Computation**

The example tested the potential of ASBM for applying famous Euclidean TSP called ATT532 city's problem. Since the stability of searching behavior depends on the gains of attractive-repulsive potential function and the gains of inflowing-outflowing energy, larger gain leads unstable with numerical computation. Furthermore the searching velocity in the stable area has non-linear relation with system parameters.However, a convenient parameter derives ASBM into good searching behavior. As shown in the comparison to the conventional studies, the proposed ASBM obtains 2.24% near optimal solution within 100 hours. Although the searching velocity is too slow for a practical use, but solution is better than conventional studies [Matsuyama91] except Padberg's study.

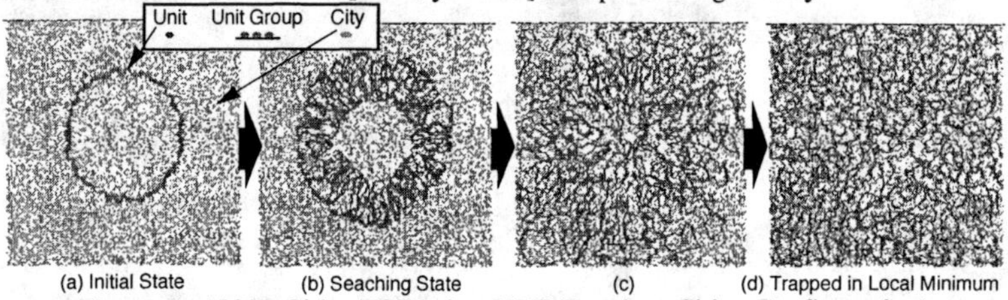


Figure 4 . 10000 Cities TSP using VPF (Random Cities Configuration)

While, for the 10000 cities problem, the searching velocity is not expornentialy growth. It costs 10th times as USA532 city's problem. Figure 4 indicates process of convergence to the 10000 random cities configuration. The reson why the computational cost is not become big, is almost computational cost is spent on field computation. Therefore, we will not face to the n! problem according to growth of number of units.

# CONCLUSIONS

For large scale combinatorial optimization, this paper proposed vibrating potential field as one approach of field technique.The VPF is computation frame work for interaction of multi-processors, and it consists of field techniques and conservation techniques of processors' states.
By computational experiment, two characteristics of VPF are shown, such as heteronomous and autonomous. The heteronomous system responds to influence from environment based on field technique. The autonomous system affects environment based on conservation technique.
For the combinatorial optimization, those two characteristics shows searching behavior on problem domain. The test problems for VPF are TSP and JSP. To adopt VPF to concete problem, the adequate potential functions are parameterized and designed for each unit that is element of VPF. The heteronomous and autonomous character gets coordination of the group of units in good condition, and optimizes its own system dynamics. The results clarified possibility for applying to large scale combinatorial optimization , because of independence from restriction of number of objects to be processed in the problem.
Advantages and disadvantages of VPF are the followings: advantages are integration of global constraints for the units, and description of recursive relationship among system and its own components. Disadvantages are difficulty of universal objective function, and large computation cost.

However, The computaion cost depends on how many processes is bundled using field, such as bundling size. The accuracy of solution will also depend on the bundling size. The problem of bundling approach is trade off among accuracy of computation and cost. We must show the relation of accuracy of computation and cost in the future work.

### References

[Suginohara 91] T.Suginohara et al.: Astrophys. J. Suppl. 75, 631 (1991).
[Husbands 91] Husbands, P., and Mill, F.: Simulated Co-Evaluation as the Mechanism for Emergent Planning and Scheduling, *Proc. of the 4th International Conference on Genetic Algorithms*, pp.264-271, (1991).
[Begchi 91] Begchi, S., Uckun, S., Miyabe, Y., and Kawamura, K.: Exploring Problem-Specific Recombination Operators for Job Shop Scheduling, *Proc. of the 4th International Conference on Genetic Algorithms*, pp.10-17, (1991).
[Matsuyama 91] Y. Matsuyama,"Self-organizing neural networks and various Euclidian traveling salesman problems", The transactions of the institute of electronics, information and communication engineers D-II, Vol.J74-D-II, No.3, pp.416-425 (1991).
[Minagawa 92] Minagawa, M., and Kakazu, Y.: Scheduling Flow-Shops with Alternative Resources: A GA Based Heuristic Approach, *Proc. of the JAPAN/USA Symposium on Flexible Automation*, vol.2, pp.1143-1146, (1992).
[Kakazu 90] Kakazu, Y., Takatori, N., and Nonaka, N.: Job Shop Scheduling by a Hopfield-type Neural Network, *Transactions on Information Processing Society of Japan*, vol.31, No.10, pp.1505-1512, (1990).
[Yokoi 92] Yokoi, H. and Kakazu, Y.: An Approach to the Traveling Salesman Problem by a Bionic Model, *Heuristics, The Journal of Knowledge Engineering, IAKE (International Association of Knowledge Engineers),* (1992).
[Yokoi 95] Hiroshi YOKOI, Takafumi MIZUNO, Masatoshi TAKITA, and Yukinori KAKAZU, Amoeba Searching Behavior Model for Traveling Salesman Problem,ASME PRESS, Intelligent Engineering Systems through Artificial Neural Networks, Vol.5 , pp705-710 (1995)

# DYNAMICAL BEHAVIOR OF ARTIFICIAL NEURAL NETWORKS WITH RANDOM WEIGHTS

**D. J. ALBERS AND J. C. SPROTT**
*Department of Physics, University of Wisconsin, Madison*

**W. D. DECHERT**
*Department of Economics, University of Houston and Department of Economics, University of Wisconsin, Madison*

**ABSTRACT**
In this paper we report a Monte Carlo study of the dynamics of large untrained, feedforward, neural networks with randomly chosen weights and feedback. The analysis consists of looking at the percent of the systems that exhibit chaos, the distrubution of largest Lyapunov exponents, and the distrubution of correlation dimensions. As the systems become more complex (increasing inputs and neurons), the probability of chaos approaches unity. The correlation dimension is typically much smaller than the system dimension.

## INTRODUCTION

In an attempt to gain insight into the general dynamics of complex systems we have constructed models using completely connected, feedforward neural networks with the output fed back to the input, to produce a time series. We chose neural networks as our general models because they are representative of a large class of functions that can represent an arbitary dynamical system on a closed and bounded domain [Hornik et al. (1990)]. We present a Monte Carlo study of the probability that deterministic systems on compact domains are chaotic and, if they are, what properties they exhibit. Our results extend previous work in [Sompolinsky et al. (1988)].

## THE STUDY OF CHAOTIC DYNAMICS

Unpredictible chaotic dynamics can arise from deterministic, non-linear systems governed by simple laws or rules. The unpredictiblility arises from a sensitive dependence on initial conditions; for the same system, the trajectories of initially nearby points exponentially diverge on average. Many methods of quantifying chaos have been proposed. For this paper we will focus on two such methods – the Lyapunov exponent and the correlation dimension. Calculation of the largest Lyapunov exponent involves perturbing the orbit a small amount, $\epsilon$, and following it and its rate of divergence from the unperturbed trajectory, thus quantifying the trajectory spreading. The attractors of chaotic systems are generally fractals with non-integer dimensions, whose value is a measure of the complexity of the dynamics. One such dimension is the correlation dimension, calculated by selecting many points on the orbit and binning the number of surrounding points within a given radius. The correlation dimension is given by the slope of this correlation sum versus radius on a log-log plot.

## DISCUSSION OF GENERAL NEURAL NETWORKS

In this paper we consider single-layer, feedforward neural networks of the form:

$$f(y) = \sum_{i=1}^{n} \beta_i \phi \left( s\omega_{i0} + s \sum_{j=1}^{d} \omega_{ij} y_j \right) \tag{1}$$

where $f : R^d \to R$. The parameters $\beta_i$ and $w_{ij}$ are real, and $s$ is the scaling factor on the weights. All the weights are randomly choosen. The $\phi$ represents a neuron or activation function which has been chosen as the hyperbolic tangent, which is general [Hornik et al. (1990)].

$$\tanh u = \frac{e^u - e^{-u}}{e^u + e^{-u}} \tag{2}$$

The dynamics we are investigating consist of a smooth map of the form:

$$x_{t+1} = F(x_t) \tag{3}$$

where $x_t \in R^m$ is a vector of state variables at time $t$. Each component function of $F$ represents a neural network with its own set of coefficients. All solutions to (1) remain bounded as is required by the squashing function (2). In this numerical experiment we use a single feedforward neural network to generate a 'time series' of scalar data. A function, $f : R^d \to R$, defines a dynamical system on $R^d$ by

$$y_t = f(y_{t-d}, y_{t-d+1}, ..., y_{t-1}) \tag{4}$$

where $y_t \in R$. Equation (4) can be written in the form (3) by mapping $R^d$ to itself:

$$(y_1, y_2, ..., y_d) \to (y_2, y_3, ..., y_{t-1}) \tag{5}$$

Such neural networks form a subset of $d$-dimensional dynamical systems. Takens has shown that systems of the form (3) are diffeomorphisms embeded in $R^d$ for some $d \leq 2n + 1$ [Noakes (1995)]. Thus, there is an open and dense set of diffeomorphisms, each one of which is topologically conjugate to a system of the form (3). These latter systems can be uniformly approximated by neural networks.

The equation used in the experiment was:

$$y_t = \sum_i^n \beta_i \tanh\left( s\omega_{i0} + s\sum_j^d \omega_{ij} y_{t-j} \right) \tag{6}$$

where the matrix of weights, $\omega$, and the vector coefficients, $\beta$, were generated using a pseudo-random-number-generator. The elements of $\omega$ are iid normal with mean zero and unit variance. The scaling parameter, $s$, is the standard deviation of the weights. The elements of $\beta$ are

$$\beta_i = \frac{u_i}{\sum_{j=1}^n u_j} \tag{7}$$

where the $u_i$ are iid uniform on $[0, 1]$. Note that the coefficients are scaled so that $\sum_{i=1}^n \beta_i = 1$. With this scaling, $f : [-1, 1]^d \to [-1, 1]$. This scaling is completely general. Without the scaling $\sum_{i=1}^n \beta_i = r$, then $f : [-r, r]^d \to [-r, r]$. The effect of scaling the inputs $y$ by $1/r$, the weights $\omega$ by $r$ and the vector $\beta$ by $1/r$, is that it maps $f : [-1, 1]^d \to [-1, 1]$, and the resulting dynamics are the same up to the scaling factor.

## NUMERICAL RESULTS

## THE FREQUENCY OF CHAOS

For this study, we consider a system chaotic if its largest Lyapunov exponent is positive. The frequency of chaos is thus the fraction of systems that have positive Lyapunov exponents. The systems are parameterized by $n$, the number of neurons, $d$, the number of input time lags, (or dimension), and $s$, the spread of the randomly chosen weights.

Figure 1 shows the percent chaotic with a 50 percent contour for $s = 8$, and various $d$ and $n$. Notice the "C" shape of the fifty-percent contour line. In figure 2, where $n = 8$ and $s$ and $d$ are varied, this "C" shaped contour line is also apparent. Both of these results suggest that $n$ and $s$ can be optimized for maximum chaos. For a given $n$ and $s$, any increase in $d$ increases the frequency of chaos, and in the limit as $d$ approaches infinity, the fraction chaotic becomes unity.

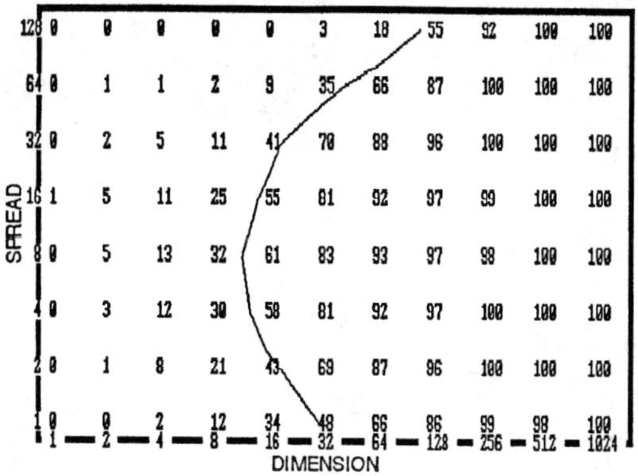


Figure 1: Frequency of chaos plot for $n = 8$, various $d$ and $s$.

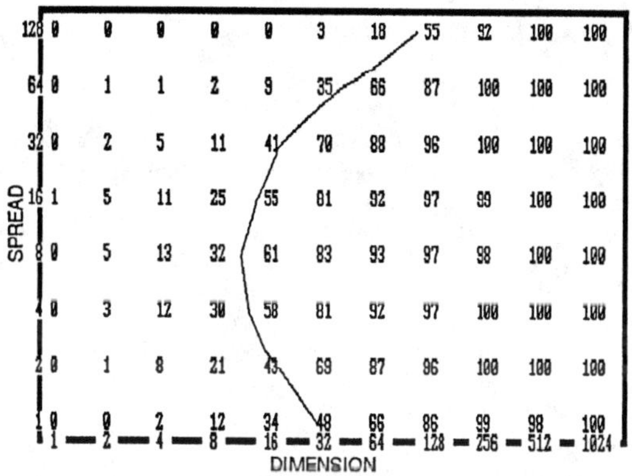


Figure 2: Frequency of chaos plot for $s = 8$, various $d$ and $n$.

The $n$ dependence evident in figure 1 is not general, but is an artifact of the $\beta$ normalization. Normalizing the $\beta$'s such that the sum of their squares is equal to $n$ causes the fraction chaotic to increase monotonically with $n$, approaching unity at large $n$.

The "C" contour in figure 2 has a fundamentally different explanation. For a given dimension, $d$, the $\tanh u$ function will approach total linearity for small values of $s$, and will approach $\pm 1$ for large values of $s$. Note that the $\beta$ normalization is dependent on $n$ whereas the $s$ parameter is independent of both $n$ and $d$, and this will affect the system in a very different way. Linear systems, (small $s$), cannot exhibit chaos. Likewise, for large values of $s$, the terms inside the activation function become large, forcing the outputs into a finite number of binary states, $\pm 1$. This produces systems that are very likely periodic since the number of states is finite and must eventually repeat.

Experiments were also done using networks with a vector output fed back to the $d$ inputs. These vector networks are much less chaotic, and the region of chaos lies at higher $n$ and $d$ and lower $s$ than in the scalar networks.

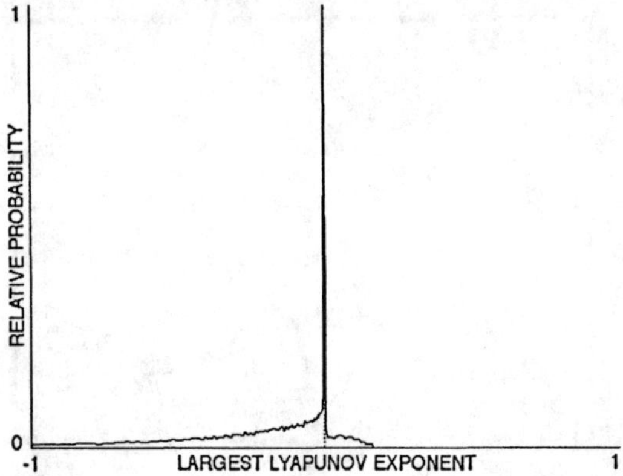


Figure 3: Lyapunov exponent distribution for $n = 8$, $d = 4$, $s = 8$.

## THE LARGEST LYAPUNOV EXPONENTS

The largest Lyapunov exponent was calculated using the numeric method described in [Wolf et al. (1984)].

As is evident from figures 1 and 2, there is a dependence on both $n$ and $s$. The Lyapunov exponent distributions for $s$ either very large or very small consists of a spike at 0. As stated in the previous section, the $n$ scaling is not general.

The Lyapunov exponent distributions at small $d$ and an arbitrary $n$ is typically represented by figure 3. The amount of the distribution that lies to the right of zero increases,and the amount to the left of zero decreases as $n$ is increased. In the limit of infinite $n$, the distributions mimic the high dimensional cases. The Lyapunov exponent spike at zero suggests that a large number of the networks are periodic limit cycles. This is not suprising since the zero Lyapunov exponent is special in that it contains all and only limit cycles, whereas the rest of the distribution contains all other behavior.

For the high dimensional networks, ($d \gtrsim 64$), a dominant feature is a narrow spike just to the right of zero. As the dimension is increased from this point for a given $n$ and $s$, this distrubution asymptotically approaches zero, and its width narrows. The explanation for this apparent asymptotic limit at zero is that the Lyapunov exponent scales with the number of time lags or dimension which has the effect of slowing the time scale. If the Lyapunov exponents are multiplied by a factor of $d$, the distributions are roughly independent of $d$.

### Correlation Dimension

We use the method of [Grassbeger and Procaccia (1983)] to determine the correlation dimension. The number of points needed to calculate a correlation dimension within 95 percent accuracy as given in [Tsonis (1992)] is

$$N_{min} = 10^{(2+.4D)} \tag{8}$$

where $D$ is taken as the attractor dimension, not the embedding dimension. We required enough points so that the Tsonis criterion was met and the deviation between iterations of the process is less that 2.5 percent. We set a minimum number of points to be correlated; if the criterita were not met, another iteration of the process was carried out, adding to the number of points correlated.

The trends in the correlation dimension match those in the largest Lyapunov exponent distribution. Note that we eliminated all cases with non-positive Lyapunov exponents before calculating the correlation dimensions.

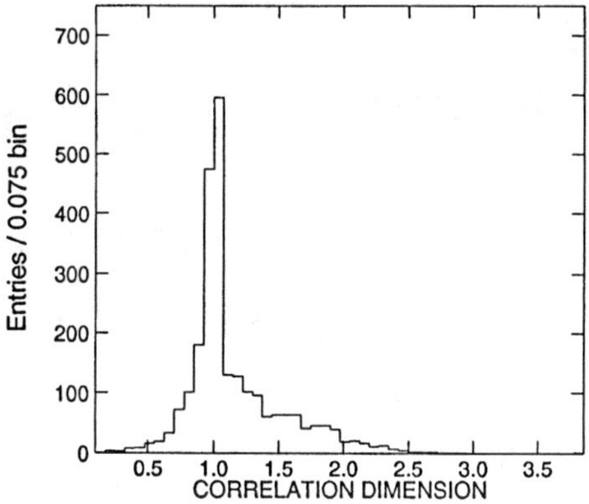


Figure 4: Correlation dimension for $n = 8$, $d = 4$, $s = 8$.

Figure 4 has several striking characteristics. The first and most pronounced is the spike at one. This spike corresponds to limit cycles. Note that this spike persists even though non-chaotic systems have been eliminated because of the numeric error inherent in digital experiments. The majority of this correlation dimension spike is due to systems which are not really chaotic and have Lyapunov exponents less than $10^{-3}$. If we require that the Lyapunov exponent be greater than $10^{-3}$ we remove the spike at one, but, nevertheless, much of the correlation dimension distribution is close to one. Removing the points with Lyapunov exponents less than $10^{-3}$ has little affect on the cluster around one, as can be seen in figure 5.

Another important feature is the tendency of the correlation dimension to be much smaller than the embedding dimension. This is an obvious feature of figure 4 where the majority of the distribution lies far below the allowed dimension of the system.

Figure 5 shows little relation between the largest Lyapunov exponent and the correlation dimension. This does not mean there is no relationship between the Lyapunov exponent *spectrum* and the information dimension, but values of the *largest* Lyapunov exponent do not yield much information about the correlation dimension [Frederickson et al. (1983)]. The largest Lyapunov exponent tends to increase with increasing correlation dimension, but this is definitely not a strict rule.

## FINAL REMARKS

The major result found is that as the dimension ($d$) and complexity ($n$) of a deterministic dynamical system increases, the fraction of systems that are chaotic approaches unity. This is not a particularly surprising result. That the correlation dimension is on average less than half the embedding dimension is much less obvious. This result is partially caused by a cluster of correlation dimensions near one, even when the cases with small Lyapunov exponents are eliminated. The majority of the systems, although complex, do not explore much of the freedom they have available. The dependence on the scaling of the weights shows that the likelihood of chaos can be maximized. Source code and additional detail can be found at http://sprott.physics.wisc.edu/neural/.

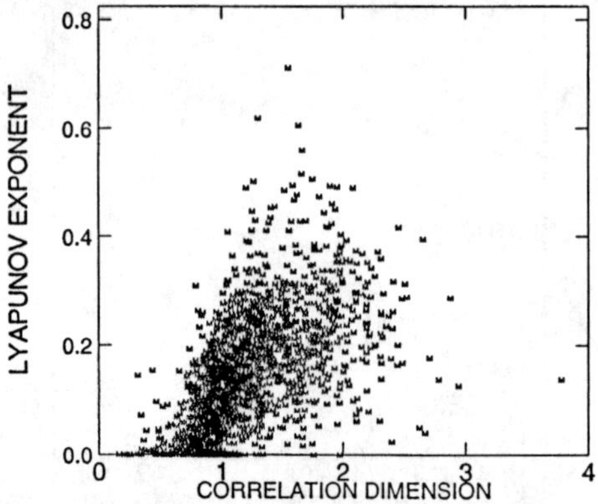


Figure 5: Correlation dimension versus Lyapunov exponent for $n = 8$, $d = 4$, $s = 8$.

ACKNOWLEDGEMENTS

D. J. Albers acknowledges Derek Wright for assistance in computing as system administrator, and the Hilldale Foundation for financial support.

# References

[Fredrickson et al. (1983)] Fredrickson, P., Kaplan, J. L., Yorke, E. D., Yorke, J. A., 1983, "The Liapunov Dimension of Strange Attractors," *Journal of Differential Equations*, Vol. 49, pp. 185–207.

[Grassbeger and Procaccia (1983)] Grassbeger, P., Procaccia, I., 1983, "Characterizaqtion of Strange Attractors," *Physical Review Letters*, Vol. 50, pp. 346–349.

[Hornik et al. (1990)] Hornik, K., Stinchocombe, M., White, H., 1990, "Universal Approximation of an Unknown Mapping and its Derivatives Using Multilayer Feedforward Networks," *Neural Networks*, Vol. 3, pp. 535–549.

[Noakes (1995)] Noakes, L., 1995, "The Takens Embedding Theorem," *International Journal of Bifurcation and Chaos*, Vol., pp. 867–872.

[Sompolinsky et al. (1988)] Sompolinsky, H., Crisanti, A., Sommers, H. J., 1988, "Chaos in Random Neural Networks," *Physical Review Letters*, Vol. 61, pp. 259–262.

[Tsonis (1992)] Tsonis, A., 1992, *Chaos: From Theory to Applications* Plenum, edition.

[Wolf et al. (1984)] Wolf, A., Swift, J. B., Swinney, H. L., Vastane, J. A., 1984, "Determining Lyapunov Exponents from a Time Series," *Physica D*, Vol. 16, pp. 285–317.

# THREE TECHNIQUES FOR EXTRACTING RULES FROM FEEDFORWARD NETWORKS

**Ismail Taha and Joydeep Ghosh** [1]
Department of Electrical and Computer Engineering,
University of Texas, Austin, TX 78712-1084
E-mail: {ismail,ghosh}@pine.ece.utexas.edu

**ABSTRACT**

Hybrid intelligent systems that combine knowledge based and artificial neural network systems typically have four phases involving domain knowledge representation, mapping into connectionist network, network training, and rule extraction respectively. The final phase is important because it can provide a trained connectionist architecture with explanation power and validate its output decisions. Moreover, it can be used to refine and maintain the initial knowledge acquired from domain experts. In this paper, we introduce three new rule extraction techniques. The first technique extracts a set of binary rules from any neural network regardless of its kind (MLP, RBF etc.,). The second technique extracts partial rules that represent the most important embedded knowledge in a trained MLP. The fidelity of the second technique is adjustable to the desired level of knowledge extraction. The third technique is a universal and comprehensive approach that extracts almost all embedded knowledge in a trained artificial neural network and represents it in a rule base format. Experimental results of implementing these three rule extraction techniques are presented.
**Keyword: Knowledge Extraction. Also: Hybrid Artificial Neural Network Systems.**

## 1 INTRODUCTION

Several researchers have investigated the design of hybrid systems that combine expert and connectionist subsystems. The main motivation of such hybrid systems is to incorporate the complementary features of knowledge based and neural network paradigms. HIA[Taha and Ghosh (1995)], KBANN[Towell and Shavlik (1994)], RAPTURE[Mahoney and Mooney (1993)] and KBCNN[Fu (1994)] are examples of these hybrid systems. The typical result is a Knowledge Based Neural Network (KBNN) system with four phases: (i) the representation phase, where initial domain knowledge is represented in a symbolic format (ii) the mapping phase, where this knowledge is mapped into an initial connectionist architecture (iii) the training phase (iv) the rule extraction phase, where the trained (and thus modified) connectionist architecture is mapped back into an updated rule-based system to provide explanation power.

Research work in the area of extracting symbolic knowledge from trained ANNs has witnessed much activity recently. KT[Fu (1994)], Subset, MofN[Towell and Shavlik (1993)], and NeuroRule[Setiono and Liu (1996)] are examples of rule extraction techniques from MLP networks. We categorize all these approaches as *Link Rule Extraction (LRE) techniques* because they all search for weighted links that cause a node (hidden or output) to be active. Then these combinations of weighted links are used to generate symbolic rules. Heuristic methods are commonly used in the *LRE* category to bound the search space for rules and to increase the comprehensibility of the extracted rules. BRAINNE, DEDEC, RuleNet and RULEX are other rule extraction techniques that extract rules from trained networks regardless the type or the structure of the neural network and hence named *Black-box Rule Extraction (BRE)* techniques. A very rich source of literature review of different rule extraction approaches can be found in[Andrews et al. (1995)].

This paper proposes three new rule extraction techniques for KBNN hybrid systems and presents their implementation results. The suitability of each technique depends on the network type, input nature, complexity, the application nature, and the required transparency level.

---

[1]This research was supported in part by ARO contract DAAH04-94-G-0417, DAAH04-95-10494, and ATP grant #442. Ismail Taha was also supported by the Egyptian Government Ph.D. Fellowship in Electrical and Computer Engineering.

## 2 RULE EXTRACTION: GAINS AND ISSUES

Extracting symbolic rules from trained ANNs is an important feature of powerful hybrid systems as it helps in: (i) Alleviating the knowledge acquisition problem and refining initial domain knowledge. (ii) Providing reasoning and explanation capabilities. (iii) Supporting cross-referencing and verification capabilities. (iv) Alleviating the catastrophic interference problem of ANNs since a different set of rules can be extracted after a network is retrained using the new environment examples. One can then examine the resulting sets of rules to find out which situation each set is suitable for. Due to these capabilities, extracting rules from trained ANNs may be essential for developing more powerful, robust, self-explanatory and self-maintained hybrid systems. Designing an efficient rule extraction module is in fact a difficult task. Based on our experience and some other reported research, we identified several factors that should be carefully considered while designing a rule extraction technique: (i) transparency of the extracted rules; (ii) granularity of the explanation feature; (iii) representation format; (iv) comprehensibility which is indicated by the number of rules and/or number of premises in each rule extracted from a trained network; (v) fidelity which is measured by the capability of the extracted rules to mimic the embedded knowledge in a trained network; (vi) Accuracy; and (vii) portability. The quality of extracted rules is improved by increasing the comprehensibility, the fidelity, and the accuracy of the rule extraction algorithm. However, to extract a comprehensive rule base from a trained ANN most, if not all, embedded knowledge in this ANN should be extracted. In this case, the comprehensibility of extracted rules will be degraded because the resulting rule base may have too many rules, some of which with many premises.

## 3 PROPOSED RULE EXTRACTION APPROACHES

The first proposed approach (BIO-RE) is categorized as a BRE technique, while the second and third techniques belong to the LRE category.

### 3.1 First Approach (BIO-RE)

It is named **Binarized Input-Output Rule Extraction (BIO-RE)** because it extracts binary rules from any neural network trained with "binary" inputs. If original inputs are not binary, they have to be binarized using Equation 1.

$$y_i = \begin{cases} 1 \text{ if } x_i \geq \mu_i \\ 0 \text{ otherwise} \end{cases} \tag{1}$$

where: $x_i$ is the value of original input $X_i$, $\mu_i$ is the mean value of $X_i$, and $y_i$ is the corresponding binarized input value of $x_i$. BIO-RE extracts rules based on the trained network input-output mapping. Some of the unique feature of BIO-RE are: (i) It does not require any information about the internal structure of the network. (ii) It can be used to extract rules from any kind of neural network (e.g. RBFs, MLPs, or Recurrent Networks). (iii) It does not require any specific training regime (supervised or unsupervised). Rules extracted by BIO-RE are represented in this format: $If$ [*Not*] *Input-Variable* [*And* [*Not*] *Input-Variable*]* $\longrightarrow Consequent_j$, where: [·] is an optional term and the [·]* means that the term [·] can be repeated 0 or n times. In terms of final rules, an extracted rule *"If* $X_1$ *And Not* $X_2$ *Then* $O_1$*"* is rewritten as *"If* $X_1 \geq \mu_1$ *And* $X_2 \leq \mu_2$ *Then* $O_1$*"*. The outline of the BIO-RE algorithm is as follows: *For a well trained neural network do:*

1. *Obtain the network output* $O(X_i) = \{o_j(x_i) \mid o_j \in \{0,1\}\}$ *corresponding to each binary input pattern* $X_i$. *If the number of input nodes is* **n** *then this conceptually requires* $2^n$ *input patterns. However, the problem specification may remove some combinations that will not occur.*
2. *Generate a truth table by concatenating each input* $X_i$ *of step 1 and its corresponding output decision* $O(X_i)$ *from the trained network. An output* $o_j(x_i)$ *is set to 1 if its corresponding output node is active (above a threshold), otherwise it is 0.*
3. *Generate the corresponding boolean function (represented in the previously described binary rule format) from the truth table of step 2.*

Any available boolean simplification method can be used to perform step 3 of the BIO-RE algorithm. We used Espresso[2] to generate the extracted rules. The BIO-RE approach is suitable when the input/output variables are naturally binary or when binarization does not significantly degrade the performance. Also the input size ($n$) should be small. Given that the above conditions are satisfied, BIO-RE has some advantages: (i) It allows the use of available logic minimization tools. (ii) It does not require any rewriting procedure since extracted rules are optimal and cannot be simplified any further. (iii) Its extracted rules do not depend on the number of layers of the trained network.

### 3.2 Second Approach (Partial-RE)

The idea underlying Partial-RE algorithm is that it first sorts both positive and negative incoming links to all hidden and output nodes in descending order into two different sets based on their weight values. Starting from the highest positive weight (say $i$), it searches for individual incoming links that can cause a node $j$ (hidden/output) to be active regardless of other input links to this node. If such a link exists, it generates a rule: $If\ Node_i \xrightarrow{cf} Node_j$, where $cf$ represents the measure of belief in the extracted rule and is equal to the activation value of $node_j$ (see Equation 3). If a node $i$ was found strong enough to activate a node $j$, then this node is marked and cannot be used in any further combinations when checking the same node $j$. Partial-RE continues checking subsequent weights in the positive set until it finds one that cannot activate the current node $j$ by itself. It is important to mention that Partial-RE assumes that all inputs are equally weighted and hence have the same effect on the next layer (i.e., have the same range of values or fall between say (0,1)). Therefore, if original inputs are not naturally normal then they need to be normalized before the training phase using Equation 2.

$$z_i = \frac{1.0}{1.0 + e^{-(\frac{x_i - \mu_i}{2.0\sigma_i})}} \tag{2}$$

where, $z_i$ ($\in (0,1)$) is the corresponding normalized input value of the original input value $x_i$ and $\sigma_i$ is the standard deviation of input feature $X_i$.

If more detailed rules are required (i.e. the comprehensibility measure $p > 1$), then it starts looking for combinations of two unmarked links starting from the first (maximum) element of the positive set. This process continues until Partial-RE reaches its terminating criteria (maximum number of premises in rule = $p$). Moreover, it looks for small combinations of positive and negative links that can activate a hidden/output node. In this case extracted rules are represented as: "$If\ Node_i\ And\ Not\ Node_g \xrightarrow{cf} Node_j$", where the link between node $i$ and $j$ is positive and between $g$ and $j$ is negative. After extracting all rules, a rewriting procedure takes place. Within this rewriting procedure any premise that represents an intermediate concept (i.e a hidden unit) is replaced by the corresponding set of conjuncted input features that causes it to be active. Final rules are written in this format: "$If\ X_i \geq \mu_i\ And\ X_g \leq \mu_g \xrightarrow{cf} Consequent_j$" (see Table 2 for examples). Experimental results show that Partial-RE algorithm is suitable for large size problems, since extracting all possible rules is NP-hard and extracting only the most effective rules is a practical alternative.

### 3.3 Third Approach (Full-RE)

Like the Partial-RE approach, Full-RE falls in the LRE category. It is notable because: (i) It extracts rules with certainty factors from trained feedforward ANNs. (ii) It extracts all possible rules that represent the semantic interpretation of the internal structure of the trained neural network that they were extracted from. (iii) It is universal since there is no restriction on the values that any input feature can take. (iv) It is applicable to any neural network node (unit) with a monotonically increasing activation function.

After examining different possible combinations of incoming links and feed-forwarding their effect on output nodes, Full-RE generates an intermediate rule in this format:

---

[2]Espresso is a software package for logic design[Ruddel and Sangiovanni-Vincentelli (1985)].

$If\ [\ (c_1 \cdot X_1 + c_2 \cdot X_2 + \cdots + c_n \cdot X_n) >= (\theta_j + \Delta)] \xrightarrow{cf} Consequent_j$
where: $c_i$ is a constant represents the effect (or the importance) of the $i^{th}$ input ($X_i$) on $Consequent_j$. If node $j$ is in the layer above node $i$ then $c_i$ represents the weight value $w_{ji}$ of the link between these two nodes. Since Full-RE is capable of extracting rules from networks trained with continuous inputs, then different values of $X_i$s that satisfy the intermediate rule, have to be computed then all possible rules can be generated. Full-RE performs this step by discretizing each input feature $X_i \in (a_i, b_i)$ into $k$ intervals: ($X_i = \{a_i, d_{i,1}, \cdots, d_{i,k-1}, b_i\}$, where: $d_{i,l-1}$ and $d_{i,l}$ are the lower and upper boundary values of interval $l$ of input $X_i$ respectively. After extracting its intermediate rules, Full-RE uses The Chi2[Liu and Setiono (1995)] algorithm[3], a powerful discretization tool, to compute discretization boundaries of input features. If Full-RE finds more than one discretization value of an input $X_i$ that can satisfy the intermediate equation (i.e., the equation has more one feasible solution) then it chooses the minimum of these values if the corresponding effect parameter $c_i$ is positive, or it chooses the maximum of them if $X_i$ has a negative effect parameter $c_i^-$. The decision of selecting the minimum or the maximum discretization value of an input $X_i$ is taken based on the sign of its corresponding effect parameters $c_i$ of the intermediate rule. However, all selected discretization values should satisfy the intermediate equation and the boundary constraints of all input features. The Full-RE method can be summarized in the following steps: *For each hidden node j in a well trained MLP do:*

1. *Consider the equation* $\sum_{i=1}^{n} w_{ji} \cdot x_i \geq \lambda_j$, *where* $\lambda_j$ *is high enough to make node j active. If activation function of node j is sigmoid then* $\lambda_j = \theta_j - \ln(\frac{1}{\Delta} - 1)$.
2. *Given the discretization boundaries of each input feature, consider the Linear Programming problem of* Minimizing $w_{j1}X_1 + w_{j2}X_2 + \cdots + w_{jn}X_n$ *such that:* $w_{j1}X_1 + w_{j2}X_2 + \cdots + w_{jn}X_n \geq \lambda_j$, *and* $X_i = \{a_i, d_{i,1}, \cdots, d_{i,k-1}, b_i\}\ \forall_i, i = 1 \cdots n$

Full-RE solves this LP problem by examining the discretization boundaries of $X_i$s that determine the feasible solution(s) and satisfy all given constraints. Note that: any linear programming tools can also be used to solve this standard LP problem[4]. Full-RE computes certainty factors of extracted rules based on[5] Equation 3.

$$cf = \begin{cases} \frac{1}{1+\exp^{(\sum_{i=1}^{n} w_{ji} \cdot x_i - \theta_j - \Delta)}} & \text{if act(j) is a sigmoid} \\ \sum_{i=1}^{n} w_{ji} \cdot x_i - \theta_j - \Delta & \text{if act(j) is a linear threshold} \\ 1 & \text{if act(j) is a hard limiting and } \sum_{i=1}^{n} w_{ji} \cdot x_i \geq 0.5 + \Delta \\ 0 & \text{if act(j) is a hard limiting and } \sum_{i=1}^{n} w_{ji} \cdot x_i \leq 0.5 + \Delta \end{cases} \quad (3)$$

Since, activation values of hidden nodes are bounded between (0,1), Full-RE uses a simplified version of the above procedure to extract rules between hidden and output nodes where discretizing the outputs from hidden nodes is no longer required. In each rule extracted between a hidden and output node, Full-RE replaces each hidden node ($h_j$) by the left hand side of the rule(s) whose right hand side is $h_j$. Final rules extracted by Full-RE are represented in the same format of Partial-RE expect that each $\mu_i$ is replaced by one of the discretization boundaries (say $d_{i,l}$) selected by Full-RE (see Table 3). Note that there is no restriction on the number of premises in the final rules extracted by the Full-RE.

## 4 IMPLEMENTATION

We applied all three rule extraction techniques to two problems: (i) an artificial rule-based system which has six rules relating four binary inputs and four binary outputs. (ii) Iris database, a simple classification problem which contains 50 examples each of three classes. These 150 instances were divides into two subsets, the first subset, used for training, is of size

---

[3]We are thankful to Liu and Setiono for making their Chi2 source code available to us.
[4]We also used Mathematica to find out feasible solutions.
[5]Full-RE only generates a rule if its $cf$ computed by Eqn. 3 is $\geq 0.5$.

Table 1. **Rules extracted by BIO-RE technique.**

| R# | Rule Body | Class | Times Correct | Times Misfired |
|---|---|---|---|---|
| 1 | If $I_2 \leq 3.0$ $I_3 \geq 3.7$ and $I_4 \leq 1.2$ | Versicolor | 10/50 | 0/150 |
| 2 | If $I_1 \leq 5.8$ and $I_3 \leq 3.7$ and $I_4 \leq 1.2$ | Setosa | 50/50 | 5/150 |
| 3 | If $I_1 \geq 5.8$ and $I_3 \geq 3.7$ and $I_4 \geq 1.2$ | Virginica | 47/50 | 24/150 |
| 4 | If $I_1 \leq 5.8$ and $I_2 \leq 3.0$ and $I_4 \geq 1.2$ | Versicolor | 11/50 | 3/150 |

Table 2. **Rules extracted by Partial-RE technique.**

| R# | Rule Body | Class | CF | Correct | Misfired |
|---|---|---|---|---|---|
| 1 | If $I_2 \geq 3.0$ and $I_3 \leq 3.7$ | Setosa | 0.98 | 48/50 | 0/150 |
| 2 | If $I_1 \geq 5.8$ and $I_3 \geq 3.7$ and $I_4 \geq 1.2$ | Virginica | 0.99 | 47/50 | 27/150 |
| 3 | If $I_1 \leq 5.8$ and $I_2 \leq 3.0$ and $I_3 \geq 3.7$ | Versicolor | 0.74 | 16/50 | 3/150 |
| 4 | If $I_1 \leq 5.8$ and $I_2 \leq 3.0$ and $I_4 \geq 1.2$ | Versicolor | 0.72 | 1/50 | 0/150 |
| 6 | Default Class | Versicolor | 1.0 | 6/50 | 2/150 |

Table 3. **Rules extracted by Full-RE technique.**

| R# | Rule Body | Class | CF | Times Correct | Times Misfired |
|---|---|---|---|---|---|
| 1 | If $I_3 \leq 2.1$ | Setosa | 0.99 | 50/50 | 0/150 |
| 2 | If $I_3 \leq 5.1$ and $I_4 \leq 1.7$ | Versicolor | 0.97 | 49/50 | 3/150 |
| 3 | If $I_3 > 4.8$ | Virginica | 0.98 | 47/50 | 1/150 |

89 and the second is of size 61 and used for testing. Each input pattern has four continuous input features: $I_1$ = Sepal-length , $I_2$ = Sepal-width, $I_3$ = Petal-length, and $I_4$ =Petal-width. In both experiments, an MLP network was trained using the backpropagation algorithm with momentum term and regularization term $P$ which contributes to the weight update term in the backpropagation equation by $\frac{2*\lambda*w_{jk}*w_0^2}{w_0^2+w_{jk}^2}$ [Ghosh and Tumer (1994)].

### 4.1 An Artificial Binary Problem

The original rules were as follows:

*Rule #1: If A And B* $\xrightarrow{0.8}$ $O_1$ *Rule #2: If B And C And D* $\xrightarrow{0.7}$ $O_2$
*Rule #3: If Not C* $\xrightarrow{0.6}$ $O_3$ *Rule #4: If Not A And D* $\xrightarrow{0.7}$ $O_4$
*Rule #5: If B And D* $\xrightarrow{0.7}$ $O_1$ *Rule #6: If D* $\xrightarrow{0.8}$ $O_1$

These six rules were mapped by the Node Links Algorithm [Taha and Ghosh (1995)], into a network with 4 input nodes, 6 hidden nodes, and 4 output nodes. **The first experiment:** is designed to check the capability of the three techniques of extracting the original rules from the mapped network. Therefore, the network was not trained before extraction procedures were applied. (a) BIO-RE extracts the same set of binary rules but without certainty factors. (b) Partial-RE with $p = 2$ (i.e maximum 2 conditions per rule) extracts all 5 original rules with two conditions or less. On increasing $p$ to 3, rule#2 was also extracted. The certainty factors attached to each output decision were approximately the same as the original rules. (c) Full-RE extracts the same six original rules from the same untrained network. **The second Experiment:** based on the original rules, $2^4$ binary patterns were generated. After training the previous network, we applied the three approaches to the final network architecture (i.e., the adapted one): (a) Both BIO-RE and Full-RE extract the same six original rules. (b) Partial-RE extracts all the six rules plus an extra one: Rule #7: *If B* And *D* $\xrightarrow{0.74}$ $O_2$. This rule was extracted when $p = 3$.

### 4.2 Iris Classification

For the Iris classification problem, we used an MLP with 4 input, 6 hidden, and 3 output nodes. Since input features of iris are continuous and both BIO-RE and Partial-RE extracts rules from networks with binary/binarized and normalized inputs respectively, we used a binarized and a normalized version of the iris data for training and testing two network architectures both with the same number of nodes as stated above. Full-RE extracts rules from the same network structure trained with the original iris data set. Note that there is no prior knowledge for the Iris data set, so the networks were initialized randomly and then trained on samples from the data set. Table 1, 2, and 3 present the rules extracted by BIO-RE, Partial-RE, and Full-RE techniques respectively from their corresponding networks trained on the Iris data set. The performance of these corresponding trained networks are 72.67%, 93.33%, and 97.33% respectively. The tables also present the correct classification and the misclassification for each extracted rule. Note that Binarizing and normalizing continuous inputs may degrade the accuracy of the extracted rules (see Table 1 and 2) as well as the generalization of the corresponding trained networks.

## 5 CONCLUSION

In this paper we introduced three new rule extraction approaches. The suitability of each approach depends on the network type, inputs, complexity, the application nature, and the required transparency level. Experimental result shows that rules extracted by the three approaches are sound. All three methods are able to extract meaningful rules for the well known Iris database, where no pre-existing rules are available. Also, it shows that BIO-RE is suitable for small networks with binary inputs, both BIO-RE and Partial-RE are able to extract "certain" rules, and Full-RE is a universal rule extraction technique that can extract rules from MLPs with any kind of input (continuous, normal, binary). The rules extracted by Full-RE are more comprehensive than those extracted by the other two techniques. It is important to mention that extracting all possible rules is NP-hard and a feasible alternative is often to extract the key rules that cover most of the key concepts of the application domain. More progress is needed in determining when an adequate set of rules has been extracted. Another important issue that needs to be investigated is how the outputs of both the rule extraction and the trained ANN modules can be integrated to provide more robust decisions.

## REFERENCES


Andrews, R., Diederich, J., Tickle, A., 1995, "A Survey And Critique Of Techniques For Extracting Rules From Trained Artificial Neural Networks," Neurocomputing Research Center, Queensland University of Technology, Queensland, Australia.

Fu, L., 1994, *Neural Networks in Computer Intelligence* McGraw-Hill, Inc.

Ghosh, J., Tumer, K., 1994, "Structural Adaptation and Generalization in Supervised Feed-Forward Networks," *Journal of Artificial Neural Networks*, Vol. 1, pp. 431–458.

Liu, H., Setiono, R., 1995, "Chi2: Feature Selection and Discretization of Numeric attributes," Proc. Proceedings of the Seventh International Conference on Tools with Artificial Intelligence, pp. 388–391.

Mahoney, J., Mooney, R., 1993, "Combining Connectionist and Symbolic Learning to Refine Certainty Factor Rule Bases," *Connection Science*, Vol. 5, pp. 339–364.

Ruddel, R., Sangiovanni-Vincentelli, A., 1985, "Espresso-MV: Algorithms for Multiple-Valued Logic Minimization," Proc. Proceedings of Cust. Int. Circ. Conf. Portland.

Setiono, R., Liu, H., 1996, "Symbolic Representation of Neural Networks," *IEEE Computer*, Vol., pp. 71–77.

Taha, I., Ghosh, J., 1995, "Controlling Water Reservoirs using A Hybrid Intelligent Architecture," Proc. Intelligent Engineering Systems Through Artificial Neural Networks, Vol. 5, pp. 63–68. ASME Press.

Towell, G., Shavlik, J., 1993, "The Extraction of Refined Rules from Knowledge-Based Neural Networks," *Machine Learning*, Vol. 13, pp. 71–101.

Towell, G., Shavlik, J., 1994, "Knowledge-Based Artificial Neural Networks," *Artificial Intelligence*, Vol. 70, pp. 119–165.

# ONTOGENESIS IN NEURAL NETWORKS WITH FEEDBACK

**THOMAS S. DRANGER and ROLAND PRIEMER**
*Department of Electrical Engineering and Computer Science*
*The University of Illinois at Chicago*

**ABSTRACT:** We present the ontogenetic network, a new autonomous feedback network arising from a study of Hopfield. The ontogenetic network incorporates Hebbian learning, neural variety, cooperative specialization, and dynamic structure via a hidden echelon of developing neurons. To compare effectiveness, Hopfield's network is augmented with a like number of extra neurons, there to represent training vectors extended by tags to increase their separation in an expanded state space. Experiments with these networks used as binary associative memories show that storage capacity of the ontogenetic network may be increased by an order of magnitude over the original and augmented Hopfield networks, and that separations as small as one bit may be resolved. The paper describes a new ontogenetic network, experiments, and compares results with benchmarks used with associative memories.

Autonomous neural function contemplates independent action upon the signals received at the inputs of a neuron. Autonomy is one of the features that define the operation of neural networks. But in training, autonomy is often abandoned in favor of a successful algorithm governing the connection weights, even though autonomous synaptic adjustment is widely regarded as the mechanism in biological systems. Extensive feedback is also a feature in biological systems, but absent in many artificial networks. Hopfield's network is autonomous in both training and operation, but alas, too limited in capacity and resolution for many applications.

Ontogenetic development is recognized in behavioral science, but mechanisms of ontogenesis are not yet identified in natural systems. Nevertheless, ontogenesis can be studied in artificial neural networks. In the new ontogenetic network, simple nonlinear neurons, asynchronous operation, extensive connectivity, massive feedback, autonomous learning, and operability at any level of training are preserved. We do not require homogeneity, uniformity, symmetry or fixity.

## EXISTING THEORY FOR HOPFIELD NETWORKS

Consider a discrete time neural network with $n$ neurons and $n$ input ports, one for each neuron. A square matrix $G$ has elements $G_{i,j}$ giving the strength of the connection from the $j^{th}$ neuron to the $i^{th}$ neuron. External signals are applied to the network through the ports. Let the $n$-component vector $X$ contain the state of the $n$ neurons so that $X_i$ represents the firing rate of the $i^{th}$ neuron. Let the square matrix $G$ contain the weights of the edges between neurons. Let the $n$-component vector $R$ specify the reference signals applied to the ports. The excitation of the $i^{th}$ neuron is $S_i = R_i + \Sigma_j G_{i,j} X_j$. The vector $S$ representing all such excitations is given by $S = R + GX$. The value of $G$ is adjusted by the learning algorithm.

The neural activation functions are bimodal and time dependent. Several are shown in Figure 1. For asynchronous systems, the order of adjustment may be fixed, leading to deterministic state trajectories, or random, leading to indeterminate

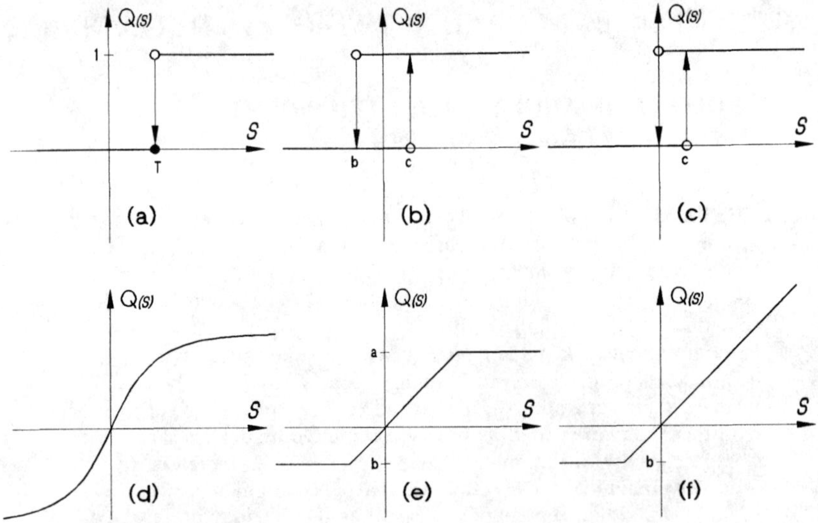


**Figure 1: Neural activation functions. Threshold (a), Hopfield (b), Modified (c), Sigmoid (d), and Bilinear (f).**

behavior. The frequency of adjustment may also vary from neuron to neuron. Let $k$ be the discrete time index. All the elements of $X(k+1)$ are the same as $X(k)$ except for one.

Suppose that a set of objects expressed as vectors $\{R1, R2, \ldots Rm\}$ contains $m$ distinct objects to be learned by a neural net. Suppose $Rq$ is one of these objects, and in turn, each object is presented to the network in training. Hebb (1949) postulated that between two natural neurons the connection strength is increased when the neurons have like states, and decreased when the neurons have unlike states. Storage algorithms are generally given by $G_{i,j} = \Sigma \Delta Gq_{i,j}$. Several autonomous storage rules used to find $\Delta Gq_{i,j}$ when $i \neq j$ are given in Table 1 for binary targets where $Wq = Rq^{T} Rq$, the inner product.

Hopfield observes that objects too close to each other (in Hamming distance) are confused in retrieval. For example, a network of $n = 100$ bivalent (on, off) neurons has an optimum Hamming distance of 45 to 55 between objects, but good performance was still observed for some objects as close together as 30. By Hopfield's criterion, severe recall error occurs when about half the objects produce well defined attractors and the rest are poorly retained. Hopfield reported that a network of $n$ neurons could store about $0.15n$ targets without severe error.

The stability of Hopfield networks has been proven absolutely by Lyapunov's

**TABLE 1: STORAGE RULES**

| HOPFIELD'S RULE | | PROPORTIONAL RULE | | COUNTERTYPE RULE | | DISTRIBUTIVE RULE | |
|---|---|---|---|---|---|---|---|
| $Rq_i$ $Rq_j$ | $\Delta Gq_{i,j}$ | $Rq_i$ $Jq_j$ | $\Delta Bq_{i,j}$ | $Rq_i$ $Rq_j$ | $\Delta Cq_{i,j}$ | $Rq_j$ $Jq_i$ | $\Delta Dq_{i,j}$ |
| 0 0 | 1 | 0 0 | 0 | 0 0 | 0 | 0 0 | 0 |
| 0 1 | -1 | 0 1 | $-Wq$ | 0 1 | 0 | 0 1 | $-1/(n-Wq)$ |
| 1 1 | 1 | 1 1 | $Wq$ | 1 1 | 1 | 1 1 | $1/Wq$ |
| 1 0 | -1 | 1 0 | 0 | 1 0 | -1 | 1 0 | 0 |

method (Cohen and Grossberg, 1983). The theorem of Abu-Mostafa & St. Jacques (1985) limits Hopfield's network to $n$ attractors, regardless of storage rules. Haykin (1994) estimates that $n/(2\ \ln(n))$ random targets can be stored with perfect recall.

**Example:** By Haykin's estimate, a network of $n = 8$ neurons can store 1.9 targets and a network of $n = 24$ neurons can store 3.8 targets. By Haykin's estimate, storage of $m = 16$ targets requires a network of 163 neurons.

## MEASURES OF PERFORMANCE

The following terms facilitate the discussion of performance in associative memories: **target** (an object in the training list), **attractor** (a stable state, a final state), **artifact** (an attractor which is not target), **probe** (an initial state used to test the response of a network), **ambiguity** (a probe which is not a target), **Hamming distance** (between two binary or bivalent vectors, the number of unequal elements), **input/output distance, I/O distance** (the Hamming distance between initial state and final state), **associative memory** (a device which, given a probe, responds with the closest or best target), **storage density** (the ratio $m/n$, which is the number of targets divided by the number of elements in each target), and **Resolution** (the ability to distinguish between similar or closely spaced objects, measured in Hamming distance). Associative memories are more effective when they have greater storage density. Better resolution of similar targets is also desirable, both for its own sake, and because greater storage density implies less distance between targets. Networks that perform better, even at low storage densities, are presumed to have better resolution. Associative memories are tested by observing the response to various inputs or probes. The most important aspect of response is the final state, which is taken as the output. Then the output, by definition, must be an attractor. We want the set of attractors to be identical with the set of targets. We also want the input/output distance to be as small as possible, but not by creation of artifacts. Five measures are used to quantify performance of associative memories. They are defined for a trained network as follows.

$N_a$ : the number of attractors.

$N_c$ : the number of attractors which are targets.

$H_e$ : the mean target identification error, equal to the Hamming distance between a probe which is a target and the output.

$H_a$ : the mean distance from probe to output.

$H_c$ : the mean distance from probe to nearest target.

$H_a/H_c$: the I/O unity check.

The first four variables are properties of $G$, the edge weight matrix. The last variable is a property of $\{R\}$, the training list. We want $N_a = m,\ N_c = m$, and $H_e = 0$. For $H_a$ to have meaning, it can be compared to $H_c$. When the I/O unity check $H_a/H_c$ deviates from unity the network is flawed. I/O unity checks near unity are inconclusive for networks which may have artifacts.

We begin by testing an augmented Hopfield network, where the number of neurons is increased from $n$, the number of elements in a target, to $n+m$, where $m$ is the number of targets in the training list. The $m$ extra neurons are used to augment each target by appending a unique tag. The tags increase the distance between targets in the expanded state space.

The state vector is now given by $X = [\ P^T\ J^T\ ]^T = [P_1 \ldots P_n\ J_1 \ldots J_m]^T$. To store a target, say $Rq$, first construct $\underline{Rq}$ which is the concatination of $Rq$ and $Jq$ where all the elements of $Jq$ are zero except for $Jq_q$ which is one. The reason for this particular construction is its similarity with the networks that will be described later. Hopfield's storage rule is applied to the augmented targets $\{\underline{Rq}\}$ to form $G$. Hopfield's activation function is used for all neurons. The network has no ports. The network is initialized by setting $P(0)$ equal to the value of the probe and allowing the network to run to the final state.

The augmented Hopfield network was tested with targets of $n = 4$ elements and $m = 1, 2, 4, 8$, and 12 targets in the training list, and also tested with targets of $n = 8$ elements and $m = 1, 2, 4, 8$, and 16 targets in the training list. The targets in each training list were distinct non-zero binary vectors chosen at random. Ten thousand training sets were generated for each combination of $n$ and $m$. In each case, the network was trained and operated to find the response to every possible probe. The probes and corresponding attractors were classified and distance between them measured so that averages could be calculated for each of the five performance measures. The results establish a benchmark for comparison of Hopfield networks with other recursive networks.

## A NEW ONTOGENETIC NEURAL NETWORK

In the new autonomous hybrid ontogenetic network, the neurons are of two classes. Neurons in one class are called protons. Neurons in the other class are called jurons. To reflect this division, let $X = [\, P^T\, J^T\, ]^T = [P_1 \ldots P_n J_1 \ldots J_m]^T$, Protons are in the first partition, also called the nucleus, where the vector $P$ represents an object of $n$ bits. Jurons are in the second partition, also called the jury, where the vector $J$ has one element for each target in a training list of $m$ vectors. The jury in vector $J$ represents a hidden internal state which does not appear in the output. The output is taken as the state of the nucleus in vector $P$.

The edge weight matrix $G$ is given an initial value which is modified by training. The storage algorithm is given by $G_{i,j} = G^0_{i,j} + \Sigma_q \Delta G_{i,j}$ where the initial value of $G$ is given in $G^0$. Also, $G$ is partitioned into four submatrices compatible with the state vector partition.

$$G = \left[\begin{array}{c|c} C_{n\times n} & B_{n\times m} \\ \hline D_{m\times n} & A_{m\times m} \end{array}\right] \qquad (1)$$

Protons have the activation function given in Figure 1(a) for T=0. Jurons have a bilinear activation function given in Figure 1(f) where $b = -1/(m+n)$. Neurons update in a fixed descending sequence as discussed in Section 3.3 for the augmented Hopfield network used to establish our benchmark. Before each round of updates, $J$ is reset to zero. The jurons update four times as frequently as the protons. A round of updates consists of resetting $J$ to zero, updating the jury four times in descending sequence, and then updating the nucleus in descending sequence.

All the elements on the main diagonal of $G$ have a fixed value of -1/24. The elements off the main diagonal of $A$ have a fixed value of -1. Ontogenesis is implemented by allowing the jurons adjust their connection weight spontaneously, one by one, each juron in its own event. These ontogenetic events are opportunities to store the state that exists in $P$ when the event is triggered. To apply the storage algorithm given above, augment each target in the training list by $Jq$ in which $Jq_q=1$, and all the other elements are zero. Then determine $\Delta G$ by the storage rules given in Table 1 for the elements of $\Delta B$, $\Delta C$ and $\Delta D$ off the main diagonal.

**Example:** Store the training list $\{Rq\} = \{[1\ 0\ 0]^T, [1\ 0\ 1]^T\}$ via the autonomous ontogenetic storage algorithm. The inner products are $\{Wq\} = \{1, 2\}$. The targets are placed in $P$ and augmented by $\{Jq\} = \{[1\ 0], [0\ 1]\}$. The augmented training list is $\{\underline{Rq}\} = \{[1\ 0\ 0,\ 1\ 0]^T, [1\ 0\ 1,\ 0\ 1]^T\}$. Then $G_{1,3} = C_{1,3} = 0 -1 +1$. $G_{1,4} = B_{1,1} = 0 +1 +0$. $G_{5,1} = D_{2,1} = 0 +0 + ½$. $G_{5,4} = A_{2,1} = -$
$1 +0 +0$. $G_{i,i} = -c +0 +0$.

$$G = \begin{bmatrix} -c & & & & \\ & -c & & & \\ & & -c & & \\ & & & -c & -1 \\ & & & -1 & -c \end{bmatrix} + \begin{bmatrix} 0 & -1 & -1 & 1 & 0 \\ 0 & 0 & 0 & -1 & 0 \\ 0 & 0 & 0 & -1 & 0 \\ 1 & -\frac{1}{2} & -\frac{1}{2} & & \\ 0 & 0 & 0 & & \end{bmatrix} + \begin{bmatrix} 0 & -1 & 1 & 0 & 2 \\ 0 & 0 & 0 & 0 & -2 \\ 1 & -1 & 0 & 0 & 2 \\ 0 & 0 & 0 & & \\ \frac{1}{2} & -1 & \frac{1}{2} & & \end{bmatrix}$$

where $c = 1/24$. The elements on the main diagonal and in submatrix $A$ have constant values given in $G^0$, the first term. The second and third terms come from application of the storage rules.

## RESULTS AND CONCLUSION

The autonomous hybrid ontogenetic network was tested by the same methods used for the augmented Hopfield network. Ten thousand training sets were generated for each combination of $n$ and $m$. The results are a great improvement over the Hopfield network. A graphic comparison is made in Figure 2 and Figure 3 between the benchmark and the new results.

Figure 2 compares the mean scores of correctly stored targets. In Figure 2(a) for $n = 4$, the benchmark performance drops off quickly to marginally acceptable performance for storage ratios $m/n = 4/4 = 1$ and larger. The ontogenetic hybrid network performs very well for the entire experiment achieving storage ratios as high as $m/n = 12/4 = 3$. In Figure 2(b) for $n = 8$, the benchmark performance drops of quickly to marginally unacceptable performance for storage ratios $m/n = 4/8 = 0.50$. The ontogenetic hybrid network performs well for the entire experiment with storage ratios as high as $m/n = 16/8 = 2$.

Figure 3 compares the mean I/O unity checks. In Figure 3(a) for $n = 4$, the benchmark I/O unity check deviates significantly from unity to values from 2.74 to 5.28, indicating increasing failure for storage ratios $m/n = 4/4 = 1$ and larger. With I/O unity checks in the range [0.67 to 1.35], there is no indication of failure in the ontogenetic hybrid network for storage ratios as high as $m/n = 12/4 = 3$. Figure 3(b) for $n = 8$, the benchmark I/O unity check deviates significantly from unity to values from 1.81 to 2.36, indicating increasing failure for storage ratios $m/n = 8/8 = 1$ and larger. With I/O unity checks in the range [0.76 to 1.62], there is at most only marginal indication of trouble in the ontogenetic hybrid network for storage ratios as high as $m/n = 16/8 = 2$.

The results of testing the new autonomous hybrid ontogenetic networks were compared to benchmarks established for Hopfield networks. The new ontogenetic network improves performance by an order of magnitude over the benchmarks. Storage densities $m/n$ were increased from about 0.2 for Hopfield's network to about 0.5 for the new augmented Hopfield networks, and then as high as 3 for the new

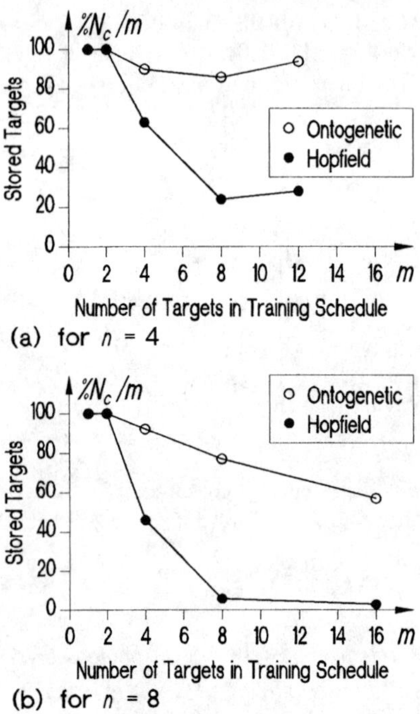


Figure 2: Comparison of stored targets in ontogenetic and Hopfield networks with nuclei of 4 protons (a), and 8 protons (b).

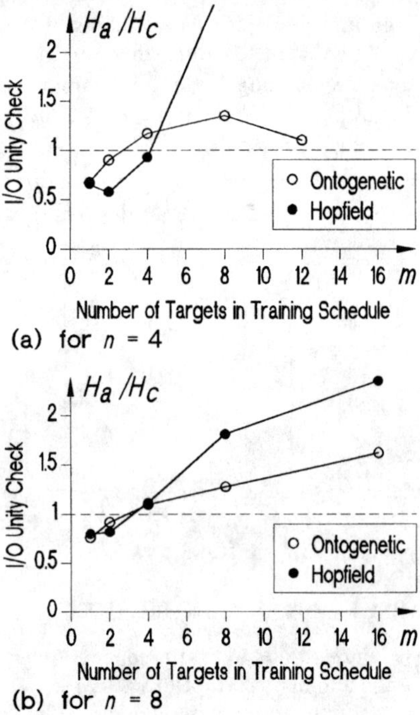


Figure 3: Comparison of mean value of $H_a/H_c$ with unity in ontogenetic and Hopfield networks with nuclei of 4 protons (a), and 8 protons (b).

autonomous hybrid ontogenetic network. The parameters of the ontogenetic networks were chosen in preliminary trials. There is no evidence to make us think that we have yet found the best parameters for these networks. Thus, we believe that even better performance than reported here is possible.

## REFERENCES

Abu-Mostafa, Y.S. and J. St.Jacques (1985), Information Capacity of the Hopfield Model, *IEEE Transactions on Information Theory* IT-31 n4, pp461-464.

Cohen, M.H. and S. Grossberg (1983). Absolute Stability of Global Pattern Formation and Parallel Memory Storage by Competitive Neural Networks, *IEEE Transactions on System, Man, and Cybernetics* SMC-13, pp813-826.

Haykin, S. (1994). *Neural Networks*, Macmillan Publishing Co., Englewood Cliffs, NJ, pp 303-307.

Hebb, D.O. (1949). *The Organization of Behavior*, John Wiley and Sons, New York, NY.

Hopfield, J.J. (Apr 1982). Neural Networks and Physical Systems with Emergent Collective Computational Abilities, *Proc. National Academy of Science, USA*, pp 2554-2558.

McCulloch, W.S. and W. Pitts (1943). A Logical Calculus of the Ideas Imminent in Nervous Activity, *Bulletin of Mathematical Biophysics* 5, pp 115-133.

# A BINARY THREE LAYERED NEURAL NETWORK WITH SWITCHED ERROR PERTURBATION AND REITERATIVE LEARNING UTILIZING THE GENERALIZATION PROPERTY

**YOHTARO YATSUZUKA AND MASARU ENOMOTO**

***Kokusai Denshin Denwa (KDD) R & D Labs.***
***2-1-15 Ohara Kamifukuoka-shi, Saitama 356 JAPAN***
yohtaro@kanri.wlb.lab.kdd.co.jp enomoto@kanri.wlb.lab.kdd.co.jp

**ABSTRACT:**
We propose a binary 3-layered neural network (BNN) using a novel learning technique in which rapid convergence can be reliably achieved by a perturbation of the unit output errors in an output layer in polarity and magnitude, and a very high generalization can also be achieved through successive training by adjusting the amount of the error perturbation after convergence.

Simulations were conducted for the proposed BNN in which 75 input units, 140 hidden units and 7 output units having a distributed representation were prepared. At 300 training cycles for 128 core input patterns under complete convergence in binary space, the rates of correct binary outputs of 98.2% for 9598 test inputs with a Hamming distance of 1 (HD1), and 91.6% for 355,040 test inputs with a Hamming distance of 2 (HD2), were achieved respectively.

A novel reiterative learning procedure by using the BNN with incremental inputs for training and test, 100 % correct binary outputs can be provided for 364,766 input samples comprising core patterns, HD1 and HD2 by using only 3% of all the input samples as training inputs.

## INTRODUCTION

Artificial multi-layered neural networks have recently been applied to static fault diagnosis problems (Chen,1994, Yamamoto,1990). The attractive features of neural networks are acquirement of expert knowledge automatically from historical data, easy maintenance and updating to solve real world problems. In applications to real time alarm correlation in telecommunication networks (Jakobson,1993), complete convergence in binary space with few training cycles within a given range and a high generalization are required for binary multi-layered neural networks.

In binary multi-layer neural networks with a back-propagation algorithm, achieving of quick and stable convergence in binary space and high generalization without overfitting and overtraining are major issues for a wide range of real applications (Reed,1993,Baum,1989,Rumelhart,1986).

A 3-layered neural network with an error perturbation learning (Yatsuzuka,1995) was proposed for avoidance of tenacious local minima, resulting in quicker and more stable convergence in binary space with insensitivity to sets of initial weights and higher generalization performance than conventional networks with a backpropagation algorithm. However, a slight degradation in generalization performance was observed.

In training multi-layered feed-forward network for the large scale of available input samples, a reiterative learning method with active data selection for fitting a desired function was proposed (Plutowski,1993). This approach provides several advantages by selecting the necessary training inputs instead of using all of the training input samples. In particular, a large amount of saving in computation is expected by preparing an appropriate selection criterion.

In this paper, we propose an improved learning method using a switched error perturbation for a binary 3-layered neural network to achieve a significantly high generalization under the condition of complete convergence in binary space, and also propose a novel reiterative learning procedure with incremental inputs for training and test, in which the property of generalization of the network with the switched error perturbation is utilized for achieving complete convergence in binary space for a huge amount of available input samples through a data base system.

In section 2, a basic configuration of the switched error perturbation training process are described, and in section 3, the reiterative learning procedure with incremental inputs for training and test is then described. Section 4 presents simulation results and discusses effects of the switched error perturbation and the performance of the reiterative learning procedure. Section 5 finally gives conclusions.

## CONFIGURATION OF THE ERROR PERTURBATION PROCESS

The basic configuration of the network with the error perturbation learning method is given in Fig.1. In the 3-layered neural network in Fig.1, it is assumed that the number of layers is L(=3), and the numbers of units are M in the output layer, H in the hidden layer and N in the input layer, respectively. In a training process, sets of training inputs and teacher patterns are prepared, and the number of training inputs is K. Weights in the multi-layered feed-forward network are basically evolved by the back-propagation algorithm (Rumelhart,1986).

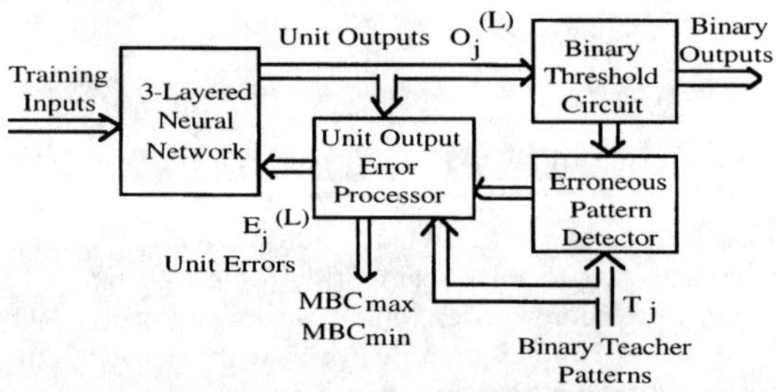

Figure 1: Configuration of a binary 3-layered neural network with the error perturbation training process.

**Error perturbation method** (Yatsuzuka,1995)
The j-th unit output error of the output layer, $Y_j^{(L)}(k)$, is defined by:

$$Y_j^{(L)}(k) = T_j(k) - O_j^{(L)}(k), \tag{1}$$

where $T_j(k)$ is a binary teacher pattern for the j-th output unit of the output layer, and $O_j^{(L)}(k)$ is the j-th unit output of the output layer for the k-th training input pattern.

In an erroneous pattern detector in Fig.1, the binary output is compared with the corresponding binary teacher pattern to determine whether bits of the binary output are correct or not[6]. In the output error processor in Fig.1, the j-th unit error of the output layer L, $E_j^{(L)}(k)$, is defined in the error perturbation method, as followings.

For the correct binary output at the j-th output unit, $E_j^{(L)}(k)$ is then given by:

$$E_j^{(L)}(k) = Y_j^{(L)}(k) - Sgn(Y_j^{(L)}(k)) \cdot D_1, \tag{2}$$

where $1 \le j \le M$, Sgn (X) is 1 for $X \ge 0$ and -1 for $X < 0$, $D_1$ is an error perturbation parameter for control of polarity and/or magnitude of the unit error of $E_j^{(L)}(k)$, and $0 \le D_1 \le 0.5$. Equation (2) indicates that when the absolute value of $Y_j^{(L)}(k)$ is equal to or less than $D_1$, the polarity of $E_j^{(L)}(k)$ is inverted from that of $Y_j^{(L)}(k)$. This means that when the unit output error of $Y_j^{(L)}(k)$ supplying the correct binary output becomes very small, the direction of updating the corresponding weight connected to the output layer is constraintly inverted. On the other hand, when the absolute value of $Y_j^{(L)}(k)$ is larger than $D_1$, the magnitude of $E_j^{(L)}(k)$ is reduced by $D_1$. Both the polarity and magnitude or only the magnitude of the unit error providing the correct binary output are controlled to avoid tenacious local minima, resulting in very quick and stable convergence in binary space.

For the wrong binary output,

$$E_j^{(L)}(k) = Y_j^{(L)}(k) - Sgn(Y_j^{(L)}(k)) \cdot D_2, \tag{3}$$

where $D_2$ is an another error perturbation parameter for control of magnitude of the unit error $E_j^{(L)}(k)$, and $0 \le D_2 \le 0.5$. The value of $D_2$ accelerates the speed of the emancipation from not only tenacious local minima but also global minima in binary space under the condition of $D_1 \neq 0.0$.

If the network is trapped into a state having tenacious local minima due to the insufficient error perturbation, the state can be easily detected (Yatsuzuka,1995). Convergence in binary space is also easily detected by the coincidence between the binary output and the corresponding binary teacher pattern.

The error signal of $\phi_j^{(L)}(k)$ for the j-th unit of the output layer (L=3) at the k-th training input pattern is given by:

$$\phi_j^{(L)}(k) = E_j^{(L)}(k) \cdot O_j^{(L)}(k) \cdot (1 - O_j^{(L)}(k)), \tag{4}$$

for a sigmoid transfer function.

**Maximum and minimum margins of bit convergence**

A margin which provides a correct binary output at the j-th output unit for the k-th training input pattern can be obtained from $|Y_j^{(L)}(k)|$ (Yatsuzuka,1995), and maximum and minimum values, $MBC_{max}$ and $MBC_{min}$, of the Margin of Bit Convergence among training input patterns and the unit outputs are defined by:

$$MBC_{max} = 0.5 - Min\,[\,|Y_j^{(L)}(k)|\,]_{j=1,M;\,k=1,K}, \tag{5}$$

and

$$MBC_{min} = 0.5 - Max[\,|Y_j^{(L)}(k)|\,]_{j=1,M;\,k=1,K}, \tag{6}$$

where Max[ $|Y_j^{(L)}(k)|$ ] and Min[ $|Y_j^{(L)}(k)|$ ] denote the detection of maximum and minimum values of magnitude of the unit output errors among output units for the correct binary output at every training cycle, respectively. $MBC_{max}$ and $MBC_{min}$ become close to $0.5-D_1$ according to the training cycles after complete binary convergence.

### Switching the error perturbation

The reduction in the margin to $0.5-D_1$ from 0.5 in the error perturbation induced a slight degradation of generalization performance, even though higher generalization performance than conventional binary 3-layered networks can be achieved due to convergence in binary space. This means that generalization can be further improved by continuous learning with $D_1$ set to zero, and maintaining convergence in binary space. Accordingly, after convergence, $D_1$ is then switched to a smaller value to absorb an abrupt variation in binary outputs and again achieve convergence without trapping of tenacious local minima, and finally is set to 0 to obtain the highest generalization performance under the condition of complete binary convergence (Enomoto,1996). Adjustment of $D_1$ is very effective in obtaining extremely high generalization performance by converging $MBC_{min}$ close to 0.5.

A simple switching procedure was adopted, that is, when the continuous convergence in binary space during 30 training cycles was obtained, the value of $D_1$ was switched to a smaller value used during 40 training cycles, and then $D_1$ is finally set to zero.

## REITERATIVE LEARNING WITH INCREMENTAL TRAINING AND TEST INPUTS

When a huge amount of input samples having clear relationships among the input patterns and the corresponding teacher patterns had been prepared in a data-base system through the acquisition of knowledge, a cross-validation method (Weiss, 1991) partitioning the input samples into sets of training and test inputs is usually applied to evaluate the convergence and generalization performance. A large amount of tedious calculations are generally required to achieve complete convergence in binary space and a reasonable generalization performance for such a huge number of training inputs, even if capture of tenacious local minima is avoided in the training process. It is also difficult to prepare an optimum network configuration and parameters that can promise successful convergence in advance.

A novel reiterative learning procedure (Enomoto,1996) is proposed for binary three-layered networks in which the switched error perturbation method is implemented to provide very quick and certain convergence in binary space and a high generalization for a huge number of input samples. The input samples are grouped into a set of core input samples and several hierarchical sets some distance away from the core input samples. At lease one core input pattern should be assigned to each teacher pattern with the distributed representation as the training input pattern. If there are no-decision states in outputs to which teacher patterns are not allocated, they are also allocated by dummy teach patterns corresponding to sparse input patterns which have a low information content in order to prevent the degradation of generalization performance due to overtraining or overfitting (Hassoun,1993).

The flow of the reiterative learning procedure with incremental inputs for training and test is given in Fig.2. In the initial conditions for reiterative learning, only the set of core input samples is assigned for the set of training inputs, and the set of input samples closest to the set of core input samples is selected as the set of test inputs.

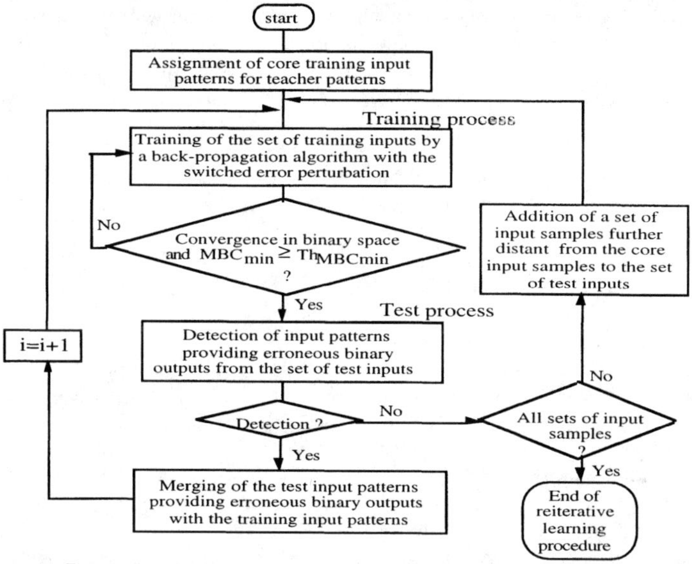


Figure 2: Re-iterative learning with incremental training inputs.

In the training process, convergence in binary space is reliably attained by the switched error perturbation method, forming the classification boundaries. To achieve reasonably high generalization performance at which the excessive increment of the training inputs can be avoided, the training process should be continued further until the $MBC_{min}$ exceeds the threshold, $Th_{MBCmin}$ after achieving convergence in binary space. The test process is then conducted for evaluating the generalization performance for the set of test inputs.

If there are test input patterns producing erroneous binary outputs, they are removed as the incremental training input patterns to the set of training inputs for the next re-iteration. If no error in the binary outputs is detected, the set of input samples further distinct from the set of core input samples is added to the set of test inputs to again conduct the reiterative learning procedure. The reiterative learning procedure with the incremental inputs for training and test is conducted until all sets of input samples provide completely correct binary outputs. It is very important that to successfully terminate the reiterative learning, the network must certainly converge in binary space, and also provide high generalization performance to reduce the number of training inputs.

The clusters of core input samples can basically form the classification boundaries around center positions at the early stage by using the error perturbation in reiterative learning, and only input samples precisely relating to the border regions are adaptively chosen from entire sets of input samples for the set of training inputs. Accordingly, the multi-layered neural network can be quickly converged with a small size set of training inputs, resulting in a large saving in calculations.

The reiterative learning procedure eventually can achieve complete convergence in binary space and 100% generalization performance for all residual input samples which are not used as training input patterns. Accordingly, the network can provide

100 % correct binary outputs for entirely specified input samples. This feature is very attractive for real applications.

Input samples, which were added into the set of training inputs due to the production of erroneous binary outputs, have more dominant features concerning the border regions for classification of input samples than other input samples remaining in the set of test inputs. The progress of the reiterative learning procedure can also asymptotically adjust the precise classification boundaries for without excess overfitting.

The achievement of complete convergence in binary space and relatively high generalization performance in the training process can effectively suppress the number of test input patterns relating to the uncertain border regions, resulting in avoidance of the excessive increment of training inputs.

## SIMULATIONS AND DISCUSSIONS

### Simulation model and conditions

N=75 bits and M=7 bits were set in all the simulations, and the distributed representation was used in the binary outputs. The initial weights were randomly chosen in the range of -0.5 to 0.5 under a uniform distribution. $D_2$ was set to 0, because $D_1$ is more dominant (Yatsuzuka, 1995).

A maximum unit output error for the set of training inputs is checked only at the first training cycle without the error perturbation since the reiterative learning procedure is conducted once, whenever a new set of test inputs is set.

If the value is below 0.8, it is almost unnecessary to utilize the error perturbation, therefore resulting in a setting of $D_1=0$. $Th_{MBCmin}$ was set to 0.42. The values of a learning factor $\alpha$ and a momentum factor $\beta$ are set under the condition of $\alpha + \beta \cong 1$ which gives a roughly constant gain for the low-pass filtering of $\Delta W_{ji}$.

Two hierarchical sets in distance, HD1 and HD2, of input samples having a Hamming distance of 1 and 2 from 128 core input samples are prepared, respectively. The sizes of the sets are 9,598 for HD1, and 355,040 for HD2.

### Effects of the switched error perturbation on convergence

Figure 3 shows the number of erroneous binary outputs in training only 128 core input samples in terms of training cycles for the switched values of $D_1$ from 0.3 to 0.0 through a buffer area having 0.1 under the conditions of H=140 hidden units, $\alpha$ of 0.1, and $\beta$ of 0.9. The number of erroneous binary outputs decreases smoothly without trapping of tenacious local minima for $D_1$=0.3, and convergence in binary space can be stably achieved at a small number of training cycles around 143.

In Fig.4, $MBC_{min}$ having a value very close to 0 implies a very small margin of unit outputs for marginally correct binary outputs due to transitions among local minima before convergence. $MBC_{min}$ then gradually increases toward 0.5-$D_1$, improving the margin along with the increase in training cycles after convergence in binary space, whereas $MBC_{max}$ also gradually decreases toward 0.5-$D_1$, decreasing the margin of unit outputs for correct binary outputs.

When $D_1$ was switched from 0.3 to 0.1, a number of erroneous binary outputs were abruptly produced during the transient region. However stable and rapid complete convergence was again achieved due to the weights slightly corrupted from optimum values, and $MBC_{min}$ started to increase again quickly along with the increase in training cycles. In this case, the switching of $D_1$ from 0.1 to 0.0 went smoothly without generating erroneous binary outputs. $MBC_{min}$ then increased rapidly to a value close to 0.5, maintaining convergence in binary space.

It was clearly observed that switching $D_1$ from 0.3 to 0.0 throughout the transient region in the error perturbation can effectively expand the maximum and minimum margins of the unit outputs for correct binary outputs toward 0.5.

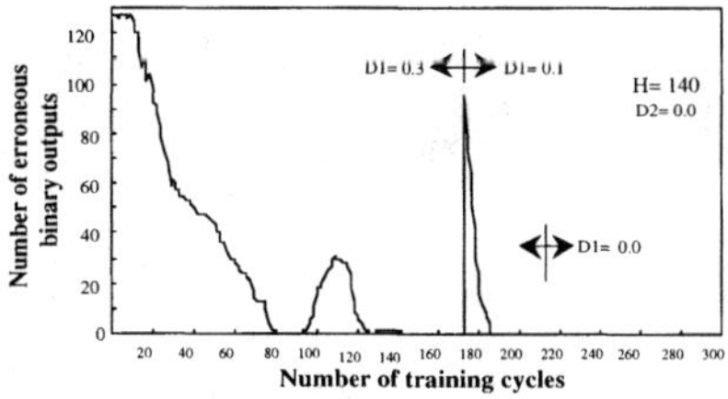


Figure 3: Number of erroneous binary outputs in terms of training cycles for 140 hidden units.

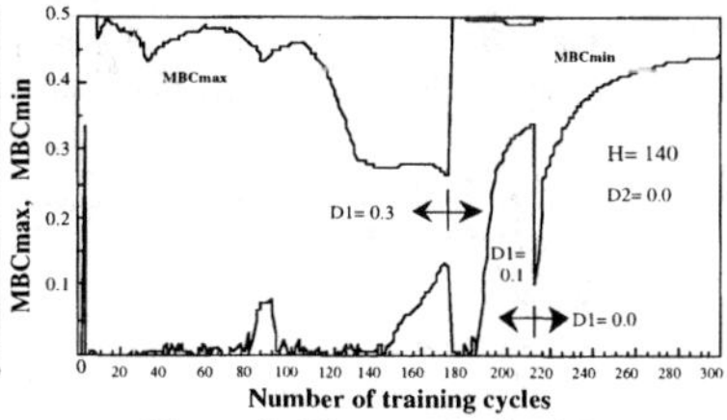


Figure 4: Maxmum and minimum margins of unit outputs providing correct binary outputs.

**Effects of the switched error perturbation on generalization**

Generalization performance for sets of test inputs having HD1 and HD2 was evaluated through simulations using the set of training inputs having only 128 core input patterns. Figure 5 shows the rates of correct binary outputs for the set of HD1 test inputs in terms of the number of hidden units for $D_1$ switched from 0.3 to 0.0 through 0.1.

It was observed in error perturbation with the fixed $D_1$ of 0.3 that even though convergence is achieved for smaller numbers of hidden units in the range of 20 and 30, the rates gradually deteriorate with an increase in the number of training cycles, indicating a tendency to overtraining. This tendency is induced, depending on the decrease in margin of $MBC_{max}$ close to 0.5-$D_1$ from 0.5. On the other hand, after switching $D_1$ to 0.0, generalization performance reaches a plateau in the switched error perturbation and is higher than that of the fixed error perturbation. For a large number of hidden units such as 100 and 140, higher and flatter generalization performance is also achieved without overtraining even with 1,000 training cycles, maintaining complete binary convergence. In conventional neural networks without the error perturbation overfitting was observed particularly due to non-convergence and the small margins of correct binary outputs. These results mean that the error perturbation can easily form the classification boundaries close to the center positions among clusters arranged by teacher patterns.

In Fig.6, the rates of correct outputs for the set of test inputs of HD1 versus the number of hidden units are also shown at 300 training cycles for the switched and fixed error perturbations. The network with 140 hidden units for $D_1$ switched from 0.3 to 0.0 through 0.1 can achieve a rate higher than 98% without overfitting. This is because the large margins of $MBC_{max}$ and $MBC_{min}$ are easily obtained in learning after binary convergence. In the network with more than 80 hidden units for a fixed $D_1$ of 0.3, the rate tends toward a slight saturation due to the smaller margin of output units for correct binary outputs, even though the network converges without trapping of tenacious local minima. In particular, in the conventional network having a number of hidden units larger than 60 without error perturbation learning, the rate gradually degrades, indicating overfitting caused by the inclined classification boundaries formed by tenacious local minima.

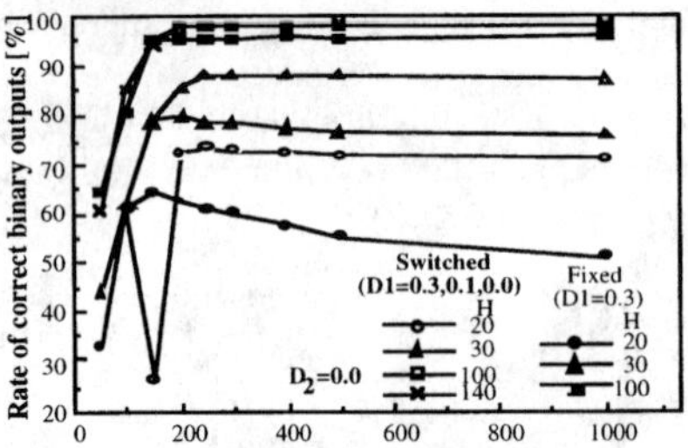


Figure 5: Generalization ability for a set of HD1 test inputs (9,598) with a Hamming distance of 1 from 128 core training input.

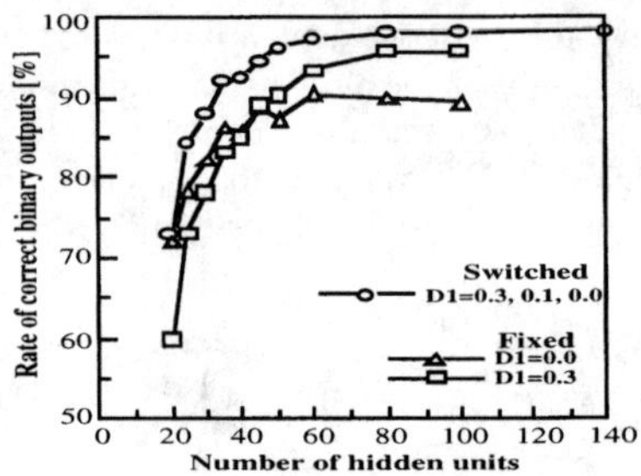


Figure 6: Generalization performance in terms of the number of hidden units at 300 training cycles for a set of HD1 test inputs

## Performance of reiterative learning with incremental inputs for training and test

Table 1 lists simulation results of the reiterative learning for the set of HD1 input samples close to the set of core input samples. Since the second reiteration observed in Table 1, only minute updating of weights is required to finely adjust the border ranges among adjacent binary outputs, and it is not necessary to use the error perturbation. Because the network has already captured global minima, therefore indicating a smaller value of maximum unit output error than 0.8. This means that tenacious local minima may not be captured along with the increase in the number of reiterations. A small value of $\alpha$ also has to be set to finely update weights, avoiding skipping over global minima, especially when training the network with a large number of hidden units for large numbers of training.

Figure 7 shows the generalization performance of the HD1 input samples remaining in the set of HD1 test inputs and the independent set of HD2 input samples in terms of the number of reiterations. It was observed that use of HD1 test inputs producing erroneous binary outputs as the incremental training inputs in the next reiterative learning gradually forms accurate classification boundaries among the binary outputs along with an increase in reiterations.

The set of training inputs had 128 core and 362 HD1 input samples during 13 reiterations, indicating only 5.03% for all of 9726 core and HD1 input samples. At this stage, the network provides a high generalization with 100% for the HD1 input samples remaining in the set of HD1 test inputs, and also provides 97.6% for HD2 input samples which are not used in the set of test inputs in the reiterative learning procedure.

A direct learning with 9726 training inputs including core and entire HD1 input samples were also conducted, and the training cycles exceeding $TH_{MBCmin}$ became 582. The reduction ratio of the total training input pattern manipulations in reiterative learning relative to that of direct learning becomes about 61.5, indicating a drastic reduction in computations. This means that the reiterative learning approach is very attractive and effective for training multi-layered neural networks on a large scale with a huge amount of input samples.

Whereas the generalization for HD2 input samples had a very high value of 99.89% in the direct learning, the generalization for HD2 input samples degraded only 2.3% from that of the direct learning in the reiterative learning procedure. This result means that an unnecessarily abrupt increment of the number of training inputs can be avoided, and this is very effective when the set of HD2 input samples is added into the set of test inputs for further progress of the reiterative learning with increments for not only training inputs but also test inputs.

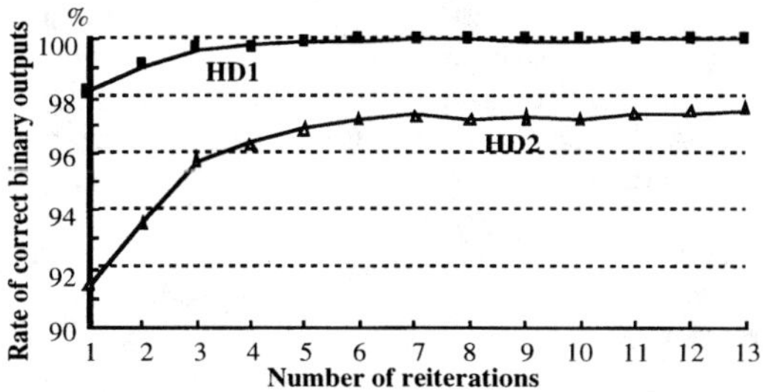


Figure 7: Generalization ability in reiterative learning for the HD1 input samples remaining in the set of HD1 test inputs and for the independent set of HD2 input samples in terms of the number of reiterations.

TABLE 1: SIMULATION RESULTS OF REITERATIVE LEARNING WITH THE INCREMENTAL INPUTS FOR TRAINING AND TEST. ( The first test increment: Tinc1, The 2nd test increment: Tinc2)

| No. of reitera -tions | Training cycles for convergence | | Training cycles ex-ceeding Th | | No. of incre-mental train-ing inputs | | No. of binary output errors per test | | Error perturbation $D_1$ | | Parameters of B.P. α | | Parameters of B.P. β | |
|---|---|---|---|---|---|---|---|---|---|---|---|---|---|---|
| | Tinc1 | Tinc2 | Tinc1 | Tinc2 | Tinc1 | Tinc2 | Tinc1 | Tinc2 | Tinc1 | Tinc2 | Tinc1 | Tinc2 | Tinc1 | Tinc2 |
| 1st | 1st:143<br>2nd:187 | 1st:403<br>3rd:504 | 262 | 596 | 128 | 9011 | 175 | 185 1 | switched (0.3,0.1,0.0) | switched (0.3,0.1,0.0) | 0.1 | 0.05 | 0.9 | 0.94 |
| 2nd | 9 | 1st: 90<br>3rd:170 | 42 | 239 | 303 | 10862 | 90 | 65 | fixed(0.0) | switched (0.3,0.1,0.0) | 0.1 | 0.05 | 0.9 | 0.94 |
| 3rd | 5 | | 22 | 206 | 393 | 10927 | 36 | 28 | fixed (0.0) | switched (0.3,0.1,0.0) | 0.05 | 0.03 | 0.95 | 0.96 |
| 4th | 8 | | 11 | 30 | 429 | 10955 | 23 | 5 | fixed (0.0) | fixed (0.0) | 0.005 | 0.01 | 0.995 | 0.99 |
| 5th | 2 | | 8 | 11 | 452 | 10960 | 12 | 0 | fixed (0.0) | fixed (0.0) | 0.005 | 0.005 | 0.995 | 0.995 |
| 6th | 4 | | 26 | | 464 | | 9 | | fixed (0.0) | | 0.005 | 0.995 | | |
| 7th | 2 | | 13 | | 473 | | 2 | | fixed (0.0) | | 0.005 | 0.995 | | |
| 8th | 2 | | 3 | | 475 | | 2 | | fixed (0.0) | | 0.005 | 0.995 | | |
| 9th | 3 | | 5 | | 477 | | 4 | | fixed (0.0) | | 0.005 | 0.995 | | |
| 10th | 2 | | 3 | | 481 | | 5 | | fixed (0.0) | | 0.005 | 0.995 | | |
| 11th | 2 | | 4 | | 486 | | 3 | | fixed (0.0) | | 0.005 | 0.995 | | |
| 12th | 2 | | 4 | | 489 | | 1 | | fixed (0.0) | | 0.005 | 0.995 | | |
| 13th | 2 | | 3 | | 490 | | 0 | | fixed (0.0) | | 0.005 | 0.995 | | |

The number of the core training inputs=128.
For the first test increment, the initial number of HD1training inputs = 0, the final number of HD1 training inputs = 362.
For the second test increment, the initial number of HD2 training inputs = 8,521, the final number of HD2 training inputs = 10,467,
For the second test increment, the initial number of HD1 training inputs = 362, the final number of HD1 training inputs = 365.

Table 1 also shows the simulation results of reiterative learning for incrementing the HD2 input samples into the set of test inputs, consequently following reiterative learning for the set of HD1 input samples as given in Table 1. The correct binary outputs for all the 364,766 input samples having core, HD1 and HD2 were achieved with an additional 5 reiterations. The progress of the reiterations with the error perturbation for incremental inputs for training and test is smoothly able to form classification boundaries among huge clusters arranged by teacher patterns.

Reiterative learning is also effective in achieving very quick and stable convergence with 100% correct binary outputs for entirely specified input samples and giving reasonably high generalization performance under the condition of a feasible amount of computations. These simulation results also indicate that the gradual increment of the number of test input samples from the set of input samples in the vicinity of core input samples to the distant set might be very effective in obtaining quick convergence with the reiterative learning procedure for a huge number of input samples.

## CONCLUSIONS

A switched error perturbation learning with a back-propagation algorithm in a binary 3-layered neural network and a reiterative learning procedure with incremental inputs for training and test were proposed. It was concluded through computer simulations that switched error perturbation is very effective in providing rapid and stable convergence in binary space and achieving very high generalization.

It was also found that the reiterative learning procedure with switched error perturbation can achieve high speed convergence with a very small size of training inputs and provide 100% correct binary outputs for a huge number of specified input samples by utilizing high generalization performance, while drastically reducing the number of required computations.

## REFERENCES


Reed R., (1993). Pruning Algorithms, IEEE Trans Neural Networks, vol.4, no.5, Sept, 740-747.

Baum E. and Haussler D., (1989). What Size Net Gives Valid Generalization?, Neural Computation,
vol.1, 151-160.

Chen H., Rinde P.B., and et. al, (1994). Expert Prediction, Symbolic Learning, and Neural Networks, IEEE EXPERT, vol.4, no.5, Dec, 21-27.

Jakobson G. and Weissman M.D., (1993). Alarm Correlation, IEEE Network, Nov, 52-59.

Yamamoto Y. and Venkatasubramanian V., (1990). Integrated approach using neural networks for fault detection and diagnosis, Proc. International Joint Conference on Neural Networks, 317-326.

Yatsuzuka Y., (1995). An Error Perturbation for Learning and Detection of Local Minima in Binary 3-layered Neural Networks, Proc. of ICNN'95, vol.1, Nov. 63 - 68.

Plutowski M. and White H., (1993). Selecting Concise Training Sets from Clean Data, IEEE Trans. on Neural Network, Vol.4, No.2, March, 305-318.

Rumelhart D.E., McClelland J.L., et al., (1986). Parallel Distributed Processing, MIT Press, Cambridge.

Enomoto M. and Yatsuzuka Y., (1996). Improvement of Generalization Performance of Binary 3-layered Neural Networks by Switched Error Perturbations, National Convention Record of IEICE. (in Japanese), Japan, No.D-2, March, D21.

Weiss S.M. and Kulikowski C.A., (1991). Computer Systems that learn, Morgan Kaufmann Publishers, Inc..

Enomoto M. and Yatsuzuka Y., (1996). Reiterative Learning with Incremental Training Inputs by Using Generalization Performance of Binary 3-layered Neural Networks with Switched Error Perturbations, National Convention Record of IEICE. (in Japanese), Sept. D-2.

Hassoun M.H., (1993). Dynamic Associative Neural Memories, Associative Neural Memories Theory and Implementation, Oxford University Press, 3-27.

# MODULAR LEARNING IN A MULTIAGENT ENVIRONMENT

**Jari VAARIO**
Nara Women's University
Department of Information and Computer Sciences
Kita-Uoya, Nishi-machi, Nara 630, JAPAN
**Kanji UEDA**
Kobe University, Department of Mechanical Engineering
Rokkodai Nada, Kobe 657, JAPAN

***ABSTRACT:***
In this paper modular learning in a multiagent environment is explored by a computer simulation. The aim of the simulation is to let each agent to learn individually its basic behaviors, and then collective behaviors through communication. The experimental task consists of four agents that try to enclose a common target from four directions simultaneously. Each agent is controlled by three learning modules implemented by a stochastic learning automaton (SLA), each taught individually by a simple reinforcement principle. The agents learn first to follow the target and then to attack. The switch from following behavior to attack behavior should occur simultaneously with the other agents. The timing of this is learned based on information exchanged between the agents. The results show that the defined task with some simplifications could be learned using the modular learning approach.

## INTRODUCTION

In multiagent systems it is necessary to combine the individual and group learning [Tan (1993)]. In many applications the situation learning is necessary. For example, in traffic conditions there are several situations where one must learn to select one's own action based on what actions the others select. It is impossible to learn all these situations beforehand and by not actually taking part on the situations.

This same phenomenon exists also in the biological world. Many predators, for example wolves, hunt collectively adjusting their behavior according to the prey and other predators. This is also the example taken in this paper. We explore a multiagent system behavior in a situation where several predators try to capture the prey by attacking it from four directions simultaneously. An engineering task could consists of robots that should cooperate in order, for example, to push a large box collectively towards a target area.

This paper reports some preliminary results of simulating multiagent systems with a communication capability. It will be shown that the problem is extremely complex, because of several combinations between learning modules and communication that can arise. The selected approach is simple. The attention is on the learning of individual behaviors and communication to synchronize the behaviors. The same problem from the evolutionary point of view has been analyzed in [Grefenstette (1992)]. However, the novelty of the selected approach is to divide the task into subtasks with each controlled by a control module taught for that particular subtask, and then to have another module to learn when to switch each subtask on/off.

In the following we first describe the selected task and its complexity. This is followed by the description of the modular structure of control system, and by the results of the simulation. At the end of the paper the results are discussed with suggestions of future work.

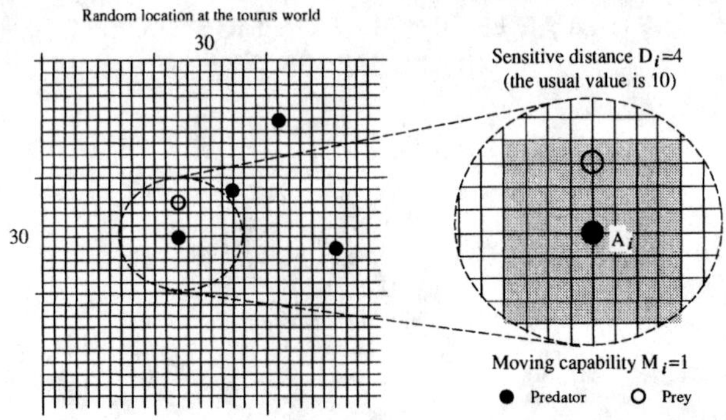


**Figure 1. The initial situation with a prey and four predators.**

## THE PROBLEM DEFINITION

The task is to capture a target by surrounding it by several agents simultaneously from four directions in the world of a torus (no edges). The environment consists of four predators and one prey ($Agents = \{A_{Predator_1}, A_{Predator_2}, A_{Predator_3}, A_{Predator_4}, A_{Prey}\}$). Each agent is defined by its position, moving capability and sensitive distance ($A_i(t) = \{pos_i(t), M_i, D_i\}$). At any given time the predators and prey position can not be same ($pos_i^{predator}(t) \neq pos^{prey}(t)$). The initial situation is illustrated in Figure 1.

### Task decomposition

The task to surround the prey from four directions simultaneously could be divided into the following subtasks. First, the predators should be capable of positioning themselves relatively to the prey so that they can simultaneously attack the prey (*positioning mode*, see Figure 2). This implies, that when a single predator has positioned itself, it must wait until all others have reached their positions with each at a different side of the prey. During this waiting time it must maintain its own relative position to the prey (*following mode*). When all predators have been positioned, they should attack simultaneously (*attacking mode*, see Figure 3). For the timing of the attack a communication mechanism is necessary (*communication mode*).

These are remotely related to the division made by Mataric [Mataric (1995)] in the context of general adaptive behavior. In our research a specific task has been given for the agents and we don't try to generalize the behaviors. Although it would be interesting to decompose the task automatically, or using a bottom-up approach (see, for example, [Vaario (1994); Vaario and Shimohara (1995)]), these are not, however, the scope of this paper.

### Simplifying conditions

A general solution for the above problem is very difficult. In the current simulation the following simplifications have been applied.

First, the simulator provides the information of the current region, *ie.*, whether the predator is at the north, east, south, or west side of the prey. In addition to this information each predator has its own region where to position, and the direction to the region from its current position. A general implementation where the predator decides itself the region to position by using the position information of other predators and the prey, or merely plain sensor information, was considered too complex at the first implementation stage.

Second, at the attack behavior the relative position to the prey is given directly, rather than using sensor information.

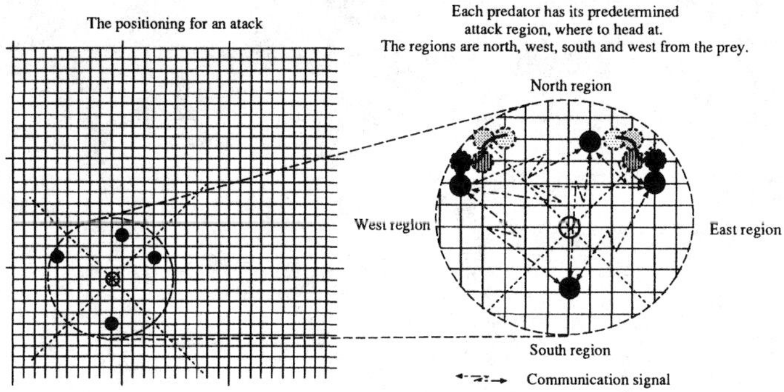


**Figure 2. Each predator will position itself relatively to the prey.**

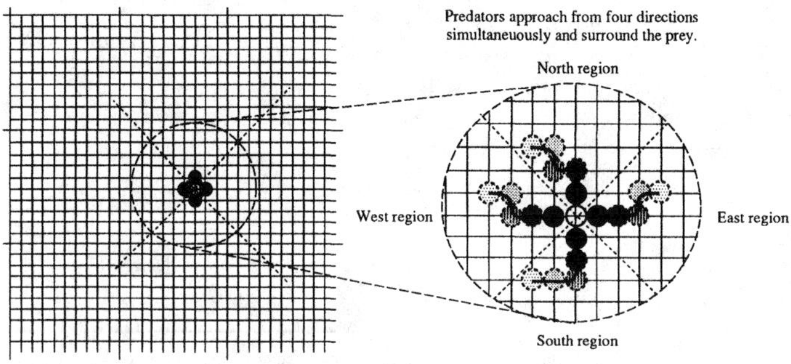


**Figure 3. Each predator approaches the prey from its predetermined direction.**

Third, an external evaluator is used. The used reinforcement learning should based on the local information received through the sensors, rather than global information.

Fourth simplification is related to the environment model. The world is restricted to a 30x30 grid, and the sensitive distance for 10 units. That means that at each time the predator is capable of seeing 1/9 of the whole area. In addition, there is only one prey (where from to select) and no obstacles in the environment.

Fifth simplification is that the prey is not capable of sensing its predators and thus is not capable of escaping. It just moves randomly one cell by time into the direction of not occupied cell.

**Communication schemas** There are, at least, two possible communication schemas: a simple signal, or a complex information exchange. In the following these two schemas are considered.

*The simple communication schema*

When the predator is at the correct location it starts to send a signal and simultaneously listen to the similar signals from the other predators. When it detects signals from all other predators, the mode is changed to the attack mode.

The signal transmission is controlled by the location of predator. If the predator loses the attack region, the signal transmission is stopped, and the mode is switched back to the following mode. Similarly, if the signals are not received from all other predators, the mode is switched back to the following mode.

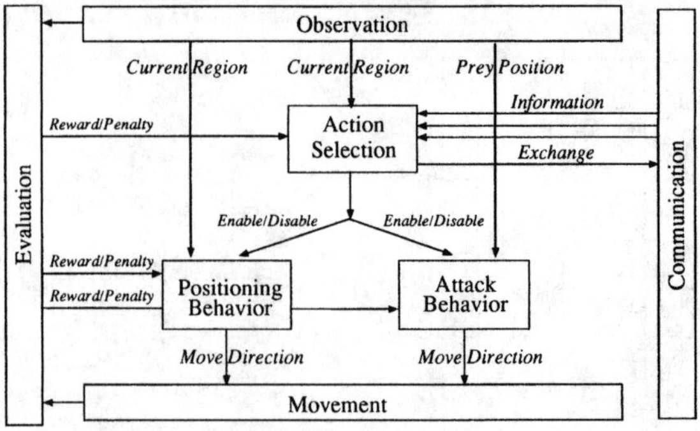


**Figure 4. An overview of the modules of a predator.**

*The complex information exchange schema*

Each agents exchanges the information of the prey position ($Inf_i(t) = \{rx_i^{prey}(t), ry_i^{prey}(t)\}$, $(|\ rx_i^{prey}(t)\ | \leq D_i, |\ ry_i^{prey}(t)\ | \leq D_i)$). The receiving partner then learns to make use of this information for deciding when to attack. This schema was thought to be more interesting, because it involves learning of the complex information and it is biologically more plausible.

## MODULAR STRUCTURE

The actual implementation of the predator model consists of three independent learning modules each modeled by a stochastic learning automaton (see Figure 4). The modules are *Action selection*, *Positioning behavior*, and *Attacking behavior*. The *Action selection* module takes care of the communication mode. The *Positioning behavior* module takes care of the positioning and following modes (it could have been more appropriate to divide these modes into separate modules). The *Attack behavior* module takes care of the attack mode.

The input for each module are as follows.

- *Action selection*: 'Current Region' ($Region = \{north, east, south, west, unknown\}$) and possible communications from all other predators ($Com_i = \{Inf_{i\oplus 1}, Inf_{i\oplus 2}, Inf_{i\oplus 3}\}$);
- *Positioning behavior*: 'Current Region' ($Region$); and
- *Attack behavior*: 'Prey Position' ($pos_{prey}(t)$).

The output from *Action selection* is to enable a particular behavior module ($\alpha^{action} = \{positioning, attack, comm\}$). The output from the behavior modules is the movement direction ($\alpha^{move} = \{forward, backward, left, right, no\_move\}$).

### Learning at modules

There are various approaches to implement learning at each module. Here the selected approach was a stochastic learning automaton (SLA) [Narendra and Thathacher (1989)].

**SLA principle** Based on the rewards and penalties determined by the evaluator, each SLA module is adjusted as follows.

The SLA is represented by the quintuplet $\{\Phi, \alpha, \beta, A, P\}$. $\Phi = \{\Phi[0], \Phi[1], \ldots, \Phi[s]\}$ is a set of states, $\alpha = \{\alpha[1], \alpha[2], \ldots, \alpha[r]\}$ is a set of actions, $\beta$ is a reinforcement $\{0, 1\}$, $A$ is a reinforcement schema, and $P$ is a set of probability vector ($P = \{P[1], P[2], \ldots, P[s]\}, P[i] = \{p[i,1], p[i,2], \ldots, p[i,k]\}$). At discrete time $t = n$, an action $\alpha(n) = \alpha[i]$ is selected statistically from the action set $\alpha$ according to the probability vector $p[i,j]$ of state $\Phi(n) = \Phi[j]$.

Thus each module has the following SLA defining them: *Action selection*: $SLA1 = \{\Phi = \{Region \cup \alpha^{action}(t-1)\}, \alpha = \alpha^{action}(t), \beta = \{0,1\}, A = A1, P\}$; *Positioning behavior*: $SLA2 = \{\Phi = \{Region\}, \alpha = \alpha^{move}(t), \beta = \{0,1\}, A = A2, P\}$; and *Attack behavior*: $SLA3 = \{\Phi = \{pos_{prey}\}, \alpha = \alpha^{move}(t), \beta = \{0,1\}, A = A3, P\}$.

**Reinforcement schema** Evaluator evaluates the action, and sends the reinforcement signal to SLA. The reward and penalty are determined for each module as follows

*Action selection (A1)*:

$$
\begin{aligned}
&if(distance(t) < distance(t+1))\ point{+}{=}\ 2\ else\ point{-}{=}\ 1\\
&if(distance(t+1) = 1)\ point{+}{=}\ 2\ else\ point{-}{=}\ 1\\
&if(dist\_target\_region(t) < dist\_target\_region(t+1))\ point{+}{=}\ 1\ else\ point{-}{=}\ 2\\
&if(region(t+1) = target\_region)\ point{+}{=}\ 2\ else\ point{-}{=}\ 1\\
&if(region(t+1) = unvisible)\ point{-}{=}\ 2\ else\ point{+}{=}\ 1\\
&if(point \geq 0)\ \beta(t+1) = 1\ else\ \beta(t+1) = 0
\end{aligned}
$$

*Positioning behavior (A2)*:

$$
\begin{aligned}
&if(dist\_target\_region(t) < dist\_target\_region(t+1))\ \beta(t+1) = 1\ else\ \beta(t+1) = 0\\
&if(region(t+1) = target\_region)\ \beta(t+1) = 1
\end{aligned}
$$

*Attack behavior (A3)*:

$$
\begin{aligned}
&if(distance(t) < distance(t+1))\ \beta(t+1) = 1\ else\ \beta(t+1) = 0\\
&if(distance(t+1) = 1)\ \beta(t+1) = 1
\end{aligned}
$$

**Learning** According to the reinforcement ($\beta$), the probability vector of current state is updated by the following the reinforcement schema.

$$
\begin{array}{ll}
if\beta(t) = 1\ (reward) & if\beta(n) = 0\ (penalty)\\
p[i,j](t+1) = (1-a)p[i,j](t), \quad \forall i \neq i^* & p[i,j](t+1) = b/(r-1) + (1-b)p[i,j](t), \ \forall i \neq i^*\\
p[i^*,j](t+1) = p[i^*,j](t) + a(1-p[i^*,j](t)) & p[i^*,j](t+1) = (1-b)p[i^*,j](t)
\end{array}
$$

where a and b are reward and penalty parameters ($0 < a < 1, 0 < b < 1$), and $i^*$ is the selected action. Thus, if selected action produced a reward, its propability is increased and the propabilities of other actions are decreased, and *vice versa*.

## SIMULATION RESULTS

The simulation was done by a simulator implemented in C++ with a X windows interface. A snapshot of the simulation is shown in Figure 5. The values for the simulations were reward coefficient $a = 0.1$; penalty coefficient $b = 0.02$; sensitivity distance $D_i = 10$ units; and moving capability $M_i = 1$ unit.

Three simulations were conducted, and the results are shown in Figure 5. The graph shows the number of execution steps required to capture the prey as a function of trials with cumulative learning.

**Case 1** Without any communication between the predators.

**Case 2** With the complex communication schema between the predators.

**Case 3** With the complex communication schema between the predators and, in addition, one predator capable to sense the whole environment.

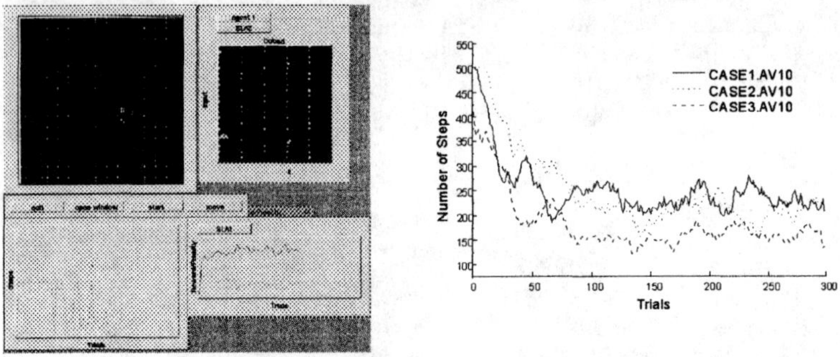


**Figure 5. A snapshot of the simulator and the summary of three simulations.**

**Discussion** The results show surprising little effect of communication. Some agents attacked too early, or not at all. This is, most likely, due to the fact that positioning and following mode were implemented by the same learning module. The position information of other agents should be exchanged only when they are in *following mode*, but not in *positioning mode*. Also the type information exchanged between agent should be reconsidered. The relative position of other agents might not be appropriate, but, for example, whether the other agents are ready for attack.

The reinforcement schema for *Action selection* is difficult to determine. It is assumed that a better schema could produce better results. This a subject of the future work.

## CONCLUSION

In this paper we divided the learning of a multiagent system into subtasks. Each subtask was taught independently, but still as an integrated part of the whole system. The selection of an appropriate subtask has been learned by another independent module receiving communication as its input.

Although the current results of communication based syncronizing of attack are debatable, the idea of learning communication based behavior control is interesting. An interesting experiment would be to learn the type of communication. The agents could have various communication methods that they try and those giving better results would be reinforced.

The simulation was conducted with simplified conditions. In the future work the simplifications are removed one by one. The most crucial simplification to be removed is to use real sensor information for the learning modules. This is planned to be implemented by utilizing Khepra simulator, and then verified by using real robots.

## REFERENCES

Grefenstette, J. J., 1992, "The Evolution of Strategies for Multiagent Environment," *Adaptive Behavior*, Vol. 1, pp. 65–90.

Mataric, M., 1995, "Designing and understanding adaptive group behavior," *Adaptive Behavior*, Vol. 4, pp. 51—80.

Narendra, K., Thathacher, M. A. L., 1989, *Learning Automata: an introduction* Prentice-Hill.

Tan, M., 1993, "Multi-Agent Reinforcement Learning: Idependent vs. Cooerative Agents," Proc. Proc. of 10th ICML, pp. 330.

Vaario, J., 1994, "Modeling Adaptative Self-Organization," In Brooks, R., Maes, P., editors, *Artificial Life IV*, pp. 313—318. The MIT Press.

Vaario, J., Shimohara, K., 1995, "On Formation of Structures," In Morán, F., Moreno, A., Merelo, J. J., Chacón, P., editors, *Advances in Artificial Life*, pp. 421–435. Springer.

# A GENERAL AUTO-ASSOCIATIVE MEMORY WITH SAMPLE LEARNING CAPABILITY

Hongchi Shi, Yunxin Zhao†, and Xinhua Zhuang

Department of Computer Engineering & Computer Science
University of Missouri-Columbia, Columbia, MO 65211

†Beckman Institute and Department of Electrical & Computer Engineering
University of Illinois at Urbana-Champaign, Urbana, IL 61801

*ABSTRACT:*
In this paper we design a general auto-associative memory with sample learning capability. We start by defining a new concept called supporting function to replace the concept of energy function. We then formulate the information recalling process as a dynamic system using the supporting function and derive the condition for convergence of an arbitrary state to a desired stable state. Finally, we devise a sample learning algorithm using Rosenblatt's perceptron rule. The effectiveness of the sample learning algorithm is demonstrated by some experiments in noisy handwritten digit recognition.

## INTRODUCTION

In the early 1980s, Hopfield proposed a biological model for memory, known as Hopfield auto-associative memory, which consists of a massively interconnected agglomerate of neurons (Hopfield, 1982). The instantaneous state of the system is defined by the collective firing status of each individual neuron. The computational capability of such an associative memory system is dictated by the interconnection matrix which is learned through a learning algorithm. The Hopfield memory uses the Hebbian learning algorithm, leading to a symmetric interconnection matrix, which fits in the energy function theory. Although the Hopfield memory has several attractive features (Sejnowski and Stanton, 1990), its performance is poor in terms of storage capacity, attraction, and spurious memories (McEliece, et. al., 1987; Cottrell, 1988; Dembo, 1989; Floreen and Orponen, 1989; Venkatesh and Psaltis, 1989, Bruck and Roychowdhury, 1990; Zhuang and Huang, 1994). To enhance the performance of the Hopfield associative memory, many alternative learning algorithms have been proposed (Kohonen and Ruohonen, 1973; Personnaz, et. al., 1986; McElicec, et. al., 1987; Cottrell, 1988; Dembo, 1989; Floreen and Orponen, 1989; Venkatesh and Psaltis, 1989; Hassoun and Youssef, 1989; Sejnowski and Stanton, 1990; Bruck and Roychowdhury, 1990; Jiang, et. al., 1990; Kam, et. al., 1990; Casasent and Telfer, 1992; Alkon, et. al., 1994; Zhuang and Huang, 1994; Zhang, et. al., 1995). Most recently proposed learning algorithms generate an asymmetric interconnection weight matrix for the associative memory. Introducing asymmetric interconnection matrix would certainly represent an important move to get around the main drawback of Hopfield auto-associative memory. But there does not exist any theory for auto-associative memory with asymmetric interconnection weights. Furthermore, none of the learning algorithms proposed so far can adaptively control the shape of basin of attraction and learn from variant samples. As well recognized, "learning from samples" would surely shape each basin of attraction more appropriately for better generalization.

In this paper we design a general auto-associative memory with sample learning capability. We start by defining a new concept called supporting function to replace the concept of energy function used in the Hopfield associative memory. We formulate the information recalling process as a dynamic system using the supporting function, explore its stability, and show each stable state is a fixed point of the system. We then study the condition for convergence of an arbitrary state to a desired attractor and describe a highly effective sample learning algorithm using Rosenblatt's perceptron rule. Finally, we present experimental results to demonstrate the effectiveness of the sample learning algorithm for the general auto-associative memory.

## SUPPORTING FUNCTION

A general auto-associative memory of size $n$ consists of $n$ interconnected neurons. We denote the $n$ neurons by $U_1, U_2, \cdots, U_n$. Each neuron $U_i$ $(i = 1, 2, \cdots, n)$ has a firing status $v_i$ which could be $+1$ or $-1$ to indicate its "firing on" or "firing off" status, respectively. The vector $v = (v_1, v_2, \cdots, v_n)^{\mathrm{T}}$ is called a firing state vector of the memory system.

There are consistency constraints on firing status assignments. Let $w_{ih}$ be the weight of the link from $U_h$ to $U_i$ which represents how the firing status at neuron $U_h$ influences the firing status at neuron $U_i$. If neuron $U_h$ firing on lends a high support to neuron $U_i$ firing on, the weight $w_{ih}$ should be large and positive. If the constraints are such that neuron $U_h$ firing on means that neuron $U_i$ firing on is highly unlikely, then $w_{ih}$ should be small and negative. No specific restrictions are placed on the magnitude of $w_{ih}$. The $n \times n$ matrix $(w_{ih})$ is called interconnection weight matrix. It does not need to be symmetric. But we do neutralize the diagonal terms, i.e., those self-supporting weights, by assuming $w_{ii} = 0$, $i = 1, 2, \cdots, n$.

Define the support on neuron $U_i$ firing on from another neuron $U_h$ having a firing status $v_h$ by $w_{ih}v_h$. Define the support on neuron $U_i$ firing on from the rest $n-1$ neurons $U_h$, $h = 1, 2, \cdots, i-1, i+1, \cdots, n$, having respective firing status $v_h$, $h = 1, 2, \cdots, i-1, i+1, \cdots, n$, by $\sum_{h \neq i} w_{ih} v_h$ denoted as $q(i|v)$, i.e.,

$$q(i|v) = \sum_{h \neq i} w_{ih} v_h \tag{1}$$

In other words, each neuron $U_i$, if firing on, will receive a support $q(i|v)$ from other neurons having various firing status.

Define the support on $U_i$ having a firing status $v'_i$ from the rest $n-1$ neurons having firing status $v_h$ for $U_h$, $h = 1, 2, \cdots, i-1, i+1, \cdots, n$ by the product $v'_i q(i|v)$. Let $v'$ denote another firing state vector, i.e., $v' = (v'_1, v'_2, \cdots, v'_n)^{\mathrm{T}}$. Define the support on the firing state vector $v'$ from the firing state vector $v$ by the sum $\sum_{i=1}^{n} v'_i q(i|v)$ or the inner product $< v', q(v) >$, where $q(v)$ is called supporting vector and defined as

$$q(v) = (q(1|v), q(2|v), \cdots, q(n|v))^{\mathrm{T}} \tag{2}$$

The inner product defines a supporting function of two variates $v$ and $v'$ each an $n$-dimensional vector. We denote the supporting function by $S(v'|v)$, i.e.,

$$S(v'|v) = < v', q(v) > \tag{3}$$

The supporting function $S(v'|v)$ measures consistency or correlation between the firing state vector $v'$ and the supporting vector $q(v)$.

## STABILITY AND RECALLING ALGORITHM

A firing state vector $v$ is considered stable if it receives the maximum support from itself, i.e.,

$$\forall v' \neq v, \; S(v'|v) < S(v|v) \tag{4}$$

In other words, a stable state vector $v$ attains the maximum consistency with its own supporting vector $q(v)$.

Define the Hamming distance $d(v', v)$ between $v'$ and $v$ by

$$d(v', v) = \#\{i : v'_i \neq v_i, \; i = 1, 2, \cdots, n\} \tag{5}$$

and the unit Hamming ball $H_1(v)$ centered at $v$ by

$$H_1(v) = \{v' : d(v', v) \leq 1\} \tag{6}$$

Obviously, $H_1(v)$ contains $n+1$ firing state vectors. Let $v^{(i)}$ denote a firing state vector that differs from $v$ only in the $i$-th component. Then, $H_1(v) = \{v, v^{(i)} : i = 1, 2, \cdots, n\}$. The following theorem holds.

**Theorem 1** (*Stability*) A firing state vector $v$ is stable iff

$$\forall v' \in H_1(v),\ v' \neq v,\ \ S(v'|v) < S(v|v) \tag{7}$$

**Remark**: Condition (7) in Theorem 1 can be shown equivalent to

$$\forall i,\ v_i q(i|v) > 0 \tag{8}$$

Motivated by (7), we propose the following dynamic system as a recalling rule to iteratively enhance the stability condition, starting at an arbitrary initial firing state vector $v^0$:

$$v^{s+1} = \arg \max_{v' \in H_1(v^s)} S(v'|v^s),\quad s \geq 0 \tag{9}$$

Let $i_0$ be an index determined by

$$i_0 = \arg \max_i \{|q(i|v^s)| - q(i|v^s)v_i^s\} \tag{10}$$

Then $v^{s+1}$ can be explicitly computed by using the following formula:

$$v_i^{s+1} = \begin{cases} q(i|v^s)/|q(i|v^s)| & \text{if } i = i_0 \text{ and } q(i_0|v^s) \neq 0 \\ v_i^s & \text{otherwise} \end{cases} \tag{11}$$

The following theorem states that each stable firing state vector comprises a fixed point of dynamic system (9).

**Theorem 2** (*Fixed Point*) Assume $v$ is stable. Then $v$ is a fixed point of the dynamic system defined by (9), i.e., $v^s = v$ implies $v^{s+1} = v$.

The following proposition gives a distance constraint on two stable states.

**Proposition 1** Assume $v^p$ and $v^q$ are two different stable state vectors. Then $d(v^p, v^q) \geq 2$.

## ERROR-REDUCTION CONVERGENCE

In this section, we investigate the behavior of dynamic system (9) and derive the condition for error-reduction convergence of an arbitrary state to a desired stable state. The meaning of error-reduction convergence will become clear after Theorem 3. We begin by defining a set $V(v', v)$ of state vectors as:

$$V(v', v) = \{v'' : v_i'' = v_i \text{ if } v_i' = v_i\} \tag{12}$$

Obviously, both $v'$ and $v$ are in $V(v', v)$. Furthermore, $v'' \in V(v', v)$ implies $V(v'', v) \subseteq V(v', v)$. The following lemma states the condition for a state to get closer to a desired state after one recalling step.

**Lemma 1** Suppose $v$ is a desired stable state vector. If the state vector $v^s$ with $d(v, v^s) = r > 0$ satisfies

$$\forall v' \in H_1(v), v' \neq v,\ S(v'|v^s) < S(v|v^s) \tag{13}$$

then $v^{s+1} \in V(v^s, v)$ and $d(v, v^{s+1}) = r - 1$.

Now we are ready to derive the condition for error-reduction convergence of a state to a desired stable state in finite steps.

**Theorem 3** (*Finite Error-Reduction Convergence*) Assume that $v$ is a stable state vector. For an arbitrary starting state vector $v^0$, if

$$\forall v' \in V(v^0, v), \forall v'' \in H_1(v), v'' \neq v,\ S(v''|v') < S(v|v') \tag{14}$$

then the recalling sequence $\{v^s, s \geq 0\}$ converges to $v$ in finite steps.

**Remark**: Condition (14) in Theorem 3 can be shown equivalent to

$$\forall v' \in V(v^0, v), \forall i,\ v_i q(i|v') > 0 \tag{15}$$

Theorem 3 establishes the basis for a capability of learning from samples. Imagine $v^0$ represents an erroneous sample of $v$. Condition (14) should be learned that assures errors in $v^0$ are reduced each time the recalling rule is applied and the desired stable state $v$ is finally recovered.

## SAMPLE LEARNING ALGORITHM

In many applications, a capability of learning from samples appears critical. For example, in handwritten digit recognition application, the same digits are written differently at different times or by different people. An associative memory would be more robust if trained by variant samples so that the shape of basin of attraction around each desired attractor could be statistically adapted. In the following, we present such a sample learning algorithm, which is the first of its kind ever been developed for associative memories.

Suppose we have $m$ desired attractors $v^1, v^2, \cdots, v^m$ and $v$ is a sample of the desired attractor $v^k$ and $v$ differs from $v^k$ in $r$ components. We want to associate $v$ to $v^k$. By the remark that follows *Finite Error-Reduction Convergence Theorem*, the associative memory will correct all errors in $v$ in $r$ steps if $v_i^k q(i|v') > 0$ for any $v' \in V(v, v^k)$. Since $V(v, v^k)$ contains $2^r$ states, associated with sample $v$ there are as many as $2^r$ conditions to be learned at each $U_i$. In practice, a few state vectors from $V(v, v^k)$ are chosen when $r$ is large.

Suppose each $v^k$ admits $l$ samples $v^k, v^{k,1}, \cdots, v^{k,l-1}$. The conditions under which each sample $v^{k,j}$ converges to $v^k$ are as follows:

$$\begin{array}{c} v_i^k \sum_{h\neq i} w_{ih} v'_h > 0,\ v' \in V(v^{k,j}, v^{k,0}) \\ k = 1, 2, \cdots, m \text{ and } j = 0, 1, \cdots, l-1 \\ i = 1, 2, \cdots, n \end{array} \tag{16}$$

which can be learned using the Rosenblatt perceptron rule (Rosenblatt, 1959). The sample learning algorithm proceeds as follows.

1. For each $v^{k,j}$, randomly select $p$ state vectors from $V(v^{k,j}, v^k)$, $j = 1, 2, \cdots, l-1$. There are a total of $(l-1)p$ state vectors selected for each $v^k$. Those plus the given $l$ samples $v^k, v^{k,1}, \cdots, v^{k,l-1}$ are used as a $v'$ in the learning condition $v_i^k q(i|v') > 0$. As $k$ varies, there are a total of $ml + m(l-1)p$ such conditions to be learned at each neuron.
2. Order those $M = ml + m(l-1)p$ state vectors (arbitrarily) as $x^0, x^1, \cdots, x^{M-1}$ and form a learning sequence $x^{(0)}, x^{(1)}, \cdots, x^{(t)}, \cdots$ with $x^{(t)} = x^{t\,\%\,M}$, where % denotes the modulus operation. By the arrangement, $x^{(t)}$ belongs to some $V(v^{k,s}, v^k)$.
3. At each neuron $U_i$, learn the $n-1$ interconnection weights as follows:

$$\left\{ \begin{array}{l} w_{ij}^{(0)} = \text{arbitrary value}, \\ w_{ij}^{(t+1)} = \left\{ \begin{array}{ll} w_{ij}^{(t)} + v_i^k x_j^{(t)} & \text{if } x^{(t)} \in V(v^{k,s}, v^k) \text{ and } v_i^k \sum_{h\neq i} w_{ih}^{(t)} x_h^{(t)} \leq 0 \\ w_{ij}^{(t)} & \text{otherwise} \end{array} \right. \\ j = 1, 2, \cdots, i-1, i+1, \cdots, n \end{array} \right. \tag{17}$$

## EXPERIMENTAL RESULTS

The proposed general auto-associative memory (GAM) highly outperforms the Hopfield auto-associative memory in terms of storage capacity, attraction, and spurious memories (Zhuang, et. la., 1996). To demonstrate the sample learning capability of GAM, we conducted experiments with ten handwritten digits. We intend to demonstrate that GAM can associate various samples of handwritten digits and perform significantly well in the presence of noise.

Figure 1 shows a total of 30 handwritten digit samples used in our experiments. Each digit has three samples and each sample is a $32 \times 25$ binary image. Thus, the memory size $n$ is 800, the number of desired attractors $m$ is 10, and the number of samples for each desired

Table: 1: Hamming Distances of samples 2 and 3 from sample 1

| Digit $k$ | 0 | 1 | 2 | 3 | 4 | 5 | 6 | 7 | 8 | 9 |
|---|---|---|---|---|---|---|---|---|---|---|
| $d(v^{k,1}, v^k)$ | 73 | 91 | 83 | 90 | 94 | 67 | 95 | 63 | 94 | 85 |
| $d(v^{k,2}, v^k)$ | 73 | 80 | 97 | 97 | 99 | 63 | 97 | 87 | 88 | 96 |

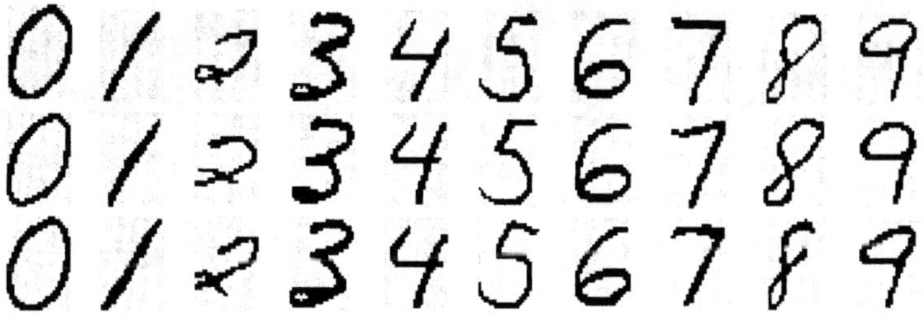

Figure: 1: Three samples of ten handwritten digits

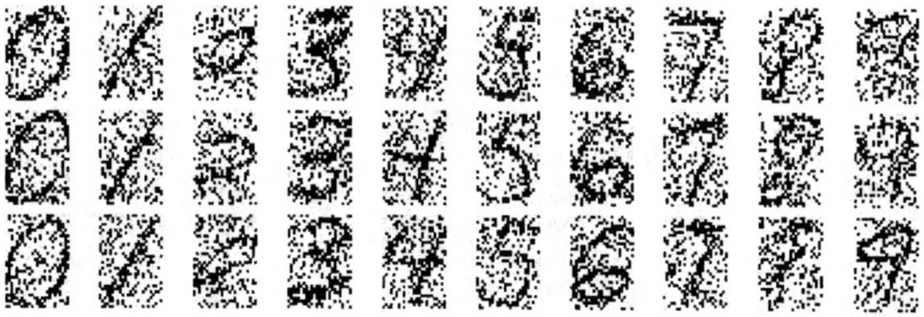

Figure: 2: Samples of handwritten digits with 25% random noise

attractor $l$ is 3. The Hamming distance between samples of each pattern is quite large, as seen in Table 1.

The GAM is trained by the sample learning algorithm so the last two samples of each digit are associated to the first one. As explained in the description of the sample learning algorithm, $p$ random states from each $V(v^{k,j}, v^k)$ along with $v^k$ and $v^{k,j}$ are chosen for training, where $j = 1, 2$ and $k = 0, 1, \cdots, 9$. After training, we present samples corrupted by random noise to the memory. Figure 2 shows the samples with 25% random noise. We then compute the percentage of correct convergence. To obtain a more objective evaluation of GAM, we randomly generate 20 sets of noisy samples at each noise level and calculate the average percentage of correct convergence. The results are shown in Table 2. The correct convergence rate significantly increases as $p$ increases at even a high noise level, which means the intermediate states along the path from $v^{k,j}$ to $v^k$, $j = 1, 2$, participating in the learning process help the convergence of corrupted samples to their desired digits

## CONCLUSION

We have introduced a general auto-associative memory with sample learning capability. We have explored its stability condition and developed a recalling algorithm. We have designed a sample learning algorithm and demonstrated its effectiveness through experiments.

Table: 2: Correct convergence rates for samples with random noise

| $p$ | Rate of correct convergence | | | | |
|---|---|---|---|---|---|
| | 5% noise | 10% noise | 15% noise | 20% noise | 25% noise |
| 0 | 99.17 | 98.00 | 95.00 | 87.83 | 77.33 |
| 20 | 100.00 | 99.00 | 97.50 | 92.00 | 82.83 |
| 40 | 100.00 | 99.67 | 98.67 | 96.00 | 84.83 |

## REFERENCES


Alkon, D. T., Blackwell, B. T., Barbour, G. S., Werness, S. A., and Vogl, T. P. (1994). Biological plausibility of synaptic associative memory models, *Neural Networks*, 7(6/7):1005–1017.

Bruck, J., and Roychowdhury, V. P. (1990). On the number of spurious memories in the Hopfield model, *IEEE Transactions on Information Theory*, 36(2):393–397.

Casasent, D. and Telfer, B. (1992). High capacity pattern recognition associative processors, *Neural Networks*, 5:687–698.

Cottrell, M. (1988). Stability and attractivity in associative memory networks, *Biological Cybernetics*, 58:129–139.

Dembo, A. (1989). On the capacity of associative memories with linear threshold functions, *IEEE Transactions on Information Theory*, 35(4):709–720.

Floreen, P. and Orponen, P. (1989). Counting stable states and sizes of attraction domains in Hopfield nets is hard, In *Proceedings of the 1989 International Joint Conference on Neural Network*, volume 1, pages 395–399.

Hassoun, M. H. and Youssef, A. M. (1989). A high-performance recording algorithm for Hopfield model associative memories, *Optical Engineering*, 27(1):46–54.

Hopfield, J. J. (1982). Neural network and physical systems with emergent collective computational abilities, *Proceedings of National Academy of Sciences USA*, 79:2554–2558.

Jiang, X., Hegde, M., and Naraghi-Pour, M. (1990). Neural network design using linear programming and relaxation, In *Proceedings of IEEE ISCAS 90*, New Orleans, pages 1090–1093.

Kam, M., Chow, J. C., and Fischel, R. (1990). Design of the fully connected binary neural network via linear programming, In *Proceedings of IEEE ISCAS 90*, New Orleans, pages 1094–1097.

Kohonen, T. and Ruohonen, M. (1973). Representation of associative data by matrix operators, *IEEE Transactions on Computers*, 22:701–702.

McEliece, R. J., Posner, E. C., Rodemich, E. R., and Venkatesh, S. S. (1987). The capacity of the Hopfield associative memory, *IEEE Transactions on Information Theory*, 33(4):461–482.

Personnaz, L., Guyon, I., and Dreyfus, G. (1986). Collective computational properties of neural networks: New learning mechanisms, *Physical Review. A*, 34(5):4217–4227.

Rosenblatt, R. (1959). *Principles of Neurodynamics*, Spartan Books, New York.

Sejnowski, T. J. and Stanton, P. K. (1990). Covariance storage in the hippocampus, In *An Introduction to Neural and Electronic Networks*, Academic Press.

Venkatesh, S. S. and Psaltis, D. (1989). Linear and logarithmic capacities in associative neural networks, *IEEE Transactions on Information Theory*, 35(3):558–568.

Zhang, B., Zhang, L., and Wu, F. (1995). Programming based learning algorithms of neural networks with self-feedback connections, *IEEE Transactions on Neural Networks*, 6(3):771–775.

Zhuang, X. and Huang, Y. (1994). Design of Hopfield content-addressable memories, *IEEE Transactions on Signal Processing*, 42(2).

Zhuang, X., Shi, H., and Zhao, Y. (1996). A general auto-associative memory model, In *Proceedings of the 1996 International Conference on Neural Networks*, Washington.

# Rapid and Reliable Learning through Data-Driven Initialization

**RALF SALOMON**
*AI Lab, Computer Science Department, University of Zürich*
*Winterthurerstrasse 190, 8057 Zurich, Switzerland*
*FAX: +41-1-363 00 35; Email: salomon@ifi.unizh.ch*

***ABSTRACT***:
Neural networks are universal function approximators and, thus, ideally suited for regression tasks. This paper shows that learning is easy, if the desired mapping function is convex or unimodal. But learning gets extremely slow or even fails, if the mapping function is multimodal. Unfortunately, increasing the number of connections does not relief the problem, because a larger network either yields poor generalization or causes even more problems to the learning algorithm. To overcome these problems, this paper suggests a new data-driven initialization procedure for feed-forward networks, which yields rapid learning and very accurate approximation. For multimodal functions, the proposed initialization procedure speeds up learning by several orders of magnitude. Furthermore, learning behavior is almost insensitive to the number of hidden units for reasonably sized networks.

## INTRODUCTION

Regression is a typical application domain for neural networks, where a network has to learn a given mapping function $f$. In this domain, the network has to mimic a system $S$ that maps $p$-dimensional input vectors $\vec{x}$ to $q$-dimensional output vectors $\vec{y}$. The system $S$ is a black box in the sense that the mapping function $f : \vec{x} \rightarrow \vec{y}$ is unknown. The task of a learning algorithm is to find an appropriate approximation $\tilde{f}$ of the mapping function $f$. Training is usually done on a set of $s$ *samples* $\{(\vec{x}_i, \vec{y}_i)\}$, $1 \leq i \leq s$.

Among other things, neural networks have two major advantages. First, they can learn from examples and, second, they can generalize unseen data reasonably well. The theory (Cybenko, 1989; Hornik, 1989) has shown that multilayer feed-forward networks are universal function approximators. One way to prove this important property is to consider pairs of hidden units, where both units have similar weight vectors. Hidden units with similar weight vectors have very similar activation patterns, and the difference of these two similar hidden units is equivalent to using one hidden unit with a sharp bell-shaped activation function. Then, a linear output unit can approximate any given mapping function with any desired accuracy by summing up *sufficiently* many hidden units with bell-shaped activation functions. Thus, in a theoretical sense, a neural network is a universal function approximator.

While this result is of great theoretical significance, it does not provide any leverage or guidance for practical applications. In practical applications, one has to solve many other problems concerning the experimental setup before the issue whether a particular

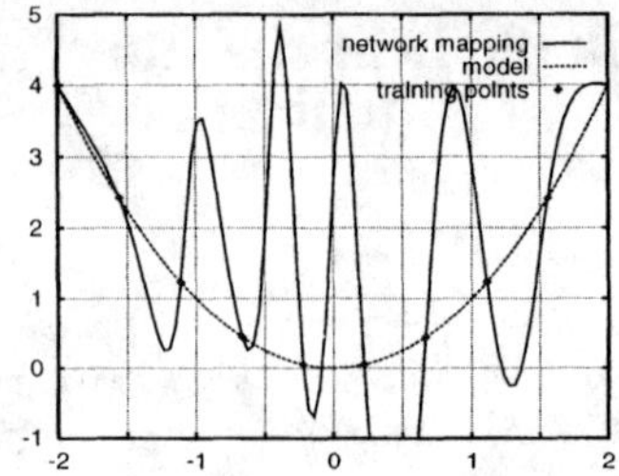


**Figure 1: An underconstrained network, which has a too high high VC dimension with respect to the number of training patterns, learns all training points perfectly, but its approximation $\tilde{f}$ is different from $f$.**

network can approximate the given function becomes relevant. First of all, the theory by no means implies that a learning procedure (e.g., backpropagation) is able to find the approximation. Furthermore, the initial conditions, such as the network's topology, the initial weights, and the learning parameters strongly influence the overall performance.

A major issue of the design process is to determine the number of connections. A network must have a minimal size in order to yield the desired mapping accuracy. Training time can be reduced significantly by using more than the minimal number of connections. However, a larger network has a higher VC dimension and, thus, a higher probability of generalizing poorly, unless further constraints are imposed to reduce its associated VC dimension (see, for example, Mostafa, 1990, 1992). An example of an underconstrained network, which has a too high high VC dimension with respect to the number of training patterns, is presented in Figure 1. The network has learned all training points perfectly, but its approximation $\tilde{f}$ is very different from the desired mapping function $f$.

Normally, it is hard to determine an appropriate number of hidden units or connections from the data itself without using any further a priori knowledge. This is mostly done in a trail and error process. In addition, dynamic approaches have been developed to avoid overlearning and to increase generalization performance. Examples are the optimal brain damage approach (LeCun, 1990), the skeletonization approach (Mozer, 1989), or the widely-used cross-validation heuristic, which schedules the learning rate and stops learning after no decrease of the generalization error is achievable.

In order to relief the problems described above, this paper proposes a new initialization method that exploits the information contained in the training data. This approach cuts the learning time by several orders of magnitude, it is universally applicable, results in a very accurate approximation, and is especially suitable, if the network should have only a few parameters. Especially, if the function is multimodal or very rough, the proposed approach cuts down learning time and makes real-world problems tackable where learning has to be performed under real-time conditions.

To make further considerations easier, the vectorial mapping function can be decomposed into a set of $q$ mapping functions $\{f^i : \vec{x} \rightarrow y^i\}$, $1 \leq i \leq q$, with $y^i$ denoting the $i$ th component of $\vec{y}$. In this case, an algorithm has to find a set of $q$ mapping functions $\tilde{f}^i$. Without loss of generality, the input space can be reduced to only one dimension. Thus, as has been done in other papers (e.g., Ji, 1990), this paper considers only the case $f : x \rightarrow y$, and consequently, indices for $f$ and the components of $\vec{x}$ and

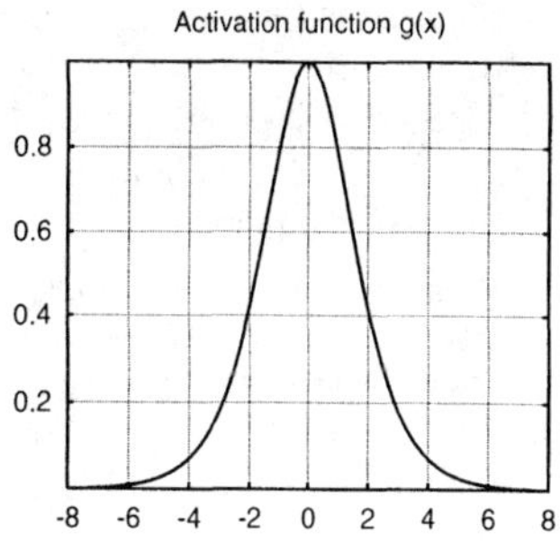


**Figure 2: The proposed activation function $g(x) = 4e^{-x}/(1+e^{-x})^2$ of the hidden units.**

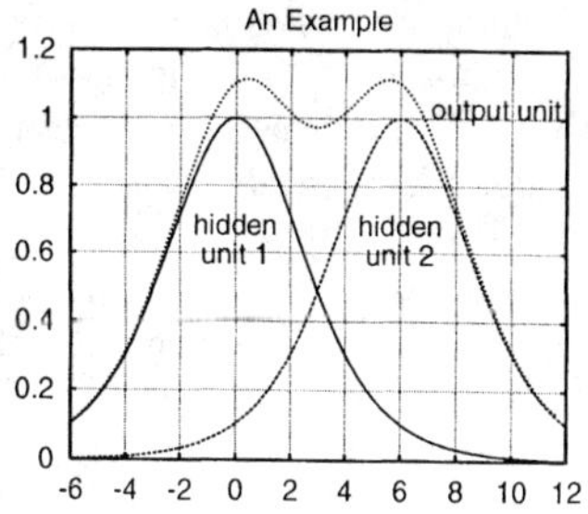


**Figure 3: An example with two hidden units and two different training patterns at $x_t = 0$ and $x_t = 6$.**

$\vec{y}$ are omitted. Furthermore, assume for the sake of simplicity that all sample points are ordered such that $i < j \rightarrow x_i < x_j$.

## DATA DRIVEN INITIALIZATION

First of all, according to the theory (Cybenko, 1989; Hornik, 1989), the hidden unit's activation function is changed to

$$g(x) = 4\,\frac{e^{-x}}{(1+e^{-x})^2}\,. \tag{1}$$

Basically, this function is the prime of the logistic function $f(x) = 1/(1+e^{-x})$ and a plot is presented in Figure 2.

Second, the procedure estimates the mean distance between two adjacent hidden units $\Delta x = (x_{\max} - x_{\min})/\#\_of\_hidden\_units$, assuming that all patterns are uniformly distributed over the input domain $x$. Third, the procedure selects one training pattern $(x_t, y_t)$ for each hidden unit $j$. The choice of the training pattern can be done randomly or deterministicly. In the latter case, the procedure selects, for example, each third training pattern. For each hidden unit, the procedure calculates the hidden unit's parameters such that the maximum of the bell-shaped activation function is at $x_t$ and determines the width according to $g(x_t + 0.5\,\Delta x) = 0.5$ (see below). Both design decisions result in the following two constraints: $w^h_j\, x_t = \text{bias}^h_j$ and $g(0.5\,\Delta x w^h_j) = 0.5$ (compare also the formula for the netinput $\text{net}_j$ as presented in the next section). The input weight $w^h_j$, the $\text{bias}^h_j$, and the corresponding output weight $w^o_j$ as well as the output $\text{bias}^o$ are determined as follows

$$\begin{aligned} w^h_j &= \frac{3.5}{\Delta x} + \delta \\ \text{bias}^h_j &= -\frac{3.5\,x_t}{\Delta x} + \delta \\ w^o_j &= y_t \\ \text{bias}^o &= 0 \quad , \end{aligned} \tag{2}$$

where $\delta$ denotes small random numbers $|\delta| < 0.1\,\Delta x$. Adding small random numbers in (2) avoids problems, if the same training pattern is chosen for more than one hidden unit. The calculation of $\Delta x$ assumes that the network has less hidden units than training patterns. The opposite case, i.e., more hidden units than training patters, is not recommended, since the VC dimension of such a network would be too high. The generalization to $p$-dimensional inputs is straightforward and omitted due to space limitations. Furthermore, its is straightforward to apply the initialization procedure to standard feed-forward networks; the weight vectors of two hidden units are set to similar values and the output unit calculates the difference of these two hidden units.

The example shown in Figure 3 has two training points $(0, 1)$ and $(6, 1)$. For a good initial approximation, it is desirable that the activation function yields the value 0.5 in the middle of both training points. This can be achieved by the constraint $g(0.5\,\Delta x\, w_j^h) = 0.5$, which leads to $0.5\,\Delta x\, w_j^h \approx 1.75$ and $w_j^h \approx 3.5/\Delta x$.

It is important to realize that the proposed initialization procedure is independent of the chosen training algorithm and it does not substitute the rtaining algorithm proper.

## METHODS

All regression experiments presented in this paper were carried out with the following functions: $f_1(x) = x^2$, $f_2(x) = \sin(2\pi x)$, and $f_3(x) = e^{-(x-1)^2} + e^{-(x+1)^2}$. A thorough discussion of several issues of function $f_3$ can be found in (Ji, 1990).

All experiments were carried out with standard feed-forward networks (Rumelhart, 1986), with one input unit, one hidden layer with $h$ hidden units, and one linear output unit. The netinput of each hidden unit is the weighted sum $\text{net}_j = \sum_{k=1}^{1} w_{jk}^h u_i + \text{bias}_j^h$, where $u_i$ denotes the activation of the input unit $i$. The activation function of each hidden unit is the logistic function $f(x)$ in the standard case and $g(x)$ in case of the data-driven initialization respectively. The output of the linear output unit is given by $o = \sum_{i=1}^{h} w_i^o h_i + \text{bias}^o$. Training is always done with standard backpropagation and online updating. Suitable values for the learning rate were obtained by previous experiments. Using more advanced learning algorithms such as the self-adaptation mechanism proposed in (Salomon, 1992) yields further speed ups but are beyond the scope of this paper, since the qualitative behavior is unchanged. For comparison purposes, all figures use the normalized mean squared error $\sqrt{\sum_{i=1}^{s}(t_i - o_i)^2}/\sqrt{s}$. All training and test patterns are equally distributed and the ratio of training and test patterns is always 1:10. To adopt a real-world situation, all training points have been deisturbed by adding $(0, \sigma)$-gaussian random numbers with $\sigma \in (0, 0.1)$.

## RESULTS

Most results are omitted due to space limitations. For the parabola $f_1(x)$, we used $x \in (-1.0, 1.0)$, 20 training points, initial weights $w_{ij} \in (-0.1, 0.1)$, and three to five hidden units. In all cases, the neural network finds an appropriate approximate very easily, but the standard approach is *very sensitive* to the initialization of the weights. The data-driven initialization procedure speeds up learning for more than one order of magnitude and there is no overlearning observable, as long as the number of training points is significantly larger than the number of connections. Additional experiments on other unimodal (convex) functions and other simple sigmoidal functions have shown the same qualitative behavior.

If the mapping function is multimodal, however, standard learning drastically slows down. Using, for example, the sine function $f_2(x)$ in the interval $x \in [-1.0, 3.0]$ with 80 training patterns, a network has to approximate four full periods of the sine function.

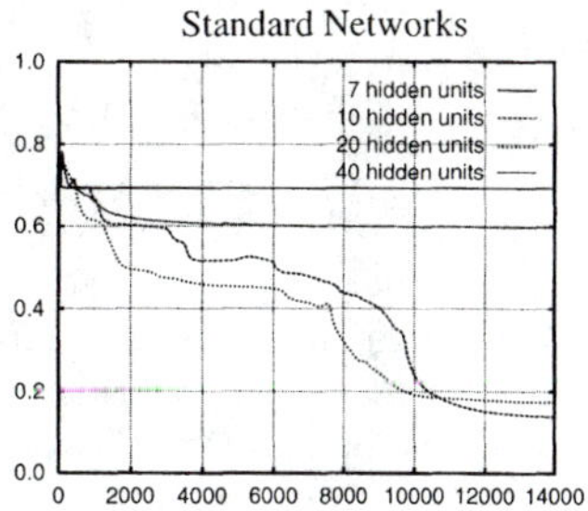


**Figure 4: Error vs. epochs of standard networks with 7, 10, 20, and 40 hidden units respectively, when approximating $f(x) = \sin(2\pi x), x \in [-1.0, 3.0]$.**

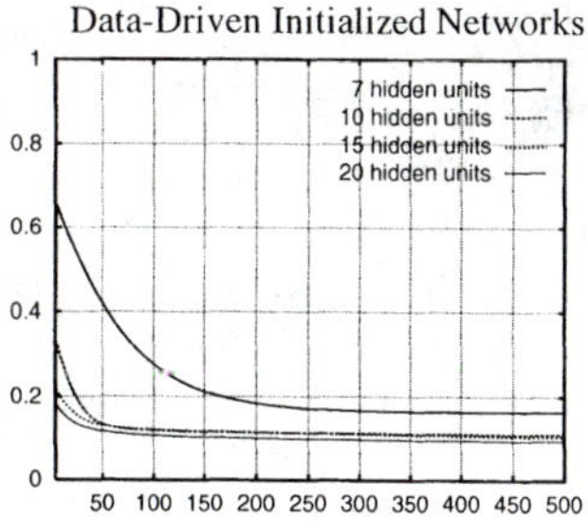


**Figure 5: Error vs. epochs of data-driven initialized networks with 7, 10, 15, and 20 hidden units, when approximating $f(x) = \sin(2\pi x), x \in [-1.0, 3.0]$.**

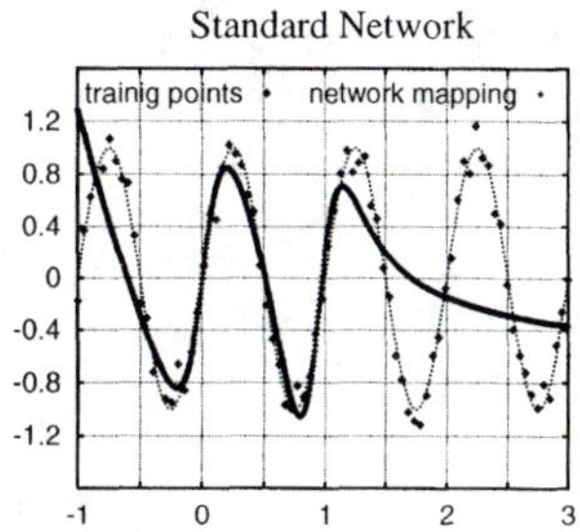


**Figure 6: Final mapping of a standard network with 40 hidden units after 10,000 epochs.**

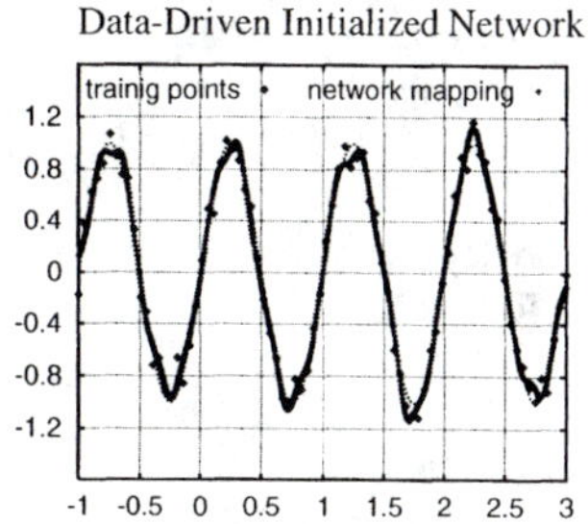


**Figure 7: Final mapping of a data-driven initialized network with 20 hidden units after 200 epochs.**

Figure 4 shows the learning behavior of standard feed-forward networks with 7, 10, 20, and 40 hidden units respectively. It can be seen that only networks with 10 or 20 hidden units yield a suitable approximation within 14,000 epochs; using 7 or 40 hidden units is without any success. By contrast, using the data-driven initialization procedure leads to much different learning behavior. Figure 5 presents the resulting performance of data-driven initialized feed-forward networks with 7, 10, 15, and 20 hidden units respectively. It can be seen that the number of hidden units has only a small effect on the required training time. In most cases, only 50 to 100 training epochs are necessary to achieve a better mapping than using standard feed-forward networks. Figures 6 and 7 show the final mapping of a standard feed-forward network with 40 hidden units after 10,000 epochs and the mapping of a data-driven initialized network with 20 hidden units after 200 epochs respectively. It can be seen that the proposed initialization procedure yields a much better approximation in a much short time. Furthermore, Figure 6 shows that a standard feed-forward network approximates well in only a subdomain of $x$, whereas the data-driven initialization leads to a good approximation in the whole domain.

The same qualitative behavior is observable when approximating function $f_3$. Here, the standard approach has even more problems and the proposed initialization leads to an easy design and training process

In another experiment, we used the proposed initialization method for the initialization of a Kohonen feature map. The feature map serves as a navigation tool in an autonomous robot.

## CONCLUSIONS

Experiments presented in this paper show that feed-forward networks do not have problems to approximate convex or sigmoidal functions. But feed-forward networks have extreme difficulties to approximating multimodal mapping functions. Learning becomes *very slow* and very sensitive to both the number of hidden units and the weight initialization. Mostly, a standard feed-forward network does not achieve a reasonable approximation within 14,000 epochs; successful learning is rather an exception.

By exploiting information contained in the training data, the proposed data-driven initialization procedure speeds up learning by several orders of magnitude. Furthermore, the new initialization method reliefs the problem of finding an appropriate number of hidden units; learning and generalization performance is much less sensitive to the number of connections. In addition, there is no overlearning observable as long as the number of connections is smaller than the number of training patterns.

The next step will be to apply the data-driven initialization procedure to real-world problems.

**Acknowledgements:** This work was supported in part by a Human Capital and Mobility fellowship of the European Union, grant ERBCHBICT941266.

## REFERENCES

Abu-Mostafa, Y.S. (1990). Learning from Hints in Neural Networks, *Journal of Complexity*, **6**, pp. 192-198.

Abu-Mostafa, Y.S. (1992). Hints and the VC Dimension, *Neural Computation*.

Cybenko, G. (1989). Approximation by Superpositions of a Sigmoidal Function, *Mathematics of Control, Signals, and Systems* Springer-Verlag.

Hornik, K., Stinchcombe, M., White, H. (1989). Multilayer Feedforward Networks are Universal Approximators. *Neural Networks* **2**, pp. 359-366.

Ji, C., Snapp, R.R., Psaltis, D. (1990). Generalizing Smoothness Constraints from Discrete Samples, *Neural Computing*, **2**, pp. 188-197.

Le Cun, Y., Denker, J.S., Solla, S.,A. (1990). Optimal Brain Damage, in *Advances in Neural Information Processing Systems 2*, D. Touretzky (ed.), pp. 598-605, Morgan Kaufmann Publishers.

Morgan, N., Bourlard, H. (1990). Generalization and Parameter Estimation in Feedforward Nets: Some Experiments. in David S. Touretzky (ed.), *Advances in Neural Information Processing Systems 2*, pp. 630-637, Morgan Kaufmann Publishers.

Mozer, M.C., Smolensky, P. (1989). Skeletonization: A technique for trimming the fat from a network via relevance assessment, in David S. Touretzky (ed.), *Advances in Neural Information Processing Systems 1*, pp. 107–115, Morgan Kaufmann Publishers.

Rumelhart, D.E., McClelland, J.L., and the PDP Research Group (1986). *Parallel Distributed Processing: Explorations in the Microstructure of Cognition*, volume I. MIT Press

Salomon, R., van Hemmen, J. (1992). A new learning scheme for dynamic self-adaptation of learning-relevant parameters, in I. Aleksander and J. Taylor (eds.) *International Conference on Artificial Neural Networks (ICANN-92)*, pp. 1047-1050, North-Holland.

# A SIMPLE HEURISTIC FOR DYNAMIC NETWORK DESIGN

**ANTONY HAMMITT AND ERIC B. BARTLETT**
*Iowa State University*

***ABSTRACT:***
Researchers have long known that it is difficult and somewhat undesirable to predetermine the number of nodes to use in each layer of artificial neural network models. Recent research has demonstrated the viability of our new approach to implementing an on-line dynamic node architecture scheme on a wide variety of real-world test problems. The new approach utilizes a simple heuristic to determine to what layer a node is to be added. This includes determining which of the available but currently unused inputs is the best one to add to the network. When using this approach, the target cost must be determined with care and some methods for determining the optimal cost are given. Examples are provided where this approach is able to determine the smallest viable architecture for a particular problem regardless of whether the network initially has too many or too few nodes to learn the problem.

## INTRODUCTION

Research on artificial neural networks (ANNs) has been progressing at a rapid pace in recent years. Advances in training algorithms as well as the ready availability of fast computers has made neural network problems much more tractable than they were even two years ago (Baba, 1989),(Moller, 1993),(Bartlett, 1994a).

While scaled conjugate gradients (Moller, 1993) have allowed researchers to explore a greater variety of problems, the user still has to choose the architecture of the networks such that the problem can be learned. Typically this means using a network that is larger than is truly necessary in order to learn the mapping in a reasonable time. However, using a large network can cause the network to generalize poorly even though it achieves the target cost. Generalization is usually better with smaller networks, but they are either hard to train or fail to learn the subtleties of the training data. This dilemma has lead researchers to develop several different kinds of dynamic network architecture (DNA) approaches (Bartlett, 1994b,c), each of which have drawbacks resulting form the complexity of the techniques used to determine where to add or delete nodes.

Our new dynamic node architecture approach uses a set of simple rules about how to modify the architecture of an ANN during the training process in such a way as to minimize the number of nodes used to map the data set. It is often the case that the network needs more nodes to learn the problem than it needs to recall accurately on the data, much in the same way that calculus is hard to learn but easy to do after it is learned. Thus the network can likely be reduced in size and still perform adequately. The user must decide on this desired level of precision, since a larger network will have to be used to map all of the features in the data (Baum and Haussler, 1989), (Hornik, Stinchcomhbe and White, 1989). The RMS error is typically used as a measure of precision or, inversely, cost.

The next section describes the algorithm and methodology of the DNA approach. Section 3 describes some results that demonstrate the viability the approach. Section 4 contains concluding remarks and proposed future work.

## TECHNICAL DISCUSSION

### Neural Network Architecture

In this paper we use the same type of networks used in (Moller, 1993) which were trained using scaled conjugate gradients. The network uses a layers of nodes, feedforward interconnected with a constant node in order to bias the input of the node, as outlined in (Lippmann, 1987).

The nodes use the transfer function given in Equation 1where $x_j$ is the output of node $j$ in layer $n$ and the $x_i$'s are either the outputs of the nodes in the layer below, or the model inputs. There are $J(n)$ nodes in layer $n$.

$$x_j(\vec{x}_i) = tanh\left( \sum_{i=1}^{J(n-1)} (w_{ji} x_i) + bias_j \right) \tag{1}$$

### Dynamic Network Architecture Approach

The dynamic network architecture heuristic used in this paper is a simplified approach developed over the past few years (Bartlett, 1994a,b,c). As mentioned in the introduction, the approach uses simple rules that were developed based on our experience of what worked well in a variety of situations.

In order to use this approach with real world data, it is important to determine which of the possible model inputs are the most useful and thereby rank the inputs. This is because the inputs of a physical system are not likely to be independent, so some of the information in an additional input is usually redundant. Eliminating the redundant inputs leads to a more robust model since the model is less likely to be confused by noise (Hammitt and Bartlett, 1995). The DNA approach can eventually find a minimal set of inputs if the inputs are not ranked, but ranking the inputs reduces a factorial search to a linear search, if the ranking is perfect. If the ranking is not perfect, then the algorithm has to add and then delete inputs until those that help training are found.

Nodes are added to the network by connecting the node to all of the nodes in the layer below it and connecting all of the nodes in the layer above it to the new node. Inputs are added based on which input is likely to be the most helpful to the network, based on the initial ranking.

Nodes are deleted by first recalling the network over the training data set in order to determine the average nodal output. Then the nodes are suppressed one at a time during recall, substituting the average output for the real output of the node. The cost is recomputed and the node whose deletion caused the smallest increase in RMS error is deemed the most useless node. If the increase in the cost is small, then the node can be deleted without harming the network performance substantially. If an input is the most useless node, then it is eliminated and moved to the end of the ranking. The approach is outlined as follows:

1. Start with a network which uses the most important input and one node in each of the processing layers (typically).
2. Train the network until the training algorithm stops making progress. If the network reaches the target cost go to step 5.
3. If there are any available inputs, add the most important unused input, if all of the inputs have been used, then add a node to the first hidden layer and go to step 4.
4. Retrain the network until either the target cost is reached or the algorithm stops making progress.
5. If the target cost is reached, save the current weights and delete the least important node from the network unless it is the only node in that layer. If it is the only node in the layer, then stop the algorithm.

6. If the target cost is not reached then try to delete a node. If none of the nodes can be deleted without raising the cost significantly, then go to the next step, otherwise go back to step 4.
7. If the cost is reduced significantly from what it was last time, add another node to the same layer that a node was last added to, this is regarded as a success. If the cost was not reduced significantly, the training is unsuccessful, add a node to the layer above the layer that the last node was added to, if there is such a layer If the next layer would be the output layer, then try to add an input and go back to step 4.

This procedure can be used to develop a network that has a minimal architecture and maps the intended function to within the target cost. The procedure can be also used with a network that is initially very large, and then gets pruned, but that approach is much slower than using an initially small network.

Before reaching the target cost, the algorithm is allowed to delete a node from a layer if there would be at least two nodes left in that layer. When the network reaches the target cost, the algorithm is allowed to delete a node as long as one node remains in the layer.

Since the behavior of the network is dependent on the specified target cost, it is important to determine what an appropriate cost should be. One approach is to set the target cost initially high, run the algorithm and then reduce the target cost in successive steps until a reasonable model is found. Another method is to set the target cost to approximately the level of the uncorrelated noise in the output, this prevents the algorithm from trying to learn the noise in the data.

## RESULTS

The first example is a simple two input one output transcendental function that is useful as a baseline function approximation benchmark, taken from (Hammitt and Bartlett, 1995). The inputs are randomly sampled in the range from 0 to 1, and the output also ranges from 0 to 1, 250 training patterns were used. For both of the training cases, the function is trained so that the model output is within 2.5% of the actual output. The training algorithm is capable of mapping the function much more precisely, but requires more hidden nodes, so for the sake of brevity, the target cost is left relatively high. The function is given in formula 2 below and plotted in Figure 1.

$$z = \left[\tfrac{1}{2} + \tfrac{1}{2} cos(2\pi x + \pi)\right] * \left[\tfrac{1}{2} + \tfrac{1}{2} cos(2\pi y + \pi)\right] \quad (2)$$

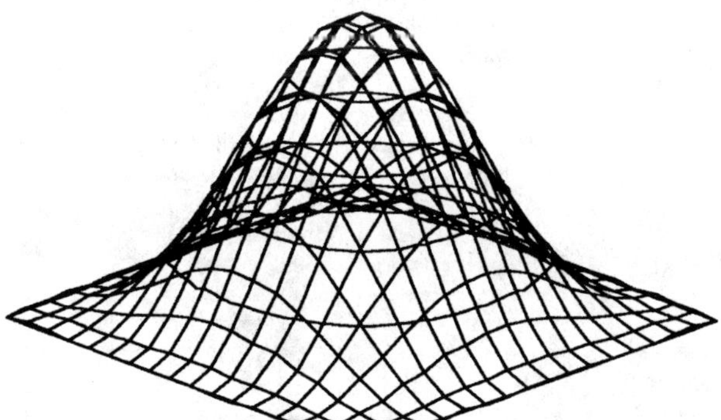

Figure 1. A representation of the cosine function

**Table 1:Small Initial Architecture Cosine Results**

| Architecture | Progress | Action |
|---|---|---|
| 1x1x1 | successful | added input 2 |
| 2x1x1 | unsuccessful | added hidden node |
| 2x2x1 | successful | added hidden node |
| 2x3x1 | reached target | deleted input 1 |
| 1x3x1 | unsuccessful | added input 1 |
| 2x3x1 | reached target | stopped |

For the first training case, the algorithm was started with only one of the inputs, and one node in the hidden layer. The DNA algorithm was stopped when the same architecture successfully mapped the output to within the target precision twice with the same architecture, and a smaller architecture was shown to be insufficient. All of the examples were trained on a DEC 3000/600 workstation. The training time for the first case was less than a minute, the results are illustrated in Table 1.

For the second case, the algorithm was started with an architecture which was much larger than was shown to be necessary by the first case. The goal was still to map the desired output to within 2.5% on average, but also to see if the DNA algorithm could come up with the same architecture as before. The algorithm took much longer to converge this time since training the larger networks takes longer and because many more steps were required. Total training time was about five minutes, the results are illustrated in Table 2.

As can be seen by these two simple cases, the DNA algorithm was able to determine what the smallest viable architecture was regardless of the size of the initial network. In the next example, the DNA algorithm was applied to some simulated chemical process data.

**Table 2: Large Initial Architecture Cosine Results**

| Architecture | Progress | Action |
|---|---|---|
| 2x10x1 | reached target | deleted hidden node |
| 2x9x1 | reached target | deleted hidden node |
| 2x8x1 | reached target | deleted hidden node |
| 2x7x1 | reached target | deleted hidden node |
| 2x6x1 | reached target | deleted hidden node |
| 2x5x1 | reached target | deleted hidden node |
| 2x4x1 | reached target | deleted input 1 |
| 1x4x1 | unsuccessful | added hidden node |
| 1x5x1 | unsuccessful | deleted hidden node |
| 1x4x1 | unsuccessful | added input 1 |
| 2x4x1 | reached target | deleted hidden node |
| 2x3x1 | reached target | deleted input 2 |
| 1x3x1 | unsuccessful | added hidden node |
| 1x4x1 | unsuccessful | deleted hidden node |
| 1x3x1 | unsuccessful | deleted hidden node |
| 1x2x1 | unsuccessful | added input 2 |
| 2x2x1 | unsuccessful | added hidden node |
| 2x3x1 | reached target | stopped |

The second example uses data generated from the kinetic equations describing the operation of a constantly stirred tank chemical reactor. The system has four inputs that describe the tank inputs and reagents and one output that quantifies the quality of the product. We generated 1250 training patterns. This data is pathological in the sense that the first input is exactly the negative of the second input. The task would therefore be to see if the DNA algorithm can learn to model the output using only one of the first two inputs. According to one of our input ranking schemes which uses information theoretic correlations of the inputs to the output, the fourth input is the most useful, with the other inputs being ranked in reverse numerical order. The model is to be trained to map the output to within 5% of the actual value, which is approximately the level of uncorrelated noise in the output variable. The results of the DNA algorithm are given below in Table 3, training the models took about 45 minutes.

Once again the algorithm stopped when the same architecture reached the target cost twice. This example shows the viability of the DNA algorithm for real world problems.

**Table 3: Constantly Stirred Tank Reactor Results**

| Input # | Architecture | Progress | Action |
|---|---|---|---|
| 4 | 1x1x1 | successful | added input 3 |
| 4,3 | 2x1x1 | successful | added input 2 |
| 4,3,2 | 3x1x1 | unsuccessful | added hidden node |
| 4,3,2 | 3x2x1 | successful | added hidden node |
| 4,3,2 | 3x3x1 | successful | added hidden node |
| 4,3,2 | 3x4x1 | reached target | deleted input 2 |
| 4,3 | 2x4x1 | unsuccessful | added input 1 |
| 4,3,1 | 3x4x1 | unsuccessful | added hidden node |
| 4,3,1 | 3x5x1 | unsuccessful | deleted input 1 |
| 4,3 | 2x5x1 | unsuccessful | added hidden node |
| 4,3 | 2x6x1 | reached target | deleted hidden node |
| 4,3 | 2x5x1 | unsuccessful | added input 1 |
| 4,3,1 | 3x5x1 | reached target | deleted hidden node |
| 4,3,1 | 3x4x1 | reached target | stopped |

## CONCLUSIONS AND FUTURE WORK

The DNA algorithm has shown itself to be applicable to a wide range of problems, both with the typical test cases and with industrial data. Most of the industrial data that we have obtained is confidential due to the proprietary nature of the processes. Selecting the next input based on the initial ranking may not be the best on-line approach, so we will be developing methods to improve the technique.

The scaled conjugate gradient training algorithm is a very fast optimization method, but it has the drawback of using only a bias and a single type of transfer function. Some different training algorithms have been developed (Baba, 1989), (Bartlett, 1994a), and we have developed as yet unpublished advances for more flexible training algorithms based on quasi-random search techniques.

The DNA approach can be extended to recurrent networks, and to use sparsely connected networks. The algorithm was much more easily implemented due to taking a modular approach to the internal representation of the nodes and weights. This allowed the node to be shut off in an efficient manner and greatly simplified the pruning of nodes.

## REFERENCES


Baba, N., A New Approach for Finding the Global Minimum of Error Function of Neural Networks, *Neural Networks*, 2, 367-373 (1989).

Bartlett, E. B., {A Stochastic Training Algorithm for Neural Networks, *Neurocomputing*, 6, 31-43 (1994).

Bartlett, E. B., Dynamic Node Architecture Learning: An Information Theoretic Approach, *Neural Networks*, 7, 129-140 (1994).

Bartlett, E. B., A Dynamic Node Architecture Scheme for Layered Neural Networks, *Journal of Artificial Neural Networks*, 1 (2),224-245 (1994).

Baum, E. B., and Haussler, D., What Size Net Gives Valid Generalization, *Neural Computation*, 1, 151-160 (1989).

Hammitt, A. M. and Bartlett, E. B., Determining Functional Relationships from Trained Neural Networks, *Mathematical and Computer Modelling*, 22 (3), 83-103, (1995).

Hornik, K., Stinchcomhbe, M., and White, H., Multilayer Feedforward Networks are Universal Approximators, *Neural Networks*, 2, 359-366 (1989).

Lippmann, R. P., An Introduction to Computing with Neural Nets, *IEEE ASSP Magazine*, April 1987, 4-22.

Moller. M.F., A Scaled Conjugate Gradient Algorithm for Fast Supervised Learning, *Neural Networks*, 6, 525-533 (1993).

# Noisy Recurrent Neural Networks

Oluseyi Olurotimi and Soumitra Das
Dept. of Electrical and Computer Engineering, MS 1G5
George Mason University
Fairfax, VA 22030
Phone: (703)993-1603
E-mail: seyi@bass.gmu.edu; soumitra@bass.gmu.edu

*ABSTRACT*

In this paper we examine the effect of noise on the general recurrent neural network (RNN) model. We outline results establishing uniform boundedness of moments of the neuron states over time. For practical usefulness we subsequently consider the RNN as a stochastic estimator of noisy dynamic patterns. We give bias and variance measures for this estimator. This has significant practical implications, since neural network design is nonminimal in the sense that several different networks can be constructed to solve the same problem. The results in this paper allow the user to quantitatively evaluate given RNN's for noise performance. In addition, the designer can use these results to constrain the design space so that the achieved design satisfies performance specifications whenever possible. We demonstrate the application of these results to a trajectory recall problem.

## 1. Introduction

Recurrent neural networks (RNN) are increasingly being proposed for use in many engineering applications. For real applications, it is important to consider the presence of noise in the system signals or in the RNN. It is prudent to anticipate that the presence of noise may alter system behavior in a significant way from what is observed in the noise-free case. There is a need for useful, design-oriented guidelines for RNN applications.

Past work on RNN that include noise has been mostly concerned either with improving learning and generalization [1], or with stochastic optimization [2, 3, 4]. In some cases, the class of RNN studied was further restricted to RNN stable-attractor networks with symmetric weight matrices

In this paper we present analytic results for the typical RNN presence of noise. To obtain our RNN model, we perturb the well-known first-order additive form with noise:

$$\dot{x_i} = -a_i x_i + g_i(\sum_{j-1}^{n} w_{ij}x_j) + I_i(t) + \sqrt{T_i}\eta_i, \; i = 1, 2, \ldots, n \quad (1)$$

where $a_i$ is a time constant, $w_{ij}$ is a weight connecting the $j^{th}$ neuron to the $i^{th}$ neuron and is not necessarily symmetric, $I_i(t)$ is an input to $i^{th}$ neuron, and $-g_{max} < g_i(\cdot) < g_{max}$ is a sigmoid or logistic function. $\eta_i$ is zero-mean, unit variance white noise. $T_i$ is a nonnegative noise scaling factor or magnitude. We do not assume control of this factor.

When the set of differential equations driven by white noise is transformed into a set of stochastic differential equations[5], we obtain

$$dx_i = (-a_i x_i + g(\sum_{j=1}^{n} w_{ij}x_j) + I_i(t))dt + \sqrt{T_i}dW_i(t), \; i = 1, 2, \ldots, n \quad (2)$$

where $i = 1, 2...n$, and $W_i$ is the Wiener process or the Brownian motion perturbing the $i^{th}$ neuron.

## 2. Main Results

In this section we will outline the main results. The results can be viewed as belonging to two classes. The first class deals with general well-behavedness of the stochastic systems with respect to existence and uniqueness, and boundedness of moments. The second class deals with the stochastic RNN as an estimator of time-varying patterns. In other words, if the deterministic RNN has been trained to follow a particular time-varying pattern, the stochastic RNN can be viewed as an estimator of that pattern. Consequently, we can derive bounds and expressions for estimator characteristics such as the bias and variance. This then allows us to make design-relevant statements about parameter requirements for good RNN design. We will state only the results and omit the proofs due to space considerations.

### 2.1. Generic properties

**Proposition 2.1.1** *Consider a noisy RNN described by the equation (2). If the initial condition $x_i(0)$ is independent of $W_i(t)$ and $Ex_i(0)^2 < \infty$, then the trajectory $x_i(t)$ is pathwise unique, that is, if there exists two trajectories $x_{i1}(t)$ and $x_{i2}(t)$ for an initial condition $x_i(0)$, then $P(\sup_{[0,T]} |x_{i1}(t) - x_{i2}(t)| = 0) = 1$.*

**Proposition 2.1.2** *The first moment, $Ex_i(t)$, of the trajectory $x_i(t)$ of the noisy RNN is uniformly bounded for all time.*

**Lemma 2.1.1** *$Cov(x_i g(\sum_{j=1}^{n} w_{ij}x_j))$ is uniformly bounded for all time.*

**Proposition 2.1.3** *The second moment, $Ex_i^2(t)$, of the trajectory $x_i(t)$ of the noisy RNN is uniformly bounded for all time.*

### 2.2. Stochastic RNN as an estimator of time-varying patterns

The earlier results established the fact that the stochastic variations of the trajectory traced by the noisy RNN will be uniformly bounded for all time. Although these boundedness results are an important step in the analysis of noisy RNN, they are generally not precise enough to be useful from the practical RNN design viewpoint. By considering the stochastic RNN as an estimator of time-varying patterns, we are able to derive more practical design and evaluation measures for noisy RNN. In particular, we are able to relate practically relevant quantities such as the bias and variance of this estimator to actual RNN parameters. This information can subsequently be used in the design, selection, and evaluation of RNN solutions.

#### 2.2.1. RNN stochastic estimator definition

Consider that a deterministic RNN models the time-varying pattern $\chi_i(t),\ i = 1, 2, \ldots, n,\ t \geq 0$ according to

$$\dot{\chi}_i(t) \quad = \quad -a_i\chi_i(t) + g_i(\sum_{j=1}^{n} w_{ij}\chi_j(t)) + I_i(t),\ i = 1, 2, \ldots, n \tag{3}$$

Now suppose that we construct a similar RNN driven by white noise. This noisy RNN can be represented by the equation (2). When a deterministic RNN is used in an environment containing input noise, the user has implicitly *chosen* to interpret $x_i(t)$ defined in the equation (2) as an estimate of $\chi_i(t),\ i = 1, 2, \ldots, n,\ t \geq 0$.

Often in practice, we are only interested in the output of a subset of the neuron units, or of specific combinations thereof. The following definition is therefore useful.

**Definition 2.2.1 (RNN output matrix)** *The deterministic RNN model defined by equation (3) has associated with it a constant output matrix $\mathbf{P}$, such that, $\boldsymbol{\kappa}(t) = \mathbf{P}\boldsymbol{\chi}\ (t)$ is considered the deterministic RNN "output". Similarly, the noisy RNN model defined by equation (2) has associated with it a constant output matrix $\mathbf{P}$, such that, $\mathbf{y}(t) = \mathbf{P}\mathbf{x}(t)$ is considered the noisy RNN "output". If the desired output is m-dimensional, $\mathbf{P}$ is $m \times n$.*

**Definition 2.2.2 (Stochastic RNN estimator)** *Suppose the deterministic, bounded, time-varying pattern $\chi_i(t)$, $i = 1, 2, \ldots, n$, $t \geq 0$ is realized by a deterministic RNN according to the equation (3) and suppose the stochastic pattern $x_i(t)$, $i = 1, 2, \ldots, n$, $t \geq 0$ is realized by a stochastic differential equation according to the equation (2). We say that $x_i(t)$ is an estimator of $\chi_i(t)$, $i = 1, 2, \ldots, n$, $t \geq 0$. We also refer to $x_i(t)$ and the corresponding stochastic differential equation as a stochastic RNN estimator. Let the output matrix be* $\mathbf{P}$. *The bias $b_i(t)$ of the state of each neuron of this estimator is*

$$b_i(t) = Ex_i(t) - \chi_i(t), \; i = 1, 2, \ldots, n$$

*We denote the vector of biases by* $\mathbf{b}(t)$. *The mean squared error (as a function of time) for each neuron of this estimator is[8]*

$$E(x_i(t) - \chi_i(t))^2 = \sigma_i^2(t) + b_i^2(t), \; i = 1, 2, \ldots, n$$

*where $\sigma_i^2(t)$ is the* variance *of the state of neuron $i$ of the estimator, $\sigma_i^2(t) = E(x_i(t) - Ex_i(t))^2$*

**Proposition 2.2.1 (RNN estimator bias)** *Consider the stochastic RNN estimator of Definition 2.2.2 Let $\mathcal{W}$ be the $n \times n$ matrix with $(i,j)$ element $w_{ij}$, $\mathbf{A}$ the $n \times n$ diagonal matrix with $i^{th}$ entry $a_i$, and $\mathbf{\Delta(t)}$ the $n \times n$ diagonal matrix with $i^{th}$ entry $0 < \alpha_i(t) \leq g_i'(\cdot)$. Let $\mathbf{b}(t)$ be the vector of biases at time $t$. Let $\lambda_{min}[\mathbf{M}]$ and $\lambda_{max}[\mathbf{M}]$ denote the smallest and largest pointwise eigenvalue of $\mathbf{M}$ respectively (we shall also alternately denote these by $\lambda_{\mathbf{M}_{min}}$ and $\lambda_{\mathbf{M}_{max}}$ respectively when more convenient), then*

$$\begin{aligned} \|\mathbf{Pb}^{zi}(t)\| &\leq \|\mathbf{Pb}_0\| e^{\frac{1}{2}\int_0^t \mu_{\mathbf{P}}(\tau)\lambda_{\mathbf{M}_{max}}(\tau)d\tau}, \; t \geq 0 \\ \mathbf{M}(t) &= \mathbf{P}^T\mathbf{P}\left(-\mathbf{A} + \mathbf{\Delta}(t)\mathcal{W}\right) + \left(-\mathbf{A} + \mathbf{\Delta}(t)\mathcal{W}\right)^T \mathbf{P}^T\mathbf{P} \end{aligned} \tag{4}$$

*where $\mathbf{b}^{zi}(t)$ is the part of the bias $\mathbf{b}(t)$ associated with the initial bias $\mathbf{b}(0) = \mathbf{b}_0$*[1]*, $0 < \mu_{\mathbf{P}}(t) < \infty$, and $\mathbf{P}$ is a user-specified output matrix. Furthermore if*

$$\lambda_{max}\left[-2\mathbf{A} + \mathbf{\Delta}(t)\mathcal{W} + \mathcal{W}^T\mathbf{\Delta}(t)\right] < 0, \; t \geq 0$$

*then an upper bound on the bias is given by*

$$\lim_{t\to\infty} \|\mathbf{b}(t)\| \leq \frac{\nu\sqrt{n}(g_{max} - g_{min})}{\lambda}$$

*where $0 < \nu < \infty$, $0 < \lambda < \infty$ are associated with the rate of decay of a state transition matrix associated with the estimator.*

**A remark on the eigenvalue tests of Proposition 2.2.1** Although the presence of the time-varying diagonal matrix $\mathbf{\Delta}(t)$ suggests a search in an $n$-dimensional compact cube to determine the suitability of the eigenvalues of $\mathbf{M}(t)$ ($\mathbf{P} = \mathbf{I}$), practical considerations simplify this considerably. Specifically, the results of the Proposition suggest that one should look for large negative eigenvalues in order to achieve small biases. Ideally then, we would like systems for which small perturbations ($b_i(t)$ small) around the nominal trajectory yield good matrices. From the (omitted) details of the proof, we have

$$\alpha_i(t) = \frac{g_i\left(\sum_{j=1}^n w_{ij}Ex_j(t)\right) - g_i\left(\sum_{j=1}^n w_{ij}(Ex_j(t) - b_j(t))\right)}{\sum_{j=1}^n w_{ij}b_j(t)} \approx g_i'\left(\sum_{j=1}^n w_{ij}Ex_j(t)\right) \approx g_i'\left(\sum_{j=1}^n w_{ij}\chi_j(t)\right)$$

This implies only a one-dimensional search along $t$. This is because for a given design, we *know* the $\chi_i(t)$, $i = 1, 2, \ldots, n$, $t \geq 0$ that we used in training. Since after training, we also *know* $w_{ij}$, $i = 1, 2, \ldots, n$, $j = 1, 2, \ldots, n$, then for each (sampled) point in time $t = t_k$, we uniformly compute *all* $n$ quantities $\alpha_i(t_k)$, and the pointwise eigenvalue can be computed. Note that this process is equivalent to a necessary condition in the sense that a matrix that fails the test under this small perturbation condition will generally not yield a good estimator. On the other hand, the matrices of good estimators will generally pass this perturbation test.

[1] The superscript $zi$ indicates that this is the unforced, homogeneous, or "zero-input" solution.

**Proposition 2.2.2** *Consider the stochastic RNN estimator of Definition 2.2.2. Then*

$$\sigma_i^2(t) \leq \frac{K}{2(a_i - \alpha_i(t)w_{ii})} \tag{5}$$

*where $\sigma_i^2(t)$ is the variance of the estimator. $K$ is a finite constant, $0 < \alpha_i(t) < g_i'(\cdot)$.*

### 2.3. Results summary and further remarks

The above results give practical guidelines for selecting between competing RNN designs for the same task. Equivalently, it guides the designer in constraining certain parameter ranges whenever possible in order to achieve improved noise performance.

We summarize as follows:

1. **Minimum bias:** For good bias performance, an RNN with the lowest values of the maximum pointwise eigenvalue of

$$\mathbf{M}(t) = \mathbf{P}^T\mathbf{P}\left(-\mathbf{A} + \mathbf{\Delta}(t)\mathcal{W}\right) + \left(-\mathbf{A} + \mathbf{\Delta}(t)\mathcal{W}\right)^T \mathbf{P}^T\mathbf{P}$$

   should be chosen.
2. **Minimum variance:** An RNN with the lowest value of

$$\beta_i(t) = -2a_i + 2\alpha_i(t)w_{ii}$$

   on the output neurons should be chosen.
3. **Uniformity of tests:** It can also be shown that the minimum variance condition is suitable for the bias condition as well. In fact, we see that since the variance condition tests the diagonals of the matrix of the bias condition, then a complementary relationship exists by Gerschgorin's theorem.

We shall denote the time average of $\beta_i(t)$ by $\bar{\beta}_i(t)$. An eigenvalue of $\mathbf{M}(t)$ is denoted $\lambda[\mathbf{M}(t)]$, and the maximum pointwise eigenvalue of $\mathbf{M}(t)$ is denoted $\lambda_{max}[\mathbf{M}(t)]$. The time average of the maximum pointwise eigenvalue of $\mathbf{M}(t)$ is denoted $\bar{\lambda}_{max}[\mathbf{M}]$.

## 3. Using the results to evaluate RNN designs

Design of RNN is nonminimal in the sense that several RNN's can be constructed to generate similar trajectories. However, the performance of these RNN's may vary significantly in the presence of noise. Certain RNN designs may not be robust to noise. As summarized above, our previous work relates the bias and variance of the RNN considered as an estimator, to RNN parameters such as weights and decay constants. In this section, we will present a design procedure for evaluating and selecting an RNN based on predicted bias and the variance performance. This same procedure can be used by a designer in constraining the design space in order to obtain better designs.

Let the trajectory to be learned be defined as $F(\chi(t)) = 0$, where $\chi_i(t)$ is the $i^{th}$ element of $\chi(t)$. $\chi_i(t)$ is the desired output of $i^{th}$ neuron unit and $F(\cdot)$ is a known continuous and bounded function. For example, if the trajectory to be learned is circular[6], we may have $\chi_1(t) = \sin t$ and $\chi_2(t) = \cos t$, and $F(\chi(t)) = F(\chi_1(t), \chi_2(t)) = 1 - \chi_1^2(t) - \chi_2^2(t) = 0$. We train the RNN such that the actual output $x_i(t)$ of the $i^{th}$ output unit follows $\chi_i(t)$. Note that the actual trajectory obtained from the RNN is $F(\{x_i(t), i \epsilon S_o\}) = 0$, where $S_o$ represents a set of indices for output units. Due to the presence of noise the actual trajectory will exhibit random deviations from its nominal path. Let the variance of the system trajectory $F(\{x_i, i \epsilon S_o\})$ be denoted by $\sigma_F^2(t)$. Then we can say that for some $F_1$, $\sigma_F^2(t) = F_1(\{\sigma_i^2, i \epsilon S_o\}$. Using this and considerations from summary points 2 and 3 above, it is possible in principle to find a function of $\bar{\beta}_i$, $F_2(\{\bar{\beta}_i, i \epsilon S_o\})$ such that in general, decreasing values of $F_2(\bar{\beta})$ leads to decreasing values of the variance of the trajectory. $F_2(\bar{\beta})$ then serves as a design or evaluation measure of performance.

Based on this problem-dependent measure of performance $F_2(\bar{\beta})$ we present a procedure for robust RNN design in the presence of noise.

| RNN | $F_2(\bar{\beta})$ | $\bar{\lambda}_{max}[\mathbf{M}]$ |
|---|---|---|
| 1 | -7.6357 | .4300 |
| 2 | -1.8035 | .6599 |
| 3 | 0.2663 | .8748 |

Table: 1: Table of performance evaluation measures for recurrent neural network design example.

### 3.1. Design procedure

There are several algorithms for training an RNN. The following design procedure is compatible with typical techniques for trajectory learning in RNN. Since we consider the RNN an estimator, the goal is to design a network which has a low variance and a low bias. The procedure is outlined below.

Step 1: Find the weights (by any technique) $\mathcal{W}$ such that the error is below a predefined error goal.

Step 2: Compute $F_2(\bar{\beta})$.

Step 3: Save $\mathcal{W}$ and $F_2(\bar{\beta})$ as $\mathcal{W}^1$ and $F_2^1(\bar{\beta})$.

Step 4: Repeat steps 1 and 2.

Step 5: Save new $\mathcal{W}$ and $F_2(\bar{\beta})$ as $\mathcal{W}^2$ and $F_2^2(\bar{\beta})$.

Step 6: Compare $F_2^1(\bar{\beta})$ and $F_2^2(\bar{\beta})$. If $F_2^2(\bar{\beta}) < F_2^1(\bar{\beta})$, set $F_2^1(\bar{\beta}) = F_2^2(\bar{\beta}), \mathcal{W}^1 = \mathcal{W}^2$.

Step 7: Is stopping condition (e.g. allowed training time) satisfied? If yes, stop. Otherwise go to Step 4.

## 4. Example: Noisy RNN Bessel function generation

We illustrate the method by considering a set of three different RNN all trained to autonomously generate a Bessel function.

When the trained RNN are perturbed by white noise, the outputs will correspond to (different) *estimates* of the original stored temporal pattern.

Since the system trajectory $F(\mathbf{x})$ is $x_6(t)$ in this problem, the variance $\sigma_F^2(t) = \sigma_F^2(x_6(t))$. Since $\sigma_F^2(t) = \sigma_6^2$, the measure of performance for variance $F_2(\bar{\beta}) = \bar{\beta}_6$. The output matrix $\mathbf{P}$ has elements $P_{66} = 1$, $P_{ij} = 0$ elsewhere. We can compute $\bar{\lambda}_{max}[\mathbf{M}]$ and $F_2(\bar{\beta})$ for each RNN.

### 4.1. RNN designs

We used Pearlmutter's algorithm [6] for training the RNN's . We arbitrarily defined the networks to have 6 fully connected neurons: no external input, 5 hidden and 1 output units. The output neuron of the networks had index number 6. We used a hyperbolic tangent for the sigmoidal function, i.e., $g_i(x) = \tanh(x)$. Target training output was generated using Bessel's differential equation of the first kind of order 1. The function was sampled at 1001 evenly spaced points in the arbitrarily chosen interval $0 \leq t \leq 100$. We used an adaptive learning strategy for faster learning. To train the time constant $a_i$, we used gradient descent in $log(a_i)$ as suggested by Pearlmutter [7]. Error goal was set at .01.

We selected three networks with different weight matrices and with similar qualitative performance in the absence of noise. We computed measures of performance $F_2(\bar{\beta})$ and $\bar{\lambda}_{max}[\mathbf{M}]$ respectively for the networks. These were observed to be different from each other. Table 4.1 summarizes the results.

Notice that $F_2(\bar{\beta})$ for RNN #1 was the lowest. $\bar{\lambda}_{max}[\mathbf{M}]$ was also lowest for RNN 1. Hence, we expected that RNN #1 would have the best performance followed by RNN #2 and RNN #3.

### 4.2. RNN performance

The Bessel functions generated in the absence of noise by the three RNN are shown by the solid lines in Figure 1*(a) - (c)*. The dashed line in the plots represents the target Bessel function.

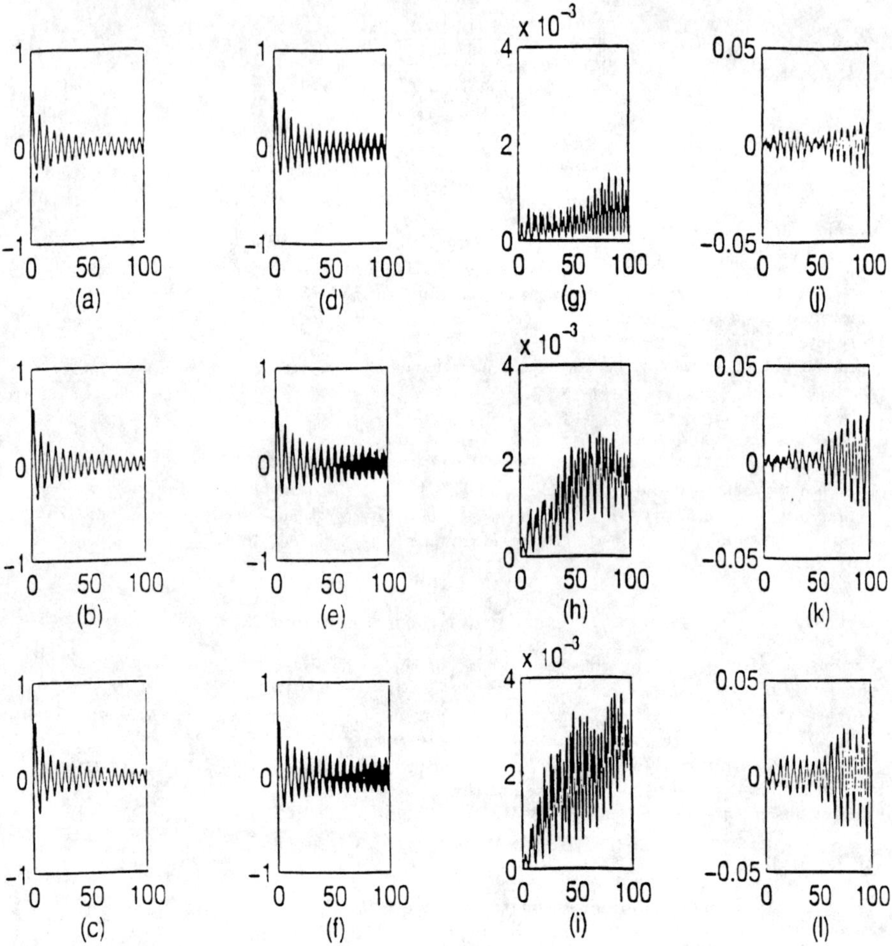


Fig. 1: *(a).* RNN #1 zero-noise output (solid line); desired output (dashed line). *(b).* RNN #2 zero-noise output (solid line); desired output (dashed line). *(c).* RNN #3 zero-noise output (solid line); desired output (dashed line). *(d).* RNN #1 noisy output. *(e).* RNN #2 noisy output. *(f).* RNN #3 noisy output. *(g).* RNN #1 variance. *(h).* RNN #2 variance. *(i).* RNN #3 variance. *(j).* RNN #1 bias. *(k).* RNN #2 bias. *(l).* RNN #3 bias.

We observe that the Bessel functions generated by these networks are almost identical even though the weight matrices of the networks are different. From observing these deterministic responses alone, we see that one cannot make a distinguishing performance evaluation.

The networks were then perturbed by white noise with a scaling factor $T_i$ of 0.0001. The noisy outputs are shown in Figure 1*(d) – (f)*. We observe that the noisy Bessel function generated by RNN #1 is least corrupted. The variances of the Bessel function are shown in Figure 1*(g) – (i)*. Now it is more clear that the best performing network in the presence of noise is RNN #1, followed by RNN #2 and RNN #3 respectively. This supported our original expectation based on the performance measures observed in Table 4.1 above before the testing commenced.

Figure 1*(j) – (l)* shows the bias of the noisy RNN's as estimators for the Bessel function. We expected, as the Figure confirms, that the performance would follow that observed for the variance.

Further plots are omitted due to space constraints.

## 5. Conclusion

We have presented some useful relations concerning the behavior of recurrent neural networks in the presence of noise. In particular, we have presented results on uniform boundedness of the dynamic systems driving important statistical measures of noisy RNN performance. For practical application purposes, the noisy RNN is viewed as a nonlinear estimator of the time pattern it was trained on. We thus give estimates of the bias and variance of this estimator as a function of RNN design parameters. This enables users to evaluate different RNN designs, and enables designers to limit parameter search, whenever possible, to regions of parameter (weight) space yielding acceptable noise performance.

We illustrated our results of noisy RNN performance for trajectory generation. Using three different noisy RNN's generating a Bessel function, we computed measures of performance for these RNN's. Based on these measures, we predicted successfully the expected RNN performance.

## References


[1] Kam Jim, Bill G. Horne & C. Lee Giles, "Effects of noise on convergence and generalization in recurrent neural networks," in Advances in Neural Information Processing Systems 7, Morgan Kaufmann, 1995, 649-656.

[2] Eugene Wong, "Stochastic Neural Networks," Algorithmica 6(1991), 466-478.

[3] Ronald A. Iltis & Peo-yih Ting, "Computing Association Probabilities Using Parallel Boltzmann Machines," IEEE Transactions on Neural Networks 4 (March 1993), 221-233.

[4] Bernard C. Levy & Milton B. Adams, "Global optimization with stochastic neural networks," Proceedings of the First IEEE International Conference on Neural Networks (June 1987), 681-689.

[5] Eugene Wong & Bruce Hajek, "Stochastic Processes in Engineering Systems", Springer-Verlag, 1985.

[6] Barak A. Pearlmutter, "Learning State Space Trajectories in Recurrent Neural Networks," Neural Computation 1 (1989), 263-269.

[7] Barak A. Pearlmutter, "Temporally Continuous vs. Clocked Networks", in Neural Networks in Robotics (Eds. Kenneth Y. Goldberg and George A. Bekey), Kluwer Academic (1992), 237 - 252.

[8] Bernard W. Lindgren, *Statistical Theory*, Macmillan Publishing Company, 1976.

# Modeling the Amacrine Cells in the Primate Retina for Contrast Enhancement and Noise Reduction of Images Provided to Artificial Vision Systems

David Enke and Cihan Dagli

Smart Engineering Systems Lab, University of Missouri at Rolla
236 Engineering Management, Rolla, MO 65409-0370

denke@umr.edu http://www.umr.edu/~denke
dagli@umr.edu http://www.umr.edu/~dagli

## ABSTRACT

*Although many of the neurons and neural connections within the human visual system have been studied extensively, this knowledge is rarely utilized in artificial vision systems. This paper examines the first processing region of the primate visual system, the retina. Previously, a retina model incorporating the interactions occurring within its first layer, involving the photoreceptors, bipolar, and horizontal cells, has proven effective for edge detection and noise reduction. Now, a model for the second layer of the retina, involving two different types of amacrine cells, is developed and tested. Simulations with gray scale images illustrate how the addition of the model amacrine cell interactions acting on the first layer outputs has the desired property of enhancing regions of low contrast, while providing gain control and noise reduction for the output response of the retina model.*

## INTRODUCTION

Recent discoveries in neuroscience are setting the stage for a renewed interest in the modeling of biologically inspired systems. No where is this more evident than with the modeling of the neural processes occurring within the human visual system. Not only is this recent research offering new insights with regard to the way the retina and the cortical areas process visual information, but advances in computer processing speeds, object oriented programming, and the use of parallel computing are making it possible to once again consider image processing models based on natural biological vision. In particular, the cells and neural interactions within the first layer of the retina have recently been modeled and shown to be successful in performing edge detection and noise reduction of an input image [Enke and Dagli, 1996]. This modeling of the retina will now continue with a examination of the second layer cells and interactions, with the amacrine cells being paramount in the following discussion. Although the functions and characteristics of these cells are somewhat speculative, models will be developed based on the knowledge that is presently available on, or hypothesized about, these cells. Simulations of the resulting retina architecture will then be made to demonstrate the performance of the model. Finally, it should be mentioned that the discussion will be taken from an engineering perspective, focusing on the retina model's overall function with regard to image processing, and less on the biologically significance of each individual cell model.

## THE RETINA NEURONS

The retina, which is located in the rear of the eye but considered by many to be part of the brain [Gregory, 1966], contains a unique network of neurons and photoreceptors. Although containing numerous feedforward and feedback connections, Figure 1 presents a simplified view of the primate retina shown with two distinct layers. Although this division is an obvious oversimplification, it suffices for the following discussion which is to focus on the image processing characteristics performed within this region of the brain.

Layer One: The first layer of processing within the retina involves the interactions between the photoreceptors, bipolar, and horizontal cells. Essentially, the bipolar cells detect edges by identifying contrast differences between the information that is provided to them by the photoreceptors and the horizontal cells, each receiving signals from a different region of the input intensity distribution, and subsequently providing both center and surround graded responses, respectively, to the bipolar cell. The modeling of these cells, along with the aforementioned connections, has been shown to perform edge detection and noise reduction of an input image [Enke and Dagli, 1996]. This first layer retina model also provides outputs responses that are physically similar to the characteristics of a Difference of Gaussian (DOG) filter, believed to have a response characteristics similar to the output of the retina [Croner and Kaplan, 1995], as well as the Laplacian of Gaussian (LOG) filter [Marr and Hildreth, 1980], often used for image processing applications.

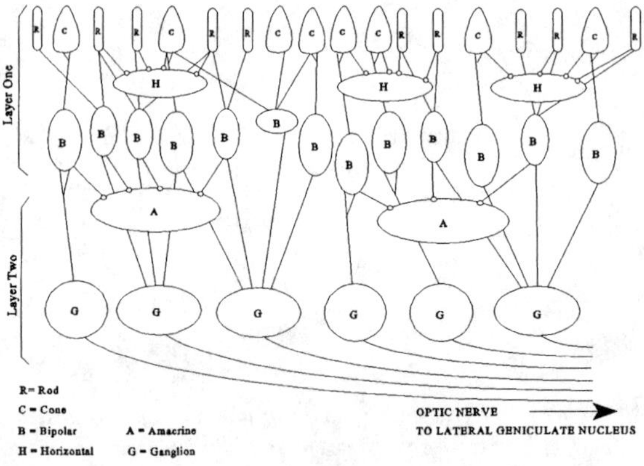


Figure 1: Neurons and Photoreceptors in the Retina

Unfortunately, although contrast detection of the layer one model has the desired property of identifying edges, it can highlight some edges at the expense of others. Often these inconsistencies result from poor lighting conditions or an improper sampling of the image, such that edge and intensity differences will be more pronounced in some regions of the acquired image. Therefore, even though the model is able to detect the lower contrast edges, their weak intensities may make them useless for any later image processing stages. Fortunately, the cells in the second layer of retina may overcome some of these difficulties by enhancing regions of low contrast, while adjusting the overall gain of the retina, allowing for a more consistent level of edge detection.

Layer Two: The second layer of retina cells, as shown in Figure 1, consist of the amacrine and ganglion cells. The first major difference between the two retina layers, besides the individual cell types, is that both the amacrine and ganglion cells fire all-or-nothing action potentials, and not graded responses like the layer one cells [Smith, 1995]. This appears to be necessary since a few types of amacrine cells, as well as the ganglion cells, are known to extend their dendrites over a large region of the retina. As such, the firing of action potentials is necessary since graded responses degrade considerably as the length of signal transmission increases. With regard to cell characteristics other than cell firing, the ganglion cells are similar to the bipolar cells in that they have a response that is influenced by the location of the cell's inputs with regard to its receptive field. The major difference involves the method in which the ganglion cells receive their inputs. Within the fovea, or higher acuity region of the retina, the bipolar cells appear to have a one-to-one relationship with the ganglion cells [Krebs and Krebs, 1989], often forming the cell's central region. Lateral support is provided to the ganglion cells by the amacrine cells. Unfortunately, less in known about the amacrine cells, making the modeling of these cells difficult.

Although 24 different amacrine cells have been identified in the retina, they generally fall into three types [Kolb, et al, 1992]. Type I amacrines have narrow dendritic fields and often act as a connection, or go between, for the bipolar and ganglion cells [Kolb, 1994]. At times, the connection of these cells with the bipolar cells may also be made through an electrical gap junction. It is believed that the main functions of these small field cells is to provide surround support for the ganglion cells [Kolb, 1994] and to sharpen the bipolar cell responses, ultimately allowing for an easier activation of the ganglion cells [Masland, 1988].

A second type of amacrine is more infrequent then the first type and usually spreads large lateral distances across the retina, with the level of spread increasing as these cells reach into the periphery [Kolb, et al, 1992]. The dendrites of these cells can also connect directly to the cell bodies of the first type of amacrines. Finally, a third type has characteristics that are a combination of the first two, they are numerous and have large lateral spreads, with these cells appearing mainly in the periphery. They also have a clearly defined axon for which to transmit their signal, which is not always apparent with the first two types of amacrines. These cells, like many of the other observed amacrines, also possess the characteristic of synapsing with the bipolar cells, and then back to themselves, in a form of reciprocal connection [Kolb, 1994], a result of their dual synaptic nature [Dowling, 1987].

Although the general functions of the amacrine cells are believed to be fairly well understood, the method in which they execute these functions is unknown. Since many amacrine cells do not have axons to transmit their signals, it can be hypothesized that they perform any necessary functions by utilizing the feedback to the bipolar cells, making connections at the triad of amacrine, ganglion, and bipolar cells. Figure 2 illustrates the possible interactions that are occurring at the triad of connections. For instance, a large field amacrine cell may perform a local enhancement of the input signal by initially summing bipolar signals within its predefined dendritic field. After the large field amacrine cell responds to this information, it will then send a signal back to each of the connected bipolar cells. This signal will be a locally normalized version of the feedforward signal previously sent by the bipolar cell. Since the amacrine cells fire action potentials, the feedback signal will also be filtered to enhance regions of low contrast, while controlling the level of much larger bipolar responses. This process will essentially perform a gain control for the retina, further limiting the range of the bipolar cell responses.

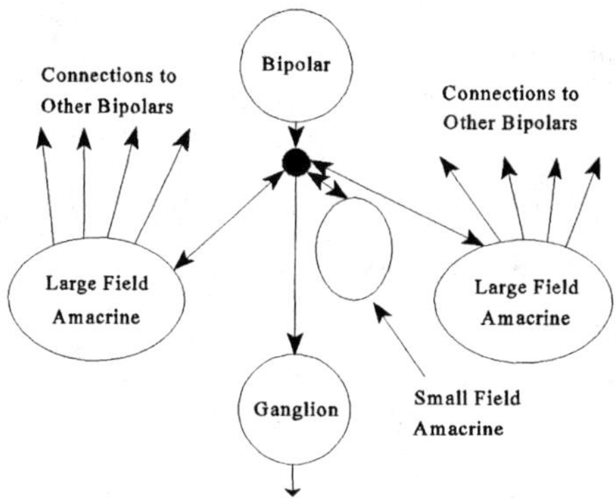


Figure 2: Triad of Connections

Since each bipolar cell will be providing input to multiple large field amacrine cells, it is possible that as a result of feedback each bipolar cell would receive feedback signals from numerous amacrine cells. Although the method the bipolar cells would use to process this information is unknown, it is possible that the type I amacrines, the ones that act as a go between for the bipolar and large field amacrine cells, and have the smallest dendritic fields, could act on this information before passing it on to the ganglion cells. Not only would these cells be able to respond to the feedback information, but they may also be providing the ganglion cells with a surround response, although in the fovea it is more likely that these cells are simply acting on the incoming information, since any additional center-surround filtering would only further reduce the details desired from this region. Finally, the resulting signal is sent on to the ganglion cell where it may be further filtered by the action potential response of this cell.

## MODELS OF AMACRINE CELL INTERACTIONS AND RESPONSES

The following section gives the static, algorithmic equations that are used to model the amacrine cell characteristics that were described in the previous section. As mentioned earlier, the modeling and performance of the first layer of the retina can be found elsewhere [Enke and Dagli, 1996].

As an initial attempt at modeling the interactions in the second layer of the retina, the interactions previously described and illustrated in Figure 2 will be modeled. As such, each bipolar cell will be paired with a corresponding ganglion cell. Each ganglion cell will likewise have both a type I (small field) and type II (large field) amacrine cell associated with it. Once again, the large field amacrine cells, which may have spans that are very similar to the small field cells in the fovea, will provide a local normalization and gain control, while the small fields cells will act on the feedback from the large field amacrine cells. For both of the amacrine cells a sigmoid function will be used as an activation function. Although a step function would define an idealized action potential, it may be more advantageous to utilize a sigmoid function since neurons that fire action potentials signify their strength of response by changing the rate of action potential firing, essentially providing a form of graded response. A sigmoid activation function will provide this advantage while still allowing the flexibility of modifying its slope, allowing the cell characteristics to vary between a pure linear response and a step function.

To model the large field amacrine cells, Equation (1) will be used to sum the incoming bipolar information that is contained within the receptive field neighborhood span of the cell. Within the equation, $a_l(x,y)$ represents the input summation of the large field amacrine cell at the retina grid location $(x,y)$ within the original two dimensional grid of bipolar cell responses. The two summations, along with neighborhood span $n_{al}$, dictate the number of bipolar cell responses that will be included in the final large field amacrine cell activation level. The term $b(i,j)$ represents the bipolar cell activation at location $(i,j)$ within the neighborhood grid. The value of $R_{al}$ is calculated from equation (2), and is used to perform an averaging of the incoming bipolar cell signals. Just as with the horizontal cell activations in layer one [Enke and Dagli, 1996], when equation (1) is used during simulations it will be necessary to constrain its use to only those receptors which can be completely surrounded by the predefined neighborhood.

Since the large field amacrine cell will be providing a filtered, and at times enhanced version of each initial bipolar cell signal, the response of the cell will depend upon which dendrite initially received the signal. Therefore, the eventual feedback signal sent to the triad of connections will be reliant upon both the large field amacrine cell activation and the level of the initial bipolar cell response that appeared on that particular connection. Equation (3) will be used to define the final large field amacrine cell activations, each of which is sent back to the corresponding triad of connections. Within this equation, $a_{l(x,y)}(i,j)$ represents the feedback response of the large field amacrine cell to the bipolar cell at location $(i,j)$ within the original cell neighborhood. The value of $net_{al(x,y)}(i,j)$ represents the net feedback activation of the large field amacrine cell at location $(x,y)$, with regard to the bipolar cell signal originating from location $(i,j)$ within the neighborhood grid. Equation (4) gives the formula that is used to calculate this net feedback activation, with the term $a_l(x,y)$ being defined by equation (1). The parameter $S_{al}$ represents the slope of the sigmoidal response, while $T_{al}$ represents a threshold, or shifting parameter that can be used to adjust the sigmoid to respond to a particular region of net activations. In other words, depending upon the value of the threshold, the amacrine cell may only respond well to net values between a small region, such as 0.4 and 1, with the value of the slope determining how the final responses will vary within this range (whether in a linear manner or more like a step function).

$$a_l(x,y) = \frac{1}{R_{al}} \sum_{i=x-n_{al}}^{x+n_{al}} \sum_{j=y-n_{al}}^{y+n_{al}} b(i,j) \tag{1}$$

$$R_{al} = \frac{1}{(2n_{al}+1)^2} \tag{2}$$

$$a_{l(x,y)}(i,j) = \frac{1}{1+e^{-S_{al}(net_{al(x,y)}(i,j)-T_{al})}} \tag{3}$$

$$net_{al(x,y)}(i,j) = \frac{a_l(x,y)}{b(i,j)} \tag{4}$$

$$net_{as}(x,y) = \frac{1}{R_{as}} \sum_{i=x-n_{as}}^{x+n_{as}} \sum_{j=y-n_{as}}^{y+n_{as}} a_{l(i,j)}(x,y) \tag{5}$$

$$R_{as} = \frac{1}{(2n_{as}+1)^2} \tag{6}$$

$$a_s(x,y) = \frac{1}{1+e^{-S_{as}(net_{as}(x,y)-T_{as})}} \tag{7}$$

After the large field amacrine cell calculations, it will then be the job of the small field cells to receive this information at the triad of connections. For instance, if the span of the small field amacrine cell is one, then a 3x3 neighborhood grid for this cell will require it to act on the feedback inputs provided by 9 different large field amacrine cells, each of which had the corresponding bipolar cell within, or only one pixel away, from its center. Given the neighborhood information, and the large field amacrine cell inputs, the net activation of the small field amacrine cells would be very similar to equation (1), and is subsequently given below in equation (5). The value of $R_{as}$ would also perform a function similar to equation (2), and is given below in equation (6). Within equation (5), $net_{as}(x,y)$ represents the net activation of the small field amacrine, with the small field neighborhood size defined by $n_{as}$. The term $a_{l(i,j)}(x,y)$ is defined by equation (3). Since the small field amacrine cell would fire an action potential, its net activation would also be acted upon by a sigmoidal function. Equation (7) gives the final response of the small field amacrine, $a_s(x,y)$, at location $(x,y)$, with the parameters $S_{as}$ and $T_{as}$ representing the slope and threshold of the small field amacrine, respectively.

Finally, in an attempt to initially examine the contrast enhancement and gain control characteristics of the small and large field amacrine cells, the ganglion cells will not be modeled at this time, but considered to take on the filtered response provided by the corresponding located small field amacrine cells. Although another sigmoidal function could be used for the ganglion response, its output would not add any extra information at this point. The characteristics of the ganglion cells will be considered at a later time.

## SIMULATIONS OF THE AMACRINE CELL INTERACTIONS

To test the response of the second layer amacrine cell interactions, modeled by equations (1-7), and used in conjunction with the first layer of retina processing, the image of a face, containing both high and low spatial frequencies, was provided to the network. Figure 3 illustrates the original 200 x 200 pixel image of a face that was captured from a camera and a 256 level gray scale video card. (Note: for the simulation, the parameters of the first retina layer were set to Q = 1, P = 0.01, $\theta_b = 0.02$, n = 1, and $\theta_f = 1.0$). Figure 4 illustrates a normalized version of the bipolar response provided by the first layer of the retina model.

As seen in Figure 4, although the first layer of the retina model is able to detect edges at varying levels of contrast, certain edges are not as clear. For instance, the nose, mouth, and chin are all at lower intensity levels, which could result in a vision system having a difficult time identifying these particular edges. Also, notice the varying intensity levels of bipolar cell responses when the borders of the hair are compared to the previously mentioned lower contrast regions. Finally, it is worth mentioning that Figure 4 gives a grid of bipolar cell responses after they have been normalized. This was done for illustration purposes

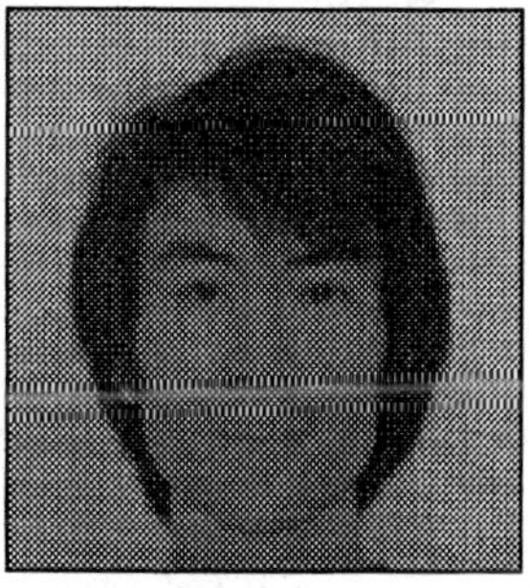

Figure 3: Original Face Image

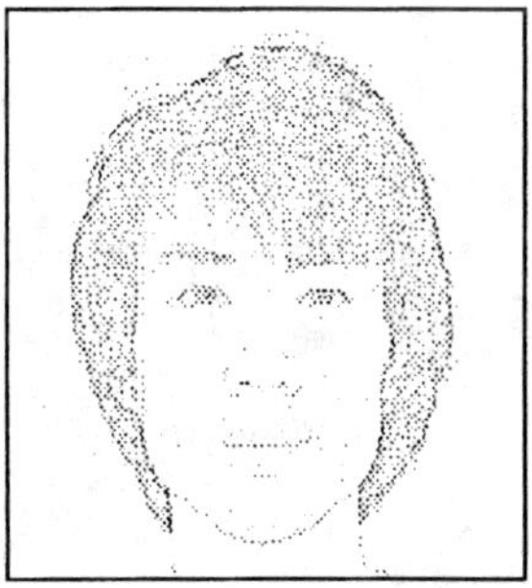

Figure 4: Bipolar Response

Figure 5: Short Amacrine, $S_{al} = 3$

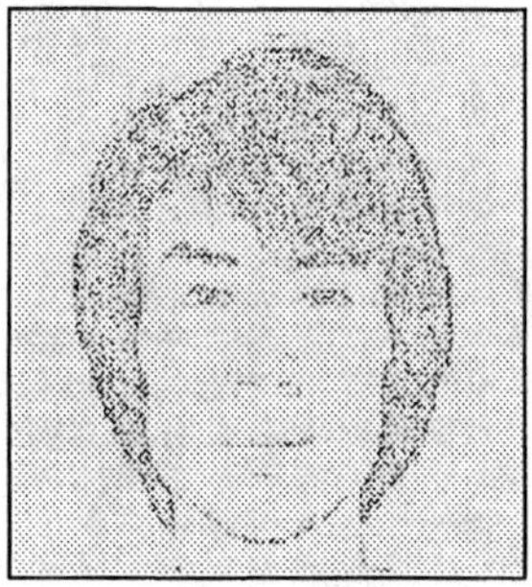

Figure 6: Short Amacrine, $S_{al} = 7$

to highlight the edge detection and contrast variations produced by the first layer of the retina. In actuality, the bipolar responses are low, resulting in a global normalization magnifying the intensity level differences.

To test the model amacrine cell interactions, the second layer parameters where set to $n_{al} = 1$, $n_{as} = 1$, $T_{as} = 0.6$, $T_{al} = 0.4$, and $S_{as} = 4$. Both $R_{al}$ and $R_{as}$ were calculated using equations (2) and (6), respectively. Figures 5 and 6 give the short field amacrine responses for long field amacrine sigmoidal slope values of $S_{al} = 3$ and 7, respectively. As illustrated in the figures, the second layer interactions of short and long field amacrine cells has the effect of enhancing the regions of low contrast, and subsequent bipolar response, while also providing gain control by limiting the range, or variation in the resulting output responses. The bipolar cell response normalization is also no longer necessary as a result of the second layer model, other than for viewing the bipolar signals. What also becomes evident from Figures 5 and 6 is how the levels of noise decrease as the slope of the long field amacrine sigmoid is increased. Remember, increasing the slope of the sigmoid causes the cell's response to come closer to approximating a step function, or pure action potential. Initially, the long field amacrine cells tend to locally normalize most points, including the noisy regions. By increasing the slope, these noisy regions, which were initially enhanced but still relatively small, are decreased closer to zero by the sigmoid function. Changing the threshold and slope parameters of the short field amacrine cell can also have a small, but similar effect on the noise, although the most noticeable differences are found at the large field amacrine cell level.

## CONCLUSIONS

Previously, a model of the layer one photoreceptors, bipolar, and horizontal cells in the retina has proven to be an effective technique for the detection of edges and the reduction of noise of an input gray scale image [Enke and Dagli, 1996]. Although this algorithm has been effective, it was incomplete in that it did not compensate for the small responses or large variations in gain that resulting from the layer one output responses. By modeling aspects of the amacrine cell interactions, and adding this layer to the overall retina model, it is now possible to not only provide a form of normalization for the layer one outputs, but to do so in such a way as to enhance regions of low contrast, while lowering the total gain of the resulting output signals. The flexibility of modifying the response characteristics of the amacrine cells, through the slope and threshold parameters, also gives the user flexibility in determining tradeoffs between noise reduction and edge detection. Although the functions and interconnections of the amacrine cells is somewhat speculative, and the models have been oversimplified and not directly compared with physiological data, equations (1-7) do offer an effective algorithm for the contrast enhancement of edge detected images. This technique should prove advantageous in a distributed environment such as parallel computing or electronic implementation.

## REFERENCES


1.) Croner, L.J., and E. Kaplan, "Receptive Fields of P and M Ganglion Cells Across the Primate Retina," *Vision Research*, Vol. 35, No. 1 (1995): 7-24.

2.) Dowling, J.E., *The Retina*, The Belnap Press of Harvard University Press, 1987.

3.) Enke and Dagli, "Modeling Biological Visual Processes for Improved Contrast Enhancement and Edge Detection of Artificial Vision Systems," *Proceedings of SPIE - Applications and Science of Artificial Neural Networks II*, Vol. 2760 (1996): 346-357.

4.) Gregory, R.L., *Eye and Brain*, McGraw-Hill Book Company, 1966.

5.) Kolb, H., K.A. Linberg, and S.K. Fisher, "Neurons of the Human Retina: A Golgi Study," *The Journal of Comparative Neurology*, Vol. 318 (1992): 147-187.

6.) Kolb, H., "The Architecture of Functional Neural Circuits in the Vertebrate Retina," *Investigative Ophthalmology and Visual Science*, Vol. 35, No. 5 (1994): 2385-2404.

7.) Krebs, W., and I.P. Krebs, "Quantitative Morphology of the Central Fovea in the Primate Retina," *The American Journal of Anatomy*, Vol. 184, No. 3 (1989): 225-236.

8.) Marr, D., and E. Hildreth, "Theory of Edge Detection," *Proceedings of the Royal Society of London, Series B*, Vol. 207 (1980): 187-217.

9.) Masland, R.H., "Amacrine Cells," *Trends in Neuroscience*, Vol. 11 No. 9 (1988): 405-410.

10.) Smith, R.G., "Retina", in *The Handbook of Brain Theory and Neural Networks* (Edited by M.A. Arbib), The MIT Press (1995): 816-820.

# Section 2 : Learning Algorithms and Training

# TRAINING OF ARTIFICIAL NEURAL NETWORKS VIA A BARRIER FUNCTION METHOD WITH $L_1$ NORM

**Theodore B. Trafalis and Tarek A. Tutunji**
*School of Industrial Engineering*
*University of Oklahoma*

**Abstract :**
Logarithmic barrier function methods are applied to the training of supervised feedforward neural networks. The $L_1$ norm (least absolute deviation) is used as the error criterion and a nonlinear constrained optimization problem is derived. Then, the barrier error is defined and the search direction is found using the recursive prediction error method (RPEM).

## INTRODUCTION

Artificial neural networks (ANN) training can be defined as a nonlinear programming problem. Different optimization methods such as gradient descent, conjugate gradient and quasi-Newton algorithms [Battiti (1992), Charalambous (1991), Watrous (1987), Bernard (1992), Chen et al. (1990), Kasparian 1994)] have been used in solving training optimization problems in ANN.

Barrier function methods form a class of interior point methods (IPM's) which are used for solving linear and nonlinear constrained optimization problems. Barrier function methods have already been used for training ANNs with good success [Tutunji (1996)].

In previous work [Trafalis and Tutunji (1994,1995)], we have developed barrier function methods for minimizing the least square error ($L_2$ norm) with box constraints on the weights. Here, we propose to use logarithmic barrier methods to minimize the least absolute error ($L_1$).

By using the $L_1$ norm as the error criterion we develop natural constraints that are incorporated directly to the objective function to form a barrier error function. Another advantage for using the least absolute deviation is robustness -- the estimated parameters are insensitive to changes in the input data.

## LOGARITHMIC BARRIER FUNCTION METHODS

Barrier functions are used to transform a constrained problem into an unconstrained problem or into a sequence of unconstrained problems. If the optimal solution occurs at the boundary of the feasible region, the procedure moves from the interior of the feasible region to the boundary. Consider the following constrained optimization problem:

$$\min f(w) \quad (1)$$
$$\text{s.t. } c(w) \le 0, \quad w \in W$$

where w is a vector and c is a vector function whose components are $c_1, c_2, \ldots, c_m$. Here, f, $c_1, \ldots, c_m$ are continuous functions on $E_n$, and the set W is a nonempty subset of $E_n$. Then the constrained problem can now be converted to a sequence of unconstrained problems which results in the barrier error:

$$\min \beta \ (w, \mu^k) = f(w) - \mu^k \sum_{i=1}^{m} \log(-c_i(w)) \quad (2)$$

where $\mu^k$ is a sequence of monotonically decreasing positive scalar parameters, the barrier parameters, which go to zero as k goes to infinity. Notice when $\mu^k$ becomes zero the barrier objective function becomes equal to the original objective function of the constrained optimization problem. For more details see Bazaraa et al. (1993).

## RECURSIVE PREDICTION ERROR METHOD

Recursive Prediction Error Method (RPEM) is a Newton-type algorithm which approximates the inverse Hessian iteratively [Soderstrom (1989)]. In this paper, we will use the RPEM update to approximate the inverse Hessian of the barrier error as follows :

$$P^{k+1} = \{P^k - P^k (\nabla\beta)^{k+1} \left( (\nabla\beta^{k+1})^T P^k (\nabla\beta)^{k+1} + \lambda \right)^{-1} (\nabla\beta^{k+1})^T P^k\} / \lambda \quad (3)$$

where $P^k$ is the matrix approximation of the inverse Hessian, $\nabla\beta^k$ is the gradient of the barrier error at iteration k, and $\lambda$ is the forgetting factor. The weights update is given as -- where $\alpha$ is the step length:

$$w^{k+1} = w^k - \alpha P^k \nabla\beta^k \quad (4)$$

## TRAINING AS A LINEAR PROGRAMMING PROBLEM

In this section we consider the training optimization problem as a linear programming problem. Consider a 2-layer neural network with an identity activation function as shown in Figure 1. This is the linear associator network. For simplicity, we consider one output node. The $L_1$ error associated with this network is:

$$\sum_{p=1}^{n} |E_p| = \sum_{p=1}^{n} \left| d_p - \sum_{i=1}^{m} x_{ip} w_i \right| \quad (5)$$

where p is the pattern, $d_p$ is the desired output, $x_{ip}$ are the inputs, and $w_i$ are the weights. We can form an equivalent linear programming problem as follows:

$$\min \sum_{p=1}^{n} e_p$$

$$\text{s.t. } e_p \geq d_p - \sum_{i=1}^{m} w_i x_{ip} \qquad p = 1,\ldots,n \qquad (6)$$

$$e_p \geq -d_p + \sum_{i=1}^{m} w_i x_{ip}$$

where $e_p = |E_p| = \left| d_p - \sum_{i=1}^{m} x_{ip} w_i \right|$

Then, we can construct the logarithmic barrier optimization problem as:

$$\min \beta(e, w, \mu) = \sum_{p=1}^{n} \left( e_p - \mu \log(e_p - d_p + \sum_{i=1}^{m} w_i x_{ip}) - \mu \log(e_p + d_p - \sum_{i=1}^{m} w_i x_{ip}) \right) \qquad (7)$$

Taking the derivatives with respect to $e_p$ and $w_i$ gives:

$$\frac{\partial \beta}{\partial e_p} = 1 - \mu \left( \frac{1}{e_p - d_p + \sum_{i=1}^{m} w_i x_{ip}} + \frac{1}{e_p + d_p - \sum_{i=1}^{m} w_i x_{ip}} \right) \qquad (8)$$

$$\frac{\partial \beta}{\partial w_i} = -\mu \sum_{p=1}^{n} \left( \frac{x_{ip}}{e_p - d_p + \sum_{i=1}^{m} w_i x_{ip}} - \frac{x_{ip}}{e_p + d_p - \sum_{i=1}^{m} w_i x_{ip}} \right) \qquad (9)$$

Now, we can use the logarithmic barrier algorithm with the RPEM to update the parameters and solve the logarithmic barrier optimization problem -- equation (7).

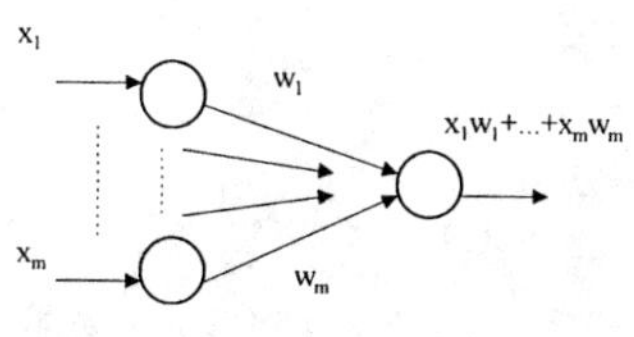


**Figure 1: Linear Associator Network**

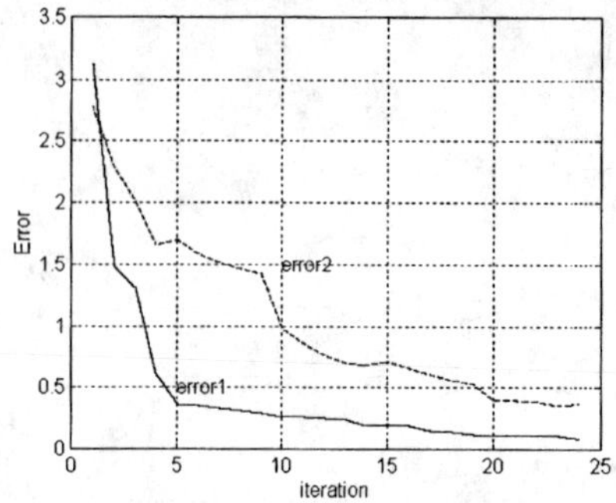


**Figure 2: Error Convergence for a Linear Associator Problem**

*Example 1* Consider a linear associator with the following two inputs $x_1 = [-0.35, 0.5]$ and $x_2 = [0.45, 0.85]$ -- each with two patterns and the desired output is [2 1]. This network can be modeled as the following linear programming problem:

$$
\begin{aligned}
\min\ & e_1 + e_2 \\
\text{s.t.}\ & e_1 \geq 2 - (-0.35w_1 + 0.45w_2) \\
& e_1 \geq -2 + (-0.35w_1 + 0.45w_2) \\
& e_2 \geq 1 - (0.5w_1 + 0.85w_2) \\
& e_2 \geq -1 + (0.5w_1 + 0.85w_2)
\end{aligned} \tag{10}
$$

The problem was changed to an equation similar to (7) and the barrier algorithm was used. The optimum variables found were w = [ -1.9445 2.2922 ] and e = [ 0.2943 0.0206] respectively. Note that the errors $e_1$ and $e_2$ are defined as variables simultaneously with the weights $w_1$ and $w_2$. Figure 2 shows the decrease of the errors over iteration number using the barrier algorithm with $L_1$ norm where error1 is the square error and error2 is the $L_1$ barrier error as described next:

$$
\begin{aligned}
\text{error1} &= (2 - (-0.35w_1 + 0.45w_2))^2 + (1 - (0.5w_1 + 0.85w_2))^2 \\
\text{error2} &= e_1 - \mu \times \{ \log(e_1 - 2 + (-0.35w_1 + 0.45w_2)) + \log(e_1 + 2 - (-0.35w_1 + 0.45w_2)) \\
&\quad + e_2 - \mu \times \{ \log(e_2 - 1 + (0.5w_1 + 0.85w_2)) + \log(e_2 + 1 - (0.5w_1 + 0.85w_2))
\end{aligned}
$$

## NEURAL NETWORK TRAINING WITH THE LEAST ABSOLUTE ERROR

A neural network training problem can be modeled using the $L_1$ norm as follows:

$$
\min \sum_{p=1}^{n} |E_p| = \min \sum_{p=1}^{n} |d_p - f_p(v, w)| \tag{11}
$$

where p is the number of patterns used, $d_p$ is the desired output, and $f_p(v,w)$ is the calculated output with $f_p$ as the sigmoidal function and v & w as the hidden and output weight variables respectively ( for simplicity we used only one output neuron, however, the results can be easily expanded to include multiple outputs).

The above optimization problem can be formulated as a nonlinear programming problem with linear objective and nonlinear constraints as follows:

$$\min \sum_{p=1}^{n} e_p \tag{12}$$

$$\text{s.t.} \quad e_p \geq d_p - f_p(v,w) \qquad p = 1, \ldots, n$$
$$e_p \geq -d_p + f_p(v,w)$$

Then, we can represent the above equation in vector form in the following way:

$$\min c^t s \tag{13}$$
$$\text{s.t.} \quad g(s) \leq 0$$

$$\text{where } s = \begin{bmatrix} e \\ v \\ w \end{bmatrix}, \quad e, v, w \text{ are vectors,} \quad c = \begin{bmatrix} 1 \\ 0 \\ 0 \end{bmatrix},$$

$$g(s) = \begin{bmatrix} g_1(s) \\ \ldots \\ g_n(s) \end{bmatrix}, \quad \text{and} \quad g_p(s) = \begin{bmatrix} d_p - f_p(v,w) - e_p \\ -d_p + f_{pj}(v,w) - e_p \end{bmatrix}$$

where the entries **1** and **0** in the vector c are themselves vectors with appropriate dimensions. The optimization problem (13) can be solved using the logarithmic barrier method. First, we construct the logarithmic barrier problem

$$\min \beta(e, v, w, \mu) = \sum_{p=1}^{n} \left(e_p - \mu \log(e_p - d_p + f_p(v,w)) - \mu \log(e_p + d_p - f_p(v,w)\right) \tag{14}$$

Now, we can calculate the gradients of equation (14) and use the RPEM to update the variables e,v, and w to decrease the barrier error function [Tutunji (1996)]. This algorithm has been applied to the traditional XOR problem with success. However, this algorithm is still under development. The two challenging issues of this algorithm are: finding a good updating formula for the barrier parameter and keeping feasibility while convergence.

Using the barrier method with the $L_1$ norm has promising results and can give important insights for developing other interior point algorithms such as the primal-dual.

## CONCLUSIONS

The logarithmic barrier function method can be used successfully for training ANN. Using the $L_1$ norm gives natural constraints that are used to form the barrier error. Preliminary computational results using a linear associator show the behavior of the barrier method. However, in order to develop this algorithm for solving large problems we need to find an appropriate updating formula for the barrier parameter and need to maintain feasibility. A primal-dual approach strongly related to the above approach is described in Trafalis and Couellan (1996).

## REFERENCES


Barnard, E. (1992), "Optimization for Training Neural Nets", *IEEE Transactions on Neural Networks*, 3(2), 232-240.

Battiti, R. (1992), "First-and-Second-Order Methods for Learning: Between Steepest Descent and Newton's Method", *Neural Computation* , 4, 141-166.

Bazaraa, M.S., Sherali, H.D., and Shetty, C.M., (1993*), Nonlinear Programming Theory and Algorithms* , Wiley, NY.

Charalambous, C., "Conjugate Gradient Algorithm for Efficient Training of Artificial Neural Networks"

Chen, S., Cowan, C., Billings, S., and Grant, P. (1990), "Parallel Recursive Prediction Error Algorithm for Training Layered Neural Networks", *International Journal of Control*, vol. 51, No. 6, 1215-1228.

Kasparian, V., Batur, C., Zhang, H., and Padovan, J. (1994), "Davidon Least Squares-Based Learning Algorithm for Feedforward Neural Networks", *Neural Networks*, Vol. 7, No.4, pp 661-670.

Soderstrom, T. and Stoica, P. (1989), *System Identification* , Prentice Hall International (UK), Englewood Cliffs, NJ. Trafalis, T. B. and Tutunji, T. A. (1994), "A Quasi-Newton Barrier Function Algorithm for Artificial Neural Networks with Bounded Weights", *Intelligent Engineering Systems Through Artificial Neural Networks*, 4, ASME Press, Editors: Dagli, Burke, Fernandez, and Ghosh, pp. 161-166.

Tutunji, T. A. (1996), " Logarithmic Barrier Functions and Newton-type Methods with Applications to Neural Network Training ", *Phd Dissertation*, School of Industrial Engineering, University of Oklahoma.

Trafalis, T.B. and N. P. Couellan (1996), " Neural Network Training via a Primal-Dual Interior Point Method for Nonlinear Programming" , *submitted to World Congress of Neural Networks '96.*

Trafalis, T. B. and Sieger, D. B. (1993),"Training of Multilayer Feedforward Artificial Neural Networks by an Algorithmic Barrier Function Adaptation Scheme", *Intelligent Engineering Systems Through Artificial Neural Networks*, 3, ASME Press, Editors: Dagli, Burke, Fernandez, and Ghosh, pp. 167-173.

Trafalis, T. B. and Tutunji, T. A. (1994), "A Quasi-Newton Barrier Function Algorithm for Artificial Neural Networks with Bounded Weights", *Intelligent Engineering Systems Through Artificial Neural Networks*, 4, ASME Press, Editors: Dagli, Burke, Fernandez, and Ghosh, pp. 161-166.

Trafalis, T. B. and Tutunji, T. A. (1995), "A Stochastic Logarithmic Barrier Function Method for Neural Networks Training with Bounded Weights", *Intelligent Engineering Systems Through Artificial Neural Networks*, 5, ASME Press, Editors: Dagli, Akay, Chen, Fernandez, and Ghosh pp. 167-172.

Watrous, R. L. (1987), "Learning Algorithm for Connectionist Networks: Applied Gradient Methods of Nonlinear Optimization", *IEEE First International Conference on Neural Networks*, San Diego, CA, 619-628.

# RISK-SENSITIVE AND ADAPTIVE FILTERING BY NEURAL NETWORKS WITH LONG- AND SHORT-TERM MEMORIES

James Ting-Ho Lo
Department of Mathematics and Statistics
University of Maryland Baltimore County
Baltimore, MD 21228
e-mail: lo@math.umbc.edu

**Abstract**

The 30 year old optimal nonlinear filtering problem was recently solved with a synthetic approach inspired by the development of artificial neural networks (ANNs). Continuing the same approach, this paper proposes the use of a risk-sensitive estimation criterion for synthesizing a neural filter that is robust toward environmental parameters not completely observable from the measurements. If there are also completely observable environmental parameters, this paper proposes using the nonlinear and linear weights of the neural filter as its long- and short-term memories respectively with the former determined in an a priori off-line training and the latter adapted on-line to the completely observable environmental parameters, both the off-line training and on-line adaptation being performed with respect to risk-sensitive criteria.

## INTRODUCTION

A solution of the general optimal nonlinear filtering problem that was intensively studied since the celebrated Kalman-Bucy filters were published was recently obtained by a synthetic approach using recurrent artificial neural networks (ANNs) [4, 6]. This paper is concerned with the corresponding robust and/or adaptive optimal filtering problems. Generally speaking, an optimal filter should be adaptive to completely observable environmental parameters and robust to those not completely observable.

When there are only environmental parameters that are not completely observable, we propose in this paper to use the standard risk-sensitive estimation error criteria for robust filtering. These criteria emphasize greater errors in an exponential manner and thereby induce robust performances. It is known that for linear systems, $H^{\infty}$-optimal filters can be derived by minimizing the standard risk-sensitive criteria [1, 7, 2].

When completely observable and not completely observable environmental parameters coexist, we propose a risk-sensitive adaptive neural filter comprising three pretrained ANNs serving as a signal estimator, a measurement predictor and a weight transformer. The nonlinear and linear weights of the signal estimator and measurement predictor are used as their long- and short-term memories (LASTMs) respectively, the former being determined in a priori trainings. The linear weights of the measurement predictor are adapted on-line and then transformed into the linear weights of the signal estimator by the weight transformer.

To prevent extreme values of the not completely observable environmental parameters from skewing the measurement predictor's linear weights adapted from the limited measurements collected on-line, Huber's estimation error criterion is used in the on-line adaptation of these linear weights.

## RISK-SENSITIVE FILTERING

A signal process to be estimated is described by the vector equation: For $t = 0, 1, \ldots,$

$$y(t+1) = f(y(t), \ldots, y(t-p+1), x(t), \ldots, x(t-q+1), \epsilon(t)), \tag{1}$$

where the vector-valued function, $f$, and the integers, $p$ and $q$, are given; $x(t)$ is a known input vector at time $t$; the components of $\epsilon(t)$ are not completely observable environmental parameters with given joint probability distributions over time; and the initial state, $(y_0, \ldots, y_{-p+1}, x_{-1}, \ldots, x_{-q+1})$, is a random vector with given probability distribution (reflecting the relative frequencies of the actual initial states of the system, (1), in operations). A measurement $z(t)$ of the vector output $y(t)$ is made available at time $t$, that satisfies

$$z(t) = h(y(t), \ldots, y(t-p_1+1), x(t), \ldots, x(t-q_1+1), \xi(t)), \tag{2}$$

where the components of $\xi(t)$ are not completely observable environmental parameters with given joint probability distributions over time. We note that time-varying parameters and/or coefficients of the signal and the measurement processes can be included in the vector $x(t)$.

For synthesizing a neural filter that is robust toward $\varepsilon$ and $\xi$, we propose to use the standard risk-sensitive estimation error criterion,

$$J(t) = \frac{1}{\lambda} \ln E \exp\left[\lambda \sum_{\tau=1}^{t} \|y(\tau) - \hat{y}(\tau)\|^2\right], \tag{3}$$

where $\hat{y}(t)$ denotes an estimate of $y(t)$ and $\lambda$ is a positive number. If the functions $f$ and $h$ are linear, the risk-sensitive filtering problem has been solved for the risk-sensitive criterion $J_{1,2}$ in [1, 7]. It was pointed out in the same papers that the obtained linear risk-sensitive filter is identical to the $H^\infty$-optimal filter for the same linear signal and measurement processes. (1) and (2).

It has tentatively been proven that if an optimal estimate of $y(t)$ with respect to the above $J(t)$ exists at each $t$, then the optimal estimate can be approximated to arbitrary accuracy with respect to the same error criterion by a multilayer perceptron with interconnected neurons (MLPWIN), under some regularity conditions. An MLPWIN is an MLP with at least one hidden layer of fully interconnected neurons, which feed their own activation levels to one another after one unit time delay. Based on this existence proof, a risk-sensitive neural filter can be synthesized to an arbitrarily small deviation from an optimal risk-sensitive estimator by trainning MLPWINs with the risk-sensitive training criterion,

$$J_T(w) = \frac{1}{\#S} \sum_{\omega \in S} \exp \left[ \lambda \sum_{\tau=1}^{T} \| y(\tau,\omega) - \hat{y}(\tau,\omega,w) \|^2 \right],$$

where $w$ denotes the weights of the MLPWIN under training; $S$ is a collection of exemplary realizations $\omega$ of the sequences, $(x(\tau,\omega), y(\tau,\omega), z(\tau,\omega))$, $\tau = -B$, $-B+1$, ..., $T$, $\omega \in S$; $B$ is an integer greater than or equal to the greater of $p$ and $q$; $\#S$ is the number of elements in the set $S$; $\lambda$ is a positive number; and $\hat{y}(\tau,\omega,w)$ is defined to be the output vector of the MLPWIN with its weights equal to the variables $w$, right after feeding $(x(s,\omega), z(s,\omega))$, $s = -B+1$, $-B+2$, ..., $\tau$ consecutively one at a time, at the input terminals of the MLPWIN. The positive integer $T$ and the collections $S$ should be selected and generated such that $S$ reflects the operating environments sufficiently.

## RISK-SENSITIVE ADAPTIVE FILTERING

Let us now consider robust and adaptive filtering for signal and measurement processes with both completely and not completely observable environmental parameters.

A signal process to be estimated is described by the vector equation: For $t = 0, 1, \ldots,$

$$y(t+1) = f(y(t), \ldots, y(t-p+1), x(t), \ldots, x(t-q+1), \theta(t), \epsilon(t)), \quad (4)$$

where the vector-valued function, $f$, and the integers, $p$ and $q$, are given; $x(t)$ is a known input vecotr at time $t$; the components of $\epsilon(t)$ are not completely observable environmental parameters with given joint probability distributions over time; $\theta(t)$ denotes a vector-valued completely observable environmental parameter; and the initial state, $(y_0, \ldots, y_{-p+1}, x_{-1}, \ldots, x_{-q+1})$, is a random vector with given probability distribution (reflecting the relative frequencies of the actual initial states of the signal, (4), in operations. A vector-valued measurement $z(t)$ of the vector output $y(t)$ is made available at time $t$, that satisfies

$$z(t) = h(y(t), \ldots, y(t-p_1+1), x(t), \ldots, x(t-q_1+1), \theta(t), \xi(t)), \quad (5)$$

where the components of $\xi(t)$ are environmental parameters not completely observable with given joint probability distributions over time.

For robust and adaptive filtering for the above signal and measurement processes, we propose a risk-sensitive adaptive neural filter that comprises three pretrained ANNs serving as a signal estimator, a measurement predictor, and a weight transformer respectively. The nonlinear weights of the signal estimator and measurement predictor are determined in a priori trainings using different types of robust training criterion. The linear weights of the measurement predictor are adapted on-line and are transformed into the linear weights of the signal estimator by the weight transformer. In the following, we will describe the three component ANNs of the proposed risk-sensitive adaptive neural filter in more detail.

### Signal Estimator

The signal estimator is an MLPWIN, whose weights $w$ are divided into the long-term memory and the short-term memory. The former consists of the nonlinear weights $u$ and is determined in an a priori training using the equations, (4) and (5). These equations are simulated to generate a priori training data as follows: Select a set $\Theta$ of examplary values, $\theta_1, \ldots, \theta_N$, of $\theta$. Using a pseudo-random number generator to simulate the initial state and the stochastic sequences $\epsilon$ and $\xi$, generate a collection $S_{\theta_n}$ of exemplary realizations $\omega$ of the sequences, $(x(\tau, \theta_n, \omega), y(\tau, \theta_n, \omega), z(\tau, \theta_n, \omega))$, $\tau = -B, -B+1, \ldots, T$, $\omega \in S_{\theta_n}$, for each selected exemplary value $\theta_n$. Here $B$ is an integer greater than or equal to the greater of $p$ and $q$. The positive integer $T$, the set $\Theta$ and the collections $S_{\theta_n}$ should be selected and generated such that the union $S := \cup_{n=1}^{N} S_{\theta_n}$ reflects the operating environments sufficiently. $S$ is called a priori training data. Then the nonlinear weights $u$ are determined by minimizing or reducing the following a priori training criterion by the variation of both $u$ and the variables $V(\theta) := (v(\theta_1), \ldots, v(\theta_N))$:

$$J(u, V(\theta)) = \frac{1}{\#S} \sum_{n=1}^{N} \sum_{\omega \in S_{\theta_n}} \exp C_1 \tag{6}$$

$$C_1 := \lambda \sum_{\tau=1}^{T} \| y(\tau, \theta_n, \omega) - \hat{y}(\tau, \theta_n, \omega, u, v(\theta_n)) \|^2 ,$$

where $\#S$ is the number of elements in the set $S$; and $\hat{y}(\tau, \theta_n, \omega, u, v(\theta_n))$ is defined to be the output vector of the MLPWIN with its linear weights $v$ equal to the variables $v(\theta_n)$, right after feeding $(x(s, \theta_n, \omega), z(s, \theta_n, \omega))$, $s = -B + 1, -B + 2, \ldots, \tau$ consecutively one at a time, at the input terminals of the MLPWIN's input terminals. Since the optimization of $J(u, v(\theta))$ is done off-line, as much computation as required to produce an optimal or very good solution can be performed. The values of the variables, $u, v(\theta_1), \ldots, v(\theta_N)$, that result from the a priori training, are denoted by $u^*, v^*(\theta_1), \ldots, v^*(\theta_N)$ and called the best-fit values of these variables.

This theorem has tentatively been proven: If an optimal risk-sensitive estimate exists for each value of $\theta(t)$ at each time $t$, it can be approximated, to any desired degree of accuracy with respect to the same risk-sensitive criterion, by the output of an MLPWIN with all its nonlinear weights independent of $\theta(t)$, for each value of $\theta(t)$ and at each time $t$, under some regularity conditions. This theorem justifies the use of the proposed risk-sensitive adaptive filter.

### Measurement Predictor

The measurement predictor is an MLPWIN, which inputs $z(t-1)$ and $x(t)$ and outputs an estimate $\hat{z}(t)$ of $z(t)$ at time $t$ during its operation. Its main function is to extract information about $\theta(t)$ from the on-line measurements and provide the information in the form of the measurement predictor's linear weights to the weight transformer. To prevent extreme values of $\varepsilon(t)$ and $\xi(t)$ from skewing the linear weights of the measurement predictor adapted with limited on-line measurements, the large prediction errors of the measurement predictor's outputs should be de-emphasized. Hence, robust error functions in statistics such as the absolute deviation or Huber's error function must be used [3]. Using only Huber's error function for illustration, the nonlinear weights $u_z$ of the measurement predictor are determined off-line by minimizing the following a priori training criterion by the variation of $u_z$ and $V_z(\theta) := (v_z(\theta_1), \ldots, v_z(\theta_N))$,

$$H(u_z, V_z(\theta)) = \frac{1}{T(\#S)} \sum_{n=1}^{N} \sum_{\omega \in S_{\theta_n}} \sum_{\tau=1}^{T} \sum_{i=1}^{m} C_2 \qquad (7)$$

$$C_2 := \rho\left(z_i(\tau, \theta_n, \omega) - \hat{z}_i(\tau, \theta_n, \omega, u_z, v_z(\theta_n))\right),$$

where Huber's error function $\rho(c) := c^2$, if $|c| \le M$ and $\rho(c) := 2M|c| - M^2$ if $|c| > M$ for the scalar variable $c$ and some prechosen positive number $M$; and $\hat{z}(\tau, \theta_n, \omega, u_z, v_z(\theta_n))$ is defined to be the output vector of the MLPWIN with its nonlinear and linear weights equal to the variables $u_z$ and $v_z(\theta_n)$, right after feeding $(x(s, \theta_n, \omega), z(s-1, \theta_n, \omega))$, $s = -B+1, -B+2, \ldots, \tau$ consecutively one at a time, at the input terminals of the MLPWIN. We note that $H(u_z, V_z(\theta))$ is twice differentiable except where $c = \pm M$. Since the optimization of $H(u_z, V_z(\theta))$ is done off-line, as much computation as required to produce an optimal or very good solution can be performed. The values of the variables, $u_z, v_z(\theta_1), \ldots, v_z(\theta_N)$, that result from the a priori training, are denoted by $u_z^*, v_z^*(\theta_1), \ldots, v_z^*(\theta_N)$ and called the best-fit values of these variables.

During the operation of the measurement predictor, $u_z^*$ is used as the nonlinear weights and only $v_z$ is adapted. This is done by minimizing the adaptive training criterion,

$$I(v_z, t) := \frac{1-\gamma}{1-\gamma^t} \sum_{\tau=1}^{t} \sum_{i=1}^{m} \gamma^{t-\tau} \rho\left(r_i(\tau) - \hat{z}_i(\tau, u_z^*, v_z)\right), \qquad (8)$$

where $\gamma$ is a positive number less that or equal to 1; and $\hat{z}(\tau, u_z^*, v_z)$ denotes the output of the MLPWIN with its nonadaptively adjustable weights equal to $u_z^*$ and its adaptively adjustable weights equal to $v_z$.

## Weight Transformer

Any mapping ANN can be used as the weights transformer. Supposing an MLP is used here, we denote its weights by $w_m$ and denote the outputs of the MLP subject to inputs $v_z$ by $\gamma(w_m, v_z)$, where the argument $w_m$ indicates the outputs' dependency on the MLP's weights. Using $(v_z^*(\theta_n), v^*(\theta_n))$, $n = 1, \ldots, N$, as the training data set of exemplary input/output pairs, the weights $w_m$ of the MLP are first determined by minimizing

$$Q(w_m) = \sum_{n=1}^{N} \left\| v^*(\theta_n) - \gamma(w_m, v_z^*(\theta_n)) \right\|, \qquad (9)$$

by the variation of $w_m$. Using the resulting minimizer $w_m^o$ as the initial quess, we obtain the optimal linear weights $w_m^*$ by minimizing, by the variation of $w_m$,

$$J_m(w_m) = \frac{1}{\#S}\sum_{n=1}^{N}\sum_{\omega\in S_{\theta_n}} \exp C_3$$

$$C_3 := \left[\lambda\sum_{\tau=1}^{T}\left\|y(\tau,\theta_n,\omega) - \hat{y}(\tau,\theta_n,\omega,u^*,v(\theta_n,w_m))\right\|^2\right].$$

## CONCLUSION

Although it has been assumed that the equations for the signal and measurement processes available to facilitate the discussion, these equations are not necessary for our method to be applied. If the equations are unavailable, experimental data can be used instead.

If the quality of a priori training data (or mathematical models) is not very good, the proposed method may still work, because the adaptation of the linear weights is expected to have great correction power, in view of the success of the conventional LMS and RLS for a linear combiner, an extremely crude "ANN".

If the a priori data is not enough for determining all the nonlinear weights, some nonlinear weights of the signal estimator and measurement predictor can be included in their respective sets of adaptively adjustable weights $v$ for adjustment in the on-line adaptation.

If the data quality is very bad, we can carry out the proposed a priori training, and then use the resulting nonlinear weights as the initial nonlinear weights in a special on-line training with different learning rates for different layers of weights.

## REFERENCES


[1] K. Glover and J. C. Doyle. State-space formulae for all stabilizing controllers that satisfy an h-infinity norm bound and relations to risk-sensitivity. *Systems Control Letters*, 11:167–172, 1988.

[2] B. Hassibi, A. H. Sayed, and T. Kailath. H-infinity optimality of the LMS algorithm. *IEEE Transactions on Signal Processing*, 44:267–280, 1996.

[3] P. Huber. *Robust Statistics.* John Wiley and Sons, New York, 1981.

[4] J. T. Lo. Neural network approach to optimal filtering. Invited paper presented at the 1992 World Congress of Nonlinear Analysts, Tampa, Florida, Aug. 1992.

[5] J. T. Lo. Synthetic approach to optimal filtering. *IEEE Transactions on Neural Networks*, 5:803–811, September 1994.

[6] J. L. Speyer, C.-H. Fan, and R. N. Banavar. Optimal stochastic estimation with exponential cost criteria. In *Proceedings of the 31st Conference on Decision and Control*, New York, New York, 1992. IEEE.

# STABILITY ANALYSIS OF PARAMETRIC UNCERTAIN NEURAL NETWORKS

**Anke Meyer–Bäse**[1]
Department of Electrical Engineering and Computer Science & University of Florida
Gainesville, Florida 32611
**Henning Scheich**[2]
Federal Institute for Neurobiology& D–39118 Magdeburg, Germany

## ABSTRACT

The stability of the equilibrium points of parametric uncertain neural networks is analyzed. The uncertainty arrises from the inexact knowledge of the synaptic contribution to the neural state. Via the presented method we are able to analyze a large group of inexact modeled neural networks. We also show the consequences of the stability analysis on the neural net parameters.

## 1 INTRODUCTION

Real–world neural systems are parametric uncertain interconnected neural systems. Parametric uncertainty can arrise in neural networks from inexact knowledge of system quantities such as neural constants, the parameters of lateral inhibition or synaptic contribution terms. The uncertainties may be expressed as deviations around a nominal parameter. The stability analysis of such systems regarding the equilibrium points is possible via a Lyapunov–based approach, if certain assumptions are satisfied. The main characteristics of this method are the linearization of a nominal system, which results in a linear subsystem with higher order nonlinear terms.

The method is interesting and useful for designing neural networks in technical applications and for explaining phenomena in neurobiological networks where parameters can not always be given explicit.

## 2 THE CLASS OF NEURAL NETWORKS WITH PARAMETRIC UNCERTAINTIES

This section defines the network of differential equations characterizing laterally inhibited neural networks. We consider a laterally inhibited network with a nonlinear synaptic contribution term which contains a parametric uncertainty.

The general neural network equations describe the temporal evolution of the STM (activity modification). For the $j$th neuron of a $N$–neuron network the equation is:

$$\dot{x}_j = -c_j x_j + \sum_{i=1}^{N} D_{ij} f(x_i) + g(x_j, \alpha) u_j \tag{1}$$

where $x_j$ is the current activity level, $c_j$ is the time constant of the neuron, $g(x_j, \alpha)$ is the contribution of the external stimulus term, $f(x_i)$ is the neuron's output, $D_{ij}$ is the lateral inhibition term and $u_i$ is the external stimulus.

[1] anke@cnel.ee.ufl.edu
[2] scheich@ifn-magdeburg.de

This system will be subject to our analysis considerations regarding the stability of their equilibrium points.

## 3 EXPONENTIAL STABILITY OF NONLINEAR PARAMETRIC UNCERTAIN SYSTEMS

We show in this section that it is possible to determine the exponential stability of the equilibrium points of this class of parametric uncertain neural networks. The idea is to use the feedback linearization and a Taylor series of the system about the nominal parameter to get a linear system plus nonlinear perturbations. Supposing further that the uncertainty satisfies a structure matching condition we can find a Lyapunov function for the perturbed system provided that the matching condition holds and the residual nonlinearities can be bounded from above.

In this sense we apply in this paper the results of this valuable analysis tool on the dynamics of laterally inhibited networks.

The system which is here to be considered is of the form [SSR86]

$$\dot{\mathbf{x}} = -\mathbf{Cx} + \mathbf{Df}(\mathbf{x}) + \mathbf{g}(\mathbf{x}, \alpha)\mathbf{u} \tag{2}$$

where it is assumed that $\mathbf{f}(\mathbf{x})$ and $\mathbf{g}(\mathbf{x}, \alpha)$ are $C^\infty$ vector fields defined on a dense submanifold $\mathbf{M} \in \mathbf{R}^\mathbf{n}$. In addition, $\mathbf{x}$ is defined on an open set $\mathbf{U}$ of $\mathbf{R}^\mathbf{n}$. The vector $\alpha$ is the unknown parameter vector. It is assumed that $\mathbf{f}$ and $\mathbf{g}$ are smooth vector fields for every $\alpha \in B_\alpha \in \mathbf{R}^\mathbf{p}$ where $\mathbf{B}_\alpha$ is an admissible set of unknown parameter vectors. Without loss of generality, we also assume that $\mathbf{f}(\mathbf{0}) = \mathbf{0}$. The nominal parameter vector $\alpha_\mathbf{n}$ is assumed known and the perturbations about $\alpha_\mathbf{n}$ are represented as:

$$\alpha = \alpha_\mathrm{n} + \delta\alpha \tag{3}$$

Furthermore the vector field $\mathbf{g}$ is analytic in $\alpha$ about $\alpha = \alpha_\mathbf{n}$.

We have to find a diffeomorphism $\mathbf{\Phi}(\mathbf{x})$ on $\mathbf{M}$ and nonlinear functions $\mathbf{a}(\mathbf{x})$ and $\mathbf{b}(\mathbf{x})$ such that the nominal system is rendered linear in the coordinates $\mathbf{z} = \mathbf{\Phi}(\mathbf{x})$ [Vid93]. That is:

$$\frac{\partial \mathbf{\Phi}(\mathbf{x})}{\partial \mathbf{x}} [-\mathbf{Cx} + \mathbf{Df}(\mathbf{x}) + \mathbf{g}(\mathbf{x}, \alpha_\mathrm{n})\mathbf{a}(\mathbf{x})]_{x=\Phi^{-1}(z)} = \mathbf{Az} \tag{4}$$

$$\frac{\partial \mathbf{\Phi}(\mathbf{x})}{\partial \mathbf{x}} [\mathbf{g}(\mathbf{x}, \alpha_\mathrm{n})\mathbf{b}(\mathbf{x})]_{x=\Phi^{-1}(z)} = \mathbf{B} \tag{5}$$

with the nonlinear feedback

$$\mathbf{u} = \mathbf{a}(\mathbf{x}) + \mathbf{b}(\mathbf{x})\mathbf{v} \tag{6}$$

This results in the linear plant

$$\dot{\mathbf{z}} = \mathbf{A}\mathbf{z} + \mathbf{B}\mathbf{v} \tag{7}$$

where $(\mathbf{A}, \mathbf{B})$ are in the Brunovsky canonical form. Supposing $\mathbf{z} = \mathbf{\Phi}(\mathbf{x}) = \mathbf{x}$, we get the following equations:

$$\mathbf{A} = -\mathbf{C} \tag{8}$$
$$\mathbf{B} = \mathbf{I} \tag{9}$$

and

$$\mathbf{a}(\mathbf{x}) = -\mathbf{g}^{-1}(\mathbf{x}, \alpha_{\mathbf{n}})\mathbf{D}\mathbf{f}(\mathbf{x}) \tag{10}$$
$$\mathbf{b}(\mathbf{x}) = -\mathbf{g}^{-1}(\mathbf{x}, \alpha_{\mathbf{n}}) \tag{11}$$

The input $\mathbf{v}$ will not only be utilized to stabilize the nominal linear system (which has all its poles at the origin), but it will be used to exponentially stabilize the system once the parametric uncertainty is introduced.

The Taylor series expansion about the nominal parameter $\alpha_n$ is

$$\dot{\mathbf{x}} = -\mathbf{C}\mathbf{x} + \mathbf{D}\mathbf{f}(\mathbf{x}) + \mathbf{g}(\mathbf{x}, \alpha_{\mathbf{n}})\mathbf{u} + \left(\frac{\partial \mathbf{g}}{\partial \alpha}\mathbf{u}\right)\Bigg|_{\alpha=\alpha_n} \delta\alpha + \frac{1}{2}\delta\alpha^T \left(\frac{\partial^2 \mathbf{g}}{\partial \alpha^2}\mathbf{u}\right)\Bigg|_{\alpha=\alpha_n} \delta\alpha + O(||\delta\alpha||^3) \tag{12}$$

The system equation of the perturbed system is given by

$$\dot{\mathbf{z}} = -\mathbf{C}\mathbf{z} + \mathbf{v} + \left(\frac{\partial \mathbf{g}}{\partial \alpha}\mathbf{a}(\mathbf{x}, \alpha_{\mathbf{n}}) + \frac{\partial \mathbf{g}}{\partial \alpha}\mathbf{b}(\mathbf{x}, \alpha_{\mathbf{n}})\mathbf{v}\right)\Bigg|_{\substack{\alpha=\alpha_n \\ x=\Phi^{-1}(z)}} \delta\alpha + O(||\delta\alpha||^2) \tag{13}$$

In the above equation all terms are known, since they are evaluated about $\alpha = \alpha_{\mathrm{n}}$ and if the parameter vector appears linearly in (2), then the expansion will be exact to first order in $\delta\alpha$.

Supposing that $\mathbf{A}$ is a Hurwitz matrix and that a matching condition [SSR86] is fullfilled that guarantees that the uncertain nonlinear terms lie in the range space of the input, we can rewrite (13) as

$$\dot{\mathbf{z}} = -\mathbf{C}\mathbf{z} + \mathbf{v} + \eta(\mathbf{z}, \mathbf{v}) \tag{14}$$

with

$$\eta(\mathbf{z}, \mathbf{v}) = \left(\frac{\partial \mathbf{g}}{\partial \alpha}\mathbf{a}(\mathbf{x}, \alpha_{\mathbf{n}}) + \frac{\partial \mathbf{g}}{\partial \alpha}\mathbf{b}(\mathbf{x}, \alpha_{\mathbf{n}})\mathbf{v}\right)\Bigg|_{\substack{\alpha=\alpha_n \\ x=\Phi^{-1}(z)}} \delta\alpha + O(||\delta\alpha||^2) \tag{15}$$

We must now assume the existance of a function $\rho(\mathbf{z},t)$ such that $||\mathbf{v}|| \leq \rho(\mathbf{z},t)$ and $||\eta|| \leq \rho(\mathbf{z},t)$.

From the above assumption and the the triangle inequality, one obtains

$$\rho(z,t) = \left[1 - ||\frac{\partial \mathbf{g}}{\partial \alpha}\mathbf{b}(\mathbf{x},\alpha_n)||_{\substack{\alpha=\alpha_n \\ x=\Phi^{-1}(z)}} \overset{max}{\alpha} ||\delta\alpha||\right]^{-1}$$

$$\cdot \; || \left(\frac{\partial \mathbf{g}}{\partial \alpha}\mathbf{a}(\mathbf{x},\alpha_n)\right) || \Big|_{\substack{\alpha=\alpha_n \\ x=\Phi^{-1}(z)}} \overset{max}{\alpha} ||\delta\alpha|| \tag{16}$$

It is important to state, that everything on the right hand of the equation is known *a priori* except for $\alpha^{max}||\delta\alpha||$ which is easily determined once one decides on a relative parameter deviation to consider.

We make now the last meaningful assumption before we state the theorem. The inequality

$$||\frac{\partial \mathbf{g}}{\partial \alpha}\mathbf{b}(\mathbf{x},\alpha_n)||_{\substack{\alpha=\alpha_n \\ x=\Phi^{-1}(z)}} \overset{max}{\alpha} ||\delta\alpha|| < 1 \tag{17}$$

must hold at the initial conditions $\mathbf{z}^0 = \mathbf{\Phi}(\mathbf{x}^0)$ and $t = t_0$.

The set of admissible initial conditions $S$ is then defined as

$$\mathbf{S} = \{\mathbf{z}^0 \in \mathbf{R}^n \, \Big| ||\frac{\partial \mathbf{g}}{\partial \alpha}\mathbf{b}(\mathbf{x},\alpha_n)||_{\substack{\alpha=\alpha_n \\ x^0=\Phi^{-1}(z^0)}} \overset{max}{\alpha} ||\delta\alpha|| < 1\} \tag{18}$$

The following theorem [SSR86] gives now the control law for $\mathbf{v}$ and the stability result for the full system.

**Theorem:** The equlibrium $\mathbf{z} = \mathbf{0}$ of the system (14) is exponentially stable with respect to initial conditions lying in $\mathbf{S}$ (defined in (18)) if $\mathbf{v}$ satisfies

$$\mathbf{v} = \begin{cases} -\rho(z,t), & \mathbf{z}^T\mathbf{P} \geq 0 \\ \rho(z,t), & \mathbf{z}^T\mathbf{P} < 0 \end{cases} \tag{19}$$

where $\mathbf{P}$ is the unique symmetric positive definite solution to the Lyapunov equation

$$\mathbf{PA} + \mathbf{A}^T\mathbf{P} + \mathbf{Q} = \mathbf{0} \tag{20}$$

with $\mathbf{Q}$ a given symmetric positive definite matrix [PH93].

The Lyapunov function for the linear system $\mathbf{A}, \mathbf{B}$ is $V(\mathbf{z}) = \mathbf{z}^T\mathbf{P}\mathbf{z}$. $V(\mathbf{z})$ is also a Lyapunov function for the system (14) with respect to the set $S$ in (18). The matrix $\mathbf{P}$ is given by:

$$p_{ij} = \begin{cases} \frac{1}{2c_i}, & i = j \\ 0, & i \neq j \end{cases} \tag{21}$$

This means that the equilibrium of (13) is globally exponentially stable with the control law in (19) for a given relative uncertainty.

## 4 DESIGN OF STABLE COMPETITIVE NEURAL NETWORKS WITH PARAMETRIC UNCERTAINTY

We design in this section a competitive neural network with parametric uncertainty that is able to store a desired pattern as a stable equilibrium. The neural network has to fullfill some initial conditions for guaranteeing stability.

The theoretical implications are illustrated in an example of a two neuron network.

*Example:* Let $N = 2$, $c_i = A$, $B_j = B$, $D_{ii} = \gamma > 0$, $D_{ij} = \beta < 0$, $g(x_i) = (x_i - \alpha)^2$ and the nonlinearity fullfilling $f(0) = 0$ in equation (1).

The linearized system is:

$$z_i = -Az_i + v_i \tag{22}$$

and the nonlinear feedback is given by:

$$u_i = \frac{\gamma f(x_i) + \beta f(x_j)}{(x_i - \alpha)^2} + \frac{1}{(x_i - \alpha)^2} v_i \tag{23}$$

The Lyapunov function of the system is:

$$V(\mathbf{z}) = \frac{1}{2A}(z_1^2 + z_2^2) \tag{24}$$

The exponential stability of the system can be guaranteed with respect to the initial condition lying in:

$$\mathbf{S} = \{\mathbf{z}^0 \in \mathbf{R}^2 \mid \frac{1}{||x_0 - \alpha_n||} \max_{\alpha} ||\delta\alpha|| < 1\} \tag{25}$$

## 5 CONCLUSIONS

We presented in this paper a procedure for linearization and stabilization of interconnected neural networks exhibiting parametric uncertainty. The chosen approach is based on a Lyapunov method and it supposes a structure matching condition on the uncertain nonlinear terms. The main advantage is that we are able to give the domain of the uncertainties for guaranteeing exponential stability of an equilibrium point. The method is interesting and useful for designing neural networks in technical applications and for explaining phenomena in neurobiological networks where parameters can not always be given explicit.

## REFERENCES


P. C. Parks und V. Hahn. *Stability Theory.* Prentice Hall, 1993.

M. W. Spong und H. Sira-Ramirez. *Robust Control Design Techniques for a Class of Nonlinear Systems. Proc. 1986 Amer. Contr. Conf.*, Seiten 1515–1522, August 1986.

M. Vidyasagar. *Nonlinear Systems Analysis.* Prentice Hall, 1993.

# BACKPROPAGATION LEARNING USING POSITIVE WEIGHTS IN OPTOELECTRONIC NEURAL NETWORKS

**W.M.D. BRADLEY AND R.J. MEARS**
*Cambridge University Engineering Dept., Trumpington Street, Cambridge CB2 1PZ, UK*

***ABSTRACT:***
Implementation of neural networks with *in situ* learning is hampered by the problems associated with representing both positive and negative synaptic weights. In this paper, it is concluded that the introduction of inhibitory neural units, combined with a simple modification of the backpropagation learning rule, leads to a learning ability that competes with momentum-aided backpropagation, using positive weights only and not itself requiring momentum terms. A multilayer optoelectronic architecture is proposed.

## INTRODUCTION

A major obstacle against the optical implementation of neural networks is the issue of signal polarity. In traditional electronic implementations, operations such as adding and multiplying, are readily accomplished with current mode circuits. In optoelectronics, however, the unidirectional nature of optical interconnect suggests that new representations of the signals are required.

In our work, we aim to implement a multilayer optoelectronic network with *in situ* learning. In this, the idea of replacing the standard network implementation with one consisting of paired sets of signals and devices is to be avoided, if possible, due to the additional physical complexity involved. Also, although biasing the signals (so that they are within the positive range) works well for summing nodes, optical multiplication of inputs (which can be accomplished extremely easily with passive spatial light modulators) typically requires unbiased positive multiplicands.

On the other hand, it has been shown in (Bródka and Macukow, 1995) and in (Dickey and DeLaurentis, 1993) that, using only twice as many neural units, it is possible to find solutions using positive weights only; but experience shows that finding that solution via steepest descent learning is often difficult or even impossible.

As it stands, this type of algorithm still attempts to search for solutions with negative weights, but the enforced saturation in a practical optical system precludes this. It is thought that the attraction towards unattainable solutions beyond the physical range leads to an arresting of the learning process at this point.

In this paper, we use an alternative approach to the backpropagation algorithm, combined with the use of inhibitory neural units, to give learning rates that typically surpass those associated with standard BP, but which accounts for the desire to use positive signals and weights.

## DELTA LEARNING IN A POSITIVE SYNAPTIC FIELD (ALPHA-SQUARED MODEL)

Consider the mapping $\alpha_{ji}^2 = w_{ji}$,, where $w_{ji}$ is an element of the weight matrix. For real values of $\alpha_{ji}$, the weight matrix will contain positive values only. If gradient descent is performed within the alpha space (as opposed to the weight space), then the search will continue to be restricted to the positive weight space.

Now consider the generalised delta rule, using gradient descent, applied to $\alpha_{ji}$, (Rumelhart and McClelland, 1986)

$$\Delta_p \alpha_{ji} \propto -\frac{\partial E_p}{\partial \alpha_{ji}}$$

With $E_p$ defined as the mean squared error between the desired and actual outputs, this leads to the following weight update rule:

$$\Delta_p w_{ji} = 4\eta w_{ji} \delta_{pj} o_{pi} \left(1 + \eta \delta_{pj} o_{pi}\right)$$

This transformation of the gradient descent into the alpha domain has the effect of warping the cost surface in a non-linear fashion. The low weight regions are stretched compared to the rest of the positive space; the slopes are made gentler, leading to minima close to the edge of the positive range becoming less attractive to a gradient descent search.

Moreover, slopes well into the positive range are effectively made steeper, leading to accelerated learning within this region.

This relatively complocated learning rule may be difficult to implement in practice, but it is possible to use a simple approximation, which assumes that, as learning progresses, the second order term becomes small. This assumption was demonstrated in simulation, showing that learning performance was not seriously impacted, and was in fact improved in many cases.

The simplified learning rule is given by:

$$\therefore \Delta_p w_{ji} = 4\eta w_{ji} \delta_{pj} o_{pi}$$

This means that the effect may be achieved by making the weight updates proportional to the existing weight value. Such learning could be realised simply if the current synaptic weight can be read into the input of a multiplier.

## INHIBITORY THRESHOLDING DEVICES

In (DeLaurentis and Dickey, 1994), DeLaurentis obtained an approximation theorem for neural networks based on sigmoidal thresholding functions. He postulated that any network based on positive sigmoids ('excitatory') and with positive and negative weights, could be approximated with another network, using only positive weights and a mixture of excitatory and inhibitory neurons. That is not to say that gradient descent will be able to find an optimum solution, due to the possible introduction of local minima. However, it seems reasonable that the introduction of such units will increase the probability of reaching a solution.

On this basis, inhibitory neurons (Figure 1), biased into the positive range, were introduced into the hidden layer of the simulation.

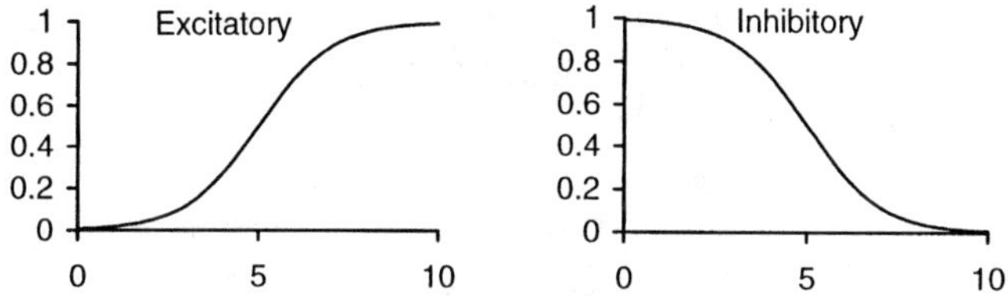


**Figure 1. Excitatory and inhibitory sigmoid functions, biased into the positive range.**

## SIMULATION RESULTS

An investigation of the proposed learning rule was undertaken and compared with standard backpropagation. A simple binary input/output 8-16-3 pattern classifier was simulated.

Training was performed using random presentations of five randomly generated 8 bit binary inputs, with fixed targets of 3 bit binary outputs. Four learning models were compared:

1. Standard backpropagation (BP)
2. Standard backpropagation with positive weights only
3. Alpha-Squared
4. Simplified Alpha-Squared

Each of the models was tested with various combinations of adaptive biasing, momentum and inhibitory neurons.

The condition for successful convergence was a mean squared error of 0.1. Learning coefficients ($\eta$) ranged between 0.5 and 0.75. Momentum coefficients were 0.9 for BP and 0.3 for the other schemes.

The results for the simulations are shown in Figure 2. It is evident that the introduction of inhibitory neurons into the hidden layer has a dramatic effect on the schemes operating with positive weights. Adaptive biasing did not generally improve things, but adding momentum enabled the restriction to positive weights to compete with standard BP.

The success of the Alpha-Squared model is also clear – with inhibitory neurons included, the algorithm converged on a solution in times close to those obtained with BP with momentum; with fine-tuning, this technique can be made to outperform it. In general, momentum and bias did not have a significant effect.

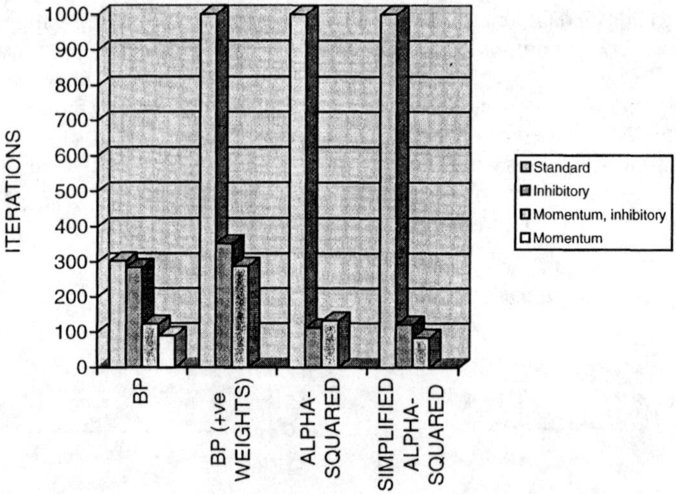


**Figure 2. Simulation of the learning scheme: the effects of using inhibitory neurons with positive weight and the improved convergence speed using the alpha-squared model are observed.**

Finally, the slight deviation from true steepest descent learning that was allowed in the simplified Alpha-Squared model, led to even faster convergence. This is doubtless due to pseudo-random variations from the ideal overcoming local attractors in the cost function. This simple model has extremely attractive properties – introducing inhibitory and bias units enabled convergence in 85 iterations.

The variance of the convergence was typically high for the positive weight models; in general, the alpha-squared models performed better.

The simulation performed using only 8 hidden units, instead of 16, showed a predictable corresponding reduction in performance of all the algorithms.

## IMPLEMENTATION OF AN OPTICAL LEARNING MULTILAYER NETWORK

Figure 5 shows a possible neural unit with optical interconnection. The cell takes the form of a ferroelectric liquid crystal based smart pixel, providing summing nodes at the photodiodes for the weighted signals from the adjoining layers, as well as thresholding.

The synapses perform simple multiplication and weight storage and update tasks. For standard backpropagation, only $\delta_{pj}o_{pi}$ needs to be calculated. The simplified version of the alpha-squared rule requires $w_{ji}\delta_{pj}o_{pi}$, so that it must, in this case, be possible to read the existing weight value before update.

Figure 3 shows a simplified schematic for the proposed optical system.

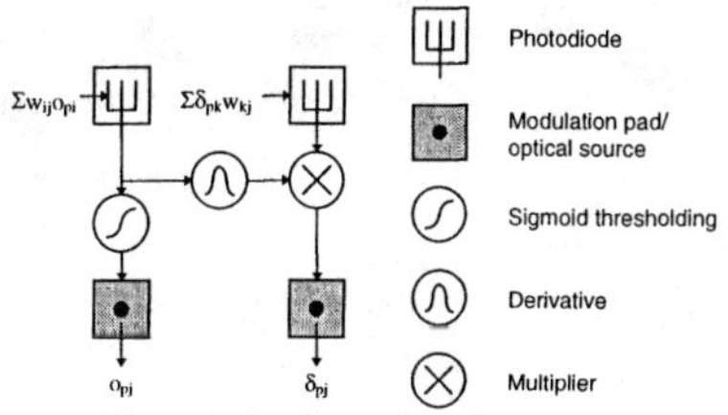


**Figure 5. Configuration of neural pixel.**

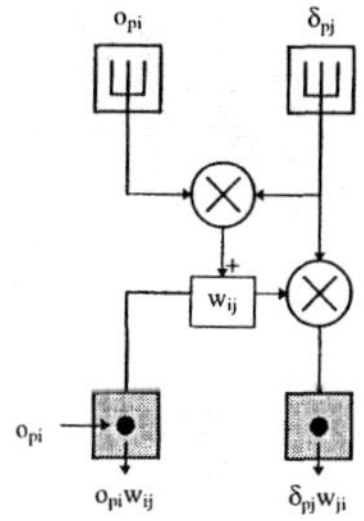


**Figure 4. Possible synapse layout.**

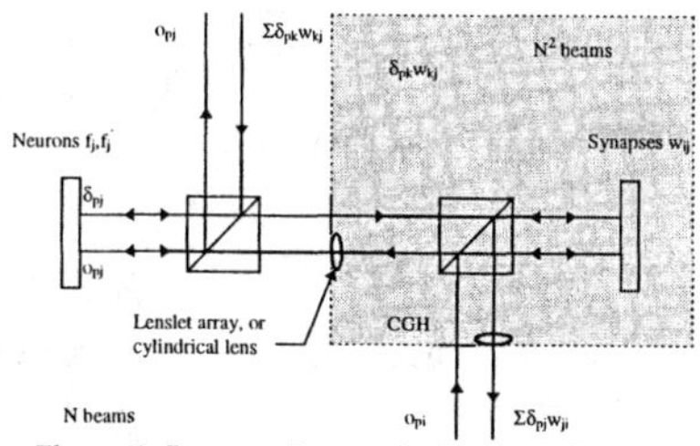


**Figure 3. Proposed cascadeable neural layer.**

## EFFECTS OF NON-LINEAR LIQUID CRYSTAL RESPONSE ON LEARNING

In this section, we shall be concerned with the effects of using liquid crystal based thresholding units, with a $SIN^2$ transmission function.

The biased bipolar derivative signal may reside within the central linear region of the $SIN^2$ function; the effects on the error backpropagation will therefore be ignored. It is assumed here that the synapses are also linear.

If the full range of the liquid crystal is employed, then it happens that the shape of the sigmoid function is retained with a scaling modification alone (squashing by a factor of 1.66). For networks with *in situ* learning, however, the fact that the derivative is **not** squashed means that there is an observed deviation from the ideal.

Nonetheless, standard backpropagation is known to be quite insensitive to such variations, as long as the general shape of the neural function is observed. Simulations using the squashed sigmoid revealed that learning rate and convergence was, in fact, improved in almost all cases.

Initial learning (central region in Figure 6) is retarded, leading to slower early network specialisation. As convergence is reached, more of the neurons approach saturation. The high value of the derivative at this stage is reasonably well matched to the ideal response, but with an elevated learning rate.

True steepest descent is not observed, but many of the problems of gradient descent algorithms are circumvented, to an extent, again by the random deviation from the ideal.

The full simulation was run using the squashed sigmoid as the thresholding function, but keeping the derivative response linear.

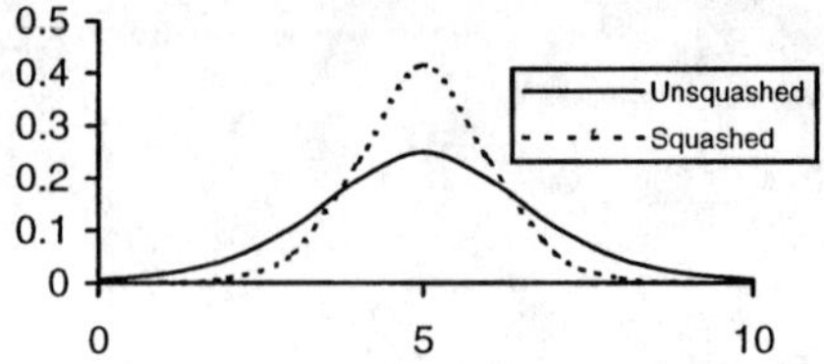


**Figure 6. Unsquashed delta function.**

The results revealed that, not only was the proposed scheme up to five times faster than standard backpropagation, (convergence was frequently observed after as few as 25 presentations), but the solution was found much more consistently, with variances that were more than an order of magnitude smaller.

## CONCLUSIONS

The tests have shown that inhibitory neurons can have a dramatic effect on learning within the positive weight plane. Such units are extremely easily constructed, based on the excitatory devices. If the simplified Alpha-Squared model can be implemented in the synapses, then very high performance is possible. A network of equivalent size can give results comparable to the standard BP momentum learning algorithm that uses positive and negative weights. Momentum terms, which are difficult to implement in hardware, are not required.

An optical system taking advantage of these results is possible using liquid-crystal on Silicon smart SLMs. Learning is improved by the non-linear response caused by the use of FLC SLMs, an effect that may be emulated in faster SEED-based devices.

## REFERENCES

Bródka, S., Macukow, B., (1995). Neural networks with unipolar weights and normalized thresholds, *Institute of Physics Conference Series*, Vol.139, 463-466

DeLaurentis, J.M., Dickey, F.M., (1994). "A convexivity-based analysis of neural networks", *Neural Networks*, Vol.7, No.1, 141-146

Dickey, F.M., DeLaurentis, J.M., (1993). "Optical neural networks with unipolar weights", *Optics Communications*, Vol.101, No.5-6, 303-305

Rumelhart, D.E., McClelland, J.L., (1986). Parallel Distributed Processing, Volume 1: Foundations, Seventh printing, The MIT Press, Cambridge MA

# Neural Networks in the Context of Autonomous Agents: Important Concepts Revisited

**RALF SALOMON**
*AI Lab, Computer Science Department, University of Zürich*
*Winterthurerstrasse 190, 8057 Zurich, Switzerland*
*FAX: +41-1-363 00 35; Email: salomon@ifi.unizh.ch*

***ABSTRACT***:
Artificial neural networks have been successfully used in many technical applications. They are also important tools for the control of autonomous agents. The major goal of research on autonomous agents is to study intelligence as the result of a system environment interaction, rather than understanding intelligence on a computational level. In contrast to other applications, autonomous agents might not distinguish between a learning and a performance phase; they have to continuously learn while they are behaving in their environment. Thus, a neural network for autonomous agents should feature incremental learning, should not exhibit overlearning and should not suffer from a high VC dimension. The review presented in this paper reveals that most existing models are not ideally suited for autonomous agents. The main goals of this paper are (1) to discuss the autonomous agents perspective, (2) to identify important properties of neural networks for autonomous agents, and (3), very important, to initiate new research on better suited network models.

## INTRODUCTION

The field of autonomous agents is an important application domain for neural networks. Autonomous agents are embodied systems that behave in the real world without any human control. An overview of special issues involved in autonomous agent design can be found in Maes (1991). One major goal of research on autonomous agents is to study intelligence as the result of a system environment interaction, rather than understanding intelligence on a computational level. It is believed that the employment of (biologically plausible) neural networks eventually leads to intelligent behavior.

Even though neural networks have been successfully applied to many technical applications, standard design methodology can usually not be applied to autonomous agents. A key aspect of learning in autonomous agents is that an agent has to learn continuously without distinguishing between a learning and a performance phase.

As already mentioned, the employment of (biologically plausible) neural networks should lead to intelligent behavior. To this end, this paper analyzes to which extent widely-used artificial neural networks feature biological neural network properties, such as incremental learning or fault tolerance. It turns out that most models lack these important properties.

This paper also discusses the frame-of-reference problem (FOR problem) (Clancey 1991) and it revisits widely-used technical terms, such as supervised and unsupervised learning, generalization and so forth; in the context of autonomous agents, these technical terms need new interpretations. This paper then analyzes how to design ideal network models for autonomous agents and it points to current approaches that pursue research in the right direction. This paper concludes with a brief summary.

## THE AUTONOMOUS AGENT FRAMEWORK

In most technical applications, such as system control, speech recognition, pattern recognition, or stock market predictions, network design is as follows. By means of some technical utilities, the designer collects a sufficiently large set of training patterns. Then, the network is trained off-line[1] until a precision criterion is met. In order to assure generalization, the network's quality is monitored by using cross validation techniques. Once the network is trained, it is considered a module and is incorporated into the application.

The usual neural network design methodology is not applicable to autonomous agents. An agent has to adapt to continuously changing environments, it has to determine the pattern's relevance at runtime by behaving in its environment, the sensor readings, which provide environmental information, are very noisy and, in addition, the sensor readings dynamically change due to the agent's interaction with the environment. Especially for the first two reasons, it is not appropriate to preselect a set of training patterns. Conversaly, an agent has to continuously learn without maintaining an infinitely growing training set. Among other reasons, the maintenance of a training set would be far too memory and time consuming.

The designer's option of training a neural network off-line can be seen as a major reason for the lack of network models that fit the requirements of autonomous agents.

## BIOLOGICAL BRAINS VERSUS ARTIFICIAL NEURAL NETWORKS

The development of artificial neural networks is mainly inspired by biological brains (see, for example, Hubel, 1988). Without doubt, a natural brain consists of millions of neurons, which are interconnected by axons and dendrites. Their ability of working inherently parallel is a key source of the brain's power. Furthermore, it is known that biological brains behave very fault tolerant. The destruction of several hundred neurons in a human brain is virtually without any effect. Probably, the most important property of a brain is its ability to learn in an incremental fashion. A biological brain learns complex concepts stepwise. Once it has acquired a set of skills, it can add new concepts without the need of re-learning already learned concepts. The features described above give natural brains a very high efficacy.

Most existing artificial neural network models, such as multi-layer perceptrons (MLPs), feature maps, or oscillatory models, do not feature biological properties. After training an MLP, the training procedure can remove those hidden units that have low saliency, e.g., LeCun et al. (1990) or Mozer (1989). But once an almost minimal network has been obtained, removing a few more hidden units drastically changes the MLP's input output mapping. Also, MLPs behave very sensitive to removing particular connections. Each hidden unit plays an important if not essential role in the network's behavior. Furthermore, the parallelism of MLPs is limited within single

[1] Off-line training should not be mixed up with updating mechanisms, such as on-line and batch mode. Off-line training refers to a separate training phase as opposed to continuous training, while is done during operation.

layers. An efficient parallel implementation of MLPs is very limited and requires tremendous programming effort (Müller, 1995).

A further problem of most neural network models concerns the network's generalization performance with respect to the number of connections. Each artificial network has an associated VC dimension (see, for example, Mostafa, 1992), which predicts the probability that the network generalizes poorly. The problem is that a larger network has a higher VC dimension and, thus, a higher probability of generalizing poorly unless further constraints are imposed to reduce the network's VC dimension (Mostafa, 1992; Simard et al. 1992). The importance of the VC dimension is the source of a huge amount of research that is devoted to the evolution of appropriately sized networks. This research can be classified as pruning algorithms (LeCun, 1990; Mozer, 1989; Prechelt, 1995), growing approaches (Kwok, 1995), or hybrid methods (Salomon, 1994; Schiffmann, 1990). All approaches consider a training set that has to be known in advance. This problem does not occur in biological brains. From the very beginning, the brain's size is very large (and rather shrinking) and even with an initially small amount of experience, animals and humans generalize well. Moreover, the requirement of well-sized artificial networks prohibits incremental learning, unless the network would allocate new nodes for unseen data. But pure pattern storage leads to poor generalization.

Kohonen feature maps (Kohonen, 1989) are useful for standard applications. As their key property, feature maps transform high-dimensional data into low-dimensional patterns, while preserving the underlaying topology. The self-organization process, however, needs to be guided by a neighborhood function, which has to be specified by the designer *in advance*. Furthermore, the weight-update mechanism requires a scheduling of the update strength in order to evolve stable feature maps with respect to the given training set. Kohonen feature maps are limited to static training sets, because the presentation of only some training patterns destroys the already learned mapping.

Oscillatory networks and hebbian networks/associative memories lack the ability of incremental learning, since the set of training patterns has to be determined prior to learning. These models are less sensitive to noisy data, but they cannot deal with translations and rotations without further enhancements. For example, a translational shift of one pixel is normally considered as an entirely new pattern, since both patterns have a very large Hamming distance.

## THE FRAME-OF-REFERENCE PROBLEM AND ITS IMPLICATIONS

The frame-of-reference problem (FOR problem) (Clancey 1991) basically states that the transfer of concepts from one context to a different context is very problematic, especially if the new context cannot be precisely described in advance. In other words, a concept that describes a situation well in one context can fail, if applied to the same problem but in a different context. An example of the FOR problem is the object-avoidance behavior of insects. For an observer it appears that insects, for example, avoid obstacles like stones while they are moving on their path. The observer can describe this behavior by an obstacle-avoidance module. But this does not imply that the insect's brain has also such an obstacle-avoidance module. Often, an observer is not aware that s/he transfers a concept from one context to another. The remainder of this section shows that the neural network community has often trapped into this pitfall. Many classifications of learning schemes are rather arbitrarily and valid only from the designer's perspective; from an agent's perspective, the semantics of technical terms need to be revisited.

For the neural network community, it is very important whether a learning scheme is supervised or unsupervised. Some learning schemes, such as self-associative back-propagation, are in between and, thus, called self-supervised. Generally, the term "supervised" indicates that a teacher is involved, whereas the term "unsupervised" indicates that a system learns without the presence of any teacher. Both learning schemes are classified from the designer's perspective.

Most designers are guided by the idea that unsupervised agents cannot use any supervised learning procedure. Consequently, they limit themselves to the use of unsupervised learning algorithms and neural networks without hidden units, since unsupervised learning procedures, such as Hebbian learning, are not applicable to networks with hidden units. But what do these terms mean from an agent's perspective? Even though an agent certainly needs some feedback about the quality of its behavior, which can be provided by reinforcement signals, reflexes, or the presence of a value system, an agent acts unsupervised if it behaves on its own in its environment and acts supervised if an external teacher pre-programs its actions or intervenes while it is operating. In other words, from the agent's perspective, supervised/unsupervised concerns the way, *how* it acquires information and *not* the way it updates its internal network parameters.

The following examples explain this (agent) perspective. An agent that uses backpropagation does not care whether it uses on-line or off-line updating, it cares about *how* it collects the patterns. If using backpropagation, the agent needs a set of training patterns, and unsupervised would mean that the agent would have gathered the training patterns by itself. This interpretation of supervised/unsupervised contradicts the common sense, since backpropagation is considered supervised. However, the re-interpretation is essential for the development of new neural network algorithms that fit the requirements of autonomous agents. Also, this reinterpretation suggest that unsupervised autonomous agents can use supervised learning procedure to overcome, for example, the limitations of neural networks without hidden units, which are limited to linear separable problems.

Other approaches that allow more complicated network structures currently require additional mechanism that are to be pre-programmed in order to allow the application of unsupervised learning algorithms. These additional mechanisms can be implemented by additional reflexes or they require a separation into a learning and a performance phase. In order to apply Hebbian learning, the DAC architecture (Pfeifer, 1992), for example, connects the hidden units with an additional collision detectors.

The term "generalization" is a second example of the FOR problem. The neural network community considers generalization as the network's ability to map similar input patterns to similar output patterns. Here, similarity is often expressed by rather arbitrary metrics, such as Euclidian distance or Hamming distance. From an agent's perspective, similarity has a different meaning. Moreover, "similarity" cannot be defined a priori for all agents. It is to be specified for a particular agent doing a particular task in a particular environment. Its meaning also depends on the agent's sensors, its environment, its embodyment, its actions etc. Thus, standard neural network terminology can be applied only to a very limited extend.

Similarly, terms, such as batch, on-line, or incremental, have other meanings from the agent's perspective. For an agent, an incremental learning scheme allows for learning concepts step by step without repeating everything it has already learned. Unfortunately, (most) existing network models require the presentation of *all* patterns; if presenting only a few patterns too often, other concepts get lost.

## IDEAL NEURAL NETWORK MODELS FOR AUTONOMOUS AGENTS

Typically, an autonomous agent behaves in the following scenario. Initially, it starts to explore its environment and during its lifetime, it compiles more and more experience. This scenario makes clear that (1) an agent determines the pattern's relevance at runtime, (2) it is not appropriate to select a set of training examples for off-line training, and (3) an agent needs a neural network that learns incrementally, since the agent acquires its experience over time.

First of all, the ideal network model would feature incremental learning properties. This is one of the key features current neural network models lack. As already discussed, it does not make sense, to repeatedly present one pattern to a backpropagation network or a Kohonen feature map; normally, *all* patterns have to presented. Using the so called on-line mode does not overcome this problem. For an autonomous agent, on-line means that it updates its network weights *while* it is behaving. Also, it is not adequate, if the agent stores all ever seen patterns in order to apply a full epoch for each new seen input pattern. The resulting computational complexity would be infeasible. Furthermore, the semantic of a particular pattern can have changed and also the pattern might be erroneous due to noise (see below).

Second, the ideal model would not suffer from problems caused by a too high VC dimension. Generally, the network needs to be oversized. Even if the network is well sized with respect to the final amount of perceived information, it will be oversized most of the time. Since *each* model has an associated VC dimension, the *effective* VC dimension needs to be bounded by imposing additional constraints.

Third, the network model for autonomous agents should be plastic. That is, during its life time, the appropriate actions associated with a particular sensory information might change. This dynamical change can be caused by a changed environment (perhaps due to the agent's interaction with its environment) or by a dynamical change of the electrical characteristics of the sensors.

In summary, to overcome current limitations of network models for autonomous agents, new models are to be developed that feature steady incremental learning and do not suffer from a high VC dimension.

## CURRENT APPROACHES

This section discusses some approaches that try to overcome the problems already discussed.

The use of evolutionary algorithms (EAs) (Baeck, 1993; Goldberg, 1989; Rechenberg, 1973; Schwefel, 1995) is one approach towards the automated design of useful control architectures for autonomous agents. EAs, such as genetic algorithms (GAs) or the evolution strategy (ES), maintain a population of individuals as opposed to only one candidate. In each generation, the algorithm generates different offspring. By applying evolutionary-inspired operators, such as mutation and crossover, offspring differ from the parents. In the evaluation phase, the algorithm assigns a fitness value to each population member and selects those members as parents for the next generation that have the highest fitness.

The evolutionary approach is certainly one step in the right direction. Successful applications of EAs to the evolution of different control architectures, such as Braitenberg vehicles (Braitenberg, 1984), have been presented in Cliff (1992), Floreano (1994, 1996), Harvey (1996), and Nolfi (1995, 1995a). However, the power of these schemes is limited, since they require a large amount of time. For example, the evolution of Braitenberg controllers is studied in (Floreano, 1994). A Braitenberg controller for the Khepera$^{TM}$ robot has approximately 20 connections. For such a small problem, a GA converges after 50 to 100 generations when maintaining a population of approximately 80 to 100 individuals. That is, the GA needs about 4000 to 10000 evaluations until convergence. Even for these very small system, the whole experiment takes about two and a half days, since the evaluation is to be done in a real robot. Other experiments on the evolution of controllers for similar tasks, such as homing or picking up pegs, require a similar amount of time.

It can be expected that the application of evolutionary approaches to more complex systems with many more parameters involved turns out to be infeasible. The major deficiency of those evolutionary approaches is that the parameters/weights of the neural network are evolved by the (very time consuming) evolutionary process and not by a self-adapting process due to the agent's interaction with the environment. Thus, the evolutionary approach should be used to evolve the controller's topology as has been done in part in Harvey (1996). The actual network weights should be adapted by proper learning mechanisms during the agent's lifetime as has been studied in Nolfi (1995a). The control architectures studied in Nolfi (1995a) feature two networks. The first network decides about the next actions, whereas the second network works as a teacher in situations, in which the robot behaves poorly.

The large search space is a main reason for the time consuming evaluation process of most evolutionary approaches. When evolving neural networks, for example, the genome has to encode all weights as well as the whole structure of the network. Eggenberger (1996) proposes a new approach that does not code a specific topology for a neural network. Among other things, the genome encodes growing processes and cell-cell interactions. Typically, the genome is very short and duplicated to all units present in a network. The actual topology is dynamically evolved, since all units process the program encoded in the genome. As a consequence, the evolved topology not only depends on the genome, but also on the actual sensory information of the agent. Thus, this approach combines evolutionary processes and dynamical self-adaptation mechanisms. Furthermore, this approach is computationally very efficient, since the genome length reflects only the network's complexity and not its's size.

Growing cell structures (Fritzke, 1994) are also a step towards the right direction. Growing cell structures consist of a number of cells, which explicitly maintain their neighborhood relationships. For each pattern, the algorithm assigns an error to those cells that are close to the pattern and it "moves" all these cells Kohonen-like towards the presented input patterns. After a certain number of steps, the algorithms splits all those cells that have accumulated a high error value. This dynamical growing process leads to an adaptation of the underlaying pattern distribution. Over time, the model dynamically grows up to an appropriate size and does not suffer from overlearning. However, growing cell structures suffer from the same problem as Kohonen feature maps; they need a dynamically scheduled update function in order to obtain a stabilized internal pattern representation and they require a prespecified training set as well. Potentially, the growing-cell-structure model can be enhanced to overcome this problem.

Radial basis functions (RBFs) seem to be most appropriate for autonomous agents. RBFs are inspired by biological receptive fields. However, RBFs are typically initialized by the results of a cluster algorithm that estimates the best centers and standard deviations. Unfortunately, the cluster algorithm requires a preselected fixed set of training patterns, which, in turn, prevents RBFs from incremental learning. Also, RBFs are sensitive with respect to overlearning. As a new research avenue, autonomous agent research suggests to develop RBF models that can start with large networks. A high VC dimension can be bounded by imposing additional continuity constraints. One of the striking features of RBF networks is that the unit's response is local with respect to the presented pattern. A learning scheme should be developed that adapts only those units that are close to the presented pattern. By using only those hidden units that have been suitably updated by previously presented patterns, such an approach would allow the agent to continuously learn during its lifetime.

## CONCLUSIONS

This paper has shown that most artificial models do not feature the same properties as biological networks. Then, it has been argued that the frame-of-reference problem is a major source of not developing neural network models that fit the requirements of autonomous agents. The ideal neural network model for autonomous agents would feature incremental learning and would not suffer from a too high VC dimension. This paper has also discussed several approaches that try to overcome current deficiencies of existing artificial neural network models. Examples of such approaches are evolutionary algorithms, growing cell structures, and radial basis function networks. However, non of these models is yet ideal for autonomous agents. Since autonomous agents are an important tool in research on artificial intelligence, future research should develop appropriate neural network models, perhaps by combining those approaches that have been considered a step in the right direction.

**ACKNOWLEDGEMENTS:** This work was supported in part by a Human Capital and Mobility fellowship of the European Union, grant number ERBCHBICT941266. Many thanks to Alex Riegler for helpful discussion.

## REFERENCES

Abu-Mostafa, Y.S. (1992). Hints and the VC Dimension. *Neural Computation*,

Bäck, T., Schwefel, H.-P. (1993). An Overview of Evolutionary Algorithms for Parameter Optimization. *Evolutionary Computation* 1(1), pp. 1-23, MIT Press.

Braitenberg, V. (1984). *VEHICLES, Experiments in synthetic psychology*, MIT Press.

Clancey, W. (1991). The Frame of Reference Problem in the Design of Intelligent Machines. In Lehn (ed.), *Architectures for Intelligence. The 22nd Carnegie Mellon Symposium on Cognition*, 257–423.

Cliff, D., Husbands, P., Harvey, I. (1992). Evolving Visually Guided Robots. In J.-A. Meyer, H.L. Roitblat, and S.W. Wilson (eds.), *From animals to animats 2*, pp. 374-383. MIT Press.

Eggenberger, P. (1996). Cell Interactions as a Control Tool of Developmental Processes for Evolutionary Robotics. To appear in *From Animals to Animats IV: Proceedings of the Third International Conference on Simulation of Adaptive Behavior.*

Floreano, D., Mondada, F. (1994). Automatic Creation of an Autonomous Agent: Genetic Evolution of a Neural-Network Driven Robot. In D. Cliff, P. Husbands, J. Meyer, and S.W. Wilson (eds.), *From Animals to Animats III: Proceedings of the Third International Conference on Simulation of Adaptive Behavior*, MIT Press.

Floreano, D., Mondada, F. (1996). Evolution of Homing Navigation in a Real Mobile Robot. To appear in *IEEE Transactions on Systems, Man and Cybernetics.*

Fritzke, B. (1994). Supervised Learning with Growing Cell Structures. In J.D. Cowan, G. Tesauro, J. Alspector (eds.), *Advances in Neural Information Processing Systems 6*, pp. 255-262.

Goldberg, D.E. (1989). *Genetic Algorithms in Search, Optimization and Machine Learning*, Addison-Wesley Publishing Company.

Harvey, I., Husbands, P., Cliff, D., Thompson, A., and Jakobi, N. (1996). Evolutionary Robotics: the Sussex Approach. In R. Pfeifer and R. Brooks (eds.), *Robotics and Autonomous Systems, Special Issues on "Practice and Future of Autonomous Agents".*

Hubel, D.H. (1988). *Eye, Brain and Vision*, The Scientific American Library.

Kohonen, T. (1989). *Self-Organization and Associative Memory*, Springer-Verlag.

Kwok, T.-Y., Yeung, D.-Y. (1995). Constructive Feedforward Neural Networks for Regression Problems: A Survey. Tech. report HKUST-CS95-43, Hong Kong University of Science and Technology.

Le Cun, Y., Denker, J.S., & Solla, S.,A. (1990). Optimal Brain Damage. In D. Touretzky (ed.), *Advances in Neural Information Processing Systems 2*, pp. 598-605, Morgan Kaufmann Publishers.

P. Maes (ed.), (1991). *Designing Autonomous Agents: Theory and Practice from Biology to Engineering and Back*, MIT Press.

Mozer, M.C., Smolensky, P. (1989). Skeletonization: A technique for trimming the fat from a network via relevance assessment. In D.S. Touretzky (ed.), *Advances in Neural Information Processing Systems 1*, pp. 107–115, Morgan Kaufmann Publishers.

Müller, U., Gunzinger, A., Guggenbühl, W. (1995). Fast Neural Net Simulation with a DSP Processor Array. *IEEE Transactions on Neural Networks*, **6**, pp. 203-213.

Nolfi, S., Parisi, D. (1995). Evolving non-trivial behaviors on real robots: An autonomous robot that picks up objects. In *Proceedings of the Fourth Congress of the Italian Association for Artificial Intelligence, Firenze*, Springer-Verlag.

Nolfi, S., Parisi, D. (1995a) Learning to Adapt to Changing Environments in Evolving Neural Networks. Tech. report 95-15, Institute of Psychology, National Research Council, Rome, Italy,

Pfeifer, R., Verschure, P. (1992). Beyond Rationalism: Symbols, Patterns and Behavior. *Connection Science* **4**

Prechelt, L. (1995). Adaptive Parameter Pruning in Neural Networks. Technical report TR-95-009, International Computer Science Institute, Berkeley, CA, ftp://ftp.icsi.berkeley.edu.

Rechenberg, I. (1973). *Evolutionsstrategie*, Frommann-Holzboog.

Salomon, R. (1994). A Hybrid Method for Evolving Neural Network Topologies. In C.H. Dagli, B.R. Fernández, J. Ghosh, and R.T.S. Kumara (eds.), *Proceedings of the Artificial Neural Networks in Engineering*, pp. 147-152, ASME Press.

Schiffmann, W., Mecklenburg, K. (1990). Genetic generation of backpropagation trained neural networks. In *International Conference on Parallel Processing in Neural Systems and Computers (ICNC)*, pp. 205-208, Elsevier Science Publishers B.V.

Schwefel, H.P. (1995). *Evolution and Optimum Seeking*, John Wiley and Sons.

Simard, P., Victori, B., Le Cun, Y., Denker, J. (1992). Tangent Prop – A formalism for specifying selected invariances in an adaptive network. In D.S. Touretzky (ed.), *Advances in Neural Information Processing Systems 4*, pp. 895–903, Morgan Kaufmann Publishers.

# A Study on State Grouping and Opportunity Evaluation for Reinforcement Learning Methods

**Wenwei Yu, Hiroshi Yokoi and Yukinori Kakazu**
*Complex System Engineering Dept., Hokkaido University,*
*North-13, West-8, 060, SAPPORO, JAPAN,*
*E-mail: yu@complex.hokudai.ac.jp*

***ABSTRACT:***
In this paper, we propose the State Grouping scheme for coping with the problem of scaling up the Reinforcement Learning Algorithm to real, large size application. The grouping scheme is based on geographical and trial-error information, and is made up with state generating, state combining, state splitting, state forgetting procedures, with corresponding action selecting module and learning module. Also, we discuss the Labeling Based Evaluation scheme which can evaluate the opportunity of the state-action pair, therefore, use better experience to guide the exploration of the state-space effectively. Incorporating the Labeling Based Evaluation and State Grouping scheme into the Reinforcement Learning Algorithm, we get the approach that can generate organized state space for Reinforcement Learning, and do problem solving as well. We argue that the approach with this kind of ability is necessary for autonomous agent, namely, autonomous agent can not act depending on any pre-defined map, instead, it should search the environment as well as find the optimal problem solution autonomously and simultaneously. By solving the large state-size 3-DOF and 4-link-manipulator problem, we show the efficiency of the proposed approach, i.e., the agent can achieve the optimal or sub-optimal path with less memory and less time.

## INTRODUCTION

It's difficult to scale up the learning scheme to real application with large problem space because of the problem of slow transfer of delayed reward and the problem of exponentially increasing state space.

For the first problem, the Experience Re-use is favored to cope with it. The concept and approach of Experience Re-use can be found in (Moore and Atkeson, 1993) and (Peng and Williams, 1993) etc..

For the second problem, emphasis was put on the organization of state space. In (Singh, 1991), abstract environment model in which high resolution state serves as the intention for low resolution state space has been studied. Kaebling's HDG Learning(Kaelbling, 1993) performed problem solving basing on Landmark Network given, but just leave the Hierarchy Forming to other Map Constructing task.

From the standpoint of both autonomous agent and RL method, the state space prepared in advance to include every detail is not expected. The more natural and effective way may be that, the agent probes and memorizes the important experience for finishing task, so that it needs not face the possible initially memory cracking and use only reasonable memory during life time. The Grouping Scheme suggested here then provides the possibility, contributing to the state organization of RL.

## LABELING-BASED EVALUATION

Many researchers try to find the evaluation measurement within the frame of RL. But as it was pointed out in (Sutton, 1988), RL methods based on TD try to take advantage of the information provided by the temporal sequence of states, which is usually helpful, though, can also be misleading. If misleading information is transferred, that will be disastrous, especially in the case that long sequence of action is necessary and the reward can only be obtained at the end of the sequence.

In (Yu and Kakazu, 1996), Labeling Based Evaluation scheme was suggested, in which the step number from start state in the state space is recorded as the label of the state, this Label is maintained through forward and backward refresh processes, then used to actively guide the Experience Re-use, thereby, improve the performance of space search. Although being combined with the RL in a tight way, this scheme can explicitly monitor the search process of RL, so as to avoid affecting reciprocally. The evaluation is also made in two direction, forward evaluating and backward evaluating. These four processes can guide the exploration effectively. See (Yu and Kakazu, 1996) for referring.

## GROUPING SCHEME

Usually, the agent can not and need not search and experience all the environment for finishing it's task during it's life time. Also, even those experienced states can not be memorized entirely, because of limited resource. Furthermore, those experienced states can not be equated with each other, they should be organized instead. So, we give out the process collecting state and organizing states into grouped state graph.

In the sections below, we will introduce each parts of the scheme.

### State Structure

Some important attributes of one State are listed in Tab. 1 with simple explanation and bytes each attribute needs. *Dim* means the degree of freedom of problem space.

### State Grouping

Grouping condition should be decided based on the nature of Group. Referring to the concept "*Cell*" of (Latombe, 1991), the Group should be geometrically simple and noncritical. So, we set the boundary of the Group as regular polyhedron, and consequently, set first grouping condition as geometrical constraint in the configuration space. Another condition is Same_Bad_Action constraint. A bad action at one state is decided by agent only if the agent failed to act at that state. As for the measure of the Group, possible unit number included by grouped state( UZ ) and possible Difference of Label of two children( DL ) are assigned to agent.

**Table 1: Attributes of State**

| Attr- Name | Explanation | Bytes Need |
|---|---|---|
| Scope | Scope of State | Dim×2×sizeof(int) |
| LinkNum | number of linkage | sizeof(int) |
| Links | link pointer of graph | LinkNum×sizeof(pointer) |
| Q | Q corresponding to Links | LinkNum×sizeof(float) |
| ExpedDir | record good or bad experienced actions | sizeof(char) |
| VisitedNum | times of being visited | sizeof(int) |
| PathNum | path step when visited within one trial | sizeof(int) |
| Label | used by Evaluation | sizeof(int) |

```
Algorithm Grouping:            Grouping( psState )
    While (psNeighbor := Search_Configuration_Space( psState )) != NULL  do
        if check-Grouping-Condition( psState, psNeighbor ) = TRUE
            UZ := Calculate-UZ( psState, psNeighbor )
            DL := Calculate-DL( psState, psNeighbor )
        if Possible_Grouping_Num != 0
            psBest := Select-Best-Candidate( psNeighbors )
        psNew.scope := merge_scope( psBest.scope, psState.scope )
        Accept-Linkage( psNew, psState )
        Accept-Linkage( psNew, psBest )
        Label := min( psState.Label, psBest.Label )
        VisitedNum := max( psState.VisitedNum, psBest.VisitedNum )
        Delete-State( psBest ), Delete( psState )
```

**Figure 1: Pseudo Code of Grouping**

Pseudo code of State Grouping Process is listed in Fig. 1. Here, the function Accept_Linkage makes the child inherit the linkages of parents. The label minimization can play kind of chance-sharing between states.

### State Splitting

If an action at one State failed, agent should split the group, for preserving the nature of a group. Again, UZ is used to decide the split direction, in accordance with the requirement of effective map forming. We omit the pseudo here, due to limited space.

### State Forgetting

In this approach, for coping with large size problem, some states experienced, need not to be bad, should be forgotten, so that, agent can avoid overloading itself and its attention can be paid to current or promising throughout search.

Here, we again utilize Label as one of the measurement deciding what should be forgotten. Two kinds of Forgetting are included in this approach, one is for state forgetting between trials, the other is for state forgetting within one trial. We omit the pseudo here, due to limited space.

### Action Selection

We just adapt the simulation annealing-based action selection mechanism, which is regularly used in Reinforcement Learning methods, to our state group. Trying to keep consistence with the original scheme, two phases are included.

First phase: weighted sum of $e^Q$ of each face of state group regular polyhedron, $Bid_{action}$, is calculated for action selection, then, Roulette Selection Method is used to decide the action of next step.

$$Bid_{action}(i) = \frac{\sum_j w_j \times e^{Q_j}}{\sum_j w_j} + e^{Bonus} \times unshaded\_area$$

where: $i$ = the number of face of group polyhedron in configuration space;

$j$ = the number of other state linking with current state;

$w_j$ = overlapping area of current state and state $j$, in direction $i$;

$Q_j$ = Q value of link $j$; *Bonus*=Bonus of exploration;

*unshaded_area* =the area not overlapped by linked state.

Second phase: within the decided face, select spot to begin with. This is done by first calculating the $Bid_{Spot}$ for each spot, then applying Roulette Selection to select.

### Frame of the Approach

Main control procedure is listed in Fig. 2. Grouping does not necessarily happen, the *StateGroupRate* is set to control the grouping rate. Labeling Based Evaluation is incorporated into the Grouping, Splitting, as well as main procedure.

```
FrameWork
    While Trial_Num < PRESET_NUM || Satisfied-with-Result do
        psState := psStart
        do until psState = psEnd
            if Time = FORGET_TIME
                State_Forgetting_in_Trial
            if Random_Data < StateGroupRate
                Grouping( psState )
            Action := ActionSelect( psState )
            do until Exec-Action( psState, Action ) Success
                Split( psState )
                Action := ActionSelect( psState )
            if Reach-New-Spot
                psState := GenerateState( Spot_Coordinate )
            Label Maintenance and Opportunity Evaluation
```

**Figure 2: Pseudo Code for Frame Work**

## EXPERIMENT SETTING AND RESULTS DISCUSSION

We will do next discussion based on the result concerning with one setting of the 3-DOF problem shown in Fig. 3. In Fig. 3, obstacles depicted by quadrilaterals distribute in the environment, rod represented by a line with one small circle at the head will search the path from start pose to end pose.

Suppose that the map size is 20×20(unit), distance interval is 1(unit) and rotate interval is 10(degree). 4-link-manipulator is mainly used to show space characteristic. One problem setting and solution is depicted in Fig. 4.

First, we discuss the mutual influence of State Grouping scheme and Labeling Based Evaluation scheme expressed by time and space characteristics.

### Labeling Evaluation with different Backup

The experiment with only State Grouping and without label scheme has been done, but as the result of ordinary QL in (Yu and Kakazu, 1996), convergence was failed to be reached. That is because that without the Experience Re-use, the search of action sequence proceeds almost randomly in large problem space at initial stage. Labeling-Based is helpful to convergence in the Grouped state space.

On the other hand, State Grouping scheme can accelerate the label improvement while the agent keep experiencing in the configuration space, e.g., Grouping processes the Label in an active way. Also, because of the state grouping, the opportunity can be transferred along a condensed labeled chain. Fig. 5 and Fig. 6 can be used to elucidate this point. Fig 6 shows the data from (Yu and Kakazu, 1996), i.e., with no State Grouping included in. *Backup* means backup number of back propagation. The performance with back up equaling to 50 over-performed that of

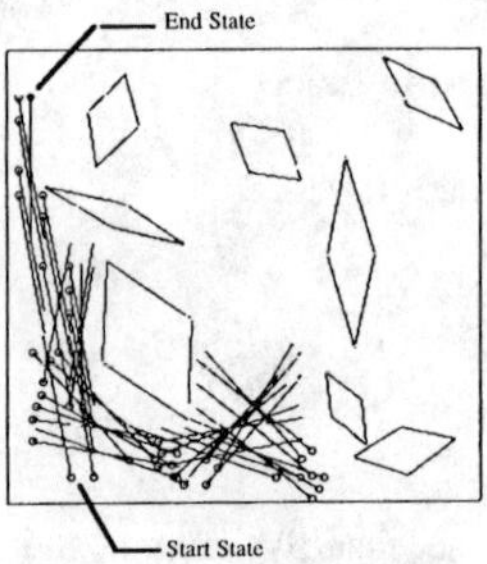


**Figure 3: Problem Setting**

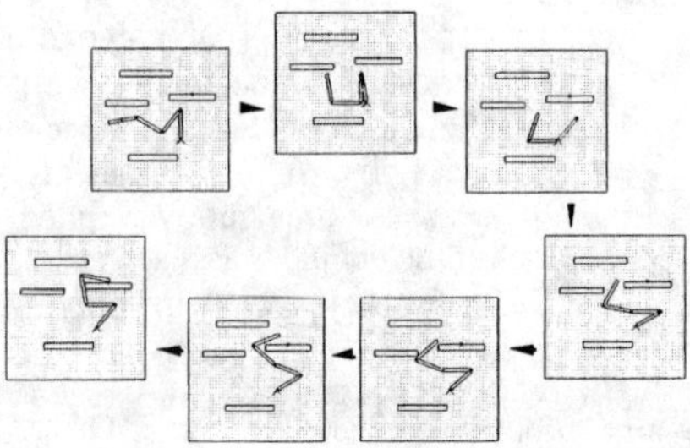

**Figure 4: Action Sequence of 4-link manipulator**

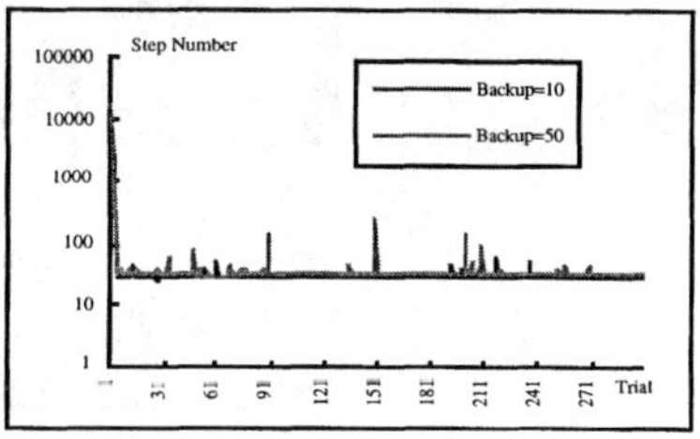


**Figure 5: Comparing Grouping+Labeling Backup=10 with Backup=50**

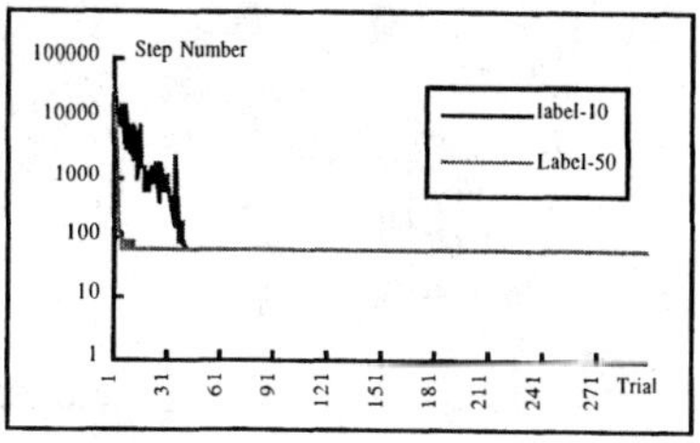


**Figure 6: Comparing Labeling, Backup=10 with Bakcup=50**

with backup equaling to 10. The agent just wandered in broad unexplored space, i.e., 10 is not enough for the deepening search. Whereas in Fig 5, the performance with 10 is not worse than the one with 50. This can indicate the effectiveness of grouped state.

**With and without Forget Step**

In Fig. 7, the left vertical axis is state number, and right vertical axis is the logarithm of step number. With Forget Process, the real state number(RSN) diminished while trial continues and experienced state number(ESN) keeps rising. Although, the intense change of step number(SN) near the trial 30 may denote that exploitation is disturbed by state forgetting, the converging tendency was not affected by the forgetting. This property makes it possible to adapt the approach to the problem with larger problem space.

Also Fig. 8 and Fig. 9 show two maps with the θ axis denotes the rotate angle, from 0 to 360 degree. Fig 8 is the map resulted in by the approach without forgetting step. While Fig. 9 is the resulting map corresponding to the procedure of Fig. 7 with some superfluous particulars neglected, while the important states get saved.

The maps depicted in Fig. 8 and Fig. 9 are not ideal, the block of group can be reasonably big, but actually not. This is because of lacking of search strategy for map and better grouping or merging strategy.

**Time Space Characteristic**

Grouping scheme takes effect basing on the state structure, thus, as shown in Tab. 1., the memory needed by one state certainly exceeds memory requirement of other scheme. Suppose the size of *float* to be 4, that of *int* to be 2, the size of one *pointer* to be 4, also average link number, shown by the result of experiment, to be 3.2. Memory Requirement of one of each scheme can be calculated:

$Mem_{Gouping} = 46.6$(byte) $Mem_{Label} = 26$(byte) $Mem_{other} = 24$(byte)

In Tab. 2, we used the maximum memory utilized by Grouping+Labeling scheme, i.e., 1573 states for comparison. The real time characteristic of each scheme is also shown in Tab. 2.

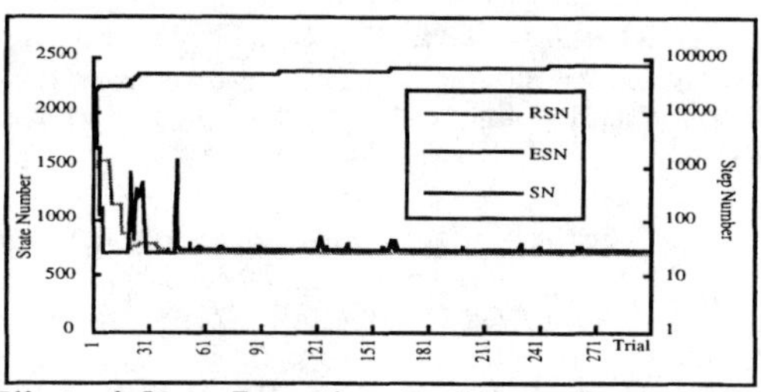


**Figure 7: The Effect of State Forgetting, Real State Number decreases although Experienced State Number keep increasing. SN means Step Number**

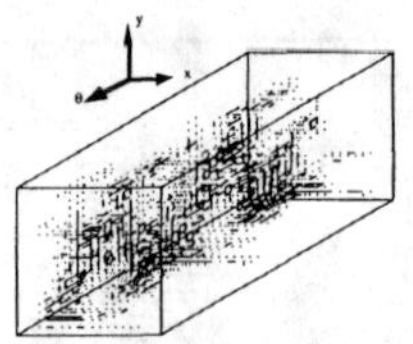

**Figure 8: Map without Forget Step**

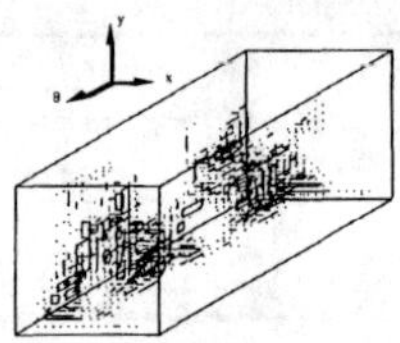

**Figure 9: Map with Forget Step**

**Table 2: Comparison of Real Time and Space Characteristic**

| | *Ordinary QL* | *Fixed Threshold* | $Q_{old}/n$ | *Label-Based* | *Grouping-Label* |
|---|---|---|---|---|---|
| Trial Num | over flow | not conver (91) | 68 | 19 | 5 |
| Step Number | over flow | not conver(227916) | 263932 | 42922 | 23584 |
| Real Time | over flow | not conver(810secs) | 867secs | 260secs | 636secs |
| Time / Step | / | (3.55msec/step) | 3.28msec/step | 6.05msec/step | 26.96msec/step |
| Memory | 345600 | 345600 | 345600 | 374400 | 73302 |

**Table 3: Space Characteristic Shown through the result of 4-link Application**

| | *Ordinary RL* | *with collect, without group* | *with collect, with group* |
|---|---|---|---|
| max state number | $36^4$=1,679,616 | 18539(7.2) | 7039(6.3) |
| max memory used(byte) | $36^4$×26=43,670,016 | 82.6×18539=1,531,321 | 76.4×7039=581,421.4 |
| steady state number | 1,679,616 | 8211(5.5) | 2714(3.6) |
| steady memory used(byte) | 43,670,016 | 69.0×8211=566,559 | 76.4×2714 |

In Tab. 3, a comparison concerning the data about 4-link-manipulator is carried out, showing further the space characteristic of suggested approach. Here, maximum number and number when the solution keeps in steady state are used as the attributes of comparison. The value in bracket reveals the average state-link number which is necessary for calculating real byte number. Even in a higher degree problem space, the grouping scheme makes the search and learning possible, i.e., the agent can find its solution without exhausting memory resource.

## CONCLUSION

In this paper, we propose the approach for coping with the curses of dimensionality. This approach first groups the experience states according to their geometrical constraint and action effect. Forgetting Mechanism can make the approach keep track with the important part and avoid being overloaded, Labeling Based Evaluation plays an important role both in Experience Re-use, and other part of Grouping scheme.

## REFERENCES


Connell Jonathan H., Mahadevan Sridhar, (1993). Rapid Task Learning for Real Robots, *in Robot Learning* edited by Jonathan H. Connell & Sridhar Mahadevan, Kluwer Academic Publishers.

Kaelbling Leslie Pack, (1993). Hierarchical Learning in Stochastic Domains: Preliminary Results, Machine Learning 93, *Proceedings of the Tenth International Conference*.

Latombe Jean-Claude, 1991, *Robot Motion Planning*, Kluwer Academic Publishers.

Moore Andrew W., Atkeson Christopher G., (1993). Memory-Based Reinforcement Learning: converging with less data and less real time, in *Robot Learning*, Kluwer Academic Publishers.

Peng Jing and Williams R. J., (1993). Efficient Learning and Planning Within the Dyna Framework, *From Animals to Animates 2*.

Singh Satinder P., (1991). Scaling Reinforcement Learning Algorithms by Learning Variable Temporal Resolution Models, *ML91, Proceedings of 8th International.*

Yu Wenwei and Kakazu Yukinori, (1996). *1996 Japan-U.S.A. Symoposium on Flexible and Automation, Boston.*

# ESTIMATING THE SOLVABILITY OF PATTERN CLASSIFICATION PROBLEMS

S. HARING[1], J.N. KOK[2], and M.A. VIERGEVER[3]

[1] *Department of Computer Science, Utrecht University*
*P.O. Box 80089, 3508 TB Utrecht, The Netherlands*
*E-mail s.haring@cs.ruu.nl*

[2] *Department of Computer Science, Leiden University*
[3] *Image Center Utrecht, Utrecht University Hospital*

***ABSTRACT:***
Training of feed-forward networks is sensitive for learning-parameters, network architecture, etc. So if a feed-forward network fails to solve a pattern classification problem one never knows if the cause of the failure is in the network or in the problem itself. Hence a well-founded indication of the problem's solvability is a helpful tool in engineering with feed-forward networks.

We present a method to estimate the solvability of a pattern classification problem. The method is based on the estimation of the Bayes error-rate of the problem, and intergrates classical theories into a novel application. The proposed algorithm is tested on a number of benchmarks and proved to be reliable. We argue that the approach is feasible if many patterns are available and if the dimension of the patterns is relatively low.

## 1 INTRODUCTION

When one, as an engineer, attempts to solve a problem, one would like to have an indication of how solvable the problem is, before putting too much effort in it. The indication can support in deciding when to stop trying to improve results; if a known minimum is reached, one is sure that no further work will yield a better solution. For pattern classification, the bound on the error-rate that can be achieved is such an indication. When feed forward networks are applied to pattern classification such an indication is important; feed-forward networks are powerful algorithms, which depend very much on parameters settings. There is a vast number of choices for the engineer to make (like the number of units, number of layers, connection-scheme, transfer-functions, learning-rules, etc.). The performance of feed-forward networks strongly relies on these decisions. Without knowledge of a well-founded estimate of a minimum error-rate, one hardly knows whether further fine-tuning of the algorithm would eventually yield better results.

The theoretical bound on the error-rate of a pattern classification problem is called the *Bayes error-rate* (Duda and Hart 1973). In this paper we propose a method to estimate the Bayes error-rate of a pattern classification problem; the method is straightforward and combines algorithms that have been known for decades to a novel application. First we use *Parzen window estimation* (Parzen 1962) to make estimates of the probability density functions that underly the data. To calculate the Bayes error-rate we need to integrate these functions; for that we apply *Monte Carlo integration* (Press, Teukolsky, Vetterling, and Flannery 1988). The algorithm is very convenient and almost independent of parameters. One inputs a pattern classification problem to the algorithm, and the algorithm outputs a reliable estimate of the smallest reachable error-rate of the problem. So when one wants to solve pattern classification problems, the method can be of great help.

# 2 FUNDAMENTALS OF THE APPROACH

Before we discuss our methods and results we briefly introduce the three basis concepts that our method relies on; i.e. the Bayes error-rate, Parzen window estimation, Monte Carlo integration. Although the applied techniques are quite standard the combination of these techniques is not. Therefore we discuss these techniques briefly. The reader that is familiar with all three theories can skip the next paragraphs.

## 2.1 Bayes error

We consider a pattern as a real valued vector $\vec{x}$ consisting of observables about some entity. It is unknown to which class the entity belongs; there are $n$ possible classes $\omega_i$, $i = 1, \ldots, n$. It is reasonable that the value of $\vec{x}$ gives information about the class that the entity belongs to; but generally there is overlap between classes and therefore $\vec{x}$ is ambiguous. Assume that there exists a function for every class $\omega_i$ that describes the probability that $\vec{x}$ has a specific value. This function is the *class conditional probability density function* for $\vec{x}$: $p(\vec{x}|\omega_i)$. Assume furthermore that the a priori probabilities $P(\omega_i)$ of all classes are known. The probability that the entity belongs to class $\omega_i$ is straightforwardly determined with the Bayes rule:

$$P(\omega_i|\vec{x}) = \frac{p(\vec{x}|\omega_i)P(\omega_i)}{p(\vec{x})}, \quad \text{where } p(\vec{x}) = \sum_{i=1}^{n} p(\vec{x}|\omega_i)P(\omega_i)$$

The Bayes decision theory states that $\vec{x}$ should be classified according to the class with the highest probability. This decision rule guarantees the theoretical minimal probability of wrongly classified patterns. The probability of an error (that is the fraction of miss-classified patterns) is as follows:

$$P(\varepsilon|\vec{x}) = 1 - \max_i(P(\omega_i|\vec{x})), \quad \text{and} \quad P(\varepsilon) = \int P(\varepsilon|\vec{x})p(\vec{x})d\vec{x} \tag{1}$$

In Figure 1 the used concepts are visualized. The Bayes error-rate corresponds to the size of shaded area in the figure; the central theme of this paper is to try to estimate the size of this area. *There are no regions, other than the regions given by Bayes decision theory that yield a lower probability of an error.* So the Bayes error-rate is a bound an the error-rate of a pattern classification problem.

The Bayes error-rate is a theoretical result. For real-world data the probability density functions will not be known. If the probability density functions would be known, the optimal, unbeatable classifier is available. However, it is possible to estimate the Bayes error-rate, using estimates of the probability density functions. Estimation of the probability density functions is the issue of the next section.

## 2.2 Parzen window estimation

A probability density function can be estimated on the basis of a number of training samples. Parametric methods assume a model of the data and adapt the parameters of the model using the training samples; the drawback of such methods is that they rely on a model that describes the data. A standard non-parametric approach is Parzen window estimation. A windowing function is centered at each training sample; the estimate is formed by the the superposition of all the windowing functions.

An appropriate windowing function $\varphi$ is positive definite, and its volume is 1. The windowing function has a parameter $h$ that relates to the width of the window, this

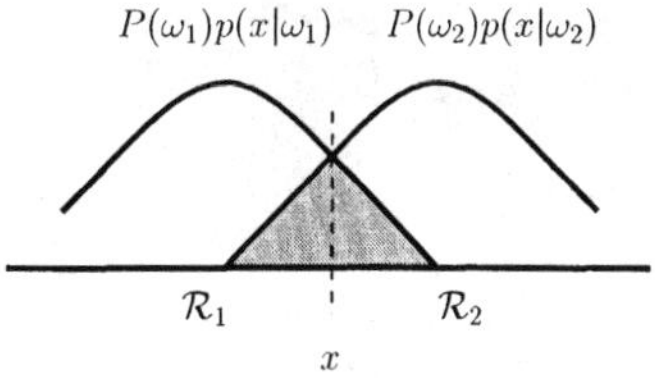


Figure 1: Bayes decision theory states the decision regions $\mathcal{R}_1$ and $\mathcal{R}_2$. The Bayes error-rate corresponds to the size of the shaded area.

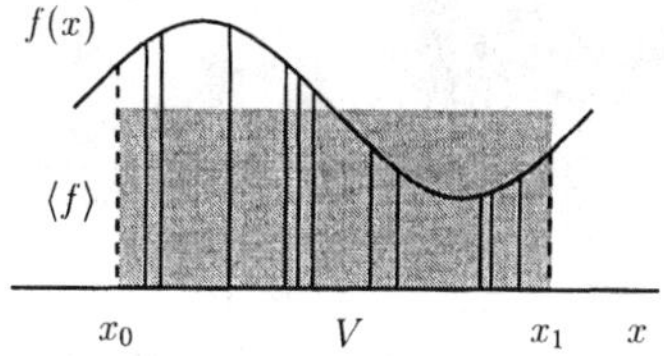


Figure 3: The surface of the shaded area, i.e. $\int_{x_0}^{x_1} f(x)\,dx$, is estimated using random samples within this area. Monte Carlo integration is based on the observation that the integral is approximated by $V\langle f \rangle$, with $V = x_1 - x_0$.

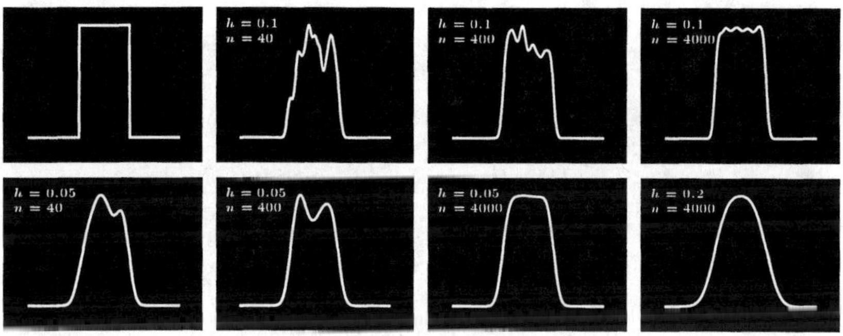


Figure 2: Parzen window estimations from training patterns that were drawn from the probability density function on the left top, various values of $h$ and $n$ are applied.

| problem name | #instances | dimension | max. accur. | Bayes error-rate |
|---|---|---|---|---|
| `breast-cancer` | 398 | 5 | 0.94 | $0.92 \pm 0.01$ |
| `letter-recognition` | 20.000 | 17 | 0.80 | $0.79 \pm 0.01$ |
| `pima-indians` | 768 | 8 | 0.76 | $0.77 \pm 0.01$ |

Table 1: Presented are, from left to right, the name of the pattern classification problem (available via `ftp ics.uci.edu:/pub/machine-learning-databases`), the number of instances, the dimension, the best reported accuracy on the problem, and finally the estimate of the Bayes error-rate.

parameter sets the resolution of the window. Typically the normalized Gaussian is used as a windowing function, but other functions may also be considered. Suppose that $n$ training samples $\vec{x}_i$ are available, $i = 1, \ldots, n$. The Parzen window estimate of the probability function that underlies the data is as follows:

$$\hat{p}(\vec{x}) = \sum_{i=1}^{n} \varphi\left(\frac{\vec{x} - \vec{x}_i}{h}\right)$$

In Figure 2 this idea is illustrated for various values of $h$ and $n$. It has been proven that if the number of training samples reaches infinity, the estimate converges to the "real" probability density function (given some constraints on the value of $h$). Unfortunately, in practice the number of training samples is always limited. The relation between the number of samples, the value of $h$, and the reliability of the estimate is not known.

If all the training patterns belong to the same class, then the estimate approximates the class conditional probability function. If we have such a function for each class, then we can estimate the Bayes error-rate by integrating over the areas of the functions, in accordance to Eq. 1. The first step that we take, is that we estimate the class conditional probability function with Parzen window estimation. Now we need an integration method with which we actually can determine areas underneath the functions. Since most applications will involve multi-dimensional patterns, the method must by able to integrate multi-dimensional functions.

### 2.3 Monte Carlo integration

The straightforward numerical method to integrate a multi-dimensional function is Monte Carlo integration. The function is evaluated at random sample points, and the integral is estimated on the basis of these samples. Suppose that we have $n$ random samples $\vec{x}_i$, $i = 1, \ldots, n$, drawn from a subspace with a volume $V$. Monte Carlo integration is based on the observation that the integral of a function $f$ in the subspace can be estimated as follows:

$$\int f dV \approx V\langle f \rangle \pm V\sqrt{\frac{\langle f^2 \rangle - \langle f \rangle^2}{n}}$$

where the angle brackets denote the average value:

$$\langle f \rangle = \frac{1}{n}\sum_{i=1}^{n} f(\vec{x}_i) \qquad \langle f^2 \rangle = \frac{1}{n}\sum_{i=1}^{n} f^2(\vec{x}_i)$$

The error estimate of the method gives only an indication, and defines no definite upper and lower bounds of the result. (See Figure 3 for an illustration of the idea of Monte Carlo Integration.)

So Monte Carlo integration of a multi-dimensional function is not difficult; the function is evaluated some times, and the average of the function is multiplied with a volume. The method does not work well if the functions are strongly peaked. Furthermore, the number of samples that is necessary to obtain a reasonable estimate, may be large.

## 3 INTEGRATION OF BASIC CONCEPTS

We have combined the three described methods to a novel application to estimate Bayes errors. Suppose that we have a pattern classification problem with $m$ different classes, in each class $\omega_i$, $i = 1 \ldots m$, there are $n_i$ training patterns $\vec{x}^{\,i}_j$, $j = 1, \ldots, n_i$. Training pattern $\vec{x}^{\,i}_j$ is the $j$-th pattern of the $i$-th class. The total number of patterns $n$ is $\sum_{i=1}^{m} n_i$. First we determine the minima and maxima along each dimension; this gives us the integration volume. Repeatedly a random sample $\vec{x}$ is drawn from the volume. The probability of an error is calculated in the sample point. According to Eq. 1 the a priori probability of each class is required, and the values of the class conditional probability density functions at $\vec{x}$. These are determined as follows:

$$\hat{P}(\omega_i) = \frac{n_i}{n}, \quad \hat{p}(\vec{x}|\omega_i) = \frac{1}{n_i}\sum_{j=1}^{n_i} \varphi\left(\frac{\vec{x}^{\,i}_j - \vec{x}}{h}\right)$$

The estimated probability of an error is as follows:

$$\hat{P}(\varepsilon) = 1 - \max_i \int \hat{P}(\omega_i)\hat{p}(\vec{x}|\omega_i)d\vec{x}$$

We estimate $P(\varepsilon)$, by repeatedly drawing samples $\vec{x}$; that is what Monte Carlo integration essentially is. After many samples we have an estimate of the Bayes error.

# 4 TESTING THE APPROACH

Beforehand it is very uncertain whether the proposed approach will be fruitful. May be Parzen window estimation requires too many training patterns to be of practical value. Or perhaps the window width influences the results very much, etc. There are a lot of aspects that have to be scrutinized before we feel that we can rely on our method. Step by step we have checked all bottlenecks using an artificial pattern classification problem with a known Bayes error-rate. From the tests we learn that our method typically needs in the order of a thousand training samples per class, if the dimension of the problem is about five. Furthermore, the number of Monte Carlo iterations that is required is roughly a hundred times the number of training samples. Fortunately, the method appeared to be not very dependent on the width of the windowing function; a whole range of widths allowed for equal—and correct—results. The major drawback of the method seems to be the fact that so many patterns are required to yield a reliable estimate. Moreover, if the dimension of the patterns increases to more than ten, an unrealistic amount of training patterns is necessary.

So our method is typically applicable for problems with a lot of training patterns, which have a small dimension. These conditions are for example satisfied in image processing; if pixels in images are to be classified there are many training patterns since there are many pixels in an image. Another field where usually a lot of patterns with a small dimension are available is time-series prediction.

# 5 BENCHMARK RESULTS

We have done a benchmark study to test our method. Description of the used data-sets are available at the UCI repository of machine learning databases and domain theories (`ftp ics.uci.edu:/pub/machine-learning-databases`). The data-sets that we chose form the extensive list contained numerous, continuously valued, patterns. For each benchmark the best reported error-rate was given. We have investigated whether our method could satisfyingly predict the best results on the benchmarks. Results are presented in Table 1.

## 6 DISCUSSION AND CONCLUSION

From the results we learn that we could successfully predict the lowest achievable error-rate in the majority of cases. One could doubt the usefulness of our approach since it seems to predict a result that can be achieved with a classifier. So why not look at the results of the classifier immediately? The reason is in the fact that the performance of the classifier is dependent on the settings of parameters that play a role during learning. Our approach is almost independent of parameters; the only parameters to decide on are the window width and the number of Monte Carlo iterations. We have shown already that the performance hardly depends on the value of the window width. The Monte Carlo integration method returns an error on the result; this error can be used to tune the number of required iterations.

Furthermore, one might comment that if the Bayes error-rate can satisfyingly be estimated with our method, then the classifier that is implicitly used in our algorithm will approximate the optimal classifyer. We make use of estimates of the a posteriori probabilities, only if these estimates approximate the real probability density functions the result of the algorithm will satisfy. So why not use the Bayes decision rule immediately, and assign patterns to the class with the highest a posteriori probability? The answer is that the classifier would be very time-consuming; for any pattern that is to be classified we have to evaluate a lot of windowing functions. However, it is no problem that the algorithm is time-consuming for the task it is designed for: to estimate the Bayes error-rate. This needs to be done only once.

Concluding we would like to stress that we have integrated three classical theories in a new algorithm to estimate the Bayes error-rate of a pattern classification problem. The algorithm is model free in the sense that it makes no assumptions about the characteristics of the data, and almost independent of parameters. One simply inputs a pattern classification problem to the algorithm, and a reliable estimate is output of the smallest reachable error-rate of the problem. This number is an indication of how solvable the problem is. Therefore, it provides the engineer with a target error-rate, and it thus prevents that too much effort is put in trying to improve results.

In further work we give attention to some issues that are left aside in this paper. For example how to speed-up the algorithm. The proposed algorithm is extremely time-consuming. A run on a benchmark problem takes typically about a few hours on a regular workstation. Although we have argued that this is is tolerable, acceleration of the algorithm is desired. There are at least three methods to increase the speed of the algorithm: (i) via alternative window functions; (ii) using multiple processors; (iii) with a vector computer. Another issue that is covered in the full paper is how the algorithm can facilitate the selection of good features for a pattern classification problem.

## References


Duda, R. and P. Hart (1973). *Pattern Classification and Scene Analysis.* John Wiley and Sons.

Parzen, E. (1962). On estimation of a probability density function and mode. *Annual Mathematical Statistics 33*, 1065–1076.

Press, W., S. Teukolsky, W. Vetterling, and B. Flannery (1988). *Numerical Recipes in C.* Cambridge University Press.

# THE INTRODUCTION OF PRINCIPLES FROM ELEMENTARY QUANTUM THEORY IN A HOPFIELD-STYLE NETWORK AND THE SOLUTION OF CONSTRAINT SATISFACTION AND VARIABLE BINDING PROBLEMS

**PHILIP VAN LOOCKE**
Lab for Applied Epistemology, University of Ghent
Senior research associate at the Belgian NFSR

***Abstract:***
This paper describes the Hopfield-style neural network model QNET in which some principles are included that are inspired on elementary quantum theory. The units of this network are allowed to have amplitudes in different frequencies instead of one single activation value. The amplitudes of the units appear in the updating rule of every unit, but the squares of the amplitudes determine the probabilities with which one unit observes another one firing in a particular frequency. The connections between the units remain simple scalars. If the connection matrix of the schema model is used, then all memories are attractors in all frequencies. In many situations where a schema model converges to a spurious mixture of different solutions, QNET gives different meaningful solutions in different frequencies. With higher order connections, the model can solve the variable binding problem of the schema model in an elegant way.

## INTRODUCTION

About a decade ago, the schema model has been introduced in order to explain how subconceptual units are able to organize cognitive processes (Rumelhart *et al.*, 1986).[1] Meanwhile, this model and related Hopfield-type models have been applied successfully to problems in which a large number of soft constraints had to be fulfilled. The model that is proposed in the present paper takes these models as its point of departure. However, in the present model, the state of a unit i is not characterized by a single activation value, but by r different numbers $a_i^1, a_i^2, \ldots, a_i^r$. The f-th number $a_i^f$ is called the amplitude of unit i in frequency f. In other terms, the state of a unit is composed of a discrete number of frequencies. Since this is not the only property of the present model that reminds us of elementary quantum theory, we have called our model QNET.

QNET has been constructed with two aims. First, it had to find all solutions that are compatible with a given input state. If different solutions are equally compatible with an initial state, traditional Hopfield-type models often converge to a spurious solution. Second, QNET was developed to solve the variable binding problem in the context of the schema model. For the sake of concreteness, consider the sentence 'John gives a red rose to Mary', and suppose that a verb or a relation is represented by a number of units that is equal to its number of arguments (e.g. 'to give' is represented by three units 'to give: subject', 'to give: object' and 'to give: receiver). Then, the pattern of activation that results if every unit of the sentence is put active is not an appropriate representation of the sentence. For instance, such a pattern does not specify if John is the subject of the sentence rather than Mary or the rose. Nor does it specify to which person or object the predicate 'red' refers, and so on. In a QNET network, units can be bound if they are active in the same frequency.

## THE UPDATING RULE OF QNET

Like the schema model, QNET has a randomly asynchronous updating rule. At every time step, a randomly chosen unit i updates its activation. To this end, it makes use of signals that it receives from the other units of the network. Then, three steps follow that remind us of elementary quantum theory:

(a) Consider a unit j that sends information to i. Unit j sends information to i concerning one frequency only. The chance that it sends information concerning the f-th frequency is given by $(a_j^f)^2$. The information that it sends is the real number $(a_j^f\text{-}nl)$, where *nl* is a network constant that corresponds to a noise level.

(b) Unit i updates its amplitude in the f-th frequency in accordance with the information that it has received about this frequency. Suppose that units $j_1,j_2,\ldots,j_q$ have sent information concerning the f-th frequency. The contributions from these units are weighted by the strengths of the connections through which they are communicated and, subsequently, these contributions are summed. Then, a term that is proportional to the bias in unit i is added (the coupling constant of the biases is denoted c). Hence, for every frequency f about which i receives information, i calculates $net_i^f$, with:

$$net_i^f = \Sigma_{\, j \text{ in } \{j1,\ldots,jq\}} w_{ij}.(a_j^f - nl) + c.b_i \qquad (1)$$

These values are multiplied by a uniform parameter μ and added to the amplitudes $a_i^f$, unless the latter would become negative:

$$\text{if } a_i^f(t)+\mu.net_i^f(t)>0 \text{ then } a_i^f(t+1)=a_i^f(t)+\mu.net_i^f(t) \qquad (2a)$$

$$\text{if } a_i^f(t)+\mu net_i^f(t)<0 \text{ then } a_i^f(t+1)=0 \qquad (2b)$$

(c) After the amplitudes of one or more frequencies in i have been adapted in accordance with (2a) and (2b), they are multiplied with a factor $((a_i^1)^2+\ldots+(a_i^r)^2)^{(-1/2)}$. This ensures that the sum of the squares of the amplitudes in every unit always equals 1, so that the probabilistic rule in (a) is meaningful. The normalization also ensures that amplitudes $a_i^f(t+1)$ that have become larger than 1 in step (b) are pushed below 1 again.

Point (a) reminds us of the quantum mechanical rule that the chance to observe a system in a particular state is determined by the square of the amplitude corresponding to this state. Like in quantum theory, the squared amplitudes determine observation probabilities, but the non-squared amplitudes govern the interactions, as is expressed in points (a) and (b). Obviously, also point (c) is familiar from quantum theory.

The fact that the squares of the amplitudes appear in (a) is vital to the QNET model. Suppose that one frequency f has acquired a clear dominance in a unit. Then, the squares in point (a) ensure that this unit communicates virtually only information concerning this frequency to other units. Hence, if the unit at issue has a strongly positive connection with another unit, it will stimulate the latter to become active in the same frequency. Conversely, if it is connected significantly negatively with a unit, it will disencourage a non-zero amplitude in frequency f in the latter unit. As another consequence, if some noise is present in the other frequencies of the unit, this noise will not propagate through the network.

## CONCURRENT SEARCH PROCESSES IN DIFFERENT FREQUENCIES

In order to illustrate how QNET works, we consider a network with 15x15 units in which ten patterns are stored. The first three patterns are shown in Figure 1; the remaining seven patterns are random patterns. The Hopfield learning rule is used to determine the connection matrix. Suppose that, in each of first three frequencies, the initial amplitude pattern corresponds with a distorted version of a different memorized pattern. More specifically, suppose that the first pattern of Figure 1 is distorted 75 successive times by a random clip, and that this pattern is initialized in the first frequency. A similar procedure is followed to initialize the next two frequencies. In frequencies 4 to 6, the mirror patterns of the patterns shown in frequency 1 to 3 are initialized (this increases the basins of attraction of patterns to be recognized, see Van Loocke, 1996). The squares of the amplitudes in every unit are normalized to one. Figures 2a, 2b and 2c show the amplitude patterns after 0, 450 and 2250 updates in six frequencies (for this run, the total number of frequencies in which a unit can be active has been put equal to six). We observe that in the first three frequencies, the three pure patterns are restored, and that in frequencies 4 to 6, the pure mirror patterns are retrieved. Depending on the input patterns, the same pattern may be retrieved in different frequencies. Alternatively, the same input pattern can be shown in every amplitude. Then, if different solutions exist, QNET realizes each of them in a different frequency. In such a situation, the original Hopfield model often converges to a spurious state (I refer to Van Loocke, 1996 for illustrations)

Figure 1. Three first patterns stored in a network with 225 units

Figure 2a. Initial amplitude patterns in six frequencies

Figure 2b. Amplitude patterns after 450 updates

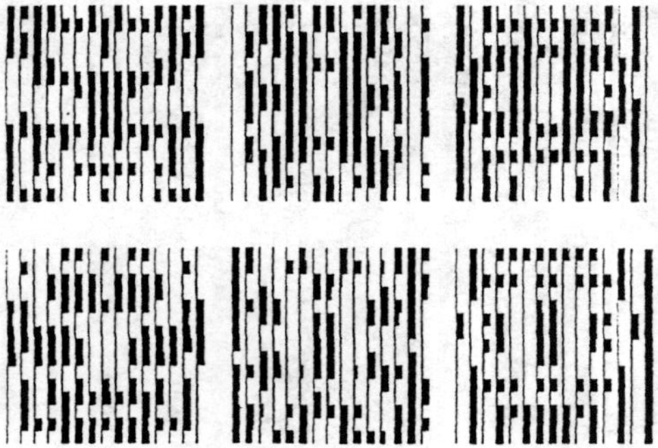

Figure 2c. Amplitude patterns after 2250 updates

## BINDING VARIABLES IN A DISTRIBUTED PATTERN

In QNET, all patterns of lowest energy are attractors in all frequencies, and different patterns may be found simultaneously in different frequencies. Once a pattern has been found in one or other frequency, QNET may still have to indicate how units that participate in this pattern are bound. For instance, in the example of the introduction, once a sentence has been retrieved in a frequency, one may wonder how the units of the sentence are bound. In order to allow a QNET model to infer this information, a second type of connection matrix has to be provided. While QNET is examining how units are bound in such a pattern, the updating operations can be confined to units that participate in this pattern. Because the second connection matrix operates on a smaller scale than the one that is responsible for the storage of the patterns themselves, we have called the second matrix the matrix of fine-grained connections (Van Loocke, 1996).

The fine-grained connection matrix contains connections of higher order. More specifically, connections that bind arguments of verbs or of relations with m arguments are of order m+1. For instance, consider the sentence 'John is larger than Mary'. Suppose that it is represented by the units 'John', 'Mary', 'larger: first argument' and 'larger: second argument'. Then, the unit 'John' sends aligning communication to 'larger: first argument' on condition that 'Mary' is active in some frequency. We remind of the fact that a unit is active if its amplitude is significantly larger than $nl$ in at least one frequency. Suppose that 'John', 'Mary', 'larger: first argument' and 'larger: second argument' are represented by units 1, 2, 3 and 4, respectively. Suppose that unit 1 ('John') has been selected for an update, and that it receives information from unit 4 ('larger: subject'). The chance that this information concerns the f-th frequency remains given by $(a_4^f)^2$. The connection $w_{41;2}$ from unit 4 to unit 1 as conditioned by unit 2 appears in the updating rule of unit 1 in the following way:

$$a_1^f\,(t) = a_1^f\,(t-1) + e.w_{41;2}\,(a_4^f\,(t-1)-nl)\;.\;\Sigma_g\,(a_2^g\,(t-1)-nl)^2 \qquad (3)$$

Unit 1 may receive such contributions from different relational units, along with contributions from non-relational units. Notice that the factor behind $a_4^f-nl$ differs from zero only if unit 2 is active in at least one frequency.

## DISCUSSION

The binding problem of connectionism can be solved if a network includes units that show different makers or amplitudes instead of a single activation value. Shastri and Ajjanagadde (1993) suggest that, from a physiological point of view, such markers correspond to temporal variables. This is consonant with the fact that the importance of temporal variables has been suggested regularly in recent physiological literature. This holds, for instance, for color vision, texture, audition, pain, taste and so on (an overview can be found in Cariani, 1995). Also the possibility that temporal variables help solving the binding problem has been put forward at different instances in this literature (Engel *et al.*, 1991). If it is assumed that temporal variables furnish markers, then different possibilities remain. For instance, the makers may be realized by frequencies or by phases. In the latter case, units are bound if they fire in synchrony (this option is defended in Shastri and Ajjanagadde, 1993). We emphasize that QNET is a model on an abstract level; the terminology 'frequencies' in the context of the QNET-units has been chosen because of the fact that QNET has been inspired on elementary quantum systems. The present study of QNET does not predict a particular neuro-physiological realization of these 'frequencies'. Rather, we studied the computational properties of QNET on an abstract level that allows for different physical realizations.

It must also be clear that QNET is not a quantum computer in the sense of Deutsch (1985), nor does it predict quantum interactions in the brain of the type that Penrose (1994) has in mind. QNET is a classical model that is inspired partially on some ideas that also occur in elementary quantum theory, but its simulation or its physical realization does not depend, for instance, on the use of SQUIDS. A quantum computer would operate in many parallel worlds. QNET operates in many parallel frequencies, but all of them belong to the same world.

## REFERENCES


Cariani, P. (1995), As if time really mattered. Temporal strategies for neural coding of sensory information, *CCAI: Journal for the Integrated Study of Artificial Intelligence, Cognitive Science and Applied Epistemology*, 12, 157-219

Deutsch, D. (1985), Quantum theory, the Church-Turing principle and the universal quantum computer, *Proceedings of the Royal Society*, A400, 97-117

Engel, A., Koenig, P., Kreiter, A. Gray, C., Singer, W. (1991), Temporal coding by coherent oscillations: a potential solution to the binding problem, in Schuster, H., Singer, W. (eds.), *Nonlinear dynamics and neural networks*, Weinheim

Mc Clurkin J., Zarbock, J., Optican, L. (1993), Neurons in primate visual cortex multiplex information about red/green, blue/yellow, and black/white opponencies using temporal codes, *Neuroscience Abstracts*, 19, 1576

Penrose, O. (1994), *Shadows of the mind*, Oxford: Oxford University Press

Rumelhart, D., Smolensky, P., Mc Clelland, J., Hinton, G. (1986), Schemata and sequential thought processes in PDP models, in: Mc Clelland, J., Rumelhart, D. (eds.), *Parallel distributed processing: explorations in the microstructure of cognition*, vol 2, MIT Press

Shastri, L., Ajjanagadde, V. (1993), From simple associations to systematic reasoning: a connectionist representation of rules, variables and dynamic bindings using temporal synchrony, *Behavioral and Brain Sciences*, 16, 417-494

Van Loocke, Ph. (1991), Study of a neural network with a meta-layer, *Connection Science*, 4, 367-379

Van Loocke, Ph. (1994a), *The dynamics of concepts: a connectionist model*, Heidelberg/New York/Berlin: Springer Verlag

Van Loocke, Ph. (1996), Parallel search processes in different frequencies and variable binding in a quantum mechanically inspired extension of the schema-model, submitted to Connection Science

# ITERATIVE TRAINING TO IMPROVE NEURAL NETWORK METAMODELS OF SIMULATED SYSTEMS

**SANIYE BURCU OZSERIM AND ALICE E. SMITH**
*Department of Industrial Engineering*
*University of Pittsburgh*
*Pittsburgh, PA 15261*
*aesmith@engrng.pitt.edu*

**LT. COL. ROBERT A. KILMER**
*Center for Strategic Leadership*
*Army War College*
*Carlisle, PA 17013-5049*
*kilmerr@csl-emh1.army.mil*

***ABSTRACT:***
This paper explores two alternatives to the traditional supervised training method of initially selecting the complete training sample randomly from the range of possible values. Both alternative methods have the same central idea of using more patterns where the input/output relationship is complex. One method estimates complexity by curvature of the response surface, while the other, "iterative training," estimates complexity by training errors. This latter approach requires virtually no additional computational effort and is shown to generally result in more precise models. The approaches are illustrated on constructing a neural network metamodel of a discrete event stochastic simulation of an inventory system.

Kilmer [3-6] used artificial neural networks to construct empirical metamodels of computer simulations. The problem that he successfully addressed was how to perform basic simulation tasks, such as prediction and comparison of alternatives, with neural approximations of a computer simulation. The main theme of this paper is to investigate two alternative approaches to selecting training patterns for the same purpose, i.e., to approximate computer simulations. This problem was selected because obtaining the training and testing pairs for a stochastic computer simulation is computationally costly, thus motivating the need to form the most accurate neural network model from a limited data set. In both new training approaches, learning is iterative. A subset of the training data is used to develop an initial neural network. The complexity of the relationship the network needs to learn is estimated by either looking at the response surface on a regional basis or by identifying regions with large training errors. The assumption is that a more complex region should be allocated more training examples, and the succeeding iterations add training patterns disproportionately from the complex region(s).

This general theme has been the subject of previous research. Ludik and

Cloete [8], explored the effect of various training strategies for backpropagation training using simple recurrent and temporal autoassociation neural networks. They demonstrated that training on increasingly more complex combined subsets reduces the number of weight updates to reach a given error criterion, suggesting that easier subtasks should be learned first. A similar approach was used by Sun et al. [10] and they also observed faster convergence and better error performance. Wann et al. [11] proposed a method to partition input patterns into groups based on distance from the class border. By using combinations of these groups, they were able to construct a variety of training sets including typical and border sets (i.e., those near class boundaries). Cheung, Lustig and Kornhauser [2] analyzed the backpropagation learning process in three stages split according to the behavior of the errors produced. These works suggest that examination of the training errors on certain observations can be used to improve the supervised neural network learning process beyond that achievable by simple batch training.

## RESEARCH APPROACH AND RESULTS

The inventory simulation problem used was taken from Law and Kelton [7]. Kilmer [3-5] also studied this problem and it is his results using uniformly generated training patterns to which we compare. Note that each simulation "run" or "replication" results in one value of the response variable, total cost, and this response depends on both the values of the input variables, reorder point ($s$) and reorder quantity ($d$), and on the particular values taken on by the many random variables inherent in the simulation. At the beginning of each month, the inventory level is reviewed and a decision is made on how many items to order from the supplier. Figure 1 shows the expected value (mean of 10 replications) of the simulation over 420 combinations of $s$ and $d$. This figure is useful when comparing the training patterns selected by the two alternative methods presented this paper. Note that the response surface reaches a minimum in the lower left quadrant, and this quadrant also contains the greatest changes in the response. The following are the parameters of the system as described in Kilmer [3].

$$\text{D (Demand)} = \begin{cases} 1 & \textit{with probability } 1/6 \\ 2 & \textit{with probability } 1/3 \\ 3 & \textit{with probability } 1/3 \\ 4 & \textit{with probability } 1/6 \end{cases}$$

$t_d$ (time between demands)~exp($1/\lambda$=0.1 month)
$I$ = inventory level ($I_o$ = 60)
$s$ = reorder point
$d$ = reorder quantity when $I = s$

$t_o$ = time between order decisions = 1 month
$t_L$ = time lag for delivery (U [0.5 to 1] month)
$k$ = setup cost = \$32.00
$i$ = incremental cost per item ordered = \$3.00
$h$ = holding cost = \$1.00 / month
$u$ = underage cost = \$5.00 / month
$n$ = time horizon = 120 months

We compare a total of eight networks. The first four are trained on the expected output (response) over ten simulation runs while the next four are trained on each of the ten responses. Each network is trained from the same sized set of training patterns (25 input/output pairs) and the network architectures (2-2-2-1), learning parameters, termination criteria (to 0.10 or 5000 epochs) and initial weights are identical. These architecture and training parameter choices were identical to those in Kilmer [3-4], who established them after experimentation. We compare the final error over a separate test set of 17 observations, where these observations include both interpolations (11 points) and extrapolations, i.e., values beyond any of those in the training set, (6 points). The first network is trained as reported in Kilmer [3-4] with training patterns evenly distributed throughout the variable ranges, and training done in one iteration on the whole training set of 25. This is the traditional method of training, using a randomly chosen set in one batch. The second network uses an initial training set of 16 patterns then adds 9 patterns in a second iteration, chosen by response surface curvature analysis. The third network uses the iterative approach based on training error with a first iteration of 16 patterns and a second iteration of 9 patterns. The fourth network also uses the iterative approach with a first iteration of 16 patterns, a second iteration of 4 patterns and a third iteration of 5 patterns. The fifth through eighth networks are the same, but trained on individual replications rather than mean responses.

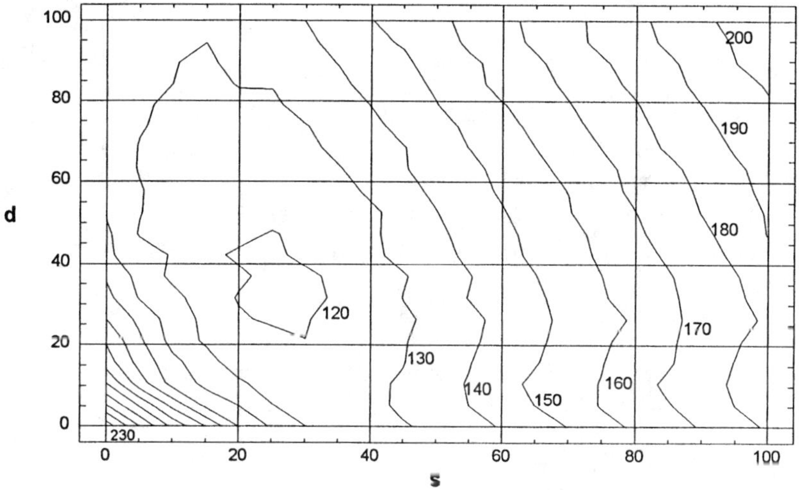


Figure 1: Simulation expected response (total cost) to s and *d* over 420 simulation runs.

**Experiments on Using Expected Values as Targets**

Network 1 - Uniformly Selected: Figure 2 shows the training (stars) and testing (squares) patterns for the network trained on 25 equally distant patterns. Note that these are the testing patterns for all networks studied in this paper.

Network 2 - Response Surface Method: A measure of regional curvature was used as an approximation to complexity. That is, a linear (planar) response surface is assumed to be simple, while curvature indicates that the region may be harder to learn. Regions are taken to be equally sized in each input dimension. We constructed four regions by dividing both $s$ and $d$ in half ($2^2$=4), i.e., four quadrants. Regions are coded by the two levels of $s$ and $d$, e.g. region (1,1) is the lower left and region (2,1) is lower right. The second derivative of a simulation is impossible to evaluate since there is no explicit form. As a surrogate, response surface methodology was used to select 9 additional training patterns to add to the 16 existing uniformly distributed patterns. Curvature and interaction checks are made independently for each region using a $2^2$ factorial with 3 center points (Figure 3).

The design chosen allows the planar model to be efficiently fitted, allows checks to be made to determine whether the planar model is representationally adequate, and provides some estimate of experimental error. The least squares estimate of $\beta_0$ is the mean of all 7 observations. $\beta_1 = 1/4\ [(-1 * r_1) + (1 * r_2) + (-1 * r_3) + (1 * r_4)]$. The values; $r_1$, $r_2$, $r_3$ and $r_4$, stand for the response of the first four runs for region (1,1) respectively. Similarly, the least squares estimate of $\beta_2$ is calculated in the same way. An estimate of the experimental error variance is obtained from the 3 replications at the center. It was very approximate because there are only 2 degrees of freedom. Standard errors of coefficients are calculated by using this estimate.

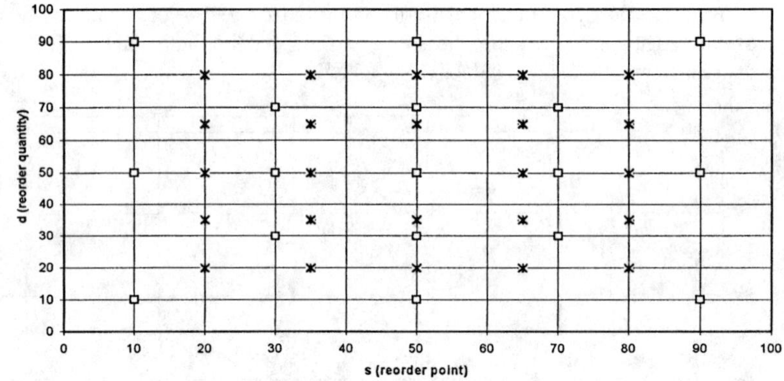


**Figure 2: Training and testing patterns for network 1.**

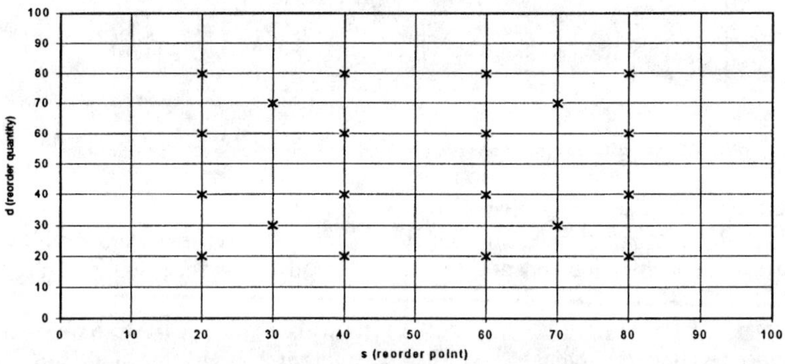


**Figure 3: Points used for response surface methodology.**

The effects of the variables are assumed to be additive. The coefficient $\beta_{12}$ of an added cross product term $x_1x_2$ in the model is used to measure the interaction between the variables. This coefficient $= 1/4\ [(1*r_1) + (-1*r_2) + (-1*r_3) + (1*r_4)]$. Another measure of planarity is provided by comparing the average of the 4 points of the $2^2$ factorial with the average at the center of the design. In [1] it is shown that if $\beta_{11}$ and $\beta_{22}$ are coefficients of the terms $x_1^2$ and $x_2^2$, this curvature measure will be an estimate of $\beta_{11} + \beta_{22}$. The estimate of the overall curvature is $\beta_{11} + \beta_{22} = 1/4\ (r_1 + r_2 + r_3 + r_4) - 1/3\ (r_5 + r_6 + r_7)$. The values $r_5$, $r_6$ and $r_7$ stand for responses of the center values. In summary, the planarity checking functions are ($\beta_{12} \pm s_{\beta12}$) and ($\beta_{11} + \beta_{22} \pm s_{(\beta11 + \beta22)}$). If both of the functions include "0" then the surface is planar, otherwise, the surface is not planar.

Table 1 shows the curvature and interaction check results. Only region (1,1) (lower left) was non-planar. Therefore 8 of the 9 additional patterns were selected randomly from this region. Since the ranges for both the interaction and curvature checks for region (1,2) (upper left) were wider than those of regions (2,1) and (2,2), 1 additional pattern was selected randomly from region (1,2) (upper left).

**TABLE 1: INTERACTION AND CURVATURE CHECK RESULTS**

| Region | Interaction check | Curvature check | Decision | # of points |
|---|---|---|---|---|
| (1,1) | (2.12, -3.45) | (10.70, 2.18) | non-planar | 8 |
| (1,2) | (1.36, -1.44) | (2.73, -1.55) | planar | 1 |
| (2,1) | (0.46, -0.94) | (1.29, -0.85) | planar | 0 |
| (2,2) | (1.29, -0.57) | (0.72, -2.13) | planar | 0 |

Network 3 - One Iteration: First, an initial 16 equally distant training patterns (4 in each region) are used to trained the network. The region with the largest training mean absolute error (MAE = Total Absolute Error / number of training points) is selected to receive the next training pattern, randomly selected from that region. Then the MAE is reduced by adding 1 to the denominator (it is now 4+1 = 5). The second additional training point goes to the region with the highest MAE, and so on until 9 points are chosen to add to the 16 to form a total set of 25. The denominator to figure the region's MAE is incremented by 1 for each additional training pattern it receives. The network continues training on this combined set until the termination criterion is met. The results of the initial training are shown followed by the selection of the 9 additional points:

| | region code | total absolute error | MAE | # points / region |
|---|---|---|---|---|
| | (1,1) | 58.46 | 14.61 | 4 |
| | (1,2) | 36.72 | 9.18 | 4 |
| | (2,1) | 22.47 | 5.61 | 4 |
| | (2,2) | 34.87 | 8.71 | 4 |
| 1 | new point from region (1,1) | (20,22) | expected new MAE in region | 11.69 |
| 2 | new point from region (1,1) | (30,21) | expected new MAE in region | 9.74 |
| 3 | new point from region (1,1) | (31,27) | expected new MAE in region | 8.35 |
| 4 | new point from region (1,2) | (36,56) | expected new MAE in region | 7.34 |
| 5 | new point from region (2,2) | (71,78) | expected new MAE in region | 6.97 |
| 6 | new point from region (1,1) | (28,33) | expected new MAE in region | 7.30 |
| 7 | new point from region (1,2) | (23,71) | expected new MAE in region | 6.12 |
| 8 | new point from region (1,1) | (37,21) | expected new MAE in region | 6.49 |
| 9 | new point from region (2,2) | (55,74) | expected new MAE in region | 5.81 |

Network 4 - Two Iterations: Network 4 was trained in two iterations. As for the third network, 16 initial equally distant patterns were used to train the initial network. Then 4 additional training patterns were added and the network continued training for the first iteration. For the second iteration, an additional 5 training patterns were selected and the network continued training on the total set of 25, as shown below.

**ITERATION 1:**

| | region code | total absolute error | MAE | # points / region |
|---|---|---|---|---|
| | (1,1) | 58.46 | 14.61 | 4 |
| | (1,2) | 36.72 | 9.18 | 4 |
| | (2,1) | 22.47 | 5.61 | 4 |
| | (2,2) | 34.87 | 8.71 | 4 |
| 1 | new point from region (1,1) | (20,22) | expected new MAE in region | 11.69 |
| 2 | new point from region (1,1) | (30,21) | expected new MAE in region | 9.74 |
| 3 | new point from region (1,1) | (31,27) | expected new MAE in region | 8.35 |
| 4 | new point from region (1,2) | (36,56) | expected new MAE in region | 7.34 |

**ITERATION 2:**

| | region code | total absolute error | | MAE | points / region |
|---|---|---|---|---|---|
| | (1,1) | 89.44 | | 12.77 | 7 |
| | (1,2) | 38.48 | | 7.69 | 5 |
| | (2,1) | 20.76 | | 5.19 | 4 |
| | (2,2) | 34.95 | | 8.73 | 4 |
| 1 | new point from region (1,1) | | (22,22) | expected new MAE in region | 11.18 |
| 2 | new point from region (1,1) | | (33,23) | expected new MAE in region | 9.93 |
| 3 | new point from region (1,1) | | (36,28) | expected new MAE in region | 8.94 |
| 4 | new point from region (1,1) | | (36,25) | expected new MAE in region | 8.13 |
| 5 | new point from region (2,2) | | (71,79) | expected new MAE in region | 6.90 |

The results of these experiments over the 17 test points are summarized in Table 2. The performance of the networks trained iteratively (networks 3 and 4) are better than the network trained on equally distant patterns (network 1). The percent improvement is the improvement in internal MAE over network 1, the conventional training strategy.

**TABLE 2: TOTAL AND INTERNAL TESTING MAE**

| | Total Testing MAE | Internal Testing MAE | Percent Improvement |
|---|---|---|---|
| Network 1 | 12.10 | 9.39 | - |
| Network 2 | 11.60 | 8.77 | 6.6% |
| Network 3 | 10.88 | 8.22 | 12.5% |
| Network 4 | 10.80 | 8.39 | 10.6% |

## Experiments on Using Simulation Replications as Targets

This section investigates networks trained on 10 replications of each pattern. Refer to Kilmer [3] for a complete discussion on expected value versus replication training.

Network 5 - Uniformly Selected: Network 5 was trained on 25 equally distant training patterns with 10 simulation replications made for each pattern. Therefore the total number of patterns to train the network was 250. The input training patterns $(s,d)$ are same as the ones used to train network 1 in the preceding section.

Network 6 - Response Surface Method: Network 6 was trained on a total of 25 training patterns, which were also selected for network 2, but the training set included each individual replication response, rather than just the mean of the 10 responses.

Network 7 - One Iteration: For network 7, 160 total patterns (replications of 16 equally distant patterns) were used for initial training. After evaluating the region absolute errors, 9 additional patterns were selected. 10 replications of these additional 9 points were added to the existing 160 points and a total of 250 were used to train the final network, as shown below.

| | region code | total absolute error | | MAE | # points / region |
|---|---|---|---|---|---|
| | (1,1) | 388.69 | | 9.71 | 4 × 10 replications |
| | (1,2) | 382.88 | | 9.57 | 4 × 10 replications |
| | (2,1) | 165.80 | | 4.14 | 4 × 10 replications |
| | (2,2) | 273.57 | | 6.83 | 4 × 10 replications |
| 1 | new point from region (1,1) | | (23,23) | expected new MAE in region | 7.77 |
| 2 | new point from region (1,2) | | (30,51) | expected new MAE in region | 7.65 |
| 3 | new point from region (1,1) | | (31,27) | expected new MAE in region | 6.47 |
| 4 | new point from region (1,2) | | (36,56) | expected new MAE in region | 6.38 |
| 5 | new point from region (2,2) | | (71,79) | expected new MAE in region | 5.47 |
| 6 | new point from region (1,1) | | (28,33) | expected new MAE in region | 5.55 |
| 7 | new point from region (1,2) | | (23,71) | expected new MAE in region | 5.46 |
| 8 | new point from region (1,1) | | (37,21) | expected new MAE in region | 4.85 |
| 9 | new point from region (2,2) | | (55,74) | expected new MAE in region | 4.55 |

Network 8 - Two Iterations: This network was like network 4, but used ten replications for each input pattern selected. The first iteration selected 4 additional points and the second iteration selected 5 more.

Iteration 1:

| region code | total absolute error | | MAE | # points / region |
|---|---|---|---|---|
| (1,1) | 388.69 | | 9.71 | 4 × 10 replications |
| (1,2) | 382.88 | | 9.57 | 4 × 10 replications |
| (2,1) | 165.80 | | 4.14 | 4 × 10 replications |
| (2,2) | 273.57 | | 6.83 | 4 × 10 replications |

| | | | | |
|---|---|---|---|---|
| 1 | new point from region (1,1) | (23,23) | expected new MAE in region | 7.77 |
| 2 | new point from region (1,2) | (30,51) | expected new MAE in region | 7.65 |
| 3 | new point from region (1,1) | (31,27) | expected new MAE in region | 6.47 |
| 4 | new point from region (1,2) | (36,56) | expected new MAE in region | 6.38 |

ITERATION 2:

| region code | total absolute error | | MAE | # points / region |
|---|---|---|---|---|
| (1,1) | 539.54 | | 8.99 | 6 × 10 replications |
| (1,2) | 600.71 | | 10.0 | 6 × 10 replications |
| (2,1) | 160.67 | | 4.01 | 4 × 10 replications |
| (2,2) | 272.12 | | 6.80 | 4 × 10 replications |

| | | | | |
|---|---|---|---|---|
| 1 | new point from region (1,2) | (22,58) | expected new MAE in region | 8.58 |
| 2 | new point from region (1,1) | (33,21) | expected new MAE in region | 7.70 |
| 3 | new point from region (1,2) | (34,57) | expected new MAE in region | 7.50 |
| 4 | new point from region (1,1) | (38,24) | expected new MAE in region | 6.74 |
| 5 | new point from region (1,2) | (41,79) | expected new MAE in region | 6.67 |

The results, as shown in Table 3, are similar to those from networks 1 through 4. The networks trained iteratively using training errors were superior, with two additional iterations the clear winner. The response surface technique did not fare well for this group of networks. Overall, the networks had lower error rates than the first four, confirming Kilmer's conclusion [3] that replication training is generally superior for small data sets.

TABLE 3: TESTING RESULTS OF NETWORKS TRAINED ON REPLICATIONS

| | Total Testing MAE | Internal Testing MAE | Percent Improvement |
|---|---|---|---|
| Network 5 | 10.20 | 7.34 | - |
| Network 6 | 10.28 | 7.53 | -2.6% |
| Network 7 | 9.70 | 7.12 | 3.0% |
| Network 8 | 9.12 | 6.27 | 14.6% |

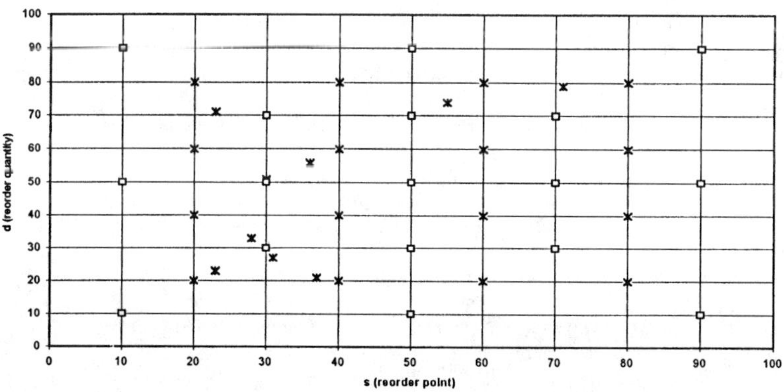


Figure 4: Typical training set selected by the iterative method.

## DISCUSSION

In both of the experiments, the regions where $s$ was between 20 and 40 and $d$ was between 20 and 80 were the ones which indicated more complexity, *viz.* regions (1,1)

and (1,2). Both of the alternative methodologies detected these regions as potential training bottlenecks, and therefore allocated them more training patterns in subsequent iterations. Figure 4 shows a typical training set selected by the iterative method. However, it is difficult and computationally expensive to approximate the simulation function by the response surface methodology. As the number of parameters grows, the complexity of approximating the second derivative increases. On the other hand, evaluation of training errors and iteratively training are easy and require hardly any additional computation. We have made further tests on the robustness of this method for larger problems [9] and have confirmed the general superiority of the iterative training method. The additional computational effort required, regardless of problem size, is minimal, which offers encouragement that this method can be applied to almost any neural network training problem with little *a priori* knowledge or experimentation.

## ACKNOWLEDGEMENT

The second author gratefully acknowledges financial support from the National Science Foundation CAREER grant DMI-9502134.

## REFERENCES

1. Box, G. E. P., W. G. Hunter and J. S. Hunter, *Statistics for Experimenters*, John Wiley & Sons, New York, 1978.
2. Cheung, R. A. M., I. Lustig and A. L. Kornhauser, "Relative effectiveness of training set patterns for back propagation," *Proceedings of the International Joint Conference on Neural Networks*, 1990, I-673-I-678.
3. Kilmer, R. A., *Artificial Neural Network Metamodels of Stochastic Computer Simulations*, unpublished Ph.D. dissertation, University of Pittsburgh, 1994.
4. Kilmer, R. A. and A. E. Smith, "Using artificial neural networks to approximate a discrete event stochastic simulation model," *Intelligent Engineering Systems Through Artificial Neural Networks, Volume 3*, (C. H. Dagli, L. I. Burke, B. R. Fernandez, J. Ghosh, editors), ASME Press, New York, 1993, 631-636.
5. Kilmer, R. A., A. E. Smith and L. J. Shuman, "Neural networks as a metamodeling technique for discrete event stochastic simulation," *Intelligent Engineering Systems Through Artificial Neural Networks, Volume 4*, (C. H. Dagli, B. R. Fernandez, J. Ghosh and R. T. S. Kumara, editors), ASME Press, New York, 1994, 1141-1146.
6. Kilmer, R. A., A. E. Smith and L. J. Shuman, "An emergency department simulation and a neural network metamodel," *Journal of the Society for Health Systems*, in press.
7. Law, A. M. and W. Kelton, *Simulation Modeling and Analysis Second Edition*, McGraw-Hill, New York, 1991.
8. Ludik, J. and I. Cloete, "Training schedules for improved convergence," *Proceedings of the 1993 International Joint Conference on Neural Networks*, 1993, 561-565.
9. Ozserim, S. B., *Improving Neural Network Simulation Metamodels Using Iterative Training*, unpublished M.S.I.E. Thesis, University of Pittsburgh, 1995.
10. Sun, X., B. Golden and E. Wasil, "The fine-tuned learning enhancement to the standard backpropagation algorithm," *Intelligent Engineering Systems Through Artificial Neural Networks Volume 5* (C. H. Dagli, M. Akay, C. L. P. Chen, B. R. Fernandez and J. Ghosh, editors), ASME Press, New York, 1995, 125-132.
11. Wann, M., T. Hediger and M. M. Greenbaum, "The influence of training sets on generalization in feed forward neural networks," *Proceedings of the 1993 International Joint Conference on Neural Networks*, 1993, 137-142.

# HIERARCHICAL NEURAL NETWORK ARCHITECTURES FOR DEPLOYMENT IN MULTI-DECISION ENVIRONMENTS


**BELUR V. DASARATHY**
*Dynetics, Inc.*
*P. O. Drawer B*
*Huntsville, Al. 35814-5050*



***ABSTRACT:***
Three different hierarchical neural network (HNN) architectures, developed for deployment in multi-decision environments and designed to respond to different, problem-environment-driven, response time (decision-speed) requirements, are presented in this study. The study estimates their relative computational demands, which in turn leads to the preferred domains of their application.


## INTRODUCTION

Neural networks (NN) have been well explored for real-time, multi-choice decision making tasks such as multi-class target recognition (Rogers et al., 1995, Botha et al., 1996). Hierarchical NN (HNN) have been proposed (Raghavan et al., 1992) to address specific objectives (feature extraction, classification etc.) with different NNs in the hierarchy. They have also been used to accomplish recognition of different object components (Nolan et al., 1993). Another example of HNN use is that of addressing special problems such as undersea events classification (Shazeer, 1992). Here, we analytically investigate the potential of HNNs for a broad spectrum of problems rather than for a specific application. In the real-world, viewing a multi-decision problem as a single multi-choice decision problem requires a relatively complicated NN (in terms of the number of nodes and hidden layers, the type of feedback, etc.) to model the complex decision surfaces separating these multiple decision domains. It is preferable instead to model the multi-decision problem as a hierarchical set of binary decision (BD) problems each of which is tackled by a far simpler BDNN (in terms of fewer number of nodes and hidden layers and so on). This is true irrespective of the specific type of NN used. Further, this construct permits definition of independent decision subspaces (features) for each BD problem. This leads to a more effective overall design than one possible with a common decision space selected to collectively address the entire multi-decision problem. Each NN can operate in a smaller optimized feature subspace and thereby be computational more efficient over those operating in a single large globally defined feature space.

The conceptual and computational issues associated with the design of this type of HNN architecture are addressed here. The analysis is independent of the specifics of the NN employed, whether it be a simple backward propagation model, or any of the other more sophisticated NNs reported in the literature. The assumption is that each NN within the overall HNN system is a BDNN only. The proposed architectures are evaluated in terms of the real-time needs and computational hardware constraints of the operational system. These software/hardware requirements are dictated by both the specific decision task (for example, target classification and/or track assignment) selected for implementation, as well as the complexity of the problem environment. The complexity of the problem environment is a function of several parameters/constraints such as dimensionality of the decision space, the extent, nature and quality of supervision in the environment, the hierarchy and number of decision choices at each level, the processing time constraints, the number of samples to be processed, requirements of physical portability of the system, and so on. Thus, this study addresses two questions: how can one reorganize a single complex NN designed to make a multi-choice decision into a hierarchy of simple BDNNs, and, how well the resulting architectures can meet the processing resource/time constraints. These constraints, defined by the problem-environment requirements, can therefore vary for different problems. For example, in a high sample density environment, minimization of the average processing time per sample (and, consequently, the total processing time) is desirable. In contrast, in a less dense environment stressed by real-time decision making requirements, minimization of the maximum processing time required for any single sample; i.e., a "worst case" design, becomes the criterion. Thus, processing-time-related constraints may require different HNN architectures. Accordingly, alternative architectures for meeting these constraints are construed and relatively evaluated in this study.

## ANALYSIS

The assumption here is that each pair of decision choices in the problem environment is (either linearly or nonlinearly) separable from one another in a correspondingly appropriate decision space and appropriate NNs can be designed for each of these BDs. Under such an assumption, the $M$-ary decision problem can be modeled either as $M$ or $Mc_2$ BD problems depending on whether each decision choice (class) is well-separable from all the other decision choices (classes) together, or each decision choice (class) is well-separable from every other decision choice (class) individually. The latter offers more flexibility and covers a wider range of practical problems. It permits a more realistic portrayal of real-world multi-decision environments. Accordingly, this study analyzes HNN architectures composed of $M(M-1)/2$ individual NNs. The number of NNs is independent of the specific functional form of the resulting decision surfaces as determined by the training of the corresponding NNs. Although this involves training of multiple NNs, instead of just one under the classical view, the training of these BDNNs is likely to be simpler and the process more likely to converge faster than would be the case of the single complex multi-decision NN. Further, not all of these NNs need be exercised in the operational phase every time such a $M$-ary decision has to be made. The number of NNs to be sequentially exercised in a specific case depends on how the process is organized relative to the ultimate decision arrived at by the system. This sequential NN activation process can be organized in several ways. One could simply use the naive and exhaustive process of activating all the NNs. Alternatively, we could visualize more sophisticated constructs, that attempt to optimize some externally specified criteria. With the overall computational demands minimization goal in view, two basic approaches can be conceived. The minimization can be attempted in a sequential manner by individually minimizing the effort required under each possible resulting decision. This leads to a decision-dependent architecture. Alternatively, one could minimize the total effort such that the computational demand is independent of the eventual decision. Both architectures have been analyzed to bring out the conditions when one or the other becomes more attractive.

**Decision-Dependent NN Architectures**

Here, the NN architecture can be viewed as a top-down tree. Under this architecture, at each node a single pair of decision choices is considered and the decision choice made by the BDNN is then paired with the next potential decision choice for consideration by the next BDNN. The nodes of the tree correspond to BDNNs, and the terminals are the decision choices (class labels) where *U* corresponds to unknown or not decidable case. The objective is to individually minimize the number of NN nodes to be exercised (depth of search or search length) for reaching each possible terminal label from 1 through *M*. The search length is constant for each decision choice and is the smallest for decision choices 1 and 2, but increases for successive decision choices. However, for the case of unknown for $M > 3$, the search length varies since the recognition of the case as not decidable can occur at different stages. This decision choice dependency feature (of increasing search length) brings out the desirability of arranging the decision choices in the order of their *a priori* probabilities, with the most likely decision choice to be tested first, and so on. While it is not feasible to determine precisely, prior to actual decision, the number of NN activations needed for a specific sample, an expected number of such activations required can be estimated using the *a priori* probabilities the different decision choices. The expected number of NN activations ($\bar{n}$) averaged over all possible label assignments, can be determined as

$$\bar{n} = (M\text{-}3) + (M+3)p_0/2 + p_1 + \sum_{i=1}^{M} ip_i \;\; ; M \geq 3 \qquad (1)$$

where $p_i$ is the a priori probability of decision choice *i*. and $n_i$ is the corresponding number of NN activations. Here, $n_0$ is not a constant, under the algorithmic architecture which has been developed to minimize $n_i$, but varies from a low of *M* to a maximum of (2*M* -3), and the relative likelihoods of each of the possible values are not analytically determinable without complex tree analysis. Such analysis is not viewed as necessary since $p_0$, (ideally equal to zero), the parameter multiplying $n_0$, is deemed to be small. Therefore, $n_0$ can be approximated by averaging the two extreme values. This average is viewed as a conservative (larger than true) estimate since the likelihoods of the lower values are greater than those of the higher values. One can now visualize the computational advantage of arranging the classes for corresponding NN node activation in the decreasing order of their *a priori* probabilities. This will minimize the value of the terms under the summation sign in eqn. (1) and hence minimizes the overall computational load. Various special cases, such as $p_0 = 0$ or *equal a priori* probabilities for all classes, or both can be construed and corresponding average computational demands for the different cases can be derived (these are not detailed here because of space constraints). It can be shown that while the average demand is of the order of 1.5 times the number of pattern classes, the maximum demand is of the order of twice the number of pattern classes. Thus, for the worst case, the computational demand is approximately a third higher than average demand.

**An Adaptive Decision-Dependent NN Architecture**

The NN architecture discussed thus far assumed that the classes were ordered as per their a priori probabilities and this ordering remained constant once the decision process was initiated for a sample. (The order could be changed for a different sample at a later time if a change in the relative values of the a priori probabilities is indicated.) However, in view of the significant impact of the relative class probabilities, one can visualize a more sophisticated, adaptive decision-dependent architecture, for organizing the inter-connects of the NNs. Instead of using the constant *a priori* probability values for pre-ordering the classes, one could at each stage determine the 'most likely pair' of classes and activate the corresponding BDNN. This determination of the 'most likely pair' is dependent on not only the class *a priori* probabilities, but also the history of the results of NN activations carried out so far on the sample under consideration, i.e., instead of *a priori* probabilities, a measure of *a posteriori* probabilities is used in determining the computational sequence. This adaptive process of identifying the next most likely classes, hastens the decision process considerably and the average number of NN node activations is likely to be reduced. However, the computational costs, of such adaptive identification of the next likely class, have to be weighed against the gains in terms of reduced NN activations. The cost of the adaptive identification process itself is dictated by the specific approach used: table look up or repeated computations. Thus, the choice of using a dynamic NN architecture will be dependent on the relative costs of NN activations (and the savings thereof) and the adaptive identification of the next most likely class. When a priori probabilities are unknown (but assumed to be equal) this approach may become all the more desirable. The different levels of trade-off involved (in determining whether or not to use the adaptive architecture and if one is to be used, which implementation is more efficient) have to be assessed in the context of a specific application and the associated parameters and constraints to identify the optimal architecture. We shall now investigate another basic alternative, a decision-independent NN architecture, before attempting development of guidelines for making the best choice for a given problem environment.

**Decision-Independent NN Architecture**

This architecture, partially inspired by the merge sort approach to sorting (Bentley, 1979), considers several pairs of classes simultaneously and reduces the number of possible classes to half that prevailing at the present level and this is repeated till only one possible decision choice is retained. The possibility of reject class is then sequentially eliminated. Under this decision-independent architecture, for the case $M=4$, the number of decision nodes passed in the decision process remains constant for all positive identification cases. For reject cases, for $M > 4$, this is not true. Unlike in the previous architecture where only one NN was exercised at each node, here $(M/2J)$ NNs are exercised at the jth level node in the tree up to $j = J = \log_2 M$. (For convenience we shall assume M to be an integer power of 2.) Thereafter, only a single NN is exercised at each of the subsequent nodes in the search path required to eliminate the possibility of rejection. Rejection can thus occur only after J levels, consisting of (M - 1) NN processes, have been traversed, i.e., at or after Mth NN activation. The expression for the average number of NN activations for this architecture can be shown to be

$$\overline{n'} = (2M - \log_2 M - 2)\ ;\ M \geq 4 \tag{2}$$

**Relative Assessment**

The difference $\delta\overline{n}$ , in the demands of the two architectures, is given by:

$$\delta\overline{n} = (M+1-\mathrm{Log}_2 M) - (p_1 + \sum_{i=1}^{M} ip_i) - (2M-\mathrm{Log}_2 M + 1)p_0/2\ ;\ M \geq 4 \tag{3}$$

Depending on whether $\delta\overline{n}$ is positive or negative, the decision-dependent or decision-result -independent architecture can be looked upon as the more desirable alternative from the viewpoint of average computational demands. This is clearly a function of the relative values of $p_i$ of the different classes and $p_0$ the reject probability. This difference between the demands of the two approaches can be analyzed for various cases, such as equal a priori probabilities of the different classes, $p_0$=0, classes as per their a priori probabilities, and the like. Corresponding conditions (parametric domains) under which one or the other architecture becomes more attractive can be derived and plotted to identify the preferred domains of the different architectures. However, due to space limitations imposed by the conference organizers, we are unable to present these varied and interesting details here. Interested readers can contact the author for such details. In the ideal case $p_0 = 0$, and equal a priori probabilities, the choice is clearly for the decision-dependent architecture. If the classes are arranged as before in decreasing order of a priori probabilities, we can establish the inequality:

$$p_0 < \frac{\left[\frac{(M^2 + M - 2)}{M} - 2\,\mathrm{Log}_2 M\right]}{[2M - \mathrm{Log}_2 M + 1]}\ ;\ M \geq 4 \tag{4}$$

i.e., as long as $p_0$ is less than the value shown, the ordered condition results in making the decision-dependent architecture more attractive than the decision-independent architecture. This automatically includes the ideal case of $p_0 = 0$. It should be kept in mind this relative evaluation has been solely from the view of minimal average computational demand.

However, if the criterion for the choice of the NN architecture is viewed differently, for example, as minimization of the maximum demand, then it can be shown that in all cases, the decision-independent architecture, irrespective of relative probabilities, has lesser maximum computational demand. Thus, in applications wherein the maximum response time should be bounded, the decision-independent architecture may become the more desirable alternative, unless the maximum imposed by the other approach is itself within acceptable limits. Further, in the general case where the classes are not necessarily ordered, another possible viewpoint, which can be of help in making the choice between these two architectures, can be construed as follows. Since, the maximum demand under the (dependent) architecture which has the lower minimum demand is higher, it is clear that there exists a point at which the two architectures have equal demand. This interesting trade-off can again be analyzed in detail but is omitted here due to page restrictions. Under the adaptive architecture, the number of NN activations required do not remain the same even for all samples allocated to a single class. This is because of the fact that paths with different search lengths are possible for the same eventual labeling. In the adaptive architecture, this variable nature is taken advantage of to adaptively direct the search and thereby optimize the search length. As expected, the naive approach and the adaptive architecture correspondingly represent the maximum and minimum demand alternatives among those considered here. The maximum demand, as expected, is lower under the decision-independent architecture than under the simple decision-dependent architecture. Under the adaptive architecture, the maximum demand is comparable to that under the decision-independent architecture while the average demand is significantly lower. Thus, if the problem environment and the computational resources permit use of the adaptive architecture with its significantly large memory requirements, then it is likely to be the best alternative among those considered here and is accordingly to be preferred for implementation.

However, this assessment of computational demands under the different architectures considered only the computational demand involved in NN activations, in terms of number of such activations. Other factors, such as overheads involved in organizing the different architectures, which are not as easily amenable to analytical assessment, were not taken into account. Thus as long as the effort of exercising the individual NNs is large compared to the overhead, the relative comparisons between the alternatives are valid. But for a simple NN with very few nodes such disregard of overheads may not always be valid. Preliminary tests have shown this is likely with regard to the decision-independent architecture, wherein the overhead can be considerable. In some cases, it distorts the picture to such an extent that even the simple-minded architecture of checking for each decision choice sequentially (without regard to the results of other NN activations) can turn out to be computationally more attractive. Therefore one has to bear in mind the cost of these overheads in deciding on a specific architecture for a given application. Also, these architectures are naturally suited for deployment in partially exposed or incompletely supervised environments (Dasarathy 1991) since they admit the possibility of unknown classes.

## REFERENCES


Bentley, J. L. (1979). An Introduction to Algorithm Design, Computer, 12(2), Feb, 66-78.

Botha, E. C., Barnard, E., and Barnard, C. J.(1996). Feature-based Classification of Aerospace Radar Targets Using Neural Networks, *Neural Networks*, 9(1), Jan, 129-142.

Dasarathy, B. V. (1991*) Nearest Neighbor (NN) Norms - NN Pattern Classification Techniques*, ISBN 0-8186-8930-7, IEEE Computer Society Press, Los Alamitos, CA.

Nolan, A. R., Wee, W. G., and Leonard, J. D.(1993). Automatic FLIR Target Recognition Using a Hierarchical Neural System, *Proc SPIE - Architecture, Hardware, and Forward-Looking Infrared Issues in Automatic Object Recognition*, 1957, Apr, 50-59.

Raghavan, S., et al. (1992). Layered Object Recognition System Using a Hierarchical Hybrid Neural Network Architecture, *Proc SPIE- Model-Based Vision Development and Tools*, 1609, Feb, 49-60.

Rogers, S. K., et al. (1995). Automatic Target Recognition Using Neural Networks, *Proc SPIE - Applications and Science of Artificial Neural Networks*, 2492, Apr, 346-360.

Shazeer, D. J. (1992). Hierarchical Neural Networks for the Classification of Undersea Events, *Proc SPIE - Applications of Artificial Neural Networks III*, 1709, Apr, 808-822.

# The Analysis and Application for Gray Correlation Theory

Chin-Ming Hong and Terng-Chiao Lin
*Graduate Institute of Industrial Education*
*National Taiwan Normal University*
*Taipei, Taiwan, R.O.C.*

Ching-Tsan Chiang
*Department of Electrical Engineering*
*Chien Hsin College of Technology and Commerce*
*Chung-Li, Taiwan, R.O.C.*

Lon-Biou Li
*Graduate Institute of Industrial Education*
*National Taiwan Normal University*
*Taipei, Taiwan, R.O.C.*

***Abstract***

This world is often modeled in terms of systems or environments, for examples: natural environment (or ecological system) and social environment. These environments and systems do not exist independently of each other. Some systems can be described as subsets of other larger systems or are dependent on other systems. One example is that the educational systems or demographic environments are within the context of a larger social environment or system. Systems and subsystems in a larger system exist many interrelationships. Changes in one system may be dependent or influenced by events in another systems. It is possible that, no isolated, independent systems exist in this world. The interrelationship of many systems are well known and understood, but some others' are poorly understood or even unknown. Methods are needed for analyzing and modeling relationships within systems. Gray system theory provides the necessary tools for creating models of systems and analyzing fuzzy relationships in those systems. This paper presents specific methods for finding measurable relationship between systems and their subsystems. The results from the California Test of Mental Maturity(CTMM) are used as independent variables in this paper, and the test scores of a midterm exams at a vocational technical school are used as dependent variables. The superiority analysis model from gray correlation theory will be used to determine whether the two are related or not. One assumption is made from this result, that is if a significant change in one variable is followed by a similar significant change in another variable, a strong correlation exists. Conversely, if a significant change in one variable is followed, by a little change or no change in another variable, then there is no significant correlation. Therefore gray correlation theory can be used as a tool to measure changing distributions in dynamic systems.

## I. Introduction

There exist some abstract systems(such as society system, economy system, agriculture system, etc.), and they can be called as gray systems because of having unknown or not completed information to describe these systems concretely [ 1, 3, 5, 13 ]. To model and describe these systems, apply logic method or proper known conception / judgment is the only way. It is the gray correlation analysis that quantitatively analyze the dynamic development among the factors of gray system. The gray correlation analysis can analyze a factor in a system(such as fertilizer, farm drug, seed, policy, water conservancy, technology, etc.) and see if it is a main factor, sub-factor, more influential factor, less influential factor, clear factor, latent factor, developmental factor or controllable factor. Using the dynamic process (historical statistical data of a system) of gray system to analyze the quantitative tendency of development can find out the relation of a system and provide the reference for calculating, making policy, controlling. According to the similar or different level of the development tendency, this method can clean and balance the relation among factors of a gray system. And this relation is named gray correlation analysis [ 1, 2, 3, 8 ].

The basic quantitative method used is a mathematical statistical method, such as recurrent analysis (include linear recurrence, multi-factors recurrence, one-factor recurrence, progressively factor recurrence, non-progressively factor recurrence), square differential analysis, main component analysis, etc. Recurrent analysis is used for one-factor, linear analysis which is more than multi-factor, non-linear analysis. There are some weaknesses [ 1, 3, 8 ] of recurrent analysis which are shown below:

1. It requires a vast amount of data to find the statistical regular.
2. Its distribution must be linear, exponential or logarithmic distribution. The linear recurrence is easy to calculate and the exponential or logarithmic distribution could be used for one-factor analysis. Recurrent analysis must have a typical distribution, not a random one.
3. Its calculation is completed. Recurrent analysis could deal with one-factor or two-factors analysis by hands. But it has to deal with multi-factor analysis by computer.
4. An unusual state can happen. The calculation of recurrent analysis includes exponential, summation, subtraction, multiplication, division, square sum, total sum for data. A mistake is caused by the error of computer calculation for transposing the relation can be occurred.

Correlation analysis has been used by the gray system theory to analyze a system for offsetting above weaknesses. It is the method for analyzing the correlation level of factors of a system. It is also the method for analyzing the quantitative tendency of dynamic process development. It is possible to use the tendency of dynamic process development of a system which is the geometric relation and the neighbor level of historical statistical data of a system to judge the correlation level. The main features of gray correlation analysis are listed below:

1. It does not require a rigid data. It is different from statistical analysis because it does require a vast amount of data and a typical distribution data.
2. Its calculation is easy. It can deal with multi-factors analysis by using computer or by hands.
3. The correlation factors analysis is to compare the geometric form of curves. If the geometric form of curves is too close, the changing tendency of their development will be close and their correlation level will be high. According to this method, an unusual state for transposing the relation will not occur.

## II. The Difference of Gray Correlation Analysis and Statistical Recurrent Analysis

The differences between gray correlation analysis and statistical recurrent analysis are compared [ 6, 7 ] in different areas which are shown below:

1. Their basic theories are different. The correlation analysis is based on the gray process of gray theory but the recurrent analysis is based on the random process of probability.
2. The correlation analysis can compare and calculate the tendency of factors of systems. The recurrent analysis calculates the relative value of factors of systems.
3. The correlation analysis does not require a vast amount of data. The recurrent analysis requires a vast amount of data.

The correlation analysis can reach the dynamic process of a system. The recurrent analysis can reach the static process of a system.

Table 2-1 The difference of gray correlation analysis and statistical recurrent analysis

| Method/Analysis | Gray Correlation Analysis | Recurrent Analysis |
|---|---|---|
| Basic Theory | Based on the gray process of gray theory | The random process of probability |
| Calculation | Compare and calculate the tendency of factors of the system | Calculate the relative value of system factors |
| Required Data | Not require a vast amount of data | require a vast amount of data |
| Reach | The dynamic process of system | The static process of system |

## III. The Model of Gray Correlation Analysis

These are 4 principles of gray correlation, and the required conditions of gray correlation and the quantitative model of gray correlation [3, 7] are listed below for measuring the quantitative level of gray correlation.

### 3-1 4 Principles of Gray Correlation

Let $X=\{Xj \mid j \in N\}$ be the subset of gray correlation factors, $X0 \in X$ be the reference order, $Xi \in X$ ( $i \neq 0$ ) be the compared order, $X0(k)$ and $Xi(k)$ ( $k=1, 2, 3,..., n$; $i=1, 2, 3,..., m$ ) be the kth point of $X0(k)$ and $Xi(k)$. If the gray correlation coefficient of Xi for X0 is real, the gray correlation level of Xi for X0 will be $r(X0,Xi)=(\sum r(X0(k),Xi(k)))/n$ and this is the average value of $r(X0(k),Xi(k))$. It needs to satisfy 4 conditions below. These 4 conditions are named 4 principles of gray correlation.

The gray correlation has to satisfy 4 principles of gray correlation: normative, coupling symmetrical, overall and near. Their mathematical definition is described as follows:

(1) Normative

$$r(X0,Xi)=1 \Leftrightarrow X0=Xi$$

$$r(X0,Xi)=0 \Leftrightarrow X0,Xi \in \emptyset$$

$$0 \leq r(X0,Xi) \leq 1$$

$$r(X0,Xi) \in (0,1)$$

It is impossible that not any relation exists in a system.

(2) Coupling Symmetrical

$$x,y \in X$$
$$r(x,y)=r(y,x)=\{x,y\}$$

If there are only two factors in a subset of gray correlation, the comparison of $r(X0,Xi)$ is symmetrical, and this is a concrete quantitative distance.

(3) Overall

$$Xj,\ Xi \in X=\{X\delta \mid \delta=0,\ 1,...n\},\ n \geq 2$$

$$r(Xj,\ Xi) \neq r(Xi,\ Xj)$$

When compare the correlation in a proper environment, then the different reference order can cause the result of comparison not satisfy the symmetrical principle.

(4) Near

When $|X0(k)-Xi(k)|$ is small, then $r(X0(k),Xi(k))$ is large.

The near principle is the limit of quantitative gray correlation.

Let k be horizontal axle and Xi(k) is vertical axle to draw a two-dimensional plane for m+1 orders of X0...Xm. If the geometric form of curve of compared order and reference order is too close, their correlation will be too high ( such as figure 3-1, there are three curves of m=2. The curve of x1 and x0 is more closer than the curve of x2 and x0. So, the correlation level of x1 and x0 is more closer than the correlation level of x2 and x0. ). According to this point and 4 principles of gray correlation, the required conditions of gray correlation and the quantitative model of gray correlation can be obtained.

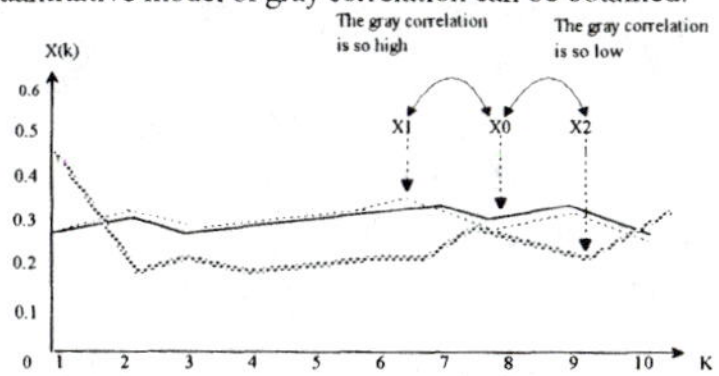


Figure 3-1: The Compared Drawing of Gray Correlation Level

**3-2 The Required Conditions of Gray Correlation**

The gray correlation needs to meet the required conditions [ 3, 7 ] below:

(1) r(X0,Xi)∈R and r(X0,Xi)∈(0,1), I=0, 1,..., m.

(2) The geometric form of two-dimension plane curve of Xi(I=1, 2, 3,..., m) and X0 is closer then r(X0,Xi) is larger.

(3) r(X0,Xi) is only related to the geometric form of two-dimensional plane curve of Xi(I=1, 2, 3,..., m) and X0, not related to its relative position in space. If there is a real $\alpha$, then the gray correlation r(X0,Xi) of X0 and Xi is equal to the gray correlation r($\alpha$X0, $\alpha$Xi) of $\alpha$X0 and $\alpha$Xi. So r(X0,Xi)= r($\alpha$X0, $\alpha$Xi).

(4) The value of r(X0,Xi) is 1 when the curve of X0 and Xi is completely equal or is completely overlapped.

If it meets the 4 required conditions above, then r(X0,Xi) is the gray correlation of Xi for X0 in the interval (1,m).

**3-3 The Quantitative Model of Gray Correlation**

In fact, the correlation is a distinction of the geometric form of curve. So the distinction value of the geometric form of curve is the scale of the correlation. There are many compared order X1, X2, ..., Xn for a reference order X0. The distinction value [ 1, 6, 8 ] of any point ( time ) in compared curve and reference curve can be obtained from the relation below.

$$\xi_i(k) = r\left(x_0(k), x_i(k)\right)$$

$$= \frac{\min_j \min_k \left|x_0(k) - x_j(k)\right| + \delta \max_j \max_k \left|x_0(k) - x_j(k)\right|}{\left|x_0(k) - x_j(k)\right| + \delta \max_j \max_k \left|x_0(k) - x_j(k)\right|}$$

$$= \frac{\Delta \min + \delta \Delta \min}{\Delta_{0,i}(k) + \delta \Delta \max}$$

The relative distinction value is the correlation coefficient of Xi for X0 in time $k$. The distinction order $\Delta_{0,j}$ is the absolute distinction value of two compared orders. It could be 1≤ i ≤m and m is a positive integer. The $\Delta$min and $\Delta$max are the minimum and maximum values of absolute distinction value of all compared orders in any point. $\delta$ is a recognized coefficient, its value is between 0 and 1. In fact, the recognized coefficient is an artificial value (it is an artificial coefficient of a quality analysis) for weakening the influence of exceeding value $\Delta$max and enhancing the obvious distinction of correlation. The recognized coefficient can one or a number between 0 and 1. Usually, it is between 0.1 and 0.5 [ 6, 8 ].

According to 4 principles of gray correlation, the model of gray correlation r(X0,Xi) is [1]:

$$r(X0,Xi)=\sum r(X0(k),Xi(k))/n$$

There are two examples to explain the gray correlation analysis and the gray correlation.

1. Example for Theory:

It is an initial value order:

X0={ 1, 1, 2, 3, 4, 5 }
X1={ 1, 1.2, 2, 2.5, 3.5, 3.5 }
X2={ 1, 1.5, 2.5, 3.4, 2.8, 2.5}
X3={ 1, 2, 1.5, 1.25, 0.9, 1.5 }

Step 1: The Distinction Order

| Order Number | 1 | 2 | 3 | 4 | 5 | 6 |
|---|---|---|---|---|---|---|
| $\Delta$1= \| X0(k)-X1(k) \| | 0 | 0.2 | 0 | 0.5 | 0.5 | 1.5 |
| $\Delta$2= \| X0(k)-X2(k) \| | 0 | 0.5 | 0.5 | 0.4 | 1.2 | 2.5 |
| $\Delta$3= \| X0(k)-X3(k) \| | 0 | 1 | 0.5 | 1.75 | 3.1 | 3.5 |

Step 2: The Min and Max Distinction of Two Level

$minmin=|X0(k)-Xi(k)|=0$

$maxmax=|X0(k)-Xj(k)|=3.5$

Step 3: The Correlation Coefficient

The minimum and maximum distinctions are used in the calculated form of correlation coefficient:

$$\zeta i(k)=(0+0.5*3.5)/(|X0(k)-Xi(k)|+0.5*3.5)=1.75/(\Delta i(k)+1.75)$$

The absolute distinction is:

$$\zeta 1(k)=(1, 0.897, 1.000, 0.778, 0.778, 0.539)$$

$$\zeta 2(k)=(1, 0.778, 0.778, 0.814, 0.593, 0.412)$$

$$\zeta 3(k)=(1, 0.636, 0.778, 0.500, 0.361, 0.333)$$

Step 4: The Correlation

The correlation curve of X0 and Xi is:

$$r1=\sum\delta 1(k)/6=0.8319$$

Then r2=0.7291 and r3=0.6014

The integrate discussion:

The correlation r1=0.8391 of X0 and X1 is the largest one. So the tendency of development of X0 and X1 is the closest one and X1 is the most influential factor for X0.

The correlation r3=0. 6014 of X0 and X3 is the smallest one. So the tendency of development of X0 and X3 is the most farthest one and X3 is the least influential factor for X0.

2. Example for Application:

If it is desired to find out the major factor which influents the growth of youths, one can apply the gray correlation analysis. Let X0(k) be the rate of bad growth, X1(k) be the rate of disqualified for bodily strength training, X2(k) be the rate of bad nutrition and X3(k) be the rate of sickliness. The percentage of relative condition value of proper age schoolgirls is listed on table 3-1.

Table 3-1 The percentage of major factor that influent the growth of students

| Order | 1 | 2 | 3 | 4 | 5 | 6 |
|---|---|---|---|---|---|---|
| Age | 11 | 12 | 13 | 14 | 15 | 16 |
| X0(k) | 31 | 45 | 46 | 45 | 51 | 47 |
| X1(k) | 36 | 33 | 32 | 38 | 36 | 42 |
| X2(k) | 38 | 38 | 25 | 13 | 10 | 17 |
| X3(k) | 7 | 8 | 9 | 7 | 5 | 7 |

Finding out the minimum and maximum distinctions of two levels on the above table:

$minmin=|X0(k)-Xj(k)|=5$

$maxmax=|X0(k)-Xj(k)|=46$

Let $\zeta=0.5$, then the correlation coefficient is:

$$\zeta 1=(\zeta 1(1), \zeta 1(2), \zeta 1(3), \zeta 1(4), \zeta 1(5), \zeta 1(6))$$
$$=(1, 0.8, 0.7568, 0.9333, 0.7368, 1)$$

At the same time:

$$\zeta 2=(0.9333, 0.9333, 0.6364, 0.5091, 0.4375, 0.5283)$$

$$\zeta 3=(0.5957, 0.4667, 0.4667, 0.4590, 0.4058, 0.4444)$$

The relative correlation is:

r1=0.8712, r2=0.6630, r3=0.4731Obviously, r1> r1> r3

The order of major factor that influent the growth of schoolgirls is the bodily strength training, nutrition and sickliness.

**IV. Superior Analysis**

The superior analysis must be applied when the reference order and the compared order are not the only one used. Let the reference order be the main order ( or main factor ) and the compared order be the sub-order ( or sub-factor ). The main order and the sub-order form the correlation matrix. According to the relation of factors of correlation matrix, the superior factors and the non-superior factors can be found[ 1, 7, 9, 10, 11 ].

If there are 5 main factors y1, y2, y3, y4, y5 and 6 sub-factors x1, x2, x3, x4, x5, x6 then there are 6 correlation at each main factor for 6 sub-factors.

The 6 correlation of first main factor for 6 sub-factors are:

r11=( the correlation of y1 and x1)=r(y1, x1)

r12=( the correlation of y1 and x2)=r(y1, x1)

r13=( the correlation of y1 and x3)=r(y1, x1)

r14=( the correlation of y1 and x4)=r(y1, x1)
r15=( the correlation of y1 and x5)=r(y1, x1)
r16=( the correlation of y1 and x6)=r(y1, x1)

Put the correlation of main factor for 6 sub-factors on a column [r11, r12, r13, r14, r15, r16 ]. And put the correlation of main factor y2, y3, y4, y5 for 6 sub-factors on columns and form a correlation matrix R which has 5 columns and 6 rows.

$$R = \begin{bmatrix} r_{11} & r_{12} & r_{13} & r_{14} & r_{15} & r_{16} \\ r_{21} & r_{22} & r_{23} & r_{24} & r_{25} & r_{26} \\ r_{31} & r_{32} & r_{33} & r_{34} & r_{35} & r_{36} \\ r_{41} & r_{42} & r_{43} & r_{44} & r_{45} & r_{46} \\ r_{51} & r_{52} & r_{53} & r_{54} & r_{55} & r_{56} \end{bmatrix} = \begin{bmatrix} \text{the 6 correlation of 1st main factor for 6 sub-factors} \\ \text{the 6 correlation of 2nd main factor for 6 sub-factors} \\ \text{the 6 correlation of 3rd main factor for 6 sub-factors} \\ \text{the 6 correlation of 4th main factor for 6 sub-factors} \\ \text{the 6 correlation of 5th main factor for 6 sub-factors} \end{bmatrix}$$

There are 2 inferiors at each element in the above matrix. The 1st inferior is the order of main factor and the 2nd inferior is the order of sub-factor. Each column is the influence of same main factor for different sub-factor and each row is the influence of same sub-factor for different main factor. According to the correlation of each column and each row, the main influence and the sub-influence can be found. The main influence factor is the superior factor. The superior main factor and the superior sub-factor are included in the superior analysis.

**V. The Application for The Gray Correlation of The School Studies Score and Ctmm**

The California Test of Mental Maturity is designed by Clark W. W. and Tiegs E. W. in 1951 and issued by California Test Bureau. It had been revised in 1957 and was divided into 12 test parts. It had been revised once in 1963 and was reduced to 7 test parts. In 1968, the psychology department of National Taiwan University had revised the test and formed the Class V of California Test of Mental Maturity ( CTTM-V ) for testing the intelligence of junior high school students to adults. It can find out the ability level of consideration of learners. There are 4 factors in the test: (1) logic, (2) number, (3) language and (4) remembrance [ 12 ].

In this paper, the score of 4 factors of CTTM-V for students of M.E. ( machine and electrical ) department of a vocational technical school is main factors and the score of mid-term examination for one student is sub-factor. It is found that 6 students are random samples in this method and they are found by the gray correlation analysis. Table 5-1, table 5-2 are the scores of 6 students:

Table 5-1 The Score List of CTTM-V of Sampled Students

| No. | 4 | 13 | 17 | 27 | 33 | 36 |
|---|---|---|---|---|---|---|
| Factor/Student | A | B | C | D | E | F |
| Logic | 72 | 58 | 72 | 42 | 82 | 62 |
| Number | 4 | 69 | 42 | 25 | 1 | 8 |
| Language | 39 | 48 | 41 | 33 | 38 | 31 |
| Remembrance | 30 | 36 | 32 | 24 | 38 | 22 |

Table 5-2 The Score List of Mid-Term Examination of Sampled Students

| No. | 4 | 13 | 17 | 27 | 33 | 36 |
|---|---|---|---|---|---|---|
| Factor/Student | A | B | C | D | E | F |
| Chinese | 79 | 76 | 85 | 78 | 84 | 89 |
| English | 60 | 62 | 60 | 56 | 72 | 86 |
| Mathematics | 61 | 81 | 69 | 82 | 61 | 75 |
| Physics | 74 | 82 | 90 | 80 | 76 | 84 |
| Basic electricity | 64 | 84 | 80 | 88 | 56 | 76 |
| Electron | 70 | 55 | 63 | 60 | 26 | 86 |

The 4 factors in table 5-1 are the main factors and are the reference order or main order for superior analysis. The 6 factors in table 5-2 are the sub-factors and are the compared order or sub-order for superior analysis. The main order and the sub-order form the gray correlation matrix, table 5-3. Figures 5-1 (a)(b) are drawn according to the statistical data of gray correlation, and the correlation coefficient $\delta$=0.5 is used to form a barograph.

Table 5-3 The Gray Correlation of CTTM-V and Examining Score

| The Gray Correlation of CTTM-V and Examining Score | | | | | | | |
|---|---|---|---|---|---|---|---|
| Factor/Object | Chinese | English | Mathematics | Physics | Basic electricity | Electron | Correlation |
| Logic | 0.73 | 0.7345 | 0.6203 | 0.6557 | 0.5775 | 0.6878 | 4.0058 |
| Number | 0.7044 | 0.7068 | 0.711 | 0.7068 | 0.7132 | 0.7165 | 4.2587 |
| Language | 0.7395 | 0.6974 | 0.7177 | 0.7167 | 0.6667 | 0.6388 | 4.1768 |
| Remembrance | 0.7596 | 0.7717 | 0.7007 | 0.7286 | 0.6627 | 0.6571 | 4.2804 |
| Correlation | 2.9335 | 2.9104 | 2.7497 | 2.8078 | 2.6201 | 2.7002 | |

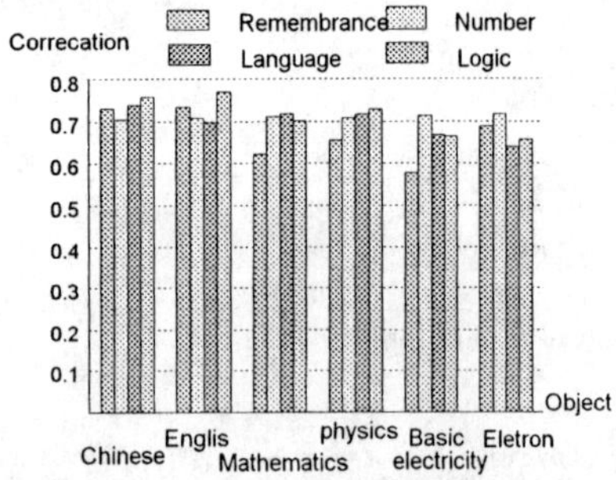


Figure 5-1 (a) The Barograph of Gray Correlation of CTTM-V and Examining Score

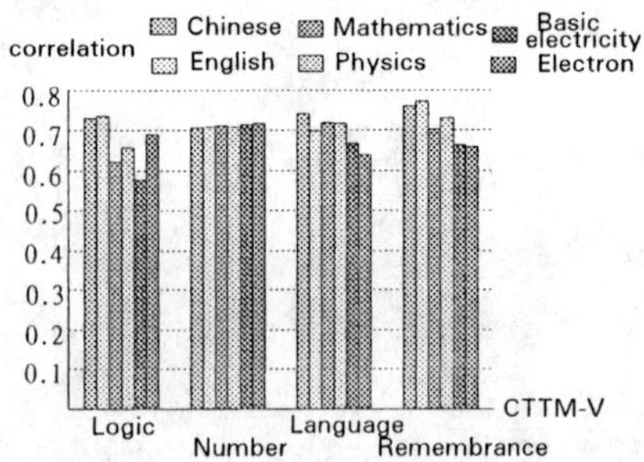


Figure 5-1 (b) The Barograph of Gray Correlation of CTTM-V and Examining Score

According to table 5-3 and figure 5-1 (a)(b), it is found that the superior main factor of influence score are remembrance, number, language and logic, and the superior sub-factor are Chinese, English, physics. Basic electricity is a bad factor. According to the score, the main factor for influencing chinese are remembrance, language. And the main factor for influencing english are remembrance, logic. These data can be refereed by their teachers. The superior main factor for influencing basic electricity is number and the bad main factor is logic.

## VI. Conclusion

1. The gray theory deals with the gray relation of factors of system easily.
2. The correlation analysis is the quantity analysis of development dynamic of factors of gray system. The correlation is the level of near of a function. It is also a measure for a dispersion function.
3. This paper is major in the gray correlation analysis and its application. So it is concentrated on the method and analysis. In the example, the data is not completely accurate and their result are not the learning tendency of students.

## VII. Reference


[1] Ju-Long Dang ( 1987 ), The Basic Method of Gray System, Hwa-Jo Science & Engineer University Publishing House.

[2] Ju-Long Dang ( 1988 ), The Gray Controlled System, Hwa-Jo Science & Engineer University Publishing House.

[3] Ju-Long Dang ( 1992 ), The Theory of Gray System, Hwa-Jo Science & Engineer University Publishing House.

[4] Ju-Long Dang ( 1992 ), The Gray Calculation & Policy Decision, Hwa-Jo Science & Engineer University Publishing House.

[5] Cang-Hwa Liu ( 1993 ), The Mineral of Gray System, Coal & Carbon Industry Publishing House.

[6] Kai-Chang Se, Ko-We Wu, Yu-Peng Hong ( 1993 ), The Relation Theory of Gray Massage, Taipei: Chang-Hwa.

[7] Ko-Shu Jang, Ye-Jeng Wo ( 1994 ), The Research for the Business & Service of Bus System: The Application of the Gray Correlation Analysis, The 2nd Discussion Conference for the Gray Theory & Application, pp. 49-55.

[8] Yen-Han Jeng ( 1991 ), Apply the Gray Theory to Discuss the Factors Analysis & Score Calculation for Decathlon of Fu-Nen Le, The newspaper of Provincial Physical Education, vol. 19, pp. 191-228.

[9] Jin-Ming Wo ( 1993 ), Apply the Dynamic Model of System to Research the Computer Simulation I: The Calculation for the Population of Students & Teachers in Vocational School, The Master Paper of Industrial Research Institute of National Taiwan Teacher University, pp. 46-75.

[10] Po-Yung Daw ( 1993 ), The Introduction of Multi-Variability Analysis ( Basic ), Fu-Cheng Publishing House.

[11] Po-Yung Daw ( 1993 ), The Introduction of Multi-Variability Analysis ( Examples ), Fu-Cheng Publishing House.

[12] The Psychology Department of National Taiwan University ( 1988 ), The Guide Manual of California Test of Mental Maturity Class V, Taipei: The Chinese Action Science Agency.

[13] Deng J. ( 1989), Introduction to Gray System Theory, The Journal of Gray System, vol. 1, pp. 1-23.

[14] Yang Bingzhai ( 1992), A New Alogorithm for G.M. (1,1), The Journal of Gray System, vol. 4, pp. 305-313.

# ERROR PHENOMENA OF BACKPROPAGATION LEARNING

**YUADKOUN DHIANTRAVAN AND ROLAND PRIEMER**

*Department of Electrical Engineering and Computer Science*
*The University of Illinois at Chicago, Chicago, Illinois*

*ABSTRACT:*
New errors are introduced. We call these errors latent errors and we investigate the latent errors in the Backpropagation (BP) algorithm. We show that the mean square latent error varies according to the correlation of the input vector sequence. With the latent error squared, we also obtain a set of weights that gives the least mean error squared for each learning iteration. This results in a new learning algorithm with a faster convergence than a conventional BP.

## INTRODUCTION

There are two kinds of errors that BP algorithm [Rumelhart, 1986] employs. The first one is written as $e(n/n-1)$, which is the error calculated with the weights from the previous iteration and the current input data. The second error is $e(n/n)$, which is calculated with the current updated weights and the current input data. The error $e(n/n-1)$ is used in updating the weights during iteration $n$. The error $e(n/n)$ is usually used in determining if the current obtained error is small enough to stop the training process or else another training iteration will be processed.

There is research that has studied the NN output error in the perspective of the training input. For example, [Waibel, 1989] studied the Time Delay NN (TDNN) in phoneme recognition which showed that the output error briefly jumped up every time more training data was introduced to the NN. [Fu, 1993] analysed the sensitivity of the output matrix in terms of the input vector, the weights, the activation function, numbers of layers and nodes and considered a new input pattern as a small perturbed version of the training input. [Wang, 1993] showed the dependency of the capability of a NN to achieve an absolute minimum training error upon the rank of the training sample matrix. [Bakker, 1993] showed that the learning speed of the adaptive linear combiner was decreased when new correlated training data was added.

This paper also relates the output errors with the input. But instead of considering the data in one input sequence at a time, we consider the data in the subsequence input vectors. This is done by employing the latent errors. The latent errors are defined as $e(n/n+i)$ where $i$ is an index that varies from $-k$ to $+l$. The error $e(n/n+i)$ is a future ($i<0$), present ($i=0$) or past ($i>0$) error with respect to iteration $n$.

To illustrate the utility of the latent errors, we model a second order autoregressive (AR) process and a first order moving average (MA) process with a NN that is composed of eight hidden nodes, five input nodes, and one output node. The results show that the latent errors depend upon the input sequence vectors.

Because the latent errors are found by using presently learned weights in other learning events, a set of weights which yields the least latent errors squared in a bound region $i$ can be obtained. That set of weights can be considered as the weights obtained from the contour of minimizing the error function. Using that set of weights will then assure that the error function is being minimized. By incorperating this new set of weights in the conventional BP algorithm, the learning process is then sped up.

**ERROR-BACKPROPAGATION**

Solving a problem with a BPNN is to determine the value of the weight coefficients. During the training phase, the input and the desired output (target) are presented to the NN. The weight coefficients are adjusted in order to give the output that best matches the target. The error between the target and the actual output is minimized along the gradient of the error squared function. As for the error to be minimized, the weights are updated iteratively in a steepest descent scheme. One way to update the weights is to update the weights after all the training patterns are presented to the NN (batch update). Another way to update the weights is to update after each training pattern is presented to the NN (pattern training) as follows

$$W_{\text{new}} = W_{\text{old}} - \eta \frac{\partial E_{\text{p}}}{\partial W} \tag{1}$$

where $W$ is the weight vector , $\eta$ is the learning rate, and $E_{\text{p}}$ is the error between the target and the actual output for the training pattern p. We use pattern training because the error phenomena will be best depicted as the training process goes from one iteration to the next.

**THE ESTIMATION AND LATENT ERRORS**

Let the scalar signal $y(n)$ be

$$y(n) = \sum_{i=1}^{\text{N}} a_i y(n-i) + u(n) \tag{2}$$

$$y(n) = \sum_{i=0}^{\text{M}} b_i u(n-i) \tag{3}$$

for the cases of AR and MA models, respectively. The scalar signal $u(n)$ is the data available to the NN. Let a set of the data points be

$$U^{\text{T}}(n) = [u(n), u(n-1), u(n-2), \ldots, u(n-\text{p})] \tag{4}$$

The estimate $y(n)$ can be modeled by a NN using the error-backpropagation algorithm. Let $W$ be the true set of weight coefficients. Hence, the error between the signal value and the estimated value is

$$e(n) = y(n) - \hat{y}(n) = y(n) - U^{\mathrm{T}}(n)W \tag{5}$$

Let $\hat{W}(n)$ be an estimate of $W$ from the data set sequence $U(0), U(1), \ldots, U(n)$, then

$$e(n/n) = y(n) - U^{\mathrm{T}}(n)\hat{W}(n) \tag{6}$$

$$e(n/n-1) = y(n) - U^{\mathrm{T}}(n)\hat{W}(n-1) \tag{7}$$

$$e(n/n+1) = y(n) - U^{\mathrm{T}}(n)\hat{W}(n+1) \tag{8}$$

Thus, eq.(8) means the error between the actual signal $y(n)$ and the estimated $y$ that is estimated from $U^{\mathrm{T}}(n)$ and the predicted $\hat{W}(n+1)$ which is given by

$$\hat{W}(n+1) = \hat{W}(n) + \eta U(n+1)e(n+1/n) \tag{9}$$

Eqs. (6), (7), and (8) above can be combined as

$$e(n/n+i) = y(n) - U^{\mathrm{T}}(n)\hat{W}(n+i) \tag{10}$$

where $i = 0$, -1, and +1, respectively. Eq.(9) can be written in general form as

$$\hat{W}(n+i) = \hat{W}(n) + \eta \sum_{s=1}^{i} U(n+s)e(n+s/n+s-1) \tag{11}$$

With eqs.(10) and (11), latent error $e(n/n+i)$ and error $e(n/n)$ are related as

$$e(n/n+i) = e(n/n) - \eta U^{\mathrm{T}}(n) \sum_{s=1}^{i} U(n+s)e(n+s/n+s-1) \tag{12}$$

The mean square of eq.(12) is given by

$$\begin{aligned} \mathrm{E}\left[e^2(n/n+i)\right] = {} & \mathrm{E}\left[e^2(n/n)\right] + \eta^2 \mathrm{E}\left[\left\{U^{\mathrm{T}}(n)\sum_{s=1}^{i} U(n+s)e(n+s/n+s-1)\right\}^2\right] \\ & - 2\eta \mathrm{E}\left[U^{\mathrm{T}}(n)e(n/n)\sum_{s=1}^{i} U(n+s)e(n+s/n+s-1)\right] \end{aligned} \tag{13}$$

When $i = 1$ eq.(13) shows that the mean square latent error depends on the correlation of two subsequence input vectors. If we rearrange eq.(12) and take the mean square, we obtain

$$\begin{aligned} \mathrm{E}\left[e^2(n/n)\right] &= \mathrm{E}\left[e^2(n/n+i)\right] + \eta^2 \mathrm{E}\left[\left\{U^{\mathrm{T}}(n)\sum_{s=1}^{i} U(n+s)e(n+s/n+s-1)\right\}^2\right] \\ &\quad + 2\eta \mathrm{E}\left[U^{\mathrm{T}}(n)e(n/n+i)\sum_{s=1}^{i} U(n+s)e(n+s/n+s-1)\right] \end{aligned} \tag{14}$$

Eq.(14) shows that the BP mean square errors depend on the latent errors and the subsequence input vectors.

## MODELING EXAMPLE AND RESULT

We study the latent errors by modeling AR and MA processes. The signal $u(n)$ is given by

$$u(n) = d_1 u(n-1) + v(n) \tag{15}$$

where $v(n)$ is white Gaussian noise with zero mean and variance of unity, and $d_1 =$ 0.1. The second order AR and first order MA processes from eq.(2) and (3), respectively, are

$$y(n) = a_1 y(n-1) + a_2 y(n-2) + u(n) \tag{16}$$

$$y(n) = b_0 u(n) + b_1 u(n-1) \tag{17}$$

where $a_1 = 0.5, a_2 = 0.3, b_0 = 0.5, b_1 = 0.9$. In eq.(4), which represents the input data, we use data points up to $n$ = (p+1) = 160. The values of $k$ and $l$ are 20 and 10, respectively. The learning rate is 0.7 and momentum is 0.9.

Because the mean of $e^2(n/n)$ depends on the subsquence input vectors (and consequencely, the mean of $e^2(n/n+i)$) as in eq.(14), we then test the NN by presenting two different set of input patterns.

1. The first set is represented by
   $U^{\mathrm{T}}(n) = [u(n), u(n-1), u(n-2), u(n-3), u(n-4)]$ and
   $U^{\mathrm{T}}(n+1) = [u(n+1), u(n), u(n-1), u(n-2), u(n-3)]$
   as a subsequence vector. The result of the first set will be represented by a dotted or a plus sign line.
2. The second set is
   $U^{\mathrm{T}}(n) = [u(n), u(n-1), u(n-2), u(n-3), u(n-4)]$ and
   $U^{\mathrm{T}}(n+1) = [u(n+5), u(n+4), u(n+3), u(n+2), u(n+1)]$
   as a subsequence vector. The result will be represented by a solid line.

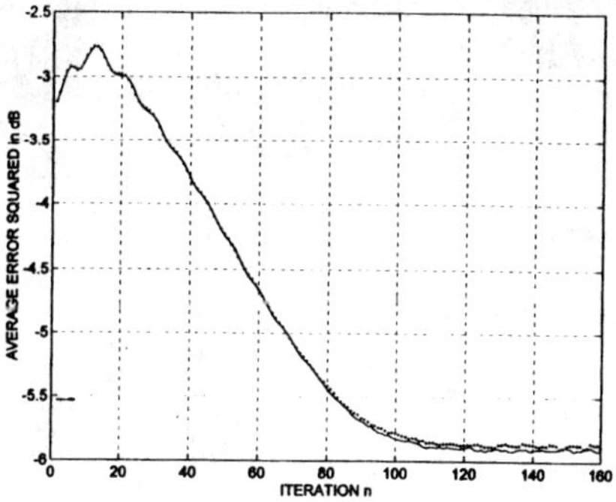


Figure 1

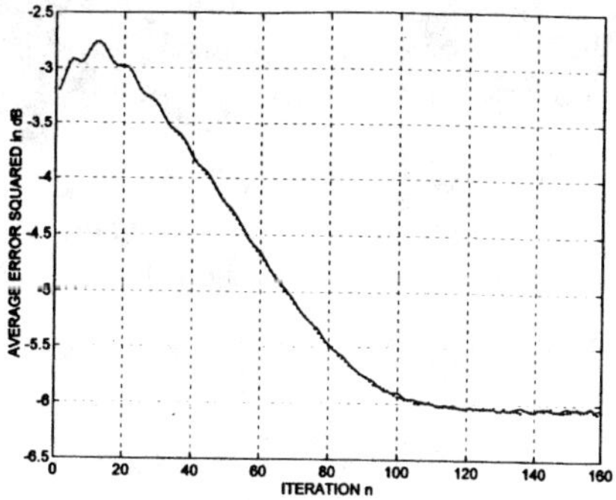


Figure 2

In the first set, two subsequence vectors have only one datapoint different, while in the second set, all five datapoints are different. Results of NN modeling AR and MA processes are obtained by taking the average of 4000 trials. They are shown in Figures 1 and 2, respectively.

Figure 3 shows the latent error squared from modeling the AR process for one trial at iterations 25 . Notice that the mean square latent errors of two different sets are not significantly different. This is because the datapoints are related by time delay. This also is the case for the output $y(n)$. In Figure 3, when $i = 0$, $E\left[e^2(n/n+i)\right]$ has a value of $0.25 \times 10^{-3}$. The least mean square latent errors with values near zero occur at $i = 1, 9, 17$. These errors are obtained by employing the weights from iteration 26, 34, and 42 in learning iteration 25 as described in eq.(10). If we replace the current weights at iteration 25 with the weights of iterations 26 or 34 or 42, the mean square error $E\left[e^2(n/n)\right]$ at iteration 25 will be reduced to be near zero or about a 3.6 dB decrease.

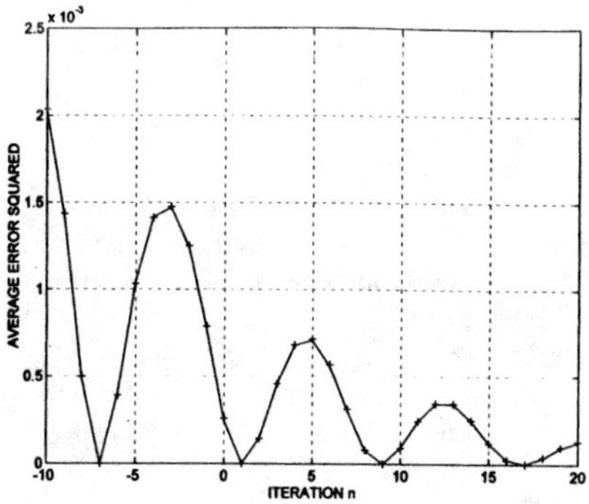


Figure 3

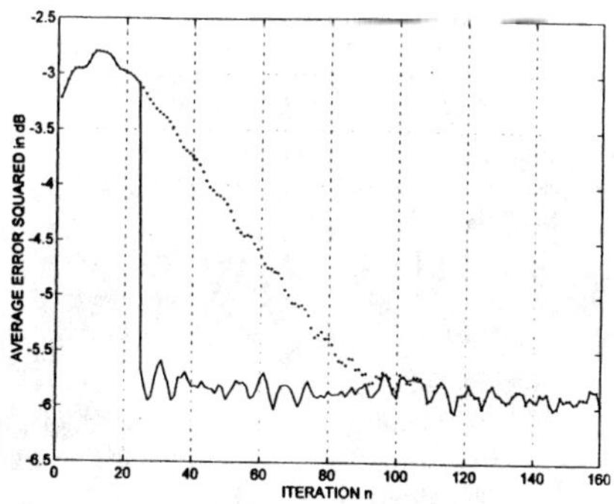


Figure 4

Also the weights at iterations 26, 34, and 42 can be considered as the weights that bring the error function of the learning process close to the contour minimum. This leads us to a new learning algorithm.

## THE NEW LEARNING ALGORITHM

A new learning algorithm is given to show how the latent errors can be incorporated into a conventional BP algorithm.

1. Specify the learning iteration $m$ where the new algorithm should start.
2. Use a conventional BP algorithm to calculate the weights around iteration $m$, i.e., $W(m+i)$ where $i$ is large enough to satisfy (3).
3. Calculate the latent errors squared around iteration $m$ with eq.(10), and find the least latent error squared $e^2(m/m+i)$ that is less than $e^2(m/m)$.
4. Replace the weights $W(m)$ with $W(m+i)$ from (2) and (3).
5. With $W(m+i)$ at learning iteration $m$, the conventional BP proceeds.

Figure 4 compares the new learning algorithm with the conventional BP algorithm. Both are averaged over 100 trials. The solid line represents result from new algorithm, the dotted line represents the result from a conventional BP algorithm. Here, we chose $m$=25, so the mean square error was brought down from -3.1 dB to -5.7 dB or a 2.6 dB decrease

## CONCLUSION

The latent errors were introduced. They are found by using presently learned weights in other learning events. We showed that the mean square latent errors depend on the correlations of the subsequence input vectors, and consequently the latent errors and the BP errors are related. We give a new learning algorithm based on the latent errors. The new learning algorithm results in a convergence rate that is faster than a conventional BP algorithm.

## REFERENCES


Rumelhart,D., Hinton, G.,and Williams,R.,(1986). 'Learning internal representation by error propagation', in Parallel Distributed Processing Vol.I, MIT Press, Cambridge, MA.

Waibel,A.,Hanazawa,T.,Hinton,G.,Shikano,K., and Lang,K.,(1989). 'Phoneme Recognition Using Time-Delay Neural Networks', IEEE Trans. on Acoustic, Speech, and Signal Processing, Vol. 37, No. 3, March.

Fu,L., and Chen,T.,(1993). 'Sensitivity Analysis for Input Vector in Multilayer Feedforward Neural Networks', IEEE International Conference on Neural Networks, San Francisco.

Wang,J., and Malakooti,B.,(1993). 'Characterization of Training Errors in Supervised Learning Using Gradient-Based Rules', Neural Networks, Vol. 6, Pergamon Press.

Bakker,R.N.,Kraaijveld,M.A.,Duin,R.P.W., and Schmidt,W.F.,(1993). 'On the Speed of Training Networks with Correlated Features', IEEE International Conference on Neural Networks, San Francisco.

# RULE SET QUALITY MEASURES FOR INDUCTIVE LEARNING ALGORITHMS

**Ralf H. Klinkenberg**
*Computer Science Department*
*University of Missouri-Rolla*
*Rolla, Missouri 65401*
*e-Mail: klinkenb@ls8.informatik.uni-dortmund.de*

**Daniel C. St. Clair**
*University of Missouri-Rolla*
*Engineering Education Center*
*St. Louis, Missouri 63121*
*e-Mail: stclair@umrgec.eec.umr.edu*

***ABSTRACT:***
Symbolic inductive learning systems that induce concept descriptions from examples are valuable tools in the task of knowledge acquisition for expert systems. Since inductive learning methods produce distinctive concept descriptions when given identical training data, questions arise as to the quality of the different rule sets produced. This work provides several techniques for comparing and analyzing rule sets. These techniques measure the accuracy, generalization, time and space complexity, and domain coverage of rule sets Based on these metrics, the performance of the rule sets generated by four different inductive learning systems is compared on six real world data sets.

## INTRODUCTION

Symbolic inductive learning algorithms learn class descriptions from examples. All information about an example must be expressible in terms of a fixed collection of *attributes* with either symbolic, discrete numeric, or continuous numeric values. Each example belongs to one of a set of mutually exclusive *classes*. The induction task is to develop rules that can determine the class of an instance from its attribute values.

Since inductive learning methods generate distinctive sets of rules when given identical training data, questions arise concerning the quality of these rule sets. The objective of this research is to develop techniques for comparing and analyzing rule sets and to apply these techniques to compare the performance of four inductive learning systems on six real world data sets. These techniques measure the accuracy, generalization, time and space complexity, and domain coverage of rule sets.

The inductive learning systems compared here are Michalski et al.'s AQ15 (1986), Quinlan's C4.5 (1993), Clark et al.'s CN2 (1989, 1991), and Janikow's GIL (1993). These systems generate different types of rule sets, i.e. unordered rule sets, ordered rule sets, and decision trees respectively. The rule set quality measures proposed in this work allow the comparison of the rule sets of these different types and thereby make a comparison of a broad class of learning algorithms possible.

## LEARNING ALGORITHMS AND RULE SET TYPES

Three of the four learning systems used here represent decision rules as expressions in **Variable-valued Logic System 1 ($VL_1$)** notation, a multi-valued logic propositional calculus with typed variables (Michalski & Larson, 1975). In $VL_1$, a *selector* relates a variable to a value or a disjunction of values using one of the relational operators <, <=, =, <>, >=, or >. A conjunction of selectors forms a *complex.* For any class, the examples given for that class in the training set are its *positive events* and all other examples are its *negative events.* A *cover* is a disjunction of complexes describing all positive and none of the negative events of a class. A cover forms the condition part of a decision rule predicting that class. The following rule contains two complexes:

*(Outlook = sunny or cloudy) and (Temperature > 60)*
*or (Windy = true) and (Temperature > 70)* => *Class(Nice)*

For each class *C*, the **AQ15** algorithm (Michalski et al., 1986) is applied in steps to generate a cover for *C*, each step producing one complex of the cover, until all positive examples are covered. Each step starts a specialization process with one selected positive example, a *seed*, and generates a *star*, a size-limited (sub)set of all maximally general complexes, which cover the seed but none of the negative examples, and then selects the best complex from the star. The AQ15 system can generate *unordered* and *ordered rules.* In case of ordered rules, the rules for class *n* assume that the rules for classes *1* to *n-1* are not satisfied. Hence the rule for the last class is a *default rule* and the rules constructed for the other classes may be less complex than in case of unordered rules. The AQ15c implementation (Wnek et al., 1995) used here can only handle discrete or discretized attribute value domains with finitely many attribute values and treats continuous numeric attribute values as discrete values.

Quinlan's inductive learning system **C4.5** (Quinlan, 1993) generates a *decision tree*, a structure that is either a *leaf*, indicating a class, or a *decision node* that specifies some test to be carried out on a single attribute value, with one branch and subtree for each possible outcome of the test. The test at a node is similar to a selector in a decision rule, but may have more than two outcomes. A decision tree can be used to classify a case by starting at the root of the tree and moving through it until a leaf is encountered. At each non-leaf node, the case's outcome for the test at this node is determined and attention shifts to the root of the subtree corresponding to the outcome of this test. When this process finally leads to a leaf, the class of the case is predicted to be that recorded at the leaf. The C4.5 inductive learning system allows to transform the decision tree to a set of *ordered rules* including a *default rule.*

The **CN2** algorithm was developed by Clark and Niblett (1989) and later modified by Clark and Boswell (1991). It produces either *ordered* or *unordered rules,* both including a *default rule.* The condition part of a rule is a $VL_1$ complex. Like AQ15, CN2 employs a star procedure to find the best complex in a pruned general-to-specific search. Besides of having a low estimated error rate, the best complex also has to pass some minimum treshold of reliability. It has to be statistically *significant* and locate a regularity unlikely to have occured by chance.

The **GIL** (Genetic-Based Inductive Learning) system (Janikow 1993) is a *genetic algorithm* for inductive learning that incorporates task specific knowledge. The traditional domain-independent genetic operators are replaced by special forms that implement the specific problem solving methodology of inductive learning. The rules are represented in $VL_1$, but GIL only considers $VL_1$ formulas using the equality relation, '='. If all attributes only take a finite number of symbolic or numeric values, this is no lack of generality, because selectors using other relational operators can be rewritten using the equality operator. For continuous numeric attributes, this is only true, if a final number of intervals is introduced to represent all possible values.

## RULE SET QUALITY MEASURES

This section proposes several metrics for comparing rule sets generated by inductive learning algorithms. These quality measures include metrics for the predictive accuracy, generalization, space and time complexity, and domain coverage of rule sets.

### Predictive Accuracy

The most widely used rule set quality measure is the predictive accuracy of a rule set on previously unseen instances. The *overall accuracy* of a rule set is the number of correctly classified test instances divided by the total number of test instances. Often not only the overall accuracy is of interest, but the *accuracies for the prediction of the distinct classes* as well. Sometimes it is desirable to have a similar accuracy on instances of all classes, while in other cases misclassifications of cases of a certain class are more costly than misclassifications of cases of another class.

While ordered rules and decision trees classify each covered instance unambiguously, several unordered rules predicting distinctive classes may cover an instance at the same time. The question of which class to predict for this instance has to be resolved. For the following experiments, the same instance classification procedure was applied to all rule sets. Each unordered rule was associated with a weight representing the number of training instances correctly classified by this rule. If a test instance was matched by several unordered rules predicting different classes, the weights of these rules were added for each class and the class with the highest count was predicted. The results of this classification heuristic do not always correspond to the results of the heuristics of the learning algorithms, but the use of this uniform heuristic allowed a better comparison of the rule sets independent of their algorithms.

### Generalization

The *generalization* measure was previously used by Gallion et al. (1993) and describes how well a rule set generalize from the set of training instances:

*Generalization = 1 - (Number of rules / Number of training instances)*

Thinking of the original training data as a set of rules, the generalization of a rule set produced by a learning algorithm presents the reduction produced by the algorithm as a percentage.

### Space Complexity

For many real world applications, the number of training instances available is very large and the space complexity of a rule set becomes a further relevant aspect as it is necessary to store the rule set and as it is desirable to have rule sets that are small enough to be intelligible for humans. The space complexity measures considered here are the *number of selectors* in a rule set, the *number of complexes* in a rule set, and the *number of rules* in a rule set including the default rule if such a rule exists.

In case of a decision tree, the number of leaf nodes corresponds to the number of rules and the attribute tests at the non-leaf nodes correspond to selectors, although for decision trees these selectors can be more complex than in case of a rule set. Attribute tests with two outcomes directly correspond to selectors, while attribute tests with $n>2$ outcomes are transformed into a chain of $n-1$ selectors in such a way that the first selector represents the first outcome of the test, the second selector the second, and so on, so that each non-leaf node in the tree becomes a chain of binary decision nodes. Each attribute test with $n$ outcomes counts as $n-1$ selectors. Since there is no real complex involved along a path in a tree, the number of complexes in a tree is set to the number of its paths, which equals the number of its leaf nodes.

**Time Complexity**

For some applications, it is relevant how long it takes to classify previously unseen instances using a particular rule set. The time complexity measures proposed here count the *average number of selector tests* and the *average number of complex tests needed to classify a previously unseen instance.*

In an unordered rule set, all rules always have to be evaluated to determine which are satisfied by an instance to be classified and hence all complexes are counted for the number of complex tests. In an ordered rule set, all rules after the first rule that matches an instance do not have to be evaluated and hence are not counted for the number of complex tests and selector tests. If an instance satisfies a complex it is tested against, all selectors of this complex are counted, since they all have to be evaluated for the instance. If an instance does not satisfy such a complex, the selectors of this complex after the first unsatisfied selector are not counted, since they do not need to be evaluated. In a decision tree, the non-leaf nodes with more than two outcomes are transformed into a chain of several binary decision nodes as described above. The number of selector tests needed to classify a particular test instance is the number of selector tests along the path in the tree followed to classify this instance.

**Domain Coverage**

The *domain coverage* of a rule set is defined as the percentage of the instances in the domain it covers, no matter whether the instances are correctly classified or not. This metric is computed as the number of test instances covered by the rule set divided by the total number of test instances. If a rule set has a low coverage, there are many instances whose class cannot be predicted by the rule set and hence the overall quality of this rule set cannot be very high. A low coverage automatically results in a low predictive accuracy, because uncovered instances are counted as not correctly classified. Therefore a high coverage is desirable, although it is no guarantee for a high accuracy. Decision trees and rule sets including a default rule always have a coverage of 100 %, because they always cover all instances.

## NUMERICAL EXPERIMENTS

This section describes the application of the proposed quality measures to compare the rule sets generated by AQ15c, C4.5, CN2, and GIL for six real world data sets. The experiments show how the variety of the proposed metrics can be used to analyze and compare the different types of rule sets generated by these algorithms.

| Table 1: Summary of the data sets used in the comparison | | | | | | |
|---|---|---|---|---|---|---|
| **Database name:** | **Cancer-loi** | **Cancer-wi** | **Lymph** | **Iris** | **Credit** | **Vote 1984** |
| No. of attributes: | 9 | 9 | 18 | 4 | 15 | 16 |
| Symbolic: | 5 | 0 | 15 | 0 | 9 | 16 |
| Discrete numeric: | 4 | 9 | 3 | 0 | 3 | 0 |
| Continuous numeric: | 0 | 0 | 0 | 4 | 3 | 0 |
| Average no. of values per attribute: | 4.56 | 9.89 | 3.28 | 30.75 | 77.93 | 2.00 |
| No. of classes: | 2 | 2 | 4 | 3 | 2 | 2 |
| Class 1: | 70.28 % | 65.52 % | 1.35 % | 33.33 % | 44.49 % | 38.62 % |
| Class 2: | 29.72 % | 34.48 % | 54.73 % | 33.33 % | 55.51 % | 61.38 % |
| Class 3: | ----- | ----- | 41.22 % | 33.33 % | ----- | ----- |
| Class 4: | ----- | ----- | 2.70 % | ----- | ----- | ----- |
| No. of instances: | 286 | 699 | 148 | 150 | 690 | 435 |
| No. of attributes with unknown values: | 2 | 1 | 0 | 0 | 7 | 16 |

**Data Sets Used In The Comparison**

The six data sets selected for the comparison are all available from the University of California at Irvine (UCI) Repository of Machine Learning Databases (Murphy & Aha, 1994). Table 1 shows the characteristics of these data sets. The **Breast Cancer Database** from the **L**jubljana **O**ncology **I**nstitute (Cancer-loi) was originally provided by Zwitter and Soklic from the University Medical Centre, Institute of Oncology, Ljubljana, Yugoslavia. The goal is to predict whether a cancer recurrence is likely or not. The **Wisconsin Breast Cancer Database** (Cancer-wi) was originally obtained from the University of Wisconsin Hospitals, Madison, Wisconsin from Wolberg (Wolberg & Mangasarian, 1990). A cancer tumor has to be classified as "benign" or "malignant". The **Lymphography Domain** (Lymph), was, like the first medical domain, originally provided by Zwitter and Soklic. The cases are labeled "normal", "metastases", "malign", or "fibrosis". The **Iris data set**, developed by Fisher (1936), lists the measurements for four characteristics of Iris flowers and the class to be predicted is the species of the Iris plant. The **Credit Approval Database** (Credit) concerns credit card applications and was submitted to the UCI Repository of Machine Learning Databases and previously used by Quinlan (1993). The **1984 United States Congressional Voting Records Database** (Vote 1984) was compiled by Jeff Schlimmer from Washington State University. It includes attributes representing the votes for each of the 435 United States House of Representatives Congressmen on 16 key votes and its class is the party affiliation, Democrat or Republican.

**Experimental Setup**

The *v-fold cross validation* technique was used to generate pairs of training and test data sets for each domain. In *v*-fold cross validation, the original data set is randomly partitioned in *v* subsets, each containing approximately the same number of instances. In each experiment, one subset is used for testing and the other *v-1* subsets for training. Here *v* was chosen as four and the results were averaged for each set of tests.

Since GIL cannot handle *unknown attribute values*, the unknown attribute values were replaced by the most common value of the corresponding attribute in the data set for GIL. The same was done for the uniform evaluation program. For the Iris and the Credit data sets, the GIL system was not able to complete the induction of a single rule set on a SUN SPARC 20 workstation within 12 hours. Both data sets contain *continuous numeric attributes* which GIL does not explicitly handle, but

**Table 2: Average predictive accuracy results [as percentage].**

| Algorithm: | AQ15c ordered rules | AQ15c unordered rules | C4.5 decision tree | C4.5 ordered rules | CN2 ordered rules | CN2 unordered rules | GIL unordered rules | Default rule |
|---|---|---|---|---|---|---|---|---|
| **Cancer-loi:** | [37.76] | [70.28] | [74.10] | [71.68] | [66.09] | [71.65] | [66.08] | [70.28] |
| No recurr.: | 75.36 | 74.33 | 94.89 | 84.63 | 90.46 | 92.98 | 81.15 | 100.00 |
| Recurrence: | 44.05 | 28.07 | 24.81 | 44.59 | 13.24 | 21.43 | 25.33 | 0.00 |
| Overall mean: | **66.10** | **60.50** | **74.10** | **71.68** | **66.09** | **72.02** | **64.32** | **70.26** |
| **Cancer-wi:** | [95.42] | [95.99] | [93.85] | [94.14] | [94.14] | [94.00] | [98.30] | [65.52] |
| Begnin: | 95.23 | 95.23 | 95.42 | 96.30 | 91.90 | 98.03 | 94.73 | 100.00 |
| Malignant: | 96.24 | 87.86 | 90.94 | 90.11 | 98.30 | 86.41 | 93.38 | 0.00 |
| Overall mean: | **95.57** | **91.28** | **93.85** | **94.14** | **94.14** | **93.99** | **94.42** | **65.52** |
| **Lymph:** | [58.11] | [81.08] | [79.05] | [76.35] | [72.30] | [79.05] | [84.55] | [54.73] |
| Normal: | 100.00 | 0.00 | 0.00 | 0.00 | 0.00 | 50.00 | 0.00 | 0.00 |
| Metastases: | 79.03 | 80.08 | 83.65 | 80.18 | 88.35 | 90.98 | 72.55 | 100.00 |
| Malign: | 69.14 | 65.43 | 75.08 | 74.37 | 71.07 | 69.99 | 66.11 | 0.00 |
| Fibrosis: | 83.33 | 50.00 | 83.33 | 83.33 | 50.00 | 33.33 | 0.00 | 0.00 |
| Overall mean: | **76.35** | **71.62** | **79.05** | **76.35** | **79.05** | **79.73** | **66.89** | **54.78** |
| **Iris:** | [50.13] | [95.34] | [94.03] | [94.03] | [98.00] | [98.00] | [94.54] | [33.33] |
| Setosa: | 92.50 | 92.50 | 100.00 | 100.00 | 100.00 | 100.00 | 97.22 | 50.00 |
| Versicolor: | 81.16 | 81.16 | 93.30 | 93.30 | 91.38 | 91.38 | 80.34 | 25.00 |
| Virginica: | 98.21 | 81.25 | 87.90 | 87.90 | 98.08 | 98.08 | 86.46 | 25.00 |
| Overall mean: | **89.94** | **84.62** | **94.03** | **94.03** | **96.68** | **96.68** | **89.37** | **25.00** |
| **Credit:** | [55.51] | [56.81] | [85.07] | [85.38] | [80.14] | [81.60] | [81.34] | [55.51] |
| + : | 12.71 | 12.71 | 84.28 | 80.89 | 78.68 | 70.81 | 86.03 | 0.00 |
| - : | 98.96 | 20.21 | 85.93 | 88.74 | 81.17 | 89.97 | 82.44 | 100.00 |
| Overall mean: | **60.59** | **16.83** | **85.07** | **85.21** | **80.14** | **81.59** | **84.06** | **55.10** |
| **Vote 1984:** | [61.38] | [88.51] | [95.86] | [95.63] | [93.79] | [93.79] | [94.92] | [61.38] |
| Republican: | 0.00 | 0.00 | 94.04 | 94.04 | 92.56 | 87.04 | 91.33 | 0.00 |
| Democrat: | 100.00 | 0.00 | 97.30 | 97.30 | 93.76 | 96.21 | 93.99 | 100.00 |
| Overall mean: | **61.40** | **0.00** | **95.86** | **95.86** | **93.10** | **93.56** | **92.88** | **61.40** |

internally represents as discrete values. The problem was solved by introducing intervals for each attribute with more than 25 distinct numeric values. As interval boundaries, the 5 % quantiles $q_0, q_5, q_{10}, \ldots, q_{95}, q_{100}$ of the particular attribute were used, where each $q_i$ is the smallest value of the attribute occuring in the data set greater than or equal to the attribute value of at least *i* % of the instances in the data set.

**Predictive Accuracy Results**

Table 2 lists the average predictive accuracy results. The average accuracy of each learning algorithm according to its own evaluation method is presented in brackets ('[', ']'), but the following comparison only considers the results of the uniform evaluation method. Since a good rule set should be better than a default rule always predicting the majority class in the training set, the average accuracy values for such a default rule are presented as well. For the default rule, the number in brackets represents the percentage of the instances of the majority class in the whole data set.

Unlike the evaluation procedure of AQ15c, the uniform evaluation procedure does not use analogical matching to cover otherwise uncovered instances. Therefore its accuracy values can significantly differ from those obtained by AQ15c. For the Credit and Vote 1984 data sets, this leads to very low accuracy results, especially for unordered rules, where no default rule keeps the accuracy from decreasing below a certain level like it happens for the ordered rule sets.

**Table 3: Average generalization and coverage results [in %].**

| Algorithm: | | AQ15c ordered rules | AQ 15c unordered rules | C4.5 decision tree | C4.5 ordered rules | CN2 ordered rules | CN2 unordered rules | GIL unordered rules |
|---|---|---|---|---|---|---|---|---|
| **Cancer-lol:** | Generalization: | 92.66 | 85.55 | 95.80 | 96.62 | 98.60 | 97.79 | 73.54 |
| | Coverage: | 100.00 | 84.63 | 100.00 | 100.00 | 100.00 | 100.00 | 90.89 |
| **Cancer-wl:** | Generalization: | 98.57 | 96.90 | 98.33 | 98.76 | 98.70 | 98.52 | 93.04 |
| | Coverage: | 100.00 | 93.71 | 100.00 | 100.00 | 100.00 | 100.00 | 98.14 |
| **Lymph:** | Generalization: | 90.99 | 86.49 | 88.29 | 90.99 | 93.69 | 91.67 | 82.66 |
| | Coverage: | 100.00 | 87.16 | 100.00 | 100.00 | 100.00 | 100.00 | 81.76 |
| **Iris:** | Generalization: | 96.60 | 94.00 | 96.89 | 95.78 | 97.33 | 95.78 | 92.00 |
| | Coverage: | 100.00 | 86.63 | 100.00 | 100.00 | 100.00 | 100.00 | 96.68 |
| **Credit:** | Generalization: | 96.38 | 93.14 | 98.65 | 98.74 | 97.73 | 97.83 | 77.45 |
| | Coverage: | 100.00 | 18.13 | 100.00 | 100.00 | 100.00 | 100.00 | 96.82 |
| **Vote 1984:** | Generalization: | 99.16 | 96.25 | 98.24 | 98.16 | 98.47 | 98.01 | 95.02 |
| | Coverage: | 100.00 | 0.00 | 100.00 | 100.00 | 100.00 | 100.00 | 98.16 |

## Generalization And Coverage Results

Table 3 lists the average generalization and coverage results. Usually a high *generalization* rate is desirable, but it does not automatically imply that the rule set has a high quality. For example, the unordered AQ15c rule sets for the Vote 1984 data set show, that a high generalization, 96.25 %, can go along with a very low coverage, 0.00 %, and a very low accuracy, 0.00 %.

A low *coverage* always implies a low accuracy. The low accuracy results of the unordered AQ15c rules on the Credit and Vote 1984 data sets shown in Table 2 were obviously due to the low coverage of the rule sets, 18.13 % and 0.00 % respectively, as listed in Table 3. While low coverage may indicate an over specialization of a rule set, high coverage does not guarantee a high accuracy. Decision trees and rule sets with a default rule cover the full domain without necessarily having a high accuracy.

## Comparison Of The Learning Algorithms

Table 4 shows the values of all rule set quality metrics for the rule sets of each algorithm averaged over all data sets. The decision trees generated by C4.5 had a slightly higher predictive accuracy than the other rule sets, but this finding is not statistically significant at the (95 % significance level). The comparatively low accuracy of the

**Table 4: Rule set quality metrics averaged over all data sets (with standard deviations).**

| Algorithm: | AQ15c ordered rules | AQ 15c unordered rules | C4.5 decision tree | C4.5 ordered rules | CN2 ordered rules | CN2 unordered rules | GIL unordered rules | Default rule |
|---|---|---|---|---|---|---|---|---|
| Predictive accuracy: | 74.99 % (13.66 %) | 54.14 % (34.10 %) | 86.99 % (8.25 %) | 86.21 % (9.37 %) | 84.87 % (10.82 %) | 86.26 % (9.03 %) | 81.99 % (12.05 %) | 55.34 % (14.63 %) |
| No. of rules: | 9.88 | 19.46 | 7.83 | 7.17 | 5.88 | 7.42 | 42.42 | ----- |
| | (5.75) | (10.28 ) | (2.96) | (1.71) | (2.98) | (2.42) | (36.75) | ----- |
| No. of complexes: | 8.88 | 19.46 | 7.83 | 6.17 | 4.88 | 6.42 | 42.42 | ----- |
| | (5.75) | (10.28) | (2.96) | (1.71) | (2.98) | (2.42) | (36.75) | ----- |
| No. of selectors: | 37.88 | 81.83 | 12.58 | 12.17 | 14.67 | 18.58 | 269.79 | ----- |
| | (31.50) | (63.32) | (5.96) | (6.03) | (11.03) | (9.35) | (323.51) | ----- |
| Complex tests: | 6.12 | 19.46 | 1.00 | 3.34 | 2.71 | 6.38 | 42.42 | ----- |
| | (5.15) | (10.28) | (0.00) | (1.25) | (1.37) | (2.35) | (36.75) | ----- |
| Selector tests: | 9.50 | 28.74 | 3.52 | 4.80 | 5.55 | 11.97 | 106.51 | ----- |
| | (6.63) | (16.86) | (1.84) | (2.04) | (3.46) | (4.99) | (118.22) | ----- |

**Table 4: Rule set quality metrics averaged over all data sets (with standard deviations).**

| | | | | | | | | |
|---|---|---|---|---|---|---|---|---|
| Generalization: | 95.63 % | 92.06 % | 96.03 % | 96.51 % | 97.42 % | 96.60 % | 85.62 | ----- |
| | (2.95 %) | (4.46 %) | (3.60 %) | (2.70 %) | (1.74 %) | (2.37 %) | (8.22) | ----- |
| Coverage: | 100.00 % | 61.71 % | 100.00 % | 100.00 % | 100.00 % | 100.00 % | 93.74 | 100.00 % |
| | (0.00 %) | (37.69 %) | (0.00 %) | (0.00 %) | (0.00 %) | (0.00 %) | (5.90) | (0.00 %) |

AQ15c rule sets is mainly due to the rule sets for the Credit and Vote 1984 data sets.

The space complexity and generalization measures in Table 4 show that AQ15c and especially GIL tend to produce more complex rule sets than C4.5 and CN2. For most data sets, the size of the GIL rule sets is a multiple of the size of the other rule sets, but obviously this has no significant impact on the accuracy of these rule sets. Furthermore the average numbers of complex and selector tests indicate that ordered rules and especially decision trees have a better time complexity than unordered rules.

## CONCLUSIONS

This work proposed a set of measures to evaluate the quality of a rule set and to compare several rule sets. Besides of the widely used predictive accuracy metric, space and time complexity metrics and two metrics called generalization and coverage were described and applied to the rule sets generated by four different learning algorithms for six data sets. These measures allowed the comparison of rule sets of distinctive type, i.e. decision trees, ordered rules, and unordered rules. The variety of measures allows the user to pick the metrics relevant for his particular application. The additional measures do not only indicate further characteristics of a rule set, but in some cases allow to find an explanation for the low predictive accuracy of a rule set, for example a low coverage indicating that the rule set is not representative for the particular domain or a low generalization indicating over-specialization.

## REFERENCES


Clark, P., and Boswell, R., (1991). "Rule Induction with CN2: Some Recent Improvements," in Y. Kodratoff, editor, *Proceedings of the Fifth European Machine Learning Conference (EWSL-91)*, pp. 151-163, Berlin, Germany, Springer-Verlag.

Clark, P., and Niblett, T., (1989). "The CN2 induction algorithm," *Machine Learning*, Vol. 3, No. 4, pp. 261-283, Kluwer Academic Publishers.

Fisher, R. A., (1936). "The Use of Multiple Measurements in Taxonomic Problems," *Annual Eugenics*, Vol. 7, Part II, pp. 179-188.

Gallion, R., Sabharwal, C., St. Clair, D., and Bond, W. E., (1993). "Dynamic ID3: A Symbolic Learning Algorithm for Many-Valued Attribute Domains," *Proceedings of the 1993 Symposium on Applied Computing (SAC '93)*.

Janikow, C. Z., (1993). "A Knowledge-Intensive Genetic Algorithm for Supervised Learning," *Machine Learning*, Vol. 13, No. 2/3, pp. 189-228, Kluwer Academic Publishers, Boston.

Michalski, R. S., and Larson, J., (1975). *AQVAL / 1 (AQ7) User's Guide and Program Description*. Report No. 731, Department of Computer Science, University of Illinois at Urbana-Champaign, Urbana, IL.

Michalski, R. S., Mozetic, I., Hong, J., and Lavrac, N., (1986). "The Multipurpose Incremental Learning System AQ15 and its Testing Application to Three Medical Domains," in *Proceedings of the Fifth National Conference on Artificial Intelligence (AAAI-86)*, Vol. 2, pp. 1041-1045, Philadelphia, Morgan Kaufmann.

Murphy, P. M., and Aha, D. W., (1994). *UCI Repository of machine learning databases*. University of California at Irvine, Irvine, CA. [http://www.ics.uci.edu/~mlearn/MLRepository.html]

Quinlan, J. R., (1993). *C4.5 : Programs for machine learning*. San Mateo, CA, Morgan Kaufmann.

Wnek, J., Kaufman, K., Bloedorn, E., and Michalski, R. S., (1995). *Selective Inductive Learning System AQ15c: The Method and User's Guide*. George Mason University, Fairfax, VA.

Wolberg, W. H. and Mangasarian, O. L., (1990). "Multisurface Method of Pattern Separation for Medical Diagnosis applied to Breast Cytology," in *Proceedings of the National Academy of Sciences*, USA, Vol. 87, December 1990, pp. 9193-9196.

# A NOVEL FRAMEWORK FOR NEURAL NET LEARNING

**TSU-SHUAN CHANG**
Department of Electrical and Computer Engineering
University of California
Davis, CA 95616

**JENQ-NENG HWANG**
Department of Electrical Engineering
University of Washington
Seattle, Washington 98195

**ABSTRACT:**
Among the main issues in neural net learning are the possibility of trapping in a local minimum, a slow learning speed, and how to obtain a neural net with a minimum number of hidden nodes. This paper presents a novel framework to deal with them by using the the concept of continuation. The intention is to lay a foundation to deal with all the issues at the same time for practical problems using ideas motivated by theoretical considerations.

## INTRODUCTION

To deal with the first issue of guaranteed convergence to a specified error tolerance, we developed in (Chang and Abdel-Ghaffar, 1992) a new learning algorithm which lets a neural net grow when trapped in a local minimum. For the second issue of getting a faster convergence, we presented in (Yu and Chang, 1995) a method to obtain an adaptive step size for learning, based upon linear bounding functions. With our better linear bounding functions newly developed in (Wang and Chang, 1996), it is expected that the previous result can be further improved. Regarding the third issue, a neural net with a guaranteed minimum number of hidden nodes is typically very difficult, if not impossible, to obtain at the present time, to our best knowledge.

In this paper, we intend to lay a foundation to deal with the three issues at the same time by using a supervised learning as an illustrative example. We will try to improve the learning speed, while keeping the system error within any specified tolerance. Although an absolute minimum number of hidden nodes can not be guaranteed, the number of hidden nodes achieves its minimum in a well-defined sense. The basic idea used is the concept of continuation, whose main purpose is to keep the weights during the learning process lying within the basin of global minimizers. By this way, an appropriate algorithm can be chosen to bring the weights back to the neighborhood of a global minimizer efficiently.

## FEEDFORWARD NEURAL NETS

Let us consider a three layer feedforward neural net composed of an input layer, a hidden layer, and an output layer, each with $N$, $X$, and $M$ nodes (neurons), respectively. Let $\bar{w}_{ki}$ denote the weight of the connection for the $i$th node in the input layer to the $k$th node in the hidden layer, and $w_{jk}$ the one from the $k$th hidden node to the $j$th output node. The $k$th hidden node has a threshold $\bar{\theta}_k$, and the $j$th output node $\theta_j$. There are $P$ distinct input patterns, where the $p$th pattern is denoted by an $N$-tuple $(i_{1p}, ..., i_{Np})$. The output of the $k$th hidden node is given by $\bar{o}_{kp} = f(\bar{net}_{kp})$, where $\bar{net}_{kp} = \sum_{i=1}^{N} \bar{w}_{ki} i_{ip} + \bar{\theta}_k$, and $f(x) = 1/(1 + e^{-x})$ is the sigmoidal function as usual. Similarly, the $j$th output is given by $o_{jp} = f(net_{jp})$, where $net_{jp} = \sum_{k=1}^{X} w_{jk} \bar{o}_{kp} + \theta_j$, $j = 1, .., M$.

The neural net is required to produce the desired output pattern $(t_{1p}, ..., t_{Mp})$, $p = 1, .., P$. The square error of the $p$th pattern is defined as $E_p = \frac{1}{2} \sum_{j=1}^{M} (t_{jp} - o_{jp})^2$, and the system error is $E = \frac{1}{MP} \sum_{p=1}^{P} E_p$. Let $W$ denote all the net parameters, $w_{jk}$'s, $\bar{w}_{ki}$'s, $\theta_j$'s, and $\bar{\theta}_k$'s. The goal is to choose $W$ such that the system error function $E(W)$ is within a specified tolerance $\epsilon$. This may be achieved by training the net through supervised learning, i.e., by presenting the input patterns, comparing the actual outputs to the desired outputs, and modifying the net parameters accordingly.

## THE CONCEPT OF CONTINUATION

Given a trained neural net with its system error strictly below the specified tolerance $\epsilon$, there always exists a neighborhood of its weights, such that a neural net with its weights in the neighborhood also has its system error within $\epsilon$. This is due to the fact that all mappings in the considered neural net are continuous. To illustrate how the concept can be used for learning, let us present one way to prune a trained neural net which has more hidden nodes than necessary.

For notational simplicity, assume the last ($X$th) hidden node is chosen to be pruned from any given pruning algorithm. Denote $W_X$ as the weights associated with the $X$th hidden node, i.e. $W_X = [\bar{w}_{X1}, .., \bar{w}_{XN}, \theta_X, w_{1X}, .., w_{MX}]$. For a given $\alpha \in [0, 1]$, let $E(\alpha)$ be the system error for the neural net (derived from a given neural net) which has the same weights as the given neural net except that its weights associated with the $X$th hidden node is multiplied by a factor $\alpha$, i.e. those weights are now $\alpha W_X$. In other words, $E(1) < \epsilon$ is the system error of the given neural net, and $E(0)$ actually represents the system error of a neural net with only $(X - 1)$ hidden nodes.

Let us now fix the weights $W_{\bar{X}}$ for the rest $(X - 1)$ hidden nodes for such a derived neural net, and denote its system error by $E(W_{\bar{X}}, \alpha)$. From the continuity property, there exists a **finite** $\delta > 0$ and $0 < \bar{\alpha}_1 < 1$ such that $E(W_{\bar{X}}, \bar{\alpha}_1) < \epsilon$ and $1 - \bar{\alpha}_1 < \delta$. If the $X$th hidden node can be indeed pruned, it seems reasonable to assume that there exists an algorithm which can bring the increased system error due to decreasing $\alpha$ back to the neighborhood of zero global minimum. More precisely, we assume that

(a) there exists an $\alpha_1 < \bar{\alpha}_1$ and $\gamma > \epsilon$, such that $E(W_{\bar{X}}, \alpha_1) < \gamma$, and

(b) there exists an algorithm which can find a weight $W_X^*$ to make

$E^*(\alpha_1) \triangleq E(W_X^*, \alpha_1) < \epsilon$.

Similarly, from the neural net associated with $E^*(\alpha_1)$, there exists an $\alpha_2 < \alpha_1$ such that $E^*(\alpha_2) < \epsilon$. By repeating the process, if we can find a sequence of $\alpha_i$'s such that $E^*(\alpha_i) < \epsilon$ and $\alpha_i - \alpha_{i+1} < \delta_i$, the sequence will reach zero in a finite number of iterations due to the **finiteness** of $\delta_i$. In other words, we have successfully pruned the given neural net to the one with one less hidden node with its system error within the specified tolerance.

By repeating the above process for each hidden node sequentially, we can reduce the number of hidden nodes one by one, until the process fails. That is, the system error of the last neural net can no longer be reduced within $\epsilon$, and thus can not be pruned any more. By this way, we can find a neural net with a minimum number of hidden nodes in the well-defined sense of continuity. Note that in practice, there is no reason to reduce the system error below $\epsilon$ before reaching the last neural net, as long as we can stay in the neighborhood of a global minimizer.

## A PRUNING ALGORITHM

In order to use the concept of continuity for pruning algorithms, we need to find an algorithm to implement (b) efficiently. That is, the algorithm must be able to quickly bring the system error from $\gamma$ back to $\epsilon$ or the neighborhood of the zero global minimum, and can also detect if this can be done or not.

In searching of such an algorithm, note that if $\gamma$ is small enough, the initial weights are expected to fall within the (convex) basin of convergence due to the continuity property. Note also that all we need is to bring the error to a specified neighborhood efficiently, there is no need to find a minimizer precisely. In other words, the convergence rate near the minimizer may not be so important except for the last neural net. To achieve a fast learning speed, an iterative algorithm should terminate with not too many iterations, and the computational requirement for each iteration should keep as low as possible. Among the good candidates considered is the way of using the basic idea of subgradient algorithms similar to that in (Chang, 1996).

To understand how the algorithm behaves, let us consider the one dimensional convex minimization problem, which has the zero global minimum. Starting from $x^0$, the next iterant $x^1$ is chosen as the intersection of the tangent line at $x^0$ and the $x$-axis. By repeating the process, $x^k$ will approach a global minimizer $x^*$ with a linear rate of convergence.

If we can keep $\gamma$ not too large, we can then use such an algorithm for neural nets. Since there is no need to use any line search or more complicated model such as a quadratic one to find the next iterant, the computational burden for each iteration is reduced considerably. Although the convergence rate is slow when near the minimizer, it does bring back the iterant to the neighborhood of the minimizer quickly from our numerical experiences and this is exactly what we want.

If the global minimum for a neural net is indeed zero, it is intuitively expected that such an algorithm can bring the system error within a given $\epsilon_e$ consecutively for (reasonable) $n_w$ times, starting from a not too large $\gamma$.. This is how Algorithm 1 stops (with $flag = 0$). If the global minimum is greater than zero, the algorithm will never converge. Such a situation is detected in Algorithm 1 (with $flag = 1$) by using criteria such as the one when the number of iterations is greater than $k_m$.

**Algorithm 1: Reaching the basin of a global minimizer**

Input: $x^0$ with dimension $dim(x^0)$, $\epsilon_e$, $n_w$, $\epsilon_g$ and $k_m$.
Output: $x^*$, and $flag$.

Step 0: Given: $k = 0$, $n = 0$, and $flag = 0$.
Step 1: Calculate the search direction $h = -grad(f(x^k))$.
Step 2: Compute the step size $\lambda_k = \beta \frac{(f(x^k) - 0.1\epsilon_e)}{h' * h}$.
Step 3: Update the iterant $x_k = x_k + \lambda_k h$. $k = k + 1$.
Step 4: If $f(x_k) < \epsilon_e$, then $n = n + 1$; else $n = 0$.
If $n > n_w$ or $(h' * h)^{0.5}/dim(h) < \epsilon_g$, or $k > k_m$, then go to step 5; else go to step 1.
Step 5: Output $x^* = x^k$. If $k > k_m$ or $((h' * h)^{0.5}/dim(h) < \epsilon_g$ and $f(x^*) > \epsilon_e)$, then $flag = 1$.

**Algorithm 2: A pruning algorithm**

Input: A neural net with weight $W$.
Output: A pruned neural net with new weight $W_n$.

Step 0: Given: All parameters needed in Algorithm 1, $\gamma$, and $flag = 0$.
Step 1: Store old weights in case for unsuccessful pruning, i.e. $W_{old} = W$.
Step 2: Select the hidden node to be pruned and set $\alpha = 0$.
Step 3: Choose next $\alpha$.
Step 4: Find $E^*(\alpha)$ with $flag$ by using Algorithm 1.
Step 5: If $\alpha < 1$ and $flag == 0$, then go to Step 3;
else if $flag == 1$, go to Step 6; else go to Step 1.
Step 6: Output the weights $W_n = W_{old}$.

Note that any existing pruning algorithm can be used to implement Step 2 in Algorithm 2. Since our purpose is to demonstrate the concept of continuation, we use a simple strategy of choosing a hidden node with the minimum norm for the weights associated with it. In Step 3, we use a simple quadratic approximation to determine $\alpha$.

The pruning algorithm was tested for one output bit XOR and PAR3 problems with two and three input bits, respectively. The desired output is 0.9 if the number of input bits equal to 1 is even; otherwise, the desired output is 0.1. The parameters used in Algorithm 1 are $\epsilon_e = 10^{-3}$, $n_w = 3$, $\epsilon_g = 10^{-6}$, and $k_m = 30$. In Algorithm 2, $\gamma = 0.1 * \epsilon_e$.

For the XOR problem, a trained neural net with four hidden nodes was obtained from the backpropagation algorithm. By applying Algorithm 2 to the neural net, we ended up with a neural net with two hidden nodes after 90 iterations. For the PAR3 problem, we started from a trained neural net with seven hidden nodes and got the final neural net with three hidden nodes after 103 iterations. Both of them obtained the final neural net with a minimum number of hidden nodes.

## A NOVEL FRAMEWORK FOR LEARNING

One way to use the concept of continuity for overall neural net learning is to apply the input patterns sequentially as follows. For a given single pattern $p$, it can be proven easily that there always exists a set of weights $W_p$ for one hidden node neural net to achieve zero system error. We can then consider training the second pattern with the first one together to achieve zero system error. Repeat the process until all the patterns are trained all together.

To implement such a process, note that whenever a new pattern arrives, there is no indication if the current neural net has enough hidden nodes. Instead of trying to learn it with the current neural net, we can always grow the net immediately and learn it in a higher dimensional space with the hope of faster and more surely convergence to the specified error tolerance. If the new pattern can be indeed incorporated by the current neural net, we then expect that a pruning algorithm such as Algorithm 2 can reduce the number of hidden nodes back to the original one.

Given a trained neural net with weight W and P patterns, we need to learn a new arrival $(P+1)$th pattern together. The subproblem defined on the current neural net is the same as before except with a modified system error function

$$E(W,\beta) = \frac{1}{M(P+1)}\sum_{p=1}^{P} E_p + \beta E_{P+1}, \quad \beta \in [0,1].$$

By choosing $\beta$, we can ensure $E(W,\beta) < \gamma$, so that those algorithms such as Algorithm 1 can be used efficiently. We can now use any growing algorithm for the subproblem to get the new weight $W^*$ such that $E^*(\beta) \triangleq E(W^*,\beta) < \epsilon$. Note that there may now have more hidden nodes in $W^*$ than those in $W$.

**Algorithm 3: A Framework for Learning**

Input: Desired output patterns associated with $P$ input patterns.

Output: A trained neural net within a specified system tolerance and has a minimum number of hidden nodes in a well defined sense.

Step 0: Given: All parameters used in Algorithm 1, and $flag = 0$.
Step 1: Find initial weight $W^*_{p_1}$ for the first chosen pattern $p_1$. Set $p = 1$.
Step 2: If $p < P$, then choose the next pattern with minimum $E(p)$ and set $p = p + 1$; else STOP.
Step 3: Set $\beta = 0$.
Step 4: Choose next $\beta$.
Step 5: Find $E^*(\beta)$ with $flag$ by using Algorithm 1.
Step 6: If $flag == 0$, then use Algorithm 2 to prune the current neural net and go to Step 8; else go to Step 7 for growing.
Step 7: Assign initial weights for new hidden node with strategies

to be explained. Go to Step 5.

Step 8: If $\beta < 1$, then go to Step 4; else go to Step 2.

The choice for $\beta$ in Step 4 is simple due to the linearity in terms of $\beta$. Since there is no need to use those algorithms in this paper for many steps, existing pruning and growing algorithms can be incorporated to the framework. In this paper, we merely use those algorithms to demonstrate our basic ideas.

Note that the idea of synergizing pruning with growing is not conceptually new. In fact, they can be blended in various ways instead of what described in Algorithm 3. What unique in Algorithm 3 is its inclusion of the continuation concept. From the construction of the algorithm, it should obtain a minimum number of hidden nodes in the well defined sense of continuity.

We can definitely use different ways to combine pruning with growing, and have various choices in many steps, such as which hidden node to be pruned, which pattern to be processed next, etc. With different choices, different continuous paths will be generated. Most likely, the path taken will lead to various neural nets with different number of hidden nodes. We need to either prove that they all have the same number of hidden nodes or choose one fixed path which is better than the other in certain sense.

Furthermore, we observe that algorithms like Algorithm 1 behaves very well when the system error is within the basin of convergence to zero global minimum, we can set the maximum number of iteration low to speed up the computation speed. Since the pruning algorithm is very reliable and fast enough, we can let the growing algorithm expand fast and achieve the specified error tolerance in higher dimensional space for higher learning speed.

## REFERENCES


Chang, T.S., Abdel-Ghaffar, K. A. S., (1992). A universal neural net with guaranteed convergence to zero system error, *IEEE transactions on Signal Processing*, 40(12), Dec, 3022-3031.

Yu, M., Chang, T.S., (1995). An adaptive step size for backpropagation using linear lower bounding functions, *IEEE transactions on Signal Processing*, 43 (5), May, 1243-1248.

Wang, X., Chang, T.S., (1996). An improved univariate global optimization algorithm with improved linear bounding functions, *Journal of Global Optimization*, 8, 393-411.

Chang, T.S., (1996). A framework of using linear programming for manufacturing scheduling, to appear at *Proceedings of the 35th IEEE Conference on Decision and Control*, Kobe, Japan, Dec.

# PREDICTING LOCAL MINIMA FOR WEIGHT TRAINING OF BACKPROPAGATION NETS

**WEI CAO AND JAMES H. BURGHART**
*Electrical Engineering Department*
*Cleveland State University*
*Cleveland, Ohio 44115*

**ABSTRACT:**

A new mathematical approach for predicting properties of local minima encountered in weight training of a backpropagation (B.P.) neural net used for pattern classification is developed. This approach is motivated by weight-training simulation results and introduces equilibrium space whose coordinates are net-output values associated with individual patterns. With an appropriate sorting procedure, the local minima reached in weight training can be predicted theoretically. An anticipated important application of this result is that it should be possible to predict the minimum number of hidden nodes needed to guarantee the solvability of a given classification problem theoretically.

## INTRODUCTION

The backpropagation (B.P.) net is one of the promising neural networks for classification problems since it can solve linearly-inseparable problems. However, because its weight training lacks global convergence, a local minimum can result [1,2,3,4]. Most research efforts [5,6,7,8] have been based on the traditional "Error Surface in Weight Space" approach. Unfortunately, the error surface of a B.P. net is high dimensional and nonlinear, limiting research results mainly to the qualitative level.

This investigation of the use of the backpropagation neural net applied to the two-class, P-pattern classification problem establishes a new approach in order to put the study of local minima on a quantitative basis [9]. A necessary condition for equilibrium has been derived. In this paper we define an L-set consisting of the collection of all local minima reached in weight training, and two of its supersets that can be established theoretically merely based on the total number of patterns, P, and pattern class memberships ($P_0/P_1$), without weight training. This implies that the local minima could be sorted out from the supersets.

This paper focuses on the development and discussion of the subset relationships. The sorting procedure and predication of local minima will be reported in [9]. An important application resulting from this investigation is to theoretically determine the minimum number of hidden nodes or layers needed to guarantee the solution of a classification problem without weight training [10].

## LOCAL MINIMA OF B.P. NETS

The multi-layer B.P. net rapidly became one of the dominant nets used for classification problem because it has the capability for solving linearly-inseparable problems, although its training process may terminate at a local minimum.

Since the TSE (Total Squared Error) of a B.P. net involves huge numbers of weights and highly nonlinear sigmoid functions, the associated error surface is high-dimensional and has a very complex non-convex shape. A special name "terraced" has been given to the error surface of B.P. net in the literature [5,6,7,8] because it is characterized by many terraces and cliffs Although many research efforts have attempted to explore the underlying properties of local minimum, the inherent weakness of using the error surface in weight space has been a hindrance to substantial quantitative results. The main issues about local minima remain open: without weight training, (1) how can we predict the number, type and properties of local minima, and (2) how can we determine the numbers of hidden layers and nodes needed to solve a particular classification problem?

We define **Weight Space** (W-space) as an R-dimensional space whose coordinates are the R weights $\ll \omega_1,\omega_2,\ldots,\omega_R \gg$ of a B.P. net, **A Local Minimum in W-space** as an R-tuple of final weights $<\omega_1,\omega_2,\ldots,\omega_R>$ when a local minimum is reached in weight training, and the **Local Minimum Set (L-set) in W-space** as the collection of all local minima in W-space.

## COMPUTER SIMULATION DATA FOR THE XOR PROBLEM USING A MULTI-LAYER 2-2-1 NET

One well-known classification problem is the XOR (exclusive OR) problem [4] using a 2-2-1 net with 9-weights. This problem was used in our preliminary investigations with hundreds of weight-training simulations starting from different initial conditions [9]. Sample data from the simulations leading to local minima are listed in Table 1.

In the top part of Table 1, the different initial weights and final weights from weight training are given. Each local minimum in weight space can be represented by a 9-tuple of final weights $<\omega_1,\omega_2,\theta_1,\omega_3,\omega_4,\theta_2,\omega_5,\omega_6,\theta_3>$. For the example, Local 1 is $<$-3.5,+11.5,+2.3,+4.8,+11.5,-1.1,-7.2,+8.0,-0.8$>$ with TSE=0.50 and Local 2 is $<$+2.7,-1.3,+9.8,-6.4,+3.7,-8.2,-7.0,+0.9,+10.6$>$ with TSE=0.50. However, it is apparently impossible to extract any valuable conclusions from the weight data.

However, if one focuses on the net output values and TSE levels associated with individual patterns rather than the weights (bottom part of Table 1), an important observation is apparent. Weights start from arbitrary real numbers and end up with arbitrary real numbers. However, although the computer algorithm is formatted in terms of decimal numbers, it is clearly apparent that the netout and the TSE values at local minima are approaching rational values (which can be written as fractional numbers) although both netout and TSE can start from arbitrary real values and can take any real values during weight training.

This fractional convergence appears not to be accidental and leads to **Equilibrium Space** (E-space) defined for a P-pattern problem as the P-dim space with coordinates $\ll netout_1,netout_2,\ldots,netout_P \gg$, the net output values associated with the P patterns. Then the **Local Minimum in E- Space** occurs when any local minimum is reached in weight training and a P-tuple of final pattern net outputs $<netout_1,netout_2,...,netout_P>$ is used to represent the local minimum in E-space. The **Local Minimum Set (L-set) in E-space** is defined as the collection of all local minima in E-space.

For the XOR Problem, each local minimum in E-space can be represented by a 4-tuple of final output values $<netout_A,netout_B,netout_C,netout_D>$. Therefore, Local 1 is $<0,1/2,1/2,1>$ with TSE 1/2 and Local 2 is $<0,1/2,1/2,1>$ with TSE 1/2. It is seen that they have the same final netout values and TSE levels, although they start from different initial weights and finish with different final weights. Thus Local 1 and Local 2 are categorized together as Local A based on E-space. Similarly, Local 3 and Local 4 are categorized together as Local B.

## FRACTIONAL CONVERGENCE PRINCIPLE

The simulation results indicated that all local minima apparently obey the underlying principle that the pattern netout values converge to fractional numbers between 0 and 1 (including 0 and 1) when a local minimum is reached, where the denominators of those fractional number are $\leq P$ and the numerators are $<P$. This is called the **Fractional Convergence Principle.** It can be shown that the total number of valid fractional numbers is $\alpha=(2+(P)(P-1)/2)$ and the TSE takes only fractional values at a local minimum as well. For instance, for XOR (P=4), the valid numerators are 1,2 and 3 and the valid denominators are 1,2,3 and 4, so there are 8 valid fractional numbers, 0, 1, 1/2, 1/3, 2/3, 1/4, 2/4 and 3/4.

Based on these observations, we define the F-set as the set of all valid fractional convergence numbers and the G-set as all possible P-tuples from the F-set. The F-set has the form {0,1, 1/2,1/3,2/3,. ..,(P-1/P)} with total number of elements $\alpha=2+(P)(P-1)/2$ and the G-set has the form $\{G^1,G^2,...,G^\beta\}$ with total number of elements $\beta=(2+(P)(P-1)/2)^P=\alpha^P$. Each element of the G-set is a P-tuple, $G^m=(netout_1^m,netout_2^m,...,netout_P^m)$ $(m=1,2,...,\beta)$. Fig.1 illustrates the geometrical representation for the F-set and the G-set. It is seen that the F-set can be formed by continuously subdividing the unit line segment between 0 and 1. It is also seen that the G-set can be formed by just looking at the corner points of the sub-blocks of a P-dimensional unit cube. The most significant meaning of the F-set and the G-set, both finite sets, is that they can be constructed without weight training, and are merely based on the number of patterns P. Table 2 lists the G-set and L-set for the 2p, 3p-line and XOR problems. The L-set is apparently a subset of the G-set.

These results are based on computer simulation results. A necessary condition for equilibrium in E-space has been derived [9]. It is called the Equilibrium Equation which, for a $P(P_0/P_1)$-pattern problem, is a 3rd-order polynomial equation

$$\Sigma netout_i^2(1-netout_i)=\Sigma netout_j(1-netout_j)^2$$

$$(i=1,..,P_0,\ j=P_0+1,..,P \text{ and } P=P_0+P_1)$$

The solution set of the Equilibrium Equation is called the E-set. And the E-surface is a graphical representation of the Equilibrium Equation as a surface in E-space. This necessary condition for equilibrium of a B.P. net in weight training implies that any local minimum reached must be a solution of the equilibrium equation. This is verified by the simulation results given in Table 3 which lists the E-set elements for the 2p, 3p-line and XOR problems. Subset relationships are shown in Fig.2 where the GE-set is the overlapping of the G-set and E-set.

## SUMMARY, CONCLUSION AND FUTURE WORK

The significant meaning of these results is that is that important sets can be established theoretically merely based on pattern number pattern membership without weight training. The development of a proper sorting procedures should theoretically allow the L-set to be sorted out from the other sets. This new approach implies that all the local minima which can be reached in weight training can be predicted without weight training. This issue will be treated in [9].

One of the important applications of these results is that the minimum number of nodes and layers which guarantee the solvability of a given classification problem could be determined theoretically without weight training. This issue will be covered in [10].

## REFERENCES


[1] R. P. Lippmann, "An Introduction to Computing with Neural Nets", IEEE Mag. ASSP, Apr.1987, pp.4-22.

[2] B. Widrow and A. L. Michael, "30 Years of Adaptive Neural Networks: Perceptron, Madaline, and Backpropagation", Proc.IEEE, V.78, No. 9, Sept. 1991.

[3] P. J. Werbos, "Backpropagation Through Time: What it Does and How to Do It", Proc. IEEE, Vol 78, No. 9, Sept. 1991.

[4] D. E. Rumelhart, G. E. Hinton, and R. J. Williams, "Learning Internal Representation by Error Propagation", Parallel, Distributed Processing: Explorations in the Microstructure of Cognition, V.1, Found. MIT Press, 1986.

[5] D. R. Hush, J. M. Salas, and B. Horne, "Error Surface for Multi-layer Perceptron", Proc. IJCNN-91, I-759, July 1991, Seattle, WA.

[6] S. C. Huang and Y. F. Huang, "Learning Algorithms for Perceptrons Using Back-Propagation with Selective Updates", IEEE Control Systems Magazine, April 1990, pp. 56-61.

[7] An Mei Chen and Robert Hechi-Nielsen, "On the Geometry of Feedforward Neural Network Weight Space". 2nd Int.Conf. on ANN. IEE. Conf. Publ n 349, 1991. London.

[8] Xiao-Hu YU, "Can Backpropagation Error Surface Not Have Local Minima". IEEE Trans. Neural Networks, V.3 Nov, 1992. p1019-1021.

[9] Wei Cao and James Burghart "Equilibrium Equation, Equilibrium Space and Equilibrium Surface", Neural Network Symposium, NASA LeRC, 1995.

[10] Wei Cao and James Burghart, "Determination of Minimum Number of Hidden Nodes to Guarantee the Solvability of B.P. Nets," Neural Network Symposium, NASA Lewis Research Center, 1996.

| | Local 1 | | Local 2 | | Local 3 | | Local 4 | |
|---|---|---|---|---|---|---|---|---|
| weight | initial | final | initial | final | initial | final | initial | final |
| $\omega_1$ | -0.214 | -3.55 | +0.002 | +2.789 | -0.437 | -8.223 | -0.346 | +9.432 |
| $\omega_2$ | +0.920 | +11.5 | -0.970 | -1.345 | -0.480 | -8.225 | +0.456 | -7.765 |
| $\theta_1$ | +0.201 | +2.37 | -0.342 | +9.876 | +0.014 | +0.711 | -0.770 | +5.928 |
| $\omega_3$ | +0.431 | +4.86 | +0.099 | -6.492 | +0.429 | +4.639 | +0.022 | +11.88 |
| $\omega_4$ | +0.879 | +11.5 | -0.762 | +3.776 | +0.466 | +4.639 | -0.549 | -6.445 |
| $\theta_2$ | -0.0111 | -1.131 | +0.267 | -8.208 | -0.008 | +2.382 | +0.333 | +2.453 |
| $\omega_5$ | -0.450 | -7.219 | -0.222 | -7.002 | -0.269 | -6.716 | -0.278 | +8.808 |
| $\omega_6$ | +0.529 | +8.05 | -0.167 | +0.908 | +0.208 | +1.304 | +0.770 | -7.354 |
| $\theta_3$ | -0.079 | -0.839 | -0.256 | +10.65 | -0.173 | -0.605 | -0.254 | -8.555 |
| $netout_A$ | 0.4803 | 0.0000 | 0.3890 | 0.0000 | 0.4432 | 0.0000 | 0.6542 | 0.0000 |
| $netout_B$ | 0.5069 | 0.5000 | 0.6443 | 0.5000 | 0.4709 | 0.6667 | 0.4894 | 0.6667 |
| $netout_C$ | 0.4834 | 0.5000 | 0.2175 | 0.5000 | 0.4676 | 0.6667 | 0.2555 | 0.6667 |
| $netout_D$ | 0.5092 | 1.0000 | 0.5003 | 1.0000 | 0.4638 | 0.6667 | 0.5677 | 0.6667 |
| TSE | 0.9958 | 0.5000 | 1.2344 | 0.5000 | 1.0050 | 0.6667 | 1.3457 | 0.6667 |
| | Local A | | | | Local B | | | |
| | <0, 1/2, 1/2, 1> TSE=1/2 | | | | <0, 2/3, 2/3, 2/3> TSE=2/3 | | | |

Table 1: Computer Simulation Samples for XOR (Multi-layer 2-2-1 Net)

| | G-set (set size) | L-set (set size) |
|---|---|---|
| 2p | (0,0)(0,1)**(1,0)**(1,1)(0,1/2)(1/2,0)(1/2,1/2)<br>(1,1/2)(1/2,1) (9) | **<1,0>** <1/2,1/2><br>(2) |
| 3p-line | (0,0,0)(0,0,1)(0,1,0)**(1,0,1)**(0,0,1/2)(0,1/2,0)<br>(0,1,1/2)(0,1/2,1)(0,1/2,1/2)(0,0,1/3)(0,1/3,0)<br>(0,1,1/3)(0,1/3,1)(0,1/3,1/3)(0,0,2/3)(0,2/3,0)<br>(0,1,2/3)(0,2/3,1).... (4096) | **<1,0,1>** <1/2,1/2,1><br><0,1/2,1/2><br><2/3,2/3,2/3><br>(4) |
| XOR | (0,0,0,0)**(1,1,0,0)**(0,0,1/2,1/2)(0,1/2,1/2,1)....<br>(0,1,2/3,2/3)(0,2/3,2/3,2/3).....<br>.....<br>(30517578125) | **<1,1,0,0>** <1/2,1/2,0,1><br><1/2,1/2,1/2,1/2><br><1/3,1/3,1/3,1><br><2/3,2/3,2/3,0><br><2/4,2/4,2/4,2/4><br>(6) |

Table 2. G-set and L-set for 2p, 3p-line and XOR Problems

| | Equilibrium Equation / E-set | L-set (set size) |
|---|---|---|
| 2p | $netout_A(1-netout_A)^2 = netout_B^2(1-netout_B)$ | <1,0> <1/2,1/2> |
| | (1,0)(1/2,1/2) ...(0.1,0.65)(0.2,0.82)...... | (2) |
| 3p-line | $netout_A(1-netout_A)^2 + netout_C(1-netout_C)^2 = netout_B^2(1-netout_B)$ | <1,0,1> <1/2,1/2,1> <0,1/2,1/2> <2/3,2/3,2/3> |
| | (1,0,1)(1/2,1/2,1)(0,1/2,1/2)(2/3,2/3,2/3) ....(0.13,0.27,0.76)(0.34,0.3,0.28) ..... | (4) |
| XOR | $netout_A(1-netout_A)^2 + netout_C(1-netout_C)^2 = netout_B^2(1-netout_B) + netout_D^2(1-netout_D)$ | <1,1,0,0> <1/2,1/2,0,1> <1/2,1/2,1/2,1/2> <1/3,1/3,1/3,1> <2/3,2/3,2/3,0> <2/4,2/4,2/4,2/4> |
| | (1,1,0,0)(1/2,1/2,0,1)(1/2,1/2,1/2,1/2) (1/3,1/3,1/3,1)(2/3,2/3,2/3,0)(2/4,2/4,2/4,2/4) ...... (0.43,0.76,0.01,0.56)..... | (6) |

**Table 3. E-set and L-set for 2p, 3p-line and XOR Problems**

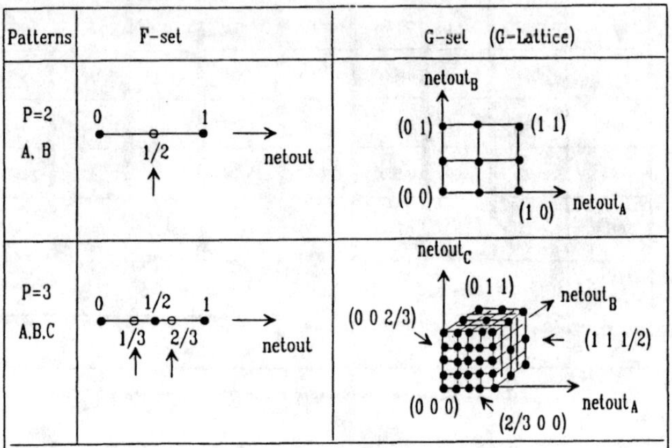


**Fig.1 Geometrical Illustration of F-set and G-set**

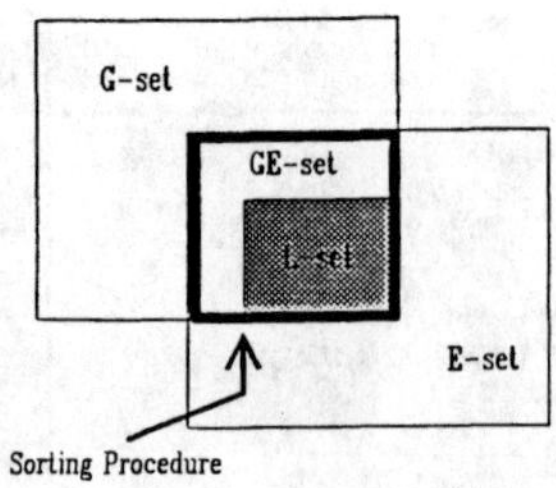


**Fig.2. Subset relationship**

# PART II : FUZZY LOGIC

# Section 3 : Fuzzy Neural Networks and Systems

# A Fuzzy-Based Concept Formation System for Categorization and Numerical Clustering

C. L. Philip Chen[1], *Senior Member, IEEE* and Yuan Lu

[1]Department of Computer Science and Engineering

Wright State University, Dayton, Ohio 45435

**Abstract**

Recently, psychologists proposed the prototype theory of concept representation, in which a concept is organized around a best example or so-called *prototype*. Most proponents of the prototype theory conceive that objects may fall in a concept to some degree rather than the all-or-none membership in the classical theory. Fuzzy-set theory is compatible with the basic premises of the prototype theory of concept representation. Concept formation is defined as a machine learning task that captures concepts through categorizing the observation of objects and also uses them in classifying future experiences. A reasonable computational model of concept formation must reflect the characteristics of human concept learning and categorization. In this paper, the design and implementation of a fuzzy-set based concept formation system (FUZZ) is presented. The main feature of the FUZZ is that the concept hierarchy is non-disjoint, in which an instance may belong to two categories in different memberships. An information-theoretic evaluation measure called *category binding* to direct searches in the FUZZ is proposed. The learning and classification algorithms of the FUZZ are also given. In order to examine FUZZ's behavior, the results of some experiments are examined.

## 1 Introduction

As a cognitive model of human learning, concept formation focuses on concept acquisition in the symbolic world. Both clustering analysis and concept formation are unsupervised processes. But these two methods also differ in several aspects. To portray the mechanism by which humans acquire and organize their concepts, a concept formation model should support the concept hierarchies with varying the levels of specificity as well as the incremental fashion of concept acquisition. On the other hand, traditional clustering methods concentrate on the numeric domains and categorize classes in the nonincremental way. Both clustering analysis and fuzzy clustering have been studied extensively for several decades. In spite of the differences, the experiences and achievements in these areas can serve as the basis of our current approach to concept formation.

Throughout this paper, we might use the terminologies such as feature, property and attribute alternatively. Similarly, class, cluster, category and concept, as well as, instance, pattern, object, example and case may be used interchangeably.

## 2 FUZZ

Similar to the extension of clustering algorithms to fuzzy clustering algorithms, our claim is that each instance can be explicitly assigned to several categories with different membership values in process of concept acquisition.

[1]The supports from Office of Naval Research, University Research Initiative Grant N00014-92-J-4096 and Air Force contract F33615-92-D-5000 are gratefully acknowledged. Any opinions, findings, and conclusions or recommendations expressed in this article are those of the authors and do not necessarily reflect the views of the funding agencies.

Here, our concern is that each instance may contribute partially to one or multiple categories with different membership values. And this partial contribution also plays a role in further learning or performance.

To determine which categories to put instances in, FUZZ employs another information-theoristic measure called *category binding*, with stimulations of COBWEB's [1] category utility and WITT's [2] category cohesion. the category binding seems to be more direct and simpler than other two measures.

UNIMEM [3], COBWEB and WITT, along with numeric clustering methods, have inspired the FUZZ system. FUZZ is a hierarchical, incremental concept formation system, with the objective of explicitly dealing with fuzzy concepts. In this section, we describe the FUZZ system in detail. Specifically, it includes the representation and organization of instances and concepts, the heuristic evaluation function *category binding*, the learning and classification algorithms, some implementation issues, and a simple detailed example.

## 2.1 Representation and organization of instances and concepts in FUZZ

FUZZ takes a sequence of instances as input, incorporating one at a time. An instance is described as set of attribute-value pairs. FUZZ supports both numeric and nominal attributes in its instance representation. To the numeric attributes, FUZZ needs information about their ranges, which is used to evaluate distances in learning. Furthermore, FUZZ allows a set of values to describe a nominal attribute.

Representation of individual concepts is a basis of any concept learning system. Many concept formation systems maintain and use the information about all attributes in the representation units. Like in COBWEB (CLASSIT [4]), it stores probabilities (mean and deviation) for each nominal (numeric) attribute.

In FUZZ, our goal is to represent fuzzy concepts. A similar representation schema is adopted to achieve this goal. A concept is defined by a set of attributes and their possible values; each attribute has a range and a *typical* range. The range and the typical range allow us to define the intentional characteristics of a concept and to capture the notion of typicality.

Let $R(a, C)$ denote the range of an attribute $a$ of a concept $C$, and $T(a, C)$ denote the typical range of attribute $a$ of concept $C$. We'll distinguish two different cases according to types of attributes:

- Attribute $a$ is numeric: $R(a, C)$ is defined as an interval $[A_{min}, A_{max}]$, which represents the possible value of attribute $a$ in all members of concept $C$. We define $T(a, C)$ as the weighted mean of attribute values of all instances in concept $C$:

$$T(a, C) = \frac{\sum_i u(i)^m X_a^{(i)}}{\sum_i u(i)^m},$$

  where $m$ is the exponential weight, $u(i)$ is the membership function of instance $i$ in concept $C$, and $X_a^{(i)}$ is the value of attribute $a$ in instance $i$. For example, in the attribute "age" of concept "Student", $R$(age,Student) may be from 15 to 30, as well as $T$(age,Student) is equal to 22.

- Attribute $a$ is nominal: $R(a, C)$ is a fuzzy set, which is constituted by all the values of attribute $a$ for instances in concept $C$ and their possibility. $T(a, C)$ represents the typical values of attribute $a$ of concept $C$. For example, the range $R$(location,Hotel) of attribute "location" of concept "Hotel" is described as

  { 0.70/highway-outlet,0.45/downtown,0.38/suburb, 0.38/close-office-center,

0.13/airport}, and $T$(location,hotel) is { 0.70/highway-outlet}.

FUZZ takes a sequence of instances and abstracts them to build a concept hierarchy in the similar manner as UNIMEM, COBWEB and CLASSIT.

## 2.2 Category binding : heuristic evaluating method in FUZZ

FUZZ uses a heuristic measure called *category binding* (CB) to guide the search. Generally, the concept formation paradigms construct categories which maximize between-category dissimilarity of instances and simultaneously minimize within-category dissimilarity of instances. Thus, in order to manifest the quality of categories, the measure of category binding should be a tradeoff between between-category dissimilarity (distance) and within-category dissimilarity (distance). Conceptually, we choose it as a contrast between between-category distance and within-category distance.

Formally, let a set of concepts $\{C_1, C_2, \ldots, C_n\}$ be a partition of the parent concept $C$. We define the category binding of this partition as:

$$CB(C_1, C_2, \ldots, C_n) = \frac{DI(C_1, C_2, \ldots, C_n)}{DW(C_1, C_2, \ldots, C_n)},$$

where $DW$ and $DI$ represent the average distance between instances within a category and the average distance between a category and all others, respectively. In addition, we can formally state $DW$ as:

$$DW(C_1, C_2, \ldots, C_n) = \frac{\sum_{k=1}^{n} \Delta_{C_k}}{n},$$

where $\Delta_{C_k}$ stands for the variance in the category $k$. Let $P_{C_i}$ denote the prototype of category $C_i$. Then, the between-category distance $DI$ is defined by:

$$DI(C_1, C_2, \ldots, C_n) = \frac{\sum_{i=1}^{n-1} \sum_{j=i+1}^{n} D(P_{C_i}, P_{C_j})}{n(n-1)/2},$$

where $D$ measures the distance between two prototypes.

Still, we can easily define the variance $\Delta$ in a category and the prototype distance $D$ by the distance of individual instances in categories. Let a set $\{I_{i,1}, I_{i,2}, \ldots, I_{i,n_i}\}$ represent the $n_i$ number of instances in the category $i$. Let $\mu_k(i)$ denote the fuzzy membership function representing the belongness of instance $I_k$ in the category $i$. Thus, we can define the distance between two individual instances $l$ and $m$ in categories $i$ and $j$ as:

$$d(I_{i,l}, I_{j,m}) = \mu_l(i)\mu_m(j) \parallel I_{i,l} - I_{j,m} \parallel,$$

while $\parallel \bullet \parallel$ can be any distance measure with various salience (weight) for each attribute. The variance $\Delta$ of a category can be defined by the average of distances of instances in the same category :

$$\Delta_{C_k} = \begin{cases} \dfrac{\sum_{l=1}^{n_k-1} \sum_{m=l+1}^{n_k} d(I_{k,l}, I_{k,m})}{\sum_{l=1}^{n_k-1} \mu_l(k) \sum_{m=l+1}^{n_k} \mu_m(k)} & \text{if } n_k > 1 \\ T_v & \text{otherwise;} \end{cases}$$

where $T_v$ is a system parameter representing the variance for a single-instance category, and $n_k$ is the number of instances in $C_k$. On the other hand, the prototype of a category is usually regarded as the average of the instances in this category. The distance between two category prototypes can be implemented by the average of distances between instances in different groups, which formally is defined as :

$$D(P_{C_i}, P_{C_j}) = \frac{\sum_{l=1}^{n_i} D(I_{i,l}, P_{C_j})}{\sum_{l=1}^{n_i} \mu_l(i)} = \frac{\sum_{l=1}^{n_i} \sum_{m=1}^{n_j} d(I_{i,l}, I_{j,m})}{\sum_{i=1}^{n_i} \mu_l(i) \times \sum_{m=1}^{n_j} \mu_m(j)},$$

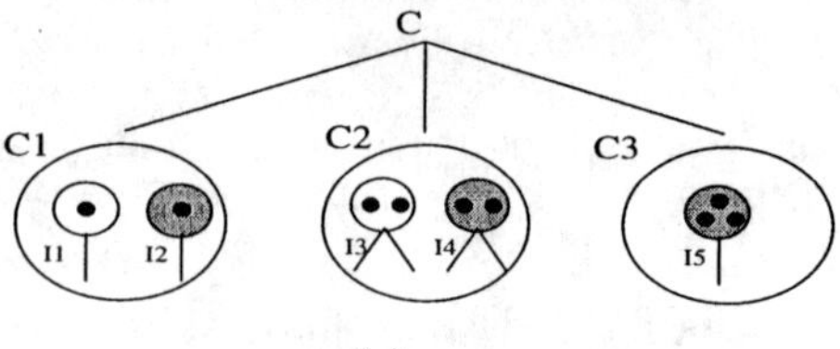


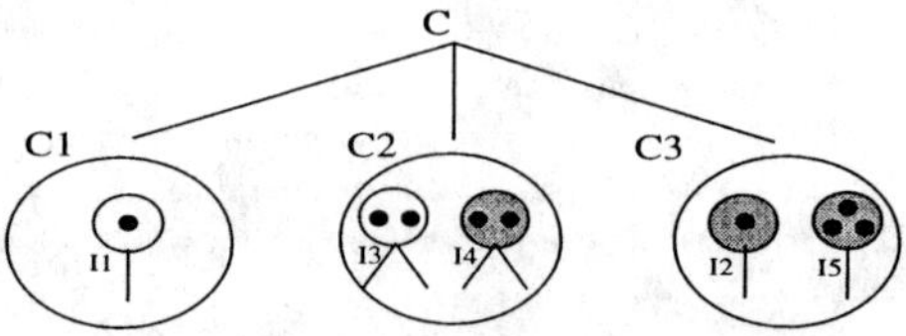


Figure 1: Two different clusterings shown in (a) and (b) illustrating the behavior of the category binding measure.

where $n_i$ and $n_j$ represent the numbers of instances in $C_i$ and $C_j$, and $P_{C_j}$ represents the prototype of category $C_j$.

As an example to explore the behavior of the category binding measure, consider two different clusterings in Figure 1. Let us define $T_v = 1$, for simplicity, in this one-celled organism domain, we define the distance of two organisms in the Hamming distance, and also no fuzzy membership function is considered. Thus, we have $d(I_1, I_2) = 1$, $d(I_1, I_3) = 2$, $d(I_1, I_4) = 3$, $d(I_1, I_5) = 3$, and so on, respectively. In Figure 1(a), clearly, we can calculate that

$$DW(C_1, C_2, C_3) = (1 + 1 + 1)/3 = 1.$$

Obviously, we can apply the above equation to get $D(C_1, C_2) = 5/2$, $D(C_1, C_3) = 5/2$, $D(C_2, C_3) = 5/2$, and

$$DI(C_1, C_2, C_3) = (5/2 + 5/2 + 5/2)/3 = 5/2.$$

Thus, $CB(C_1, C_2, C_3) = \frac{5/2}{1} = 5/2$. Similarly, in Figure 1(b), we have, $DW(C_1, C_2, C_3) = 4/3$, $DI(C_1, C_2, C_3) = 7/3$, and $CB(C_1, C_2, C_3) = \frac{7/3}{4/3} = 7/4$. According to the measure of category binding, we would expect that the clustering in Figure 1(a) is superior to one in Figure 1(b). From the experience of human categorization, we would agree with this conclusion. In summary, the category binding measure is accordance with the intuitive measure in human categorization. In next section, we utilize this meaure in FUZZ's learning algorithm.

## 2.3 Learning in FUZZ

FUZZ is an unsupervised, incremental, top-down conceptual clustering algorithm intergrating both learning (concept formation and modification) and performance (classification and predication). In the concept hierarchy, each node maintains a fuzzy concept description representing a category of instances. Concept descriptions are updated constantly in the process of incorporations of new instances. FUZZ includes four basic operators same as COBWEB and CLASSIT in the learning stage, which are : (1) placing the instance into an existing category; (2) creating a new category to accommodate the instance; (3) combining two categories into one to accept the instance; (4) splitting a category into its sub-categories.

The system considers all four operators and selects the best one depending on the measure of category binding as we discussed above. Before we step into the detailed algorithms, we need to introduce some predefined system parameters. We first define the parameter $T_f$, which is a threshold of the difference between the category binding values of two different partitions. That is, if the difference between $CB_1$ and $CB_2$ is below to this threshold, the system considers that these partitions are all acceptable.[2] Therefore, the system incorporates the instance into different categories. Another system parameter $T_c$ called *cutoff* threshold is also defined in FUZZ. The motivation of cutoff threshold is that, rather than always descends into the leaves of the hierarchy in the learning procedure, the system may halt at some high-level category. When it happens, the system decides that the instance is "similar enough" to the specific category such that no further descent is necessary, and also the some specific features of this instance are thrown away. In FUZZ, it compares the variance $\Delta_C$ of a category $C$ to $T_c$ to determine the cutoff of the further descent in the hierarchy.

We summary the FUZZ learning algorithm as the followings: (FUZZ starts at the root of hierarchy).

FUZZ $(N, I, U)$

1. IF $N$ is a leaf;

    create two children of $N$, $L_1$, $L_2$; copy the instances $I_k$ in $N$ into $L_1$ and set $\mu_{L_1}(I_k) = random(0.75, 1)$ [3];

    place $I$ into $L_2$ and set $\mu_{L_2}(I) = random(0.75, 1)$; place $I$ into $N$ and set $\mu_N(I) = U$;

    compute the center of category $N$, $P_N$; return the node $L_2$ and halt;

2. ELSE

    place $I$ into $N$ and set $\mu_N(I) = U$; compute the center of category $N$, $P_N$;

    IF $\Delta_N < T_c$; return the node $N$ and halt;

3. ELSE

    FOR each child $C_k$ of node $N$; compute the category binding (CB) of the clustering by placing $I$ in $C_k$;

    let $S_1$ be the score for the best categorization of hosting $I$ in $C_1$;

    let $S_2$ be the score for the second best categorization of hosting $I$ in $C_2$;

    let $S_3$ be the score for creating a new category $C_n$ to host $I$;

    let $S_4$ be the score for merging $C_1$ and $C_2$ into one category;

    let $S_5$ be the score for splitting $C_1$ into its children;

4. IF $S_1$ is the best score;

    IF $S_1 - S_2 < T_f$; compute $\mu_{C_1}(I)$ and $\mu_{C_2}(I)$ by using the distance $D(I, P_{C_k})$;

        call FUZZ $(C_1, I, \mu_{C_1}(I))$; call FUZZ $(C_2, I, \mu_{C_2}(I))$;

    ELSE

        call FUZZ $(C_1, I, random(0.75, 1))$;

   ELSE

    IF $S_3$ is the best score; create a new category $C_n$;

        place $I$ into $C_n$ and set $\mu_{C_n}(I) = 1$; return the category $C_n$ and halt;

   ELSE

---

[2] In current version, only two partitions are considered at same time maximally. We feel that, in practical, it is enough.

[3] The function $random(a_1, a_2)$ returns a random generated value between $a_1$ and $a_2$.

IF $S_4$ is the best score; let $C_m$ be the result of merging $C_1$ and $C_2$; call FUZZ $(C_m, I, random(0.75, 1))$;

ELSE

IF $S_5$ is the best score; split $C_1$ into its children; call FUZZ $(N, I, random(0.75, 1))$;

FUZZ performs a hill-climbing search in the space of possible partitions using category binding to evaluate and determine optimal clustering. It initializes the hierarchy using the first instance. For each subsequent instance, the algorithm begins with the root category and moves through the hierarchy, and depends on the category binding measure to select the appropriate operator. Furthermore, it uses two user-defined parameters to control if further descent in the hierarchy is needed, or multiple categories are chosen. If later case happens, FUZZ uses the distance metric to calculate the fuzzy membership value based on the method used in the fuzzy ISODATA algorithm [5]. Since FUZZ preserves the basic motivations of COBWEB, indeed, it introduces the fuzzy concept in the categories of the hierarchy, we believe that it could supply more inferential capability. More detail is dicussed in Chen's coming paper [6].

## 3 Summary

As a concept formation system, the central motivation of the FUZZ system is to introduce the fuzzy set theory explicitly in the process of concept formation. It also extends the previous works in the representation of instance and concept, in which it allows the mixed numeric and nominal attributes , multiple-valued attributes. In addition, we establish and experiment with a new evaluation function to control the construction and search of concept hierarchy. In summary, FUZZ retains the advantages of the previous concept formation system, and provides its own characteristics to model real-world concept formation.

## References


[1] Fisher, D.H. (1987). Knowledge Acquisition via Conceptual Clustering. Machine Learning, 2, 139-171, 1987. Reprinted in Shavlik, J.W., &Dietterich, T.G. (Eds) Reading in Machine Learning. San Mateo, CA.: Morgan Kaufmann, 1990.

[2] Hanson, S.J., &Bauer, M. (1989). Conceptual Clustering, Categorization, and Polymorphy. Machine Learning, 3, 343-372, 1989.

[3] Lebowitz, M. (1987). Experiments with Incremental Concept Formation: UNIMEN. Machine Learning, 2, 103-138, 1987.

[4] Thompson, K., &Langley, P. (1991). Concept Formation in Structured Domains In Fisher, D.H., Pazzani, M.J., &Langley, P. (Eds) Concept Formation: Knowledge and Experience in Unsupervised Learning, 127-161. San Mateo, CA.: Morgan Kaufmann, 1991.

[5] Bezdek, J.C. (1980). A Convergence Theorem for the Fuzzy ISODATA Clustering Algorithms. IEEE Trans. Pattern Analysis and Machine Intelligence, Jan., 1980. Reprinted in Bezdek, J.C., &Pal, S.K. (Eds) Fuzzy Models for Pattern Recognition. The IEEE, New York, 1992.

[6] Chen, C. L. P. and Y. Lu (1996). FUZZ: A Fuzzy-Based Concept Formation System that Integrates Human Categorization and Numerical Clustering. Accepted to appear in IEEE Trans. on Systems, Man, and Cybernetics.

# GENERAL PARAMETER NEURAL NETWORKS WITH FUZZY SELF-ORGANIZATION

**DAOUREN AKHMETOV AND YASUHIKO DOTE**
*Department of Computer Science and Systems Engineering*
*Muroran Institute of Technology, Mizumoto 27-1, Muroran 050, Japan*
*Fax: +81-0143-47-3374; Phone: +81-0143-44-4181*
*E-mail: dote@dote_sun.csse.muroran-it.ac.jp*

*ABSTRACT:*
A novel neural network structure formation approach and learning method which has improved training features are proposed based on compressed representation of teaching data during learning process. Acceleration of convergence rate is achieved by initial reduction of network parameter space dimensionality. Improved structure backpropagation and radial basis function networks are developed. Simulations are performed to confirm the feasibility of these networks for practical applications. Lastly, these are experimentally applied to nonlinear time series prediction problems resulting in higher accuracy and faster convergence rate relatively conventional networks. A fuzzy procedure is designed for an optimal neural network structure determination based on information available from learning stage of modified network.

## INTRODUCTION

Requirements of high quality automatic control had made it necessary to use nonlinear mathematical models for real processes description (Kosko, 1992; Narendra and Parthasarathy, 1990).

Identification systems using neural networks are widely spread now, due to their well-known ability of nonlinear process approximation to arbitrary accuracy. However, conventional methods used for network training require rather long time period and sometimes cannot be recommended for "on-line" identification, especially for substantially nonlinear systems.

Assume that a neural network, which exactly describes the plant process (reference or sample NN), may be concerned as a system $S$ characterized by the finite dimensional vector of parameters $\theta$ ( for example, network weights). The system $S$ is approximately represented with an adjusting model M $(\hat{\theta})$, where $\hat{\theta} \in R^n$ is a model parameter vector, $R^n$ is a parametric space with a metric $\rho(\theta,\hat{\theta})$. The random time sequences of input $\{u(t)\}$ and output $\{x(t)\}$ values of the plant, which possess final variances and satisfy the experimental conditions $E$, are observed.

This paper describes the problem of achieving the matching between the model $M(\hat{\theta})$ and the system $S$ by means of $\hat{\theta}$ estimation choice with an iterative method of identification $I$ and the convergence of the estimation $\hat{\theta}(t)$ to $\theta$ provided when $t \to \infty$. Thus, the solution of the identification problem in the time interval $t$ is the estimation $\hat{\theta}(t) = \hat{\theta}(t;J)$, where $J$ is an identification system described by a quadruple $\{S,M,E,I\}$.

The choice of the NN type, structure and learning algorithms sometimes does not provide the desired accuracy of parameter estimation within the finite observation intervals. In such situation the identification problem can be solved by using new constructions of models $M_K$ and estimation schemes $I_K$ of parameters. Then the parameter estimation in the time interval $t$ can be denoted as $\hat{\theta}(t) = \hat{\theta}(t;J_K)$ where $J_K$ is identification system defined by a quadruple $\{S,M_K,E,I_K\}$. The system $J_K$ is more effective in comparison with the known systems $J$ if the condition, $\rho[\hat{\theta}(t;J),\theta] > \rho[\hat{\theta}(t;J_K),\theta]$, is satisfied.

In such a situation there was proposed the general parameter method (GP) of complex system identification (Ashimov and Syzdykov, 1981; 1987). The main features of the GP approach are the high convergence rate and ability of estimation accuracy calculation in the steady state of learning algorithms. The aim of the presented paper is to investigate the possibility of implementing the GP approach with the artificial neural networks (ANN).

## GENERAL PARAMETER METHOD

Recurrent procedures of identification are based on adjusting model schemes. The structure of the adjusting model is chosen as

$$\hat{x}(t) = \hat{F}\{\tilde{u}(t),\hat{\theta}(t-1)\} \tag{1}$$

where $\hat{x}(t)$: model output; $\tilde{u}(t)$: input signal; $\hat{F}\{.\}$ is a plant operator estimate; $\hat{\theta}(t) = [\hat{\theta}_1(t)\hat{\theta}_2(t)...\hat{\theta}_N(t)]^T$: model parameter vector that has to be estimated by using some identification technique; $t$ =0,1,2,...: discrete time; N: plant dimensionality. Using the general parameter approach the model (1) is represented by:

$$\hat{x}(t) = \hat{F}\{\tilde{u}(t),\hat{\theta}(0),\tilde{\beta}(t-1)\} \tag{2}$$

where $\hat{\theta}(0)$: vector of the initial model parameters; $\tilde{\beta}(t) = [\beta_1(t)\beta_2(t)...\beta_M(t)]^T$: general parameter (GP) vector; $M$: dimensionality of the GP-vector$(1 \le M < N)$. The number of model parameters to be adjusted in the case (2) is reduced to increase the identification algorithm convergence rate. For fixed values of signal/noise ratio the quadratic form $S(t)$ of the general parameter vector error may be considered as an identification accuracy measure.

## GENERAL PARAMETER RADIAL BASIS FUNCTION NETWORKS (GP RBFN)

Recently theoretical results have been developed to prove approximation abilities of

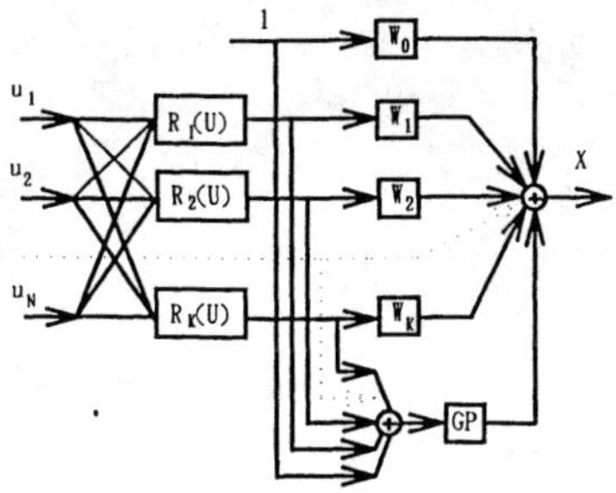


**Figure 1: Simplest GP–RBFN structure.**

neural networks (Hornik, *et al.*, 1989; Cybenko, 1989). There is trade-off between network approximation ability and learning procedure convergence rate (Poggio and Girosi, 1990). Let consider one of typical representatives of MNN, namely the radial basis function networks.

As soon as RBFN is linear on its weights, the GP method may be implemented in a straightforward manner. The simplest GP-RBFN structure is shown in Fig.1. The equation describing GP-RBFN for a single output network is

$$x = \sum_{j=1}^{K} (w_j^0 + \beta) exp[-\sum_{i=1}^{N} (u_i - c_j)^2 / \delta_j^2] + (w_o^o + \beta) \tag{3}$$

where $w_j^0$ : fixed initial values of network weights; $\beta$: scalar general parameter adjusting with the following algorithm

$$\beta = \beta + \gamma_\beta (x^* - x) \times \{ \sum_{j=1}^{K} exp[-\sum_{i=1}^{N} (u_i - c_j)^2 / \delta_j^2] + 1 \} \tag{4}$$

where $\gamma_\beta$ : adjusting gain; $x^*$ : desired network output.

## GENERAL PARAMETER BACKPROPAGATION NETWORK (GP-BPNN)

Let consider a one-layer sigmoid activation function neural network which is described by (5).

$$\begin{cases} \hat{x}_i = \sum_{j=1}^{p} (\hat{w}_{ij}^0 + \beta_i) u_j \\ y_i = f(\hat{x}_i) \\ i = \overline{1,q} \end{cases} \tag{5}$$

where $\hat{w}_{ij}^0$ : fixed initial values of network weghts; $\tilde{\beta} = [\beta_1\ \beta_2 \ldots \beta_q]^T$: general

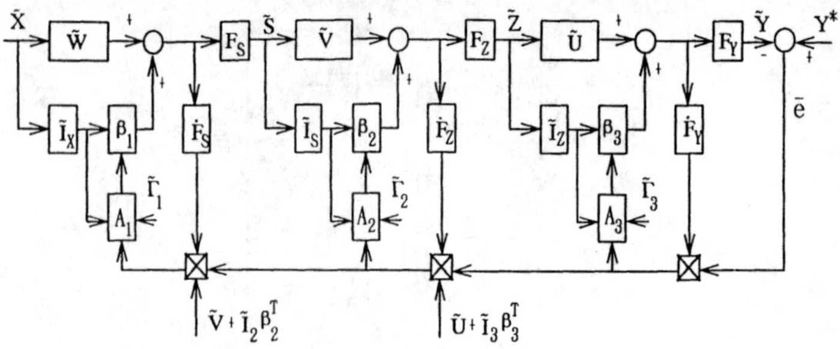


Figure 2: Three layer GP-backpropagation neural network.

parameter vector. Iterative procedure of parameter $\beta_i$ identification:

$$\beta_i = \beta_i + \gamma_i e_i f(x_i)[1 - f(x_i)] \sum_{j=1}^{p} u_j, \quad i = \overline{1,q} \tag{6}$$

where $\gamma_i$: adjusting algorithm gain with its convergence.

The developed approach may be implemented with the multilayer network structure as represented in Fig.2. The following designations are accepted: $\tilde{X}, \tilde{Y}$: input and output vectors; $\hat{S}, \tilde{S}, \hat{Z}, \tilde{Z}, \hat{Y}$: internal variable vectors of NN; $\tilde{W}, \tilde{V}, \tilde{U}$: weight matrixes; $F_S, F_Z, F_y$ : nonlinear operators with $F_S = diag\{ f_i^S(.)\}$; $f_i(.)$: sigmoidal function; $\tilde{I}_i = [1\,1 \ldots 1]^T$ : units array of i-th layer input space dimensionality; $\tilde{\beta}_i$: general parameter vector of i-th layer; $A_i$: adjusting algorithms.

Recurrent procedure of general parameter adjusting in matrix form is

$$\begin{cases} \tilde{\beta}_1 = \tilde{\beta}_1 - \tilde{\Gamma}_1 \dot{F}_S (\tilde{V} + \tilde{I}_2 \tilde{\beta}_2^T) \dot{F}_Z (\tilde{U} + \tilde{I}_3 \tilde{\beta}_3^T) \dot{F}_Y \tilde{e} \tilde{X}^T \tilde{I}_X \\ \tilde{\beta}_2 = \tilde{\beta}_2 - \tilde{\Gamma}_2 \dot{F}_Z (\tilde{U} + \tilde{I}_3 \tilde{\beta}_3^T) \dot{F}_Y \tilde{e} \tilde{S}^T \tilde{I}_S \\ \tilde{\beta}_3 = \tilde{\beta}_3 - \tilde{\Gamma}_3 \dot{F}_Y \tilde{e} \tilde{Z}^T \tilde{I}_Z \end{cases} \tag{7}$$

where $\tilde{\beta}_i = [\beta_{i1}\, \beta_{i2} \ldots \beta_{ij}]^T$: i-th layer vector general parameter with dimensionality $j$ equal to i-th layer unit number; $\tilde{\Gamma}_i = diag\{\gamma_{ij}\}$: gain matrix; $\dot{F}_S = diag\{ \dot{f}_j(S)\}$: nonlinear matrix operator with elements corresponding to activation function derivatives; $\tilde{e} = \tilde{Y}^* - \tilde{Y} = [e_1\, e_2 \ldots e_q]^T$ : output error vector; $\tilde{I}_x, \tilde{I}_S, \tilde{I}_Z, \tilde{I}_1, \tilde{I}_2, \tilde{I}_3$ : unit arrays of appropriate dimensionality.

## FUZZY SELF-ORGANIZING OF GPNN STRUCTURE

The steady state expectation and variance of the general parameter provide information about network approximation quality. Both values may be easily calculated and used for decision making purposes. The fuzzy self-organizing network structure determination approach was proposed which results in quasi-optimal structure network with desired approximation ability. The generalized representation of fuzzy rules is as follows

$$R_i : if\ E\{\beta\}\ is\ E_i\ and\ D\{\beta\}\ is\ D_i\ then\ St\ is\ St_i \quad (8)$$

where i : rule number, $E\{\beta\}$ , $D\{\beta\}$ : general parameter average and variance correspondingly, $St$ : structure characteristic of neural network to be corrected.

## SIMULATION RESULTS AND EXPERIMENTS

Simulations were performed to justify the proposed network structures and adjusting procedure. The adjusting model was represented by two different structure neural networks. In the first case it was a conventional network with individual parameter adjusting. In the second case the combined GP method was used, when initial adjusting steps were performed by correction of $\beta$. After GP procedure steady state achieving, the identification system was switched to individual adjusting of weights for increasing estimation accuracy .

The efficiency of both of methods was determined by using the following measure of algorithm convergence rate:

$$\chi(t) = \frac{E\{\|\tilde{r}(t)\|^2\}}{\|\tilde{r}(0)\|^2} \quad (9)$$

where $\tilde{r}(t) = \tilde{w} - \tilde{w}(t)$: parameter error vector ; E – expectation symbol; $\|.\|$: Euclidean norm. The simulation results show that the proposed method is characterized by higher convergence rate compared with conventional approaches, particularly during initial learning periods.

The convergence and approximation features of conventional and modified (GP) networks were compared by usage of real vibration data from boring machine. Fig.3 shows approximation abilities of conventional radial basis network and GP-RBFN after finite number of adjusting steps. The proposed method is superior to conventional one because of less prediction error for short training time.

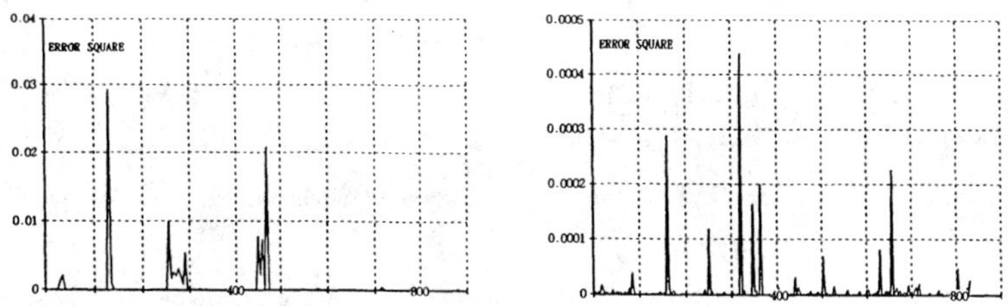


a) b)

**Figure 3: Prediction with RBFN**
**a) conventional RBFN, b) GP-RBFN.**

## CONCLUSION

The novel structures and learning procedures of radial basis function and backpropagation neural networks are developed based on general parameter approach of complex system identification. The main feature of the latter is artificial reduction of adjusting model parameter space dimensionality at least during some learning periods, especially at initial steps when approximation error is large.

Simulation was performed to confirm the feasibility of the proposed technique. But even in the worst case of the full initial uncertainty GP provides fast convergence to some parameter space domain which includes solution point. As a result, the network is characterized by certain approximation accuracy, sufficient for some problems and unsufficient for another one. The mapping quality of the considered neural networks became better after modification, especially for short time learning periods. Therefore, proposed approach may be recommended in some cases for real-time nonlinear mapping problems.

The fuzzy self-organizing GPNN structure determination approach based on learning period information analysis was proposed which results in quasi-optimal network structure with desired approximation abilities.

## REFERENCES

Akhmetov, D., and Dote, Y. (1995) Novel "on-line" identification procedure using artificial neural network, Proc.of the IFAC Workshop on On-Line Fault Detection and Supervision in the Chemical Process Industries, pp.90-97, Newcastle, U.K.

Ashimov, A.A. and Syzdykov, D.Zh. (1987) Identification by the general parameter method. In: *Reference book on the automatic control theory* (A.A.Krasovsky, Ed.), pp.263-271. Nauka, Moscow. (In Russian).

Cybenko, G. (1989). Approximation by superpositions of a sigmoidal function. *Math.Control Signals Syst.,* **Vol.1**, pp.303-314.

Hornik, K., Stinchcombe, M. and White, H. (1989). Multilayer feedforward networks are universal approximators.*Neural Networks,* **No 2**, pp.359-366.

Kosko, B. (1992) *Neural Networks and Fuzzy Systems: a Dynamical Systems Approach to Machine Intelligence.* Prentice Hall, Englewood Cliffs, New Jersey.

Moody, T.J. and Darken, C.J. (1989). Fast learning in networks of locally tuned processing units. *Neural Computation.* **Vol.1**, pp.151-160.

Narendra, K.S. and Partasarathy, K. (1990). Identification and control of dynamical systems using neural networks.*IEEE Trans. on Neural Networks.* **1,1**, pp.4-27.

Poggio, T. and Girosi, F. (1990). Extensions of a theory of networks for approximation and learning: dimensionality reduction and clustering. *Artificial Intelligence Lab. Memo.* **No** 1167. MIT, Cambridge, MA.

Syzdykov D. Zh., Ashimov A.A. (1981). The General Parameter Method of Complex System Identification. YIII IFAC Congress,Kyoto,Japan.

# AN ANN ARCHITECTURE FOR THE FUSION OF FUZZY PATTERN RECOGNITION SYSTEMS

**M. E. ULUG**

Intelligent Neurons Inc., 1537 E. Hillsboro Blvd., 342, Deerfield Bch., FL 33441, intneurons@aol.com

***ABSTRACT:***

In most problems there is numerical data that describes the system and there is linguistic data that represents human judgment and experience. We achieve a consensus between these two types of data by fusing the outputs of two fuzzy pattern recognition systems. This is done in the testing program of a fuzzy neural network. In pattern recognition for each input vector the network provides a spectrum of "object to a class belongingness" with respect to all classes. The output spectrums of these two ANNs are treated as two fuzzy premises and aggregated using a disjunction operator. The resulting fuzzy conclusion is a combined spectrum. When new information becomes available the "event representation," an AI technique, is used to modify the combined spectrum (Ulug 1995). The system architecture is single layer and uses a learning method that substantially reduces the training time of fuzzy ANNs. A military application that can benefit from the fusion of numerical and linguistic pattern recognition systems is also described.

## INTRODUCTION

The human reasoning involves dealing with many imprecisions, ambiguities, and uncertainties. For ANNs to function more like humans, it is necessary to build into their design fuzzy set theory (Zadeh, 1965) and fuzzy logic (Bellman and Zadeh, 1976). In this paper we describe such a fuzzy ANN intended for the fusion of numerical and linguistic pattern recognition systems.

The training times of fuzzy neural networks using the backpropagation learning rule are exceptionally slow. Moreover, convergence of the network cannot always be guaranteed. This is because the target values represent the membership values of the training samples. Therefore, they are dynamic. In Section 2 we describe a single layer orthogonal neural network and an associated learning method that resolves this problem by substantially reducing the training time of fuzzy ANNs. In Sections 3 and 4 we describe the input and output fuzzification used with this architecture.

In Section 5 we describe a military application that can benefit from the fusion of numerical and linguistic pattern recognition systems. Mainly we propose to improve the accuracy of ballistic missile identification by combining the target recognition and situation assessment systems. In Section 5 we also describe the event representation technique for incorporating the new event data into the combined target recognition/situation assessment spectrum.

In Section 6 we provide system parameters and give training times for our fuzzy orthogonal neural network.

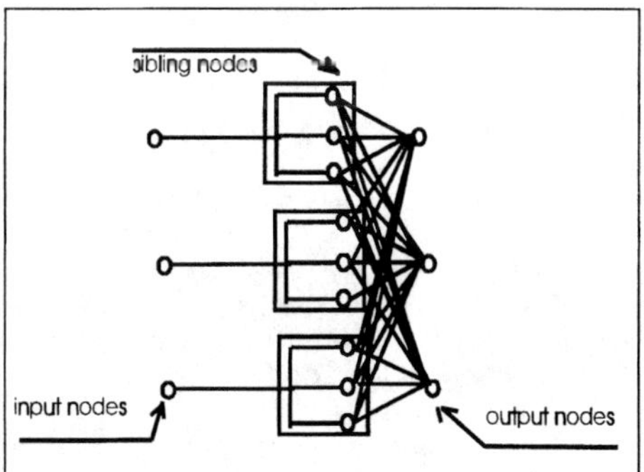


**Figure 1: Orthogonal Neural Network**

## FUZZY ORTHOGONAL NEURAL NETWORKS

In our fuzzy ANN (Ulug, 1993), (Ulug, 1992), FONN, an Orthogonal Neural Network, ONN, is placed between the input and output fuzzification layers. An ONN consists of n input signals, [n.N] sibling nodes, and m output nodes or classes. As shown in Fig. 1 each input signal is connected to N sibling nodes where the Fourier series processing is done. Because of this processing, an ONN can map any $L_2$ mapping function between input and output vectors within a mean square error accuracy without using hidden layers or the backpropagation rule (Ulug, 1992). We will prove this by referring to the following result: "Given a function $f:[0,1]^n \rightarrow \mathbf{R}^m$, we say that f belongs to $L_2$ (Royden, 1971) if each of f's coordinate functions is square-integrable on the unit cube. If a vector function of a vector variable is in $L_2$, then each of its components can be approximated by its Fourier series to any degree of accuracy in the mean squared error sense (Dunford and Schwartz, 1966), (Wylie, 1966)."

For 1<=k<=[(N-1)/2], where N is the number of sibling nodes per input signal (an odd number), the following functions are used in the sibling nodes:

$$\varphi_1(x_i) = \frac{1}{\sqrt{2}} \quad (1)$$

$$\varphi_{2k}(x_i) = \sin(k\pi x_i) \quad (2)$$

$$\varphi_{2k+1}(x_i) = \cos(k\pi x x_i) \quad (3)$$

It can be shown that the functions given above are orthonormal for -1<=x<=1 (Spiegel, 1962). In training we are using a modified version of the Widrow and Hoff rule (Widrow and Hoff, 1960). One of the advantages of using the Fourier series expansion is that the synaptic link values can be calculated and used for initialization. The use of this technique, called **Coefficient Initialization**, results in very short training times (please see Section 6) for fuzzy neural networks (Ulug, 1994), (Ulug, 1993), (Ulug, 1992). For Coefficient Initialization we use the following algorithm:

(i) Calculate the coefficient of the constant term, $c_1$, assuming an activation of f(x)=[(ea)/n] as follows:

$$c_1 = \frac{e.a.\sqrt{2}}{n} \quad (4)$$

where a = the actual target value assigned to the output nodes, e = percentage of activation achieved before training, n = number of input signals,

(ii) Set the coefficients of the other nodes, $c_2$ to $c_N$, to zero.

The architecture of FONN is shown in Fig. 2. The neural network is very similar to the one described above and depicted in Fig. 1 except that an input fuzzifier is placed before the sibling nodes and output fuzzifier is placed before the output nodes. The membership function for the input vector of the pattern i, $\mathbf{x}_i$, is calculated using the following equation which is a simplified version of the introduced by S. K. Pal (Pal and Mitra, 1992).

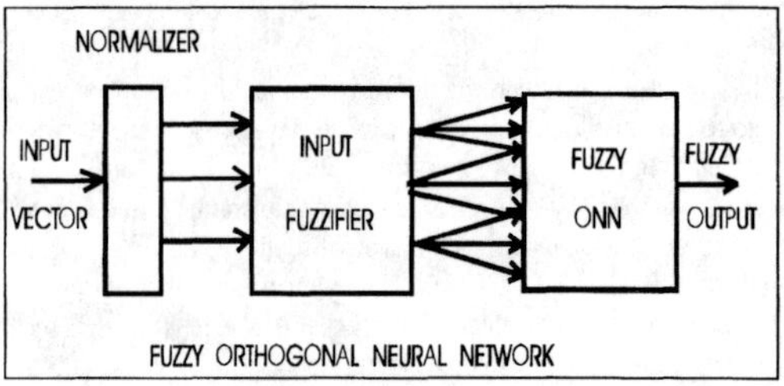


**Figure 2: Fuzzy ONN**

$$\mu(x_i) = \frac{1}{1 + \sqrt{\sum_{j=1}^{n}[x_{ij} - mean_{kj}]^2}} \quad (5)$$

where $mean_{kj}$ = j'th component of the mean vector of the k'th class, for k=1, ... ,m, and $x_{ij}$ = j'th component of the i'th sample.

Here we calculate a membership value for all the training samples of all the classes with respect to a given mean vector. We then repeat the process for all the mean vectors. In classification of an unknown sample, we calculate the vectorial differences between the vector representing the sample and all the mean vectors. The vectors used in Equation 5 are shown in Fig. 3. In Fig. 3 we show only two classes. For clarity we did not show all the vectors.

## INPUT FUZZIFICATION

For the fuzzification of input signals we used the following technique in our software system: (i) we normalized the input signals, and (ii) we placed seven 2-dimensional π functions at the entrance to the ONN.

Our π functions were given by the following equations (Pal and Mitra, 1992):

$$\pi(x;c;r) = 2(1 - \frac{|x-c|}{r})^2 \quad (6)$$

$$\pi(x;c;r) = 1 - 2(\frac{|x-c|}{r})^2 \quad (7)$$

The Equations 6 and 7 are for r/2<=|x-c|<=r and for 0.0<=|x-c|<=r/2 respectively

otherwise

$$\pi(x;c;r) = 0.0 \quad (8)$$

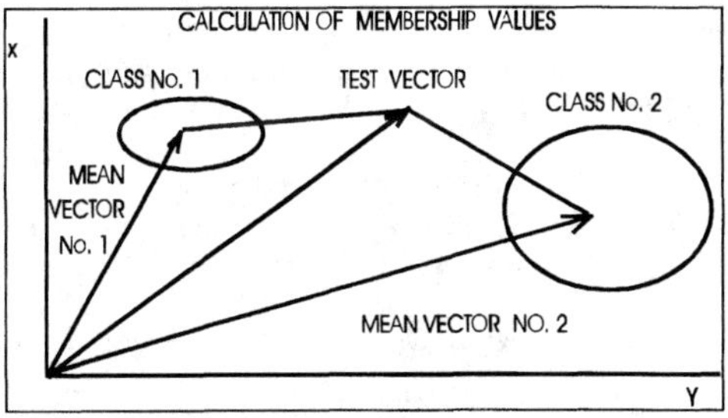


**Figure 3: Calculation of Membership**

where x = distance from origin, c = center of the π function, r = radius of the π function.

After normalizing the input signal, we placed the medians of our π functions as follows: Extremely low (EL) = 0.125, Very low (VL) = 0.250, Low (L) = 0.375, Midpoint (MD) = 0.5, High (H) = 0.625, Very high (VH) = 0.750, Extremely high (EH) = 0.875.

## OUTPUT FUZZIFICATION

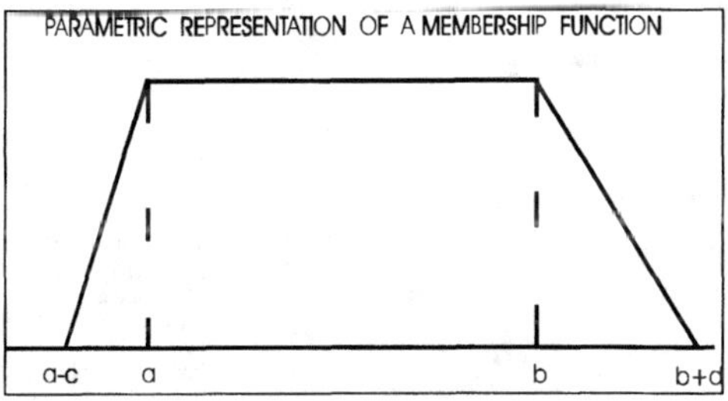


**Figure:4 Parametric Representation**

For the output fuzzification we used parametric representation (Bonissone and Decker, 1986), (Zadeh, 1965) shown in Fig.4. For the fuzzification of the output signals we use the triangular norms and conorms of Schweizer and Sklar (Schweizer and Sklar, 1963) to handle conjunctions and disjunctions respectively. A user could choose uncertainty granularity by selecting a 5, 9, or 13-element term set (Bonissone and Decker, 1986).

## APPLICATION

We are now going to describe a fusion application by giving an example. In this example we consider an automatic target recognition, ATR, system having four stages. These are as follows: 1. Data and sensors, 2. Preprocessing and segmentation , 3. Feature selection and extraction, 4. Classification. We assume that a fuzzy ANN will be used in the fourth stage. In situation assessment, SA, up coming events are predicted based on all relevant input data such as "increased communication activities," "troop movements," "economic tensions," etc.. We assume that in SA there is a database feeding linguistic data into a second fuzzy ANN. Using AI we combined the outputs of the ATR and SA systems in the testing program of our fuzzy ANN.

The event representation is used to modify the SA spectrum when new information becomes available. Following that the corresponding components of the spectrums ATR and AS (two fuzzy premises) are aggregated using a disjunction operator resulting in a combined spectrum (fuzzy conclusion).

The training program of a SA is carried out using data collected over a long period of time. In addition, a typical training program is not executed in real-time. When a new event occurs, there is no past data to use for training and usually there is no time to wait for the training program to run. Because of this, we incorporated the new event data into the combined ATR/SA testing program which is executed in real-time and does not require past data (Ulug 1993).

Using AI we aggregate an event certainty with an "object to a class belongingness" of the SA spectrum using either a conjunction or disjunction operator (Ulug, 1993). When we use a conjunction operator, we divide the aggregate by the certainty of the event to find a conditional certainty in a similar way to conditional probability. In both types of operators (see Appendix A) the user specifies his or her confidence level on the occurrence of the event by selecting one of the following descriptions: Optimistic, average, or pessimistic.

In both systems we used four classes and 10-dimensional input vectors. In SA ANN linguistic input vectors were used. Each component of this vector has two elements. The first one is a linguistic expression such as "increased communication activities," "troop movements," "economic tensions," etc.. Using templates these expressions are mapped into input nodes. The second element is a term set that expresses the magnitude of the, for examples, troop movements in a fuzzied form. As a result a typical linguistic vector is as follows (see Section 3): MD H EH EL EL EL MD VH VH MD. In our tests we used a 9-element term set. As illustrated in Figure 4 the elements of this set, (a, b, c, d), are as follows: **impossible** (0 0 0 0), **extremely unlikely** (0.01 0.02 0.0075 0.05), **very low chance** (0.1 0.18 0.06 0,05), **small chance** (0.22 0.36 0.05 0.06), **it may** (0.41 0.58 0.09 0.07), **meaningful chance** (0.63 0.8 0.05 0.06), **most likely** (0.78 0.92 0.06 0.05), **extremely likely** (0.98 0.99 0.05 0.01), **certain** (1 1 0 0). The values of (a, b, c, d) were derived from an adaptation of the results of psychological experiments on the use of linguistic probabilities (Beyth-Marom, 1982)

We aggregate the ATR (premise #1), SA (premise #2), and new events (premise #3) spectrums. First using the rule #1 we aggregate the premises #2 and #3 to obtain a modified SA spectrum. We name the conclusion of rule #1 premise #4. We then aggregate the premises #1 and #4 using rule #2. The conclusion of this rule is the combined spectrum of the ATR and SA systems. The following simple examples illustrate this process: A ballistic missile has been fired towards Japan. The following spectrum is offered by ATR system:

**PLOT OF BELONGINGNESS OF MISSILE**

| BELONGINGNESS | CLASS NO. | PREMISE #1 |
|---|---|---|
| 0.1698 | 1 (China) | \|*** |
| 0.3439 | 2 (N. Korea) | \|****** |
| 1.0000 | 3 (S. Korea) | \|***************** |
| 0.1329 | 4 (Russia) | \|** |

The spectrum of SA system is as follows:

**PLOT OF BELONGINGNESS OF MISSILE**

| BELONGINGNESS | CLASS NO. | PREMISE #2 |
|---|---|---|
| 0.1915 | 1 (China) | \|*** |
| 0.9709 | 2 (N. Korea) | \|***************** |
| 0.5188 | 3 (S. Korea) | \|********* |
| 0.1141 | 4 (Russia) | \|** |

At the time of these analyses the following two events occurred. First, CNN reported that North Korea has just signed a nuclear inspection agreement with the USA. When the program asked the user whether the Class 2 was affected by a new event, the analyst assigned **extremely unlikely** with pessimistic confidence level to North Korea. Second, CIA reported that there has been an increased communication activity among the dissidents of South Korea's military. This time the analyst assigned **most likely** with optimistic confidence level to South Korea. The program aggregated the CNN report with Class 2 and CIA report with Class 3 using conjunction. The new SA spectrum is as follows:

**PLOT OF BELONGINGNESS OF MISSILE**

| BELONGINGNESS | CLASS NO. | PREMISE #4 |
|---|---|---|
| 0.1915 | 1 (China) | |*** |
| 0.5000 | 2 (N. Korea) | |********* |
| 0.6061 | 3 (S. Korea) | |********** |
| 0.1141 | 4 (Russia) | |** |

The modified SA spectrum (Premise #4) is then aggregated with the ATR spectrum (Premise #1) using the pessimistic disjunction operator $S_3(a,b)=\max(a,b)$. The resulting combined spectrum (fuzzy conclusion) is as follows:

**PLOT OF BELONGINGNESS OF MISSILE**

| BELONGINGNESS | CLASS NO. | FUZZY CONCLUSION |
|---|---|---|
| 0.1915 | 1 (China) | |*** |
| 0.5000 | 2 (N. Korea) | |********* |
| 1.0000 | 3 (S. Korea) | |****************** |
| 0.1329 | 4 (Russia) | |** |

**TEST RESULTS**

We used 15 sibling nodes per input nodes. There were four classes and input vectors were ten dimensional. The training rate was $\eta=0.0075$ and the coefficient momentum term was $\alpha=0.1$. On the average after 50 cycles the error was reduced to $10^{-7}$. We used the coefficient initialization technique described in Section 2 with e=0.975. This resulted in a very rapid training period of about a 15 second per class using a 75 MHz Pentium with a 32 bit code. However, because the target values were dynamic, in testing we could not use the constant terms. To compensate for this loss we divided the output by (1-e)=0.025.

**CONCLUSIONS**

The architecture of a fuzzy ANN that is intended for fusion of the outputs of numerical and linguistic pattern recognition systems is described. The details of a learning method that substantially reduces the training times of fuzzy ANNs are also given. By means of an example it is shown that the accuracy of ballistic missile identification may be improved by combining the ATR and AS systems, and using the event representation technique.

**APPENDIX A**

| Confidence | Conjunction | Disjunction |
|---|---|---|
| Pessimistic | $T_1(a,b)=\max(0,(a+b-1))$ | $S_3(a,b)=\max(a,b)$ |
| Average | $T_2(a,b)=a.b$ | $S_2(a,b)=a+b-a.b$ |
| Optimistic | $T_3(a,b)=\min(a,b)$ | $S_1(a,b)=\min(1,(a+b))$ |

where a and b are two fuzzy numbers between zero and one.

**REFERENCES**

Bellman R. E. and Zadeh L. A. (1976), "Local and Fuzzy Logic", Mem. No. ERL-M584, University of California, Berkeley.

Beyth-Marom, (1982), "How Probable is Probable? A Numerical Taxonomy Translation of Verbal Probability

Expressions," Journal of Forcasting, Vol. 1, pp.275-269.

Bonissone, P.P., Decker, K.S., (1986) "Selecting Uncertainty Calculi and Granularity: An Experiment in Trading-off Precision and Complexity," in Uncertainty in Artificial Intelligence, 2217-247, North Holland.

Dunford, N., and Schwartz, J. T. (1966); "Linear Operators," Part 1, Third Printing, Wiley, NY.

Khan E., (1993), "Neural Network Based Algorithms for Rule Evaluation and Defuzzification in Fuzzy Logic Design," Proceedings of WCNN93, Portland.

Pal K.S. and Mitra S. (1992), "Multilayer Perceptron, Fuzzy Sets, and Classification", IEEE Transactions on Neural Networks, Vol. 3, No. 5.

Royden, H.L. (1971); "Real Analysis," p. 111, The Macmillan Company, N.Y.

Schweizer B., Sklar A. (1963), "Associative Functions and Abstract Semi-Groups," Publicatione Mathematicae Debrecen, Vol. 10, pp.69-81.

Spiegel, M.R. (1962); "Theory and Problems of Advanced Calculus," pp. 301-320, McGraw-Hill Book Company, New York, NY.

Ulug, M.E., (1993), "A Single Layer Fuzzy and Fuzzy-Like ANN for Parallel Electro-Optical Implementation," Proceedings of 1993 International Neural Networks Society, WCNN 93, Portland, Oregon.

Ulug, M.E., (1992), "A Rapid Learning Orthonormal Neural Network for Pattern Recognition," Proceedings of 1992 IEEE/RNNS Symposium on Neoro-informatics and Neurocomputing, Rostov-on-Don.

Ulug, (1995), M.E.,"Event Representation in Fuzzy Neural Network with Qualitative Inputs," Proceedings of ICNN95 IEEE International Conference on Neural Networks, Perth, Australia, Dec. 1995.

Ulug, M.E., (1994) ,"A Single Layer Fast Learning Fuzzy Controller/Filter," Proceedings of the IEEE World Congress on Computational Intelligence, June 26-July 2, 1994, Orlando, Florida.

Wylie, Jr., C. R. (1966), "Advanced Engineering Mathematics," McGraw-Hill Book Co., New York.

Widrow, B., and M.E. Hoff (1960); "Adaptive Switching Circuits," 1960 IRE [16] Zadeh L.A. (1965), "Fuzzy Sets", Information and Control, 8, 338-353WESCON Convention Rec., New York: IRE, 96-104.

Zadeh, L. A., (1965), "Fuzzy Sets," Information and Control, 8, 338-353, Westcon Convention Rec., New York: IRE, 96-104.

# Classification of Eddy-Current Signals using Fuzzy-Logic

Hartmut Ewald, Michael Stieper
Hochschule Wismar, 23952 Wismar, Germany

Abstract:
Eddy-current systems are commonly used in automated defect inspection to detect cracks in cars, power and aircraft industry. The 'damage tolerance concept' assumes that defects such as cracks are tolerable until they reach a defined limit. This concept requires very sensitive nondestructive methods for detecting and quantifying these defects (cracks). The complex eddy-current signal from an infinitely long crack can be classified with the help of the fuzzy logic (crack width and crack depth).

## Introduction

Eddy-current systems are commonly used in automated defect inspection to detect cracks in cars, power and aircraft industry, etc. The 'damage tolerance concept' assumes that defects such as cracks are tolerable until they reach a defined limit. This concept requires very sensitive nondestructive methods for detecting and quantifying these defects (cracks). The eddy-current method is a very sensitive method for the detection of surface defects, the surface being the zone of interaction between the eddy-current and the material. The eddy-current crack signal has to enable a judgement to be made with regard to the width and the depth of the cracks. This process of

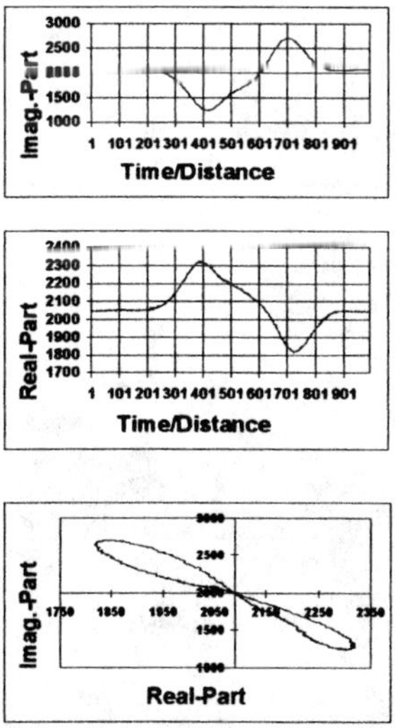


**Figure 1:** Eddy-current 'crack'-signals in the time domain plane and impedance plane

classification in production lines should occur on-line, e.g.: for detecting cracks in the bore holes. The eddy-current signal is complex and the real- and imaginary parts can be captured while the probe is moving over the surface.
Typical eddy-current crack signals, generated by a special differential eddy-current (MDX) probe, are shown in Figure 1 (time domain plane and impedance plane). This probe is characterized by an asymmetric field distribution (Mook, 1994). Each Crack produces two peaks. The distance between the two peaks corresponds to the width of the crack, and the amplitude to the depth.

## Feature extraction

Eddy-current systems have used various methods of classification of this 'error'-signal. In order for the classification process to produce a good result it is necessary to use a fit classification method in conjunction with a powerful method for the feature extraction. In eddy-current systems approaches which have been used for feature extraction include:

- The eddy-current loop (in the complex impedance plane) is developed as a Fourier series, the coefficients of the series being the features (Brow, 1985).
- The eddy-current loop is interpreted as a time function which is Fourier-transformed, the amplitudes and phases being the features (Mensching and Stegemann, 1993).
- From the eddy-current loop simple geometric values are calculated: loop length and loop width.

In another paper the features of the cracks-signals were calculated using the data from the binary 'eddy-current image' (C-Scan) and the data from the original (analog) eddy-current signal (Güttinger and Wezel, 1993).

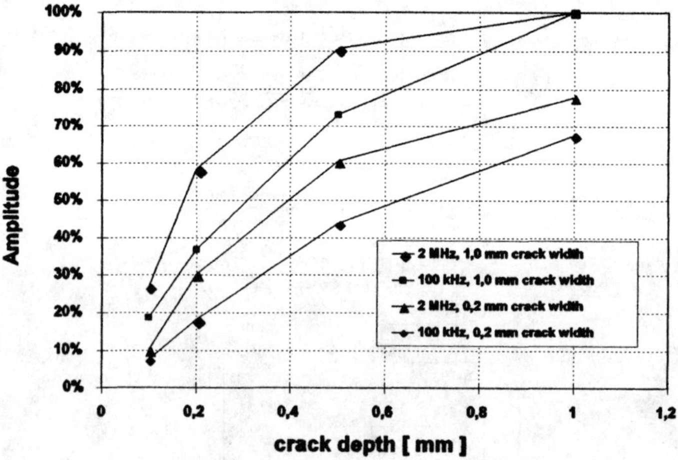


**Figure 2**: Amplitude of eddy-current signal versus crack depth for different frequency

Many well-know methods from the binary image processing of 2-dimensional objects can be used for the feature extraction, such as radius- and line-segment coding.
The amplitude versus the width of the cracks and the depth for two different frequencies are shown in Figure 2 (for every frequency the value is normalized on the highest amplitude). The interpretation of eddy-current signals from the Figure 1 is difficult, because the response of the eddy-current amplitude for most parameters (conductivity, permeability, measurement frequency) is nonlinear If the cracks are deeper the change in amplitude is smaller (skin effect).
In our case the form and the amplitude of the probe signal are dependent on the width and the depth of the cracks . All influences are functions of frequency . Determining the depth of cracks the amplitude of the eddy-current signals is only possible if the width is known. In our case the measured eddy-current signal is interpreted as a complex time function, so that we can use the complex Fast Fourier Transformation for the feature extraction. The first 10 amplitudes of the real and imaginary parts are the relevant features (each feature has two elements: real and an imaginary part).
The following section contains a description of the classification of eddy-current signals, which were produced by artificial and infinitely long cracks, using fuzzy-logic. The investigation was carried out for three groups of different width on nonferrous material (eg. aluminum) and each group had four depths .

## Classification of eddy-current data

In the first steps we analysed the structure of the eddy-current data and the possibility of classification with help from the classical methods of pattern recognition. From the host of classical approaches available in the field of multivariate data analysis the hierarchic clustering was chosen (with various methods such as single linkage, complete linkage etc.) because this technique does not require the measured values to have any special distribution.

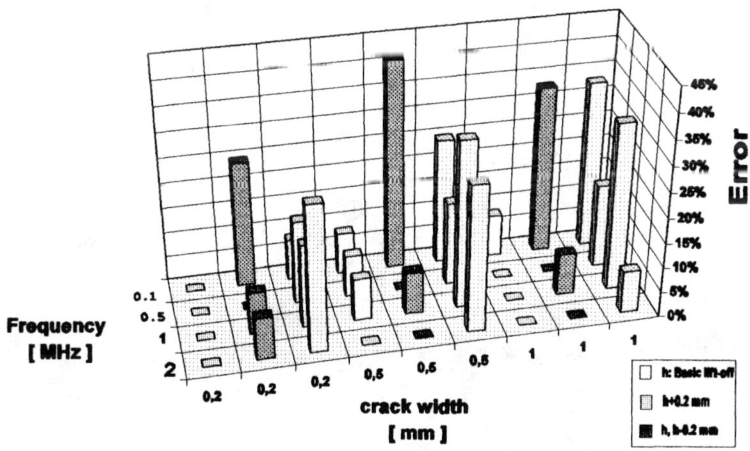


Figure 3: Hierarchical Clustering versus the measurement frequency and the crack width

Furthermore, the estimation of other important influences of the classification result are applicable, such as:

- Measurement frequency and lift-off effect.
- The eddy-current signal area for the Fast Fourier Transformation.
- The number of the features.
- Angle between the crack and the direction the probe is moving.

The results of this investigation were then used for the development of the Fuzzy-sets (rule based fuzzy classificator). In Figure 3 the results of the hierarchic clustering as a function of the measurement frequency and the lift-oft effect are represented .
Only with the constant lift-off of the probe and the high measurement frequency is an error-free assignment available. With the single linkage method we obtained the best results using the classical methods. With 4 -10 features and at a high frequency we obtained a error-free classification. An error of +/- 2,5 % on the starting point of the eddy-current signal has no influence of the result of the Fast Fourier Transformation. With our special MDX probe the assignment of cracks are angle independent after 70 degrees.
If the information in the eddy-current signal is increased by using a second measurement frequency then it is not possible to classify the width using classical methods; but on the other hand, new assignment features become available: we can achieve a classification into two groups with only four features which is independent of the width, i.e.: groups of low depth cracks and deep cracks.

## Fuzzy clustering

In the second stage the classification of eddy-current signals using fuzzy-logic is investigated. The fuzzy-logic is realized by a hardware or a software solution. For both methods of pattern recognition using fuzzy-logic (the fuzzy clustering and the rule-based fuzzy classification system) we have implemented a software solution for different hardware: standard PC and digital signal prozessor board.
The input signals of the classificator are the values from the fast Fourier transformed

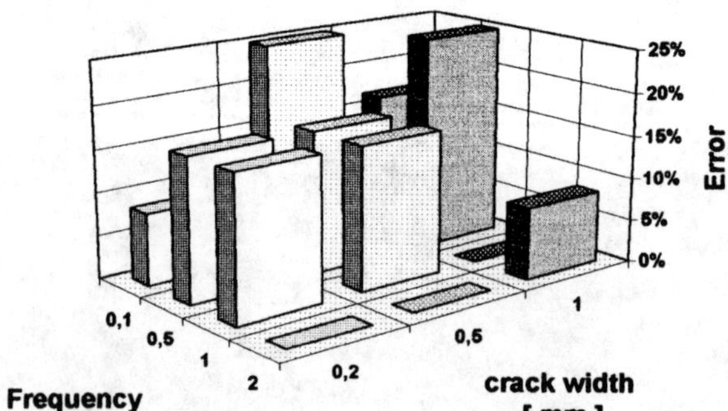


**Figure 4**: Fuzzy-clustering versus frequency for three different groups of cracks

eddy-current signal.

An important characteristic of the fuzzy clustering in comparison the classic distance methods is that object (crack signals) is a member of class k (crack group k) and a member of class m (crack group m). The degree of membership is controlled by a weight-factor w: If w =1 then the assignment is 'sharp'. In our implemented system we use the Fuzzy-C-means algorithm with the Euclid distance. The results of fuzzy clustering of eddy-current signals as a function of the frequency are shown in Figure 4. An eddy-current signal was here assigned to the crack width when the degree of membership was higher than 50%. If the weight-factor $w < 1.5$ then the results from the fuzzy clustering are similar to the assignment from the classical distance methods. If the number of features is decreased down to four then the classification result is stable.

In the rule-based classificator, the first step of fuzzification of the inputs (feature of the objects) occurs. For fuzzyfication we used the trapezoid function. The fuzzy sets are generated automatically from sets of eddy-current data. The distribution of each feature is calculated by a data set of m signals and is approximated by a trapezoid function. (the fuzzy sets may be edited online).

The rule-based inference machine realized the minimum-, the algebraic- and the fuzzy-AND-operator. By the fuzzy-AND-classification the inference mechanism can be controlled by a weight-factor d. In the defuzzyfication method we used the center of gravity. Figure 5 show the fuzzy classification results with the rule-based version.

By using the fuzzy-AND-classificator and with the weight factor d=0.5 all signals can be assigned to the classes with a membership higher then 60%.

Using the rule-based classification system the fuzzy sets may be edited online. In practice, the eddy-current testing system, particularly in production lines, the processing time is important. The processing time reached in the fuzzy-clustering versus the number of objects and the number of clusters for two hardware configurations

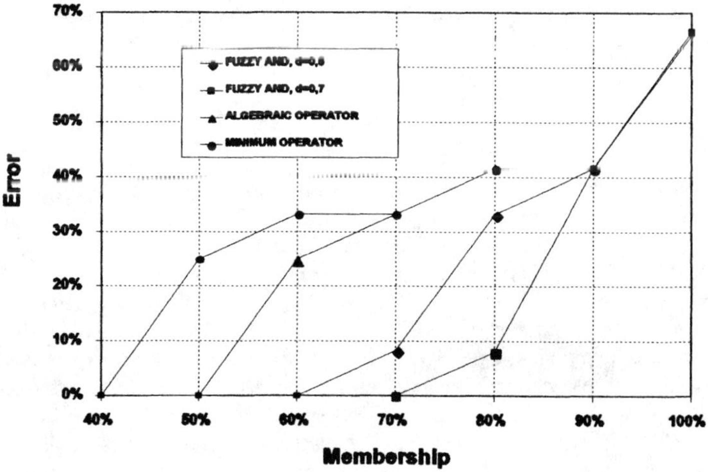


**Figure 5:** Rule-based fuzzy logic classificator results of eddy-current 'crack'-signals

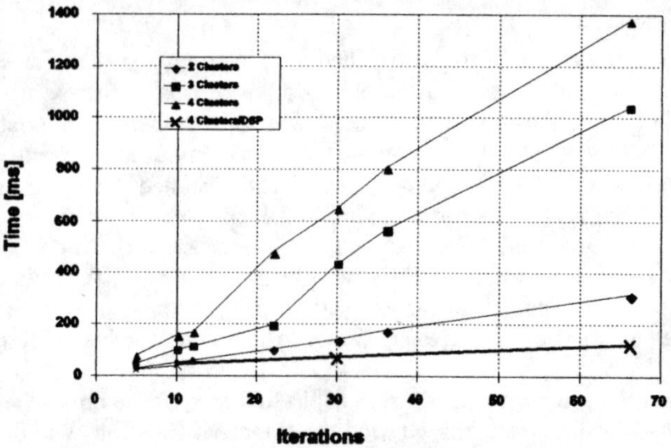


**Figure 6:** Processing time versus Iterations for different number of clusters (groups)

(standard PC/486/DX2-66 and digital signal prozessor board TMS 320C31/40) is shown in Figure 6.
With a rule-based classification system we obtained processing time which was substantially higher (factor 10) by comparing the number of parameters and the number of clusters (for example 4).

## Summary

The extraction of the eddy-current signals' features is the first and the most important step. The amplitudes of the complex Fast Fourier Transformation are used as features. The main components are the input vectors for all the classification methods.
A rule-based fuzzy-logic-classification guaranties better results than a fuzzy-cluster-means-algorithm (FCM) because the classification results can be increased step-by-step online (fuzzy sets and rules).

## References

Bedzdek, J.C., (1981). Pattern Recognition with Fuzzy Objective Function Algorithmus. Plenum Press, New York

Brow, St.D., (1985).Template Matching - An Approach for machine Sorting of Eddy-Current Data. Materials Evaluation (43), Nov, 1553-1661

Güttinger, T., Grotz, K., Wezel, H., (1993). Eddy-Current Imaging. Materials Evaluation (51) , Apr, 444-451

Mensching, L., Stegemann, D., (1993). Klassifikation fouriertransformierter Wirbelstromsignale mit Neuronalen Netzen. DGZfP Annual Conference Proceedings, Mai 1994 Garmisch-Patenkirchen, Germany

Mook, G., (1994). Aspekte der Bildgewinnung und der -verarbeitung in der Wirbelstromprüfung. DGZfP Annual Conference Proceedings, Mai 1994, Timmendorfer Strand, Germany

# COMPARISON OF THE PERFORMANCE OF THE NEURAL NETWORK- AND FUZZY LOGIC-BASED DECISION UNITS FOR A ROBOT SAFETY SYSTEM

**JOZEF ZURADA and ANDREW L. WRIGHT**
*Computer Information Systems, University of Louisville*

*ABSTRACT*:
This paper presents the results from the development and simulation of neural network- and fuzzy logic-based, real-time decision units for a robot safety system implemented at a high level of data processing. The neural network decision unit accepts on input a map with potential collision zones with an intruder, robot's velocity, and its steering angle. It produces one of the following safety decisions: 'move as intended', 'slow down', or 'emergency stop'. The safety decision is based on the computation of the scalar product between two vectors: the virtual repulsive force vector produced by an obstacle and the robot's velocity vector. The fuzzy logic decision unit accepts on input three variables: the magnitude of the virtual repulsive force vector, the magnitude of the robot's velocity vector, and the angle between the two vectors and generates the output robot's velocity, as a percentage of the intended velocity. The performance of both units is compared. The results from computer simulation indicate that the fuzzy logic-based decision unit is more robust than its neural network counterpart, which uses a crisp logic.

## INTRODUCTION

Robots are used widely in the industrial, manufacturing and military environments. They are controlled by computer hardware and software. Faults in robot control are difficult to predict. Moreover, possible aberrant motion of a robot may be dangerous to personnel engaged in its installation, programming, debugging, and maintenance. As a result, appropriate safety measures have to be designed to avoid human accidents or injuries.

Recently, Zurada et al. (1995, 1996) presented a robot safety system based on neural networks. The system contains the neural network detection and decision units implemented at an intermediate and high level of data processing, respectively. The detection unit combines local sensor maps into one integrated map of the environment that contains potential collision zones with an intruder. This map along with the robot's velocity and its steering angle is passed to the neural network decision unit - a classifier. The classifier produces safety decisions which may subsequently be passed to the robot control unit.

Although the decision unit is able to generalize very well and produces 97% of correct safety decisions in real-time of less than 6 ms (measured on the Pentium 90 MHz processor), its operation is based on a crisp logic. The major disadvantage of this crisp approach is a sudden change in the decision. For example, the neural network is trained in such a way that the values of the scalar product $S = 0.69$ and $S = 0.71$ generate safety decisions 'move as intended' and 'slow down', respectively. If this approach is applied to the real world manufacturing environments, it may cause false alarms that could halt or slow down production in situations when this is not warranted. Furthermore, the neural network decision unit does not provide any information as to the recommended speed of a robot, especially when the 'slow down' decision is issued. Also, it produces 3% incorrect (unsafe) decisions, and requires training.

These events prompted the authors to develop and test by simulation a fuzzy logic-based decision unit which should be more robust, will not have to be trained, and will not misclassify. The use of a fuzzy logic controller making robot safety decisions at the highest level of data processing was suggested by Graham (1992). Also, Beom et al. (1992) and Fukuda et al. (1992) used fuzzy logic concepts to determine the robot's velocity and its steering angle.

## THE NEURAL NETWORK-BASED DECISION UNIT

The neural network decision unit is implemented at the high level of data processing. The decision unit accepts the following randomly generated input: a map containing potential collision zones encoded as belief values into grids, the magnitude of a robot's velocity vector, and robot's steering angle. The neural network decision unit produces one of the following safety decisions: 'move as intended', 'slow down', or 'emergency stop'. These decisions are based on computation of the scalar product $S$ of two vectors, i.e., the virtual repulsive force vector $F$ generated by the obstacle and robot's velocity vector $V$. The network is trained by a human expert who heuristically determines thresholds for safety decisions.

The idea of having obstacles conceptually exerting virtual forces onto a mobile robot was first suggested by Khatib (1985). Borenstein et al. (1989) used the combination of certainty grids and virtual force fields (VFF). The VFF algorithm scans a small square window around or in front of the robot. Each cell inside the window contains a numerical value within the range [0,1] expressing support for the hypothesis that there is an intruder in the cell. Also, each cell in the window applies a hypothetical repulsive force $F$ to the robot, pushing the robot away from the cell. In order to compute the resultant virtual repulsive force $F$, the individual repulsive force vectors must be computed and accumulated vectorially. It is assumed that the robot moves with a certain velocity and steering angle toward the target. It is reasonable to increase the strength of the repulsive forces when the robot moves toward an obstacle, and reduce it when the robot moves along, or away from, the obstacle. This can be achieved by computing the scalar product $S$ of two vectors $F$ and $V$, multiplying the magnitudes of these two vectors by the cosine of the angle $\theta$ between the two vectors. The scalar product $S = |F||V|\cos\theta$ is the largest when a robot faces an obstacle ($\theta$=0°), equal to zero when the robot moves along the obstacle ($\theta$=90°), and is negative when the robot moves away from the obstacle ($\theta$>90°). The $S$ is a function of five variables: $S$=f($B,D,OS,V,\theta$), where $B$ is the belief in an obstacle in the sensed area, $D$ is the distance to an obstacle, $OS$ is an obstacle's size, and $V$ and $\theta$

were defined earlier. The scalar product $S$ appears very difficult to handle. As a result, the value of $S$, based on which a safety decision is made, is determined heuristically by a human expert. After studying many cases, the following decision lines were selected for the network's training: $S$=0.7 and $S$=1.4. The reader is referred to Zurada et al. (1996) for details.

The decision unit is implemented and tested by simulation. The network is trained for 1200 training patterns which constitute a training set. During training, the neural network accepts on input a training pattern that consists of a map with robot position and collision zones, the magnitude of the robot's velocity vector $V$ and its steering angle $\psi$. All these values are generated at random. The standard algorithm for training a two-layer feedforward neural network with error back-propagation is used. The neural network contains 124 and 3 neurons in each of the hidden and output layers, respectively.

The performance of the decision unit is tested using 300 test cases. None of them have previously been used for training. They are presented on input one at a time. The neural network decision is determined by the largest of its three outputs. Among 300 test cases, 25 misclassifications occurred. In nine cases, the neural network produced a 'slow down' decision while 'move as intended' was expected, and in five cases it generated an 'emergency stop' decision while 'slow down' was to be produced. In four cases, the neural network produced a 'move as intended' decision while 'slow down' was expected, and in seven cases it produced a 'slow down' decision while 'emergency stop' was to be generated. Thus, it appears that the neural network has the tendency to produce more conservative, that is, safer decisions. When safety is considered, one may assume that those fourteen 'safer' decisions mentioned above are correct decisions. This assumption would actually increase the reliability of the system to 97%. It is also worth noting that in neither case the neural network generated an extreme error, i.e., producing a 'move as intended' decision while 'emergency stop' was expected, or producing an 'emergency stop' while a 'move as intended' decision is justified. In addition, the response time of the neural network decision unit measured on a Pentium 90 MHz microprocessor, which might be used as a robot controller, is 5.6 ms.

## FUZZY LOGIC APPROACH

Fuzzy logic (Zadeh, 1992) is a type of multi-valued logic. Fuzzy set theory uses a concept of membership functions and degree of memberships which are defined as follows. If $X$ denotes a universe of objects, $x$ denotes an individual element from that universe, $A$ is a subset of $X$, and $x \in A$, then $\mu_A(x)$=[0,1] is called the membership function of $A$. The value assigned by the membership function for a given $x$ is called the degree of membership. Membership functions are frequently of a triangular or trapezoidal shape, for simplicity. Fuzzy logic allows the uncertainty to be directly represented in the inference process by assigning each proposition, as well as the actual implication operator, to a real number from within the interval [0,1] to indicate its degree of truth.

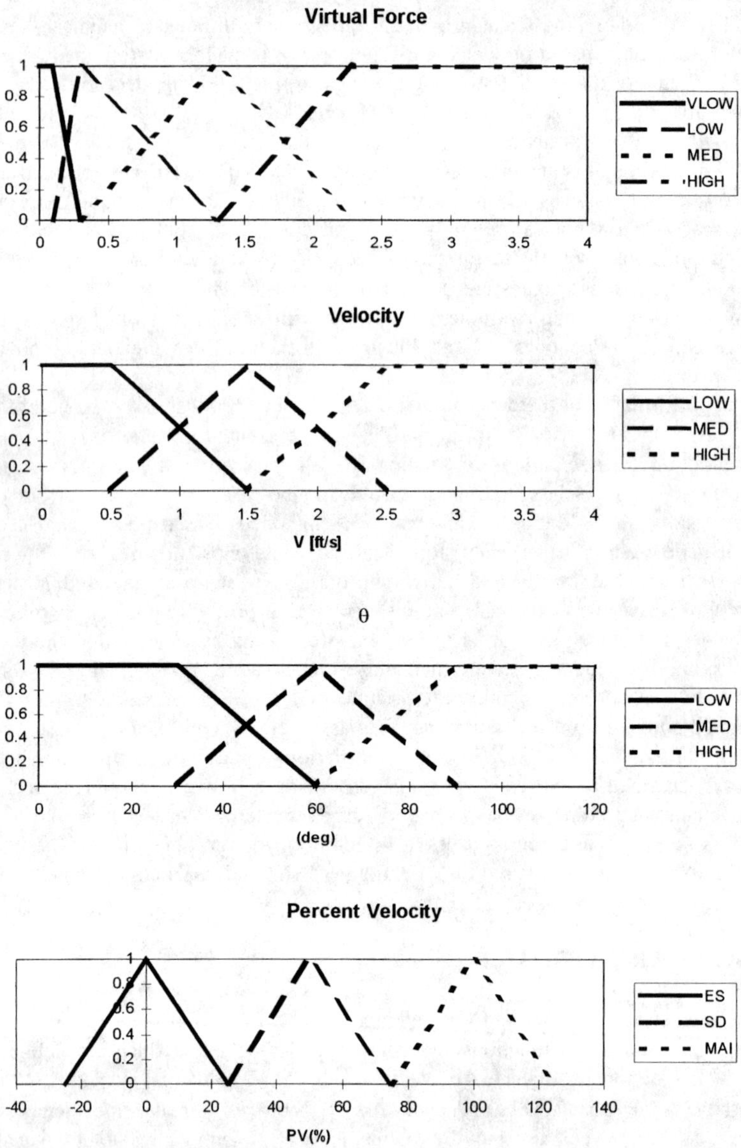


**Figure 1: Membership functions**

The fuzzy logic-based decision unit accepts on input three variables: the magnitude of the virtual repulsive force, $F$, of an obstacle; the magnitude of the robot's intended velocity, $V$; and the relative angle, $\theta$, defined as the difference between the robot's steering angle, $\psi$, and the position of an obstacle, $\alpha$. The decision unit produces one output variable, $PV$, which is a percentage of the input velocity, $V$. The membership function for $PV$ has the following fuzzy categories of safety decisions: 'emergency stop', 'slow down', and 'move as intended' that are mapped into the crisp $PV$ value which may subsequently be passed to the robot control unit. The membership functions for the four variables are shown in Figure 1. These triangular and/or trapezoidal membership functions and the fuzzy rules were constructed based on human/expert reasoning using the Windows-based software TILShell 3.0.1 introduced by Togai Infralogic. It is interesting to note that during testing, it was decided that more membership functions were required for the virtual force than the other variables because more information was embedded in this variable (i.e., belief, distance, obstacle size). The current system contains 24 fuzzy rules. By directly utilizing the calculated virtual force, the number of input variables and, therefore, the number of rules have been decreased significantly. For defuzzification, the center of gravity method is used. To obtain the crisp response $PV = 0\%$ and $PV = 100\%$, the membership function for $PV$ crosses 0 at $PV = -25$ and $PV = 125$, respectively.

## COMPARISON

The original neural network was trained and 300 simulated test cases were run. The test cases were equally divided between the three possible safety decisions, 100 of each. The neural network achieved 97% accuracy compared to the responses of the teacher after training. Unfortunately, some of the erroneous responses given by the neural network were less conservative than those of the teacher, indicating possibly unsafe decisions. In addition, the thresholds used to determine the safety categories are crisp, resulting in abrupt changes in the decisions. The fuzzy logic approach gives as output a percentage of intended velocity, allowing for a more robust response. In order to compare these more meaningful responses of the fuzzy system to those of the neural network teacher, the output axis was divided into 5 categories, with 50% in the center ('Slow Down'). The fuzzy responses were considered in general agreement with the teacher if they were within one category on either side of the teacher's response. This approach allows the percent output of the fuzzy system to be more fairly compared to the teacher's response. The fuzzy system was in general agreement in 294 of the 300 cases, although the actual outputs were distributed over the interval [0, 100]. In the six cases where the fuzzy system was not in agreement with the teacher, it made a more conservative response than the teacher indicated. At no time did the fuzzy rule base provide a less safe decision. See Table 1 for a listing of some representative test cases involving the teacher, the trained neural network, and the fuzzy system. Note, that the fuzzy system outputs are more robust than those of the other methods.

Table 1: Comparison of the performance of the neural network and fuzzy logic decision units.

| *F* | *V* | *θ* | Teacher Decision | Neural Net. Decision | Fuzzy Sys. *PV* |
|---|---|---|---|---|---|
| 0.24 | 0.3 | 16.9 | MAI | MAI | 100.0 |
| 0.37 | 1.2 | 18.9 | MAI | MAI | 90.8 |
| 0.52 | 2.8 | 71.0 | MAI | SD | 84.7 |
| 0.68 | 0.4 | 47.9 | MAI | MAI | 72.5 |
| 0.38 | 2.8 | 45.0 | SD | SD | 64.6 |
| 0.86 | 1.1 | 39.1 | SD | MAI | 55.8 |
| 0.63 | 2.2 | 35.9 | SD | SD | 42.5 |
| 1.20 | 1.2 | 21.2 | SD | SD | 29.2 |
| 1.69 | 1.0 | 20.2 | ES | ES | 14.0 |
| 2.31 | 2.9 | 29.9 | ES | ES | 0.0 |

## CONCLUSIONS

This paper presented some initial results from the simulation of the fuzzy logic-based decision unit for robot safety. These results were compared with the results generated by the neural network-based decision unit proposed by Zurada et al. (1996). The results produced by the fuzzy logic-based system are more robust, never produce unsafe decisions, and are generated without training. The fact that the output from the fuzzy system is a percentage rather than a simple safety decision is a significant improvement over the neural network approach. However, a more fair comparison to the neural network could be accomplished if it were retrained to remove its crisp decision thresholds. Future work will consider this and the refinement of the fuzzy decision unit by adding additional membership functions and rules.

## REFERENCES


Beom, H.R. and Cho, H.S., 1992, A sensor-based obstacle avoidance controller for a mobile robot using fuzzy logic and neural network. In *Proceedings of IEEE/RSJ International Conference on Intelligent Robots and Systems*, **2**, pp. 1470-1475.

Borenstein, J., and Koren, Y., 1989, Real-time Obstacle Avoidance for Fast Mobile Robots, *IEEE Transactions on Systems, Man and Cybernetics*, **19**(5), 1179-1187.

Fukuda, T. and Shibata, T., 1992, Hierarchical intelligent control for robotic motion by using fuzzy, artificial intelligence, and neural network. In *Proceedings of IEEE International Conference on Neural Networks*, **1**, pp. 269-274.

Graham, J.H., 1992, A fuzzy logic approach for safety and collision avoidance in robotics systems. In *Ergonomics of Hybrid Automated Systems III* by P. Broedner and W. Karwowski (eds.), (Elsevier Science Publishers B.V.).

Khatib, O., 1985, Real-time Obstacle Avoidance for Manipulators and Mobile Robots. In *Proceedings of the IEEE International Conference on Robotics and Automation*, St. Louis, MO, March 25-28, pp. 500-505.

Zadeh, L.A., 1992, Knowledge representation in fuzzy logic. In *An Introduction to Fuzzy Logic Applications in Intelligent Systems* by R.R. Yager and L.A. Zadeh (Eds.) (Kluver Academic Publishers), 1-26.

Zurada, J. and Graham, J.H., 1995, "A neural network-based robot safety system". In *Proceedings of the IEEE International Conference on Man, Systems, and Cybernetics*, October, pp. 2922-2927.

Zurada, J. and Graham, J.H., 1996, Intelligent neural network-based decision unit for robot safety, *Journal of Intelligent & Fuzzy Systems*, (in press).

# APPLICATION OF FUZZY NEURAL HYBRID NETWORKS TO STATE PREDICTION

Faridoon Shabani Nadipuram R. Prasad Howard A. Smolleck
The Klipsch School of Electrical and Computer Engineering
New Mexico State University
Las Cruces, NM 88003

**ABSTRACT**
In this paper, a fuzzy neural state predictor is presented. The fuzzy neural predictor uses the past estimated states and measurement vectors as entrance parameters, and it infers the next estimated states. The design of the fuzzy neural predictor is described. The results show how the fuzzy neural predictor obtains an accurate predict of the state. An example is presented to illustrate the performance and applicability of fuzzy neural predictor.

## INTRODUCTION

The task of determining the actual values of the state variables using past history is called **state prediction**.

The stability of a power system is an important task in its planning and its operation. This stability may change under different operating conditions and for different contingencies, i.e., faults or disturbances (perturbation) that may occur on the power system. Since disturbances are a fast phenomena, the distortion of the complex voltage must be corrected quickly. In order to accomplish this, the complex voltage (state) must be predicted ahead of time to take the appropriate control action and maintain the stability. Many of the inherent problems associated with automatic generation control (AGC) could be solved with an accurate state prediction algorithm [1].

Intensive research efforts have been placed on power system state prediction [2-5]. The standard solution toward state prediction is to use a Kalman predictor or one of its many variants [6]. This is optimal in the sense that the expected squared error in the estimated vector $\hat{\mathbf{x}}_k$ is zero. However, the system perturbation and observation noise variables must be from a zero mean, gaussian, white noise distribution and the observer must supply the variables of the system perturbation and the observation noise.

The robustness of the method has made it very popular in the control literature, but it requires the system and observation noise variances to be known in advance. One cannot obtain these values simply by looking at past observations, since the sequence $\mathbf{Z}_k$ is the result of both the system perturbation and the observation noise [7, 8].

To summarize, if the variances of the system and observation noise are available, the Kalman predictor is an attractive estimation techniques. However, for situations where these variables are not available, we present a heuristic predictor that works even in the absence of knowledge about system and observation noise.

In this paper, we present a new approach by using a fuzzy neural technique for state prediction in power systems. The approach is illustrated for a sample system.

## DESIGN OF THE FUZZY NEURAL STATE PREDICTOR

The basis for our approach is a variant of the Kalman update state estimation [9], given by the recurrence relation

$$\hat{\mathbf{x}}_k = \alpha\hat{\mathbf{x}}_{k-1} + (1-\alpha)\mathbf{z}_{k-1} \qquad k = 1, 2, \ldots, m \qquad (1)$$

where

$\mathbf{z}_{k-1}$ is the $(k-1)\times 1$ past measurement vector,

$\hat{\mathbf{x}}_{k-1}$ is the $(k-1)\times 1$ past estimated state vector,

$\hat{\mathbf{x}}_k$ is the $k\times 1$ future estimated state vector, and

the value of $\alpha$ can be determined by a knowledge of the system and error variances.

In Equation 1, "+" represents the union of two input variables $\alpha\,\hat{\mathbf{x}}_{k-1}$ and $(1-\alpha)\,\mathbf{z}_{k-1}$ and $\hat{\mathbf{x}}_k$ is the output variable. This is in the MISO fuzzy neural system form as show in Figure 1.

## SIMULATION RESULTS

The authors simulated the fuzzy neural predictor for the 6-bus network given in Figure 2 using the "C" language on an IBM PS/2 system.

We have two inputs $X_1$ and $X_2$ and one output y, where $\alpha$, $\hat{\mathbf{x}}_{k-1}$ and $\mathbf{z}_{k-1}$ are given in Table 1 and Figure 3. The I/O relation for 6-bus network (magnitude and angle) is shown in Figures 4 and 5.

Figure 3 illustrates fuzzy sets for the error and $\alpha$. The shapes and ranges of the fuzzy labels **low**, **medium** and **high** are specified according to the standard assumption that they are all trapezoidal(or triangular)-shaped, normal fuzzy sets. The value of $\alpha$ is in the range [0,1] and the error is expressed in per-unit, where the per-unit value is defined as $|$ error/base value $|$ scaled to fit the response range [0,1]. The absolute value is adequate since there is no reason, in general, to suppose that positive and negative errors should induce different behavior for $\alpha$.

### Creating the fuzzy model

Figure 6 shows the fuzzy rules created for the I/O relationships. Three fuzzy numbers are defined for $X_1$ and $X_2$ with term set (Low, Medium, High). The triangular membership functions are defined for each fuzzy number. The combinations of input variables $X_1$ and $X_2$ are divided into eighteen areas. These rules give two rough values for output variable y (True, False).

**Fuzzy neural network prewiring and preweighting**

Figure 7 shows a structured neural network prewired with fuzzy rules. In this diagram, each circle represents a neuron. This neural network corresponds to a fuzzy model as follows:

. The output of the neurons in second layer correspond to antecedent propositions.

. The output of the neurons in third layer correspond to the fuzzy rules.

. The output of the neurons in fourth layer correspond to the consequent proposition.

. The output of the neurons in fifth layer provide defuzzification strategies.

In the fuzzy reasoning operation, the rules were assigned a weight of 0.5 that indicates a rule as having the same probability of being true as false.

**Fuzzy neural network after learning**

As teacher data set, I/O samples data for 6-bus are generated in Tables 3 and 4.

Tables 5 and 6 summarizes the changes in the fuzzy rules occurring before and after learning.

At the time of wiring, the link weights on the fourth layer were set to 0.5. After training, rule optimization can be done by deleting unimportant rules as indicated by small weights. Tables 7 and 8 shows the rules after optimization.

Finally, the prediction value can be calculated by adding the estimated value and measured value. The output (defuzzification) is calculated by the centroid method. Table 9 shows 6-bus prediction results after 500 iterations. From the results obtained it may seen that the desired and predicted values are the same.

Figure 8 shows the learning rate through the training process. Each time the error decreased the learning rate was increased. Conversely, whenever the error increased by more than the old-to-new ratio the learning rate was decreased.

Figure 9 shows the sum squared error. Note that while the error curve has a few small bumps, all increases in error are extremely small. This is so because new weights that caused an increase in error greater than the old-to-new error ratio were rejected.

## CONCLUDING REMARKS

The following critical points can be derived from the above results:

1. Using all the possible generated rules (a total of 18) the fuzzy neural network produced the desired prediction values.
2. The computed output values of fuzzy neural network (or fuzzy system) using a lesser number of rules (a total of 7) produced similar results. This is extremely advantageous as it will take less time (as well as less memory) to compute the output.
3. Further reduction in number of rules is possible but with noticeable errors.

## REFERENCES


[1] Erkmen, I., and Adanir, T., A short-term load prediction algorithm for dynamic economic dispatch, Proceeding of the 7th Mediterranean Electrotecnical, pp. 885, 1994.

[2] Mcclelland, E. C., and Van Horne, P. R., Fast voltage prediction using a knowledge based approach, *IEEE Transactions on Power Apparatus and Systems*, Vol. PAS-102, No. 2, pp. 315-318, February 1983.

[3] Schwartzenberg, J. W., Nwankpa, C. O., and Sundaram, A., Prediction of distribution system disturbances, IEEE Annual Power Electronic Specialists Conference, Vol. 2, pp. 1077-1082, 1994.

[4] Wvong, M. D., Prediction of power system stability by catastrophe theory, IEEE International Symposium on Circuits and Systems, Vol. 4, pp. 974-977, 1991.

[5] Yang, C., and Hsu, Y., Estimation of line flow and bus voltages using decision trees, IEEE Transactions on Power Systems, Vol. 9, No. 3, pp. 1569-1574, August 1994.

[6] Simutis, R., Havlik, I., and Lubbert, A., A new approach to state estimation and prediction exemplified by alcohol formation in beer brewing, *Journal of Biotechnology*, Vol. 24, pp. 211-234, 1992.

[7] Sorenson, H. W., Least-squares estimation: from Gauss to Kalman, *IEEE Spectrum*, vol. 7, pp.63-68, July 1970.

[8] Wallace, J. N., and Clarke, R., The application of Kalman filtering estimation techniques in power station control systems, *IEEE Trans. Automat. Contr.*, Vol. AC-28, pp. 416-427, Mar. 1983.

[9] F. L. Lewis, *Optimal Estimation*, John Wiley: New York, 1986.

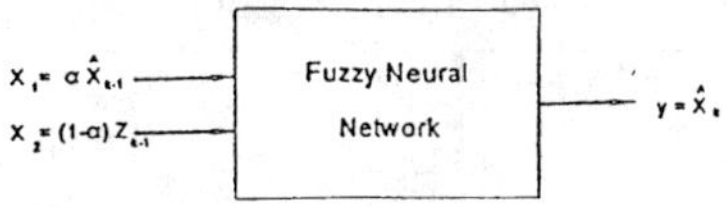


Figure 1. Block Scheme of the Equation 1.

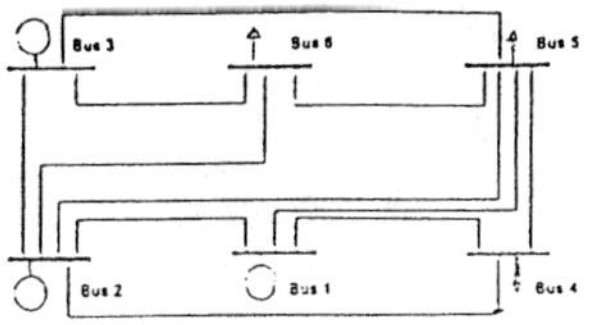


Figure 2. 6-Bus Network.

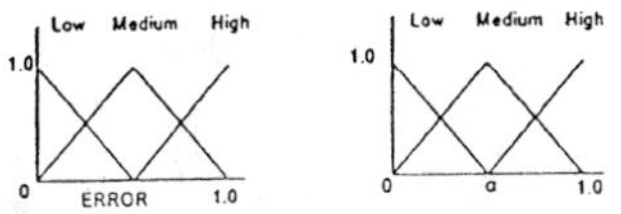


Figure 3. Membership Function for Error and α.

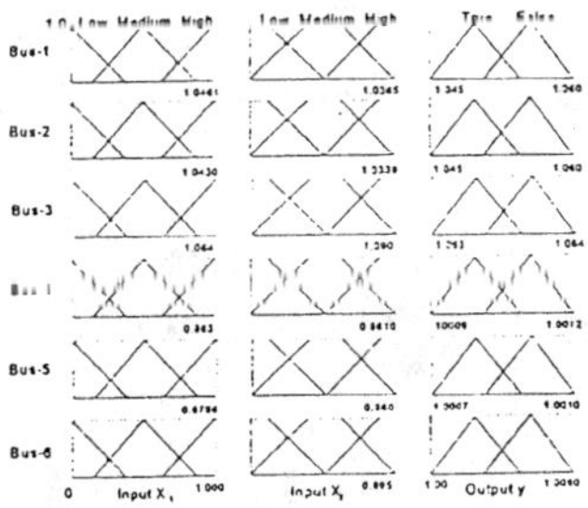

Figure 4. 6-bus Input-output Variables (Magnitudes).

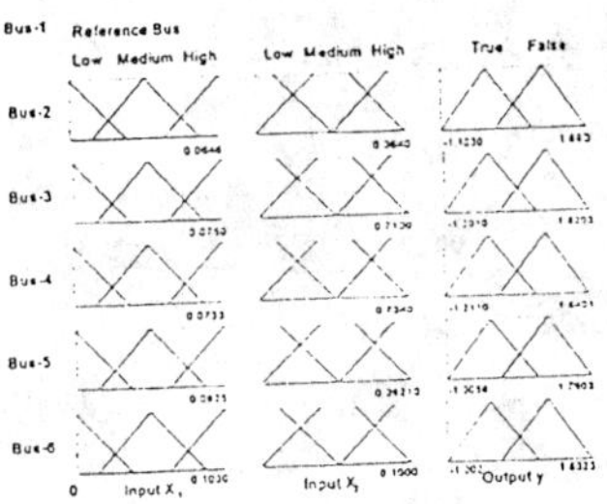

Figure 5. 6-bus Input-output Variables (Angles).

Rule 1: if ($x_1$ is low) and ($x_2$ is low) then y is true.
Rule 2: if ($x_1$ is low) and ($x_2$ is low) then y is false.
Rule 3: if ($x_1$ is low) and ($x_2$ is medium) then y is true.
Rule 4: if ($x_1$ is low) and ($x_2$ is medium) then y is false.
Rule 5: if ($x_1$ is low) and ($x_2$ is high) then y is true.
Rule 6: if ($x_1$ is low) and ($x_2$ is high) then y is false.
Rule 7: if ($x_1$ is medium) and ($x_2$ is low) then y is true.
Rule 8: if ($x_1$ is medium) and ($x_2$ is low) then y is false.
Rule 9: if ($x_1$ is medium) and ($x_2$ is medium) then y is true.
Rule 10: if ($x_1$ is medium) and ($x_2$ is medium) then y is false.
Rule 11: if ($x_1$ is medium) and ($x_2$ is high) then y is true.
Rule 12: if ($x_1$ is medium) and ($x_2$ is high) then y is false.
Rule 13: if ($x_1$ is high) and ($x_2$ is low) then y is true.
Rule 14: if ($x_1$ is high) and ($x_2$ is low) then y is false.
Rule 15: if ($x_1$ is high) and ($x_2$ is medium) then y is true.
Rule 16: if ($x_1$ is high) and ($x_2$ is medium) then y is false.
Rule 17: if ($x_1$ is high) and ($x_2$ is high) then y is true.
Rule 18: if ($x_1$ is high) and ($x_2$ is high) then y is false.

Figure 6. Fuzzy Rules: 6-bus network (Magnitudes and Angles).

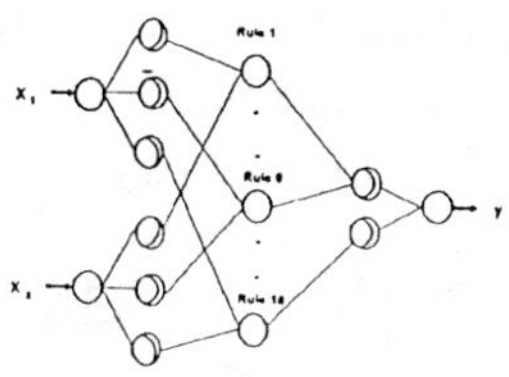

Figure 7. Neural Network Prewired with Fuzzy Rules.

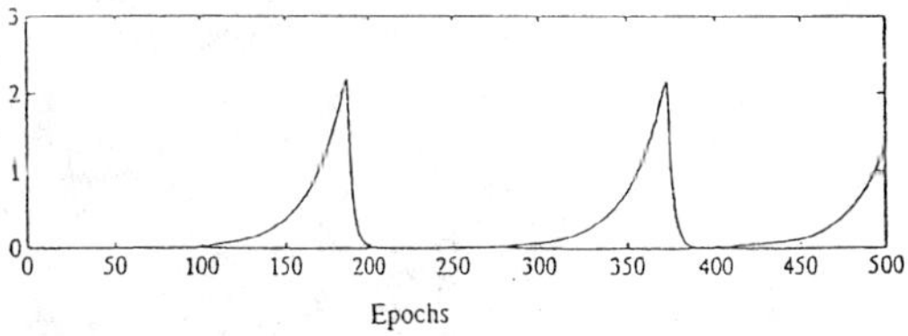


Figure 8. Fuzzy Neural Network Learning Rate.

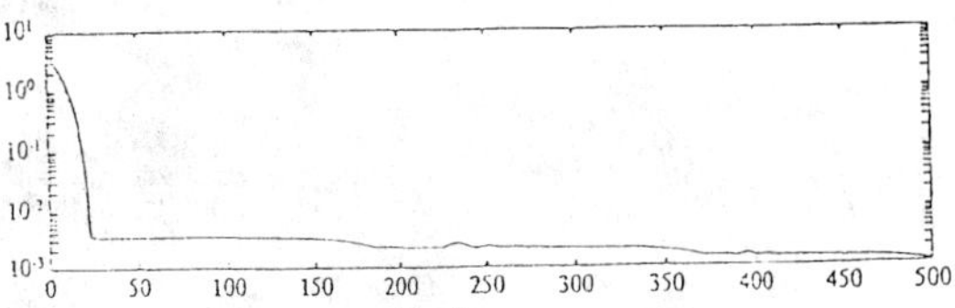


Figure 9. Fuzzy Neural Network Error.

Table 1. 6-bus Voltage Magnitudes and Angles.

| | Past Estimated Voltage | | Measurement Voltage | |
|---|---|---|---|---|
| Bus | Magnitude | Angle | Magnitude | Angle |
| 1 | 1.0461 | 0.0000 | 1.0365 | 0.0000 |
| 2 | 1.0430 | -0.0647 | 1.0339 | -0.0640 |
| 3 | 1.0640 | -0.0752 | 1.0900 | -0.0710 |
| 4 | 0.9830 | -0.0733 | 0.9813 | -0.0734 |
| 5 | 0.9796 | -0.0927 | 0.9791 | -0.0921 |
| 6 | 1.0000 | -0.1032 | 0.9952 | -0.1000 |

Table 3. Fuzzy-Neural-Samples Data: 6-bus network (Magnitudes).

| Samples | $X_1$ | $X_2$ | Desired Output (Magnitudes) Bus-1 | Bus-2 | Bus-3 | Bus-4 | Bus-5 | Bus-6 |
|---|---|---|---|---|---|---|---|---|
| Sample-1 | 0.0 | 0.0 | 1.05 | 1.05 | 1.07 | 1.00 | 1.00 | 1.00 |
| Sample-2 | 0.2 | 0.7 | 1.05 | 1.05 | 1.07 | 1.00 | 1.00 | 1.00 |
| Sample-3 | 0.2 | 0.9 | 1.05 | 1.05 | 1.07 | 1.00 | 1.00 | 1.00 |
| Sample-4 | 0.7 | 0.2 | 1.05 | 1.05 | 1.07 | 1.00 | 1.00 | 1.00 |
| Sample-5 | 0.7 | 0.7 | 1.05 | 1.05 | 1.07 | 1.00 | 1.00 | 1.00 |
| Sample-6 | o.7 | 0.9 | 1.05 | 1.05 | 1.07 | 1.00 | 1.00 | 1.00 |
| Sample-7 | 0.9 | 0.2 | 1.05 | 1.05 | 1.07 | 1.00 | 1.00 | 1.00 |
| Sample-8 | 0.9 | 0.7 | 1.05 | 1.05 | 1.07 | 1.00 | 1.00 | 1.00 |
| Sample-9 | 1.0 | 1.0 | 1.05 | 1.05 | 1.07 | 1.00 | 1.00 | 1.00 |

Table 4. Fuzzy-Neural-Samples Data: 6-bus network (Angles).

| Samples | $X_1$ | $X_2$ | Desired Output (Angles) Bus-1 | Bus-2 | Bus-3 | Bus-4 | Bus-5 | Bus-6 |
|---|---|---|---|---|---|---|---|---|
| Sample-1 | 0.00 | 0.00 | -0.0000 | -0.0646 | -0.0750 | -0.0733 | -0.0925 | -0.1030 |
| Sample-2 | 0.02 | 0.07 | -0.0000 | -0.0646 | -0.0750 | -0.0733 | -0.0925 | -0.1030 |
| Sample-3 | 0.02 | 0.09 | -0.0000 | -0.0646 | -0.0750 | -0.0733 | -0.0925 | -0.1030 |
| Sample-4 | 0.07 | 0.02 | -0.0000 | -0.0646 | -0.0750 | -0.0733 | -0.0925 | -0.1030 |
| Sample-5 | 0.07 | 0.07 | -0.0000 | -0.0646 | -0.0750 | -0.0733 | -0.0925 | -0.1030 |
| Sample-6 | 0.07 | 0.09 | -0.0000 | -0.0646 | -0.0750 | -0.0733 | -0.0925 | -0.1030 |
| Sample-7 | 0.09 | 0.02 | -0.0000 | -0.0646 | -0.0750 | -0.0733 | -0.0925 | -0.1030 |
| Sample-8 | 0.09 | 0.07 | -0.0000 | -0.0646 | -0.0750 | -0.0733 | -0.0925 | -0.1030 |
| Sample-9 | 0.10 | 0.10 | -0.0000 | -0.0646 | -0.0750 | -0.0733 | -0.0925 | -0.1030 |

Table 5. Changes in the Weights of Rules (Magnitudes).

| Rules | Weighs Before Learning Bus 1-6 | Weights After Learning Bus-1 | Bus-2 | Bus-3 | Bus-4 | Bus-5 | Bus-6 |
|---|---|---|---|---|---|---|---|
| Rule 1 | 0.5 | 0.92 | 1.00 | 1.00 | 1.00 | 1.00 | 0.98 |
| Rule 2 | 0.5 | 0.02 | 0.06 | 0.02 | 0.00 | 0.03 | 0.00 |
| Rule 3 | 0.5 | 1.00 | 0.92 | 0.98 | 0.98 | 1.00 | 1.00 |
| Rule 4 | 0.5 | 0.00 | 0.02 | 0.02 | 0.00 | 0.02 | 0.00 |
| Rule 5 | 0.5 | 1.00 | 1.00 | 1.00 | 0.98 | 0.94 | 1.00 |
| Rule 6 | 0.5 | 0.03 | 0.00 | 0.00 | 0.00 | 0.00 | 0.02 |
| Rule 7 | 0.5 | 0.87 | 0.70 | 0.66 | 0.93 | 0.95 | 0.52 |
| Rule 8 | 0.5 | 0.02 | 0.03 | 0.00 | 0.00 | 0.00 | 0.00 |
| Rule 9 | 0.5 | 0.50 | 1.00 | 1.00 | 0.50 | 0.52 | 1.00 |
| Rule 10 | 0.5 | 0.00 | 0.02 | 0.03 | 0.03 | 0.01 | 0.00 |
| Rule 11 | 0.5 | 1.00 | 1.00 | 1.00 | 1.00 | 1.00 | 1.00 |
| Rule 12 | 0.5 | 0.00 | 0.03 | 0.00 | 0.02 | 0.00 | 0.02 |
| Rule 13 | 0.5 | 0.50 | 0.50 | 0.50 | 0.15 | 0.50 | 0.00 |
| Rule 14 | 0.5 | 0.05 | 0.04 | 0.10 | 0.01 | 0.00 | 0.02 |
| Rule 15 | 0.5 | 0.50 | 0.50 | 0.50 | 0.15 | 0.50 | 0.00 |
| Rule 16 | 0.5 | 0.02 | 0.04 | 0.01 | 0.30 | 0.22 | 0.02 |
| Rule 17 | 0.5 | 1.00 | 1.00 | 0.50 | 1.00 | 1.00 | 1.00 |
| Rule 18 | 0.5 | 0.00 | 0.00 | 0.00 | 0.00 | 0.02 | 0.00 |

Table 6. Changes in the Weights of Rules (Angles).

| Rules | Weights Before Learning Bus 1-6 | Weights After Learning Bus-1 | Bus-2 | Bus-3 | Bus-4 | Bus-5 | Bus-6 |
|---|---|---|---|---|---|---|---|
| Rule 1 | 0.5 | 0.00 | 0.97 | 1.00 | 0.98 | 1.00 | 1.00 |
| Rule 2 | 0.5 | 0.00 | 0.04 | 0.02 | 0.03 | 0.00 | 0.00 |
| Rule 3 | 0.5 | 0.00 | 0.98 | 1.00 | 0.98 | 1.00 | 0.97 |
| Rule 4 | 0.5 | 0.00 | 0.00 | 0.04 | 0.01 | 0.00 | 0.00 |
| Rule 5 | 0.5 | 0.00 | 1.00 | 1.00 | 1.00 | 0.96 | 1.00 |
| Rule 6 | 0.5 | 0.00 | 0.03 | 0.00 | 0.06 | 0.02 | 0.00 |
| Rule 7 | 0.5 | 0.00 | 0.90 | 0.86 | 1.00 | 0.97 | 0.83 |
| Rule 8 | 0.5 | 0.00 | 0.00 | 0.02 | 0.05 | 0.06 | 0.00 |
| Rule 9 | 0.5 | 0.00 | 1.00 | 1.00 | 0.95 | 1.00 | 0.56 |
| Rule 10 | 0.5 | 0.00 | 0.01 | 0.05 | 0.00 | 0.00 | 0.00 |
| Rule 11 | 0.5 | 0.00 | 0.95 | 1.00 | 0.98 | 1.00 | 1.00 |
| Rule 12 | 0.5 | 0.00 | 0.04 | 0.00 | 0.01 | 0.01 | 0.00 |
| Rule 13 | 0.5 | 0.00 | 0.01 | 0.03 | 0.50 | 0.08 | 0.50 |
| Rule 14 | 0.5 | 0.00 | 0.00 | 0.02 | 0.06 | 0.02 | 0.02 |
| Rule 15 | 0.5 | 0.00 | 0.50 | 0.40 | 0.15 | 0.50 | 0.50 |
| Rule 16 | 0.5 | 0.00 | 0.00 | 0.04 | 0.03 | 0.09 | 0.00 |
| Rule 17 | 0.5 | 0.00 | 0.98 | 0.65 | 1.00 | 1.00 | 0.50 |
| Rule 18 | 0.5 | 0.00 | 0.03 | 0.05 | 0.02 | 0.00 | 0.50 |

Table 7. Rules Optimization after Training (Magnitudes).

| Rules | Weights After Learning Bus-1 | Bus-2 | Bus-3 | Bus-4 | Bus-5 | Bus-6 |
|---|---|---|---|---|---|---|
| Rule 1 | 0.92 | 1.00 | 1.00 | 1.00 | 1.00 | 0.98 |
| Rule 3 | 1.00 | 0.92 | 0.98 | 0.98 | 1.00 | 1.00 |
| Rule 5 | 1.00 | 1.00 | 1.00 | 0.98 | 0.94 | 1.00 |
| Rule 7 | 0.87 | 0.70 | 0.66 | 0.93 | 0.95 | 0.52 |
| Rule 9 | 0.50 | 1.00 | 1.00 | 0.50 | 0.52 | 1.00 |
| Rule 11 | 1.00 | 1.00 | 1.00 | 1.00 | 1.00 | 1.00 |
| Rule 17 | 1.00 | 1.00 | 0.50 | 1.00 | 1.00 | 1.00 |

Table 8. Rules Optimization after Training (Angles).

| Rules | Weights After Learning Bus-1 | Bus-2 | Bus-3 | Bus-4 | Bus-5 | Bus-6 |
|---|---|---|---|---|---|---|
| Rule 1 | 0.00 | 0.97 | 1.00 | 0.98 | 1.00 | 1.00 |
| Rule 3 | 0.00 | 0.98 | 1.00 | 0.98 | 1.00 | 0.97 |
| Rule 5 | 0.00 | 1.00 | 1.00 | 1.00 | 0.96 | 1.00 |
| Rule 7 | 0.00 | 0.90 | 0.86 | 1.00 | 0.97 | 0.83 |
| Rule 9 | 0.00 | 1.00 | 1.00 | 0.95 | 1.00 | 0.56 |
| Rule 11 | 0.00 | 0.95 | 1.00 | 0.98 | 1.00 | 1.00 |
| Rule 17 | 0.00 | 0.98 | 0.65 | 1.00 | 1.00 | 0.50 |

Table 9. Desired and Predicted Voltage: 6-bus network.

| | Desired Voltage | | Predicted Voltage | |
|---|---|---|---|---|
| Bus | Magnitude | Angle | Magnitude | Angle |
| 1 | 1.0500 | 0.0000 | 1.0500 | 0.0000 |
| 2 | 1.0500 | -0.0646 | 1.0500 | -0.0646 |
| 3 | 1.0700 | -0.0750 | 1.0700 | -0.0750 |
| 4 | 1.0000 | -0.0733 | 1.0000 | -0.0733 |
| 5 | 1.0000 | -0.0925 | 1.0000 | -0.0925 |
| 6 | 1.0000 | -0.1030 | 1.0000 | -0.1030 |

# Design and Application of Fuzzy-Logic-based Gray Prediction Controller

Chin-Ming Hong and Chih-Tang Chiang
*Graduate Institute of Industrial Education*
*National Taiwan Normal University*
*Taipei, Taiwan, R.O.C.*

Ching-Tsan Chiang
*Department of Electrical Engineering*
*Chien Hsin College of Technology and Commerce*
*Chung-Li, Taiwan, R.O.C.*

***ABSTRACT:***

This paper proposes a newly-designed strategy of gray prediction control based on fuzzy logic. The conception of the proposed control algorithm is that the prediction step size of gray predictor in the Gray Prediction Controller (GPC), which can be adjusted dynamically by fuzzy inference mechanism during different system response periods in order to produce superior control effect. The proposed controller is called Fuzzy-Logic-based Gray Prediction Controller, (FLGPC). The simulation results show the superiority of the proposed FLGPC in many cases. In addition, this paper applies the FLGPC to high-speed response systems, such as DC servo motor speed control, in order to verify its practicability.

## 1. Introduction

The Gray System theory [1-2], was proposed by Deng Julong, and can deal with poor, incomplete, or uncertain problems in the systems. Now, the Gray System theory has been successfully applied to many fields in China, such as history, economics, agriculture, medicine, geography, seismology, industry, and industrial control, etc.. In 1986, Cheng Biao applied the Gray Prediction Controller (GPC) successfully to control an industrial process and obtained a good control effect [3]. From that moment, the applications of GPC in industrial control fields grew rapidly [4-7]. The conception of GPC is that a gray model (GM) predictor in GPC is used to predict the future system response and then the decision-making mechanism adopts this prediction information to generate the control strategy. Thus, the system behavior can be controlled before going wrong, and the gray prediction control is called "prior control".

From the simulation and experimental results, it can be found that the GPC has good performance only when tuning the prediction step size properly. Hence, when the plant is changed, the prediction step size in GPC must be adjusted, too. For this reason, this paper proposes a novel self-tuning strategy which based on fuzzy reasoning technique, which can produce a suitable prediction step size for GPC, in order to work well, called Fuzzy-Logic-Based Gray Prediction Controller (FLGPC) [7]. The time-variant prediction step size in FLGPC is adjusted dynamically by a fuzzy inference mechanism according to the error of controlled plant's output and the change in error of controlled plant's output. The rules of fuzzy rule base in FLGPC can be generated by systematically modifying the design of the GPC based on simulation results and experimental results. From the results, it is obvious that the proposed FLGPC is superior to conventional fixed-prediction-step-size GPCs including a GPC with a small prediction step size and a GPC with a large prediction step size in many cases.

## 2. Gray Prediction Control

### 2.1 Gray Model

In the Gray System theory, modeling a system requires only the system's output signals. In addition, the gray modeling method requires much less sampling data (four samples are enough) to generate a highly precise gray forecasting model. The gray model, GM(n, h), is the protagonist of the Gray System theory, and it is a dynamic model which contains a group of differential equations, where the parameter, n, stands for the order of the differential equations, and the parameter, h, stands for the number of variables. So, the GM(1,1) model stands for a kind of first-order and single-sequence linear dynamic forecasting model, and it is one of the most commonly used for gray forecasting in gray model. In this paper, the gray predictor in the proposed FLGPC just adopts the GM(1,1) model. According to the Gray System theory, the irregular original data can be transformed to the more regular data by using data generating technique. It is possible that the generated data can be used to construct a system model by establishing the so-called gray differential equation. The Gray System theory constructs the gray model out of these gray differential equations.

In the Gray System theory, any stochastic process is regarded as a gray process. It can weaken the randomness of the process by data generating technique [8]. There are many data generating operations which have been proposed in the Gray System theory. Here, two of the most essential of them are described as follows: one is the Accumulated Generating Operation (AGO), the other is the Inverse Accumulated Generating Operation (IAGO). The AGO is the most popular data generating technique which transforms the irregular original data to more regular data (or called generated data), and the IAGO is a method which restores the generated data to the original data. The AGO and IAGO are necessary to establish the GM(1,1) model. The techniques of AGO and IAGO, and the process of establishing the GM(1,1) model are summarized in the following section.

### 2.2 AGO, IAGO, and GM(1,1)

Summarizing up the results of reference [8-9], the modeling procedure of GM(1,1) is briefly described as follows:

Let $x^{(0)}$ be a nonnegative original data sequence, that is

$$x^{(0)} = \left\{x^{(0)}(1), x^{(0)}(2), \ldots, x^{(0)}(n)\right\}, \tag{1}$$

taking AGO on $x^{(0)}$, we obtain a first order AGO sequence $x^{(1)}$, denoted as

$$x^{(1)}(k) \equiv AGO \circ x^{(0)}(k)$$
$$= \sum_{i=1}^{k} x^{(0)}(i) \qquad \forall k \in \{1,2,\cdots,n\} \tag{2}$$

define $z^{(1)}$ as the sequence obtained by applying MEAN generating operation to $x^{(1)}$,

$$z^{(1)}(k) = \frac{1}{2}\left\{x^{(1)}(k) + x^{(1)}(k-1)\right\} \qquad \forall k \in \{2,3,\ldots,n\} \tag{3}$$

The equation

$$x^{(0)}(k) + az^{(1)}(k) = b \tag{4}$$

is called a gray differential equation of GM(1,1) and the equation

$$\frac{dx^{(1)}}{dt} + ax^{(1)} = b \tag{5}$$

is said to be an whitening equation of the gray differential equation.

The parameter vector $\hat{a}$ can be expressed as follows:

$$\hat{a} = \begin{bmatrix} a \\ b \end{bmatrix} = \left(\mathbf{B}^T\mathbf{B}\right)^{-1}\mathbf{B}\mathbf{y}_N \tag{6}$$

where

$$\mathbf{B} = \begin{bmatrix} -Z^{(1)}(2) & 1 \\ -Z^{(1)}(3) & 1 \\ -Z^{(1)}(4) & 1 \\ \vdots & \vdots \\ -Z^{(1)}(n) & 1 \end{bmatrix} \tag{7}$$

and

$$\mathbf{y}_N = \left[x^{(0)}(2), x^{(0)}(3), \ldots, x^{(0)}(n)\right]^T \tag{8}$$

here, the coefficients a and b are called the developing coefficient and the gray input, respectively. The solution of the whitening equation (5) is

$$\hat{x}^{(1)}(n+p) = \left(x^{(0)}(1) - \frac{b}{a}\right)e^{-a(n+p-1)} + \frac{b}{a} \tag{9}$$

where the parameter, p, is the prediction step size of gray model.

Taking IAGO on $\hat{x}^{(1)}$, the corresponding IAGO sequence $\hat{x}^{(0)}$ is denoted as $\hat{x}^{(0)} = IAGO \circ \hat{x}^{(1)}$, and can be expressed by the following relation,

$$\hat{x}^{(0)}(n+p) = \hat{x}^{(1)}(n+p) - \hat{x}^{(1)}(n+p-1)$$
$$= \left(x^{(0)}(1) - \frac{b}{a}\right)\left(e^{-a(n+p-1)}\left(1-e^{a}\right)\right) \tag{10}$$

where $\hat{x}^{(0)}(n+p)$ is the prediction value of $x^{(0)}(n+p)$.

## 3. Design of the Fuzzy-Logic-based Gray Prediction Controller

The block diagram of proposed FLGPC is shown in Fig. 1. It consists of three main components, that is: 1) the fuzzy inference mechanism, 2) the modified gray predictor, and 3) the decision-making mechanism.

In Fig. 1, the sensor senses the output signals of plant, y(n), during each sampling period. The sampling data are delivered to the fuzzy inference mechanism and the modified gray predictor, respectively. The inputs of the fuzzy inference mechanism, error, e, and change in error, ce, are derived according to the result of comparing the output signal, $y^{(0)}(n)$, with the desired control object, R.

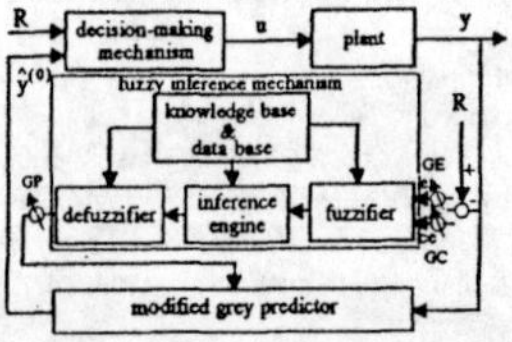


Fig. 1 Block diagram of the proposed FLGPC

The fuzzy inference mechanism is a policymaker which determines the prediction step size for the modified gray predictor. On the other hand, the sampling data can be collected by data collector and constructs a sequence out of these sampling data, then, it can be fed back to the modified gray predictor. The modified gray predictor, which is based on GM(1,1) model, predicts the p-step ahead system output, $\hat{y}^{(0)}(n+p)$, according to the prior information, $y^{(0)}(n)$, where, the prediction step size, p, is produced from the fuzzy inference mechanism. The decision-making mechanism produces the control signal, u, according to a control strategy based on the difference between the desired control object, R, and the prediction value, $\hat{y}^{(0)}$. Because the decision-making mechanism produces the control signal, u, according to the prediction information of system, the system behavior can be controlled before going wrong. More details of the FLGPC will be described in the following section.

### 3.1 The fuzzy inference mechanism

The proposed FLGPC will adopt the fuzzy inference mechanism [10] to determine the prediction step size for the modified gray predictor. The design of fuzzy inference mechanism is discussed as follows in detail.

(1) variable definition

A fuzzy rule base consists of a set of IF-THEN rules in the form of

**IF** ***conditions*** **THEN** ***conclusions***

The fuzzy inference mechanism in our FLGPC has two input variables, e (k) and ce (k), defined as follows:

$$e(k) = R - y^{(0)}(k) \tag{11}$$

$$\begin{aligned} ce(k) &= e(k) - e(k-1) \\ &= y^{(0)}(k-1) - y^{(0)}(k) \end{aligned} \tag{12}$$

where $R$ = set point

$y^{(0)}(k)$ = the present process output

$y^{(0)}(k-1)$ = the last process output

$e(k)$ = the error between the present process output and set point

$e(k-1)$ = the error between the last process output and set point

$ce(k)$ = the change in error of process

In addition, this paper will define the input fuzzy sets including the error and change in error of controlled system's output, and the output fuzzy sets which are the prediction step size. These fuzzy sets are necessary to construct a rule base. The input variables are error, e, and change in error, ce, and its input linguistic sets are assigned in Table 1. The output variable is the prediction step size, p, and its output linguistic sets are assigned in Table 2.

Table 1 The input linguistic sets

| Fuzzy Set | Linguistic Meaning |
|---|---|
| PB | Positive Big |
| PM | Positive Medium |
| PS | Positive Small |
| ZE | Zero |
| NS | Negative Small |
| NM | Negative Medium |
| NB | Negative Big |

Table 2 The output linguistic sets

| Fuzzy Set | Linguistic Meaning |
|---|---|
| VS | Very Small |
| SM | Small |
| MD | Medium |
| LG | Large |
| VL | Very Large |
| HU | Huge |
| VH | Very Huge |

Now, this paper will define the membership functions of the above linguistic sets. The shape of the linguistic sets are determined according to our knowledge and experience of relations between the prediction step size and the GPC's performance. For the simplicity of implementation, the triangular shapes are adopted here. The membership functions for the error, change in error, and the prediction step size are shown in Fig. 2~4, respectively.

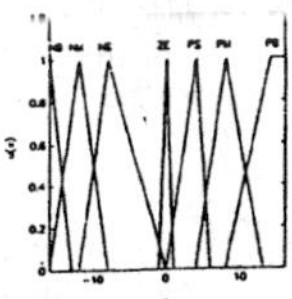

Fig. 2 The membership function of 'error'

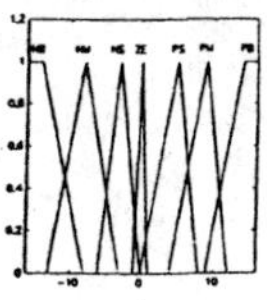

Fig. 3 The membership function of 'change in error'

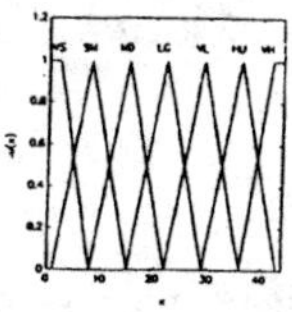

Fig. 4 The membership function of 'prediction step size'

(2) Inference rules

In the conventional GPC adopting the different prediction step size can produce the distinct control effect. When the prediction step size in GPC is small, it can produce a fast system response, that is, the rise time of system is short, but it may lead to a large overshoot and oscillation. On the contrary, when the prediction step size in GPC is large, the system response is slower and without any overshoot. Only when the GPC has an appropriate prediction step size, it does have a good system response. According to the preceding conceptions, the prediction step size of GPC can be adjusted dynamically by a fuzzy inference mechanism in order to produce better performance. The inference rules of the proposed FLGPC are established according to the distinct control effect of GPC due to different prediction step size.

The tuning rules of prediction step size in the fuzzy rule base can be constructed as follows:

Rule 1 If *error is PB* then *step-size is VS.*

Rule 2 If *error is PS* and *change in error is NB* then *step-size is LG.*

Rule 1 implies a general condition. When the output of plant is distant from the set point, then this scheme must provide a very small prediction step size for the modified gray predictor in order to produce a fast response. Rule 2 implies that when the output of plant is close to the set point, this scheme must provide a larger prediction step size for the modified gray predictor in order to restrain the overshoot and oscillation. The other rules are shown in Table 3.

Table 3 Rule base for FLGPC

| e \ ce, p | NB | NM | NS | ZE | PS | PM | PB |
|---|---|---|---|---|---|---|---|
| NB | VH | VH | VH | SM | SM | MD | MD |
| NM | VH | VH | HU | MD | MD | VL | VL |
| NS | HU | HU | VL | MD | LG | HU | VL |
| ZE | VL | VL | LG | LG | LG | HU | HU |
| PS | LG | VL | LG | MD | HU | VH | VH |
| PM | MD | LG | SM | SM | HU | VH | VH |
| PB | VS | VS | VS | VS | VH | VH | VH |

(3) Fuzzy inference engine

The inference rules in the rule base are reasoned by fuzzy inference engine. Here, the fuzzy inference engine adopts the method of sup-min compositional operation. The reasoning process of sup-min compositional operation is described as follows: Let $X$ and $Y$ denote the universe of discourse of input and output, respectively. Suppose $R \in X \times Y$ be a fuzzy relation representing the implication $A \to B$, where $A \in X$ and $B \in Y$ are the fuzzy sets defined in the input space and output space, respectively. Now, give the fuzzy relation R, and the input fuzzy set $A'$, then the output fuzzy set $B'$ can be inferred:

$$\begin{aligned} \mu_{B'}(y_j) &= \bigvee_{x_i \in X_i} \left(\mu_{A'}(x_i) \wedge \mu_R(x_i, y_j)\right) \\ &= \max_{x_i \in X_i}\left(\min_{x_i \in X_i}\left(\mu_{A'}(x_i), \mu_R(x_i, y_j)\right)\right), \quad \forall y_j \in Y \end{aligned} \tag{13}$$

(4) Defuzzification

To determine the crisp value of prediction step size, $p$, we must defuzzify the fuzzy set B'. There are many methods of defuzzification which can be used, such as the max. criterion method, the mean of maximum (MOM) method, and the center of gravity (COG) method, and so on. This paper will adopt the widely-used COG method. The COG method can be expressed as follows

$$p = \frac{\sum_{k=1}^{n} p_k \cdot \mu_{B'}(p_k)}{\sum_{k=1}^{n} \mu_{B'}(p_k)} \tag{14}$$

here, we assume that the support of $B'$ is a discrete set, i.e., $\mathrm{supp}(B') = \{p_1, p_2, \ldots, p_n\}$.

**3.2 Modified Gray Predictor**

The forecasting technique of gray predictor adopts the popular GM(1,1) model in the proposed FLGPC. However, the gray forecasting model GM(1,1) can only predict the nonnegative sequences or the non-positive ones. But the response sequences of control system may be one of the positive values, negative values, or both. Therefore, it must transfer the sequence containing both positive values and negative ones to the relative positive sequence by means of data mapping technique in order to make the gray predictor work well. For this reason, the conventional gray forecasting modeling must be modified. The modified gray predictor can work on any type of sequences well. In this paper, two data mapping techniques are proposed, one is referred as the Mapping Generating Operation (MGO), the other is Inverse Mapping Generating Operation (IMGO). The MGO is defined as follows:

Let $x_0$ be an original sequence,

$$x_0 = \{x_0(1), x_0(2), \ldots, x_0(n)\} \tag{15}$$

$x_m$ be the MGO image sequence of $x_0$,

$$x_m = \{x_m(1), x_m(2), \ldots, x_m(n)\} \tag{16}$$

where,

$$\begin{aligned} x_m(k) &= MGO \circ x_0(k) \\ &\equiv b^{\gamma \cdot x_0(k)} \qquad k = 1, 2, \ldots, n \end{aligned} \tag{17}$$

where $b$ and $\gamma$ are positive constants selected by designer.

Similarly, the IMGO can also be defined as follows:

$$x_m(k) = IMGO \circ x_m(k) \equiv \frac{1}{\gamma}\log_b^{x_m(k)} \tag{18}$$

Thus, the conventional gray predictor may be modified to a new type of gray predictor, and the block diagram is shown in Fig. 5.

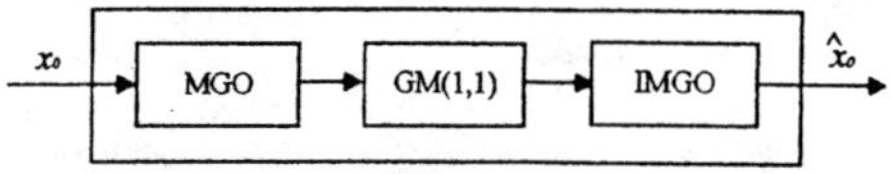


Fig. 5 Block diagram of the modified gray predictor

### 3.3 Decision-making mechanism

The decision-making mechanism is one of the important parts in both GPC and the proposed FLGPC. It is a policymaker which determines the control strategy and produces the control signal based on the information containing both of the desired control goal, R, and the prediction information which is produced by gray predictor. Because the decision-making mechanism adopts the system prediction information to produce the control strategy which controls the plants, the proposed control scheme can control the system behavior before going wrong. Here, the decision-making mechanism takes the Proportional-integral (PI) control as an example to control many different types of the dynamic systems. And it works according to comparing the distance between the desired control goal, R, and the prediction value of system response, $\hat{y}^{(0)}$. In the next section, this paper will illustrate the performance of control systems that the feedback signal using the prediction information, $\hat{y}^{(0)}(n+p)$, is better than those which using the prior data of system output,. $\hat{y}^{(0)}(n)$. In addition, this paper is also going to show that the proposed FLGPC with adjustable prediction step size, p, may yield a response superior to those of GPC.

## 4. Simulation Results

In this section, according to the principle of proposed FLGPC described previously, this paper will make simulations in order to verify the proposed scheme. For this purpose, this paper simulates three distinct control algorithms. They are a GPC with a small prediction step size, a GPC with a large prediction step size, and the proposed FLGPC. In order to obtain more general conclusions, this paper tests the following plants containing the second order linear stable case [11], the third order linear stable case, the third linear unstable case, the first order time delay system [12], the second order nonlinear case [13], and the square-wave-tracking case ( The plant of case 2 is adopted). These plants are described as follows:

(1) the second order linear stable plant

$$y'' + 6y' + 5 = 100u_s$$

(2) the third order linear stable plant

$$y''' + 24y'' + 80y' = 800u_s$$

(3) the third order linear unstable plant

$$y''' + 19y'' - 20y' = 100u_s$$

(4) the first order time delay plant

$$y' + y = u_s(t-0.2)U_s(t-0.2)$$

$$U_s = \begin{cases} 1 & t > 0 \\ 0 & t \le 0 \end{cases}$$

(5) the second order nonlinear plant

$$y'' + \left(|y'| - 1.5\right)y' + y = u_s$$

(6) the square-wave-tracking case

$$y''' + 24y'' + 80y' = 800u_t$$

where $u_S$ is step function

$u_t$ is square wave function

From Fig. 6 to 11, the proposed FLGPC obviously improves the transient response of system, and produces shorter rise time, shorter settling time, lower overshoot, and more accuracy than the other control algorithms. In addition, from the simulation result of the square-wave-tracking case, the modified gray predictor in FLGPC can successfully work with a sequence containing both positive values and negative ones, and shown in Fig 11. That is, the modified gray predictor in the proposed FLGPC can work on any control system case well. The simulation results show that the proposed FLGPC can produce a superior performance in many cases.

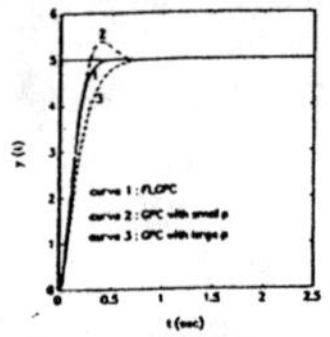

Fig. 6 Output response of the second order linear stable plant

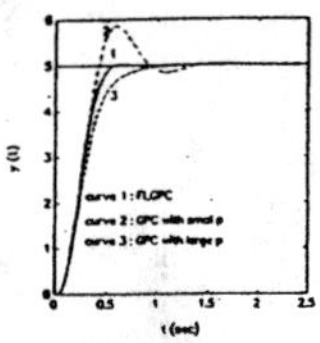

Fig. 7 Output response of the third order linear stable plant

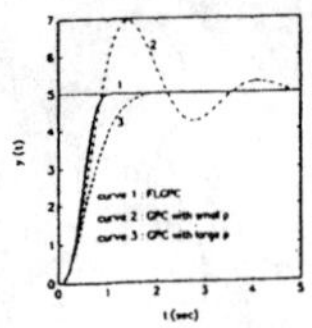

Fig. 8 Output response of the third linear unstable plant

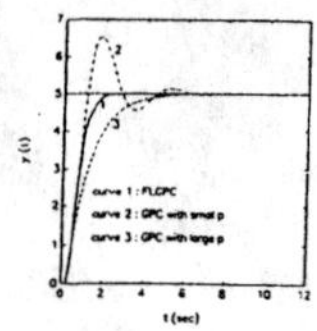

Fig. 9 Output response of the first order time delay system plant

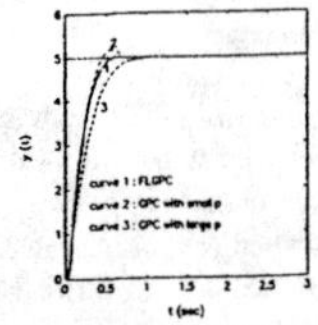

Fig. 10 Output response of the second order nonlinear case plant

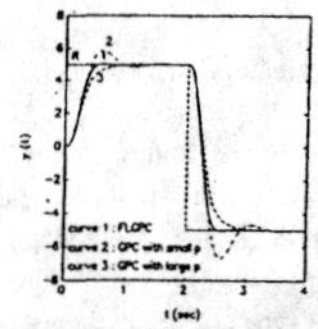

Fig. 11 Output response of the square-wave-tracking case

## 5. Application of FLGPC in DC servo motor speed control system

To verify the practicability of the proposed FLGPC, this paper applies FLGPC to high speed response system, such as DC servo motor speed control system. In this section, the implementation of the proposed FLGPC in DC servo system is described, including system configuration and experimental results.

### 5.1. Configuration

The block diagram of the DC servo motor speed control system is shown in Fig. 12. In this system, most of the process units are implemented by software program except the driver and DC servo plant. The control strategy of decision-making mechanism in FLGPC can adopt many control policies, here, the popular PI speed control strategy is adopted because it has good performance and convenient computing. In previous description, it is necessary that the gray predictor can only operate when a sequence, which includes four samples at least, is established by data collector. Here, the sequence form adopts the fixed-dimensional model which is suggested by D. Yi and R. Yang [4] for its great precision and simple computing. Every sequence consists of four samples in any time and recorded in computer's memory. In other words, the sequence dimension in our system is four. These sequence provides the system behavior information which the modified gray predictor required in order to predict the future system behavior.

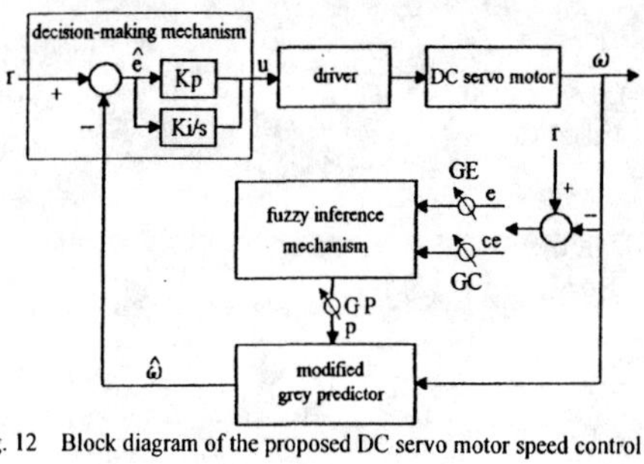


Fig. 12 Block diagram of the proposed DC servo motor speed control system

### 5.3 Experimental Results

In this experiment, the specification of DC servo motor, is shown in Table 4. The sampling period, T, is 5 ms in our system. Four set points of the rotor speed are tested in this experiment, they are 1000 rpm and 1500 rpm in normal speed control, 1000 rpm and 1500 rpm in reverse speed control. All of the system information including sampling data, the prediction value from the modified gray predictor, and the control signal, u, from the decision-making mechanism are stored in a disk file, and the experimental results are shown in Fig. 13-16. From that experimental results, whenever in the normal speed control or the reverse speed control, the proposed FLGPC, can provide the system with a good performance including a shorter rise time, lower overshoot and oscillation, and high accuracy. That is, this paper can successfully apply the proposed FLGPC to the DC servo motor speed control and produces a superior control effect.

Table 4 The specification of DC servo motor

| ITEM | SPECIFICATION |
|---|---|
| type | permanent magnet |
| power | 90 W |
| voltage | 30 V |
| current | 4 A |
| speed | 3000 rpm |
| pole | 2 poles |
| rotor resistance | 5.7 ohms |

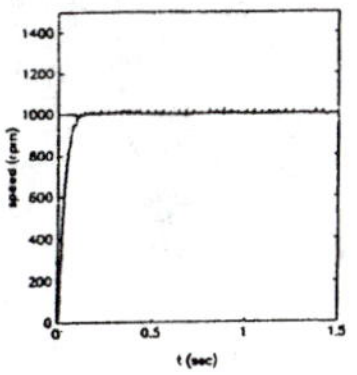

Fig. 13 Output response of normal speed control in 1000 rpm

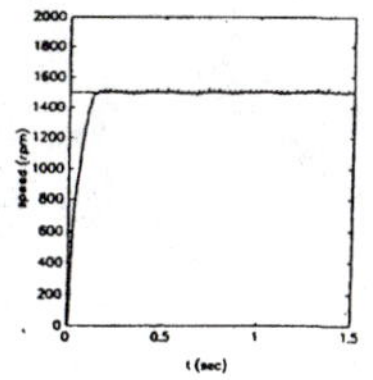

Fig. 14 Output response of normal speed control in 1500 rpm

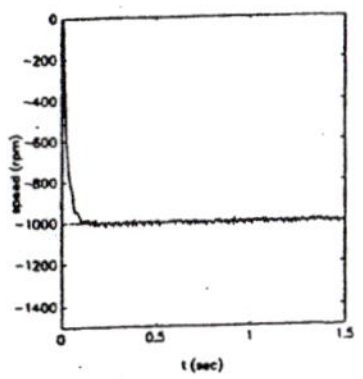

Fig. 15 Output response of reverse speed control in 1000 rpm

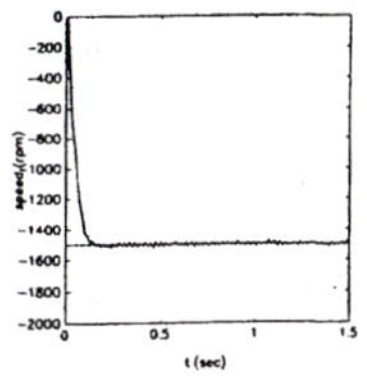

Fig. 16 Output response of reverse speed control in 1500 rpm

## 6. Conclusions

This paper proposes a novel type of fuzzy-logic-based gray prediction controller developed by the fuzzy logic control theory and the Gray System theory. Since the conventional gray predictor can only predict a nonnegative sequence or non-positive sequence, and it can not satisfy the control system requirement. For this reason, this paper modifies the conventional gray predictor in order to predict a sequence containing both positive values and negative ones. The simulation results show that the modified gray predictor can work well with any type of sequences. In addition, by comparing the control response of the proposed controller with those of the conventional GPC in many cases, the simulation results show that the performance of the proposed FLGPC is superior to the others. Furthermore, this paper also implements the proposed FLGPC on a personal computer for DC servo motor speed control in order to verify its practicability. The experimental results show that the proposed controller can produce a good performance for DC servo motor speed control. Areas of future research are focused on the design of strategies to automatically tune the membership functions of the fuzzy inference mechanism to make the more superior performance of FLGPC.

References


[1] J. L. Deng, "Control problems of Gray Systems," *Systems and Control Letters*, Vol. 5 pp. 288-94, 1982.

[2] --, "Introduction to Gray System Theory," *The Journal of Gray System*. Vol. 1, pp. 1-24, 1989.

[3] B. Cheng, "The gray control on industrial process," *Journal of Huangshi College*, Vol. 1, pp. 11-23, 1986 (in Chinese).

[4] D. Yi & R. Yang, "Gray Predictor Controller for DC Speed Control System," *The Journal of Gray System*. Vol. 2, pp. 189-215, 1990.

[5] C. C. Kung and S. C. Wang, "Gray Fuzzy Sliding Mode Controller Design," *Processdings of The 2nd Natl Conf. on Fuzzy & Appl., Taipei, Taiwan, R. O. C.*, pp. 60-65, 1994.

[6] Y. P. Huang, C. C. Huang, and C. H. Huang, "Determination of the Preferred Fuzzy Variables and Applications to the Prediction Control by the Gray Modeling," *Processdings of The 2nd Natl. Conf. on Fuzzy Theory & Appl., Taipei, Taiwan, R. O. C.*, pp. 406-10, 1994.

[7] C. M. Hong, S. C. Lin, and C. T. Chiang, "Control of Dynamic Systems by Fuzzy-based Gray Prediction Controller," *The Journal of Gray System*. Vol.7, No. 1, pp. 23-44, 1995.

[8] J. L. Deng, The Essential Methods of Gray System. *Huazhong University of Science and Technology Press*, Wuhan, 1987 (in Chinese).

[9] --, Gray Systems control, *Huazhong University of Science and Technology Press*, Wuhan, 1985 (in Chinese).

[10] L. A. Zadeh, "Fuzzy sets," *Information and Control*, Vol. 8, pp. 338-353, 1965.

[11] B. C. Kuo, Automatic control systems, Ed. 6, *Prentice-Hall: New Jersey*, 1991.

[12] Hao Ying, William Siler and James J. Buckley, "Fuzzy control theory: a nonlinear case," *Automatica*, Vol. 26, No. 3, pp. 513-520, 1990.

[13] M. M. Gupta and J. Qi, "Design of fuzzy logic controllers based on generalized T-operators," *Fuzzy sets and Systems*, Vol. 40, pp. 473-489, 1991.

# Section 4: Fuzzy Control

# FUZZY LOGIC PROCESSING AND DYNAMIC ALARM HANDLING FOR REAL-TIME MACHINE HEALTH MONITORING

**TIM D. SEIFERT and JOHN P.H. STEELE**
*Laboratory for Intelligent Automated Systems*
*Division of Engineering, Colorado School of Mines*

***ABSTRACT:***
Dynamic alarm handling is an important component of any monitoring system that is used to assess the condition of equipment in dynamic environments. This paper identifies problems with conventional methods used to set alarm thresholds, filter nuisance alarms, and determine the condition of the equipment. An alternative monitoring approach, using fuzzy logic and statistical methods, is presented for condition monitoring of hydraulic systems.

System Health Assessment and Real-Time Prediction (SHARP) technology provides remote on-line monitoring of machine condition to assess and express the operational condition of the equipment to the operator and management in real-time. As a result, management will be able to plan and prepare for the repair of equipment prior to failure, thus avoiding unnecessary down time.

In order to accomplish the above goals, relevant information must be extracted from the system to be monitored. Therefore, the sensor data are reduced and classified into fuzzy states based on knowledge of the system. Data reduction decreases the necessary computer resources for archiving and aids in pattern recognition. Classifying the information into fuzzy states allows the application of linguistic expert knowledge of a to assess its condition. Finally, dynamic alarming evaluates the system's apparent condition and relates it to its dynamic operating environment to determine the condition of the system (Steele et al., 1995).

## ALARM HANDLING FUNDAMENTALS

Alarm handling is a critical component in any monitoring system. It is responsible for taking the signal values from the system and assigning alarm states which reflect whether or not the system is performing as expected. In general, alarms can be generated from any set of measurable variables on a system. Typically, the variables consist of analog values measured by transducers or digital values that change state. When these values exceed an acceptable value or change state they are flagged as alarms. These alarms are then related to events in the system that caused the change. Once an event has been determined from a set of alarms, the event that occurred is then presented to the operator for remediation efforts.

Using conventional methods, transforming a variable into an alarm state is essentially a three stage process. First, alarm thresholds have to be set for each individual variable. As a result, an alarm state can be inferred based on the measured value relative to the threshold. The next stage in traditional alarm processing involves filtering and suppressing nuisance alarms from the entire system set of alarms. The third stage in alarm handling is alarm processing. From the alarms that are not filtered from the input data stream, the status of each component and the machine as a whole can be determined through a causal network of events.

**Traditional Alarm Threshold Setting**

Generally, a system's performance will be acceptable if all the components of that system are operating within their respective acceptable limits. (Lee, 1991) The limits are flagged as alarm thresholds and are used to alert an operator of a process or a component that has changed from one state into another. Typically, these states are defined by either acceptable performance or unacceptable performance. Defining the threshold values, however, is not always straightforward. Typically, all the variables in the system being monitored are not fully understood and therefore make setting the alarm thresholds more difficult. As a result, comfortable, experience-based estimates are often used as initial thresholds and then tested in operation. This method of limit definition has some characteristics that may not be acceptable. If these limits produce alarms when the system is performing acceptably, the limits are adjusted until the so-called "nuisance alarms" are reduced. This can result in an iterative cycle with the resulting alarm limits losing their "expert" meaning over several iterations. The net result is estimated alarm thresholds that have little meaning and are not repeatable on other systems.

In order to obtain repeatable results with significant meaning, Statistical Process Control (SPC) methods can be used to set the alarm thresholds. SPC is a collection of statistical methods that have been developed to provide an indication of whether a system is performing as expected. For a given measurable variable of the system, the mean ($\mu$) and standard deviation ($\sigma$) of a data sample are key components in SPC. By holding the system in steady state while its performance is acceptable, data can be taken and its mean and standard deviation calculated for the sample. By definition, 99.75% of the collected data would lie within the expected performance limits of $\mu \pm 3\sigma$. Therefore, using alarm thresholds of $\mu \pm 3\sigma$ would provide meaningful upper and lower thresholds.

Although Statistical Process Control methods can be employed to set the thresholds for alarms, it has limitations that must be addressed. SPC derived alarm limits are generally correct for steady state operations, but are too restrictive during transient periods of the system such as start-up. As a result, nuisance alarms will be generated during these times (Lee, 1991). Alarm suppression and filtering can be employed to remedy these weaknesses.

**Alarm Suppression and Filtering**

In order to eliminate nuisance alarms, alarm suppression and filtering can be employed prior to alarm processing. "Toggling" nuisance alarms cross the edge of an alarm threshold repeatedly due to unexpected signal noise. Discrete signals with bad connections are also susceptible to this type of toggling (Shoop et al., 1992). "Redundant" nuisance alarms occur when monitored variables of the system are directly related, e.g., a low pump input pressure and low pump flow rate caused by a restricted pump input line. Nuisance alarms need to be removed prior to alarm processing for two reasons. During times of system disturbance and in systems with many variables to monitor, many alarms are generated that consume CPU time and disk resources which are valuable to a real-time alarm handler. Additionally, nuisance alarms are cumbersome to operators and can bury the important alarms that carry information and require further action. Alarm suppression and alarm filtering are employed to reduce nuisance alarms.

**Traditional Methods of Alarm Processing**

A traditional alarm processor needs to determine the health of the system being monitored based on the individual alarms that are generated. An alarm processor characteristic critical to assessing system health involves determining what has happened, which is accomplished by event recognition.

In order to conclude which event has generated the set of alarms in the data stream, it is necessary to define the relationship between alarms and events. Each event that can occur in a system has a set of characteristic alarms that may be present in the alarm data stream. In practice, a system alarm set in the input stream will not typically match the defined characteristic alarm sets exactly (Dijk, 1992). To determine which event can explain the alarms, strong or weak causal relationships can be used.

***Strong Causal Relationships***

The use of strong causal relationships is one method of relating system events to the alarms that are generated by the system. The principle of this method relies on a pattern recognition technique for alarm explanation. A relationship between the events and the alarms must be defined. By using expert knowledge, each system event is related to a set of characteristic alarms for that particular event. Next, the alarm input stream is searched to determine if the characteristic alarm set for any event is contained within the input stream. If the characteristic alarm pattern is within the alarm input stream the event is concluded to have occurred (Dijk, 1992).

There are, however, several issues that need to be addressed in the used of strong causal relationships. This method assumes that if an event occurs, it will produce its complete set of characteristic alarms. As a result, any false or missing alarms in the characteristic set would result in no solution and no event will be concluded. Therefore, strong causal relationships are very sensitive to false and missing alarms and can leave alarms uncovered.

***Weak Causal Relationships***

Weak causal relationships can be introduced by relaxing the event-alarm relationships. In this relationship, events are not required to produce all of the characteristic alarms for the particular event. First, the relationship between the events and alarms must be defined, just as when using strong causal relationships for event recognition. Next, the alarm input stream is searched with a pattern recognition algorithm. The search starts by taking all defined alarm sets that contain at least one alarm in the generated set and flags those solutions as possible events that occurred. The flagged characteristic alarm sets are then compared to the generated alarms and close matches are identified.

Inherently, weak causal relationships require further processing. Since this method essentially makes allowances for missing and false alarms, multiple solutions can be concluded for a generated alarm set. The possibility for multiple solutions for a given alarm set requires a criterion to be used to determine the actual event that has caused the generated alarms. Using Occam's razor as a criterion, the minimal set of events causing the given set of alarms is a solution (Dijk, 1992). If multiple events are determined, the most likely faulty component is the component associated with the smallest number of inconsistencies in the overall system diagnosis.

## APPLICATION TO SHARP

The architecture of SHARP is constructed to enable an alternative approach to the traditional alarm handling methods just outlined. A Mamdani-style (Asai, 1995) fuzzy logic inference engine is used to analyze incoming sensor data and conclude event alarms. The fuzzy logic inference system classifies the state of each sensor variable and establishes a rule-based environment for inferring system health. This eliminates the need for traditional causal networks and fragile, crisp thresholds on each sensor. A defuzzified event alarm state is then passed on to G2, an expert system shell, for alarm filtering (Gensym, 1995). G2 presents the operating condition of monitored components and equipment in a graphical environment to the operator.

### Fuzzy Logic, SPC, and Alarming in SHARP

As mentioned previously, SPC methods can be effective for setting alarm thresholds in a steady state system. However, SHARP will typically be monitoring systems in a dynamic environment. Therefore, only a small portion of the SPC techniques is borrowed for use in the fuzzy inference system. Specifically, SPC methods can be used to define the membership functions used to fuzzify the system variables. By holding the system in steady state at various performance levels, the expected levels of each variable can be assessed. In other words, the sensor values at which the membership functions should be centered upon and each function's width can be determined by using methods borrowed from SPC. For example, taking a sample of temperature data when the system is at "normal" operating conditions would produce a calculated mean and standard deviation for the data. The membership function that represents the "normal" temperature should be centered at a value equal to the sample mean and have a width equal to six times the standard deviation. This can then be repeated for various levels of operation if necessary, e.g., very low, low, high, and very high. The net result is sensor values that have been transformed into fuzzy states with true meaning.

To accommodate alarming in dynamic environments, the linguistic approach in fuzzy logic will be utilized. Rules can be created which can reason about the operating environment and change the alarm severity accordingly. An example rule for a hydraulic circuit might look like: *IF the ambient_temperature is high and the fluid_temperature is high, THEN pump_cavitation_alarm_severity is slightly_decreased.*

### Alarm Filtering in SHARP

G2, the expert system shell, provides an excellent means for suppressing and filtering nuisance alarms. This is a result of the rule based and object oriented nature of G2.

Rules can be written which suppress toggling alarms caused by alarms originating from both analog and discrete values. Suppression rules would determine if any particular alarm has been toggled within a given time frame and might look like: *IF any object alarm toggles more than three times within a five second window, THEN suppress it until no alarms are seen for five seconds.*

Filtering out redundant nuisance alarms is achieved gracefully due to the object oriented nature of the programming environment. If the alarms are created and defined in the same hierarchy that represents the actual system being monitored, subsystem alarms are easily identified and suppressed. For a system class of pressure sensors, an example rule might look like: *IF any manifold pressure sensor is in alarm, THEN suppress all subsystem pressure sensor alarms.*

## RESULTS

The fuzzy logic and SPC technique discussed previously has been implemented in a simulation environment in our laboratory. A high pressure hydraulic power unit (HPU), including valves for loading and starving the pump, has been set up to simulate the cavitation of a hydraulic pump in heavy equipment. The measurable system variables include motor speed, hydraulic flow, pump inlet and outlet temperatures, and pump inlet and outlet pressures. Data were taken at various motor speeds and valve settings while the pump was operating at two different performance levels: acceptable and unacceptable. This data was then grouped into the appropriate performance level: acceptable (no cavitation) and unacceptable (cavitation), for further analysis.

Further analysis entailed creating the fuzzy membership functions and rules to determine cavitation of the pump and can be summarized as follows: First, each set of sensor values was statistically examined at each performance level to obtain a mean and standard deviation. For each sensor variable, and each performance level, a membership function was created that was centered at the sensor mean and spread six standard deviations in width. Once all the membership functions were established, a single sensor variable was identified with the greatest distinction between the different performance levels and was used to write fuzzy rules that concluded the corresponding performance level. Next, the fuzzy system was tested to evaluate its performance. The sensor value distributions were statistically re-evaluated for the fuzzy inference system's performance levels of correct and incorrect misclassifications. A single variable was identified with the greatest distinction between the different performance levels and used to write fuzzy rules that concluded the corresponding performance level. This process was repeated until there were essentially no distinctions in sensor value distributions between performance levels.

The resulting fuzzy logic inference system, depicted in Figure 1, classified 85.34% of the data sets correctly and produced the correspondingly correct cavitation alarm state. The incorrect alarm states consisted of no missing alarms and 18.11% false alarms. Using a desired alarm state of 1.0 for pump cavitation and a desired alarm state of 0.0 for no pump cavitation, this system produced a total sum squared error of 2.48 and a root mean squared error of .070.

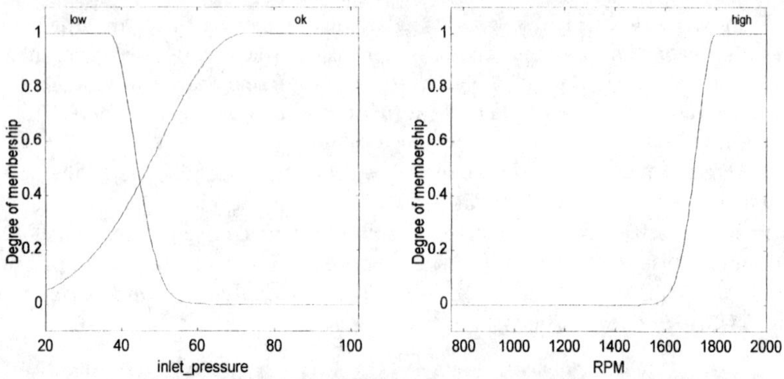


1. If (inlet_pressure is low) then (cavitation_alarm is severe)
2. If (inlet_pressure is ok) then (cavitation_alarm is off)
3. If (inlet_pressure is not ok) and (RPM is high) then (cavitation_alarm is severe)

**Figure 1. Membership functions and fuzzy rules used to determine cavitation.**

## CONCLUSIONS

A method for determining fuzzy membership function placement, shape and fuzzy rules has been presented for inferring the alarm status of a cavitating hydraulic pump. This method utilizes fuzzy logic and SPC fundamentals and can be used for any process where sensor information is available for the desired operating conditions. The resulting fuzzy logic inference engine only requires the use of dominant sensor variables and is therefore more computationally efficient than adaptive techniques. Additionally, this method enhances the developer's expert knowledge about the system by identifying cause and effect relationships in the system.

## REFERENCES


Asai, K., (1992). *Fuzzy Systems for Information Processing*, IOS Press, Inc., Burke, VA.

Dijk, Harold E., (1992). AI-Based Techniques for Alarm Handling, *Electric Power and Energy Systems*, Vol. 14, No 2/3, April/June 1992, pg. 131-137.

Gensym, (1995). G2 Guide: User's Guide Version 4.0, Gensym Corporation, 125 CambridgePark Drive, Cambridge, MA 02140.

Lee, J. L., (1991). Application of Statistical Process Control Methods to the Establishment of Process Alarms in a Distributed Process Control System, *ISA Transactions*, Vol. 30, No 1, 1991, pg. 59-64.

Shoop, Alvin H., Silverman, Steven, (1992). A Real-Time Alarm Processor, *Electric Power and Energy Systems*, Vol. 14, No 2/3, April/June 1992, pg. 108-113.

Steele, John P.H., Archuleta, Michelle, Brown, Galen, Seifert, Tim, (1995). SHARP - System Health Assessment and Real-Time Prediction, *Predictive Maintenance Technology National Conference Proceedings*, 1995, T9-T15.

# Inverse-Model Compensation Using Fuzzy Modeling and Fuzzy Learning Schemes

**P.J. Costa Branco[*] and J.A. Dente[**]**

*Instituto Superior Técnico (I.S.T.) CAUTL- Laboratório de Mecatrónica.*
*Av. Rovisco Pais, 1096 Lisboa Codex Portugal (Fax Nº 351-1-8417167)*
** pbranco@alfa.ist.utl.pt * edentepc@alfa.ist.utl.pt*

***ABSTRACT:***

This paper presents a learning control system that uses fuzzy modeling and fuzzy learning schemes to design an inverse-model compensation controller. It displays interesting features such as generalisation and learning capabilities that we can use to compensate non-linear terms that affect system dynamics. The controller is implemented in an experimental system composed by an electro-hydraulic actuator. The paper shows experimental results demonstrating the controller capability in learn new knowledge and compensate unmodeled uncertainties of the actuator improving its performance in trajectory following.

## INTRODUCTION

Most of actual systems present uncertainties at a structural level as too at the parameter level, they are multidimensional and have high nonlinearities. So, there are difficulties in the application of control techniques, as for example the adaptive control and the predictive control that use linear or linearized models of the system to be controlled.

The use of same mathematical model for all operation region does not permit the controller to modify the functional relation between the control variables in the manner to cancel new system operating modes that can deviate the system's response from the defined trajectory. So, it is important the integration of an *intelligent component* in the production system as increase in its flexibility level. Here, the *intelligent component* means the capability to extract relations from the process, change them to improve the control performance and manifesting reasoning and learning abilities.

In this paper, we exploit the potentialities of fuzzy models and fuzzy learning schemes to design an inverse-model compensation control scheme. We present a learning control system which is based on the method of the feedback-error-learning method introduced in [1] and [2]. In our scheme, the learning process does not utilizes the error signal from the feedback loop. The feedback command is monitored by the rules in the estimated inverse fuzzy model and it is expected that the feedback signal tends to zero as learning proceeds. Use of existing knowledge of the system being controlled is incorporated into the model, facilitating the model design and helping to accelerate the initial controller actuation.

To investigate the performance of the learning controller, it is applied to position control of an electro-hydraulic actuator. The actuator contains nonlinear characteristics as a dead-zone and hysteresis effects. These nonlinearities are dominant in the low performance of an existing coarse position controller. So, it is a good solution the adittion of a signal to the controller in order to compensate the nonlinearities. The controller is implemented in a PC with a 80386 CPU and an interface with A/D (analog to digital) and D/A (digital to analog) converters. At each control cycle, the learning process builts the inverse fuzzy model and generates a compensation signal from the extracted inverse model to the current controller. The results show that the learning

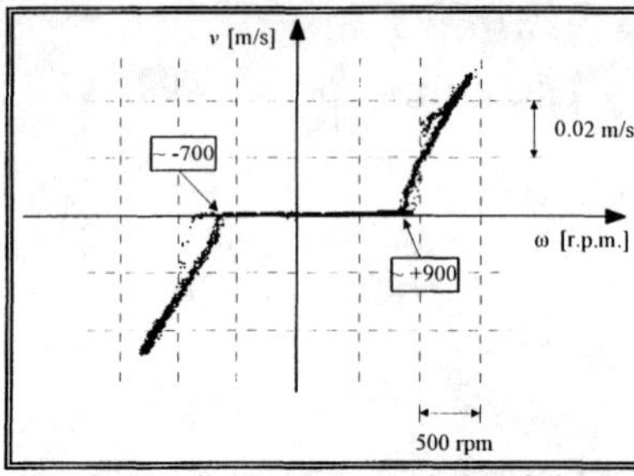

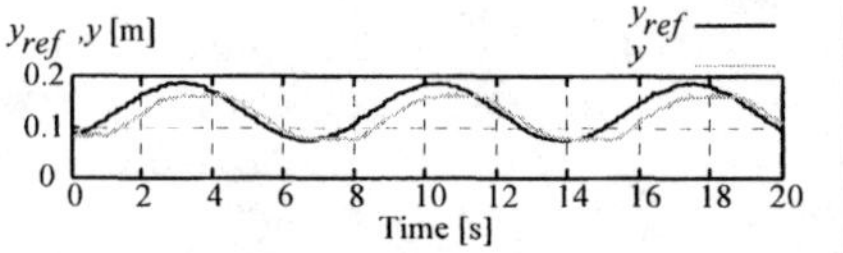


**Fig. 1: Non-linear characteristic of the hydraulic circuit. Fig. 2: Reference signal ($y_{ref}$) evolution and the position measured ($y$) for a closed-loop control using only a proportional controller. This figure shows the nonlinearity effects in the control system.**

controller progressively reduces the trajectory errors, it generalises the acquired knowledge for any new trajectory and learn new relations when are changes in system dynamics.

The paper is organized as follows. In section II the experimental system is described. Section III describes the fuzzy learning approach to extract the actuator model. In Section IV we explain the learning controller for inverse-model compensation. At last, Section V presents the experimental results with the run-time learning controller.

## EXPERIMENTAL SYSTEM CHARACTERIZATION

The experimental system is an electro-hydraulic actuator composed by a commercial electrical drive that controls the speed of a permanent-magnet motor connected with a pump that sends fluid to a hydraulic circuit to move a linear piston. Following, we resume the componnents of the experimental system:

- an electrical drive system made by an inverter with IGBT's and current control;
- a synchronous permanent-magnet motor (P.M. Motor) operated by the commercial drive system (220V/ 1.2Nm/ ±3000 rpm);
- a hydraulic pump with fixed displacemment;
- the hydraulic actuator controls a linear piston with 0.2m as maximal displacemment;
- the speed control system of the P.M. motor. Its control system is composed by a proportional+integral (PI) controller setted in factory for the motor;

If we see the experimental system as a *black-box* and make some simplifications, we can interpret the piston position signal ($y$) being a function of the reference signal ($y_{ref}$), the motor speed ($\omega$), and the linear speed of the piston ($v$). The relation is presented in (1).

$$y = f(y_{ref}, \omega, v) \tag{1}$$

An experimental training data, composed by the variable set of relation (1), is used for the learning process. To assure that the training data covers a significant part of the system working domain, we use a proportional controller in a closed-loop to control the piston postion. The actuator presents an assymetric nonlinear behaviour caused by a nonlinear chacracteristic in the hydraulic circuit. Figure 1 displays the non-linear characteristic obtained from the actuator. It relates the motor speed ($\omega$) with the piston speed ($v$). The curve shows an assymetric dead-zone localized between -700 r.p.m. and +900 r.p.m, and a hysteresys effect out of the dead-zone.

As shown in figure 2, the piston follows the reference ($y_{ref}$) with an assimetric time-delay caused by the nonlinearitie. When the motor speed is into the dead-zone, the hydraulic system stays desacoupled from the electrical part. In this situation, the fluid debited by the pump is near zero and the piston stops. So, into the dead-zone there is not command signal to the hydraulic circuit. Out of the dead-zone, the pump debits more fluid in one direction than to the other causing the piston move faster to one side than to the other. This can be seen by the inclination angle of the curves out of the dead-zone

## FUZZY-LOGIC APPROACH FOR LEARNING THE INVERSE MODEL

**Overview of the Fuzzy Learning Process**

In this section, we overview the learning algorithm introduced by Wang in [3] and used in the learning controller. The learning process considers each variable equally partitioned by symmetric membership functions of triangular type. The learning method uses the *t*-norm *max* to select the degree to which two fuzzy sets match and extract the conclusion part of each rule.

The model uses rules with the form of expression (2) where $R^{(l)}$ is the *l*th rule, $x_j$ the chosen variables expressing the system's condition and $y$ the considered output variable. The symbols $A_j^{(l)}$ are membership functions and $\omega^{(l)}$ is the rule consequent part being a real number (*fuzzy singleton*).

$$R^{(l)}\text{: IF } x_1 \text{ is } A_1^{(l)} \text{ and } x_2 \text{ is } A_2^{(l)} \ldots \text{ and } x_m \text{ is } A_m^{(l)} \text{ THEN } y \text{ is } \omega^{(l)} \quad (2)$$

The firing degree of each rule is measured by expression (3). The membership values of each variable, defined for every linguistic term of rule *l*, are combined using the *algebraic product operator* (*).The variables denoted by $x_1^{'}, x_2^{'} \ldots, x_m^{'}$ are the numerical values of $x_1, x_2 \ldots, x_m$ and the expressions $\mu_{A_1^{(l)}}(.), \ldots, \mu_{A_m^{(l)}}(.)$ are the membership functions attributed to the fuzzy sets $A_1^{(l)}, \ldots, A_m^{(l)}$.

$$\mu_{A_1^{(l)} \times \ldots \times A_m^{(l)}}(x_1^{'}, \ldots, x_m^{'}) = \mu_{A_1^{(l)}}(x_1^{'}) * \ldots * \mu_{A_m^{(l)}}(x_m^{'}) \quad (3)$$

Each variable is quantized by a number of triangular and symmetric fuzzy sets. The working domain is partitioned in subspaces constituting each one a rule. The rules are represented as a fuzzy implication with a degree given by (4). In that expression, the operation "*" is the *inference product*.

$$\mu_{A_1^{(l)} \times \ldots \times A_m^{(l)} \to B^{(l)}}(x_1^{'}, \ldots, x_m^{'}, y^{'}) = \mu_{A_1^{(l)}}(x_1^{'}) * \ldots * \mu_{A_m^{(l)}}(x_m^{'}) * \mu_{B^{(l)}}(y^{'}) \quad (4)$$

To extract from learning data the function that models the system, we proceed as follows. For each input $x_i^{'}$ and output $y_i^{'}$ coming from the sensors, their different grades in each attributed membership function are calculated. To each variable are attributed *n* grades corresponding to the number of membership functions or the number of linguistic terms of each one. The grade set of every variable constructs a vector with *n* possibilities. Then, for all variables, it is chosen a final degree corresponding to the highest degree in all membership functions. The fuzzy implication degree is calculated by the product operation rule as described before. That rule with highest implication degree is chosen to define the pre-defuzzified output value $\omega^{(l)}$.

An inference method, based on the centroid-defuzzification formula, combines all rule contributions in a weighted form (5), where *Y* is the output of the model, $\omega^{(l)}$ is the pre-defuzzified output, $\mu(R^{(l)})$ is the rule firing degree, and *c* the total number of extracted rules.

$$Y = \sum_{l=1}^{c} \mu(R^{(l)}).\omega^{(l)} \Big/ \sum_{l=1}^{c} \mu(R^{(l)}) \quad (5)$$

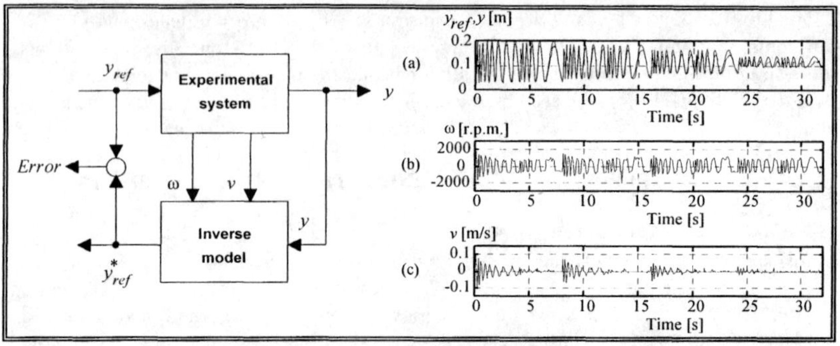


**Fig. 3: Diagram used to obtain the reference position error. Fig. 4: Learning set. (a) Reference and position signal. (b) Motor speed signal. (c) Piston speed signal.**

**Extracting the Inverse Model of the Actuator**

The anterior learning process is used to extract the inverse relation (6) of the actuator. To verify the performance of the extracted fuzzy inverse-model, the reference piston position ($y_{ref}$) and the reference signal estimated by the fuzzy model ($y^*_{ref}$) are compared as described in figure 3.

$$y_{ref} = f^{-1}(\omega, v, y) \tag{6}$$

Two different experimental data sets are used: one for the learning phase and other to test the generalisation capability of the extracted model. The learning examples are acquired imposing a sinusoidal reference for piston position with amplitude ranging from 0 a 0.2m and the frequency among 0 and 1Hz. Figure 4 shows the learning set obtained from the experimental system with the coarse position controller. Figure 4(a) shows the reference signal evolution and the measured piston position. Figure 4(b) presents the motor speed and figure 4(c) shows the piston speed signal.

In figure 5 we show first the extraction of the direct relation $y = f(\omega, v, y_{ref})$.The direct fuzzy model is extracted by the learning process using the training data set. After, we use the testing set with examples not presented in the learning phase to measure the model performance. The signals in figure 5(a) show the good approximation of the model. Although, as we can see by the error evolution in figure 5(b), there are some errors. They originate from noise interference in the learning algorithm, from incomplete information supplied to the learning process, and even from the discrete nature of the fuzzy model.

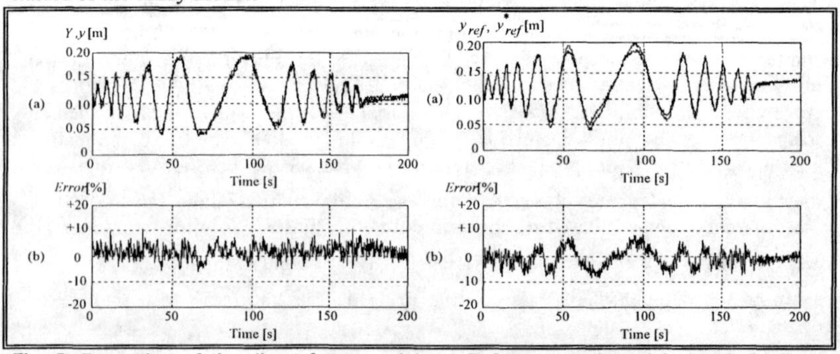


**Fig. 5: Extraction of the direct fuzzy model. (a) Reference and model signal. (b) Error evolution. Fig. 6: Extraction of the inverse fuzzy model. (a) Reference and model signal. (b) Error evolution.**

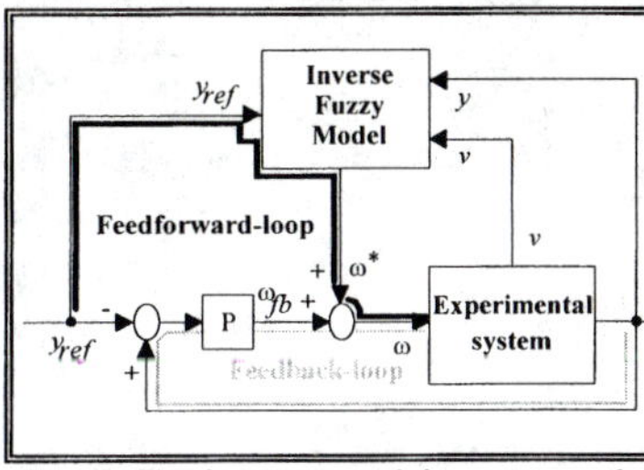


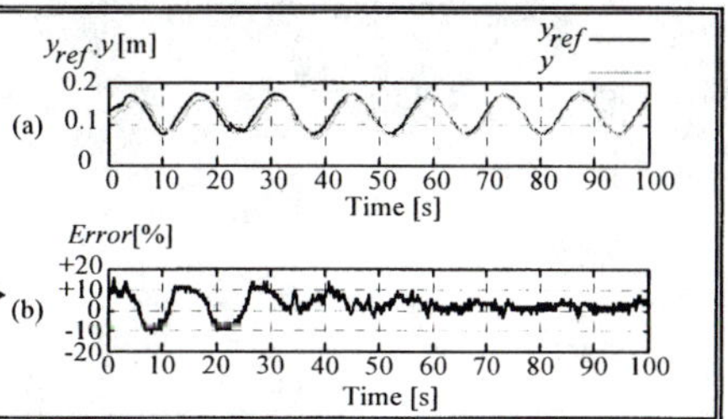


**Fig. 7: The inverse-model compensation diagram. Fig. 8: (a) Compensation of the actuator nonlinearities by the fuzzy learning scheme. (b) Error signal evolution.**

Figure 6 presents the results when we use the learning process to extract the inverse relation $y_{ref} = f^{-1}(\omega, v, y)$. The error signal in figure 6(b) shows high values near the regions where the piston changes its direction. This can be seen too in figure 6(a) where the model presents high prediction errors when the piston changes its direction. This happens because the pump speed changes its signal and passes through the dead-zone. As there is no inverse relation into the dead-zone, the model deteriorates and generates high errors in its prediction of the $y_{ref}$.

## THE FUZZY LEARNING CONTROLLER

The learning controller is based on the feedback-error-learning scheme proposed by Kawato in [1,2]. Two differences distinguish the proposed controller from the original one: the error signal is not need for the learning scheme as the fuzzy learning algorithm automatically extracts the rules by an unsupervised manner; the other difference is that the feedback-error-learning neural network studies assume no a priori knowledge of the plant being controlled. As is rarely the case that no knowledge is available for a real control problem, we effectively use the a priori knowledge in the fuzzy model.

The diagram in figure 7 shows the implemented inverse-model compensation scheme. It includes the proportional controller (P) and the addition of feedback ($\omega_{fb}$) and feedforward ($\omega^*$) commands. The total command $\omega$ fed to the actuator (7) is the sum of the feedback command $\omega_{fb}$ generated by the proportional controller, and the feedforward command $\omega^*$ which is calculated by the fuzzy inverse-model.

$$\omega = \omega_{fb} + \omega^* \tag{7}$$

The inverse model receives the desired trajectory $y_{ref}$ and more two signals from the actuator, the piston speed $v$ and the piston position $y$, generating the feedforward command $\omega^*$. The learning process does not uses the error signal from the feedback loop. The feedback command is monitored by the rules in the inverse model and it is expected that the feedback signal tends to zero as learning proceeds.

## EXPERIMENTAL RESULTS

To describe the effectiveness of the proposed controller, we present some experimental results on following a trajectory. The proportional controller is maintained constant during the learning process and the fuzzy inverse-model is initialised by the rules extracted from the training set, as described in section III.

The results in figure 8 illustrate the actuation of the inverse model in compensate the nonlinearitie. Until 17 seconds, only the feedback-loop of the proportional controller is actuating. After, the feedback and feedforward loops are present, and the control and learning cycles are executed in real-time decreasing the trajectory errors as shown in figure 12(b).

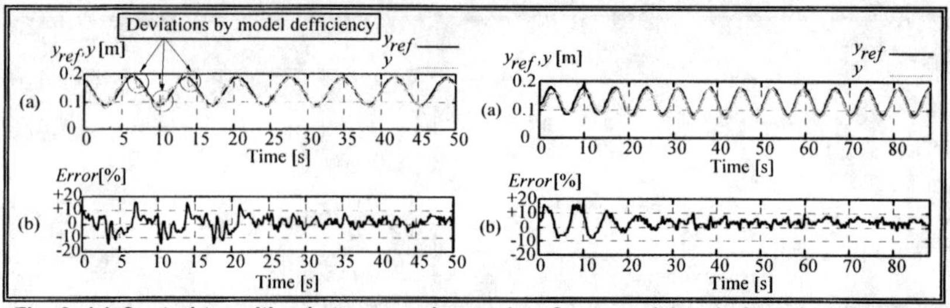


**Fig. 9: (a) Control transition between an incomplete fuzzy model and after when it begins to acquire knowlege and corrects the model. (b) Error signal evolution. Fig. 10: (a) Learning and compensation for a new system dynamic caused by load increase. (b) Error evolution.**

Figure 9 illustrates the situation where the initial fuzzy model is incomplete by information fault on the learning set. Incomplete means that there are rules that were not extracted. When the model needs to activate these rules, it fails and cause great deviations from the desired trajectory, as shown in Figure 9(a). As the system proceeds in the learning and control cycles, these rules are extracted completing the model. Figure 9(a) displays the transition between the two anterior situations. Until about 20 seconds, the control system presents deviations from the trajectory. After that, additional examples begin to be used by the learning process to complete the model and the trajectory error decreases, as shown in figure 9(b).

The main feature of the control system is its real-time learning capability to compensate changes in actuator dynamics. An example is presented in figure 10 where the load applied to the actuator is increased. Until 10 seconds the system is controlled only by the proportional controller with this load. The initial fuzzy model was extracted using the anterior training set which uses a lower load applied to the piston. After, the learning loop is connected to the system and the model starts to change itself to learn and compensate the new dynamics. The sensor signals coming from the actuator describe the new dynamic condition imposed by the increase of the load. So, as the fuzzy learning process modifies the initial knowledge accumulated in the inverse model, the position error decreases as seen in figure 10(b).

## CONCLUSION

This paper presented the capabilities of a fuzzy learning controller, based on the feedback-error-learning method, in compensate the unstructured uncertainties of an experimental electro-hydraulic actuator. The method used heuristic knowledge and numerical information to extract its inverse fuzzy model. The learning controller avoids the *black-box* approach usually present in the neural network based schemes. The experimental results showed that the control system has the capability of learn relations not incorporated in the initial model, change itself to characterise a new process dynamic, and has generalisation capability.

## REFERENCES

[1] Myamoto, H., Kawato, M., Setoyama, T., and Suzuki, R. (1988). *"Feedback-error-learning neural network for trajectory control of a robotic manipulator"*, **Neural Networks** 1:251-265.

[2] Kawato, M., Uno, Y., Isobe, M. and Suzuki, R. (1988). *"Hierarchical neural network model for voluntary movement with application to robotics,"*, **IEEE Control System Magazine**, pp. 8-16, April.

[3] Wang, L. X.; Mendel, J.; (1992) *"Generating Fuzzy Rules by Learning From Examples"*, **IEEE Trans. on Syst. Man and Cybernetics**, vol. 22, no. 6, July.

# MODEL REFERENCE ADAPTIVE FUZZY CONTROLLER DESIGN USING GENETIC ALGORITHMS

**WEN-RURY HWANG, SALEH ZEIN-SABATTO, AND JOHN G. KUSCHEWSKI**
Department of Electrical and Computer Engineering
Tennessee State University, Nashville, TN 37209-1561

## ABSTRACT

A methodology for combining genetic algorithms (GAs) with fuzzy algorithms to create fuzzy controllers for model reference adaptive control systems is presented. Using GAs, optimal or near optimal fuzzy rules and membership functions can be designed without a human operator's experience or a control engineer's knowledge, although such information can be useful for initial designs. This genetic/fuzzy approach involves searching the encoded fuzzy rule and membership function parameter spaces using a fitness function that is defined in terms of a system performance criterion. We demonstrate this approach in an application where a GA adapts the fuzzy rules and membership functions of a fuzzy controller at the same time for a model reference adaptive tracking system.

## INTRODUCTION

There have been a number of successful fuzzy control system studies in which the controller designs are based entirely on the linguistic equivalent of a human operator's experience or a control engineer's knowledge [1]. However, this experience or knowledge may be necessary but not sufficient for the design of optimal fuzzy control systems.

To address this concern, it is appealing to consider an adaptive fuzzy control system equipped with evolutionary search procedures known as genetic algorithms (GAs). GAs are randomized, parallel, and global search methods based on natural selection and natural genetic operators such as crossover and mutation. Often used as a basis for parameter search techniques, (e.g. searching for optimal PID controller parameters [2]), they manipulate parameter set codings to find near optimal solutions to performance optimization problems. GAs have been quite successfully applied to a variety of problems [3, 4] and are an increasingly important tool in the solution of difficult function optimization and machine learning problems. A thorough introduction to GAs and a list of possible application areas can be found in [5, 6].

This paper presents a methodology for designing an optimal or near optimal Model-Reference Adaptive Controller (MRAC) by combining fuzzy and genetic algorithms (MRAF-GA). We illustrate this methodology by applying it to tracking system design.

# MRAF-GA CONTROL SYSTEMS

## Structure

A traditional Model-Reference Adaptive Control (MRAC) system has three components: a plant $G_p(s)$ with unknown parameters, a reference model $G_m(s)$ that specifies the desired plant response, and a controller with adaptable parameters equipped with a parameter adaptation mechanism [7]. The block diagram of the Model-Reference Adaptive Fuzzy-GA (MRAF-GA) control system is shown in Figure 1.

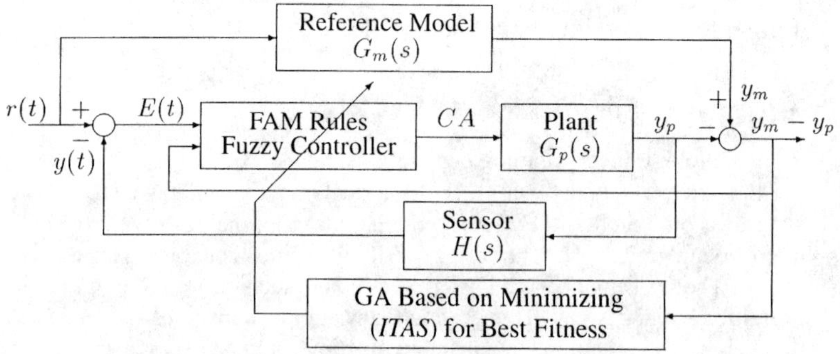


Figure 1: A Model-Reference Adaptive Fuzzy-GA (MRAF-GA) control system.

## Encoding Parameters and Fitness Measures

In this paper, fuzzy rules, which relate input variables such as $E(t) = r(t) - y(t)$ and $y_m(t) - y_p(t)$ to an output variable such as $CA$ (control action), have 49 parameters that range over 7 fuzzy levels. Each parameter is encoded into 3 binary bits that cover the range 0 to 7. The range of the membership functions (e.g. $\mu_{LN}$, $\mu_{MN}$, $\mu_{SN}$, $\mu_{ZE}$, $\mu_{SP}$, $\mu_{MP}$, $\mu_{LP}$) for the input signal $E$ is encoded into 8 binary bits that cover 0.01 to 2.57, and is represented by $M_E$. The range of the membership functions for the input signal $y_m(t) - y_p(t)$ is also encoded into 8 binary bits that cover 0.01 to 2.57, and is represented by $M_{CE}$. The $M_E$ and $M_{CE}$ are the adaptive ranges of fuzzy variable membership functions of triangular shape for the input signals $r(t) - y(t)$ and $y_m(t) - y_p(t)$. Thus, the total number of bits in the string describing each member of the population is $49 \times 3 + 8 + 8 = 163$.

In an analogy to biology, each encoded parameter can be thought of as a gene, where the parameter's values correspond to the gene's traits. The string produced by concatenating all the encoded parameters forms a genotype, an individual member of population. Evaluation of each string is based upon a fitness measure that is problem dependent. In control system applications, a performance criterion often used is a weighted combination of various performance characteristics such as rise_time, settling_time, overshoot, and the ITAE (the integral of time multiplied by absolute error). We use the performance criterion $J$ = ITAE for MRAF-GA control system applications. The MRAF-GA controllers track a reference model system output which makes the controlled system outputs smooth. The fitness measure is defined as $fitness = \frac{1}{1+J}$; thus one can see that since $J$ is non-negative, minimizing $J$ maximizes fitness. Other aspects of our genetic algorithm are: binary parameter encoding, stochastic remainder reproduction, simple crossover, uniform mutation, crossover probability, $P_c = 0.80$, mutation probability, $P_m = 0.02$, total population size is 100, total chromosome length is 163, and maximum generations is 1000.

# APPLICATION: MRAF-GA TRACKING SYSTEMS

We now present examples of MRAF-GA tracking systems that illustrate the effectiveness of the genetic algorithm design procedure.

## Description and Results For a Stable System

Consider the control structure shown in Figure 1, and let the desired second-order reference model $G_m(s)$ be:

$$G_m(s) = \frac{15(s+4)}{(s^2+12s+52)}. \tag{1}$$

The reference model generates an ideal plant response that the adaptation mechanism should seek when adjusting the controller parameters. Hence, the choice of the reference model is a part of the overall control system design.

In traditional MRAC systems, the controller contains a number of adjustable parameters. After choosing a control law and an adaptation law, the adaptation mechanism adjusts the controller parameters so that optimal tracking is asymptotically achieved. In MRAF-GA systems, the fuzzy rules and membership function parameters are the controller parameters and the adaptation law is GA evolution, a search for optimal fuzzy rules and membership function parameters. The optimal fuzzy rules and membership function parameters should make the plant output $y_p(t)$ identical to the reference model output $y_m(t)$. In our example, two types of inputs, $r(t) = 4U(t)$ and $r(t) = 4\sin(6t)$, are used as reference inputs. The reason for choosing these two types of inputs is that they are typical tracking system reference inputs.

In the first simulation, we chose a second order plant characterized by the following transfer function:

$$G_{p1}(s) = \frac{10(s+2)}{(s^2+0.1s)} \tag{2}$$

with feedback sensor transfer function:

$$H(s) = \frac{0.4}{(s+2)} \tag{3}$$

The resulting $y_m(t)$ and $y_p(t)$ for different generations are shown in Figure 2 for $r(t) = 4U(t)$ and in Figure 3 for $r(t) = 4\sin(6t)$.

It is clear that with $r(t) = 4U(t)$, the plant output $y_p(t)$ becomes nearly identical to the model reference output $y_m(t)$ at generation 749 as shown in Figure 2. Similarly, with $r(t) = 4\sin(6t)$, $y_p(t)$ becomes nearly identical to $y_m(t)$ at generation 772 as shown in Figure 3. From these responses, we see that $M_E$ changes dramatically close to the highest boundary value of 2.57 and $M_{CE}$ decreases close to the lowest boundary

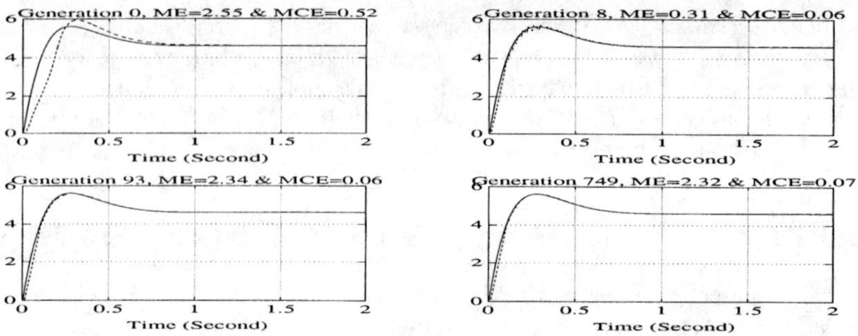


Figure 2: Resulting $y_m(t)$ (solid line) and $y_p(t)$ (dashed line) for generation 0, 8, 93, and 749 for $G_{p1}(s)$ with $r(t) = 4U(t)$.

value of 0.01 for $r(t) = 4U(t)$. For $r(t) = 4\sin(6t)$, $M_E$ and $M_{CE}$ decrease close to the lowest boundary value of 0.01.

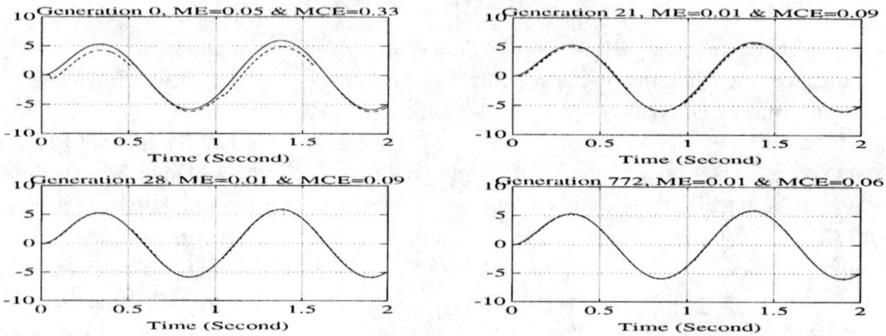


Figure 3: Resulting $y_m(t)$ (solid line) and $y_p(t)$ (dashed line) for generation 0, 21, 28, and 772 for $G_{p1}(s)$ with $r(t) = 4\sin(6t)$.

The performance index $J$ versus generation with inputs $r(t) = 4U(t)$ and $r(t) = 4\sin(6t)$ was plotted. From these plots, we see that although in both cases the best results are obtained after generation 700, results almost as good are obtained before generation 50.

## Description and Results For an Unstable System

The performance of a MRAF-GA tracking system on an unstable, non-minimum phase, plant is now evaluated. We choose

$$G_{p2}(s) = \frac{10(s+2)}{(s^2 - 0.1s)}. \tag{4}$$

The resulting $y_m(t)$ and $y_p(t)$ for different generation are shown in Figure 4 for $r(t) = 4U(t)$, and in Figure 5 for $r(t) = 4\sin(6t)$.

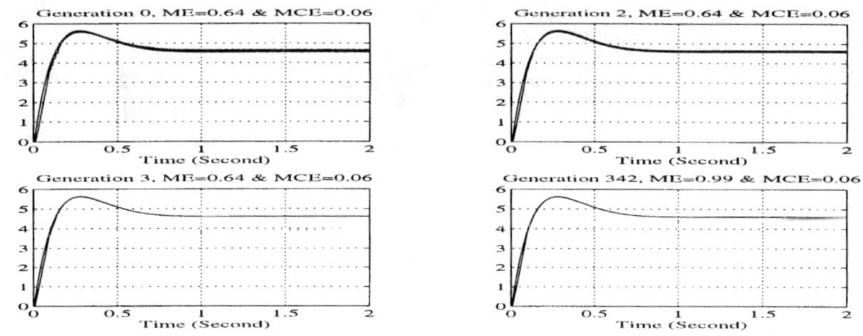


Figure 4: Resulting $y_m(t)$ (solid line) and $y_p(t)$ (dashed line) for generation 0, 2, 3, and 342 for $G_{p2}(s)$ with $r(t) = 4U(t)$.

It is clear that with $r(t) = 4U(t)$ the plant output $y_p(t)$ becomes nearly identical to the model reference output $y_m(t)$ at generation 342, as shown in Figure 4. Similarly, with $r(t) = 4\sin(6t)$ the plant output $y_p(t)$ becomes nearly identical to the reference model output $y_m(t)$ at generation 223.

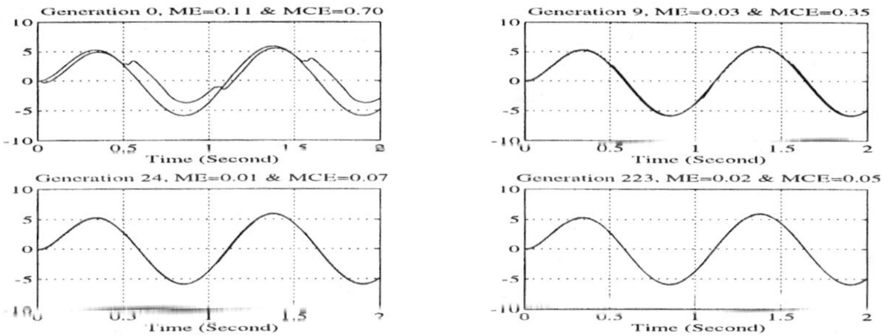


Figure 5: Resulting $y_m(t)$ (solid line) and $y_p(t)$ (dashed line) for generation 0, 9, 24, and 223 for $G_{p2}(s)$ with input $r(t) = 4\sin(6t)$.

The performance index $J$ versus generation with $r(t) = 4U(t)$ and $r(t) = 4\sin(6t)$ was plotted. Although the best result is obtained at generation 342 with $r(t) = 4U(t)$, a very good result is achieved by generation 10. Similarly, with $r(t) = 4\sin(6t)$, a very good result is achieved by generation 50 even though the best result occurs at generation 223.

# DISCUSSION

In this paper, we presented the elements of a procedure for designing near optimal model reference adaptive fuzzy logic controllers (MRAFs) using genetic algorithms (GAs). The use of GAs for evolutionary search results in a near optimal MRAF-GA by finding near optimal fuzzy rules and membership functions. In addition, using GAs for designing a FLC (Fuzzy Logic Controller) requires much less effort than traditional methods that are based on a priori information. Although a previous methodology that fixes the domain of the membership functions and finds the fuzzy rules can obtain good results [8, 9], adapting the domain of the membership functions and finding the near optimal fuzzy control rules at the same time yields improved performance in a fewer number of generations. The success in the MRAF-GA control system application shows that the combination of genetic and fuzzy algorithms can result in a successful control methodology.

## ACKNOWLEDGMENTS

This work is supported by NASA Lewis Research Center under NASA grant No. NAG3-1471. This effort is also a part of research activities conducted in the Center for Neural Engineering (CNE) at Tennessee State University. The CNE is supported by the U.S. Navy Grant No. N00014-92-J-1372. The authors wish to thank both NASA and ONR for their support of this research.

## REFERENCES

[1] W. R. Hwang, C. W. Tao, W. E. Thompson, and R. Paz. A comparison of a fuzzy controller with an optimal pd controller for a nonlinear control system. In *The 36th Midwest Symposium on Circuits and Systems*, Detroit, MI, August 1993.

[2] W. R. Hwang and W. E. Thompson. An intelligent design based on genetic algorithms. In *The 32nd IEEE Conference on Decision and Control*, Marriott Rivercenter, San Antonio, TX, December 1993.

[3] M. Pearce, R. Arkin, and A. Ram. The learning of reactive control parameters through genetic algorithms. In *IEEE/RSJ International Conference on Intelligent Robots and Systems*, pages 130–137, Raleigh, NC, July 1992.

[4] K. Kristinsson and G. A. Dumont. System identification and control using genetic algorithms. *IEEE Transactions on Systems, Man, and Cybernetics*, 22(5):1033–1046, 1992.

[5] D. E. Goldberg. *Genetic Algorithms in Search, Optimization, and Machine Learning.* Addision-Wesley Publishing Company, Inc., Reading, MA, 1989.

[6] Z. Michalewicz. *Data Structures + Genetic Operators = Evolutionary Programs.* Springer-Verlag, Berlin Heidelberg, 1992.

[7] J. E. Slotine and W. Li. *Applied Nonlinear Control.* Prentice-Hall, Inc., 1991.

[8] W. R. Hwang and W. E. Thompson. Design of intelligent fuzzy logic controllers using genetic algorithms. In *Proceedings of the Third IEEE International Conference on Fuzzy Systems*, pages 1383–1388, Orlando, FL, June 1994.

[9] W. R. Hwang and W. E. Thompson. A methodology for designing an optimum fuzzy tracking using genetic algorithms. In *Proceedings of the SPIE International Symposium on Optical Engineering in Aerospace Sensing: Signal Processing, Sensor Fusion, and Target Recognition*, volume 2232, pages 97–104, Orlando, FL, April 1994.

# DESIGN OF A FUZZY CONTROLLER BASED ON FUZZY CLOSED-LOOP SPECIFICATIONS

**Cezary Kolodziej[1] and Roland Priemer[2]**
Electrical Engineering and Computer Science Department
The University of Illinois at Chicago
Chicago, Illinois

ABSTRACT

Very often the control knowledge of an expert is not complete or, even worse, it consists of contradictory rules. Thus, the performance of a closed-loop system based on this knowledge does not meet requirements. We give a theorem about obtaining the closed-loop system fuzzy relation using the fuzzy relation of the controller and the fuzzy relation of the plant. Based on this theorem and using a fuzzy rule model of the plant and a desired fuzzy rule model of the closed-loop system, we develop a complete design procedure to obtain the fuzzy controller.

**Keyword: Fuzzy relations, fuzzy control.**

## 1 INTRODUCTION

For the last 20 years, fuzzy control theory has emerged as a fruitful approach to a wide variety of control problems [Bardossy and Duckstein (1995); Hellendoorn and Palm (1994); Jamshidi et al. (1993); King and Mamdani (1977); von Altrock (1995); Stenz and Kuhn (1995); Takahashi et al. (1994); Yamakawa (1989); Yen et al. (1995)]. The design of a fuzzy controller is usually based on control knowledge of an expert or on an adaptive process, both of which rely on experience to achieve acceptable performance.

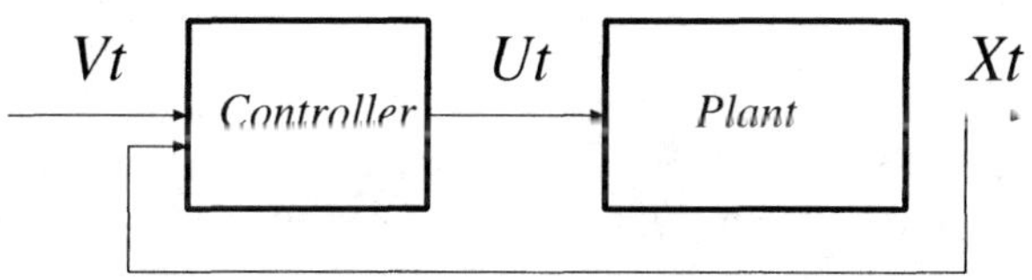


Figure 1. Block diagram of a closed-loop system.

Very little work concerned with the analysis of a fuzzy closed-loop control system, shown in Figure 1, has been reported [Tong (1980); Chen and Chen (1993)]. Furthermore, there is no widely accepted method of closed-loop fuzzy relation calculation. Tong [Tong (1980)] has studied a fuzzy closed-loop system, but his closed-loop system fuzzy relation is not only a function of the fuzzy relations of the closed-loop system components but also the initial conditions of the fuzzy state variables. This increases the dimension of the fuzzy relation of the closed-loop system, and makes it very difficult to analyze the stability of a system by fuzzy

[1]ckolodzi@eecs.uic.edu
[2]priemer@eecs.uic.edu

relations. For these reasons Chen *at el* [Chen and Chen (1993)] have proposed an analytical model of a fuzzy closed-loop control system that does not increase the dimension of the closed-loop fuzzy relation over that of the open-loop system, and it is based on the linguistic fuzzy model of the process to be controlled and on the fuzzy relation matrix of a fuzzy controller. The major drawback of Chen's method is that it was derived just for the particular case when there is no external input to the fuzzy controller, and the linguistic model of the plant has the particular form cosisting of two fuzzy propositions in the antecedent and a single fuzzy proposition in the consequent of a fuzzy rule. This limits the applicability of the method not only to error based control strategies but also to very simple (low-order) plants.

First, the paper gives a new method for closed-loop fuzzy relation calculation using the fuzzy relation of a process and the fuzzy relation of a fuzzy controller of any dimension with an external input. Then, based on the proposed method we give an algorithm that synthesizes the fuzzy relation of a fuzzy controller based on the desired fuzzy relation of a closed-loop system and the fuzzy relation of the plant. This algorithm can be used to design a fuzzy controller that meets performance objectives that are inherently included in fuzzy rules of a closed-loop system.

## 2 FUZZY RELATION OF A CLOSED-LOOP SYSTEM

The block "Plant" in Figure 1 is a fuzzy model of a plant to be controlled. It is given in terms of rules such as

$$\begin{array}{l} p_1 : \textit{If } X_t \textit{ is } A_1 \textit{ and } U_t \textit{ is } B_1 \textit{ then } X_{t+1} \textit{ is } C_1 \\ \vdots \\ p_n : \textit{If } X_t \textit{ is } A_n \textit{ and } U_t \textit{ is } B_n \textit{ then } X_{t+1} \textit{ is } C_n \end{array} \tag{1}$$

where $p_i$ represents $i$th fuzzy rule, $A_i$ and $B_i$ represent linguistic values in fuzzy propositions of the antecedents of the rules, and $C_i$ represents linguistic values in fuzzy propositions of the consequents of the rules. Then, the equation

$$X_{t+1} = X_t \circ U_t \circ R_p \tag{2}$$

is the fuzzy relational equation of the open-loop system, where $X_t \in F\left(\bar{X}\right)$ and $X_{t+1} \in F\left(\bar{X}\right)$ are fuzzy state variables at times $t$ and $t+1$, respectively. The fuzzy relation $R_p \in F\left(\bar{X} \times \bar{U} \times \bar{X}\right)$ can be obtained with

$$R_p = \bigvee_{i=1}^{n} \{X_{t_i} \wedge U_{t_i} \wedge X_{t+1_i}\} \tag{3}$$

where $n$ is the number of fuzzy rules. Here, $U_t \in F\left(\bar{U}\right)$ is the input fuzzy variable, both $\bar{X}$ and $\bar{U}$ represent the universes of discourse, and $\circ$ is a max-min composition operator.

The block "Fuzzy Controller" in Figure 1 is the fuzzy controller in terms of rules such as

$$\begin{array}{l} c_1 : \textit{If } X_t \textit{ is } A_1 \textit{ and } V_t \textit{ is } D_1 \textit{ then } U_t \textit{ is } B_1 \\ \vdots \\ c_m : \textit{If } X_t \textit{ is } A_m \textit{ and } V_t \textit{ is } D_m \textit{ then } U_t \textit{ is } B_m \end{array} \tag{4}$$

where $X_t$ and $U_t$ are defined as before, $V_t \in F\left(\bar{V}\right)$ is an external input fuzzy variable and the $D_i$ are fuzzy values representing linguistic values. From the above linguistic description of the fuzzy controller, the fuzzy relational equation of the controller is given by

$$U_t = X_t \circ V_t \circ R_c \tag{5}$$

where $R_c = F \in \left(\bar{X} \times \bar{V} \times \bar{U}\right)$ is obtained with

$$R_c = \bigvee_{i=1}^{m} \{X_{t_i} \wedge V_{t_i} \wedge U_{t_i}\} \tag{6}$$

The fuzzy relational equation of the closed-loop system can be written as follows

$$X_{t+1} = X_t \circ (X_t \circ V_t \circ R_c) \circ R_p \tag{7}$$

However, equation (7) is of higher order than we would expect according to Figure 1. The increased order of the fuzzy relational equation of the fuzzy closed-loop system is a major drawback for fuzzy description. We would prefer to have a method to calculate a fuzzy relational equation of the closed-loop system which does not artificially increase the order of the fuzzy relational equation. The other disadvantage of (7) is that it cannot be easily used to analyze the stability of the closed-loop system. To apply the already developed methods for stability analysis of fuzzy systems [Chen and Chen (1993); Kiszka et al. (1985)] we require the fuzzy relational equation of a closed-loop system in the form of (2). Therefore, we prove the following theorem.

**Theorem :** Any fuzzy system described by (2) and coupled with a fuzzy controller given by (5) can be described by the fuzzy relation given by

$$X_{t+1} = X_t \circ V_t \circ R_l \tag{8}$$

The matrix $R_l \in F\left(\bar{X} \times \bar{V} \times \bar{X}\right)$ is a closed-loop fuzzy relation calculated as follows

$$R_l\,(i,j,l) = \bigvee_{k=1}^{m} \{R_c\,(i,k,l) \wedge R_p\,(i,j,k)\} \tag{9}$$

where the $R_l\,(i,j,l)$ are elements of the 3-dimensional closed-loop system fuzzy relation, and $m$ is the number of elements in $\bar{U}$.

**Proof :** Suppose a fuzzy system is described by the fuzzy relation

$$X_{t+1} = X_t \circ U_t \circ R_p \tag{10}$$

where $X_t$ and $X_{t+1}$ are the discrete time universes of discourse given by

$$\begin{aligned} X_t &= [x_1, x_2, ..., x_n] \\ X_{t+1} &= [x_1, x_2, ..., x_n] \end{aligned} \tag{11}$$

Thus, $X_{t+1}$ and $X_t$ can be rewritten as follows

$$\begin{aligned} X_t &= \{x_t\,(i)\}_{i=1}^{n} \\ X_{t+1} &= \{x_{t+1}\,(j)\}_{j=1}^{n} \end{aligned} \tag{12}$$

$U_t$ is the discrete time universe of discourse given by

$$U_t = [u_1, u_2, ..., u_m] \tag{13}$$

or

$$U_t = \{u_t(k)\}_{k=1}^{m} \tag{14}$$

And, $R_p$ is a 3-dimensional fuzzy relation of a fuzzy process.

The algorithm to calculate $X_{t+1} = \{x_{t+1}(j)\}_{j=1}^{n}$ is given by

$$X_{t+1} = X_t \circ (\bigvee_{k=1}^{m} u_t(k) \wedge R_p(i,j,k)) \tag{15}$$

By further extension, we get

$$X_{t+1} = \bigvee_{i=1}^{n} x_t(i) \wedge (\bigvee_{k=1}^{m} u_t(k) \wedge R_p(i,j,k)) \tag{16}$$

A 3-dimensional fuzzy controller description has the form given by

$$U_t = X_t \circ V_t \circ R_c \tag{17}$$

where $U_t$ and $X_t$ are defined as before, and $V_t$ is an external input. To have the same scale as for $X_t$ and $X_{t+1}$ ( Since we describe the output of the closed-loop control system in terms of $X_t$ and $X_{t+1}$, we want to use the same scale to describe $V_t$, and then be able to compare it with $X_t$. ) we describe $V_t$ similarly to $X_t$ and $X_{t+1}$. Thus,

$$V_t = [v_1, v_2, ..., v_n] \tag{18}$$

or

$$V_t = \{v_t(l)\}_{l=1}^{n} \tag{19}$$

And, $R_c$ is a 3-dimensional fuzzy relation.

Using (17), we can calculate the controller output with

$$\{u_{t+1}(k)\}_{k=1}^{m} = X_t \circ \{\bigvee_{l=1}^{n} (v_t(l) \wedge R_c(i,k,l))\} \tag{20}$$

By further extension, we can write

$$\{u_{t+1}(k)\}_{k=1}^{m} = \bigvee_{i=1}^{n} [x_t(i) \wedge \{\bigvee_{l=1}^{n} (v_t(l) \wedge R_c(i,k,l))\}] \tag{21}$$

When we connect all closed-loop system components together, we obtain the closed-loop fuzzy relation given by

$$X_{t+1} = X_t \circ (X_t \circ V_t \circ R_c) \circ R_p \tag{22}$$

Using the extended notation, (22) can be rewritten to become

$$\{x_{t+1}(j)\}_{j=1}^{n} = X_t \circ (\bigvee_{k=1}^{m} \{u_t(k) \wedge R_p(i,j,k)\}) \tag{23}$$

Thus, we get

$$\{x_{t+1}(j)\}_{j=1}^{n} = \bigvee_{i=1}^{n} x_t(i) \wedge (\bigvee_{k=1}^{m} \{u_t(k) \wedge R_p(i,j,k)\}) \tag{24}$$

Substituting (21) into (24) gives

$$\{x_{t+1}(j)\}_{j=1}^{n} = \bigvee_{i=1}^{n} x_t(i) \wedge (\bigvee_{k=1}^{m} \{\bigvee_{i=1}^{n} x_t(i) \wedge [\bigvee_{l=1}^{n} (v_t(l) \wedge R_c(i,k,l))] \wedge R_p(i,j,k)\}) \tag{25}$$

By switching the second and third max-operators, we get

$$\{x_{t+1}(j)\}_{j=1}^{n} = \bigvee_{i=1}^{n} x_t(i) \wedge (\bigvee_{i=1}^{n} x_t(i) \wedge \bigvee_{k=1}^{m} [\bigvee_{l=1}^{n} (v_t(l) \wedge R_c(i,k,l))] \wedge R_p(i,j,k)) \tag{26}$$

This simplifies to

$$\{x_{t+1}(j)\}_{j=1}^{n} = \bigvee_{i=1}^{n} x_t(i) \wedge (\bigvee_{k=1}^{m} [\bigvee_{l=1}^{n} (v_t(l) \wedge R_c(i,k,l))] \wedge R_p(i,j,k)) \tag{27}$$

Further, we can write

$$\{x_{t+1}(j)\}_{j=1}^{n} = \bigvee_{i=1}^{n} x_t(i) \wedge (\bigvee_{l=1}^{n} v_t(l) \wedge [\bigvee_{k=1}^{m} \{\wedge R_c(i,k,l) \wedge R_p(i,j,k)\}]) \tag{28}$$

On the other hand, we know that the fuzzy closed-loop system can be defined by the fuzzy relation given by

$$X_{t+1} = X_t \circ V_t \circ R_l \tag{29}$$

where $R_l$ is the closed-loop system fuzzy relation.

Using the extended notation, (29) becomes

$$\{x_{t+1}(j)\}_{j=1}^{n} = X_t \circ (\bigvee_{l=1}^{n} \{v_t(l) \wedge R_l(i,j,l)\}) \tag{30}$$

Thus, we get

$$\{x_{t+1}(j)\}_{j=1}^{n} = \bigvee_{i=1}^{n} x_t(i) \wedge (\bigvee_{l=1}^{n} \{v_t(l) \wedge R_l(i,j,l)\}) \tag{31}$$

If we compare (31) with (28), we get the final result

$$R_l(i,j,l) = \bigvee_{k=1}^{m} \{R_c(i,k,l) \wedge R_p(i,j,k)\} \tag{32}$$

Graphically, we can visualize the closed-loop system fuzzy relation calculation as that of obtaining the minimum overlapping of two 3-dimensional arrays through U projections in each component of the fuzzy relation. Hence, it can be applied to higher order systems.

## 3 SYNTHESIS OF A FUZZY CONTROLLER.

Now we assume that the fuzzy rules of a closed-loop system and the fuzzy rules of a plant are given and we want to synthesize the fuzzy relation of a fuzzy controller. Thus, the controller will meet closed-loop system performance specifications of an expert.

An algorithm that is based on equation (9) consists of the following steps.

1. Based on the fuzzy rules of a plant and equation (3), obtain the fuzzy relation of $R_p$.
2. Based on the fuzzy rules of a closed-loop system and equation (3), obtain $R_l$.
3. Initialize the fuzzy relation of the controller, $R_c$, with grades of membership values representing "everything" for the given fuzzy input sets and fuzzy output sets of the designed fuzzy controller.
4. Do the following for all elements in the fuzzy relation of the initialized fuzzy controller, i.e.

$$\begin{array}{ll}\textbf{If } \{R_c(i,j,k) \wedge R_p(i,j,k)\} > R_l(i,j,l) \textbf{ then,} & R_c(i,j,k) = R_l(i,j,l) \\ & \textbf{otherwise } R_c(i,j,k) \textbf{ is not changed}\end{array} \tag{33}$$

5. Based on synthesized fuzzy controller relation, obtain the fuzzy rules of the fuzzy controller.

## 4 CONCLUSION

A new method for closed-loop system fuzzy relation matrix calculation has been presented. It does not artificially increase the dimension of the fuzzy relation of the closed-loop system. The obtained fuzzy relational equation of a closed-loop system is in the same form as the open-loop system fuzzy relational equation, and therefore it is possible to apply stability criteria developed for fuzzy open-loop systems. Then, based on the given theorem, we gave an algorithm to synthesize a fuzzy controller using fuzzy rules of a fuzzy closed-loop system and fuzzy rules describing the plant to be controlled. This is a great advantage over design procedures developed elsewhere, where trial and error methods are used to achieve acceptable performance instead of predefined closed-loop system performance criteria. This way of describing closed-loop system behavior is easily understandable by experts, and it can speed up the design process. Despite its advantages, the presented algorithm needs further development to explicitly include available stability criteria to insure that the overall closed-loop system stability is not only based on experience of a designer but also on systematic stability analysis.

## REFERENCES


Bardossy, A., Duckstein, L., 1995, *Fuzzy Rule-Based Modelling with Application to Geophysical, Biological and Engineering Systems* CRC Press, New York.

Chen, J., Chen, L., 1993, "Study on stability of closed-loop control systems," *Fuzzy Sets and Systems*, Vol. 57, pp. 159–168.

Hellendoorn, H., Palm, R., 1994, "Fuzzy systems technologies at Siemens R & D," *Fuzzy Sets and Systems*, Vol. 63, pp. 245–269.

Iskandar, M., Mizutani, Y., Suzuki, A., Ishizeki, M., 1994, "Stabilizing control of power system usig fuzzy control," *Electrical Engineering in Japan*, Vol. 114, pp. 33–45.

Jamshidi, M., Vadiee, N., Ross, T., 1993, *Fuzzy Logic and Control* PTR Prentice Hall, Englewood Cliffs, New Jersey 07632.

Kawakita, Y., Ohsawa, Y., Arai, K., 1994, "Power-System Stabilizing Control by SMES Using Fuzzy Techniques and Neural Networks," *Electrical Engineering in Japan*, Vol. 114, pp. 9–16.

King, P., Mamdani, E., 1977, "The application of fuzzy control systems to industrial processes," *Automatica*, Vol. 3, pp. 235–242.

Kiszka, J., Gupta, M., Nikiforuk, P., 1985, "Energetic stability of fuzzy dynamic systems," *IEEE Transactions on Systems, Man, and Cybernetics*, Vol. 15, pp. 783–792.

Kitauchi, Y., Taniguchi, H., 1994, "Development of fuzzy exitation control system," *Electrical Engineering in Japan*, Vol. 114, pp. 47–57.

Stenz, R., Kuhn, U., 1995, "Automation of a batch distilation column using fuzzy and conventional control," *IEEE Transactions on Systems Technology*, Vol. 3, pp. 171–176.

Takahashi, T., Kitou, M., Asai, M., Kido, M., Chiba, T., Kawakami, J., Matsui, Y., 1994, "A new voltage equipment using fuzzy inference," *Electrical Engineering in Japan*, Vol. 114, pp. 18–32.

Tong, R., 1980, "Some properties of fuzzy feedback systems," *IEEE Transactions on Systems, Man, and Cybernetics*, Vol. 10, pp. 327–330.

von Altrock, C., 1995, *Fuzzy Logic and NeuroFuzzy Applications Explained* Prentice Hall PTR, Englewood Cliffs, New Jersey 07632.

Yamakawa, T., 1989, "Stabilization of an inverted pendulum by a high-speed fuzzy logic controller hardware system," *Fuzzy Sets and Systems*, Vol. 32, pp. 161–180.

Yen, J., Langari, R., Zadeh, L., 1995, *Idustrial Applications of Fuzzy Logic and Intelligent Systems* IEEE Press, IEEE, Inc, New York.

# FUZZY LOGIC CONTROL SYSTEM FOR THREE DIMENSIONAL LINE FOLLOWING FOR A MOBILE ROBOT

**TAYIB SAMU, NIKHIL KELKAR AND ERNEST L. HALL**
University of Cincinnati
Center for Robotics Research, ML # 72
Cincinnati, OH 45221
Phone: (513) 556-2730
FAX: (513) 556-3390
Email: elhall@uceng.uc.edu

**ABSTRACT:**
Automated Guided Vehicles (AGV's) have many potential applications in manufacturing, medicine, space and defense. The purpose of this paper is to describe exploratory research on the design of a modular autonomous mobile robot controller. The controller incorporates a fuzzy logic approach for steering and speed control, a neuro-fuzzy approach for ultrasound sensing (not discussed in this paper) and an overall expert system. The advantages of a modular system are related to portability and transportability, i.e. any vehicle can become autonomous with minimal modifications. A mobile robot test-bed has been constructed using a golf cart base. This cart has full speed control with guidance provided by a vision system and obstacle avoidance using ultrasonic sensors. The speed and steering fuzzy logic controller is supervised by a 486 computer through a multi-axis motion controller. The obstacle avoidance system is based on a micro-controller interfaced with six ultrasonic transducers. This micro-controller independently handles all timing and distance calculations and sends a steering angle correction back to the computer via the serial line. This design yields a portable independent system in which high speed computer communication is not necessary. Vision guidance is accomplished with a CCD camera with a zoom lens. The data is collected by a vision tracking device that transmits the X,Y coordinates of the lane marker to the control computer. Simulation and testing of these systems yielded promising results. This design, in its modularity, creates a portable autonomous fuzzy logic controller applicable to any mobile vehicle with only minor adaptations.

**Keywords:** vision guidance, ultrasonic sensing, mobile robots, intelligent machines, three dimensional imaging, fuzzy logic.

## 1.0 Introduction

An excellent introduction to the mathematical analysis of mobile robots can be found in [1]. Even though the visualization and recognition of image information for the guidance of mobile robot have been studied for years[2-7], the design of a mobile system is a challenging task. The challenge of designing an intelligent controller is in determining what information to measure and how to use this information in a manner that will satisfy the performance specifications of the machine. Several fuzzy logic approach for modeling control system for vehicles have been studied in the past. A fuzzy logic controller[8] that guarantees stability of a control system for a computer-simulated model car and advanced fuzzy logic application for automobiles application was discussed in [9].

The main aspect of intelligent control addressed in this paper is the design of the controller for a mobile robot using fuzzy logic. The design specifications were to build a robot which could follow a line, avoid obstacles, and adapt to variations in terrain. The adaptive capabilities of a mobile robot depend on the fundamental analytical and architectural designs of the sensor systems used. The mobile robot provides an excellent test bed for investigations into generic vision guided robot control since it is similar to an automobile and is a multi-input, multi-output system. An algorithm was developed to establish a mathematical and geometrical relationship between the physical 3-D ground coordinates of the line to follow and its corresponding 2-D digitized image coordinates. This relationship is incorporated into the vision tracking system to determine the perpendicular distance and angle of the line with respect to the centroid of the robot. The information from the vision tracking system was used as input to a closed loop fuzzy logic controller to control the steering and the speed of the robot. Section 2 presents the model for the vision tracking system. In Section 3, the fuzzy logic controller is related. Results of simulation of model and conclusion are respectively presented in Sections 4 and 5.

## 2.0 Vision Guidance

The purpose of the vision guidance is to guide the robot to follow the line using a digital CCD camera. To do this, the camera needs to be calibrated. Camera calibration is a process to determine the relationship between a given 3-D coordinate system (world coordinates) and the 2-D image plane a camera perceives (image coordinates). More specifically, it is to determine the camera and lens model parameters that govern the mathematical or geometrical transformation from world coordinates to image coordinates based on the known 3-D control field and its image. The CCD camera digitizes the line from 3-D coordinate system to 2-D image system. Since the process is autonomous, the relationship between the 2-D system and the 3-D system has to be accurately determined so that the robot can follow the line. The objective of this section is to show how a model was developed to calibrate the vision system so that, given any 2-D image coordinate point, the system can mathematically compute the corresponding ground coordinate point. The X and Y (the Z is constant) coordinates of two ground points are then computed from which the angle and the perpendicular distance of the line with respect to the centroid of the robot are determined.

The vision system was modeled by the following equations.

$$xPI = A_{11}\, x_g + A_{12} y_g + A_{13} z_g + A_{14}\,. \qquad (1)$$

$$yPI = A_{21}x_g + A_{22}y_g + A_{23}z_g + A_{24} \qquad (2)$$

where $A_{nm}$ are coefficients, xPI and yPI are x and y image coordinates, and $x_g$, $y_g$, and $z_g$ are the ground coordinates. In transforming the ground coordinate points to the image coordinate points the following transformation operations occur on the points: scaling, translation, rotation, perspective, and projective. Solving for the transformation parameters to obtain the image and ground coordinate relationship is a difficult task. Fortunately, in the model equations given above, the transformation parameters are imbedded into the coefficients. To compute the coefficients, a calibration device was built to obtain 12 data points. With the 12 points, a matrix equation was yielded as shown below:

$$XPI = CA_{1k} \qquad (3)$$

$$YPI = CA_{2k} \qquad (4)$$

where

$$C = \begin{bmatrix} xg_1 & yg_1 & zg_1 & 1 \\ xg_2 & yg_2 & zg_2 & 1 \\ . & . & . & . \\ . & . & . & . \\ . & . & . & . \\ xg_{12} & yg_{12} & zg_{12} & 1 \end{bmatrix}; XPI = \begin{bmatrix} xPI_1 \\ xPI_2 \\ . \\ . \\ . \\ xPI_{12} \end{bmatrix}; YPI = \begin{bmatrix} yPI_1 \\ yPI_2 \\ . \\ . \\ . \\ yPI_{12} \end{bmatrix}; A_{1k} = \begin{bmatrix} A_{11} \\ A_{12} \\ A_{13} \\ A_{14} \end{bmatrix}; A_{2k} = \begin{bmatrix} A_{21} \\ A_{22} \\ A_{23} \\ A_{24} \end{bmatrix}$$

Equations (3) and (4) consist of 12 linearly independent equations and four unknowns, the least-square regression method is applied to yield a minimum mean-square error solution for the coefficients. Below are the equations for the solution:

$$A_{1K} = (C^TC)^{-1}C^TXPI \qquad (5)$$

$$A_{2K} = (C^TC)^{-1}C^TYPI \qquad (6)$$

Given an image coordinate xPI and yPI, and z ground coordinate (the z coordinate of the points with respect to the centroid of the robot is maintained constant since the robot is run on a flat surface in this model) the corresponding $x_g$ and $y_g$ ground coordinates are computed as indicated by the following matrix equations.

$$\begin{pmatrix} x_g \\ y_g \end{pmatrix} = Q^{-1}B \qquad (7)$$

where

$$Q=\begin{bmatrix} A_{11} & A_{12} \\ A_{21} & A_{22} \end{bmatrix}; B=\begin{bmatrix} xPI - A_{14} - (A_{13} z_g) \\ yPI - A_{24} - (A_{23} z_g) \end{bmatrix}$$

Note that equation (1) and (7) can be modified to accommodate the computation of $z_g$ when an elevation of the ground surface is considered.

The image processing of the physical points is done by the Iscan tracking device, which returns the centroid of the brightest or darkest region in a computer controlled windows and returns its X and Y coordinates. Two points on the line are windowed and their corresponding coordinates are computed as described above. From the computed x and y ground coordinates of the points, the angle of the line with respect to the centroid of the robot is computed from simple trigonometric relationship. In the next section, we shall show how the angle of the line just computed is used with other parameters to model the steering control of the robot with a fuzzy logic controller.

**3.0 Fuzzified Steering and Speed Controller**

The mobile robot is designed to maneuver an obstacle course. The width of the course is 10 feet. In controlling the robot to follow the painted line of the course, the angle of the line and that of the front wheel with respect to the centroid of the robot along the direction of motion (y-axis) are considered. Also considered is the perpendicular distance of the line to the centroid with respect to the x-axis. Both the distance and the angle need to be minimized. If only the angle is controlled, the robot might draw closer to the line eventually and strays out of bounds and the same is true for minimizing only the distance error.

A dual level fuzzy logic controller was modeled to control the robot. The first level of fuzzy system has two inputs, $\theta_{error}$ and $d_{error}$, and one output, $\theta_{steering}$. The second level of the fuzzy system has one input, $\theta_{steering}$ and one output, the speed of the robot. $\theta_{error}$ is the difference between the wheel and the line angle. The convention for steering the wheel is that positive angles are to the right and negative angles to the left. Four cases are considered in computing the $\theta_{error}$. The cases are: If angle of the wheel, $\theta_w$, is positive while the angle of the line, $\theta_l$, is negative then the error angle is computed as:

$$\theta_{error} = -(\theta_w + |\theta_l|) \tag{8}$$

The steering wheel needs to be turned to the left, i.e. $\theta_{error}$ has a negative value.

2. If $\theta_w$ is positive, and $\theta_l$ is also positive, the error angle is:

$$\theta_{error} = \theta_l - \theta_w \tag{9}$$

The direction of the steering is dependent on the sign of $\theta_{error}$.

3. If $\theta_w$ is negative, and $\theta_l$ is positive, then the error angle is:

$$\theta_{error} = |\theta_w| + \theta_l \tag{10}$$

The sign of $\theta_{error}$ is positive, as such, the direction is right.

4. If $\theta_w$ is negative, and $\theta_l$ is also negative, then the error angle is:

$$\theta_{error} = \theta_w + \theta_l \qquad (11)$$

The direction of steering will depend on the sign of $\theta_{error}$.
The algorithm to calculate $\theta_{error}$ is coded in C++. The steering wheel has a limit of ±20 degrees turning angle. As a result, $\theta_{error}$ is bounded by ±20. If the magnitude of $\theta_{error}$ is greater than 20, it is reduced to 20. $\theta_l$ is computed from the vision guidance model discussed in the previous section while $\theta_w$ is obtained from the encoder.

The difference between the required perpendicular distance from the centroid to the line, $d_r$, and the actual perpendicular distance of the line to the centroid, $d_l$ is $d_{error}$.

$$d_{error} = d_r - d_l \qquad (12)$$

where $d_r$ is five feet because the width of the obstacle course is ten feet. If $d_{error}$ is negative, it means that the robot is closer to the left line of the course and if $d_{error}$ is positive, the robot is nearer to the right line of the course. The $d_{error}$ and $\theta_{error}$ are combined by the first level of the fuzzy logic system to produce a suitable steering angle, $\theta_{steering}$.

The second level of the fuzzy system takes $\theta_{steering}$ as input and determines the most appropriate speed of the robot. If the magnitude of $\theta_{steering}$ is big, the robot has to slow down to turn gently. And if $\theta_{steering}$ is small, the robot moves with its normal speed. In the next sub-section, the rules and the membership functions for the fuzzy system are discussed.

**Fuzzy Inference System**

Fuzzy inference is the actual process of mapping from a given input to an output using fuzzy logic[10]. Fuzzy logic starts with the concept of a fuzzy set. A fuzzy set is a set without a crisp, clearly defined boundary. It can contain elements with only a partial degree of membership. The MATLAB Fuzzy Logic Toolbox was used to build the fuzzy inference system. There are five parts of the fuzzy inference process: fuzzification of the input variables, application of the fuzzy operator (AND or OR) in the antecedent, implication from the antecedent to the consequent, aggregation of the consequents across the rules, and defuzzification.

**Fuzzify Inputs:**

In the first level of the fuzzy system we have two inputs, $\theta_{error}$ and $d_{error}$. These inputs are resolved into a number of different fuzzy linguistic sets. For $\theta_{error}$, there are five fuzzy sets: left, soft-left, center, soft-right, and right while $d_{error}$ also has five sets: far-left, left, middle, d1(right), and far-right. These inputs are taken and the degree to which they belong to each of the appropriate fuzzy sets is determined via membership functions. A membership function is a curve that defines how each point in the input space is mapped to a membership value (degree of belongingness). A Gaussian distribution curve membership function was used in our system. The $d_{error}$ is on a scale of -3' to 3' while $\theta_{error}$ is on -20 to +20 degrees. The fuzzy sets for each input has its own sub-scale. For example, the fuzzy set far-left is mapped between the values of -3' and -1.5'. Figure 1 shows the membership function plot for the inputs $\theta_{error}$ and $d_{error}$ has similar plot. A fuzzy operator is applied to obtain one number that represents the result of the antecedent for a given rule.

**Fuzzify Outputs:**

The output of the first level of the fuzzy system is $\theta_{steering}$. It has nine fuzzy linguistic sets. The sets are: extreme-left, far-left, soft-left, left, straight, right, soft-right, far-right, and extreme-right. $\theta_{steering}$ is on a scale of -20 to +20 degrees. The Gaussian distribution curve membership function was used on all the sets to map the output. The membership function plot for the output is similar to figure 1.

**Fuzzy Rules and Defuzzification Process:**

If-then rules are generated to make conditional statements on the inputs and output to obtained a defuzzified output. First, the outputs of each rule are unified or aggregated by joining the parallel threads. All the fuzzy sets that represent the output of each rule are combined into a single fuzzy set and serves as an input to the defuzzification process. Defuzzification has several computational methods. The centroidal and bisector types were tested and bisector was selected. In Figure 2, a surface plot for the inputs and the output of the system is shown. Twenty five rules were generated for the first level. Some of the rules are given in Table 1.

The second level of the fuzzy system has one input and one output. The input is $\theta_{steering}$ and the output is the speed of the robot. Three fuzzy sets were defined for the input and they are too-sharp, normal and too-sharp. $\theta_{steering}$ is on a scale of -20 to +20 degrees. The output has two fuzzy sets, slow and fast and is on a scale of 0 to 5 miles per hour. A Gaussian distribution curve membership function is used for both the input and the output. Three rules were generated for this level and are listed in Table 2. The input-output plot is shown in Figure 3.

## 4.0 Results

The dual level fuzzy system controller was simulated with a MATLAB M-file. The model is also being implemented on the mobile robot, which is scheduled to take part in a nation wide obstacle avoidance and path following competition in July. The inputs to the M-file are $\theta_{error}$ and $d_{error}$. The outputs are $\theta_{steering}$ and the speed of the robot. The results obtained from the simulation are given below. Seventeen cases were considered. Results of the simulation are shown in Table 3. Appendix A has the code for the M-file. Note that the results appear reasonable for both angle and speed.

## 5. 0 Conclusion

The design of a modular autonomous mobile robot fuzzy-logic-based controller for following a line along an obstacle course has been presented. From the results obtained in the MATLAB simulation and the preliminary testing of the model on the robot, it is convincing that the model presented can be reliably and successfully implemented permanently on the robot. No system is perfect in the onset; as such the necessary refinements and modifications will be made to make the E-Z-Go golf cart a competitive mobile robot. Future considerations will include a total vision and ultrasonic fuzzy control system with an additional input from the ultrasonic signals, which will influence the decision making for obstacle avoidance.

## 6.0 References

[1] P. F. Muir and C. P. Neuman, 'Kinematic Modeling of Wheeled Mobile Robots,' **Journal of Robotic Systems**, 4(2), 1987, pp. 281-340

[2] E. L. Hall, **Robotics: A User-Friendly Introduction**, Holt, Rinehart, and Winston, New York, NY, 1985, pp. 23.

[3] Z. L. Cao, S. J. Oh, and E. L. Hall, "Dynamic omnidirectional vision for mobile robots," **Journal of Robotic Systems**, 3(1), pp. 5-17, 1986.

[4] Z. L. Cao, Y. Y. Huang, and E. L. Hall, "Region Filling Operations with Random Obstacle Avoidance for Mobile Robots," **Journal of Robotics Systems**, 5(2), 1988, pp. 87-102.

[5] S. J. Oh and E. L. Hall, "Calibration of an omnidirectional vision navigation system using an industrial robot," **Optical Engineering**, Sept. 1989, Vol. 28, No. 9, pp. 955-962.

[6] R. M. H. Cheng and R. Rajagopalan, "Kinematics of Automated Guided Vehicles with an Inclined Steering Column and an Offset Distance: Criteria for Existence of Inverse Kinematic Solution," **Journal of Robotics Systems**, 9(8), 1059-1081, Dec. , 1992.

[7] M. P. Ghayalod, E. L. Hall, F. W. Reckelhoff, B. O. Mathews and M. A. Ruthemeyer, "Line Following Using Omnidirectional Vision," Proc. of SPIE Intelligent Robots and Computer Vision Conf., SPIE Vol. 2056, Boston, MA, 1993.

[8] Kazuo Tanaka, "Design of Model-based Fuzzy Controller Using Lyapunov's Stabilty Approach and Its
Application to Trajectory Stabilization of a Model Car," **Theoretical Aspects of Fuzzy Control**, John Wiley & sons, Inc, pp.31-50, 1995. 2nd IEEE conference on fuzzy system, San Francisco, CA, 1993

[9] C. V. Altrock et al., Advanced fuzzy logic control technologies in automotive applications. Proceedings of 1st IEEE international Conference on Fuzzy Systems, pp. 835-842, 1992.

[10] J.-S. Roger Jang and Ned Gulley, **Fuzzy Logic Toolbox For Use with MATLAB**, The MathWorks Inc. 1995.

## Tables

Fuzzy System Rules Tables:

**Table 1.**

1. If (Distance-Error is far-right) and (Angle-Error is left) then (Steering-Angle is extre-left) (1)
2. If (Distance-Error is far-right) and (Angle-Error is soft-left) then (Steering-Angle is far-left) (1)
3. If (Distance-Error is right) and (Angle-Error is left) then (Steering-Angle is extre-left) (1)

.

.

25. If (Dis_Error is far-left) and (Ang_Error is right) then (Steer_Angle is extre-right) (1)

**Table 2.**

1. If (steering-angle is too-sharp) then (speed is slow) (1)
2. If (steering-angle is normal) then (speed is fast) (1)
3. If (steering-angle is too-sharp) then (speed is slow) (1)

**Table 3. Results from simulation**

***Input Variables into the fuzzy controller:*** ***Ouput Variables form the fuzzy controller:***

| Remarks: | Distanee Error:(de) | Angle Error: (ae) | Steering angle: (st) | Speed of the robot: (speed) |
|---|---|---|---|---|
| *Case 1:Ideal* | 0 | 0 | 0 | 3.5591 |
| *Case 2:* | 2 | 0 | -5.2 | 3.2751 |
| *Case 3:Extreme* | 3 | 0 | -10 | 2.5057 |
| *Case 4:Extreme* | -3 | 0 | 10 | 2.5057 |
| *Case 5:Extreme* | 0 | 20 | 19.6 | 1.4679 |
| *Case 6:Extreme* | 2.5 | 20 | -19.6 | 1.4679 |
| *Case 7:Extreme* | -3 | 20 | 19.6 | 1.4679 |
| *Case 8:* | -1.5 | -1 | 10 | 2.5057 |
| *Case 9:* | 1.5 | 10 | -13.6 | 1.9084 |
| *Case 10:* | -2 | 13 | 10.8 | 2.356 |
| *Case 11:* | -2.8 | 5 | 13.6 | 1.9084 |
| *Case 12:* | 1.8 | -20 | -19.6 | 1.4679 |
| *Case 13:* | -2.2 | -15 | 10.8 | 2.356 |
| *Case 14:* | 2.6 | 15 | -19.2 | 1.4811 |
| *Case 15:* | 0 | -12 | -10 | 2.5057 |
| *Case 16:Extreme* | 3 | 19 | -19.6 | 1.4679 |
| *Case 17:* | 1.2 | -8 | -13.6 | 1.9084 |

## Figures

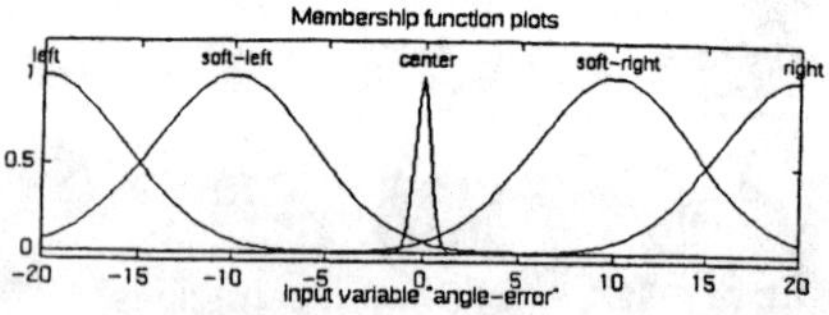


Figure 1. Membership Function plot for input angle-error

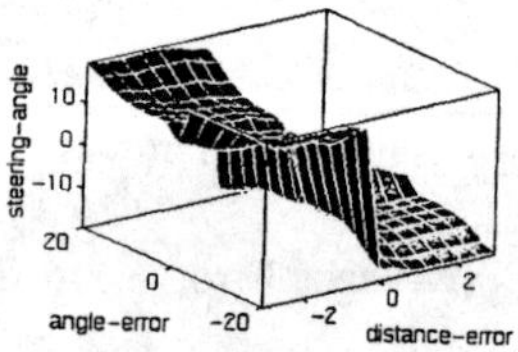


Figure 2. Input-output Surface plot

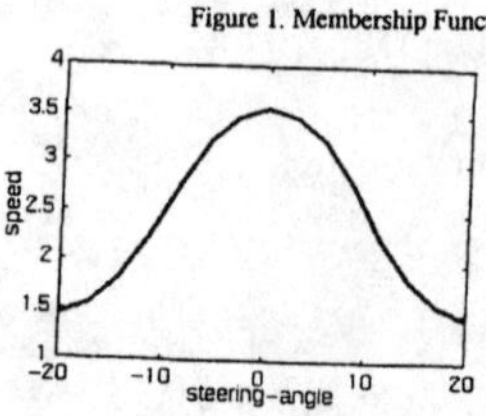


Figure 3. Input-output plot

# ROBOT CONTROL USING FUZZY LOGIC AND NEURAL NETWORK

**BEHESHTEH SAYEH**
*Mac Murray College*
*Jacksonville Illinois 62650*

**Abstract**

This paper explores the application of three control algorithms (based on fuzzy logic, neural network, and neurofuzzy logic) to a problem in robot control. The robot is considered in this work moves forward toward a track. The track is located on the center of a road. Once the robot reaches the track, it moves toward the north direction. The fuzzy controller has twelve rules, and performs very efficiently for all test runs, where the robot is on the road. It fails when the robot is on the shoulder close to the road. The neural network approach is based on the cluster formation of the training data. The training data space coordinates consist of the location, direction, and steering angle of the robots. When the clustering method is applied to the data, total of five clusters are formed. Given the coordinates of the robot, the minimum-distance classifier is used for obtaining the steering angle for controlling the robot. The test runs show that the neural network controller is successful in most difficult situations, but not as efficient as the fuzzy controller. It also suffers from chattering when reaches to the center track. The last controller is the neurofuzzy controller, where the rules and membership functions are obtained from a training set by applying the neural network. This nerofuzzy controller has only five rules. The simulation runs show that this controller is successful for most cases. It does not suffer from the chattering of the steering angle. However it covers the longest distance before reaching to the center track when compared with the other two controllers. From this work we conclude that the fuzzy logic method works the best given the fuzzy rules. In the absence of these rules, the neural network can be used to generate a more compact fuzzy rules.

## INTRODUCTION

Fuzzy logic concept was first lunched by the work of Zadeh [1]. Zadeh proposed the method to enhance the use of artificial intelligence(AI) techniques in certain areas such as speech recognition. Recently, the fuzzy logic has been applied to several control applications, such as process control and robotics [2].

Fuzzy control is the first successful industrial application. A fuzzy controller is able to control systems that previously could only be controlled by skill operators. Fuzzy logic is very effective in nonlinear process control because it models the experience of a human operator, rather than the process itself. However, the success of the fuzzy logic all depends on having a good set of rules in hand.

In this work, a technique is developed for generating the fuzzy rules. This is based on the formation of cluster centers when applied to a training data, drawn from the manual control of the robot. The clustering is achieved by using a neural network model called LYAM (Lyapunov Associative Memory) [3]. The LYAM system consists of a set of differential equations. The initial state of the system is chosen randomly over the training data. The clustering process is then performed by the dynamics of the system. When the system is an equilibrium state, this state gives the cluster centers coordinates.

In literature, there are other neural network models which are used in generating the fuzzy rules [4-7]. In this work, the LYAM system is considered because of its simplicity of operation. It has also shown reliable clustering results in applications such as signal sorting [8].

In generating the fuzzy rules using the neural network, it is found that a controller could be designed based solely on the classification results. This is further explored in the neural network controller section. The control signal is issued based on the fact that which cluster is closer, in Euclidean sense, to the current position of the robot.

**FUZZY LOGIC CONTROLLER**

Fuzzy logic has been applied to several control applications, such as process control and robotics. Fuzzy control is the first successful industrial application [2]. A fuzzy controller is able to control systems that previously could only be controlled by skilled operators.

Auto-steering Control of A Robot

This example of auto-steering control of a robot presents a control application of fuzzy logic, neural network classifier, and the neurofuzzy controller. First the problem is stated and the three controllers are applied.

We consider a robot (adopted from [2]) that moves forward toward a track. Once reached the track, it moves toward the north direction. Figure 1 shows the robot and the road with the corresponding track. The position of the robot is determined by two linguistic variables, the direction angle $\alpha$ and the distance from the left shoulder x. The steering direction of the robot movement $\beta$, is determined by the angle of the front wheel. For a given initial robot position within a specified area, the goal is to drive the robot toward the center of the track, where $x = 0.5$. The desired final position is for the robot driving north on the center track.

Let us assume the ranges of the linguistic variables x, $\alpha$, and $\beta$, are as follows:

$$0 \leq x \leq 1, \quad 0^0 \leq \alpha \leq 360^0, \quad -5^0 \leq \beta \leq +5^0.$$

Notice that the rotation of the steering angle $\beta$ is limited to only 5 degrees. A positive value of $\beta$ represents a right turn of the front wheel, and a negative value represents a left turn.

To each of the above linguistic variables, a set of values are assigned as shown in Table 1. A range of numerical values can be assigned to each linguistic value. In Figure 2, each graph is called a membership function, indicating the degree to which an input value belongs to a particular linguistic value. Such a degree of membership ranges from 0 to 1. The value 0 indicates no membership, and the value 1 represents full membership. A value between 0 and 1 represents partial membership. For example, $x = 0.1$ belongs to fuzzy value L with degree 1 and to C with degree 0.

Similar to an expert system design, a set of rules must be defined. In general, each rule produces some output linguistic values based on some input values. For example, in the robot case, some of the rules can be defined as

$$\text{If } (\alpha = S \text{ and } x = L) \text{ then } \beta = TL$$
$$\text{If } (\alpha = S \text{ and } x = R) \text{ then } \beta = TR$$

These rules can be extended to consider all possible values for $\alpha$, generating 12 rules. Table 2 shows all the rules for our robot.

For given input values of x and $\alpha$, the fuzzy controller should determine an output value of $\beta$, the steer angle. First, for each input value the controller determines the membership degrees of its corresponding linguistic values. Next, for each rule, the minimum of the membership degrees of its antecedents is chosen as a membership degree for the rule's consequent. When there is more than one membership degree for a consequent, the maximum degree is chosen for the consequent. Hence, at this point, a membership degree is assigned to each linguistic output value. To compute the output value for the steering angle $\beta$, defuzzification is performed. A simple defuzzification rule is

$$\beta = \frac{\sum_{i=1}^{n} c_i r_i}{\sum_{i=1}^{n} r_i},$$

where n is the number of output variables, $c_i$ is the value of $\beta$ at which the ith output membership function is maximum, and $r_i$ is the maximum membership value of the inputs for ith output variable.

Simulation Results

A computer program, written in Microsoft C++, is used to simulate the trajectory of the robot and its fuzzy logic controller with 12 rules.

Robot is assumed to be a point with structure, and its trajectory is computed in x (distance from the left shoulder, see Figure 1) and y (distance along the road) by the following increments:

$$\Delta\alpha = \beta$$
$$\Delta x = A\cos\alpha$$
$$\Delta y = A\sin\alpha,$$

where Δ refers to the difference between the present value and the future value. Parameter A is a constant controlling the trajectory interval.

Series of test runs have shown that the fuzzy logic controller is very efficient and effective. A challenging control action is shown in Figure 3, where the robot starts at x = 0.125 and $\alpha = 270^0$, pointing south. However, the fuzzy controller fails in the test run shown in Figure 4, where the robot is placed outside of the road, heading west. This failure is expected since the rules are not defined in this domain

**NEURAL NETWORK CLASSIFIER**

In this section, a classification technique is used for generating clusters which in turn will be used to control the robot. Cluster centers are formed on a training data, drawn from the manual control of the robot.

Pattern clustering, in general, is a challenging task with many unsolved problems. Neural networks have shown excellent results, but they are very complicated systems with many tuning parameters. The LYAM (Lyapunov Associative Memory) system offers a compact set of governing equations with only a single tuning parameter (λ, gain factor or vigilance) in the simplest case [9].

The LYAM system operates in many modes by simply changing its parameters. The system exhibits the characteristics of the Kohonen network when the recall process is very slow with respect to the learning process. On the other hand, if the recall process is very fast with respect to the learning process, the LYAM system is considered as a content-addressable memory. Finally, the LYAM system exhibits its full dynamics and relates to the operation of an ART-like network. The classification capability of the LYAM system in regard to the AKH networks, is summarized in Reference [8].

The "Kohonen-like" weight ($w_{ij}$) dynamics is given below for the gated operation and for the virtual-batch operation.

$$\tau\left(\frac{dw_{ij}}{dt}\right) = x_j(1-x_j)(u_i - w_{ij})\left(\gamma - \sum_{k=1}^{M} x_k\right)$$

where τ is fixed in time and it may safely be unity. The parameters $x_j = f(r^2 - \| U - W_j \|)$, $\| . \|$ is the Euclidean norm, $f(\alpha) = 1/(1 + e^{-\lambda\alpha})$ is the logistic function, r is the radius, λ is the gain factor, and $u_i$ are the elements of the input vector with the dimension N. The parameter M is the number of clusters.

For true batch operation, the weight dynamics is

$$\tau\left(\frac{dw_{ij}}{dt}\right) = \sum_{p=1}^{P}\left[x_{jp}(1-x_{jp})(u_{ip} - w_{ij})\left(\gamma - \sum_{k=1}^{M} x_{kp}\right)\right],$$

where P is the number of input samples. Note that this is unsupervised classification of patterns with known number of clusters, M. The following algorithms describes the process of the true-batch operation.

<u>Neural Network Controller Design</u>

The neural network clustering technique, reviewed earlier, is used to control the robot discussed earlier. In what follows, we discuss how a clustering technique can be used in controlling the robot.

Let us consider a hypothetical situation, where the input/output data ( x and α as inputs and β as an output in our robot control problem) are shown by crosses in Figure 8. The collection of there points clearly cluster into two distinct classes. The center of each cluster is shown by a solid dot. This clustering is achieved by using the LYAM network discussed earlier. Given the input/output date set, the network issues the cluster center coordinates.

In order to issue the control signal (β) for a given input (x and α), the Euclidean distance of the input to each cluster center is computed. The β component of the center with the smallest distance is then chosen as the control signal to the robot.

In order to apply the above technique, a training data set is collected by manual driving the robot simulator. We try to cover all possible positions that robot may encounter during the control process. An example of the training data set is shown in Figure 5. There are four test runs shown with initial positions as: (1) far left pointing north, (2) far right pointing north, (3) far left pointing south, and (4) far right pointing south. The direction angle $\alpha$, the distance from the left shoulder x, and the steer angle $\beta$ are recorded from a initial position to the final position on the center track.

Since the direction angle has discontinuity at $0^0$ approaching from $360^0$, $\cos(\alpha)$ and $\sin(\alpha)$ are used for the clustering technique. The clustering technique is applied to the input data along with the control signal (the steering angle). The clustering task is then to consolidate these data into clusters. Its outcome will be the coordinates of the cluster centers. The neural network controller compares the input components of the cluster centers with a given input. The steer angle component of the closest center will then be used as a control signal steering the robot in a correct direction.

Using the LYAM dynamics, with parameters, $u_1 = x$, $u_2 = \cos(\alpha)$, $u_3 = \sin(\alpha)$, $u_4 = \beta$, $N = 4$, $M = 5$, $P = 23$, $\lambda$ = variable, $r = 0$, $w(0)$ = uniform distributed random numbers over the training data region, the algorithm results in five clusters tabulated in Table 3.

Once the robot reaches the center track, the neural network controller has to switch back and forth between Cluster 2 and Cluster 3 in keeping the robot on the track heading north. Certainly, this chattering is not desirable in practice. The chattering will be reduced when the fuzzy logic is applied to this controller. This will be developed in the next section as a neurofuzzy controller.

Test runs are performed on the neural network controller with initial robot positions used earlier for the fuzzy controller. The simulation results are shown in Figures 3 and 4. The controller performs relatively well in comparison with fuzzy controller. The robot drives longer distance before reaching the center track when compared with that of the fuzzy controller. In the test run, shown in Figure 3, where the initial position is at far left pointing south, the neural network controller almost failed. It swings into the shoulder in order to bring the robot back on the road. In all simulation tests, the chattering of the robot is not shown because the small distance intervals ($\Delta x$ and $\Delta y$).

**NEUROFUZZY LOGIC CONTROLLER**

Fuzzy logic is very effective in nonlinear process control because it models the experience of a human operator, rather than the process itself. However, the success of the fuzzy logic all depends on having a good set of rules in hand.

Neurofuzzy logic controller refers to a fuzzy logic controller where the rules and the membership functions are obtained from a neural network. The fuzzy logic engine is similar to that of the fuzzy controller developed earlier. The neural network used is the LYAM clustering system discussed earlier. The clusters coordinates are considered rules, and the cluster shapes generate membership functions. The clusters are bell-shape type defined the membership functions as

$$f_i(x) = \frac{2}{1+e^{\lambda_1 (x-x_i)^2}}$$

$$g_i(\alpha) = \frac{2}{1+e^{\lambda_2 (\cos\alpha_i - \cos\alpha)^2}}$$

$$h_i(\alpha) = \frac{2}{1+e^{\lambda_3 (\sin\alpha_i - \sin\alpha)^2}}$$

where $\lambda_1$, $\lambda_2$, and $\lambda_3$ are constants, controlling the line-width of each function. The parameters $x_i$, $\cos(\alpha_i)$, and $\sin(\alpha_i)$ are ith cluster centers components, where i = 1, 2, 3, 4, 5. For example, x membership functions are illustrated in Figure 6.

Given the cluster coordinates (see Table 3), the five rules are:

*If* [(x belongs to $f_i$) AND ($\cos(\alpha)$ belongs to $g_i$) AND ($\sin(\alpha)$ belongs to $h_i$)]
*then* $\beta_i$, for i = 1, 2, 3, 4, and 5.

For the above rules, the control signal $\beta$ is computed by the following defuzzification method.

$$\beta = \frac{\sum_{i=1}^{5} r_i \beta_i}{\sum_{i=1}^{5} r_i},$$

where $r_i = \min\{f_i(x), g_i(\alpha), h_i(\alpha)\}$.

Simulation Results

All the test runs for the fuzzy controller and the neural network controller, are applied to the neurofuzzy controller developed in this chapter. In these test runs, the robot drives very smoothly as compared to the neural network controlled robot, where some chattering is observed. However, for some test runs, the neurofuzzy controller fails to steer the robot into the center track.

## CONCLUSION

The objective of this work was to explore the application of three control algorithms (based on fuzzy logic, neural network, and neurofuzzy logic) to a problem in robot control. The fuzzy controller had twelve rules, and performed very efficiently for all test runs, where the robot was on the road. It failed when the robot was on the shoulder close to the road.

The neural network approach was based on the cluster formation of the training data. The training data space coordinates consisted of the location, direction, and steering angle of the robots. When the clustering method was applied to the data, total of five clusters were formed. Given the coordinates of the robot, the minimum-distance classifier was used for obtaining the steering angle for controlling the robot. The test runs showed that the neural network controller was successful in most difficult situations, but not as efficient as the fuzzy controller. It also suffered from chattering when reached to the center track.

The last controller was the neurofuzzy controller, where the rules and membership functions were obtained from a training set by applying the neural network. This nerofuzzy controller had only five rules. The simulation runs showed that this controller was successful for most cases. It did not suffer from the chattering of the steering angle. However it covered the longest distance before reaching to the center track when compared with the other two controllers.

From this work we conclude that the fuzzy logic method works the best given the fuzzy rules. In the absence of these rules, the neural network can be used to generate a more compact fuzzy rules.

## REFERENCES


[1] L.A. Zadeh, "Fuzzy Sets," Information and Control, pp. 338-353, 1965.

[2] M.R. Zargham, *Computer Architecture, Single and Parallel Systems,* Prentice Hall, 1996.

[3] M.R. Sayeh, A. Ragu, and H.H. Szu, "Design of Unsupervised Classifier," IJCNN, Seattle, WA, pp. II417-II422, 1991.

[4] J.S.R. Jang, "ANFIS: Adaptive Network Based Fuzzy Inference System," IEEE Tansactions on System, Man, and Cybernetics, Vol. 23, No. 3, 1993.

[5] A. Kawamura, N. Watanabe, H. Okada, and K. Asakawa, "A Prototype of Neuro-fuzzy Cooperation System," IEEE International Conference on Fuzzy Systems, 1992.

[6] C.T. Lin and C.S.G. Lee, "Real-time Supervised Structure/parameter Learning for Fuzzy Neural Network, IEEE International Conference on Fuzzy Systems, 1992.

[7] S.K. Halgamuge, W. Poechmueller, A. Pfeffermann, P. Schweikert, and M. Glesner, "A New Method for Generating Fuzzy Classification Systems Using RBF Neurons with Extended RCE Learning," IEEE International Conference on Fuzzy Systems, 1994.

[8] M.R. Sayeh, R. Athinarayanan, and M.R. Zargham, "Operation of An Associative Memory," Pattern Recognition, Vol. 27, No. 12, pp. 1815-1821, 1994.

[9] R. Athinarayanan and M.R. Sayeh, "Limit Cycles and stability of Critical Points of LYAM System," Intelligent Engineering Systems Through Artificial Neural Networks, Volume 5, Editors: C.H. Dagli, M. Akay, C.L.P. Chen, B.R. Fernandez, and J. Ghosh, ASME Press, pp. 103-108, 1995.

Table 1. Linguistic Variables

| | |
|---|---|
| *Distance x* | *Input Variable* |
| L | left side of the track |
| C | center of the track |
| R | right side of the track |
| *Direction Angle α* | *Input Variable* |
| N | north |
| W | west |
| S | south |
| E | east |
| *Steer Angle β* | *Output Variable* |
| TR | turn right |
| ST | straight |
| TL | turn left |

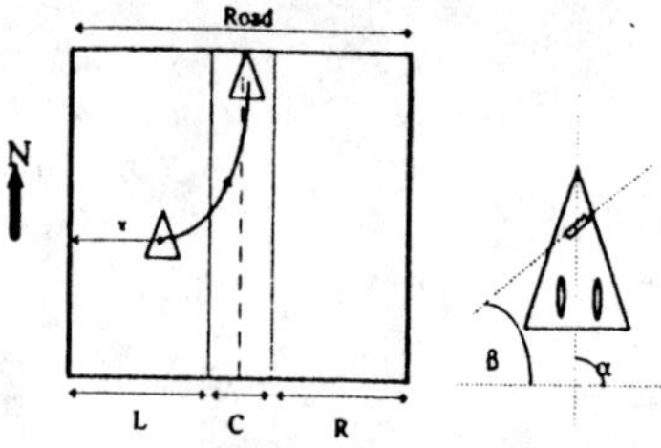


Figure 1. Illustration of the Parameters L, C, R, x, α, and β.

Table 2. Fuzzy Logic Rules for Robot Control

| | L | C | R |
|---|---|---|---|
| S | TL | TL | TR |
| E | ST | TL | TL |
| N | TR | ST | TL |
| W | TR | TR | ST |

Table 3. Coordinates of the Cluster Centers

| | x | cos(α) | sin(α) | β |
|---|---|---|---|---|
| Cluster 1 | 0.229489 | 0.484068 | 0.661413 | 1.244918 |
| Cluster 2 | 0.616462 | -0.605429 | 0.707852 | 0.765869 |
| Cluster 3 | 0.424693 | 0.481618 | 0.596129 | -0.777801 |
| Cluster 4 | 0.188879 | 0.737439 | -0.270011 | -1.899304 |
| Cluster 5 | 0.764772 | -0.565251 | 0.679077 | -0.789129 |

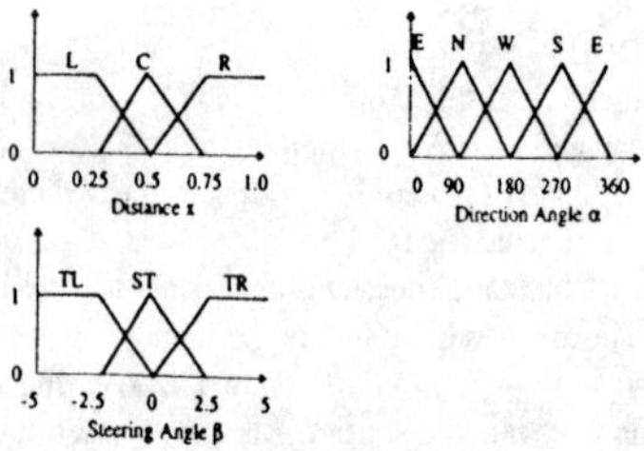


Figure 2. Membership Functions

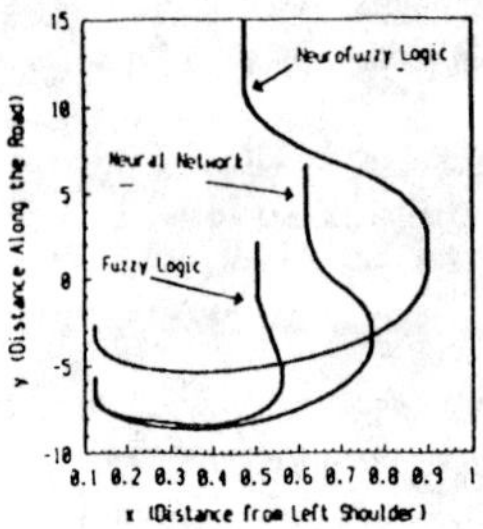


Figure 3. Trajectories of the Robot starting at x = L, α = S for all three controllers

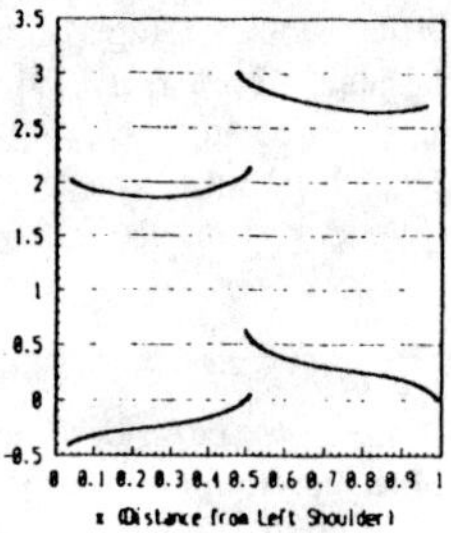


Figure 5. Trajectories of Manual Control of the Robot at Four Different Starting Points

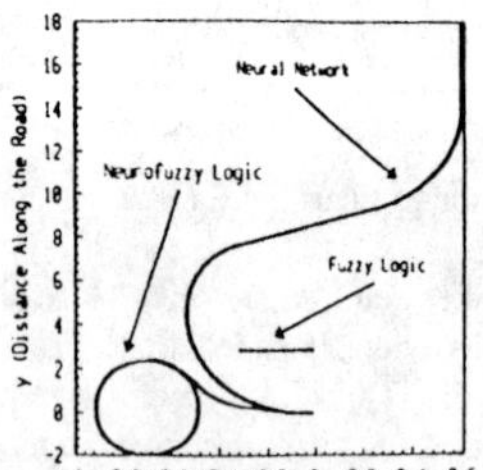


Figure 4. Trajectories of the Robot starting at x = off L, α = W for all three controllers

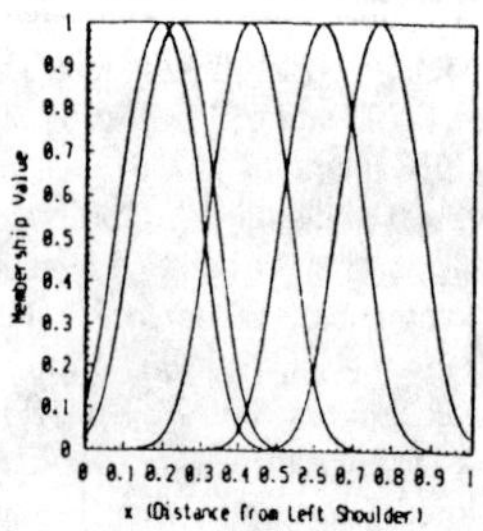


Figure 6. Membership Functions of x

# FUZZY LOGIC CONTROL OF A FOUR-LINK ROBOTIC MANIPULATOR IN A VERTICAL PLANE

**YUZHOU ZHOU AND ROBERT A. MCLAUCHLAN**
*Mechanical Engineering and Industrial Engineering Department*
*Texas A&M University, Kingsville, Texas 78363*

**RAJAB CHALLOO AND S. IQBAL OMAR**
*Electrical Engineering and Computer Science Department*
*Texas A&M University, Kingsville, Texas 78363*

***ABSTRACT:***
Fuzzy logic is used to implement two-dimensional control of the end point of a four-link, mechanical linkage following a straight line in a vertical plane. A non-algorithmic, model free approach has been developed that relies on a fuzzy rule base to evaluate the required axis motion to result in the desired two-dimensional endpoint motion. This scheme does not require solution of the complex nonlinear inverse kinematics equations to arrive at joint control commands. This is in contrast with traditional robot controllers. The fuzzy rule-based control gives fast execution speed because fuzzy rules perform simple addition and multiplication to evaluate the required axis motion. Since there is no interaction between the fuzzy rules governing the motion of the individual manipulator joints, the evaluation of the control signals for each joint can be performed independently and simultaneously.

## INTRODUCTION

The rapid expansion in recent years in the area of robotic systems has given rise to new and challenging problems. These systems are now expected to perform tasks that require a high degree of dexterity and adaptability to various and uncertain task conditions (Blauer and Belanger, 1987).

A prerequisite to the successful execution of a complex manipulation task is the availability of data that describe the task geometry with sufficient accuracy throughout its progression. The term task geometry as used here refers to the instantaneous position, orientation, and velocity of the manipulator itself and to the shape, position and orientation of manipulated objects.

### Multiple-Degree-of-Freedom Robotic Systems

The majority of robot systems in use today have no kinematic redundancy, so that they are constrained to reach a specified end position in only one geometric orientation. Such systems suffer from their lack of flexibility, and sensitivity to disturbance. A good geometric orientation for force control is also hard to obtain for these systems. A robotic system can reach a specified end position in many geometric orientations, if additional degrees of freedom are provided to the mechanical structure of a robot manipulator. This gives a capability to a robot system to perform tasks with dexterity

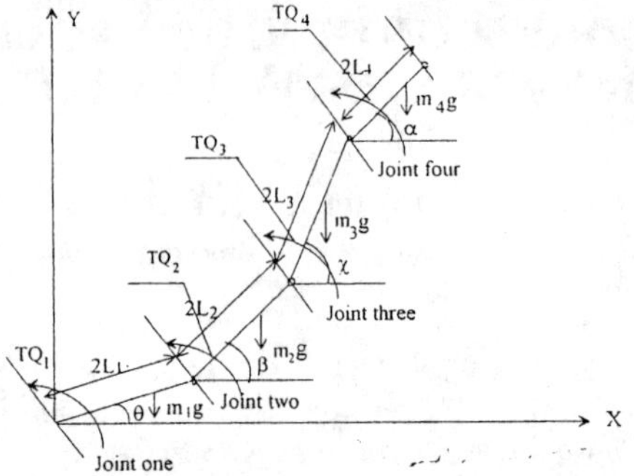


**Figure 1: Four-Link, Open-Chained Manipulator System**

and adaptability as a function of events. It is driven by the interaction of the robotic system with a dynamically changing environment (Franke and Nedungadi, 1991).

**Control Method**

Let us assume that the manipulator of Fig. 1 is in the vertical plane. Our control objective here is to make the end point of the robot manipulator track a given desired trajectory, in a time domain $[0, t_f]$. However, this objective is in general very difficult to achieve. This is largely due to the highly nonlinearity character of the system.

The system is even harder to control when the unknown system parameters have time-varying effects such as payload changes, mechanical wear, aging, etc. This type of system change, as well as the disturbance is extremely difficult to manage. Many methods and techniques have been proposed, to control this multiple link system more efficiently and overcome this kind of difficulty. Kuc, Nam, and Lee, 1991, for example, used an iterative learning technique, which was comprised a feedforward learning controller and a linear feedback controller.

By using the method of conservation of total mechanical energy introduced in Slotine 1981, one can show simply the stability of simple proportional-derivative position controllers for robot manipulators. This approach can also be systematically exploited to design trajectory controllers for these systems.

Spong 1987 used the feedback linearization technique to transform a nonlinear system to a controllable linear system. Liu, Iberall, and Bekey, 1989 used a Neural Network approach to learn the relationship between object primitives and a useful set of grasp modes of a multiple-degree-of-freedom articulated robot hands. The fuzzy logic method was used by Franke and Nedungadi 1991 to control a redundant, multiple-link system.

All of the methods proposed in these references could fall into two categories: conventional control and intelligent control. Intelligent control is completely different from conventional control, and it is our interest in this project. Intelligent control, as used here refers to the control of a complex dynamical system by a neural network or a fuzzy logic method.

Fuzzy Associative Memory (FAM) elements and the relationship between specific performance measures and FAM rules can be determined more easily than for neural network systems. Both neural network and fuzzy logic approaches could be used to control the system in Figure 1. The Fuzzy logic approach has been selected for use in the results discussed herein.

**Dynamic Equations of Motion**

Using appropriate expressions for the kinetic energy T and for the gravity force on each link and applying Lagrange's equations of motion yields for the joint coordinates θ (link 1), β (link 2), ψ (link 3), and α (link 4):

$$\frac{d}{dt}\left[\frac{\partial T}{\partial \frac{dq_i}{dt}}\right] - \frac{\partial T}{\partial q_i} = Q_{q_i} \tag{1}$$

in which, $1 \le i \le 4$, and $q_1 = \theta$, $q_2 = \beta$, $q_3 = \psi$, $q_4 = \alpha$. Substituting into equation (1) and carrying out the appropriate operations yields four coupled, second order, nonlinear ordinary differential equations. A Maple program has been used to change this sytem of differential equations to state variable form by solving for the second derivatives of θ, β, ψ, and α explicitly. The outputs of the program are translated into FORTRAN statements, which are then used in four function subprograms in a Runge-Kutta fourth order numerical integration package (Zhou 1994).

## FUZZY LOGIC RULES AND DEFUZZIFICATION

**Fuzzy Logic Rules**

The fuzzy rule base for the four-link, planar manipulator system is derived from the forward kinematics equation. A XY rectangular coordinate system is chosen in which x is the X coordinate and y is the Y coordinate of the end point of the four-link system,

$$x = x(\theta,\beta,\psi,\alpha) = 2l_1\cos\theta + 2l_2\cos\beta + 2l_3\cos\psi + 2l_4\cos\alpha \tag{2}$$
$$y = y(\theta,\beta,\psi,\alpha) = 2l_1\sin\theta + 2l_2\sin\beta + 2l_3\sin\psi + 2l_4\sin\alpha$$

and the virtual displacements of x, y are

$$\delta x = \frac{\partial x}{\partial \theta}\delta\theta + \frac{\partial x}{\partial \beta}\delta\beta + \frac{\partial x}{\partial \psi}\delta\psi + \frac{\partial x}{\partial \alpha}\delta\alpha$$
$$\delta y = \frac{\partial y}{\partial \theta}\delta\theta + \frac{\partial y}{\partial \beta}\delta\beta + \frac{\partial y}{\partial \psi}\delta\psi + \frac{\partial y}{\partial \alpha}\delta\alpha \tag{3}$$

Thus, $\delta x = A_1\delta\theta + A_2\delta\beta + A_3\delta\psi + A_4\delta\alpha$ and $\delta y = B_1\delta\theta + B_2\delta\beta + B_3\delta\psi + B_4\delta\alpha$ in which the $A_i$ and $B_j$ are functions of $\theta,\beta,\psi,\alpha$, and $l_i$, $1 \le i,j \le 4$.

**Fuzzy Membership Functions**

In encoding information, δx ( *A* or δθ) can be divided into seven fuzzy subsets using triangular membership functions. Therefore, given any variable value inside the fuzzy set domain, its degree of membership in each of the fuzzy subsets can be calculated. Experience has shown that to use fuzzy rules effectively, a 25% overlap between every two adjacent fuzzy functions is appropriate (Kosko 1992).

**Fuzzy Encoding**

There are two methods to encode a set of Fuzzy Associative Memory rules, correlation minimum encoding and correlation-product encoding (Kosko 1992). Correlation-product encoding is used for the work reported in this paper. For multi-antecedent FAM rules like, if x is A and y is B, then z is C, or in another form, A and B → C. The following decompositional inference is proposed, [A AND B → C] → [A → C] AND [B → C]. This statement is true for bivalent logic only if A is true and B is true. The above equation still holds in the sense of continuous-valued logic, for general fuzzy-set inputs A' and B'. Thus, for correlation product encoding, if A is activated by $a_i$ and B is activated by $b_i$, then C is activated by $w_i$, in which, $w_i = \min(a_i, b_i)$.

**Defuzzification**

There are two ways to implement defuzzification, maximum-membership defuzzification and fuzzy centroid defuzzification. Maximum-membership defuzzification has two fundamental problems: (a) the mode of the output fuzzy set is not unique, and (b) the maximum-membership scheme ignores the information in much of the wave-form of the output fuzzy set, the fuzzy centroid defuzzification method is chosen.

## SIMULATION RESULTS

The simulated system has the four-link, planar manipulator system with the following parameter values: $[m_1, m_2, m_3, m_4] = [1, 1, 1, 1]$ kg and $[l_1, l_2, l_3, l_4] = [0.17, 0.17, 0.17, 0.17]$ m. The time domain for all of the simulations is $[t_o, t_f] = [0, 20]$ sec. The time step size used is 0.02 sec.

Figures 2 and 3 give graphs for the output of the control used for the two line cases shown of the six considered (Zhou 1994). The lines were chosen randomly, and the initial and final specifications for the end positions are given by random locations on each line. The error in both x and y are shown as functions of time for each case.

## CONCLUSIONS

A non-algorithmic, model free approach has been developed that relies on a fuzzy rule base to evaluate the required axis motion to result in the desired two-dimensional end-point motion of a four-link, planar robotic manipulator. This scheme does not require solution of the complex non-linear inverse kinematics equations to arrive at joint control commands. This is in contrast with traditional robot controllers. The fuzzy rule based control gives fast execution speed because the fuzzy rules perform simple additions and multiplication to evaluate the required axis motion. Since there is no interaction between fuzzy rules governing the motion of the individual manipulator joints, the evaluation of the control signals for each joint can be performed independently and simultaneously which leads to a very straight forward parallel microprocessor implementation. The fuzzy rules used in this project could also be extended to the control of higher DOF, redundant, etc. planar manipulator systems.

The simulated system in this project is a redundant planar manipulator. Redundant manipulators can have an infinite number of joint motions that lead to the same end-effector trajectory. It is desired that the kinematic component of a redundant manipulator control scheme must generate a set of joint angle trajectories, from the

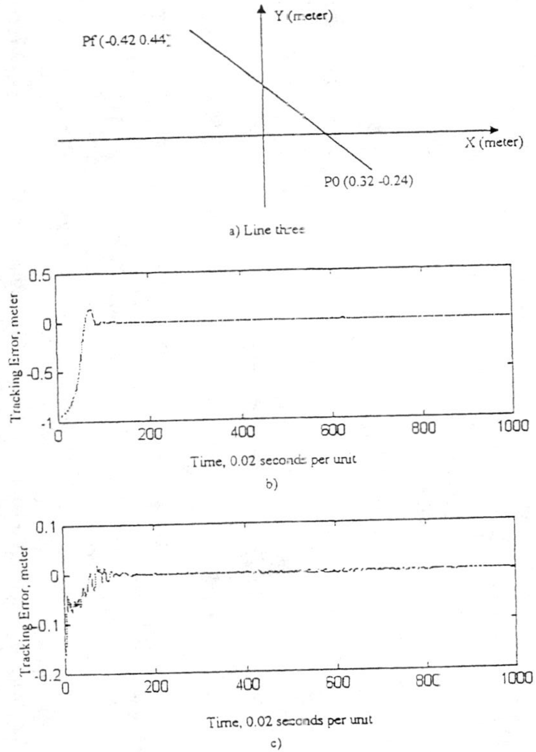


**Figure 2. Tracking both errors in X and Y for line three**
**a) Line to track, b) Tracking error in X,**
**c) Tracking error in Y**

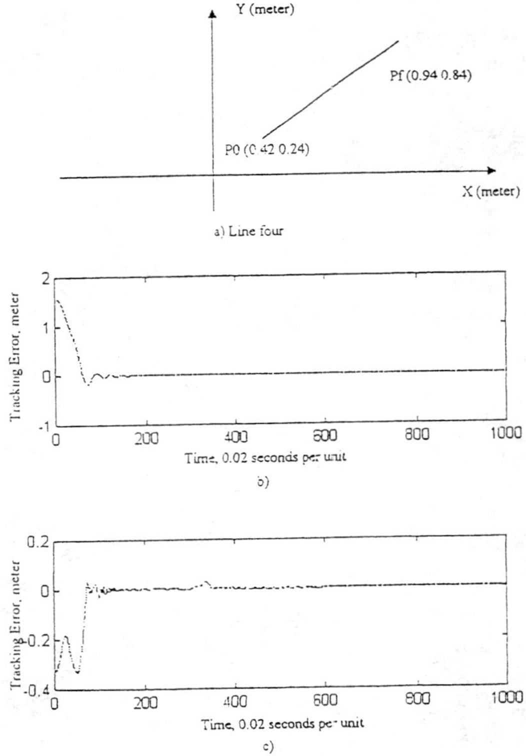


**Figure 3. Tracking both errors in X and Y for line four**
**a) Line to track, b) Tracking error in X,**
**c) Tracking error in Y**

infinite set of possible trajectories, which causes the end-effector to follow a desired trajectory while satisfying additional constraints, such as collision avoidance, maximum mechanical advantage, singularity avoidance, sensor motor torque minimization, elbow control, optimal joint movement and cyclicity of motion. Developing techniques to simultaneously achieve end-effector trajectory control while meeting additional task requirements are known as the redundancy resolution problem (Seraji 1993).

Since redundancy is an important evolutionary step toward versatile manipulation, much research has been devoted to redundancy resolution and related areas in recent years. Most redundant manipulator control schemes resolve the redundancy through local or global optimization of various performance criteria (Oh, Orin, and Bach, 1984, Baillieul 1985, Egeland 1987, Sciavicco and Sicilao, 1988, Anderson and Angeles, 1989). H. Seraji proposed a better method, the configuration control approach (Seraji 1989), and used this control scheme to control a 7-DOF robot arm (Seraji 1993). If we could extend our FAM control scheme to the redundancy resolution problem, the computational efficiency resulting from the parallel implementation should be better than the configuration control scheme approach. In order to do so, we need additional fuzzy rules for the existing FAM bank. Thus, finding a method of composing new FAM rule bases for redundancy resolution problems and testing them by simulation and/or by use with a real system are recommended for our future research work.

## REFERENCES


Anderson, K. and Angeles, J., (1989). "Kinematic Inversion of Robotic Manipulators in the Presence of Redundancies," *International Journal of Robotics Research*, **8** (6), 80-97.

Baillieul, J., (1985). "Kinematic Programming Alternatives for Redundant Manipulators," *Proc. IEEE Int. Conference on Robotics and Automation*, St. Louis, Mar., 80-97.

Blauer, M. and Belanger, P., (1987). "State and Parameter Estimation for Robotic Manipulators Using Force Measurements", *IEEE Transactions on Automatic Control*, **AC-32** (12), Dec., 1055-1066.

Egeland, O., (1987). "Task-Space Tracking with Redundant Manipulators," *IEEE Transactions on Robotics and Automation*, 1987, **RA-3** (5), 471-475.

Franke, E. and Nedungadi, A., (1991). "Fuzzy Logic Control of Telerobot Manipulators", *SOAR 1991 Proceedings*, 248-253.

Kosko, Bart, (1992). *Neural Networks and Fuzzy Systems*, Prentice-Hall, Inc.

Kuc, T., Nam, K. and Lee, J., (1991). "An Iterative Learning Control of Robot Manipulators", *IEEE Transactions on Robotics and Automation*, **RA-7** (6), Dec., 835-841.

Liu, H., Iberall, T., and Bekey, G., (1989). "Neural Network Architecture for Robot Hand Control", *IEEE Control System Magazine*, **CSM-9** (3), Apr., 38-43.

Oh, S.Y., Orin, D., and Bach, M., (1984). "An Inverse Kinematic Solution for Kinematically Redundant Robot Manipulators," *Journal of Robotic Systems*, **1** (3), 235-249.

Sciavicco, L. and Siciliao, (1988). "A Solution Algorithm to the Inverse Kinematic Problem for Redundant Manipulators," *IEEE Transactions on Robotics and Automation*, **RA-4**, 403-410.

Seraji, H., (1989). "Configuration Control of Redundant Manipulators: Theory and Implementation," *IEEE Transactions on Robotics and Automation*, **RA-5**, 472-490.

Seraji, Homayoun, (1993). "Motion Control of 7-DOF Arms: The Configuration Control Approach", *IEEE Transactions on Robotics and Automation*, **RA-9** (2), Apr., 125-139.

Slotine, J., (1981). "Putting Physics in Control--the Example of Robotics", *IEEE Control Systems Magazine*, **CSM-8** (6), Dec., 12-17.

Spong, M., (1987). "Modeling and Control of Elastic Joint Robots", *Journal of Dynamic Systems, Measurement and Control*, **DSMC-109**, Dec., 310-319.

Zhou, Y., (1994). "Fuzzy Control Simulation with a Planar, Four-Link Robotic Manipulator," M.S. Research Project Report (Mechanical Engineering), Department of Mecnanical and Industrial Engineering, Texas A&M University - Kingsville.

# APPLICATION OF THE FUZZY STATE FUZZY OUTPUT FINITE STATE MACHINE TO THE PROBLEM OF RECOVERY FROM VIOLATIONS OF ONTOLOGICAL ASSUMPTIONS

**JANOS L. GRANTNER**
*Department of Electrical and Computer Engineering*
*Western Michigan University, Kalamazoo, MI 49008-5066, USA*

**GEORGE FODOR**
*ABB Industrial Systems AB, ISY/AMC*
*S-721 67 Vasteras, Sweden*

**DIMITER DRIANKOV**
*Department of Computer and Information Science*
*Linkoeping University, S-581 83 Linkoeping, Sweden*

**MAREK J. PATYRA**
*Department of Electrical and Computer Engineering*
*University of Minnesota, Duluth, MN 55812-2496, USA*

***ABSTRACT:***
A programmable logic controller (PLC) carries out a control algorithm under violations of the ontological assumptions (VOA) when the plant does not meet one or more unstated but essential assumptions used in the design of the control algorithm. This paper presents a recovery technique based on the theory of Fuzzy State Fuzzy Output Finite State Machines (FSFO FSM). The appeal of this approach is that the only known recovery results today are for fuzzy controllers. A common drawback of most linguistic models is that they are essentially static and thus not suitable to model the sequential behavior of a PLC program. A FSFO FSM can act as a sequential machine during the normal course of control, yet it possesses all the required linguistic properties during a VOA.

In the case of modern programmable controllers for large industrial process control applications, the modeling assumptions in the control algorithms are extended by additional assumptions about the complex environment in which a given programmable logic controller (PLC) is embedded. These additional assumptions refer to segments of the plant that are not directly controlled by the PLC, but which interact with other segments under direct control, and assumptions on the dynamic behavior of other controllers interacting with the given PLC. Those assumptions that are implicit in a control algorithm, but are essential for its validity are referred to as ontological assumptions. The control objectives of the PLC can be achieved only if the ontological assumptions are not violated during the execution of the control algorithm. The PLC executing a particular control algorithm cannot detect itself a violation of an ontological assumption (VOA). Ontological control is a novel type of supervisory control (Driankov and Fodor, 1993). The Ontological Controller supervises a control algorithm in order to detect when it encounters a situation in which the ontological assumptions are violated. In addition, when it is possible, the OC moves the control algorithm into a state from which it can recover and achieve its control objective in spite of a VOA. In this paper, a recovery technique based on the theory of Fuzzy State Fuzzy Output Finite State Machines (FSFO FSM) is presented. The appeal of this approach is that the only known recovery results today are for fuzzy controllers (Driankov and Fodor, 1993). However, a common drawback of most linguistic models is that they are essentially static and thus not suitable to model the sequential behavior of a PLC program. A FSFO FSM can act as a sequential machine during the normal course of control, yet it possesses all the required linguistic properties during a VOA.

## FSFO FSM MODEL

The model of FSFO FSM that is implemented by a Boolean automata based on two-valued logic is given by the formulas (1), where X and Z stand for a finite set of fuzzy inputs and outputs, respectively, R* is the composite linguistic model (3), and $\circ$ is the operator of composition. Each state of the Boolean FSM is characterized by an overall linguistic model $R_s$, or a set of linguistic sub-models with multiple-input-single-output (MISO), and multiple-input-multiple-output (MIMO) systems

$$\begin{aligned} Z &= X \circ R^* \\ R^* &= G(R_S) \\ z_c &= DF(Z) \\ X_B &= B(X) \\ Y &= f_y(X_B, y) \end{aligned} \tag{1}$$

A fuzzy state is defined by a crisp (Boolean) state and a state membership function

$$S_{F_k} : S_k, g_{S_k} \tag{2}$$

where $S_{F_k}$ stands for fuzzy state k, $S_k$ represents crisp state k, and $g_{S_k}$ is the state membership function associated with $S_k$. G stands for the matrix of state membership functions, $X_B$, Y, and y are two-valued Boolean input and state variables, respectively. B stands for a Fuzzy-to-Boolean transformation algorithm to map a change in the status of a fuzzy variable into state changes of a finite set of Boolean variables. The $z_c$ crisp values of the fuzzy outputs are obtained by computing the DF defuzzification strategy. On the basis of the concept of fuzzy state, the FSM stays in a number of crisp states simultaneously, to a certain degree in each. One of these states is referred to as a dominant state for which the state membership function is a 1 (full membership). The concept of the fuzzy FSM based on a Boolean FSM and the State Membership Functions, in this context, have been introduced in (Grantner and Patyra, 1993), and (Grantner, Patyra, and Stachowicz, 1994), respectively.

For each fuzzy state of the FSFO FSM model, a $R_i^*$ composite linguistic model is created from the finite set of $R_{S_i}$ overall linguistic models (i=1,..,p). Let the FSFO FSM be in fuzzy state $S_{F_k}$, then

$$R_k^* = \max[\min(\beta_1^k, R_{S_1}), \min(\beta_2^k, R_{S_2}), \ldots, \\ \ldots, \min(\beta_k^k, R_{S_k}), \ldots, \min(\beta_p^k, R_{S_p})] \tag{3}$$

where $\beta_1^k, \beta_2^k, \ldots, \beta_p^k$ stand for the degrees of state membership function $g_{S_k}$, and $R_{S_1}, R_{S_2}, \ldots, R_{S_p}$ are the overall rules in crisp states $S_1, S_2, \ldots, S_p$, respectively. With (3), a SISO system is assumed.

In adaptive systems $R_k^*$ is not stored in memory, it is dynamically created by computing (3), instead. By modifying the $\beta$ degrees of the state membership functions on-line, new $R^*$ composite linguistic models can be created under real-time conditions. The switch over between composite linguistic models is determined by the state transients of the FSFO FSM.

The state transients of the FSFO FSM are specified by means of a sequence of changes at the fuzzy inputs and outputs. Those changes are mapped into the corresponding sequence of changes of the Boolean input and output variable sets, respectively, using the B algorithm. The Boolean input/output sequence specification is passed to an algorithm (Arato, 1987) (Klein, 1993), and the Boolean FSM will then be automatically synthesized. A fuzzy logic controller hardware accelerator of pipeline architecture has been presented in (Grantner and Patyra, 1994).

The theory of FSFO FSM will be used in the sequel to model the changes in the control algorithm of a PLC recovering from a VOA.

## ONTOLOGICAL DESYNCHRONIZATION

The architecture of an Ontological Controller (OC) capable to detect violations of ontological assumptions (VOA) in the context of a desynchronization from the execution of a goal path was shown in (Fodor, 1995). However, the OC is unable to recover from a VOA by itself (Fodor, 1995). The concept of state is defined in the OC by (4):

$$S_i = (y_i, u_{i,j}) \ (i,j=1,\ldots,n) \tag{4}$$

where $y_i$ stands for a Boolean formula (referred to as "plant formula") showing what condition is true at a given time in the controlled plant. Each relevant plant situation has a corresponding plant formula in some state. A control action denoted as $u_{i,j}$ is executed when $y_i$ is true. The expected outcome of this action is the plant changes such that at the next time instance $y_j$ will be true. However, if some external action (disturbance) occurs, the expected change in the plant does not take place but a new plant state, $y_k$, will materialize instead of $y_j$. The disturbance is considered as an external action and, if known in advance, is denoted as $u_{i,k}^{ext}$ where the indices point to the respective two plant formulas before and after the external action. The new state can be denoted as some $S_k = (y_k, u_{k,l})$. The control then proceeds in a succession of states. The states that can materialize from an arbitrary state $S_i$ by external actions (disturbances) are referred to as "collaterals to $S_i$" and the set of collaterals is denoted as $K_{\sim}(S_i)$.

An example is given below to illustrate the problem of ontological desynchronization. A position controller has the goal to move a radar antenna into a desired position 'p'. The controller has two inputs: (1) the angle 'x' of the radar antenna given by a mechanical sensor and (2) the angle 'y' of the radar antenna given by an electronic sensor. The measurement y is much more accurate than x, however the range $\varepsilon_2$ of y is much smaller than the range $\varepsilon_1$ of x. Thus the positioning algorithm has two steps: first a regulator $R_1$ brings the radar in the range $\varepsilon_2$ of the electronic sensor, then a regulator $R_2$ brings the radar in the desired position p. The position error is $\varepsilon$=q-p, where q can be either x or y, depending on the current range. By using (4), state $S_1$ can be given as $S_1=(y_1=|\varepsilon<\varepsilon_1|, u_{1,2}=$'$R_1$ acting'). The other states are shown in Fig. 1. A disturbance can bring the radar outside the range $\varepsilon_1$, when a state transition from $S_1$ to $S_6$ occurs, i.e. $K_{\sim}(S_1)=\{S_6\}$, or a disturbance outside $\varepsilon_2$ makes the transition from $S_3$ to $S_7$, i.e. $K_{\sim}(S_3)=\{S_7\}$.

A violation of ontological assumptions (VOA) occurs at a state $S_j = (y_j, u_{j,k})$ if $u_{i,j}$ is executed in state $S_i$, but the expected $y_j$ does not materialize in the plant, although no external action has occurred. It has been shown in (Fodor, 1995) that a VOA manifests always as a state transition from $S_i$ to a state in $K_{\sim}(S_j)$ where $S_j$ is the consecutive (next expected) state to $S_i$. This type of transition is referred to as an ontological desynchronization. Its has also been shown in (Fodor, 1995) that for certain state sets, a VOA has the effect such that the controller enters a cycle. This cycle is repeated indefinitely such that the ontological desynchronization will be repeated as well. For the system above, let us consider the following ontological violation: the mechanical part has developed an error consisting of an angle offset between the two sensors such that when x=r then y=r+offset. This kind of error is a violation of the assumption that the two measurements are always aligned and thus there are no measurements or other means to verify it. This error has the effect that after the plant formula for the state $S_2$ materializes, (i.e., $|x-p|<\varepsilon_1$), the state $S_3$ will not materialize since now $|y+\text{offset}-p|>\varepsilon_1$ and thus the next state will be $S_7$ instead. However, no disturbance has occurred that would correspond to the transition to $S_7$. The transition from $S_2$ to $S_7=K_{\sim}(S_3)$ is the ontological desynchronization. After $S_7$, the next state materialized will be $S_0$, then $S_1$ and then $S_2$. At $S_2$, the relation $|y+\text{offset}-p|>\varepsilon_1$ is still satisfied so the controller will jump again to $S_7$ and the cycle repeats as predicted by the theory.

The recovery from this situation requires to determine a state that corresponds to the current plant situation and which has a control action that can "break" the cycle. For fuzzy controllers, there can be found recovery solutions by adding extra rules to the rule base (Driankov and Fodor, 1996). However, for controllers such as PLCs that have discrete states, the problem of recovery becomes more difficult since there is no state with acceptable properties in the state space of the controller that can materialize at a VOA. Thus the recovery algorithm suggested in this paper relies on techniques that can accommodate fuzzy states and yet produce Boolean outputs based on dynamically modified linguistic models.

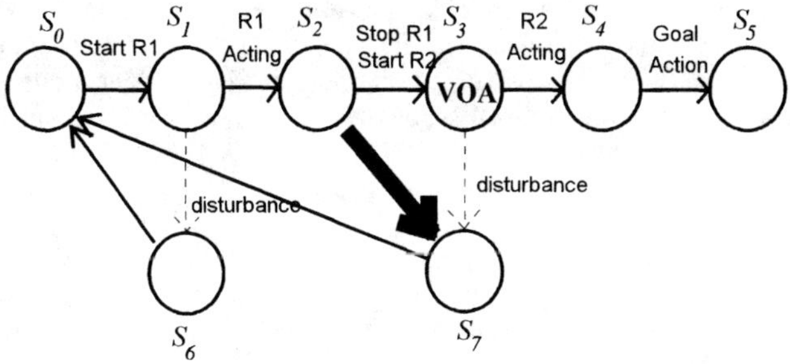


**Figure 1: The state set of a position controller.**

## RECOVERY USING FSFO FSM

For the sake of clarity, the use of the FSFO FSM will be outlined for a single state transition at a VOA. At design time, each plant formula is associated with linguistic intervals. After an ontological desynchronization is detected, the FSFO FSM uses the following information: (i) the relevant fuzzy input X to the FSFO FSM is the fuzzified plant formula $y_i$ of the last state materialized as expected, (ii) the fuzzy output set Z consists of the fuzzified plant formula of the expected next state, and in addition, the fuzzified plant formulas of the collateral states of the expected next state, (iii) the state membership function degrees are set such that the expected next state (dominant state) is assigned the degree of 1.

Due to a VOA, a fuzzified plant formula leading to a collateral state as next state will materialize instead, however, the expected dominant model has a less than 1 degree in that state. The Ontological Controller has detected the ontological desynchronization, and it will attempt to dynamically tune the β degrees of state membership functions using fuzzy set theory methods in order to recover from the VOA. The defuzzified result $z_c$, by means of a look-up table, will be given the following meaning: (i) if the value of $z_c$ is zero, then the control action of the collateral state will be executed and, in effect, the recovery is not possible, (ii) if the value is 1, then the control action of the expected next state will be executed and, in effect, the recovery is possible.

The intervals used for the plant formulas of the state sets are, in fact, Boolean. However, fuzzification makes the boundaries between states continuous such that a state preserves its boundaries beyond its Boolean limits. Thus fuzzification "balances" the fuzzy boundaries of the states against the degree of the ontological violation and, if it is possible, helps to select a control action to recover from the VOA.

## CONCLUDING REMARKS

The use of the FSFO FSM proposed in this paper departs from the traditional use of fuzzy logic in control. Traditionally, the advantage of using fuzzy logic has been associated with properties such as faster execution, lower memory requirement, better man-machine interface, etc., which have been compared to ones yielded by more traditional approaches. For ontological control the FSFO FSM approach represents such a recovery method for sequential controllers that, until now, has no known alternatives. The existence of such a method opens a way to applications that can detect and recover from violations of ontological assumptions such as unmanned vehicles, complex control systems, and alike.

## REFERENCES


Antsaklis, P.J., Passino, K.M., (1994). *An Introduction to Intelligent and Autonomous Control*, Kluwer Academic Publishers, Boston/Dordrecht/London.

Arato, P., (1987). Specification and realization of logic control procedures on the basis of prescribed input-output changes, *Per. Pol. El. Eng.*, Budapest, Hungary, Vol. 31, No. 3-4, 59-153.

Driankov, D., Fodor, G., (1993). Ontological Real-Time Control, Proc. and Plenary Talk, First European Congress on Fuzzy and Intelligent Technologies, Aachen, Germany, Sept.

Driankov, D., Fodor, G., (1996). Fuzzy control under violations of ontological assumptions, invited plenary talk, FLAMOC'96 Proceedings, Sydney, Australia, January 15-18, 109-115.

Fodor, G., (1995). Ontological Control: Description, Identification and Recovery from Problematic Control Situations, PhD Thesis, Dept. of Computer Science, University of Linkoeping, Sweden.

Grantner, J., Patyra, M., (1993). VLSI implementation of fuzzy logic finite state machines, IFSA'93 World Congress, Proceedings, Seoul, Korea, July 4-9, Vol. II, 781-784.

Grantner, J., Patyra, M., (1994). Synthesis and analysis of fuzzy logic finite state machine models, WCCI'94, Proceedings of the FUZZ-IEEE'94, Orlando, FL, June 26-29, Vol. I, 205-210.

Grantner, J., Patyra, M., Stachowicz, M., (1994). Intelligent fuzzy controller for event-driven real-time systems and its VLSI implementation, in the book *Fuzzy Control Systems* (Eds. A. Kandel, G. Langholz), CRC Press, Boca Raton, FL, USA, 161-179.

Klein, M., (1993). Automatisierung des Schaltwerkentwurfs auf der Basis einer zustandsfreien Spezification", MS Thesis, University of Karlsruhe, Germany.

# PART III : EVOLUTIONARY PROGRAMMING

# Evolving and Optimizing Braitenberg Vehicles

**RALF SALOMON**
*AI Lab, Computer Science Department, University of Zürich*
*Winterthurerstrasse 190, 8057 Zurich, Switzerland*
*FAX: +41-1-363 00 35; Email: salomon@ifi.unizh.ch*

***ABSTRACT***:
This paper presents a practical application of the evolution strategy (ES) to the evolution and optimization of Braitenberg vehicles. Braitenberg vehicles are a special class of autonomous agents. Autonomous agents are embodied systems that behave in the real world without any human control. One major goal of research on autonomous agents is to study intelligence as the result of a system environment interaction, rather than understanding intelligence on a computational level. Braitenberg vehicles are controlled by a number of parameters, which are mostly determined by hand in a trial and error process. This paper shows that a simple ES evolves Braitenberg vehicles very efficiently. Other research has used genetic algorithms (GAs) for very similar tasks. A comparison of both approaches shows that the ES approach is much more efficient. Since autonomous agents are very important in the field of new AI, this research field should spend more attention to evolution strategies.

## INTRODUCTION

Autonomous agents are autonomous, self-sufficient embodied systems (Brooks, 1986) that do useful work. Autonomous means that the agent operates without any human control. Self sufficient means that the agent can maintain its internal energy level over a long time; typically, the agent gets a reward after doing useful work. Autonomous agents are equipped with sensors, such as infrared sensors, and effectors, such as motors or grippers, to actively behave in their environment. One major goal of research in autonomous agents is to study intelligence as the result of a system environment interaction, rather than understanding intelligence on a computational level. Thus, autonomous agents are very important in the field of new AI. For an overview of special issues involved in autonomous agent design see, for example, Maes (1991).

The next section describes the *Khepera*$^{TM}$ robot that is widely used for research on autonomous agents. In order to achieve the goal of intelligent behavior, an agent has to perform different tasks, such as exploring the environment, moving around, and avoiding obstacles. Typically, the agent's behavior is controlled by a neural network, which coordinates the sensor readings with the agent's actions. This paper describes a simple neural control architecture called Braitenberg vehicle (Braitenberg, 1984).

Standard neural network training cannot be applied to Braitenberg vehicles. Rather than working with a fixed, predefined set of training patterns, an autonomous agent determines the pattern's relevance at runtime while it behaves in its environment. Furthermore, the sensor readings are very noisy and dynamically changing due to the agent's interaction with the environment.

Figure 1: The Khepera robot.

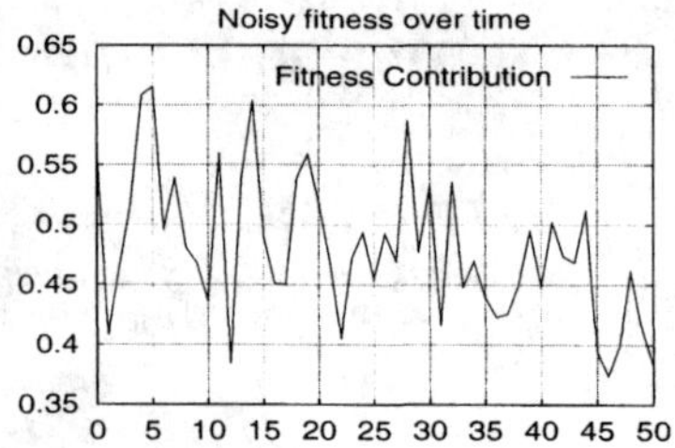


Figure 2: Extremely noisy fitness contributions $f_t$.

For the reasons discussed above, the evolution strategy (ES) (Schwefel, 1995) is an ideal candidate for the evolution and optimization of Braitenberg vehicles. The fourth section describes the experimental setup and the fitness function that is designed such that the evolved agent favors straight movements while it avoids obstacles. The fifth section reports some experiments obtained by applying a (3,6)-ES to the evolution of Braitenberg vehicles. The results show that such a simple ES evolves reasonable controllers within 30 generations, which take approximately one and a half hours. Other research (Floreano, 1996; Nolfi, 1995) that uses genetic algorithms (GAs) (Goldberg, 1989) takes more than two and a half days. The obtained speedup is of great practical relevance. A comparison of ES and GAs can be found in Baeck (1993).

## THE KHEPERA ROBOT

The Khepera robot (cf. Figure 1) is 55 mm in diameter and 32 mm high. The robot is equipped with eight infrared and eight ambient light sensors as well as two motors, which can be controlled independently. Khepera's sensors and motors are controlled by a Motorola 68331 micro controller. Khepera can be connected to a workstation or it can operate autonomously by using an on-board controller. For experimental purposes, the robot was controlled from a workstation in all experiments reported in this paper, and the ambient light sensors were not used. A detailed description of the robot and its electrical parts can be found in the *Khepera Users Manual*.

The motors can be controlled independently of each other by sending commands (i.e., function calls) to the robot. Valid speed values are in the range $[-40, 40]$; inside the program, these values are normalized such that they are in the range $[-1, 1]$. To avoid problems caused by the floating-point-to-integer conversion, [0,1]-equal-distributed random numbers are added to the motor speeds at each time step. For example, by setting both motors to the same speed but with different signs, the robot spins on the spot. The robot interprets the speed settings as commands. Internal PID controllers take care of the robot's dynamics. However, rapidly changing motor commands induce additional dynamics, which can cause problems for the fitness evaluation. By means of attached wheel encoders, the robot measures the motor's real speed, which can differ from the command setting in certain situations.

Khepera is equipped with eight infrared proximity sensors (cf. Figure 3). The sensor readings are of type integer and the values are in the range $[0, 1023]$. Within the program, the sensor readings are normalized such that the values are in the range $[0, 1]$. The sensors give reasonable input values for object distances between 10 mm and 60 mm. A sensor value of 1023 indicates that the robot is very close to the object.

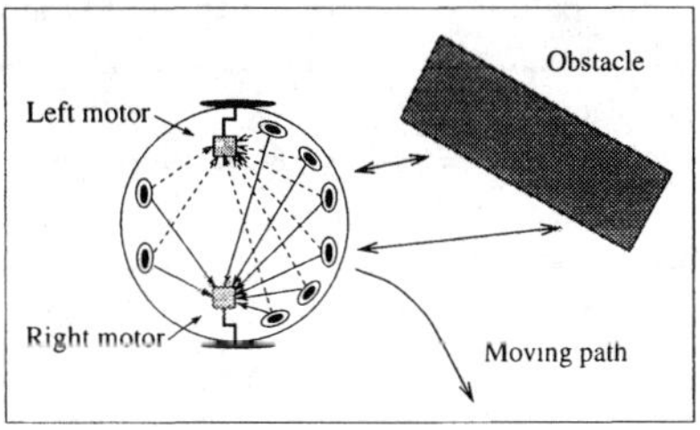


**Figure 3: Controlling a Braitenberg type-3C vehicle.**

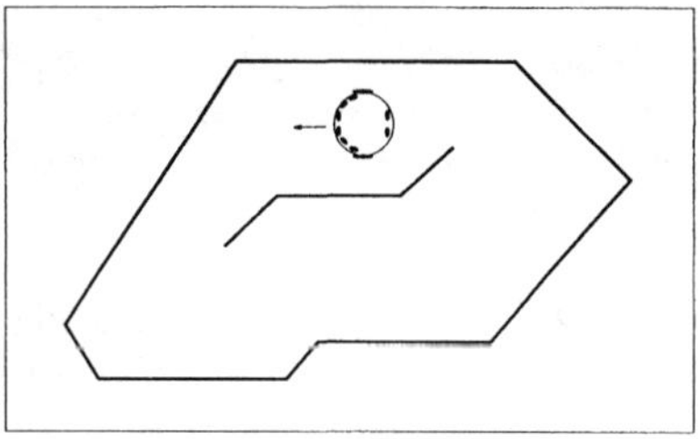

**Figure 4: The arena with the starting position.**

A major problem with Khepera is that the sensor readings are extremely noisy, which causes several problems for the fitness evaluation of a given controller, i.e., the weights of the Braitenberg network. This effect is visualized in in Figure 2, which shows how the fitness contributions $f_t$ are dynamically changing under constant speed and constant environmental conditions. Furthermore, the sensor readings are subject to, for example, the material the object is made of, the object's surface, and the ambient light. During a long series of experiments, the environment cannot be kept constant.

## BRAITENBERG VEHICLES

As outlined in the introduction, an autonomous agent should not sit somewhere while doing nothing, since it would consume energy and would eventually die. Rather, the agent has to *operate* in its environment. Thus, moving around while avoiding obstacles is a key issue in autonomous agent research. Braitenberg (1984) has proposed a simple architectures for such tasks. Figure 3 shows a control architecture inspired by a Braitenberg type-3c vehicle. The main idea is that a sensor with a high proximity activation accelerates the motor on the sensor's side whereas this sensor slows down the motor on the opposite side. Depending on the activation of all proximity sensors, the robot does a particular move/turn. The activation of the left and right motor $M^l$ and $M^r$ are calculated as follows

$$M^l = \sum_{i=1}^{8} IP_i w_i^l + w_0^l \,, \quad M^r = \sum_{i=1}^{8} IP_i w_i^r + w_0^r \,, \tag{1}$$

where the weights $w_i^l$ and $w_i^r$ connect the proximity sensors $IP_i$ with both motors. The "idle" weights $w_0^l$ and $w_0^r$ are responsible for the robot's forward motion in the absence of any obstacle.

Even though a Braitenberg type-3c architecture is simple and straight forward, finding appropriate weights is not easy. The main problem is to determine the weights $w_i^l, w_i^r, w_0^l$, and $w_0^r$ such that the robot is moving around while it avoids obstacles. This is mostly done in a trial and error process. Standard neural network training procedures cannot be used, since it is not reasonable to determine a sufficient training set. Thus, the evolution strategy is an ideal candidate for this optimization task.

## THE EXPERIMENTAL SETUP

The experimental setup has been chosen as close as possible to setups proposed in other research (Floreano, 1996; Nolfi, 1995). Figure 4 shows the arena, in which the robot has to move. The arena is of size 60x45 cm and the walls are made from wood with a height of 3 cm. The width of the corridors is chosen such that always at least one proximity sensor has a less-than-maximum value.

In order to evolve good Braitenberg vehicles, the fitness function has to incorporate the motor speeds and the distance to obstacles. Such a fitness function, however, would allow for *useless* robots that are spinning on the spot with a high speed, since they would have a high fitness. Therefore, the fitness function is enhanced by a third term that favors straight movements by penalizing turns (Floreano, 1996; Nolfi, 1995).

The fitness is measured as follows. At each time step $t$, both motor speeds $V_l$ and $V_r$ as well as all eight proximity sensors $IR_i$ are measured. Then, the speed of the robot's center $V_t = (V_l + V_r)/2$, the penalty term $\Delta v_t = |V_l - V_r|$, and the sensor with the highest activation $\hat{IP}$ is calculated. The fitness contribution $f_t$ for step $t$ is then

$$f_t = V_t(1 - \sqrt{\Delta v_t})(1 - \hat{IP}_t)\,. \tag{2}$$

Finally, the total fitness is the sum over $t_{\max}$ (e.g., 240) time steps

$$F = \sum_{t=1}^{t_{\max}} f_t = \sum_{t=1}^{t_{\max}} V_t(1 - \sqrt{\Delta v_t})(1 - \hat{IP}_t)\,. \tag{3}$$

Khepera's sensor readings can be obtained every 100 ms. That means, the robot moves with constant motor speeds for 100 ms. Then, the controller receives the new sensor readings and calculates new motor speeds for the next time step. Meanwhile, the individual fitness component $f_t$ is calculated and added to the total fitness.

## EXPERIMENTS

This section reports some typical results when evolving Braitenberg vehicle by means of a $(3,6)$-ES with self-adaptation of the step size (Schwefel 1995). A $(3,6)$-ES generates 6 new offspring per generation and selects the 3 fittest individuals as parents for the next generation. Due to the extremely noise fitness evaluation, a *non-elitist* selection scheme was chosen to eliminate spurious candidates. In all experiments, the ES generates two offspring without crossover, two with uniform recombination, and two with intermediate recombination, and the initial standard deviation was set to 0.5. Over a long time, the environmental conditions cannot be hold constant; this phenomenon is intrinsic to real-world applications. For each series of experiments, six to eight runs were performed and all presented figures are typical with respect to these runs. Fitness evaluation was done over 240 time steps (approx. 24 seconds).

One main problem is to find a first Braitenberg controller that exhibits an even tiny reasonable behavior. To help guide the evolution process, all weights were initialized with very small random numbers $[-0.5, 0.5]$. Furthermore, the first series of experiments exploit the morphology of the robots, i.e., the left and right part of the controllers are constrained to be equal $w_0^l = w_0^r$ and $w_i^l = w^r_{(14-i)\bmod 8+1}$. This constraint leads to a reduced search space of nine parameters.

A typical run of the evolution of a *constrained* Braitenberg controller can be seen in Figure 5. This figure shows the average fitness as well as the fittest agent. In the first few generations, most vehicles are sitting at the initial position or crashing backwards into the wall resulting in a negative fitness. After a few more generations, some agents are moving forward, but after some steps, got stuck at the wall. Such a short forward

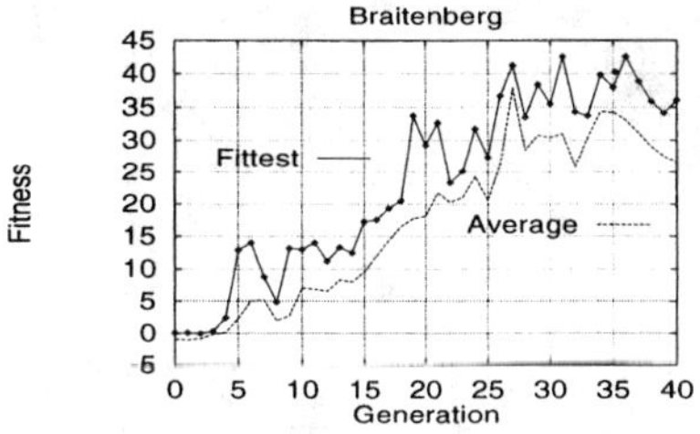


**Figure 5: A typical run of constrained controller.**

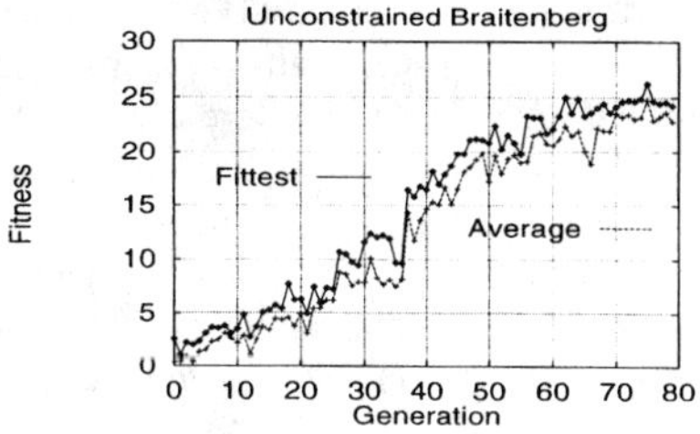


**Figure 6: A typical run of a unconstrained controller.**

movement results in a small positive fitness, which is a first step towards a useful agent. After about eight to ten generations, the fittest agents are moving slowly inside the corridor. But sometimes, they are hitting the walls again. After hitting a wall, good controllers set the motor speeds to negative values, which cause the robots to back up from the wall, and after turning, they continue moving inside the corridor. After 30 generations, the fittest agents perform up to three complete laps in the arena. Even though both controller sides are constrained, identical weights do not ensure total symmetric behavior. Tolerances in the electrical characteristics of the motors and sensors impose an unsymmetric behavior. Consequently, such controllers have to find a good compromise. As a result, Braitenberg vehicles with constrained controllers move forward in an almost straight line and turn, if they approach an obstacle.

Figure 6 shows a typical run of the evolution of an *unconstrained* Braitenberg controller. Such a controller has 18 parameters and the ES has to evolve both sides of the controller, i.e., the connections for both motor activations $M^l$ and $M^r$. Thus, the evolution of an unconstrained controller requires more time, and also, the final performance is lower. The unconstrained controller develops a different survival strategy. From the very beginning, the controllers have different bias weights $w_0^l$ and $w_0^r$, which causes the robot to turn in small circles. Circling around results in a small positive fitness. During the next generations, this circling process is preserved, but the radius becomes larger and larger. After about 30 generations, one controller side improves its behavior so that it prevents the robot from crashing into the wall, and after about 80 generations, most robots perform two complete laps in the arena. The behavior of the evolved controllers can be interpreted as wall following. This wall-following behavior is not encoded in the fitness function; rather, it is emerged.

## DISCUSSION

This paper has discussed the practical application of the evolution strategy to the evolution and optimization of Braitenberg vehicles. Braitenberg vehicles are autonomous agents with a simple control architecture, which is typically implemented as a neural network. Autonomous agents are very important tools in New Artificial Intelligence, since they study intelligence as the result of a system environment interaction, rather than understanding intelligence on a computational level.

In the experiments, two different controller types, constrained and unconstrained controllers, were investigated. The constrained control architecture is inspired by the biological observation that most animals are of symmetric shape. The experiments indicate that the evolution process can benefit from an exploitation of physical matters.

A constrained controller has less parameters, which accelerates the evolution process, but, at the same time, limits the number of potential solutions. Consequently, both controller types develop different survival strategies.

Further investigations have shown that a controller reacts very sensitive to parameter changes; the controller's parameters exhibit high epistasis. Consequently, all parameters have to be adapted simultaneously. The ES obeys this requirement, by applying mutation to all parameters at the same time. It is suspected that the observable epistasis is the main reason for the inefficiency of GA-based approaches presented in Floreano (1996) and Nolfi (1995). That research reports that GAs with a population of 80 individuals need about 50 to 100 generations (approx. 66 hours). Such a drastic slow down of GAs under the influence of epistasis has also been observed when optimizing widely-used test functions (Salomon, 1996). Salomon (1996) has indicated that the independence of the parameters to be optimized is an essential prerequisite for GAs and that the GA's performance significantly degrades under epistasis.

Even though we pointed to other GA-based research, the main focus of this paper is the application of the ES to the evolution of Braitenberg vehicles. A comparison across remote research would be problematic, since not all parameters of such a real-world application can be replicated. Our own additional control experiments with GAs yielded roughly the same performance as reported in Floreano (1996) and Nolfi (1995); the GA needs approximately ten times more fitness evaluations than the ES.

Further research will be devoted to more complex control architectures for object avoidance, navigation, and manipulation as well as other neural controllers, which are designated to enhance the robots capabilities/competences

**ACKNOWLEDGEMENTS:** This work was supported in part by a Human Capital and Mobility fellowship of the European Union, grant number ERBCHBICT941266.

## REFERENCES

Bäck, T., Schwefel, H.-P. (1993). An Overview of Evolutionary Algorithms for Parameter Optimization, *Evolutionary Computation* 1(1), pp. 1-23.

Braitenberg, V. (1984). *VEHICLES, Experiments in synthetic psychology*, MIT Press.

Brooks, R. (1986). A robust layered control system for a mobile robot, in: *IEEE Journal of Robotics and Automation*, **2**, pp. 14-23.

Floreano, D., Mondada, F. (1996). Evolution of Homing Navigation in a Real Mobile Robot, to appear in: *IEEE Transactions on Systems, Man and Cybernetics*.

Goldberg, D.E. (1989). *Genetic Algorithms in Search, Optimization and Machine Learning* Addison-Wesley Publishing Company.

*Khepera Users Manual*, Laboratoire de microinformatique, Swiss Federal Institute of Technology (EPFL) 1015 Lausanne, Switzerland.

P. Maes (Ed.) (1991). Designing Autonomous Agents: Theory and Practice from Biology to Engineering and Back, The MIT Press, Cambridge, Massachusetts.

Nolfi, S., Parisi, D. (1995). Learning to Adapt to Changing Environments in Evolving Neural Networks, Tech. Report 95-15, Institute of Psychology, National Research Council, Rome, Italy.

Salomon, R. (1996). Performance Degradation of Genetic Algorithms under Coordinate Rotation. To appear in *Fifth Annual Conference on Evolutionary Programming EP'96*.

Schwefel, H.P. (1995). *Evolution and Optimum Seeking* John Wiley and Sons, Inc, New York.

# OPTIMAL SAMPLING FOR GENETIC ALGORITHMS

**Brad L. Miller**
Dept. of Computer Science
University of Illinois
at Urbana/Champaign
bmiller@uiuc.edu

**David E. Goldberg**
Dept. of General Engineering
University of Illinois
at Urbana/Champaign
deg@uiuc.edu

**ABSTRACT**

In many real-world environments, a genetic algorithm designer is often faced with choosing the best fitness function from a range of possibilities. Fitness functions differ primarily based upon the speed, accuracy, and cost of a fitness evaluation. An important type of fitness function is the *sampling fitness function*, which utilizes sampling in order to reduce the noise of fitness evaluations. The accuracy and speed of a sampling fitness function are directly related to the *sample size*, which is the number of samples used by the sampling fitness function to evaluate an individual chromosome. The optimal sample size denotes the sample size that maximizes the performance of a genetic algorithm within a fixed time bound. In this paper, a domain independent lower bound of the optimal sample size is derived, and a sample size pruning method is described.

## INTRODUCTION

Genetic algorithms (GAs) are increasingly being used to solve real-world problems (Davis, 1991; Winter, Périaux, Galán, & Cuesta, 1995). Implementing a GA for a real-world environment is difficult for several reasons. First, the GA must identify a solution within a bounded amount of computational time. For many domains, this entails that the GA will not be allowed to run until convergence, but instead will be halted after a fixed amount of time and the best individual in the ending population selected as the final solution. Second, noise is often present in the fitness functions. An important subclass of noisy fitness functions are *sampling fitness functions*, which utilize sampling to reduce the noise from fitness evaluations. An entire range of possible fitness functions can be generated by simply changing the sample size of a sampling fitness function, with each different sample size producing a fitness function with unique speed and accuracy characteristics. It is difficult for a GA implementor to determine which sample size will maximize GA performance, as each different sample size produces a fitness function that is pareto-optimal in terms of speed and accuracy. This paper develops a lower bound for the optimal sample size, and describes a pruning method that can greatly reduce the sample size search space.

Sampling can be used in any domain whose fitness function has a random noise component. One application is image registration, where sampling can be used to efficiently estimate the difference between two images (Grefenstette & Fitzpatrick, 1985). Image registration has been used in such fields as aerial photography and medical imaging. Another possible application for genetic algorithms employing sampling is numerical integration, specifically the Monte Carlo method (Dahlquist & Björk, 1974), where the integral of a function is estimated based upon randomly sampled points. The accuracy of the integral estimate is directly based on the number of samples. Monte Carlo techniques are used in such diverse fields as telecommunications, particle physics, conflict modeling, and queuing problems. A thorough understanding of how GAs can use Monte Carlo sampling techniques will open up new application areas for genetic algorithms.

### Overview

Section 2 provides background information on some research upon which this research is based. It discusses sampling fitness functions, population sizing, and the onemax domain. Section 3 presents a GA computational requirements model, and uses it in deriving the lower bound for the optimal sample size. Section 4 discusses some experimental results, which are compared with the behavior predicted by convergence models. Lastly, section 5 presents some general conclusions drawn from this research.

## BACKGROUND

This section gives some background information on research areas relating to this paper. First, noisy fitness functions and sampling fitness functions are examined. Next, the concept of how to adequately size a population for a computationally bound GA is introduced. Lastly, the onemax domain is briefly discussed.

### Sampling Fitness Functions

A noisy fitness function returns a fitness value $f_i'$ for an individual $i$ based upon the sum of the real fitness of the individual $f_i$ and a random noisy component. For the purposes of this paper, we assume that the noise is randomly drawn from an unbiased (mean of zero) Gaussian distribution with a variance of $\sigma_E^2$:

$$f_i' = f_i + rnd[N(0, \sigma_E^2)] \tag{1}$$

As the noise component is randomly drawn each evaluation, subsequent evaluations of the same individual will return different noisy fitness values $f_i'$. A sampling fitness function seeks to reduce the amount of noise by taking the mean of multiple noisy fitness evaluations of an individual. The evaluation of individual $i$ by a sampling fitness function using a sample size of $n$ can be expressed as:

$$f_{i,n}^* = \frac{1}{n}\sum_{j=1}^{n} f_{i,j}' \tag{2}$$

where $f_{i,j}'$ is the $j$th noisy fitness evaluation of individual $i$. The sampling fitness $f_{i,n}^*$ will have a distribution of $N(f_i, \sigma_E^2/n)$, as it is distributed as the mean of $n$ samples from the normal distribution $f_{i,j}'$, which is $N(f_i, \sigma_E^2)$.

The speed and accuracy of the sampling fitness function therefore directly depend on the sample size. For a noisy fitness evaluation with variance $\sigma_E^2$ and computational cost $\beta$, the sampling fitness function has a variance and computational cost of $\sigma_E^2/n$ and $\beta n$, respectively. Thus by simply changing the sample size, an entire range of fitness functions can be generated, each of which is pareto-optimal (Srinivas & Deb, 1995; Goldberg, 1989) in terms of speed and accuracy. Thus any of the sampling sizes could result in a GA performing optimally within a given time-bound, as none of the sample sizes are absolutely better than any of the other sample sizes in terms of speed and accuracy of the resulting sampling fitness function.

### Population Sizing

The correct size for a population is very important for a computationally bound GA. If the population size is too large, valuable time is wasted processing redundant individuals, thereby reducing the number of generations processed within the time-bound, which in turn reduces the final solution quality. If the population size is too small, the GA can prematurely convergence to a sub-optimal solution, again reducing the final solution quality. In Goldberg, Deb, and Horn (1992), the question of how to correctly size populations for different domains and noise levels was addressed. Their population sizing model, including the noise component, can be expressed as:

$$N = \frac{2c\chi^k}{d^2}\frac{(m-1)}{m}(\sigma_F^2 + \sigma_N^2) \tag{3}$$

where $N$ is the population size, $c$ is the scaling factor based on a given level of confidence that the population will not prematurely converge, $\chi$ is the cardinality of the alphabet, $k$ is the largest building-block length, $d$ is the minimum signal difference between two competing building blocks, $m$ is the number of independent building blocks in the chromosome, $\sigma_F^2$ is the fitness variance of the initial population, and $\sigma_N^2$ is the noise variance of the noisy fitness function. For the purposes of this paper, the terms to the left of the sum of the population and noise variances will be grouped together and represented as the population sizing coefficient $\Gamma$. The population sizing coefficient $\Gamma$ is thus determined solely on the confidence factor and domain characteristics, and is independent of both the population fitness variance and the fitness function noise variance. The population sizing equation then becomes:

$$N = \Gamma(\sigma_F^2 + \sigma_N^2) \tag{4}$$

Equation 4 is used to size the population for different levels of fitness function noise in this paper. By generalizing equation 3, the model derived in this paper can used for any population sizing model that can be expressed in the form of equation 4, so the results in this paper are not tightly tied to the model derived in Goldberg, Deb, and Horn (1992). Within this paper, we use equation 3 to calculate the population sizing coefficient $\Gamma$, but other methods could also be used.

### Onemax Domain

This paper uses the onemax domain, also known as the counting ones or bit-counting problem, to verify the accuracy of the selection scheme models. The onemax is a simple domain, in that all alleles are uncorrelated. This domain is used as it is easy to demonstrate, and model, the affects of different sample sizes on GA performance. A thorough understanding of how sampling affects GA performance in the onemax domain is an important step towards predicting how sampling affects GA performance in other domains. This subsection reviews the onemax domain, and presents the convergence model for the onemax domain.

Within the onemax domain, the real fitness of an individual is the number of one bits in the chromosome. The optimal chromosome is simply the chromosome containing all one bits. The population mean fitness and variance for the onemax domain are given by $\mu_t = lp_t$ and $\sigma_t^2 = lp_t(1-p_t)$, respectively (Mühlenbein & Schlierkamp-Voosen, 1993; Thierens & Goldberg, 1994), where $l$ is the chromosome length and $p_t$ is the proportion of correct alleles in the population at generation $t$. The proportion of correct alleles in the initial population $p(0)$ is assumed to be 0.5. In Miller and Goldberg (1995), a performance model for the onemax domain with noisy fitness functions is derived:

$$\frac{dp}{dt} = I \frac{p(1-p)}{\sqrt{lp(1-p) + \sigma_N^2}} \tag{5}$$

where $I$ is the selection intensity (Mühlenbein & Schlierkamp-Voosen, 1993; Miller & Goldberg, 1995), and $\sigma_N^2$ is the fitness function noise variance. Equation 5, taken together with the assumption that half of the alleles are correct in the initial population ($p(0) = 0.5$), can be numerically solved to predict the convergence of the onemax population with noisy fitness functions. This model is used in this paper to predict the performance of a GA utilizing a noisy sampling fitness function for different sample sizes, by simply setting the noise variance of the model to the noise variance of the sampling fitness function for the sample size of interest.

## LOWER BOUND OF OPTIMAL SAMPLE SIZE

### GA Computational Model

The primary goal of this paper is to bound the optimal sample size that maximizes GA performance within a fixed amount of time. To do so, we must model the computational requirements of the GA over time. Fitzpatrick and Grefenstette (1988) developed a model for the total time $T$ required by a GA:

$$T = (\alpha + \beta n)GN \tag{6}$$

where $G$ is the total number of generations, $N$ is the population size, and $n$ is the sample size of the sampling fitness function. The variable $\alpha$ represents the fixed amount of GA overhead time per individual per generation, which includes the costs of selection, recombination and mutation, but excludes the cost of fitness evaluations. The variable $\beta$ represents the cost of a single fitness evaluation (a sample), so $\beta n$ is simply the total computational cost of a sampling fitness function with a sample size of $n$. This model does ignore the cost of initializing the population, but this is reasonable as the runtime costs dominate. If the population is sized according to equation 4, equation 6 becomes:

$$T = (\alpha + \beta n) G \Gamma (\sigma_F^2 + \sigma_N^2) \tag{7}$$

For our purposes, we are interested in determining the ending generation of a time-bounded GA for different sample sizes. Each different sample size results in a different sampling fitness function noise variance, as $\sigma_N^2 = \sigma_E^2/n$. This in turn changes the population size required by equation 4. The different sample sizes thus not only affect the $\beta n$ component of equation 6, but also affect the population size. If the computational bound $T$ is given, along with the relevant domain dependent variables $\alpha, \beta, \sigma_E^2, \sigma_F^2$, and $\Gamma$, the ending generation as a function of the sample size $g^*(n)$ can be calculated as follows:

$$\begin{aligned} g^*(n) &= \frac{T}{(\alpha + \beta n) N} \\ &= \frac{T}{(\alpha + \beta n)\Gamma(\sigma_F^2 + \sigma_N^2)} \\ &= \frac{T}{(\alpha + \beta n)\Gamma(\sigma_F^2 + \sigma_E^2/n)} \end{aligned} \tag{8}$$

The ending generation is important, as the performance of the GA at the ending generation $g^*(n)$ can be predicted using a convergence model for the domain. The predicted performance for different sample sizes can thus be compared to identify the optimal sample size for a given computational bound $T$.

#### Domain Independent Lower Bound

To determine the lower bound for the sample size, we make a critical assumption: the improvement in the convergence rate resulting from increased sample sizes (resulting in lower noise variance) is ignored. As this entails that GA runs for all sample sizes will have the same convergence rate, the optimal sample size is the sample size that maximizes the ending generation $g^*$. Determining the optimal sample size for the ending generation is easily determined by solving for the sample size value $n$ that makes $\frac{dg^*}{dn} = 0$, where $g^*(n)$ is defined in equation 8. Deriving from equation 8

$$\frac{dg^*}{dn} = -\frac{\beta T}{\Gamma(\sigma_F^2 + \sigma_E^2/n)(\alpha + \beta n)^2} + \frac{\sigma_E^2 T}{n^2 \Gamma(\sigma_F^2 + \sigma_E^2/n)^2 (\alpha + \beta n)} \tag{9}$$

Setting $\frac{dg^*}{dn} = 0$ and solving for $n$ yields a lower bound for the sample size $n_{lb}$:

$$n_{lb} = \sqrt{\frac{\alpha}{\beta}} \sqrt{\frac{\sigma_E^2}{\sigma_F^2}} \tag{10}$$

This lower bound balances the competing forces of population sizing and sampling costs, but ignores the convergence rate improvements for larger sampling sizes. It is easy to see that $n_{lb}$ is a lower bound, as if the performance increase due to faster convergence is also factored in along with the performance increase due to decreased population sizes at higher sample sizes, the performance advantages at $n_{lb}$ then outweighs the performance detriment due to the larger population size at $n_{lb}$. Thus the optimal sampling size must be at least as large as the lower bound ($n_{lb} \leq n^*$).

## EXPERIMENTAL RESULTS

These experiments, using the onemax domain, verify that the lower bound on the optimal sample size is valid by testing if the experimentally determined optimal size occurs after the predicted lower bound. The experimental results are averaged over 50 runs, and plotted against the performance predicted by the noisy convergence models for the onemax domain (equation 5). To predict the performance, the noise variance parameter of the onemax convergence model is simply set to $\frac{\sigma_E^2}{n}$, where $n$ is is the sample size, and used to determine the performance at the predicted ending generation. The performance metric is the average population fitness of the ending generation for each sample size. This metric was also used in the study by Fitzpatrick and Grefenstette (1988), as the fitness of the final population can be considered to represent the knowledge obtained by the GA.

For each GA run, the following parameters are user-specified: $n, \alpha, \beta, \sigma_E^2, \sigma_F^2, I, T$, and $\Gamma$. The stopping criterion for the GA is the ending generation calculated by equation 8, rounded down to the nearest generation, so as to not exceed the GA time bound $T$. The chromosome length is $l = 100$, and half of the alleles in the initial population are assumed to be correct ($p(0) = 0.5$). This assumption also leads to a population variance of $\sigma_F^2 = l\,p(1-p) = 25$. The population size is set according to equation 4, with the population sizing coefficient determined according to the population sizing model of Goldberg, Deb, and Horn (1992), resulting in $\Gamma = 8$. The selection mechanism is binary tournament selection, with a corresponding selection intensity of 0.5642 (Thierens & Goldberg, 1994). The crossover mechanism is uniform crossover (Syswerda, 1989). As done in other studies (Thierens & Goldberg, 1994; Miller & Goldberg, 1996; Miller & Goldberg, 1995), two rounds of crossover are performed in each generation in the following manner: the mating pool is selected by the selection scheme, crossover is used to produce a new pool of offspring, and then crossover is run again on the new offspring to produce the next generation. The creates a new generation that is more normally distributed than if one crossover per generation is used. This makes the selection intensity convergence model assumption of a normally distributed population more accurate. However, this is not critical, as Miller and Goldberg (1996) shows that for tournament selection, two crossovers per generation only has a small improvement on the accuracy of the predicted convergence performance of the GA as compared to one crossover per generation.

A sample experimental run is shown in figure 1. In figure 1, the top smooth continuous plot is the performance predicted using the onemax convergence model described above. The stair-shaped continuous plot is a result of the ending generation being rounded down to the nearest generation, as the GA can not run for a fractional number of generations and is not allowed to exceed the GA time bound. The points represent the experimentally determined mean performance of 50 runs for each sample size. They closely agree with the predicted truncated performance plot.

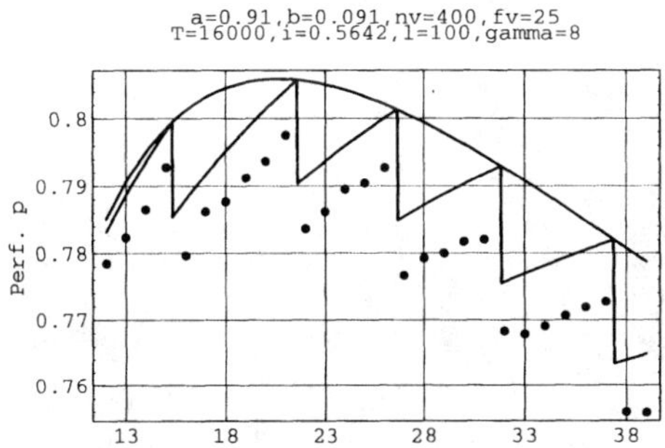


**Figure 1. Lower Bound for Optimal Sample Size.**

Although for brevity's sake only one figure is shown here, the lower bound was shown to be valid for a wide range of $\frac{\alpha}{\beta}$ and $\frac{\sigma_E^2}{\sigma_F^2}$ settings. In all cases, the experimentally determined performance was very close to the predicted performance. The experimentally determined optimal sample size in all cases matched the predicted optimal sample size. Another interesting result of the experiments is that the stair-shape of the truncated performance plot can be exploited to prune sample sizes not on outside stair edges from consideration, as it can be shown that they are dominated by the sample size at the rightmost edge of each stair. This pruning method is a result of the inability of a GA to run for a fractional number of generations, and is independent of the domain.

## CONCLUSIONS

This research derived, and experimentally verified, a lower bound for the optimal sample size. It is important to note that the lower bound for the sample size, equation 10, is independent of the domain. As practically all sampling domains are characterized by the competing forces of faster convergence rates, smaller population sizes, and higher sampling costs for increased sample sizes, equation 10 is broadly applicable to most domains for determining a lower bound on the sample size. It has the added advantage that the convergence equation for a particular domain does not have to be determined, as the lower bound ignores the effect of different sampling sizes on the convergence rate. This is an important benefit, as in many domains it is difficult to determine an exact convergence model, due to many factors, including allele correlation, non-uniformly scaled building blocks, and lack of sufficient domain knowledge. Another interesting result of this research is that truncation of the ending generation in order to complete the GA within the time alloted can be exploited to prune sample sizes from consideration. This research showed that only the rightmost sample size of each truncation "stair" needs to be examined, as it can be proved that the performance resulting from the rightmost samples of each stair dominates the performance resulting from the other samples on the same truncation "stair." This pruning technique can drastically reduce the number of samples that need to be examined to identify the optimal sample size for a given domain.

## ACKNOWLEDGMENTS

This work was supported under NASA Grant No. NGT 9-4. This effort was also sponsored by the Air Force Office of Scientific Research, Air Force Materiel Command, USAF, under grant numbers F4960-94-1-0103 and F49620-95-1-0338. The U.S. Government is authorized to reproduce and distribute reprints for Governmental purposes notwithstanding any copyright notation thereon.

The views and conclusions contained herein are those of the authors and should not be interpreted as necessarily representing the official policies or endorsements, either expressed or implied, of the Air Force Office of Scientific Research or the U.S. Government.

### References

Dahlquist, G., & Björk, Å. (1974). *Numerical methods* (N. Anderson, trans. ed.). Englewood Cliffs, NJ: Prentice-Hall, Inc.

Davis, L. (Ed.) (1991). *Handbook of genetic algorithms.* New York: Van Nostrand Reinhold.

Fitzpatrick, J. M., & Grefenstette, J. J. (1988). Genetic algorithms in noisy environments. *Machine Learning, 3*, 101–120.

Goldberg, D. E. (1989). *Genetic algorithms in search, optimization, and machine learning.* New York: Addison-Wesley.

Goldberg, D. E., Deb, K., & Horn, J. (1992). Massive multimodality, deception, and genetic algorithms. *Parallel Problem Solving from Nature, 2*, 37–46.

Grefenstette, J. J., & Fitzpatrick, J. M. (1985). Genetic search with approximate function evaluations. In Grefenstette, J. J. (Ed.), *Proceedings of an International Conference on Genetic Algorithms and Their Applications* (pp. 112–120). Hillsdale, NJ: Lawrence Erlbaum Associates.

Miller, B. L., & Goldberg, D. E. (1995, November). *Genetic algorithms, selection schemes, and the varying effects of noise* (IlliGAL Report No. 95009). Urbana, IL: University of Illinois at Urbana-Champaign.

Miller, B. L., & Goldberg, D. E. (1996). Genetic algorithms, tournament selection, and the varying effects of noise. *Complex Systems*, *9*(3), 193–212.

Mühlenbein, H., & Schlierkamp-Voosen, D. (1993). Predictive models for the breeder genetic algorithm: I. Continuous parameter optimization. *Evolutionary Computation*, *1*(1), 25–49.

Srinivas, N., & Deb, K. (1995). Multiobjective optimization using nondominated sorting in genetic algorithms. *Evolutionary Computation*, *2*(3), 221–248.

Syswerda, G. (1989). Uniform crossover in genetic algorithms. *Proceedings of the Third International Conference on Genetic Algorithms*, 2–9.

Thierens, D., & Goldberg, D. (1994). Convergence models of genetic algorithm selection schemes. *Parallel Problem Solving fron Nature, 3*, 119–129.

Winter, G., Périaux, J., Galán, M., & Cuesta, P. (Eds.) (1995). *Genetic algorithms in engineering and computer science.* Chichester: John Wiley and Sons.

# GESTURE-BASED PROGRAMMING, PART 1: A MULTI-AGENT APPROACH

**RICHARD M. VOYLES**
*Robotics Ph.D. Program*
*Carnegie Mellon University*

**PRADEEP K. KHOSLA**
*Electrical and Computer Engineering*
*Carnegie Mellon University*

***ABSTRACT:***
Gesture-Based Programming is a paradigm for the evolutionary programming of rapidly deployable manipulation systems by human demonstration. The goal is to provide a more natural environment for the user and to generate more complete and successful programs by focusing on task experts rather than programming experts. What makes it unique from other programming by human demonstration approaches are the same things that make it evolutionary: a composable knowledge base of expertise agents and a facility for supervised practice after initial training. Prior knowledge of previously acquired skills (sensorimotor expertise) facilitates the interpretation of "gestures" during training and then provides closed-loop control of execution during run-time. This paper, part one of two, presents a high-level description of the system as well as descriptions of capabilities we've demonstrated on a PUMA robot and Utah/MIT hand. The companion paper provides a detailed account of one method of acquiring, matching, and even transforming sensorimotor expertise.

## INTRODUCTION

Programming robotic systems is difficult not only for novices but experts alike. Planning operation sequences and collision-free paths while maintaining realistic cycle times can be hard enough, but add in the complexities of robot/environment contact, uncertainty, and exception handling and the expertise required becomes daunting. As a result, robotic technologies have had minimal impact on most manufacturing domains (in terms of the number or robots versus number of humans employed), particularly as product life cycles and batch sizes decrease. Ironically, these are the conditions under which robots were originally touted as being most effective (Hartley, 1983).

Once implemented, many robotic applications fail to meet expectations or prove insufficiently robust for the desired level of autonomy. A primary reason for this is that programming difficulties maintain a layer of insulation between the system's use and it's development. Because programming requires so much expertise, unskilled users, by definition, cannot be programmers and experienced programmers are too "valuable" to be users. As a result, there is a gap between personnel needs and expertise.

Numerous attempts have been made to ease the discomfort of robot programming with varying degrees of success. Behavior-based and multi-agent systems (Brooks, 1986) seek to modularize software for easier programming by programming experts. Visual programming environments like Chimera/Onika (Stewart et al, 1993)(Gertz, 1994) and commercial packages such as MatrixX/SystemBuild and LabView capitalize on modular, reconfigurable software blocks by iconifying them within "software assembly" environments. These visual environments allow programming at a much higher level, hiding many details of the particular implementation, and lessening the burden of programming expertise.

Despite these advances, a fairly high level of expertise is still required to program and interact with robots. For example, Onika allows novice users to easily build simple pick-and-place robotic applications in a matter of minutes (Gertz, 1994). But, it is imperative that the user understands and consciously considers the importance of via points, collision avoidance, grip selection, and the dynamic effects of transport in order to for the application to be successful. Since these characteristics are inherent in all manipulative tasks and are adeptly mastered by humans, they are usually handled unconsciously. Hence the distinction between task experts and programming experts. Programming experts are trained to bring these subconscious requirements up to the conscious level in order to transcribe them as required by the particular programming environment. Visual programming does little for the bulk of robotic applications that involve substantive contact with the environment, uncertainty, and complex sensing.

A more recent approach to human/robot interaction is the field of *learning by observation* (Kang and Ikeuchi, 1996)(Kuniyoshi et al, 1989). By forcing the robot to observe a human interacting with the world, rather than forcing the human to interact with a *textual representation* of the robot interacting with the world, a more natural, "anthropocentric" programming environment results. But these systems are mostly kinematically-based and operate off-line; if the robot misinterprets the desired trajectory, the whole sequence must be re-taught. (More on learning by observation later.)

It is instructive to look at the inroads robots have made into manufacturing to see how these problems can be mitigated. Two of the most common industrial applications are spray painting and spot welding. Both of these use natural programming methods such as "lead-through teaching" (Todd, 1986) (i.e. human demonstration) and point-teaching, respectively, that are easily mastered by semi-skilled workers. Lead-through teaching, in particular, allows anyone who can perform a "valid" task to program the robot to perform the task by demonstrating it. As such, programming changes can be made reactively or proactively by a task expert, not by calling in a robot programming expert. Tasks such as spray painting involve *skill transfer* from the teacher (i.e., programmer) to the robot. Lead-through teaching provides an intuitive mechanism for a task expert to accomplish this. Unfortunately, it is limited to simple tasks because of its reliance on pure kinematic replay.

## GESTURE-BASED PROGRAMMING OVERVIEW

If one views programming as a form of teaching, textual programming is clearly unnatural for humans . Despite a human's ability to master some task -- from high-level planning of sequence and trajectory, to grasp selection, to robust, low-level execution -- there is no guarantee that human will be able to describe the task in syntactically correct prose in a foreign (programming) language. Humans are more effective at

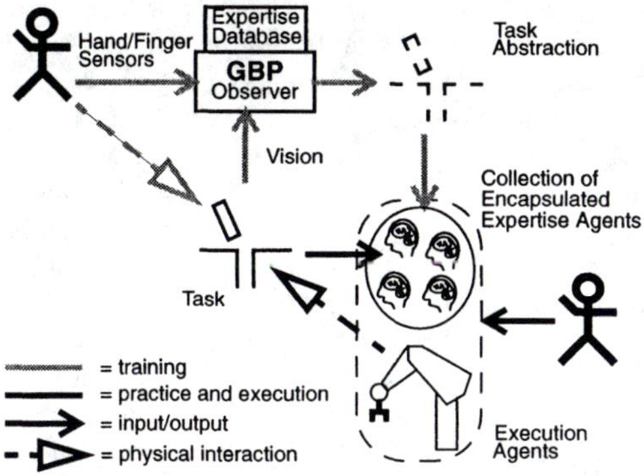


**Figure 1: Gesture-Based Programming system overview.**

teaching by demonstration and practice and, although there are no guarantees here, either, it has been proven empirically to be the most successful technique between humans (Patrick, 1992). But teaching by demonstration is not guaranteed to work and one of the reasons is it assumes a basic set of *a priori* capabilities. Imagine trying to teach calculus to someone who doesn't know algebra. Gesture-Based Programming (GBP) addresses all of the above issues by providing a more natural, demonstration-based programming paradigm that allows both demonstration and practice phases and relies on an underlying skill base of robust, semi-autonomous primitives from which new tasks can be assembled.

GBP begins by observing a human demonstrate the task to be programmed. (See the stick figure on the left of Figure 1.) Observation of the human's hand and fingertips is achieved through a sensorized glove. The modular glove system senses hand pose, finger joint angles, and fingertip contact conditions. Objects in the environment are sensed with computer vision while a speech recognition system extracts "articulatory gestures." (Gestures will be described later.) Primitive gesture classes are extracted from the raw sensor information and passed on to a gesture interpretation network. These autonomous multi-agent networks extract the demonstrator's intentions with respect to the system's skill base to create an abstraction of the task. In other words, the system is not merely remembering everything the human does, but is trying to understand -- within its scope of expertise -- the subtasks the human is performing ("gesturing"). These primitive capabilities in the skill base take the form of *encapsulated expertise agents* -- autonomous agents that encode sensorimotor dexterity.

The output of the GBP system is the executable program for performing the demonstrated task on the target hardware. This program consists of a network of encapsulated expertise agents of two flavors. The primary agents implement the primitives required to perform the task and come from the pool of primitives represented in the skill base. The secondary set of agents includes many of the same gesture recognition and interpretation agents used during the demonstration. These agents perform on-line observation of the human to allow supervised practicing of the task, if desired. (Stick figure on far right of Figure 1.)

As mentioned above, the human model for teaching by demonstration most often involves a practice phase. The reason for this is that passive observation of a task rarely provides accurate parametrization for the trainee's deduced task "model" (in this case, the model is represented by the collection of primary encapsulated expertise agents) and sometimes the deduced model is wrong (e.g. missing or incorrect encapsulated expertise agents). Incorporating gesture recognition and interpretation agents into the executable provides an intuitive way for the demonstrator, or another user, to fine tune the operation of the program without having to demonstrate the entire task over again. Because all our agents are implemented as autonomous software modules, these observation agents can easily be disabled without recompiling the program to prevent unauthorized modification.

In the real world, it will not be possible to represent most useful tasks with one network of encapsulated expertise agents. Therefore, it is necessary to segment the demonstration into a series of discrete subtasks (e.g. a grasping subtask followed by a manipulation subtask). Each subtask will be embodied by a network as described above. In this case, the executable program will consist of a sequence of networks rather than one network.

### What are gestures?

We define gestures as "imprecise events that convey the *intentions* of the gesturer." There are many gestural modalities including *symbolic gestures*, such as the pre-shape of a dextrous grasp, *motion gestures*, such as transporting or manipulating an object, *tactile gestures*, such as fingertip contact force, and *articulatory gestures*, such as "oops!' when an error is made or model-based information such as "tighten the *screw*." Other gestural modes exist but this is the subset we consider for our GBP system. Because gestures are imprecise events, they are context dependent; in isolation they convey little information. Yet, when combined with the state of the system and the environment at the time the gesture occurred, perhaps augmented with some past history of gestures, the intention becomes interpretable.

In effect, gestures form an "alphabet" from which "words" are formed when combined with state information. The state information is also represented by characters in the gestural alphabet. These gestural words are strung together by the gesturer into "sentences" which form the basis for interpretation. The gestural alphabet is like a linguistic alphabet in the sense that characters are parametrized. For instance, in English, the letter "a" has many sounds including the long "a" as in "save," the short "a" as in "sad," and the silent "a" as in "read." This is an implicit parametrization. In contrast, the gestural characters are explicitly parametrized by either a single integer or floating-point argument. An example can be found in (Voyles and Khosla, 1995b).

Gesture recognition is handled by individual gesture recognition agents for each gestural mode. Each recognition agent is made up of two subordinate agents, one general, one domain-specific. The general agent is the gesture agent in Figure 2. It recognizes the instance of a gesture and the type of gesture within the gesture class (mode). The domain specific agent grabs the relevant state information associated with the gesture type for the given domain and appends it to the gestural word.

## PROJECT STATUS

### Learning by Observation

GBP is a direct extension of the real-time control work by Voyles for the functional demonstration of Kang and Ikeuchi's Learning by Observation system (Kang and

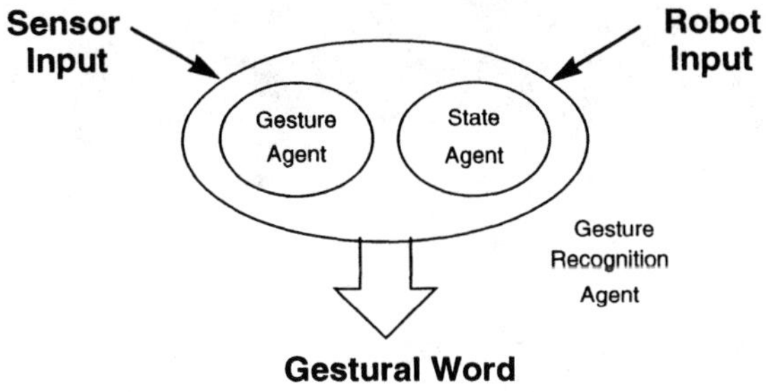


**Figure 2: Structure of the gesture recognition agent.**

Ikeuchi, 1996). The human is instrumented with a sensorized glove but only kinematic information is reported. As the human demonstrates simple tasks, hand and wrist trajectories are recorded and a multi-baseline stereo machine looks for changes in the pose of detected objects. These data streams are automatically segmented to detect pre-grasp, grasp, and manipulation phases of motion and then mapped from the human hand to the target robot hand. A PUMA and Utah/MIT hand demonstrated lifting, transporting, and re-depositing objects as well as low-tolerance peg-in-hole insertions.

While these demonstrations were impressive when successful, getting them to succeed was difficult. Based solely on kinematic trajectories, duplication by the robot demanded accurate calibration of the workspace since no feedback was provided. The robot was incapable of reacting to even the most insignificant of error conditions. In fact, optimization of the kinematic grasp to fit the surfaces of the manipulated objects prevented the exertion of grasping forces. We had to command trajectories that penetrated the object surfaces in order to hold them stably. These experiences demonstrated the need for a skill-based approach to programming by human demonstration. They also suggested reconfigurable agents for the implementation of those skills.

**Gesture Interpretation Networks**

An example gesture-based interaction system is shown in Figure 3 and described more fully in (Voyles and Khosla, 1995a). This is not a programming environment but an application scenario utilizing tactile gestures. The robot executes one of four polygonal cyclic trajectories: rectangle, triangle, cross, or pick-and-place. Each trajectory has parameterized height, width, and thickness. The user selects and tunes the trajectories with tactile gestures (nudging the end effector) sensed by a force/torque sensor or trackball. The recognition agents segment gestures from the sensory data streams and pass gestural words on to the interpretation agents. Each interpretation agent contains a fuzzified model of a single hypothesis which gestures either support or refute. The agents individually build a confidence value in their respective hypotheses and vote within agent pools, each pool being winner-take-all. Some of these agents were also applied to a cable harnessing task as described in (Voyles and Khosla, 1995b).

## CONCLUSIONS

We have presented our concept for a gesture-based programming system as well as brief portrayals of work we have completed on individual components of that system. Although the complete programming environment is not yet finished, the success of

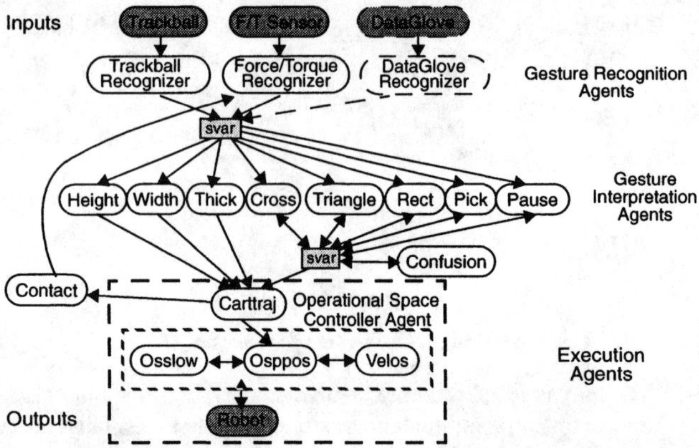


**Figure 3: Multi-agent interpretation network for tactile gestures.**

our preliminary work leads us to believe it will provide a much more natural and successful programming paradigm for certain classes of contact-based tasks by allowing task experts, rather than programming experts, to specify the operations. Part 2 of this series describes one method of acquiring expertise agents for the skill-base.

## ACKNOWLEDGMENTS

Mr. Voyles was supported in part by a National Defense Science and Engineering Graduate Fellowship, a DOE Integrated Manufacturing Predoctoral Fellowship and by Sandia contract AL-3019.

## REFERENCES

Brooks, R.A., (1986). "A Robust Layered Control System for a Mobile Robot," *IEEE Journal of Robotics and Automation*, v.RA-2, n.1, March, pp. 14-23.

Gertz, M.W, (1994). *A Visual Programming Environment for Real-Time Control Systems*, Ph.D. Thesis, Department of Electrical and Computer Engineering, Carnegie Mellon University.

Hartley, J., *(1983). Robots at Work*, IFS Publications, Ltd.,Kempston, Bedford, UK, chapters 5, 8.

Kang, S.B., and K. Ikeuchi, (1996). "Toward Automatic Robot Instruction from Perception -- Temporal Segmentation of Tasks from Human Hand Motion," *IEEE Trans. on Robotics and Auto.*, Mar.

Kuniyoshi, T., M. Inaba, and H. Inoue, (1989). "Teaching by Showing: Generating Robot Programs by Visual Observation of Human Performance," *Proc. of 20th Intnl. Symp. on Ind. Robots*, pp. 119-126.

Patrick, J., (1992). *Training: Research and Practice*, Academic Press, San Diego, CA.

Stewart, D.B., R.A. Volpe, and P.K. Khosla, (1993). "Design of Dynamically Reconfigurable Real-Time Software Using Port-Based Objects," CMU Robotics Institute tech. rpt., CMU-RI-TR-93-11.

Todd, D.J., (1986). *Fundamentals of Robot Technology*, John Wiley and Sons, chapters 3, 7.

Voyles, R.M. and P.K. Khosla, (1995a). "Tactile Gestures for Human/Robot Interaction," *Proc. of IEEE/RSJ Conf. on Intelligent Robots and Systems*, Pittsburgh, PA, v. 3, pp. 7-13.

Voyles, R.M. and P.K. Khosla, (1995b). "Multi-Agent Gesture Interpretation for Robotic Cable Harnessing," *Proc. of IEEE Conf. on Systems, Man, and Cybernetics*, Vancouver, B.C., pp. 1113-1118.

# GESTURE-BASED PROGRAMMING, PART 2: PRIMORDIAL LEARNING

**RICHARD M. VOYLES**
**J. DANIEL MORROW**
*Robotics Ph.D. Program*
*Carnegie Mellon University*

**PRADEEP K. KHOSLA**
*Electrical and Computer Engineering*
*Carnegie Mellon University*

***ABSTRACT:***
In part one of this two-part series, we described our Gesture-Based Programming paradigm for programming by human demonstration. This paradigm depends on a pre-existing knowledge base of capabilities, collectively called "encapsulated expertise," that comprise the real-time sensorimotor primitives from which the run-time executable is constructed as well as providing the basis for interpreting the teacher's actions during programming. In this paper we present a technique based on principal components analysis, augmentable with model-based information, for learning and recognizing sensorimotor primitives. We describes simple applications of the technique to a mobile robot and a PUMA manipulator. The mobile robot learned to escape from jams while the manipulator learned guarded moves and accommodations that are composable to allow flat plate mating operations. While these initial applications are simple, they demonstrate the ability to extract primitives from demonstration, recognize the learned primitives in subsequent demonstrations, and combine and transform primitives to create different capabilities.

## INTRODUCTION

Gesture-Based Programming (GBP) is a paradigm for programming rapidly deployable systems by human demonstration. A fundamental assumption on which the paradigm is based is the availability of a knowledge base of sensorimotor dexterity -- precompiled capabilities the target system already "knows." There are three basic mechanisms by which these primitives can come into existence. The first is "clairvoyance." The user (or automatic planner), acting as a god, hand codes a primitive that he/she *knows* is useful and composable. An example of this is a guarded move. The second mechanism is through assembly of lower-level primitives (i.e. Morrow and Khosla, 1995). Using a calculus analogy, algebraic primitives may form the basis for calculus, but arithmetic primitives form the basis for algebra. Finally, there is learning.

In this paper, we present a learning approach for extracting sensorimotor primitives from teleoperated manipulations that is a natural extension of an autonomous sensor calibration technique called *Shape from Motion* (Voyles et al, 1995a). Calibration is, afterall, a form of learning. One wants to "learn" the transformation from the sensor's input space to the output space. Most traditional calibration techniques (such as least-squares) do not pose it as a learning problem, however. Shape from Motion calibration extracts the "shape" of an input/output mapping that results from random "motion" of the input through the sensor space. The result is calibration of the sensor with little or no knowledge of the loads applied to the sensor. It became apparent that Shape from Motion could be applied to other types of learning problems, as well.

Shape from Motion is based on the extraction of an eigenspace representation of the input/output mapping by SVD to derive principle components. In the case of force/torque sensor calibration (described in Voyles et al, 1995a), the principle components analysis (PCA) is constraint-directed, so model-based information can be incorporated into the learning process. Because the underlying representation (a set of eigenvectors) is linear, it is limited to learning linear, or at least piece-wise linear, interactions. The drawback is the difficult task of segmenting the observations into linear, learnable subtasks. The advantage is the representation provides not only a mechanism to learn the primitives, but natural mechanisms for identifying primitives in subsequent observations and combining and transforming existing primitives into new primitives. These are abilities many other learning approaches have difficulty demonstrating.

## PRIOR WORK

We are interested in developing a primitive representation which supports acquisition, integration, and transformation of sensorimotor primitives. Previous work in learning robot skills is relevant to this effort because it provides potential primitive acquisition techniques and because primitives learned by techniques other than primordial learning may also be used in the execution phase of our system. The idea is to relate primitives to recurring subgoals of complex tasks and to reuse primitives in the construction of complex task strategies, whatever the origin of the primitives may be.

Yang et al (1994) have applied hidden markov models to skill modelling and learning from telerobotics. The skills learned are manipulator positioning tasks without force or vision feedback. Pook and Ballard (1992) performed similar work.

Along these lines of skill acquisition by supervised learning is ALVINN, an artificial neural network for vehicle steering (Pomerleau, 1992). ALVINN has demonstrated robust mastery over a wide range of vehicles and road types by observing a human drive the vehicle on the target road type. For the purpose of comparison, Hancock and Thorpe (1995) developed ELVIS, a PCA-based vehicle steerer to operate on the same vehicles. The success of ELVIS provided additional impetus to complete this work.

Several researchers have applied reinforcement learning methods to the recovery of a peg insertion skill. Simons et al (1982) learn how to interpret a force feedback vector to generate corrective actions for a peg insertion which has an aligned insertion axis. Gullapalli et al (1992) learned close tolerance peg insertion using a neural network and a critic function, but the result is specific to hole location.

## PRIMORDIAL LEARNING

*Primordial Learning* refers more to a way of thinking than to a specific algorithm; it refers to learning fundamental interactions, or mappings, with no prior knowledge. It's like an infant that learns gross motor control by flailing. The infant is aware of "sensor and actuator ports," but has no comprehension of their connection to the body or to the outside world. Some might say primordial learning is non-parametric learning. Some might call it unsupervised learning. The implementations in this paper can be called principle components analysis. Yet, all these terms, broad and narrow, are, in one way or another, inaccurate in describing our work in its entirety, from autonomous sensor calibration to mobile robot behaviors, hence the new terminology.

**Mobile Robot Primitives**

The infant analogy provided not just the impetus for the name, but the impetus for the first application. Somewhat as a lark, the challenge was made to create a mobile robot that could learn to "crawl" before the newborn infant of the first author could do so. The term "crawl" was interpreted loosely to mean "purposefully motivate" so as not to exclude the creation of a wheeled vehicle as opposed to a much more mechanically complex legged vehicle.

This sort of robot learning has been suggested for mobile robots before and prior work and detailed results from our experiments can be found in Voyles, et al (1995b). To summarize, we applied primordial learning to the MK-V, a three-wheeled, non-holonomic mobile robot. Sensors include wheel encoders, a compass, bump sensors and motor current sensors. The drive and steering motors have three-valued -- forward-off-reverse -- commands with no closed-loop controllers on velocity or position. For this initial experiment there are no redundant groups of very similar and correlated sense elements such as a visual retina (as in Hancock and Thorpe, 1995) or sonar ring.

The shape from motion technique allows the robot to develop an internal representation of teleoperated interaction to produce "meaningful" externalized behavior. We do not use an explicit objective function or provide a reinforcement signal, but the learned result is explicitly dependent on the teleoperated interactions presented by the teacher. For training, we teleoperated the robot in a cluttered environment allowing it to wander while bumping into obstacles and jamming the wheels.

The training data consists of a matrix of input/output vectors sampled periodically during teleoperation. The input/output vector is composed of the actuator commands concatenated to the sensor data. The mean of this vector, **a**, over the entire sequence was subtracted out to normalize the data values. Next, the matrix was batch-processed using SVD to extract the eigenvectors and the largest $n$ eigenvectors were selected using the largest ratio of adjacent singular values as the threshold for $n$. This test assumes there is a group of "significant" eigenvectors, $\mathbf{e}_i$, with similar singular values and then there's a bunch of noise with small singular values. Examining the ratios of adjacent, ordered singular values indicates the dividing line between the groups. For run-time operation, the new sensor image, **x**, is projected onto the eigenspace as described in Hancock and Thorpe (1995):

$$\mathbf{v} = \mathbf{a} + \sum_{i=1}^{n} ((\mathbf{x} - sensor(\mathbf{a})) \bullet sensor(\mathbf{e}_i))\,\mathbf{e}_i \qquad (1)$$

Using this technique, the real robot successfully achieved both wandering behavior and escaped stalls and collisions without highly redundant or correlated sensors and with no algorithmic structure imposed on the learned result. Plots demonstrating the MK-V learned when to steer and reverse direction in order to free itself appear in Voyles, et al (1995b).

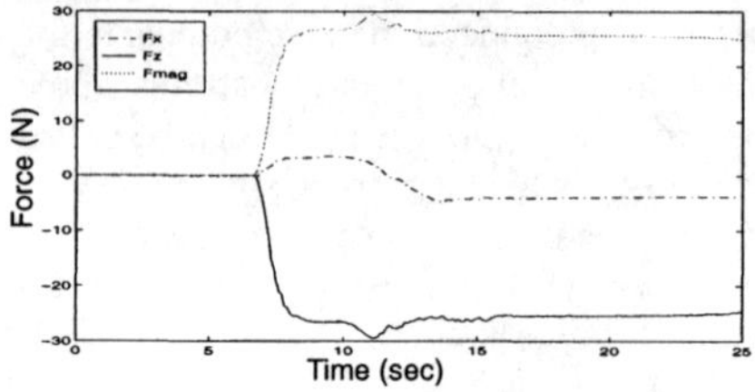


**Figure 1: Autonomous operation of the learned, one-dimensional guarded move.**

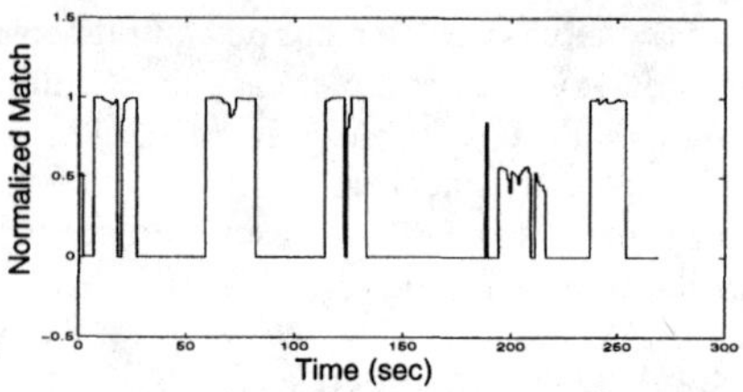


**Figure 2: Goodness of match determined by automatic skill identification algorithm.**

## PUMA Manipulator Primitives

While the application on the MK-V mobile robot started out as a lark, learning sensorimotor primitives for robot manipulation skills has long been a serious goal of our lab. Automatic generation of usable primitives is a necessary technology demonstration for the credibility of GBP. Because the skill base can incorporate primitives from a variety of sources, it is not necessary to generate all primitives automatically. Nonetheless, it is important to show proof-of-concept of the end-to-end system.

**Guarded Move.** As a first demonstration, we attempted to learn a one-dimensional guarded move. Although this is a fairly trivial primitive that can easily be hand-coded, it is not as obvious how to robustly identify it during human demonstration. This is the strength of an integrated approach like shape from motion primordial learning; it provides primitive identification and transformation as well as basic learning.

To learn *zguard* -- a guarded move along the z-axis -- we teleoperated a PUMA robot with a 6-DoF force/torque sensor so that it came into contact with a table while moving along the approach (z) axis of the end effector. We repeated this one-dimensional guarded move ten times, logging data only during the guarded move, not during the retraction phase when the end effector was moved away from the surface. The robot was controlled by a cartesian velocity controller, teleoperation input was provided by a 6-DoF trackball, and operator feedback was visual as well as a graphical display of the real-time force/torque components.

The input/output vector from which the data matrix for training was generated consisted of the useful data -- 6 measured force components, the total force and torque magnitudes, and 6 commanded cartesian velocities -- plus some irrelevant data -- 3 cartesian position elements and 9 cartesian orientation elements. We used the same algorithm for extraction of the eigenvectors as used on the MK-V.

The output of the training was a primitive of one eigenvector that performed very well. Figure 1 shows the measured force components as the guarded move primitive autonomously acquires a hard surface. 25N was the target force threshold applied during training and this can be varied by scaling the output portion of the extracted eigenvector. The primitive worked equivalently in all trials performed regardless of region of workspace, orientation of the end effector, compliance of the surface, or external perturbations. Analysis of the components of the eigenvector support this. The primitive is looking at the *z*-component of force as well as the total magnitude of the force.

To identify instances of the primitive from demonstration data, we run the training algorithm on windowed batches of run-time data and examine the parallelism between the extracted eigenspace and the eigenspace representation of the primitive. The dot product gives a quantified measure of eigenvector parallelism that we compare to a high threshold for presence or absence of the primitive.

To test the recognizer, we randomly moved the robot around the workspace, occasionally coming into contact with surfaces. The resulting goodness-of-fit is plotted in Figure 2. The identification algorithm accurately picked out every instance of a guarded move in z and rejecting every instance of random motion or random contact, including guarded moves along other axes. The occasional dropouts to zero are inconsequential as this fitness measure is but one input to the gesture interpretation network of the GBP system.

**Edge-to-Surface Alignment.** The guarded move is a "move until force threshold" operation. Movement is an explicit part of the goal. Alignment operations, on the other hand, only move in order to accommodate. Since no explicit motion is part of the primitive, it is difficult to teleoperate. What we'd like to do is use the previously learned guarded move to bring the edge and surface together while we teleoperate only the alignment primitive. The problem is the learned result would include both primitive operations, since there is no way to separate the sensor stream into "guarded move" and "alignment" portions. However, the eigenspace representation provides a convenient mechanism for separating the two after learning, if they are truly elemental and decoupled.

Learning the *yroll* alignment task -- alignment of an edge to a surface by accommodating rotationally around the y-axis -- was accomplished in the same manner as the *zguard* primitive with the exception that *zguard* was running during the teleoperation phase. This complicated the eigenvector extraction, as well. We expect the *zguard* eigenvector to appear in the new set of trained vectors and, in fact, it does. Using our standard ratio of eigenvalues test, only one eigenvector is extracted during training and its dot product with *zguard* is 0.9993, indicating it is the *zguard* primitive.

To extract the alignment primitive, we must look to the next significant group of eigenvectors. Ignoring the *zguard* eigenvector produces a set of four eigenvectors that very nicely implement the desired accommodation operation. Combining the *zguard* and *yroll* primitives by superposition produced behavior that reliably pressed a line against a table. Simple superposition of eigenvectors allows for decomposition and recomposition of the primitives at least in this case of decoupled operations.

**Surface-to-Surface Alignment.** The above decoupled primitives can be combined to produce more complex behavior. Combining *zguard*, *xroll*, and *yroll* produces a "skill" that allows two surfaces to be pressed together. An experimental run of these three primitives is shown doing just that in Figure 3 and Figure 4.

## CONCLUSIONS

We have described shape from motion primordial learning, a learning technique based on principle components analysis that has been successfully applied to extracting and identifying robotic sensorimotor primitives from teleoperated actions of both mobile robots and manipulators. By incorporating model-based information, we have also applied it to nearly autonomous calibration of force/torque sensors with equal or greater accuracy than the traditional least-squares technique (data in Voyles et al, 1995a).

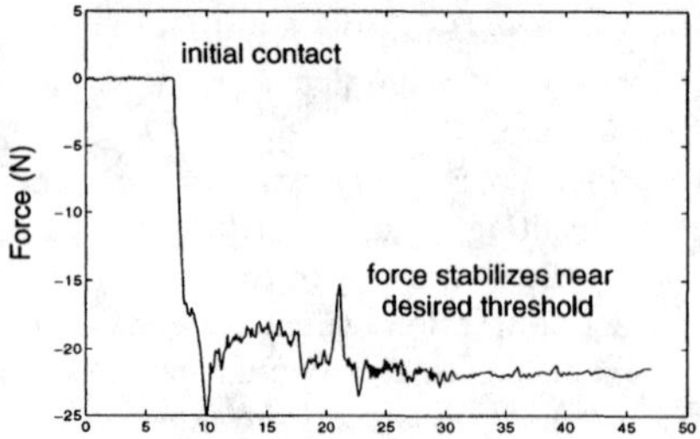


**Figure 3: Force in z-axis as zguard, xroll, and yroll cooperate to press one surface against another.**

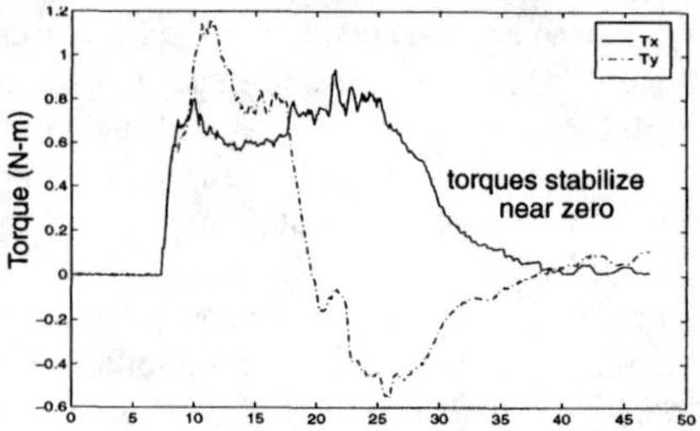


**Figure 4: Torque around x- and y-axes as zguard, xroll, and yroll cooperate.**


## ACKNOWLEDGEMENTS

Mr. Voyles was supported in part by a National Defense Science and Engineering Graduate Fellowship. Mr. Morrow was supported in part by a DOE Computational Science Graduate Fellowship. Both Mr. Voyles and Mr. Morrow were further supported, in part, by DOE Integrated Manufacturing Predoctoral Fellowships and Sandia contract AL-3019.

Trident Robotics and Research, Inc. provided components for constructing the MK-V and retrofitting the PUMA robots for low-level control.


## REFERENCES


Gullapalli, V., R. Grupen, and A. Barto, (1992). "Learning Reactive Admittance Control," *Proc. of IEEE International Conference on Robotics and Automation*, Nice, France.

Hancock, J. and C. Thorpe, (1995). "ELVIS: Eigenvectors for Land Vehicle Image System," *Proc. of IEEE/RSJ International Conf. on Intelligent Robots and Systems*," Pittsburgh, PA, pp 35-40.

Morrow, J.D. and P.K. Khosla, (1995). "Sensorimotor Primitives for Robotic Assembly Skills," *Proc. of the IEEE International Conference on Robotics and Automation,* v. 2, pp. 1894-1899.

Pomerleau. D., (1992). *Neural Network Perception for Mobile Robot Guidance*, Ph.D. Thesis, Carnegie Mellon University, Pittsburgh, PA.

Pook, P.K. and D.H. Ballard, (1993). "Recognizing Teleoperated Manipulations," *Proc. of the IEEE International Conference on Robotics and Automation*, v. 2, pp. 578-585.

Simons, J., H. Van Brussel, J. De Schutter, and J. Verhaert, (1982). "A Self-Learning Automaton with Variable Resolution for High Precision Assembly by Industrial Robots," *IEEE Transactions on Automatic Control*, Vol AC-27, No. 5.

Voyles, R.M., J.D. Morrow, and P.K. Khosla, (1995a). "Shape from Motion Approach to Rapid and Precise Force/Torque Sensor Calibration," *Proc. of Dynamic Systems and Control Division*, ASME Winter Annual Meeting, San Francisco, CA., v. 57-1, pp. 67-73.

Voyles, R.M., J.D. Morrow, and P.K. Khosla, (1995b). "Shape from Motion Decomposition as Learning Approach for Autonomous Agents," *Proc. IEEE Intl. Conf. on Sys., Man, and Cyber.,* pp 407-412.

Yang, J., Y. Xu, and C.S. Chen, (1994). "Hidden Markov Model Approach to Skill Learning and Its Application to Telerobotics," *IEEE Trans. on Robotics and Automation*, Vol. 10, No. 5, pp. 621-631.

# CONSTRAINED GENETIC PROGRAMMING WITH *lil-gp*

Cezary Z. Janikow[1]
Department of Mathematics and Computer Science
University of Missouri – St. Louis
janikow@radom.umsl.edu

***ABSTRACT:***

Genetic Programming refers to a class of genetic algorithms using parameterized tree representation, specifically determined by a set of labels and arities allowed. Such representation is more natural for many problems, but unfortunately the space of trees is usually much larger than the space of sought solutions. Constrained genetic programming addresses this issue. It uses data typing and semantic information to constrain the representation space so that only constraint-valid solutions can be represented. This paper describes the methodology as implemented in *lil-gp*.

## PRELIMINARIES

Evolutionary algorithms, such as genetic algorithms (GAs) (Davis, 1991)(Goldberg, 1989)(Holland, 1975), provide search mechanisms simulating nature. A GA uses a population of chromosomes coding individual potential solutions to the problem at hand. These chromosomes undergo a simulated evolution, facing Darwinian selective pressure proportional to chromosomes' qualities as problem solutions. Chromosomes also interact with each other via crossover in producing new solutions, and they are subjected to mutation.

Genetic programming (GP) (Kinnear, 1994)(Koza, 1992) uses tree-like representation, often utilized to represent computer programs. Such representation space is usually much larger than the sought solution space. This problem has recently received some attention. For example, Montana has proposed to use data type information to avoid generating invalid programs (Montana, 1995). Our approach, developed independently (Janikow, in press), allows for processing more arbitrary constraints, provides a more systematic means, and uses more powerful crossover.

*lil-gp 1.0* is a tool (Zonker) for developing GP applications. Its implementation is based on the standard GP *closure* property (Koza, 1992), which states that every function can call any other function and that any terminal can provide values for any func-

1. This research was supported through NASA/JSC grant NAG 9-847.

tion argument. This property leads to enlarging the representation space.

Almost any application imposes some constraints on the placement of functions and terminals in program trees. Except for some very specific constraints coded with *ad-hoc* approaches (Kinnear, 1994)(Koza, 1992), up to very recently such were a nuisance, leading to invalid or redundant chromosomes. For moderately and heavily constrained problems, this may lead to a serious waste of evolutionary resources. The objectives of constrained genetic programming (CGP) is to provide means to specify syntactic and semantic constraints, and to provide systematic mechanisms to enforce them (Janikow, in press). Because *lil-gp* does not allow explicit set definitions, while CGP was based on such, the resulting *CGP lil-gp* implementation uses its own constraint specification language. Such constraints are transformed (removing redundancies) into a minimal normal form. The normal form is stored in a special form allowing for efficient processing (within the same complexity). This paper briefly presents the methodology.

## GP AND ITS SEARCH SPACE

GP operates on trees, which are here interpreted as computer programs, with internal nodes corresponding to function calls (with user-defined interpretations), branches corresponding to function arguments, and leaves corresponding to atomic values (constants or functions of no arguments). Therefore, the search space is the space of computer programs, given some elements of the computer language.

GP evolves such trees from use-defined functions and terminals. Since all programs are evolved from the primitive elements, these must be sufficient to build the sought program. This assumption is called the *sufficiency* principle.

The space of all possible programs (trees), given the function and terminal sets, can be abstracted to two dimensions. The first is the space of uninstantiated tree templates (only those that can be instantiated). To control complexity, all GP applications restrict the size of possible trees. Thus, the dimension of tree templates is bounded.

A given template can be instantiated with functions and terminals of appropriate arity. In fact, arity is the only restriction imposed while instantiating trees. The absence of other restrictions is called *closure*. A given template can usually be instantiated in a number of ways. Thus, template instances provide another dimension.

The two-dimensional space of templates and instances is the representation space that will be explored by GP. Every element (tree) of that space must have a mapping to the solution space, to interpret the evolved solution (*i.e.*, to provide Darwinian pressure). However, the representation space is usually much larger than the solution space. This has traditionally been dealt with by providing *many-to-one* mappings. An obvious question is whether this redundancy has any implications on the search. Moreover, *closure* can make it difficult to interpret some trees. Traditionally, such trees have been considered invalid and would evaluate very low to ensure their extinction. Another question is what implications this might have on the search. CGP provides mechanisms to reduce the representation space, hopefully so that a *one-to-one* mapping will be found. It also allows studying the raised questions.

## CGP METHODOLOGY

### Constraint-Specification Language

*lil-gp* does not provide for explicit (neither extensive nor even intensive) set specifications. Instead, sets are defined by functions generating random elements, which are provided by the user. As mentioned before, the user provides two sets of functions: those for labeling internal nodes (type I) and those for labeling leaves (no arguments, type II).

In CGP, the notion of type compatibility has been defined, which specifies compatible types. With explicitly defined sets, compatibilities can be automatically computed. However, in *lil-gp* no explicit sets are provided, and all compatibilities must be given by the user as constraints of type ***T***.

***Definition 1*** *Define the following **T** constraints:*
*i)* $T^{Root}$ *- the set of functions which return data type compatible with the problem specification.*
*ii)* $T_*^*$ *-* $T_i^j$ *is the set of functions compatible with the jth argument of* $f_i$*.*

In terms of a labeled tree, $T^{Root}$ is the set of functions which, according to data types, can possibly label the root node. $T_i^j$ is the set of functions that can possibly label the *j*th child node of a node labeled $f_i$. Then, the above restricts *closure*, thus reduces the effective representation space.

*T* specifications reduce both the space of program templates (and thus structures) and the space of program instances. They can be seen analogous to data types and function prototypes available in high level languages.

***Example 1*** Assume $F_I = \{f_1, f_2, f_3\}$ (functions of type I) with arities 3, 2, and 1, respectively, and $F_{II} = \{f_4, f_5, f_6, f_7\}$ . Assume that the problem specifications state that a solution program should compute a real number, and integers are compatible with reals, while booleans are not compatible with either. Assume $f_1$ takes boolean and two integers, respectively, and returns a real. Assume $f_2$ takes two reals and returns a real. Assume $f_3$ takes a real and returns an integer. Assume $f_4$ reads an integer. Assume the other type II functions generate random constants of type boolean, integer, and real, respectively. These assumptions are expressed with the following *T* specifications:

$$T^{Root} = \{f_1, f_2, f_3, f_4, f_6, f_7\}\ ,\ T_3^1 = \{f_1, f_2, f_3, f_4, f_6, f_7\}$$
$$T_1^1 = \{f_5\}\ ,\ T_1^2 = \{f_3, f_4, f_6\}\ ,\ T_1^3 = \{f_3, f_4, f_6\}$$
$$T_2^1 = \{f_1, f_2, f_3, f_4, f_6, f_7\}\ ,\ T_2^2 = \{f_1, f_2, f_3, f_4, f_6, f_7\}$$

However, syntactic fit does not necessarily mean that a function *should* call another function. One needs additional specifications based on program semantics. These are provided by means of *F* specifications, which further restrict the space of program templates. These are additional constraints, thus they will override those provided by $T$ specifications.

***Definition* 2** *Define the following* ***Fspecs*** *(semantic constraints):*
*i)* $F^{Root}$ *– the set of functions disallowed at the Root.*
*ii)* $F_*$ – $F_i$ *is the set of functions disallowed as direct callers to* $f_i$.
*iii)* $F_*^*$ – $F_i^j$ *is the set of functions disallowed as* $arg_j$ *to* $f_i$.

***Example* 2** Continue Example 1. Assume that we know that the sensor reading function $f_4$ does not provide the solution to our problem. We also know that a boolean value (generated by $f_5$) cannot be the answer (this information is redundant). Also assume that for some semantic reasons we wish to exclude $f_3$ from calling itself. These constraints are expressed with the following $F$ constraints (the other sets are empty):

$$F^{Root} = \{f_4, f_5\}$$
$$F_3 = \{f_3\}$$

**Transformation of the Constraints and Mutation Sets**

Given the above language, which can be used to express some problem constraints (those expressible with the language), an obvious issue is that of possible redundancies, or that of the existence of sufficiently minimal specifications.

Transformation necessary to generate the minimal specifications have been detailed in (Janikow, in press). Because $T$ and $F$ specifications are redundant, the first step is to merge them to remove such redundancies while preserving complementary information. Some $F$ specifications are also redundant. These redundancies are removed as well. Finally, some severe constraints may prohibit the existence of valid finite trees – this is also checked. The resulting constraints are expressed in a special normal form, which is guaranteed to express the same information as the user-supplied constraints.

***Example* 3** Constraints of Example 1 and Example 2 have the following normal form:

$$F^{Root} = \{f_4, f_5, f_6\}$$
$$F_1^1 = \{f_1, f_2, f_3, f_4, f_6, f_7\}, F_1^2 = \{f_1, f_2, f_5, f_7\}, F_1^3 = \{f_1, f_2, f_5, f_7\}$$
$$F_2^1 = \{f_5\}, F_1^2 = \{f_5\}$$
$$F_3^1 = \{f_3, f_5\}$$

Assuming that valid finite trees exist (it is possible to determine whether valid trees restricted by GP's size exist – this feature is being added), the normal form is expressed in so called *mutation sets*. A pair of mutation sets is computed for each argument of every function, plus one extra pair for the root. For the argument $j$ of the function $f_i$, they list functions of type I (denoted $\mathcal{F}_3^1$) and type II (denoted $\mathcal{T}_3^1$) which can label the child node. For the root, these are functions that can label the root node.

***Example 4*** Here are selected examples of mutation sets generated for Example 3:

$$\mathcal{T}_{Root} = \{f_7\},\ \mathcal{F}_{Root} = \{f_1, f_2, f_3\}$$

$$\mathcal{T}_3^1 = \{f_4, f_6, f_7\},\ \mathcal{F}_3^1 = \{f_1, f_2\}$$

## Constrained Initialization, Mutation, and Crossover

The most important property of *CGP lil-gp* is that it does not use its constraints to determine that a tree is invalid after generating the tree. Instead, an invalid tree will never be generated. This is accomplished with the help of the mutation sets. First, the population is initialized with constraint-valid trees, each generated using the following procedure: generate the root node and label it with a random element of $\mathcal{T}_{Root} \cup \mathcal{F}_{Root}$; if the function has arguments, then build all subtrees recursively, following the same method but using the appropriate mutation sets rather than those for the root.

Mutation is implemented similarly, but it starts from a random node and uses mutation sets determined by its parent. Crossover first determines the destination location in one of two parents. Based on the parent of that node, appropriate mutation sets are retrieved. The second parent is traversed, marking and counting all nodes appearing in the sets, after which a random marked node is selected and moved (along with its subtree) to the first parent. In (Janikow, in press) we have shown the following properties:

- the initialization procedure is guaranteed to adhere to all constraints,
- given constraint-valid parents, both mutation and crossover are guaranteed to produce constraint-valid offspring,
- the new operators can be implemented with the same complexity levels as those of the unconstrained operators.

***Example 5*** Assume the mutation sets of Example 4. Assume mutating *parent1* as in Figure 1. Assume the node *N* is selected for mutation. It is the 1st child of a node labeled with $f_3$. Thus, $\mathcal{T}_3^1 = \{f_4, f_6, f_7\}$ and $\mathcal{F}_3^1 = \{f_1, f_2\}$. If the current mode is to grow the tree, then the mutated node will be randomly labeled with either $f_1$ or $f_2$. If the current node is to generate a leaf, then label *N* with either $f_4$, $f_6$, or $f_7$.

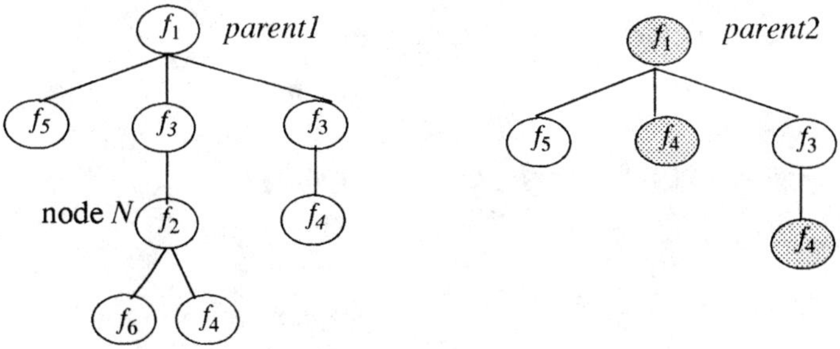


**Figure 1: Illustration of mutation and crossover.**

***Example 6*** Assume mutation sets of Example 4. Assume *parent1* and *parent2* as in Figure 1. Assume the node $N$ is selected for replacing with a subtree of *parent2*. It is the 1st child of a node labeled with $f_3$. Then, $\mathcal{T}_3^1 = \{f_4, f_6, f_7\}$ and $\mathcal{F}_3^1 = \{f_1, f_2\}$, and only the subtrees with the shaded roots can be used to replace $N$. Crossover would select a random element from so marked set of nodes, and move the corresponding subtree.

## SUMMARY

We have presented the CGP methodology applied to the widely used GP toolkit *lil-gp*. This methodology differs from the standard CGP methodology by not being able to deal directly with sets, as *lil-gp* does not use explicit set definitions. In the resulting *CGP lil-gp*, the user specifies constraints given a fixed language. These constraints are handled by built-in generic mechanisms transforming them into a minimal normal form, generating only valid programs, and ensuring that mutation and crossover will not create invalid programs from valid ones. In other words, the evolution is closed to operate on valid programs only. This dramatically reduces evolution complexity. Moreover, the mechanisms introduce only minor overhead. Initial experiments, reported in (Janikow, in press), indicate that constraining the search may indeed be very beneficial under certain circumstances.

## REFERENCES


Lawrence Davis (ed.). *Handbook of Genetic Algorithms*. Van Nostrand Reinhold, 1991.

David E. Goldberg. *Genetic Algorithms in Search, Optimization, and Machine Learning*. Addison Wesley, 1989.

John Holland. *Adaptation in Natural and Artificial Systems*. University of Michigan Press, 1975.

Cezary Z. Janikow. *"A methodology for processing problem constraints in Genetic Programming". Computers and Mathematics with Applications,* to appear.

Kenneth E. Kinnear, Jr. (ed.). *Advances in Genetic Programming*. The MIT Press, 1994.

John R. Koza. *Genetic Programming*. The MIT Press, 1992.

David J. Montana. *"Strongly Typed Genetic Programming". Evolutionary Computation*, Vol. 3, No. 2, 1995.

Douglas Zongker & Bill Punch. *"lil-gp 1.0 User's Manual"*. zongker@isl.cps.msu.edu.

# SOLVING LARGE SCALE BLOCK STACKING PROBLEM: A DISTRIBUTED APPROACH

**MASAAKI MINAGAWA**
Faculty of Social Information, Sapporo Gakuin University
Bunkyodai-11, Ebetsu, 069, Japan
minagawa@earth.sgu.ac.jp

**YUKINORI KAKAZU**
Faculty of Engineering, Hokkaido University
Kita-13, Nishi-8, Sapporo, 060, Japan
kakazu@complex.hokudai.ac.jp

***ABSTRACT*:**
A distributed approach to large scale block stacking problem is shown. A block is assumed to be an autonomous agent which transmit messages of motion request and activates with each other. Conflicts among agent motions are resolved according to motivation value calculated considering agent's goal dependency, goal position occupation, and mobility. An activated agent determines its destination according to outputs of ANN based behavior function trained using Evolutionary Programming technique. Experimental results show applicability and scalability of the approach up to the case of 200 blocks.

## INTRODUCTION

To solve the block stacking problems of large scale, even if the size of relevant block states is small concerning particular local block motions, due to the existence of the Frame Problem (Crockett,1994), the conventional AI-based search methods requires keeping and tracking costly huge state space. Furthermore, as the problem has nonlinear nature (Sacerdoti,1990) , any single strategy is hard to be effective. As well as careful description of block states, operators, conditions, to guide search process effectively, several heuristics have been investigated (Tate,1990). To date, some novel approaches were made to solve the problem using evolutionary approaches (Koza,1992; Hirano,1992; Minagawa,1996; Hama et al. 1994), but these resulted in limited success with respect to solutions of practical size problems.

Recent studies discussed the problem from the point of view of computational complexity. It is pointed out that complexity of search in the block stacking problem is on the same level as for sliding block puzzle problem, and classic NP-hard problems like the traveling salesman problem and bin packing (Chenoweth 91). Gupta and Nau reported that the problem of finding shortest-length plan is proved to be NP-hard (Gupta and Nau,92) even if the table large enough to hold all the blocks is assumed. They also asserts that the complexity of the problem results from existence of what they call "deadlock" situations where some critical steps are required to resolve the goal conflict.

Another obstacle in finding solutions of the problem is the existence of "arbitrariness among block states transitions (i.e. several ways to reach the goal state )". Before moving a block, the decision have to be made both on "which block to be moved" and "which position to go". This decision is implicitly driven ( or motivated ) by final goal attainment. So, how to focus planner's attention in the problem space is a crucial problem (Whitehead et al.,1990).

Our objective here is to find solutions of large scale block stacking problems. Starting from solving small size problems, we gradually increase the problem size while acquiring domain-dependent heuristics automatically.

First, we view the problem as a cooperative task by multi-agents considering the block to be an autonomous agent . This seems to be a natural solution; e.g. the sliding block puzzle problem was successfully solved by using a distributed approach (Drogoul and Dubreuil, '92). The agent transmit messages of motion-activation/goal-attainment request and activate motions with each other. Conflicts among agent motions are resolved according to motivation values (Parker,93) which is calculated considering agent's goal dependency, goal position occupation, and mobility (Ribeiro,93).

Second, to eliminate necessities for costly search-tree traverse, the states of blocks are represented using a cellular coordinate rather than conventional propositional representation.

Third, an activated agent determines "one of the promising stack positions to go" according to the outputs of ANN based behavior function. The connection weights of ANN are modified using Evolutionary Programming (hereafter EP) technique which is a real-valued, mutation-based evolutionary computation algorithm ( Angeline,94; Back,91; DeJong,93; Fogel,90; Fogel,94). For this purpose, we just use simple feedforward network with fixed-structure rather than attempting to evolve problem-oriented network structures (Mandischer, 93 ; Roberts,93; Utecht,94 ). In our experiments, the proposed methodology worked effectively up to the case of 200 blocks example.

## THE PROBLEM DESCRIPTION

We describe the problem $P$ using 5-tuples:

$$P = (Q, \Delta, q_0, q_G, E) \tag{1}$$

where $Q$ is the set of finite states and $\Delta$ is the state transition function. An initial state and a goal states are described as $q_0, q_G$ $(\in Q)$ respectively. $E$ is the environmental reward. The function $\Delta$ maps i-th state $q_i$ to j-th state $q_j$ ( obtained by moving one of the existing top blocks ) .

$$q_j = \Delta(q_i) \quad (q_i, q_j \in Q) \tag{2}$$

Assuming that each block is an autonomous agent, the state transition is obtained through local actions by the agents. Here we define $\Delta$ as:

$$\Delta = (\Psi, \Sigma) \tag{3}$$

where $\Psi$ is the mapping to determine an agent to be activated according to motion requests by other agents, and $\Sigma$ determines destination of the activated agent.
Denoting that $\Delta$ being applied n times as $\Delta^n$ , the problem is to find $\Delta$ such that

$$q_G = \Delta^n(q_0) \quad (n < L) \tag{4}$$

where $L$ is a large enough number until reaching the goal.

To describe the state space, we use cellular coordinate form rather than well known conventional propositional form ( see Fig.1 ) . First, let $A$ be the set of agents:

$$A = \{ \alpha_i \mid i = 1, 2, \ldots, N_B \} \tag{5}$$

where $\alpha_i$ is $i$-th agent and $N_B$ is the number of agents. By saying "large scale" we assume that $N_B \geqq 100$. We define the state $q$ $(q \in Q)$ as:

$$q = \{ u_i \mid i = 1, 2, \ldots, N_B \}$$
$$u_i = (a_i, h_i, v_i) \quad (a_i \in A,\ h_i, v_i \in I) \tag{6}$$

where $h_i$ and $v_i$ indicates horizontal position and vertical position of $i$-th agent respectively.

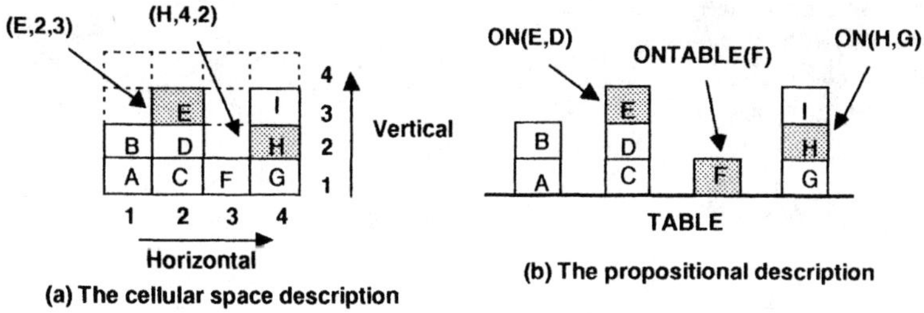


**Figure 1: The state descriptions of the blocks world.**

We define the set $S$ as follows and call it "stack set".

$$S = \{ s_i \mid i = 1,2,..,N_S \}$$
$$s_i = \{ a_{ij} \mid a_{ij} \in A , j = 1,2,..,N^i_{SB} \} \quad (7)$$

$N_S$ is the number of stack sets and $N^i_{SB}$ is the number of agents in $i$-th stack set. Each $s_i$ corresponds to $i$-th stack of agents and the first element corresponds to the agent on the table:

$$s_i = \begin{cases} \phi & ( if\ N^i_{SB} = 0 ) \\ a_{i1} & ( if\ N^i_{SB} = 1 ) \end{cases} \quad (8)$$

According to this, completion of the task is examined as follows. If we let $S^G$ be the goal stack set , the $i$-th stack of height $y$ will be completed if

$$for\ \forall j\ ( j = 1,..,y,\ y \le N^i_{SB} ) \quad a_{ij} = a^G_{ij} \quad (9)$$

We consider special stack sets with respect to given goal state; we call the stack set "buffer stack" such that the relation $s_i = \phi$ holds in the given goal state. As seen in Fig.2, the buffer stack (indicated as Buf) will be used as temporary place for the blocks motions .

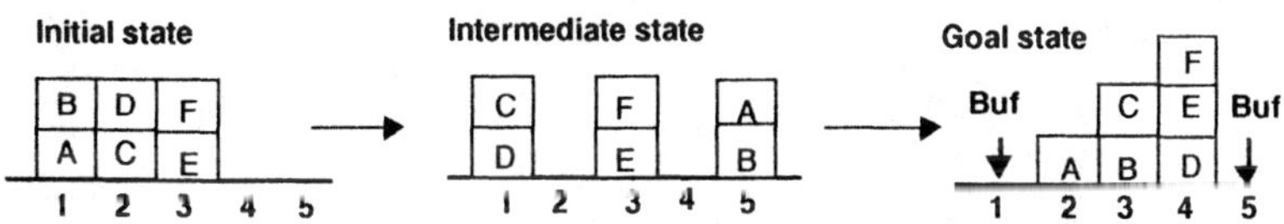


**Figure 2: The buffer stack.**

If $N_S$ is quite large ( i.e infinite table capacity ) , we may easily get "trivial solution " ( unstack every incorrectly located blocks onto the table and then correctly restack again ) (Chenoweth,91) , but we limit the value of $N_S$ so that the buffer stack size becomes relatively small ( e.g. less than ten ). This limitation adds difficulties to the solution: sophisticated unstack/restack operations are required before goal attainment.

## THE APPROACH

As the number of blocks increases, traversing the search tree is combinatorially explosive. So, we propose the distributed approach to localize decisions. To do this, we assume that each block to be an autonomous agent which can transmit request signals and activate motions with each other. The steps

(1) transmit goal attainment request signal
(2) transmit motion request signal
(3) calculate motivation value
(4) determine an agent to be activated
(5) determine destination stack of the activated agent

will be repeated until either (a) the given goal is attained, or (b) predetermined maximum number of motion steps ( $L$ ) is exceeded.

## Activation through motivation calculation

Since individual goal attainment is relevant to both current and goal positions of other agents, we describe the mapping $\Psi$ as:

$$\Psi : q \times q_G \rightarrow a \quad ( a \in A ) \tag{10}$$

and, to evaluate activation priority, we introduce "motivation". To carry out janitorial service simulation, Parker proposed a reactive architecture for cooperative agents, and in (Parker93) motivation is calculated considering inputs such as sensory feedback, inter-agent communication etc. In our approach, motivation values are calculated considering positional relations between agents such as (a) depend-on relation, (b) occupied-by relation, and (c) blocked-by relation ( see Fig.3 ). Ribeiro proposed multi-agent based approach and calculated "energy (activation level)" based on the relations (a) and (c) (Ribeiro,93). This work is suggestive, but practicality was not demonstrated.

The *depend-on* relation is defined only in the goal state and specifies goal attainment dependency. Starting from the "top" agent of each stack set, motivation values with respect to this relation will be transmitted and accumulated. This calculation will not be carried out to goal-attained agents.

Goal position of the most motivated agent in the relation above has significance with respect to individual goal attainment. So, we define *occupied-by* relation on the agent which occupy the goal position of the most motivated agent. As "deadlock situations" are considered in (Gupta and Nau, 92), this relation considers critical steps to resolve conflicts among the agents.

The *blocked-by* relation is given based on mobility ( i.e. easiness of starting motion ). Starting from the "bottom" of individual stack set ( goal attained agent will be excluded ), to clear the top, agents which are blocked by upper agents send motion activation signals.

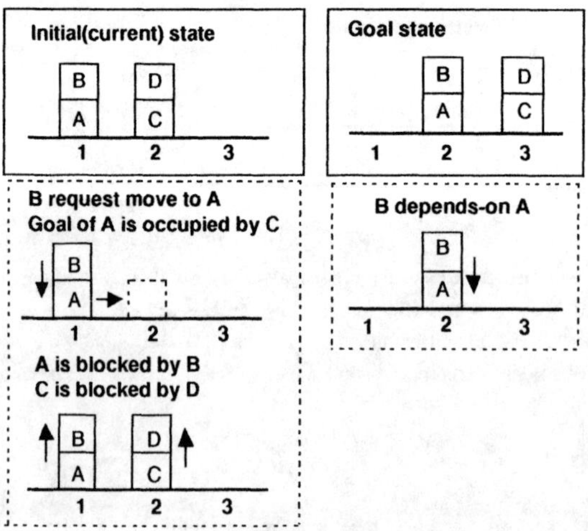


**Figure 3: The request signals for motivation calculation.**

The motivation value $m(\alpha)$ of agent $\alpha$ will be accumulated through these calculations. After this calculation, we calculate *i*-th agent's activation probability and select an agent to be activated according to the following equation:

$$Prob(\alpha_i) = \frac{exp(m(\alpha_i)/T)}{\sum_k exp(m(\alpha_k)/T)} \tag{11}$$

## Destination determination through ANN behavior function

The second problem is to determine "where to go". The mapping $\Sigma$ gives a destination stack number:

$$\Sigma : q \times q_G \rightarrow h \quad (1 \leq h \leq N_S) \tag{12}$$

We use simple feedforward ANN to do this (see Fig.4) and we have seven binary inputs consists of

(1) attainability of individual goal
(2) empty stack (table)
(3) goal position of stack below
(4) completed stack
(5) attainability of goal by other non-activated agents
(6) buffer stack
(7) bottleneck position.

The goal position of the agent receiving the highest motivation value is defined as the *bottleneck position* ; i.e. the position satisfies

$$\exists i, j \ \ such\ that\ \ max(m(a_{ij})) \tag{13}$$

We have three outputs, each of them corresponds to (a) go signal, (b) no-go signal and (c) don't-know signal. Output value $\xi_i$ is given by familiar sigmoid function and we select an output such that

$$\max_i(\xi_i) \tag{14}$$

Among stack positions where "go" signal output was obtained, we select a destination stack which gives maximum strength of "go" signal. When all the outputs indicate "don't know" signal, destination stack will be determined randomly.

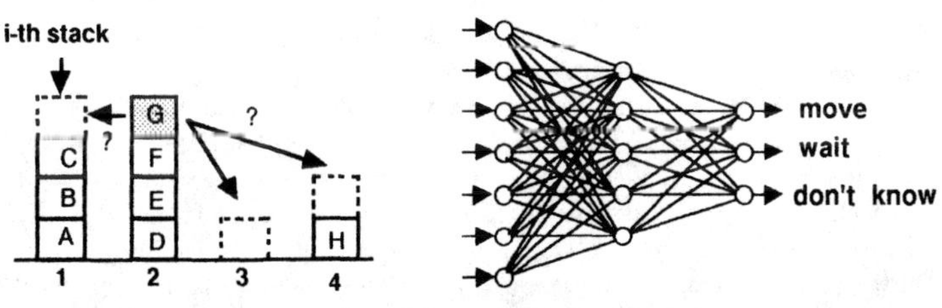


**Figure 4: Determination of a destination stack.**

## Training ANN

The connection weights of ANN are modified using EP technique which is a real-valued, mutation-based evolutionary computation algorithm (Angeline,94; Back,91; DeJong,93; Fogel,90; Fogel,94). First, we introduce a distance evaluation in the solution space. Assuming that block $\alpha_i$ is to be moved, we calculate the number of step motions required to

$\delta_1$ : clear the top of $\alpha_i$

$\delta_2$ : move $\alpha_i$ to another stack

$\delta_3$ : clear the given goal position of $\alpha_i$

$\delta_4$ : satisfy the goal position ( if needed )

$\delta_5$ : move to the goal position

And then, we calculate

$$D = \sum_{i=1}^{N_B} \sum_{j=1}^{5} \delta_{ij} \quad (15)$$

where $\delta_{ij}$ represents $j$-th $\delta$ value of $i$-th agent. $D$ will be used to estimate distance between the current state and the given goal state.

Within the population of the sets of weights of ANNs, selection, reproduction ( before the start of each successive generation, ANNs having below-average performance will be deleted and replaced by above-average ones ) and mutation will be carried out. Fitness of $i$-th network $F_i$ is given by

$$F_i = \left( \frac{E_{max} - E_i}{E_{max} - E_{min}} \right)^2 \quad ( E_i = D_i + f_i ) \quad (16)$$

where $f_i$ is the number of step motions until goal attainment.

In a network, connections to be mutated are selected according to uniform distribution and connection weights are modified (mutated) using the following equation ( Gaussian distribution )

$$w = w + c \cdot N(\mu, \sigma^2)$$

$$c = \begin{cases} \dfrac{F_{max} - F_i}{F_{max}} & (success) \\ 1 & (failure) \end{cases} \quad (17)$$

## EXPERIMENTS

In our experiments, we solved the problem up to the size of 200 blocks. Due to the existing problem complexity explained before, we started from simple examples and applied the training results ( i.e. evolved network ) to scaled up problems.

Fig.5 shows training example (left) and obtained result(right). Starting from randomly initialized weight set, the networks were trained ( Before this, 8 blocks example was used ). "Buf" indicates the location of the "buffer stack". There are 18 blocks and 5 buffer stacks. In this experiment, (1) population size = 100, (2) mutation rate = 0.05, (3) $\mu = 0$, (4) $\sigma = 0.3$.

The graph (right) shows the number of successful networks ( Success ), the number of step motions required to complete the task ( Motions ), and average goal distance ( Ave. Distance ). The training was terminated when the success rate exceeded 90% ( i.e. 90 successful ANNs ). The graphs show "smooth" network population fitness changes. It may be expected that the environmental reward was properly given concerning these training examples.

The example in Fig.6 were tried using the training results above. The number of blocks is 200 and blocks in the goal state are aligned alphabetically in horizontal direction. As well as scalability, this test example examines dependency of the trained network performance on block arrangement pattern. The table (below) shows $N_{buf}$, $L$, success rate and required motions steps. Minimal motion steps in this trial was 1400. ( Note that plan optimality is not discussed here; feasibility of the approach is stressed. )

## CONCLUSIONS

We showed a solutions of large scale block stacking problem. We viewed the problem as a cooperative task by multi-agents. The states of blocks were represented using cellular coordinates. As

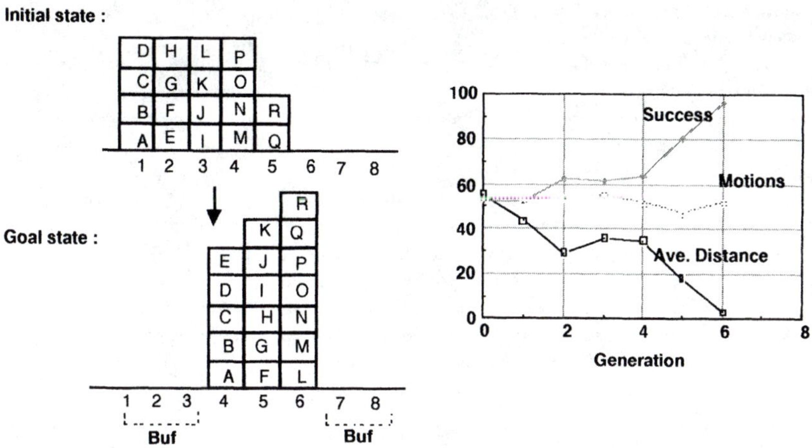


Figure 5: The training example and the result.

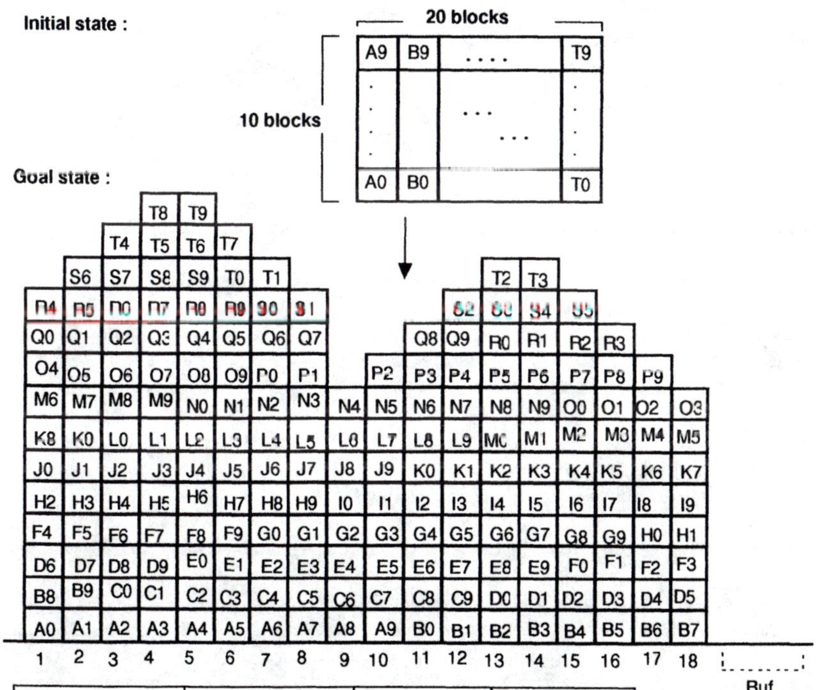


| No. of buffer stacks | Motion step limitation ( *L* ) | Success rate (%) | Motion steps |
|---|---|---|---|
| 7 | 2000 | 65 | 1400 |

Figure 6: The test example ( 200 blocks ) and the result.

shown in the problem space descriptions, this representation eliminated necessities for keeping and tracking huge state space.

Each block was assumed to be an autonomous agent which transmit messages of motion request and activate motions with each other. Conflicts among agent motions were resolved according to motivation value calculated considering agent's goal dependency, goal position occupation, and mobility. To attain the given goal state collectively, an activated agent determined "one of the promising stack positions to go" according to the outputs of ANN based behavior function. The connection weights of ANN were modified using mutation based EP technique. The experimental results showed applicability and scalability of the approach up to the case of 200 blocks.

The proposed methodology worked fairly well , but plan optimality was not considered here. To find a shortest plan, the proposed methodology needs further improvements. The number of buffer stacks ( including "no buffer" case ) may become an important issue. Including feasibility study of constructing the state transition function ( i.e. mapping ) Δ by using more sophisticated ANN architectures ( or other learning mechanisms ) , these remained as our future work.

## REFERENCES


Angeline,P.J. et al. (1994). An Evolutionary Algorithms that Constructs Recurrent Neural Networks, *IEEE Trans. on Neural Networks*, Vol.5, NO.1

Back,T. et al. (1991). A Survey of Evolution Strategies, *ICGA-91* , Morgan Kaufmann

Chenoweth,S.V.(1991). On the NP-hardness of Blocks World , *Proc. of AAAI-91* ,Vol.2

Crockett,L.J.(1994). The Turing Test and the Frame Problem, Ablex

DeJong,K.,Spears,W.(1993).On the State of Evolutionary Computation, *ICGA-95* , Morgan Kaufmann

Drogoul,A. and Dubreuil,C.(1992). Eco-problem-solving Model: Results of the N-puzzle, *Decentralized A.I.-3* , Elsevier

Fogel,D.B. and Atmar,J.W.(1990). Comparing Genetic Operators with Gaussian Mutations in Simulated Evolutionary Processes Using Linear Systems, *Biological Cybernetics*, 63

Fogel,D.B.(1994). An Introduction to Simulated Evolutionary Optimization, *IEEE Trans. on Neural Networks*, Vol.5, No.1

Gupta, N. and Nau, D. S.(1991). Complexity Results on Blocks-World Planning, *Proc. of AAAI-91*, Vol.2, AAAI Press

Hama, K. et al. (1994). A Distributed Approach to Block Stacking Problem Based on Evolutionary Learning, *Intelligent Systems Through Artificial Neural Networks*, ASME Press

Hirano, H.(1992). Solving the Block Stacking Problem by Genetic Algorithms , *Interface*, No.2

Koza,J.R.(1992). Genetic Programming, MIT Press

Mandischer,M.(1993). Representation and Evolution of Neural Networks, *Artificial Neural Nets and Genetic Algorithms*, Springer

Minagawa,M.(1996). Solving Nonlinear Block Stacking Problem by Operator Oriented Genetic Algorithm, *Social Information* , Vol.5, No.2

Parker,L.E.(1993). Adaptive Action Selection for Cooperative Agent Teams.*From Animals to Animats 2*, MIT

Ribeiro,F.,et al.(1993). Dynamic Selection of Action Sequences, *From Animals to Animats 2*, MIT

Roberts,S.G. and Turega,M.(1995). Evolving Neural Network Structures:An Evaluation of Encoding Techniques,*Artificial Neural Nets and Genetic Algorithms*,Springer

Sacerdoti,E.D.(1990). The Nonlinear Nature of Plans", *Readings in Planning* , Morgan Kaufmann.

Tate,A. et al.(1990). A Review of AI Planning Techniques, *Readings in Planning*, Morgan Kaufmann

Utecht,U. and Trint,K.(1994). Mutation Operators for Structure Evolution of Neural Networks, *PPSN-3*, Springer

Whitehead,S.D. and Ballard,D.H.(1990). Active Perception and Reinforcement Learning, *Proc. of the Seventh Intl. Conf. on Machine Learning*, Morgan Kaufmann

# Optimum Performance of Genetic Algorithms and Derivatives of Simulated Annealing for Combinatorial Optimisation

**Dr J.N.Carter** email: j.n.carter@ic.ac.uk
Centre for Petroleum Studies, Imperial College(ERE), London, SW7 2BP, UK.

**Abstract:**
It is generally accepted that Genetic Algorithms can out-perform many other algorithms on hard combinatorial optimisation problems. However, a number of authors have reported results suggesting that sometimes Genetic Algorithms perform significantly worse than simpler techniques, such as Simulated Annealing or one of its derivatives. These apparently conflicting views may have arisen through differing extents of tuning.

In this paper we investigate the performance of a tuned Genetic Algorithm, with the best tuned performance of three derivatives of Simulated Annealing, ie Great Deluge Algorithm, Record to Record Travel and Threshold Acceptance. Results are obtained for two practical problems encountered in industry.

## INTRODUCTION

The Genetic Algorithm is a flexible optimisation method that can be easily applied to many problems. GAs using conventional mutation and crossover are particularly well suited to solving problems whose feasible solutions can easily be represented by binary strings, eg (1001 ... 101).

GAs to solve combinatorial problems, which are not easily mapped to binary strings, require specifically designed representations and non-conventional mutation and crossover operators. This makes GAs difficult to use as a "black box" optimisation scheme.

Some studies have shown that GAs can perform badly, in comparison with simpler derivatives of Simulating Annealing, when applied to combinatorial problems. In particular, Sinclair (1993) has reported results on the turbine problem, used in this paper, that show that, when these algorithms are used as "black box" optimisers, the Great Deluge Algorithm and Record-to-Record Travel out-perform the GA.

In this paper we investigate the impact of tuning these algorithms. We show that, while un-tuned SA derivatives may out-perform an un-tuned GA, superior performance can be expected from a well-formulated, tuned GA.

## TEST PROBLEMS

### Turbine Balancing Problem

Balancing the blades of a hydraulic turbine is a real-life combinatorial optimisation problem which can be posed as a Quadratic Assignment Problem (QAP).

It has been treated using a number of different techniques by the operations research community (Laporte, 1988; Mosevich, 1986).

A "Francis" hydraulic turbine is constructed from a central shaft and an outer annulus. Between these are fixed a number (typically 20) of curved blades. As the water flows from the outer edge to the centre the turbine turns, rotating the shaft which is connected to an electric alternator. These machines can be very large with diameters of 10m and blade masses of about 18 tonnes. Blade masses can vary by ±5%. As the position of the blades is fixed, the system is balanced by welding the blades in an order that minimises the distance of the centre of gravity of all the blades from the centre of the shaft axis. Final balancing is achieved by adding lead weights to the outer annulus.

The Turbine problem is to specify the positions of each of the turbine blades which minimises the distance of the centre of gravity of the blades from the centre of the shaft axis. Using some simplifying assumptions (Laporte, 1988) the objective function can be written as

$$\left(\sum_{i=1}^{N}\sum_{j=1}^{N} M_i M_j \cos(\theta_i - \theta_j)\right)^{\frac{1}{2}} \times 10^2 \qquad (1)$$

where $M_k$ is the mass of blade $k$, $\theta_k$ is the angle of blade $k$, relative to some fixed reference, $\theta_k \in \{2\pi(l-1)/N; l = 1..N\}$, and $N$ is the number of blades. The final factor in the objective function is a scaling factor. The masses of the blades used in this paper have been published previously (Carter, 1995).

To code the turbine problem, we label each blade and allow the algorithms to find the best order to weld these into the turbine. In practice, pairs of blades on opposite sides of the turbine will be welded into position in successive operations. This is done to keep the system balanced. This suggests that a good coding could be based on the weld order.

### Cutting-Stock Problem

The one-dimensional Cutting-Stock problem is another industrial problem which has received much attention from the operations research community (Haessler, 1991). The problem is to cut a given list of lengths of cloth (or other materials) from as few rolls of stock as possible. Each of the un-cut rolls is the same standard length. Each piece of cloth to be cut, must be cut from one roll, but more than one piece may be cut from the same roll. The objective is to minimise both the number of rolls and the amount cut from the final roll (Sinuany, 1994). In this form the problem is equivalent to the bin packing problem, for which there are many references (eg Falkenauer, 1992).

For our problem, each of the rolls of stock material is 150 units in length and the lengths to be cut can be found in a previously published paper (Carter, 1995). The objective function used is the number of rolls cut minus one, plus the fraction of the final roll cut. For our best solution, 44 rolls are used and 29 units are cut from the final roll, so the objective function value is $(44-1)+29/150 = 43.193333$.

As for the turbine problem, we label each length of cloth required and use an ordered list of these labels as the string (gene) for the algorithms. The list is interpreted as the input to a greedy algorithm. In this, each time we have a new length to cut, we try to cut it from each of the cut rolls and use a new un-cut roll only if the length cannot be cut from an old roll. The Greedy Coding means that the same solution may have several string representations, while some (non-optimal) feasible solutions are not represented. However, we are guaranteed that the minimum can be found.

## THE OPTIMISERS

All three algorithms to be examined, Threshold Accepting (TA), Great Deluge Algorithm (GDA) and Record-to-Record Travel (RRT), are developments of the Simulated Annealing (SA) algorithm (Dueck, 1990, 1993). The current solution is subject to a small random mutation to produce a trial solution. This is accepted, becoming the new current solution, if it has a better function value. If the trial solution has a worse function value than the current solution, then SA accepts it on a probabilistic basis. The three proposed algorithms make this decision in a deterministic way.

### Threshold Accepting

This is the most simulated annealing like of the three. If the trial function value is not more than $\Delta$ worse than the current function value, then it is accepted (in SA this would be a probabilistic decision based on the difference in function values). $\Delta$ is a user defined quantity, and plays a similar roll to the 'cooling schedule' in SA. As the number of function evaluations ($n$) increases then the size of $\Delta$ is gradually reduced. In principal the variation of $\Delta$ as a function of $n$ can be quite arbitrary. The method used here, as suggested by Dueck *et al.* , is that $\Delta$ is multiplied by a constant factor $\alpha$ after $M$ trial acceptances.

In this study we have considered four parameters to optimise over: the initial value of $\Delta$; values for $\alpha$ and $M$; and the number $m$ of Swap Mutations (Oliver, 1989), which randomly selects two elements in the ordered string and exchanges their positions.

### Great Deluge Algorithm

In this algorithm we define a function level ($L$), which is independent of the current function value. A trial is accepted whenever its function value is better than the function level. Every time a trial is accepted then the function level is moved towards the best function value found. The idea is that the current solution moves around on its fitness landscape, but avoids a continuously rising water level.

In this study we have considered three parameters to optimise over: the gap between the initial current function value and the function level, $L$; the factor $\alpha$ by which the function level is changed; and the number $m$ of swap mutations.

### Record-to-Record Travel

This is the simplest of the three algorithms. The function level $L$ is kept at a fixed distance $\Delta$ from the best function value currently found. Every time a new 'record' solution is found then $L$ is changed. The only parameters optimised in this study are: the size of $\Delta$; and the number $m$ of swap mutations.

### Genetic Algorithm

The crossover operator used is HTBX (Poon, 1995), which is described below, and has been optimised previously on the problems in this paper (Carter, 1995).

In a typical combinatorial problem a list of objects ($a$,$b$,$c$...) has to be placed in some order so as to optimise some function which depends on the order of the objects. A GA crossover operator has to take a pair of solutions and combine them in such a way as to produce valid offspring. In many problems the objects have some property that would enable them to be ranked, e.g. their weight. So if object $d$ was the 6th heaviest, then we could ask for the 6th heaviest object rather than object $d$. Also if this object had already been used we could ask for the next heaviest object, which is then looked up. This is the principle employed by the HTBX operator.

Our GA employs the HTBX crossover operator, with the Swap mutation operator, Bakers (Baker, 1987) selection scheme and single solution elitism.

The configuration parameters for the GA are:

- $m$, the mutation rate — expected number of mutations per string.
- $N$, the population size — $N$ is the number of strings available at the start of a breeding cycle.
- $B$, the selection probability parameter — when parents are selected, the selection is limited to the $B$ highest ranking solutions in the population. Baker's ranking scheme (Baker, 1987) is used to set the selection probabilities for these $B$ solutions. This is essentially the ($\mu$,$\lambda$) algorithm described by Bäck (1991, 1995).
- $c$, the crossover rate — $c$ is the probability that a pair of parents breed rather than clone themselves.
- $r$, the replacement parameter — the highest $r$ ranking solutions are copied directly into the next generation. $r = 1$ is "elitism" and $r = N - 2$ is "steady-state".

## RESULTS

To study the effect of changing the configuration parameters, for both the GA and the other algorithms, a large number of cases have been tested. In each test, 100 runs of the algorithm were performed, with a run terminating after $2^{15}$ function evaluations. The average performance of each test was then assessed. A small selection were then run for four times the number of function evaluations, to check the performance. For each selected case the figure shows the mean (over 100 runs) of the best know solution, against number of function evaluations. The table gives the results after $2^{17}$ function evaluations.

### Turbine Balancing Problem

Figure 1 shows the results, plotted against number of function evaluations, obtained from four test cases chosen on the basis of their performance after $2^{15}$ function evaluations (no random search is given since its performance was very poor):

⋆ Tuned TA ($\delta = 0.03$, $\alpha = 0.1$, $M = 400$, $m = 3$)
◇ Tuned GDA ($L = 0.2$, $\alpha = 0.01$, $m = 2$)

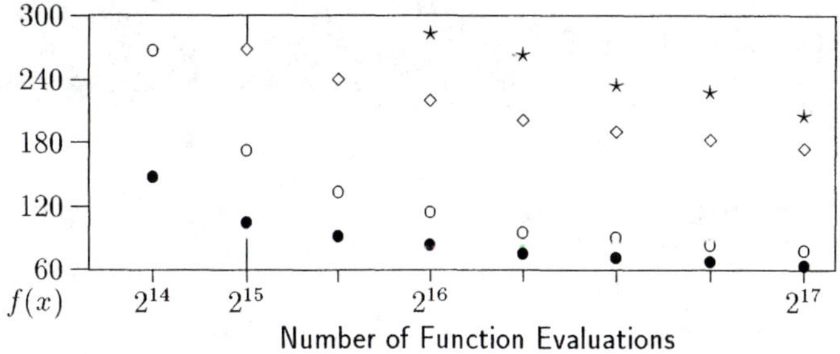


| | | |
|---|---|---|
| ★ | 206 | 119 |
| ◇ | 174 | 92 |
| ○ | 78 | 45 |
| • | 63 | 30 |
| | $f(x)$ | $\sigma$ |

Figure 1: Results for Turbine balancing Problem

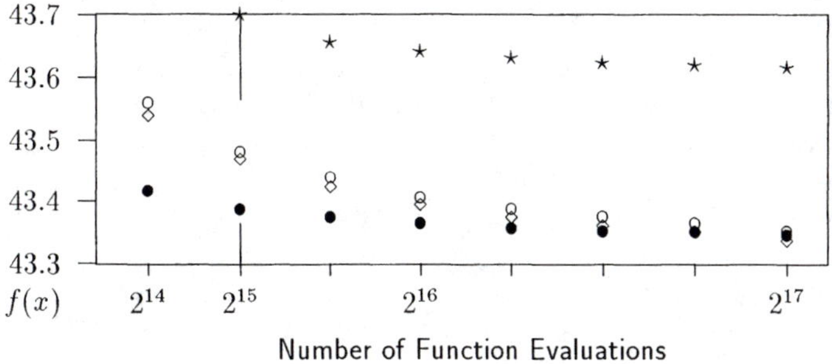


| | | |
|---|---|---|
| ★ | 43.616 | 0.112 |
| ○ | 43.350 | 0.068 |
| • | 43.344 | 0.063 |
| ◇ | 43.336 | 0.071 |
| | $f(x)$ | $\sigma$ |

Figure 2: Results for Cutting Stock Problem

∘ Tuned RRT ($\delta = 0.002$, $m = 1$)
• Tuned GA with ($m = 0.2$, $N = 64 + 1$, $B = 13$, $c = 1.0$ and $r = 1$).

The results show that the optimised GA out performs the optimised versions of the other algorithms.

## Cutting Stock Problem

Figure 2 shows the results, plotted against number of function evaluations, obtained from four test cases, the table at the end gives the final mean optimum and the standard deviation:

⋆ Tuned TA ($\delta = 1.0$, $\alpha = 0.005$, $M = 400$, $m = 1$)
⋄ Tuned GDA ($L = 1.0$, $\alpha = 0.2$, $m = 2$)
∘ Tuned RRT ($\delta = 0.01$, $m = 2$)
• Tuned GA with ($m = 0.2$, $N = 64 + 1$, $B = 65$, $c = 1.0$ and $r = 51$).

Again, the results show that the optimised GA out performs the tuned versions of the other algorithms.

## CONCLUSIONS

We conclude that, for combinatorial problems, a GA can be tuned to perform very well, and, in particular, better that the other algorithms considered in this paper. Therefore, when selecting an algorithm to use, we must recognise that there is a trade-off between the effort required to optimise the algorithm's formulation and the performance of that algorithm. For problems where it is important to get as close as possible to the optimal solution, it will be worthwhile expending the extra effort required to choose the most appropriate representation, crossover operators and parameter values for the GA.

## REFERENCES


Bäck, T, (1995). Generalized Convergence Models for Tournament and ($\mu$,$\lambda$) Selection, *Proceedings of the Sixth International Conference on Genetic Algorithms*, 2, ed LJ Eshelman.

Baker, JE, (1987). Adaptive Selection Methods for Genetic Algorithms, *Proceedings of the First International Conference on Genetic Algorithm and Their Applications*, 101, ed JJ Grefenstette.

Carter, JN, (1995). Optimal Performance of Genetic Algorithms and Stochastic Hill-climbers for Combinatorial Optimisation, *Intelligent Engineering Systems Through Artificial Neural Networks*, 5, 399.

Dueck, G, and Scheuer, T, (1990). Threshold Accepting: A General Purpose Optimization Algorithm Appearing Superior to Simulated Annealing, *J. Computational Physics*, 90, 161.

Dueck, G, (1993). New Optimisation Heuristics: The Great Deluge Algorithm and the Record-to-Record Travel, *J. Computational Physics*, 104, 86.

Falkenauer, E, (1992). A Genetic Algorithm for Bin Packing and Line Balancing, *Proceedings of the 1992 IEEE International Robotics and Automation Conference*, 1186

Haessler, RW, and Sweeney, PE, (1991). Cutting Stock Problems and Solution Procedures, *European Journal of Operational Research* 54, 141.

Laporte, G, and Mercure, H, (1988). Balancing Hydraulic Turbine Runners: A Quadratic Assignment Problem, *European Journal of Operational Research* 35, 378.

Mosevich, J, (1986). Balancing Hydraulic Turbine Runners — A Discrete Combinatorial Optimisation Problem, *European Journal of Operational Research* 26, 202.

Oliver, IM, Smith, DJ, and Holland, JRC, (1989). A Study of Permutation Crossover Operators on the Travelling Salesman Problem, *Proceedings of the Second International Conference on Genetic Algorithm and Their Applications*, p224 ed JJ Grefenstette.

Poon, PW, and Carter, JN, (1995). Genetic Algorithm Crossover Operators for Ordering Applications, *Computers and Operations Research* 22, 135.

Sinclair, M, (1993). Comparison of the Performance of Modern Heuristics for Combinatorial Optimization on Real Data, *Computers and Operations Research* 20, 687.

Sinuany-Stern, Z, and Weiner, I, (1994). The One Dimensional Cutting Stock Problem Using Two Objectives, *Journal of the Operational Research Society* 45, 231.

# DIRECTED MUTATION APPLIED TO NONLINEAR PROGRAMMING

**Prof. Francisco Antonio Pereira Fialho - Dr. Eng.**
Universidade Federal de Santa Catarina, CCE - EGR, Campus Universitário, Trindade, Florianópolis, Santa Catarina, CEP 88041-800

***Abstract:***
Mathematical functions describe virtual "n" dimensional worlds exhibiting the most strange topology one can imagine. To find optimum points means to climb the highest mountains of these fantasy realms, taking care by not being deceived by false maximums. Usually, inside these kingdoms, there are forbidden zones places that must be avoided. Most of the times one become lost inside these prohibited territories, even without noting it. The search for good and fast algorithms to guide the traveler in his journey toward the pot of gold hidden in these heights tells a never ending story where, everyday, a new record is established. Like in a Olympic game, the yesterday impossible is inevitably beaten by some ignominious ignorant that doesn't know that such a record could not be overthrown. This paper tells the story of this search and suggests some approaches to enhance the performance of nowadays optimization algorithms.

## INTRODUCTION

Engineering problems can be roughly understood as optimization issues. In order to understand a system performance, all variables related to it must be identified, and the constrains, signaling that we are dealing with events happening in a real world, must also be enumerated. The goal is to find the best combination of factors that drives the system under study toward its best operation state. To do so, one must travel inside the Space of all Possible Solutions, and search for the prize winner.

In fact, variables and constrains choices, are consequence of a decision making process occurring in the establishment of the problem. In other words, we are talking about fuzzy functions that must respect fuzzy constrains.

Because we prefer flying through space we will talk, most of the time, about maximums. It is understood that maximums and minimums depends only on what side of Alice's mirror we are.

## TALKING ABOUT IMAGINARY REALMS

The first step to solve our optimization problem is to identify the place we need to visit in order to search for the winner. A fuzzy definition would guide us to a multidimensional fractal universe. In this article, for simplification, Solution Space will be taken as the virtual world defined by some traditional objective function, constrained by a well behaved set V. The formulation of the problem is complete when we have:

- an objective function $f: R^n \rightarrow R$
- a viable region $V = \{ x \in R^n \mid g_i(x) \leq 0 , i = 1, \ldots , m\}$

## CLIMBING HILLS

In a broad classification, one can say that there are three general optimization procedures: Random, Numerical, and Enumerative methods. Numerical methods encompasses two big families of solution approaches, depending on the need or not to differentiate f. Virtually nothing is known on how to distinguish a local from a global optimum. Two strategies can be explored. First we can search for several optimums, departing from different points, choosing the best one. Second we can perturb a local candidate, observing if this procedure will lead us toward a better solution.

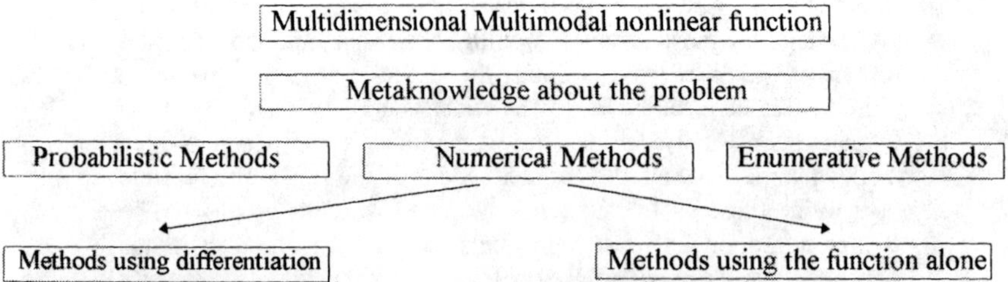


Gradient conjugated methods
    Fletcher-Reeves algorithm
    Polak-Ribiere algorithm
Quasi Newton methods
    Davidson-Fletcher-Powel algorithm
    Broyde-Fletcher-Goldfarb-Shame algorithm
etc.

Powell method
Uphill multidimensional simplex
Combinatorial maximization
etc.

Figure 1 Searching methods

**Simulated annealing**

Simulated annealing is a computational technique derived from statistics mechanics, that has, as a goal, to find solutions near the global minimum for big optimization problems.

Suppose a system is identified by a set of positions of its components If the system is in thermal equilibrium for a temperature t, there is a probability $\pi_T(s)$ that the system is in a configuration *s* that depends on the energy E(s) associated to this configuration. Boltzmann distribution writes:

$$\pi_T(s) = \frac{e^{\frac{-E(s)}{kT}}}{\sum_{w \in S} e^{\frac{-E(s)}{kT}}},$$

The metaphor that inspired this method came from metallurgical annealing process. The idea is that, at high temperatures, molecules behave in a random way traveling both to high as to low levels of energy. If temperature is gradually reduced the probability of traveling toward highest energy levels decrease.

This technique is being applied successfully to global optimization of multimodal function. The problem is the time needed to arrive at the solution and the complexity of choosing parameters and strategies for the temperature decreasing process.

**COMPUTATIONAL INTELLIGENCE (SOFT COMPUTING)**

Artificial Intelligence deals with symbol manipulation. Recently, several methods, based on nature metaphors, were developed. Computational intelligence (Bezdek, 1994), incorporates Neural Network, Fuzzy Systems, and Evolutionary Computing.

Neural networks are, at the present state of the art, extremely simple models of natural neural systems, searching for solutions in a bottom up approach. Fuzzy logic tries to imitate human reasoning when dealing with human environment. Evolutionary computing reinforces evolution as a method for wise optimization. Direct mutation explores the idea that evolution can occur not only gradually but jumping for a new equilibrium state as verified in modern biology and anticipated by Jean Piaget.

**Evolutionary computing**

Evolutionary computing comprises Evolutionary Programming (Fogel et all, 1966), Evolutionary Strategies (Rechenberg, 1994), and Genetic Algorithms. Genetic Algorithms are applied also in Genetic Programming (Koza, 1994), and Classifier Systems (Wilson, Goldberg, 1988). Cellular Automata is another correlated field.

**Genetic Algorithms**

Developed by John Holland, his collaborators, and students at Michigan University, Genetic Algorithms are searching methods based in a natural selection (genetics) metaphor.

The first step needed for GA utilization is the definition of the Solution Space. Next it is necessary to code the solutions. Usually a alphabet A = {0, 1, #} is used, where symbol # is a don't care. The search is carried out by a complete population (50 to 200 individuals, typically) that evolves in accordance with individual fitness to compete in the space defined. Stochastic operators are used instead of deterministic ones. The most simple GA works with only three operators: reproduction, crossover, and mutation.

Selection copies individuals fitted to the environment. Recombination mixed genes hoping obtaining better individuals and mutation favors that all gene repertoire will be available for future generations. Without selection, GA could be classified as a random method. Selection and recombination (crossover) realizes a local search in best individuals surroundings. A fundamental dilemma is in balancing exploration, exploitation. Recombination exploits while mutation explores.

One kind of classifiers mentioned by Holland, the *Echo*, was used by Dowkins (1986) to simulate a true 'biological run'. An agent has *chromosomes* and a *gene reservoir* that comes from an existing in the environment '*source of genes*'. The iterations between agents is determined by simple tests sequences performed over *labels* (attack, defend, couple); *conditions* (fight, exchange, marry) and *enzymes* that specifies the capacity of an agent to capture other environment resources.

**Genetic Programming**

Genetic programming is leaded by the works of Koza and Rice, from Stanford. Each chromosome represents a variable size computational tree, coding for a program, generally written in LISP. Genetic Operators (Reproduction, Crossover, Mutation, etc.) are successively applied at this population creating new generations of probable solutions. In this way new algorithms are build in a natural selection process.

The search is for an initial set of algorithms, or eventual solutions that, in a way similar to that used for the 23 aminoacids that forms the basis of all the infinite protein structures, be able to build all the algorithms, or all the eventual solutions for our problems.

**Evolutionary Strategies**

Evolutionary strategies were first developed in Germany as an efficient method for optimization of multimodal non differentiable functions. In the simplest scheme, a father generates only one son through a mutation based on a Gaussian distribution with media zero and variable variance, in such a way that big mutations occur less frequently then small ones. Usually 20% of the mutations generated result in sons that are best fitted then their parents. In these cases, sons replace fathers in the next generation.

**Evolutionary Programming**

Typically, in evolutionary programming, there is a population of N individuals that are copied to a temporary population that suffers variable mutations. A tourney is so realized and the best of these 2N individuals are chosen to compose the next generation.

**Cellular Automata**

A cellular automata is a finite state machine surrounded by other identical cellular automata. All cells can change their state in accordance with local rules and the state of its neighbors. These rules can be stochastic or deterministic (Casti, 1994) (Whitley, 1989) (Forrest, 1990).

## STATE OF THE ART IN HYBRID METHODS

Through the classic use of gradient as a pointer to the way to follow (downhill or uphill climbing), a population of initial points evolve simultaneously toward a optimal one desires to be the global one.

In order to avoid convergence to local optimums, and inspired in the simulated annealing paradigm, random perturbations move these points, guiding to a mixed G-PA-GPF (genetic, gradient, and simulated annealing) method. Cursi and Cortes (1996) have shown that this hybrid technique is faster and more accurate, when compared with the simple genetic algorithm or other methods.

Downhill Climbing (DHC) algorithm considers an external and an internal loop. The internal loop contains a very efficient technique for finding local minimum. The external loop will force the external loop in investigating different regions of the Solution Space. DHC has four parameters that must be defined: the objective function, the starting point, the maximum and minimum step sizes. Usually one establishes that the maximum step size is half the search space length.

## DIRECT MUTATION

Genetic Algorithms paradigm is sustained by a biological metaphor. Natural selection seems to provide a robust answer for the challenges a specie must overcome in order to survive. To imitate it seems a good approach for solving practical problems. Populations of individuals, in their struggle for survival, behave as living beings, succeeding in finding answers to environment challenges.

Any nonlinear problem defines a virtual world where staying alive means to look for the highest peaks inside these fantasy realms, where one can breeze a pure atmosphere, avoiding measles and other venomous places. A population is then defined, struggling for survival in this artificial environment. The winner phenotype is a candidate for being the champion we were looking for in our quest.

Nevertheless, in order to keep update with the most recent researches in evolution, we must consider novel theories that, although confirming that natural selection remains a very important mechanism for explaining environment adaptation, points that development can also occur in steps or even as a consequence of a random genetic drift. To keep the relation between GA and the biological metaphor it is necessary to review GA fundamentals.

Recent research in genetics signals to mutations occurring, also, as a non-random process, as consequence of environment menaces. For instance, the transposon which determines erythromycin resistance in bacteria will generate an appropriate mutation only when needed, and stress induced somatic mutation in plants may get into the germ line (Endler and McLellan, 1988).

The idea, behind direct mutation proposition, is that mutations can occur as consequence of phenotypic information given, for instance, by an external loop, like the one specified in DHC, or based in the metaknowledge about the problem granted by some expert system, or fuzzy decision making processor.

The proposal of this article is to join DHC techniques with G-PA-GPF methods, adding direct mutation as a concept that reestablishes the commitment with the biological metaphor implicit in the Computational Intelligence paradigm.

Gradient method gives the flexibility for fast adaptation while simulated annealing avoids deceiving choices. Genetic operators acts as memories of successful trials that must be preserved for future generations. An external loop is then defined. An airplane flying above this world, capturing other regions that must be explored and forcing individual to migrate for these promising lands to harvest for the big prize.

Mutation, in this context, preserves its exploration characteristic, letting for the other genetic operators the non trivial task of exploiting the near environment. Direct mutation can justify also the several artifices used for speeding up GA convergence

## CONCLUSION

Artificial computer tricks, which disregard the biological metaphor, are normally employed, for speeding up genetic algorithm or taking into account an ever changing environment. The present article proposes a way of taking into account changes in the environment through a new mechanism we called directed mutation which is applied to all members of a population, so preserving the biological similarity embedded in the genetic algorithm conception.

Ecology is teaching us that diversity is fundamental for survival. A robust solution must consider that 'bad elements' can hide the key for facing up some new challenge in a ever changing environment.

Ecological niches that compete only between themselves can make full use of parallel processing techniques. These niches are defined by an external loop that, based in cues given by the metaknowledge one has about the problem being solved, forces the migration, by the use of this direct mutation operator, of members of the original population toward these new regions of Solution Space. Elitist crossing between the best individuals of these niches can also be investigated.

Diversity can be implemented coding and using GA in different ways. One can think in using Cellular GA automatons where each individual, controlled by one processor, change its possible states using our G-PA-GPF strategy.

Finally we point out the urgent need of taking in consideration that we are dealing with fuzzy objective functions constrained by fuzzy conditions. Penalties for members of a population traveling in forbidden zones must be reduced to avoid precious genetic material losses.

There is much to be done. Fortunately this article is just a more tale in the never ending story about find optimums in crazy non stationary fractal worlds.

## REFERENCES


Pogu, M., Cursi, J. E. S., (1994). Global optimization by random perturbation of the gradient method with a fixed parameter, *Publication du Service de Mathématiques de l'ECN.*

Nelder, J. A., Mead, R., (1965). A simplex method for function minimization, *Computer Journal*, 7, 308-313.

Geman, S., Hwang, C. H. (1986). Diffusions for global Optimization, Siam, J. *Control and Optimization*, vol. 24, no. 5, 1031 - 1043.

Bezdek, J. C., (1994). What is computational intelligence, In: Zurada, J. M. et all, *Computational Intelligence. Imitating Life*, IEEE press.

Fogel, L. J., Owens, A. J., Walsh, M. J., (1966). *Artificial Intelligence Through Simulated Evolution*. John Wiley.

Rechenberg, I., (1994). Evolution Strategy. In: Zurada, J. M. et all, *Computational Intelligence - Imitating Life*. IEEE Press.

Koza, J. R., (1994). *Genetic Programming II. Automatic Discovery of Reusable Program.* MIT Press.

Wilson, S. W., Goldeberg, D. E., (1989). A Critical Review of Classifier Systems: In: *Procedures* Third International Conference Genetic Algorithms.

Holland, J., (1975). *Adaptation in Natural and Artificial Systems.* University of Michigan Press, 1975.

Dowkins, R. (1988). *The Blind Watchmaker*. London: Penguin Books.

Casti, J. L., (1994). *Complexification*. HarperPerennial.

Forrest, S. (ed.), (1995). *Emergent Computations: Self Organizing, Collective, and Cooperative Phenomena in Natural and Artificial Computing Networks*, MIT Press.

Toffoli, T., (1995). Cellular Automata. In: Arbid. M. A., *The Handbook of Brain Theory and Neural Networks*, MIT Press.

Whitley, D., (1993). Cellular Genetic Algorithms. In: *Procedures* Fifth International Conference Genetic Algorithms.

Cursi, J. E., Autrique, L. (1993). Solution numerique d'un probleme d'identification comprenant des equations couplees a l'aide de processus stochastiques, *Rapport Interne*, Laboratoire d'Automatique de Nantes, jul.

Cursi, J. E., Souza, B. (1996). Genetic Algorithms in Problems of Continuous Nonlinear Programming. Thesis. Programa de Pós Graduação em Engenharia de Produção. UFSC.

Endler, J. A., McLellan, T. (1988). The process of evolution toward a newer synthesis. *Ann. Re. Evol. Syst.*, v. 19, p. 395 - 421.

# GENETIC ALGORITHM-BASED METHOD FOR SOLVING AN OVERDETERMINED SYSTEM OF EQUATIONS CORRUPTED BY MEASUREMENT NOISE

ANNA L. BUCZAK, JOSEPH J. BARRETT
*AlliedSignal, Inc., 101 Columbia Road, Morristown, NJ 07962-1021*

***ABSTRACT:***
This paper describes a novel method for prediction of atmospheric temperature profiles. The profiles are described by an overdetermined system of equations corrupted by measurement noise. The goal is to predict the temperatures in four range bins with reasonable accuracy. Solving the equations by a classical linear least squares fit results in predicted temperature error over 30K for small amount of noise in the measurements themselves. Since this result is physically implausible, we developed a genetic algorithm based method and we incorporated into the fitness function correction terms based on our experience with the atmospheric temperature profiles. The results improved dramatically. The technique devised is tested on a simulated data set and the results are described.

## INTRODUCTION

Many real world systems are described by overdetermined systems of linear equations. The most common method for solving this type of problems is the least squares fit [Pre-95]. Solutions of that kind are satisfactory, as long as the measurements contain very limited amounts of noise. For certain measurements, as those resulting from integral equations, noise is more pronounced and it is impossible to find the solution to the system by exact analytical computations. Methods that currently exist for solving the problem include, the linear regularization (also called Phillips-Twomey) method [Pre-95] and Backus-Gilbert method [Pre-95]. Linear regularization maximizes the smoothness of the solution, while Backus-Gilbert method maximizes the stability of the solution.

The problem of unwanted oscillations often found in numerical solutions of integral equations is very pronounced in radiometric measurements of temperature. In those measurements even a small amount of noise causes the least squares method to generate physically invalid answers, e.g. when the amount of noise is very pronounced (5K) the least squares solution will contain negative temperatures in Kelvin! In order to infer useful information from those noisy measurements, information about atmospheric behavior has to be somehow introduced when solving those equations. The method that we developed is based on genetic algorithms and atmospheric behavior information is incorporated into the fitness function. In this way, we are able not only to incorporate into the solving mechanism the smoothness constraint (as in linear regularization) but also the atmospheric behavior information.

The approach we are using for determining atmospheric temperature is based on the observation that the relationship between radiometric measurements and range bin temperatures can be described by an overdetermined system of linear equations. We are using radiometric data from an atmospheric model (FASCODE) for determining the matrix describing this relationship. Next we are solving the equations for simulated data sets by the genetic algorithm.

## MEASUREMENT CONCEPT

The atmosphere emits infrared radiation in selective wavelength bands. These wavelength bands correspond to the infrared-active molecular transitions of the various atmospheric constituents such as carbon dioxide, water vapor, ozone, etc. The magnitude of the infrared radiation depends on the strength of the molecular transitions, the molecular concentration and the temperature of the atmosphere [Wol-93]. A wavelength-tunable infrared radiometer can be used to measure the emission from atmospheric $CO_2$. Since the infrared emission is temperature dependent, the radiometer's signal is related to atmospheric temperature. The FASCODE atmospheric computer model, developed by the U.S. Air Force Phillips Laboratory, provides high spectral resolution line-by-line calculations of atmospheric transmittance and radiance [Tho-93]. This program was used to determine the radiance and absorption coefficients for atmospheric $CO_2$.

The radiometer integrates all the atmospheric radiation along the line of sight (LOS) that passes through the spectral filter in front of the detector. The LOS path length varies with the angular position of the spectral filter in the radiometer. The radiometer signal for a horizontal path at a constant height and constant temperature is

$$S(\theta) = Const. \sum_j N_j T_j(\theta) \int_0^{\infty} e^{-K_j R} dR \tag{1}$$

where $N_j$ and $K_j$ are the spectral radiance and absorption values for the $j$-th wavelength interval as determined from the FASCODE atmospheric model. The filter transmission function is represented by $T_j(\theta)$ and $R$ is the range variable. The constant term in Eq. (1) contains parameters involving the radiometer that are not relevant to the analysis presented in this paper.

In order to determine the atmospheric temperature at different ranges in front of the radiometer, we divide the path along the LOS into range bins (Figure 1) such that the sum of the signals from all the range bins is equal to the total integrated signal of the radiometer. The signal contribution from the $i$-th range bin is determined by (1) integrating the signal out to the $R_i$ range, (2) integrating the signal out to the $R_{i-1}$ range and (3) taking the difference between these two values. This procedure will yield the signal contribution from the $i$-th range bin and includes the attenuation of the signal along the path up to the $i$-th range bin. The result is

$$\begin{aligned} S_i(\theta) &= Const. \sum_j N_j T_j(\theta) \left[ \int_0^{R_i} e^{-K_j R} dR - \int_0^{R_{i-1}} e^{-K_j R} dR \right] \\ &= Const. \sum_j \frac{N_j T_j(\theta)}{K_j} \left[ e^{-K_j R_i} - e^{-K_j R_{i-1}} \right] \end{aligned} \tag{2}$$

Equation (2) was used to calculate the signal from each range bin for each angular position of the radiometer's spectral filter.

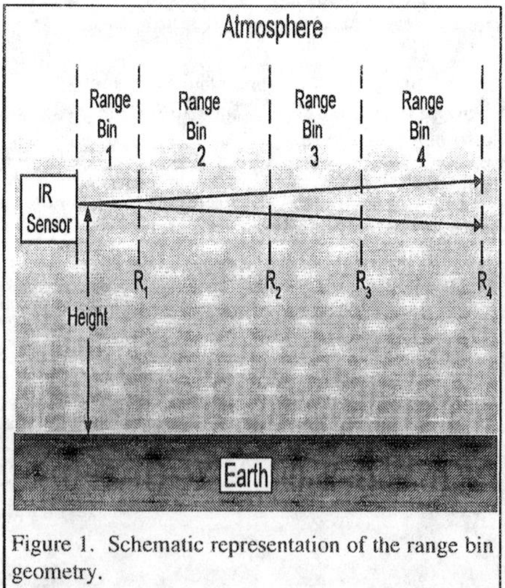


Figure 1. Schematic representation of the range bin geometry.

## MATRIX FORMULATION

For small temperature variations, the radiometer signal is nearly linearly dependent on temperature. We can create a matrix whose columns correspond to the fraction of the signal in each range bin and whose rows correspond to the various filter angles. For a given filter angle θ, the fraction of the signal is calculated by normalizing the sum of contributions (obtained from Eq. (2)) from all range bins to one. The range bin boundaries, $R_1$, to $R_4$, were chosen to be 2, 7, 20 km and to space, respectively. The filter angles corresponding to rows 1 through 9 of the matrix are 0°, 10.8°, 21.6°, 25.2°, 28.8°, 32.4°, 36.0°, 39.6° & 43.2°, respectively. The relationship between the range bin temperatures and the signal temperatures is given by the equation

$$Tsig = A \cdot Tbin \tag{3}$$

where the $A$ matrix is

$$A := \begin{bmatrix} 0.98205 & 0.011451 & 0.005642 & 0.000857 \\ 0.962291 & 0.024916 & 0.011593 & 0.001201 \\ 0.80832 & 0.119379 & 0.065298 & 0.007003 \\ 0.699668 & 0.17884 & 0.107938 & 0.013554 \\ 0.566698 & 0.244609 & 0.164151 & 0.024541 \\ 0.440556 & 0.294386 & 0.224146 & 0.040912 \\ 0.345992 & 0.316708 & 0.275492 & 0.061808 \\ 0.28979 & 0.313983 & 0.310244 & 0.085983 \\ 0.254193 & 0.301143 & 0.331761 & 0.112903 \end{bmatrix} \tag{4}$$

From Eq. (3) we can calculate the average range bin temperatures by the linear least squares fit:

$$Tbin = [A^T \cdot A]^{-1} \cdot A^T \cdot Tsig \tag{5}$$

In matrix $A$ (Eq. (4)), for the 0° filter angle (row 1), 98.2% of the signal comes from the first range bin (0 to 2 km) while only 0.08% comes from range bin 4 (20 km to space). On the other hand, for the 43.2° filter angle (row 9), 25.4% of the signal comes from range bin 1 and 11.3% comes from range bin 4. If we assume an input temperature vector, *Tsig*, of 300K for all the filter angles, then equation (5) calculates the average range bin temperatures at exactly 300K. However, if we add some measurement noise on the order of ±0.3K to the input temperatures, then the exact temperature solution for the range bin average temperatures yields physically unrealistic results. Typical results for the input temperature vector, *Tsig*, and the output temperature vector, *Tbin*, are shown below:

$$Tsig = [300.1\ \ 300.3\ \ 299.8\ \ 299.9\ \ 299.8\ \ 299.8\ \ 300.3\ \ 299.9\ \ 299.7]^T \tag{6}$$

$$Tbin = [300.3\ \ 284.7\ \ 327.0\ \ 258.0]^T \tag{7}$$

where the superscript T represents the transpose of the vectors. For slight variations in the *Tsig* vector of only ±0.3K, the exact solution for the output range bin temperatures, *Tbin*, vary by 69K between range bins 3 and 4. Since this result does not make sense from a physical point of view, we explored the use of genetic algorithms to solve for the range bin average temperatures.

## GENETIC ALGORITHMS

Genetic algorithms (GA) are search algorithms based on the mechanics of natural selection and natural genetics. They efficiently utilize historical information to obtain new search points with expected enhanced performance. In every generation, a new set of artificial individuals is created, using the information from the best of the old generation. Genetic algorithms combine the survival of the fittest from the old population with a randomized information exchange which helps to form new individuals with higher fitness. There are three basic genetic algorithm operators: reproduction, crossover and mutation. Those operators combined with the proper fitness function definition constitute the main body of genetic algorithms. Genetic algorithms are described in detail in [Gol-89, Mic-92].

## METHOD DESCRIPTION

The scheme that we are proposing for finding the most physically plausible solution to an overdetermined system of linear equations is based on genetic algorithms. In order to be able to obtain a better solution using genetic algorithms than solving the system of equations by the least squares method, information about the physical behavior of the system has to be incorporated into the fitness function of the genetic algorithm. This information is incorporated in the form of penalties for solutions which do not conform to the underlying physical reality. We incorporated three constraints:

1. when the radiometric measurements are close, the resulting range bin temperatures are close.
2. the parametric equations hold as closely as possible.
3. the range bin temperatures obtained at a given time step should be close to the range bin temperatures from the immediately preceding step. The underlying range bin temperature is a smooth function of time and adding this term to the fitness function assures a better smoothness.

The first two constraints are mutually contradictory: constraint 1 drives all range bins to the same temperature, and constraint 2 causes the range bin temperatures to be conform to the least squares solution. The exact least squares solution is just what we want to avoid and incorporating constraint 1 and 3 allows the genetic algorithm to do so. The fitness function of the genetic algorithm uses three coefficients, one for each of the constraints. These coefficients are the weighting factors, reflecting the importance of each constraint to be fulfilled.

The fitness function has the following form:

$$fitness = 1 - error \tag{8}$$

where

$$error = c1 * err1 + c2 * err2 + c3 * err3 \tag{9}$$

In Eq. (9) constraint 1 is implemented as *err1,* constraint 2 as *err2* etc. The coefficients *c1, c2, & c3* were set with the help of our knowledge of underlying physical reality. Important are not the values of the coefficients themselves but their ratio since the values of the fitness function are later scaled to be non-negative thus allowing all individuals to compete for a place in the mating pool. The coefficients *c1, c2,* and *c3* that we are using have the values 0.3, 40 and 2, respectively.

The genetic algorithm uses binary encoding for each of the four chromosomes, each corresponding to one range bin temperature. The population size is 101 and each GA is run 3 times for 600 generations. The results from the best run are chosen as the range bin temperatures. Probability of crossover is constant and is set to 0.6. In order to improve convergence of GA we introduced non-uniform mutation that allows the mutation probability for the bits located on the left side of the chromosome to decrease as the population ages, and the mutation probability for the bits at the right side of the chromosome to increase. This allows GA to explore more of the space in earlier generations, and to perform better fine tuning in later generations. The particular function used for changing the mutation probability is described in the earlier work of one of the authors [Buc-95].

## DATA SETS AND RESULTS

The matrix formulation of Eq. (4) is based on experimental radiometric data. The matrix designed in this fashion will be used in our later work in processing experimental infrared data. When working with experimental data, the exact knowledge of range bin temperatures is usually not available. This is the reason to test the method first on a simulated data set in which we know all the range bin temperatures. The data consist of four simulated sets. Each set corresponds to one of the four possible combinations of temperature 300K in three range bins and temperature 295K in the fourth range bin. Each data set consists of 10 series of nine radiometric measurements each. When simulating data in this fashion ±0.1 and ±0.3 uniform noise was added to the calculated radiometer temperatures.

Range bin temperatures are calculated by both the least squares fit of Eq. (5) and the genetic algorithm. Examples of two solutions are shown on Figures 2 and 3. On those figures the first four curves represent GA solution for each of the consecutive range bins, while the last four curves represent the least squares solution for the same range bins. The curves corresponding to GA solution are flat, meaning that the result from one case to the next does not change drastically. In the curves corresponding to the least squares solution, the changes are very pronounced from one measurement to another. This causes the curves to contain sharp peaks.

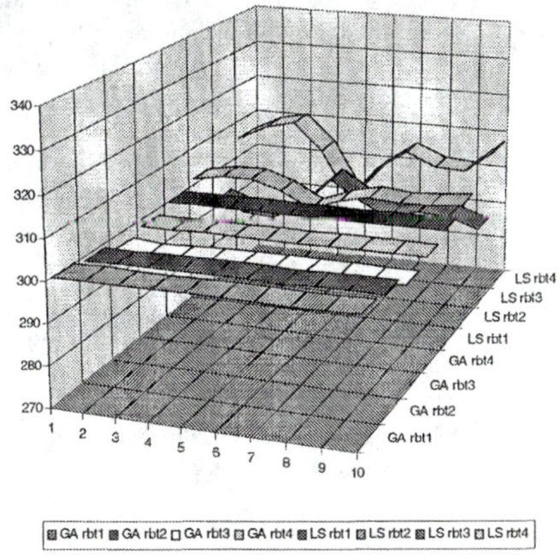


Figure 2. GA and least squares solution for range bin 3 temperature 295K, and other range bin temperatures 300K with 0.1 noise.

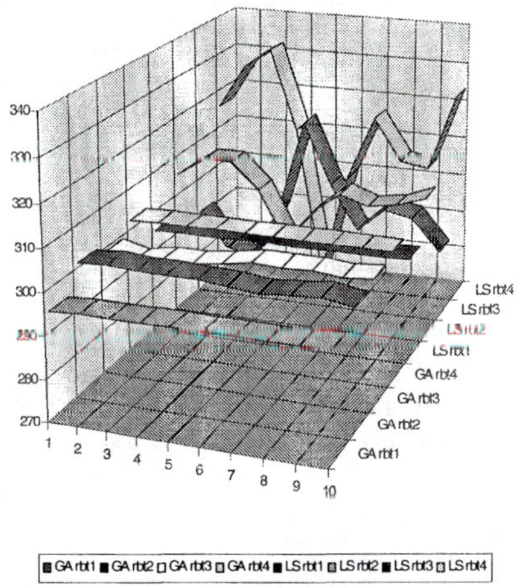


Figure 3. GA and least squares solution for range bin 1 temperature 295K, and other range bin temperatures 300K with 0.3 noise.

Figures 4 and 5 show the error of the calculations. In this case the first four curves represent GA error for each of the consecutive range bins, while the last four curves represent the least squares error for the same range bins. Those curves for ±0.1 noise do not exceed 2.8K, even for the last range bin. Least squares solution for ±0.1 uniform noise introduced have an error up to 10.8K. For data with ±0.3 noise, the maximum GA error is 3.7K, while the maximum error of least squares fit is 32.4K. The same characteristics of flat results for GA, and considerably varying results for the least squares solution are encountered.

It is worth noting that for all four data sets both least squares and GA solution for the first range bin gives excellent results for each amount of noise. For GA solution the maximum error is 0.1 and 0.3 for ±0.1 noise and for ±0.3 noise respectively. For least squares solution this result is even better: 0.1 and 0.2 error. The reason can be found by looking at the matrix A of Eq. (4): the percent contribution from the first range bin is very high in each radiometric measurement: in five out of nine measurements this contribution largely exceeds 50%. With such a high contribution the error for this range bin temperature will be minimal, while errors for other range bins that contribute much less to the radiometric measurements will be much larger.

Setting $c1$ to zero in Eq. (9) causes the minimum and maximum of the four range bin temperatures to be very different, thus giving improper results. Setting $c3$ to zero removes the smoothing of temperatures from one step to the next, resulting in a much noisier solutions from step to step. Setting both those parameters to zero, changes the result to the least square solution. It is interesting to note that both parameters $c1$ and $c3$ are small in comparison to $c2$, and still their inclusion changes the results so significantly.

The comparison between genetic algorithm results and least squares fit shows the immense improvement obtained using genetic algorithms for three last range bins. For last range bin with ±0.3 noise GA error is almost 10 times smaller than least squares error. This is due to the fact, that GA do not blindly calculate the least squares solution, but are driven towards the correct solution by using the knowledge of the underlying physical phenomena, as incorporated into the fitness function.

**CONCLUSIONS**

A novel method for solving an overdetermined system of linear equations describing the relationship between radiometric temperatures and range bin temperatures was developed. The method performing

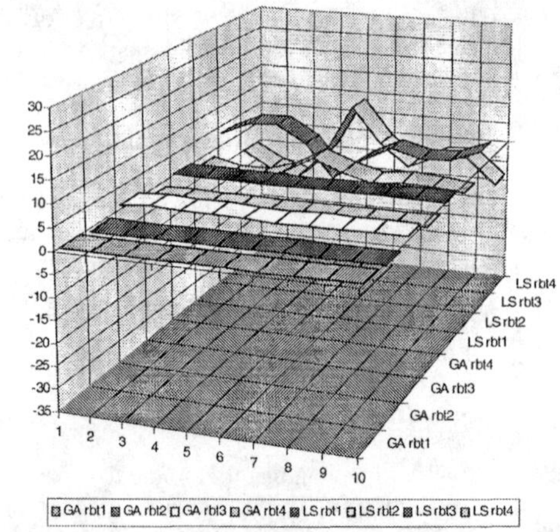


Figure 4. GA and least squares solution fit calculation errors for range bin 2 temperature 295K, and other range bin temperatures 300K with 0.1 noise.

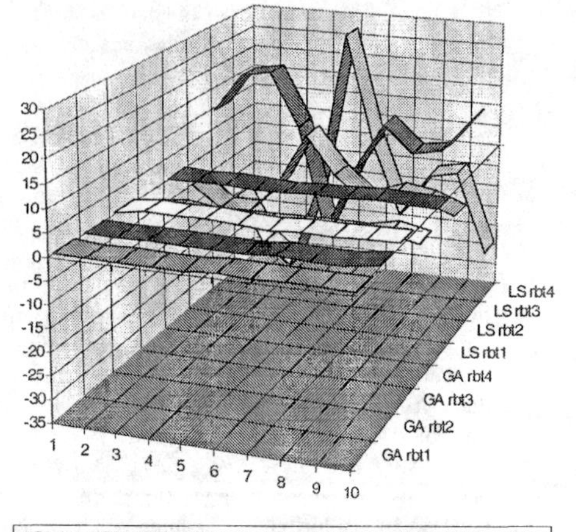


Figure 5. GA and least squares solution errors for range bin 4 temperature 295K, and other range bin temperatures 300K with 0.3 noise.

the prediction of atmospheric temperature profiles is based on genetic algorithms and results in solutions which are physically acceptable. This is a tremendous improvement over least squares solutions that even for small noise in radiometric measurements (on the order of ±0.3K), may give results over 30K off. The method developed works extremely well for radiometric measurements. However it can be adapted to any problems described by an overdetermined system of linear equations. For other types of measurements the terms introduced into the fitness function will differ since the underlying physical reality will differ.

This paper shows that GA can help in solving problems without a solution in the exact mathematical world. All that is needed is some knowledge of the underlying physical problem and the incorporation of this knowledge into the fitness function of the genetic algorithm. The GA method developed for solving the overdetermined system of linear equations can be easily generalized to other type of measurements.

**REFERENCES**


[Buc-95] A. L. Buczak, R. E. Uhrig, "Information Fusion by Fuzzy Set Operations and Genetic Algorithms", *Simulation, Special Issue on Application of Computational Intelligence*, Vol. 65, No. 1, July 1995.

[Gol-89] D. Goldberg, *Genetic Algorithms in Search, Optimization and Machine Learning*, Addison-Wesley Publishing Company, Inc., 1989.

[Mic-92] Z. Michalewicz, *Genetic Algorithms + Data Structures = Evolution Programs*, Springer-Verlag, 1992.

[Pre-95] W. H. Press, S. A. Teukolsky, W. T. Vetterling, B. P. Flannery, "Numerical Recipes in C", Cambridge University Press, 1995.

[Tho-93] M. E. Thomas, D. D. Duncan "Atmospheric Transmission" in *The Infrared and Electro-Optical Systems Handbook, Volume 2 Atmospheric Propagation of Radiation*, Ed. by F. G. Smith, Infrared Information Analysis Center, Environmental Research Institute of Michigan and SPIE Optical Engineering Press, Bellingham, WA 1993.

[Wol-93] W. L. Wolfe, "Radiation Theory" in *The Infrared and Electro-Optical Systems Handbook, Volume 1 Sources of Radiation*, Ed. by G. J. Zissis, Infrared Information Analysis Center, Environmental Research Institute of Michigan and SPIE Optical Engineering Press, Bellingham, WA 1993.

# OPTIMIZING THE GROWTH OF PECVD SILICON NITRIDE FILMS FOR SOLAR CELL APPLICATIONS USING NEURAL NETWORKS AND GENETIC ALGORITHMS

**SEUNG-SOO HAN, LI CAI, AJEET ROHATGI AND GARY MAY***
*School of Electrical and Computer Engineering*
*Georgia Institute of Technology*
*Atlanta, GA 30332-0250*
* *Office: (404) 894-9420, FAX: (404) 894-5028, E-mail: gary.may@ee.gatech.edu*

***ABSTRACT*:**
Silicon nitride films grown by plasma-enhanced chemical vapor deposition (PECVD) are useful as anti-reflection coatings in solar cells. PECVD systems are controlled by RF power, pressure, gas flow rates, reactant composition, and substrate temperature. The wide variety of operating conditions and the complex nature of plasmas make tailoring $Si_3N_4$ film properties very challenging. In this study, silicon nitride PECVD modeling using neural networks is investigated. The deposition of $Si_3N_4$ is characterized by a central composite design, and data from this experiment is used to train optimized neural networks using the back-propagation algorithm. Using these process models, the effect of deposition conditions on film properties is studied. Conditions which optimize effective carrier lifetime and hydrogen concentration are then determined using genetic algorithms to search the response surface described by the neural process models. Optimal conditions predicted by the genetic algorithms are used to grow improved films, and excellent agreement is obtained between predicted film qualities and experimental data. Deposition experiments are carried out in a Plasma Therm 700 series PECVD reactor.

## INTRODUCTION

Silicon nitride films grown by plasma-enhanced chemical vapor deposition (PECVD) are useful as anti-reflection coatings in solar cells and can be beneficial in improving cell efficiency (Johnson et. al., 1983). These films also enhance cell performance by passivating the device surface by introducing significant amounts of atomic hydrogen during the PECVD process (Michiels et. al., 1990). In addition, achieving low surface recombination velocities and low interface trap densities using PECVD nitrides has been reported (Chen et. al., 1993). Furthermore, the PECVD process possesses a number of other qualities which are attractive from a manufacturing standpoint, including high throughput, good uniformity and thickness control, and excellent reproducibility (Hezel and Jaeger, 1989).

Silicon nitride film properties are determined by the composition of the plasma, which are in turn controlled by the deposition variables. Due to the complex nature the plasma, it is very difficult to quantify from first principles the exact relationship between input factors (such as substrate temperature, pressure, RF power, and gas flows) and critical output parameters (such as effective lifetime and positive charge). Although empirical models for plasma-based and other semiconductor processes have been developed using response surface methods (May et. al., 1991), process models derived from neural networks have recently been shown to offer advantages in both learning accuracy and generalization capability (Himmel and May, 1993).

This paper discusses the use of neural networks to develop models for PECVD of silicon nitride. To characterize the process, a central composite experiment consisting of a $2^{6-1}$ fractional factorial design augmented by twelve axial points and three center points was performed (Box et. al., 1978). Data from these 47 experiments was used to train neural network models of effective carrier lifetime and positive charge concentration using the back-propagation algorithm. Since the development of optimal neural process models is complicated by the fact that back-propagation neural networks contain several adjustable parameters, the effect of these factors on network performance was also investigated, and parameter sets which minimized the training and prediction error of the $Si_3N_4$ PECVD models were determined (Han et. al., 1994). The neural process models were then used in conjunction with genetic algorithms to determine the optimal condition for depositing PECVD silicon nitride. The results were compared with the experiment data, and excellent agreement was obtained.

## EXPERIMENTAL METHODOLOGY

Silicon nitride films were deposited in a Plasma-Therm 700 series batch reactor operating at 13.56 MHz using ammonia ($NH_3$), silane ($SiH_4$), and nitrogen as feed gases. Deposition conditions were varied using a central composite circumscribed design over the ranges shown in Table 1. Three-inch (100) p-type silicon wafers with a resistivity of 2.0 Ω-cm were used as substrates. The wafers were thoroughly cleaned, and each was then cut into four pieces and subjected to PECVD. During deposition, $SiH_4$ was diluted to 2% in nitrogen. After depositing approximately 0.05 μm of silicon nitride, the samples underwent annealing at 350 °C for 20 minutes.

To measure positive charge, capacitors were made by evaporating 0.5 μm of aluminum on top of the nitride layer and defining the gate region using standard photolithographic techniques. Charge concentration was obtained by high frequency C-V measurements. Effective lifetime was measured using a laser photoconductive decay tester. Based on data obtained from these experimental trials, neural network models were trained and PECVD process output characteristics were subsequently analyzed.

## NEURAL NETWORK PROCESS MODELING

The complex nature of plasmas has limited the success of PECVD modeling from a fundamental physical standpoint. Recently, neural networks have emerged as an attractive alternative to physically-based models and statistical methods (May, 1994).

## TABLE 1: RANGES OF DEPOSITION PARAMETERS

| Parameter | Range |
|---|---|
| Substrate Temperature | 200 - 400 °C |
| Pressure | 0.6 - 1.2 torr |
| RF Power | 20 - 40 watts |
| $NH_3$ Flow | 1 - 1.4 sccm |
| $SiH_4$ Flow | 180 - 260 sccm |
| $N_2$ Flow | 0 - 1000 sccm |

Feed-forward neural networks used for semiconductor process modeling are trained via *error back-propagation* (BP). The input layer of neurons in this BP neural network corresponds to the six adjustable input parameters varied in the PECVD experiment. The output layer corresponds to the two deposition variables being modeled. The network also incorporates one "hidden" layer of neurons which can be viewed as representing fundamental, yet not directly controllable plasma properties such as electron temperature or reactive species concentration.

In the BP algorithm, performance is influenced by both the number of hidden layers and the number of neurons in each layer. Since it has been shown that a network containing one hidden layer can encode any arbitrarily complex input-output relationship (Irie and Miyake, 1988), the number of layers has been fixed at three. The number of hidden neurons is then varied to determine the optimal network structure. The performance of BP networks also depends on learning rate, momentum, and training tolerance (Kim and May, 1993). Learning rate determines convergence speed, but the network may settle too far from the actual minimum value of the error surface if it is too large. Smaller rates can ensure stability, but result in longer training time. Momentum is designed to prevent the training algorithm from settling in local minima and also increases the speed of convergence. This term tends to keep the weight changes going in the same direction. Another important factor is network training tolerance, which specifies an acceptable tolerance for the accuracy of the neural outputs. Smaller tolerances increase learning accuracy, but can result in less generalization capability and longer training time.

In optimizing the models, hidden neurons, learning rate, momentum, and training tolerance were considered. Initially, neural PECVD models were obtained using a default network structure and set of learning parameters. These "rough" models were then refined by varying the values of the four critical parameters according to a $2^{4-1}$ fractional factorial design and analyzing the results using the *RS/Discover* statistical analysis software package. Network learning ability and prediction capability were quantified by their mean-squared error. Prediction error was determined by using the trained networks to predict the nitride film properties for ten of the original 47 experimental runs selected at random. To minimized both training and prediction error, the following performance index (*PI*) was implemented:

$$PI = K_1\sigma_t^2 + K_2\sigma_p^2 \qquad (1)$$

where $\sigma_t$ is training error and $\sigma_p$ is prediction error. $K_1$ and $K_2$ represent the relative importance of each performance measure (values chosen here were $K_1 = 1$, and $K_2 = 10$). Optimization was then performed using the simplex search algorithm to minimize the performance index (Nash, 1979). Optimal training parameters, training error, and prediction error for each PECVD process model are shown in Table 2.

**TABLE 2: OPTIMIZED NEURAL NETWORK PARAMETERS**

| PECVD Response | Hidden Neurons | Learning Rate | Momentum | Tolerance | Training Error | Prediction Error |
|---|---|---|---|---|---|---|
| Effective Lifetime ($\mu$s) | 6 | 0.05 | 0 | 0.01 | 10.54 | 16.09 |
| Positive Charge ($cm^{-2}$) | 10 | 0.05 | 0 | 0.13 | 0.372 | 0.556 |

## RECIPE SYNTHESIS

The next objective was to develop a systematic methodology to design specific sets of PECVD process conditions (or process "recipes") to achieve desired process objectives. These process objectives are specific locations on multidimensional response surfaces which geometrically depict the variation of process output characteristics with respect to its input variables. Recipe generation is achieved by employing the optimal neural process models in conjunction with genetic algorithms (GAs) as a response surface exploration scheme (Goldberg, 1989). GAs refer to a family of computational models inspired by evolution. These algorithms encode potential solutions to a problem on a chromosome-like data structure and apply recombination operators to these structures so as to preserve critical information. They operate through a simple cycle of four stages: (1) creation of a "population" of strings; (2) evaluation of the "fitness" of each string; (3) selection of most "fit" strings; and (4) genetic manipulation (crossover and mutation) to create the new population of strings.

For the synthesis of PECVD recipes from neural process models, the desired output characteristics of the PECVD silicon nitride film to be produced are reflected by the following fitness function ($F$):

$$F = \frac{1}{1 + \sum_r |K_r(y_d - y)|} \tag{2}$$

where $r$ is the number of responses, $y_d$ are the desired responses, and $y$ are the outputs dictated by the current choice of inputs and $K_r$ is a constant which represents the relative importance of the $r^{th}$ process response. Optimization was stopped after a fixed number or iterations or when $F$ was within a predefined tolerance. The probabilities of crossover and mutation for the genetic algorithms were set to 0.6 and 0.01, respectively. A population size of 100 was used in each generation. Each of the five process input parameters were coded as 40-bit string, resulting in a total chromosome length of 200 bits. The *elitist roulette wheel selection* method was used. Maximization of $F$

continued until a final solution was selected after 500 generations.

## RESULTS & DISCUSSION

Effective lifetime and positive charge as were chosen as response variables because: (1) long effective lifetime results in high solar cell efficiency (Uematsu et. al, 1989); and (2) high positive charge concentration not only improves the surface passivation of a solar cell (leading to longer effective lifetime), but also results in stronger inversion and low sheet resistance in MIS inversion cells (Shengliang et. al., 1987). Investigation of the changes in a few of these critical film properties as a function of substrate temperature, RF power, pressure, and gas flow was conducted, and a summary of key results follows.

### Effect of Deposition Parameters

Figure 1 shows effective carrier lifetime as a function of temperature and RF power. Increasing the substrate temperature increases the lifetime. Effective lifetime is a measure of the quality of the bulk and surface passivation. High effective lifetime means better passivation. At lower temperatures, bulk lifetime is not affected. Positive charge stems from dangling bonds, the presence of which can be reduced by increased atomic hydrogen. An increase in deposition temperature results in less bonded hydrogen, which therefore increases the positive charge density. The amount of positive charge in silicon nitride film is closely related to the surface quality - higher concentration yields better surface passivation and therefore, longer lifetime. Our models confirmed a correlation between effective lifetime of a sample and positive charge concentration in the silicon nitride film.

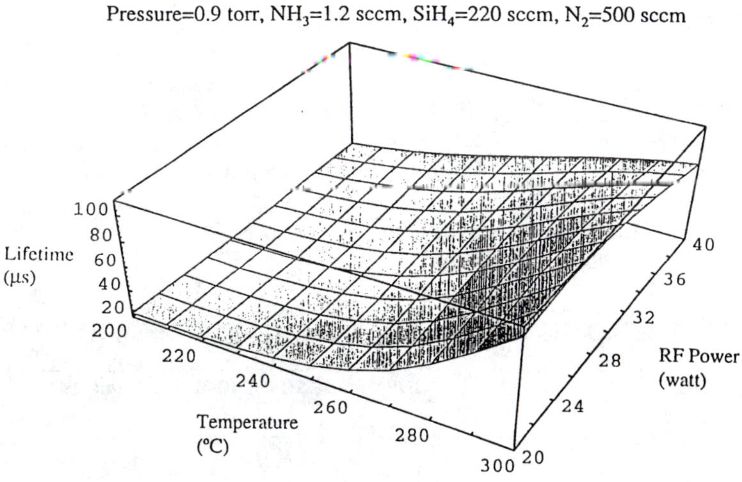


*Figure 1: Effective lifetime vs. temperature and RF power.*

RF power determines the level of ion bombardment. High RF power increases bombardment by more energetic ions, causing a greater number of collisions as ions are accelerated through the plasma sheath. Increased ion bombardment also aids the removal of reaction by-products from the substrate surface. With increasing power, therefore, films that are denser and more homogeneous are deposited, resulting in better passivation. Figure 2 shows positive charge concentration as a function substrate temperature and RF power. A comparison of Figures 1 and 2 shows the correlation between positive charge and the effective lifetime.

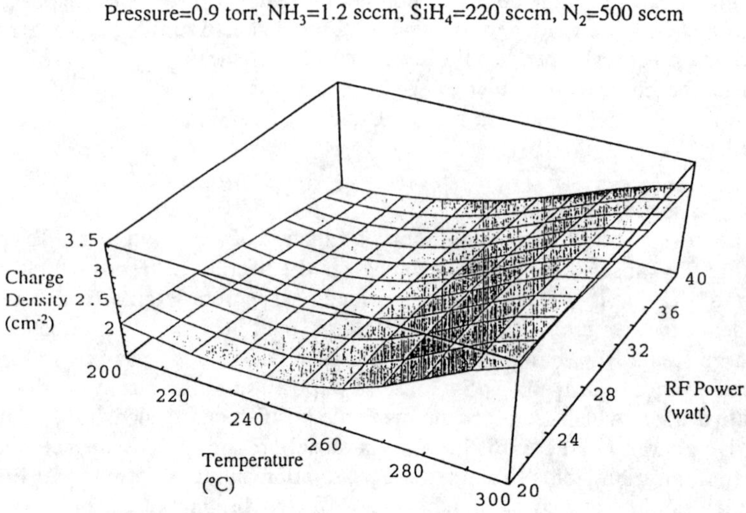


*Figure 2: Positive charge vs. temperature and RF power.*

**Recipe Synthesis and Optimization**

The genetic algorithm based search procedure provided optimized deposition recipes for effective lifetime and positive charge concentration. The optimal PECVD deposition conditions for optimal positive charge were slightly different from those for the effective lifetime, requiring slightly lower RF power, higher pressure, and larger gases flow rates. Experiments using these recipes were performed to verify the GA predictions. The lifetime and charge from the modeling predictions and experiment results are shown in Table 3, along with the percent error between the predictions and experimental values. Error is less than 7% in each case, representing good agreement.

**TABLE 3: PREDICTED AND EXPERIMENTAL LIFETIME AND CHARGE**

| PECVD Response | Prediction | Measurement | Error |
|---|---|---|---|
| Effective Lifetime (μs) | 170 | 160 | 6.3% |
| Positive Charge ($cm^{-2}$) | $5.3 \times 10^{12}$ | $5.1 \times 10^{12}$ | 3.9% |

## SUMMARY & FUTURE WORK

Properties of PECVD silicon nitride films have been modeled using back-propagation neural networks. Genetic algorithms were then employed as a recipe synthesis method to optimize the deposition conditions for effective lifetime and positive charge. Deposition using the optimized conditions was performed, and experimental results were consistent with model predictions. Although the recipes generated are useful to obtain films optimized for a single criterion, they can lead to unacceptable results for the remaining film characteristics. It is desirable to grow films with the best combination of all desired qualities. Since this will involve some processing trade-offs, future work will involve deriving a recipe which optimizes both responses.

## ACKNOWLEDGEMENT

The authors would like to thank NASA (Contract No. NAS7-918), JPL, ARPA, NREL and Sandia for support of this research.

## REFERENCES

Box, G., Hunter, W., and Hunter, J., *Statistics for Experimenters*, New York: Wiley, 1978.

Chen, Z., Sana, P., Salami, J., and Rohatgi, A., "A Novel and Effective PECVD $SiO_2$/SiN Antireflection Coating for Si Solar Cells," *IEEE Trans. Elec. Dev.*, vol. 40, no. 6, pp. 1161-1165, June, 1993.

Goldberg, D., *Genetic Algorithms in Search, Optimization & Machine Learning*, Addison Wesley, 1989.

Han, S., Ceiler, M., Bidstrup, S., Kohl, P., and May, G., "Modeling the Properties of PECVD Silicon Dioxide Films Using Optimized Back-Propagation Neural Networks," *IEEE Trans. Comp. Pack, & Manufac. Tech.*, vol. 17, no. 2, June, 1994.

Hezel, R., and Jaeger, K., "Low-Temperature Surface Passivation of Silicon for Solar Cells," *J. Electrochem. Soc.*, vol. 136, p. 518, 1989.

Himmel, C., and May, G., "Advantages of Plasma Etch Modeling Using Neural Networks Over Statistical Techniques," *IEEE Trans. Semi. Manufac.*, vol. 6, no. 2, pp. 103-111, May, 1993.

Irie, B., and Miyake, S., "Capabilities of Three-Layered Perceptrons," *IEEE Internat'l Conf. on Neural Networks*, 1988, pp. 641-648.

Johnson, C., Wydeven, T., and Donohoe, K., "Plasma-Enhanced CVD Silicon Nitride Antireflection Coatings for Solar Cells," *Solar Energy*, vol. 31, no. 4, pp. 355-358, 1983.

Kim, B., and May, G., "An Optimal Neural Network Process Model for Plasma Etching," *IEEE Trans. Semi. Manufac.*, vol. 7, no. 1, pp. 12-21, Feb., 1993.

May, G., Huang, J., and Spanos, C., "Statistical Experimental Design in Plasma Etch Modeling," *IEEE Trans. Semi. Manufac.*, vol. 4, no. 2, pp. 83-98, May, 1991.

May, G., "Manufacturing ICs the Neural Way," *IEEE Spectrum*, vol. 31, no. 9, pp. 47-51, Sept., 1994.

Michiels, P., Verhoef, L., Stroom, J., Sinke, W., van Zolingen, R., Denisse, C., and Hendriks, M., "Hydrogen Passivation of Polycrystalline Silicon Solar Cells by Plasma Deposition of Silicon Nitride," *Proc. 21st IEEE Photovoltaic Specialists Conf.*, 1990, pp. 638-643.

Nash, J., *Compact Numerical Methods for Computers: Linear Algebra and Function Minimization*, John Wiley & Sons, 1979.

Shengliang, Z., Enke, L., and Quanxi, J., "Research on the Charge Stored Effect in Silicon Nitride for Improving Performance of MIS/IL Solar Cells," *Fifth International Conference on Energy Options - The Role of Alternatives in the World Energy Scene*, p.107-110, 1987.

Uematsu, T., Kokunai, S., Saitoh, T., Iwashita, T., and Tokuyama, T., "Improvements in Surface Passivation for High-Efficiency, Thin Silicon Solar Cells," *Optoelectronics - Devices and Technologies*, vol. 4, no. 1, p. 67-73, 1989.

# HYDRAULIC SYSTEM PARAMETER IDENTIFICATION USING GENETIC ALGORITHM METHODS

**RUTH BOOK**
*Department of Agricultural Engineering*
*University of Illinois at Urbana-Champaign*

*ABSTRACT:*
A simple, real-coded genetic algorithm identifying the unmeasured parameters of a fluid power system replaces the trial and error method used in the development of a nonlinear valve-motor system model. A tournament selection scheme with crossover and mutation is successful in determining a combination of parameters which optimizes the match of simulation results to the prototype data, with a reasonable number of function evaluations. Fitness is evaluated by comparing frequency and magnitude of simulated to actual pressure response. The automation of parameter identification using the genetic algorithm improves the performance of the system model.

## INTRODUCTION

In the development of typical computer models of hydraulic systems, simulated solutions are compared to actual data for validation of the model. Generally, there are a number of parameters called for in the equations which can not be (or have not been) measured in the actual test, and must be selected by the engineer. There are many ways to go about determining these values, including trial and error manual methods and computerized optimization techniques.

The work done in this study involves the application of a simple genetic algorithm (GA) to optimize the selection of unknown parameters in a nonlinear dynamic model. The model was developed for a hydraulic valve-motor system operating the swing function of a log loader, and eventually became part of a more complex model which included the load sensing pump and related dynamics (Book and Goering, 1995). Prototype data was used for validation, and the parameter identification was done manually. Figure 1 shows a comparison of these simulation results to the actual data. The genetic algorithm is the method of choice in this work due to its

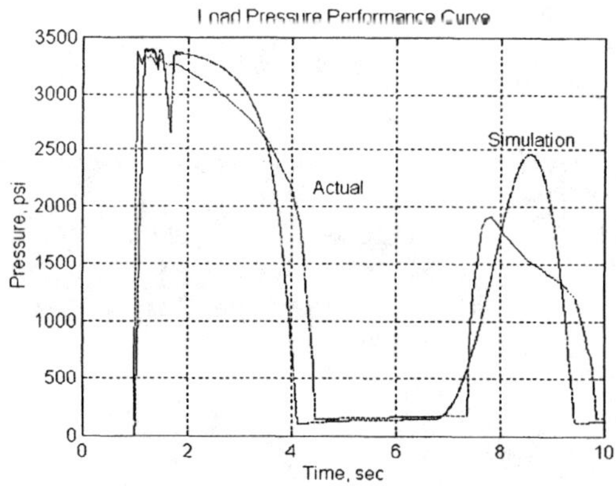


**Figure 1: Manual parameter identification performance.**

reported advantages in finding global rather than local optima, and the relative ease of application to existing models (Goldberg, 1994). The main goal is to establish whether the use of a simple genetic algorithm is superior to the manual method of optimizing parameters in the hydraulic system. Factors to be considered in the determination of superiority include the quality of the solution found, the amount of time required to find the solution, and the ease of implementation.

## RELATED WORK

Typically, selection of values for the unknown parameters in fluid power simulations is done using calculations to get into a reasonable range and then fine tuning by manually comparing and correcting the simulation results (Book and Goering, 1995), (Mai and Dransfield, 1989), (Zeiger and Akers, 1986). This manual method can take weeks, or months, due to the large number of possible solutions involved. Also, good solutions can be missed if multiple optima exist in the problem.

More advanced techniques have been used for parameter identification in fluid power models. A Rosenbrock algorithm was used to optimize six parameters by Kim et al. (1988). Krus et al. (1992) developed a modified Complex algorithm, a method closely related to genetic algorithms in its randomization of the parameters in an initial population, or complex, of possible solutions. The Krus procedure required 490 simulations to optimize seven sizing variables in the design of a hydraulic system, although convergence was nearly reached after approximately 200 simulations. Krus et al. predicted that the genetic algorithm approach , while being more robust, would likely require a greater number of function evaluations.

## ALGORITHM DEVELOPMENT

The choice of software for this work was based on what had been used for the fluid power model: SIMULINK, a dynamic simulation package which works with MATLAB (The MathWorks, 1994), on a 100mHz Pentium PC.

The challenge when automating the optimization was to develop a reliable way for the computer to assign a fitness value to each set of simulated results. Five critical locations along the pressure profile were identified, and these five pressure and time coordinates were used to define the matching errors in the two peak heights ($\Delta h_1$ and $\Delta h_2$), the width of each pressure pulse ($\Delta w_1$ and $\Delta w_2$), and the frequency of the pulses ($\Delta f$). A single fitness number was assigned to each simulation run (Equation 1), lower being better fitness. Scaling was done to give approximately equal penalties to both pressure and time errors, and a penalty was assigned when a feature was not found.

The seven parameters to be optimized were assigned ranges of reasonable values (Table 1). Each range was discretized into ten values, and coded as a number from 1 to 10. Each parameter was assigned a fixed location in a 7-character string, or chromosome. Features that are likely to interact form "building blocks" which are best positioned adjacent so that successful combinations are not broken up during the

$$Fitness = \frac{|\Delta h_1|}{100} + \frac{|\Delta h_2|}{200} + \Delta f^2 + \Delta w_1^{\,2} + \Delta w_2^{\,2} \qquad (1)$$

**TABLE 1: CONSTRAINTS ON PARAMETER VALUES**

| Parameter | Range |
|---|---|
| equivalent fluid bulk modulus (psi) | 4,000 - 49,000 |
| motor coefficient of friction (in-lb-sec) | 0.01 - 1.0 |
| orifice coefficient of discharge | 0.5 - 0.95 |
| meter-out opening area ($in^2$) | 0.05 - 0.14 |
| moment of inertia ($in\text{-}lb\text{-}sec^2$) | 26 - 35 |
| internal motor leakage coefficient ($in^3/sec/psi$) | 0.00001 - 0.001 |
| external motor leakage coefficient ($in^3/sec/psi$) | 0.00001 - 0.001 |

crossover operations (Goldberg, 1989). In the manual optimization, interactions had been observed between the bulk modulus, the motor coefficient of friction, the orifice coefficient of discharge and the meter-out area, so these were positioned together. The leakage parameters were assumed to act together, and were kept adjacent. The total search space was thus defined to be a matrix of seven variables, each having ten possible values, creating a possible $10^7$ different chromosomes.

With the specifics of the problem defined, the remainder of the GA development involved the application of standard operators. Binary tournament selection without replacement was chosen (Goldberg and Deb, 1991) with probability $p$=1 of the better individual in each tournament being selected. Single-point crossover between the selected chromosomes was utilized to generate a new population for each generation (Goldberg, 1989). These reportedly were some of the more effective simple operators developed to date. A mutation rate of 0.07 was chosen, through experimentation, to retain some diversity in the population and avoid premature convergence.

Two methods of convergence were incorporated, to reduce the likelihood of the search running longer than necessary.

- Compare the lowest fitness value found from one generation to the next, and terminate if this has not changed in five generations.
- Compare the average population fitness value for each generation against the minimum found in the entire search, and terminate if these values become sufficiently close.

## RESULTS AND DISCUSSION

The genetic algorithm was very successful in identifying sets of parameters which improved the fit of the simulated results to the actual data. Figure 2 shows one such simulation, which is representative of all of the solutions found in several trials of the GA. The main improvement in performance lies in the height of the secondary pressure peak. Minor variations in pulse width were observed.

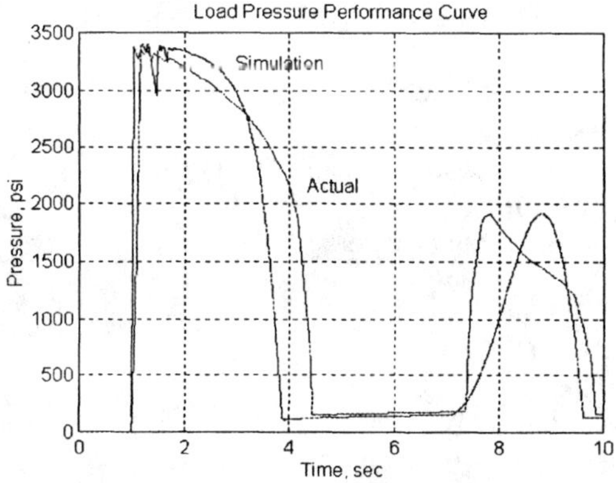

**Figure 2: GA Parameter identification performance**

| TABLE 2: VARIATIONS IN OPTIMA | | |
|---|---|---|
| **Optimization Method** | **Best Coded String** | **Fitness Value** |
| Manual | [1 7 4 4 5 6 7] | 6.1805 |
| 7-Parameter GA, trial 1 | [1 7 4 5 2 4 8] | 0.9574 |
| 7-Parameter GA, trial 2 | [1 2 2 4 10 7 10] | 0.9204 |
| 6-Parameter GA | [* 7 4 5 3 5 9] | 0.8207 |

Several trials were performed with the final GA configuration to get some idea of variability. The bulk modulus parameter was consistently driven by the GA to 4,000 psi, but there seemed to be a wide variety of optima, all producing approximately the same pressure performance, depending on combinations of the other six parameters. In addition, it was discovered that a 6-parameter GA, holding the bulk modulus constant at its optimum value as determined by the 7-parameter GA, not only ran in minutes instead of hours, but also was able to produce an even better solution. Table 2 presents a short comparison of these results.

As the GA progresses through its generations, the average fitness quickly drops to about 30% of its original level within four generations, and then begins to fluctuate (Figure 3). At the same time, the minimum fitness value found in each generation is getting better and better. At convergence, the overall rate of improvement of both the average and the minimum fitness has decreased to zero. Further optimization on the solution at this point would require the addition of some other "hillclimbing" technique. This hybridization was not done for the fluid power optimization, because the solutions found were already so much better than what had been found manually.

The main problem encountered in this work was the length of time required for the function evaluations. Convergence generally occurred in 13 - 28 generations, requiring 650 - 1400 calls to the SIMULINK model. With the full 7-parameter optimization problem, this took upwards of 12 hours. The 6-parameter version (with bulk modulus held constant) took approximately 30 minutes for convergence. It was evident that wide variations in the bulk modulus slowed the SIMULINK model considerably. Overall, implementation of the genetic algorithm was relatively easy, however. Proper choice of software made it possible to use the hydraulic system model for function evaluation without any modifications. The simple GA operators were well documented in the literature and easy to program into MATLAB. With the existence of this genetic algorithm as applied to a fluid power model, applications to future models will require only a small amount of tailoring.

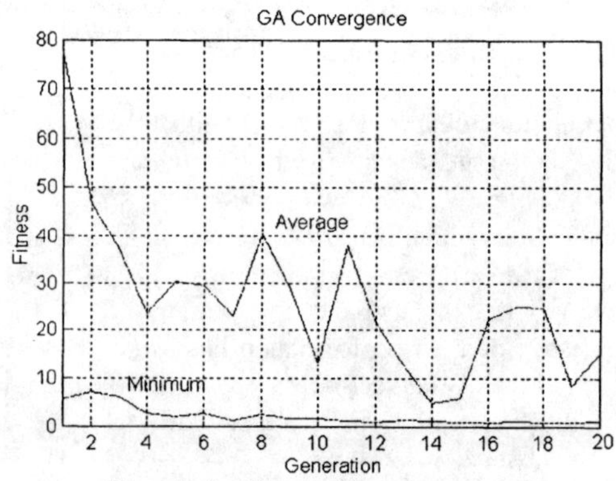

**Figure 3: Genetic algorithm convergence.**

## EXTENSIONS AND RAMIFICATIONS

The number of function evaluations required for the genetic algorithm in this problem was indeed 1.3 to 2.8 times higher than that found for the modified Complex algorithm, as predicted (Krus, et al, 1992). However, this comparison is not necessarily valid, since the two optimization problems were not the same, and it is not possible to compare the relative quality of the solutions found. As Krus points out, there is no guarantee that the solutions found are the absolute best possible. To get a direct comparison, both methods would need to be implemented on the same problem.

Also, it is important to note that in any fluid power optimization problem, there is usually more than one data trace to be matched. This work has concentrated on the single pressure trace which formed the baseline for the manual optimization work. However, there exists a flow trace for the same test which would need to be verified. In addition, existing test data with other configurations of the hydraulic components would need to be evaluated.

## SUMMARY AND CONCLUSIONS

The genetic algorithm easily outperformed the manual search in the quality of solution found. The speed with which the GA found this solution was several orders of magnitude faster than that of the manual method, even allowing for time to install and tailor the GA to the application. Approximate run time of the genetic algorithm as applied to the fluid power model was 12 hours.

Only one of the seven parameters to be identified, the effective bulk modulus of the system, was consistently driven to one optimum value. Several different optima with equivalent fitness values were found in successive trials, with varying combinations of the other six parameters, indicating that there was a complicated relationship between these parameters with multiple "valleys" of optimum fitness. This illustrates one of the strengths of genetic algorithms: the ability to locate global optima in situations where there are many local ones.

A six-parameter GA implemented with the bulk modulus optimum value held as constant was able to find an even better solution in a much shorter time. This suggests the possibility of designing the genetic algorithm to first identify parameters which are consistently optimized to one value, fixing these values, and then rerunning the GA as a fine-tuning method.

The number of function evaluations required by the genetic algorithm was quite high. However, the ease of application, due to the simplicity of the programming of the genetic operators and the ability of the GA to use the existing fluid power model without modification, made the GA an extremely attractive alternative to the manual method of parameter identification.

As a result of this work, the following conclusions are drawn:

1. A simple genetic algorithm produces a superior quality solution to the parameter identification problem in a much shorter time than the manual method with no modifications to the fluid power model needed.
2. Excessive run time can be minimized using a set of two criteria for termination of the search.
3. Fine tuning can be accomplished by re-applying the genetic algorithm to the problem with consistently optimized parameters held fixed.

**REFERENCES**


Book, R., & Goering, C.E. (1995). Hydraulic valve-motor system simulation. Presented at the *Agricultural Equipment Technology Conference,* Chicago, Illinois, 1-4 November, Paper No. AETC95109. ASAE, 2950 Niles Rd., St. Joseph, MI 49085-9659 USA.

Goldberg, D.E. (1994). Genetic and evolutionary algorithms come of age. *Communications of the ACM* 37(3): 113-119.

Goldberg, D.E. (1989). *Genetic algorithms in search, optimization and machine learning.* Reading, Massachusetts: Addison-Wesley Publishing Company, Inc.

Goldberg, D.E., & Deb, K. (1991). A comparative analysis of selection schemes used in genetic algorithm. *Foundations of Genetic Algorithms.* Morgan Kaufmann Publishers, Inc., San Mateo, California: 69-93.

Kim, S.D., Cho, H.S., & Lee, C.O. (1988). Stability analysis of a load-sensing hydraulic system. *Proceedings of the Institution of Mechanical Engineers, Part A: Power and Process Engineering* 202(A2): 79-88.

Krus, P., Jansson, A., & Palmberg, J-O. (1992). Optimization Based on Simulation for Design of Fluid Power Systems. ASME Paper No. 92-WA/FPST-8. New York: ASME.

Mai, V.K.L., & Dransfield, P. (1989). Load-sensitive pump control performance. *Mechanical Engineering Transactions - Institution of Engineers, Australia* ME14(3):149-157.

The MathWorks, Inc. (1994). *SIMULINK User's Guide For Microsoft Windows.*

Zeiger, G., & Akers, A. (1986). Dynamic analysis of an axial piston pump swashplate control. *Proceedings of the Institution of Mechanical Engineers, Part C* 200(C1): 49-58.

# Testing Trainable Software Components by Combining Genetic Algorithms and Backpropagation Algorithms

**JUNDA CHEN AND DAVID C. RINE**
*School of Information Technology and Engineering,*
*George Mason University, Fairfax, Virginia 22030*

***ABSTRACT:***
This paper presents an effective adaptive method for off-line controller software component testing. Each component is represented at the abstract level by means of a set of fuzzy logic rules and is represented at the concrete level by means of fuzzy membership functions. The method investigated uses a combination of a genetic algorithm (GA) and a neural network (NN) with an error back-propagation algorithm (BP). Component testing consists of two functionalities: *generating* fault scenarios and using the scenarios to *locate* erroneous fuzzy elements (EFEs) within the component. *Generating* fault scenarios involves training the NN and recalling the scenario. In this GA-BP combination, the GA searches for fault scenarios (training examples for the NN), while BP is used to train the NN. After being trained, the NN recalls candidate fault scenarios for an unseen environment. EFEs are *located* by examining the accumulated absolute error difference generated by the fuzzy logic (FL) component. Effectiveness of this combination is determined by the number of EFEs uncovered under new environments. Experimental results on comparing GA-BP and GA-only methods are presented.

## INTRODUCTION

This paper presents an effectively adaptive method for off-line controller software component testing. Each component is represented at the abstract level by means of a set of fuzzy logic (FL) rules and is represented at the concrete level by means of fuzzy membership functions (MBFs). The method investigated uses a combination of a genetic algorithm and a neural network with an error-back-propagation algorithm. Component testing consists of two functionalities: *generating* fault scenarios and using the scenarios to *locate* erroneous fuzzy elements within the component at the concrete level. A scenario is a sequence of states where each state is a vector of environmental and vehicular attributes (simply called an *environment*), e.g. road grade $v$, cruising speed $y_d$, road friction $\mu$, and vehicular mass $m$. In this paper, a *fault scenario* is a scenario used to uncover a fault in the operation of a vehicle's controller software component. Finding a fault scenario by duplicating the circumstances that caused a fault in vehicular operation, in particular in the cruise controller software component, is a time consuming and tedious task. Duplicating these circumstances is necessary before efficiently and effectively adapting the component can be carried out. Hence, there is a need to automatically generate test scenarios, especially under a time-limited situation.

Principles of reusable software components are: information hiding, abstraction, encapsulation, portability, and modularity. A component satisfying these principles is also called an *abstract data type* (ADT). An ADT has an abstract specification layer interfacing with the ADT user and a concrete body layer containing current internal representation details. For example, specification of a cruise controller software module (component) may contain high level rules, such as

> *"if actual speed is less than desired speed and the vehicle is accelerating, then press the accelerator a little."*

This specification is implemented in terms of a crisp code body. To reuse this component for different engine models and road conditions, we usually take advantage of ADT information hiding by only modifying the body by hand with tools. However, such manual modification of representation level code and parametric numerical values is often inefficient or impractical. To address problems such as unreachability and continuously evolving application environments, we present a training framework as an alternative to traditional software component reuse after extensive and programmer-intensive modifications (Rine, 1993; Chen and Rine, 1995; Rine, Ahmed, and Chen, 1996). We argue that the fuzzy rules and the MBFs of a FL component are analogous to the specification and the body of software components specified as ADTs. In addition, numerous studies show that it is desirable and possible to automatically adapt the FL component by mainly adjusting the MBFs in conjunction with modification of the fuzzy rules (Karr, 1993; Lin and Lee, 1994). Hence, we call this FL-based adaptive ADT a *trainable software component*. In this paper we only deal with the front-end phase of the training framework, testing. Using adaptation algorithms, we may automatically search for a fault scenario which is suitable for representing a current faulty environment. There are related works where genetic algorithms and neural network algorithms are used in testing and diagnosis for various applications (Polucarpou, 1995; Rodríguez, 1996; Schultz, 1993; Simpson, 1994; Thomas, 1994).

## ADAPTATION ALGORITHMS FOR TESTING TRAINABLE COMPONENTS

Our problem is to find an effective method for off-line testing of trainable software components used to develop controllers. A FL controller with a trainable software component is shown in (Figure 1). Each trainable component is defined in two levels, the user level with fuzzy rules (also called fuzzy associative memory (FAM)) (Table 1) and the implementation level with the MBFs (Figure 2). Let E, $\Delta$E, and $\Delta$U denote fuzzy linguistic variables of error, change of error, and change of control respectively. E, $\Delta$E, and $\Delta$U have fuzzy linguistic values of Negative (N), Small Negative (SN), Zero (Z), Small Positive (SP), and Positive (P). Each fuzzy linguistic value at the abstract end-user level is represented at the concrete level by a given triangular (saw-toothed) MBF. Each $\Delta u_{ij}$ denotes the right hand side of a fuzzy rule. Fuzzy membership functions, $MBF_0$, $MBF_1$, and $MBF_2$, represent fuzzy linguistic variables E, $\Delta$E, and $\Delta$U respectively.

*Genetic algorithms (GAs)* are used to maintain a best set of effective fault scenarios (solutions). Each candidate solution is represented as a linear structure (chromosome)

$$Y = \{ y_0, y_1, y_2, \dots y_{N-1} \} \quad (1)$$

where $y_i$ is a gene and $N$ is the size of a chromosome. Let *popsz* be the population size of

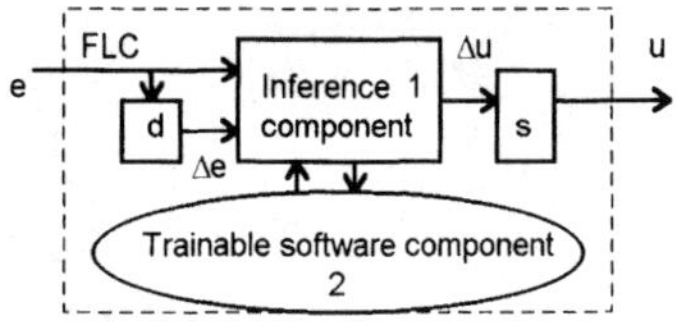


**Figure 1. A fuzzy logic controller.**

**TABLE 1: A FAM AS USER SPECIFICATION**

| ΔU | E | | | | |
|---|---|---|---|---|---|
| **ΔE** | **N** | **SN** | **Z** | **SP** | **P** |
| **N** | N | N | SN | SP | SP |
| **SN** | N | N | SN | SP | SP |
| **Z** | SN | SN | Z | SP | SP |
| **SP** | SN | SN | SP | P | P |
| **P** | SN | SN | SP | P | P |

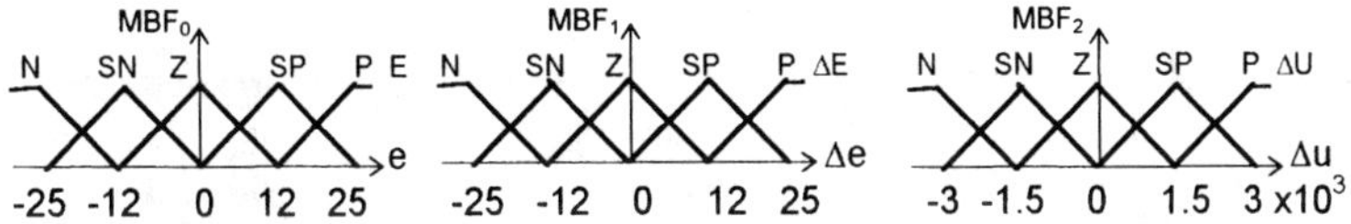


**Figure 2. MBFs for a two-input one-output as concrete implementation body.**

chromosomes. A population of candidates, $Y_0, \ldots, Y_{popsz\text{-}1}$, is manipulated by GA operations of mutation, crossover, and selection. Let $G$ represent a combination of these operations. In each generation, $G(Y_0, \ldots, Y_{popsz\text{-}1}) \Rightarrow Y_0', \ldots, Y_{popsz\text{-}1}'$, selection operation which determines a new set of chromosomes, is based on the evaluation of a fitness function, $f(Y) = \Sigma e^2$, where $e$ denotes the difference between the desired performance and the actual performance of $Y$.

*Neural networks and their training algorithms (NNs)* are used to fine tune, generalize, and store candidate solutions from a large set of numeric inputs by adjusting weights of network edges. A multi-layer feedforward neural network can be trained by an *error back-propagation algorithm (BP)* (Figure 3). Let $X = \{x_0, x_1, \ldots, x_{Ni\text{-}1}\}$ represent an input with $N_i$ attributes (input nodes). Let $Y_d = \{y_0, y_1, \ldots, y_{No\text{-}1}\}$ represent a desired solution with $N_o$ attributes (output nodes). Each training cycle includes a feedforward phase in which a candidate solution $Y$ is generated and an error back-propagation phase with edge weight updating. After being trained, the network is expected to come up with a reasonable $Y'$ efficiently for any new $X'$. It is also expected that if $X'$ is close to a given $X$ in the training examples, then $Y'$ is also close the $Y_d$ of $X$.

The purpose of *testing* is to find effective *fault scenarios* used to locate the possible *erroneous fuzzy elements (EFEs)* inside the component at the concrete level. An *EFE* is defined as a miss-used element in a FL component, such as an un-tuned linguistic value. One wishes to have the following functionalities during testing.

- Identify the new environment.
- Generate an applicable fault scenario.
- Execute the fault scenario for duplicating the problem causing the fault.
- Locate the EFEs of the component.

To achieve these functionalies, we combine a GA and a NN with BP (GA-BP) in the following four steps. (1) We use a GA to search training examples (Figure 4). A training example consists of a *circumstance* (environment $env_j$ and software component at abstract level $fam_k$) and a *solution* (fault scenario $fs_i$). This saves a trainer's effort in finding these examples. A GA is effective in finding an optimal solution in a complex environment in the expense of time. (2) We train a 3-layer network using a BP algorithm with the training

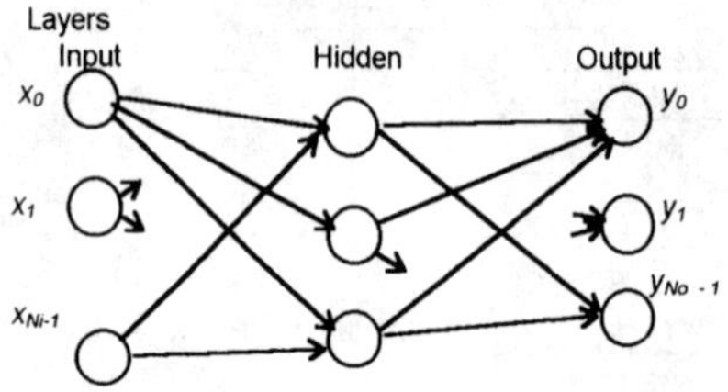

Figure 3. A 3-layer feedforward neural network.

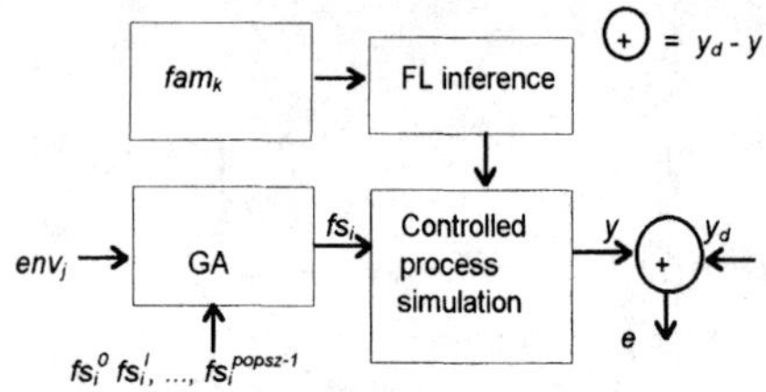

Figure 4. Finding a training example via a GA.

exmples found. (3) We use this trained NN to generate (recall) a solution under a new circumstance. This method is expected to be more efficient than running a GA for each new circumstance. (4) The fault scenario is executed for locating the EFEs. A EFE causes a significant error when it is used.

## EFFECTIVENESS OF THE TESTING METHOD

A test is the application of a fault scenario. The effectiveness of a test is measured by counting the number of EFEs uncovered in the fuzzy software component by executing the fault scenario. A testing method may apply a set of tests. The effectiveness of a testing method is measured by the effectiveness of each of its tests.

A test in this paper is one of two types: GA-only and GA-BP. First, let us consider a testing method with GA-only tests. Under any new circumstance $c$, the GA-only testing method needs a searching time $t_c(GA)$ to find an effective fault scenario.

Second, let us consider a testing method with GA-BP tests. For the GA-BP testing method, the method needs searching time for $n_T$ training examples plus NN training time $n_T * t_c(NN)$. $t_c(NN)$ is the time spent for each training example during training. Fault scenario recalling time $t_r(NN)$ through the NN can be ignored. Our experimentation with these two testing methods shows that the numbers of EFEs found by these two methods individually are approximately the same. Hence, the more times $n_N$ the NN trained by the GA-BP method is used, the more effort (time) can be saved in generating fault scenarios to be used in the GA-BP tests. The total effort (time) per circumstance $c$ using this GA-BP testing method, $(n_T * t_c(NN) + n_T * t_c(GA)) / n_N$, will be less than the total effort (time) per circumstance $c$ using the GA-only testing method $t_c(GA)$. This can be shown mathematically as follows:

$$(n_T * t_c(NN) + n_T * t_c(GA)) / n_N < t_c(GA) \quad (2)$$

$$n_T (t_c(NN) / t_c(GA) + 1) < n_N \quad (3)$$

When equation (3) is satisfied, the GA-BP method takes less effort than the GA-only method.

## EXPERIMENTS

We implemented a program in C++ on a 486 PC to carry out the testing method. The controlled process is simulated according to a vehicle mathematical model,

$$dx / dt = Ax + Bu + Rv \quad (4)$$

$$y = x \quad (5)$$

where $A = -2ky_d / m$, $B = 1 / m$, $R = \mu\, g \sin v_0 - g \cos v_0$. Here $v$, $y_d$, $\mu$, and $m$ are the same meanings as defined in the Introduction. $v_0$ is the starting $v$ and $g$ is gravity. The actual speed $y$ can be obtained by solving this differential equation. Each trainable component has a FAM similar to that specified in Table 1. There are four FAMs and one set of MBFs for interpreting all FAMs (Figure 2). There are ten environment descriptions. An environment is defined by the min-max value ranges of its attributes $(v, y_d, \mu, m)$. A fault scenario *fs* is a sequence with two states, $st_0$ and $st_1$. Each state is described by an environment instance. The GA parameters are as follows. Crossover rate is 0.6. Mutation rate is 0.001. Population size is 10. Maximum number of generations is 500. Structure (bit string) length is 160 bits (20 bits per variable in the fault scenario). There is a total of 40 fault scenarios generated. A fault scenario is selected as a training example if it causes a significant error, e.g. $\Sigma e^2$ is 500.0. The NN parameters are as follows. Number of layers is 3. Numbers of nodes per layer are 33 (input layer), 60 (hidden layer), and 8 (output layer). In the input layer, 8 nodes are for environment descriptions, and 25 nodes are for representing FAM. In the output layer, there is one node per variable in each *fs*. Learning rate for updating all edge weights is 0.02. Activation for each node is a sigmoid function, $f(x) = 1 / (1 + e^{-x})$. Inputs to the input layer are linearly normalized to the value range [-1, 1], while outputs from the output layer are linearly denormalized from the value range [0, 1]. After the NN is trained, a set of 40 new circumstances are presented to the NN for recalling fault scenarios. Fault scenarios obtained by using NN and GA respectively are shown in Table 2 and 3. Only 3 *fs*'s for each testing method are shown for demonstration. These fault scenarios for the same circumstance by different method are very close (observation 1). Finally, we execute these fault scenarios. During a composition rule inference, if there are $n$ rules executed, a $|\Delta e|$ is accumulated for each of these $n$ rules. The fuzzy rule usage

**TABLE 2. A SET OF FAULT SCENARIOS GENERATED FROM A NN**

| *circum* | $st_0$ | | | | $st_1$ | | | |
|---|---|---|---|---|---|---|---|---|
| *or fs#* | $v$ | $y_d$ | $\mu$ | $m$ | $v$ | $y_d$ | $\mu$ | $m$ |
| 1 | 0.699 | 1.600 | 0.046 | 9200.6 | 0.6889 | 1.6788 | 0.081 | 6441.8 |
| 2 | 0.69576 | 1.2782 | 0.0015 | 9336.6 | 0.7086 | 2.63798 | 0.0015 | 6757.66 |
| 3 | 0.68687 | 0.9596 | 0.0672 | 9458.75 | 0.7099 | 1.61428 | 0.0877 | 6969.93 |

**TABLE 3. A SET OF FAULT SCENARIOS FOUND BY A GA**

| *circum* | $st_0$ | | | | $st_1$ | | | |
|---|---|---|---|---|---|---|---|---|
| *or fs#* | $v$ | $y_d$ | $\mu$ | $m$ | $v$ | $y_d$ | $\mu$ | $m$ |
| 1 | 0.6867 | 2.2085 | 0.0314 | 10076.4 | 0.6126 | 2.84487 | 0.10429 | 6346.48 |
| 2 | 0.69269 | 1.0803 | 0.0005 | 10100.0 | 0.6204 | 3.26656 | 0.0015 | 7491.85 |
| 3 | 0.69188 | 1.1488 | 0.0882 | 10026.8 | 0.7694 | 1.69279 | 0.0918 | 7554.63 |

**TABLE 4. A COMPARISON OF THE MOST SIGNIFICANT *EFES* UNCOVERED**

| *circum* | Using GA-BP method | | | | Using GA-only method | | | |
|---|---|---|---|---|---|---|---|---|
| *or fs#* | *e&de->du* | *uc* | *e_val* | *e_rate* | *e&de->du* | *uc* | *e_val* | *e_rate* |
| 1 | ZE&PO->SP | 1 | 2.268681 | 2.268681 | ZE&PO->SP | 1 | 2.866558 | 2.866558 |
| 2 | ZE&PO->SP | 1 | 1.943881 | 1.943881 | ZE&PO->SP | 1 | 1.744166 | 1.744166 |
| 3 | ZE&PO->ZE | 1 | 1.632775 | 1.632775 | ZE&PO->ZE | 1 | 1.826837 | 1.826837 |

statistics are shown in Table 4, where *uc* is the usage counter, *e_val* is the $\Sigma|\Delta e|$, and *e_rate* is *e_val / uc*. It is indicated in Table 4 that both methods can find out the same (number of) EFEs with very close *e_rate*s (observation 2). These two observations show that the GA-BP method is as effective as the GA-only method. In our experiment, $t_c(NN)$ = 9.7 sec. and $t_c(GA)$ = 24 sec. From equation (3), if $n_N > (9.7 / 2.4 + 1) * n_T$ or $n_N > 1.404 * n_T$, then GA-BP method is more effecitient than GA-only method.

## CONCLUSION AND FUTURE WORKS

An effective testing method is presented which combines GA and NN to test FL based trainable software components. A program is implemented to automatically adapt to new circumstances, to generate fault scenarios for an unseen circumstance, and to locate the erroneous fuzzy elements. This work is the first phase under our training framework to develop reusable software components (Chen and Rine, 1995). The next phase of the training framework is to adapt the fuzzy software component based on the testing result.

## REFERENCES


Chen, J., and Rine, D.C., (1995). A training framework to develop reusable software components by combining adaptation algorithms, *Proceedings of Artificial Neural Networks In Engineering (ANNIE'95) Conference*, 159-165, Sponsored by the Office of Naval Research, St. Louis, MO.

Karr, C.L., (1993). Fuzzy control of pH using genetic algorithms, *IEEE Trans. on Fuzzy Systems,* vol. 1, no.1, 46-53.

Lin, C.-T., and Lee, C.S.G., (1994). Supervised and unsupervised learning with fuzzy similarity for neural network-based fuzzy logic control systems, in *Fuzzy Sets, Neural Networks, and Soft Computing* (R.R. Yager, and L.A. Zadeh, Eds.), Van Nostrand Reinhold, 85-125.

Polucarpou, M.M., and Helmicki, A.J, (1995). Automated fault detection and accommodation: a learning systems approach, *IEEE Trans. on Neural Networks,* vol., 25, no. 11, 1447-1458.

Rine, D.C., (1993). Software perfective maintenance: including retrainable software in software reuse, *Information Sciences: International Journal*, vol. 75.

Rine, D.C., Ahmed, M., and Chen, J., (1996). A reusable software adaptive fuzzy controller architecture, *Proceedings of the International Conference on Software Applications,* Sponsored by the ACM, Feb. 15-17.

Rodríguez, C., Rementería, S., Martín, J.I., Lafuente, A., Muguerza, J., and Pérez, J. (1996). A modular Neural Network approach to fault diagnosis, *IEEE Trans. on Neural Networks,* vol., 7, no. 2, 326-340.

Schultz, A., Grefenstette, J., and De Jong, K., (1993). Test and evaluation by Genetic Algorithms, *IEEE Expert*, Oct. 9-14.

Simpson, W.R., and Sheppard, J.W., (1994). *System Test and Diagnosis,* Kluwer Academic Publishers.

Thomas, M.D., and Flockton, S.J., (1994). Fault detection in a mechanical system by neural networks. *Proceedings of SPIE* - The International Society for Optical Engineering, vol. 2361, 55-58.

# VLSI PLACEMENT USING GENETIC ALGORITHMS

**K. W. Leavitt, A. R. Marudarajan, and A. M. Nouhi**,
Electrical and Computer Engineering Department, California State Polytechnic University, Pomona, California

**H. N. Riley**
Computer Science Department, California State Polytechnic University, Pomona, California

## Abstract

In this paper, an attempt is made to find solutions to three VLSI placement optimization problems using standard and non-standard genetic techniques. The authors chose the VLSI placement problem because the size of its search space makes the problem intractable for even relatively small numbers of modules, and because sub-optimal layout must therefore be acceptable in VLSI design. Genetic algorithms were chosen due to their inherent parallelism, stochastic search capabilities, and the ability to define constraints within the objective function. GA techniques applied include proportional selection and crossover, as well as two non-standard approaches; master race initialization and sigmoidal mutation. Candidate solution populations were evolved for three progressively larger VLSI layout configurations to analyze the scalability of the GA techniques. The results of these runs are compared with a Hopfield neural network-derived optimization of the same VLSI configurations. The results of this research show that genetic algorithms are particularly well suited for finding acceptable solutions to very complex VLSI placement problems.

**Indexing Terms**: Genetic Algorithms, Optimization, VLSI Placement, Evolution

**I. Characterization of the VLSI Placement Problem**. There are two key design optimization problems in VLSI circuit layout, area minimization and cell placement. The cell placement problem may be simplified to optimizing the placement of $n^2$ modules in an n x n matrix of slots such that the total length of the interconnect lines is minimized. Other constraints may be imposed such as an interconnect distribution function and maximum lengths for specific connections. Considering only total interconnect length, the complexity of this problem is a function of the number of possible placements. Since order matters here, the problem is actually one of permutation rather than combination, and complexity is given by O(n!). Since the complexity function is $n^2!$, the problem becomes intractable for even small values of n. As an example, for a 4 x 4 matrix with a total of 16 slots, there are over 20 trillion possible placement permutations, and 10 x 10 placements are not uncommon. While it may not be practical to determine a globally optimal placement, a sub-optimal placement may be sufficient for purposes of the VLSI cell placement problem. (Date, Seki and Hayashi)

VLSI problems of successively larger scale were chosen for this effort. A 4x4 configuration analyzed (M.R Zargham, 1996), as well as 5x5 and 6x6 layouts were optimized using GA techniques to ascertain their scalability. The complexity of these problems is proportional to the size of their search spaces, which for the 4x4, 5x5 and 6x6 problems are 2.09E13, 1.55E25 and 3.72E41 respectively.

**II. Genetic Algorithm Fundamentals**. Genetic algorithms offer a stochastic search mechanism capable of finding near optimal, and sometimes globally optimal solutions to combinatorial optimization problems such as VLSI cell placement. GA's borrow evolutionary techniques from the natural world such as selection, crossover and mutation to evolve generations of potential solutions to a given problem. To apply a GA techniques, one needs only to encode potential solutions in strings (or chromosomes), and to evaluate and rank solutions produced by the GA against an objective function. Properly implemented, the Simple Genetic Algorithm (SGA) will take a random population of solutions and evolve successive generations of progressively better solutions to the problem through the following steps:

Step 1.) Initialize a random population of strings (potential solutions to the given problem).

Step 2.) Assign each string a fitness value based on its performance of the objective function. If a sufficient solution is produced within this population, then we are done. Otherwise;

Step 3.) From this population, select the parents of the next generation. The probability of any string's selection will be a function of its fitness value.

Step 4.) Create the strings for the next generation by crossing the genetic structure of the selected parent strings.

Step 5.) If the genetic content of strings within the new population surpasses a pre-established threshold of internal similarity, mutate (randomly exchange) a fixed number of genes.

Step 6.) Return to step 2.

Selecting appropriate parameter variables is critical to the performance of the GA in optimization. The VLSI optimization problem may be viewed as an optimization of an underlying graph that represents the circuit topology. A graph may explored in two ways: depth first and breadth first. The depth first search process explores nodes in the inverse order of their creation. The breadth first search process explores nodes in the order of their creation. The GA implements a combination of these approaches by balancing the explorative elements of evolution (crossover and mutation) with the exploitative (selection). Too much stasis, and the population stagnates on a sub-optimal solution. Too much dynamism, and the population degrades to random noise.

**III. Genetic Algorithms vs. Other Optimization Techniques**. Hopfield neural networks (Hillier and Lieberman, 1990) have also been successfully applied in optimization problems. These are recurrent nets because of the existence of feedback paths within the network. They are fully connected networks in that every unit receives input from every other unit. Each unit computes the weighted sum of its inputs to generate a net input. Similarly, each of these units applies a non-linear activation function such as hyperbolic tangent function on its net input in order to produce its output that is sent to all the other units. Hopfield has shown that for a given network, there is an energy function (or Lyapunov function) that depends on the values of the weight matrix. In particular, if the network outputs are mapped to the decision variables of an optimization problem; and, if the weights are chosen so that the energy of the network can be identified with variables of the optimization problem; then the network can solve the problem by minimizing the energy function and finding the associated stable output values of the units in the network. However, there are generally three problems associated with Hopfield networks:

1. The size of the weight matrix increases quadratically with problem size.
2. The network may get stuck in a local minimum
3. The network may even oscillate without converging to the optimal solution.

While Hopfield nets have been applied to optimize small VLSI placements (M.R Zargham, 1996), these problems diminish the utility of this approach as the scale of the problem grows. As a result, this approach does not guarantee a solution for a large size problem.

Simulated annealing (Kirkpatrick, Gelatt and Vecchi, 1983), as it is applied to the optimization of a circuit placement problem, uses an algorithm developed in statistical mechanics to find the lowest-energy configuration of the atoms and molecules of a solid material in thermal equilibrium. The mathematical model used is based on the experiment of growing a material into a single crystal structure, a process that requires a very elaborately controlled thermal annealing. The growth process starts with first melting the material, then lowering its temperature very slowly, and spending a relatively long time at temperatures in the vicinity of the freezing point in order to achieve a highly ordered configuration having the lowest energy. Finding the low-temperature state of such a material system with minimum energy is very similar to those encountered in combinatorial optimization problems like an optimized placement in a VLSI circuit . It is indeed this similarity that led to the development of simulated annealing process. However, in the absence of an obvious physical temperature equivalent in this type of problems, a new system parameter, called an effective temperature, is defined. The equivalent of system energy is the cost function of the system that is calculated as a function of the effective temperature. The simulated annealing process consists of first melting the system being optimized at high effective temperature, then lowering the temperature in small decrements until the system freezes and no further changes occur. Simulated annealing techniques are very similar to those of GAs, in that they implement evolutionary processes modeled accurately by stochastic means (Kirkpatrick, Gelatt and Vecchi, 1983).

Branch and bound (Brassard and Bratley , 1996) is yet another useful technique for pruning the search space. In this method, at each node a bound is calculated on the possible value of any solutions that may lie on farther on in the graph. If the bound shows that it is worse than the best solution found so far, then there is no need to explore that part of the graph further. On many occasions, the calculated bound guides one to choose which open path looks most promising so that it can be explored first. Since a VLSI cell placement problem is a kind of assignment problem, it is possible to calculate bounds by adding the smallest values in each row or column of the cost matrix. Philosophically, branch and bound is an enumeration technique that underscores the divide and conquer principle. Since it is only a heuristic, there is no guarantee that it would converge very quickly to an optimal solution. Since the computational complexity of a branch and bound algorithm is proportional to the scale of the problem, its utility diminishes as the scale of the problem grows. Conversely, the random mechanism inherent in a GA prevents it from exhaustive searches of all nodes of a search tree. At the same time, its exploitative mechanisms keep the GA focused on only the most productive search space areas. As a direct result of these factors, The GA can produce very strong solutions after exploring extremely small percentages of a vast search space

## IV. Methodology for Genetic Algorithm Implementation of the VLSI placement problem.

The GA was developed to optimize VLSI placement problems of any scale, and was run on three layouts; 4x4, 5x5 and 6x6. The VLSI GA was based on the SGA, and contains the following steps:

**Step 1.) Population Initialization.** The population size chosen for the VLSI GA was 30, the lower bound of a small genetic population. Each chromosome consisted of N genes (where N = 16, 25 or 36 in the problems we chose), each locus representing its respective slot in the circuit layout. The alleles (genetic instances) consisted of random placements of the numbers one through 16 with no duplicates. These represented a specific module placement in each slot. As an example, a chromosome such as 4, 11, 6, 1, 12 ... would represent module 4 in slot 1, module 11 in slot 2, module 6 in slot 3, etc. The user also has the option of initializing the first generation with the best 30 chromosomes produced in previous runs. This tends to significantly reduce the average and minimum string lengths, but mutation quickly neutralizes this advantage. The goal here is to produce even shorter chromosome lengths before the mutation operation kicks in.

**Step 2.) Fitness Assessment**. The objective function chosen for this problem is quite simple; sum the length of interconnects between modules in a given chromosome, and you have its fitness. This is accomplished by establishing the number of connections between any two modules and multiplying this number by the distance between the slots in which those modules reside, as contained in a look-up table. Several other calculations are performed in this subroutine such as minimum and maximum population fitness and a maximum cutoff length for crossover selection. If one or more chromosomes are evaluated at a pre-established acceptable level, the algorithm terminates.

K. Shahookar and P. Mazumder (1991) evaluated a scoring (or objective) function which measured not only a string's total interconnect length, but the distribution of interconnects as well. The scoring function penalized all channels with a wiring density greater than one standard deviation above the population mean, thus encouraging more uniform wiring distributions. This ability to perform this type of constrained optimization is a key strength of the GA.

**Step 3.) Parent Selection.** Proportional selection was implemented as the mechanism for parent selection. That is, a string with a fitness value in the upper 25% of the population range will have a 75% probability of being selected as a parent at least once. The user may specify a minimum percentage cutoff value as a selection limitation criteria. The default value of 75% will reject any chromosome whose length lies within the longest 25% of lengths in its generation. This way, each chromosome competes within the standards set by its own population. A random number is generated between this minimum percentage (maximum length) cutoff value and a length slightly greater than the minimum chromosome length within the population. As an example, parent candidate one is randomly selected. If it has a length less than or equal to the previously mentioned random number, it qualifies as a parent. If not, another selection is made. Parent 2 must meet the same selection criteria plus one more; it must not be parent 1.

The user may also establish a maximum number of times any one parent may be selected for crossover. The higher this index, the more quickly the building blocks of very short strings will come to dominate future generations (high exploitation). The lower this index, the more strings of longer length will be allowed to crossover and propagate into future generations (high exploration)

**Step 4.) Offspring Generation.** The crossover operator chosen was similar to that implemented by Cohoon and Paris (1986) in that of the two parents selected for crossover, one is designated as a "target" parent, and the other as a "passing" parent. A random crossover mask is generated to split the passing parent into two sub-strings. For a parent of length N, the length of each section $N_s$ is between 2 and N-2. This insures a minimal schema (two genes) remains intact from each parent, and thus building blocks within the population will be given some shelter from the disruptive effects of crossover. The genetic content of the left sub-string is analyzed and exchanged between parents. While the genes extracted from each parent are the same, the sequence is likely to be unique, and it is the sequence which determines the strength or weakness of a particular schema, and thus the likely strength of the offspring.

Figure 1 depicts this crossover technique for two parents with N = 16 and the mask chosen at 6. The first six genes in parent 1 are re-arranged to their respective sequence in parent 2 for the creation of offspring 1. The same 6 genes in parent 2 are re-arranged to their respective sequence in parent 1 for the creation of offspring 2.

| | | | | | | | | |
|---|---|---|---|---|---|---|---|---|
| 5 | 12 | 7 | 4 | | 8 | 4 | 7 | 5 |
| 1 | 8 | 9 | 11 | | 12 | 1 | 9 | 11 |
| 15 | 3 | 14 | 16 | | 15 | 3 | 14 | 16 |
| 10 | 6 | 13 | 2 | | 10 | 6 | 13 | 2 |
| | | | | | | | | |
| 11 | 2 | 8 | 4 | | 11 | 2 | 5 | 12 |
| 7 | 5 | 13 | 15 | | 7 | 4 | 13 | 15 |
| 16 | 12 | 1 | 10 | | 16 | 1 | 8 | 10 |
| 3 | 9 | 14 | 6 | | 3 | 9 | 14 | 6 |

Parents 1 & 2 Offspring 1 & 2

**FIGURE 1**

**PARENTS AND OFFSPRING FOR N = 16 AND MASK VALUE 6.**

Note that the specific genes exchanged are the same for both parents, but the sequence is unique. Thus parental strengths and weaknesses are passed to the next generation

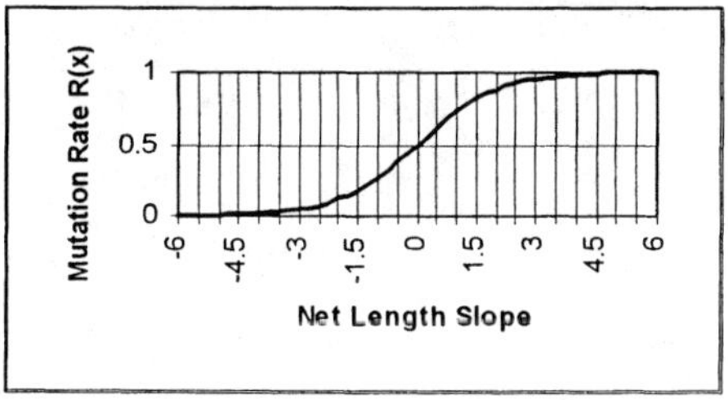


**FIGURE 2**
Sigmoidal Mutation with

$$\text{Mutation Rate } R(x) = \left( \frac{1}{1 + e^{-\sigma x}} \right)$$

In practice, we used a sigma value of 3, and only applied sigmoidal mutation if the net length slope was negative over the past two generations. The full mutation rate was applied if the slope was zero or positive.

**Step 5.) Mutation.** As building blocks start to dominate populations, future generations evolve to better average and minimum net lengths. However, as always, exploitation comes at the expense of exploration, and stronger populations start to become in-bred. Once all 30 strings in a population contain the same gene at a particular locus, no amount of crossover will ever change this, and the population stagnates at a local (sub) optimum. Our VLSI GA tracks a similarity index which ranges from 1 to 100 and represents the lack of diversity within each population. A population with all possible genes represented at least once in every locus across all chromosomes will have a similarity index of zero. A population with every locus dominated by a single gene (30 instances of the same chromosome) will have a similarity index of 100. Once the similarity index of a sequence of generations crosses a pre-set threshold, one of two mutation factors is activated.

The simplest of these mechanisms is a preset mutation rate between 1 and 100%. After the similarity threshold is crossed, a percentage of the genes within the population is randomly mutated. This was implemented through switching the loci of two or more genes within strings to preserve overall genetic content. A small mutation rate continues to explore the search space in the genetic vicinity of the last populations, and a high mutation rate has the effect of resetting the population to a random content.

The problem with the flat percentage mutation approach is that the population may be genetically reset regardless of the progress it may be making toward the production of shorter net lengths. After observing the performance of this flat percentage mutation rate, we implemented a sigmoidal variation which applies mutation more judiciously. Using sigmoidal mutation, the rate $R$ is set as a function of the local slope of the net length curve and a modulation factor $\sigma$. The intent is to apply mutation more gradually if the populations are evolving toward a better solution. Figure 2. depicts the sigmoidal variation of the mutation factor as a function of the fitness slope.

After a new population is established through selection, crossover and mutation where applicable, the algorithm returns to step 2, where the genetic operators are cyclically applied, and building blocks compete to produce stronger future generations. The strings of highest fitness value from each generation are stored to a database for statistical analysis.

**V. Results**. The GA was run on the 4x4 case for a total of 500K populations, varying all parameters including initialization technique, mutation rate, similarity threshold and minimum acceptable string fitness. When the population was initialized with random gene placement, this resulted in average chromosome lengths of around 105 units; more than twice the minimum results achieved by Zargham. For 30 - 40 generations, (dependent upon where the mutation threshold was set) the average and minimum net lengths would drop exponentially, while the population similarity indices increased. Once the similarity threshold was breached, the mutation factor kicked-in and average and minimum net lengths quickly returned to their random levels. The GA cycle then repeated itself in a different area of the problem search space. This cycling is depicted in figure 3.

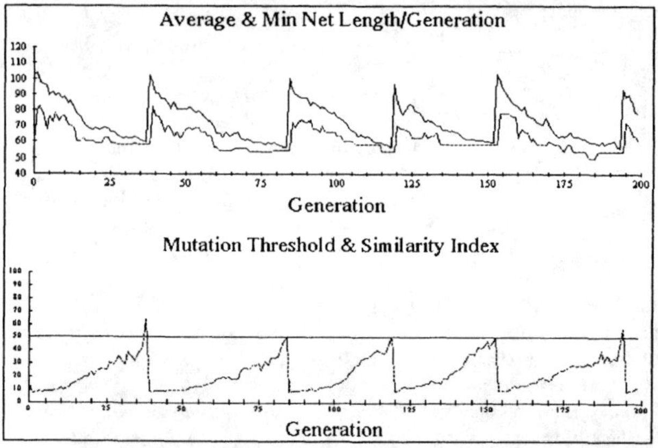


**FIGURE 3**

4x4 VLSI average and minimum net length curves, and the similarity index curve for a 200 generation run with a mutation threshold of 50%.

Note the strong negative correlation between the net length curves and the similarity index. Also notable is the fact that the average net length curve is often still decreasing when the similarity index breaches the mutation threshold.

After cycling through 500K generations of chromosomes, the GA had evaluated 15 million candidate solutions. Of these, 35 unique solutions had total net lengths of 46. This result was very unusual, given the stochastic mechanisms of the GA. It is highly unlikely there would exist this many independent solutions exactly four units longer than Zargham's solution, yet none shorter than 46. Upon running Zargham's solution through the GA net length evaluation routine, it was discovered that his published solution of 42 was incorrect, and that in fact 46 is minimal for this configuration. Zargham has confirmed that this was due to a typo in the publisher's representation of his 4x4 configuration.

Zargham's Hopfield neural network simulation produced a minimal configuration (adjusted for publication error to 46 units in length) in 149 iterations. Clearly, this is an improvement in efficiency over the GA, in terms of both coding and run-time brevity. However, the GA was able, not only to duplicate the global optimum produced by the NN, but it also produced 34 other unique and equally optimal solutions. These solutions could in turn, be optimized for additional criteria such as layout distribution and area minimization. The optimal solutions produced by the GA are contained in appendix A. It is noteworthy that in every optimal solution, the GA positioned modules 1, 2, 3 and 4 in adjacent slots. Due to the high level of interconnectivity between these cells, they constitute a high fitness value schema of short defining length, or a building block.

While the evaluation of 15 million possible solutions to this problem seems like a substantial effort, consider that this represents only $.717 \times 10^{-6}$ % of the search space for the 4x4 problem. In light of this efficiency, the scalability of the GA to problems with substantially larger search spaces has very significant implications, as demonstrated by our results on the larger problems.

The 5x5 and 6x6 placement problems were also run for 500K generations. At 15 million strings each, these runs examined $.97 \times 10^{-18}$ % and $.4 \times 10^{-34}$ % of the respective search space of each design problem. All parameters were also varied throughout these iterations. It is noteworthy that the optimal solutions produced for both problems resulted from runs initialized by a master race of previous best results, and with sigmoidal mutation implemented. The GA produced optima for the 5x5 and 6x6 problems with total net lengths of 1878 units and 1995 units respectively. These are likely sub-optimal solutions, as substantial progress (2-5 units) was still being made with each new run.

**VI. Conclusions**. This research has demonstrated both the strengths and weaknesses of genetic algorithms in optimization. Due to their coding complexity, and stochastic mechanisms, they are generally too slow for applications requiring real-time decisions and are inefficient at producing solutions to relatively small optimization problems. Mature algorithms such as backtracking, branch and bound, and Hopfield Neural Network simulators offer better approaches to these types of problems. However, as optimization problems grow in size and complexity, these same weaknesses become the GA's strengths. Stochastic approaches to crossover and mutation prevent the GA from stalling on sub-optimal solutions, or oscillating around a single area of a problem's search space. Inherent parallelism allows a GA to propose and evaluate thousands of schema simultaneously, and quickly move toward the most promising solutions to a problem of vast size and complexity. The objective function determines this direction, and the criteria it implements may range from simple to compound.

This research effort has also demonstrated the effectiveness of two techniques which further enhance the GA's effectiveness in solving extremely large scale problems with multiple inter-dependent variables. Master-race initialization affords a method of accelerating the exploitative mechanisms of a GA. Sigmoidal mutation provides a method of delaying the disruptive effects of mutation while successive generations are still making progress toward a globally optimal solution.

**Acknowledgments**. The authors wish to thank Prof. R.H. Cockrum, Chairman, Electrical and Computer Engineering Dept., College of Engineering, California State Polytechnic University, Pomona for his encouragement and support in the development of this paper. We also thank all members of the ANNIE organizing committee and referees.

## REFERENCES

M.R Zargham, Computer Architecture, Single and Parallel Systems, Prentice Hall, 1996

J.P. Cohoon and W.D. Paris, *Genetic Placement*, IEEE International Conference on Computer-Aided Design, pp. 422-425, 1986

H. Date, M. Seki and T. Hayashi, *LSI Module Placement Methods Using Neural Computation Networks*, Hitachi Research Laboratory, Hitachi Ltd.

K. Shahookar and P. Mazumder, *VLSI Cell Placement Techniques*, ACM Computing Surveys, Vol 23, No. 2, June 1991

G. Brassard and P. Bratley, Fundamentals of Algorithms, Prentice-Hall, 1996

S. Kirkpatrick, C. D. Gelatt, P. P. Vecchi, *Optimization by Simulated Annealing*, Science, 13 May, 1983, Volume 220, Number 4598

F. S. Hillier and G. J. Lieberman, Introduction to Operations Research, McGraw-Hill (Fifth Edition), 1990

### Appendix A: Optimal Placements Produced by the GA for each VLSI Problem

Optimal placements consist of the series of module-slot placements for the chromosome and the total net length of the placement solution. Each column represents one slot in the 4x4 matrix, while each number represents a cell placement within that slot.

Optimal solutions produced for the 4x4 layout (n=4):

| 1 | 2 | 3 | 4 | 5 | 6 | 7 | 8 | 9 | 10 | 11 | 12 | 13 | 14 | 15 | 16 | NetLength |
|---|---|---|---|---|---|---|---|---|---|---|---|---|---|---|---|---|
| 5 | 6 | 7 | 16 | 9 | 13 | 14 | 15 | 10 | 12 | 4 | 3 | 8 | 11 | 1 | 2 | 46 |
| 11 | 12 | 13 | 15 | 8 | 10 | 9 | 16 | 1 | 2 | 6 | 14 | 4 | 3 | 7 | 5 | 46 |
| 13 | 12 | 10 | 11 | 15 | 6 | 9 | 8 | 16 | 7 | 1 | 4 | 14 | 5 | 2 | 3 | 46 |
| 13 | 12 | 11 | 10 | 15 | 6 | 8 | 9 | 16 | 7 | 3 | 4 | 14 | 5 | 2 | 1 | 46 |
| 16 | 7 | 6 | 5 | 15 | 10 | 9 | 8 | 14 | 11 | 1 | 2 | 13 | 12 | 4 | 3 | 46 |
| 5 | 6 | 7 | 14 | 9 | 13 | 16 | 15 | 10 | 12 | 4 | 3 | 8 | 11 | 1 | 2 | 46 |
| 12 | 10 | 9 | 8 | 13 | 7 | 6 | 5 | 11 | 15 | 1 | 2 | 14 | 16 | 4 | 3 | 46 |
| 2 | 3 | 7 | 6 | 1 | 4 | 5 | 9 | 14 | 15 | 16 | 8 | 11 | 13 | 12 | 10 | 46 |
| 5 | 6 | 7 | 16 | 9 | 13 | 15 | 14 | 10 | 12 | 1 | 4 | 8 | 11 | 2 | 3 | 46 |
| 11 | 13 | 12 | 8 | 15 | 6 | 10 | 9 | 16 | 7 | 4 | 3 | 14 | 5 | 1 | 2 | 46 |
| 5 | 6 | 7 | 16 | 13 | 9 | 14 | 15 | 12 | 10 | 3 | 2 | 11 | 8 | 4 | 1 | 46 |
| 8 | 6 | 7 | 5 | 9 | 15 | 16 | 14 | 10 | 12 | 1 | 4 | 11 | 13 | 2 | 3 | 46 |
| 10 | 12 | 13 | 11 | 9 | 16 | 15 | 14 | 6 | 7 | 2 | 1 | 8 | 5 | 3 | 4 | 46 |
| 6 | 9 | 10 | 8 | 7 | 13 | 12 | 11 | 5 | 14 | 4 | 1 | 16 | 15 | 3 | 2 | 46 |
| 1 | 4 | 7 | 5 | 2 | 3 | 6 | 14 | 8 | 10 | 9 | 16 | 11 | 12 | 13 | 15 | 46 |
| 2 | 3 | 6 | 5 | 1 | 4 | 7 | 16 | 8 | 10 | 9 | 15 | 11 | 12 | 13 | 14 | 46 |
| 3 | 4 | 12 | 13 | 2 | 1 | 11 | 15 | 8 | 9 | 10 | 14 | 5 | 6 | 7 | 16 | 46 |
| 7 | 6 | 5 | 9 | 16 | 12 | 10 | 8 | 15 | 13 | 3 | 2 | 14 | 11 | 4 | 1 | 46 |
| 7 | 6 | 5 | 9 | 16 | 12 | 8 | 10 | 15 | 13 | 1 | 2 | 14 | 11 | 4 | 3 | 46 |
| 7 | 6 | 5 | 9 | 16 | 12 | 10 | 8 | 15 | 13 | 1 | 2 | 14 | 11 | 4 | 3 | 46 |
| 13 | 12 | 11 | 8 | 15 | 6 | 10 | 9 | 16 | 7 | 3 | 4 | 14 | 5 | 2 | 1 | 46 |
| 8 | 9 | 10 | 11 | 5 | 6 | 7 | 12 | 3 | 2 | 14 | 13 | 4 | 1 | 16 | 15 | 46 |
| 15 | 13 | 12 | 11 | 14 | 9 | 10 | 8 | 16 | 6 | 4 | 3 | 5 | 7 | 1 | 2 | 46 |
| 10 | 12 | 13 | 11 | 8 | 16 | 15 | 14 | 9 | 5 | 1 | 2 | 6 | 7 | 4 | 3 | 46 |
| 11 | 7 | 6 | 5 | 12 | 10 | 9 | 8 | 13 | 16 | 4 | 3 | 15 | 14 | 1 | 2 | 46 |
| 8 | 6 | 7 | 5 | 9 | 15 | 16 | 14 | 10 | 11 | 2 | 1 | 12 | 13 | 3 | 4 | 46 |
| 14 | 13 | 12 | 11 | 15 | 9 | 10 | 8 | 16 | 6 | 2 | 3 | 5 | 7 | 1 | 4 | 46 |
| 9 | 6 | 7 | 5 | 10 | 15 | 16 | 14 | 12 | 13 | 1 | 4 | 8 | 11 | 2 | 3 | 46 |
| 5 | 7 | 6 | 8 | 12 | 16 | 10 | 9 | 13 | 15 | 1 | 4 | 11 | 14 | 2 | 3 | 46 |
| 10 | 12 | 13 | 11 | 8 | 16 | 15 | 14 | 9 | 7 | 2 | 1 | 5 | 6 | 3 | 4 | 46 |
| 9 | 6 | 7 | 5 | 10 | 15 | 16 | 14 | 8 | 12 | 1 | 4 | 11 | 13 | 2 | 3 | 46 |
| 15 | 13 | 12 | 11 | 16 | 9 | 10 | 8 | 7 | 6 | 2 | 1 | 14 | 5 | 3 | 4 | 46 |
| 5 | 6 | 7 | 11 | 8 | 9 | 10 | 12 | 3 | 2 | 14 | 13 | 4 | 1 | 16 | 15 | 46 |
| 15 | 13 | 12 | 11 | 16 | 9 | 10 | 8 | 14 | 6 | 3 | 2 | 5 | 7 | 4 | 1 | 46 |
| 15 | 13 | 12 | 11 | 16 | 9 | 10 | 8 | 14 | 7 | 3 | 2 | 5 | 6 | 4 | 1 | 46 |

Best solution produced for the 5x5 layout:

| 1 | 2 | 3 | 4 | 5 |
|---|---|---|---|---|
| 19 | 21 | 6 | 4 | 17 |
| **6** | **7** | **8** | **9** | **10** |
| 8 | 23 | 25 | 2 | 15 |
| **11** | **12** | **13** | **14** | **15** |
| 12 | 10 | 14 | 13 | 11 |
| **16** | **17** | **18** | **19** | **20** |
| 16 | 3 | 1 | 24 | 22 |
| **21** | **22** | **23** | **24** | **25** |
| 5 | 18 | 20 | 7 | 9 |

| NetLength |
|---|
| 1878 |

Best solution produced for the 6x6 layout:

| 1 | 2 | 3 | 4 | 5 | 6 |
|---|---|---|---|---|---|
| 35 | 24 | 29 | 14 | 3 | 7 |
| **7** | **8** | **9** | **10** | **11** | **12** |
| 34 | 5 | 36 | 19 | 27 | 31 |
| **13** | **14** | **15** | **16** | **17** | **18** |
| 26 | 11 | 23 | 16 | 17 | 28 |
| **19** | **20** | **21** | **22** | **23** | **24** |
| 33 | 21 | 1 | 8 | 32 | 12 |
| **25** | **26** | **27** | **28** | **29** | **30** |
| 4 | 6 | 13 | 2 | 18 | 9 |
| **31** | **32** | **33** | **34** | **35** | **36** |
| 15 | 30 | 25 | 22 | 20 | 10 |

| NetLength |
|---|
| 1995 |

## Appendix B: Connectivity matrices for the 5x5 and 6x6 placement problems

These matrices (available from the primary author at keith.leavitt@trw.com) represent the connectivity of each VLSI circuit considered. The first row and column represent the cells 1-25 (or 1-36) and each entry represents the number of connections required between the cells denoted by the row and column headers. Note that the matrices are upper triangular. The 4x4 connectivity is discussed in the Zargham reference.

```
ID,1,2,3,4,5,6,7,8,9,10,11,12,13,14,15,16,17,18,19,20,21,22,23,24,25
1,0,1,0,2,0,3,0,4,0,5,0,6,0,7,0,8,0,9,0,10,0,11,0,12,0
2,0,0,1,0,2,0,3,0,4,0,5,0,6,0,7,0,8,0,9,0,10,0,11,0,12
3,0,0,0,1,0,2,0,3,0,4,0,5,0,6,0,7,0,8,0,9,0,10,0,11,0
4,0,0,0,0,1,0,2,0,3,0,4,0,5,0,6,0,7,0,8,0,9,0,10,0,11
5,0,0,0,0,0,1,0,2,0,3,0,4,0,5,0,6,0,7,0,8,0,9,0,10,0
6,0,0,0,0,0,0,1,0,2,0,3,0,4,0,5,0,6,0,7,0,8,0,9,0,10
7,0,0,0,0,0,0,0,1,0,2,0,3,0,4,0,5,0,6,0,7,0,8,0,9,0
8,0,0,0,0,0,0,0,0,1,0,2,0,3,0,4,0,5,0,6,0,7,0,8,0,9
9,0,0,0,0,0,0,0,0,0,1,0,2,0,3,0,4,0,5,0,6,0,7,0,8,0
10,0,0,0,0,0,0,0,0,0,0,1,0,2,0,3,0,4,0,5,0,6,0,7,0,8
11,0,0,0,0,0,0,0,0,0,0,0,1,0,2,0,3,0,4,0,5,0,6,0,7,0
12,0,0,0,0,0,0,0,0,0,0,0,0,1,0,2,0,3,0,4,0,5,0,6,0,7
13,0,0,0,0,0,0,0,0,0,0,0,0,0,1,0,2,0,3,0,4,0,5,0,6,0
14,0,0,0,0,0,0,0,0,0,0,0,0,0,0,1,0,2,0,3,0,4,0,5,0,6
15,0,0,0,0,0,0,0,0,0,0,0,0,0,0,0,1,0,2,0,3,0,4,0,5,0
16,0,0,0,0,0,0,0,0,0,0,0,0,0,0,0,0,1,0,2,0,3,0,4,0,5
17,0,0,0,0,0,0,0,0,0,0,0,0,0,0,0,0,0,1,0,2,0,3,0,4,0
18,0,0,0,0,0,0,0,0,0,0,0,0,0,0,0,0,0,0,1,0,2,0,3,0,4
19,0,0,0,0,0,0,0,0,0,0,0,0,0,0,0,0,0,0,0,1,0,2,0,3,0
20,0,0,0,0,0,0,0,0,0,0,0,0,0,0,0,0,0,0,0,0,1,0,2,0,3
21,0,0,0,0,0,0,0,0,0,0,0,0,0,0,0,0,0,0,0,0,0,1,0,2,0
22,0,0,0,0,0,0,0,0,0,0,0,0,0,0,0,0,0,0,0,0,0,0,1,0,2
23,0,0,0,0,0,0,0,0,0,0,0,0,0,0,0,0,0,0,0,0,0,0,0,1,0
24,0,0,0,0,0,0,0,0,0,0,0,0,0,0,0,0,0,0,0,0,0,0,0,0,1
25,0,0,0,0,0,0,0,0,0,0,0,0,0,0,0,0,0,0,0,0,0,0,0,0,0
```

```
ID,1,2,3,4,5,6,7,8,9,10,11,12,13,14,15,16,17,18,19,20,21,22,23,24,25,26,27,28,29,30,31,32,33,34,35,36
1,0,1,0,0,8,0,0,0,0,0,0,0,0,0,0,0,0,3,0,0,2,0,9,7,0,6,0,4,0,0,18,0,0,2
2,0,0,1,0,0,0,0,0,20,0,0,5,8,0,0,6,0,5,0,0,0,8,3,0,0,0,0,0,0,0,0,0,0,0,0
3,0,0,0,1,0,0,9,0,0,2,0,0,0,0,0,0,0,6,0,0,0,0,0,2,0,0,0,3,0,0,4,0,0,5,0
4,0,0,0,0,1,0,0,4,0,0,0,0,0,0,19,0,0,0,0,0,11,0,0,0,0,0,3,0,0,0,0,0,5,0,0,0
5,0,0,0,0,0,1,0,0,0,0,20,0,0,0,0,0,2,0,3,0,0,0,12,0,0,0,0,0,0,0,0,0,0,0,11,6
6,0,0,0,0,0,0,1,0,0,0,0,0,0,1,0,0,3,0,0,0,7,0,0,9,0,8,0,5,0,13,0,0,0,0,0,2
7,0,0,0,0,0,0,0,1,0,0,0,5,0,0,0,0,0,0,0,0,0,0,5,0,0,0,0,0,5,0,0,0,0,3,0,0
8,0,0,0,0,0,0,0,0,1,0,0,0,0,0,0,0,0,5,0,0,17,0,0,0,2,0,17,0,0,0,0,13,0,0,0,7
9,0,0,0,0,0,0,0,0,0,1,0,0,0,0,7,0,3,0,0,1,0,0,0,0,0,0,0,6,0,0,7,0,5,0,0,0
10,0,0,0,0,0,0,0,0,0,0,1,0,0,0,0,0,0,0,0,2,0,7,0,0,0,0,0,0,0,5,0,0,0,0,0,0
11,0,0,0,0,0,0,0,0,0,0,0,1,0,0,0,14,0,11,0,0,18,0,19,0,0,9,3,0,0,0,0,0,0,7,0,5
12,0,0,0,0,0,0,0,0,0,0,0,0,1,0,0,0,0,0,0,0,0,0,0,0,0,0,16,0,0,0,5,0,0,0,0,0
13,0,0,0,0,0,0,0,0,0,0,0,0,0,1,0,0,0,0,0,0,0,0,0,0,4,0,0,0,0,0,0,0,0,0,0,0
14,0,0,0,0,0,0,0,0,0,0,0,0,0,0,1,0,0,8,0,0,0,4,0,0,0,0,0,0,7,0,0,2,4,0,0,8
15,0,0,0,0,0,0,0,0,0,0,0,0,0,0,0,1,0,0,0,0,3,0,0,0,0,6,0,0,0,1,0,0,0,0,0,0
16,0,0,0,0,0,0,0,0,0,0,0,0,0,0,0,0,1,0,6,0,0,0,0,5,0,0,0,8,0,0,0,0,0,0,0,3
17,0,0,0,0,0,0,0,0,0,0,0,0,0,0,0,0,0,1,0,0,0,0,4,0,0,0,0,0,0,0,0,6,0,0,0,0
18,0,0,0,0,0,0,0,0,0,0,0,0,0,0,0,0,0,0,1,0,0,4,0,0,0,0,0,0,0,6,0,0,0,0,0,0
19,0,0,0,0,0,0,0,0,0,0,0,0,0,0,0,0,0,0,0,1,0,0,0,0,0,0,18,0,9,0,4,0,0,6,0,6
20,0,0,0,0,0,0,0,0,0,0,0,0,0,0,0,0,0,0,0,0,1,0,0,0,9,0,0,0,0,0,0,9,0,0,0,0
21,0,0,0,0,0,0,0,0,0,0,0,0,0,0,0,0,0,0,0,0,0,1,0,0,0,0,0,6,0,0,0,0,6,0,2,0
22,0,0,0,0,0,0,0,0,0,0,0,0,0,0,0,0,0,0,0,0,0,0,1,0,0,0,0,0,0,0,0,0,0,0,0,0
23,0,0,0,0,0,0,0,0,0,0,0,0,0,0,0,0,0,0,0,0,0,0,0,1,0,0,0,0,0,0,0,9,0,0,0,14
24,0,0,0,0,0,0,0,0,0,0,0,0,0,0,0,0,0,0,0,0,0,0,0,0,1,0,0,0,9,0,0,0,5,0,0,0
25,0,0,0,0,0,0,0,0,0,0,0,0,0,0,0,0,0,0,0,0,0,0,0,0,0,1,0,0,0,5,0,0,0,6,0,0
26,0,0,0,0,0,0,0,0,0,0,0,0,0,0,0,0,0,0,0,0,0,0,0,0,0,0,1,0,0,0,0,0,3,0,1,9
27,0,0,0,0,0,0,0,0,0,0,0,0,0,0,0,0,0,0,0,0,0,0,0,0,0,0,0,1,0,0,19,0,0,7,0,0
28,0,0,0,0,0,0,0,0,0,0,0,0,0,0,0,0,0,0,0,0,0,0,0,0,0,0,0,0,1,0,0,0,0,0,0,0
29,0,0,0,0,0,0,0,0,0,0,0,0,0,0,0,0,0,0,0,0,0,0,0,0,0,0,0,0,0,1,0,0,5,0,0,15
30,0,0,0,0,0,0,0,0,0,0,0,0,0,0,0,0,0,0,0,0,0,0,0,0,0,0,0,0,0,0,1,0,0,0,0,0
31,0,0,0,0,0,0,0,0,0,0,0,0,0,0,0,0,0,0,0,0,0,0,0,0,0,0,0,0,0,0,0,1,0,0,0,0
32,0,0,0,0,0,0,0,0,0,0,0,0,0,0,0,0,0,0,0,0,0,0,0,0,0,0,0,0,0,0,0,0,1,0,0,6
33,0,0,0,0,0,0,0,0,0,0,0,0,0,0,0,0,0,0,0,0,0,0,0,0,0,0,0,0,0,0,0,0,0,1,0,0
34,0,0,0,0,0,0,0,0,0,0,0,0,0,0,0,0,0,0,0,0,0,0,0,0,0,0,0,0,0,0,0,0,0,0,1,0
35,0,0,0,0,0,0,0,0,0,0,0,0,0,0,0,0,0,0,0,0,0,0,0,0,0,0,0,0,0,0,0,0,0,0,0,1
36,0,0,0,0,0,0,0,0,0,0,0,0,0,0,0,0,0,0,0,0,0,0,0,0,0,0,0,0,0,0,0,0,0,0,0,0
```

# DESIGN OF 2-DIMENSIONAL CROSSOVER OPERATORS FOR THE FACILITIES LAYOUT PROBLEM

**SULUNG AND K. C. CHAN**
*School of Mechanical and Manufacturing Engineering*
*The University of New South Wales, Sydney 2052, Australia*

***Abstract:***
Genetic algorithms (GAs) have successfully been applied to the facilities layout problem in a number of previous attempts. This paper proposes three two-dimensional crossover operators based on the partial matched crossover (PMX) operators, namely PMX-HV, PMX-2PT, and PMX-1PT. The PMX-HV operator selects a horizontal or vertical cutting edge randomly for crossover. The PMX-2PT selects two points along the boundary of the plant to establish a cutting edge. The PMX-1PT starts from one point along the boundary of the plant and travels randomly to generate a cutting edge. The PMX operator and the three new operators were applied to a large 36 location plant. Extensive experiments were performed and the performance of these operators was evaluated and compared. The performance of these three new operators was not as good as the PMX operator due to the significant reduction of number of possible crossing boundaries.

## INTRODUCTION

The facilities layout problem, or quadratic assignment problem, is a field of study concerning the assignment of machines/facilities to locations within a plant. It can be formulated as an optimisation problem with the objective of minimising materials handling costs. Plant layout belongs to the Non-Polynomial hard (NP-hard) class of problems believed to be unsolvable in polynomial time. Heuristics are usually involved in solving this type of problems and numerous such algorithms are available in the literature.

Genetic Algorithms are heuristic search and optimisation techniques that imitate the natural selection and biological evolutionary process (Goldberg, 1989). GAs combine the notion of survival of the fittest, random and yet structured search, and parallel evaluation of nodes in the search space. GAs have been successfully applied to various optimisation problems, such as the travelling salesman problem (TSP) (Oliver, Smith, and Holland, 1987) (Jog, Suh, and Gucht, 1989), a problem similar to the facility layout problem. In a more recent attempt, GA was applied to solve the layout problem and very good results were achieved for medium and large sized plants (Chan and Tansri, 1994). Among a number of crossover operations studied, the PMX crossover operator was found to be the best for the problem.

Although the PMX operator was very successful for the layout problem, PMX operates on one-dimensional strings. However, the facilities layout problem is a 2-dimensional problem and it is envisaged that new operators which can work on 2-dimensional data structures representing layouts may be more appropriate.

In the quantitative approach of facilities layout, the objective function is to minimise the total materials handling cost. The objective function (Chan and Tansri, 1994) is defined as:

$$\text{Min} \sum_{i}^{n} \sum_{j}^{n} f_{ij} d_{p(i)p(j)} C_{ij} \qquad (i>j) \tag{1}$$

where $f_{ij}$ = frequency of material flow between machine $i$ and machine $j$,
$d_{ij}$ = Euclidean distance between location $i$ and location $j$,
$p(i)$ = the location to which machine $i$ is assigned,
$C_{ij}$ = cost of material handling between machine $i$ and machine $j$,
$n$ = number of machines (or manufacturing cells, or departments).

## GENETIC ALGORITHMS FOR PLANT LAYOUT

### String Representation

The formulation of a genetic algorithm requires representation of solutions. Classical genetic algorithms use binary string representation (Goldberg, 1989). However, binary strings are not naturally suitable for the plant layout problem. Hence, decimal integer representation was used. For example, a string 3-2-4-5-1-6 for a 6-location plant means machine 3 is placed in the first location, machine 2 in the second location, and machine 4 in the third location, and so on. This string representation was originally used to represent routes for the travelling salesman problem. Although this string representation is one dimensional, it may be used to represent a 2×3 plant, or a 1×6 plant, or other plant structures. This string representation was used in the PMX crossover operations. However, a layout would be more suitably represented by a 2 dimensional data structure. Therefore, an array type of structure was used for the three modified PMX operators, as discussed in the next section. For example, a data structure $\begin{bmatrix} 3 & 2 & 4 \\ 5 & 1 & 6 \end{bmatrix}$ is used to represent a 2×3 plant unambiguously.

### The Crossover Operators

There are three basic genetic operations used to generate new strings from generation to generation. They are reproduction, crossover and mutation. Reproduction is a biased selection process to determine which strings will be reproduced in the new population. Strings with higher fitness values (better solutions) have better chances of being selected than those with lower fitness values. Hence, the strong continues to live while the weak dies off. The crossover operator is the most important operator in genetic algorithms. It is the "mating" operator that allows production of new strings through partial information exchange between pairs of strings. The crossover operators proposed are based on PMX (Goldberg, 1989).

PMX-HV

Unlike a route for a travelling salesman problem, a layout is a 2 dimensional structure and it could be more naturally represented as an array. The three proposed PMX-HV, PMX-2PT and PMX-1PT crossover operators are developed specifically for the 2 dimensional layout structure. The main difference between the PMX and the three modified PMX crossover operators is the way that crossover boundaries are generated. For the PMX-HV crossover, the boundaries are selected by the following procedures:

1. horizontal (H) or vertical (V) boundary type is chosen at random,
2. for H boundary type, select two horizontal lines randomly, Fig. 1;
   for V boundary type, select two vertical line randomly, Fig. 2,

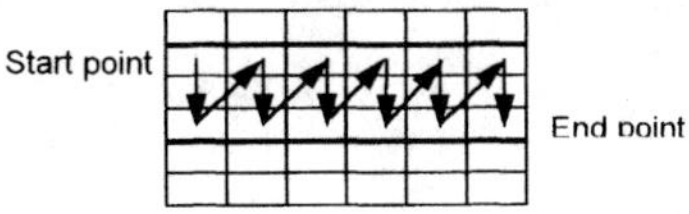


Figure 1: Horizontal boundary

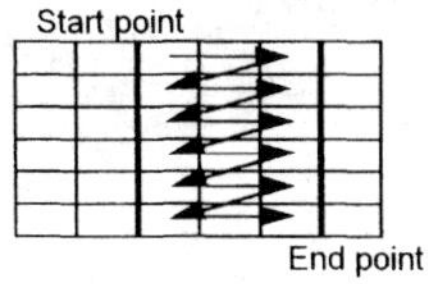


Figure 2: Vertical boundary

After the boundaries are created, crossover is then performed as in PMX, either column-wise (top to bottom) within horizontal boundaries or row-wise (left to right) within vertical boundaries. An example with vertical boundaries is shown in Fig. 3.

PMX-2PT

The PMX-2PT crossover uses the same 2 dimensional representation as the PMX-HV. The PMX-2PT is designed to allow boundaries be created by joining two random points, one at the top and one at the bottom, so that more variety and higher flexibility can be achieved in generating boundaries. To simplify the process, only one boundary is generated and one side, which is chosen randomly, will undergo the usual PMX operation. As shown in Fig. 4, two points (the starting and end points) are randomly selected and a boundary is created by joining the nearest intersections along the straight line joining the two points. Having defined the boundary, the PMX crossover will be applied row-wise to one side, which is chosen randomly, of the plant.

PMX-1PT

The PMX-1PT crossover operator is the most flexible among the three operators being proposed. The PMX-1PT operator selects one point randomly at the top and then travels down the plant with moves to the left and right randomly, until the bottom boundary is reached, Fig. 5. Then the usual PMX operation is performed on one side of the boundary.

## EXPERIMENTS

The performance of the proposed crossover operators: PMX-HV, PMX-2PT, and PMX-1PT, was tested on a 36 (6×6) location plant. The programs were originally written in Borland C (ANSI) on a Pentium PC. After the programs were successfully debugged they were uploaded to Silicon Graphics workstations with 16 MB of RAM. A very large number of experiments were conducted using over 30 workstations in parallel for days. Due to page limits, the frequency and cost charts are not shown in the paper, but can be obtained from the authors. The experimental results presented are for the following GA parameters: 1) number of generations: 500; 2) population size: 2000; 3) crossover probability: 0.9; 4) mutation probability: 0.005; and 5) fitness function: $y = x^{-\log x} * 10^{12}$. All the experiments were repeated 20 times. Their performance was compared with that of the PMX operator. As

Layout A

| | | |
|---|---|---|
| 3 | 1 | 5 |
| 9 | 7 | 4 |
| 2 | 8 | 6 |

Layout B

| | | |
|---|---|---|
| 4 | 6 | 7 |
| 8 | 9 | 1 |
| 5 | 3 | 2 |

Layout A'

| | | |
|---|---|---|
| 8 | 6 | 5 |
| 7 | 9 | 4 |
| 2 | 3 | 1 |

Layout B'

| | | |
|---|---|---|
| 4 | 1 | 9 |
| 3 | 7 | 6 |
| 5 | 8 | 2 |

Figure 3: PMX-HV crossover

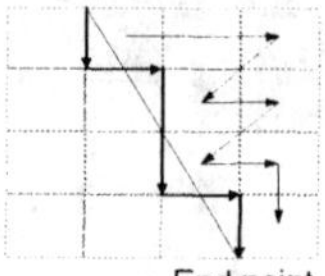


**Figure 4: PMX-2PT crossover**

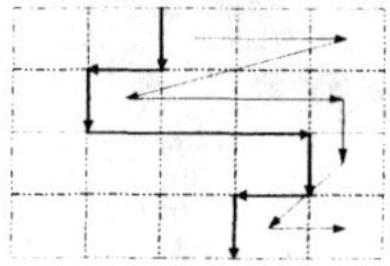


**Figure 5: PMX-1PT crossover**

measures of performance, the following data was recorded: minimum cost of each generation, maximum, minimum and average fitness values and time taken.

**Results**

Table 1 shows the statistical analysis of the best layout generated in each run by these four crossover operators. The PMX produced the lowest cost solution but has the largest variance within its population at generation 500.

Fig. 6 shows the average fitness curves. Both the PMX-2PT and PMX-1PT converged to a fixed value very quickly. It appeared that these two operators would get trapped at local minima and were unable to escape from them. Although the average fitness values for the PMX-HV were not as good the PMX-2PT or the PMX-1PT, the PMX-HV appeared to be able to avoid being trapped at local optima. However, the PMX was the best operator and continued to find better solutions.

Fig. 7 shows the average minimum costs for 20 runs. The PMX-2PT and PMX-1PT were not able to produce any obvious better solutions from the initial population; while the PMX-HV was able to produce better offspring. However, none of the three proposed operators performed better than the PMX.

Fig. 8 compares the fitness ranges between the PMX-1PT and the PMX operators. The fitness gap between the maximum and minimum of PMX is much greater than the PMX-1PT and widens over generations. This shows the desirable characteristic of PMX in producing a diverse population. Classical GA theory dictates the requirement of diverse population after convergence in order to climb out of local optima.

The time taken to run the experiment is almost a linear relationship with population sizes, Fig. 9. It is surprising that PMX took the second longest time to complete. We would expect it to be the fastest to run due to its simpler (one-dimensional) structure. PMX-HV took the longest time to complete. The most flexible operator, PMX-1PT, had the shortest running time.

**TABLE 1: COST STATISTICS (BEST LAYOUTS AT-END-OF-GENERATION)**

| | PMX | PMX-HV | PMX-2PT | PMX-1PT |
|---|---|---|---|---|
| Mean | 16527 | 17168 | 17982 | 18043 |
| Median | 16627 | 17170 | 18005 | 18010 |
| Standard Deviation | 513.79 | 313.34 | 229.13 | 193.96 |
| Minimum | 15513 | 16506 | 17456 | 17771 |
| Maximum | 17331 | 17599 | 18290 | 18375 |

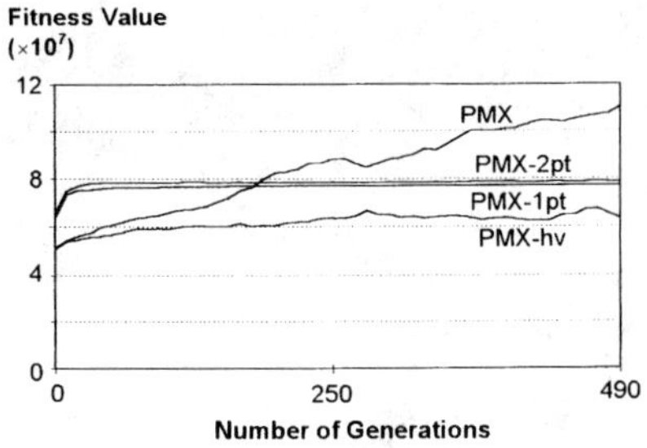


**Figure 6: Average fitness value**

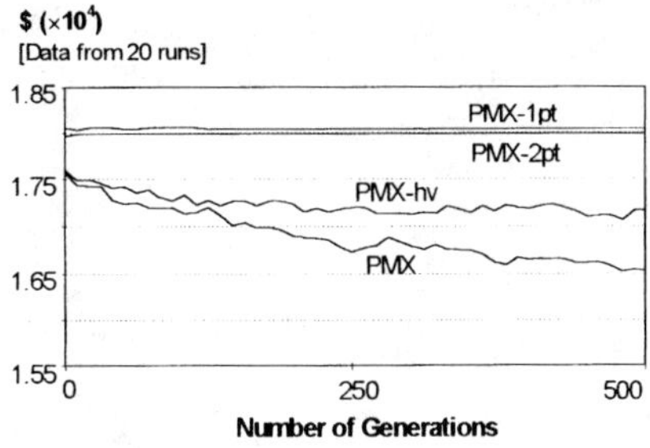


**Figure 7: Average minimum cost**

## LESSONS LEARNT AND CONCLUSIONS

Although the PMX crossover was an effective operator for the facility layout problem, it was originally designed for the travelling salesman problem with one dimensional strings used to represent travelling routes. We thought that direct implementation of PMX operator may not achieve the best results for the layout problem since it was not tailor-designed for the problem. Unlike a travelling route, a layout is a 2 dimensional structure. Therefore, a 2 dimensional array type of structure should be used to represent layouts, to which genetic operations are performed. We envisaged that the spatial relationships between facilities could be better preserved from one generation to the next with a 2 dimensional structure and operators. Therefore, this study was aimed to design three 2 dimensional crossover operators based on the PMX. They are the PMX-HV, PMX-2PT, and PMX-1PT. The proposed data structure and modified PMX operators were tested on a 36 (6×6) location plant. Extensive numerical experiments were performed using Silicon Graphics workstations. However, the results were disappointing. All the proposed operators did not perform well, and in fact, they are worse than the original PMX operator.

Trying to understand why this new approach does not work well, we examined the pools of solutions over generations and noticed that the 2 dimensional crossover operators were not able to produce a large variety of solutions and converged relatively fast. This can be seen in Fig. 8. The main reason was the number of boundaries that could be created was significantly reduced by using the 2 dimensional operators. Hence, the variety of offspring was reduced. Take the 36 (6×6) location plant as an example. The PMX has $\binom{37}{2}$ choices in creating boundaries; the PMX-HV has $\left(2 \times \binom{7}{2}\right)$ choices, the PMX-2PT has (7 × 7) choices; and the PMX-1PT has $7^6$ choices. This also explains why the PMX takes a bit longer to run than some of the modified PMXs. Due to the limited number of crossing boundaries and uniformity of solutions, the amount of gene shuffling during crossover, and consequently, the computing time, was greatly reduced.

The lesson learnt in this study was: while tailor-making genetic algorithms, may be an effective approach in solving certain specific problems, we must be careful in the design to ensure that the capability of the GA in generating a diversified pool of solutions is not significantly reduced. A balance between specific and generic formulations may be the key to build powerful genetic algorithms. A GA is not always as simple as many people think or want.

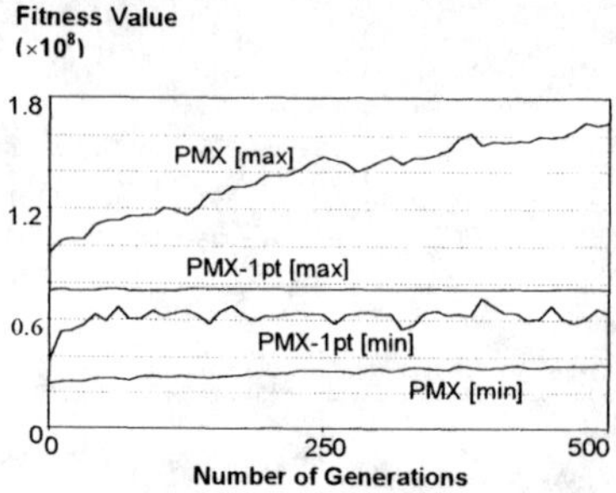


**Figure 8: Maximum and minimum fitness values of PMX and PMX-1PT**

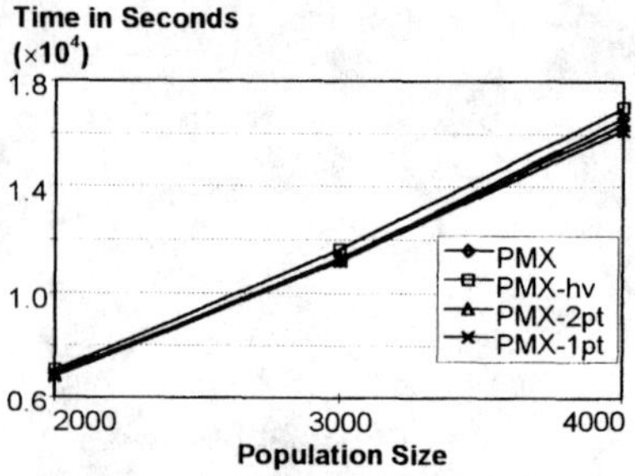


**Figure 9: Time taken for various population sizes for 500 generations**

Further research is currently undertaken to investigate:

- if higher mutation rate would improve the overall performance of the GA using the 2 dimensional PMXs;
- if the 2 dimensional PMXs have better future in much larger plants,
- the effect of flexibility in creating boundaries, Fig. 10, and
- the effect of multiple crossing zones, Fig. 11.

## REFERENCES

Chan, K. C. and Tansri, H., (1994). A Study of Genetic Crossover Operations on the Facilities Layout Problem, *Computers and Industrial Engineering, An Int. J.*, (26), No. 3, July, 537-550.

Fogel, D. B., (1995). *Evolutionary Computation, Towards a New Philosophy of Machine Intelligence*, IEEE Press.

Goldberg, D. E., (1989). *Genetic Algorithms in Search, Optimization, and Machine Learning*, Addison Wesley.

Jog, P., Suh, J. Y. and Gucht, D. V., (1989). The effects of population size, heuristic crossover and local improvement on a genetic algorithm for the travelling salesman problem. *Proc. of the 3rd Int. Conf. on Genetic Algorithms*, 110-115.

Kaku, B. K., Thompson, G. L., and Morton, T. E., (1991). A Hybrid Heuristic for the Facilities Layout Problem, *Computers Opns. Res.*, (18), No. 3, 241-253.

Michalewicz, Z., (1994). *Genetic Algorithms + Data Structures = Evolution Programs*, 2nd Edition, Springer-Verlag.

Oliver, I. M., Smith D. J. and Holland, J. R. C., (1987). A study of permutation crossover operations on the travelling salesman problem. *Proc. of the 2nd Int. Conf. of Genetic Algorithms*, 224-230.

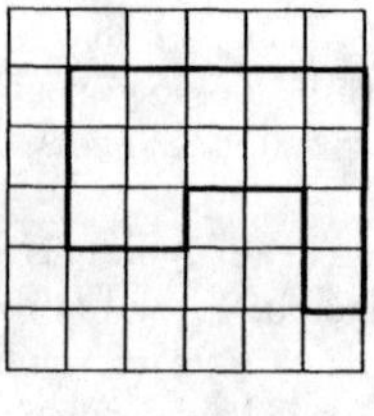

**Figure 10: Higher boundary creation flexibility**

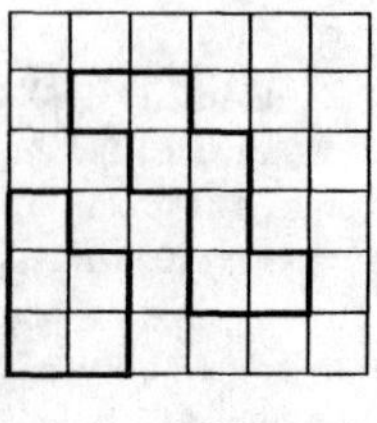

**Figure 11. Multiple crossing zones**

# SPATIAL LAYOUT PLANNING USING EVOLVED DESIGN GENES

**John S. Gero and Vladimir A. Kazakov**
Key Centre of Design Computing,
Department of Architectural and Design Science,
The University of Sydney, NSW 2006 Australia.
e-mail:{john,kaz}@arch.su.edu.au

## ABSTRACT

Spatial layout planning is an NP-complete optimization problem. In this paper we apply to it an extension to the standard genetic algorithm based on an analog with the genetic engineering of natural systems. This allows not only the solution to evolve but also evolves the genetic representation at the same time producing 'evolved genes'. In spatial layout problems these evolved genes map onto spatial compact clusters of activities (spaces). The evolution of these genes can be viewed as a form of learning of problem-specific knowledge related to the structure of the problem being solved. The resulting search process proceeds in two stages. During the first stage these clusters of activities are produced and optimally arranged with respect to each other. During the second stage this 'large-scaled layout' is improved by rearranging the activities within each such cluster. It is shown that the genetic engineering extension to genetic algorithms produces results faster than the standard GA. Further, it is demonstrated that the evolved genes can be reused in a class of related problems further increasing the efficiency of the method.

**Keyword: Enhanced Genetic Algorithms, Engineering Optimization, Genetic Engineering.**

## 1 INTRODUCTION

The spatial layout planning (SLP) problem arises in various domains such as architectural design, location-allocation problems, backboard wiring in electronics, scheduling, statistical data analysis etc. We will use the formalization of the SLP problem as a quadratic assignment problem (Liggett, 1985) to find a one-to-one mapping $\rho$

$$\rho : \{M \rightarrow N\}, \quad j = \rho(i), \quad i \in M, \quad j \in N$$

of the discrete set $M$ with $m$ activities, onto another discrete set $N$ of $n$ locations $m \leq n$, such that the overall cost of the layout $I$ is minimal

$$I = \sum_i f_{i\rho(i)} + \sum_i \sum_j q_{ij} c_{\rho(i)\rho(j)} \rightarrow \min$$

where $f_{i,j}$ is the fixed cost of assigning element $i \in M$ to element $j \in N$; $q_{ij}$ is measure of interaction of elements $i, j \in M$; and $c_{ij}$ is measure of distance between elements $i, j \in N$. This problem is NP-complete. Computationally, SLP problems of size more than 20 are not, currently, practically solvable to guaranteed optimality. In engineering applications it is often required to solve a number of similar SLP problems. The existing computational methods are not designed as extractors and re-users of the knowledge about the solutions of such group of problems, they do not become "smarter" after solving a few similar problems, each new problem is solved from scratch.

## 2 GENETIC ALGORITHMS

Randomized heuristic methods have proven very successful recently in solving combinatorial optimization problems. One of the most popular among such methods are genetic algorithms (GAs) - a family of search methods which model Darwinian evolutionary theory. The search space consists of character strings (genotypes) composed of the elements of a given alphabet. The objective is to adapt the population to the given fitness landscape. Typically GAs start by generating a random population. Then the successive populations are generated recursively using the crossover and mutation operators. Evolution proceeds until the population stabilizes. Selection for reproduction is carried out stochastically based on the defined fitness ("survival of the fittest"). Similar to the evolution of natural organisms the genetic material of the population is not accessible and is controlled indirectly by the evolution process. Hence the overall efficiency of the search is low. We employ an order-based genetic representation - the length of the genotype is fixed ($n$), the j-th letter from the alphabet $M$ in the i-th position corresponds to the placement of the j-th activity in the i-th location. Each letter from the alphabet must be used once and only once in any valid genotype. We use a simple 1-point crossover where the front half of the offspring's genotype is copied from the corresponding parent exactly and the rear half from the other parent is copied with minimal changes which make it valid. The mutation is a pair-wise exchange of the genes.

## 3 THE GENETIC ENGINEERING BASED EXTENSION OF GAS

This is the computational model of Darwinian evolution supplemented with practice of the genetic engineering (Gero and Kazakov, 1995). Genetic engineering is based on the assumption that there is a link between the fitness performance of a set of phenotypes in a population and some structure in their genotypes and that this structure is not present in the genotypes of those phenotypes which do not exhibit that fitness performance. Genetic engineering is a well developed process in the genetics of natural organisms. It is assumed that there is an external force in evolution - a genetic engineer - who firstly studies the dynamic changes in the genetic structure of the population during the evolution and identifies the genetic features (called evolved genes) directly linked to the beneficial features of the organism. Then he intervenes in the reproduction by making sure that the evolved genes are not damaged by crossover and mutation. Finally he checks the population before each reproduction and makes sure that the population is genetically healthy by changing genotypes which lack evolved genes in such a way that they acquire evolved genes with minimal changes to the genetic material. It is clear that since more powerful and direct means of control are introduced into the search its overall efficiency is higher than the efficiency of a method based on Darwinian evolution alone. The evolved genes once identified for some species can be re-applied to enhance the evolution process of similar species. This reuse of the learned information obviously leads to computational savings.

## 4 IDENTIFICATION OF EVOLVED GENES

We analyse the genetic structure of the population after the execution of some preset number of generations. It is carried out in the following way: the fittest 10% of the population and the most unfit 10% of the population are separated; then an attempt is made to find the genotypic features (characteristic substring, patterns, etc.) which distinguish the former from the latter groups; the identified genetic features are declared as newly evolved genes. Then statistical tests are carried out to ensure that the presence of these evolved genes in the genotype is beneficial.

## 5 EVOLVED GENES IN SLP PROBLEMS

It turns out that in many SLP problems the evolved genes are the groups of activities which are naturally "attracted" to each other, that is, whose gene's placement within the genotype as a closed group is beneficial for the fitness. Although one can simply test this hypothesis experimentally, the intuitive reason for such a phenomenon to occur is clear. It is the presence of a subgroup among the activities such that the layout cost due to their interaction depends more strongly on the average distance between activities within the group than the rest of the layout cost does on the average distance between the remaining activities. It is clear that the average distance between activities within the group changes less than the average distance of all activities if the group is scattered along the genotype. We assume here that the distance matrix $|c|$ possesses "natural" properties: $c_{ij}$ is a monotonically decreasing function of $j - i$, triangular inequality holds, etc.

## 6 TYPICAL GA'S SEARCH

The solution to many SLP problems proceeds in two stages (let us emphasize that we consider here a generic GA process with or without genetic engineering enhancement). During the first stage crossover plays a major role which leads to macrosteps in the search space. The genetic regularities (evolved genes) which emerge here are the gene clusters described above. One can interpret this process as packaging of such clusters and ordering them optimally with respect to each other. It yields a population shifted into the basin of attraction of one of the problem's minima. During the second stage mutations are the major search tool which primarily affect the order of genes within previously evolved genes clusters. This yields local search within the already approached minimum.

## 7 GENETIC ENGINEERING GA FOR SLP PROBLEMS

Assuming that this two-step pattern in GA search is a general feature of many SLP problems one can design the following particular version of genetic engineering based GA. It proceeds in two steps. During the first step the genetic engineering analysis is performed. Its aim is the identification of evolved genes (beneficial contiguous or non-contiguous gene clusters). Then statistical tests are carried out to ensure that the genotypes with evolved genes are fitter than the ones without. The reproduction operators of GAs remain the same except the choice of the crossover point is restricted to the areas of the genotypes which are not taken by the evolved genes and the mutation swaps are confined to these areas. Before reproduction the population's genotypes are checked for the presence of the evolved genes. The genotypes which lack them are subjected to "genetic surgery" - the components of the corresponding evolved gene are gathered (pushed) together using the minimal number of pair-wise gene swaps (one can view it as directed extended macromutation). The yield of this stage of the algorithm is a "large-scale" structure of the optimal layout - the evolved genes (compactly spatially placed groups of activities) and their spatial arrangement with respect to each other.
During the second step of the algorithm the problem is decomposed into a number of quasi-independent problems of optimally arranging the activities within the evolved genes. Hence, a number of standard GAs (for each one of the evolved clusters) run in parallel. The genetic analysis is not carried out during this stage.

## 8 NUMERICAL RESULTS

As a testbed we used the problem of the placement of 19 office departments into a four-floor building (Liggett, 1985) and the problem of assigning 19 hospital departments to 19 possible locations (Elshafei, 1977). The result of the run of the standard GA which converges to the best solution for office example (averaged over 10 such runs from different initial random seeds) is

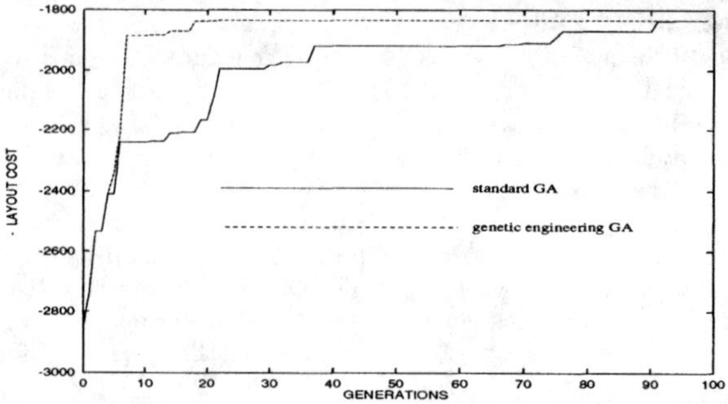


**Figure 1. The best fitness vs. generation number for the standard GA and genetic engineering based GA in the office building example. The results are averaged over 10 runs with different initial random seeds, which converge to the best solution found.**

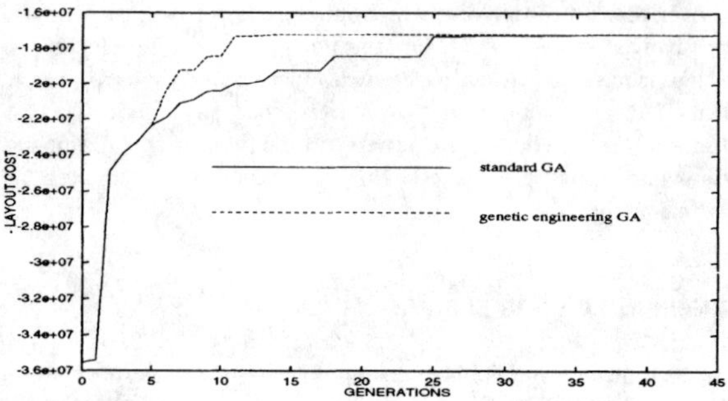


**Figure 2. The best fitness vs. generation number for the standard GA and genetic engineering based GA in the hospital example. The results are averaged over 10 runs with different initial random seeds, which converge to the best solution found.**

presented in Figure 1 (bold line). The population size was 200, probability of crossover 0.6 and the probability of mutation 0.01. We use elitist GA with a generation gap of 3. On average, it was necessary to run the GA 13 times in order to get the best solution. The results of the genetic engineering based GA are presented in Figure 1 with the dotted line. On average just two runs were required with different initial seeds to find the layout with the minimal cost. The results of the numerical simulations of the hospital problem are presented in Figure 2. The bold line denotes the results of using the standard GA. The dotted line denotes the results of the genetic engineering based GA. On average it took 26 runs of the standard GA from different initial random seeds to find the layout with the minimal cost. The genetic engineering based GA needed on average only three runs with different initial seeds in order to find this layout. The corresponding dependence of the best layout cost on the generation number, averaged over 10 successful runs using genetic engineering based GA, is shown in Figure 2 with the dotted line.

In order to demonstrate the possibility of re-using the evolved genes in a similar problem we consider the problem of placing the same 19 activities of the office example into the floors of a two-storey building. This building is obtained by placing the two upper levels of the four-storey building of the original problem next to the two first levels (Gero and Kazakov,

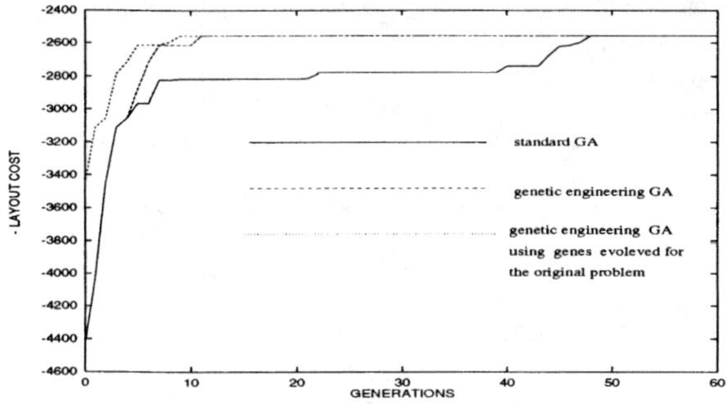


**Figure 3. The best fitness vs. generation number for the standard GA and genetic engineering based GA in the modified office building example. The results are averaged over 10 runs with different initial random seeds, which converge to the best solution found.**

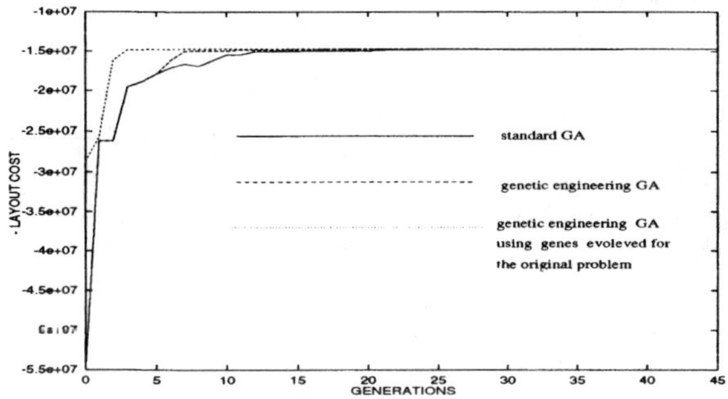


**Figure 4. The best fitness vs. generation number for the standard GA and genetic engineering based GA in the modified hospital example. The results are averaged over 10 runs with different initial random seeds, which converge to the best solution found.**

1996 ). The results of the optimization are shown in Figure 3. We run the standard GA and genetic engineering based GA beginning without the previously evolved genes and then using the previously evolved genes. We can see that the re-use of the evolved genes yields about a 20% savings in terms of the number of generations.

The simulations were also carried out for the hospital problem with a modified distance matrix (the average change of its component was about 6%), (Gero and Kazakov, 1996). The results (standard GA, genetic engineering based GA with and without the previously evolved genes) are shown in Figure 4. Again, we can see computational savings when re-using the previously evolved complex genes.

## 9 CONCLUSIONS

The approach presented here has some intuitive appeal for designers because of the direct mapping between the genes and their expression in a layout. It has a further appeal since the evolved genes readily map on to groups of activities which need to located next to or near each other. Such information is hard to see intuitively in a large scale problem. Another advantage of this approach which has not been seen in other approaches is the ability to reuse the evolved genes in related problems. The evolved genes represent problem-specific knowledge that has been learned by the system which is now externalised and available for reuse in similar problems.

In conclusion, we have developed a genetic engineering based extension of GAs for layout planning problems. We demonstrated that in a number of layout planning problems the solution process can be roughly divided in two stages. During the first stage some activities are aggregated in compact groups and the objective is to find the optimal placement of such groups. During the second stage the optimal location of the activities within these groups is sought. The first stage corresponds primarily to non-local search in the original search space (since even pair-wise permutation of two aggregated activities corresponds to the macrostep in the placement space of non-aggregated activities). The second stage sooner or later becomes local search. We demonstrated significant computational advantages of such an extension. We have also shown that such optimal (natural) aggregation of activities can be used in a range of similar problems with significant computational benefits.

## 10 ACKNOWLEDGMENTS

This work is directly supported by a grant from Australian Research Council, computing resources are provided through the Key Centre of Design Computing.

## REFERENCES

Elshafei, A. N. (1977). Hospital layout as a quadratic layout problem, *Operations Research Quarterly*, **28**: 167-179.

Gero, J. S. and Kazakov, V. (1995). Evolving building blocks for genetic algorithms using genetic engineering. *Proceedings of the IEEE Conference on Evolutionary Computing*: 340-345.

Gero, J. S. and Kazakov, V. (1996). Evolving design genes in space layout planning problems. submitted to *Artificial Intelligence in Engineering.*

Liggett, R.S. (1985). Optimal spatial arrangement as a quadratic assignment problem, *in* Gero, J. S. (ed.), *Design Optimization*, Academic Press, New York: 1-40.

# ASSEMBLY LINE BALANCING USING A GENETIC ALGORITHM WITH REPAIR ROUTINE

**K. C. CHAN AND R. TANTONO**
*School of Mechanical and Manufacturing Engineering*
*The University of New South Wales, Sydney 2052, Australia*

***Abstract:***
This paper presents the development of a genetic algorithm for the assembly line balancing problem. In the implementation, decimal integer encoding was used for data representation, and two crossover operators: one point crossover and uniform crossover, were used. Due to the strict precedence and cycle time constraints, the chance of obtaining a feasible solution at random during initialisation was extremely small. Therefore, an efficient procedure was developed to generate feasible initial solutions at a random but guided manner. Because of the same constraint reason, it was also very likely to produce infeasible children during crossover and mutation processes. A repair routine was developed to convert infeasible children to feasible ones. The proposed genetic algorithm, in conjunction with the initialisation method and repair routine, were tested on a 21 task problem and very satisfactory results were achieved.

## INTRODUCTION

An assembly line is a production line where material moves through a sequence of workstations, in which one or more assembly tasks are performed. The production rate of the assembly line is governed by the slowest or the bottleneck workstation. The other workstations may have to slow down or wait to keep in pace with the bottleneck workstation. The primary consideration in assembly line balancing (ALB) is to reduce or eliminate the idle time of individual workstations so that smooth and efficient production can be achieved.

Assembly lines are widely used for high volume production of goods and services, and a substantial amount of research effort has been devoted in the past to the efficient design and operation of assembly system. Most research has focused on the development of enumerative and heuristic methods for minimising the number of workers required for a desired production rate, or minimising the cycle time for a given number of workstations.

In this paper, the single model deterministic ALB is investigated. The task time is assumed to be constant. Other external factors such as machine layout, station length, tools, safety, delays, etc. are assumed to have no effect in the problem model.

The assembly line balancing problem falls into the NP (non - polynomial) hard class of problems. Recent studies are directed towards the development of computer-efficient algorithms or heuristic techniques. However, numerous classical operation research methods are also capable of solving the assembly line balancing problems. Excellent reviews of other methods for solving the assembly line balancing problem can be found in (Bhattacharjee and Sahu, 89) (Hoffmann, 90).

This paper presents the use of genetic algorithms (GAs) to solve the assembly line balancing problem. Genetic algorithms are heuristic search and optimisation techniques whichuse a process similar to biological evolution to improve an initial set of feasible solutions through an iterative process (Goldberg, 89) (Michalewicz, 94). The details of the development are provided in the following sections.

## A GENETIC ALGORITHM FOR ASSEMBLY LINE BALANCING

### Representation

A coding system was defined to represent possible solutions of an ALB problem. In this coding system, a solution is represented by a string. Each element of the string is a decimal integer. The number in the $i^{th}$ place in the string is the workstation to which the $i^{th}$ task (in alphabetical order) is assigned. For the example given in Fig. 1, the string representation is 1-1-1-2-2-3-2-3-4. The numbers shown next to the tasks are the task times.

### Initialisation

In most genetic algorithms, initial solutions are generated randomly without any constraints. In our initial attempts, this approach was adopted. However, it was found that the chance to generate a feasible initial solution is extremely small due to the presence of cycle time and precedence constraints. For a small 11 task problem, the probability of generating a feasible solution is in the order of 0.0001. Therefore, a more efficient initialisation method was developed. In this method, only feasible tasks, which follow the precedence constraints, will be available for random selection and workstation assignment. The following two rules are followed in string initialisation until all tasks are assigned:

1. If there exist any tasks which meet both the precedence and cycle time constraints, select one randomly. The selected task is then either added to the current workstation or assigned to a new workstation at random or according to prescribed probabilities.
2. If there only exist tasks which meet the precedence constraints only, select one randomly and put it in a new workstation.

### Objective Functions

For ALB problems, a number of goals need to be achieved. One possible goal is to obtain the lowest idle time in order to maximise the utilisation of machines and operators. Other objectives are to minimise the number of workstations, to have the highest line efficiency (minimum balance delay), to obtain the minimum cycle time for the available number of workstations, to minimise the labour cost per unit and the inventory, set-up and idle time costs. All these objectives of ALB can be considered towards a common goal which is to have a "well balanced" assembly line. Only two objectives are investigated in this study. They are :

1. to minimise the number of workstations for a given cycle time, and
2. to minimise the balance delay.

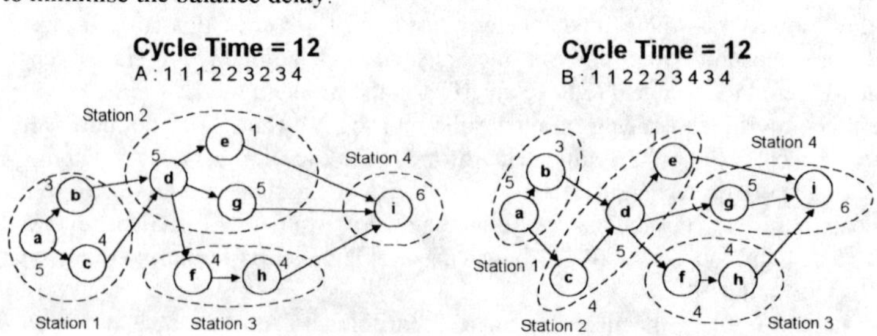


**Figure 1: Schematic diagram of two feasible solutions.**

### Fitness Function

The fitness value of each string in the population determines how likely it will survive to the next generation, or will be selected for mating. Since the objectives in this study are formulated as minimisation problems, better or fitter solutions are those whose objective values are low. Therefore, the fitness function should have an inverse relationship with the objective function. In order to make the fitness function more sensitive and biased in selecting fitter solutions for crossover, instead of a linear function the following function is employed:

$$y = x^{-\log(x)} \times \text{constant} \tag{1}$$

where $y$ is the fitness value and $x$ is the objective function value.

### Mating Pool Selection

A number of methods can be used in selecting individuals for the mating pool. Despite the wide range of methods, the underlying principle is to select individuals according to probabilities in proportion to their fitness values. The conventional selection method, spinning roulette wheel method (Goldberg 89), is employed in this research.

### Crossover and Mutation

Two crossover operators were investigated in this research. They are the conventional one point crossover operator and the uniform crossover. The one point crossover operator is the simplest crossover operation, in which a random point is chosen along two strings drawn randomly from the population, both strings are separated at the same point, and then the front part of one string is joined to the back part of the other string, and vice versa, to form two children. The uniform crossover requires individual gene swapping between two strings in the mating pool. Genes in the same positions are swapped according to a user-defined probability, $p$. In our experiments, $p$ was set to 0.5. Therefore, roughly 50% of the genes would be swapped. $p$ is a parameter which can be used to control the convergence behaviour and performance of the GA.

The mutation operator involves the random change of assignments of tasks to workstations. The workstation number is "mutated" to either the workstation number immediately before or after it. In other words, the particular gene in the string is changed by adding or subtracting one according to a certain mutation probability.

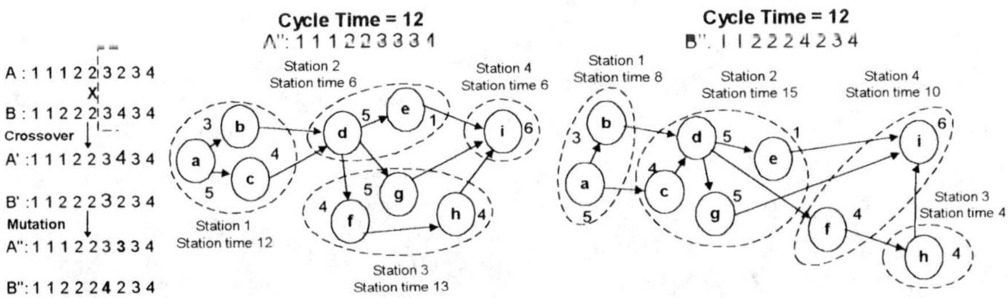


Figure 2: Two infeasible solutions after crossover and mutation.

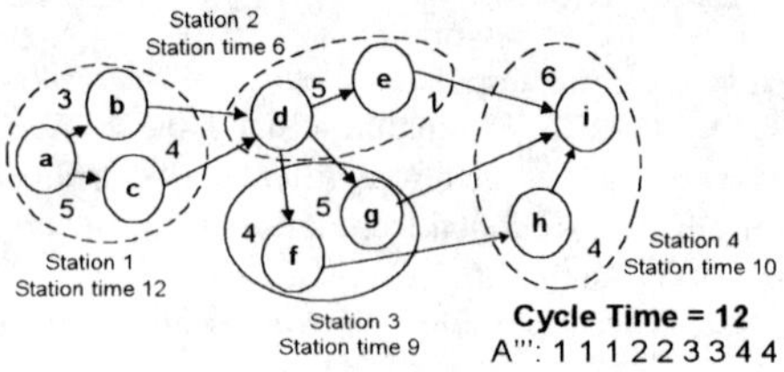


**Figure 3: Repair of the first child (A" to A'").**

## Repair Routine

Due to the precedence and cycle time constraints, infeasible solutions are often generated by the crossover and mutation operations. One approach is to use a penalty function to reduce the fitness values of the infeasible solutions, thus reducing their chances to be reproduced to the next generation. Another approach is to use a repair routine to change infeasible solutions to feasible ones. In this research, a repair routine was developed.

Let us consider the following example. Two feasible solutions of a 9-task problem are coded as 111223234 and 112223434, as shown in Fig. 1. After undergoing crossover and mutation operations, the new strings become 111223334 and 112224234 respectively, Fig. 2. They are clearly violating the defined constraints. One string has one station that exceeds the given cycle time (station number 3 has a station time of 13 seconds which is exceeding the given cycle time of 12 seconds), and the other string violates both constraints (station number 2 has a station time of 15 seconds, and station 4 violates the precedence relationships).

The repair routine will correct the arrangement of the task elements in the string so that it becomes feasible. Firstly, the routine examines each station following the precedence relationship (from left to right). If one station time is greater than the cycle time, the routine will discard the last task element and assign it to the next station (the station contains the task's immediate successor). If one station has a task that violates the precedence relationship, the routine will assign the last task element to the previous workstation (the station contains the task's immediate predecessor). The routine performs the above operations iteratively until a feasible solution is obtained. After the repair routine, the final feasible solutions for the above example are shown in Figs. 3 and 4.

## EXPERIMENTS AND RESULTS

The performance of genetic algorithms in any application area can only be determined by experimentation. Parameters like population size, crossover and mutation probabilities can only be set based on previous experience and on a trial-and-error basis. A large number of tests have been carried out to evaluate the proposed approach. In this paper, we only present

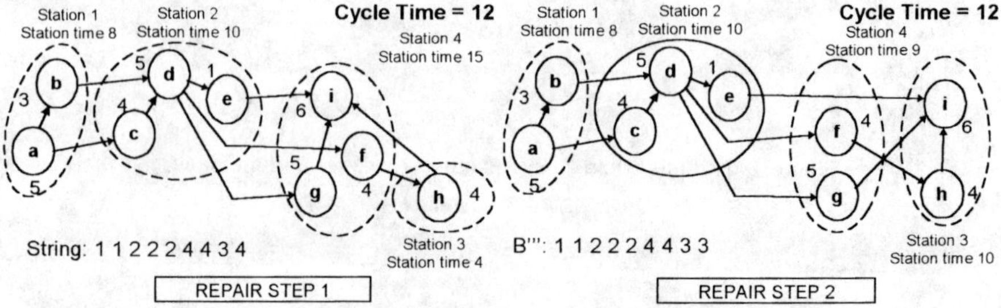


**Figure 4: Repair of the second child (B" to B'").**

the results of the 21 task problem proposed by Mitchell (Baybars, 86). This problem has been widely used by assembly line balancing researchers in testing their algorithms. The objective was to minimise the number of workstations. Both the one point crossover and uniform crossover operators were tested. A Pentium 100 Mhz PC with 16 MB of RAM computer was used for the experiments. The software was written in the Borland C++ language.

**Experimental Settings:**

| | | |
|---|---|---|
| • Cycle time | : | 35, adopted from (Baybars, 86) |
| • Crossover Operator | : | Single point and uniform crossovers |
| • Fitness Function | : | $Y = (X)^{-\log(X)}$ 10E2 |
| | | X = the total number of stations |
| • Number of generations | : | 1000 |
| • Population size | : | 20, 30, 40, 50, 80, 100, 200 |
| • Probability of crossover | : | 0.7, 0.8, 0.9, 0.95, and 0.99 |
| • Probability of mutation | : | 0.001, 0.005, 0.01, 0.02, and 0.05 |
| • Number of simulations | : | 100 |

**One Point Crossover**

Fig. 5 shows the experimental results of varying the crossover probability for a fixed mutation probability of 0.001 and population size of 50. Fig. 6 shows the results of varying the mutation probability for a fixed crossover probability of 0.95 and population size of 50. According to the experiments using all combinations of parameters shown in Section 3.1, the genetic algorithm performs better with the combination of a crossover probability of 0.95, a mutation probability of 0.05, and a population size of 50. No improvement was observed using either higher crossover or mutation probabilities. The maximum fitness value of 146.34 (4 workstations formed) was obtained in each of the 100 trials. The average running time is 6.62 minutes per run. The minimum number of workstations found previously by other researchers was also 4 (Baybars, 86). Therefore, the genetic algorithm was successful in achieving similar results. The best solution found by the genetic algorithm is shown in Table 1.

**Uniform Crossover**

The performance of the uniform crossover operator is similar to the one point crossover. Even though both crossover operators can achieve similar final results, a general observation is that

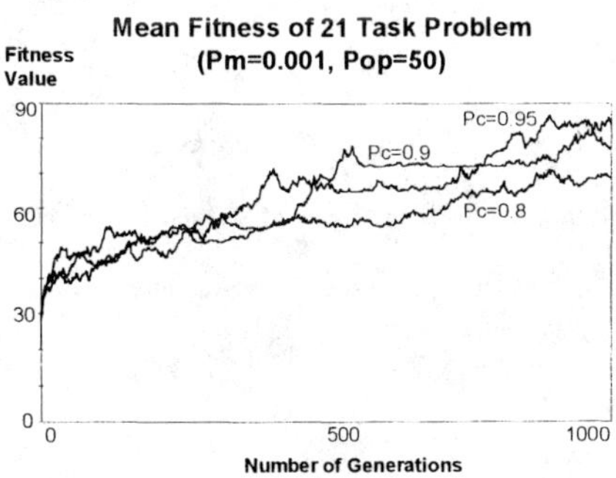


**Figure 5: Mean fitness values for different crossover probabilities**

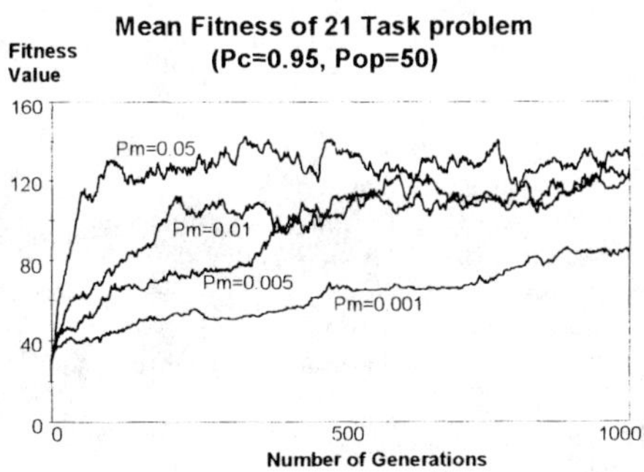


**Figure 6: Mean fitness values for different mutation probabilities**

| TABLE 1: ONE OF THE BEST WORK ASSIGNMENTS | | |
|---|---|---|
| Workstation | Assignments | Station time |
| 1 | { 1, 2, 3, 4, 5} | 30 |
| 2 | {6, 7, 14, 21} | 22 |
| 3 | {8, 9, 10, 11, 12, 13, 15, 18} | 35 |
| 4 | {16, 17, 19, 20} | 21 |

the uniform crossover is more stable, not as oscillating as the one point crossover. The uniform crossover tends to converge and improve steadily and can maintain the good solutions until the end of the experiments. The best solution found was also 4 workstations, using the following parameters: crossover probability of 0.95, mutation probability of 0.05, population size of 100, and 1000 generations. The best fitness value found was the same as that found by the one point crossover operator, 146.34. The minimum number of workstations (4) was also found in each of the 100 trials.

## DISCUSSION AND CONCLUSION

This paper has presented the development of genetic algorithms for the simple assembly line balancing problem. One of the difficulties in the implementation was the initialisation of strings. Due to the precedence and cycle time constraints that must be satisfied, the chance to obtain a feasible initial solution at random was extremely small. Therefore, a heuristic procedure was developed to generate initial strings at a random but guided manner so that only feasible solutions would be generated. The strings were encoded in decimal integers representing the workstation numbers to which tasks are assigned. Two crossover operators were tested: the conventional one point crossover and uniform crossover operators. Again due to the strict precedence and cycle time constraints, the crossover operations frequently generate infeasible children from feasible parents. To ensure every generation will consists of feasible solutions only, a repair routine was developed to convert infeasible children to feasible children. This approach worked well with both crossover operators on a 21 task problem, used by many assembly line balancing researchers. Comparable results were achieved for the minimisation of number of workstations. Although both crossover operators achieved similar results, the uniform crossover was more stable and improved more steadily over generations. This research is continuing in a number of areas including: refinement of the GA, multiple objectives, and mixed-model assembly line balancing.

## REFERENCES

Arcus, L. A. (1966). COMSOAL - A Computer Method of Sequencing Operations for Assembly Lines, The International Journal of Production Research, 4, No. 4.

Baybars, I (1986a), An Efficient Heuristic Method for Simple Assembly Line Balancing Problem, International Journal for Production Research, 24, 149 - 166.

Bhattacharjee, T. K. and Sahu. S. (1989), A Comparative Study of Heuristic General Assembly Line Balancing Techniques, Asia Pacific Journal of Operation Research, 6, 63 - 76.

Goldberg, D. E. (1989), Genetic Algorithm in Search , Optimization, and Machine Learning, Reading, MA: Addison - Wesley.

Hoffmann, T. R. (1990), Assembly Line Balancing : A Set of Challenging Problems, International Journal of Production Research, 28, No. 10.

Michalewicz, Z (1994), Genetic Algorithms + Data Structures = Evolution Programs, Berlin: Spriger - Verlag.

Tonge, F. M. (1960), Summary of A Heuristic Line Balancing Procedure, Management Science, 7, No.1.

# USING GENETIC ALGORITHMS TO DESIGN COMBINATIONAL LOGIC CIRCUITS

**CARLOS A. COELLO COELLO AND ALAN D. CHRISTIANSEN**
*Department of Computer Science, Tulane University, New Orleans, LA 70118, USA*

**ARTURO HERNÁNDEZ AGUIRRE**
*Department of Computer Science, Tulane University, New Orleans, LA 70118, USA*

***ABSTRACT:***
We introduce a method, based on a Genetic Algorithm (GA) approach, to design combinational logic circuits. This problem is quite difficult for a traditional GA, but we have overcome these difficulties and have implemented a computer program that can automatically generate high-quality circuit designs. We describe the important issues to consider when solving this circuit design problem: the importance of the representation scheme, the encoding function, and the definition of the fitness function. We compare the solutions produced by our system against those generated by a human designer. We also show that our representation approach, when compared to a standard binary encoding, produces better performance both in terms of quality of solution and in terms of speed of convergence.

## INTRODUCTION

Design is usually considered to be an activity requiring considerable human creativity and knowledge. Even the definition of the term *design* itself is quite elusive, since it can be interpreted in several different ways depending on the task to be performed. Although there have been many attempts at developing programs for automated design, such programs are notoriously difficult to build.

In the research reported in this paper, we seek a computer-based tool that can make the design process less tedious for the human designer without sacrificing quality of the design produced. In this paper, we limit our focus to combinational logic circuits, which contain no memory elements. Such circuits contain no feedback paths.

## PREVIOUS WORK

A general search technique inspired by natural evolution, called the *genetic algorithm* (Holland, 1992), has been widely used for optimization tasks (Goldberg, 1989) and is known to be a very powerful tool in certain domains. In our current work we wish to find a way to use the genetic algorithm (GA) as a design tool, with particular emphasis in the design of combinational circuits. Louis (1993) is one of the few sources found in the literature to address the use of GAs for the combinational logic design problem. In his dissertation (Louis, 1993) Louis combines knowledge-based systems with the genetic algorithm, making use of a genetic operator called *masked crossover* that adapts to the encoding, being able to exploit information unused by classical crossover operators. His results, although very encouraging for certain examples, do not seem to have solved the combinational circuit design problem completely. Koza (1992) has used genetic programming to design combinational circuits. He has designed, for example, a two-bit adder, using a small set of gates (AND, OR, NOT), but his emphasis has been on generating functional circuits rather than on optimizing them. In fact, this is also the case of Louis' research, where the main focus was to provide an easier way to generate functional designs using the GA rather than in optimizing a functional design according to certain metrics.

In the work reported here, we are interested not only in producing functional designs, but also in optimizing them according to certain metrics. This is a quite complicated task for the GA, because we must deal with two difficult problems at the same time.

## STATEMENT OF THE PROBLEM

The problem of interest to us consists of designing a circuit that performs a desired function (specified by a truth table), given a certain specified set of available logic gates. The complexity of a logic circuit is a function of the number of gates in the circuit. The complexity of a gate generally is a function of the number of inputs to it. Because a logic circuit is a realization (implementation) of a Boolean function in hardware, reducing the number of literals in the function should reduce the number of inputs to each gate and the number of gates in the circuit—thus reducing the complexity of the circuit. The algebraic method used to minimize functions is tedious and error prone. Its success depends on our ability to recognize the application of a theorem or a postulate during the minimization process. Such recognition may not be obvious. Furthermore, there is no general set of rules to aid that recognition.

Two popular minimization techniques are the *Karnaugh Map* (Karnaugh, 1953), which is based on a graphical representation of Boolean functions, and the *Quine-McCluskey Procedure* (Quine, 1955; McCluskey, 1956), which is a tabular method. Both of these methods are mechanical in nature, and their efficiency depend on the designer's abilities.

In this work, we compare the designs produced by a GA with those generated by a human designer using Karnaugh maps. The comparison is in many ways unfair because of differing capabilities of man and machine. For example, a human designer tends to use only the gates NOT, AND, OR and has more difficulties using XOR because the Karnaugh Map does not support the identification of XOR as well as it supports "seeing" simple product terms. Our overall measure of circuit optimality is the total number of gates used, regardless of their kind. This is approximately proportional to the total part cost of the circuit. Obviously, we perform this analysis for only fully functional circuits.

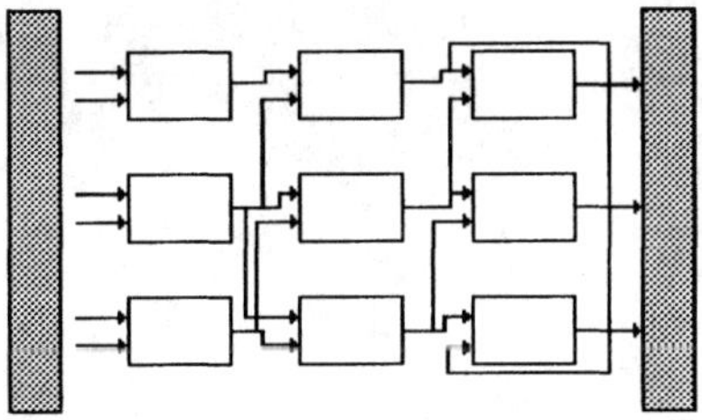


**Figure 1: A gate in a two-dimensional template, gets its second input from either one of the two gates in the previous column.**

**Example 1: A Two-Bit Adder**

We want to find the combination of five possible types of gates (AND, NOT, OR, XOR and WIRE) so that our circuit performs the two-bit addition of its inputs, producing a three-bit sum. Our objective is to find a functional design that minimizes the use of gates other than WIRE (essentially a logical no-operation). The circuit can be represented as a two-dimensional array of gates $S_{i,j}$. The index $j$ indicates the *level* of the gate, where gates closest to the inputs have the lowest values of $j$. (Level numbers increase from left to right in Figure 1.) For fixed $j$, the index $i$ ranges over gates that are "next to" each other in the circuit, but such gates are not directly connected to each other. Each gate can get its inputs from any row (any gate in the next smaller level) in the matrix.

**Example 2: A Two-Bit Multiplier**

This problem is similar to the previous one, but in this case we want to design a circuit that performs the two-bit multiplication of its inputs, producing a four-bit result. The set of gates available is the same as in the previous example, and again, each gate can get its inputs from any gate in the previous level according to our matrix representation.

## USING THE GENETIC ALGORITHM

The first interesting aspect of this problem is the encoding of solutions. Each circuit is encoded in the following way: *<input1><input2><gate_type>*, where *input1* and *input2* can be any of the gates or inputs at the previous column in the matrix representing the circuit (see Figure 1), and *gate_type* is the gate that we will be using for that particular position of the array. A chromosome is formed with as many triplets of this kind as needed, according to the size of the matrix that the user wants to use to represent the circuit. In our experiments, both binary and floating representation were used, and the order chosen for the gates was column-order, starting from the position $S_{1,1}$ (the top leftmost position), followed by the gate below it ($S_{2,1}$), and so on.

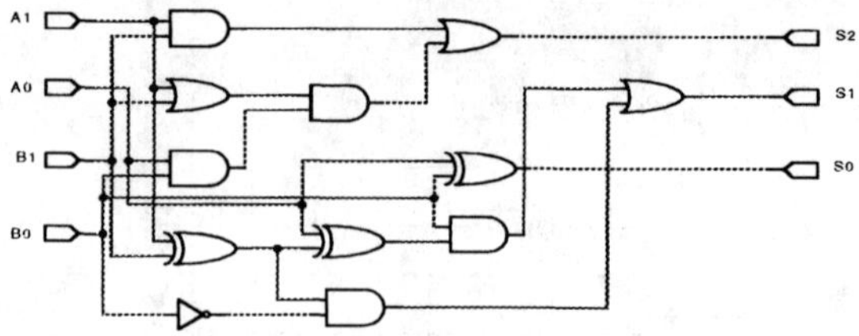


**Figure 2: A 2-bit adder generated by a human designer**

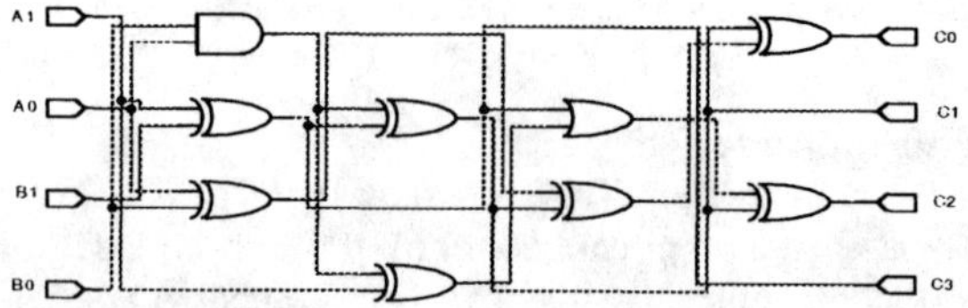


**Figure 3: A sample 2-bit adder generated by our GA-based approach using binary representation**

We also experimented with an alphabet of cardinality *n*, where *n* can be defined by the user and will be normally taken as the number of rows allowed in our circuit, according to the matrix encoding adopted in this problem. This representation allows the manipulation of shorter strings, it decreases the complexity of the decoding task, and it provides better solutions (Coello, 1996).

Another difficulty is the development of a good fitness function. Our approach compared the output produced by the circuit generated by the GA with the desired values according to the truth table, on a bit-per-bit basis. Our fitness function works in two stages. At the beginning of the search, only validity of the circuit outputs is taken into account, but once functional solutions start appearing in the population, we switch to a second fitness function which tries to minimize the number of gates by rewarding those circuits that use more WIREs.

## COMPARISON OF RESULTS

We compared, for both examples, our results with those produced by an experienced human designer. Populations of 3000 chromosomes were used in all tests, and the GA was run for 200 generations.

### Example 1: A Two-Bit Adder

Figure 2 shows a functional circuit generated by a human designer using Karnaugh maps. In this example, a fitness of 48 or higher indicated a functional circuit. To solve this problem, we first used a traditional GA with binary representation, and the solution generated is shown in Figure 3. This circuit has only 9 gates, saving 3 from the human solution previously shown. However, we were convinced that we could do better using a higher cardinality alphabet, so we decided to experiment with a floating point representation that used the integers from 0 to 3, to represent 4 rows of inputs. The solution found was quite efficient, since only 7 gates were required (see Figure 4).

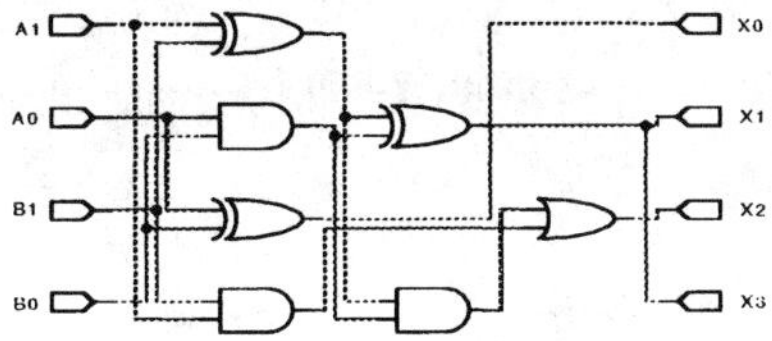


**Figure 4: A sample 2-bit adder generated by our GA-based approach**

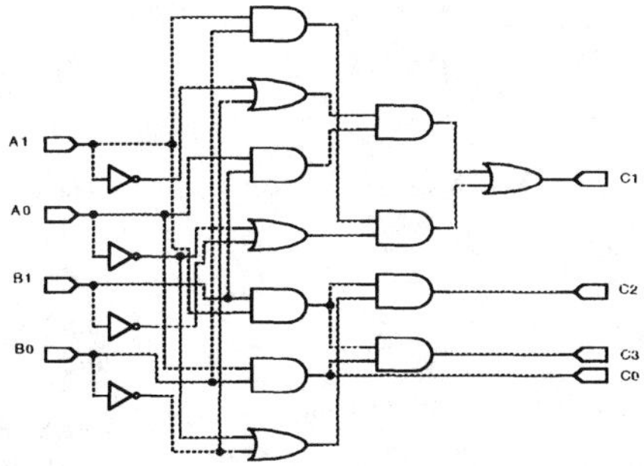


**Figure 5: A 2-bit multiplier generated by a human designer**

**Example 2: A Two-Bit Multiplier**

This is a more complicated problem in which we compared again our solution with the circuit generated by a human designer (see Figure 5). Once more, the solution generated by the GA using binary representation was somewhat disappointing, because it uses one more gate than the solution generated by a human (see Figure 6). Then, we decided to try a GA with an alphabet of cardinality 8 (using the integers from 0 to 7). With only 7 gates, we generated a circuit equivalent to the one produced by a human designer using 16 gates. In this context, a fitness of 64 represents a functional circuit, and any increment above that indicates the number of WIREs used. For the GA with binary representation, the maximum fitness achieved was only 79, whereas for the GA with our alternative representation, we could find a design with a fitness of 89, as can be seen in our results.

## CONCLUSIONS AND FUTURE WORK

We have demonstrated a fully implemented genetic algorithm approach to the combinational logic circuit design problem, and we have compared the results obtained by our system to circuit designs produced by experienced human designers. The results for our two simple multiple-output logic functions are quite encouraging. The automated designs use fewer gates than the human designs, and can be obtained in a few minutes by our program. In the future, we will determine how our GA-based approach scales to larger circuits. We are interested in building a system that can automatically design a wide range of combinational and sequential logic circuits. We are also interested in exploring optimization tradeoffs involving number of levels in the circuit, total number of gates, number of gate inputs, number of integrated circuit packages, part costs, etc. We believe that a GA-based approach has great potential to provide a practical tool for assisting designers of logic circuits.

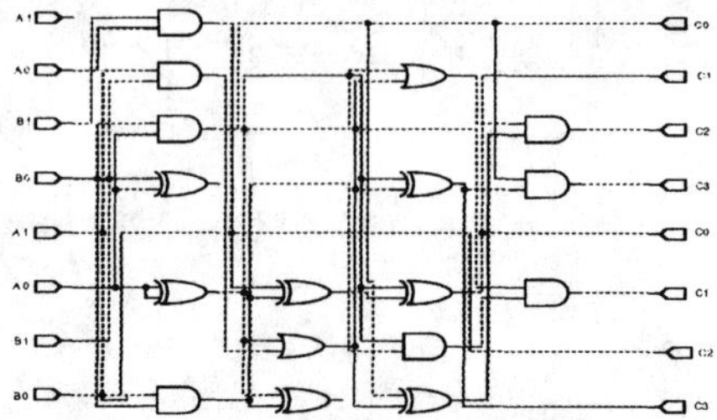


**Figure 6: A 2-bit multiplier generated by a GA using binary representation**

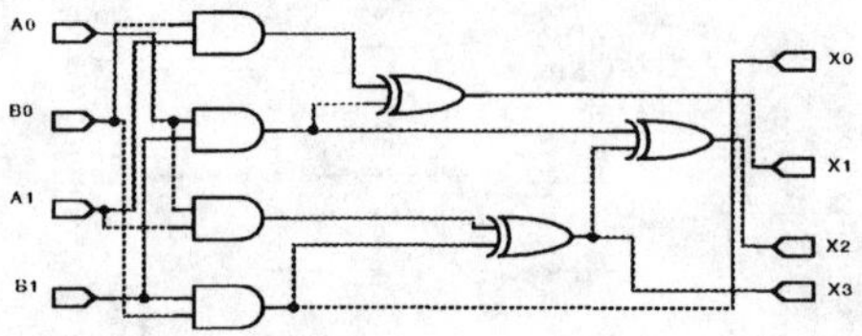


**Figure 7: A 2-bit multiplier generated by a GA using our approach**

## REFERENCES


Coello, C. A. (1996). An Empirical Study of Evolutionary Techniques for Multiobjective Optimization in Engineering Design. Ph.D. Dissertation, Department of Computer Science, Tulane University, New Orleans, LA.

Goldberg, D. E. (1989). Genetic Algorithms in Search, Optimization and Machine Learning. Reading, Mass. : Addison-Wesley Publishing Co.

Holland, J. H. (1992) Adaptation in Natural and Artificial Systems. An Introductory Analysis with Applications to Biology, Control and Artificial Intelligence. MIT Press, Cambridge, Massachusetts.

Karnaugh, M. (1953) A map method for synthesis of combinational logic circuits. *Transactions of the AIEE, Communications and Electronics* 72 (I):593-599.

Koza, J. R. (1992) Genetic Programming. On the Programming of Computers by Means of Natural Selection. The MIT Press.

Louis, S. J. (1993) Genetic Algorithms as a Computational Tool for Design. Ph. D. Dissertation, Department of Computer Science, Indiana University.

McCluskey, E. J. (1956) Minimization of boolean functions. *Bell Systems Technical Journal* 35 (5):1417-1444.

Quine, W. V. (1955). A way to simplify truth functions. *American Mathematical Monthly* 62(9):627-631.

# Application of a Modified Genetic Algorithm to Parameter Estimation in the Petroleum Industry

**M.D.Bush** email: mdbush@lon.amoco.com
EPTG, Amoco Services Ltd, Amoco House, West Gate, London, W5 1XL, UK.

**Dr J.N.Carter** email: j.n.carter@ic.ac.uk
Centre for Petroleum Studies, Imperial College(ERE), London, SW7 2BP, UK.

**Abstract:**
In the petroleum industry, large numerical simulators are used to approximate the flow of oil, water and gas in the reservoir rocks. It is possible to estimate, from geological information, the likely ranges of the parameters in the numerical model. However, to obtain better estimates of the parameters, the output of the simulator is matched to the measured history of the reservoir, by iteratively adjusting the model inputs until the error between the model outputs and the measured history is minimised. It often happens that the history can be matched by more than one set of parameters. The challenge is to identify all possible matches in as few function evaluations as possible.

To allow us to search the whole of the parameter space, whilst minimising the number of function calls, we use the following modifications to a GA: sampling theory to determine the chromosome length and the nearest approach to an optimum; a non standard method to encode the parameters into binary strings; modified niching and breeding strategies.

Using a simple reservoir model, we compare the results of our approach, with those of a standard GA. The modified GA shows significant gains compared to the standard GA.

## INTRODUCTION

In the petroleum industry use is made of large numerical simulators to estimate the flow of oil, water and gas through the rocks that form the reservoir. For a typical reservoir a single simulation may take several hours CPU, even on a supercomputer. When a simulator is used to evaluate the object function for a GA, the costs of these evaluations are high both in terms of time and money.

When we try to match our simulation model to the measured history of a reservoir, by adjusting some of the model parameters, it often happens that more than one optimum exists, ie we can find more than one set of parameters which reproduce the measured history of the reservoir. In general we do not know in advance how many optima there are. The challenge is to identify all the optima, using as few function evaluations as possible.

Genetic Algorithms are one possible way of approaching this problem. They allow the global parameter space to be explored using only function evaluations. The various niching strategies that have been proposed would allow the identification of multiple optima. A potential draw-back is the slow convergence of the GA. This difficulty can be alleviated by stopping the GA before it has converged and switching to local search methods.

In this paper we discuss the various modifications that we have found to be of value in optimising our models, whilst minimising the total number of function calls. These involve the use of a non-standard encoding method, modified niching and breeding strategies, and the use of sampling theory to determine chromosome length.

Finally we demonstrate the performance gain of our GA, compared with a standard GA, on a simple oil reservoir problem.

## PETROLEUM ENGINEERING ASPECTS

### History Matching

How fluids flow through rocks, is commonly studied by making use of large numerical simulators. In such a simulation the oil reservoir will be modelled on a mesh represented by discrete grid cells. There may be several hundred thousand cells, each of which has several parameters associated with it. Parameters include: the ratio of fluid to solids (porosity), how easily the fluid can flow through the rock (permeability) and the ratios of different types of fluid (saturation).

Because of the location of the rocks, thousands of metres below ground and spread over many kilometres, direct measurement of the parameters is impossible. A few of the parameters may be estimated from measurements made at wells drilled into the reservoir. The other parameters can only be inferred from experiments on the system. These experiments usually take the form of either injecting or extracting fluid, and measuring the response at a number of points.

The process of obtaining the actual parameter values from the measured response is solved by the following iterative method. A set of parameter values is chosen, then the simulation is run and the results compared to the measured response (the 'history'). This process is repeated with a new set of parameter values until a match between the simulation and the history is obtained. In general a smaller set of meta-parameters are modified, rather than the model parameters themselves (in this case three meta-parameters rather than 4000 model parameters). The difficulty with this process is that inordinately large amounts of computing resources are needed. A single simulation, for a single set of parameters, can take several hours, even on a supercomputer. To obtain a satisfactory history match many simulations have to be run.

### Reservoir Model

The model used is a vertical cross section between an injection well and a production well with an intervening fault (figure 1). The reservoir has six geological layers alternating between shale and sand. If there is no loss of permeability at the fault, e.g. due to clay smearing, then the transmissibility of fluids across the fault will rise and fall as the throw on the fault is changed. The model was developed as part of an European Union research project (REP, 1996).

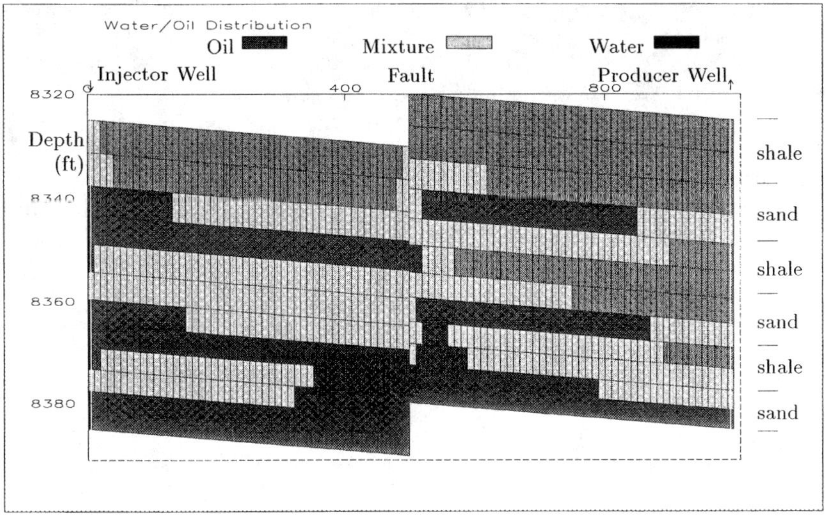


Figure 1: Distribution of Oil and Water Within the Reservoir

The simulation model consists of 1200 cells, the wells are situated in columns 1 and 100 of the model, the fault is exactly half way between the wells. At the fault the section containing the injector well is lower (down-thrown) relative to the section containing the producer well. However, at the positions of the wells the reservoir depths are equal. The sand and shale have constant properties throughout the reservoir. There is no gas cap and the initial water saturation is uniform. The parameters to be history matched are the sand permeability $K$(sand), the shale permeability $K$(shale) and the throw on the fault, everything else is kept constant. The production policy is that both wells are completed across all layers. The producer is held at constant bottom hole pressure and the oil production rate is monitored. Water is the injected fluid and the injection well is controlled on fixed bottom hole pressure again.

To obtain a "measured" history the following approach was used. First a set of reservoir parameters (158.6mD, 2.0mD, 10.3 feet) was selected in an arbitrary manner; then the model was run to simulate three years of production, the oil production rate (in stock tank barrels per day - STB/d) was noted at the end of each six month period. A standard deviation for the error of each measurement was assumed, and a random error added/subtracted from each recorded value to obtain the "measured" history. The values are given in table 1, figure 1 shows the distribution of oil and water at the end of the history period for this "truth" model.

## GENETIC ALGORITHM

### Determination of Chromosome Length

It is well understood that the number of binary bits, is related to the number of functions calls need to locate an optimum. The chromosome therefore should be as short as possible, particularly when function calls are expensive. To determine the optimum length of a chromosome we have used sampling theory.

From sampling theory it is well known that the smallest object that can be reliably detected, has a length equal to twice the sampling interval being used. Consequently, if you can specify the smallest structure containing an optima that would be of interest, you can specify the maximum chromosome length.

Because it is important in our problems to minimise the total of number of function cells, and it is known that GAs converge only slowly to the final optima, we use the GA only to locate areas likely to contain an optimum. Having done this, we switch to a local search method.

To further enhance the search efficiency, we start with a short chromosome, allowing only large structures to be identified. In subsequent generations, the chromosome is lengthened so allowing smaller structures to be identified. There is a danger in this process that small isolated structures might be missed. However, from a physical understanding of the problem such structures would not be expected.

In the case reported in this paper initially a $(3 \times 4)$ bit chromosome was used, then after ten generations this was increased to $(3 \times 8)$ bits per chromosome, and new random individuals introduced into the population to provide examples of the new low-significant bits.

### Problems of Ghost Clusters

In our initial studies using a two parameter analytical model, with two known optima, we used a simple graphical method to determine if a potential optimum area had been located. That is to say we plotted the positions of all the solutions generated throughout a run. A typical plot is shown on the left hand side of figure 2. What we saw were four clusters rather than the expected two. The real clusters where centred on (50,100) and (100,50), the ghost clusters where centred on (50,50) and (100,100). Clearly breeding between solutions in different real clusters was taking place. Examining only the final populations showed that it was mainly composed of solutions from the real clusters, so our niching scheme was working. However, the ghost clusters represented wasted effort.

The ghost clusters are generated when parents from different real clusters are mated and the children inherit the high order bits for the first parameter from one parent and the high order bits for the second parameter from the other parent. This tendency was easily removed by changing the way the chromosome was constructed, from concatenation to inter-leaving.

$$\begin{array}{ll} \text{P1} \quad a_1a_2a_3a_4 \\ \text{P2} \quad b_1b_2b_3b_4 \end{array} \left\} \begin{array}{lll} \text{concatenation} & \Rightarrow & a_1a_2a_3a_4b_1b_2b_3b_4 \\ \text{inter-leaving} & \Rightarrow & a_1b_1a_2b_2a_3b_3a_4b_4 \end{array} \right.$$

This simple modification gave the results shown on the right hand side of figure 2.

### Modified Niching and Breeding Strategies

During our experiments we were forced to develop complicated breeding and niching strategies, which are described below, to allow us to find optima whilst minimising the total number of function calls. Our problem arose because we

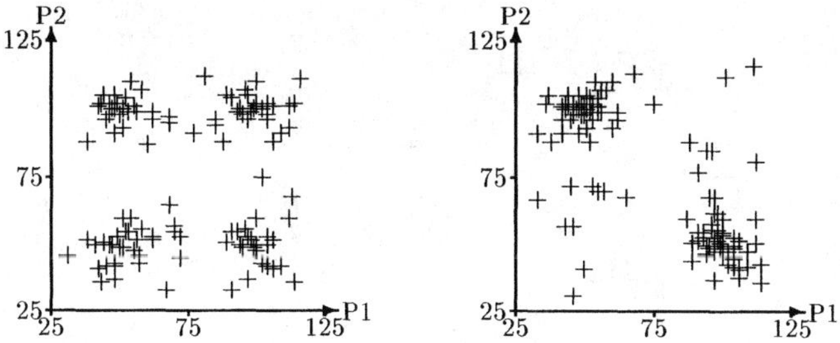


Figure 2: Ghost Clustering

tried to force rapid convergence, whilst trying to avoid premature convergence. We had difficulty in maintaining diversity with the small populations used.

Our first modification was not to allow any individual to die out. Hence, all previous generated individuals were available for breeding at each generation. So after each breeding cycle the size of the 'total' population increases. This allows us to maintain diversity in the parent population.

From this total population we generate a breeding population of unique individuals. The size of the breeding population is less than the size of the child population, this is similar to the ($\mu$,$\lambda$) method of Bäck (1995). The breeding population is selected as follows: rank the total population according to function values; copy the best individual; select the next best individual with a 50% probability; repeat the last step until the breeding population is filled.

We now generate a parent population, of size equal to that of the required child population. Using the concept of shared function values (Goldberg, 1987) and linearly scaled fitness functions (Goldberg, 1989), we randomly clone members of the breeding population. The final stage in the breeding cycle is to pair-off the parents, perform crossover and mutation. Pairing off is achieved by randomly choosing a parent from the parent population. Each of the remaining individuals is assigned a weighting depending on its distance from the selected parent, with nearer individuals having higher weights. A random choice is then made for the second parent. A random $k$-point crossover is used to produce two children ($k$ is strongly biased towards low values), finally mutation may be applied. The size of the parent population has now been reduced by two, and the process repeated.

**Results**

To demonstrate the improvements that our modifications have achieved we present the results of trying to match the reservoir model described earlier. We compare the results of a simple GA, with our version. Both cases start with 50 individuals in the breeding population, after ten generations the chromosome size is increased and the breeding population doubled. After a further 20 generations the GA is stopped, a total of 2500 function evaluations. The simple GA uses concatenated chromosomes with an elitist generational replacement policy, and random $k$-point crossover. The comparison is made by searching the final total population in each case for clusters, and recording the best individuals.

| True History | True + Noise = Measured History | | Simple GA | Modified GA Cl. 1 | Modified GA Cl. 2 |
|---|---|---|---|---|---|
| | | $K$(sand) | 156.8 | 164.7 | 149.0 |
| | | $K$(shale) | 11.84 | 12.12 | 8.16 |
| | | throw | 29.6 | 28.9 | 13.6 |
| 552.4 | 571.3 | 6 months | 547.1 | 565.4 | 550.3 |
| 564.1 | 559.7 | 12 months | 557.1 | 574.9 | 561.5 |
| 552.3 | 554.2 | 18 months | 545.7 | 557.3 | 555.1 |
| 557.4 | 545.4 | 24 months | 536.5 | 554.3 | 555.2 |
| 498.4 | 499.9 | 30 months | 486.5 | 500.2 | 509.5 |
| 354.6 | 355.1 | 36 months | 364.8 | 350.6 | 346.7 |
| | | $\sum(\text{error})^2$ | 1.71 | 0.48 | 1.05 |

Table 1: Cluster Positions and Production Data (STB/d)

The results for the two methods are given in table 1, the objective function is a sum of error squared. The simple GA developed just a single cluster, while the modified GA developed two clusters, the first of which is related to the cluster identified by the simple GA. This cluster is clearly not related to the parameters of the 'real' reservoir. However, the local optimum in this region has been shown, by other work, to be the global optimum. The second cluster is close to the point that generated the history, and the associated local optimum is closer still.

## CONCLUSION

We have shown that GAs can be usefully applied to problems where the aim is to identify regions likely to contain optima. We can then use standard methods to improve estimates within these regions. Our modifications to the GA has given substantial improvement over the simple GA.

Further work needs to be done to improve the early identification of real clusters. Once a cluster has been identified, then the GA needs to be persuaded to search elsewhere. This is particularly difficult given that the number of optima and their distribution in parameter space is unknown.

## REFERENCES


Bäck, T, (1995). Generalized Convergence Models for Tournament and ($\mu$,$\lambda$) Selection, *Proceedings of the Sixth International Conference on Genetic Algorithms*, 2, ed LJ Eshelman.

Goldberg, DE, and Richardson, J, (1987). Genetic Algorithms with Sharing for Multimodal Function Optimisation, *Proceedings of the Second International Conference on Genetic Algorithms*, 41, ed JJ Grefenstette.

Goldberg, DE, (1989). *Genetic Algorithms in Search and Machine Learning*, p77, publisher Addison-Wesley.

Final Report to the European Union on the Reservoir Engineering Project, (1996), to be published.

# APPLICATION AND EVALUATION OF GENETIC PROGRAMMING FOR AIMPOINT SELECTION

**CAREY SCHWARTZ, CHARLES KEYES , ERIK VANBRONKHORST**
*Naval Air Warfare Center Weapons Division, China Lake, CA*

**ABSTRACT:**
We have applied Genetic Programming techniques to the determination of ship aimpoints from an input vector. The problem we have studied reduces to mapping a multidimensional input vector to a single real output describing one coordinate of the aimpoint. The multidimensional input vector is obtained by segmenting an image of the ship and performing spatial averages. We have found that Genetic Programming techniques can be used to provide a solution to the problem which maps the inputs to the outputs within the quantization level of the problem. For purposes of comparison we have developed artificial neural network and classifier solutions to the same problem based upon C4.5.

## INTRODUCTION

A number of problems of military interest are such that while a solution is desired, a technique which guarantees an optimally performing solution is not known. Problems in this class include the design of non-linear control systems, route planning for long range weapon systems, and many areas in the field of automatic target recognition. A partial listing techniques employed to attack these problems include feed forward neural networks (Rosenblatt, 1962), adaptive resonance theory neural networks (Carpenter and Grossberg, 1987), fuzzy logic (Zadeh, 1979), and genetic algorithms (Holland, 1975;1989). However, in some of the techniques, particularly neural networks, it is difficult to discern the "rules" that have been developed to solve the problem thereby providing little or no insight into the solution.

Recently a new paradigm has emerged, genetic programming (GP) which, if it can be successfully applied to a problem, addresses this concern. GP is closely related to genetic algorithms (GA) and borrows considerably from the machinery of GA. Details regarding GP can be found in Koza (1992; 1994). GP not only provides a solution to the problem, the solution is in the form of a computer program that can be used to extract rules, perform sensitivity analysis, or even be employed to generalize. The information extracted from a GP program is similar to that obtained from a classification tree generated by systems such as ID3 (Quinlan, 1979) or C4.5 (Quinlan, 1993). However, unlike the classification tree approaches, there is no requirement to pre-assign class memberships or be constrained solely by the elementary comparison operators utilized by ID3 and C4.5.

On the surface GP is an attractive alternative technique for generating solutions to the types of problems previously described. In this paper, a comparative study is undertaken to assess the performance of GP on a real world aimpoint selection problem, which has been previously solved using a feed-forward neural network. The performance of GP is compared against the neural network and a solution generated using Quinlan's C4.5 classification system.

## STATEMENT OF PROBLEM

Autonomous aimpoint selection for an anti-ship missile is a complex process that is discussed elsewhere (McCauley, 1993). We have used a database containing 192 digitized ship profiles and their designated aim-points as inputs for this study. Briefly, a ship's signature is obtained by filtering an IR image for vertical and horizontal edges and then generating a closed contour. A broadside view of the ship with the bow oriented to the left of the image is assumed. Any variance in perspective introduces noise by foreshortening the ship's profile. A robust solution is desired which provides the correct aimpoint given an input vector containing averaged ship heights on a fixed spatial grid. The data set was previously used in pursuing a three-layer feed-forward neural network solution (McCauley, 1993). McCauley showed that a three-layer neural network could successfully compute aimpoints from discretized silhouette data within specified error tolerances. His work provides a basis of comparison for examining the performance of neural networks, genetic programming, and decision.

## NEURAL NETWORK APPROACH

McCauley previously used a three-layer (20-10-2) feed-forward neural network trained using back propagation. The inputs consisted of the 20 normalized height samples and the output layers yielded the X and Y aimpoint (McCauley, 1993). McCauley used 163 randomly selected ships from the database and we have extended the data set to include 192 cases. In this study training was accomplished with back propagation with nominal values of 0.1 for the gain (learning rate) and 0.9 for the momentum term (Rumelhart, Hinton, and Williams, 1986).

Results of the aimpoint computations for three different feed-forward neural network architectures have been summarized elsewhere (Schwartz, 1996). We examined several three-layer architectures with varying numbers of hidden units. Training of the neural network was stopped when the mean squared error computed over the training patterns reached 0.0001 or reached a local minimum where no further reduction in error was achieved. The results show that the choice of network architecture for this problem is important for the aimpoint computation. McCauley's previously reported 20-10-2 architecture correctly computed aimpoints for the subset of 163 ships in his work but failed to converge to a 100% correct solution for the complete set of 192 ships. Numerous unsuccessful attempts at randomly re-initializing the initial neuron weights were attempted to escape the premature convergence reached after 150,000 training epochs. As in McCauley's work, neuron weights were updated after the presentation of each input pattern rather than a single update per epoch. The fixed sequential order of presentation of the ship patterns to the network was not varied to maintain consistency with McCauley's previous work.

A more optimal architecture was found consisting of 20 input neurons, 15 neurons in the hidden layer and the 2 output neurons corresponding to the aimpoint coordinates. This configuration resulted in a network which converges to a 100% correct solution, independent of the randomly selected initial weight settings, after approximately 50,000 training epochs. We also examined a 20-5-2 configuration for completeness and found little change in the performance and considerable decrease in training epochs relative to the 20-10-2 architecture. The key point here is that a feed-forward neural network is able to develop a mapping of the input 20 vector to a two vector representing the coordinates of the aimpoint.

## GENETIC PROGRAM (GP) APPROACH

GP is an extension of the genetic algorithm technique developed by John Holland (1975). GPs are distinguished from genetic algorithms in that the structure of the population are computer programs rather than a strings (Kinnear, 1994). It is convenient to think of a computer program as a tree with the program's operands residing on leaf nodes and its operands on the remaining nodes. The number of branches and depths of the branches of any individual may change as the individual moves from generation to generation. Programs move from generation to generation based upon a measure of fitness as in a genetic algorithm. Programs that show a higher degree of fitness receive greater opportunity for reproduction into successive generations than programs which show little propensity for solving the problem at hand. Koza provides in-depth discussion on reproduction, recombination, and mutation of individual programs in Koza, 1989. The process of forming new programs terminates after an exact solution to the problem has been determined or a specified number of generations of programs has been developed. In the course of this work we used our own implementation of the GP algorithm as well as the lil-gp v1.0 GP kernel (Punch and Zongker, 1995). All code was run on an eight processor Silicon Graphics Challenge XL.

The representation of the ship aimpoint problem within the GP paradigm is identical to the Neural Network representation save that we have decoupled the calculation of the X and Y coordinates of the aimpoint. This approach required that two programs be evolved, independent of one another, to compute the X and Y aimpoints. We examined the issue of decoupling the calculation of both coordinates of the aimpoint for the neural net architecture and found no measurable difference in training times or aimpoint error relative to the 20-10-2 configuration. Thus we assumed that the GP system suffers no cross coupling penalty for independent computation of the aimpoint coordinates.

Initially, this work focused on computing the X coordinate of the aimpoint. In these GP runs only 100 of the available 192 input vectors were used as training data to reduce the complexity of the problem and minimize CPU cycles. The genetic programming paradigm, without automatically defined functions (ADFs), is controlled by 19 control parameters which must be specified by the user (Koza,

1994). Typical control parameters employed in this portion of the study are presented elsewhere (Schwartz, 1996). The form of the fitness function initially selected for minimization is

$$\sum |y_i - Y_i|^{1/n} \tag{1}$$

with $y_i$ the calculated result of the GP for the ith input state and $Y_i$ the known coordinate (X or Y) for the ith state. Initially, n was set to 1/2 so that a conventional L2 norm was calculated corresponding to the same error measure used by the feed-forward neural network calculations. Details can be found in Schwartz, et. al. (1996).

Our initial results were disappointing. In examining the best performing programs it was observed that the genetic programs were attempting to create floating point constants from the available input terminals, which were added to the data set. We did this by augmenting the input vector to include two random floats. Multiple runs were made using different initial populations. Table 1 presents a summary of these results. It is of interest to note that the error is close to being the product of quantization error squared and the number of test cases.

**TABLE 1. SUMMARY OF THE RESULTS OF SEVENTEEN INDEPENDENT GP RUNS ATTEMPTING TO PRODUCE THE X COORDINATE OF THE AIMPOINT USING THE CONTROL PARAMETERS OF TABLE 2 AND THE INPUT VECTOR OF SHIP PROFILES FOR VARIOUS NORMING POWERS OF THE ERROR.**

The columns represent the minimum L2 error, maximum L2 error, average L2 error, and standard deviation of the L2 error associated with the best of breed individual in each run. The best of breed individual is defined to be the individual with the smallest norm observed over all generations of the run.

| N | Minimum Error | Maximum Error | Average Error | Standard Deviation |
|---|---|---|---|---|
| 1/2 | 0.251087 | 0.544061 | 0.416981 | 0.098336 |
| 1 | 0.298956 | 0.595493 | 0.435997 | 0.08960674 |
| 2 | 0.287563 | 0.594307 | 0.457337 | 0.09984284 |
| 3 | 0.396101 | 0.682000 | 0.504005 | 0.07267541 |
| 4 | 0.293819 | 0.627787 | 0.509942 | 0.09182974 |
| 5 | 0.309464 | 0.634111 | 0.483494 | 0.08534475 |

We note that altering the exponent in the norm has a discernible but small effect upon the average best of breed and the best of breed. If we examine the mean square error associated with the 100 cases treated using GP and the 192 cases employing neural networks, we see that there is a factor of 10 difference of performance in favor of the neural networks. In addition, the neural network employs roughly 40000 epochs in contrast to the 300 generations of 200 individuals allowed to the GP. While it is possible that allowing the GP a larger number of generations or increasing the width of the population to evolve might result in performance equivalent to the neural network, we are unable to afford the CPU cycles at this time.

An alternative scheme for converting the aimpoint selection problem to an optimization problem is to require the GP to return the aimpoint such that the returned aimpoint is within the quantization error of the problem, i.e. 0.05 corresponding to the division of the normalized X axis into twenty bins. If the GP returns an aimpoint that is within the quantization error the program receives a score of 1 (a hit) and otherwise receives a score of 0. The goal of the GP is then to maximize the score of hits. The lil-gp kernel was used exclusively in an effort to compute the X coordinates utilizing ADFs, ephemeral constants, and a function set consisting of the arithmetic operators, and a four argument IFLTE (If Less Than or Equal function. Numerous runs, corresponding to a multitude of configurations, were made over a period of months in an attempt to achieve success in computing X aimpoint coordinates. Population sizes were varied from 200 to 10,000 with evolution allowed to occur up to 4000 generations before terminating a run. Each configuration was run numerous times with different initial populations. Various selection schemes were tried (fitness-proportionate, tournament, greedy over selection) with no significant variation in results. Automatically defined functions were implemented utilizing argument lists containing up to five parameters. Multiple ADFs were permitted using differing scopes of visibility (which ADFs can invoke other ADFs). A systematic approach was initiated where the number of ADFs and corresponding parameter lists were modified in a manner that moved toward more complexity, but was abandoned due to combinatorial explosion, run times that exceeded weeks, and an absence of improvement over earlier baseline experiments. The best program ever generated correctly computed 166 X aimpoint coordinates. This program utilized two ADFs, with the first ADF using two arguments and the second ADF using four parameters. The population size in which the program was generated consisted of 2000 individuals and was found in generation 40. Evolution for this population was allowed to continue until 4000 generations had been created with no additional improvement realized.

## RESULTS USING DECISION TREES

Quinlan's follow on to ID3 (Quinlan, 1979) is a set of computer programs collectively called C4.5 (Quinlan, 1993), which build decision trees and production rules for classifying individual items into classes based upon attribute value descriptions of the items. The C4.5 classification system can accommodate continuous valued attributes such as the normalized elevation samples in our data set, which makes the system an attractive approach in determining target aimpoints from the sampled silhouette data.

The aimpoint problem described can be recast as a classification problem by partitioning the designated aimpoints into bins according to specified error criteria. Aimpoints sharing the same bin can be considered members of a class with the objective of the classification problem to identify the class membership of a ship based upon its sampled silhouette data. Once the class membership of a ship has been determined, table lookup provides the aimpoint coordinates shared by all members of the class. This transformed, supervised classification problem, differs from the neural network and genetic programming approaches in that this pre-assigned class membership information is provided to the decision tree system.

The aimpoints were grouped into 24 classes according to the two-dimensional 0.05 by 0.05 bin they fall into using a simple gridding scheme. System parameters for C4.5 were set to their default values with the exception of the -m parameter specifying the minimum number of cases which can be assigned to a class. Our simple gridding scheme forced classes with singleton members requiring the -m parameter to be set to 1 rather than the default value of 2. The decision tree generated by C4.5 was then able to correctly classify 191 of the 192 ships into their assigned aimpoint classes.

## COMPARISON OF TECHNIQUES

Any comparison of the performance of neural networks, GP, and C4.5 on the aimpoint problem must account for the differences in problem representation, expended computing effort, correctness, and robustness. The problem for the neural network and GP is one of finding a functional mapping from the input data to a computed aimpoint. The C4.5 representation is quite different. The C4.5 representation is a classification problem where the system assigns class membership to an individual using predetermined classes. This distinction is important when comparing results from the different systems. The C4.5 representation benefits from the additional information presented to the system in the form of the predetermined classes. The problem of assigning class membership based upon ship signature is easier than computing an actual aimpoint from the signature. The C4.5 system is required to generate 1 of only 24 outputs corresponding to the predetermined classes for each input, while the neural network and GP systems must produce a single coordinate in the infinite domain of real valued numbers. A GP representation analogous to the C4.5 representation is possible but is not nearly as convenient as the C4.5 implementation since the desired multi-valued output would have to be artificially interpreted via a wrapper, or require multiple programs to be evolved, with each program solution functioning as a single group classifier. Requiring the GP system to recognize an encoding scheme used by a wrapper to identify class membership adds unnecessary complexity to the problem.

The computational effort afforded to and expended by each of the three machine learning techniques also provides a basis for comparison. Clearly, C4.5 requires the least computation of the three methodologies in the development of the classification trees, including the off-line computations required for clustering the input data into equivalence classes. A significant portion of the C4.5 computational budget is spent in measuring the information entropy within the data set so that compact trees can be constructed. The neural network solution required more computational effort as measured by the number of computations performed than C4.5, but significantly less than GP. Once an architecture has been specified, the computations required for determining the weights of the network are isolated on the single network being developed. GP in contrast, apportions the available computing power to numerous candidate solutions in its parallel exploration of the search space. A qualitative comparison of the computational effort required by neural networks and GP would be difficult for a problem in that both paradigms reached acceptable conclusions but is impossible when either methodology fails to culminate with an acceptable conclusion. The search for an acceptable GP solution appears to be more severely constrained by the computing resources available for this work than either the neural network or C4.5 approaches.

Correctness, or the number of aimpoints correctly computed, is the figure of merit which ultimately distinguishes the different approaches. The best neural network solution correctly computed all of the 192 aimpoints making up the data set for this problem. C4.5 provided a solution yielding 191 correct aimpoints. The best GP solutions computed 166 of the 192 X coordinates and 170 of 192 Y coordinates. Ancillary statistical data such as mean square error reflects the same correctness relation as the number of correctly computed aimpoints with the neural network having the smallest error and the GP solutions having the greatest error.

## CONCLUSION

Caution must be exercised when drawing conclusions from this work for extrapolation to other problems. The C4.5 system provides a solution that provides insight into the components of the ship signatures that distinguish one ship from another. From the preliminary noise studies however, the C4.5 solutions appear to be incapable of generalizing to handle perturbations in the input data. The GP approach was unsatisfactory for this application. We attribute this conclusion to the difficulty in developing a configuration for the GP leading to an acceptable solution. Multiple control parameters including population size, unsuccessful termination criteria, reproduction and cross over rates, selection mechanisms (proportionate, greedy, tournament) each affect the behavior of the GP operation. Choices for the basis functions, automatically defined functions, and ephemeral constant ranges also add to the configuration complexity. Koza has provided techniques for evolving the form and function of ADFs that might resolve these issues (Koza, 1995). Insight into which parameters to change when the GP system is not providing acceptable results is difficult to discern from the incorrect programs being generated. Perhaps a formal design of experiments methodology such as Taguchi methods would provide direction in reconfiguring the GP system (Phadke, 1988). From the results obtained from C4.5, it may be useful to increase the probability of selection of certain basis functions such as the IFGT (if greater than) operator for inclusion into programs in the initial generation. The C4.5 classification trees are replete with decision operators indicating that the GP programs should perhaps be biased to utilize conditional operators to a greater extent. However, we are somewhat reticent in moving in this direction since part of our attraction to the GP paradigm is in the autonomous manner in which the evolutionary process discovers this type of information.

The potential benefits afforded by GP solutions guarantee that it will remain a tool to be used on future problems. It's relative poor performance on this problem compared to neural networks and C4.5 is disappointing but is not to be construed as a general dismissal of GP for the class of complex problems which have no existing solution or sub-optimal solution. The ultimate goal is to find solutions for these type of problems coupled with an understanding of how the solutions work.

## ACKNOWLEDGMENTS

We are thankful to Howard McCauley for making available the data used in this project as well as the results of his neural network calculations and the Office of Naval Research for financial support.

## REFERENCES

Carpenter, G. and Grossberg, S., (1987). A massively parallel architecture for a self-organizing neural pattern recognition machine, *Computer Vision, Graphics Image Process.* (37), 54-115.

Carpenter, G. and Grossberg, S., (1987). ART 2: Self-organization of stable category recognition codes for analog input patterns, *Appl. Optics* (26), 4919-4930.

Holland, J., (1975). Adaptation in Natural and Artificial Systems. Ann Arbor, University of Michigan Press.

Holland, J., (1989). *Genetic Algorithms In Search, Optimization & Machine Learning*. New York, Addison-Wesley.

Kinnear K., ed., (1994). *Advances in Genetic Programming*. Cambridge, MIT Press.

Koza, J., (1989). *Genetic Programming: On the Programming of Computers by Means of Natural Selection*. Cambridge, MIT Press.

Koza, J., (1992). Hierarchical genetic algorithms operating on populations of computer programs, in Proceedings of the 11th International Joint Conference on Artificial Intelligence, Volume I. New York, Morgan Kaufman.

Koza, J., (1994). *Genetic Programming: Automatic Discovery of Reusable Programs*. Cambridge, MIT Press.

Koza, J., (1995). Evolving the architecture of a multi-part program in genetic programming using architecture altering operations, in *Proceedings of the Fourth Annual Conference on Evolutionary Programming*, ed. by J. McDonnell, R. Reynolds, and D. Fogel. Cambridge, MIT Press.

McCauley, H., (1993). Aimpoint Selection - A Heterogeneous Neural Network Approach, in *IEEE Visual Signal Processing and Communications, Workshop Proceedings*, ed. by K. Ngan.

Phadke, M., (1988). *Quality Engineering Using Robust Design*. New York, Prentice-Hall.

Punch, B. and Zongker, D, (1995). lil-gp Genetic Programming System, version 1.0, Michigan State University.

Quinlan, J, (1979). Discovering Rules by Induction from Large Collections of Examples, *Expert Systems in the Micro-Electronic Age*. Edinburgh, Scotland, Edinburgh University Press.

Quinlan, J., (1993). *C4.5 Programs For Machine Learning*. New York, Morgan Kaufmann.

Rosenblatt, F., (1962). *Principles of Neurodynamics*. New York, Spartan.

Rumelhart, D., Hinton, G., and Williams, R., (1986). Learning Internal Representations by Error Propagation, in *Parallel Distributed Processing, Vol. 1: Foundations*, ed. by D. Rumelhart and J. McClelland. Cambridge, MIT Press, 318-362.

Schwartz, C., Keyes, C., and VanBronkhorst, E., (1996). "Application and Evaluation of Genetic Programming for Aimpoint Selection" in the Proceedings of SPIE Conferrence 2824, Denver, CO.

Spector, L, (1995). Evolving Control Structures With Automatically Define Macros, *AAAI Symposium On Genetic Programming, Working Notes*.

Teller, A., (1994). The evolution of mental models, in *Advances In Genetic Programming*, ed. by K. Kinnear Jr. Cambridge, MIT Press, 199-220.

Zadeh, L., (1979). A theory of approximate reasoning, in *Machine Intelligence 9*, ed. by J. Hayes, D. Michie, and L. Mikulich. New York, Halstead Press, 149-194

# JOB SHOP SCHEDULING PROBLEM REVISITED : AN EFFICIENT UNIFICATION OF GENETIC ALGORITHM WITH SIMULATED ANNEALING

**SABYASACHI GHOSHRAY**
*School of Computer Science & Department of Electrical Engineering*
*Florida International University, Miami Florida, 33133 USA*
*E-mail : ghoshray@fiu.edu*

***Abstract:***
Job shop scheduling problems have been investigated in this research. The three broad categories of search methods have been analyzed and a newer genetic search method is proposed based on domain specific recombination. Also, in this research a hybrid genetic algorithm is proposed which efficiently unifies the salient features of simulated annealing and simple genetic algorithm to create a modified genetic algorithm (MGA). The generation of chromosomes are based on various permutations and combinations of job orders, process plans and allocated resources. The proposed MGA shows different measures of performance as well as different convergence time based on three types of chromosome representations. It has been found that the most complex chromosome representation that is based on the actual schedule has the slowest convergence time, but it provides near optimal solution for the scheduling problem.

## INTRODUCTION

Within an organization scheduling pertains to establishing the timing of specific resources of the organization. It relates to the use of equipment, facilities, and human services. A job shop is a process-organized manufacturing facility. Scheduling tasks are largely a function of the volume of system output. A job shop produces goods or parts, which have one or more alternative process plans. Each process plan consists of a sequence of operations which require resources and have certain predefined durations on machines. The main characteristic of job shop is that a great diversity of jobs to be performed within a given schedule; thus, the problem in job shop scheduling lies in selecting a sequence of operations together with an assignment of start/end times and resources for each operation. In this regard, the goal of job shop scheduling is to minimize service time and maximize machine use while satisfying other predefined constraints. Here the main consideration to be taken in account are the cost of having idle machine and labor capacity, the cost of carrying in process inventory, and the need to meet certain order completion due dates. The problem lies in the fact that these considerations tend to conflict with each other, because one can have a low cost idle machine and labor capacity by providing only a minimum amount of machinery and man power. This results in a considerable waiting time which leads to large in-process inventories and difficulty in meeting completion due dates. On the other end of the spectrum, we can have an excess of machine and labor capacity that ensures no waiting time and completion by due date. The drawback of this scenario is the cost increase due to idle machine and labor capacity. Therefore, optimizing different constraints and making economic compromises among the considerations outlined before, lay the foundation for solving job shop scheduling problem. Let us present an overview of some of the existing job shop scheduling problems and try to ascertain their inherent drawbacks.

## BACKGROUND OF JOB SHOP SCHEDULING

The plethora of job shop scheduling algorithms can be broadly classified into three classes. *Brute-force or "blind" search algorithms* are of no use in complex scheduling problems due to time and space requirements. These algorithms aim at producing near-optimal solutions by maintaining a population of potential solutions which undergoes some unary and binary transformation ("mutations" and "crossovers") under a selection scheme biased towards fit individuals (Dejong,1988). The *heuristic search methods* can solve optimization problems of lower polynomial complexity, but they fail when the search space is either huge or multidimensional. However, when the exponential complexity of these methods are reduced, these methods can produce near optimal solutions. But the drawback lies in the fact that the solution becomes less flexible by virtue of the restraint on the search space. Here in comes the efficiency of *random search and optimization methods*, such as genetic algorithms. Since we are after "the best" solution, we can view the task as an optimization process, which can be efficiently performed for larger spaces by genetic algorithms (GA). GAs are stochastic algorithms whose search methods model some natural phenomena, such as genetic inheritance and Darwinian strife for survival.

Let us now explain job shop scheduling in detail. A job shop is a facility that produces goods according to prespecified process plans and under several domain-dependent and common-sense constraints. The scheduling of orders in a job shop is a very complex process which can have a large number of possible combinations, each satisfying a number of preselected criteria. While selecting a particular combination, two things have to be constantly verified, one: the flexibility of the scheme, and two: the optimization of cost. A typical problem has three distinct search spaces: (i) order permutations, (ii) process plans, and (iii) resources.

The following description of job shop scheduling is due to Michalewicz (Miehalewicz, 1992). Assume there are three orders, $0_1$, $0_2$, and $0_3$. For each order, the parts and the number of units to be produced are:

$0_1$: 30x part a; $0_2$: 45x part b; $0_3$: 50x part a.

Each part has one or more alternative process plan, such as:

a: plan $\#1_a$ ($opr_2$, $opr_7$, $opr_9$); b: plan $\#2_a$ ($opr_1$, $opr_3$, $opr_7$, $opr_8$); etc.

where terms $opr_i$ denote the required operations to be performed. Each operation requires some items on one or more machines, these can be as follows:

$opr_1$ : ($m_1$, 10) ($m_3$ 20); $opr_2$ : ($m_2$ 20 ); etc.

Finally, each machine has its setup time necessary for changes in operation:

$m_1 : 3;$ $m_2 : 5;$ $m_3 : 7.$

The job shop problem enjoyed some interest in the GA community. Our research has found that selecting simple chromosomes is not a suitable method to solve job-shop scheduling problems. It is necessary therefore, to incorporate the complexity of the problem both in the operators and in the chromosome structures and subsequently to look for the best fit chromosome. Therefore, genetic algorithms (GA) are to be developed where the chromosome representations are based on the problem specific-knowledge (Bagchi et al, 1994). In this context, comes the uniqueness of our method in which, we use the problem-specific knowledge in creating both the operators and chromosomes and also combine the features of both simulated annealing (SA) and genetic algorithms (GA).

## THE PROPOSED METHOD

The method described here is based on our earlier research to develop a solution for the job shop scheduling problems (Ghoshray, 1996). The user of a GA system has to provide a set of processing functions, a fitness function. The GAs also require that the problem states to be defined as chromosomes. We have investigated the performance of the proposed MGA based on three different chromosome representations. The first type of chromosome deals with priority of orders. That means each chromosome consists of only 3 elements, each element representing a particular order. In the second type of chromosomes, the priority of order are combined with process plans. The degree of complexity increases in the third type of chromosome, as the order priority, process plans and resources are combined together. These representations are depicted in Table 1.

**TABLE 1: DIFFERENT CHROMOSOME REPRESENTATIONS**

| | | |
|---|---|---|
| order 1 | order 2 | order 3 |
| order 2 | order 3 | order 1 |
| **(a) Simple representation of Chromosomes using only order in the permutations** | | |
| order1 plan 1 | order 2 plan 2 | order 3 plan 3 |
| order 1 plan 3 | order 3 plan 1 | order 2 plan 2 |
| **(b) representation of Chromosomes that involve combinations of both plans and orders** | | |
| order 1 oper 1 oper 4 oper 6 res. 1 res. 2 res. 3 | order 2 oper 3 oper 2 oper 1 res. 2 res. 3 res. 4 | order 3 oper 7 oper 2 oper 1 res. 2 res. 3 res. 3 |
| order 2 oper 2 oper 3 oper 5 res. 2 res. 3 res. 4 | order 3 oper 1 oper 4 oper 5 res. 5 res. 6 res. 4 | order 1 oper 1 oper 4 oper 8 res. 1 res. 2 res. 3 |
| **(c) the most complex of all the chromosomes that involve representation of Chromosomes that involve all possible permutations thereby representing an actual schedule** | | |

**Chromosome Representation**

In the first representation, the ordering of elements in the chromosome represents the scheduling priority. The chromosome does not depend on the process plans, resources or start or stop times. This type of chromosome representation is very simple and they converge very quickly as the MGA has to search only the possible permutations on the set of orders. Here they suffer from several disadvantages. First, since the chromosome is not a direct representation of the schedules, the offspring chromosomes has to be converted to legal schedules; therefore, the savings in time by quick convergence gets offset by the additional step necessary to come up with the optimal solution. In addition, since the search space does not contain process plan and resource selections, the efficiency of the MGA is severely constrained. The second chromosome representation involves enlarging the search space by adding all possible process plans. Therefore, the chromosome is a quasi-representation of a typical schedule in the sense that it lists the combination of order and plan in order of priority. In this type of chromosome, the convergence time towards optimatlity increases as the representations become more complex. In addition, the search space is still constrained as it does not involve t he selection of resources. The third type of representation addresses all the drawbacks of the previous two types of chromosome representations. Since the chromosomes themselves represent the subject of optimization, no extra step is necessary to convert the chromosomes to legal schedules after the convergence. Also, the fitness function can be applied directly to the chromosomes.

If $0_i$ represents the ith order from the set of orders, $p_j$ represents any specific plan in an order and $m_k$ and $op_l$ represents respectively the $k^{th}$ plan and $l^{th}$ operation. We can have the three types of chromosome representations as follows:

$(0_i)(0_{i+3})(0_{i-5})$ represents a chromosome representing a schedule with $i^{th}$, $i+3^{th}$, and $i-5^{th}$ orders in priority.

$(0_i m_k)(0_{i-2} m_{k+1})(0_{i+1} m_{k-3})$ represents the second type of chromosome described above in which the ith order with $k^{th}$ plan will be the 1st in the que for the scheduling. Finally,

$(0_i(op_l{:}m_k)(op_{l-3}{:}n_{k-5})(op_{l+5}{:}m_{k+1}))$

$(0_{i+3}(op_{l+1}{:}m_{R=k-5})(op_l{:}m_{k-1})(op_{l+2}{:}m_{k-3}))$

$0_{i2}(op_l{:}m_k)(0p_{l-2}{:}m_k)(op_{l+1}{:}m_{k-4}))$is an example of the most complex of all the chromosome representations in which the chromosome itself is the optimized schedule.

The MGA ( Modified Genetic Algorithm) that we have used has the following algorithm:

```
begin MGA
Get any one initial population applying f_S1 = x ∈ S_i
          While f_eval (x) = True do
                    Begin
                              Get p, q, r, s
                              y = f_SZ (p, x)
                              z = f_RP (v, y)
```

$z' = f_m (r, z)$
$t = x \cup z'$
$x = f_{Rd} (s, t)$

end
output the present population
end MGA

$f_{sl}$ is the function that randomly selects the initial population from the search space
p is the parameter that randominzes the selection function $f_{sz}$
q, r, s randomizes the reproduction function $f_{rp}$, the mutation function $f_m$, the reduction function $f_{rd}$ respectively.
A detailed description of this algorithm can be found in Ghoshray, 1995.

Determining a suitable fitness function is the most important step in assessing the effectiveness of the MGA. In this case the legality as well as the efficiency of the schedule corresponding to the reported chromosome will determine the fitness of the chromosome. The schedule is evaluated based on the measurement of the utilization efficiency of the schedule and the measurement of the overall response time of the schedule. If j is the job order where there are n number of jobs and, *o* is the operation in a particular process where m is the number of total operations, and l is the number of machines in the overall operation, we calculate the fitness as follows:

$$\frac{\sum_{j=1}^{n} \sum_{0=1}^{m} t_{tpt} (j.0)}{1 * [ t_{end} - t_{start}]} \times 100$$

where t = total time, $t_{end}$ = end time of the schedule and $t_{start}$ = start time of the schedule

## RESULTS AND DISCUSSIONS

In this section, we are going to have a discussion on the comparative performances of the proposed genetic algorithm, simple genetic algorithm and simulated annealing. The probability of selection is calculated as $p_i = eval(v_i) / F$ in which F is the total fitness of the population calculated as $F = \sum_{1}^{20} eval(V_i)$. Let us now define two characteristics that is used to compare the performances of different genetic algorithms. Accuracy ( $\mu$ ) is defined as the value of the best found chromosome in the proposed algorithm relative to the value of the global optimum found in the original GA. Therefore, $\mu = f_{best} / f_{global}$. Imprecision ($\lambda$ ) is defined as the standard deviation of the optimal vector found. This measure evaluates the closeness of individual components selected in the found optimal chromosome. Therefore,

$$\lambda = \sqrt{\frac{\sum_{i=1}^{\#element} (index_i)^2}{\#elements - 1}}$$

Our experiments have shown the values of accuracy μ for different algorithms are as follows: 95.71 for GA, 96.47 for SA and 98.45 for the proposed MGA. The values of imprecision λ for different algorithms are as follows: 2.243 for GA, 2.108 for SA and 2.012 for the proposed MGA.

## CONCLUSION

This paper has established the drawbacks of the classical GA in solving job shop scheduling problem. We have also presented a refined genetic algorithm by efficiently combining simple genetic algorithms with simulated annealing. The empirical results show that modified genetic algorithms provide efficient search heuristics for solving job shop scheduling problems. The model proposed has a short run time compared to that of SGA and the results produced have been qualitatively correct. Also, the results show that MGA provides less imprecision. However, the main difficulty remaining is in the efficient modeling of selection function $f_{s_1}$ that randomly selects the initial population from the search space. We hope to address this issue in more detail in a future publication. Our forthcoming research will also focus on developing more efficient parallel selection algorithms by using fractal methods.

## REFERENCES


Bagchi, S., Uckun, S., Miyabe, Y., & Kawamura, K., (1994).Exploring Problem Specific Recombination Operators for Job Shop Scheduling, *Third International Conference on Genetic Algorithms*, p.p.10 -17.

Davis, L., (Editor)(1987). Genetic Algorithms and Simulated Annealing, Morgan Kaufmann Publishers, Inc., Los Altos, CA.

Dejong, K.A., (1988). Learning with Genetic Algorithm: An Overview, *Machine Learning*, vol.3.

Ghoshray, S.,Yen, K.K., & Andrian, J, (1995). Modified Genetic Algorithm By Efficient Unification Wth Simulated Annealing, *International Conference on Artifiical Neural Networkds and Genetic Algorithms*. April 24-28.

Ghoshray, S.(1996).Modified Genetic Algorithms to Solve Job Shop Scheduling Problems, *ICAFS-96 Second International Conference on Application of Fuzzy Systems and Soft Computing*, June 25-27.

Kirkpatrick, S., Gelatt, C. D. & Vechi, M. P.(1983). Optimization by Simulated Annealing, *Science 220*, pgs. 671-680.

Koza, J. R., (1994). Genetic Programming II Automatic Discovery of Reusable Programs. *The MIT Press*.

Michalewicz, Z.,(1992). Genetic Algorithms + Data Structures = Evolution Programs, *Springer Verlag*. pp. 204-205.

# PART IV : PATTERN RECOGNITION

# BUFFERED RESET LEADS TO IMPROVED COMPRESSION IN FUZZY ARTMAP CLASSIFICATION OF RADAR RANGE PROFILES

STEPHEN GROSSBERG, MARK A. RUBIN AND
WILLIAM W. STREILEIN
*Department of Cognitive and Neural Systems*
*Boston University*
*Boston, MA 02215*

**Abstract**

Fuzzy ARTMAP has to date been applied to a variety of automatic target recognition tasks, including radar range profile classification. In simulations of this task, it has demonstrated significant compression compared to $k$-nearest-neighbor classifiers. During supervised learning, match tracking search allocates memory based on the degree of similarity between newly encountered and previously encountered inputs, regardless of their prior predictive success. Here we investigate techniques that buffer reset based on a category's previous predictive success and thereby substantially improve the compression achieved with minimal loss of accuracy.

**Introduction.** Fuzzy ARTMAP (Carpenter et al., 1992) is a neural network for supervised learning and recognition of $N$-dimensional analog patterns. In response to labeled training inputs, it generates (hyper)rectangular "internal-category boxes" in the $N$-dimensional space of pattern vectors.[1] Newly-encountered patterns are matched to the closest category box.[2] If the box label[3] is the same as the label of the new pattern, and the new pattern falls inside the box defined by the weights, the weights remain unchanged. If the new pattern falls outside the box, the weights are adjusted by the minimum amount necessary to encompass the pattern. In either case, the amount of information that the network must store remains constant.

If the labels do not match, the node associated with the category box[4] is reset; the box in question is no longer considered as a possible category for the current input pattern. The next-best-matching box is examined, and if that does not match, the next. If match with previously-learned categories is not achieved, then a new category is selected, and the information stored by the network increases by $2N$ real numbers.[5] This process is called *mismatch reset*.

Nodes may be reset even if their labels match the labels of training vectors coded to them. A vigilance parameter governs the maximum size to which an internal-category

[1] Throughout this paper we employ the complement-coded form of fuzzy ARTMAP, in which there are two weight vectors per category. The information needed to specify completely the dimensions of each box are the two $N$-dimensional weight vectors specifying, respectively, the vertices closest to and farthest from the origin.

[2] "Closeness" is defined by an $L_1$-like norm.

[3] *I.e.*, the label of the training vectors previously matched to this box.

[4] *I.e.*, the neural unit which computes the degree of closeness of input vectors to the box defined by the weights.

[5] Not counting the label of the new box (Rubin, 1995b).

box can grow. For greatest code compression the vigilance parameter is typically set at a baseline value of zero, which sets no restriction on maximal box size. However, when a node is reset by a predictive error, the vigilance is raised by the minimum amount needed to disallow the box that would have been created if the new category had been coded to that box. When the vigilance parameter is nonzero, a node is reset if the input pattern which is matched to it would cause the category box to grow beyond the size allowed by the vigilance parameter. This type of reset is called *vigilance reset*.[6]

The net effect of these processes is to cover the input space with internal-category boxes in an adaptive manner, with the sizes of the boxes covarying with the sizes of the partition of the underlying pattern space into categories. The classifier which is thus created concentrates its memory resources to make fine distinctions between test input patterns in regions of pattern space where fine distinctions are needed, and conserves memory in regions of pattern space which do not require such close attention.

When compared with other pattern recognition algorithms, such as $k$-nearest-neighbors ($k$NN), fuzzy ARTMAP yields comparable classification accuracy with greatly reduced memory requirements (Rubin, 1995b; Gjaja, 1996). Nevertheless, important applications of pattern recognition algorithms—e.g., mobile artificial vision and missileborne automatic target recognition systems—impose extreme constraints on both memory storage and processing time, and it is natural to examine the basic fuzzy ARTMAP algorithm to see whether even greater memory efficiency can be achieved.

**Success-based-buffering.** During fast learning, fuzzy ARTMAP does not compute the cumulative predictive success of the categories that it learns. Statistical factors have been incorporated into fuzzy ARTMAP using slow learning, among other mechanisms (Bradski and Grossberg, 1995; Carpenter et al., 1995; Markuzon, 1995), to enable internal categories which more frequently code training patterns to have a larger say in determining the category membership, or probabilities of category membership, of a test pattern. These techniques reduce category proliferation in response to datasets with high noise or strongly-overlapping probability distributions and therefore lead to improved code compression.

The present work focuses on improving code compression in response to high-dimensional inputs with largely *non*overlapping underlying distributions. Since node reset ultimately leads to the generation of new nodes, we focus on how to modify reset to be sensitive to the cumulative statistics of the training process. Specifically, nodes that have previously led to predictive success are *buffered* against being reset, to a degree commensurate with the degree of their success; that is, to the degree that they code training vectors that have been assigned to them during training without being reset. This operation thus uses concepts from reinforcement learning to modulate the process of recognition learning (Grossberg, 1982, 1987).

Buffering against reset is defined stochastically. With each node $i$ is associated a number $f_i$, such that $0 \leq f_i \leq 1$, where $f_i$ is the probability of avoiding reset of the $i^{th}$ node. When a condition occurs which would, in the absence of buffering, certainly result in reset, reset will in fact occur with probability $1 - f_i$. With probability $f_i$ reset does not occur, and the input pattern is coded to the node in question. We let $f_i = f(R_i)$, where the confidence measure $R_i$ depends only on the coding history of node $i$, and function $f$ is independent of $i$. Several forms for $f$ are considered below.

The simplest choice for $R_i$ is $R_i = x_i \equiv (y_i^+ - y_i^-)/(y_i^+ + y_i^-)$, where $y_i^+ \equiv$ ( # of successful codings of node $i$) and $y_i^- \equiv$ (# of resets of node $i$), so $y_i^+ + y_i^- =$ (total # of attempts at coding node $i$). Function $R_i$ embodies the idea that a higher relative

[6] For a more detailed description of the operation of fuzzy ARTMAP, see Carpenter et al. (1992).

reward rate better buffers the recognition category against reset. Alternatively, one could use $z_i \equiv (y_i^+)/(y_i^+ + y_i^-)$, which is related $x_i$ by $x_i \equiv 2(z_i - .5)$.[7]

**Testing with simulated radar range profiles. Functional form of $f$.** Radar range profiles are one-dimensional images of radar targets (Smith and Goggans, 1992; Borden, 1993; Hudson and Psaltis, 1993). Fuzzy ARTMAP has been successfully applied to range profile automatic target recognition (ATR), using simulated range profiles. In these applications it has achieved accuracy comparable to $k$NN classifiers, while yielding significant storage compression[8]

We first performed tests on small range profile datasets (4 targets, 2 wavelengths) using three functional forms for $f$: $f_0(x_i) = (1 + tanh(sx_i))/2$, $f_1(x_i) = [tanh(sx_i)]^+$ and

$$f_2(x_i) = \begin{cases} p_0 + (x+1)(p_0 - p_1), & x < 0 \\ p_1 + x(p_2 - p_1), & x \geq 0 \end{cases}$$

where $p_0$, $p_1$, and $p_2$ are constant parameters between 0 and 1. These simulations indicated that the probability-of-reset-avoidance function $f_0$ with the slope parameter $s$ at 2 works particularly well; i.e., gives high accuracy and/or compression in most cases, with no cases of extremely poor performance in either regard. It was also found that buffering against mismatch resets, in addition to vigilance resets, degrades performance relative to unbuffered fuzzy ARTMAP in both compression and accuracy, for all functional forms of $f$ and parameter ranges tested. All subsequent work uses $f = f_0$, $s = 2$, and only buffers vigilance resets, unless otherwise noted.

**Varying the baseline degree of buffering.** Function $f(x_i) = (1+tanh(2x_i))/2$ gives a probability-of-reset-avoidance of 1/2 when a node has, up to the current point in training, had as many success as failures, i.e., $y_i^+ = y_i^- \Rightarrow x_i = 0$. A more general function would allow shifts in the baseline degree of buffering, by shifting the value of $x_i$ at which the $tanh$ function sweeps up from $\approx 0$ to $\approx 1$. Specifically, consider $f(x_i, b) = (1 + tanh(2[x_i - b]))/2$ or, equivalently, $F(z_i, b) = f(x_i(z_i), b) = (1 + tanh(4[z_i - .5 - b/2]))/2$. Keeping in mind that $-1 \leq x_i \leq 1$, we see that large positive values of $b$ correspond to no buffering, since the argument of the $tanh$ function is then $\ll 0$. Choosing $b = 0$ corresponds to the standard buffering previously done, in which nodes which are successful half the time have a 50% chance of avoiding reset. Large negative values of $b$ correspond to 100% buffering, since the argument of the $tanh$ function is $\gg 0$. This case is called maximum buffering.

Typical results of buffering with $b$ decreasing from $\gg 0$ (no buffering) through 0 (standard buffering) to $\ll 0$ (maximum buffering) are presented in Table 1. The number of category nodes decreases as $b$ decreases, reaching the number of targets with maximum buffering. This is the case even with Gaussian noise (snr=3) added.

**Circle-in-the-square benchmark.** Is buffering a universally useful tool? The range profiles considered in the above sections are relatively high-dimensional, since each has at least (30 range bins) $\times$ (2 wavelengths) = 60 components. It turns out that certain low dimensional datasets whose statistical distributions are not separable are not helped by buffering. For example, consider the circle-in-the-square problem.

[7]In a hardware implementation, $y_i^+$ and $y_i^-$ could be stored as an additional layer of weights, along the lines of the instance-counting weights of Markuzon (1995), which (in our notation) record $y_i^+ + y_i^-$.

[8]Except where otherwise noted, all range profile simulations have 30 range bins of size 2/3 m, have 2 or more wavelengths evenly spaced from 1.35 cm to 1.65 cm, use 3 training epochs, have 21 training views evenly spaced in azimuth from $-5°$ to $5°$ with respect to target heading and have at least 1000 test views randomly chosen in the same azimuthal range. All training and test views are in the plane of the target. For details of the simulation and classification procedures see Rubin (1995a,b).

| buffering | none | $b = 0$ | $b = -1$ | $b = -40$ |
|---|---|---|---|---|
| error (# nodes) | 43.7% (18) | 47.2% (10) | 46.6% (5) | 44.9% (4) |
| | 5.85% (259) | 5.85% (259) | 8.63% (78) | 8.73% (36) |

Table 1: Success-based buffering with different values of the shift $b$. Range profile datasets. First line: 4 targets (2 wing positions, 2 wing lengths), 2 wavelengths, Gaussian noise with snr=3 for both training and testing. Second line: 36 targets (6 wing positions, 6 wing lengths), 11 wavelengths, 10 sequential views of each test target (ART-EMAP, stage 3), no noise added.

| buffering | none | $b = 0$ | $b = -1$ | $b = -40$ |
|---|---|---|---|---|
| error (# nodes) | 6.9% (20) | 14.2% (5) | 10.6 (20) | 23.0% (3) |

Table 2: Success-based buffering with different values of the shift $b$. Circle-in-the-square dataset: 1000 training samples, 1 training epoch.

In this benchmark, a network is presented, for both training and testing, with two-dimensional inputs representing the locations of points chosen at random from the unit square. The points belong to one of two classes, depending on whether or not they are on the interior or exterior of a circle whose center lies at the center of the unit square and whose area is one-half that of the square (Wilensky, 1990; Carpenter et al., 1992). As illustrated in Table 2, buffering, notably maximum buffering, while still capable of yielding savings in the number of category nodes, tends to cause significant degradation in accuracy.

The challenge thus becomes to devise a buffering rule which is self-adjusting—*i.e.*, invokes buffering when the training data has the characteristics of the range profile dataset, but reduces to the usual fuzzy ARTMAP when the data are more like circle-in-the-square.

**Competitive buffering.** Buffering need not depend only upon the success rate of each node taken separately. An additional source of information is the *competitiveness* of the input which has tentatively been assigned to the $i^{th}$ node. Competitiveness means the degree to which, in the presence of each input, the activation of the winning $i^{th}$ node constitutes the main source of activation. Here, the confidence $c_i$ is given by $c_i \equiv T_i/(\beta + \Sigma_k T_k)$ where $\beta$ is a constant and $T_i$ is the Weber-law activation function employed by fuzzy ARTMAP (Carpenter et al., 1992). A very competitive winning node thus satisfies $c_i \approx 1$ when $\beta \ll 1$, since then $T_i \approx \Sigma_k T_k$. In contrast, an uncompetitive winning node satisfies, $c_i \approx 1/(\#\text{ allocated nodes})$. Since $c_i$ is nonnegative, we use it in the role of $z_i$,as the argument of a sigmoidal probability-of-reset-avoidance function: $g(c_i, b) = (1 + tanh(4[c_i - .5 - b/2]))/2$.

In both range profile and circle-in-the-square trials, this competitive form of buffering produces little effect of any kind, even for small $\beta$. The degree of competitiveness, as defined by $c_i$, is too small in general to make a noticeable difference. Since the sum in the denominator scales with the number of allocated nodes, the entire term is rescaled by $\mathcal{N} = \#$ allocated nodes $+1$, and the confidence measure for competitive buffering is $C_i \equiv \mathcal{N}T_i/(\beta + \Sigma_k T_k)$

With small $\beta$, ($\beta = .001$), the results for both range-profile and circle-in-the-square datasets are qualitatively the same as for success-based buffering with large $b$; that is, good for range profiles, not so good for circle-in-the-square.[9]

[9]Similar results are obtained using forms for the confidence measure which join properties of node

| | range profiles | | | | | circle-in-the-square | | | | |
|---|---|---|---|---|---|---|---|---|---|---|
| $\log_{10}(\beta)$ | n. b. | -2 | 0 | 5 | $\beta = \mathcal{N}$ | n. b. | -2 | 0 | 5 | $\beta = \mathcal{N}$ |
| error (% ) | 43.7 | 44.8 | 43.5 | 47.0 | **43.8** | 6.9 | 22.5 | 22.8 | 6.5 | **8.4** |
| # nodes | 18 | 6 | 7 | 18 | **10** | 20 | 4 | 4 | 20 | **9** |

Table 3: Competitive buffering with various fixed values of $\beta$, and self-scaled ($\beta = \mathcal{N}$). ("n. b." = no buffering.) Range profiles: 4 targets (2 wing positions, 2 wing lengths), 2 wavelengths, Gaussian noise with snr=3 for both training and testing. Circle-in-the-square: 1000 training samples, 1 training epoch.

| buffering | none | success-based | | self-scaled competitive |
|---|---|---|---|---|
| | | $(b = 0)$ | $(b = -\infty)$ | $(\beta = \mathcal{N})$ |
| error (# nodes) | 21.3% (10) | 22.8% (5) | 22.8% (3) | 22.0% (6) |

Table 4: Buffering procedures applied to *xpatch*-generated range profiles. 3 targets, one wavelength at 1.5 cm, 60 range bins 2/3 m wide, 3 training epochs, 21 training views evenly spaced from $-90°$ to $90°$ with respect to target heading. Testing on 1020 views from $-5°$ to $5°$ with respect to target heading, none coinciding with training views. All training and testing views in plane of the target.

**Self-scaled competitive buffering.** As was seen earlier, the amount of success-based buffering can be controlled by changing the value of $b$. The amount of competitive or competitive success buffering can also be adjusted, by changing the value of $\beta$. Small values of $\beta$ lead to large amounts of buffering and a decreased number of category nodes. Conversely, as $\beta$ is increased, the degree of buffering is decreased, the number of nodes generated increases, and the negative impact on the accuracy of the circle-in-the-square is decreased. From the examples studied so far, there is no obvious fixed value to choose for $\beta$ which gives good results in a variety of situations. There is, however, a natural dynamically changing value at to which set $\beta$; namely, $\beta = \mathcal{N}$. This choice realizes a compromise when tested on datasets summarized in Table 3; that is, it achieves less compression than maximum, but accuracy (in the circle-in-the square case) somewhere between the unbuffered (accurate) and highly-buffered (inaccurate) values.

**Buffering algorithms applied to range profiles simulated using *xpatch*.** *xpatch* is a sophisticated electromagnetic-scattering simulator developed under U. S. Department of Defense sponsorship (Volakis, 1994). In Table 4, small dataset (3 target) range profile simulations generated using *xpatch* are classified using various buffering approaches and show the expected pattern of decrease in the number of nodes.

**Discussion.** The tests of various buffering algorithms on various datasets described above show that, in certain classes of application domains, including the range-profile ATR application, the standard fuzzy ARTMAP approach may generate more internal nodes than necessary. Code compression can be improved by buffering with little, if any, impact on accuracy.

At the same time, the buffering algorithms may cause significant degradation of accuracy if applied to certain datasets. For example, even the self-scaling forms pro-

---

competition and predictive success, which are called competitive success buffering; namely, $C_i^{\alpha} \equiv \mathcal{N}(T_i(1+\alpha z_i))/(\beta + \Sigma_k T_k(1+\alpha z_k))$ and $C_i^{\pm} = \mathcal{N}(T_i[1+\alpha^+ y_i^+ - \alpha^- y_i^-]^+)/(\beta + \theta \Sigma_k T_k[1+\alpha^+ y_k^+ - \alpha^- y_k^-]^+)$ for various values of the parameters $\alpha$ and $\alpha^{\pm}$. These forms are suggested by neural models of how reward focuses attention upon competing recognition codes (Grossberg, 1982, 1987). Results presented below for competitive buffering use $C_i$.

posed here perform poorly on the two-spirals benchmark (Lang and Witbrock, 1989; Carpenter et al., 1992). These algorithms can, at present, only be used confidently in conjunction with a validation set to determine their suitability. This is not an uncommon situation in neural network applications; networks trained by backpropagation, for example, must generally have such parameters as learning rate and number of training epochs determined by such a procedure (Masters, 1993). A self-adjusting buffering mechanism would be preferable. It remains to be seen whether such a general purpose method can be developed.

**Acknowledgements.** M. A. R. would like to thank Brett Borden, Natalya Markuzon and Jim Williamson for helpful discussions. This research was supported in part by Advanced Research Projects Agency and Office of Naval Research grants ONR N00014-95-1-0657 (S. G.), ONR N00014-95-1-0409 (S. G., M. A. R.), and ONR N00014-96-1-0659 (M. A. R., W. W. S.).

**References.**

Borden, B. (1993). Problems in airborne target recognition. *Inverse Problems*, **10**, pp. 1009-1022.

Bradski, G., and Grossberg, S. (1995). Fast-learning VIEWNET architecture for recognizing three-dimensional objects from multiple two-dimensional views. *Neural Networks*, **8**, pp. 1053 -1080.

Carpenter, G. A., Grossberg, S., Markuzon, N., Reynolds, J. H., & Rosen, D. B. (1992). Fuzzy ARTMAP: a neural network architecture for incremental supervised learning of analog multidimensional maps. *IEEE Transactions on Neural Networks*, **3**, pp. 698-731.

Carpenter, G. A., Grossberg, S. and Reynolds, J. H. (1995). A fuzzy ARTMAP nonparametric probability estimator for nonstationary pattern recognition problems. *IEEE Transactions on Neural Networks*, **6**, pp. 1330-1336.

Gjaja, M. (1996) Neural networks for learning and prediction with applications to remote sensing and speech perception. Ph.D. thesis, Department of Cognitive and Neural Systems, Boston University.

Grossberg, S. (1982). The processing of expected and unexpected events during conditioning and attention: A psychophysiological theory. *Psychological Review*, **89**, pp. 529-572.

Grossberg, S. (1987). *The Adaptive Brain*, Vol. I. Amsterdam:North-Holland.

Hudson, S. and Psaltis, D. (1993). Correlation filters for aircraft identification from radar range profiles. *IEEE Transactions on Aerospace and Electronic Systems*, **29**, pp. 741-748.

Lang, K. J., and Witbrock, M.J (1989). Learning to tell two spirals apart. In *Proceedings of the 1988 Connectionist Models Summer School*, pp. 52-59.

Markuzon, N. (1995). ARTMAP-PI: a neural network architecture for probabilistic estimation of outcome and its application in complex medical data sets. In *World Conference on Neural Networks—Washington, D. C.: 1995 International Neural Network Society Annual Meeting*, July 17-21, Vol. 2, pp. 186-190. Mahwah, NJ:Lawrence Erlbaum Associates.

Masters, T. (1993). *Practical Neural Network Recipes in C++*. Boston:Academic Press, Inc.

Rubin, M. A. (1995a). Issues in automatic target recognition from radar range profiles using fuzzy ARTMAP. In *World Conference on Neural Networks—Washington, D. C.: 1995 International Neural Network Society Annual Meeting*, July 17-21, Vol. 1, pp. 197-202. Mahwah, NJ: Lawrence Erlbaum Associates.

Rubin, M. A. (1995b). Application of fuzzy ARTMAP and ART-EMAP to automatic target recognition using radar range profiles. *Neural Networks*, **8**, pp. 1109-1116.

Smith, C. R. and Goggans, P. M. (1992). Radar target identification. *IEEE Antennas and Propagation Magazine*, **35**, pp. 27-38.

Volakis, J. L. (1994). XPATCH: a high-frequency electromagnetic-scattering prediction code and environment for complex three-dimensional objects. *IEEE Antennas and Propagation Magazine*, **36**, pp. 65-69.

Wilensky, G. (1990). Analysis of neural network issues: scaling, enhanced nodal processing, comparison with standard classification. DARPA Neural Network Program Review, Oct. 29-30.

# SPATIO-SPECTRAL IMAGE ANALYSIS USING CLASSICAL AND NEURAL ALGORITHMS

**SOPHIA ROBERTS**
*Department of Computer Science*
*Florida State University*
*Tallahassee, FL 32306-4019*

**GALEN R. GISLER, JAMES THEILER**
*Astrophysics and Radiation Measurements Group*
*Los Alamos National Laboratory*
*Los Alamos, NM 87545*

*ABSTRACT:*

Remote imaging at high spatial resolution has a number of environmental, industrial, and military applications. Analysis of high-resolution multi-spectral images usually involves either spectral analysis of single pixels in a multi- or hyper-spectral image or spatial analysis of multi-pixels in a panchromatic or monochromatic image. Although insufficient for some pattern recognition applications individually, the combination of spatial and spectral analytical techniques may allow the identification of more complex signatures that might not otherwise be manifested in the individual spatial or spectral domains. We report on some preliminary investigation of unsupervised classification methodologies (using both "classical" and "neural" algorithms) to identify potentially revealing features in these images. We apply dimension-reduction preprocessing to the images, cluster, and compare the clusterings obtained by different algorithms. Our classification results are analyzed both visually and with a suite of objective, quantitative measures.

## INTRODUCTION

High resolution imaging of the Earth's surface in multiple spectral bands is becoming increasingly routine from both airborne and satellite platforms. An agricultural business may probe the health of its crops or a food-exporting country may utilize remote sensing to measure the extent of land devoted to a specific crop in another country. Environmentalists may monitor remotely the success of reforestation, waste remediation efforts or the deterioration of natural vegetation in particular areas. Industrial concerns may employ remote surveillance of natural resources. National security analysts may observe the construction of new military facilities at known installations as well as at unusual locations.

According to Richards, "the purpose of acquiring remote sensing image data is to be able to identify and assess, by some means, either surface materials or their spatial properties" [Richards (1994)]. Techniques for assessing remote sensing data usually involve either spectral analysis of single pixels in a multi-spectral or hyper-spectral image or spatial analysis of multi-pixel features in a panchromatic or monochromatic image. Often these two types of analyses are quite distinct from each other: for example, the intelligence community has traditionally relied on exclusively spatial analysis, while the environmental and agricultural communities have focussed on purely spectral analysis. But for some applications, it may be advantageous to combine both approaches. Signals that are too subtle to identify in one domain may become more readily recognizable using both domains.

## APPROACH

The usual approach for interpreting multispectral images (such as those provided by Landsat) involves analyzing the spectral signature of each of the pixels independently. But this neglects spatial aspects of the image, including "texture" [Jain and Dubes (1975)] and the general tendency for nearby areas on the ground to be similarly constituted. For example, if a pixel is surrounded by a region of pixels identified as water, then that enhances the likelihood that the pixel in question is also water. A system incorporating spatial as well as spectral information can potentially provide more reliable guesses as to what is on the ground.

When ground truth is not available, an unsupervised clustering technique is advantageous, because it reduces the number of characteristic spatio-spectral signatures that need to be identified. In this study, we will apply a variety of classical [Hartigan (1975); Späth (1975); Jain and Dubes (1975)] and neural [Carpenter and Grossberg (1988); Carpenter and Grossberg (1991); Carpenter et al. (1991); Kohonen (1991)] algorithms to cluster multi-spectral images using both purely spectral and spatio-spectral embeddings of the image data.

A data set is a single seven-band Landsat image. Our aim is to cluster the pixels within a given image. Each pixel provides a separate input pattern, and is represented as a vector with $M \times (Q+1)$ components, where $M = 7$ is the number of spectral bands and $Q$ is the number of neighboring pixels in the embedding. For the purely spectral embedding, $Q = 0$; for a spatio-spectral embedding that uses a $3 \times 3$ spatial patch for each pixel, we have $Q = 8$.

Particularly with the spatio-spectral embeddings, one very quickly runs up against the curse of dimensionality. We have attempted to ameliorate this problem by using a straightforward linear technique to reduce the dimensionality. Principal components analysis (PCA) provides linear coefficients for a rotation of the input vectors such that the projection of the rotated data into the first $m$ components maximizes the variance in the data.

## NEURAL CLUSTERING ALGORITHMS

Adaptive Resonance Theory (ART) is a real-time self-organizing neural network that uses competitive, unsupervised learning to classify patterns [Carpenter and Grossberg (1988); Carpenter and Grossberg (1991); Carpenter et al. (1991)]. ART networks have two fully-connected layers: the F1 layer or "feature representation field" accepts the input patterns and the F2 layer or "category representation field" represents the categories learned by the network. The two layers are connected via bottom-up and top-down weights. The bottom-up weights connecting each node in the F1 layer to every node in the F2 layer encode the input pattern while the top-down weights connecting each node in the F2 layer to every node in the F1 layer encode learned expectations [Carpenter and Grossberg (1988)].The version of ART utilized in this approach is ART 2-A [Carpenter and Grossberg (1991); Carpenter et al. (1991)], which embodies the basic design of ART 2 [Carpenter and Grossberg (1991)] but permits the use of analog input data necessary for this application.

In ART networks, a vigilance parameter, $\rho$, a sort of variable error criterion [Carpenter and Grossberg (1988)], sets the coarseness of similarity allowed in feature categorization; that is, it controls the degree of similarity to which input patterns must conform to be placed in a particular category in the F2 layer. Normally, a vigilance value that is too high constrains the cluster evolution resulting in large numbers of smaller clusters, whereas vigilance values that are too low produce the opposite effect, resulting in small numbers of very large clusters. Thus, establishing the vigilance range for a particular data set and application is critical to obtaining useful cluster results.

When using spectral data only, the dimension of the input vector is equal to $M$,the number of spectral bands. Because the spectral signatures of ground materials may be very similar and therefore difficult to differentiate, a very high vigilance value of 99.5 percent was applied and produced useful cluster results. However, when combining spatial and spectral data, the dimension of the input vector increases to $M \times (Q+1)$, where again $M$ is the number of spectral bands and $Q$ is the spatial area surrounding each pixel. Use of a high vigilance value at this point results in an excess of clusters from which it is now more difficult to extract critical or concise cluster information. Therefore, when using combined spatial and spectral image data vigilance values in the range of 50 to 60 percent were applied for training and produced workable cluster results.

Though our emphasis has been on the ART architecture, we have also done some preliminary comparisons with Kohonen's self-organizing map (SOM) [Kohonen (1991)].

## CLASSICAL CLUSTERING ALGORITHMS

As a benchmark, we compared our neural clusterings with some classical algorithms, most of which are well described in texts [Hartigan (1975); Späth (1975); Jain and Dubes (1975)]. The Leader algorithm is the most ART-like, in that it takes a single pass at the data, and identifies each data point as either belonging to an existing category or being a prototype (a "leader") for a new category. The Leader algorithm also requires a "vigilance" parameter to determine how closely an existing data point needs to be to a leader in order to be classified into that leaders category. Unlike ART, the prototypes do not adapt with further presentation of data. A variant of the leader algorithm, Max-a-min, makes multiple passes through the data. The first passes identifies the "leaders" as those points farthest away from the other leaders, and the last pass collects points into the category for which the leader is closest. Unlike ART or Leader, the Max-a-min algorithm takes as input the number $K$ of desired clusters.

Perhaps the most famous classical algorithm, $k$-means, is not actually a clustering algorithm, but a scheme for iteratively improving an existing clustering. Improvement is defined explicitly in terms of minimizing a global cost function which is the same as the compactness measure that we describe below. Although it is traditional to begin with a "random" clustering and to use $k$-means to achieve a final clustering our approach is to begin with the clustering given by the Leader algorithm.

Space, unfortunately, does not permit us to describe the details, but we have also devleoped an algorithm we call contig-$k$-means which works on the same principle as the $k$-means algorithm, but the cost function that it attempts to minimize is a weighted sum of the compactness and the (negative of) the contiguity.

## COMPARATIVE MEASURES

Since we do not (and often, cannot) compare our clustering results to ground truth, it is difficult to quantify whether a given clustering is any better or worse than any other clustering. This is a necessary consequence of the nature of the problem. Still, it is imperative that we seek, to as great an extent as possible, objective and quantitative measures which can be used for comparing different approaches.

Perhaps the most straightforward measure of cluster quality is just an evaluation of how "compact" each cluster is. Compactness is computed as the sum of squared distances from each data vector to the center of its cluster. In fact, it is just this quantity which the $k$-means algorithm attempts to minimize. We report the square root of this quantity divided by the standard deviation of the data, so it is a measure that ranges between zero and one.

We also considered a measure which explicitly takes the spatial aspect of the clustering into account. This measure quantifies the "contiguity" of the clustering image by counting the fraction of adjacent pixel pairs which are identified as being in the same cluster. This value varies between zero (for a checkerboard, for instance) and one (for all pixels the same).

A third diagnostic, the entropy, measures how evenly distributed the data is among its clusters. If there are $N$ total data points and $N_i$ points in the $i$-th cluster, then the entropy is

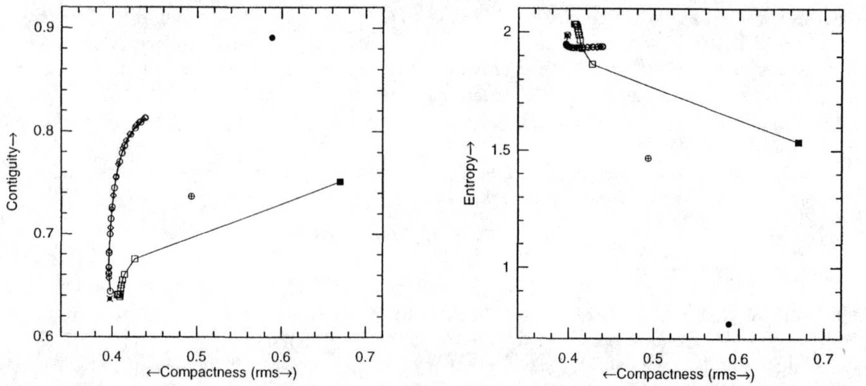


**Figure 1.** (a) **In this plot of compactness (as measured by root-mean-square of distances from points to their cluster centers) versus contiguity (which measures the tendency of adjacent pixels to be in the same cluster), each clustering is a single point. A variety of algorithms were used to cluster the pixels in the Hatch, New Mexico, image into $K = 9$ categories. In general, one prefers compact clusters and categories which form nearly contiguous regions; that is, one prefers clusterings toward the upper left in this plot. The solid square is the leader algorithm, and it is used as an initial point for the $k$-means algorithm, successive iterations of which are shown as open squares. The solid circle is the max-a-min algorithm, and the crossed diamond in the lower left corner is Kohonen's self-organizing map. The circled plus sign in the middle is the clustering produced by the ART algorithm. The line of clusterings indicated by open circles and open diamonds correspond to the modified $k$-means algorithm at different parameter values $\lambda$ ranging from $\lambda = 0$ (ordinary $k$-means) to $\lambda = 2000$.** (b) **In this plot, the same clusterings are shown, but with entropy versus compactness.**

defined by $S = -\sum_i (N_i/N)\log(N_i/N)$. For instance, if the points are evenly divided among $n$ clusters, then the entropy is $\log n$. At the other extreme, if all the points are in one cluster, and all the other clusters are empty, then the entropy is zero.

## ANALYSIS OF RESULTS

Although we looked at Landsat images from a number of different geographical regions in New Mexico, we will report here results only for an image which includes the village of Hatch, some of the surrounding countryside, and a small segment of the Rio Grande river. Fig. 2(a) shows this scene as an ordinary greyscale image.

Our first wave of comparisons are between algorithms, and Fig. 1 illustrates the tradeoff of compactness and contiguity, two properties that we would "like" to have in a clustering. It is interesting that successive iterations of the $k$-means algorithm produced clusterings with decreasing rms (more compactness), but at a cost of contiguity. This is also evident from visual inspection of the clustered images; qualitatively, those images seem more "mixed up." This illustrates that trying to improve one metric can lead to the degradation of another. Not surprisingly, the modified $k$-means algorithm that included a weighted term for increasing contiguity produced clusterings with larger contiguity, and (though only slightly) less compactness. However, compactness and continuity are not the only measures of quality — indeed, the ultimate measure will depend on the application at hand — so optimizing those measures may be at the expense of some other, more useful, but still unspecified, quantity.

We have also used PCA to reduce the dimension of this still purely spectral image, and some results are shown in Table 1 for the ART clustering. Since both the dimension $m$, and the number $K$ of clusters varies, it is difficult to make quantitative comparisons.

| Algorithm | Compactness | Entropy | Contiguity | Clusters | Vigilance |
|---|---|---|---|---|---|
| Original Landsat image, $m$=7 | | | | | |
| ART | 0.493007 | 1.46791 | 0.736733 | 9 | 99.5% |
| Kohonen SOM | 0.397199 | 1.98671 | 0.63702 | 9 | |
| Leader | 0.6698 | 1.53581 | 0.751214 | 9 | 97.5% |
| $k$-means | 0.397943 | 1.98803 | 0.644128 | 9 | |
| Contig-$k$-means, $\lambda = 200$ | 0.407339 | 1.93471 | 0.767682 | 9 | |
| Max-a-min | 0.588268 | 0.763005 | 0.890817 | 9 | |
| Spectral dimension-reduced image, $m$=2 | | | | | |
| ART | 0.58298 | 2.54718 | 0.455475 | 18 | 60% |
| Leader | 0.640351 | 1.30408 | 0.773245 | 9 | 99.5% |
| $k$-means | 0.351972 | 1.86678 | 0.667263 | 9 | |
| Contig-$k$-means | 0.425397 | 1.47298 | 0.828742 | 9 | |
| Max-a-min | 0.49572 | 1.00513 | 0.852693 | 9 | |
| ART, $m = 3$ | 0.352945 | 3.17245 | 0.454283 | 31 | 55% |
| ART, $m = 5$ | 0.878115 | 0.363317 | 0.929868 | 30 | 60% |
| Spatio-spectral dimension-reduced image, $m$=2 | | | | | |
| ART, $m = 3$ | 0.442424 | 3.14658 | 0.5333 | 33 | 55% |
| Leader | 0.626686 | 1.41603 | 0.839017 | 9 | 99.25% |
| $k$-means | 0.332666 | 1.98129 | 0.741429 | 9 | |
| Contig-$k$-means | 0.362035 | 1.78707 | 0.830537 | 9 | |
| Max-a-min | 0.497689 | 1.07366 | 0.89128 | 9 | |

**Table 1. Comparison of compactness, entropy, and contiguity for various clusterings of the original Landsat image of Hatch, New Mexico, also shown are clusterings on a dimension reduced image, and on a dimension reduced image of the spatio-spectral embedding.**

ART Cluster results are produced in a pixel color coded image, where pixels belonging to the same cluster have the same color. Sample color coded image results are reproduced here as grey scale images and shown in Figures 2(b) and (c). Figure 2(a) is the original Landsat image. The dimensionality of the original image was reduced to $m = 3$ and clustered using spectral data only and then both spectral and spatial data, as illustrated in Figures 2(b) and (c) respectively. Training of a subset of pixels from the dimension-reduced Landsat image was achieved at a vigilance of 55 percent. Clustering of spectral data only yielded 31 clusters. Clustering using combined spectral and spatial data yielded 33 clusters.

Inspection of the visual cluster results obtained using both spectral and spatial data reveal a more diverse classification of the ground materials present in the image. Further inspection and comparisons to actual ground truth of the image reveal that the clustering obtained not only identified features recognizable using the spectral data only, but also identified other features which may be later used to make stronger judgments as to actual texture and terrain of the ground materials.

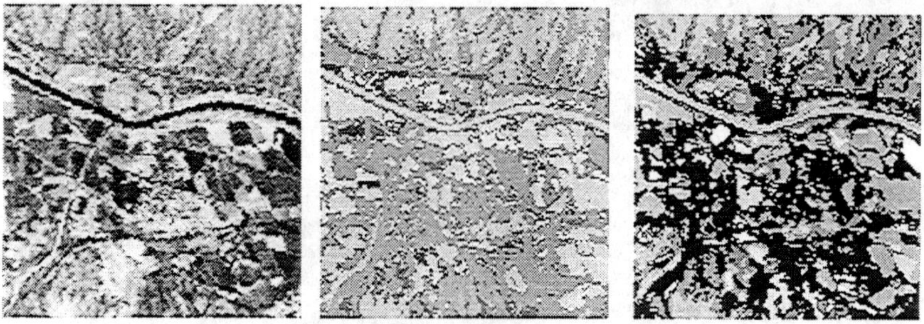

Original Landsat Image (a) | Di-Reduced with Spectral Data (b) | Di-Reduced with Spectral and Spatial Data (c)

**Figure 2. ART Cluster Results for spectral and spatio-spectral embeddings of the image data, dimension-reduced to $m = 3$.**

## CONCLUDING REMARKS

We have presented results of unsupervised clustering techniques for the purpose of identifying and clustering features of high resolution remote sensing images using combined spatial and spectral pixel data.

This continues to be a work in progress. Consequently, several issues currently under study include the determination of if the ART network is the most viable tool for the task, the continuing comparative analysis of the ART cluster results to those obtained from other clustering techniques, and development of an adaptive system for real-time detection of anomalies in multi-spectral images. Issues regarding training time and network scalability are also currently under study.

## REFERENCES


Carpenter, G. A., Grossberg, S., 1988, "The ART of Adaptive Pattern Recognition by a Self-Organinzing Neural Network," *IEEE*, Vol., pp. 77 - 88.

Carpenter, G. A., Grossberg, S., 1991, "ART 2: Self-Organization of Stable Category Recognition Codes for Analog Input Patterns," Proc. Pattern Recognition by Self-Organizing Neural Networks, chapter 12, pp. 397 - 423. MIT PRESS, Cambridge, Massachusetts.

Carpenter, G. A., Grossberg, S., Rosen, D. B., 1991, "ART 2-A: An Adaptive Resonance Algorithm for Rapid Category Learning and Recognition," *Neural Networks*, Vol. 4, pp. 493 - 504.

Hartigan, J. A., 1975, *Clustering Algorithms* John Wiley & Sons, New York.

Jain, A. K., Dubes, R. C., 1975, *Algorithms for Clustering Data* Prentice Hall, Englewood Cliffs, NJ.

Kohonen, T., 1991, "The Self-Organizing Map," *Proc. IEEE*, Vol. 78, pp. 1464–1480.

Richards, J. A., 1994, *Remote Sensing Digital Image Analysis, An Introduction* Springer-Verlag, New York, 2nd. edition.

Späth, H., 1975, *Cluster Analysis Algorithms for Data Reduction and Classification of Objects* John Wiley & Sons, New York.

where $Y$ indicates the class, $C_Y$ represents the cost of misclassification, $P_Y$ is the *a priori* probability, $pdf_Y$ is the probability density function.

Without a loss of generality this method will be detailed for one-dimensional patterns, ignoring misclassification costs and *a priori* probabilities. The problem of pattern classification is now reduced to estimating the conditional probability density functions from a set of "training patterns" belonging to known classes. The probability density function for a given class can be estimated as a convolution of all the patterns of the training set which belong to this class with a Gaussian of fixed kernel width (σ) and performing the appropriate normalization. The (conditional) probability density function for a given one-dimensional unlabeled pattern ($x$) based on a set of labeled training patterns ($x_i$) belonging to a known class is given by estimation from (Masters, 1994)

$$pdf(x) = \frac{1}{N\sqrt{2\pi\sigma}}\sum_{i=1}^{N} e^{-D_i^2}, \tag{2}$$

where $N$ is the number of samples of the training set belonging to the class, $x_i$ is a particular pattern known to belong to class A, $\sigma$ is the Gaussian kernel width (assumed to be unity), and

$$D_i^2 = \frac{(x - x_i)^2}{2\sigma^2}. \tag{3}$$

Estimating the appropriate kernel width for each class is crucial for the performance of this scheme and is often done by trial and error.

A different insight into probabilistic or statistical pattern classification can be gained by considering the first-order Taylor series expansion of the exponent in equation (2). For patterns with P independent dimensions this leads to:

$$pdf(\vec{x}) = \frac{1}{\left(\sqrt{2\pi}\right)^P \prod_{i=1}^{P}\sigma_i}\left[1 - \sum_{i=1}^{P}\frac{(x_i - \bar{x}_i)^2 + VAR(x_i)}{2\sigma_i^2}\right], \tag{4}$$

where $x_i$ is the average and $VAR(x_i)$ is the variance of the points $x_i$.

The above formula adds a refreshing intuitive insight to pattern classification by a PNN. Determining the most likely class for a given pattern $x$ is similar to simply deciding for which class the pattern has the smallest "variance corrected Euclidean distance" from the mean pattern of that class.

The intuitive interpretation of equation (4) is illustrated in figure 1 for two dimensions. The two classes (class A and class B) of two-dimensional patterns shown in this figure are centered around (0, 0) and (3, 2) respectively. Both classes consist of a random Gaussian distribution of points with unity σ in the x and y directions for class A, and a σ equal to 1/2 and 1/3 for the x and y directions for class B patterns. The boundary separating class A patterns from class B patterns using a naive smallest distance from the mean criterion is labeled "Euclidean separation." A Bayesian separation between the two classes based on the Gaussian pdf expression leads to an elliptical separation region and is indicated on the figure. The Taylor series expansion for the pdf estimation for the classification boundary is labeled "variance corrected Euclidean separation" and matches closely the Bayesian separation for the proper choice of the Gaussian kernel width. While point C is closer to class B in the Euclidean sense it is more likely to belong to class A when the variances of class A and class B are accounted for.

# IMAGINET: A NOVEL PROBABILISTIC NEURAL NETWORK FOR RAPID MULTISCALE IMAGE CLASSIFICATION

**MARK J. EMBRECHTS** (embrem@rpi.edu)
*Department of Decision Sciences and Engineering Systems*

**VIJAY SANKARAN, RUSSELL P. KRAFT, and DON L. MILLARD**
*Center for Integrated Electronics and Electronics Manufacturing*
*Rensselaer Polytechnic Institute, Troy, NY 12180-3590, USA*

**ABSTRACT:**
This paper presents a novel interpretation of Bayesian image classification and implements this interpretation into a simple neural network paradigm. IMAGINET is a novel probabilistic neural network for image recognition that has single-pass training and fast execution. Only the average patterns and the pixel variances for each pixel and class have to be stored in memory. IMAGINET is very flexible and can take neighborhood information into account. The results for a solder-joint image classification problem show that IMAGINET has a superior performance when compared to standard backpropagation and a self-organizing map.

## 1. INTRODUCTION

Rapid automated classification of raw pixelized images remains a challenge for computational intelligence because:
(a) Memory requirements are extensive, and training is slow.
(b) Algorithms are sensitive to shifts and rotations of the image.
(c) A feedforward network trained with the backpropagation algorithm (BP) does not rely on neighborhood information, and it is therefore hard to argue that such image classifiers could be considered to be intelligent at all.

In this paper the theory of probabilistic neural networks (PNNs) is extended to a novel neural network paradigm for image classification (IMAGINET) that (a) can retain neighborhood information, (b) trains and executes quickly, (c) is not memory intensive, (d) is not extremely sensitive to shifts in the positioning of the image, and (e) does not have to resort to wavelet or Fourier preprocessing. Training of IMAGINET is fast and consists of one or more passes where prototype images are constructed by averaging the training set images and a convolution with one or more Gaussian kernels of predetermined width, $\sigma$.

## 2. PROBABILISTIC NEURAL NETWORKS: A BRIEF REVIEW

Probabilistic neural networks were introduced by Donald Specht (Specht, 1990). PNNs are a straightforward -- but nevertheless clever -- neural network implementation of statistical pattern recognition. Statistical pattern recognition is equivalent to determining the class to which the pattern most likely belongs in the Bayesian sense (Bishop, 1995). This is also the class (Y) for which a given pattern ($\vec{x}$) has the highest figure of merit (FOM) according to:

$$FOM_Y(\vec{x}) = C_Y . P_Y . pdf_Y(\vec{x}) \tag{1}$$

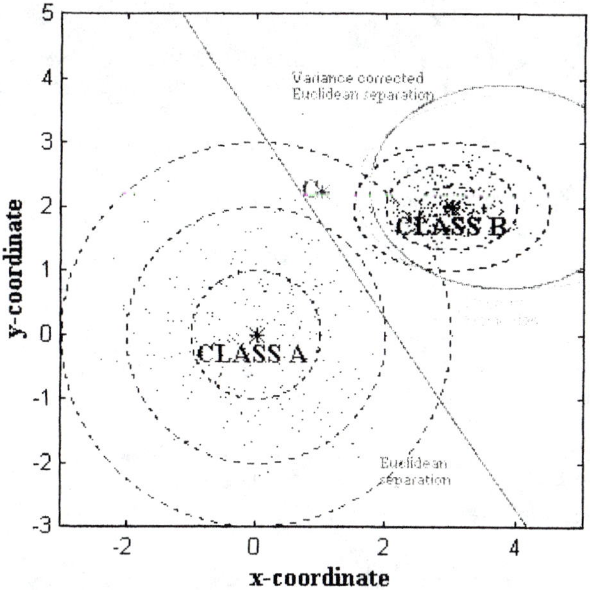


**Figure 1. Illustration of boundary separating class A patterns from class B patterns using a naive smallest distance from the mean criterion (straight line, labeled Euclidean separation) and Bayesian pattern classification (most elongated elliptical region). The variance corrected Euclidean separation is labeled on the other elliptical region.**

## 3. PNNs FOR IMAGE CLASSIFICTION

We will now build on our naive interpretation of statistical pattern recognition to extend PNNs to image classification problems. This method will be framed in the next section as a simple and fast novel neural network paradigm for image classification: IMAGINET. The paradigm will then be extended to include neighborhood information in section 5.

The extension of PNNs to two-dimensional gray scale pixel images is straightforward when the naive interpretation from the previous section is kept in mind: each pixel is considered as an independent dimension. Rather than assigning unity if a pixel is occupied for a particular pattern and calculating the variance over a continuous scale, the variances of the gray scale levels are tabulated for each pixel. "Training" a PNN for two-dimensional images is as simple as calculating the average image for each class (normalized by taking the *a priori* distribution into account). Image classification with a PNN is nothing more than determining the "variance corrected Euclidean distance" (with appropriate scaling to accommodate the choice of the Gaussian kernel width) between a "test image" and the average image for each class according to (4) and selecting the class with the smallest variance corrected Euclidean distance. Casting such a PNN as a neural network paradigm parallels Specht's implementation and the pros and cons of PNNs are reflected as well: fast training and slow recognition.

## 4. IMAGINET

The neural network implementation of IMAGINET deviates from the traditional PNN approach by relaxing the variance corrected Euclidean distance measure in the following manner:

(a) The layer of neurons of IMAGINET contains as many neurons as there are image classes and the weights of each neuron correspond to the average pixel value for each class of images. The neurons do not have a transfer function and act as only summation operators.

(b) The bias contains the lumped variance measure modified by the particular choice for the Gaussian kernel widths. The lumped variance measure is a combination of the individual pixel variances.

(c) The neural network is executed in a competitive mode by relaxing the Euclidean distance to a dot product similarity measure so that the neuron with the largest response corresponds to the class with the smallest variance corrected distance from the mean.

Under the assumption that the Gaussian kernel widths are equal for all classes finding the class with the maximal pdf now reduces to finding the class for which the variance corrected Euclidean distance (VCED) is minimal according to:

$$VCED = \sum_{i=1}^{P} \frac{(x_i - \bar{x}_i)^2}{2\sigma_i^2} + \sum_{i=1}^{P} \frac{VAR(x_i)}{2\sigma_i^2} \tag{5}$$

By rescaling the variables to the appropriate Gaussian kernel the minimal Euclidean distance measure is almost equivalent with a dot product maximal similarity measure and finding the class with the maximal pdf can now be approximated with finding the class for which the variance corrected similarity (VCS) is maximal according to:

$$VCS \approx \vec{\mathbf{x}}' \bullet \vec{\mathbf{x}}'_{AV} - \sum_{i=1}^{P} \frac{VAR(x_i)}{2\sigma_i^2} \tag{6}$$

The primes in the notation indicates that these variables have been scaled with respect to the Gaussian kernel width. Figure 2 illustrates the implementation of this strategy for two-dimensional patterns and two possible classes in a competitive neural network. Note that the weights for this network are just the averages and that the variance correction is lumped in a bias for each class. The normalization factor of equation (4) was not considered so far in this implementation to keep the explanation of the network implementation as simple as possible.

The implementation of IMAGINET with different Gaussian kernels for the classes (general case) is straightforward. The normalization factor in front of the expression for the pdf in (4) does not cancel out in the competitive network anymore, but could be included by subtracting unity from the bias and incorporating an appropriate transfer function (i.e., a multiplier).

IMAGINET does a one-pass training by storing the average images as weights and determines the class based on the variance corrected similarity measure. Some crude approximations (first-order Taylor series expansion) were made to be able to cast IMAGINET as a simple competitive neural network. The *a priori* probabilities and cost functions could easily be included by modifying the transfer function to the neurons appropriately. IMAGINET has the obvious advantage over a traditional PNN: only the average patterns, scaled for each class by the corresponding Gaussian kernels, have to be stored in memory (rather than all the patterns as was the case for the PNN).

While the neural network implementation of IMAGINET now shows the advantage of fast training and fast classification, it still suffers from the same potential fundamental flaws as the PNN when it comes to image classification:

(a) The PNN is not inherently tolerant for image translations, rotations or shifts.
(b) A neural network trained with raw pixel-based images suffers from the curse of dimensionality (Bishop, 1995). I.e., a 64x64 image has 4096 independent variables. Most neural network applications require 3 to 10 sample images for each independent variable. The required number of images for training might therefore be excessive for most practical implementations.
(c) Smart image recognition should incorporate neighborhood information.
(d) Classification is sensitive to the choice of the Gaussian kernel widths.

There are several ways to mitigate some of the above deficiencies by applying appropriate preprocessing techniques:

(a) Any combination of translation, rotation and shift invariance can be accommodated by applying a Fourier-polar-Fourier transformation (Fuchs and Haken, 1988).
(b) The curse of dimensionality can be alleviated by reducing the number of independent variables: i.e., considering selected two-dimensional (2-D) Fourier components or block pixel averaging.
(c) 2-D Fourier transforms and block pixel averaging do actually account for neighborhood information.
(d) Several authors make suggestions for the approprate choice of the Gaussian kernel withs with statistical preprocessing techniques (Silverman, 1986).

IMAGINET does not exclude Fourier-based preprocessing per se. However a different approach was attempted to account for neighborhood information and to mitigate small shifts in the centering of the images. Rather than storing average images in the network the average images are convoluted with a Gaussian.

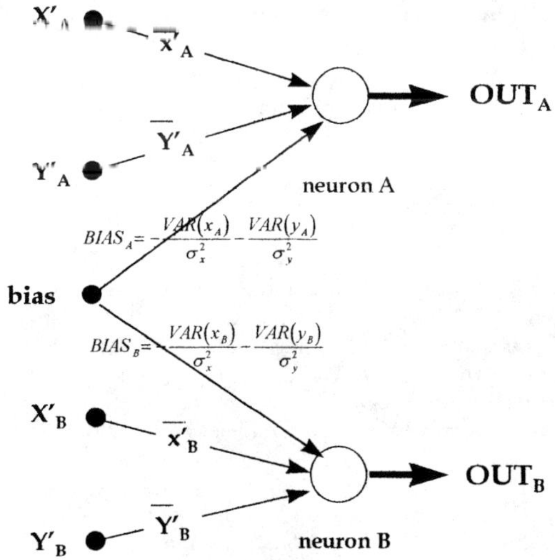


**Figure 2. Implementation of IMAGINET as a competitive neural network for two-dimensional patterns which can either belong to class A or class B.**

## 6. SOLDER JOINT CLASSIFICATION WITH ANNs

The performance of IMAGINET was compared with a standard BP trained network for a set of 600 solder thruhole joint images. The solder joint image data base consists of 256 gray-scale 64x64 pixelized images which can belong to three classes (too little solder, good solder joints, and excess solder). The number of solder joints for each class is about equal. Images were split up in an equal number of patterns for training and testing. Details about this solder joint classification problem were reported earlier by the authors (Embrechts et al., 1995).

A backpropagation network trained on raw images could classify 64 percent of the images correctly. By training a BP neural network on the lowest 7x7 frequency components of a two-dimensional Fourier spectrum of the solder joint images, 78 percent of the joints could be classified successfully.

For the experimental determination of the performance of the IMAGINET on raw images the average image for the three classes and their variance were computed for 100 training images per class. All Gaussian kernel widths were arbitrarily set equal to unity. Different Gaussian blurs were applied to each class average and their variance corrected Euclidean distance from the test images were computed. The correct classification percentages are tabulated in Table I. From the results, it can be observed that the pdf approximation technique of IMAGINET provides about 78 percent overall accuracy.

**Table I. Results of the multiscale implementation of IMAGINET**

| Data set | Blur9 | Blur7 | Blur5 | Blur3 | None | Sharp1 | Sharp2 | Sharp3 |
|---|---|---|---|---|---|---|---|---|
| Thru-II | 60.7 | 69.7 | 76.8 | 76.1 | 77.7 | 77.6 | **78.1** | 75.3 |

## CONCLUSIONS

This paper provides a novel insight into statistical pattern classification and implements this interpretation in a novel neural network for image classification: IMAGINET. IMAGINET has one-pass training, performs well on raw pixelized images and - unlike the PNN - has extremely fast execution with modest memory requirements.

## REFERENCES


Bishop, C.M., (1995). *Neural Networks for Pattern Recognition,* Clarendon Press.

Embrechts, M.J. , Harsson, L.E., Kraft, R. P. , Millard, D.L., , and Sankaran, V., (1995). Preprocessing for the Inspection of Printed Wiring Boards with Neural Nets, in *Intelligent Engineering Systems through Artificial Neural Networks, Vol. 5: Fuzzy Logic and Evolutionary Programming,* ASME Press.

Masters, T., (1994). *Advanced Algorithms for Neural Networks: A C++ Sourcebook,* John Wiley.

Fuchs, A., and Haken, H., (1988). Pattern Recognition and Associative Memory as Dynamical Processes in a Synergetic System, *Biological Cybernetics*, **60**, 107-109.

Silverman, B.W., (1986). *Density Estimation for Statistics and Data Analysis*, Chapman & Hall.

# AN AUTONOMOUS BOOLEAN NEURAL NETWORK APPROACH FOR IMAGE UNDERSTANDING

**ESSAM A. EL-KWAE AND MANSUR R. KABUKA**
*Center for Medical Imaging and Medical Informatics*
*University of Miami*
*1150 NW 14 Th. St., Suite 301, Miami , FL 33136*

***ABSTRACT:***
The Boolean Neural Network (BNN) has been successfully applied in many image processing tasks. In this paper, a method to incorporate rules into the BNN is proposed. The concept of ROA (Radius of Attraction) is used to incorporate more complex rules into the BNN which are difficult and inefficient to incorporate in traditional production systems. Then, a Rule-Based BNN approach is proposed as an autonomous approach for image understanding. This approach integrates unsupervised image segmentation, feature extraction, image labeling, and a task-specific knowledge-base to extract information from images.

The approach was applied to the analysis of colored driver licenses (Kabuka, Waly, and Sauer, 1995). The target was to extract the driver personal information. Results of testing are provided in the paper.

## INTRODUCTION

Neural Networks present a powerful alternative to conventional classifiers due to their inherent parallelism and fault tolerance. In spite of that, ANNs are faced with some problems as complex architecture, slow learning, and non-guaranteed convergence. Choosing an ANN architecture for a certain application is a complex task usually done by trial and error. In addition, there are many problems in practice where descriptive methods are more suitable. Descriptive methods are often based on classification rules that map the input vectors to output categories. The classification rules, in this case, can be stored in a knowledge-base.

In this paper, incorporating rules into the BNN is suggested. The BNN is shown to be ideally suited to incorporate domain knowledge. Then, a Rule-Based BNN approach is proposed as an autonomous approach for image understanding. The approach integrates unsupervised image segmentation, feature extraction, image labeling, and a task-specific knowledge-base to extract information from images.

The rest of the paper is organized as follows: An overview of the BNN is given. Then, encoding knowledge into the BNN is discussed followed by the proposed image understanding approach. Results of applying this approach for extracting driver information from driver licenses are provided followed by conclusions.

## OVERVIEW OF THE BNN

The BNN developed in (Hussain and Kabuka, 1992) can be used to synthesize any boolean function. It was used to solve optimization problems (Bhide, John, and Kabuka, 1993). A feature recognition BNN has been developed in (Bhide, John, and

Kabuka, 1994). In (Gazula and Kabuka, 1995), the BNN was used for the design of classifiers. In the medical field, the BNN was used for supervised and unsupervised segmentation of MR Brain images (Bhide, 1994), for labeling MR Brain images (Li, Bhide, and Kabuka, 1994)), and for detection of malignant regions in 3D MR breast images (El-Kwae and Kabuka, 1996).

The BNN consists of layers of neurons fully connected in a feedforward structure. Layers in the BNN have simple perceptron connections. While learning, the BNN converges in a single iteration because patterns are encoded directly into the weights of their corresponding neurons. Thus, weight calculations are simple and no weight update algorithm is needed. After learning, a new training pattern can be applied without unlearning those already learned. The concept of ROA (Radius of Attraction) is used to classify noisy and distorted patterns. Every exemplar node is associated with a programmed threshold $\Theta_k$ and a ROA $r_k$. The programmed threshold is then decremented by the ROA to get the actual threshold.

$$\Theta_k = \Sigma_i a_{ik} w_{ik} - r_k \qquad (1)$$

This leads to all patterns that differ from the stored pattern with a hamming distance of ROA to be classified as belonging to the class of that exemplar. The ROA builds hyperspheres of radius $r_k$ with the class programmed threshold as center. It was proved in (Gazula and Kabuka, 1995) that the number of patterns from $R^n$ enclosed by an ROA r defined over a certain neuron are:

$$^n\Lambda_r = \Sigma^r_{i=0} \, ^nC_i \qquad (2)$$

In (Bhide, 1994), a generalization of the BNN for handling scalar weighted functions was given. The scheme is called the Fixed Weight Coding Scheme (FWC). Input data is quantized into g gray scales represented by m bits where $m=\log_2 g$. One layer of preprocessing is added to the BNN whose weights are allowed to have fixed values from $2^0$ to $2^{m-1}$. Inputs to next layers are not binary but the BNN still have binary external inputs and integer weights. The ROA in this case is based on the Euclidean Distance and is called the EROA (Euclidean Radius Of Attraction).

## KNOWLEDGE INCORPORATION INTO THE BNN

Incorporating knowledge into connectionist form has several advantages. One is the inherent parallel rule inferencing and the other is the trainability of the networks. Several connectionist knowledge representation methods exist, e.g (Towell and Sahvlik, 1992).

Decision rules generally take the form: IF condition(s) THEN action

Where conditions(antecedents) and actions(consequence) are boolean functions. Knowledge described in this form is easy to understand and can be directly used to construct the knowledge-base for expert systems. Domain knowledge is translated into the BNN by defining the topology, connection weights, and thresholds. This idea is best illustrated by examples.

**Example 1: A Conjunction/Disjunction Rule**

IF ($a_1$) AND ($a_2$) AND ...... ($a_n$) THEN C

Only a single neuron is required as shown in Figure 1, where n is the number of conditions in the antecedent. C equals 1 (and the neuron fires) only when all $a_i$ are 1 and is 0 otherwise. To encode a disjunction rule, the threshold is changed to 1.

$$C = \begin{cases} 1 & \text{If} \quad \sum_i W_i \, a_i \geq T \\ 0 & \text{Otherwise} \end{cases} \qquad (3)$$

$$\text{Where,} \ T = \sum_i W_i = n \qquad (4)$$

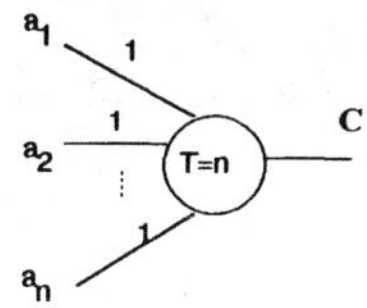


Figure 1. BNN Implementation of A Conjunction Rule.

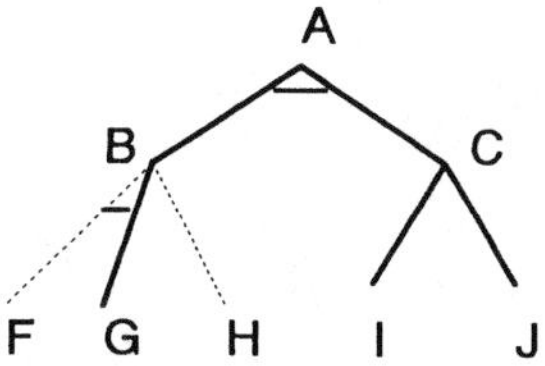


Figure 2.a AND/OR Graph.

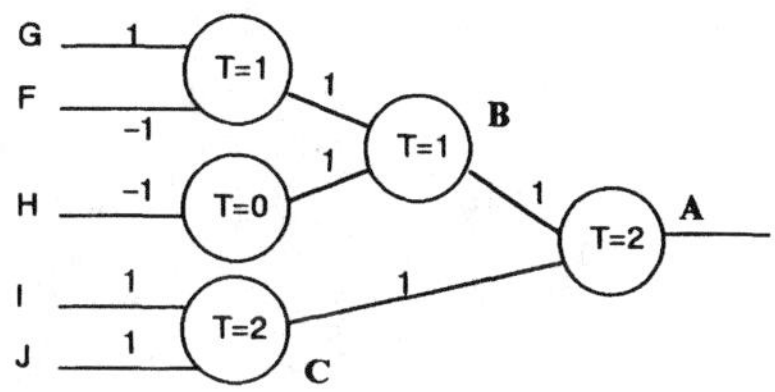


Figure 2.b. BNN implementation.

**Example 2: Encoding A set of rules**

IF (B) AND (C) THEN A

IF (NOT H) OR ((NOT F) AND (G)) THEN B

IF (I) AND (J) THEN C

This example is similar to that in (Towell and Shavlik, 1992). Figure 2.a. illustrates the AND/OR Graph and Figure 2.b. illustrates the BNN implementation.

The ROA can be used to encode more general rules such as the majority rule, m out-of-n rule, and the all-but-one rule, thus leading to a reduction in the number and size of the rules in the system and easier maintainability of the knowledge-base.

**Example 3: Encoding the majority rule**

IF (majority of input conditions is true) THEN C

If number of input conditions is 3, then 4 rules are required. For larger n, the task is quite tedious and error prone. On the other hand, encoding the majority rule into the BNN requires only a single neuron. Weights and thresholds are set as example 1 above, the ROA is set to flow(n/2).

## THE PROPOSED IMAGE UNDERSTANDING APPROACH

The proposed image understanding approach (Figure 5) consists of three components: the P-BNN (Partitional BNN) for clustering (Bhide, 1994), the CS-BNN (Constraint-Satisfying BNN) for labeling (Li, Bhide, and Kabuka, 1994) and the KB-BNN (Knowledge-Base BNN).

The P-BNN (Figure 3) was developed in (Bhide, 1994). It is based on a variation of the K-means algorithm. The algorithm segments the image into regions starting with a number of clusters equal to twice or thrice the number of clusters eventually desired. Clusters are then reduced by merging close ones. This allows for the formation of hyperelliptical clusters which model the data better than the hyperspherical clusters that are a characteristic of most K-means based algorithms.

Results from the clustering stage are passed to the CS-BNN (Figure 4) for labeling. The labeling process can be viewed as a constraint satisfaction problem.

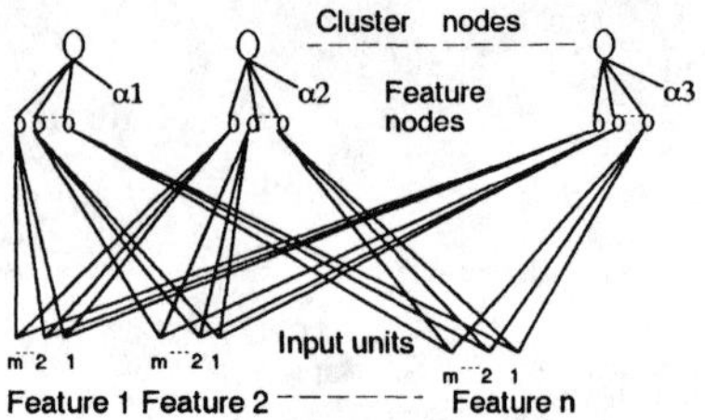


Figure 3. Network for the P-BNN.

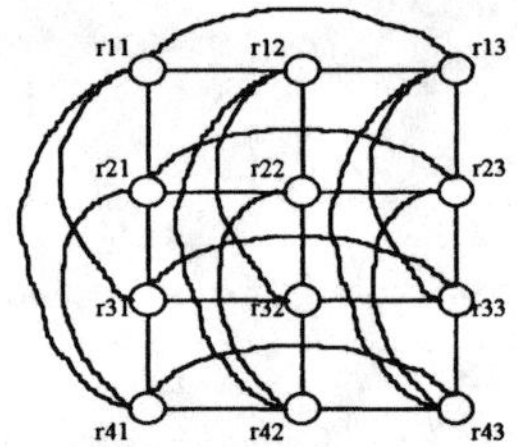


Figure 4. : CSBNN neuron array.

Some of the constraints involved are: single label for each region, two adjacent regions cannot have same labels, two labels may be known to be adjacent, and region property constraints like pixel intensity, and shape.

In (Li, Bhide, & Kabuka, 1994), the CS-BNN was used for image labeling. Each neuron represents the pairwise combination of a region and a label. Thus, for a n-region, m-label problem, the network will have a total of n x m neurons with mn(m+n-2)/2 interconnections. The Single label constraint can be enforced by the horizontal inhibitory connections which have weights of -1. To enforce the Region adjacency constraint, a Region Adjacency Matrix (RAM) n x n is developed. An entry of -1(0) corresponds to two adjecent(non-adjecent) regions respectively.

The KB-BNN then checks the consistency of the obtained labels. If inconsistent, it generates corrective feedback to the P-BNN to improve on the clustering. This cycle of segmentation, labeling, and consistency checking is repeated until consistent labels are assigned. If unsatisfactory results are obtained, tuning of the knowledge-base might be needed. Rules may be added, deleted, or modified to reflect the environment change or correct any failures.

**TEST RESULTS**

To test the proposed Image Understanding approach, a collection of 8 Florida driver licenses were scanned at a resolution of 610 x 611 pixels and stored in true RGB color, 24-bit/pixel. The goal was to extract the drivers' personal information.

To establish a base for comparison, the original image (Figure 9.a.) was scanned and input to the OmniPage® OCR software. It was not able to recognize any characters because of the background and the map contour superimposed on the image. Then, the image was segmented by the P-BNN. The text cluster, Figure 9.b., was input to the OCR. Results for the lines of interest are shown in Figure 6.

A set of rules were then extracted with all color and neighborhood information. These rules were then incorporated into the KB-BNN. An example of the rules used:

IF ITEM1="CLASS" AND ITEM2="A or B or C or D or E" AND ROW=8 THEN ROW_8_OK

For the P-BNN, the number of final clusters was determined to be 3, a text cluster (Figure 9.b.), edges and Florida map contour cluster (Figure 9.c.), and a background cluster (Figure 9.d.). After segmentation, row and column projection were done to extract different characters. A threshold on the space between characters was used to group characters into items. To recognize the extracted characters, a feature extraction BNN defined in (Bhide, John, and Kabuka, 1994) was used. Training characters were extracted from the test images. Results of the first iteration of the image understanding system are shown in Figure 8. These results are better than the OCR results given before.

A phase of rule application was done to correct errors before any reclustering was needed. For example: In row 0, item 2, the letter B was recognized as 8 and I as 1. A simple rule that the first row is always recognized as alphabets easily corrected this problem. The same can be said about the height field in row 3 , item 3, in which 0 was recognized as Q. Also, in row 7, item 1, SAFE was recognized as SAFF. This problem is simply solved by lowering the threshold of the neuron responsible for detecting this word by 1. i.e. putting ROA=1. Results of corrections are shown in Figure 7. In the following phases, the KB-BNN issued reclustering commands to the P-BNN component until label consistency was attained.

## CONCLUSION

In this paper, knowledge incorporation into the BNN was presented. The BNN proved to be ideally suited for knowledge representation. The concept of ROA was used to encode complex rules which are inefficient in traditional production systems. Then, an autonomous image understanding approach was proposed. The approach goes through cycles of segmentation and labeling followed by a rule-based BNN to check on the labeling results. Simple rules specific to the application were shown to have a great effect on improving labeling results. The proposed approach was tested on extracting driver information form scanned driver licenses. The approach is highly adaptive and can be used for a plethora of applications by altering the KB-BNN.

## REFERENCES


Bhide, S., (1994). Image Segmentation with Boolean Neural Networks, *M.Sc. thesis*, Univ. of Miami.

Bhide, S., John, N., Kabuka, M. R., (1993). A Real-Time Solution for the Traveling Salesman Problem using a Boolean Neural Network, *Proc. of the 1993 IEEE Int'l Conf. on Neural Networks* , San Fransisco, CA, 1096-1103.

Bhide, S., John, N., Kabuka, M. R., (1993). A Boolean Neural Network approach for the Traveling Salesman Problem, *IEEE Trans on Computers*, 42(10), Oct, 1271-1278.

El-Kwae, E., Bianchi, M., Kabuka, M.R., (1996). Detection of Malignant Regions in 3D MR Breast Images Using a Boolean Neural Network, *Proc. of the 13th Conf. on Computer Application in Radiology (SCAR/96)*, Denver, Colorado, Jun, 304-309.

Gazula S., Kabuka, M. R., (1995). Design of supervised classifiers using Boolean Neural Networks, *IEEE Trans on Pattern Analysis and Machine Intelligence*, Dec.

Hussain, B., Kabuka, M. R., (1992). Neural Net transformation of arbitrary Boolean functions, *SPIE 1992 Int'l Symposium on Optical and Applied Science: Neural and Stochastic methods in Image and Signal Processing* (Invited).

Hussain, B., Kabuka, M. R. , (1994). A novel feature recognition neural network and its application to character recognition, *IEEE Trans on Pattern Analysis and Machine Intelligence*, 16(1), Jan.

Kabuka, M. R. , Waly, S., Sauer, F., (1995). Automatic Data Entry Using Scanning and Color Segmentation of Driving Licenses, *11Th Int'l Symp. on the Creation of Electronic Patient Record Systems and Global Conference on Patient Cards*, Orlando, FL.

Li, X., Bhide, S., M. R. Kabuka, (1994). Labeling of MR Brian Images Using a Boolean Neural Network, *Proc. of the 12th Conf. on Computer Application in Radiology*, NC, Jun.

Towell, G., Shavlik, J. W., (1992). Interpretation of Artificial Neural Networks: Mapping Knowledge-Based Neural Networks into Rules, *In Advances In Neural Information Processing Systems 4*, Moody, J. E., Hanson, S. J., and Lippmann, R. P., eds., Morgan Kaufmann Publishers, 977-984.

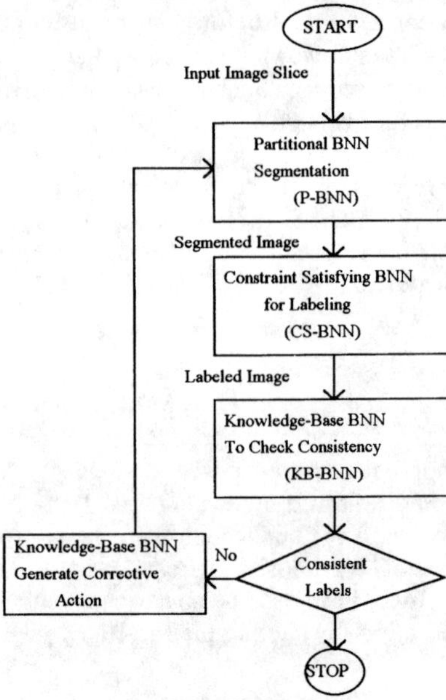


Figure 5. : Block diagram for Image Understanding Approach.

| Row | Item1 | Item2 | Item3 |
|---|---|---|---|
| 0 | SHIHIDE | BHID | |
| 3 | | | 5-06 |
| 4 | 08-8-192 | | |
| 7 | SAFE | | -7-09-1993 |

Figure 7. Refining results by KB-ANN.

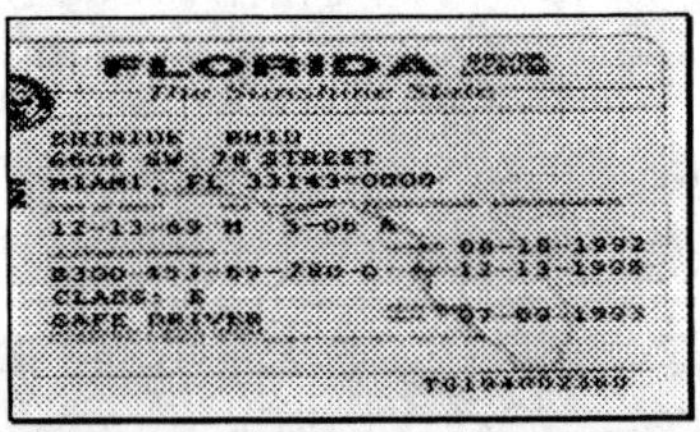

Figure 9.a. The Original Image.

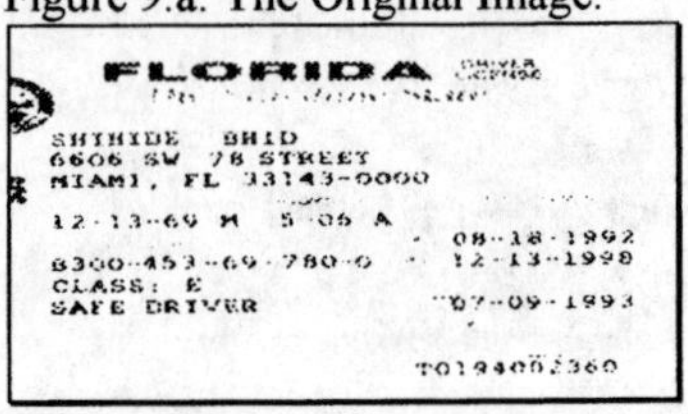


Figure 9.b. Text Cluster.

SHIHIDE BHID
6606 SW 78 STREET
MIAN1 ~ Ft 3 3 ~ 4 3 ~0000
~ ~ ~ 1 3 ~69 ~5 -06 A
08~ 18- I992
B300 ~ 453~9 ~7 R0 12
~` 1 3~I998
CLASS. E
SAP E DR IVER ~7 ~Q'- ~ DD 3

Figure 6. Applying OCR Software to Segmented Image.

| Row | Item 1 | Item 2 | Item 3 | Item 4 |
|---|---|---|---|---|
| 0 | SH1H1 DE | 8H1D | | |
| 1 | 6606 | SW | 78 | STREE T |
| 2 | MII | F | 3143- 0Q00 | |
| 3 | 2-1369 | M | 5-Q6 | A |
| 4 | 68-8- 192 | | | |
| 5 | B3O045 3-69-8O | 23-1998 | | |
| 6 | CLASS | E | | |
| 7 | SAFF | DRIVER | -7-69- 1993 | |

Figure 8. Results of iteration 1 of the image understanding approach.

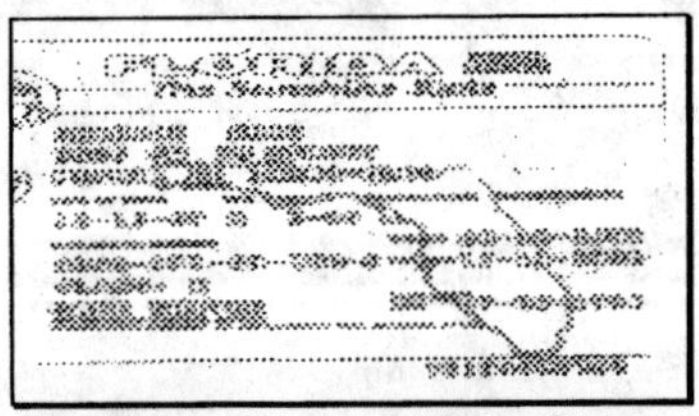

Figure 9.c. Edges and Map Contour.

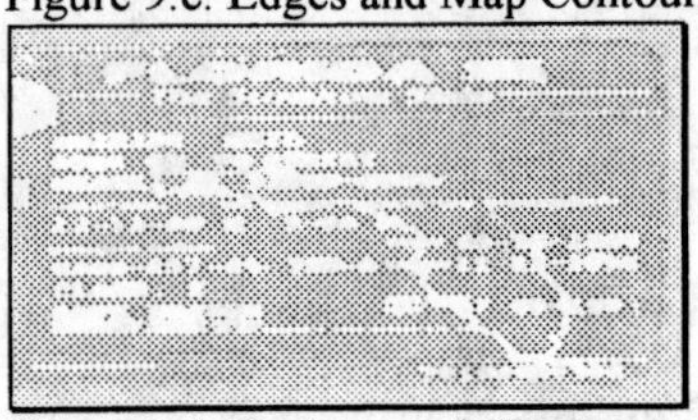

Figure 9.d. Background Cluster.

# USING IMMUNOLOGICAL PRINCIPLES IN ANOMALY DETECTION

**DIPANKAR DASGUPTA**
Dept. of Mathematics & Computer Science
University of Missouri-St. Louis
St. Louis, MO 63121

**Abstract**

In this paper, a method for anomaly detection is described which is based on principles of the natural immune system. Compared to existing methods, this method has the advantage of not requiring prior knowledge about all possible failure modes of the monitored system. Instead it uses the normal behavior pattern to generate a set of detectors which can detect any deviation that exceed allowable variation from the defined normal. An example using Mackey-Glass equation is presented and the performance of the proposed anomaly detection algorithm is reported.

## INTRODUCTION

The problem of detecting anomalies and novelties can be viewed as finding non-permitted deviations of a characteristic property in the system of interest. The detection of anomaly is an important task in many diagnostic or dynamical systems. In safety-critical applications, it is essential to detect the occurrence of unnatural events as quickly as possible before any significant performance degradation results. This can be done by continuous monitoring of the system for changes from the normal behavior patterns.

There have been a number of techniques suggested in the literature for detecting anomalies and faults in monitored systems. They include model-based methods, knowledge-based expert systems, pattern recognition and cluster analysis etc. Recently, neural network-based methods have been increasingly used for anomaly and novelty detections. Specifically, Multi-Layer Perceptrons (MLP) [Roberts and Tarassenko, 1994] and Adaptive Resonance Theory (ART) [Hwarng and Chong, 1994; Caudell and Newman, 1993] have received the most attention in such applications.

However, most existing methods require either prior knowledge about various anomaly conditions [Kozma et al., 1994] or precise theoretical models of the monitored system or the process. But a robust method should detect any unacceptable (unseen) change rather than looking for specific known novel activity patterns. This paper proposes a new anomaly detection method that addresses some of these deficiencies.

## AN ANOMALY DETECTION ALGORITHM

This approach is inspired by the information-processing properties of natural immune system. The immune system uses learning, memory, and associative retrieval to solve pattern recognition problems (of its defense mechanism). Animal immune systems are capable of distinguishing virtually any foreign cell or molecule from the body's own cells - this is known as the self - nonself discrimination problem. In the immune system, T cells have receptors on their surface that can detect foreign proteins (antigens). During the generation of T cells, receptors are made by a pseudo-random genetic rearrangement process. Then they undergo a censoring process, called negative selection, in the *thymus* where T cells that react against self-proteins are destroyed, so only those that do not bind to self-proteins are allowed to leave the thymus. This complex censoring process is very important in self-nonself discrimination. Our artificial immune system [Forrest et al., 1994; D'haeseleer et al., 1996] is an abstract of the complex chemistry of antibody/antigen recognition in natural immune systems.

The basic principle of our algorithm is as follows:

- Define *Self* as a normal pattern of activity or stable behavior of a system/process which need to be monitored. In particular, we build up a database of normal behavior of each process of interest, then logically split the pattern sequence to represent as a multiset $S$ of equal-size strings of length $l$ over a finite alphabet.

- Generate a set $R$ of *detectors*, such that each detector fails to match any string in $S$. We use a partial matching rule, in which two strings match if and only if they are identical for at least $r$ contiguous positions, where $r$ is a suitably chosen parameter (as described in [Forrest et al., 1994]).

- Monitor new observations (of $S$) for changes by continually matching the detectors against the representative of $S$. If any detector ever matches, a change (or deviation) must have occurred in system behavior.

In our implementation, we reduce the anomaly detection problem to the problem of detecting whether or not a string has changed (actually matched using negative selection), where a change (or match) implies a shift in the normal behavior patterns. During preprocessing, analog data is sampled from a moving time window and each data value in the window is encoded in binary ($m$ bits) string. Each window, therefore, is a sequence of data items (Win_Size) . A historical database is then constructed from a succession of windows sliding along the time series (Win_Shift) to capture the semantics (regularities) of the normal behavior of the system. The number of such windows needed for the normal database will depend on how much inherent variability there is in the monitored process. So as long as the time series data pattern maintains similar behavior, collected strings should be sufficient to define normal behavior of the system. This collection of strings for windows is our self ($S$). We then generate strings that do not match any of the strings in $S$ to become member of the detector set. However, there exist many different ways of generating set of detectors in the complement space (refer [D'haeseleer et al., 1996; Forrest et al., 1994] for detailed description).

## EXPERIMENTS

We have tested the feasibility of the immune system-based algorithm for anomaly detection on a number of test problems including tool breakage detection in milling operations, and some sensory data. In this paper, we report the experiments on Mackey Glass series [Mackey and Glass, 1977]. It has been used as a task for detecting any change from previous samples presented (as used in [Caudell and Newman, 1993]).

The normal data set with Mackey Glass series was generated using $\tau = 30$, and the novel pattern was generated using $\tau = 17$ (as used in [Caudell and Newman, 1993]). In these experiments, 2000 samples are taken for generating the self and different test databases (see figure 1).

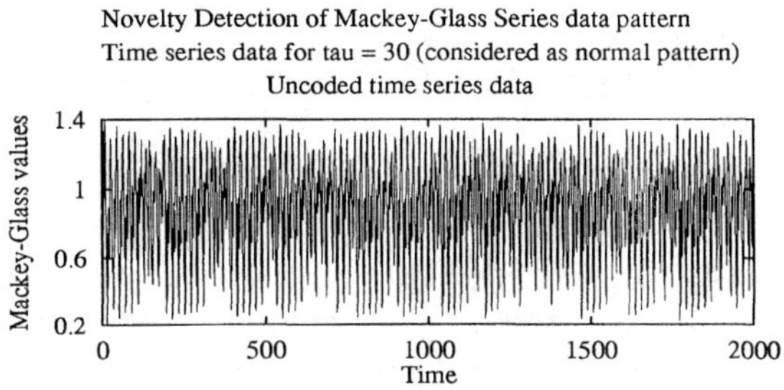


Figure 1: Mackey-Glass time series data considered as normal pattern.

In all experiments, we set $m = 5$ for binary encoding of data (i.e. the interval number it belongs to). Results of the experiments are shown in Table 1. Two different window sizes were considered with various parameter settings in each case. In first set of experiments, self strings were generated using non-overlapping sliding windows, and in the second set, overlapping windows were used. We generated diverse set of detectors in such a way that they do not match each other by $r$-contiguous bit rule.

Figure 3 shows a typical result demonstrating that the proposed algorithm can easily detect the change in data pattern by with a small set of detectors. In particular, it could detect changes during the time period 1000 and 1500. This suggests that the detection of gradual change can be monitored with a suitable detector set.

## DISCUSSIONS

In this paper, we have proposed and demonstrated a method for anomaly detection based on an negative selection algorithm inspired by the natural immune system. The preliminary results of this anomaly detection method are encouraging and the method appears to have significant advantages over other approaches.

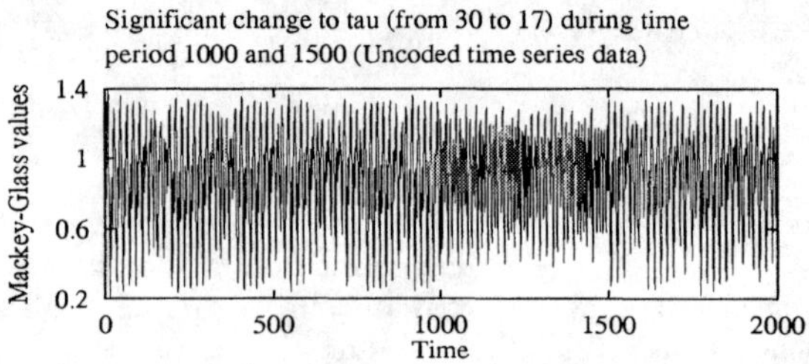


Figure 2: Shows significant changes in time series data pattern.

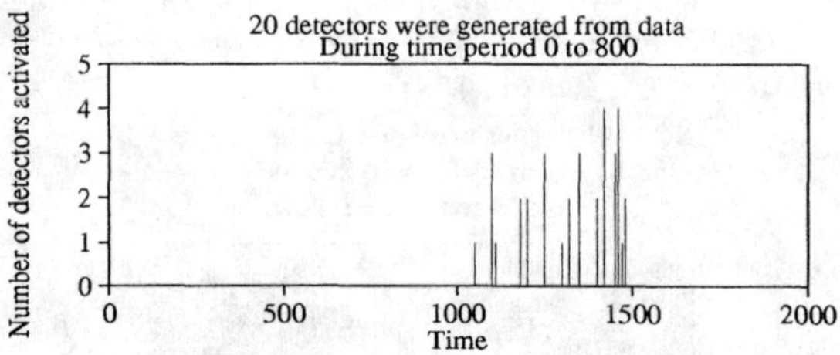


Figure 3: The vertical lines indicate the activation of detectors indicating the changes in data pattern. The detectors are only activated during the time period when the series gets distorted as in figure 2.

Since it does not look for any particular (or known) anomaly, rather it can detect any shift from the normal pattern.

Recently, neural network methodologies (in particular, ART networks) were increasingly used to the problem of fault and anomaly detection [Caudell and Newman, 1993; Hwarng and Chong, 1994]. Our results agree qualitatively with those of ART. However, the similarities and differences of our immune system based model and the ART network for anomaly detection are as follows:

- Both ART networks and our immune system based detection algorithm are biologically inspired techniques and have capability of pattern recognition, but they use different set of principles for recognition and learning.

- The ART network tries to recognize/classify an input pattern in the space defined by the training data set (actual encoded space of data or signals), but on the contrary our detection algorithm is based on negative selection, recognizing patterns in the complement space as novel.

- ART's vigilance threshold ($\rho$) provides a degree of sensitivity in discriminating input patterns, and forming clusters (or classes). A $\rho$ value near zero gives a network with low discrimination, and a value near one gives a network with high discrimination (forming many clusters). An optimal vigilance value creates a resonable number of clusters with minimum er-

| Encoding parameters | $r$ | $N_r$ | Anomaly Detection Mean(Std. dev) | Success |
|---|---|---|---|---|
| Win_size =4 | 9 | 30 | 13.24 (2.86) | 100% |
| Win_shift=4 | | 40 | 18.35 (3.91) | 100% |
| Self_length, | | 50 | 20.65 (4.36) | 100% |
| $l = 20$ | 10 | 30 | 6.10 (2.83) | 98% |
| Self_size, | | 50 | 9.62 (2.93) | 100% |
| $N_s = 500$ | | 70 | 13.66 (3.87) | 100% |
| Win_size =6 | 10 | 20 | 8.34 (3.12) | 100% |
| Win_shift=2 | | 30 | 11.22 (4.10) | 100% |
| Self_length, | | 40 | 15.98 (5.09) | 100% |
| $l = 30$ | 12 | 30 | 6.84 (2.66) | 98% |
| Self_size, | | 50 | 9.34 (4.65) | 100% |
| $N_s = 998$ | | 70 | 15.24 (5.11) | 100% |

Table 1: Reports the results for anomaly detection of Mackey-Glass time series (as in figure 2). Here $r$ is the matching threshold and $N_r$ is the number of detectors generated for monitoring novel patterns. Columns 4 and 5 give the success rate of detections in two different set of experiments (averaged over 50 runs) where column 4 shows the number of detection (average and standard deviation) of novel patterns in successful runs.

ror. A similar effect is also noticed while choosing matching threshold, $r$ $(1 \leq r \leq l)$ in our algorithm.

- In an ART network, a cluster (or class) is represented in its neural memory as a template. A template is an abstraction of the patterns in a cluster which are formed and modified during the learning (training) process. So the number of templates increases (for a fixed parameter setting) according to the diversity in input patterns. In our algorithm, the effect of such diversity in data is opposite, also there is an indication that for an independent self, $S$ (a training data set), the size of a detector set remains almost unchanged for a fixed probability of success (or failure) [Forrest et al., 1994].

- During testing (or monitoring) new data patterns in an ART network, the existence of novel patterns are determined by examining the frequency and the degree of template formation and recoding. Activation of any detector (or a threshold number) determines the change in the behavior pattern of data in our implementation.

- Recognizing an input pattern is a global decision over the ART network (while determining the class memberships of object represented by the input patterns), whereas the recognition of a novel pattern by detectors in our algorithm is a decentralized local decision, which appears to be an important advantage over the ART methodology. This property results in generating a quick response to any changes, and may be very useful in monitoring safety-critical systems.

We have considered the Mackey-Glass equation as an example of time series data; other methodologies (in particular, ART networks) in anomaly detection also used this example [Caudell and Newman, 1993]. However, nonperiodic nature of the Mackey-Glass dynamics restricts its use for continuous monitoring test. There are many potential application areas in which this method may be useful including the tool breakage detection [Dasgupta and Forrest, 1996; Dasgupta and Forrest, 1995], monitoring wafer processing, fault detection etc.

This research was initiated by the author at the University of New Mexico, Department of Computer Science, Albuquerque, NM 87131. The author would like to specially thank Prof. Stephanie Forrest for encouraging to carry out this work.

## References

Caudell, T. P. and Newman, D. S. (1993). An Adaptive Resonance Architecture to Define Normality and Detect Novelties in Time Series and Databases. In *IEEE World Congress on Neural Networks*, pages IV166–176, Portland, Oregon.

Dasgupta, D. and Forrest, S. (1995). Tool breakage detection in milling operations using a negative-selection algorithm. Technical Report Technical Report No. CS95-5, Department of Computer Science, University of New Mexico.

Dasgupta, D. and Forrest, S. (1996). Novelty detection in time series data using ideas from immunology. In *ISCA 5th International Conference on Intelligent Systems*, Reno, Nevada.

D'haeseleer, P., Forrest, S., and Helman, P. (1996). An immunological approach to change detection: algorithms, analysis, and implications. In *Proceedings of IEEE Symposium on Research in Security and Privacy*, Oakland, CA.

Forrest, S., Perelson, A. S., Allen, L., and Cherukuri, R. (1994). Self-Nonself Discrimination in a Computer. In *Proceedings of IEEE Symposium on Research in Security and Privacy*, pages 202–212, Oakland, CA.

Hwarng, H. B. and Chong, C. W. (1994). A Fast-Learning identification system for SPC: An Adaptive Resonance Theory approach. *Intelligent Engineering Systems Through Artificial Neural Networks*, 4:1097–1102.

Kozma, R., Kitamura, M., Sakuma, M., and Yokoyama, Y. (1994). Anomaly Detection by neural network models and statistical time series analysis. In *Proceedings of IEEE International Conference on Neural Networks*, Orlando, Florida.

Mackey, M. C. and Glass, L. (1977). Oscillation and Chaos in Physiological Control Sysyems. *Science*, 197:287–289.

Roberts, S. and Tarassenko, L. (1994). A Probabilistic Resource Allocating Network for Novelty Detection. *Neural Computation*, 6:270–284.

# A Hybrid Neural Network Based Pattern-Recognition Engine for Out-door Electronic Nose Application


B. Yang[◇], M. C. Carotta[◇], G. Faglia[♭], M. Ferroni[◇]
V. Guidi[◇], G. Martinelli[◇] and G. Sberveglieri[◇♭]
[◇] *INFM, Department of Physics,*
*University of Ferrara, via Paradiso, 12; I-44100 Ferrara, Italy*
[♭] *INFM, Department of Chemistry and Physics for Materials,*
*University of Brescia, via Valotti, 9; I-25133 Brescia, Italy*



August 6, 1996



**Abstract**

Construction of artificial nose for identification of toxic gases is gaining high priority of microsystem development. In this paper, a robust quantifier of chemical gas/odour, based on an adaptive resonance theory network(ART2) and a multi-layer perceptron(MLP) is proposed as pattern-recognition engine of an out-door electronic nose. The ART2 subsystem is expected to be a vigilant system, which predicts nonstationary gas/odorous environments even with unpredictable interfering odour, while a MLP subsystem is employed for quantitative identifications of gas/odorous mixtures. Unlike these single networks, the hybrid system is expected to have capability to identify in an environment in the presence of interfering patterns, which are not learned in the design stage, moreover to have better noise and fault tolerance, and therefore intends to be more reliable and robust in an out-door ambient. The applicability of the network was demonstrated experimentally in identifications of $CH_4$ and $C_2H_5OH$.


## 1 Introduction

An electronic nose must be capable of identifying low concentrations(PPB-PPM) of chemical signals in an odorous environment with detection times of seconds. The receptors in biological nose are replaced by sensors(e.g. metal oxide films)that respond to a broad range of chemical vapors or odour. As its name implies, the electronic nose is a emulation of the human olfactory system, it aims to do exactly the same as the human nose but with considerable fewer sensors and pattern-recognition(PARC) engine.

Improvement of performances of an electronic nose can be achieved from both sensor selectivity and PARC techniques. Selectivity in gas-sensing materials can, in principle, be improved by the use of an array of gas sensors with partially overlapping sensitivities and appropriate PARC techniques, which were previously concentrated on the use of conventional methods, such as partial least squares and partial model building. There are several disadvantages to the use of these classical methods. First, these techniques only well behaved, i.e., quadratic and power law. Secondly, these techniques often assume a multivariate normal distribution that may not be appropriate and lead to considerable errors.

Many problems of pattern recognition are far more easily solved by humans than by any supercomputers. Human reasoning may be somewhat fuzzy in nature such that to tackle the uncertain or ambiguous data so often encountered in real life. Hence, to enable to work in real-life situations in a manner more like humans, one may incorporate different neural networks, which are adaptive and fast, and hold both noise and fault tolerance. Recently artificial neural network(ANN) has been successfully used in replacing conventional PARC techniques in chemical gas/odour identifications[1, 2, 3, 4].

The multilayer perceptron(MLP) is the most practically used network in electronic nose approach, it predicts gas/odour concentration precisely for a well-controlled ambient. The feature of an out-door ambient is that the gas/odour species are non-stationary and with interfering gas. Therefore the MLP basically can not manage properly this situation because it require a strict forecasting of the odorous environment at the design stage. If presented with strongly distorted or interfering signals in a real operation condition, the network will output wrong results.

The aim of this work is to investigate an robust ANN system for an out-door electronic nose, i.e., selecting an ANN architecture which not only classifies sensor signals correctly in stationary condition, but also works well in non-stationary odorous environment even with interfering gas/odour. In section 2 we propose the principle of the hybrid network, while the performance of the network is demonstrated in section 3. In section 4 we give conclusions.

# 2 A hybrid neural network for an out-door electronic nose application

The hybrid neural network for an out-door electronic nose application is shown in Fig.1. It consists of a MLP and a ART2 subsystem. A MLP consists of a network of units, which are connected by the lines known as weights or links. The input variables are applied to the $L$ input units. These variables are multiplied by a matrix of parameters $w_{lm}(l = 1, .., L; m = 1, .., M)$ corresponding to the first layer of links. Here $M$ is the number of units in the middle, or hidden, layer. Similarly the signal in the hidden layer are multiplied by another matrix of parameters $w_{mn}(n = 1, .., N)$, $N$ is the number of output units. Then the output value is

$$y_n = f[\sum_{m=0}^{M} w_{mn} f[\sum_{l=0}^{L} w_{lm} x_l]] \tag{1}$$

where $f$ is generally chosen as a sigmoid function. According to the error back-propagation process these weights $w_{lm}$ and $w_{mn}$ are iteratively changed to minimize the difference between the ideal output $t^q$ and actual output $\overline{y}(\overline{x}, \overline{w})$.

As we discussed before[5], the MLP is very precise in a well-controlled gas/odour ambient, however it can not manage well non-stationary odorous environment due to its inherent characters. It actually fails to solve the stability-plasticity dilemma, i. e., learning a new pattern erases or modifies previous training. In order to solve the problem, ART2 is proposed to by Carpenter and Grossberg[6].

ART2 networks self-organize stable recognition categories in response to arbitrary sequences of analog(gray-scale, continuous-valued)input patterns, as well as binary input patterns. The ART2 network self-organizes in response to newly input patterns. The basic architecture of the ART2 network is shown in Fig.2. The ART2 network consists of an attentional subsystem and an orienting subsystem. Through a hypothesis testing cycle, the input patterns are recognized by the attentional subsystem, which consists of a feature detect (F1), a category representation (F2), and a set of bottom-up and top-down LTM's. The connections between the neurons in the F1 layer and the corresponding neurons in the layer F2 layer forms the LTM. In the hypothesis testing cycle, the F1 layer receives an input pattern and extracts its features using a nonlinear signal function. When the activity of a neuron in the F1 layer is large enough, it generates an excitatory signal, which are multiplied by the bottom-up LTM and sent to the neurons in the F2 layer. The neuron, in the competitive F2 layer, is activated when it achieves largest input, whereas the other neurons are inhibited. The activated neuron associates a pattern through the top-down LTM at the F1 layer. If the associated pattern does not match the input pattern, the orienting subsystem sends a reset signal to the F2 layer so that the activated neurons may be strongly inhibited and may not activated again until the hypothesis testing cycle ends. After that, the input pattern is reinstated and generates an output to the F2 layer again. When the associated pattern approximately matches the input pattern, the hypothesis testing cycle ends.

The principle of the network application is as following: the long-term memories(LMT) of ART2 serve a suitable interface between well-controlled network design and a real situation. The typical gas/odour patterns at the design stage are memorized in the LMTs. In the out-door situation, if the gas/odour environment are similar to these at the design stage, a MPL is set to the forward mode for quantification tasks. On the other hand, if the gas/odour environment are significantly different from those at the design stage, the MPL is inhibited in order to avoid giving misleading identification results.

# 3 The neural system performance

The performance of the neural system is demonstrated on quantifications of methane and ethyl alcohol in the individual component cases at various humidity together with an influencing pattern of $CO$. Experimentation is as in the following.

An array consisting of five tin dioxide thin film sensors is used to measure methane and ethyl alcohol. These thin-film sensors were deposited by sputtering through the rheotaxial growth and thermal oxidation(R.G.T.O.) technique over alumina substrates[7]. The sensing films were obtained through deposition of a 360 nm thick Sn layer followed by 30 hours thermal annealing in oxidizing environment. In order to improve the selectivity toward specific gases, catalyst depositions of either Pt, or Pd, or In, or Ti-W layers (7.5 nm in thickness) have also been done. Prior to measurements, the temperature of the sensor array had been kept constantly at $500^0$ C for two weeks by a Pt heater in order to achieve reliable measurements. 1000-5000 ppm of $CH_4$ and 100-500 ppm of $C_2H_5OH$ were introduced in the test chamber at various humidity from 20% to 50%. The conductance changes have been monitored by a volt-amperometric technique[7].

In order to simulate a strong noise(or interfering pattern) situation, 500 ppm $CO$ is introduced in the test chamber. These measurements, including the $CO$ signal, are presented to the ART2 network for classification tests.

The hybrid neural system consists of a ART2 network and a MLP network. The ART2 network has 5 inputs take the signals from the five sensors; 4 neurons are chosen for the F2. After several comparison studies, the vigilance parameter is set to be 0.96. The MLP network is with five input neurons, five hidden neurons and two output neurons. The SNNS simulator is used for the training procedure[8].

The ART2 subsystem is supplied with two typical input patterns, one is from $CH_4$ and another from $C_2H_5OH$ measurements. These two patterns are memorized in the LTMs through the training procedure. Then the total experimental results, which are measurements at various humility, are presented to the network for identification; in addition the $CO$ signal is also included. The ART2 starts real-time learning, and assigns correctly $CH_4$ and $C_2H_5OH$ in corresponding categories at different humidity. When receiving the $CO$ signal, the ART2 does not match it with the stored patterns, then a new category is automatically created for the pattern without modifying the old memories, and at same time the ART2 inhibits the quantification of the MLP. Figure 3(a) presents a typical classification result from the ART2.

When an known pattern arrives, such as $CH_4$ or $C_2H_5OH$ in this study, the ART2 enables the MLP to processing this pattern. On the other hand, if a interfering pattern is presented, such as $CO$ here, the ART2 inhibits the MLP in order to avoid misleading results. Figure 3(b) shows the identification results from the MLP network. As we can see, the vigilant ART2 system successfully inhibited the MLP system from producing misleading result, which the MLP did not learn in the design stage.

# 4 Conclusions

In this paper we proposed a hybrid neural network for application of out-door electronic nose, i.e., suitable for a non-stationary gas/odour ambient. The ART2 subsystem is expected to identify new patterns, while the MLP subsystem produces quantification results of gas/odour ambient. The applicability was successfully demonstrated in a experiment for identification of $CH_4$ and $C_2H_5OH$ with interfering gas $CO$.

Figure 1: The neural-network system for a nonstationary odour applications

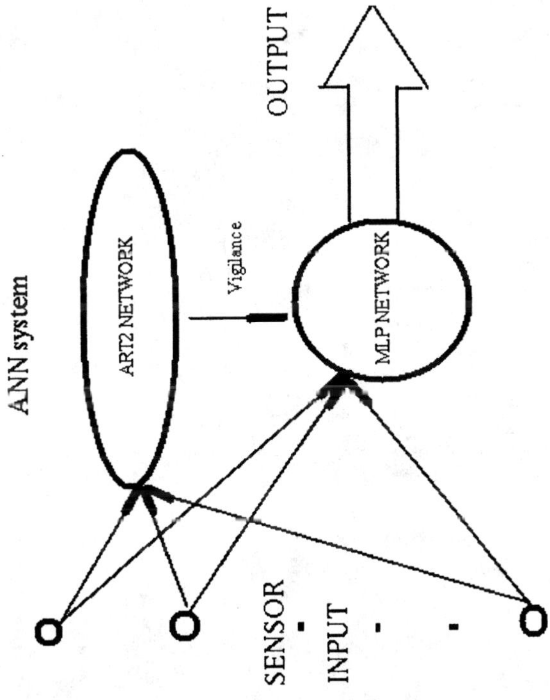

Figure 2: The ART2 network architecture[6]. The signal processing is performed by a pair of connected neurons between the $F_1$ and $F_2$ layers. Nodes $w_i, x_i, u_i, v_i, p_i$ and $q_i$ characterize the short term memory(STM) activities in the $F_1$ layer. Nodes *rec* characterize the STM activities in the $F_2$ layer. Details are described in [6].

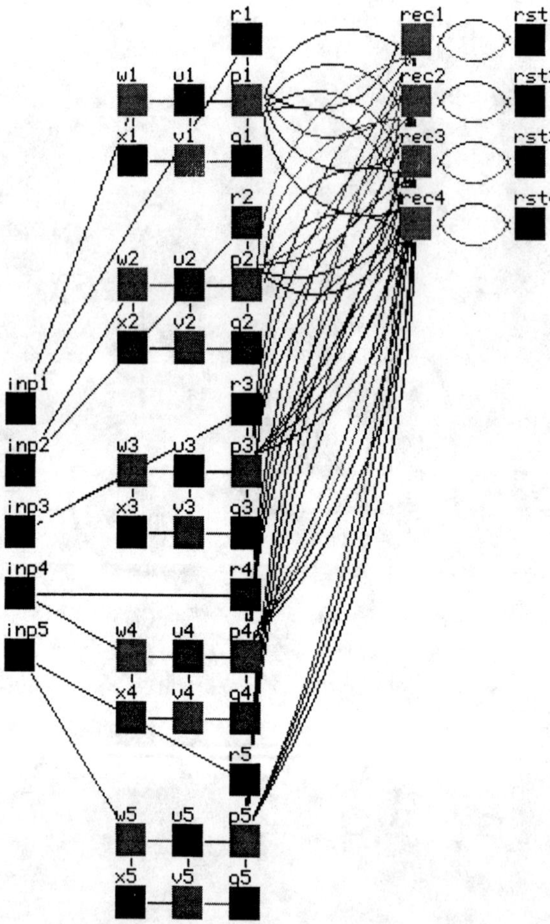

Figure 3: The MLP neural network performance on the $CH_4S$ and $C_2H_5OH$ quantifications with interfering pattern $CO$. a: the classification results from the ART2 subsystem; b: the quantification results from the MLP network, the gas concentration range(0 to 5000ppm) is normalized to [0,0.5].

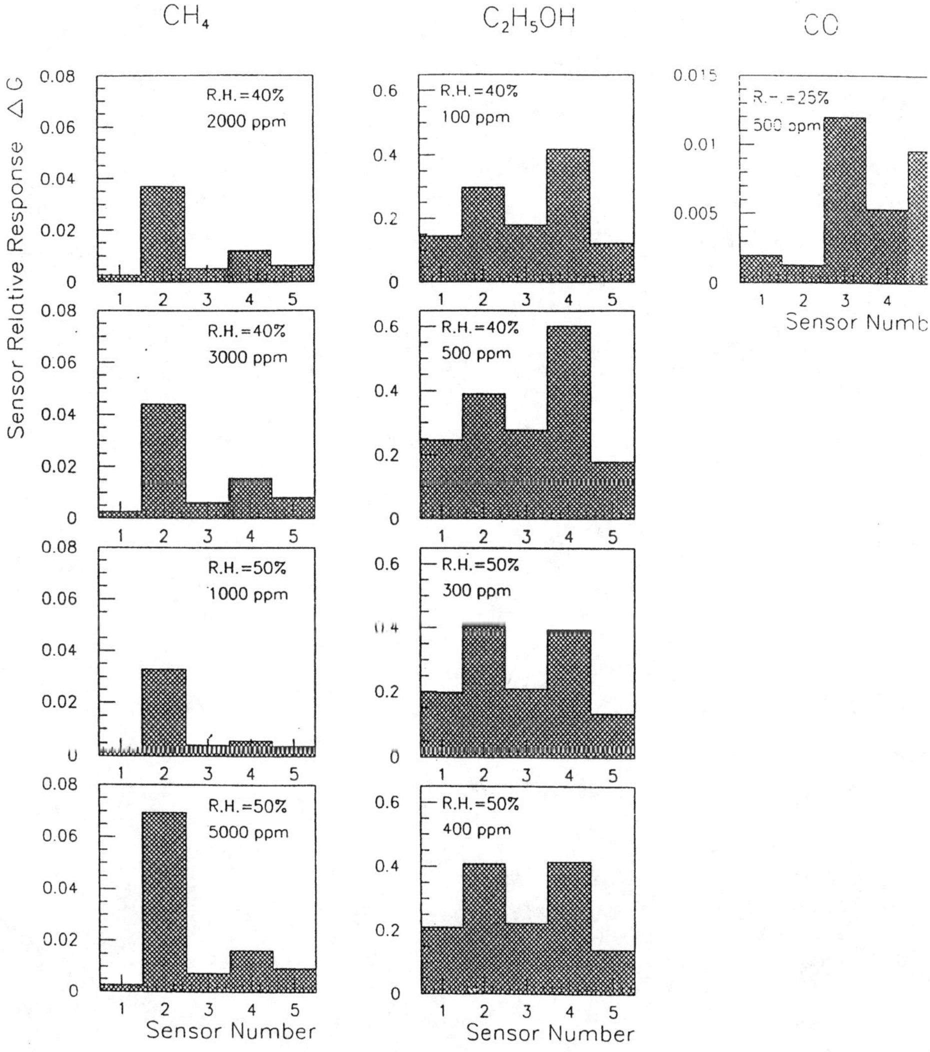

Figure 3: The MLP neural network performance on the $CH_4S$ and $C_2H_5OH$ quantifications with interfering pattern $CO$. a: the classification results from the ART2 subsystem; b: the quantification results from the MLP network, the gas concentration range(0 to 5000ppm) is normalized to [0,0.5].

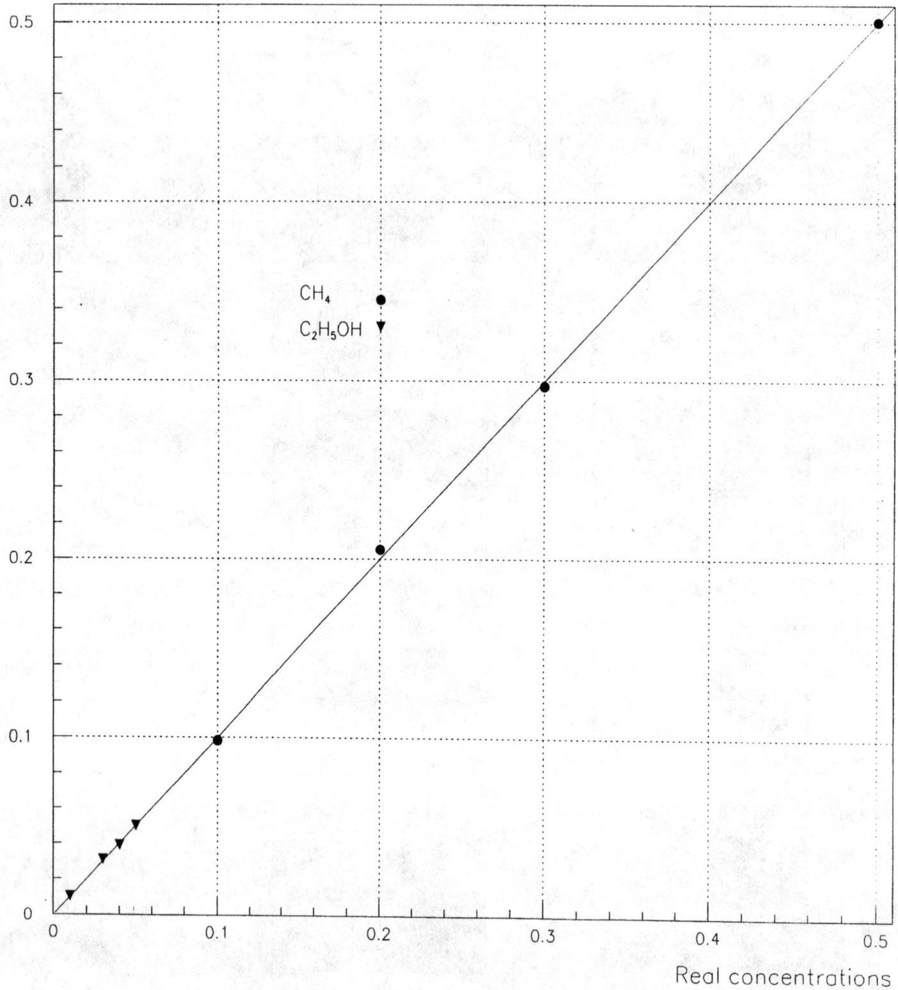

# References


[1] J. W. Gardner ,E. L. Hines and M. Wilkinson *Meas. Sci. Technol.*, **1**(1990)446

[2] H. Sundgren , F. Winquist , I. Lukkar and I. Lundstrom *Meas. Sci. Technol.*, **2**(1991)464

[3] S. W. Moore , J. W. Gardner , E. L. Hines , W. Gopel and U. Weimar, *Sensors and Actuators B*, **15-16**(1993)344

[4] T. Nakamoto , A. Fukuda , and T Moriizumi., *Sensors and Actuators B*, **10**(1993)85

[5] B. Yang, M. C. Carotta, G. Faglia, M. Ferroni, V. Guidi, G. Martinelli, and G. Sberveglieri, The adaptive resonance theory network approach to chemical gas classifications, *Submitted to Sensors and Actuators B*

[6] G. A. Carpenter and S. Grossberg, ART2: Self-organization of stable category recognition codes for analog input patterns, *Applied Optics*, 23(1987)4919-4930.

[7] A. D'Amico and G. Sberveglieri 1995 *Sensors for Domestic Applications* (Singapore: World Scientific)

[8] A. Zell et al., *SNNS Stuttgart Neural Network Simulator, Version 4.1*, Stuttgart 1996.

# Qualification of shape characteristics by using Neural Nets


Sanjay Goel
General Electric Corporate Research and Development
Schenectady NY-12309
goels@crd.ge.com

Mark Embrechts
Rensselaer Polytechnic Institute
Troy, New York 12205


## Abstract


A new methodology which uses a neural network to evaluate the quality of an airfoil by interpreting the distribution of flow parameters on a quasi-3D surface has been developed. The training data was generated by Computational Fluid Dynamics (CFD) analysis of randomly perturbed airfoil. Heuristic rules based on Fluid boundary layer analysis were used for evaluating the flow parameters on the airfoil. Two different heuristic measures of quality are used 1) Diffusion which is a measure of losses due to local pressure changes in the flow distribution, and 2) Deviation which is a measure of losses due to non-uniformity of acceleration of the flow. These two parameters were used as the outputs of the neural network.

One dimensional Fourier analysis was used to parametrically represent the shape characteristics of the velocity distribution. The Fourier coefficients generated from the transformation were used as the inputs to the Neural Network. Various configurations of one and two hidden layer feed-forward networks were investigated. Backpropagation algorithm was used for the training which was considered complete when a training error of less than 2 percent and a test error of 10 percent was obtained. The trained neural network was used for evaluation of the airfoils during the design optimization. Data from a high pressure steam turbine airfoil and a low pressure gas turbine airfoil were used for network training. The results of the network training and airfoil optimization using the trained networks are described in the paper.


## Introduction

A large number of problems which do not have a closed form solution are solved iteratively by using analysis codes to predict the performance. In this process the problem parameters are changed successively in each iteration and the performance of the resultant solution is evaluated until a satisfactory solution is obtained [1,2]. Solution procedure for these problems can be automated by coupling optimization techniques with analysis codes, which analyze these problems and can predict the performance of a candidate solution. Optimization techniques are driven by a numerical value of the performance parameter predicted by the analysis code. The numerical value of the performance parameter is computed by post processing of the analysis results. In case the solution is in the form of an arbitrary geometric shape, the visual perception of the shape quality must be translated to a numerical score using analytic methods. This transformation is non-trivial and often relies on heuristics which are imprecise and subjective. This paper discusses a generic methodology based on neural networks for quality characterization of shape based solutions in optimization problems.

Turbine airfoil design problem exhibits both of the characteristics described above, it does not have a closed form solution and it's solution is in the form of a velocity distribution curve. Solution procedure for airfoil design can been automated by coupling optimization techniques with CFD codes which analyze the performance of candidate solutions[3,4]· Burgreen and Baysal[6] have used direct design techniques for wing design and stationary nozzle design in which the ratio of lift to drag coefficients is used as the evaluation criteria. Huddleston and Mastin[5] have coupled direct optimization techniques with a CFD code for design of a supersonic planer nozzle and a planar supersonic forebody simulator. They use in inverse approach in which they define an ideal solution and use the error norm between the ideal and the actual solution as a measure of quality. In turbomachinery design a direct approach in which the solution quality is defined explicitly is preferred since the aerodynamically ideal solution may not be topologically feasible. Goel et. al. [3,4,5] have developed a direct optimization technique in which the airfoil quality is determined heuristically from the CFD solution. Even though heuristic criteria are based on physical theories, they are imprecise, and often require significant amount of trial-and-error in order to obtain a robust criteria for evaluation. The approach described in this paper provides an algorithm which is precise and eliminates subjectivity.

## Airfoil Design

The turbine airfoil design process is a shape optimization problem which involves slicing the blade shown in Figure 1 on quasi-3D surfaces, designing each section independently and stacking the sections together to obtain a smooth radial geometry. There are two distinct steps in the optimization i.e. section optimization and radial stack optimization. Both the section and stack optimization problem are shape evaluation problems; the focus of this investigation is on the design of airfoil sections. A closed form solution is not obtained for airfoil section design, so, numerous airfoil geometries need to be evaluated to obtain the desired flow characteristics. During a typical airfoil design, the user starts with an estimate of the airfoil shape, which may not meet all the design criteria, and iteratively changes the blade shape until all the design requirements are satisfied.

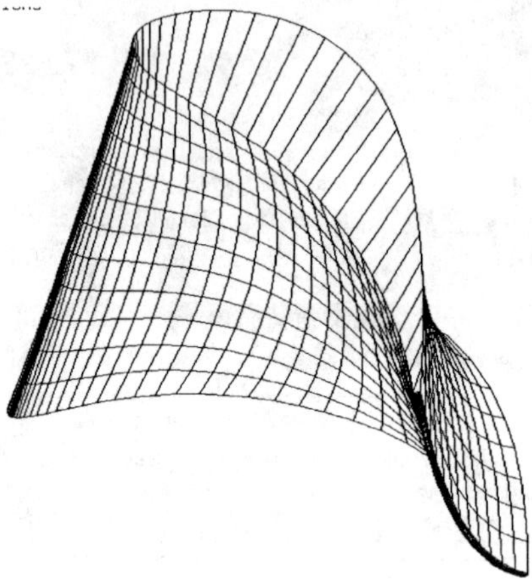

**Figure 1** Wire frame model of an airfoil in 3d.

The airfoil optimization process, shown in Figure 2 is an iterative process with four discrete steps i.e. Updating the airfoil geometry (or starting from an initial design), analyzing the airfoil flow field using CFD analysis, Computing the quality of the airfoil by interpreting the flow field and making changes to the parametric model of the airfoil.

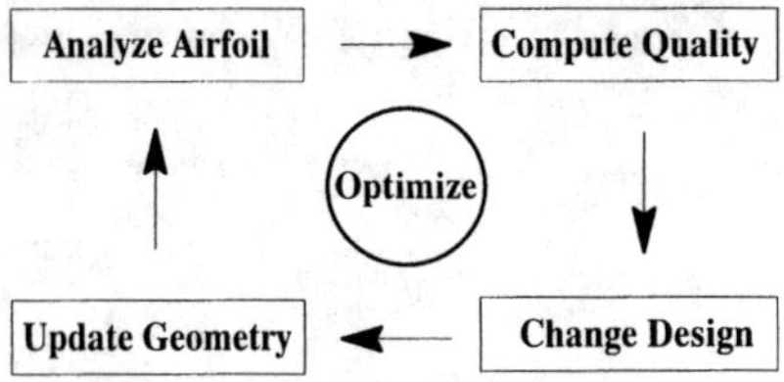


**Figure 2** Quasi-3d section optimization.

Mathematically the problem can be stated as:

Minimize: $F(x_i),\ i = 1 \ldots n$

subject to: $G(x_i) \leq 0\ i = 1 \ldots m$

$x_{imin} \leq x_i \leq x_{imax}\ i = 1 \ldots n$

To formulate airfoil optimization as a standard optimization problem three primary issues need to be addressed 1) parametric representation of the design variables 2) evaluation of the CFD solution 3) definition of the constraint boundaries. The subsequent sections address these issues in greater depth. Evaluating the CFD solution is the primary focus in this paper.

**Parameteric Representation:**

A critical element in the success of any shape optimization scheme is the ability to generate a great variety of physically realistic shapes. In the proposed design scheme, the blade geometry is represented by high order Bezier polynomials to obtain a smooth continuous geometry. The Bezier geometry is represented by control points which are perturbed to modify the physical geometry of the airfoil. These control points are used as the design variables during the optimization process. In addition to the Bezier control points geometric parameters like stagger angle, wedge angle, nose radius etc. are defined as shown in Figure 3. These parameters are essential for defining heuristics for the design optimization. since designer heuristics are defined in terms of these parameters.

**Evaluation of the CFD solution**

The CFD solution for an airfoil is in the form of Mach number curve (Figure 4) obtained by a CFD blade-to-blade analysis. In traditional design procedure the designer uses his expertise to visually examine the plot, correlates the plot to the airfoil quality and makes suitable design changes. The process of visually examining the solution and correlating it to quality needs to be automated for design optimization. Translating the designer heuristics based on visual perception, to a completely analytic method is not easy since the subjectivity of the designer is difficult to capture. Some guidelines are available in literature for defining an acceptable Mach number distribution; however these guidelines represent a small fraction of the designer's evaluation criteria[3,4].

There have been numerous attempts to analytically estimate the characteristics of the flow solution in both 2D and 3D. Goel et al. [3,4,5] have estimated the quality of Mach number distribution by defining two parameters i.e., Diffusion and, Deviation.

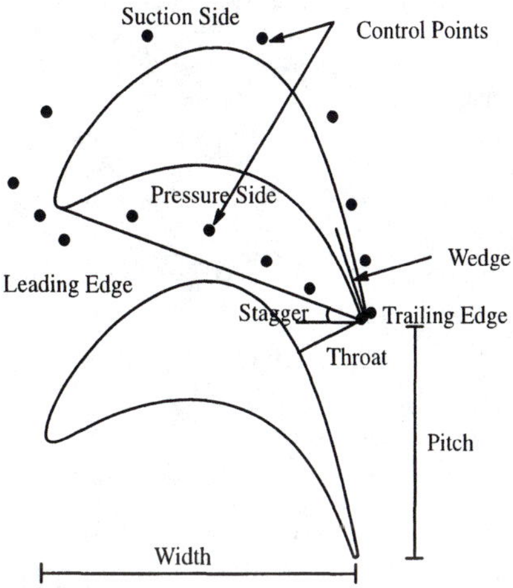


**Figure 3** Parametric representation of airfoil section

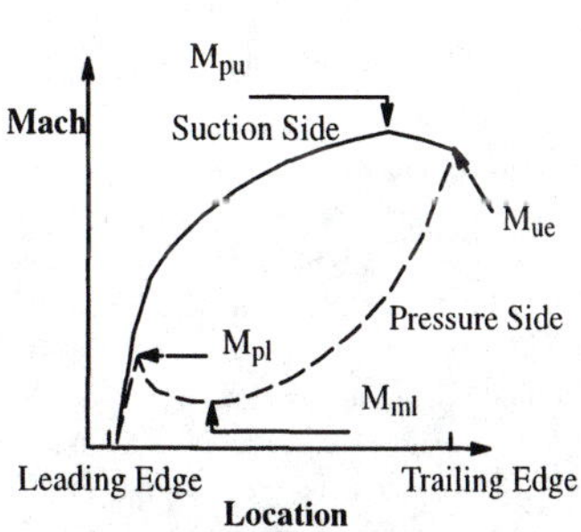


**Figure 4** Sample Mach number distribution.

Diffusion of flow leads to a loss of velocity head and may lead to separation of the boundary layer resulting in increased viscous losses and increased pressure drag. The diffusion parameter measures the cumulative velocity drops in the flow field and thus measuring the diffusion effect. A flow may be non-diffusing, but may have an uneven acceleration along the airfoil surface. This can again cause increased drag on the airfoil. The deviation parameter measures this uniformity by fitting polynomials to the Mach number distribution and computing an error norm of the deviation of the actual Mach number distribution from the fitted distribution. Such methods even though effective are highly problem dependent and require considerable amount of work for each problem solved.

The neural network based method is generic in nature and can be used to automatically recognize an arbitrary shape and associate characteristics based on the domain by providing suitable training data. This method is used for characterization of Mach number curve and compared to the analytic method used in references [3,4].

## Shape Recognition

To determine the characteristics associated with a specific shape, the prediction model must be able to distinguish one shape from another. Numerous statistical shape classification techniques have been used for character recognition and are described in the literature. More recently neural networks have been used for such problems and have been shown to work better than conventional statistical techniques[8].

A shape recognition problem can be formally represented as a nonlinear mapping problem between the input shape parameters and the shape characteristic parameters. Assume a function $f$, which maps from an n-dimensional input space $R^n$ into an m–dimensional output space $R^m$. Let vectors $X$ and $Y$ represent the points in $R^n$ and $R^m$, respectively. Given a set of input/output pairs $\{X_i, Y_i\}$, where $i = 1...k$, $X_i \in R^n$, $Y_i \in R^m$, and $Y_i = f(X_i)$, then one can predict $Y_j$ for any $X_j$, where $X_j \not\subset X$ and $Y_j \not\subset Y$ [8]. Using this formal representation a neural network can be trained to learn the map in order to predict characteristics based on the shape.

One characteristic that is often desired in pattern recognition problems is invariance of the network output to changes in the position and size of the input pattern or image. One method involves including in the training set several examples of such exemplar transformed in size, angle, and position, but with a desired response that depends only on the original exemplar. For shapes represented by non-parametric data, it may also be necessary to preprocess the data to extract features which are representative of the geometry. One way of characterizing a shape is to use Fourier transform which allows a shape to be described as a weighted sum of functions of different frequencies. The weights of the terms in the Fourier transform are used as shape characteristics. Shape can also be characterized by defining moment invariant features which are a set of non-linear functions invariant to translation, scale and, orientation and are defined on geometrical moments of the image. Grace and Span [9] have compared the two methods for character recognition and have found the Fourier transforms to perform better than the moment invariant features for noisy images.

In the airfoil optimization process the Mach number distribution computed by using the CFD analysis, is in the form of [x,y] data pairs along the airfoil surface. Unlike the problem of character recognition, the relative magnitudes of the Mach number at different location are critical for airfoil design, so, invariance to scaling is undesirable. Invariance to rotation is also undesirable since the slope of the distribution is important. Invariance to translation is however required and is obtained by translating the data to the origin. Subsequent sections explain the data generation and transformation in greater detail.

**Training Data**

Input data for training the network is generated by a blade-to-blade CFD analysis of sample airfoils generated by random perturbations of an airfoil shape. The Mach number distribution data is transformed using Fourier analysis. The Fourier series was truncated at order 30 giving 61 coefficients including the constant coefficient. The constant coefficient was ignored and 60 inputs were used in the training. Deviation and Diffusion are computed for each sample using the heuristic based analytical method. The Fourier coefficients were normalized over the entire data set between 0.1 and 0.9. The output parameters of diffusion and deviation were also normalized between 0.1 and 0.9. Three-quarters of the data was used for training and one-quarter was used for testing the network

The training for the pressure and suction sides is treated independently without loss of generalization or accuracy. The current work is focused primarily on the characterization of the suction side Mach number distribution which is the critical part of the airfoil design. Two different turbines were selected for the investigation 1) High pressure steam turbine for a nuclear power unit and 2) Low pressure gas turbine for power generation. 125 samples were obtained for the stream turbine; 100 samples were used for training, and 25 samples were used for testing. 250 were generated for the gas turbine; 200 samples were used for training, and 50 samples were used for testing.

**Architecture**

Figure 5 shows a schematic diagram of a feed-forward neural network with N layers of neurons each containing $n_i$ nodes, where i is the layer index. The first layer consists of fanout input units, and the neurons of the last layer represent the outputs. The neurons are interconnected as shown in the figure, and the strength of these interconnections are referred to as network weights; $w^K_{ij}$ denotes the strength of connection between the i-th and the j-th neuron of the K-th layer.

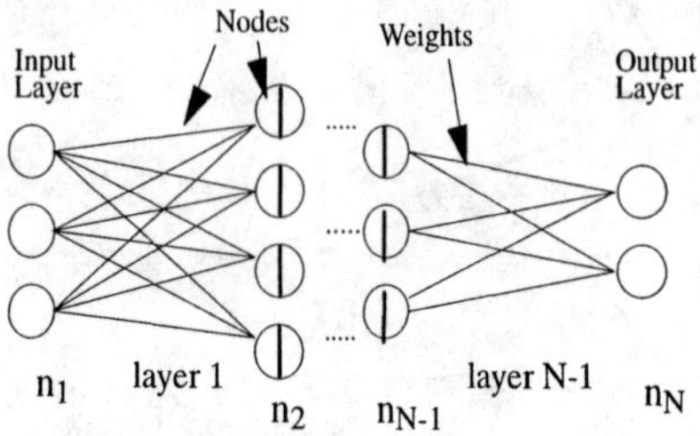


**Figure 5** Schematic diagram of a neural network.

The outputs of nodes in one layer are transmitted to the nodes in the next layer through links that amplify or attenuate these outputs through weights factors. The net input to each neuron is the sum of the weighted outputs of the nodes of the prior layer. Each node is activated in accordance with the input to the node, the activation function of the node, and the bias to the node. The activation function is a normal sigmoidal function i.e. $[1+e^{-x}]^{-1}$.

At the start of network training, the weights are randomly initialized. The network is then exposed to a set of training data which includes a number of input vectors and the corresponding outputs. As the training proceeds the weights are incrementally adjusted with the Backpropagation algorithm until the error between the network predicted output and and the actual known output for all patterns drops below a threshold. In each adjustment, weights leading to the output nodes are adjusted in proportion to the difference between the output node's actual output and its desired or target output; the weights leading into hidden nodes are in turn adjusted in proportion to their contributions to the higher nodes until the training is complete. At the end of the training phase, the weights are frozen; then, inputs for which the outputs are not known can be fed into the stabilized network for characterization[10].

**Training**

A proper network architecture for back propagation training with 60 inputs and two outputs was determined by trail-and-error. Initially the complete set of data was presented to train a single hidden layer layer network with number of nodes in the hidden layer varying from 15 to 300 without much success. Network convergence was obtained by presenting the network with a small set of data and gradually increasing the size of the training set in increments of 5 to 15 data points as the training progressed.

## Results

Two different sets of data, one from a high pressure steam turbine and, the other from a low pressure gas turbine were used for training and testing the Neural Networks. Detailed optimization results from the steam turbine test are described in this section.

**High pressure steam turbine**

One hundred data samples were generated for training a single hidden layer network. The neural network training information is summarized in Table 1. The 3 layer 60-30-2 configuration had the best generalization and the 4 layer 60-25-25-2 configuration had the least training error. The training objective is to predict the evaluation parameters of diffusion and deviation so, generalization is more critical and thus the network configuration of 60-30-2 was selected.

The search path comparing the analytical method and the neural network are shown in Figure 6. The analytic method and the trained network follow the same search path for 90 iterations and then the search paths diverge. Figure 7 shows the initial and the final design obtained by using the Neural Network approach and the Figure 8 shows the initial and final design obtained using the analytic approach. The designs, though different, are equally good. The neural network based approach did well when coarse changes were being made to the design as seem from the large steps in Figure 6, however when the design was being fine tuned the neural network approach was not very sensitive. So the neural network based approach worked very well for the steepest part of the optimization path but was not sensitive enough to make small changes for fine tuning their performance. two approaches is quite similar.

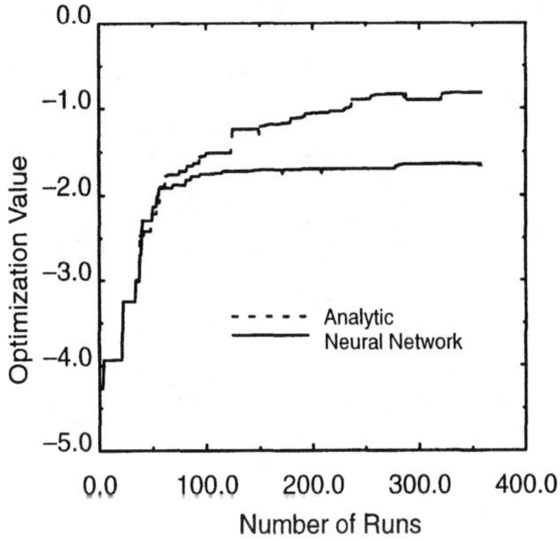


**Figure 6** Comparison of Search Paths.

Figures 7-11 show the optimization results obtained using the conventional heuristic based design evaluation and the neural network based evaluation technique. Figures 7-8, 8-9, and 9-10 show the optimization results for the tip, pitch and the hub sections of the airfoil respectively. The optimization results are quite similar using both techniques with the conventional technique outperforming the Neural Network based technique.

| Case # | Sample Size | | Layers | Architecture | Step Size | momentum | Training Error | Average Prediction Error Per Pattern |
|---|---|---|---|---|---|---|---|---|
| | Train | Test | | | | | | |
| 1 | 100 | 23 | 4 | 60-15-15-2 | 0.5→0.10 | .6→.4 | .0072 | .11 |
| 2 | 100 | 23 | 3 | 60-30-2 | 0.5→0.05 | .6 | .0112 | .08 |
| 3 | 100 | 23 | 4 | 60–25–25–2 | 0.5→0.15 | .6→0.4 | .0027 | .1422 |

**Table 1:** Network training for different architectures for Nuclear Turbine.

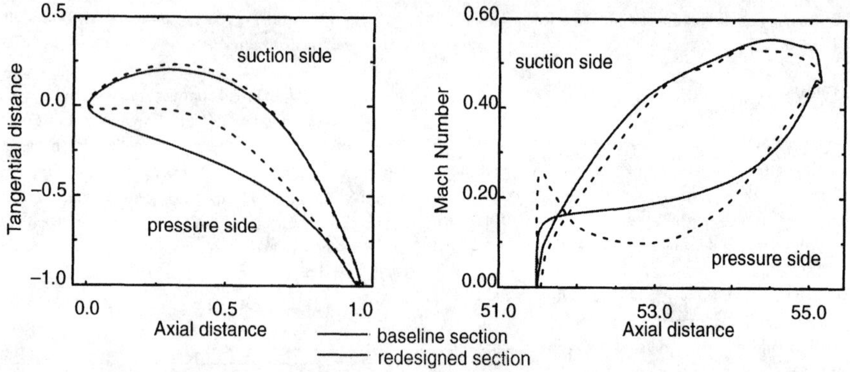


Figure 7 Airfoil geometry and Mach number distribution change at tip section (Neural Net)

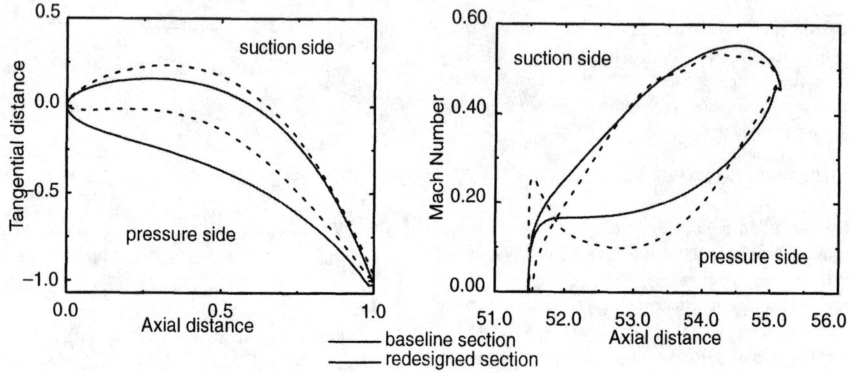


Figure 8. Airfoil geometry and Mach number distribution at the tip section (Analytic)

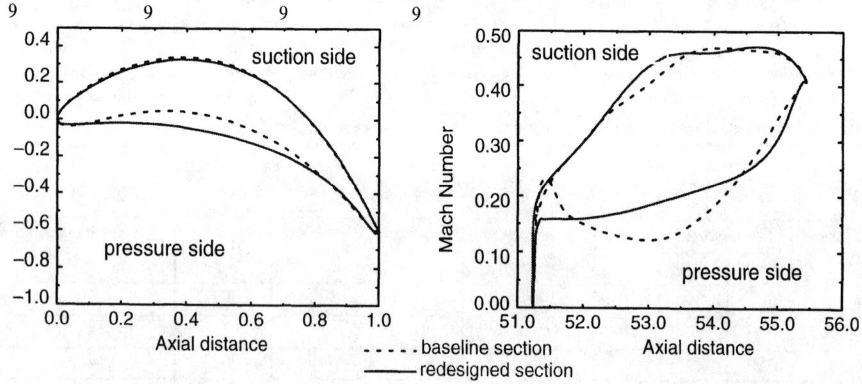


Figure 9. Airfoil geometry and Mach number distribution at the pitch section (Neural Net)

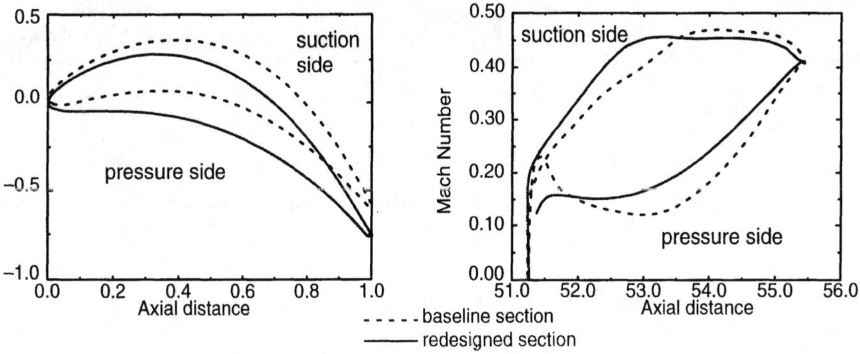


Figure 10. Airfoil geometry and Mach number distribution at the pitch section (Analytic)

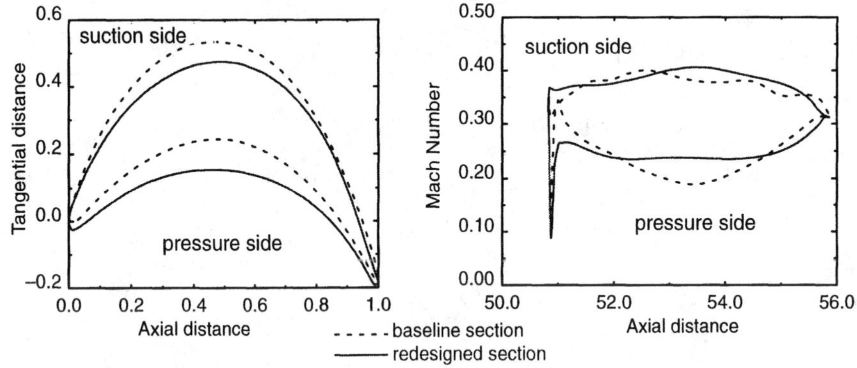


Figure 11. Airfoil geometry and Mach number distribution at the pitch section (Neural Net)

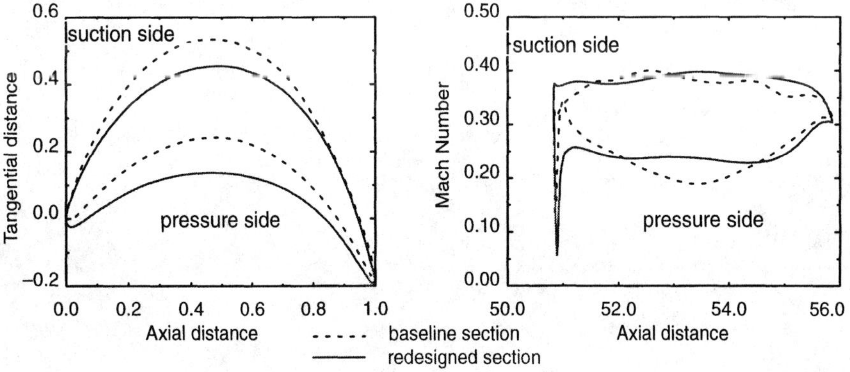


Figure 12. Airfoil geometry and Mach number distribution at the pitch section (Analytic)

| Case # | Sample Size | | Layers | Architecture | Step Size | momentum | Training Error | Average Prediction Error Per Pattern |
|---|---|---|---|---|---|---|---|---|
| | Train | Test | | | | | | |
| 1 | 200 | 50 | 3 | 60-20-2 | 0.6→0.20 | .5 | .0043 | .034 |
| 2 | 200 | 50 | 3 | 60-40-2 | 0.6→0.2 | .5 | .007 | .031 |

Table 2: Network training for different architectures for Gas Turbine

**Low pressure gas turbine**

Table 2 shows the results for a low pressure gas turbine. Two hundred data sets were used for training and 50 data sets were used for testing the network. Two different neural network configurations were used for training. Both of these preformed very well for both training and testing. This test was performed to verify the generality of the approach for use in other turbines. The airfoil designed with this approach showed similar results.

## Conclusions

A generic methodology based on neural networks for quality characterization of shape based solutions in optimization problems has been developed. A neural network was trained to predict the aerodynamic performance of turbine airfoils. The raw CFD data was processed using Fourier analysis prior to presenting it to the feed forward Neural Network. The Neural Network was trained using a Backpropagation algorithm and was used for airfoil optimization. The results were very close to those obtained using the analytic technique based on design heuristic.

The advantage of Neural Network based approach is it's domain independence which allows generic use of the approach for shape based evaluation problems. The analytic method in contrast is very a very subjective trial-and -error approach.

## References


1 Goel S., Gregory B. and Cherry D., "Knowledge-Based System for the Preliminary Aerodynamic Design of Aircraft Engine Turbines," *Proceedings of the SPIE – The International Society for Optical Engineering*, Vol. 1963, pp. 160-170, 1993.

2 Tong S.S., Powell D., and Goel S., "Integration of artificial intelligence and numerical optimization techniques for the design of complex aerospace systems", *AIAA 1992 Aerospace Design Conference*, February 3–6, 1992, Irvine, California.

4 Goel S., Lamson S. H., "Knowledge-Based System for Detailed Blade Design of Turbines", *ASME 1995 International Mechanical Engineering Congress and Exposition*, Nov 12-17 1995, San Francisco, California

5. Huddleston D.H., and Mastin C.W., "Optimization Methods Applied to Aerodynamic Design Problems in Computational Fluid Dynamics", *7th International conference of Finite Element Methods in Fluid Flow*, 1987, pp. 899-907.

6 G.W. Burgreen, and O. Baysal, "Three–Dimensional Aerodynamic Shape Optimization of Wings Using Sensitivity Analysis", *AIAA 32nd Aerospace Sciences Meeting & Exhibit*, Reno, Nevada, January 10–13, 1994.

7 Farrin G., "Curves and Surfaces for Computer Aided Geometric Design - A Practical Guide", Academic Press Inc.

8 Lee H. and Hajela P., "Prediction of Turbine Performance using MFN by Reducing Mapping Nonlinearity", *Artificial Intelligence and Object-Orinted Approaches for Structural Engineering*, Edited by G.H.V. Topping and M. Papadrakakis, pp. 99-105, Civil-Comp Press, Proceedings of the 2nd International Conference on Computational Structures Technology, Athens, greece, Aug. 30 - Sept. 1, 1994.

9 A.E. Grace and M. Spann, Comparison between Fourier-Mellin descriptors and moment based features for invariant object recognition using using Neural Networks, Pattern Recognition Letters 12, October 1991, pp. 635–643.

10 M. Chester, "Neural Networks - A tutorial", Prentice Hall, Englewood Cliffs, New Jersey.

# IDENTIFICATION OF THE WELL TEST INTERPRETATION MODEL USING THE EXTENDED HAVNET NEURAL NETWORK

**EDWARD A. MAY AND CIHAN H. DAGLI**
***Smart Engineering Systems Laboratory, University of Missouri-Rolla***
***1870 Miner Circle, Rolla, MO 65409-0370***

***ABSTRACT:***
The well test is a classic example of the inverse problem. Based solely on the pressure response to a change in flow rate, the model (assumed physical system) which most closely matches the response must be selected. These response models can be visualized as a family of parametrically related curves. Some of the difficulties that must be overcome in a neural network solution are invariance in both pressure and time, non-uniqueness, and both random and systematic noise.

This paper demonstrates a neural network approach to selection of the model based on test response. Invariance in the pressure term is removed using by converting the response to a syntactic metric, based on the magnitude of the slope. The Hausdorff-Voronoi (Havnet) neural network is then employed for pattern matching. Time invariance is overcome by using an extended version of the Havnet network which allows multiple aspects of a given class to be learned.

Several examples of reservoir models are tested with various levels of noise are used to evaluate the performance of the network and different degrees of overlap in the definition of the syntactic metrics are investigated. The results show good identification ability when noise is low, however, performance drops as the amount of noise in the data increases. The optimal degree of syntactic overlap is surprising, showing the best response occurs with smaller amounts of overlap.

**Keywords: Pattern Recognition**, Artificial Neural Networks, Classification, Data Analysis, Noise Sensitivity, Syntactic Pattern Recognition.

## INTRODUCTION

The well test is one of the primary diagnostic tools used in the evaluation of the productive capacity of oil and gas wells. It is a very important tool as it one of the few ways that engineers can actually "see" into a reservoir located many hundreds or thousands feet below the earth's surface. The results of the test can directly affect the profitability of the well and aid in decision making about the future of the well and the need for additional wells in a given reservoir.

Our understanding of the reservoir is limited to analytical models which have been derived to explain the response of wells over time. These models are based on equations of fluid flow and heat transfer, and include not only the reservoir, but also inner and outer boundary conditions. While the reservoir model itself can tell us the maximum potential of the well, the boundary conditions are perhaps even more important. From these boundary conditions the engineer can determine the size of the reservoir and if the formation was damaged by the drilling process.

The well test is a classic example of the inverse problem. An input signal is applied to the reservoir, in the form of a constant flow rate, and the pressure response over time is measured. From this response the model must be inferred (Horne, 1995). If the model selected is incorrect, predictions of future response are also likely to be incorrect. Selection of the model is further complicated by a high degree of similarity of test responses between dissimilar models and the presence of noise in the test response.

Neural networks of various types, including feed-forward networks using backpropagation (Al-Kaabi, 1990), a hybrid approach combining symbolic and ANN (backpropagation) methods (Allain, 1992), and a network incorporating the sequential predictive probability method (Anraku, 1993) have been implemented to attempt to identify the proper well test interpretation (WTI) model. Although these implementations have proven successful, there is a need to improve the time it takes to perform the analysis and train the network. A more recent work investigated the use of a higher order neural network (Kumoluyi, 1994) in order to remove the pressure and time invariance. In all these cases the full range of response was provided. However, since a typical test only provides a small window of the overall reservoir response, the network must be able to handle responses where early or late time data are missing and where the key features occur at different times, and with different magnitudes, during the test.

One approach to accomplish these goals is to use a syntactic metric to describe the pressure response. The data is divided into sections of equal time and syntactic metric is used to represent the data in that period. A matrix of the syntactic metrics over time is constructed and presented to a neural network for pattern classification. In order to account for horizontal (time) shifting and missing data in the early or late time, the network is trained with several aspects of each class. The input pattern is then compared to each aspect of the class and the winning class is determined by selecting the class containing the aspect with the highest degree of match.

## DATA PREPARATION

Well test data from very dissimilar models often gives highly similar responses. Figure 1(a-c) shows the similarity of pressure response for the three reservoir models used in this study. One common way of representing the data is the derivative curve (Bourdet, 1983). This representation acts to enhance key features on the data and to eliminate the scaling problem, as shown in Figure 1 (d-f). The derivative is defined as:

$$t\left(\frac{\delta p}{\delta t}\right) \qquad (1)$$

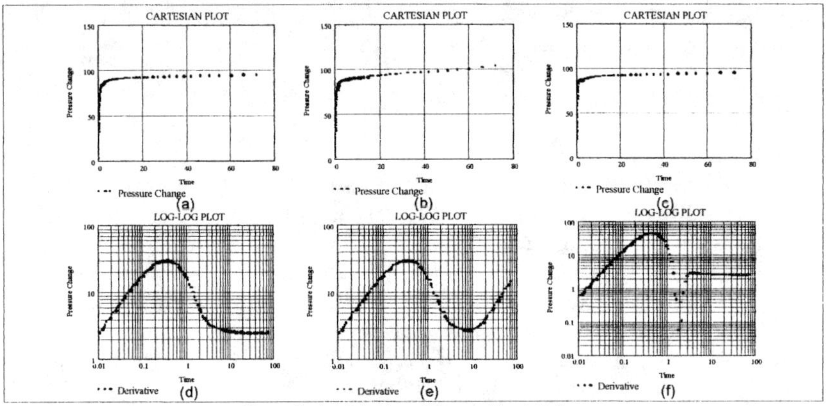


**Figure 1 - Analytical models used for training.**

The nature of this calculation acts to enhance any noise which might be present in the data. For this reason the derivative should be evaluated over a larger interval than simple numerical differentiation. A numerical differentiation with respect to the logarithm of time, using data points which are separated by a differentiation interval of at least 0.2 of a log cycle, is recommended (Horne, 1995) (Koederitz, 1994). Hence the numerical differentiation becomes:

$$\left(\frac{\delta p}{\delta \ln t}\right)_i = \left[\frac{\ln(t_i/t_{i-k})\Delta p_{i+j}}{\ln(t_{i+j}/t_i)\ln(t_{i+j}/t_{i-k})} + \frac{\ln(t_{i+j}t_{i-k}/t_i^2)\Delta p_i}{\ln(t_{i+j}/t_i)\ln(t_i/t_{i-k})} - \frac{\ln(t_{i+j}/t_i)\Delta p_{i-k}}{\ln(t_i/t_{i-k})\ln(t_{i+j}/t_{i-k})}\right] \tag{2}$$

subject to:

$$\ln(t_{i+j}) - \ln(t_i) \geq 0.2$$
$$\ln(t_i) - \ln(t_{i-k}) \geq 0.2$$

In well testing the derivative is always a positive number. Noise can also result in negative derivatives being calculated. Therefore a second test is performed. If the calculated derivative is negative, increasing the differentiation interval by one data point each time, until the calculated derivative become positive. This allows the data to be used as is, without filtering the noise.

The next phase in data preparation is to convert the derivative data into syntactic representation (Fu, 1982). The data is first broken into eight section of equal log time. The factor of eight was arbitrarily chosen to provide a small enough matrix to provide fast calculational time and yet detailed enough to capture the significant features. The log-log slope of the data points within each region is calculated and the syntactic membership is determined. A variable degree of overlap of the syntactic groups was incorporated in order to allow greater generalization. The various syntactic groups and their cutoff values are shown in Figure 2.

The final step in data preparation is to convert the syntactic representation into a matrix suitable for input to the Havnet neural network. The rows of the matrix, from top to bottom, represent the fuzzy classes shown in Figure 2, from left to right. The columns, from left to right, represent the equal spaces in log time.

## NEURAL NETWORK DESIGN

### Architecture Description

The neural network used in this study is the Hausdorff-Voronoi network, or Havnet, introduced by Rosandich (Rosandich, 1994). The network gets its name from the use of the Hausdorff distance as a measure of similarity between patterns, and because it employs a learned version of the Voronoi surface to perform the comparison. The neural network behaves as a binary pattern classifier. The Havnet network takes as inputs two dimensional binary patterns, employs feed-forward processing, and produces an analog output value. One output is generated by each node, with the analog value indicating the level of match between the input pattern and the class represented by that node. The Havnet neural network consists of three layers, the plastic layer, the Voronoi layer, and the Hausdorff layer. The plastic layer contains neurons with the weights that are trained during the learning process, the Voronoi layer serves to measure the distance between individual points in the input and learned patterns, and the Hausdorff layer uses information from the Voronoi layer to compute the overall level of similarity between the input pattern and the learned pattern. Figure 5 shows a detailed diagram of the architecture for a single node. The node is shown in a configuration for one-dimensional inputs for reasons of clarity. In the actual network, the input pattern, plastic layer, and Voronoi layer are all two-dimensional.

### Network Learning

Learning in the Havnet architecture is conducted off-line and in a supervised manner by presenting examples of each class to the network during a training phase. The network is informed *a priori* of the class to which each training pattern belongs.

During the learning phase, each training pattern is presented to the network in sequence, and the appropriate node is trained using the equations above. The learning rate determines the magnitude of the effect that each training pattern has on the trained weights. A learning rate of 1 was used in this study which allows for one-pass training.

### Recognition

When a pattern is presented to the network for recognition, the response of a node n to an input pattern $a_m$ is determined by computing, in sequence, the output of the plastic

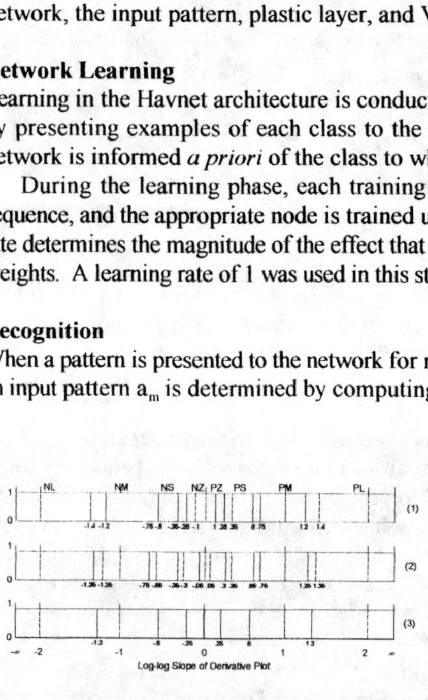


Figure 2 - Syntactic Classification Schemes

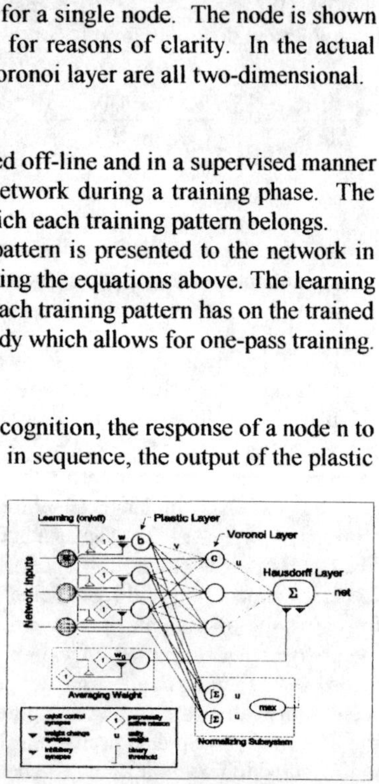


Figure 3 - Havnet Node Structure

layer and the Voronoi layer. The results for all nodes are then passed to the Hausdorff (output) layer. The value of the output layer represents the degree of match of the input data set to the trained class. In this study the winner was considered to be the class with the highest output value.

**Extended Havnet Architecture**

One of the strengths of the Havnet architecture is the ability to train different aspects. The plastic layer of the Havnet neural network is expanded to include several aspect representations for each learned object rather than the single representation used previously, and the learning process is modified to allow for the self-organization of these aspect representations. The recognition process of the expanded network is also modified to incorporate the multiple-aspect object representation.

In the case of this study, aspects are used to train the network on various windows of data. These include cases where the early time and late time data are truncated or missing, and the principal features, therefore, lie at different points along the time axis.

## RESULTS AND CONCLUSIONS

The Havnet neural network was trained with synthetic data from the three models shown in Figure 1(d-f). These represent the following analytical models:

- Homogeneous, infinite, with wellbore storage and skin
- Homogeneous, closed outer boundary, with wellbore storage and skin
- Dual porosity (fractured), infinite, with wellbore storage and skin

To account for the time shifting phenomena and the variability of reservoir parameters the network was trained with 9 aspects of each of the three classes (27 total training sets) covering the expected range of parameters, as shown in Table 1. All of the training sets were noise free.

**TABLE 1: TRAINING SET PARAMETERS**

| Parameter | Expected Range | Training Values |
|---|---|---|
| Permeability | .01 - 5000 md | .01, 10, 500, 5000 |
| Storage Coefficient | 0.0 - 0.5 rvb/psi | 0.0, 0.25, 0.5 |
| Skin Factor | 0.0 - 20 | 0.0, 5.0, 20.0 |
| Fracture Permeability | 1000 - 100000 md | 1000, 10000, 100000 |
| Fracture Porosity | 0.05 - 0.45 | 0.05, 0.20, 0.45 |

Testing was performed using 100 synthetic data sets from the three models using different parameters from those trained, but still within the range of expected values. Noise was introduced at 5 increasing levels, using a normal distribution. The degree of noise was controlled by increasing the standard deviation of the distribution from 0 to 0.2. This results in data very similar to real-world situations. In addition, the tests were repeated using the three syntactic groups shown in Figure 2.

The results are summarized in Table 2. It can be seen that the network was readily able to identify the noise-free models, but had difficulty as the amount of noise increased. Also, the performance improved as the amount of overlap in the syntactic groups was decreased, with the best performance occurring with no overlap. In the best case the network was able to recognize 100% of the noise-free data, and 50% of the noisy data, although the performance still decreases as the noise level increases.

**TABLE 2: RESULTS**

| **Syntactic Data Set —>** | | **(a)** | **(b)** | **©** |
|---|---|---|---|---|
| **Noise(Std dev)** | **# of Test Data Sets** | | **% correct** | |
| 0.0 | 20 | 75 | 85 | 100 |
| 0.05 | 20 | 60 | 65 | 80 |
| 0.10 | 20 | 25 | 40 | 65 |
| 0.15 | 20 | 15 | 25 | 35 |
| 0.20 | 20 | 0 | 5 | 20 |
| Average | | 34 | 44 | 60 |

The extended Havnet architecture has proven its ability to classify the data in a noise-free environment. Use of the Havnet with one-pass training, small matrix size, and multiple aspects, has the potential to offer a computationally fast solution. Based on these results, future work will focus on a better data representation in order to remove the time invariance and improve the noise handling ability. Noise filtering of the data prior to calculation of the derivative should improve the noise handling capability, although the degree of filtering must be limited to avoid distortion of key features. Time invariance can be addressed with a variety of feature extraction techniques. Another column in the Havnet matrix will likely be used to store yes/no answers to prompted questions, providing some *a priori* knowledge which will be necessary to discriminate between some of the more complicated and similar analytical models. Future work will also use real data rather than synthetic data.

## REFERENCES


Al-Kaabi, A.O., Lee, W.J., (1990). Using Artificial Neural Networks to Identify the Well Test Interpretation Model, paper SPE 20332 presented at the 1990 SPE Petroleum Computer Conference, Denver CO, June 25-28.

Allain, O.F. , Houzé, O.P., (1992). A Practical Artificial Intelligence Application in Well Test Interpretation, paper SPE 24287 presented at the 1992 SPE European Petroleum Computer Conference, Stavenger, Norway, May 25-27.

Anraku, T., Horne, R.N., (1993). Discrimination Between Reservoir Models in Well Test Analysis, paper SPE 26426 presented at the 1993 SPE Annual Technical Conference and Exhibition, Houston, TX, October.

Bourdet, D., Ayoub, J.A., Whittle, T.M., Douglas, A.A., and Pirard, Y.M., (1983), A New Set of Type Curves Simplifies Well Test Analysis," *World Oil* (May 1983), pg. 95-106.

Fu, K.S. (1982). *Syntactic Pattern Recognition with Applications*, Prentice Hall, Englewood Cliffs, NJ.

Horne, R.N. (1995). *Modern Well Test Analysis*, 2nd Edition, Petroway, Inc., Palo Alto, CA.

Koederitz, L.F. (1994). *Notes on Well Test Analysis*, Revised Edition - December 1994, Lecture notes for PE ENG 341, University of Missouri-Rolla, pg. 3.18-3.20.

Kumoluyi, A., Daltaban, T.S., Archer, J.S., Koncar, N., Jones, A.J., (1994). Well Reservoir Model Identification Using Higher Order Networks, *Intelligent Engineering Systems Through Artificial Neural Networks*, ASME Press. New York, NY, pg. 467-475.

Rosandich, R.G. (1994). Artificial Vision: Three Dimensional Object Recognition Using Neural Networks, Ph.D. Dissertation, University of Missouri-Rolla (1994).

# A NEURAL NETWORK BASED CLASSIFICATION OF BONGARD PATTERNS

**LAKSHMI N. SRINIVASA, ERNEST L. HALL[1] AND SAM ANAND[2]**
Center for Robotics Research,
Department of Mechanical, Industrial and Nuclear Engineering,
ML#72, University of Cincinnati,
Cincinnati, OH 45221-0072

**ABSTRACT**
Artificial neural networks provide a powerful method for machine classification. This paper presents an automated vision based classification of screws. The problem of classification of screws in our case is similar to that of classification of Bongard patterns and falls in the category of non-linear problems which require an exclusive-or (X-OR) type decision rule. A multilayer feedforward neural network implemented with an error back propagation algorithm is known to solve this X-OR type of problem. A two layer feedforward network was implemented with a generalized delta rule. Input to the neural network was the diameter and angle of orientation of the screws. Training of the network was accomplished on a training sample of 24 screws. A 100% classification was achieved. The significance of this work is that the multilayer feedforward networks are shown to be powerful classifiers even for linearly non-separable patterns.

**Keywords:** Pattern Recognition, Artificial Neural Networks, Bongard Patterns, Classification of Screws, Backpropagation Algorithm, Exclusive-OR type, Image Processing, Manufacturing Applications, Multilayer Feedforward NN

## 1. INTRODUCTION

Many important manufacturing decisions can be made based on automated measurements. Such automated procedures can often result in improved product quality or decreased costs. In recent years there has been an increase in activity to solve optimization and pattern recognition problems using neural networks. One reason is that both linearly and non-linearly separable patterns can be classified. Many of the manufacturing problems encountered are either optimization or pattern recognition problems. In this paper, a manufacturing application has been considered which is a pattern recognition problem.

[1] Director, Center for Robotics Research, Author for Communication, e-mail: elhall@uceng.uc.edu
[2] Director, Computer Aided Manufacturing Laboratory, ML#72, University of Cincinnati, Cincinnati, OH-45221-0072

Bongard [1] has listed a set of geometric problems that vary from easy to difficult to solve by single layer perceptrons. These problems involve simple geometric objects like circles, cubes, triangles, pentagons, etc., that vary in size such as large, small, and in orientation such as right, left, oriented vertically, oriented horizontally, etc. Humans have excellent ability to solve many of these recognition problems. However, a particularly interesting problem is not so easily solved by humans. This problem involves the recognition of certain screw heads which have a certain diameter and angle of the screw slot.

The top portion of a screw is generally a circle with a line segment running along the diameter of the circle. Extraction of the features such as orientation, diameter of the circle and length of the segments in the geometric objects can be used as features for representing the data.

Multilayer feedforward network implemented with error back propagation algorithm has been shown to solve the X-OR problems. Gupta [2] and Colin Wang [3] demonstrated the application of supervised backpropagation in shape classification and automated visual inspection. Kashyap and Chellappa [4] proposed a Circular Autoregressive (CAR) model to represent the shape of screws. Dubois and Glanz [5] applied the CAR model to classify artificial boundary shapes. Taioun Kim [6] developed a novel curvature and circular fitting algorithm to represent circular boundaries for classifying boundary defects using neural networks. These methods were concerned more with the representation schemes whereas we avoid such difficult representation schemes in our shape classification of screws. Tou and Gonzalez [7], Rumelhart and McClelland [8], and Richard P. Lippmann[9] provide an in depth overview of pattern recognition schemes and neural networks. A simple Bongard problem involving the classification of screws is shown in Figure. 1.

## 2. DESCRIPTION OF THE PATTERN RECOGNITION PROBLEM

The problem of classification of screws belongs to the class of problems which are characterized by non-linear decision boundaries. The problem can be illustrated as shown in Figure 1. This shows the top view of several screws. Figure 2. shows the screws fixed loosely on a pallet. A set of screws belonging to two plates are to be classified. The top view of a screw will look like a circle with a line segment on the diameter of the circle. If a robot were to pick up the set of screws and fix them on plate 1 or plate 2, then it needs a decision making scheme to classify the screws. The two dimensional plot of the orientation and diameters of the screws is shown in Figure 3. From the 2D plot of the angle versus diameter of the screws, it is seen that a nonlinear decision boundary is required to classify the screws belonging to each plate. The non-linear boundary can be described as two linear boundaries and the general form of the decision rule may be called an exclusive-or (X-OR) since patterns can lie on either side of the two decision boundaries .

A vision system was used for capturing images of the screws. Preprocessing and extraction of features will be performed to compare the ideal to an actual case. A multilayer feedforward neural network was implemented with the generalized delta rule (error back propagation algorithm). This was found to provide good classification of the sample patterns for generating the nonlinear decision boundaries. A two layer network was implemented with a generalized delta rule in ANSI C on a

Sunsparc workstation for classifying the Bongard patterns. The set of screws belonging to plate 1 was assigned class 1 and the rest of the screws belonging to plate 2 were assigned class 2.

## 3. IMAGE ACQUISITION AND PREPROCESSING

The experimental setup for the classification of screws consisted of the screws fixed loosely on a pallet. Figure 4. shows the block diagram of the experimental set up used for classification of the Bongard patterns. This top view would reveal the top portion of the screw and the angle of orientation making the measurement easier for classification.

The images of the screw heads were captured by a camera and then digitized. The processing is done on the computer using Khoros, an image processing tool. Preprocessing and feature extraction will be done on the images to extract the diameter and the angle of orientation of each screw. The angle of orientation was normalized to lie between 0 and 1 by dividing the angles by 90 degree. Table I shows the diameter, angle of orientation, normalized angle of each screw and their respective class. As the classes were known beforehand, a supervised learning scheme was used for training the network.

## 4. MULTILAYER FEEDFORWARD NETWORK IMPLEMENTED WITH ERROR BACK PROPAGATION TRAINING ALGORITHM

The error back propagation training algorithm is a supervised learning scheme for the classification of linearly non-separable patterns. The fundamental concept by which backpropagation works is by calculation of the output, layer by layer, and feeding the output of the previous layer as the input to the next layer neurons, propagating the error (using gradient descent search procedure) backwards layer by layer, updating the weights layer by layer till it reaches the input layer. This procedure is repeated until the actual outputs converge to the desired values.

A diagram illustrating the architecture of a multilayer feedforward network is shown in Figure 5. The error back propagation training algorithm is described in detail in Lippmann [9]. A two layer perceptron is used for our purpose. The network for our purpose consisted of three to four neurons in the hidden layer and one neuron in the output layer. A sigmoidal function described in equation (1) is used as the activation function

$$f(w'x) = 1/1 + \exp(-\lambda w'x) \tag{1}$$

The root mean square error in equation (2) was used as a measurement of the distance between the desired result and the computer output.

$$\text{RMSE} = \sqrt{(C_i - Y_i)^2 / n \times m} \tag{2}$$

where $C_i$ is the desired output, $Y_i$ is the computed output, $n$ is the number of patterns an $m$ is the number of neurons in the output layer.

## 5. SIMULATION RESULTS

The extracted features of the 24 patterns (screws) was used as input to the input layer of the multilayer feedforward network with a bias input of +1. Three experiments were performed in order to study the behavior of this algorithm. On the first hand, the weights were set at random small values. The learning rate, number of hidden neurons, and the number of epochs were varied to see the results. As a second experiment, weights in the hidden layer were initialized to [1.0,-0.4,-0.3,-0.2,-0.1,0.1,0.2,0.3,0.4]. The weights in the output layer were set to [1.0,0.1,0.5]. The number of hidden layer neurons, learning rate and the number of epochs were varied and the results were tabulated in Table II. The inputs to the neural network are given in the Table I. The desired result was set as 0 for class 1 and 1 for class 2 set of screws. If the neural network outputs a value in between [0, 0.5] then the pattern is classified as class 1. The patterns are classified as class 2 if the output value ranges in between [0.5, 1.0].

The first experiment gave a bad result by mis-classifying 4 pattern vectors. The second experiment gave a good result by classifying all elements of the two pattern classes. In the second experiment, the weights for the hidden layer was chosen to lie between -0.5 to +0.5 whereas in the first experiment, the weights were randomly generated for all the layers. With 8 neurons in the hidden layer, the multilayer feedforward network mis-classified 4 out of the 24 screws whereas in experiment 2, the same network classified all the screws correctly with just 3 neurons in the hidden layer. The same network architectures were applied to X-OR problem and the results were compared. The first network (the network with random initial weights) was able to solve the X-OR problem correctly with 8 neurons in the hidden layer. The second network (which started with initial weights of 1, -0.4,-0.3,-0.2,-0.1,0.1,0.2,0.3,0.4,0.5 in the hidden layer) was able to solve the X-OR problem correctly with just 3 neurons in the hidden layer. A two dimensional plot was drawn to show the convergence of the output with the desired output. The rms-error starts with a high value for 0.01423 for 200 iterations and reaches a stable low value of 0.0001 from 2000 iterations. Figure. 6. shows the decrease of the rms-error as the number of iterations increase.

In the third experiment, the diameter of the screws were normalized to zero mean and unit variance in order not to bias the network. The network was able to classify the patterns with four neurons in the hidden layer. Jack-knifing was performed on the patterns. This was done by setting aside one pattern for testing and using the rest for training the network. This was repeated for the 24 combinations. It was found that 6 patterns were misclassified during the testing phase. It was also found that all these six patterns were of class 1. When the networks were fed with normalized angle, diameter (without any normalization) and bias input, 4 patterns were misclassified during the testing phase. When 4 patterns were tested, leaving aside 20 patterns for training, the network classified the 4 patterns for both the normalized diameter input as well as the diameter input without normalization. From the third experiment, it was concluded that the normalization of the diameter of the screws to zero mean and unit variance did not significantly improve the network's performance.

## 6. CONCLUSION AND RECOMMENDATIONS

A vision based classification of screws similar to that of the Bongard patterns was presented in this paper. In the method described, the representation problem was overcome by fixing the screws on a pallet. Further research work also needs to be done on the representation of screws to achieve robust classification of screws. A multilayer feedforward neural network was implemented with error back propagation training algorithm. This method provided 100% correct classification. The disadvantage of the generalized delta rule is that the weights might settle at a local minima for complex problems since there is no clear cut theoretical foundation for initializing the weights. Although, the neural network performed well, more research needs to be done into the theoretical aspects of selecting weights and speeding the convergence of this algorithm. The main theoretical aspects that are to be studied in the future are 1) increase in speed of learning the samples, 2). determine the number of layers and neurons required for any problem and 3) decreasing the size of the network without any degradation in performance.

## REFERENCES


1. Bongard, M., *Pattern Recognition*, Spartan Books, New York, 1970.
2. Gupta, Lavit, Sayeh, M.R., and Tammana, Ravi., "A Neural Network Approach To Robust Shape Classification", *Pattern Recognition*, Vol. 23, No. 6, pp. 563-568, 1990.
3. Colin Wang, David J. Cannon, Soundar R.T. Kumara, and Guowen Lu, "A Skeleton and Neural Network-Based Approach for Identifying Cosmetic Surface Flaws", *IEEE Transactions on Neural Networks*, Vol. 6, No. 5, pp. 1201-1211, Sept. 1995.
4. Kashyap, R.L., and Chellappa, R., "Stochastic Models for Closed Boundary Analysis: Representation and Reconstruction", *IEEE Transactions on Information Theory*, Vol. 27, No. 5, pp. 627-637, 1981.
5. Dubois, Swan R. and Glanz, Filson H., "An Autoregressive Model Approach to Two-Dimensional Shape Classification", *IEEE Transactions on Pattern Analysis and Machine Intelligence*, Vol. 8, No. 1, pp. 55-66, 1986.
6. Taioun Kim, "Automated Visual Inspection and Classification by Neural Networks in Manufacturing", Ph.D. Dissertation, The Pennsylvania State University, 1995.
7. Julius T. Tou and Rafael C. Gonzalez, *Pattern Recognition Principles*, Addison-Wesley Publications, 1974.
8. Rumelhart, D.E., and McClelland, J.L., *Parallel Distibuted Processing: Explorations in the Microstructure of Cognition Vol 1. Foundations.* MIT Press (1986).
9. Richard P. Lippmann, *An Introduction to Computing with Neural Nets*, IEEE ASSP Magazine, p4-22, April 1987.

# Figures and Tables

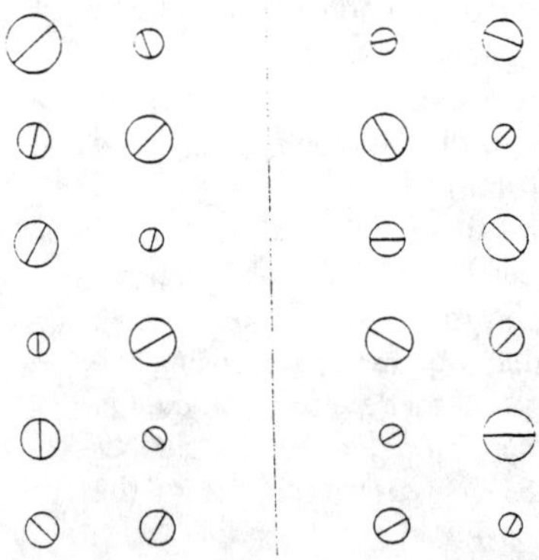


Figure 1. Bongard Patterns

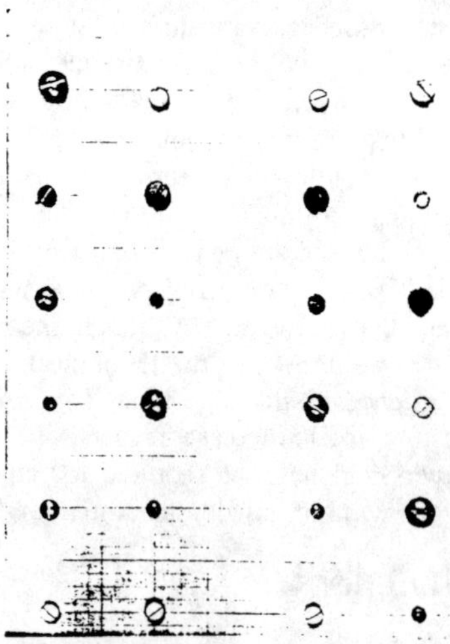


Figure 2. Top View of Screws fixed on a pallet

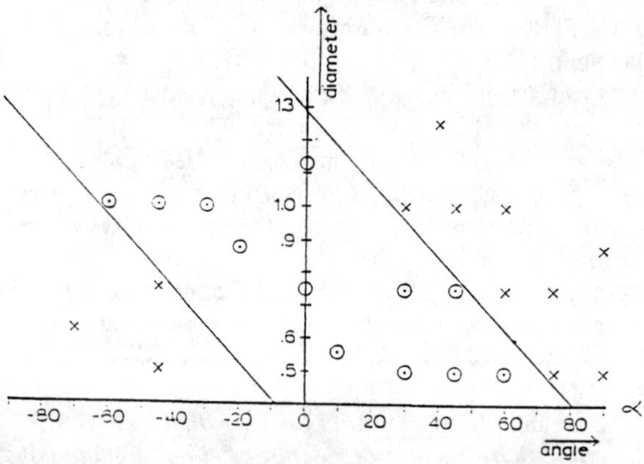


Figure 3. Two dimensional plot of angle of orientation Vs diameter. The classes are separated by two linear decision boundary or one non-linear decision boundary

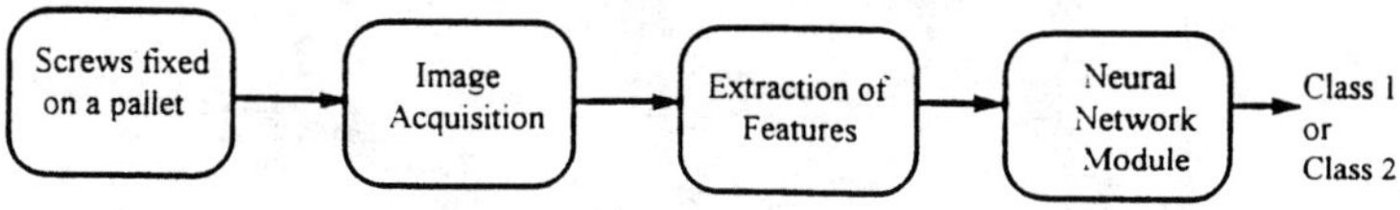


**Figure .4. Block diagram of the experimental setup for classification of screws.**

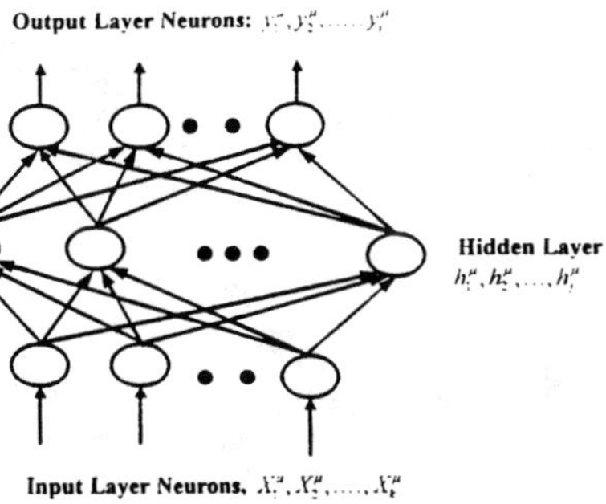


**Figure 5. Multilayer feedforward network implemented with Error Back Propagation Training Algorithm**

**Table I**
**Inputs to the Two layer network**

| Pattern Number | Bias Input | Diameter of screw | Angle of Orientation | Normalized Angle | Class |
|---|---|---|---|---|---|
| 1 | 1.0 | 0.5 | -45 | -0.5 | 0 |
| 2 | 1.0 | 0.5 | 30 | 0.33 | 1 |
| 3 | 1.0 | 0.5 | 45 | 0.5 | 1 |
| 4 | 1.0 | 0.5 | 60 | 0.66 | 1 |
| 5 | 1.0 | 0.5 | 75 | 0.833 | 0 |
| 6 | 1.0 | 0.5 | 90 | 1.0 | 0 |
| 7 | 1.0 | 0.55 | 10 | 0.11 | 1 |
| 8 | 1.0 | 0.6 | -70 | -0.77 | 0 |
| 9 | 1.0 | 0.75 | -45 | -0.5 | 0 |
| 10 | 1.0 | 0.75 | 0.0 | 0.0 | 1 |
| 11 | 1.0 | 0.75 | 30 | 0.33 | 1 |
| 12 | 1.0 | 0.75 | 45 | 0.5 | 1 |
| 13 | 1.0 | 0.75 | 60 | 0.667 | 0 |
| 14 | 1.0 | 0.75 | 75 | 0.833 | 0 |
| 15 | 1.0 | 0.85 | 90 | 1.0 | 0 |
| 16 | 1.0 | 0.9 | -20 | -0.22 | 1 |
| 17 | 1.0 | 1.0 | 60 | -0.667 | 1 |
| 18 | 1.0 | 1.0 | -45 | -0.5 | 1 |
| 19 | 1.0 | 1.0 | -30 | -0.33 | 1 |
| 20 | 1.0 | 1.0 | 30 | 0.33 | 0 |
| 21 | 1.0 | 1.0 | 45 | 0.5 | 0 |
| 22 | 1.0 | 1.0 | 60 | 0.667 | 0 |
| 23 | 1.0 | 1.1 | 0.0 | 0.0 | 1 |
| 24 | 1.0 | 1.3 | 40 | 0.44 | 0 |

**Table II**

**Best Answers for case 1(case 1 has random intial weights in the hidden layer) and case 2 (case 2 has fixed intial weights in the range [-0.5,0.5]) compared eta indicates learning rate. HN is the number of hidden layer neurons X mark in the column indicates misclassification**

| No of Hidden Neurons | Desired Class | Results of Case one | Results of Case 2 | RMS Error |
|---|---|---|---|---|
| Case 1= 8HN | 0 | 0 048 | 0 0038 | Case 1=0 021 |
| Epochs=2000 | 1 | 0 621 | 0.996 | Case 2=0 0001 |
| eta=0.5 | 1 | 0 468 X | 0 998 | |
| Case 2=3HN | 1 | 0 382 | 0 989 | |
| Epochs=2000 | 0 | 0 322 | 0.123 | |
| eta=0.5 | 0 | 0 217 | 0 001 | |
| | 1 | 0 933 | 0.965 | |
| | 0 | 0 034 | 0 003 | |
| | 0 | 0 808 X | 0.083 | |
| | 1 | 0 905 | 0.998 | |
| | 1 | 0 631 | 0.998 | |
| | 1 | 0 528 | 0.802 | |
| | 0 | 0 479 X | 0.008 | |
| | 0 | 0 287 | 0.0008 | |
| | 0 | 0 172 | 0.0004 | |
| | 1 | 0 968 | 0.998 | |
| | 1 | 0 615 | 0.9392 | |
| | 1 | 0 991 | 0 996 | |
| | 1 | 0 992 | 0.999 | |
| | 0 | 0 699 X | 0.148 | |
| | 0 | 0 295 | 0.001 | |
| | 0 | 0 186 | 0.0005 | |
| | 1 | 0 773 | 0.998 | |
| | 0 | 0.106 | 0.0005 | |

## Table III

**Table showing the decrease of rms-error as the number of epochs increase with learning rate kept constant**

| No of epochs | learning rate | rms-error |
|---|---|---|
| 200 | 0.5 | 0.01423 |
| 400 | 0.5 | 0.0029 |
| 600 | 0.5 | 0.0014 |
| 800 | 0.5 | 0.000878 |
| 1000 | 0.5 | 0.000545 |
| 2000 | 0.5 | 0.0001 |

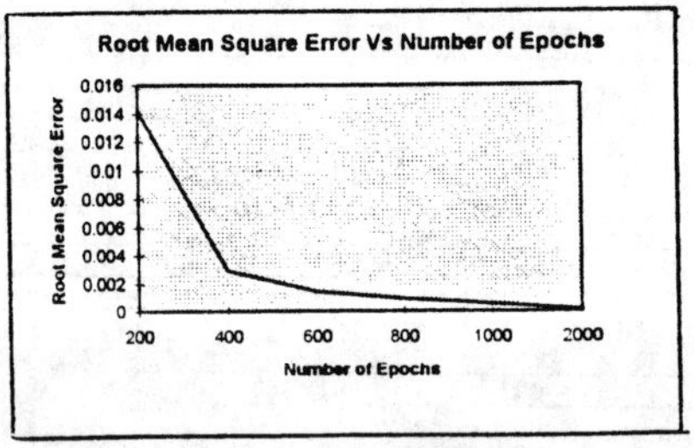


**Figure 6. Graph showing the decrease of RMS-Error with increase in the number of iterations**

# DETERMINATION OF CARBONATE LITHOFACIES USING HIERARCHICAL NEURAL NETWORKS

**H. YANG AND H. C. CHEN**
*Department of Computer Science, The University of Alabama*

**J. H. FANG**
*Department of Geology, The University of Alabama*

***ABSTRACT:***
The primary task of geological and engineering characterization of a petroleum reservoir is to determine various lithofacies of the reservoir rocks from the examination of cores taken from wells. However, cores are not always available, and the characterization must depend on well logs. We have devised a hierarchical scheme of using two kinds of neural networks to perform lithofacies determination from log data. The first network is that of backpropagation, which is used to determine the lithofacies from a log suite, and the second one, a recurrent network, is used to examine their vertical sequence. The well-log and core data from two wells in Appleton field, southwest Alabama were used for training and testing. This approach has achieved a comparable level of accuracy as those obtained from the core data. The results of using hierarchical networks showed a remarkable improvement compared to when backpropagation alone was used.
**Keywords: pattern recognition, petroleum reservoir, lithofacies, recurrent neural network, backpropagation, hierarchical scheme.**

## INTRODUCTION

Broadly speaking, there are two kinds of petroleum reservoirs: clastic and carbonate. The former is composed of sandstones and the latter, limestone or dolomite. As the first step in petroleum reservoir characterization, lithofacies (i.e., rocktypes) of the reservoir must be determined.

Conventional practice of clastic and/or carbonate lithofacies determination from cores is still the mainstream of reservoir geologists, but a method not requiring cores, that is, using well-log data alone, is highly desirable because core recovery is not always 100%, and, moreover, it is very costly to take cores. Well logs are continuous recording of geological properties of subsurface rock formations at depths. Cores provide direct observations, whereas well logs only provide indirect information which must be transformed to yield lithofacies.

One may argue at this juncture that lithofacies obtained from logs should be called electrofacies, the term coined by Wolf and Pelissier-Combescure [12]. We are of the opinion that "electrofacies" is a misnomer, because the facies determined via logs are not related to neither electronic nor electric properties of the rocks. It is, therefore, more appropriate to designate it as logfacies [6]. Logfacies are herein defined as the collective set of log responses, reflecting both the rock and fluid properties which allow discrimination between beds or sedimentary units. Logfacies are often equivalent to lithofacies, especially when they are calibrated with cores; thus, logfacies may be construed as surrogate for lithofacies.

Compared to clastic lithofacies, carbonate lithofacies show only subtle changes in logfacies, therefore, the task of determining the latter is more demanding [3, 5]; a single-stage determination with backpropagation is not always successful or satisfactory. Because of this difficulty, we have devised a hierarchical network system, combining the first-stage simple backpropagation architecture, coupled with the second-stage, "backpropagation through time", originally proposed by Werbos [11]. The networks used and their hierarchy are described in the next section.

## HIERARCHICAL NEURAL NETWORKS

### Backpropagation Neural Network

The Artificial Neural Networks (ANNs) have been widely used in the field of geology where in many cases conventional computing algorithms and statistical methods are not adequate [1, 2, 4, 8-10]. Among those applications, Backpropagation Neural Network (BPNN), a popular paradigm of ANN architecture, has generated many successful demonstrations of its power despite of its simplicity. The BPNN is a layered feed-forward network. A set of input is fed from outside of the network into the input layer; inside the network each node transforms the sum of its weighted inputs to the next layer with a differentiable activation function. The calculation is carried out for every node at each layer till the network produces output for each node in the output layer. The computed output is then compared with the target output or desired output from the training set, and the error is backpropagated to modify the weights on the network. This weight-changing procedure is called "backpropagation training". The objective of training a BPNN is to adjust the weights so that, when given proper input data, the network will produce the desired outputs. Mathematically, it is a steepest-descent algorithm for finding a set of weights which minimizes the squared errors of the outputs.

In this paper we employed a 3-layer BPNN with 5 nodes (we have 5 types of log measurements) in the input layer, 10 nodes in the hidden layer, and 6 nodes (we classify the lithofacies into 6 classes) in the output layer. The number of the hidden-layer nodes was chosen by experiments. After the network was trained, we simply used the same feedforward procedure as in the training process with the weights obtained from the training to calculate the output. The result obtained from the feedforward process at the output layer, denoted by $y_i$, $i = 1..6$, was then filtered by a threshold function:

$$y_i = \begin{cases} 1; & when \quad y_i > 0.8 \\ 0; & when \quad y_i < 0.2 \end{cases} \tag{1}$$

**Recurrent Backpropagation Neural Network**

A simple BPNN is still not good enough to determine the lithofacies. Werbos[11] proposed a scheme known as "backpropagation through time", where the input is a time-series set of data. These time-series data will be denoted by [*y(t), t=1..n*]. The output can be determined based on any past input or even past output within the network itself. This network involves retaining experiences of the past in memory so that it can be applied for time-series predictions [7]. Therefore, the network has to go beyond the regular non-linear input-output mapping and pattern generalization of BPNN.

For the sake of simplicity, we assume that the output *d(t)*, at the current time *t* depends on the inputs: *y(t-2), y(t-1)* and *y(t)*. The architecture of the network is depicted in Figure 1. The mathematical form of this network can be defined as:

$$d_i(t) = F\left[\sum w_{ij}.y_j(t) + \sum w'_{ij}.y_j(t-1) + \sum w''_{ij}.y_j(t-2)\right] \tag{2}$$

where $w_{ij}$, $w'_{ij}$ and $w''_{ij}$ are weights of the network and $F$ is an activation function. The part of the algorithm that feedforwards the input and backpropagates the errors remains the same, except that now more interconnections are considered. This BPNN implies that the output at time *t* is governed by the actual states at time *t, t-1, t-2* and system evaluation functions. Many recognition algorithms involve this kind of "time-series" analysis where the pattern recognition at time *t* can be more accurate if what has happened at earlier times can be taken into consideration. In our research, we substitute time with depth, since the vertical sequence at various depths plays an important role in determining the lithofacies. In addition, the output from the previous time period is also used as part of the input to the current time period. As a result, the accuracy of the lithofacies classification is improved. This network recurrently uses the previous output, thus this scheme is called "recurrent backpropagation neural network" (RBPNN). The details of the complete network system will be discussed in the next section.

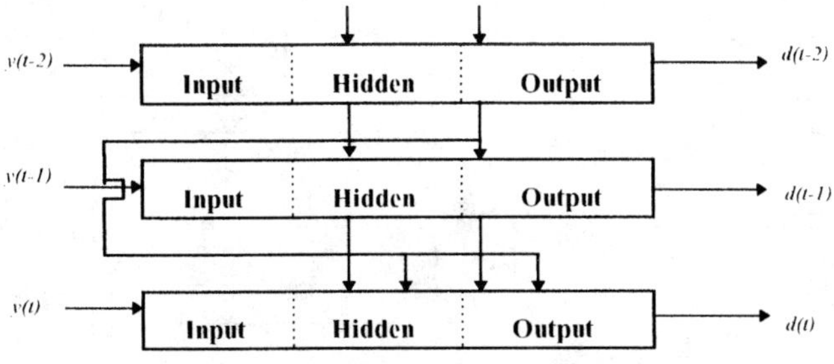


**Figure 1: Architecture of a basic recurrent neural network**

**Hierarchical Neural Network**

To solve our lithofacies problem, we employ a hierarchical neural network system as shown in Figure 2. The architecture of each BPNN is a variant of a 3-layer backpropagation network with 5 nodes in the input layer, 10 nodes in the hidden layer, and 6 nodes in the output layer. These BPNNs determine the lithofacies from well-log data (neutron, density, sonic, resistivity and gamma-ray) at each depth. Then the RBPNN, with 18 nodes at the input layer, 10 nodes at the hidden layer, 6 nodes at the output layer, is used to examine the vertical sequence with a lag of 1 foot. We use the output from the RBPNN in the previous depth and two outputs from BPNN (at the current depth and next depth) to determine the lithofacies at the current depth. Mathematically, the equation used can be defined as:

$$d(t) = F_1[W'd(t-1)] + F_2[Wy(t)] + F_3[W''y(t+1)] \tag{3}$$

where $W$, $W'$ and $W''$ are the weight matrices for the respective RBPNNs, and $F_1$, $F_2$ and $F_3$ are the corresponding activation functions used in these RBPNNs. The procedure is repeated for the determination of lithofacies of the subsequent depths.

## SOURCE OF DATA

The cores and logs we used for this study came from two wells (#6247 and #3986) in Appleton field, located in north-central Escambia County, Alabama. The field produces oil from the Smackover carbonate formation at a subsea depth of approximately 12,900 feet. The Smackover reservoir in Appleton field includes a variety of carbonate lithofacies.

Core examination revealed six lithofacies: (1) dolostone, (2) grainstone, (3) dolomitic limestone, (4) boundstone, (5) laminite, and (6) anhydrite. The suite of the logs is composed of: (1) gamma-ray, (2) neutron, (3) density, (4) sonic, and (5) resistivity.

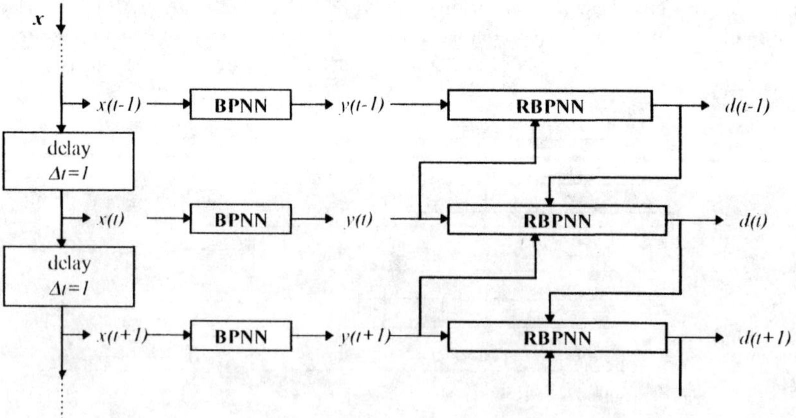


**Figure 2: Structure of the hierarchical neural network**

## RESULTS AND DISCUSSION

Lithofacies of the whole core from well 6247 were described using a modified form of the Dunham classification. The network was trained on the data from well 6247 and then tested on well 3986. Subsequently core chips from this well were described and classified using the same lithofacies defined for well 6247. One lithofaicies had to be added, for there is no massive anhydrite in the core from well 6247.

Because only core chips were available from well 3986-B, the accuracy of lithofacies identification is less than for well 6247. Some lithofacies, such as boundstone, are easy to determine in cores, but difficult to identify in core chips.

The amount and quality of core and well-log data available for this study are typical for the Smackover Formation in Alabama, thus it provides an appropriate test of the network's ability to predict lithofacies from well-log data. Of 156 1-foot cored intervals (12,873-13,028) in well 3986-B, 130 were correctly identified by the network. In fact, 51 intervals identified as probably boundstone from the core-chip examination, were identified by the network as boundstone. If this attribution is correct, then the network actually performed better in this interval. In 23 intervals, lithofacies determination by the network was a near miss; in these intervals the network predicted dolomitic limestone whereas the core chips were identified as grainstone. This grainstone interval was partially dedolomitized, so the network identified the lithfacies correctly. If the training set had included some partially dedolomitized grainstone, then the network might have been able to distinguish between partially dedolomitized grainstone and other dolomitic limestone. In 11 intervals, the network's second choice was dolomitic limestone, which would have been correct. In summary, the network predicted 130 of 156 (83%) intervals correctly, and if correct second choices are included, the success rate rises to 90%, while an implementation of conventional statistical multivariate cluster analysis shows only approximately 60% accuracy (which doesn't provide a second best choice).

As stated previously, The network predicted lithofacies better than 80% in well 3986-B. This success ratio would probably rise with a larger training set, because some of the rock types in the core from well 3986-B were confusingly different from those in the training set from well 6247. For instance, all of the grainstone in the training set was completely dolomitized, whereas some of the grainstone in the predicted well had been significantly dedolomitized (up to 1/2 or more of the rock). Also, no thin sections were made from the core chips from well 3986-B. Thin section examination makes identification of lithofacies more reliable.

## CONCLUSIONS

Results from this study allow the following conclusions to be drawn:

1. ANNs are efficient and efficacious in determining lithofacies using only well-log data. This task has traditionally been performed with statistical multivariate cluster analysis and discriminant function [3, 6]. Compared to the conventional approach, the new approach is more powerful.

2. A simple BPNN is inadequate to classify complex patterns; hierarchical networks are needed to yield better results. We have demonstrated this with real-world examples.
3. The veracity of our approach should be tested with additional examples.

## ACKNOWLEDGEMENT

The first author would like to thank the Alabama DOE/EPSCoR Program (DE-FC02-91ER75678) for awarding him the Young Investigator Enhancement Award. All the authors acknowledge the financial support from the same program.

## REFERENCES

[1] Baldwin, J. L., Bateman, A. R. M., Wheatley, C. L., (1990). "Application of neural network to the problem of mineral identification from well logs", *The Log Analyst*, V. 3, 279-293.

[2] Baldwin, J. L., Otte, D. N., Wheatley, C. L., (1989). "Computer emulation of human mental process: application of neural network simulations to problems in well log interpretation", *Society of Petroleum Engineers SPE* paper 19619, 481-493.

[3] Busch, J. M., Fortney, W. G., Berry, L. N., (1987). "Determination of lithology from well logs by statistical analysis", *SPE Formation Evaluation*, V. 2, 412-418.

[4] Davis, J. C., (1986). "Statistics and data analysis in geology", 2nd ed., *John Wiley*, New York, 646p.

[5] Doveton, J. H., (1986). "Log analysis of subsurface geology", *John Wiley*, New York, 273p.

[6] Gill, D., Shomrony, A., H. Fligelman, H., (1993). "Numerical zonation of log suites and logfacies recognition by multivariate clustering", *The American Association of Petroleum Geologists Bulletin*, V. 77, 1781-1791.

[7] Grauger, C. W. J., Newbold, P., (1986). "Forecasting Economic Time Series", *Academic Press Inc.*, Florida.

[8] Raiche, A., (1991). "A Pattern Recognition Approach to geophysical inversion using neural nets", *Geophysical Journal International*, V. 105, 629-648.

[9] Rogers, S. J., Chen, H. C., Kopaska-Merkel, D. C., Fang, J. H., (1995). "Predicting Permeability from Porosity Using Artificial Neural Networks", *The American Association of Petroleum Geologists Bulletin*, V. 79, no. 12, 1786-1797.

[10] Rogers, S. J., Fang, J. H., Karr, C. L., Stanley, D. A., (1992). "Determination of Lithology from well logs using a Neural Network", *The American Association of Petroleum Geologists Bulletin*, V. 76, no. 5, 731-739.

[11] Werbos, P. J., (1990). "Backpropagation Through Time: What It Does and How to Do it", *Proceedings of the IEEE*, V. 78, no. 10, 1550-1560.

[12] Wolf, M., Pelissier-Combescure, J., (1982). "FACIOLOG - Automatic electrofacies determination", *SPWLA Annual Logging Symposium*, Paper FF, 23p.

# GBF NETWORK ARCHITECTURES FOR ROBOT VISION

**JOSEF PAULI**
*Christian-Albrechts-Universität, Institut für Informatik, Preusserstrasse 1-9, D-24105 Kiel, Germany*

***ABSTRACT:***
A versatile robot manipulator is based on techniques of computer vision and neural network learning. For grasping objects four principal tasks have to be done in a cycle — detect the desired object and the grasping fingers in the images, evaluate the spatial relationship wrt. grasping stability, choose a more stable grasping pose (if possible), and move the manipulator to it. In this work we focus on two subproblems thereof, specifying the hand–eye coordination and evaluating the grasping situation. These procedures are based on (nonlinear) functions which are not known a priori and therefore have to be learned. We uniformly approximate the required functions by means of networks of Gaussian Basis Functions (GBF networks). Modifying the number of GBFs and/or the accessory size of the gaussian support changes the quality of the learned function. We use GBF networks both to specify the hand–eye coordination and to recognize the grasping situation and show how various network configurations influence the accuracy and therefore the usefulness of the function approximations. All experiments are carried out in real world environments using an industrial robot manipulator and the computer vision system KHOROS.

## INTRODUCTION

The motto of William of Occam, a scholastic of the middle ages, reads as follows: It's vain to do with more what can be done with less. This principle, known as Occam's Razor (Blumer et al., 1990), is very much alive in neural network learning. Several theoretically oriented papers study the aspects of learnability and derive lower and upper bounds on the sample size versus net size needed such that a function approximation of a certain quality can be expected (Baum and Haussler, 1990). A sample is a set of input–output pairs ( examples) characterizing the desired function. Most recently, the relationship between sample compression and probably approximately correct learning (PAC–learnability) has been discussed (Floyd and Warmuth, 1995). The term PAC–learnability characterizes a function as learnable if and only if a learning algorithm can be formulated which produces with a certain expenditure of work a probably approximately correct function. Based on pre–specified levels for probability and correctness of a function approximation, we are interested in discovering the most simply and efficiently applicable representation (Rissanen, 1984).

An approximation scheme leading to a function which is close to the sample data and meets smoothness constraints is known under the term regularization. It was shown that regularization principles can be implemented with neural networks consisting of an input layer, a layer of hidden nodes and an output layer (Poggio and Girosi, 1990). Each hidden node is defined by a so–called support basis function acting as a window in the input space. In our application we use GBF networks whose basis functions are multidimensional symmetric Gaussians. On account of applying Occam's Razor to GBF networks it is desirable to discover the minimum number of GBFs to reach a critical quality for the function approximation. Our work treats that problem from a practical point of view by doing real world experiments in vision based robotics. We show the relationship between net size and/or support size of a GBF network on the one hand and the quality of the function approximation on the other hand. Various configurations of GBF networks are applied for eye–hand coordination and for evaluation of grasping situations. Taking the experimental results into account we can configure appropriate GBF networks for vision based robot grasping.

## EYE–HAND COORDINATION

For grasping an object the end–effector of the robot manipulator has to be moved into a stable grasping pose. The desired pose (position and orientation) must be extracted from visual information which will be produced by two cameras. The camera system is arranged in an appropriate position and orientation for watching the scene (no physical connection to the robot). Taking stereo images and detecting the target object in the two images result in two twodimensional positions representing the centers of gravity (two 2D–vectors). The two positions are defined in the coordinate systems of the two cameras and will be combined in a single vector (4D–vector). On the other hand, the end–effector moves within a 3D working space which is defined in the basis coordinate system of the robot (the position of the end–effector is a 3D–vector). Thus we need a function transforming the object positions from the coordinate systems of the cameras to the cartesian coordinate system of the robot (4D–vector $\Longrightarrow$ 3D–vector).

The procedure to acquire that function, which determines the eye–hand coordination, is as follows. We make use of a training sample for learning a GBF network. First the set of GBFs must be configured, and second the combination factors of the GBFs are computed. We configure the set of GBFs by simply selecting certain elements from the training sample and using the input parts (4D–vectors) of the selected examples to define the centers (of the GBFs). The combination factors for GBFs are computed with the pseudo inverse technique, leading to least square errors between pre-specified and computed output values.

The prerequisite for running the learning procedure is the existence of a training sample. To get it, we take full advantage of the robot dexterity. The end–effector moves in the working space systematically, stops on equidistant places, and 3D–positions of the end–effector are carefully recorded. These 3D-vectors are simply provided by the control unit of the robot. Furthermore, at each stopping place an SSD–based (sum of squared distances) recognition algorithm detects the end–effector bend in the stereo images (see Figure 1) and the two twodimensional positions are combined to a 4D-vector. Alternative striking features, e.g. the end–effector tip, could be detected as well. All pairs of 4D–/3D–vectors are used as training sample for the desired eye–hand coordination.

Based on image coordinates of the end–effector bend the GBF network has to estimate its 3D position in the robot basis coordinate system. The mean 3D position error should be as low as possible. The main question of interest is: How many GBFs and which support sizes are needed to get a certain quality for the eye–hand coordination? To answer that question four experiments have been carried out. In the first and second experiment we applied two different numbers of GBFs exemplarily. The third experiment shows the effect of doubling the image resolution. Finally, the fourth experiment takes special care for training the combination weights of the GBFs. In all four experiments we systematical increase the GBF support size and measure the mean position error.

For each experiment we take a training sample. The working space of the end–effector underlying the sample is cube–shaped of maximum 300 millimeters (mm) side length. The GBFs will be spread over a subspace of 4D–vectors according to certain stopping places of the end–effector. That is, the 4D image coordinates resulting from the end–effector bend position at a certain stopping place are used for defining the center of a Gaussian. The following experiments differ wrt. the size and the use of the training sample. The evaluation of the resulting GBF network is based on a testing sample. It consists of input–output pairs from the same working space as above, but definitely the robot fingers moved in discrete steps of 20 mm. It is assured, that training and testing sample differ essentially and have only a small number of elements in common.

In the first experiment the manipulator moved in discrete steps of 50 mm through the working space resulting in a training sample of 7x7x7 = 343 elements. Every second example is used for defining a GBF (4x4x4 = 64 GBFs), and the whole training sample for computing the combination weights of the GBFs. The image resolution is set to 256x256 pixel. Figure 2 shows in curve (a) the course of mean position error for systematically increasing the support. As the GBFs become more and more overlapped the function approximation improves, and the mean position error decreases to a value of about 2.2 mm.

The second experiment differs from the first in that the manipulator moved in steps of 25 mm. Thus the training sample consists of 13x13x13 = 2197 examples and every second example is used for defining a GBF (7x7x7 = 343 GBFs). Figure 2 shows in curve (b) that the mean position error converges to 1.3 mm.

In the third experiment the same configuration was used as before, but the image resolution is doubled to 512x512 pixels. The accuracy of detecting the finger bend in the images increases, and the mean position error decreases further. Figure 2 shows in curve (c) the convergence to error value 1.0 mm.

The fourth experiment takes special care of both the training of weights and the testing of the resulting GBF network. It is obvious, that at the border of the working space there is only a one–sided overlap between GBFs. Therefore the quality of the function approximation can be improved, if a specific subset

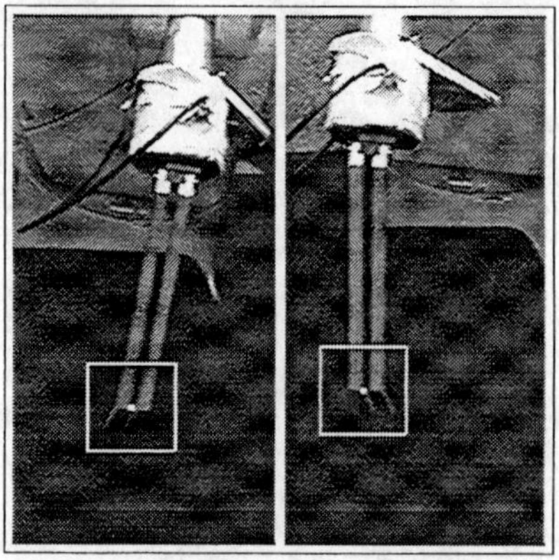

Figure 1: Detecting end–effector in stereo images.

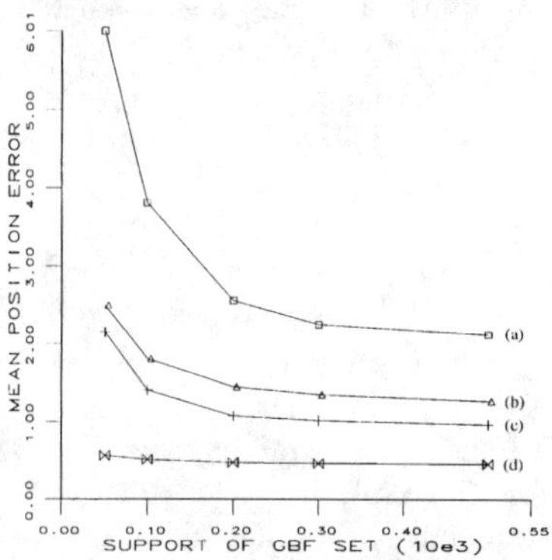


Figure 2: Courses of mean position error.

of 4D-/3D vectors located at the border of the working space will not be taken into account. In this experiment, the 343 GBFs are spread over the original working space as before, but for computing combination factors and testing the GBF network an inner working space of 250 side length is used. Figure 2 shows in curve (d) that the mean position error decreases to a value of 0.5 mm.

We used GBF networks to learn the mapping from image coordinates in stereo images into coordinates of a robot manipulator. The main advantage is that extrinsic and intrinsic camera parameters don't have to be determined. Rather we configure a GBF network appropriately, depending on the required accuracy (e.g. 1 mm). In order to approach the manipulator to a target object, we detect the target in stereo images, compute the centers of gravity, put the image coordinates into the GBF network, and compute the relevant robot coordinates.

## EVALUATION OF GRASPING SITUATION

Having the manipulator near the object we must fine-tune the pose of the robot fingers in order to stably grasp the target. Therefore the spatial arrangement between target and fingers have to be recognized and evaluated wrt. grasping stability. According to our vision based approach the image depiction of that arrangement is used to draw conlusions about the grasping stability. Figure 3 shows three images each depicting a target object, two bended robot fingers, and some other objects. On the left and the right the grasping situation is unstable because the horizontal part of the two parallel fingers is behind respectively in front of the target. The grasping situation in the middle image is most stable.

The grasping situations will be evaluated by applying a specific function to the contents of the relevant image area. To learn that evaluation function a GBF network is used. The input nodes describe the relevant image area, the hidden nodes represent a set of typical grasping situations, and the only output node computes the grasping stability. There are many approaches known in the computer vision literature for describing image contents. In this application we prefer

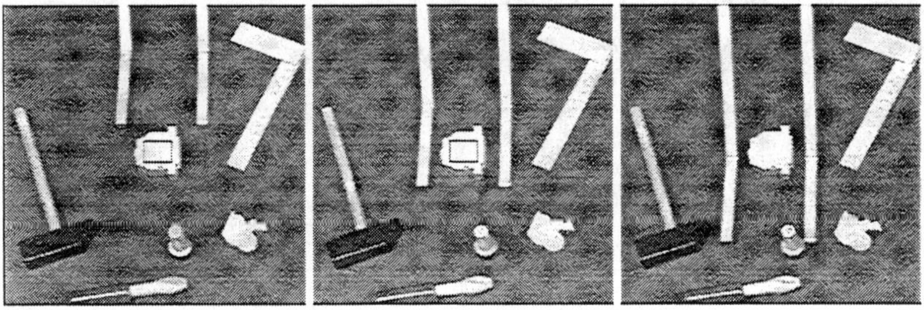

Figure 3: Three typical arrangements of target object and grasping fingers.

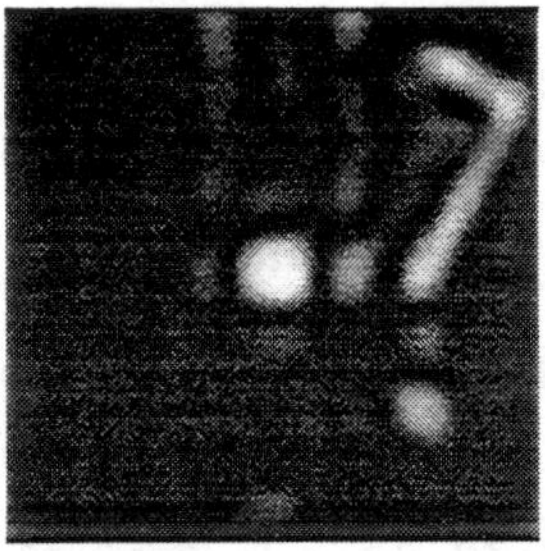

Figure 4: Response of Gabor filter.

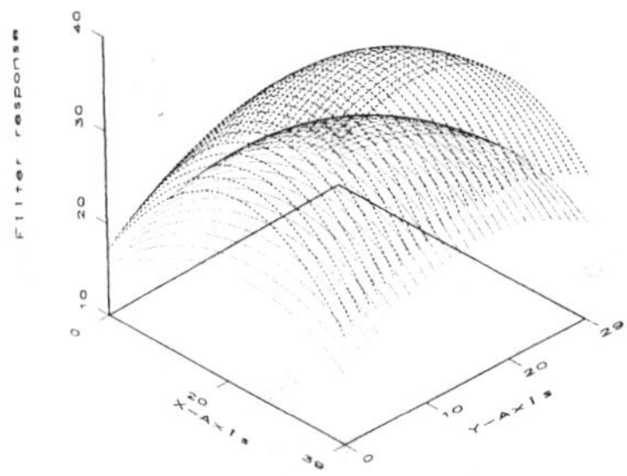

Figure 5: Overlay of two filter response patterns.

to use the response of a simple filter operation and avoid image segmentation. Thus we don't have to bridge the problematic gap between photometric gray level edges and geometric surface discontinuities (Maxwell and Shafer, 1994).

In line with this concept it would (!) be possible to take the raw gray levels and use directly the appearance patterns of grasping situations. Unfortunately these patterns are large–sized and so the efficiency of recognition is low. Therefore, we are interested in an image operator which concentrates the contents of a large image area into a smaller patch. The Gabor wavelet filter can be applied for this purpose (Pauli et al. 1995). For example, Figure 4 shows the response of an adequately parameterized Gabor filter applied to the left image in Figure 3. Rather than using appearance patterns we take so-called response patterns of a pre–defined size for evaluating the grasping situation. Specific arrangements consisting of the fingers and the target result in specific filter response patterns. For example, Figure 5 shows the overlay of two response patterns which result by applying the Gabor filter to the left and the middle image in Figure 3 and selecting the response of the (black) outlined rectangular area.

Unlike the simple application of GBF networks in eye–hand coordination, the dimension of the evaluation function (for grasping situations) is extremely high. Definitely, the dimension of the input space is equal to the number of pixels of the response patterns (the pre–defined size typically has several hundred pixels), and furthermore the GBFs have to be defined according to this dimension. The input space of the GBF network is the set of all possible response patterns of the pre–defined size. Each hidden node only responds significant for a certain subset of these patterns. The factors combining the GBFs encode values of grasping stability assigned to typical grasping situations. The output node computes the definite grasping stability for the grasping situation put into the input layer.

The approach for learning the evaluation function is as follows:
First, we take example images containing various grasping situations. Especially in our experiment the robot fingers will be moved step by step to the most stable grasping situation and step by step moved off afterwards. The movement is photographed in 25 discrete steps (Figure 3 shows three images thereof). Every second image will be used for training and the images in between for testing.
Second, we apply the Gabor filter to the training images and extract the rectangular area (of pre-specified size) describing the grasping situation. This training sample of response patterns is used for learning the evaluation function.
Third, according to the approach for learning a GBF network we have to cluster the response patterns with regard to similarity. The ISODATA clustering procedure is used for this purpose (Schalkoff, 1992, pp. 109-125).
Finally, we determine appropriate combination factors of the GBFs using the pseudo inverse technique. For that a set of pre-specified stability values must be assigned to the training sample. Considering the order in which the examples of grasping situations have been photographed, we define that the course of stability values should take the form of a bell–shaped parabolic curve. Therefore, the course of stability value for the ordered set of training examples increases systematically until the maximum is reached and decreases afterwards.

According to this approach a GBF network can be configured representing an evaluation function. Four experiments have been carried out by taking different numbers and/or support sizes of the GBFs. Figure 6 shows in curve (a) and (b) the course of stability values if we take six GBFs and a small respectively large support size. Alternatively, the curves (c) and (d) in Figure 7 depict the courses for 13 GBFs and a small respectively large support size. Curve (d) depicts the best approximation of the evaluation function.

Using such a GBF network for evaluating the grasping situation, the robot system automatically controls the manipulator in order to reach the optimal grasping stability.

## CONCLUSION

Our approach of vision based robotics uses GBF networks both for eye–hand coordination and for the evaluation of grasping situations. Furthermore, GBF networks can be used to learn operators for view independent object recognition (Pauli, 1996). In numerous experiments it was demonstrated how specific network configurations influence the accuracy of the function approximation. Depending on pre-specified limits for the accuracy the GBF networks can be trained appropriately and then used for online operation.

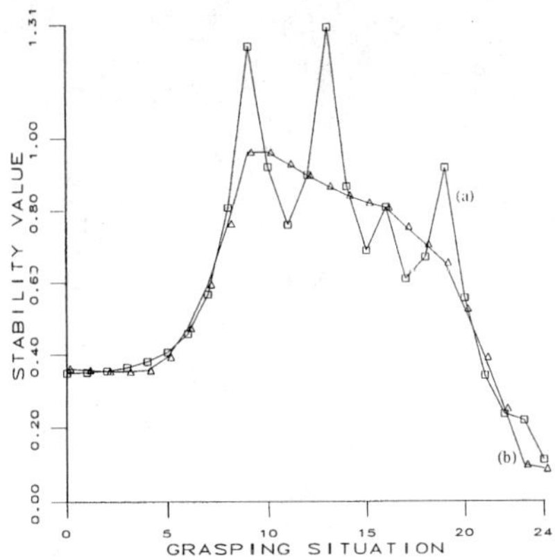


**Figure 6: Courses of grasping stability.**

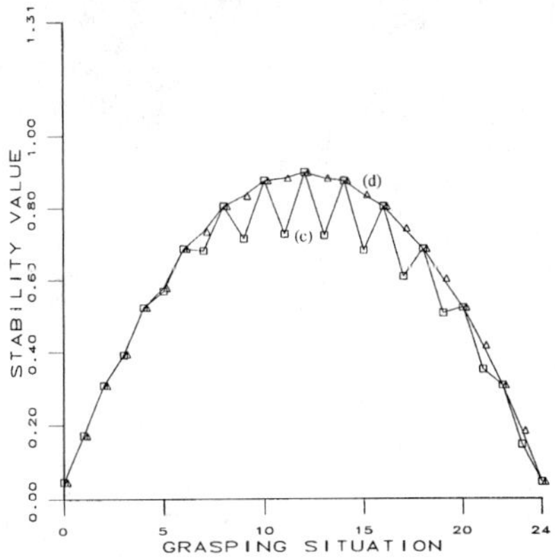


**Figure 7: Courses of grasping stability.**

## REFERENCES

Baum E., Haussler D., (1990). What net size gives valid generalization?, in Shavlik J., Dietterich T. (1990), Readings in Machine Learning, Morgan Kaufmann Publishers, 258-262.

Blumer A., Ehrenfeucht A., Haussler D. (1990). Occams' Razor, in Shavlik J., Dietterich T. (1990), see above, 201-204.

Floyd S., Warmuth M., (1995). Sample compression, learnability, Vapnik–Chervonenkis Dimension, Machine Learning Journal, Vol. 21(3), 269-304.

Maxwell B., Shafer S., (1994). A framework for segmentation using physical models of image formation, IEEE Conference on Computer Vision and Pattern Recognition, 361-368.

Pauli J., Benkwitz M., Sommer G., (1995). RBF networks for object recognition, in Krieg-Brückner B., Herwig Ch. (1990), Workshop Kognitive Robotik, Zentrum für Kognitionswissenschaften, Universität Bremen, Bericht 3/95.

Pauli J., (1996). View independent object recognition, British Machine Vision Conference, Edinburgh.

Poggio T., Girosi F., (1990). Networks for approximation and learning, Proceedings of the IEEE, Vol. 78, 1481-1497.

Rissanen J., (1984). Universal coding, information, prediction, and estimation, IEEE Transactions on Information Theory, Vol. 30(4), 629-636.

Schalkoff R. (1992). Pattern Recognition - Statistical, Structural, and Neural Approaches, John Wiley and Sons, New York.

# TUNING IMAGE FILTERS USING FEED-FORWARD NETWORKS

S. HARING[1], J.N. KOK[2], and M.A. VIERGEVER[3]

[1] *Department of Computer Science, Utrecht University*
*P.O. Box 80089, 3508 TB Utrecht, The Netherlands*
*E-mail s.haring@cs.ruu.nl*

[2] *Department of Computer Science, Leiden University*
[3] *Image Center Utrecht, Utrecht University Hospital*

***ABSTRACT:***
We present a method to automatically tune filters that are used to extract features from image data. In a standard image processing approach filters are fixed. The filters are embedded in a feed-forward network architecture. Tuning of the filter-parameters is achieved using the back-propagated error terms that are available for the input units of such a network. Details are provided for the Gaussian filter and its derivatives, but the approach is also applicable to Gabor filters, wavelets, etc. We will show that the proposed algorithm can find optimal parameter settings for filters—given a specific task.

## 1 INTRODUCTION

A central image processing task is *pixel classification*. Applications like image segmentation, texture recognition etc. are often based on pixel classification. First, features of the image pixels are extracted that are of importance for the application at interest. Following, the feature pattern representation of each pixel is input to a classifier, like a feed-forward network, to yield a classification of the pixel.

Feature extraction is usually done by convolving the image with specific filters. Performance of filters is strongly influenced by the settings of their parameters; only specific parameter-settings allow the filters to extract useful pixel properties. Parameters of a filter—like its width or orientation—are usually fixed and set manually on the basis of heuristics and/or expert knowledge. Time is saved, and better results are achieved if the filter parameters are tuned automatically to optimal values for the application at interest. In our approach image filters are incorporated in a feed-forward network, yielding automatic tuning of the filter parameters with a specific learning rule. In this manner the filter-parameters are "hidden" in the black box of the feed-forward network.

The main difference of our algorithm with respect to alternative methods to tune filter parameters is that our method is *supervised*, i.e. a task is defined that is to be accomplished. Other methods tune filters solely on the basis of the input image.

## 2 LEARNING RULE

We first derive the learning rule for a simple feature-extractor with only one parameter. The feature-extractor convolves an input image with a filter. Later we show how the learning rule can be extended to feature-extractors that (i) have more parameters and (ii) are more complex in the sense that they *combine* convolved images to one *feature-image*. (A feature-image is an image of which the grey values correspond to feature values; we can say that a feature-extractor transforms an input image to a feature-image.) The basic idea is to use the error terms that are available for the input units of the network to adapt the filter parameters.

In the following $I : \mathbb{Z}^2 \to \mathbb{R}$ denotes the input image. An $n$-parameter filter kernel with parameter vector $\vec{p}$ is defined by $F : P \times \mathbb{Z}^2 \to \mathbb{R}$, $(\vec{p}; \vec{x}) \mapsto F(\vec{p}; \vec{x})$, where $\vec{p} \in P \subseteq \mathbb{R}^n$.

A feature-image depends on the filter parameters that were applied. Hence a feature-image $J$ is a function of $\vec{p}$ like $F$; $J : P \times \mathbb{Z}^2 \to \mathbb{R}$, $(\vec{p}; \vec{x}) \mapsto J(\vec{p}; \vec{x})$.

Input image $I$ is transformed by $n$ feature-extractors to feature-images $J_i$, $i = 1, \ldots, n$. Index $i$ denotes that the feature image is a result of feature-extractor $i$. Each feature-extractor has a parameter vector $\vec{p}_i$. Repeatedly a pixel $I(\vec{x})$ is drawn from the input image; the pixel yields a pattern $\vec{y} = (J_1(\vec{p}_1; \vec{x}), \ldots, J_n(\vec{p}_n; \vec{x}))$ to be input to the feed-forward network. The output pattern $\vec{o}$ of the classifier yields the classification of the pixel. We assume that we know a target pattern $\vec{t}$ for all training pixels.

We derive the learning rule for *one* feature-extractor with *one* parameter (this architecture is illustrated in Figure 1). Hence we can leave out the indices $i$ and we may write $p$ instead of $\vec{p}$. The filter $F$ is applied to the image as follows:

$$J(p; \vec{x}) = \{I \otimes F(p; \cdot)\}(\vec{x})$$

where $\otimes$ denotes the convolution operation. Since only one feature-extractor is considered the input pattern of the classifier is a scalar: $y = J(p; \vec{x})$ The input layer consists of units denoted with an index $j$. The weights of the feed-forward network are modified using the standard back-propagation learning rule. Thus in unit $j$ a back-propagated delta-factor $\delta_j$ is available, and for the feature-extractor a delta-factor $\delta = \sum_j w_j \delta_j$. Error gradient descent with respect to $p$ gives:

$$\Delta p \quad \sim \quad -\frac{\partial \varepsilon}{\partial p} \quad = \quad \delta \frac{\partial J}{\partial p}(p; \vec{x}) \tag{1}$$

In this equation $\frac{\partial J}{\partial p}$ expresses the dependency of the feature-image with respect to the parameter of the filter; we coin this the *parameter derivative image.* The learning rule implies that, *after a pixel is drawn from position $\vec{x}$, parameter $p$ should be adapted proportional to $\delta$, and proportional to the grey value of the parameter derivative image at $\vec{x}$.*

The parameter derivative image is straightforwardly determined in the case of a simple feature-extractor:

$$\frac{\partial J}{\partial p}(p; \vec{x}) \quad = \quad \frac{\partial \{I \otimes F(p; \cdot)\}}{\partial p}(\vec{x}) \quad = \quad \left\{ I \otimes \frac{\partial F(p; \cdot)}{\partial p} \right\}(\vec{x}) \tag{2}$$

Substitution of Eq. 2 in Eq. 1 gives a complete learning rule for $p$. Note that in Eq. 3 is referred to $\frac{\partial F}{\partial p}$; this is the derivative of the filter kernel with respect to its parameter. With the learning rule we can iteratively adapt $p$ to gain optimal classification results. For every update we need a convolution, and after each update we need to process the input image with the new filter. In this way, the algorithm finds the right parameter setting of the filter for the specific application. In Section 3 we show an example of how Eq. 1 can be extended to more complex feature-extractors.

Because of the close resemblance between our learning-rule and back-propagation one might get the impression that the learning-rule is nothing more than just back-propagation. That is not so; parameters of filters are adapted, and not weights.

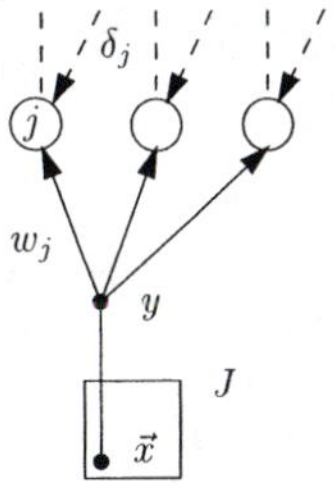


Figure 1: The architecture on which the derivation of the learning rule is based; there is one feature-image $J$ from which grey values $y = J(p; \vec{x})$ are input to the network. In each unit $j$ in the input layer there is a back-propagated delta-factor $\delta_j$ available.

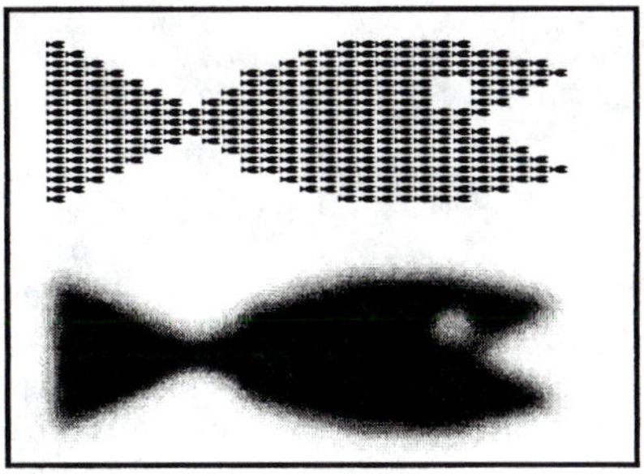

Figure 2: The effect of scale on the emergent structures in an image. The top image has a high resolution (small scale), whereas the bottom one has a lower resolution (larger scale).

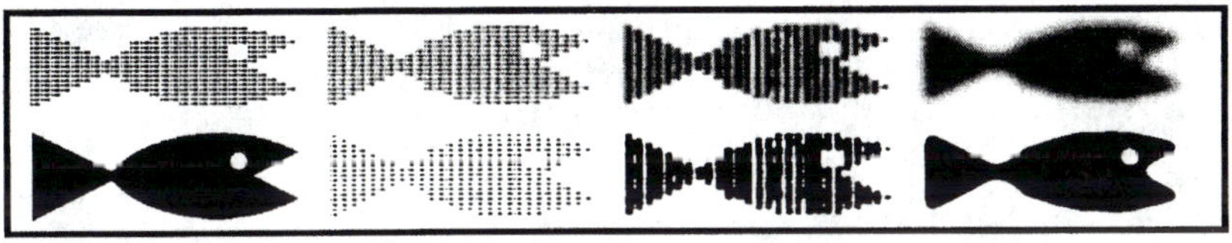

Figure 3: Left is the image, with its intended segmentation below. Then starting from left to right we see the feature-image (blurred image in this example) during learning, and below the classification made by the network. Initially the scale is too small and only the small fishes can be recognized but after a while the scale converges to a value such that the objective structure—the big fish—is found.

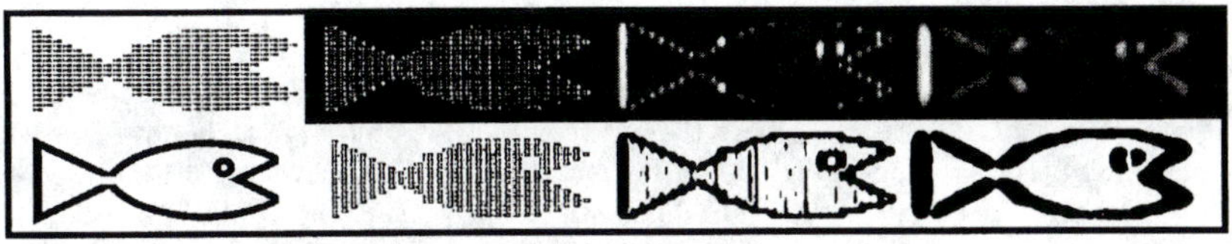

Figure 4: Left is the image, with its intended edge depicted below. Starting from left to right we see the feature-image (which expresses the squared norm of the gradient) during learning, and below the classification made by the network. One can observe that initially the edges of the individual fishes are found, but after learning the objective edge was identified.

Furthermore the filters are used as convolution kernels. Such functionality is not incorporated in a regular feed-forward network architecture, and the effect of this on the learning-rule is non-trivial.

## 3 SCALE SELECTION

The algorithm can be used to select the optimal scale for a specific image processing task (scale is the reciprocal of resolution and corresponds to the width of a filter). For example, the extent to which an image is blurred—i.e. the scale of a blurring filter—influences the structures that can be found in the image. And an edge-detector will only extract desired edges if a proper scale is applied.

We will consider two scaled feature-extractors, which follow from scale space theory (Florack, ter Haar Romeny, Koenderink, and Viergever 1993): (i) *blurring* an image—this happens by convolving the image with a Gaussian; (ii) *edge-detection*—this is done by convolving the image with derivatives of a Gaussian. For both feature-extractors we will derive learning-rules for scale $\sigma$.

**Blurring** an image with a Gaussian can be seen as a simple feature-extraction operation ("simple" in the same sense as explained in Section 2). The Gaussian removes noise; the larger the scale (or width) of the Gaussian, the more noise is removed, but also fine scale structures in the image are removed. For some applications it is required that a lot of noise is removed, and for other applications it is essential to leave small structures intact. Clearly the Gaussian has an optimal scale for a given application. Figure 2 illustrates how the emerging features in an image depend on the scale of the Gaussian. The Gaussian with scale $\sigma$ is defined by:

$$G : \mathbb{R}^+ \times \mathbb{Z}^2 \rightarrow \mathbb{R},\ G(\sigma; \vec{x}) \ = \ \frac{1}{2\pi\sigma^2} \exp\left(-\frac{\vec{x}\cdot\vec{x}}{2\sigma^2}\right)$$

To derive the learning rule for $\sigma$ we substitute the derivative of the Gaussian with respect to $\sigma$ in Eq. 3:

$$\frac{\partial G}{\partial \sigma}(\sigma; \vec{x}) = \left\{\frac{\vec{x}\cdot\vec{x} - 2\sigma^2}{\sigma^3}\right\} G(\sigma; \vec{x})$$

The parameter derivative image is the input image, convolved with this filter. We now have learning rule to adapt the scale of the Gaussian filter iteratively. How this learning rule allowed to find the coarse structure of the image in Figure 2 (i.e. the big fish) is illustrated in Figure 3. In the next section we will show how the process of segmenting medical images benefits from the learning-rule.

**Edge-detection** is done by a more complex feature-extractor. The feature that is extracted is the squared norm of the gradient of the grey-value "landscape". There is probably an edge at locations where the gradient is relatively large. The gradient is determined by convolving the image with derivatives of the Gaussian. In Figure 4 is illustrated how the edges that are found depend on the scale of the applied Gaussians; if a small scale edge-detector is applied to the top image of Figure 2, all the edges of the individual fishes will be found, whereas if a larger scale is used, the edge of the big fish can be obtained.

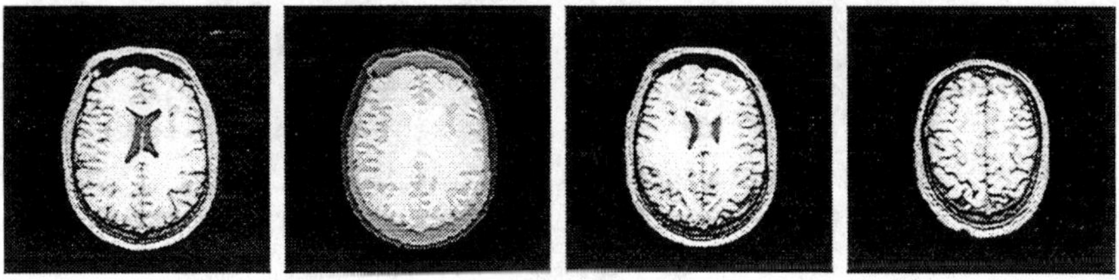

Figure 5: From left to right slices of a three-dimensional MRI of the head: the training image; the intended segmentation; and two test images.

In order to not have to write so many $\sigma$'s we use an extended definition of the convolution operator:

$$J = I \otimes G \Leftrightarrow \forall \sigma : J(\sigma; \vec{x}) = \{I \otimes G(\sigma; \cdot)\}(\vec{x})$$

Furthermore we use in the next formulae subscripts to denote partial derivatives (i.e. $G_x = \frac{\partial G}{\partial x}$).

The feature-image $J$ , which expresses the squared norm of the gradient, is as follows:

$$J = (I \otimes G_x)^2 + (I \otimes G_y)^2$$

And the parameter derivative image $J_\sigma$ is:

$$J_\sigma = 2(I \otimes G_x)(I \otimes G_{x\sigma}) + 2(I \otimes G_y)(I \otimes G_{y\sigma})$$

We now have the parameter derivative image for edge-detection, yielding a learning rule for the scale of an edge-detector. In Figure 4 is illustrated how the learning rule finds the coarse edge (i.e. the edge of the big fish).

# 4 APPLICATION

We have applied the algorithm for segmentation of medical images. The various segments are intended to the refer to the anatomical objects in the input image. In Figure 5 a slice of a three-dimensional MRI of the brain is illustrated, and besides the intended segmentation in which the various objects are depicted. We apply a multi scale (or multi-resolution) technique; the input image is convolved with various Gaussians with a different scale. In previous work we tuned the scales of the Gaussians manually (S.Haring, M.A.Viergever, and Kok 1994). With the algorithm proposed here, we can tune the scales automatically. Table 1 reveals that the automatically found scales yield better segmentation results than fixed scales. Furthermore, we need less Gaussians than before, to get these results. Note that the percentages presented in Table 1 are for the test images, which are shown on the right of Figure 5.

# 5 DISCUSSION AND CONCLUSION

The application reveals that our method (i) allows for better results, and (ii) saves time since. The method is not only applicable to scale selection; it can also be applied to many other kinds of filters like Gabor filters and wavelets. In the full paper we present results on these topics.

| | $I$ | $\sigma$ | % |
|---|---|---|---|
| manual | • | | 72 ± 4 |
| | • | 1 | 86 ± 2 |
| | • | 1, 2 | 84 ± 2 |
| | • | 1, 4 | 88 ± 2 |
| | • | 1, 8 | 86 ± 2 |
| | • | 1, 2, 4, 8 | 96 ± 1 |
| autom. | | 1.7, 7.3 | 97 ± 1 |

Table 1: The first six rows show results that are achieved with manually tuned scales. Note that in these experiments also the raw intensity ($I$) was input to the classifier as a feature. The third row shows the applied scale; in the last row the percentage of correctly classifed pixels in test images is presented. The last row shows how two automatically tuned scales outperform the manually tuned scales.

Methods that are related to ours are, amongst others, the *Adaptive-Subspace* SOM (ASSOM) (Kohonen 1995), *steerable filters* (Freeman and Adelson 1991), *Radial Basis Function* (RBF) networks (Moody and Darken 1989), and the *Neocognitron* (Fukushima 1980). The ASSOM is also used to tune filter parameters but it is an unsupervised approach based on the Self-Organizing Map. Steerable filters are less related to our work than one might expect from their name; the theory of steerable filters is about how filters with fixed orientations can be combined to new filters with arbitrary orientations. RBF networks differ from our method because in these networks the scales of Gaussian *transfer functions* are tuned, whereas in our method the Gaussians are used for *convolutions*. Finally, the Neocognitron does not tune parameters of a convolution kernel but adapts all the weights of a kernel.

Concluding, we have presented a fruitful synthesis of image processing theory and neural networks. The integration of convolution filters in a feed-forward network allows for a further benefit of the back-propagation concept. The parameters of the filters are included in the network and therefore, manual tuning of the filter's parameters is unnecessary any more. The algorithm is practical; one only needs to depict in an example image the structures that are to be found, and the algorithm will tune filters to allow for an optimal recognition of these structures, in the example image and in other images. In short, the algorithm facilitates automated setting of image filter parameters.

# References


Florack, L. M. J., B. M. ter Haar Romeny, J. J. Koenderink, and M. A. Viergever (1993). Cartesian differential invariants in scale-space. *Journal of Mathematical Image and Vision*.

Freeman, W. and E. Adelson (1991). The design and use of streerable filters. *IEEE Transactions on Pattern Analysis and Machine Intelligence 13*, 891–906.

Fukushima, K. (1980). Neocognitron: A self-organizing neural network model for a mechanism of pattern recognition unaffected by shift in position. *Biological Cybernetics 36*, 193–202.

Kohonen, T. (1995). The adaptive-subspace SOM (ASSOM) and its use for the implementation of invariant feature detextion. In *Proceedings of the International Conference on Artificial Neural Networks*, pp. 3–10.

Moody, J. and C. Darken (1989). Fast learning in networks of locally-tuned processing units. *Neural Computation 1*, 281–294.

S.Haring, M.A.Viergever, and J. Kok (1994). Kohonen networks for multiscale image seg-

# Unsupervised Cluster Discovery using Reconciliation

**LARRY K. BARRETT and PETER BOCK**
lawrence.barrett@ccmail.gsfc.nasa.gov and pbock@seas.gwu.edu
Department of Electrical Engineering and Computer Science
The George Washington University, Washington, DC 20052

***ABSTRACT:***
A new unsupervised method of deriving classes for a statistical learning system is presented. The method is based on the sequential discovery of clusters in a multi-dimensional feature-space histogram. The $R^3$ (R-cubed) algorithm iteratively recognizes, reconciles and removes parametrically defined clusters from the feature histogram, refining its parametric estimates on each pass. This process correctly identifies clusters even when they are significantly overlapped or when the histogram is corrupted by a moderate level of noise. Results from a preliminary version of the $R^3$ algorithm demonstrate its capabilities.

The problem of sensing and correctly classifying environmental inputs pre-dates recorded history, when the survival of our primitive ancestors totally depended on this capability. In that environment, external supervision (*e.g.*, the presence of a teacher) was not always possible. Thus, individuals had to develop adaptive and robust methods for classifying the elements of a complex and hostile environment or suffer the consequences of misclassification [Edelman 87]. Contemporary data classification has evolved but is still complex, with the results of misclassification no less costly. The diagnosis of tumors as either malignant or benign, the detection of explosives in luggage, and the intent of aircraft approaching a military installation (*e.g.*, a carrier group) are examples where misclassification can be extremely expensive.

Classification is the partitioning of the elements of a specified population into disjoint subsets such that the elements within each subset are more 'similar' to each other than are those between subsets [Fukunaga 90]. One of the goals of classification is therefore, to discover the underlying order or natural categories present in the overall population [Anderberg 73].

One prevalent method of extracting classification information is the discovery of clusters in a feature histogram created from training samples [Anderberg 73] [Duda and Hart 73] (*e.g.*, radial basis functions [Musavi 92]). Each cluster is presumed to represent a common class of data with a limited range of feature values. The specification of these clusters can be represented as elements of a finite mixture distribution [Titterington, Smith and Markov 85]. Because the clusters, and thus the classes they represent, can overlap each other in the feature-space, classification error is possible even when the classes are correctly specified. The overlapping of clusters also significantly increases the complexity of discovering each cluster in the feature-space as a result of mutual interference [Fukunaga 90].

Pioneering work in this area resulted from the development of a method to graphically (*i.e.*, manually) decompose a Gaussian mixture [Bhattacharya 67]. A version of this method was later automated [Gregor 69]. With the increased availability of computers, more computationally intensive methods that are aimed at discovering the unknown component density parameter values (*e.g.*, Method of Moments, Maximum-Likelihood, and EM Algorithm) have become more heavily used. The discovery of clusters has become the basis for several unsupervised machine learning methods [Matthews and Kearne 91] [Michalski and Stepp 83] [Wilson and Spann 90].

The weaknesses of existing unsupervised learning methods have helped clarify the need for specific algorithmic capabilities. For example, an unsupervised machine learning method at the sub-symbolic level should be able to support a general cluster structure and not just Gaussian clusters (as required by many techniques). The learning method must to able to determine the number of clusters in the feature histogram without being told *a priori*. It must be robust in the presence of sampling noise found in a typical histogram, without the need to corrupt the original density by filtering. Finally, the learning method must be able to locate and extract concealed clusters in the feature-space. The primary purpose of this paper is to present a new iterative cluster discovery and refinement algorithm to support unsupervised learning systems.

## APPROACH

An unsupervised learning method has been devised that provides the previously discussed algorithmic capabilities. The Recognize-Reconcile-Remove ($R^3$ or R-cubed) algorithm has been formulated to solve the unsupervised class discovery problem by combining the rigor of statistical analysis and pattern recognition with the flexibility of heuristic search and iterative refinement in order to discover the classes (*i.e.*, clusters) in a feature histogram. Its initial goal was to enable the Adaptive Learning for Image and Signal Analysis (ALISA) system to function in an unsupervised learning mode [Bock, *et al.* 92].

Once initialized, the $R^3$ algorithm iteratively executes each of its three primary processing phases: cluster recognition, cluster reconciliation and cluster removal. The cluster recognition phase is responsible for locating the next cluster in the histogram to be analyzed. In order to prevent previously discovered clusters from interfering in this process, a working copy of the histogram where all previously discovered clusters have been removed, referred to as the histogram residue, is used. The cluster recognition phase performs centroid location and parameter estimation functions. The centroid location function locates a new cluster by finding the global mode of the histogram residue. Because histogram values typically include sampling noise, a low-pass spatial filter is used to reduce the noise so that a more accurate assessment of the mode can be obtained. The filter is used only to locate the cluster centroid, not to estimate cluster parameters. The parameter estimation function approximates the values of cluster parameters. As defined, the $R^3$ algorithm does not place any restrictions on the parametric form that must be used to define a cluster. Currently, the initial implementation of the $R^3$ algorithm has been limited to the discovery of Gaussian clusters in the residue.

The second processing phase of the $R^3$ algorithm reconciles the new cluster with those that have already been discovered within the histogram. Reconciliation is the process in which the parameters of all previously discovered clusters are reassessed in light of the discovery of a new cluster in the feature-space. This step is needed to account for the fact that clusters can overlap in the feature-space, and thus, significantly affect each others statistics (*i.e.*, parameters). The cluster reconciliation phase performs conflict set creation and multi cluster parameter optimization functions. The conflict set creation function identifies those clusters that overlap, and thus, could mutually affect each others parameters. The multi-cluster parameter optimization function determines updated parameter values by performing a metric-driven search of the parameter spaces of the conflict set clusters. The goal is to find a combination of parameter values that minimizes the difference (*i.e.*, squared-error) between the restored residue and the proposed mixture density as defined by the conflict set. The $R^3$ algorithm does not place any constraints on the search process to be used. In the current version a gradient ascent search is used; however, replacement of this approach with a genetic algorithm is being explored.

The third processing phase of the $R^3$ algorithm is the generation of an updated residue from the original histogram and updated cluster parameters. An updated residue allows the cluster recognition phase to find clusters that are concealed within larger ones. The cluster removal process performs parameter validation and residue generation functions. Specifically, special checks are performed to ensure that clusters whose amplitudes approach zero (as a result of reconciliation) are removed from the list of discovered clusters. The residue generation function then produces an updated histogram residue by subtracting all discovered component densities (using updated parameter values) from the original histogram.

The halting criteria established by the $R^3$ algorithm determine when sufficient information has been extracted from the original histogram. Until the halting criteria are met, the $R^3$ algorithm iterates to discover additional clusters in the histogram residue; otherwise the algorithm terminates and all discovered clusters are reported. A generic halting criterion for cluster discovery algorithms has been defined and is being explored as part of the $R^3$ algorithm analysis. However, it is not discussed further in this paper.

## $R^3$ PERFORMANCE EVALUATION

The preliminary version of the $R^3$ algorithm has successfully discovered clusters in multi-dimensional feature histograms. The following sections describe the discovery process both when the clusters in one- and two-dimensional histograms are overlapped and when those histogram values are corrupted with noise. All clusters in the feature histograms are Gaussian in structure.

### One-Dimensional Results

The first test of the preliminary $R^3$ algorithm was to determine its ability to discover a combination of both separated and overlapped clusters in a noiseless feature histogram. A mixture density was generated from the three components. Table 1 summarizes the results of a typical cluster discovery process by presenting the actual cluster parameters (amplitude, mean, and standard deviation) used to generate the histogram, those estimated by the cluster recognition phase prior to reconciliation,

TABLE 1: ONE-DIMENSIONAL HISTOGRAM RESULTS WITHOUT NOISE

| Cluster | Actual Values | | | Pre-Reconciled | | | Reconciled | | |
|---|---|---|---|---|---|---|---|---|---|
| | Amp | Mean | Std | Amp | Mean | Std | Amp | Mean | Std |
| 1 | 0.50 | 21 | 6 | 0.50 | 21 | 6 | 0.50 | 21 | 6 |
| 2 | 0.42 | 65 | 10 | 0.52 | 63 | 12 | 0.42 | 65 | 10 |
| 3 | 0.08 | 50 | 5 | 0.01 | 49 | 2 | 0.08 | 50 | 5 |

TABLE 2: ONE-DIMENSIONAL HISTOGRAM RESULTS WITH NOISE

| Cluster | Actual Values | | | Pre-Reconciled | | | Reconciled | | |
|---|---|---|---|---|---|---|---|---|---|
| | Amp | Mean | Std | Amp | Mean | Std | Amp | Mean | Std |
| 1 | 0.50 | 21 | 6 | 0.50 | 21 | 6 | 0.50 | 21 | 6 |
| 2 | 0.42 | 65 | 10 | 0.52 | 65 | 12 | 0.35 | 67 | 9 |
| 3 | 0.08 | 50 | 5 | 0.01 | 50 | 1 | 0.15 | 52 | 7 |
| *Squared Error* | *3.889* | | | *4.063* | | | *3.886* | | |

and the same parameter estimates after reconciliation. Note that although the parameter estimates for clusters 2 and 3 deviate slightly from the known values the reconciliation phase of $R^3$ successfully corrects this discrepancy.

The second test of the $R^3$ algorithm repeated the previous experiment, but added sampling noise to the histogram. The same processing sequence was performed by $R^3$ as it discovered the three clusters in the feature histogram. Table 2 summarizes the results of this process for the histogram with sampling noise. Because the reconciled parameters do not match those of the original mixture density, the total squared error between each set of cluster definitions and the histogram with noise is provided for comparison. Note that the reconciled clusters fit the histogram as well as the original mixture density that was used to create the histogram.

**Two-Dimensional Results**

The objective of the two-dimensional test of the $R^3$ algorithm was to verify that it was able to discover both separated and overlapped clusters in a multi-dimensional feature histogram. Again, this initial test was to verify the ability of $R^3$ to successfully discover non-overlapping clusters in the histogram. For simplicity, only two clusters were included in the histogram.

A histogram containing the two clusters was input to the $R^3$ algorithm, and both clusters were sequentially recognized and removed from the histogram. Because these clusters did not overlap, the reconciliation process was not performed and the algorithm terminated. An analysis of the results showed that the parameter values for the discovered clusters *were the same as* those of the components used to generate the histogram.

$R^3$ was then applied against a histogram containing two overlapping clusters, as depicted in Figure 1. This histogram is presented as a contour plot; the highest point in this uni-modal histogram is labeled as "centroid 1". The parameter values for the overlapping clusters are presented in Table 3. Each cluster was sequentially discovered by the $R^3$ algorithm. Because the two clusters overlap, the $R^3$ algorithm

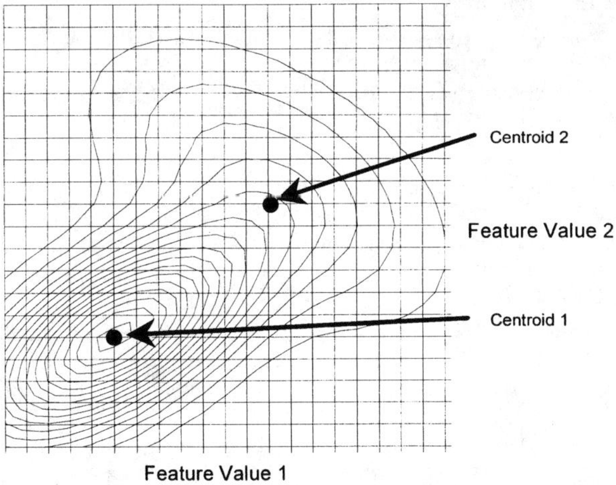


**Figure 1: Two-Dimensional Histogram with Overlapped Clusters**

**TABLE 3: 2-D OVERLAPPED CLUSTERS PARAMETERS (WITHOUT NOISE)**

| Cluster | Actual | | | | Discovered | | | |
|---|---|---|---|---|---|---|---|---|
| | Mixing Weight | Mean (x,y) | Std (x,y) | Correl | Mixing Weight | Mean (x,y) | Std (x,y) | Correl |
| 1 | 0.70 | 6,6 | 3,1 | 0.6 | 0.70 | 5,5 | 3,4 | 0.6 |
| 2 | 0.30 | 12,12 | 4,5 | -0.5 | 0.30 | 12,12 | 4,5 | -0.5 |

**TABLE 4: 2-D CLUSTERS BEFORE AND AFTER RECONCILIATION (WITH NOISE)**

| Cluster | Before | | | | After | | | |
|---|---|---|---|---|---|---|---|---|
| | Mixing Weight | Mean (x,y) | Std (x,y) | Correl | Mixing Weight | Mean (x,y) | Std (x,y) | Correl |
| 1 | 0.76 | 5,5 | 3,3 | 0.6 | 0.70 | 5,5 | 3,4 | 0.6 |
| 2 | 0.14 | 12,12 | 3,4 | -0.3 | 0.30 | 12,12 | 4,5 | -0.5 |

performed reconciliation processing. Table 3 also presents the cluster parameters reported at the conclusion of the reconciliation process.

The last test of the $R^3$ algorithm is a repeat of the two-dimensional, overlapping clusters with sampling noise altering the surface of the histogram. The standard processing sequence was performed by $R^3$ and it discovered the two clusters in the feature histogram. Table 4 provides the cluster parameter values produced by $R^3$ before the reconciliation process, as well as the values for the same parameters after reconciliation. Note that in spite of the introduction of sampling noise, the reconciled cluster parameters *are the same as* the original mixture density used to create the histogram.

**Summary of Results and Ongoing Work**

The preliminary $R^3$ algorithm successfully demonstrated its ability to discover clusters in one- and two-dimensional histograms both with and without sampling noise. Since no major conceptual changes are needed to handle higher-dimensional feature histograms, the algorithm is expected to perform equally well under those conditions.

In addition, the $R^3$ algorithm displayed its ability to discover clusters that are separated in the feature-space, even when the feature histogram was intentionally contaminated with a moderate level of sampling noise. The algorithm produced acceptable results without the need to filter, and thus corrupt, the feature histogram.

Finally, the $R^3$ algorithm was able to discover multiple clusters in uni-modal segments of the feature histogram. The reconciliation process has shown the ability to compensate for the statistical interference that occurs when two clusters overlap. This represents a significant achievement since no other technique known to the authors reports to have this capability.

An enhanced version of the $R^3$ algorithm is currently under evaluation. Preliminary results indicate that the algorithm performs as well in three and four dimensions as the preliminary version that is discussed in this paper does at lower dimensions.

## REFERENCES


Anderberg, M.R., *Cluster Analysis for Applications*, Academic Press, New York, 1973.

Bhattacharya, C.G., A simple method of resolution of a distribution into Gaussian components, *Biometrics 23*, 115-135

Bock, P., Klinnert, R., Kober, R., Rovner, R.M., and Schmidt, H., Gray-Scale ALIAS, *IEEE Transactions on Knowledge and Data Engineering*, Vol. 4, No. 2, April 1992, 109-122

Duda, R.O. and Hart, P.E., *Pattern Classification and Scene Analysis*, John Wiley and Sons, New York, 1973

Edelman, G.M., *Neural Darwinism*, Basic Books, New York, 1987

Fukunaga, K., *Introduction to Statistical Pattern Recognition*, Academic Press, San Diego, CA, 1990

Gregor, J., An algorithm for the decomposition of a distribution into Gaussian components, *Biometrics*, March 1969, pp 79-93

Matthews, G., and Kearne, J., Clustering Without a Metric, *IEEE Transaction on Pattern Analysis and Machine Intelligence*, Vol. 13, No. 2, February 1991

Michalski, R.S., and Stepp, R.E., Learning from Observation: Conceptual clustering, in *Machine Learning, An Artificial Intelligence Approach*, Morgan Kaufman, Los Altos, CA, 1983, pp 331-363

Musavi, M.T. Ahmed, W., Chan, K.H., Faris, K.B., and Hummels, D.M., On the training of radial basis function classifiers, *Neural Networks*, Vol.5, pp 595-603, 1992

Titterington, D.M., Smith, A.F.M., and Markov, U.E., *Statistical Analysis of Finite Mixture Distributions*, John Wiley and Sons, New York, 1985

Wilson, R., and Spann, M., A New Approach to Clustering, *Pattern Recognition*, Vol. 23, No. 12, 1413-1425, 1990

# APPROPRIATE SCALES WHEN USING WAVELETS FOR FEATURE EXTRACTION

L. M. BRUCE, R.R. ADHAMI, J.W. BRUCE
*University of Alabama in Huntsville*

***ABSTRACT*:**
The wavelet transform can be used effectively in conjunction with neural networks and fuzzy logic systems for pattern recognition. One method of using wavelets for the extraction of pattern features is the wavelet transform modulus-maximus (mod-max) method. This method has been shown to be effective for extracting features of sharp signal variations for the use in pattern recognition systems. However, the method assumes that the sharp signal variations are "isolated". The isolation assumption may or may not hold for many applications depending on its definition. This paper explains the motivation for the isolation assumption and the definition "isolation" for adjacent signal variations. This paper also discusses how this definition places limits on the range of scales to be used. Finally, this paper outlines the mod-max method for use with neural networks.

## INTRODUCTION

The wavelet transform is an expanding and evolving tool in the area of signal processing, particularly in the areas of compression, enhancement and recognition. The wavelet transform can be used effectively in conjunction with neural networks and fuzzy logic systems for pattern recognition. One method of using wavelets for the extraction of patterns is the wavelet-transform modulus-maximus (mod-max) method (Mallat, 1992). This multiresolution method has been used recently, along with neural networks for multiresolution recognition of handprinted characters (Wunsch, 1995). The mod-max method has been used with fuzzy logic for the automatic detection of dominant points on digital curve (Pei, 1992). The mod-max method has been used for feature extraction in an automated mammographic lesion classification system (Bruce, 1996).

In the systems by Wunsch and Laine (Wunsch, 1995), Pei and Lin (Pei, 1992) and other (Alter-Gartenberg, 1994), (Lu, 1994), (Velho, 1995), the mod-max method is used to recognize sharp signal variations. With this use of the mod-max method, the signal variations are assumed to be "isolated". The isolation assumption may or may not hold for many applications depending on its definition. This paper discussed the motivation for isolation assumption and the definition of "isolated" for adjacent, sharp signal variations (Bruce, 1996).

For a set minimum separation distance of adjacent variations, the isolation definition places limits on the uses of the mod-max method. This method utilizes various scales of the wavelet transform The isolation definition places an upper bound on the scales of the wavelet transform. The use of scales greater than this upper bound can cause the mod-max method to produce inaccurate results. That is, the mod-max method is used to extract features describing the sharp signal varieties. By using scales greater than those associated with isolated variations, the extracted features may not accurately describe the variations.

This paper is organized as follows: Section I is a brief introduction. Section II provides the mathematical background and notation for the wavelet transform mod-max method of muti-resolutional feature extraction. Section III discussed the isolation assumption and the definition of "isolated" for adjacent signal variations. Section IV outlines the use of the mod-max method with neural networks. Section V provides the concluding remarks.

## BACKGROUND

In this section, we provide a brief overview of the wavelet transform mod-max method and associated notation. For a detailed description of this method see (Mallat, 1992). This method is used for extracting features that numerically describe sharp signal variations. The method is based on the dyadic wavelet transform with the wavelet function being the first derivative of a Gaussian (Babaud, 1986), (Bruce, 1996), (Mallat, 1992). The continuous wavelet transform of a function $f(x)$ is defined by the convolution product

$$W_s f(x) = f(x) * \psi_s(x) \tag{1}$$

at scale $s$. $\psi_s(x)$ is a scaled version of the wavelet function $\psi(x)$. The scaling is defined as

$$\psi_s(x) = \frac{1}{\sqrt{s}}\psi\left(\frac{x}{s}\right). \tag{2}$$

The discrete dyadic wavelet transform (DDWT) is defined as the sequence of discrete signals

$$\left\{S_{2^J} f, \left(W_{2^j} f\right)_{1 \le j \le J}\right\}, \tag{3}$$

where the fine signal components $W_{2^j} f(x)$ is the wavelet transform at dyadic scales (powers of 2) and the coarse signal $S_{2^J} f(x)$ provides the signal components below the scale $2^J$.

The mod-max representation of a signal is obtained by detecting the local extrema of the DDWT at each scale $2^j$. The values are then set to zero if the location does not coincide with a local extrema. Thus the mod-max representation is defined as

$$\left\{ \overline{S}_{2^J} f,\ \left( \overline{W}_{2^j} f \right)_{1 \le j \le J} \right\},\tag{4}$$

where

$$\overline{S}_{2^J} f(x) = \begin{cases} S_{2^J} f(x), & \text{where } \left| S_{2^J} f(x) \right| \text{ is a local maxima} \\ 0, & \text{otherwise} \end{cases}\tag{5}$$

and

$$\overline{W}_{2^j} f(x) = \begin{cases} W_{2^j} f(x), & \text{where } \left| W_{2^j} f(x) \right| \text{ is a local maxima} \\ 0, & \text{otherwise} \end{cases}.\tag{6}$$

The mod-max representation can be used for various applications, but when being used to characterize sharp signal variations, the choice of the wavelet function is crucial. For the remainder of this paper, the wavelet function will be assumed to be the first derivative of a zero-mean Gaussian defined as

$$\psi(x) = -\frac{x}{\sigma^2} \cdot \frac{1}{\sigma\sqrt{2\pi}} e^{\frac{-x^2}{2\sigma^2}}.\tag{7}$$

The mod-max method utilizes the mod-max representation to compute the numerical descriptors of sharp signal variations. The development of the mod-max method is described in detail in (Mallat, 1992). The method is based on the fact that a sharp variation in the input signal will produce an extremum in each scale of the mod-max representation. Thus, due to a sharp signal variation, a sequence of extrema will appear in the mod-max representation. See Figure 1. This sequence will propagate from the scale $2^I$ to the scale $2^1$. At the scale $2^1$, the sequence converges to the location $x_o$ which is the location of the sharp variation. Using this sequence of extrema, the mod-max method can be used to extract features which characterize the signal variation.

One method of extracting these features is given here. Let $a_j$ be the local extrema value of the wavelet transform at scale $2^j$. By this, we mean that in a given neighborhood of $x_o$, we assume that the wavelet transform is smaller than $a_j$. Based on this assumption, three numerical descriptors of the variation located at $x_o$ may be computed. The descriptors are i) $K$, the amplitude of the sharp variation; ii) $(s_1\sigma)$, the sharpness of the variation; and iii) $\alpha$, the Lipshitz order of the variation. Note that smooth variations are modeled as singularities convolved with a Gaussian. The Gaussian function used to model the smooth variation has zero mean and a variance of $(s_1\sigma)^2$, where $\sigma^2$ is the variance of the Gaussian wavelet function.

One example method of obtaining these numerical descriptors is by minimizing the following (Mallat, 1992),

$$\sum_{j=1}^{I} \left( \log_2 |a_j| - \log_2 (K) - j - \frac{\alpha - 1}{2} \log_2 ((s_1\sigma)^2 + 2^{2j}) \right)^2,\tag{8}$$

where $a_j$ is the local extremum value of the wavelet transform at scale $2^j$. See Figure 1. The computation can be done with the steepest gradient descent algorithm. Mallat *et al* state that this analysis is valid only for isolated singularities (Mallat, 1992). However, the term "isolated" was not defined clearly. The isolation assumption was investigated and clearly defined in (Bruce, 1996).

## ISOLATION ASSUMPTION

From the previous section, we see that the mod-max method can be used to compute numerical descriptors of sharp signal variations. The mod-max method depends on the ability to accurately track wavelet transform extrema, due to a specific signal variation, across a set of given scales. The scales vary from $2^1$ to $2^I$. By incorporating information from more scales, the approximations of the numerical descriptors are of a greater accuracy. However, at large scales, adjacent signal variations begin to "blur" together. That is, the extrema due to two adjacent variations merge into one extrema. See Figure 1. Since the mod-max method needs to track extrema due to individual signal variations, the merging of adjacent extrema can cause the method to produce inaccurate results. Thus, limits need to be set on the range of scales to be used with the mod-max method.

Limits have been set on the appropriate range of scales to be used with the mod-max method (Bruce, 1996). These limits vary depending on the separation distance of adjacent signal variations ($x_o$), the variance of the Gaussian wavelet function ($\sigma$), and the truncation length of the Gaussian wavelet function ($k$). For a particular range of scales, two adjacent signal variations are defined as "isolated" when they can be accurately tracked across the mod-max representation.

In defining isolation for adjacent signal variations, three classes of isolation were formed: i) total isolation; ii) semi-isolation; and iii) non-isolation (Bruce, 1996). Total isolation refers to the case where the adjacent signal variations cause no interference. Semi-isolation refers to the case where the adjacent signal variations cause interference, but this interference does not affect the mod-max propagation method for tracking extrema across scales. Non-isolation refers to the case where the adjacent signal variations cannot be accurately characterized by the mod-max method. For two adjacent signal singularities, the range of scales for each of these classes of isolation are given in Table 1. a.

Table 1. a. provides the appropriate range of scales to use for signal singularities with the mod-max method. However, in many applications, the signal features that need to be recognized are smooth variations rather than discontinuities. By modeling smooth variations as singularities convolved with a Gaussian, the three classes of isolation were defined for smooth variations (Bruce, 1996). The Gaussian function used to model the smooth variation has zero mean and a variance of $(s_1\sigma)^2$, where $\sigma^2$ is the variance of the Gaussian wavelet function. The definition of isolation for smooth variations is given in Table 1. b.

### NEURAL NETWORK APPLICATIONS

In the previous sections, we have given an overview of the wavelet transform mod-max representation and its use for extracting feature that numerically describe sharp signal variations. This representation has been used in various applications of pattern recognition. In (Pei, 1992), the method was used to detect dominant points on digital curves. In (Alter-Gartenberg, 1994), edge primitives were extracted for image compression. In (Velho, 1995), the mod-max method was used for extracting shape features. In these applications, observations were made concerning the "inaccuracies" of the results when adjacent signal variations were too close. The limits on the range of appropriate scales discussed in Section III could be used to overcome these shortcomings.

In (Wunsch, 1995), neural networks were used to form a multiresolutional classifier for recognizing handwritten characters The use of the mod-max method with neural networks can take many forms. The information from each scale can be the input to a neural network. The extrema information across a given set of scales for each signal variation can be the input to a neural network. Or the signal variation numerical descriptors ($K$, $s_1\sigma$, $\alpha$) can be the input to a neural network system.

Whichever design of the neural network system is chosen, the assumption of isolated signal variations must be taken into consideration. If the application requires the recognition of closely spaced signal variations, an upper bound should be placed on the utilized wavelet transform scales. This bound should be set according to the ranges specified in Table 1. On the other hand, the application may require the recognition of greatly separated signal variations. This case could be related to the idea of using wavelets to analyze the coarse or global structure of the input pattern. For this situation, a lower bound should be placed on the utilized wavelet transform scales according to Table 1.

## CONCLUSIONS

The wavelet transform can be used effectively with neural networks for pattern recognition systems. One example of the use of wavelets for pattern recognition is the mod-max method. This method uses the wavelet-transform mod-max representation to extract features that numerically describe sharp signal variations. The method assumes that the characterized signal variations are isolated. This assumption places restrictions on the range of scales that are appropriate for used with the mod-max method. When using the mod-max method with neural networks to form pattern recognition systems, these limits on the applied scales must be observed.

## REFERENCES


Alter-Gartenberg, R., Huck, F.O. and Narayanswamy, R., (1994). Compact image representation by edge primitives, *CVGIP: Graphical Models and Image Processing*, vol. 56, no. 1, pp. 1-7.

Babaud, J., Witkin, A., Baudin, M. and Duda, R. (1986). Uniqueness of the Gaussian kernel for scale-space filtering, *IEEE Trans. Patt. Anal. Machine Intell.*, vol. 8, no. 1, pp. 26-33.

Bruce, L.M., (1996). The use of multiresolutional analysis for automated recognition of non-isolated patterns, Ph.D. dissertation, Dept. of Elec. & Comp. Eng., The University of Alabama in Huntsville.

Lu, J., Healy, D.M.,Jr. and Weaver, J.B., (1992). Contrast enhancement of medical images using multiscale edge representation, *Optical Engineering*, vol. 33, no. 7, pp. 2151-2161.

Mallat, S. and Zhong, S., (1992). Characterization of signals from multiscale edges, *IEEE Trans. Patt. Anal. Machine Intell.*, vol. 14, no. 7, pp. 710-732.

Pei, S. and Lin, C., (1992). The detection of dominant points on digital curves by scale-space filtering, *Pattern Recognition*, vol. 25, no. 11, pp. 1307-1314.

Velho, L., Terzopoulos, D. and Gomes, J., (1995). Constructing implicit shape models from boundary data, *Computer Vision and Image Processing*, vol. 57, no. 3, pp. 220-234.

Wunsch, P. and Laine, A.F., (1995). Wavelet descriptors for multiresolution recognition of handprinted characters," *Pattern Recognition*, vol. 28, no. 8, pp. 1237-1249.

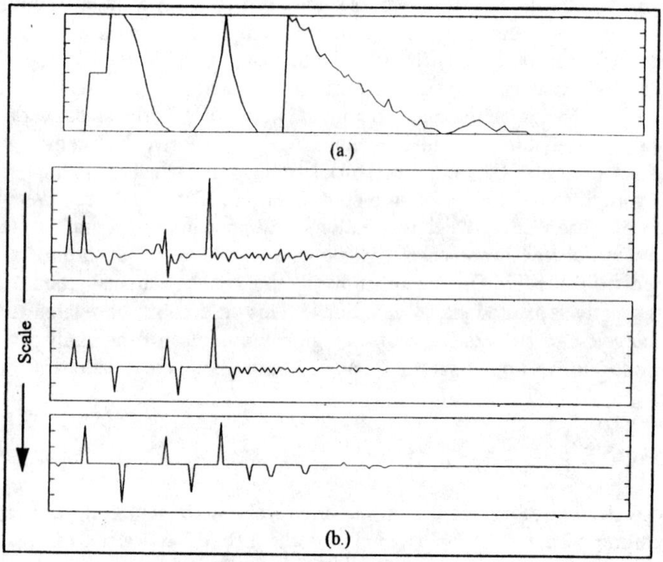


Figure 1. (a.) One-dimensional test signal, f(x),
(b.) Mod-max representation of f(x) for various scales.

| | Total Isolation | Semi-Isolation | Non-Isolation |
|---|---|---|---|
| (a.) Singularities | $s \le \frac{x_o}{2k}$ | $\frac{x_o}{2k} \le s \le \frac{x_o}{k+\sigma}$ | $s \ge \frac{x_o}{k+\sigma}$ |
| (b.) Smooth Variations | $s \le \sqrt{\left(\frac{x_o}{2k}\right)^2 - s_1^2}$ | $\sqrt{\left(\frac{x_o}{2k}\right)^2 - s_1^2} \le s \le \sqrt{\left(\frac{x_o}{k+\sigma}\right)^2 - s_1^2}$ | $s \ge \sqrt{\left(\frac{x_o}{k+\sigma}\right)^2 - s_1^2}$ |

Table 1. Summary of isolation conditions for adjacent (a.) singularities and (b.) smooth variations separated by a distance of $x_o$. The Gaussian-derivative wavelet is windowed to have length *2k*. Smooth variations have variance of $(s_1\sigma)^2$.

# A FUZZY LOGIC CLASSIFIER FOR THE DETECTION OF BOUNDED WEAK ECHO REGIONS IN METEOROLOGICAL IMAGES


**V Lakshmanan**[1]
National Severe Storms Laboratory & University of Oklahoma
Norman, Oklahoma 73069
**Arthur Witt**[2]
National Severe Storms Laboratory, Norman, Oklahoma 73069



## ABSTRACT

The presence or absence of a BWER within a storm is important for severe weather forecasting. The three dimensional structure of a BWER is often not clearly defined in radar data and a fuzzy logic approach where the various uncertainties associated with a BWER's radar profile (capping, three-dimensional structure, vertical height, reflectivity ranges, proximity to a mesocyclone, etc.) are taken into account leads to a reliable BWER detection scheme [Lakshmanan and Witt (1996)].

The BWER detection scheme, however, provided crisp output in the form of BWERs and non-BWERs while meteorologists provide a fuzzy description, identifying some structures as "marginal" BWERs. The classifier described in this paper identifies BWERs in a way similar to that of a human observer of the same weather phenomena. We discuss the limitations of classification done by thresholding a weighted sum and introduce a fuzzy logic classifier, comparing its performance with that of the weighted-sum thresholder used in [Lakshmanan and Witt (1996)].



**Keyword: Feature Identification And Classification.** Also: fuzzy logic, remote sensing


## 1 INTRODUCTION

Tornadoes and very large hail are commonly observed with supercell severe thunderstorms. One radar feature often associated with supercells is the Bounded Weak Echo Region (BWER). BWERs (see Figure 1 [Lemon (1980)]) are representative of local storms that develop in a strongly sheared environment and which tend to a steady state circulation [Browning and Donaldson Jr (1963)]. They are usually indicative of a high speed updraft and the presence or absence of a BWER within a storm is important information for a weather forecaster.

There are several reasons why the radar signature of a BWER rarely appears as clear cut as would be expected from the structure indicated in Figure 1. Since the area scanned by a radar increases the farther away one gets from the radar, a higher proportion of the storms analyzed will be located at fairly distant ranges from the radar. As the distance between the radar and the storm increases, the ability of the radar to properly sample small- scale features within the storm, such as the BWER, becomes more difficult, since the radar sampling volume becomes larger. Also, as distance increases, the lowest part of the storm that the radar can scan is at increasingly higher altitudes. Another problem can occur with rapidly moving storms. In the time it takes for the radar to scan upward through the storm, the higher altitude capping region of the BWER may have moved (with the storm) such that it no longer is located over the relative reflectivity minimum, detected at a lower altitude. There is also an error associated with vertical height as measured by the radar [Howard and Gourley (1995)] and this error varies with weather conditions.


[1] lakshman@nssl.uoknor.edu
[2] witt@nssl.uoknor.edu

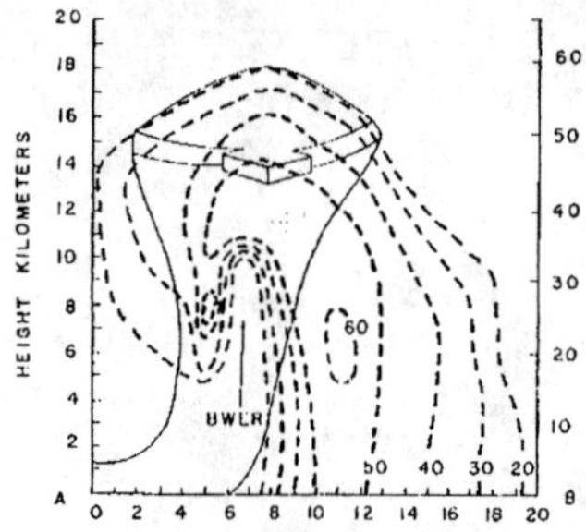


**Figure 1. Vertical cross-section of a BWER. The contours represent constant radar reflectivity.**

Performance evaluation of meteorological algorithms is commonly done using the contingency table, which is a 2×2 table defined as:

$$\begin{Bmatrix} a\,b \\ c\,d \end{Bmatrix} \tag{1}$$

where $d$ is the number of correctly detected (or forecast) events (often referred to as "hits"), $b$ the number of false alarms and $c$ the number of events not detected ("misses"). The number of correctly classified non-events, $a$, is not clearly defined in the case of rare weather events like BWERs. The critical success index (CSI) is defined on the above table as [Donaldson et al. (1975)]:

$$CSI = \frac{d}{b+c+d} \tag{2}$$

A fuzzy logic scheme that takes into account the uncertainties associated with the appearance of a BWER in a radar image has been found to work reliably [Lakshmanan and Witt (1996)]. The performance of the BWER scheme was evaluated using the critical success index (CSI) defined as in Equation 2. A meteorologist identified all the BWERs in the data, marking them either as well-formed or as marginal. The BWERs detected by the scheme were then proofed. The number of hits, misses and false alarms was computed over all the data sets with the scheme not penalized for missing marginal BWERs. From the total hits, misses and false alarms, the CSI was computed.

## 2 METHODS

The BWER detection scheme described in [Lakshmanan and Witt (1996)] accounted for the ambiguities and vagueness in the appearance of BWERs in radar images and performed reliably in tagging structures as either BWERs or non-BWERs. Meteorologists, however, also identify structures as being "marginal" BWERs. A major cause of error in the BWER detection scheme was the misidentification of marginal BWERs as BWERs and the consequent tracking of such misidentified structures through successive volume scans.

In [Lakshmanan and Witt (1996)], we used a classifier that thresholded a weighted sum of all the fuzzy criteria. The enforced linearity in the response of the classifier used seemed to cause identification of structures when a majority of criteria were satisfied. It was hoped that an optimization scheme would make the classifier "learn" which criteria were crucial. While the more important criteria did receive higher weights as a result of optimization, this did not preclude false alarms caused by structures that did not satisfy a crucial criterion and instead satisfied other equally important criteria to a high degree. This seems to be a problem inherent in using weighted sums for classification.

The performance of the scheme in both the heuristic and linear-optimal modes was comparable. This suggested that the heuristic membership functions performed about as well as the "optimal" set of membership functions generated as the result of optimization. Hence, we retained the membership functions used in the heuristic case (we find them easier to explain) and redesigned the classifier to use a fuzzy rule base where the rules were dictated by our expectation of a "textbook BWER".

## 2.1 Measures of Skill

An improvement of the performance of the fuzzy logic scheme can be had when its output matches the indecision inherent in that of a human observer i.e. when a marginal BWER is identified as a marginal BWER and not as a BWER.

Scoring the skill of such a classifier can be done based on the $3 \times 3$ table:

$$\begin{Bmatrix} a & b & c \\ d & e & f \\ g & h & i \end{Bmatrix} \tag{3}$$

in which rows represent the truth values 0, 0.5 and 1. The columns represent the output of the scheme. Thus, $h$ would represent BWERs (true value = 1) that were misidentified as marginal BWERs (output = 0.5) by the scheme.

We can then describe a modified CSI based on this table as:

$$CSI = \frac{i+e}{c+g+\frac{b+d+h+f}{2}+i+e} \tag{4}$$

where the "misses" and "false alarms" are weighted by the degree to which they are "misses" and "false alarms". The CSI of Equation 2 (where marginal BWERs are counted neither when detected nor when missed) is given in terms of this table by:

$$CSI = \frac{i}{c+g+i} \tag{5}$$

## 2.2 Generating Membership Functions

The fuzzy rules were formulated based on the properties of a region in an image that would be part of a 3D BWER structure. For further details on what these properties are, the reader is directed to [Lakshmanan and Witt (1996); Browning and Donaldson Jr (1963)]. Each of these properties is measured quantitatively within a volume scan of the radar and based on quantities carried over from preceding volume scans.

The degree to which a rule such as "The minimum reflectivity within the region is low" holds is obtained by first finding the minimum reflectivity within the region and then using a fuzzy membership function for low in the context of minimum reflectivity to read out the degree to which the rule holds. The membership function, $f(x)$ is generated using two values – the value of the variable in a "textbook case", $x_2$, and the value, $x_1$, of the variable beyond

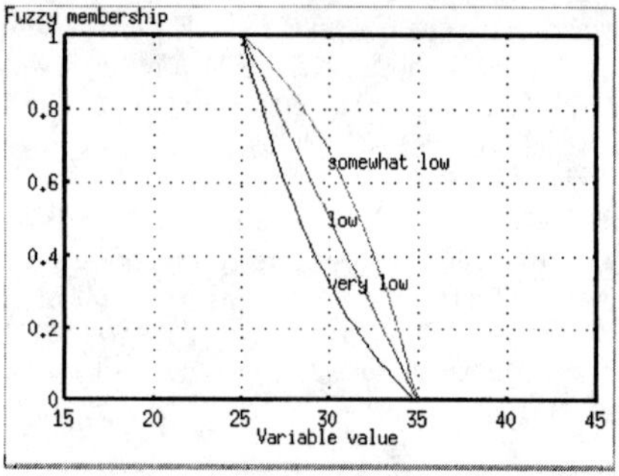


**Figure 2. The membership function generated for "low minimum reflectivity" given that minimum reflectivity in an ideal BWER would be less than $25dBZ$ and that $35dBZ$ is quite high for the minimum value.**

which the rule does not hold true – as:

$$f(x) = \begin{array}{cl} (x-x_1)/(x_2-x_1) & if \;\; 0 <= (x-x_1)/(x_2-x_1) <= 1 \\ 0 & if \;\; (x-x_1)/(x_2-x_1) < 0 \\ 1 & if \;\; (x-x_1)/(x_2-x_1) > 1 \end{array} \tag{6}$$

The membership functions for the rules "somewhat low" ($f^+(x)$) and "very low" ($f^-(x)$) are generated from $f(x)$ using:

$$\frac{1-e^{-\gamma \cdot f(x)}}{1-e^{-\gamma}} \tag{7}$$

where $\gamma$ is 1.5 for $f^+(x)$ and $-1.5$ for $f^-(x)$.

The membership function for "low minimum reflectivity" given that in an ideal BWER, the minimum reflectivity within the BWER would be less than $25dBZ$ and that $35dBZ$ is quite high for the minimum value is shown in Figure 2.

Fuzzy sets are combined using standard operators. The inverse, $\bar{A}$, of a set, $A$, has the membership function $f_{\bar{A}}(a) = 1 - f_A(a)$ where $f_A(a)$ is the membership of element $a$ in the set A. In the rest of this paper, we will represent $f_A(a)$ by $f_A$ when we talk about a relationship valid over all the elements in the universe of discourse. The membership of an element in the union of two sets ($A \cup B$) is given by the maximum of memberships in the two sets ($f_A \vee f_B$). This represents a logical OR. The membership of an element in the intersection of two sets ($A \cap B$) is given by the minimum of the two sets' memberships ($f_A \wedge f_B$) and represents a logical AND.

### 2.3 Matching Human Indecision

The only question asked in the original BWER detection scheme is: To what degree is this 3D structure a BWER? The classifier then thresholds the fuzzy answer to this question, providing a binary classification. Instead, we ask three more questions:

- To what degree is this 3D structure not a BWER?
- To what degree is this 3D structure very much a BWER?
- To what degree is this 3D structure very much not a BWER?

We then use a classifier that acts on the fuzzy outputs of the above four questions to decide which of the three categories the region belongs to. The four fuzzy measures are utilized by the classifier to tag a given structure as one of the three categories - BWERs, marginal BWERs and non-BWERs.

### 2.4 Fuzzy Rule Base

The fuzzy measures to the questions in Section 2.3 were determined using a fuzzy rule base. We will denote as $f_W$, the fuzzy measure that denotes to what degree the BWER is at the right location relative to the storm and has the right physical properties. The fuzzy measure, $f_W$, is derived by combining about twenty fuzzy sets such as the extent of capping, and the closeness to a mesocyclone. The fuzzy rule base consists of the positive rules that go into $f_W$ and the negative ones that determine $f_{no}$, the degree to which the structure is not a BWER.

The structure is a BWER to the extent $f_{yes}$ given by:

$$f_{yes} = ((f_D \wedge f_W) \vee (f_D{}^- \wedge f_W{}^+) \vee f_W{}^-)^- \tag{8}$$

where $f_D$ is the degree to which this region is associated with a BWER that occurred in a previous volume scan. The structure is very much a BWER to the extent of $f_{vyes}$ given by $(f_D \wedge f_W)^-$. The final hedge is done to reduce false alarms and is justified by the apriori likelihood that the structure is not a BWER – BWERs are rare events!

The structure is not a BWER to the extent $f_{no}$ given by:

$$f_{no} = f_{gap} \vee f_{low\_gradient} \vee f_{not\_capped} \vee \bar{f}_W \vee (\bar{f}_{edge} \wedge \bar{f}_{bounded}) \tag{9}$$

where the terms on the right are negative criteria i.e. criteria that a human observer would cite when dropping the region from consideration. It is very much not a BWER to the extent of the above criteria holding together (i.e. the intersection of all the sets on the right-hand side).

### 2.5 Fuzzy Classifier

The fuzzy classifier uses as input the fuzzy measures that are the answers to the four questions described in Section 2.3. The rule base of the classifier yields fuzzy scores for a given structure being a BWER, a non-BWER and a marginal BWER. These scores are given by:

$$f_{marginal} = (f_{yes} \wedge f_{no}) \vee (f_{vyes} \wedge f_{vno}) \tag{10}$$

$$f_{bwer} = (f_{yes} \wedge \bar{f}_{no}) \vee (f_{vyes} \wedge \bar{f}_{vno}) \tag{11}$$

$$f_{nonbwer} = (\bar{f}_{yes} \wedge f_{no}) \vee (\bar{f}_{vyes} \wedge f_{vno}) \tag{12}$$

The structure is tagged as the maximum of the three fuzzy scores, unless the maximum fuzzy score pertains to the "marginal" category. A structure that receives the maximum score for being a marginal BWER is tagged as such only if $f_{yes}$ exceeds $f_{no}$. If $f_{yes}$ does not exceed $f_{no}$, the structure is tagged as a non-BWER.

| Location | Date | b | c | d | e | f | g | h | i | CSI( 4) | CSI( 5) |
|---|---|---|---|---|---|---|---|---|---|---|---|
| **Weighted-sum thresholder** | | | | | | | | | | | |
| Fort Worth, TX | May 7, '95 | 0 | 0 | 0 | 0 | 2 | 5 | 0 | 5 | - | - |
| Lubbock, TX | Jun 2, '95 | 0 | 16 | 10 | 0 | 13 | 7 | 0 | 16 | - | - |
| Norman, OK | May 11, '92 | 0 | 2 | 0 | 0 | 3 | 0 | 0 | 0 | - | - |
| Training cases: Weighted-sum thresholder | | 0 | 18 | 10 | 0 | 18 | 12 | 0 | 21 | 0.32 | 0.41 |
| Oklahoma City, OK | Apr 21, '96 | 0 | 1 | 5 | 0 | 7 | 9 | 0 | 6 | - | - |
| Oklahoma City, OK | Jun 1, '95 | 0 | 0 | 0 | 0 | 0 | 0 | 0 | 0 | - | - |
| Test cases: Weighted-sum thresholder | | 0 | 1 | 5 | 0 | 7 | 9 | 0 | 6 | 0.27 | 0.38 |
| **Fuzzy logic Classifier** | | | | | | | | | | | |
| Fort Worth, TX | May 7, '95 | 6 | 5 | 0 | 0 | 2 | 1 | 0 | 9 | - | - |
| Lubbock, TX | Jun 2, '95 | 14 | 9 | 0 | 13 | 10 | 5 | 2 | 27 | - | - |
| Norman, OK | May 11, '92 | 5 | 15 | 0 | 2 | 5 | 0 | 0 | 0 | - | - |
| Training cases: Fuzzy logic classifier | | 25 | 29 | 0 | 15 | 17 | 6 | 2 | 36 | 0.43 | 0.51 |
| Oklahoma City, OK | Apr 21, '96 | 5 | 2 | 4 | 5 | 3 | 3 | 5 | 8 | - | - |
| Oklahoma City, OK | Jun 1, '95 | 3 | 2 | 0 | 0 | 0 | 0 | 0 | 0 | - | - |
| Test cases: Fuzzy logic classifier | | 8 | 4 | 4 | 5 | 3 | 3 | 5 | 8 | 0.38 | 0.53 |

**Table 1. Performance of the BWER Detection Scheme The letters b–i correspond to the elements of the table in Equation 3. The CSIs are computed using Equations 4 and 5.**

## 3 RESULTS AND CONCLUSIONS

Training and testing were done on the same (respective) cases used in [Lakshmanan and Witt (1996)]. Two test cases were chosen – the first containing several BWERs and inflow notches (thus necessitating differentiation) and the second a storm sequence with no BWERs (necessitating a good false alarm ratio). The scheme was scored on these test cases using the CSIs described as in Equations 4 and 5. The performance of the scheme on the training and test cases before and after introduction of the fuzzy logic classifier is shown in Table 1.

Performance improvements are noticed using either measure of skill by using the fuzzy logic classifier described in this paper. However, the success index is still quite low, mainly because of the number of false alarms. This suggests that improvements will have to be made to the rule base.

## REFERENCES


Browning, K. A., Donaldson Jr, R. J., 1963, "Airflow and Structure of a Tornadic Storm," *J. Atmos. Sci*, Vol. 20, pp. 533–545.

Donaldson, R., Dyer, R., Krauss, M., 1975, "An Objective Evaluator of Techniques for Predicting Severe Weather Events," Proc. 9th Conference on Severe Local Storms. Amer. Meteor. Soc.

Howard, K., Gourley, J., 1995, "Evaluating the Magnitude of Uncertainties in WSR-88D Measurements and Measuring This Impact on Storm Attribute Trends – An Initial Study," National Severe Storms Laboratory.

Lakshmanan, V., Witt, A., 1996, "Detection of Bounded Weak Echo Regions in Meteorological Images," Proc. 13th Int'l Conference on Pattern Recognition.

Lemon, L. R., 1980, "Severe Storms Radar Identification Techniques and Warning Criteria," National Severe Storms Forecast Center.

# A Fuzzy Technique for Flow Oriented Image Direction Computing

*D. C. Douglas Hung and Ching-Yu Austin Huang, Department of Computer and Information Science, New Jersey Institute of Technology, Newark, New Jersey 07102*

*Jane H. C. Cheng, Computer Information System, Business Division, Bloomfield College, Bloomfield, New Jersey 07003*

***Abstract:***

The flow image is direction oriented. In this paper, we present a technique to compute the *pixel-wise direction* of a flow oriented image. A fuzzy balanced direction number algorithm is applied by a circular model. In addition, we use a ridge width dependent technique to control the window size under different resolution. Several experiments are shown in the paper to display the merit of our technique.

The vector or direction technique is widely used in flow oriented image such as fingerprint amd optic flow because their particular characteristic. Mehtre [4] Srinivasan [8] computed the block direction by the major histogram within a square window size by a gray values difference method. Kawagoe [2] uses the gradient in the threshold image and computes the direction for each block. Sherlock [7] described a local ridge orientation method to get the block direction. Tao [9] presented a orientation normalized vector quantization image coding scheme. O'Gorman [5] approached a filter technique for fingerprint.

The result of gray values difference method may need iteratively to be smoothed until good quality. The block size in their methods is highly depended on the quality of direction image. When the block size is larger, their method just can get the rough singular point positions that the range may be more couple blocks.

Fuzzy theory has proved itself to be of significant importance in pattern recognition problems. Fuzzy methods are particularly useful when it is not reasonable to assume class density functions and statistical independence of features [3] [1] [6]. In the paper, a fuzzy model is applied to get a better *pixel-wise direction*. Several experiments are shown to display our approach. We will also discuss the affected results under different techniques, direction number and window sizes.

## Local Direction Computation

A flow image can be described as a set of concentric ridge flows. As well known, ridge flows are direction-oriented characteristic. Let $N$ be the number of directions used in representing directions of a ridge flow, $D = \{ d_1, ..., d_n \}$ be a set of integer directions which are represented directional orientation ranging [0, π) (shown as Fig. 1 ). To get any pixel $P_{ij}$'s direction is by computing all 8-connected pixels of neighbors $Q_{ij}$ which are the same type (ridge or valley) as $P_{ij}$ within $W$x$W$ area. Then the maximum one in the direction histogram is assigned as $P_{ij}$'s direction.

*Pixel-wise Calculation*

For each pixel $P_{ij} = (i,j)$, a $W \times W$ domain centered at $p$ is used and the domain is denoted as $\Omega_W(P)$. Given $q = (a,b) \in \Omega_W(P)$, the $N$ orientation of a line linking of $P$ and $q$ is defined by,

$$d_n(P,q) = \begin{cases} \text{Quan}\left[\tan^{-1}\left(\dfrac{b-j}{a-i}\right)\right] & \text{if } q \text{ is 8-connected to } p, \\ -1 & \text{otherwise} \end{cases} \quad (1)$$

where value -1 marks a status of invalid and Quan(•) is a quantization function which the resulted value is in { 0, 1, ..., $N$–1 }. Then the local direction $d(P)$ is set to the majority direction over the domain of $\Omega_W(P)$. If no majority direction can be found or multiple directions are found having similar high counts, a higher domain of $\Omega_{2W}(P)$ is applied. If a majority still can not be found, then $d(P)$ is set to -1 or a status of invalid.

Topologically, directional measurements obtained along a circular trajectory surrounding a core are monotonically varying from 0 to 2π. As the radii of these pseudo circular trajectories decrease, the rates of directional different increase. Differ from a core, the direction changes around a delta can be divided into 3 disconnected ranges, each of which is monotonically decreasing. The direction of a ridge line can be also defined as its $k$-slope.

*Block Direction*

After every pixel has its own direction, the block direction is then assigned by the majority direction in the $W \times W$ region, in other words, all pixels' direction in this area are the same as majority. By this way, the local noisy will be smoothed. So, for avoiding the premature and local disturbances, most of the current research use block direction to process fingerprint image.

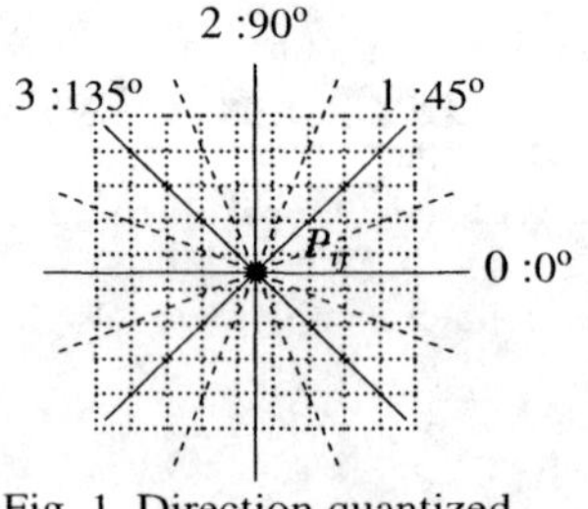


Fig. 1 Direction quantized with $N = 4$.

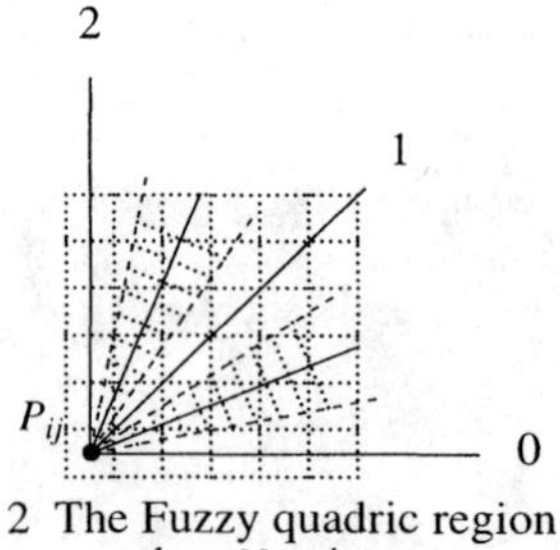


Fig. 2 The Fuzzy quadric region when $N = 4$.

*Gray Difference Method*

In the equation (2), $S_d$ is the sum of differences of gray values in a local region along the direction $D$. $f(i,j)$ and $f_d(i_k,j_k)$ are the gray values at pixels $P_{ij}$ and ${}_kP_{ij}$ respectively, where ${}_kP_{ij}$ is the $k$ th pixel along direction $D$ from $P$, $m$ is the number of pixels chosen for this computation in $\Omega_W(P)$.

$$S_d = \sum_{k=1}^{m} |f(i,j) - f_d(i_k,j_k)|, \text{for } d = 1,...,N \qquad (2)$$

The direction $d(P)$ is the minimum one in $S_d$. Alternately, the direction indicates the direction of maximum gray level uniformity in the image. The resulted direction image is the pixel-wise direction image. Although this method does not need to binary the input image, the quality of resulted image is highly depended on the size $\Omega_W(P)$, and computing the angle for every pixel is time consuming.

*Fuzzy Circular Technique*

After thresholding the gray scale image, the image is cataloged into 2 distinct level - ridge and valley. In a square region, the number of diagonal direction pixels are more than vertical and horizontal. So, we propose a circular method to compute the direction because as well known the circle has balanced number for every direction. A circle is first divided into $2N$ direction which each cone contains $\pi/N$ angle. Then it is applied on a digitized grid plan with radius as $W$. $\{{}_{ij}C_{mn}\}$ is the pixel set within the circle on the digitized area as center at $(i,j)$.

In Fig. 2, we can find that there are some pixels which are absolutely on the direction 0, 1 and 2, but most pixels are not (shown as the marked area). It is no doubt that these pixels on the direction 0, 1 and 2 should have that direction. But most of them are uncertain. Then, the fuzzy concept is allied to solve this problem. For any pixel $P_{mn} \in \{{}_{ij}C_{mn}\}$ satisfies the following fuzzy function is called absolute pixel. $\{{}_dV_{ij}\}$ is defined as the fuzzy type marker of direction of $P_{ij}$:

$$\frac{\pi}{4N} * (k-1) \le f(i,j,m,n) \le \frac{\pi}{4N} * (k+1), \text{ for } k = 0, 4, ..., 4(N-1) \qquad (3)$$

where $f(i,j,m,n)=\cos^{-1}\left(\frac{(m-i)}{\sqrt{(i-m)^2+(j-n)^2}}\right)$ The $\{{}_aV_{ij}\}$ of absolute pixel is { 0, 1, 2, ..., $N-1$ } And, for any pixel $P_{mn} \in \{{}_{ij}C_{mn}\}$ satisfies the following fuzzy function is called uncertain pixel.

$$\frac{\pi}{4N} * (k-1) \le f(i,j,m,n) \le \frac{\pi}{4N} * (k+1), \text{ for } k = 2, 6, ..., 4(N-1)+2 \qquad (4)$$

The uncertain pixel's $\{{}_zV_{ij}\}$ is { $A, B$, ... } which is a fuzzy set. So, we have the fuzzy direction marker of $P_{ij}$ when $N = 4$ direction as following:

$${}_dV_{ij} = {}_aV_{ij} \cup {}_zQ_{ij} = \{ 0,1,2,3,A,B,C,D,E,D,F,G \} \qquad (5)$$

where $A$ = { 0, 1, 2, 3 }, $B$ = { 0, 1 }, $C$ = { 0, 3 }, $D$ = { 2, 1 }, $E$ = { 2, 3 }, $F$ = { 0, 1, 3 } and $G$ = { 1, 2, 3 }. { $A$ } usually can be ignored. Then we have a fuzzy template $T$ for the image. For each ridge or valley pixel direction $D'_{ij}$ computing, equation (6) is used:

$$D' = \text{maximum number of } {}_dV_{ij}. \qquad (6)$$

Fig. 3(a) is our experimental template with $N = 4$ and $W = 6$. We can see that the pixels marked in $\{A, ..., G\}$ are those bidirectional area where are not clearly defined in the most current researches. Given a partial ridge flow as (b). After applying our technique, the result is shown in (c). A circle's center is located at $P_{ij}$. The insided □ is marked as valued neighbors for computing $P_{ij}$'s direction when the template is applied on it.

*Smoothing*
After the direction image is computed, it is necessary to smooth the image filter the local disturbance. Let $M(W,h)$ be a smoothing function which the number of major direction is above $h\%$ within $\Omega_W(P)$ for pixel $P_{ij}$. First we apply $M(1,0)$ to remove the noise which is formed around boundary. Then we use $M(6,50)$ to filter the local disturbance which is formed at digitizing process or by some small spots.

**Feasible Analysis**

There are several important variables, such as direction number $N$ and window size $W$. We would like to know what the influence with different $N$ and $W$.

*The affection of direction number*
It is noticed that the direction-crossings of a digitized direction image are convergent toward the cores. In practical, direction-crossings will merge before they reach the prospect singular point. That is, a phenomena of a prematured convergence. By the definition of singular point,this means a few extraneous cores will be induced.

We observed that such a premature will be occurred earlier if more levels are used to quantize the direction image Using Fig. 4 to show the affection of direction number. When direction number $N = 3$, the image (b) looks clearly. When $N = 4$, the noisy increases as shown in (c). When $N = 6$, the image (d) is hard to be identified. The bigger number, the more information, but with more noisy. The less number, the less information, but also with less noisy. In this proposal, we focus on detecting the singular points. It is better to reduce the noisy. To reduce the aforementioned phenomena, obtained directions are quantized into 3 orientations in most of the experiment results. If it is not 3, we will specify the number.

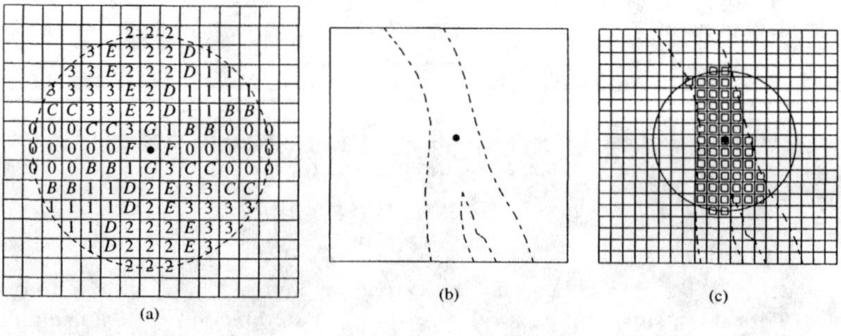

Fig. 3(a) Fuzzy template with $W = 6$ and $N = 4$; (b) A part of flow image. • indicates the pixel $P_{ij}$ to be tested. (c) the result after applied circular template.

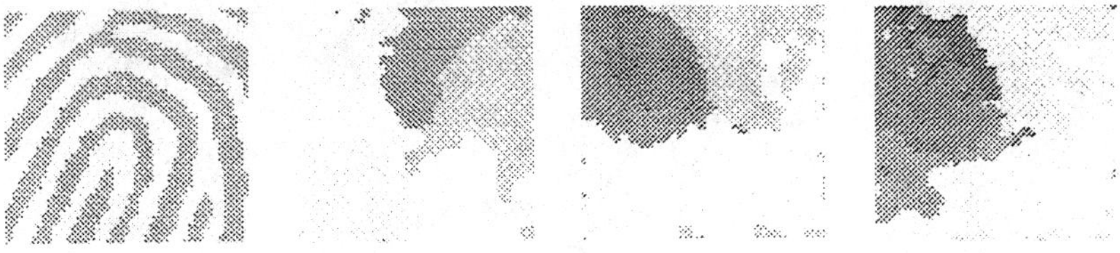

Core image (b) $N = 3$ (c) $N = 4$ (d) $N = 6$

Fig. 4 Using (a) Core image illustrates the premature of directional difference when the quantized levels $N$ is 3 in (b), 4 in (c) and 6 in (d), respectively.

*Window Automated Control*

Average width under 300 dpi is around 4 - 5 pixels, so $W$ in $\Omega_W(P)$ must be bigger than the average width. But when the window increases, the performance will be decreased, and if the window is too big, it also may cover more noisy or the ridge already changed its direction. We used a ridge width dependent method to decide the window size automatically. Using the partial ridge in Fig. 3(c) as our explanation, then apply the following procedures:

a). Locating the center of window that size is $1/2(sx \times sy)$ at the image center which is $1/4(sx \times sy)$, where $sx$ and $sy$ are the $x$ and $y$ size of the image, respectively.

b). Computing the 4 connected distance for each pixel to the other type of it in the window $1/4(sx \times sy)$, and mark the distance as the pixel's flag.

c). Accumulating each flag and sorting flags as queue according the distance, we can get the automated window size $W$ as $3\Phi$, where $\Phi = \lceil (Q_{min} + Q_{max})/2 \rceil$, $Q_{min}$ and $Q_{max}$ are the minimum and maximum distance value in the sorted queue.

According our experiments, we select $W$ as 6 under 300 dpi resolution.

## Experiment Results

Using image shown in Fig. 4(a) as our experiment source, and the results are shown in Fig. 5. (a) is the gray difference method result, (b) is computed by our method, and both using $N = 4$, $W = 6$ and $M(1,0)$. (c) and (d) are the (a) and (b) results, respectively. under applying $M(6,50)$ again.

## Conclusion

In this paper, we discussed how to compute the direction image and proposed our new fuzzy circular model to get better direction image. Also, we analyzed the affection of different number direction. The bigger number, the more noisy. The less number, the less information. In addition, The automatic window size control gives more flexibly.

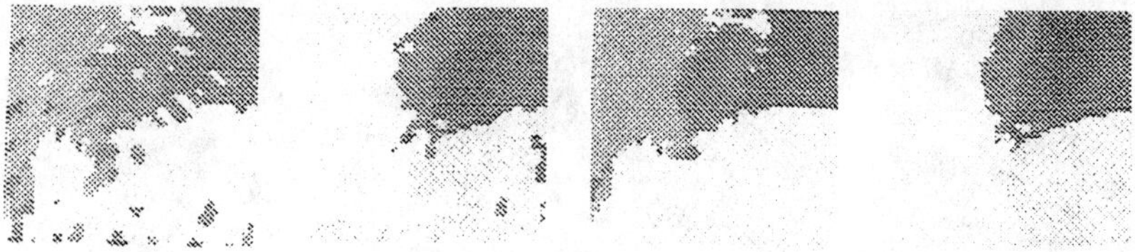

(a) (b) (c) (d)

Fig. 5 Direction results under different condition of Fig. 4(a). (a)Gray difference method result, (b) our circle model, both using $N = 4$, $W = 6$ with $M(1,0)$. (c)Gray difference method result of (a), (d) our circle model of (b), both applied $M(6,50)$.

**References**


[1] Didier Duboris and Henri Prade, "Fuzzy Sets - A Convenient Fiction for Modeling Vagueness and Possibility" *IEEE Trans. on Fuzzy Systems*, vol. 2, no. 1, pp. 16 - 21, February, 1994.

[2] Masahiro Kawagoe and Akio Tojo, "Fingerprint Pattern Classification" *Pattern Recognition*, vol. 17, no. 3, pp. 295 - 304, 1984.

[3] Hon Keung Kwan, "A Fuzzy Neural Network and its Application to Pattern Recognition" *IEEE Trans. on Fuzzy Systems*, vol. 2, no. 3, pp. 185 - 193, August, 1994.

[4] B. M. Mehtre, "Fingerprint Image Analysis for Automatic Identification" *Machine Vision and Application*, 6: pp. 124 - 139, 1993.

[5] Lawrence O'Gorman and Jeffrey V. Nickerson, "An Approach to Fingerprint Filter Design" *Pattern Recognition*, vol. 22, no. 1, pp. 29 - 38, 1989.

[6] Nikhil R. Pal and James C. Bezdek, "Measuring Fuzzy Uncertainty" *IEEE Trans. on Fuzzy Systems*, vol. 2, no. 2, pp. 107 - 118, May, 1994.

[7] B. G. Sherlock and D. M. Monro, "A Model for Interpreting Fingerprint Topology" *Pattern Recognition*, vol. 26, no. 7, pp. 1047 - 1055, 1993.

[8] V. S. Srinivasan and N. N. Murthy, "Detection of Singular Points in Fingerprint Images" *Pattern Recognition*, vol. 25, no. 2, pp. 139 - 153, 1992.

[9] Hai Tao and Zhaoqi Bian, "An Orientation Normalized Vector Quantizer for Flow-like Image Coding" *IEEE Pattern Recognition Conference*, pp. 465 - 470, 1993.

# A NEURAL NETWORK FOR DETECTING EDGES IN AN IMAGE

**ROBERT B. HEATHCOCK**
*Top of the Gulf, Inc., Mexico Beach, Florida, USA*

***ABSTRACT:***
This research paper describes a neural network architecture based on the instar and outstar networks of Adaptive Resonance Theory to enhance and detect edges in an image. The network can discern edge pixels in natural scenes and catagorize them into one of 32 outputs representing differences of 11.25 degrees. The network is built on a series of instar and outstar pairs with a competitive hard limiting network between. The instar network is organized as 32 neurons with 25 inputs each which requires a 5-by-5 subwindow of the input image. A unique weighting and summation technique are used for the instar neurons to decrease the calculation time for the network. The weighting and summation functions produce an asymmetric output so a radial basis network is used to scale and reshape. The image intensity levels are contrast enhanced by the network as it detects edges and the feedback mechanism merges pre- and post-neurally processed images. The network was developed with synthetic edge images to determine the best parameters for the network. Both synthetic and photographic images were presented to the network to test the final configuration. Piece-wise linear edge elements are correctly classified as to one of 32 angles.

## INTRODUCTION

Artificial neural networks have made significant advances in feature extraction and object recognition for the field of image processing. The network methods usually depend on classical image processing techniques for preprocessing the image to determine edges and boundaries. This paper introduces a neural network to perform edge detection and image contrast enhancement. The network is based on the instar and outstar networks developed by Grossberg and Carpenter as part of Adaptive Resonance Theory, ART (Grossberg, 1982).

The edge detecting network was trained to determine the weights associated with edges of different angles and as in an ART network the weights are modified as images are presented to it after the initial training phase. In this way the network can adapt to anomalies in specific images while maintaining the basic purpose of edge detecting. A feedback mechanism builds upon the edges detected in the instar network and a rebuilt image from the outstar network to enhance the edge detecting process. As a side benefit to these networks, a contrast enhanced image is formed.

## APPLICATION

The edge detecting artificial neural network is the beginning phase for an object recognition network. The completed network is being developed as part of a system for automatic target recognition. By incorporating a neural network edge detecting process in a neural network object recognition system, a homogenous system can be developed. Each sub-network of the system processes data and passes it to the next forward process while accepting data from the forward process as feedback.

## THE MODEL

The edge detecting neural network model is composed of two networks, an instar edge element detecting network and an outstar image rebuilding network (see Figure 1). Each instar network determines the closeness of the edge element angle represented by the different illumination intensity levels of the input sub-window. The no edge detector, NED, then determines if there is no edge passing through the sub-window. A winner-take-all network fires the winning neuron, representing the angle of the edge, or none of the neurons if no edge is present. Finally an outstar network generates a model of what an edge's intensity levels should be for merging into the next iteration of the image. The instar and outstar configuration resembles the ART networks (Carpenter 1991) (Grossberg, 1988) (Carpenter, 1987) (Carpenter, 1990).

The instar network is a 25 input by 32 output neural network which determines how close the data from the input buffer is to the weights of the network. The network accepts input from the 5-by-5 pixel sub-window and returns a correlation factor to each of 32 evenly spaced angles of 11.25 degrees. The value of 11.25 degrees is in the range of values determined by Hubel and Weisel (Hubel, 1977) in studying the visual discrimination of the macaque monkey and is supported by the window size.

The neurons directly after the photoreceptors in a biological system use a simple subtraction mechanism followed by a transfer function which is a form of distance measure. Dividing by the weight has the effects of both directing and scaling the function so that small downhill movements were in approximately the right direction and approximately the right magnitude. The process takes the form of the equation:

$$\mathbf{a}_i = \sum \Phi(\mathbf{p}_j \cdot \frac{1}{\mathbf{w}_{ij}} - \text{bias}) \tag{1}$$

Multiplying the input vector by the reciprocal of the weight returns a 1 if $\mathbf{p}_j$ is equal to $\mathbf{w}_{ij}$ and tends to either 0 or 255 as the two are less equal. The bias term shifts the result so the radial basis function returns a 1 if the two numbers are equal and a number tending to 0 when the two numbers become distant. The function returns a value near one, for inputs near the weight value, and the value tends to zero as the input looks less

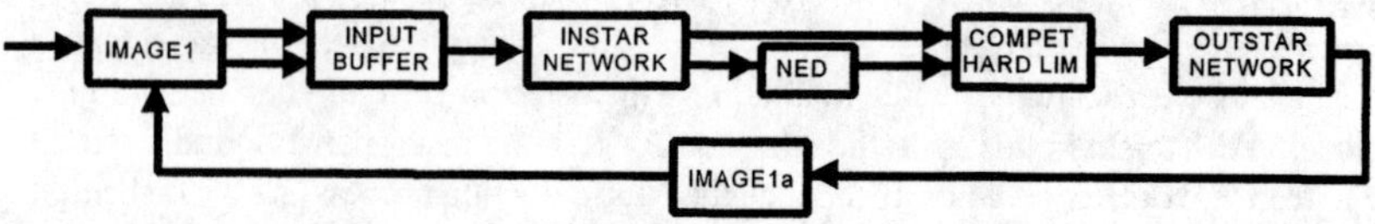


**Figure 1: Architecture of edge detecting neural network.**

like the weights. Summing the 25 numbers in the vector yields a result between zero and 25, with a larger number signifying a better match.

The learning phase for the instar network occurs just after the processing of the outstar network. By delaying the learning phase, the weights are only updated for the neuron whose input most closely matches the perceived edge angle. Although this network uses a non-standard function, the goal is to make the weight vector represent the desired vector of the edge input. Learning occurs using the standard instar learning rule for a given learning rate *lr*

$$\Delta \mathbf{w}_{ij} = lr \cdot \mathbf{a}_i \cdot (\mathbf{p}_j - \mathbf{w}_i j) \tag{2}$$

$$\mathbf{w}_{ij}^{k+1} = \mathbf{w}_{ij}^{k} + \Delta \mathbf{w}_{ij}^{k} \tag{3}$$

The weights for the network are initialized with random numbers. The image presentations cause the weights to change depending on the angle of the edge running through the center pixel. The weights were constrained to lie between a floor of 5 and a ceiling of 250. This range is 96 percent of the total gray scale yet it avoids problems associated with division by zero or near zero.

The output of the instar network determines the closeness of the input sub-window to each of the 32 angled edges represented in the weights. The no edge detector, NED, determines if any of the outputs from the instar network are close enough to an edge representation to match. If there is an edge present then there will be one measure that is close to the maximum value and two edge representations, at ± 11.25 degrees, that are close to the maximum value. The rest of the neurons' outputs will fall off sharply because of the shape of the Gaussian shaping function. If no edge is present the output values will be either random or uniform. The profile of the instar's output determines the presence of an edge going through the center of the image sub-window.

If the NED determines there is not an edge present, then the output neurons of the hard limiter are set to zero. Otherwise, the instar neuron with the largest value wins and the corresponding neuron in the hard limiter is set to one while all other neurons are set to zero.

The outstar network uses a standard 32-input by 25-output Grossberg outstar network to regenerate the image intensities. In order to guarantee that the output of the network represents the desired image sub-window, the weights are the transpose of the instar network. Because the instar and outstar share the same weights there is no need to update the weights beyond the updating done for the instar network.

## DEVELOPING THE NETWORK

The edge detecting network was derived from the ART networks and parts of other networks (Fukushima, 1975) (Fukushima, 1980) (Fukushima, 1986) (Fukushima, 1987) (Hecht-Nielsen, 1987) were incorporated, maintaining a neural network theme and some biological plausibility.

### The input image sub-window

The main considerations in developing this network were the size, transfer function, and weight updating of the instar network. The size of the instar network determines the

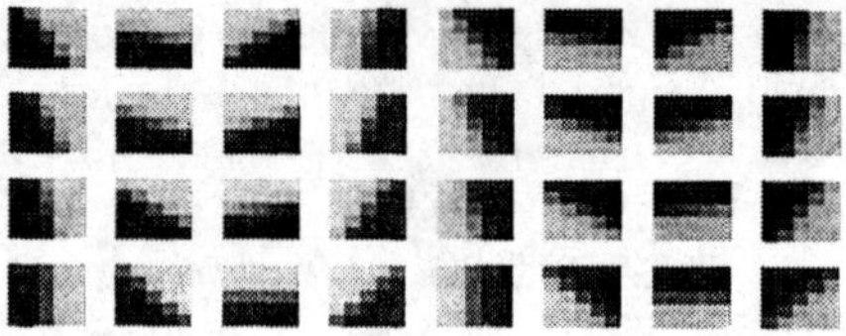

**Figure 2: Gray level representation of edge detector's instar weights.**

size of the image sub-window and the edge angle increments. The image sub-window has to be a square with an odd number of rows and columns. Image sub-windows of 9, 25, 49, and 81 pixels were tried before the 25 pixel window was chosen. A nine pixel sub-window could not determine the presence of an edge if the edge was blurred beyond the sub-window's boundary, and the sub-window was too small to determine if noise in the image was either creating or destroying edges. The 81 pixel sub-window experienced stability and convergence problems. Sub-windows of 25 and 49 performed similarly. The 25 pixel sub-window converged faster, after about 200,000 presentations versus 1.2 million, and executed eight times faster.

The angle increments determine how coarse or fine the network will differentiate the illumination intensity levels. Large increments make ambiguous determinations which will not provide good feedback. Small increments are constrained by the ability to differentiate two angles in the given size of the sub-window. Five degrees or less does not allow enough intensity differentiation for a 5-by-5 sub-window. At 10 degrees, eight of the 36 neurons have the same pixel representation. The final configuration employs a network with 32 output angles which equates to 11.25 degrees.

**Weight convergence**

The final network design was tested with the configuration and weight values determined during the development process. The edge path network was presented with 320,000 presentations of model edges based on the edge equation

$$\mathbf{I}_{ij} = k \pm s \bullet \tanh(n \bullet (i \bullet \cos(\theta) + j \bullet \sin(\theta) - d)) \tag{4}$$

The weights converged to represent the 32 angles as shown in Figure 2 which is the 5-by-5 weights for each of the 32 neurons. The weights were manually arranged for display purposes to reveal the pattern, resembling edges of the different angles, arrived after convergence.

## EXPERIMENTAL RESULTS

Data that represented random synthetic images with and without linear edges and noise was presented to the network. The noise took the form of blurring, Gaussian noise, and speckle noise. The network was able to identify all edges in a simple 5-by-5 sub-window with simple linear edges in one iteration.

Next, a synthetic three-box basis image was created. Figure 3a is the basis image, 3b is blurred by a 3-by-3 window, 3c is with 5% Gaussian noise, and 3d is with 2%

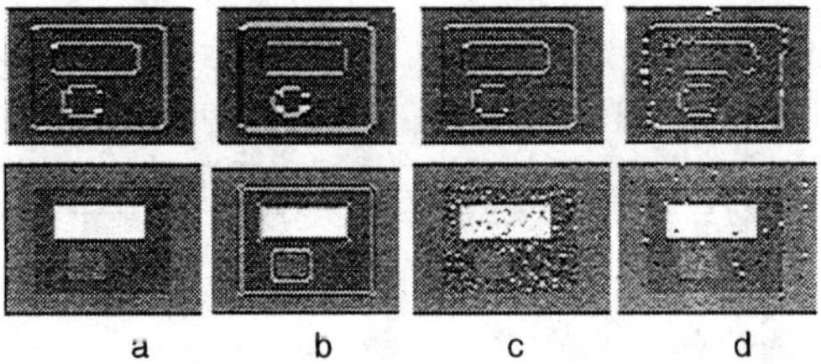


**Figure 3: Edge Element Detection of Three Boxes with (a) No Noise (b) Blurring (c) Gaussian (d) Speckle.**

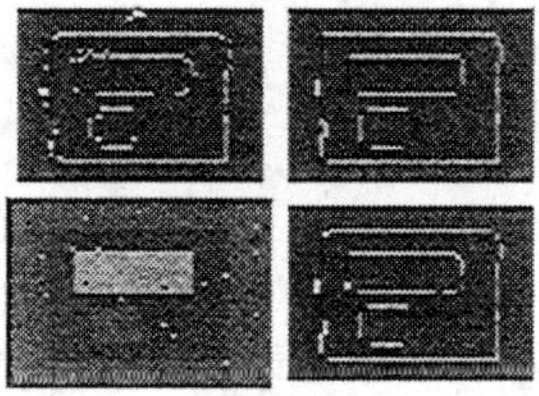

**Figure 4: Three iterations of the complete network on 3 boxes with speckle noise.**

speckle noise. The output is from a single iteration. As can be seen in Figure 3a output, corners are not properly detected since they are not piece-wise linear. The blurring causes multiple parallel edges. Gaussian noise seems to have little effect, while speckle noise has the most severe loss in performance.

Applying the feedback network to the basis figure with speckle noise is presented in the images of Figure 4. Figure 4a represents the basis image with two-percent speckle noise. Figures 4b, c, and d are the outputs for 3 iterations of the total network. Figure 4b contains many false edges and lost edges caused by the noise. Subsequent iterations quickly remove distant false edges while displaced edges and missing edges require more iterations. A further iteration closes some of the gaps while extending others.

As a test image, an aircraft from the LAX imagewas selected (Figure 5). The input image size is 81 pixels in length and 76 pixels in width for a total input neuron count of 6,156 and a total of 177,408 connections in the edge element detecting layer. Figure 6 shows the output of the network after each of four iterations. The edges of the aircraft are detected as are the shadows of the aircraft and building. Several of the containers were large enough to be detected also. Since the network is only detecting edges and not looking for features, shadows and shading created false edges along the aircraft wing and fuselage.

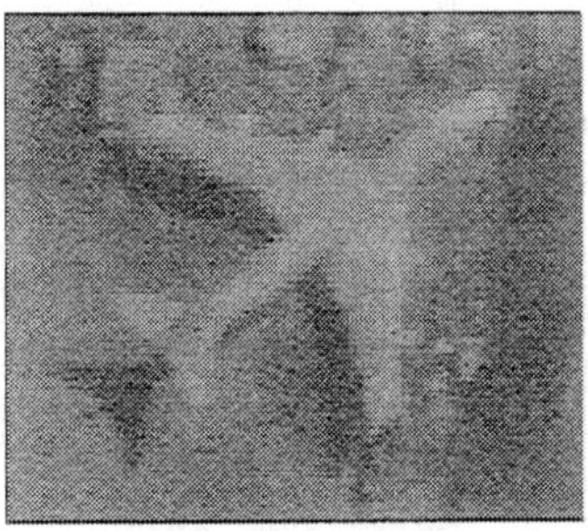

**Figure 5: Sub-image used as input to the network.**

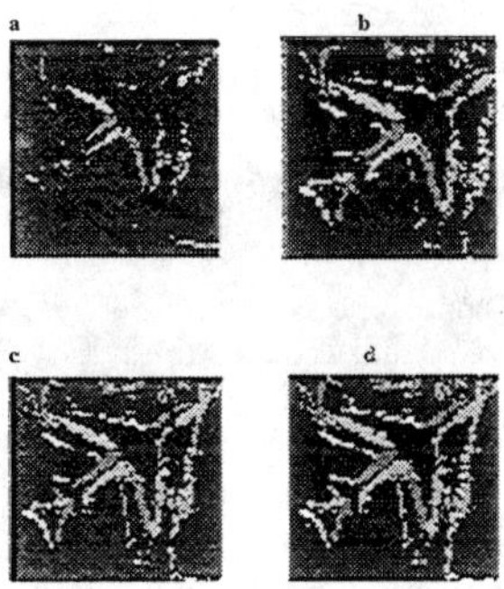


**Figure 6: Output of the edge detecting network.**

## CONCLUSION

The edge detecting network is able to detect piece-wise linear edges in an image. The network performs a good first approximation after only one iteration. Subsequent iterations of the feedback mechanism remove most of the noise in slightly noisy images and extends and connects edges. The local operation in a 9-by-9 image is very promising while implementation on a larger image with intersecting lines and speckle noise produces misclassification of some edges.

By adding additional neurons to the model, nonlinear features could be detected. The architecture of this neural network model lends itself well to a larger model of object recognition. An attentional feedback can be added just after the instar network to provide the edge classifier with contextual information. By incorporating the early stages of object recognition, i.e. edge detection, in a neural network, the final goal will be a more homogeneous model.

## NOMENCLATURE

| | | | |
|---|---|---|---|
| $k \pm s$ | range of values for the step | $d$ | pixel fraction distance the edge is from center |
| $n$ | steepness measure of the edge | $\theta$ | angle of the edge |
| $\Phi$ | Gaussian shaping function. | | |

## REFERENCES


Grossberg, Stephen, Studies of the Mind and Brain, Reidel Press, Drodrecht, Holland, 1982.

Carpenter, Gail A. and Stephen Grossberg, Pattern Recognition by Self-Organizing Neural Networks, Ed., A Bradford Book by MIT Press, 1991.

Grossberg, Stephen, "Nonlinear Neural Networks: Principles, mechanisms, and Architectures," Neural Networks, vol. 1, Pergamon Press, Inc., 1988, pp. 17-61.

Carpenter, Gail A. and Stephen Grossberg, "ART 2: Self-organization of Stable Category Recognition Codes For Analog Input Patterns," Applied Optics, vol. 26, 1987, pp. 4919-4930.

Carpenter, Gail A and Grossberg, Stephen, "ART 3: Hierarchical Search Using Chemical Transmitters in Self-Organizing Pattern Recognition Architectures," Neural Networks, vol. 3, Pergamon Press, Inc., 1990, pp. 129-152.

Hubel, D. H. and T. N. Wiesel, "Functional Architecture of Macaque Monkey Visual Cortex," Proceedings of the Royal Society of London, B 198, 1977, pp. 1-59.

Fukushima, Kunihiko, "Cognitron: A Self-organizing Multilayered Neural Network," Biological Cybernetics, (20), Springer-Verland, 1975, pp. 121-136.

Fukushima, Kunihiko, "Neocognitron: A Self-organizing Neural Network Model for a Mechanism of Pattern Recognition Unaffected by Shift in Position," Biological Cybernetics, (36), Springer-Verland, 1980, pp. 193-202.

Fukushima, Kunihiko, "A Neural Network Model for Selective Attention in Visual Pattern Recognition," Biological Cybernetics, (55), Springer-Verland, 1986, pp. 5-15.

Fukushima, Kunihiko, "Neural Network Model for Selective Attention in Visual Pattern Recognition and Associative Recall," Applied Optics, (26/23), Optical Society of America, December 1987, pp. 4985-4992.

Hecht-Nielsen, Robert, "Counter-propagation Networks," Proceedings of the IEEE First International Conference on Neural Networks, 1987.

# PART V : ADAPTIVE CONTROL

# NEURAL NETWORK CONTROLLERS: A LITERATURE REVIEW

B. FERNÁNDEZ-RODRÍGUEZ[1], N. JARAMILLO[2], N. PANDYA[3]
*Neuro-Engineering Research and Development Laboratory*
*Mechanical Engineering Department, The University of Texas at Austin*

***ABSTRACT:***
This paper presents a review of recent advances in neural networks and their applications to automatic control systems.

## NEURAL NETWORK ARCHITECTURES

Neural Network architectures are defined by the arrangement and interconnection of basic processing elements called neurons. The basic processing elements can be considered to have two components:

- *A Weighted Summer*: As shown in Figure 1, the weighted summer can be described by

$$u_k = \sum_{j=1}^{p} w_{kj} x_j \tag{1}$$

where, output $u$ is the weighted sum of the inputs $x$, and $w_{kj}$ are the synaptic weights of neuron $k$. The figure also shows the nonlinear activation function $\varphi(.)$ which acts on $u$ to give the neuron output $y$.

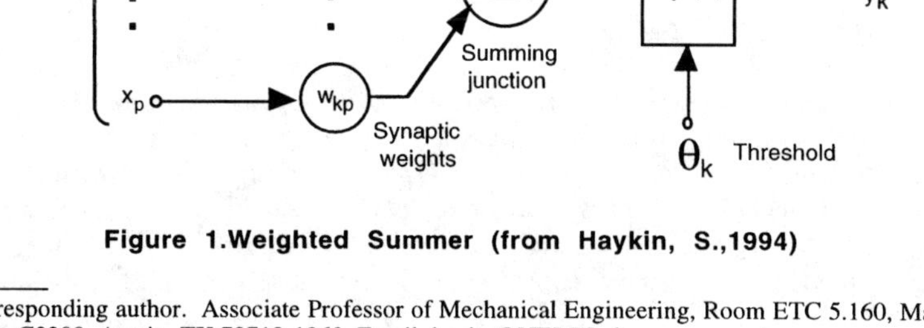


**Figure 1.Weighted Summer (from Haykin, S.,1994)**

---

[1] Corresponding author. Associate Professor of Mechanical Engineering, Room ETC 5.160, Mail Code C2200, Austin, TX 78712-1063. Email: benito@NERDLab.me.utexas.edu.

[2] Doctoral candidate. Also Profesor (on-leave) La Universidad del Zulia, Maracaibo, Venezuela. Email: nelsonj@NERDLab.me.utexas.edu.

[3] Doctoral candidate. Email: nirad@NERDLab.me.utexas.edu.

- *A Non-Dynamic Nonlinear Function*: The non-dynamic nonlinear function $\varphi(.)$, often referred to as the *activation* function gives the element output $y$ in terms of the summer output $u$.

$$y = \varphi(u) \tag{2}$$

The type of activation function used in a particular neuron significantly affects its output for different inputs. For example, adaptation algorithms such as backpropagation require smooth activation functions, whereas sharp functions such as step functions, are useful if binary outputs are required.

Individual neurons by themselves are not very powerful in terms of computation or representation. A network formed by their interconnection, however, allows the approximation of the relation between several inputs and outputs, thus making the network a function approximator. Neurons can be interconnected in several ways to give different characteristics. This section provides a brief overview of some of the important Neural Network architectures that have been developed and lists sources which can provide more detailed information about their operation and applications.

In general, neural networks can be classified as either static or dynamic. The following are characteristics of each:

*Static Feedforward Networks*: Such a neural network is composed of several interconnected layers of neurons, and the neurons in one layer receive inputs only from the neurons in the previous layer. Typically, the neurons between adjacent layers have full factorial connectivity and no feedback.

*Dynamical Networks*: The introduction of feedback in a neural network causes it to be dynamic, with several stable states. Feedback implies that the output of a particular neuron is fed back to itself as an input after a time delay. In general, the dynamic network can be represented as:

$$\begin{aligned} x(t) &= F(x(t), u(t), \theta) \\ y(t) &= G(x(t), \theta) \end{aligned} \tag{3}$$

where $x$ represents the states of the network, $u$ is the set of external inputs, and $\theta$ are the parameters of the network. $F$ and $G$ are functions which represent the structure of the network and the relation between the state variables and the outputs, respectively. Originally, the feedback network was introduced in the context of associative memory by Hopfield (1984) and later by Kohonen (1987). In associative memory, an uncorrupted pattern is stored in the recurrent network and is therefore a Lyapunov stable equilibrium point. Corrupted or noisy versions of that stored pattern lie in its basin of attraction and a dynamic system associated with a set of patterns is created.

There exist numerous neural network architectures and learning algorithms. Several papers survey the various types of networks, describing their architecture and applications. Hunt *et al.* (1992) provide a very detailed account of many of the networks that have been used in control, and this paper draws significantly from their work. Jaint *et al.* (1996) discuss the motivation behind the development of artificial neural networks, describe the basic biological neuron and the artificial computational model; and outline network architectures and learning processes. The following are some of the original schemes with applications to automatic control systems.

*The Multilayer Perceptron*: This is one of the simplest forms of static feedforward networks. The network has a layered structure, with a layer of input nodes, one or more hidden layers, and a layer of output nodes. The number of neurons in the input and the output layers is problem dependent, but the number of hidden layers and also the number of hidden neurons in each hidden layer are determined heuristically.

The presence of the hidden layers enables the network to extract higher-order statistics. Since the multilayer perceptron with sufficient hidden neurons can theoretically approximate any nonlinear function, it is often used in the control of nonlinear systems. Haykin (1993) provides a very detailed explanation on the structure, learning and applications of the multilayer perceptron.

*The Lateral Connection Neural Networks*: So far, only feedforward adjacent layer factorial connection networks have been mentioned here. However, neural networks with lateral connections between neurons in the same layer have some interesting applications. Lateral connections are vital in applications like maximum selection, competitive learning and contrast enhancement. Grossberg (1988) describes how the combination of several types of connections in a neural network can create complex architectures like analog Adaptive Resonance Theory (ART), which is a two layer neural network with feedback and lateral interaction capable of pattern classification. The idea of using encoded inputs has been effectively used by Hecht-Nielsen (1987) in the counterpropagation network (CPN). The CPN has the first layer of competitive neurons called the LVQ layer, introduced by Kohonen (1987) and a second layer of output neurons. The output of the network can thus be thought of as a weighted sum of functions that represent the encoder characteristic.

*Cerebellar Model Articulation Controller (CMAC)*: Another model that uses the encoding system to approximate nonlinear functions is the CMAC. Albus (1975 a, b) shows that the CMAC does not use competitive learning to produce an encoding of the input to reach the same general objective. Instead, these neural networks use the concept of locally tuned overlapping receptive fields, and can approximate complex nonlinearities much faster than networks which use sigmoids. A detailed account of the relationship between splines and the CMAC has been provided by Lane (1991).

*Cellular Neural Networks (CNN)*: In 1988, Chua and Yang proposed an architecture derived from cellular automata. Each layer of the CNN is two dimensional with sparse interconnections and local properties based on spatial invariant templates.

*Multilayer Recurrent Networks*: As discussed earlier, recurrent networks are nonlinear dynamic systems whose behavior depends both on the initial conditions and the inputs used. Much of the current literature dealing with recurrent nets assumes the use of only single layer networks. Kosko (1988) and Narendra (1988) have considered recurrent networks constructed by including multilayer networks in the feedback loop.

*Reinforcement Learning in Adaptive Critic Networks*: Reinforcement learning is the on-line learning through a trial-error process designed to maximize a parameter called the reinforcement signal. The principles of reinforcement learning were developed by Minsky (1961) and were applied to control applications by Waltz and Fu (1965). The idea is that if an action taken by a learning system is followed by a satisfactory response in the application at hand, then the tendency of the system to produce that particular action is strengthened or *reinforced*. Otherwise the tendency of the system to produce that action is weakened.

A reinforcement learning system called the adaptive critic described by Barto *et al.* (1983) and Sutton (1984), uses a critic to convert a primary reinforcement signal into a higher quality reinforcement signal called the heuristic reinforcement signal. Such a system is designed to learn delayed reinforcement where some evaluation function is optimized based on a cumulative reinforcement computed over a sequence of two or more time steps rather than simply the immediate reinforcement.

*Hopfield Networks*: The best known example of a dynamic associative memory network is the Hopfield Network. Hopfield published two important papers (Hopfield (1982, 1984)) detailing the discrete and continuous associative memory models. The discrete model uses asynchronous update of the neural network states (outputs) choosing a random neuron to be updated at any given instant. The energy of the discrete Hopfield model is bounded and each successive state update causes a decrease in the energy. Therefore, the noisy version of a stored memory will always converge to its uncorrupted form in a finite number of state updates (Bruck (1989)). The continuous model Hopfield net on the other hand uses a simultaneous update of the output of each neuron. Unlike the discrete model, the continuous Hopfield model can be represented as a system of ordinary differential equations and it has been shown that the continuous Hopfield model is Lyapunov stable.

*Simulated Annealing and Boltzmann Machines*: Learning in neural networks is an optimization process in which the learning algorithm tries to minimize the error function with respect to the weights of the neuron connections. Due to the nonlinear nature of the networks, there can be multiple sub-optimal local minima in the error surface in which the network could be trapped if pure gradient descent learning algorithms are used. An alternative approach, which would potentially increase the chances of finding the global minimum (if it exists) is to permit the error function to occasionally increase in certain search directions, thus allowing the network to escape the local minima. The Boltzmann machine, as described by Ackley *et al.* (1985) is an application of this idea to the continuous Hopfield network. The method is found to be statistically convergent but reaching the global minimum could theoretically require an infinite number of iterations, and *ad-hoc* stopping methods are used.

## ARTIFICIAL NEURAL NETWORKS IN CONTROL SYSTEMS

The use of neural networks in control systems can be seen as a natural step in the evolution of control methodology to meet new challenges. Looking back, the evolution in the control area has been fueled by three major needs; the need to deal with increasingly complex systems, the need to accomplish increasingly demanding design requirements, and the need to attain these requirements with less precise advanced knowledge of the plant and its environment- that is, the need to control under increased uncertainty. Today, the need to control increasingly complex dynamical systems with significant uncertainty has led to a reevaluation of the conventional control methods, and it has made the need for new methods quite apparent. It has also led to a more general concept of control, one that includes higher-level decision making, planning, and learning, which are capabilities necessary when higher degrees of system autonomy are desirable, (Antsaklis, 1990).

According to Åstrom and McAvoy (1992), at present there is an extremely strong interest in using neural networks in several control application areas including: robotics, process control, communications, vision, and aerospace. One of the first research efforts into neurocontrol was conducted by Narendra and Pathasarathy (1990). They developed generalized neural network models for both identification and control of unknown nonlinear dynamical systems. These generalized neural network models result from a unified treatment of multilayer and recurrent neural network models. They discuss direct and indirect control architectures which are based on model reference adaptive control (MRAC).

An excellent survey on the use of neural networks in control systems is given in the work of Hunt *et al.* (1992), but an extensive research has been developed in the last three years about the application of neural networks in control systems.

Willis *et al.* (1991) reviewed several applications of artificial neural networks to process engineering control systems. They discussed process modeling via the use of artificial neural networks and its potential applications to solve some process engineering problems.

Uhrig (1993) addressed the potential use of neural networks in nuclear power plants, discussing six projects related to transient diagnosis, plant monitoring, analysis of vibrations, and monitoring of performance and efficiency.

El-Gindy and Palkovics (1993) conducted a review of the existing and possible applications of artificial neural networks to vehicle systems. They analyzed control systems to guide autonomous vehicles, to navigate autonomous undersea vehicles, and control aircraft motion. They proposed that artificial neural networks can be used as a suspension controller or "*neuro suspension-controller*," and as a combined vehicle and controller emulator or "*neuro vehicle-controller*" for surface vehicle systems.

Warwick (1995) evaluated the application of neural networks for fairly simple, linear type control actions, and shows that applying a neural-network based controller could result in drastic over-parametrization compared with a self-tuning controller.

Table 1 has been taken and updated from Hunt *et al.* (1992) and shows a resume of the control applications of neural networks.

## PERSPECTIVES ON NEUROCONTROL

The area of neurocontrol is very active as demostrated by the large number of publications in scholar conferences and journals. The field is maturing to a stage where is currently facing problems that have not been solved satisfactorily by alternative techniques or no technique is available (e.g., the auto-lander problem, Miller, *et. al.* (1990)). The advent of adaptive techniques with optimization and adaptive critics (White and Sofge, 1992) poses the next step in neurocontrol. Several groups are working in this area and many more are joining their ranks. A recent conference Sian ka'an (Fernández and Lara, 1996) reunite some of these researchers and a recent NSF workshop brought them together to exchange the state-of-the-art in adaptive critics and reinforcement learning. Upcomming books will publish their findings.

## TABLE 1: SUMMARY OF APPLICATIONS

| Application | Architecture | Learning | References |
|---|---|---|---|
| Adaptive control using generic index | BP | Reinforcement learning | Barto (1988); Barto *et al.* (1983); Helferty *et al.* (1989); Anderson (1989); Porcino and Collins (1990); Guhua and Mathur (1990); Chen (1990); Franklin (1989); Lee and Berenji (1989) |
| Adaptive filtering and prediction | BP | BP learning alg. | Lapedes and Farber (1987); Casdagli (1989); Chen *et al.* (1990); Ydstie (1990); Weigand *et al.* (1990) |
| Adaptive nonlinear control | BP | BP learning alg. | Goldberg and Pearlmutter (1988); Yeung and Beckey (1989); Bassi and Beckey (1989); Josin (1988); Sekiguchi *et al.* (1990); Chen and Pao (1989); Psaltis *et al.* (1988); Guez and Selinsky (1990); Josin (1990); Grant and Zhang (1989); Sanner and Akin (1990); Chen (1989); Karsai *et al.* (1989); trouder and Merril (1989); Ciliz and Isik (1989); Bhat and McAvoy (1990); Ydstie (1990); Ungar *et al.* (1990); Savis and Tan (1989); Bhat *et al.* (1990); Akhyar and Omatu(1993); Hajek (1983); Kuschewski *et al.* (1993); Kim *et al.* (1993) |
| | | Decoupled ext. Kalman filter alg. | Puskurious and Feldkamp (1992, 1993, 1994); Feldkamp *et al.* (1993) |
| | CMAC | Delta rule | Atkenson and Reinkensmeyer (1989); Ersü and Tolle 91984); Kraft and Campagna (1990); Miller *et al.* (1987); Nordgren and Meckl (1993); |
| | Kohonen and linear system | Kohonen alg. and delta rule | Graf and Lalonde (1988); Martinez *et al.* (1988); Ritter and Schulten (1986, 1988) |
| | CPN | Kohonen alg. and delta rule | Cheok and Huang (1989) |
| Optimal control | Kohonen and linear system | Kohonen alg. and delta rule | Fu (1970) |
| Nonlinear modeling | BP | BP learning alg. | Bhat and McAvoy (1990); Bhat *et al.* (1990); Parlos *et al.* (1994) |
| Adaptive linear control | Hopfield | | Chi *et al.* (1990) |
| | ART II | | Kumar and Guez (1990) |
| | Hopfield | | Zak (1990) |
| Fault detection and control | | | Narendra and Mukhopadhyay (1992) |
| Gain scheduling | Hopfield | | Guez *et al.* (1988) |

Source: Hunt *et al.* (1992)

## REFERENCES


Ackley, D.H., *et al.* , 1985, A learning algorithm for Boltzmann machines, *Cognitive Science*, v. 9, pp. 147-159.

Akhyar, S., and Omatu, S., 1993, Self-Tuning PID Control by Neural Networks, *Proceedings of the 1993 Int. Joint Conference on Neural Networks*, v. 3, pp. 2749-2752.

Albus, J.S., 1975a, Data storage in the cerebellar model articulation controller (CMAC), *Trans. of the ASME J. of Dynamic Systems, Measurements and Control*, v. 97, pp. 228-233.

Albus, J.S., 1975b, A new approach to manipulator control: The cerebellar model controller (CMAC) *Trans. of the ASME J. of Dynamic Systems, Measurements and Control*, v. 97, pp. 220-227.

Anderson, C.W., 1989, Learning to control an inverted pendulum using neural networks, *IEEE Control Systems Magazine*, v. 9, pp. 31-37.

Antsaklis, P.J, 1990, Neural Networks in Control Systems, *IEEE Control Systems Magazine*, v. 10, n. 3, pp. 3-5.

Åström, K.J, and McAvoy, T.J, 1992, Intelligent Control: an Overview and Evaluation, *Handbook of Intelligent Control*, White, D, and Sofge, D., ed., Van Nostrand Reinhold, New York, pp.11-17.

Åström, K.J., Hang, C.C., Persson, P., and Ho, W.K., 1992, Towards Intelligent PID Control, *Automatica*, v. 28, n. 1, pp. 1-9.

Atkenson, C.G., and Reinkensmeyer, D.J., 1989, Using associative content-addressable memories to control robots, *Proceedings of the 1989 Int. Conf on Robotics and Automation*, pp. 1859-1864.

Barto, A.G, 1988, An approach to learning control surface by connectionist systems*Vision, Brain and Cooperative Computation*, Arbib and Hanson, ed., MIT Press, Cambridge, MA., pp. 665-701.

Barto, A.G., *et al.*, 1983, Learning and sequential decision making, *Learning and Computational Neuroscience: Foundations of Adaptive Networks*, Gabriel and Moore, ed., MIT Press, Cambridge, MA., pp. 539-602.

Barto, A.G., *et al.*, 1983, Neuronlike adaptive elements that can solve difficult learning control problems, *IEEE Transactions on Systems, Man, and Cybernetics*, v. 13, pp. 834-846.

Bassi, D.F., and Beckey, G.A., 1989, Decomposition of neural network model of robot dynamics: a feasibility study, *Simulation and AI*, n. 220, pp. 8-13.

Bhat, N.V., and McAvoy, T.J., 1990, Use of neural nets for dynamical modelling and control of chemical process systems, *Computers Chem. Eng.*, v. 14, pp. 573-583.

Bhat, N.V., *et al.*, 1990, Modelling chemical process via neural computation, *IEEE Control Systems Magazine*, v. 10, n. 3, pp. 24-29.

Bruck, J., 1989, Computing with Networks of Threshold Elements, D. Phil. Thesis. Stanford University.

Casdagli, M., 1989, Non-linear system prediction of chaotic time series, *Physica D.*, v. 35, pp. 335-356.

Chen, F-C., 1989, Back-propagation neural network for nonlinear self-tuning adaptive control, *IEEE Symposium on Intelligent Control*, pp. 274-279.

Chen, S., *et al.*, 1990, Non-linear system identification using neural networks, *Int. J. of Control*, v. 51, pp. 1191-1214.

Chen, V.C., and Pao, Y.H., 1989, Learning control with neural networks, *Proceedings of the 1989 Int. Conf. on Robotics and Automation*, pp. 1448-1453.

Chen, V.C., 1990, Problem solving by using reinforcement learning neural nets, *IEEE Int. Joint Conference on Neural Networks*, v. I, pp. 583-586

Cheok, K.C., and Huang, N.J., 1989, Lyapunov stability analysis for self-learning neural model with application to semi-active suspension control system, *IEEE Int. Symposium on Intelligent Control*, pp. 329-331.

Chi, S.R., *et al.*, 1990, Neural networks for system identification, *IEEE Control Systems Magazine*, v. 10, n. 3, pp. 31-34.

Chua, L.O., and Yang, L., 1988a, Cellular neural networks: Applications, *IEEE Trans. on Circuits and Systems*, v. 35, pp. 1273-1290.

Chua, L.O., and Yang, L., 1988b, Cellular neural networks: Theory, *IEEE Trans. on Circuits and Systems*, v. 35, pp. 1257-1272.

Ciliz, M.K., and Isik, C., 1989, Time optimal control of mobile robot motion using neural nets, *IEEE Int. Symposiurn on Intelligent Control*, pp. 368-373.

El-Gindy, M., and Palkovics, L., 1993, Possible Application of Artificial Neural Networks to Vehicle Dynamics aod Control: a Literature Review, *Int. J. of Vehicle Design,* v. 14, nos. 5/6, pp. 592-614.

Ersü, E., and Tolle, H., 1984, A new concept for learning control inspired by brain theory, *Proc. 9th World Congress of IFAC*, v. 7, pp. 245-250.

Feldkamp, L.A., *et al.*, 1992, Neural Control Systems Trained by Dynamic Gradient Methods for Automotive Applications, *Proceedings of the 1992 Int. Joint Conference on Neural Networks*, v. II, pp. 798-804.

Franklin, J.A., 1989, Input space representation for refinement learning control, *IEEE Int. Symposium on Intelligent Control*, pp. 115-122.

Fu, K.S., 1970, Learning control systems- review and outlook, *IEEE Transactions on Automatic Control*, v. 16, pp.210-221.

Goldberg, K., and Pearlmutter, B.A., 1988, Using a neural network to learn the dynamic of the CMU direct-drive arm II, Report CMU-CS-88-160. Carnegie-Mellon University.

Graf, D.H., and Lalonde, W.R., 1988, A neural controller for collision-free movement of general robot manipulators, *IEEE Int. Joint Conference on Neural Networks, IJCNN'88*, pp. 77-84.

Grant, E., and Zhang, B., 1989, A neural net approach to supervised learning of pole balancing, *IEEE Int. Symposium on Intelligent Control*, pp. 123-129.

Grossberg, S., 1988, Neural Networks and Natural Intelligence. The MIT Press, Cambridge, MA.

Guez, A., *et al.*, 1988, Neural network architecture for control, *IEEE Control Systems Magazine*, v. 8, pp. 22-25.

Guez, A., aod Selinsky, J.W., 1990, A neurocontroller with guaranteed performance for rigid robots, *IEEE Int. Joint Conference on Neural Networks, IJCNN'90*, pp. 347-350.

Guhua, A., and Mathur, A., 1990, Setpoint control based on reinforcement learning, *IEEE Int. Join Conference on Neural Networks, IJCNN'90*, pp. 511-514.

Hajek, M., *et al.*, 1993, Global Control of Nonlinear Dynamical Systems Using Neural Networks, *Proceedings of the 1993 Int. Joint Conference on Neural Networks*, v. 3, pp. 2819-2822.

Haykin, S., 1994, Neural networks: A Comprehensive Foundation, MacMillan.

Hecht-Nielsen, R., 1987, Counterpropagation networks, *Applied Optics*, v.26, pp. 4979-4984.

Helferty, J.J., *et al.*, 1989, A neural network learning strategy for the control of a one-legged hopping machine, *Proc. of the 1989 Int. Conf on Robotics and Automation*, pp. 1604-1609.

Hopfield, J.J., 1982, Neural networks and physical systems with emergent collective computational abilities," *Proc. of the National Academy of Sciences*, v. 79, pp. 2554-2558.

Hopfield, J.J., 1984, Neurons with graded response have collective computational properties like those of two-state neurons, *Proc. of the National Academy of Sciences*, v. 81, pp. 3088-3092.

Hunt, K.J., *et al.*, 1992, Neural Networks for Control Systems - A Survey, *Automatica*, v. 28, n. 6, pp. 1083-1112.

Jain, A.K., *et al.*, 1996, Artificial Neural Networks: A Tutorial, *Computer*, v. 29, n. 3, pp. 31-44.

Josin, G., 1988, Neural-space generalization of a topological transformation, *Biological Cybernetics*, v. 59, pp. 283-290.

Josin, G.M., 1990, Development of a neural network autopilot model for high performance aircraft, *IEEE Int. Joint Conference on Neural Networks, IJCNN'90*, pp. 547-550.

Karsai, G., *et al.*, 1989, Dynamic modelling and control of nonlinear process using neural network techniques, *IEEE Int. Symposium on Intelligent Control*, pp. 280-286.

Kasparian, V., and Batur, C., 1992, Neural Network Structure for Process Control Using Direct and Inverse Process Model, *Proceedings of the American Control Conf.*, v. 1, pp. 562-566.

Kim, J-T., *et al.*, 1993, Linearization Control of VCR Servo System by Using a Neural Network Based Feedforward Compensator, *Proceedings of the 1993 Int. Joint Conf. on Neural Networks*, v. 3, pp. 2753-2756.

Kohonen, T., 1987, Self Organization and Associative Memory, Springer-Verlag, Berlin.

Kosko, B., 1988, Bidirectional associative memories, *IEEE Trans. Syst., Man,Cybern*, v. 18, pp. 49-60.

Kraft, L.G., and Campagna, D.P., 1990, A comparison between CMAC neural network control and two traditional adaptive control systems, *IEEE Control Systems Magazine*, v. 10, pp. 36-43.

Kumar, S., and Guez, A., 1990, Adaptive pole placement for neurocontrol, *IEEE Int. Joint Conference on Neural Networks, IJCNN'90*, pp. 397-400.

Kuschewski, J.G., *et al.*, 1993, Application of Feedforward Neural Networks to Dynamical System Identification and Control, *IEEE Trans. on Control Systems Technology,* v. 1, pp.37-49.

Lane, S.H., *et al.*, 1991, Higher order CMAC neural networks- theory and practice, *Proc. American Control Conf.*, pp. 1579-1585.

Lapedes, A., and Farberr, R., 1987, Non-linear signal processing using neural networks: prediction and system modelling, *Report* LA-UR-87-2662, Los Alamos National Laboratory.

Lee, C-C., and Berengi, H.R., 1989, An intelligent controller based on approximate reasoning and reinforcement learning, *IEEE Int. Symposium on Intelligent Control*, pp. 200-205.

Martinez, T.M., et al., 1988, Three-dimensional neural net for learning visuomotor coordination of a robot arm, *IEEE Int. Joint Conference on Neural Networks, IJCNN'88*, pp. 351-356.

Miller, W.T., *et al.*, 1987, Application of a general learning algorithm to the control of robotic manipulators, *Int. J. of Robotic Research*, v. 6, pp. 84-98.

Miller, W.T, and Anderson,C.W.,1990, Challenging Control Problems, Neural Networks for Control Miller, W.T., *et al* , Eds., The MIT Press, Cambridge, MA.

Minsky, M.L., 1961, Steps toward artificial intelligence, *Proc. of the Institute of Radio Engineers*, v. 49, pp. 8-30.

Narendra, K.S., and Mukhopadhyay, S., 1992, Intelligent Control Using Neural Networks, *IEEE Control Systems Magazine*, v. 12, n. 3, pp. 11-18.

Narendra, K.S., and Parthasarathy, K., 1988, "Neural networks and dynamical systems, Part I: A gradient approach to Hopfield networks, Center Syst. Sci., Dept. of Electrical Eng., Yale University, New Haven, CT, *Tech. rep.* 8820.

Narendra, K.S., and Parthasarathy, K., 1990, Identification and Control of Dynamical Systems Using Neural Networks, *IEEE Transactions on Neural Networks*, v. 1, pp. 4-27.

Nordgren, R.E., and Meckl, P.H., 1993, An Analytical Comparison of a Neural Network and a Model-Based Adaptive Controller, *IEEE Transactions on Neural Networks*, v. 4, pp. 685-694.

Parlos, A.G., et al., 1994, Application of the Recurrent Multilayer Perceptron in Modeling Complex Process Dynamics, *IEEE Transactions on Neural Networks*, v. 5, pp. 255-266.

Porcino, D.P., and Collins, J.C., 1990, An application of neural networks to the guidance of free-swimming submersibles, *IEEE Int. Joint Conf. on Neural Networks, IJCNN'90*, pp. 417-420.

Psaltis, D., *et al.*, 1988, A multilayered neural network controller, *IEEE Control Systems Magazine*, v. 8, pp. 17-21.

Puskorius, G.V., and Feldkamp, L.A., 1992, Model Reference Adaptive Control with Recurrent Networks Trained by the Dynamic DEKF Algorithm, *Proceedings of the 1992 Int. Joint Conference on Neural Networks*, v. II, pp. 106-113.

Puskorius, G.V., and Feldkamp, L.A., 1993, Automotive Engine Idle Speed Control with Recurrent Networks, *Proceedings of the American Conirol Conference*, v. 1, pp. 311-316.

Puskorius, G.V., and Feldkamp, L.A., 1994, Neurocontrol of Nonlinear Dynamical Systems with Kalman Filter Trained Recurrent Networks, *IEEE Transactions on Neural Networks*, v. 5, pp. 279-297.

Ritter, H., and Schulten, K., 1986, Topology conserving mappings for learning motor tasks, *Neural Networks for Computing, Aip Conf. Proc.*, v. 151, pp. 376-380.

Ritter, H., and Schulten, K., 1988, Extending Kohonen's self-organizing mapping algorithm to learn ballistic movements, *Neural Computers*, pp.393-406, Springer-Verlag.

Sanner, R.M., and Akin, D.L., 1990, Neuromorphic pitch attitude regulation of an underwater telerobot, *IEEE Control Systems Magazine*, v. 10, pp. 62-67.

Savic, M. and Tan, S-H., 1989, A new class of neural networks suitable for intelligent control, *IEEE Int. Symposium on Intelligent Control*, pp. 418-423.

Sekiguchi, M., *et al.*, 1989, Behaviour control for mobile robot by multi-hierarchical neural network, *Proc. of the 1989 Int. Conf on Robotics and Automation*, pp. 1578-1583.

Sutton, R.S., 1984, Temporal credit assignment in reinforcement learning. Ph.D. Dissertation, University of Massachusetts, Amherst, MA.

Troudet, T., and Merril, W., 1989, Neuromorphic learning continuous valued mappings in presence of noise, *IEEE Int. Symposium on Intelligent Control*, pp. 312-319.

Uhrig R.E., 1993, Use of neural networks in nuclear power plants, *ISA Transactions*, v. 32, pp. 139-145.

Ungar, L.H., *et al.*, 1990, Adaptive networks for fault diagnosis and process control, *Computers Chem. Engineering*, v. 14, pp. 561-572.

Waltz, M. D., and Fu, K. S., 1965, A heuristic approach to reinforcement learning control systems, *IEEE Trans. on Automatic Control*, v. 10, pp. 390-398.

Warwick, K., 1995, A critique of neural networks for discrete-time linear control, *International Journal of Control*, v. 61, n. 6, pp. 1253-1264.

Weigand, A.S., *et al.*, 1990, Predicting the future: a connectionist approach, *Int. J. Neural Systems*, v. 3, p. 193.

White, D.A., and Sofge, D.A., Eds., 1992, Intelligent Control: an Overview and Evaluation, Handbook of Intelligent Control, White, Van Nostrand Reinhold, New York.

Willis, M.J., *et al.*, 1991, Artificial neural networks in process engineering, *IEE Proceedings-D*, v. 138, n. 3, pp. 256-266.

Ydstie, B.E., 1990, Forecasting and control using adaptive connectionist networks, *Computers Chem. Engineering*, v. 14, pp. 583-599.

Yeung, D.Y., and Beckey, G.A., 1989, Using a context-sensitive learning network for robot arm control, *Proc. of the 1989 Int. Conf. on Robotics and Automation*, pp. 1441-1447.

Zak, S.H., 1990, Robust tracking control of dynamic systems with neural networks, *IEEE Int. Joint Conf. on Neural Networks, IJCNN'90*, pp. 563-566.

# NEURO-GAIN APPROXIMATION (A CONTINUOUS APPROXIMATION TO THE NONLINEAR MAPPING BETWEEN LINEAR CONTROLLERS)

**J. NEIDHOEFER AND K. KRISHNAKUMAR**
*Department of Aerospace Engineering*
*The University of Alabama, Tuscaloosa, AL 35497-0280*

**ABSTRACT:**
This paper presents an approach for designing a psuedo-global, nonlinear, adaptive controller. The adaptive capabilities of this controller are such that, real time adaptive control should be possible. Inspiration for this controller design approach was found largely in the methodology for "gain scheduling", however, the resultant controller is not a gain scheduler, but a continuous approximation to the nonlinear mapping between linear gains. The approach presented here is applied to the control of a Lotke-Volterra ecosystem, and also to the longitudinal control of a high performance aircraft.

## INTRODUCTION

As detailed by Shamma and Athans in [1], the general idea behind gain scheduling is to break the control design process into two steps. First, one designs a set of "locally valid" (usually linear) controllers, and second a global nonlinear controller is designed by utilizing some "logical operator" to choose which local controller is the most valid. A "third step", now often used is to employ some type of tuning technique to the "chosen" local controller. This idea is not a new one. In the 1970's Lainiotis[2] proposed using multiple Kalman filters to improve the accuracy of state estimations. Athans used multiple models with MMAC switching (a probability based logical operator) for aircraft control[3]. Much more recently Narendra, Balakrishnan, and Ciliz have proposed using multiple models with switching [4]. How neural networks might be employed is also discussed in [4]. Krishnakumar's idea of Immunized Neural Systems [5][6] is based upon the idea of creating and combining several neural networks which are locally valid in some global hypersphere of uncertainty. For fast adaptation, Krishnakumar has discussed the idea of "innate immunity" and has proposed the use of "linear controllers with fuzzy schedulers" and also that "systems known *a priori* can be stored in look up tables"[7].

The two main limitations of gain scheduling as discussed by Shamma[1], are that 1) the scheduling variable must vary slowly, and 2) the scheduling variable should capture the nonlinearities of the plant. A third disadvantage of most gain scheduling methods is that we have no way of knowing how far from the equilibrium point (around which the linear model has been created) that the linear controller will remain valid for the nonlinear system. Two advantages of modern neural networks are 1) they can generalize a function arbitrarily well utilizing a only few (relative to the whole space) representative data points, and 2) they can model highly nonlinear functions.

This paper sets forth a technique which utilizes the strengths of artificial neural systems to overcome some weaknesses (limitations) in the general gain scheduling technique. The result is a very fast adapting controller architecture suitable for real-time control. The technique is applied to a Lotke-Volterra ecosystem and an aircraft control problem

## CONTROLLER DESIGN METHODOLOGY

### Define The Hypersphere of Uncertainties

The first step in the controller design is to specify the hypersphere of uncertainties. In other words, we must specify which uncertainties the system may encounter. These are the uncertainties whose effects must be negated by the controller. These uncertainties may be grouped into several different types[11]:

Defining the hypersphere of uncertainties may be considered the single most important step in successfully implementing a neural gain approximator. This is because this method may be impractical for specific problems in which uncertainty measurements, estimations, or approximations are unobtainable.

### Define The Scheduling Variables

Once we have specified the disturbances which define our hypersphere of uncertainty, we must next define our "scheduling variables. It should become apparent from the design methodology that the controller inputs will have the same restrictions as the "scheduling variables" of contemporary "scheduling type" controllers. Thus we will term the inputs to our controller "scheduling variables". As formalized by Shamma [1], the two main restrictions on the scheduling variables are that 1) the scheduling variables must vary slowly, and 2) the scheduling variables should capture the nonlinearities of the plant. A rigorous mathematical justification for these restrictions is presented in [12]-[14].

### Define Suitable Set of Equilibrium Points

For purposes of this design methodology, the choice of equilibrium points is critical. The equilibrium points will represent the points at which we generate linear controllers, and will also be closely related to our neural net training set. These must be chosen to sufficiently capture the nonlinearities of the hypersurface.

In the abstract, this approximator was termed "pseudo-global", this terminology was used to mean that the region of controller validity is actually designed by the choice of training points (equilibrium points).

### Design a Linear Controller At Each of the Equilibrium Points

Once we have defined a suitable set of equilibrium points, we must next linearize the system and design controllers at each point. Stability augmenting, or tracking controllers should work equally well. The only restriction on the controllers is that the gains must be statistically correlated between equilibrium points. The necessity for correlation between gains arises from the fact that standard neural networks cannot generalize uncorrelated data.

An example of a way to ensure correlated data is to use pole placement in the individual controller designs and place the poles the same for each linear controller. Thus, as we move from one equilibrium point to another, the gains will change, but they will be correlated, and the response of the closed loop system will remain constant.

**Train A NN Approximator**

The key to our design process is as follows: *If multiple linear controllers designed for a given system are correlated between equilibrium points, then we can train a neural network using input/output pairs consisting of the scheduling variables (inputs) and the controller gains (outputs).* The resultant network will provide an approximation of the nonlinear, possibly multidimensional, mapping between linear gains. If NN training data is uncorrelated, the best we can hope for is for the NN to *memorize* the data. If the NN memorizes the data, then the NN *generalization* will probably be meaningless.

**Analyze the Approximator**

The main objective of this analysis is to convince ourselves that the approximator will provide a valid control law in the designed region. Since we have the ability to linearize the system, we may be able to check the eigenvalues of (A-BK). If it is unstable, two things can be tried, 1) train the NN better, or 2) increase the stability margin of the linear controllers. In the second case, it is possible to determine the maximum and mean error and the standard deviation in the approximator output. These values can then be used to calculate the stability margin necessary for the linear controllers.

**Implement the NN Approximator in the Control Loop**

Once we have successfully trained the approximator, and achieved satisfactory results in the closed loop analysis, we can implement the approximator in a closed loop control system.

## THE CONTROL OF A LOTKE-VOLTERRA ECOSYSTEM

The population dynamics of a predator-prey system can be described by the Lotke-Volterra equations [10]:

$$\dot{x}_1(t) = a\, x_1(t) + b\, x_1(t)\, x_2(t) \qquad (1)$$
$$\dot{x}_2(t) = c\, x_2(t) + d\, x_1(t)\, x_2(t)$$

where $x_1$ is the predator population and $x_2$ is the prey population. Parameter b describes the benefits the predator gains by eating the prey. Parameter d describes the loss in the prey population due to the predators. Parameters a and c describe the birth and death rates not dependent on the predator-prey relationship. For a given set of parameters, equations (1) above, have two equilibrium points, one is at the origin ($x_1 = x_2 = 0$) and the other is at $x_1 = -(c / d)$ and $x_2 = -(a / b)$.

If we assume a linear controller of the form: $u = -k * \delta$ then equations (1) above become:

$$\dot{x}_1(t) = a\,x_1(t) + b\,x_1(t)\,x_2(t) + u \qquad (2)$$
$$\dot{x}_2(t) = c\,x_2(t) + d\,x_1(t)\,x_2(t) + u$$

Now linearizing about the equilibrium point $x_1 = -(c / d)$ and $x_2 = -(a / b)$ yields:

$$\begin{bmatrix} -k_1 & -k_2 + (b*c)/d \\ -k_1 + (a*d)/b & -k_2 \end{bmatrix}$$

which we can see by inspection is a coupled system of equations whose stability is dependent upon both the choice of "k" and the parametric values.

**Define The Hypersphere of Uncertainties**

For this experiment, initial condition, and parametric uncertainties will be introduced to the system. Parametric uncertainties will be introduced by varying the system parametric values a and c. The design region for each parameter was specified as being between ±3. It is important to note that the specified design region contains an uncontrollable set, when a=c.

**Define the Scheduling Variables**

Now that we have specified the disturbances which define our hypersphere of uncertainty, we next define our "scheduling variables". For this experiment, the parameters "a" and "c" which define part of our hypersphere of uncertainties were chosen as the scheduling variables.

**Define Suitable Set of Equilibrium Points**

For this experiment, the design region is for "a" and "c" both between ±3. So equispaced training points were chosen at the integer values a={-3, -2, -1, 0, 1, 2, 3} and c={-3, -2, -1, 0, 1, 2, 3}. These points form a grid which is a discrete approximation to a plane in the hyperspace.

**Design a Linear Controller At Each of the Equilibrium Points**

Now that we have defined a suitable set of equilibrium points, we must next linearize the system and design linear controllers at each point.

**Train a NN Approximator**

The first step to accomplish this was to create a neural network training data file which used the scheduling variable as the inputs and the linear gains as outputs of NN training pairs.

For this experiment, two networks were trained each having three inputs, parameter a, parameter c, and the bias. Each NN had a single output, the first net's output was k1 and the second net's output was k2.

**Analyze the Approximator**

For this experiment the NN approximators were trying to map the nonlinear space between linear controllers (see Figure 1). The output of one of the NN approximators can be seen in Figure 2 for comparison with Figure 1.

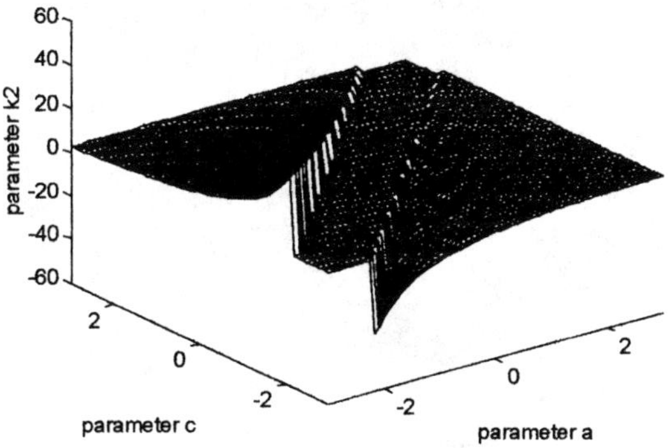


Figure 1. The Nonlinear Mapping Between Linear Controllers

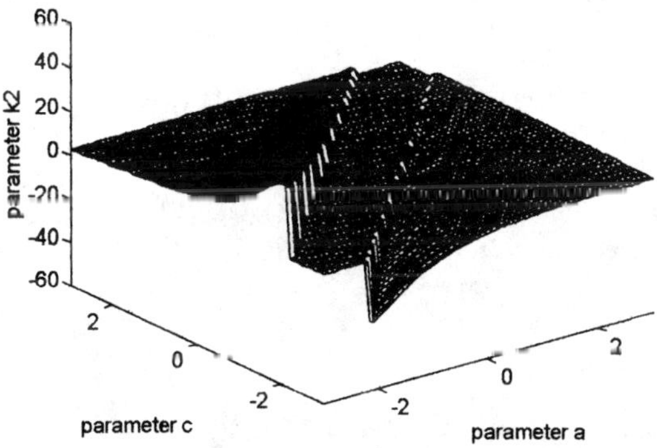


Figure 2. NN Approximation to Nonlinear Mapping Between Linear Controllers

Notice that Figure 2 is very similar to Figure 1, and thus, the NN approximators did a good job of mapping the surface. The greatest error is near where a = c. In this region, the gains go to infinity no matter where the poles are placed. We could have reduced this error by training the approximators to handle this region, i.e. used more training points near the region.

Now stability was checked over a fine (relative to the training) grid. A stable control was provided until the absolute value of a - c was less than 0.3. This was expected since when a = c the system is uncontrollable, and our nearest training points were not this close to a = c.

**Implement the NN Approximator in the Control Loop**
Now that we have analyzed our NN approximators for k1 and k2 we implement our controllers.

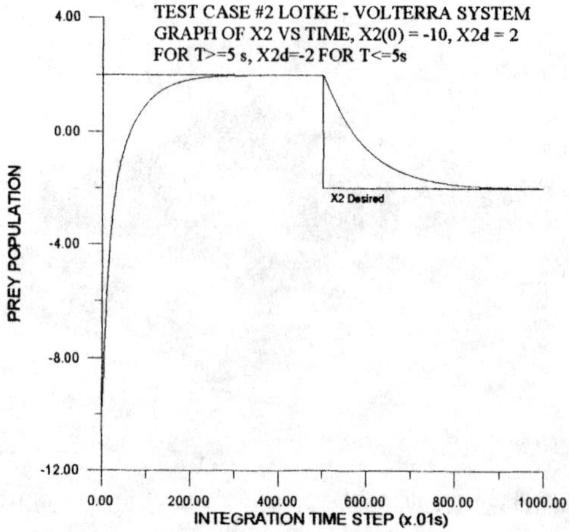


Figure 3. X2 vs. Time

Figure 3 is a simulation of the system with the approximators implemented in the closed loop. The desired condition is changed in mid-simulation.

## STABILITY AUGMENTATION IN A HIGH PERFORMANCE AIRCRAFT

A generic, state-of-the-art, high-performance aircraft model, representative of a current day fighter aircraft can be found in [15]. The model has fully nonlinear aerodynamics and kinematics. It was desired to use neuro gain approximators as a method of providing stability augmentation to the model. In order to simplify the problem, initial efforts have been aimed at improving the aircraft's short period damping only.

**Define the Hypersphere and the Scheduling Variables**
For this experiment, the hypersphere was defined using the operational envelope of the aircraft in terms of altitude and Mach number. The operational envelope is described fully in [15]. A grid of points were chosen to cover the envelope between 500 and 1000 fps and zero and 20,000 ft.. The velocity was incremented by 100 fps and altitude was incremented by 5,000 ft.

Similar to the experiment with the Lotke-Volterra system, the scheduling variables for this problem were obtained from the definition of the Hypersphere. Two scheduling variables were used. The first, was Mach number and the second was altitude.

**Design a Linear Controller at Each Point**

The nonlinear system was first linearized at each of the chosen equilibrium points using the perturbation method. The next step was to extract a simplified system which approximated the short period mode of the aircraft. Now, pole placement was used to define linear feedback gains to improve the short period qualities.

**Train a NN Approximator**

The first step to accomplish this was to create a neural network training data file which used the scheduling variable as the inputs and the linear gains as outputs of NN training pairs.

For this experiment, two networks were trained each having three inputs, Mach number, altitude, and the bias. Each NN had a single output, the first net's output was k1 and the second net's output was k2.

**Analyze the Approximator**

For this experiment the NN approximators were trying to map the nonlinear space between linear controllers. The neural gain approximators were put in a closed loop system, at off-training point equilibrium points. And the short period poles were analyzed and compared to the open loop (See Table 1). It should be noted that actual simulations of the aircraft for this controller were hard to analyze as controlling only the short period mode also affected the phugoid mode.

| EQUILIBRIUM POINT | OPEN LOOP POLES | CLOSED LOOP POLES |
|---|---|---|
| 7,000 ft, 525 fps | -1.5434 ± 2.4534i | -1.6032 ± 0.0529i |
| 9,000 ft, 550 fps | -1.5006 ± 2.5105i | -1.5, -1.7375 |
| 13,000 ft 625 fps | -1.4187 ± 2.7107i | -1.6250 ± 0.2872i |
| 17,000 ft, 650 fps | -1.2492 ± 2.6571i | -1.6152 ± 0.3425i |
| 19,000 ft, 750 fps | -1.1421 ± 2.8174i | -1.5852 ± 0.1846i |

Table 1: Open Loop Poles Compared To Closed Loop Poles

Notice that each of the cases in Table 1 are off-training points and that in all cases, there was a substantial improvement in the short period damping characteristics.

## CONCLUSIONS AND FUTURE RECOMMENDATIONS

During the course of this study, it was concluded that if a system is linearizable, then there is little advantage in using nonlinear local controllers when trying to implement a global nonlinear controller with scheduling type control laws. If we can linearize the system then we can use linear controllers along with neuro-gain approximation to derive a psuedo-global valid control law.

A drawback of this technique is that we must be able to get correlated control laws. Linearization is one easy way of doing this, however, it may also be possible to employ a similar technique using parameterized nonlinear local controllers.

The most pressing area of future research must be method justification. It is easy to conceptualize using Lyapunovs linear stability techniques along with the understanding of what a neural net is doing when it maps a higher dimensional surface, why this method is valid. Proving such a thing may be more challenging.

**REFERENCES**


[1] Shamma, J.S., Athans, M., ( 1992). "Gain Scheduling: Potential Hazards and Possible Remedies", Control Systems Magazine, Vol. 12, No. 3, June.

[2] Lainiotis, D.G.,(1976). "Partitioning: A Unifying Framework for Adaptive Systems, I: Estimation, II: Control," *Proceedings of the IEEE*, 64: 1126-1143, 1182-1197, August.

[3] Athans, M. et al., (1977). "The Stochastic Control of the F-8C Aircraft Using a Multiple Model Adaptive Control (MMAC) Method-Part I: Equilibrium Flight, " *IEEE Transactions on Automatic Control*, AC-22: 768-780.

[4] Narendra, K.S., Balakrishnan, J., Ciliz, M.(1995). "Adaptation and Learning Using Multiple Models, Switching, and Tuning", IEEE Control Systems, June.

[5] Krishnakumar, K., (1993). "Immunized Artificial Neural Systems", World Congress on Neural Networks, Portland, July 11-15.

[6] Krishnakumar, K., (1992). "Immunized Neural Control: Concepts and Initial Implementation Results", BER Report #513-177, The University of Alabama, Tuscaloosa AL.

[7] Krishnakumar, K., Satyadas, A., Neidhoefer, J., (1995). "An Immune System Framework For Integrating Computational Intelligence Paradigms With Applications To Adaptive Control", Computational Intelligence (A Dynamic System Perspective), IEEE Press.

[8] Slotine, J.J., Li, W., "Applied Nonlinear Control",(1991).Prentice Hall, Englewood Cliffs, New Jersey.

[9] Rugh, W.J. (1991). "Analytical Framework for Gain Scheduling", Control Systems Magazine, Vol. 11, No. 1, Jan..

[10] Cook, P., (1986). "Nonlinear Dynamical Systems", Prentice / Hall International.

[11] Krishnakumar, K., (1992). "Immunized Neurocontrol: Concepts and Initial Results", COGANN-92, A Workshop on Combinations of GA and NN. June 6.

[12] Shamma, J.S., (1988). "Analysis and design of gain scheduled control systems," Ph.D. thesis, Dept. of Mechanical Engineering, M.I.T..

[13] Shamma, J.S., and Athans, M., (1990). "Analysis of nonlinear gain scheduled control systems," *IEEE Trans. Auto Control*, vol. AC-35, no. 8, pp. 898-907.

[14] Shamma, J.S. and Athans, M., (1991). "Guaranteed properties of gain scheduled control of linear parameter-varying plants," *Automatica*, vol. 27, no. 3, pp. 559-564, May.

[15] Brumbaugh, R., (1994). "Aircraft Model for the AIAA Controls Design Challenge", Journal of Guidance, Control, and Dynamics, Vol. 17 No. 4. July-August.

[16] Bailey, J.E., Krishnakumar, K., Homaifar, A., (1985). "GPS-Based Automatic Landing System For The AH-64 Helicopter", BER Report No. 361-177, September.

# NEUROCONTROLLERS FOR BALL-AND-BEAM SYSTEMS

**PAUL H. EATON,**
**DANIL V. PROKHOROV,**
**DONALD C. WUNSCH, II**
*Applied Computational Intelligence Laboratory*
*Department of Electrical Engineering, Box 43102*
*Texas Tech University, Lubbock, TX 79409; http://www.acil.ttu.edu*

***ABSTRACT*:**
The ball-and-beam problem is a well known benchmark for testing new control algorithms. We deal with the off-line training of neurocontrollers to balance the ball at a fixed arbitrary location on the beam. Resulting neurocontrollers are tested on our original hardware. We record a time series of positions of the ball, and it is the only signal permitted to use for identification and control. We utilize recurrent neural networks for all modules of our designs. We obtain a sufficiently accurate neural network identification model of the system using the parallel identification method. Two neurocontrol designs are discussed. The conventional approach is based on truncated backpropagation through time. Another design uses an adaptive critic approach, which is a form of approximate dynamic programming.

## INSTALLATION FOR DATA ACQUISITION AND EXPERIMENTS

The first author designed and built the ball-and-beam setup as shown in Figure 1. The design uses movement like a cradle with a pivot point high above the actual beam. This distinguishes our design from others' (see e.g., (Godfrey and Li, 1995)). The wooden black beam on which the ball rolls is about 35 cm of length with foam bumpers at the ends, and the white ball is 2 cm in diameter. The beam has two main electrical components - the linear Charge-Coupled Device (CCD) used to determine the position of the ball, and the servo motor to change the angle of the beam. The cradle design allows the CCD to remain in the same position above the beam no matter what angle the beam is at. The CCD scans the length of the beam looking for the bright ball, and determines its position.

The servo motor is set to a desired position using a Pulsed Width Modulation (PWM) control input. The servo motor can be set by sending any control input from 0 to 255, but the hardware limits in the design reduce the allowable values to be between 50 and 140, and these numbers change the angle of the beam anywhere between 79 degrees and 103 degrees, with 90 corresponding to a flat beam. This allows for changes in the beam angle of $0.26^{\circ}$.

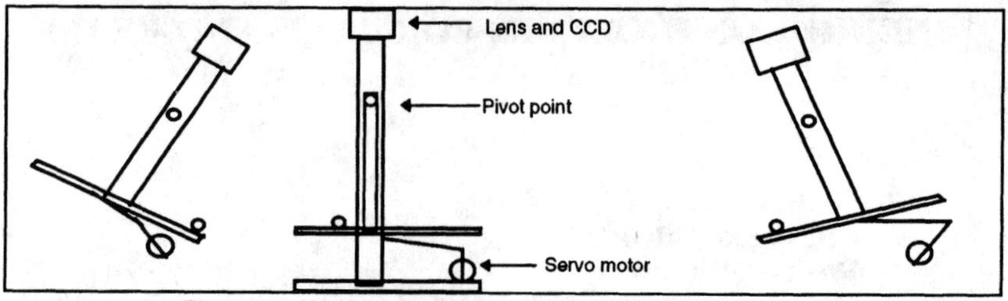

**Figure 1: Hardware setup for ball-and-beam problem.**

For the detection of the ball, a 512-element linear CCD is used. The CCD is placed behind a camera lens which focuses the image of the beam onto its elements. The ball's position can be determined down to 2 mm accuracy, which corresponds to 0.006 in units normalized by the length of the beam.

The data files are obtained by recording the position of the ball and the PWM signal at discrete intervals. The time between samples is created in the software by using the computer clock which provides 18 ticks per second, and after each tick, the ball position and PWM value are stored. The data is then normalized so the largest and smallest values for the ball position and the PWM is between 1 and -1. The last step is to organize the data for the program to read. This is done by setting the data up in three columns; the current ball position, the current PWM signal, and the ball position at the next time step (the third column is considered to be the desired output for the model given the first two columns as inputs). Each line in the file then corresponds to a different time step.

## CONVENTIONAL NEUROCONTROL

### System Identification

One of the main components of this project is the creation of a neural network model (or identification network) of the ball-and-beam systems. Since the training of the controller (and later the critic) is fairly slow, the networks cannot be trained in real-time. This makes it necessary to record many data files showing how the system behaves, and these files are then used to train a neural model which can reproduce outputs similar to the systems outputs. In this project, the model network takes two inputs, the ball position, B(t) at time (t), and the PWM(t) signal at time (t). The model will then produce an output which represents the ball position at the next time step B(t+1). Our design uses a network with 2 neurons in a hidden layer with 1 output, and the neurons in the hidden layer have recurrent connections which pass the output of the neurons from the previous time step as inputs to the current neurons.

Two different designs were tested for the model; the series-parallel training (also known as teacher forcing) and the parallel training methods, the later being shown connected to the controller in Figure 2. In the series-parallel method, both B(t) and PWM(t) are read in from a data file, and the ball position for the next time step is calculated. The error is then calculated by comparing the network's predicted position and the actual position, and this error is used to update the weights of the network.

In the parallel method of training (Narendra and Parthasarathy, 1990), the input B(t) for the next time step is connected to the output from the model network of the current time step. In the parallel configuration, the model needs only the first few ball positions to be read in from the data file, and after these few first points the ball position is determined by the previous time step. Each output is compared with the corresponding desired output ball position from the data file, and an error is determined equal to the squared difference between ouputs. This parallel method was worse than the series-parallel configuration on training data, but performs much better when the model network is being used to train a controller. While series-parallel is better for single-step ahead predictions, this type of model does not truly grasp the dynamics of the system, but must rely on its two inputs being actual hardware data, keeping previous errors from affecting current outputs.

The training of the model in this parallel configuration offers challenges. First, the optimal length of a data file is difficult to determine. If the data file is too short, the network will not grasp all of the dynamics of the system it is modeling. Conversely, if the data file is too long, it is impossible to train a model with any resemblance to the real system. Even with a good model, there will be a small amount of error between actual and modeled ball positions. With each time step, this error builds up. This can cause improper training when the model might make a correct move in predicting a change in the ball position in a time step, but be penalized because the previous time step prediction was incorrect. As an example, the model from the parallel configuration showed the sum of squared errors of 20.01, and the series-parallel model had the sum of 71.469 when tested on the same 1000 sample data file.

The training method used in this project is BackPropagation Through Time truncated at depth h (BPTT(h)) with the Node-Decoupled Extended Kalman Filtering. The BPTT(h) uses h copies of the network(s) with the same weights but different node activations corresponding to different time steps. We typically used h=10. Starting at the beginning of the data, each copy calculates its outputs. When the end is reached, the error is calculated for the last copy, and the weights are updated. The copies are then moved forward one time step and the process is repeated (Williams and Zisper, 1995).

Among many methods proposed for training recurrent neural networks, the Node-Decoupled Extended Kalman Filter (NDEKF) training stands out as an effective and powerful technique. The NDEKF is viewed as a parameter identification problem for a nonlinear dynamic system (recurrent neural network). This method adapts weights of the network in a pattern-by-pattern fashion accumulating important training information in approximate error covariance matrices and providing individually adjusted updates for each of the network's weights (Puskorius and Feldkamp, 1994).

**Training Conventional Neurocontroller**

Once we have the model network, we can now proceed with the designing and training of our system's controller. In both our conventional and the adaptive critic-based neurocontrollers, the design of the controller has been the same, but their training differs. The controller, also called the action network, is a two layer neural network with two nodes in the hidden layer and one node in the output layer. The network uses the two inputs, the current ball position (B) and the desired ball position (Bdesired), to produce an output which is a PWM signal to be used by the model. The current ball position comes from the previous time step model network and the desired ball position is read in from a file. The controller then calculates a PWM value that would best move the ball to its desired position for the next time step. The controller is shown with its connections to the model in Figure 2. Although we describe experiments on balancing around a fixed point, a neurocontroller trained in this framework is capable of moving the ball in a certain desired pattern, more complicated than a simple skyline function (see below). The output of the controller (PWM) is a sigmoid node, where PWM=-1.0 sets the beam at the highest angle possible in one direction, and PWM=+1.0 does the highest angle in the other direction. The controller is trained using BPTT(10) with a single error term corresponding to the difference between the ball position and the desired ball position at each time step. The weights of the controller are then updated using NDEKF and appropriate derivatives with respect to the controller weights. Conventional neurocontroller performance is illustrated in the next section in comparison to the neurocontroller trained using DHP (see Figure 4).

## NEUROCONTROLLER BASED ON DHP

Another approach to training a controller is to create a third neural network, the Critic, shown in Figure 3. In conventional neurocontrol, the controller is trained using an error term derived through BPTT from how well the network performed in the current time step or a series of time steps of a particular trajectory. The critic network is trained with a "look ahead" algorithm which is potentially more efficient than a single trajectory BPTT. In several types of adaptive critic designs originally proposed by Werbos (Werbos, 1992), the critic takes an input from the model network, in this case, the predicted ball position, and produces an output, which is an assessment of how well the controller is performing. Two adaptive critic designs were implemented, Heuristic Dynamic Programming (HDP) and Dual Heuristic

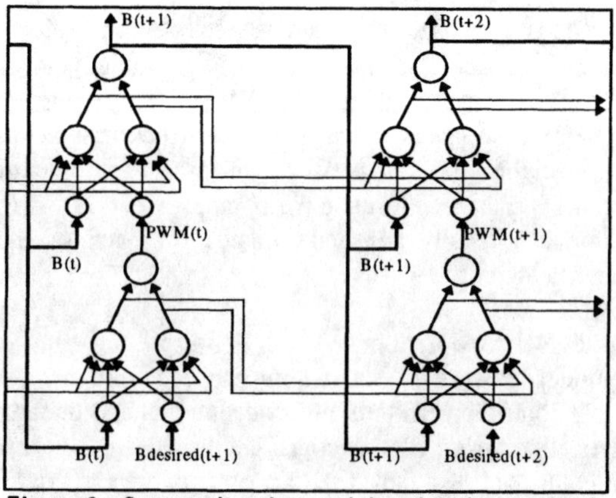


**Figure 2: Connecting the model and action network.**

Programming (DHP). Both designs output either a function J(t), called the "cost-to-go" function, or its state derivatives ∂J(t)/∂X(t). The critic is repeatedly trained until it can accurately match its target output(s) at each step, as described below. Once the critic is trained, the weights in the critic are fixed, and the controller weights are updated based on the critic output at each step (Prokhorov and Wunsch, 1995). In the HDP design, the critic produces the J(t) term directly as its output. The derivative of this term is then calculated and this derivative is used to update the controller weights. In the DHP design, the critic produces the derivative of J(t) as its output. This makes the training of the controller simpler, since this term can be used directly, but is slightly more difficult in training the critic network, as described below.

In HDP, the critic produces a function J(t) as an output. This output tries to drive the Error(t) term to zero using the equation

$$Error(t) = J(t) - \gamma J(t+1) - U(t) \tag{1}$$

where

$$U(t) = \frac{1}{2}(B(t) - Bdesired(t))^2 \tag{2}$$

Once trained, the critic network would be applied to training the controller. This is done by creating an error term by backpropagating the output of the critic network through the critic, creating the term ∂J(t+1)/∂B(t+1), which is treated as the error term for the model. This error is then backpropagated through the model, giving the necessary error term for the controller, ∂J(t+1)/∂PWM(t). When training of the critic in HDP was completed, it became apparent that this derivative term was not accurate enough for training the controller. This led to switching to the DHP design.

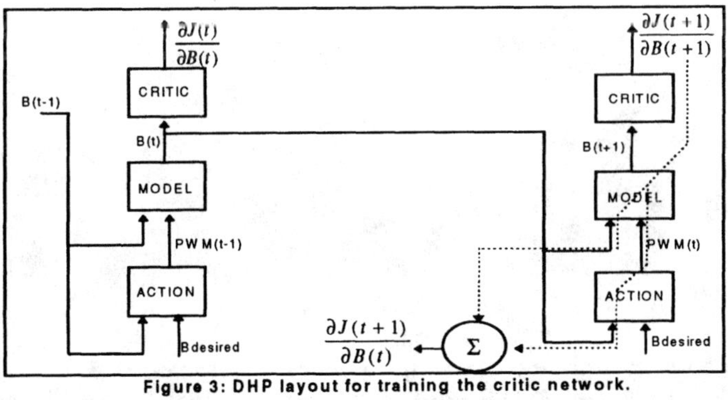


**Figure 3: DHP layout for training the critic network.**

The DHP design uses a critic whose output tries to minimize the error term:

$$Error(t) = \frac{\partial J(t)}{\partial B(t)} - \gamma \frac{\partial J(t+1)}{\partial B(t)} - \frac{\partial U(t)}{\partial B(t)} \quad (3)$$

through the model and controller. The critic network is trained using NDEKF until the critic output error converges to its minimum. This convergence of the critic is simple as it usually converges in less than 20 passes through the data sets of 300 time steps and reaches the sum of squared errors of 0.005. After the critic is trained, the training switches to updating the controller weights. The forward propagation travels through the controller, model, and critic. Necessary derivatives are then computed through BPTT(10) (it is used for both critic and controller training). The output of the critic is then used as the error term for updating the controller. As the controller reaches its minimum error for the data by moving the ball to the desired position as quickly as possible, the process is continued by training a new critic which evaluates the newly trained controller's performance. This alternating training of the critic and controller networks resulted in the DHP-based neurocontroller with better overall performance than that of the conventional neurocontroller. Balancing around multiple fixed points (skyline pattern) required adding a second input to the critic network which would read in the desired ball position for the current time step. An example of the DHP-based neurocontroller performance is shown in Figure 5. We emphasize that both Figures 4 and 5 are results of testing the neurocontrollers on the actual hardware.

## CONCLUSION

We describe off-line training of neurocontrollers for the ball-and-beam benchmark and their subsequent testing on the actual hardware. Results to date indicate better performance of the DHP-based neurocontroller than that of the conventionally trained neurocontroller. Future research will be devoted to adapting the NDEKF training procedure for simultaneous minimization of multiple error terms in the time

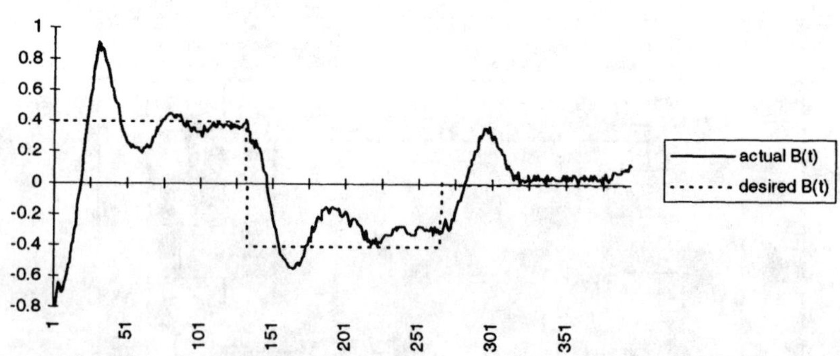


**Figure 4: Conventional neurocontroller performance on a skyline function. The data is normalized with the edges of the beam at ±1 and 0.0 corresponding to the center.**

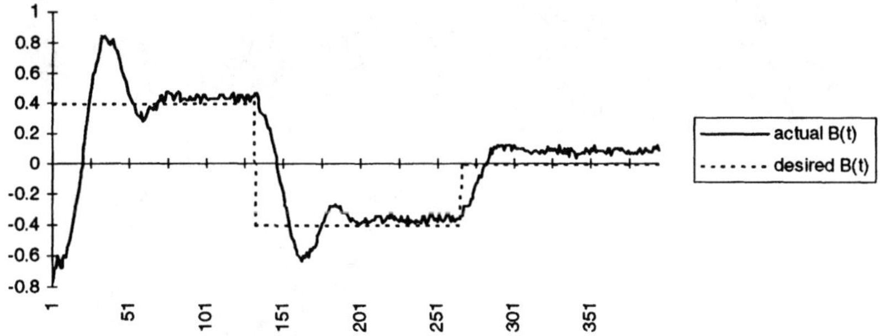


**Figure 5: Critic based neurocontroller performance on a skyline function.**

sequence to improve performance of the conventional neurocontroller. In addition, we will study neurocontrol of other ball-and-beam systems derived from ours using, e.g., an inhomogeneous friction coating on the beam's surface.

## REFERENCES


Godfrey, N., Li, H., Ji, Y., and Marcy, W., (1995). Real Time Fuzzy Logic Controller for Balancing a Beam-and-Ball System, In *Fuzzy Logic and Intelligent Systems* (H.Li, M.Gupta, Eds.), Kluwer AP. pp. 157 -154.

Narendra, K., and Parthasarathy, K., (1990). Identification and Control of Dynamical SystemsUsing Neural Networks, *IEEE Trans. on Neural Networks*, Vol. 1, No. 1.

Williams, R., and Zipser, D., (1995). Gradient-Based Learning Algorithms for Recurrent Networks and Their Computational Complexity, Ch. 13 in *Backpropagation: Theory, Architecture, and Applications* (Chauvin and Rumelhart, Eds.), LEA.

Puskorius, G., and Feldkamp, L., (1994). Neurocontrol of Nonlinear Dynamical Systems with Kalman Filter Trained Recurrent Networks, *IEEE Trans. on Neural Networks*, Vol. 5, No. 2, pp. 279-297.

Werbos, P., (1992). Approximate Dynamic Programming for Real-Time Control and Neural Modeling, In *Handbook of Intelligent Control* (D. White, D. Sofge, Eds.), Van Nostrand.

Prokhorov, D., and Wunsch, D., (1995). Adaptive Critic Designs, submitted to *IEEE Trans. on Neural Networks*, abstract is available at http://www.ieee.org/nnc. (Earlier version of this paper appeared in *Computational Intelligence: A Dynamic System Perspective* (R. Marks et. al., Eds.), IEEE Press, Nov. 1995, pp. 98-107.)

# CONTROL OF A NONLINEAR MULTIVARIABLE SYSTEM WITH ADAPTIVE CRITIC DESIGNS

**NIKITA A. VISNEVSKI AND DANIL V. PROKHOROV**[1]
*Applied Computational Intelligence Laboratory*
*Department of Electrical Engineering, Box 43102*
*Texas Tech University, Lubbock, TX 79409-3102, USA*

***ABSTRACT:***
We study neurocontrol of a nonlinear multivariable plant proposed by Narendra and Mukhopadhayay (1994). The plant's outputs should closely follow two independent reference signals. To train neural networks we utilize the Node-Decoupled Extended Kalman Filter (NDEKF) algorithm. After briefly discussing our results with conventional neurocontrol, we describe solutions of the problem based on two Adaptive Critic Designs (ACDs): Heuristic Dynamic Programming (HDP) and Dual Heuristic dynamic Programming (DHP). These designs are approximations of dynamic programming using neural networks. The DHP solution is much more successful than the one of HDP, and it is similar with results of conventional neurocontrol.

## PROBLEM FORMULATION

We consider the following nonlinear multivariable plant (Narendra and Mukhopadhayay, 1994):

$$x_1(t+1) = 0.9x_1(t)\sin[x_2(t)] + \left[2 + 1.5\frac{x_1(t)u_1(t)}{1+x_1^2(t)u_1^2(t)}\right]u_1(t) + \left[x_1(t) + \frac{2x_1(t)}{1+x_1^2(t)}\right]u_2(t)$$

$$x_2(t+1) = x_3(t)\{1+\sin[2x_3(t)]\} + \frac{x_3(t)}{1+x_3^2(t)}$$

$$x_3(t+1) = \{3+\sin[2x_1(t)]\}u_2(t)$$

$$y_1(t) = x_1(t) \qquad y_2(t) = x_2(t) \tag{1}$$

where:

$x_i(t),\ i\in(1,3)$ are state variables of the nonlinear system,
$y_i(t),\ i\in(1,2)$ are outputs of the system,
$u_i(t),\ i\in(1,2)$ are control inputs of the system.

As we can see from state equations above, time delays from $u_1$ and $u_2$ to $y_1$ are equal to one, the delay from $u_2$ to $y_2$ is two, and the one from $u_1$ to $y_2$ is three. The plant is stable for constant control inputs, but it is highly unstable for small input

[1] Authors gratefully acknowledge support from the National Science Foundation (Neuroengineering Program Grant # ESC-9413120), Ford Motor Co. and the Texas Tech Center for Applied Research. We would also like to thank Drs. Lee Feldkamp and Don Wunsch for many helpful discussions.

changes. The linearized system at $x=0$ and $u=0$ is controllable, observable and of minimal phase (Narendra and Mukhopadhayay, 1994). The goal is to develop a controller for the plant (1) to track two independent reference signals $y_1^*(t)$ and $y_2^*(t)$ as closely as possible, minimizing the following error measure over time:

$$\sum_t \gamma^t \left\{ \left[ y_1(t+1) - y_1^*(t+1) \right]^2 + \left[ y_2(t+2) - y_2^*(t+2) \right]^2 \right\} \qquad (2)$$

where $\gamma$ is the discount factor ($0<\gamma<1$), and we assume that $y_1^*(t+1) = r_1(t)$ and $y_2^*(t+2) = r_2(t)$ are specified at instant $t$.

Authors of the original paper have explored several cases for control of this plant varying from the case of fully accessible states and known plant equations to the case of only input-output data used for control. This paper deals with the first case, when state variables of the system are accessible and plant equations are known.

## TRAINING NEURAL NETWORKS

We use the feed-forward multilayer perceptron with forward method of obtaining gradients similar to those described in (Puskorius and Feldkamp, 1994), and the NDEKF algorithm for weight updates. Among many methods proposed for training neural networks, Extended Kalman Filter-based training stands out as an effective and powerful technique (Puskorius and Feldkamp, 1994). This training is viewed as a parameter identification problem for a nonlinear system - neural network. This method adapts weights of the network in a pattern-by-pattern fashion accumulating important training information in approximate error covariance matrices and providing individually adjusted updates for each of the network's weights.

## CONVENTIONAL NEUROCONTROL

Under the name of conventional neurocontrol we mean the well-known model-reference control scheme using neural networks for either identification model or controller, or for both of them. As mentioned in the Section 1, we assume the states accessible and the plant equations known. Proper training can not be started without first obtaining a stabilizing controller network. We obtain a stabilizing controller network by training for the plant equations (1) linearized around the origin and $u=0$. Our architecture for the controller network is 5×6×2 inputting $y_1(t)$, $y_2(t)$, $x_3(t)$, $y_1^*(t+1)$, and $y_2^*(t+1)$ and outputting $u_1(t)$ and $u_2(t)$. We perform one pass of training through a set of 300 vectors, with the reference signals $y_1^*(t)$ and $y_2^*(t)$ chosen randomly from the range ±1.5 for each instant $t$, where $t \in (1,300)$.

When training neurocontroller for the full system of the plant equations (1), it is crucial to correctly obtain all the necessary derivatives of the plant outputs with respect to the control inputs. We use the following nonzero derivatives: $\partial y_1(t+1)/\partial u_1(t)$, $\partial y_1(t+1)/\partial u_2(t)$, $\partial^+ y_2(t+2)/\partial u_2(t)$, $\partial^+ y_2(t+3)/\partial u_1(t)$, and $\partial^+ y_2(t+3)/\partial u_2(t)$. Here the notation "+" defines ordered derivatives described in details by P. Werbos (1990). Presence of the last two derivatives requires augmentation of the original error measure (2) by another quadratic term depending on $y_2^*(t+3) = r_3(t) = r_2(t+1)$, the shifted version of the second reference signal. Thus, our new error measure to be minimized over time is:

$$\sum_t \gamma^t \left\{ \left[y_1(t+1) - y_1^*(t+1)\right]^2 + \left[y_2(t+2) - y_2^*(t+2)\right]^2 + \left[y_2(t+3) - y_2^*(t+3)\right]^2 \right\} \quad (3)$$

Training the controller network starts with initial weights obtained for training on the linearized plant, as described above. We then perform 400 passes of the NDEKF training algorithm over an ensemble of desired trajectories $r_1(t)$ and $r_2(t)$ specified for $t \in (1,300)$. Results of testing performance of the neurocontroller are shown in Figure 1 for the following reference trajectories not used in training (Narendra and Mukhopadhayay, 1994):

$$r_1(t) = 0.75\left(\sin\left[\frac{2\pi t}{50}\right] + \sin\left[\frac{2\pi t}{10}\right]\right) \text{ and } r_1(t) = 0.75\left(\sin\left[\frac{2\pi t}{30}\right] + \sin\left[\frac{2\pi t}{20}\right]\right) \quad (4)$$

Our results are similar with those reported in (Narendra and Mukhopadhayay, 1994). However, they are obtained much faster and for a significantly smaller size of the controller network, the former advantage being attributed to the NDEKF training algorithm.

**ADAPTIVE CRITIC DESIGNS**

A family of Adaptive Critic Designs (ACD) was proposed by Werbos (Werbos, 1992) as a new optimization technique combining together concepts of reinforcement learning and approximate dynamic programming. In ACD we have two networks called Critic and Action which are connected together directly (Action-dependent designs) or through an identification model of a plant (Model-dependent designs). Model-dependent designs are considered to be more advanced, but they should have a model accurate enough to operate. As in the case of conventional neurocontrol, we use the plant equations (1) instead of a separate identification network. The Action (controller) network has plant outputs $y_1(t)$, $y_2(t)$, the state variable $x_3(t)$, and next step desired plant outputs $y_1^*(t+1)$ and $y_2^*(t+1)$ as its inputs. It outputs control signals $u_1(t)$ and $u_2(t)$ which serve as inputs for the plant. Actual and desired outputs of the plant serve as inputs for the Critic which is trained to estimate the well-known cost-to-go function from the Bellman equation of dynamic programming or some function related to it. The goal is to train the Action network to minimize this estimate over time, thereby finding an optimal control strategy.

There are three basic implementations of ACD called Heuristic Dynamic Programming (HDP), Dual Heuristic dynamic Programming (DHP), and Globalized Dual Heuristic dynamic Programming (GDHP) (Werbos, 1990). Here we use HDP and DHP and compare results obtained by both designs.

**Solution by HDP**

We introduce the utility or the single step cost function for the ACD in the following form:

$$U[t] = \frac{1}{2}\left\{ \left[y_1(t+1) - y_1^*(t+1)\right]^2 + \left[y_2(t+2) - y_2^*(t+2)\right]^2 + \left[y_2(t+3) - y_2^*(t+3)\right]^2 \right\} \quad (5)$$

where:

$y_i(t+j)$ is the $i$-th output of the system at the time instance $t+j$,

$y^*_i(t+j)$ is the value of the $i$-th reference signal at the time instance $t+j$.

In terms of specifications above our HDP design is defined in the following form:

$$\varepsilon[t]=\gamma J[t+1]+U[t]-J[t] \quad E_C=\frac{1}{2}\varepsilon^2[t] \quad \frac{\partial^+ E_C}{\partial\omega^l_{gh}}=-\varepsilon[t]\frac{\partial^+ J[t]}{\partial\omega^l_{gh}} \tag{6}$$

where:

| | |
|---|---|
| $J[t]$ | is an estimate of the cost-to-go function of the Bellman's equation of dynamic programming, |
| $\omega^l_{gh}\in W$ | is $h$-th weight of the $g$-th neuron in the $l$-th layer of the Critic, |
| $\varepsilon[t]$ | is the error term for the training process, |
| $E_C$ | is the squared error for the training procedure of the Critic, |
| $\gamma\in(0,1)$ | is the discount factor (in all our experiments $\gamma=0.5$). |

Note that $U[t]$ corresponds to a single term in the sum of the error measure (3). $J[t]$ in (6) is an estimation of the sum in (3) running to infinity.

Plant states $y_1(t+1)$, $y_2(t+2)$, and $y_2(t+3)$ and their desired values $y^*_1(t+1)$, $y^*_2(t+2)$, and $y^*_2(t+3)$ are fed into the Critic which is trained to generate estimates of cost-to-go function $J[t]$ based on this inputs. As in the Section 3, we used the same desired trajectories $r_1(t)$ and $r_2(t)$ specified for $t\in 1{,}300$. The utility function $U[t]$ is used in the training process in the error term (6). To produce the $J[t+1]$ term it is necessary to make one more step in time using $y_1(t+2)$, $y_2(t+3)$, and $y_2(t+4)$, and than restore all the state variables of the Critic, Action and plant to the values at the time $t$. Weight updates are carried out by the NDEKF algorithm.

After the Critic is trained, its weights become fixed and Action training procedure starts. The error term for this training can be written in the following form:

$$\varepsilon_i[t]=0-\frac{\partial^+ J[t]}{\partial u_i(t)} \quad E_A=\frac{1}{2}\sum_{i=1}^{NA_L}\varepsilon_i^2[t] \tag{7}$$

$$\frac{\partial^+ J[t]}{\partial u_i(t)}=\frac{\partial^+ J[t]}{\partial y_1(t+1)}\frac{\partial^+ y_1(t+1)}{\partial u_i(t)}+\frac{\partial^+ J[t]}{\partial y_2(t+2)}\frac{\partial^+ y_2(t+2)}{\partial u_i(t)}+\frac{\partial^+ J[t]}{\partial y_2(t+3)}\frac{\partial^+ y_2(t+3)}{\partial u_i(t)} \tag{8}$$

and the training itself can be expressed as below:

$$\frac{\partial^+ E_A}{\partial v^l_{gh}}=-\sum_{i=1}^{NA_L}\varepsilon_i[t]\frac{\partial^+ u_i[t]}{\partial v^l_{gh}} \tag{9}$$

where:

| | |
|---|---|
| $v^l_{gh}\in V$ | are Action's weights, |
| $\varepsilon_i[t]$ | is the error term for the i-th network's output, and $E_A$ is the squared error of the Action network, |
| $NA_L$ | is the number of outputs in the Action network. |

The Action network generates control signals $u_j$ for the plant at time steps $t$, $t+1$, and $t+2$ causing the plant to produce output signals $y_1(t+1)$, $y_2(t+2)$, and $y_2(t+3)$

which serve as Critic's inputs, and, together with reference signals at corresponding time instances, are used to obtain *U[t]*. Critic based on its inputs produces the estimate of *J[t]*. Propagating through the Critic and the plant copies at the time $t$+3, $t$+2, and $t$+1 appropriate derivatives for the learning process are obtained (8). These derivatives are used in the Action's training, where we exploit the NDEKF algorithm. Training proceeds until convergence of the Action's weights. The Critic is then trained again from randomly initialized weights and keeping the Action's weights fixed. These iterations of Critic-Action training are continued until no improvement is attained for both the networks (Prokhorov and Wunsch, 1995). Results of the procedure are shown in Figure 2 We used the same Action's architecture as in conventional neurocontrol case, and the Critic was designed as a two-layer network 6×6×1. These results are analyzed and compared to results obtained with an alternative ACD in the next section.

**Solution by DHP**

The major difference between HDP and DHP designs is that the Critic network outputs estimates of derivatives of the cost-to-go function with respect to Critic's inputs, rather than the cost-to-go function itself. The error terms for DHP Critic are:

$$\varepsilon_1[t]=\gamma\frac{\partial^+ J[t+1]}{\partial y_1(t+1)}+\frac{\partial^+ U[t]}{\partial y_1(t+1)}-\lambda_1[t] \quad \varepsilon_2[t]=\gamma\frac{\partial^+ J[t+1]}{\partial y_2(t+2)}+\frac{\partial^+ U[t]}{\partial y_2(t+2)}-\lambda_2[t]$$

$$\varepsilon_3[t]=\gamma\frac{\partial^+ J[t+1]}{\partial y_2(t+3)}+\frac{\partial^+ U[t]}{\partial y_2(t+3)}-\lambda_3[t] \tag{10}$$

where $\lambda_1[t]$, $\lambda_2[t]$, $\lambda_3[t]$ are the outputs of the Critic, estimates of derivatives of *J[t]* with respect to $y_1(t+1)$, $y_2(t+2)$, and $y_2(t+3)$, respectively.

The derivative of *J[t+1]* with respect to Critic's inputs at the time $t$ have to be obtained by propagating along all possible pathways through the Action network and the plant equations (1) of instances $t$+2 and $t$+3.

Training procedure for the Critic is described below:

$$E_C=\frac{1}{2}\sum_{i=1}^{NC_L}\varepsilon_i^2[t] \quad \frac{\partial^+ E_C}{\partial \omega^l_{gh}}=-\sum_{i=1}^{NC_L}\varepsilon_i[k]\frac{\partial^+ J[t]}{\partial \omega^l_{gh}} \tag{11}$$

where:

$\varepsilon_i[t]$ is the error term for the $i$-th network's output, and $E_C$ is the squared error of the Critic network,

$NC_L$ is the number of outputs in the Critic network.

The Action's training remains almost the same as in HDP, except for derivatives of *J[t]* with respect to $u_i(t)$, which in DHP case should be written in the following form:

$$\frac{\partial^+ J[t]}{\partial u_i(t)}=\lambda_1[t]\frac{\partial^+ y_1(t+1)}{\partial u_i(t)}+\lambda_2[t]\frac{\partial^+ y_2(t+2)}{\partial u_i(t)}+\lambda_3[t]\frac{\partial^+ y_2(t+3)}{\partial u_i(t)} \tag{12}$$

where $\lambda_i[t]$ is the $i$-th Critic's output.

As in HDP, weights of both Critic and Action networks are updated by the NDEKF algorithm.

The same alternating Critic's and Action's training was carried out as in HDP. Results are in Figure 3. The Action network has the same architectures as in the case of HDP, and the Critic is a 6×6×3 network. Our results show that the Action network trained in DHP can control the plant with almost the same quality as the conventional neurocontroller. However, the Action network trained in HDP has much worse final representation than the conventional neurocontroller (Figure 2). This is a result of significantly less accurate approximation of gradients of the cost-to-go function by the Critic in HDP. Indeed, such a Critic is directly trained to estimate the cost-to-go only. However, it is not the cost-to-go function itself but its derivatives with respect to the plant outputs that be used in the Action's training process. DHP design does not have this disadvantage because its Critic is directly trained to estimate the derivatives of the cost-to-go which are used in the Action's training procedure.

In our simulations we used small networks to show advantages of the Node-Decoupled Extended Kalman Filter algorithm over the usual gradient descent training procedure (Narendra and Mukhopadhayay, 1994), and our results can be improved if sizes of networks (especially, Critic's) are increased.

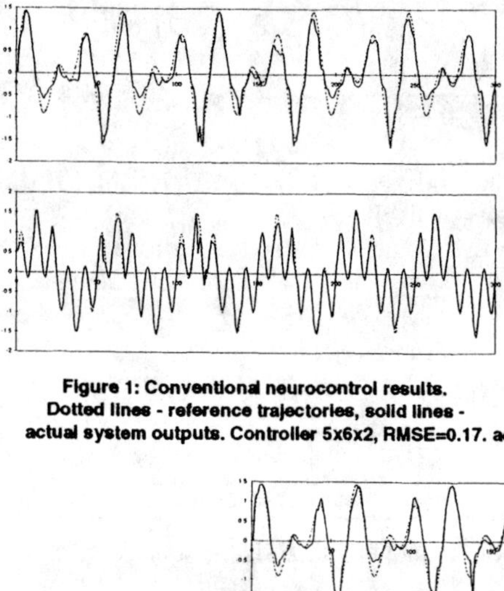

**Figure 1: Conventional neurocontrol results. Dotted lines - reference trajectories, solid lines - actual system outputs. Controller 5x6x2, RMSE=0.17.**

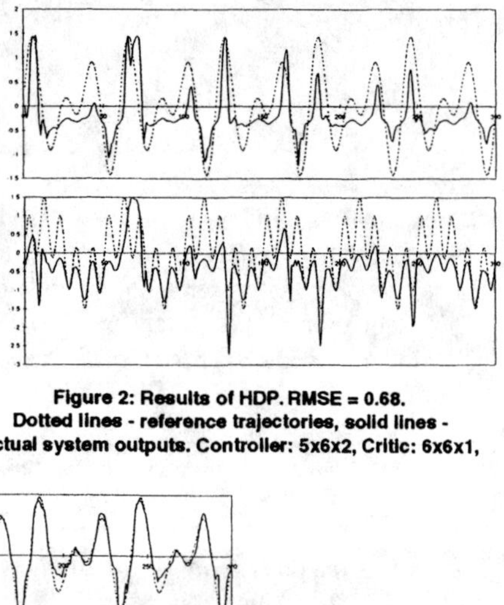

**Figure 2: Results of HDP. RMSE = 0.68. Dotted lines - reference trajectories, solid lines - actual system outputs. Controller: 5x6x2, Critic: 6x6x1,**

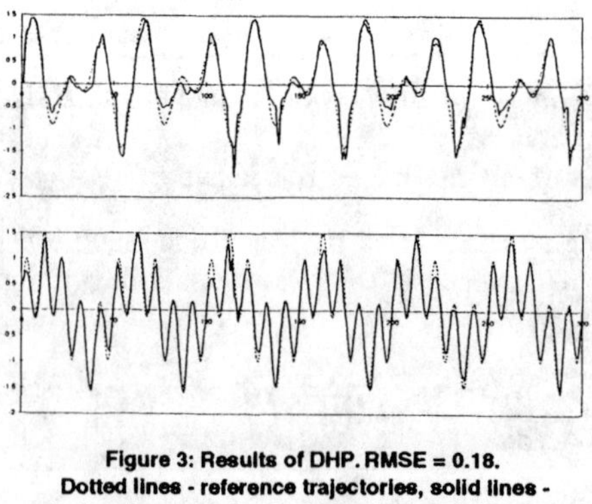

**Figure 3: Results of DHP. RMSE = 0.18. Dotted lines - reference trajectories, solid lines - actual system outputs. Controller: 5x6x2, Critic: 6x6x3,**

## CONCLUSION AND FUTURE WORK

We have discussed neurocontrol of the nonlinear multivariable plant using both conventional and adaptive critic designs. It is shown that DHP is quite competitive with one of the best methods of conventional neurocontrol whereas HDP does not seem to have good prospects in modern control. In this preliminary study we have successfully attempted the simplest case of control when the state variables are available and the plant equations are known. Future research will explore cases of adaptive critic-based control with an identification model instead of the plant equations as well as issues on training the same Critic network for multiple reference trajectories.

## REFERENCES

Narendra, K. S., Mukhopadhyay, S., (1994). Adaptive Control of Nonlinear Multivariable Systems Using Neural Networks, *Neural Networks*, Vol. 7, No. 5, 737-752.

Puskorius, G., Feldkamp, L. (1994). Neurocontrol of Nonlinear Dynamical Systems with Kalman Filter Trained Recurrent Networks, *IEEE Trans. on Neural Networks,* Vol. 5, No. 2, 279-297.

Werbos, P., (1990). Backpropagation Through Time: What it Does and How to Do It, *Proc. of the IEEE*, Vol. 78, No. 10, 1550-1560.

Puskorius, G., Feldkamp, L., (1994). Truncated Backpropagation Through Time and Kalman Filter Training for Neurocontrol, In *Proc. of the World Congress on Computational Intelligence (WCCI)*, Orlando, FL, June/July, 2488-2493.

Werbos, P., (1992). Approximate Dynamic Programming for Real-Time Control and Neural Modeling, In *Handbook of Intelligent Control* (D. White, D. Sofge, Eds.), Van Nostrand.

Werbos, P., (1990). A Menu of Designs for Reinforcement Learning Over Time, Ch. 3 in *Neural Networks for Control* (W. Miller, R. Sutton, P. Werbos, Eds.), MIT Press.

Prokhorov, D., Wunsch, D., (1995). Adaptive Critic Designs, submitted to *IEEE Trans. on Neural Networks*, abstract is available at http://www.ieee.org/nnc. Earlier version of this paper appeared in *Computational Intelligence:A Dynamic System Perspective* (R. Marks et. al., Eds.), IEEE Press, Nov. 1995, pp. 98-107.

# REAL-TIME LEARNING OF AIRCRAFT PARAMETERS USING RECURSIVE LEAST-SQUARES TO TRAIN RBF NETWORKS[1]

**J. NICHOLAS LANEMAN**
*Massachusetts Institute of Technology, Cambridge, Massachusetts, 02139*

**GERALD E. PETERSON**
*McDonnell Douglas Corporation, St. Louis, Missouri, 63166*

***ABSTRACT:***
We describe a real-time algorithm for learning aircraft parameters to be used by an adaptive controller. Learning consists of training a collection of radial basis function neural networks to approximate the incoming data stream in the least-squares sense. Only the heights of the basis functions are trained; heuristics are used to find their centers and widths. Since the heights enter the equations linearly, we employ recursive least-squares to quickly obtain the new heights when we incorporate additional data points and basis functions. In order to keep the computations manageable, we break the data stream into segments, with each segment approximated by about 10 basis functions. We illustrate the algorithm on a set of F-15 flight data.

## INTRODUCTION

Designing gains for aircraft flight controllers requires estimates of the stability and control derivatives, which are the primary components of the $A$ and $B$ matrices in a state-space model of the dynamics. Estimates of these derivatives are normally obtained from a database of aerodynamic information derived primarily from wind tunnel testing. Controller development proceeds with the gains being designed initially using the aerodynamic database and then refined using information obtained over many flight tests.

There is a desire in the aerospace community to provide a better response to changes in aerodynamic loading, control surface failures, or minor aircraft damage during flight by adapting the flight control laws in real time. However, this approach requires online estimates of the time-varying stability and control derivatives. To estimate these parameters, the aircraft dynamics must be excited in such a way that the parameters influence (are observable in) the sensor measurements. For example, a derivative with respect to $\alpha$ requires movement in the $\alpha$ dimension. During periods when accurate estimates are possible, they must be stored and made available for when accurate estimates are not possible. This storage function can be provided by a neural network.

[1]This work was supported by the McDonnell Douglas Independent Research and Development Program.

A radial basis function (RBF) neural network is a good candidate for this online storage function because the learning can be localized to the area of current flight, and because these networks can be trained quickly. Localization along the trajectory is accomplished by placing the centers of the RBFs along the flight path in parameter space. The widths of the basis functions and their spacing are chosen in advance using heuristics, *e.g.*, requiring each point along the trajectory to be supported by at least four basis functions. With the widths and centers chosen, the heights of the RBFs can be quickly determined to satisfy the least sum-of-squared errors criterion by using the recursive least-squares methodology.

Recursive least-squares (RLS) is a method for maintaining a least-squares solution as new data points are added to the training cases or as new basis functions are added to the RBF network. RLS has been used extensively in adaptive filter theory (Haykin, 1991) and adaptive control (Åström and Wittenmark, 1989). If points or basis functions are added one at a time, no matrix inversions are required by this process; only matrix multiplications and additions are necessary. However, when a basis function is added, the size of the primary matrix grows by one row and one column. We take additional steps to limit the size of this matrix and ensure the possibility of a real-time implementation.

The remainder of this paper is organized as follows. In the next section we develop the RBF network model used in our approximations. We derive the RLS equations used to incorporate new basis functions and data points into the training model, and we discuss steps taken to reduce the amount of computation required to maintain the least-squares solution. Finally, we illustrate our algorithm by approximating aerodynamic derivatives from a segment of flight data.

## TRAINING RBF NETWORKS USING RECURSIVE LEAST-SQUARES

Suppose data $\{z_i, y_i\}_{i=1}^n$ is given, where $z_i$ is the input vector $[z_{i_1}, \ldots, z_{i_\ell}]^T$ and the measurement $y_i$ is a scalar. A *radial basis function* (RBF) neural network is a model of the form

$$y \approx b_1 f_1(z) + \cdots + b_m f_m(z) \tag{1}$$

where the order of the model satisfies $m < n$ to achieve compression. Here $\{f_j(z), j = 1, \ldots, m\}$ are the radial basis functions, which are derived from a radially symmetric function $f(z)$, called the *basic* RBF (Bishop, 1995). The vector of coefficients $b = [b_1 \ \ldots \ b_m]$ must be determined so that the model fits the data in the least-squares sense.

Since the $b_j$ occur linearly in (1), we may write the model as a set of linear equations. Specifically, let the measurement data vector be $y = [y_1 \ldots y_n]^T$, and let $R = (r_{ij})$ be the $n \times m$ regression matrix defined by $r_{ij} = f_j(z_i)$. Then (1) becomes $y \approx Rb$, and we assume that the columns of $R$ are linearly independent. To fit the data in the least-squares sense, that is, to find the heights such that

$$\hat{b} = \arg\min_b \, (Rb - y)^T (Rb - y)$$

requires

$$\hat{b} = [R^T R]^{-1} R^T y \tag{2}$$

which is the standard least-squares solution to an over-determined set of linear equations (Strang, 1980).

**Adding a New Basis Function to the Model**

Given a solution in the form of (2), suppose we wish to add an additional function $f_{m+1}(z)$ to the model and calculate the new solution with a minimum of additional computation. The regression matrix becomes $R' = [R\ H]$, where

$$H = \left[\ f_{m+1}(z_1)\ \ \dots\ \ f_{m+1}(z_n)\ \right]^T$$

Let

$$\hat{b}' = \left[\ \hat{b}'_1\ \ \dots\ \ \hat{b}'_{m+1}\ \right]^T$$

be the new vector of coefficients to be determined. By (2),

$$\hat{b}' = \begin{bmatrix} R^TR & R^TH \\ H^TR & H^TH \end{bmatrix}^{-1} \begin{bmatrix} R^T \\ H^T \end{bmatrix} y \qquad (3)$$

A calculation shows that

$$\begin{bmatrix} R^TR & R^TH \\ H^TR & H^TH \end{bmatrix}^{-1} = \frac{1}{\theta} \begin{bmatrix} \theta[R^TR]^{-1} - cc^T & c \\ c^T & -1 \end{bmatrix} \qquad (4)$$

where $c = [R^TR]^{-1}R^TH$ and $\theta = (Rc - H)^TH$. Equation (4) is valid whenever $\theta \neq 0$. Substituting (4) into (3) gives

$$\hat{b}' = \begin{bmatrix} \hat{b} - \alpha c \\ \alpha \end{bmatrix} \qquad (5)$$

in which

$$\alpha = \frac{(Rc - H)^Ty}{\theta} \qquad (6)$$

Thus (4) may be used to efficiently calculate the new inverse matrix from the old, and (5) may be used to obtain the new least-squares solution.

**Adding a New Data Point to the Training Set**

Given a solution in the form of (2), suppose it is desired to add an additional data point $\{z_{n+1}, y_{n+1}\}$ to the training set and calculate the new solution with a minimum of additional computation. In this case, the regression matrix becomes $R' = [R^T\ G^T]^T$, where

$$G = \left[\ f_1(z_{n+1})\ \ \dots\ \ f_m(z_{n+1})\ \right]$$

and the output vector becomes

$$y' = \left[\ y\ \ y_{n+1}\ \right]^T$$

Let $P = R^TR$, and $P' = P + G^TG$. Starting with (2) and utilizing the matrix inversion lemma (Haykin, 1991), we reduce the solution to

$$\hat{b}' = \hat{b} + K(y_{n+1} - G\hat{b}) \qquad (7)$$

$$P'^{-1} = [I - KG]P^{-1} \qquad (8)$$

where

$$K = P^{-1}G^T[GP^{-1}G^T + 1]^{-1} \qquad (9)$$

Equations (7), and (8) may be used to find the new solution heights and the new inverse matrix from the old, all without matrix inversion.

**Finding Centers and Widths**

In our application, the data is a function of time. That is, $z_i = z(t_i)$ and $y_i = y(t_i)$ for some sequence of time values $\{t_i\}$. Therefore, the data points form a one-dimensional *trajectory* in an $\ell + 1$ space of parameter values. In order to give approximately equal treatment to each of the dimensions in parameter space, the input is scaled so that each parameter varies between -1 and +1. Also we assume the basis functions have equal widths in all dimensions, which we define as the number $\sigma$.

The basis functions must be placed near the trajectory in order to be responsive to the incoming data stream, and they should be approximately equally spaced in order to give uniform coverage. With each new data point, the Euclidean distance from the last data point is computed. We add a series of these distances to approximate the arc length along the trajectory. When the total distance exceeds the intended RBF spacing, $\Delta$, a new basis function is centered at the current $z_i$ and incorporated into the solution.

Given $\Delta$, the intended RBF spacing in parameter space, we compute the width, $\sigma$, according to a heuristic: roughly four basis functions should be non-zero for each data point in order to form accurate approximations. (This heuristic results from experiments with RBF approximations to multi-dimensional polynomials.) Ensuring such a condition requires the width of the basis functions to satisfy $\sigma \geq (3/2)\Delta$ which is easily verified in the single dimensional case.

**Segmenting the Data**

Equation (4) shows that the new $P$ matrix grows by one row and one column when a new basis function is added to the model. If steps are not taken to restrict this growth, $P$ will eventually grow too large for real-time implementation of the algorithm.

Our approach is to allow the window to grow to a fixed size, *e.g.*, 10 basis functions, store the entire window, and begin a new window. This approach inherently segments the data trajectory, with each segment approximated by 10 basis functions. To evaluate this approximation at a point on a trajectory, the segment containing this point is determined and evaluated. The problem is that for a given point, it is not obvious which segment it belongs to. The solution we adopt is to find the basis function which is closest to the given point and use the segment containing that basis function. With this approach, the only time that minimum least-squares accuracy is not obtained is when the wrong segment is determined for a point, and then a close approximation is still obtained.

## APPLICATION TO ONLINE LEARNING OF AIRCRAFT PARAMETERS

We now consider the problem of learning aircraft parameters online using an RBF network trained with RLS. We wish to approximate a portion of the surface of an aerodynamic derivative in the least-squares sense, by placing the centers of the basis functions along the trajectory and adjusting their heights to provide an optimum fit.

The variables of interest are aeronautical state inputs such as the angle of attack, $\alpha$, the angle of sideslip, $\beta$, the speed in mach units, and the deflection angles of the stabilator and the canard, $\delta_S$ and $\delta_C$. These inputs are used with the aircraft equations of motion in order to estimate the aerodynamic derivatives $C_{m_\alpha}$ and $C_{m_{\delta_S}}$, which are the rates of change of the coefficient of pitching moment with respect to $\alpha$ and $\delta_S$,

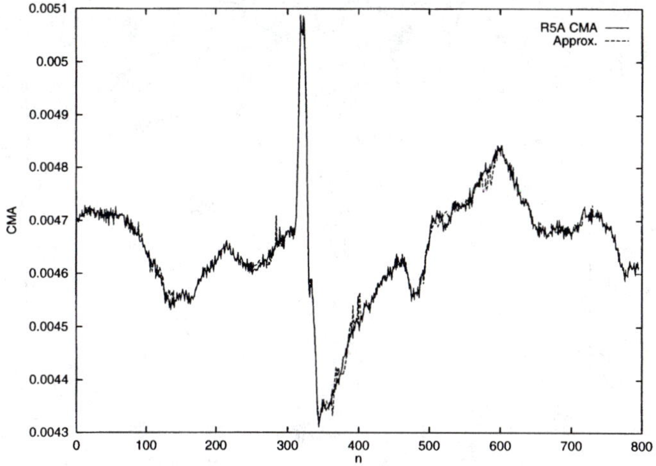


Figure 1: Segmented RLS approximation to $C_{m_\alpha}$ with $\Delta = 2$ and 10 RBFs per window, 42 RBFs total.

respectively. It is required to save $C_{m_\alpha}$ and $C_{m_{\delta_S}}$ as functions of $\alpha$, $\beta$, mach, $\delta_S$ and $\delta_C$ using a neural network.

A data set from F-15 flight tests is used to demonstrate the approach. The data is sampled at 40 Hz, and the duration of our example flight is 20 seconds. Maneuvers are being performed during this flight, but the type of aircraft movement is not important to the approach. Centers are placed along the trajectory in $(\alpha, \beta, \text{mach}, \delta_S, \delta_C)$-space using the above heuristics with $\Delta = 2$ (in normalized coordinates), and RLS is used to determine the heights so that the RBF approximates $C_{m_\alpha}$ and $C_{m_{\delta_S}}$ in the least-squares sense. The segmenting approach is used to limit the required computation time. The true value of $C_{m_\alpha}$ and its approximation in time for the flight data are shown in Figure 1. Similarly, the approximation for $C_{m_{\delta_S}}$ is shown in Figure 2.

The total range of $C_{m_\alpha}$ is the interval $[-0.035, 0.019]$ and the range of $C_{m_{\delta_S}}$ is $[-0.0136, 0.0014]$. Dividing the error by the size of these intervals shows that our approximations are accurate to less than 1.5% of the range of the approximated parameter. Evidence that these estimates are accurate enough for use by the controller comes from internal studies at McDonnell Douglas, which have shown that stability is maintained if all the derivatives are accurate to within 12% of range.

The compression ratio results are as follows. Our data set contains 797 data points, and 42 basis functions are used to approximate it as Figures 1 and 2 indicate. Thus, the compression ratio for this dataset is roughly 19:1. From these results, we expect compression ratios of 20:1 to be achievable and effective.

The average time to add a point using actual flight data on a SPARCstation 2 is 1.34 ms, and the average time to evaluate a point is 1.02 ms. Since the update rate of the aircraft controller is expected to be 12.5 ms, there is reason to believe that the RBF neural network can be trained and evaluated quickly enough to operate in real time.

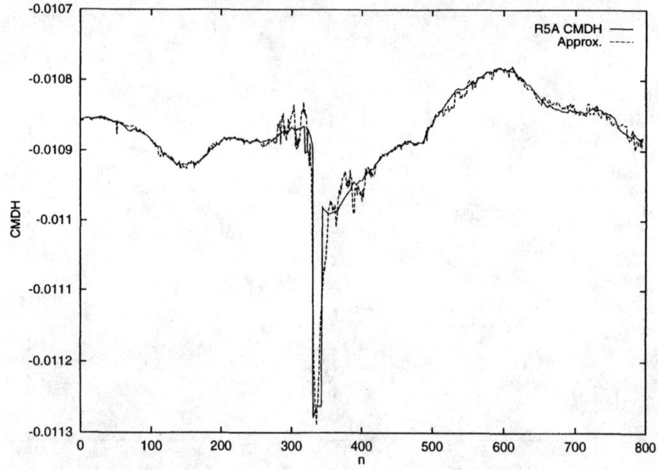


Figure 2: Segmented RLS approximation to $C_{m_{\delta_S}}$ with $\Delta = 2$ and 10 RBFs per window, 41 RBFs total.

## CONCLUSIONS

Adaptive aircraft controllers are of interest because the maintenance of excellent control is desired when the aircraft structure changes or unexpected conditions occur in flight. A quickly trained localized neural network consisting of radial basis functions can be used to provide an adaptive aircraft model for such a controller. Simulations have shown that the algorithm works in real time. Flights for the purpose of testing this concept are expected to occur in 1997.

## REFERENCES


Åström, Karl J. and Bjön Wittenmark, (1989). *Adaptive Control.* Reading, MA: Addison-Wesley Publishing Company.

Bishop, Christopher M., (1995). *Neural Networks for Pattern Recognition*, Chapter 5. Oxford, UK: Clarendon Press.

Haykin, Simon, (1991). *Adaptive Filter Theory*, Second Edition. Englewood Cliffs, NJ: Prentice-Hall, Inc.

Strang, Gilbert, (1980). *Linear Algebra and Its Applications*, Second Edition. New York, NY: Academic Press.

# A NEW ABSTRACT ARTIFICIAL NEURAL ARCHITECTURE FOR SOLVING DIFFICULT LEARNING CONTROL PROBLEMS

**EDWARD A. THOMPSON AND GEORGES A. BÉCUS**
*AsE/EM Dept., University of Cincinnati, Cincinnati, OH 45221-0070*

***ABSTRACT:***
We introduce a neural architecture designed for efficient, reliable, and adjustable solutions to difficult learning control problems. It consists of a hierarchy of command and control centers governing motor selection networks. Within the controller, internal drives, similar to biological ones, facilitate learning. The inverted pendulum control problem is used to validate the proposed architecture.

## INTRODUCTION

For the past fifty years, scientists/engineers have developed computational neuronal models which can recognize/complete complex patterns and solve difficult optimization and control problems. In spite of successes in various fields, their scope is limited as their lack of efficiency, reliability, and adjustability virtually eliminates artificial neural systems (ANS) as solutions for larger, nonacademic problems.

Three distinct learning paradigms have emerged for ANS. Supervised learning requires a teacher/critic for the entire learning; unsupervised learning requires none as the system self-organizes according to some learning algorithm. Reinforcement learning (RL) combines supervised and unsupervised learning: the system receives occasionally performance evaluation via a reward/punishment (R/P) signal but must construct its own reinforcement signal throughout most of the learning process.

In many control problems, there is no well-defined performance index (PI) and, especially in the presence of system dynamics uncertainty, not enough information to train a controller to perform satisfactorily; eliminating supervised and unsupervised learning strategies for control applications. RL is attractive but not combined with control easily. Internal PI construction for efficient/reliable learning is a difficult task complicated by a trend in ANS of using building blocks much more elaborate than the perceptron. Crude R/P signals reflect the performance of the overall system, not the elements within. RL must solve "credit assignment" and is affected by delays in R/P signals. In developing RL mechanisms, we aimed to simplify internal R/P signal construction by providing the ability to construct, through exploration, a pseudo PI.

A Command and Control Center/Internal Drive ($C^3$/ID or "Cid") ANS, designed for high efficiency/reliability/adjustability, and implementing a novel heuristic-like RL strategy for solving difficult control problems is presented. Cid does not model any specific biological neural mechanism, it emulates and implements within the controller more abstract biological phenomena as internal drives (ID), similar to hunger, thirst, etc. We build a hierarchy of command and control centers ($C^3$) to govern ID. In Cid, which may be applied to multiple task problems, each control task is considered an ID.

Single and multiple task configurations are validated on the benchmark cart-pole problem. Cid, when compared to other RL ANS, is more reliable and efficient and experiences less failures. Cid can also be user fine-tuned for improved performance.

## $C^3$/ID ARCHITECTURE AND OPERATION

The architecture of Fig. 1, for a single ID controller, has three main components: environment sensors (ES), a command and control center ($C^3$), and motor selection networks (MSN) (one per variable of interest). $C^3$ and MSN together form an ID. Information from ES is fed to $C^3$ and passed to the proper MSN, which makes decisions based on past/present information; choosing the MSN to receive attention which then selects the control action (CA) to be implemented at that time.

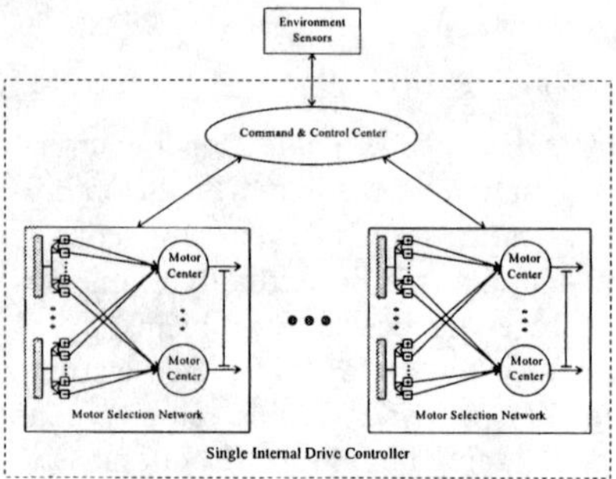


**Figure 1: $C^3$/ID architecture for a single ID controller.**

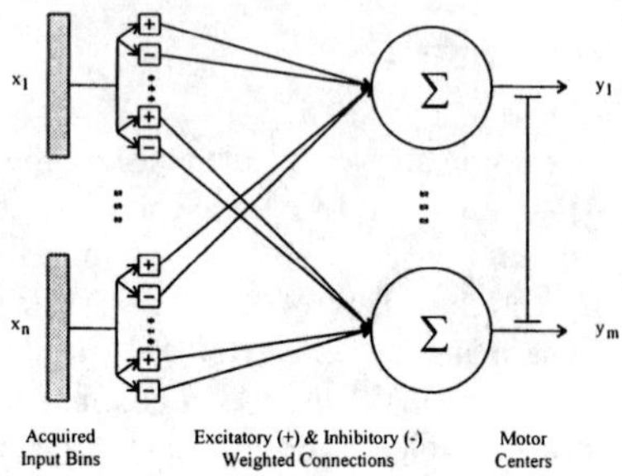


**Figure 2: Motor selection network diagram.**

Klopf (1993) envisioned that a "more advanced model of nervous system function ... would include provisions for ID and for an attention mechanism that may reside within a $C^3$". Our $C^3$ meets these goals. Its mechanism to map learning system (LS) environment and its exploration strategy can find/remember/learn to avoid failures by giving attention to appropriate MSN. In Cid, $C^3$ tasks are to keep track of state variables (SV) explored region (ER), compute SV ER centers, form input vectors to the MSN, calculate SV priorities and ID failure measures (FM), select MSN to be activated, and govern weight changes in MSN. A timing issue arising in the attention mechanism is solved by attaching timing constants to each ID FM. In the example below, a timing constant is used. ID implemented in Cid reduce MSN workload. Breaking larger control problems into smaller sub-problems by identifying ID makes for a highly efficient and reliable LS. Cid architectures for multiple ID is shown in Fig. 4. A high $C^3$ regulates ID. MSN (one per controlled variable) are ANS selecting appropriate CA for the current SV when called upon by the $C^3$. MSN are made up of acquired input bins, weighted connections, and motor centers (MC), see Fig. 2. The number of input bins for SV is user-selected. However, Cid performs bin partitioning through exploration.

## APPLICATION TO BENCHMARK POLE BALANCING PROBLEM

To demonstrate single/multiple ID controller architectures, we selected the cart-pole problem which offers nonlinearity, multiple conflicting goals, multiple SV, and a variety of data for comparison. Fig. 3 depicts the system. The pole is free to fall left/right in the plane of the figure. The goal is to balance the pole within ±12° of vertical and keep the cart within ±2.4m track bounds by applying a ±10N force to the cart cg at each time step. System dynamics, Barto (1983), simulated by Euler method with a 0.02s time step, are unknown to the controller receiving SV inputs at each time step. The only external evaluative feedback to the controller is a failure signal when the pole falls more than 12° and/or the cart exceeds track ends. Upon failure, cart-pole assembly returns to its initial state. An infinite track length version of the problem (only goal being to keep pole from falling more than 12°) is considered first to illustrate single ID controller architecture.

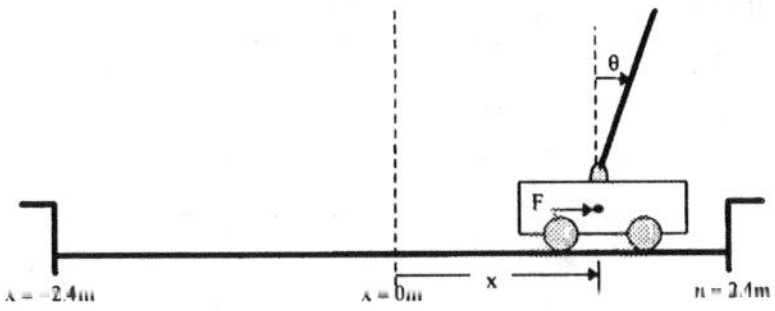


Figure 3: Cart-pole assembly to be controlled.

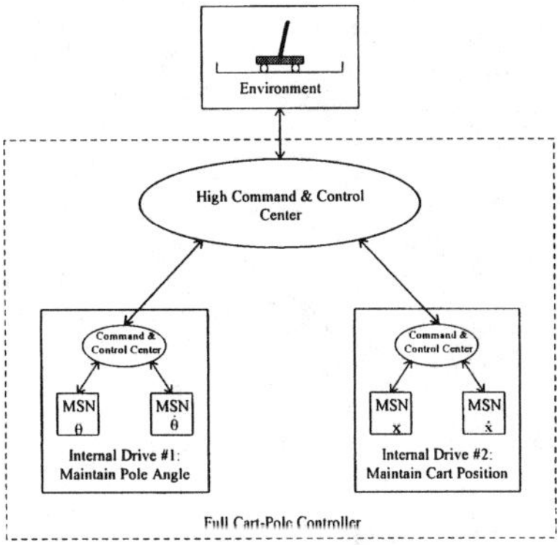


Figure 4: Architecture for finite track problem.

### Architectures

Cid architecture for the infinite-track problem, similar to that in Fig. 1, has a $C^3$ and two MSN combining in a single ID to maintain a safe pole angle. Only two SV (pole angle/angular velocity) have their own MSN, each with two MC exerting a left/right control force if activated. Each MSN has two sensor bins. Fig. 4 shows the finite track problem architecture: a high $C^3$, two regular $C^3$, and four MSN, one per SV (pole angle/angular velocity, cart position/velocity). Two ID, each with a $C^3$ and two MSN, maintain a safe pole/cart position. Each MSN has two MC (for left/right force) and two sensor bins. High $C^3$ passes ES SV measurements to lower $C^3$ and gives attention to one ID which implements the CA.

### Simulation Results

The efficiency/reliability for our controller in solving both versions of the pole balancing problem was tested by performing the following simulation 10 times for each problem. Initially, all SV are set to zero. A trial ends when a failure occurs: the pole falls more than 12° left/right of vertical (infinite track problem) and/or the cart exceeds the track ends (finite track problem). Upon failure, all MSN weights are updated and the system is returned to its initial state. The controller learns to balance the pole successfully if it avoids failure for 100,000 time steps (about 33.3

minutes of simulation time). If not, it is unsuccessful. All excitatory/inhibitory weights were initialized to ±0.1. Initial values for SV upper/lower ER bounds were ±0.0001. The learning rate constant was 0.1. For the finite track length problem, a 1.3 timing constant for the pole angle ID FM was introduced to allow the controller to handle random starts (randomly selected initial SV). We assisted the controller learning capability by enforcing symmetry for each SV ER bounds. Also, for the finite track length problem, correct track length bounds (±2.4m) were provided.

**Infinite Track Length Problem**

Cid learns to balance the pole with 100% success in an average of 2.4 trials and 1.51s; runs 1-10, Table 1. No exploration mechanism is used to find SV bounds leading to pole angles widely varied within ±12°. Runs 11-20 are another test: a simple exploration is implemented; a left/right control force is exerted during the 1st/2nd trial (CA in others are determined as usual) for accurate determination of SV bounds. The controller was 100% reliable, solving the problem in an average of 3.6 trials, 2 for exploration, 0.95s learning time, and much improved pole angles.

TABLE 1: INFINITE TRACK CONTROL PROBLEM TRAINING RESULTS.

| Run | Required Trials | Time Steps | Avg Pole Angle(°) | Pole Angle Band (°) | Avg Pole Velocity (°/s) | Pole Velocity Band (°/s) |
|---|---|---|---|---|---|---|
| 1 | 2 | 117 | 1.98 | (1.14,2.68) | 0.00098 | (-21.71,19.05) |
| 2 | 1 | 43 | -6.58 | (-6.03,-7.05) | -0.0034 | (-17.25,17.34) |
| 3 | 1 | 22 | -7.99 | (-8.5,-7.46) | -0.0041 | (-18.91,18.84) |
| 4 | 5 | 186 | -7.77 | (-7.23,-8.34) | -0.0040 | (-18.87,18.8) |
| 5 | 2 | 65 | 4.64 | (4.04,5.24) | 0.0023 | (-19.07,18.34) |
| 6 | 3 | 67 | 7.36 | (6.84,7.85) | 0.0036 | (-20.02,19.26) |
| 7 | 2 | 34 | 9.65 | (9.09,10.25) | 0.0049 | (-18.62,19.25) |
| 8 | 2 | 56 | -7.62 | (-7.04,-8.14) | -0.0039 | (-18.76,18.76) |
| 9 | 3 | 103 | 2.62 | (1.82,3.26) | 0.0016 | (-20.99,18.75) |
| 10 | 3 | 61 | 0.94 | (-0.014,1.81) | 0.00042 | (-22.43,20.47) |
| Avg | 2.4 | 75.4 | | | | |
| 11 | 3 | 43 | -0.00032 | (-0.67,0.67) | 0.000084 | (-24.82,24.81) |
| 12 | 4 | 46 | 0.00032 | (-0.67,0.67) | -0.000084 | (-24.81,24.82) |
| 13 | 3 | 36 | -0.00025 | (-0.75,0.67) | 0.000074 | (-24.69,25.38) |
| 14 | 4 | 49 | -0.00032 | (-0.67,0.67) | 0.000084 | (-24.82,24.81) |
| 15 | 4 | 58 | -0.00032 | (-0.67,0.67) | 0.000084 | (-24.82,24.81) |
| 16 | 3 | 43 | 0.00032 | (-0.67,0.67) | -0.000084 | (-24.81,24.82) |
| 17 | 4 | 50 | 0.00025 | (-0.63,0.71) | -0.000024 | (-23.56,24.87) |
| 18 | 4 | 52 | -0.00032 | (-0.67,0.67) | 0.000084 | (-24.82,24.81) |
| 19 | 3 | 48 | -0.00032 | (-0.67,0.67) | 0.000084 | (-24.82,24.81) |
| 20 | 4 | 52 | 0.00032 | (-0.67,0.67) | -0.000084 | (-24.81,24.82) |
| Avg | 3.6 | 47.7 | | | | |

**Finite Track Length Problem**

In this problem Cid learns to balance the pole with 100% success in an average time of 17.72s. Due to imposed symmetry, SV centers were zero and SV means were all nearly zero after learning. At the end of each run in Table 2, the resulting controller ability to handle random starts was tested. Initial SV values were chosen from: pole angle [–6°,+6°], pole angular velocity [–50°/s,+50°/s], cart position [–0.8m,+0.8m], and cart velocity [–0.8m/s,+0.8m/s]. This test was performed 50 times for each run. In every case, the controllers were successful. One final experiment was performed to illustrate Cid adjustability. After the controller learned to avoid failure for

100,000 time steps and all SV converged to average values of Table 2, the cart position center was moved from 0m to 1.0m by increasing upper/lower bounds for the cart position SV by 0.2m every 5 time steps over 25 time steps. Fig. 5 shows SV before, during, and shortly after this change. Only small shocks to the system SV appear as the cart moves from 0m to 1.0m. Adjustments are made manually here; done automatically with a center moving algorithm, Cid could also solve tracking problems.

TABLE 2: FINITE TRACK CONTROL PROBLEM TRAINING RESULTS.

| Run | Required Trials | Time Steps | Avg Pole Angle (°) | Avg Pole Velocity (°/s) | Avg Cart Position (m) | Avg Cart Velocity (m/s) |
|---|---|---|---|---|---|---|
| 1 | 3 | 957 | 0.0000 | -0.0005 | -0.0039 | -0.0001 |
| 2 | 3 | 930 | 0.0003 | -0.0001 | -0.0015 | 0 |
| 3 | 3 | 689 | 0.0002 | 0.0006 | 0.0026 | 0.0001 |
| 4 | 5 | 1107 | -0.0003 | 0.0003 | -0.0020 | 0 |
| 5 | 3 | 931 | 0.0002 | 0.0001 | -0.0024 | 0 |
| 6 | 3 | 1194 | 0.0001 | 0.0002 | -0.0010 | 0.0001 |
| 7 | 5 | 920 | -0.0002 | 0.0001 | -0.0022 | 0.0001 |
| 8 | 3 | 652 | 0.0003 | 0 | 0.0019 | -0.0001 |
| 9 | 5 | 1135 | 0.0001 | 0.0002 | -0.0016 | 0.0001 |
| 10 | 3 | 344 | 0.0002 | 0.0002 | 0.0003 | 0 |
| Avg | 3.6 | 885.9 | | | | |

**Comparison Of Results**

Prior to developing Cid, we examined different RL systems: Barto (1983) ASE-ACE network, Klopf (1988) ACP networks, Baird (1993) modified ACP networks, Gullapalli (1994) (SRV) reinforcement algorithm, and Berenji (1992) GARIC architecture. These ANS can learn to solve difficult nonlinear control problems; but in solving the pole balancing (ball balancing for SRV) problem, which may appear trivial, their performance is less appealing than Cid's. They require either a large amount of training time, a large number of failures before success, and/or exhibit limited reliability. A comparison of the different learning systems appears in Table 3.

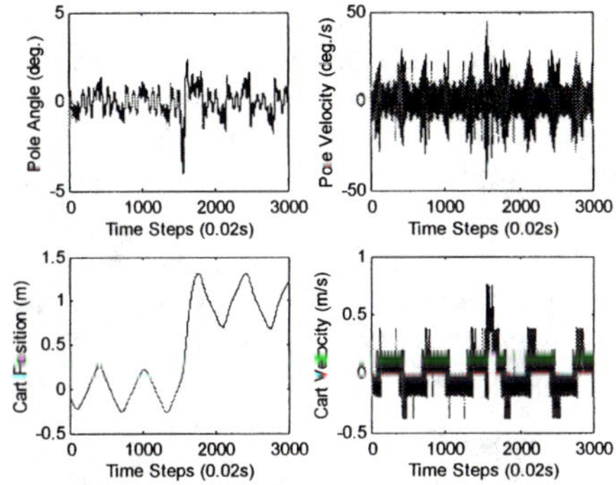

**Figure 5: State responses to cart position adjustment.**

The ASE-ACE network needs over 32 minutes of training and 70 failures before balancing the pole with a reported success rate of only 80%. The original ACP network cannot handle efficiently simultaneous multiple sensors. It is only 30% reliable with the same amount of failures and training time as the ASE-ACE network. The modified/optimal ACP networks are 100% reliable and need only a minute or so of training time but have about 70 failures. No reliability rate is given for the SRV network applied to a similar ball balancing problem. It had 700 failures before learning to solve the problem. GARIC architecture needed 322 failures before solving the pole balancing problem. Our Cid architecture is clearly the most reliable and efficient learning system of those listed. It also offers adjustability.

## CONCLUSION

TABLE 3: COMPARISON OF LEARNING RELIABILITY AND EFFICIENCY.

| Method | % Success | Avg Number of Required Trials | Avg Number of Time Steps | Avg Simulated Time (minutes) |
|---|---|---|---|---|
| ASE/ACE | 80 | 70 | 96511 | 32.2 |
| Original ACP | 30 | 77 | 75000 | 25.0 |
| Modified ACP | 100 | 71 | 4016 | 1.3 |
| $C^3$/ID | 100 | 4 | 886 | 0.295 |
| SRV | not given | 700+ | not given | not given |
| GARIC | not given | 322 | not given | not given |

Our new Cid architecture shows that ANS can find very efficient, reliable, and adjustable solutions to difficult learning control problems. Rather than develop elaborate ANS as building blocks for larger network, we believe focus should be on the organization and attention of simpler architectures. Our approach of organizing simple MSN and $C^3$ into ID proved very successful in learning to solve the pole balancing problem. Preliminary results reported here validate the approach and suggest it could handle larger, non-academic control applications with similar success. We are now working on a jet engine control problem.

Besides solving larger control problems, we will focus also on developing an exploration mechanism which can adequately explore SV operating regions and determine SV sensitivities or rates of change to the system CA. This information can be used to automatically compute appropriate timing constants leading to more robust controller performance. We would also like to further exploit the adjustability offered by our architecture. By exploring different center moving strategies, we hope to enhance the tracking capabilities of our system.

Cid may appear to require knowledge of all SV. This is not so. Measuring quantities that influence/determine success/failure is all that is needed by Cid. The example used SV, but it need not be so; even if SV are used, not all are needed as the infinite-track pole balancing problem demonstrates. It may also appear that exact (non noisy) measurements are required. Since, through the use of bins (albeit determined from measurements), measured quantities are 'fuzzified', we do not believe measurement accuracy is an issue. Further work on both issues is also in progress.

## REFERENCES


Baird, L.C., Klopf, A.H., (1993). A Hierarchical Network of Provably Optimal Learning Control Systems: Extensions of the Associative Control Process (ACP) Network, *Adaptive Behavior*, 1, 321-352.

Barto, A.G., Sutton, R.S. and Anderson, C.W. (1983). Neuronlike Elements that Can Solve Difficult Learning Control Problems, *IEEE Trans. on Systems, Man, and Cybernetics*, 13, 835-846.

Berenji, X.X. P Khedkar, P., (1992). Learning and Tuning Fuzzy Logic Controllers Through Reinforcements, *IEEE Trans. on Neural Networks*, 3, 724-740.

Gullapalli, Y.Y., Franklin, J.A. and Benbrahim, H (1994). Acquiring Robot Skills via Reinforcement Learning, *IEEE Control Systems Magazine*, 13-24, 1994.

Klopf, A.H. (1988). A Neuronal Model of Classical Conditioning, *Psychobiology*, 16, 85-125.

Klopf, A.H., Morgan, J.S. and Weaver, S.E. (1993). A Hierarchical Network of Control Systems that Learn: Modeling Nervous System Function During Classical and Instrumental Conditioning, *Adaptive Behavior*, l, 263-319.

# CHARACTERIZATION OF CONTACT MOTION KINEMATICS WITH RECURRENT NEURAL NETWORKS

**CHIDAMBAR GANESH, DAVID J. FERKINHOFF AND KAI F. GONG**
*Naval Undersea Warfare Center, Division Newport*
*Newport, Rhode Island 02841*

***ABSTRACT :***
*This paper addresses the problem of characterization of motion kinematics for enhanced tracking of undersea contacts. In particular, this work focuses on high-speed and maneuvering vehicles. A recurrent neural network architecture is employed to exploit sensor measurement features, such as the Doppler shift in received acoustic signal frequency from a maneuvering contact, to determine the motion characteristics of the vehicle. The architecture employed here, also known as the Jordan configuration, uses feedback from the output nodes in a conventional neural network to provide additional data inputs known as context nodes. Experimental results, using simulated kinematics, are presented to demonstrate the performance of the recurrent network system. A confusion matrix, designed to measure the level of correct motion classification, is evaluated to quantify performance of the system. Recommendations for future work are also presented.*

## INTRODUCTION

In this paper, we address the problem of motion characterization of a high-speed and maneuvering contact. In this context, contact motion characterization can be regarded as an augmented description of the kinematic behavior of an undersea vehicle that provides additional information necessary to perform improved tracking. This information includes detection of contact maneuvers, classification of possible maneuver type, or determination of vehicle motion phase. In particular, this work focuses on identification of the motion phase of a specific contact-of-interest. A schematic illustration of this concept is shown in Figure 1.

Previous studies have shown that adequate modeling of contact dynamics is a major consideration in underwater tracking; errors induced by differences in actual and assumed motion significantly impact trajectory estimation [1]. Assuming contact motion to be constant velocity for all tactical conditions may be unrealistic, since maneuvers typically occur fairly often in shallow-water environments. Hence, rapid detection and evaluation of maneuvers become critical to enable adaptive response with a new motion model in a timely fashion. A detailed exposition of this contact tracking concept is provided in [2], whereby information is hierarchically integrated to provide improved estimation of vehicle trajectory.

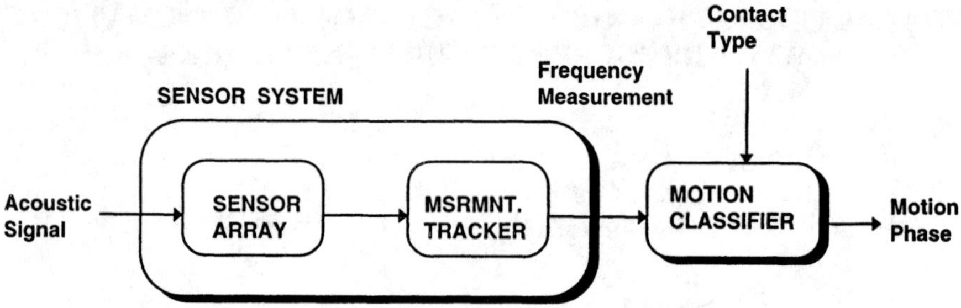


*Figure 1 : Conceptual schematic of the motion phase classifier system*

For the problem at hand, two vehicle types are considered : these are denoted A and B. A contact scenario simulator was developed and used to generate synthetic frequency measurements, which were corrupted with additive white Gaussian noise. The variance on the noise changed from one scenario to the next, and had a uniform distribution across the scenario pool with low-to-medium noise levels. Inputs to the neural network are defined to be the relative change in frequency from the initial value; they are obtained by scaling the frequency measurements as

$$f_{in}(t) = \frac{f_m(t) - f_m(t_0)}{f_m(t_0)} + 0.5$$

## APPROACH

Artificial Neural Systems (ANS), or neural networks, constitute a novel and innovative approach to signal and information processing that have achieved a fair degree of maturity in recent years. A recurrent network architecture provides context to the signal analysis by the use of feedback. The architecture employed here, known as the Jordan configuration, uses feedback from the output nodes in a conventional neural net to provide additional data inputs known as context nodes [3]. This architecture is shown in Figure 2. If $u_n$ is the input signal to the ANS at the present time $t_n$ and $y_{n-1}$ is the output response of the ANS from the **previous** time $t_{n-1}$ , then the output of the hidden layer at the present time is

$$x_n = f(u_n, y_{n-1})$$

and the net output is $y_n = g(x_n)$. Figure 2 shows that as a pattern of inputs is introduced to the network, prior inputs will affect the network output due to the feedback from the output nodes.

A system of two ANSs is developed, corresponding to contact types A and B. Each ANS is an output-feedback or Jordan recurrent network : type A has configuration (40+3)x10x3 and type B has configuration (40+2)x10x2. Here, the notation (40+3)x10x3 is interpreted as an architecture with 40 inputs together with 3 additional feedback or context inputs, 10 hidden nodes and 3 outputs. Input patterns are generated from the frequency measurement time-series using a window of 40 1-

second data samples and sliding the window 5 seconds at a time. Type A used 6 out of the total 216 scenarios and type B required 9 out of 36 scenarios for adequately learning the problem under consideration.

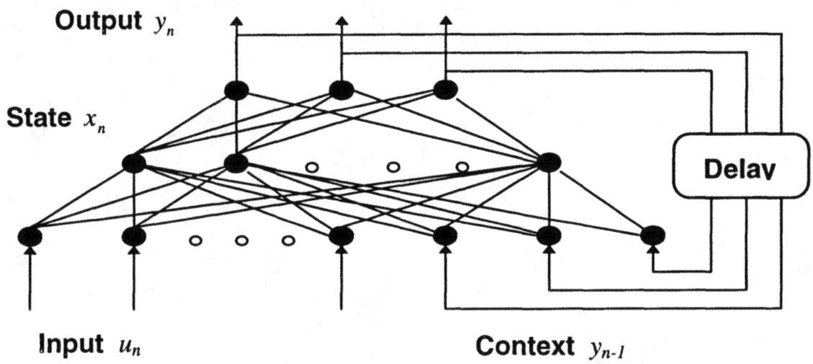


*Figure 2 : Recurrent architecture with output feedback ( Jordan configuration )*

To achieve faster training times, the learning algorithm used is based on the Levenberg-Marquadt optimization technique [4]. The weight update rule is given by

$$\Delta \mathbf{W} = \left(J^T J + \mu I\right)^{-1} J^T \mathbf{e}$$

where $J$ is the Jacobian matrix of the error vector, $\mathbf{e}$ , and $\mu$ is an adjustable scalar.

## EXPERIMENTAL RESULTS

The confusion matrix standard is applied to quantify performance of the motion phase classifier system. For contact type A, the confusion matrix is evaluated over an independent test set of 210 scenarios with 7633 individual test patterns, and is shown in Table 1. It can be seen that the ANS performance is very consistent in classifying the correct motion phase. An in-depth examination of the recurrent net performance shows that the non-zero off-diagonal entries in this matrix are due to either an early recognition or a delayed identification of the phase transition for some scenarios. For the scenarios tested, there is never a situation where motion phase is misclassified in its' entirety owing to a missed transition.

| | GROUND TRUTH | | |
|---|---|---|---|
| ESTIMATE | *Phase 1* | *Phase 2* | *Phase 3* |
| Phase 1 | 0.982 | 0 | 0 |
| Phase 2 | 0.018 | 0.925 | 0.001 |
| Phase 3 | 0 | 0.075 | 0.999 |

*Table 1 : Confusion matrix for motion phase classifier for contact type A*

A sample case-analysis is presented in Figures 3(a)-(d). The scenario geometry is plotted in Fig. 3a, and shows that the vehicle trajectory goes through all 3 phases.

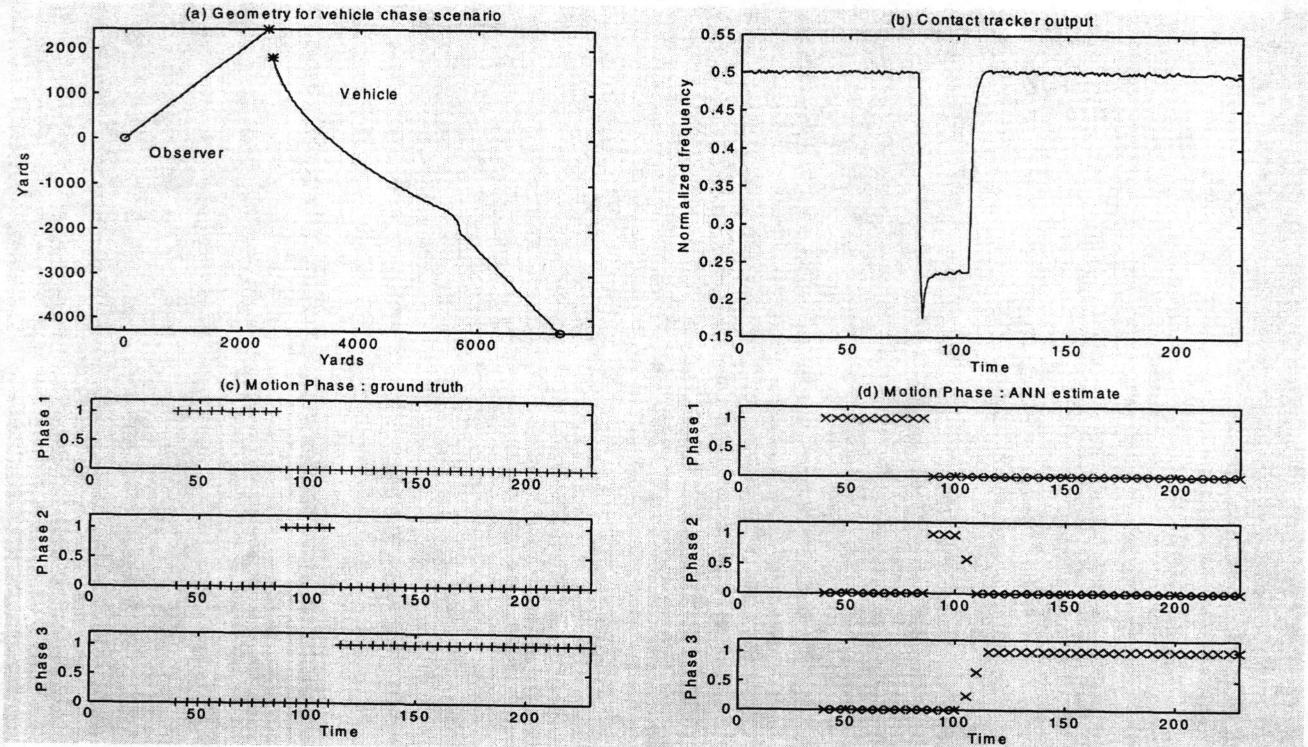


*Figure 3 : Sample case-analysis of motion phase classifier performance for contact type A.*

From the observer's perspective, this can be seen in the normalized frequency output from a measurement tracker assigned to following the vehicle (Fig. 3b). Figs. 3c and 3d show the ground truth and motion phase estimate of the ANS. There is an early recognition of transition from phase 2 to phase 3, leading to a contribution to the 'below-diagonal' (3, 2) entry of the confusion matrix. However, it is evident that this effect does not impact overall system performance.

Similar performance results were obtained for contact type B. The confusion matrix is evaluated over an independent test set of 27 scenarios with 118 patterns, and is given in Table 2. It is worth noting that owing to the close-range short-time nature of these scenarios, larger training sets and longer learning times are characteristic of type B.

| | *GROUND TRUTH* | |
|---|---|---|
| ESTIMATE | *Phase 1* | *Phase 2* |
| Phase 1 | 1 | 0.09 |
| Phase 2 | 0 | 0.91 |

*Table 2 : Confusion matrix for motion phase classifier for contact type B*

To evaluate the robustness of the motion phase classifier, a controlled-noise analysis was performed to determine system performance with increasing noise level. The results of this analysis are schematically illustrated in Figure 4 (a)-(b) for contact types A and B respectively. While the performance of both networks deteriorated with higher noise levels, there was a marked difference in the actual decrease between the contact types. This is due to the nature of the signal features being used; contact type A has a very strong signal level shift between phases, whereas for contact type B the transition between phases is marked by a significantly less prominent feature.

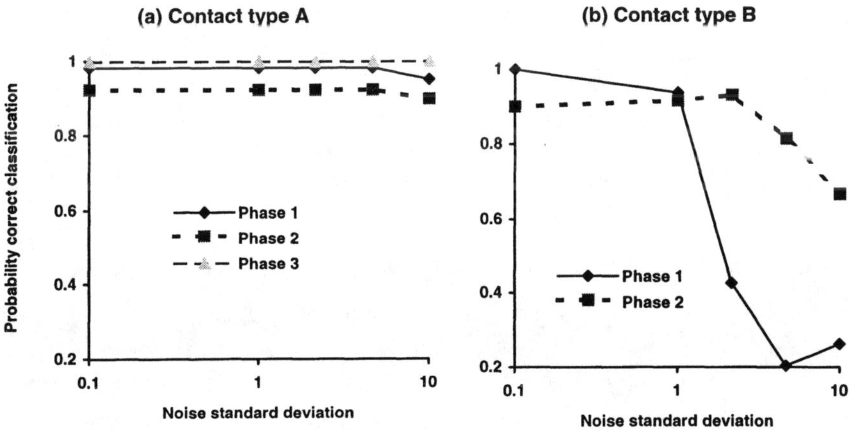


*Figure 4 : Performance analysis for contact types A and B*

## CONCLUSIONS AND DISCUSSIONS

In this work, a recurrent network architecture is employed to exploit sensor measurement features, such as the Doppler shift, to determine the motion phase of an acoustic contact. The recursion inherent in this output-feedback network allows for usage of information from recent measurement history, such as the prior motion phase, to process the information contained in the current data-window. The networks developed here perform with high accuracy over a wide range of independent test scenarios.

The confusion matrix is used to evaluate performance of the motion phase classification system for two types of contacts. This performance criterion is determined on the basis of experimental testing with synthetic data. The subsequent controlled-noise analysis demonstrates that the system is robust above and beyond the training domain. As the noise is increased to higher levels, motion classification performance deteriorates owing to a higher occurrence of false detection of phase transitions. The increased noise level results in false features that the system triggers on as indicative of motion phase changes.

The concept underlying motion phase classification of a high-speed and maneuvering vehicle is readily extended to motion characterization of any maneuvering platform. Such a characterization involves detection of contact maneuvers and classification of possible maneuver type, such as a right turn. This enhanced kinematic description provides additional information necessary to perform improved tracking, and is presently the focus of on-going investigations.

## ACKNOWLEDGMENTS

This work was supported under the Combat Control Science and Technology Project, sponsored by the Office of Naval Research, Program Managers Jim Fein (ONR-333) and Teresa McMullen (ONR-342).

## REFERENCES

1. S.C. Nardone, D.J. Ferkinhoff, S.E. Hammel and K.F. Gong, "Contact Management Model Assessment : An Approach and System Description", NUWC-NPT Technical Report 10,444, Newport, RI, 30 April 1995 (UNCLASSIFIED).

2. S.E. Hammel, D. J. Ferkinhoff and K.F. Gong, "A System and Method for Intelligent Data Fusion for Tracking Highly Dynamic Vehicles", U.S. Patent Application, Navy Case 77334, in preparation.

3. M I. Jordan, "Attractor Dynamics and Parallelism in a Connectionist Sequential Machine", *Proceedings of the Eight Annual Conference of the Cognitive Science Society*, Hillsdale, NJ, 1986.

4. M.T. Hagan and M.B. Menhaj, "Training Feedforward Networks with the Marquadt Algorithm", *IEEE Trans. on Neural Networks*, vol. 5, no 6, pp. 989-993, Nov. 1994.

# A REAL-TIME NEURAL NETWORK CONTROLLER FOR NONLINEAR PNEUMATIC VALVES

**J. F. Glenn, L. M. Schaina, S. Cheng, A. R. Marudarajan, Y. Cheng, M. S. El-Sawah and E.T. Ibrahim**
*Electrical and Computer Engineering Department, California State Polytechnic University, Pomona, California*

***ABSTRACT:***
This paper reports a recurrent neural network based model-reference adaptive controller whose objective is to control a non-linear pneumatic valve of an aircraft environment control system. Usually, such a system includes an identifier in order to estimate the sensitivity of the plant output to the control input. Our approach has eliminated the need for this identifier. In addition, the neural network may contain as few as 1 or 2 neurons. Simulation results indicate that the simplified controller performs consistently well for a general nonlinear plant that is characterized by typical nonlinearities such as relays, dead-zones, or hysteresis. These simplifications make microcontroller based real-time implementation quite promising.

The motivation for this work stems from the desire to develop a robust controller for a plant that is known to be nonlinear and time-varying. A model reference adaptive controller is promising because of its ability to force a plant to follow a linear model (Ku and Lee, 1995). When this type of adaptive controller is designed using a multilayer perceptron, the resulting controller becomes very versatile in the sense that the system can perform consistently well for a variety of nonlinear and/or time varying plants. The neural network derives much of its versatility from its distinctive features such as connectivity, density of neurons in the hidden layer, non-linear transfer functions following the hidden layer neurons, and feedback paths provided by the recurrent type architecture which allows the network to capture the temporal behavior.

The neural network approach proposed by Ku and Lee consists of two major modules: The On-line identifier and the Adaptive controller. The parameters of both of these modules have to be constantly updated using the backpropagation algorithm, a computationally intensive method that makes its real-time implementation highly prohibitive. On the other hand, our approach has eliminated the need for an on-line identifier, reducing the computational requirements by fifty percent. Thus, this simplification greatly enhances the possibilities for a real-time implementation using a single-chip microcontroller.

## DYNAMICS OF THE NONLINEAR PNEUMATIC VALVE OF THE AIRCRAFT ENVIRONMENTAL CONTROL SYSTEM

The airflow into an aircraft cockpit can be controlled by a pneumatically actuated butterfly valve. The valve dynamics can be modeled as a mass-spring-damper system with all of the nonlinear characteristics (static vs. coulomb friction, relay, hysteresis and deadband) contained in the damping term:

$$my''+\Psi\{y'\}+Ky=F_T \tag{1}$$

where $m$ is the valve piston and rod mass, $K$ is the flow force spring rate, $\Psi$ is the nonlinear characteristic function, and $F_T$ are the total forces.

The valve dynamics may also be represented as a nonlinear characteristic function and a linear transfer function in series. The linear transfer function parameters can be determined experimentally from the frequency response of the valve. The nonlinear parameters can be approximated by estimating the hysteresis in the valve, and then running simulations with a choice of other nonlinear characteristic functions.

The frequency response was determined by noting the valve angular deflection given an actuating input of a sine wave of various frequencies. These experiments were carried out with the aid of Labview, a data acquisition software. The bode plot showed that the system had two poles at $\omega_1$= .0135 rad/s and $\omega_2$= .2 rad/s. From this data the linear transfer function was approximated to be:

$$H(s)=\frac{.06}{(s+.0135)(s+.2)} \tag{2}$$

Experimentation on the valve proved the predominant nonlinearities to be hysteresis and static vs. coulomb friction (Glenn, 1996). A wide range of force values describing these nonlinearities were used in the simulation.

Given the estimate of the system poles and $F_T$ as the system input, the linear dynamics of the valve model can be estimated using equations (1) and (2):

$$y''+\Psi\{.2135y'\}+.0027y=F_T \tag{3}$$

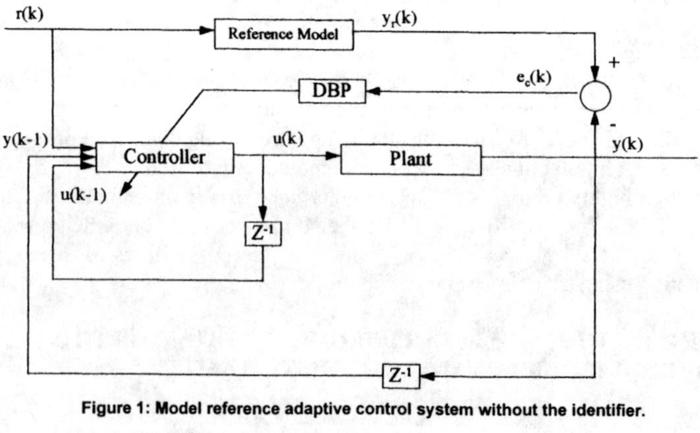


**Figure 1: Model reference adaptive control system without the identifier.**

## DESIGN OF THE NEURAL NETWORK CONTROLLER

The overall system assumes the configuration of the model reference adaptive system (Astrom, 1989) as diagrammed in figure 1. The goal of this system is to force the nonlinear valve to follow a linear model. The inputs to the adaptive controller are: the reference signal r(k), the previous value of the controller output u(k-1) and the previous value of the plant output y(k-1). The controller parameters are adjusted to minimize the square of the deviation, $e_c(k)$, between reference model and plant outputs.

The controller is realized using a recurrent neural network whose generic structure is shown in figure 2. The input-output relationship for this neural network can be described by the equations (4), (5) and (6).

$$S_j(k) = W_j^D X_j(k-1) + \sum_{i}^{n_i} W_{ij}^I I_i(k) \tag{4}$$

$$X_j(k) = \sigma\{S_j(k)\} \tag{5}$$

$$O(k) = \sum_{j}^{n_o} W_j^o X_j(k) \tag{6}$$

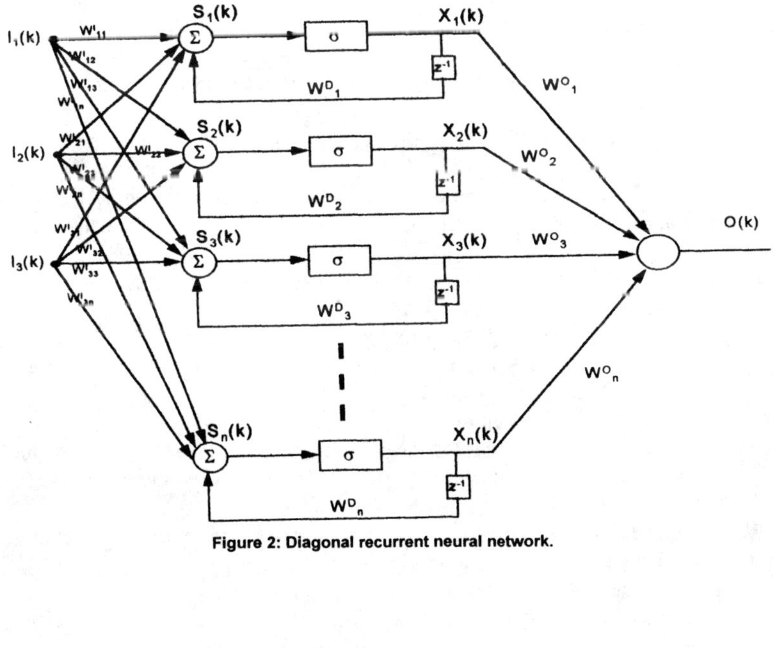


**Figure 2: Diagonal recurrent neural network.**

Here, *O(k)* is the output of the neural network, the $I_i(k)$'s are the inputs, W is the weight matrix, and $\sigma$ is the activation function. The values of $n_i$ and $n_o$ are the number of inputs and the number of neurons in the hidden layer respectively. The controller parameters are updated as shown in equation (7):

$$\begin{bmatrix} W_j^o(k+1) \\ W_j^D(k+1) \\ W_{ij}^I(k+1) \end{bmatrix} = \begin{bmatrix} W_j^o(k) \\ W_j^D(k) \\ W_{ij}^I(k) \end{bmatrix} + \eta e_c(k) y_u(k) \begin{bmatrix} X_j(k) \\ W_j^o P_j(k) \\ W_j^o(k) Q_{ij}(k) \end{bmatrix} \tag{7}$$

where

$$P_j(k) \equiv \frac{\partial X_j(k)}{\partial W_j^D} = \frac{\partial X_j(k)}{\partial S_j(k)}\left[X_j(k-1) + W_j^D P_j(k-1)\right] \; and,$$

$$Q_{ij}(k) \equiv \frac{\partial X_j(k)}{\partial W_{ij}^I} = \frac{\partial X_j(k)}{\partial S_j(k)}\left[X_j(k-1) + W_j^D Q_{ij}(k-1)\right]$$

and the sensitivity of the plant output to the control input is:

$$y_u(k) = \frac{\partial y(k)}{\partial u(k)} \tag{8}$$

The learning rate is a small positive number. The steady state value of the sensitivity parameter $y_u(k)$ is either positive or negative depending on the particular system. Usually it is positive for a valve, since a positive input will normally result in a positive steady state output. Therefore, it is possible to assume a value of +1 for the sensitivity parameter.

This simplification implies that the system does not require any on-line identifier. The computational and hardware savings significantly enhance the potential for real-time implementation of the adaptive controller using a standard single-chip microcontroller.

## SIMULATION RESULTS WITH AND WITHOUT IDENTIFIER

A nonlinear plant, $y(k) = y(k-1)/(1 + y^2(k-1)) + u^3(k)$, is to be designed to follow the responses of a first-order linear model: y (k) = 0.6 * y ( k-1) + r(k). A reference signal, r(k) = sin( 2*pi*k/25) + sin( 2*pi*k/10), is used as the input to both the model and the plant. A simulation MATLAB program was written to study the diagonal recurrent neural network in controlling the nonlinear plant. The initial values of the coefficients of the neural network are random numbers from -0.05 to 0.05.

We started the simulation with the neural network identifier. However, the combination of the controller and identifier did not converge about half of the time. In order to check the validity of the algorithm and the simulation code, a very simple plant, y(k) = u(k), was used. The identifier output, or the sensitivity function $y_u(k)$, has a theoretical value of one. When the simulation program gives negative values

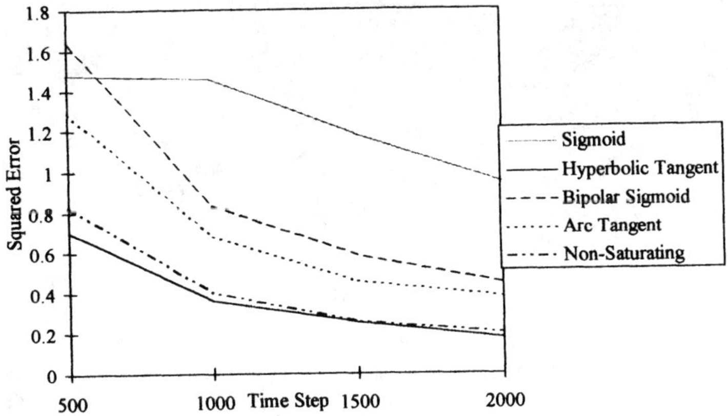


**Figure 3: The effect of learning rates on system performance.**

for this sensitivity function, the iterative process diverges. When the sensitivity function is positive, the process always converges.

All the neural network controller simulation results presented in this paper use +1 for the sensitivity function without using an identifier. The computation requirement is cut by half and the convergence of the adaptation process is very consistent.

**Activation Functions**

Five different functions were tested as the activation function σ in equation (5). These functions were Sigmoid, Bipolar Sigmoid, Hyperbolic Tangent, Arc Tangent and Non-saturating (Logarithm). Because the controller output, u(k), can be either positive or negative, the Bipolar Sigmoid is expected to be the better activation function for a neural network controller than the Sigmoid function. The Hyperbolic Tangent function is a special case of Bipolar Sigmoid with $a$=2. Thus, the Bipolar Sigmoid function can have any optimal value for $a$ if $a$ is allowed to be optimized. The mean square errors vs. the five different activation functions are shown in figure 3.

**Number of Neurons**

To determine the effect of the number of neurons in the hidden layer on the performance of the neural network controller, the number of neurons were changed from one to seven. The mean square error vs. the time step of different numbers of neurons with an arc tangent activation function is shown in figure 4. The minimum mean squared errors for different numbers of neurons are about the same, while the computations required to calculate the neural network adaptive controller are proportional to the square of the number of the neurons. Thus, the number of neurons can be limited to three or less to minimize real-time computation requirements without sacrificing the neural network performance.

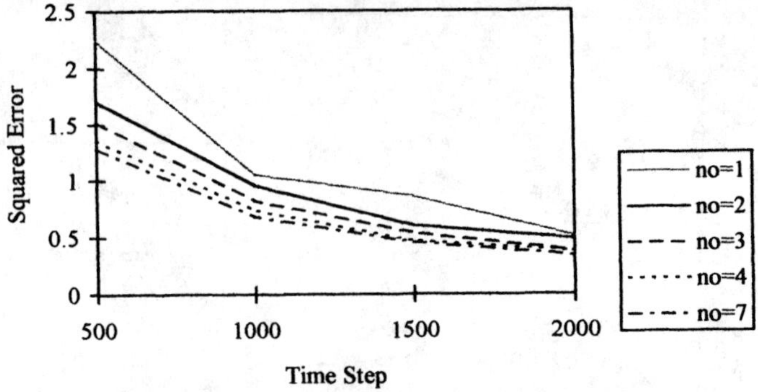


**Figure 4: Mean values for different $n_o$, arc tangent, $\eta_c$=0.01.**

## Different Nonlinearities

Four different types of hard nonlinearities were studied with the simulation program: relays, relays with overlaps, relays with overlaps and deadbands, and idealized frictional forces. The continuous-time equation of motion for the valve can be described as equation (3), where $\psi$ is the appropriate nonlinear characteristic equation. Using a sampling period of 1 second, and backward differences to approximate the derivatives, the following discrete-time equation is obtained:

$$y(k) = 2y(k-1) - \psi(0.21359[y(k-1) - y(k-2)]) - 1.0027y(k-2) + u(k) \qquad (9)$$

The open-loop response of this plant is not satisfactory because the difference between the open-loop response and the model response is too large. The results of the closed-loop responses with the four different types of nonlinearities are listed in table 1. These simulations used a step size of 0.01 and the arc tangent function as the activation function. Each controller consisted of one or two neurons.

**TABLE 1: RESULTS OF CLOSED LOOP RESPONSES**

| | Type of nonlinearity | | | |
|---|---|---|---|---|
| | **Relays** | **Relays with overlaps** | **Relays with overlaps and deadbands** | **Idealized friction forces** |
| **Parameters** | Static friction $F_s$ = 0.1 or 0.5 | Static friction $F_s$ = 0.1 or 0.5<br>Overlap = 0.25 | Static friction $F_s$ = 0.1 or 0.5<br>Overlap = 0.25<br>Deadband = 0.1 | Static friction $F_s$ = 0.1 or 1.0,<br>Coulomb friction $F_c$ = 0.1 or 0.5 |
| **Average square error** | 0.3 after 500 time steps | 0.6 after 1000 time steps | 0.5 after 1000 time steps | 0.4 after 750 time steps |
| **Comments** | Quick convergence | Quick convergence | Quick convergence | Quick convergence |

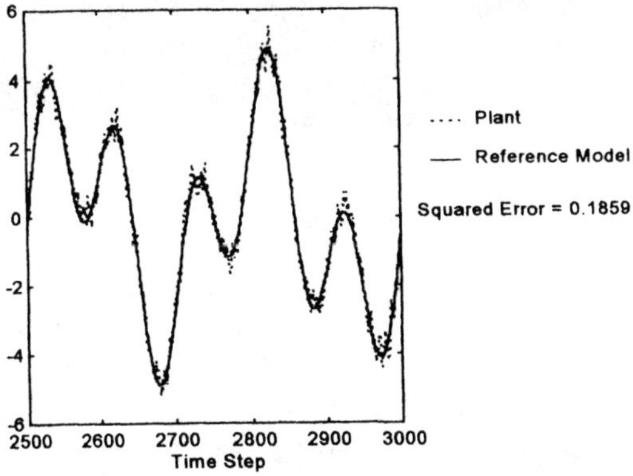


**Figure 5a: Plant response with relay, $n_o$=2, $F_s$=0.5.**

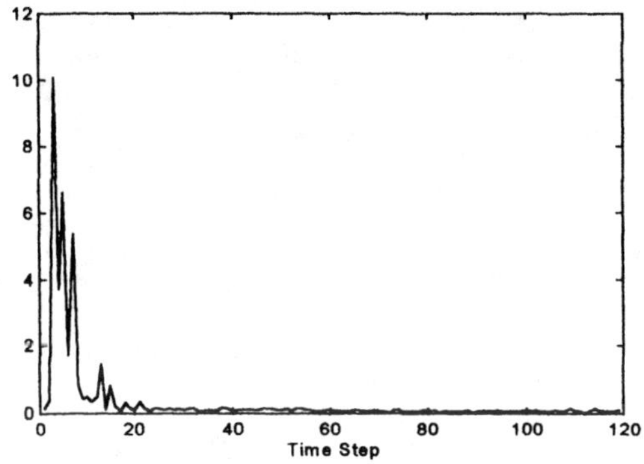


**Figure 5b: Average squared error with relay, $n_o$=2, $F_s$=0.5.**

An example of the neural network controller's responses and the average square errors to a sinusoidal reference input are plotted in figure 5. From the simulation results, one can conclude that the neural network controller with one or two neurons is capable of controlling the nonlinear dynamics.

## CONCLUSIONS

In this paper, design of a recurrent neural network based adaptive controller is presented. Simulation results have indicated that this controller performed consistently with plants that contained nonlinearities such as hysteresis, deadzone and coulomb friction. The approach suggested in this paper offers a great promise for microcontroller based real-time implementation for the following reasons:

1. The need for on-line identification is eliminated.

2. The adaptive controller with 1 or 2 neurons in the hidden layer is sufficient to force the plant to follow a linear model. This result is not a surprise because in many real world adaptive control systems, the reference model is found to be either of first order or second order.

## REFERENCES


Angstrom and Wittenmark, (1989). *Adaptive Control*, Addison-Wesley.

Cheng, Y., Karjala, T., Himmelblau, D., (1995). Identification of nonlinear dynamic processes with unknown and variable dead time using internal neural networks, *Ind Eng Chem Res*, 34, No 5.

Glenn, John, (1996). *Real-Time Adaptive Control of Nonlinear Systems Using Neural Networks*, MSEE thesis, California State Polytechnic University, Pomona.

Ku, C., Lee, K. Y., (1995). Diagonal recurrent neural networks for dynamic systems control, *IEEE Transaction on Neural Networks* 6, No 1, January.

Liang, J., Nikiforuk, P., Gupta, M., (1994). Dynamic recurrent neural networks for control of unknown nonlinear systems, *Journal of Dynamic Systems, Measurement and Control* 116, December.

Widrow and Stearns, (1985). *Adaptive Signal Processing*, Prentice-Hall.

# ROBUST VIBRATION CONTROL OF COMPOSITE BEAMS USING PIEZOELECTRIC DEVICES AND NEURAL NETWORKS

**C. P. SMYSER**
*Department of Mechanical and Aerospace Engineering and Engineering Mechanics, University of Missouri-Rolla, Rolla, Missouri*

**S. VARADARAJAN**
*Structural Dynamics Research and Analysis Corp., Los Angeles, California*

**K. CHANDRASHEKHARA**
*Department of Mechanical and Aerospace Engineering and Engineering Mechanics, University of Missouri-Rolla, Rolla, Missouri*

***ABSTRACT:***
A finite element model of a laminated composite beam with integrated piezoelectric sensors and actuators was developed. The finite element formulation, which was based on a higher order shear deformation theory and accounts for the lateral strains, is applicable for both thin and thick laminated composite beams. The reduced order model was used to develop a state space model of the system. An LQG/LTR based robust controller was designed using the state space model of the system. The performance of the controller was verified for various arbitrary initial conditions. A system of neural networks was then trained to emulate the robust controller. A robustness study including effects of tip mass, structural parameter variation, and loss of a sensor input was performed. The NN controller is shown to provide robustness and control capabilities equivalent to that of the LQG/LTR controller.

## INTRODUCTION

Advanced applications including jet fighters, concept automobiles and spacecraft have structural requirements that include high strength, light weight, high damping and low acoustical noise. Because of their light weight, low cost, and easy implementation, piezoceramic patches are widely used as sensors and actuators for vibration control, shape control and noise reduction. (Crawley and de Luis, 1987) developed analytical models for surface bonded and embedded piezoelectric actuators.

The use of neural networks for vibration control has been investigated by several researchers. (Flanders et al., 1994) used a Radial Basis Function network to minimize beam vibration. (Long, 1993) used a self similar network to control a cantilever beam system. (Chen et al., 1994) implemented neural networks as state estimators in using the Modified Independent Modal Space Control (MIMSC) algorithm for vibration control of a cantilever beam.

In this paper, an LQG/LTR based robust controller is designed using the procedures outlined by (Athens, 1986) to control the vibrations of a laminated composite beam with integrated piezoelectric sensors and actuators. The performance robustness of the neural network controller is compared to that of the LQG/LTR controller through a robustness study of various system parameter variations and different initial conditions.

## MATHEMATICAL FORMULATION

The structural system considered in the present work is a fiber reinforced laminated composite beam (length L, width b and thickness h) with integrated piezoelectric sensors and actuators, as can be seen in Figure 1. Piezoelectric materials possess anisotropic properties. The constitutive equation for a laminated beam accounting for lateral strains and shear deformation can be written as:

$$\begin{Bmatrix} \sigma_x \\ \tau_{xz} \end{Bmatrix} = \begin{bmatrix} \bar{Q}_{11} & 0 \\ 0 & \bar{Q}_{55} \end{bmatrix}_k \left( \begin{Bmatrix} \epsilon_x \\ \gamma_{xz} \end{Bmatrix} - \begin{bmatrix} d_{31} \\ 0 \end{bmatrix} E_3^{\ k} \right) \tag{1}$$

where $\bar{Q}_{11}$ and $\bar{Q}_{55}$ are the transformed coefficients.

The sensor equation can be written as:

$$\{D\}_k = [d]\,[Q]_k\,\{\epsilon\}_k + [\bar{\epsilon}]\,\{E\}_k \tag{2}$$

Treating each distributed actuator as an independent actuator, the assembled finite element equations can be written as:

$$[M]\,\{\ddot{\Delta}\} + [C]\,\{\dot{\Delta}\} + [K]\,\{\Delta\} = \{F\}_{ex} + [P_{ac}]\,\{V\}_{ac} \tag{3}$$

where $\{\Delta\}$ is a (4nx1) vector of nodal displacements, $\{\ddot{\Delta}\}$ is a (4nx1) vector of nodal accelerations, [M] is the global mass matrix, [C] is the damping matrix, [K] is the global stiffness matrix, $\{F\}_{ex}$ is a (4nx1) global force vector, $[P_{ac}]$ is a (4nxP) transformation matrix from actuator voltage to the actuator force, $\{V\}_{ac}$ is a (Px1) vector of actuator voltages, n is the number of nodes in the mesh and P is the number of independent actuators.

The sensor output equations can be written as:

$$\{q\} = [S]\,\{\Delta\} \tag{4}$$

The mode superposition method is used to transform the coupled equations of motion in the modal coordinates to a set of uncoupled equations of motion in the modal coordinates.

## LQG/LTR DESIGN

An optimal full state feedback gain matrix is calculated such that the requisite design criteria is achieved. Typical design performance requirements for vibration control problems are specified in terms of required damping for various modes of vibration, available control command and desired minimum loop gain. The minimum loop gain is specified to reflect the expected uncertainty in the system. The higher the expected uncertainty, the higher the required minimum loop gain in order to ensure performance robustness. The loop gain is particularly important at and around the natural frequency of the structure, since the vibration behavior of the system is most sensitive around these frequencies. A loop gain of more than 0 dB is desired at the natural frequencies of the structure.

## NEURAL NETWORK CONTROLLER DESIGN

Once the LQG/LTR controller design was complete, a neural network controller was developed to emulate the robust controller. The neural network (NN) controller consisted of a two feedforward neural network system as shown in Figure 2. Each network is made up of an input layer, a hidden layer and an output layer. The inputs to the first network are the sensor voltages and the controller states. Network 1 then outputs the new predicted states of the controller. The inputs to the second network are the predicted states of the controller (the output of network 1). The outputs of the second network are the actuator voltages required to control the beam. The number of neurons on the input and output layers of the networks are determined by the number of sensors and actuators used to control the beam. For the present work, only 2 sensors and actuators are used, leading to 12 states for the controller. Thus network 1 has 14 inputs and 12 outputs, and network 2 has 12 inputs and 2 outputs.

## RESULTS AND DISCUSSION

The sensors are bonded at the top and bottom surfaces of the beam, and the actuators are bonded on top of the sensors. The two collocated actuators are polarized in opposite directions in order to generate only bending moments. The material properties, dimensions, and cross-sectional data for the beam and the piezoelectric sensors and actuators are given in Table 1. The beam is clamped at one end and free at the other. Figures 3a and 3b show the open loop response, LQG/LTR controlled response and NN controlled response for initial conditions equal to the first mode of vibration and second mode of vibration, respectively. As can be seen in the figures, both the LQG/LTR controller and the NN controller successfully damp the vibrations of the beam. Figure 3c shows the responses for a distributed force of 20000 N/m$^2$ applied along the length of the beam for a duration of 0.001s. As in the previous figures, both controllers successfully damp the vibrations of the beam, with the NN controller performing slightly better than LQG/LTR controller.

**TABLE 1: MATERIAL AND GEOMETRIC DATA**

| | T300/5208 Graphite/Epoxy | G-1195 Piezoceramic |
|---|---|---|
| $E_1$ (GPa) | 174.8 | 63.0 |
| $E_2$ (GPa) | 10.3 | 63.0 |
| $G_{12}$ (GPa) | 7.17 | 24.8 |
| $\nu_{12}$ | 0.25 | 0.28 |
| $G_{13}$ (GPa) | 7.17 | -- |
| $G_{23}$ (GPa) | 6.21 | -- |
| $\rho$ (kg/m$^3$) | 1389.23 | 7600 |
| $d_{31}$ (m/volt) | -- | $-166.0 \times 10^{-12}$ |
| $d_{32}$ (m/volt) | -- | $-166.0 \times 10^{-12}$ |
| Length (m) | 1.0 | 0.1 |
| Width (m) | 0.0254 | 0.0254 |
| Thickness (m) | 0.0127 | 0.0002 |

## ACKNOWLEDGMENTS

This work is supported by the Office of Naval Research under Grant # ONR N00014-94-1200 with Dr. Thomas M. McKenna as the technical monitor. Partial support from the Missouri Department of Economic Development is gratefully acknowledged.

## REFERENCES

Athans, M. (1986). A tutorial on the LQG/LTR method, *American Control Conference*, Vol 2, 1289-1296.

Chandrashekhara, K., Agarwal, A.N. (1993). Active vibration control of laminated composite plates using piezoelectric devices: a finite element approach, *Journal of Intelligent Material Systems and Structures*, Vol. 4, 496-508.

Chen, C. I. (1994). Active vibration control using the modified independent modal space control (MIMSC) algorithm and neural networks as state estimators, *Journal of Intelligent Material Systems and Structures*, Vol 5, Jul, 550-558.

Crawley, E. F., de Luis, J. (1987). Use of piezoelectric actuators as elements of intelligent structures, *AIAA Journal*, Vol 29, 1373-1385.

Flanders, S. W., Burke, L. I., and Yalcintas, M. (1994). Alternate neural network architectures for beam vibration minimization, *Adaptive Structures and Composite Materials: Analysis and Application*, AD-Vol 45/MD-Vol 54, 293-298.

Long, T. W. (1993). A self-similar neural network for distributed vibration control, *Proceedings of the 32nd Conference on Decision and Control*, Vol 4, 3243-3248.

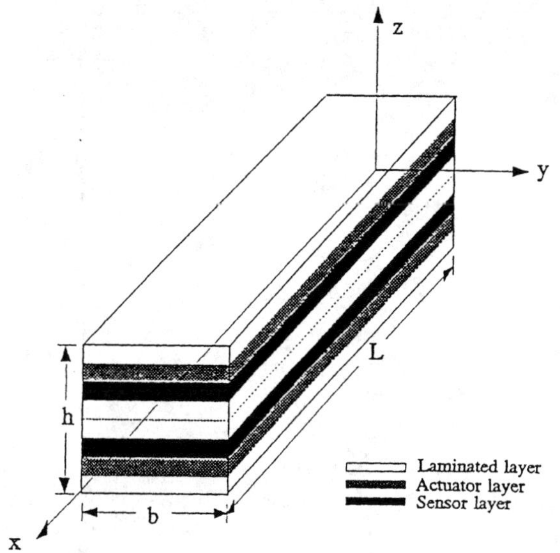


Figure 1 Laminated beam with integrated piezoelectric sensors and actuators

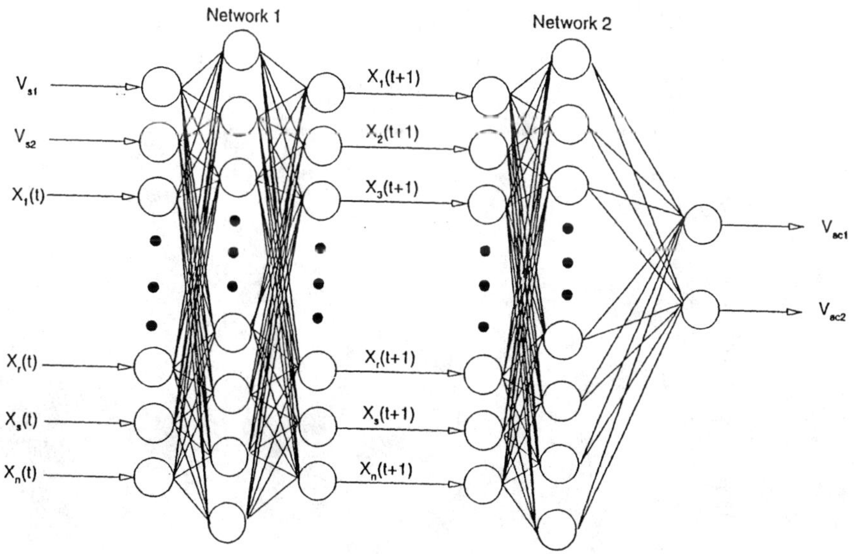


Figure 2 Architecture of neural network controller

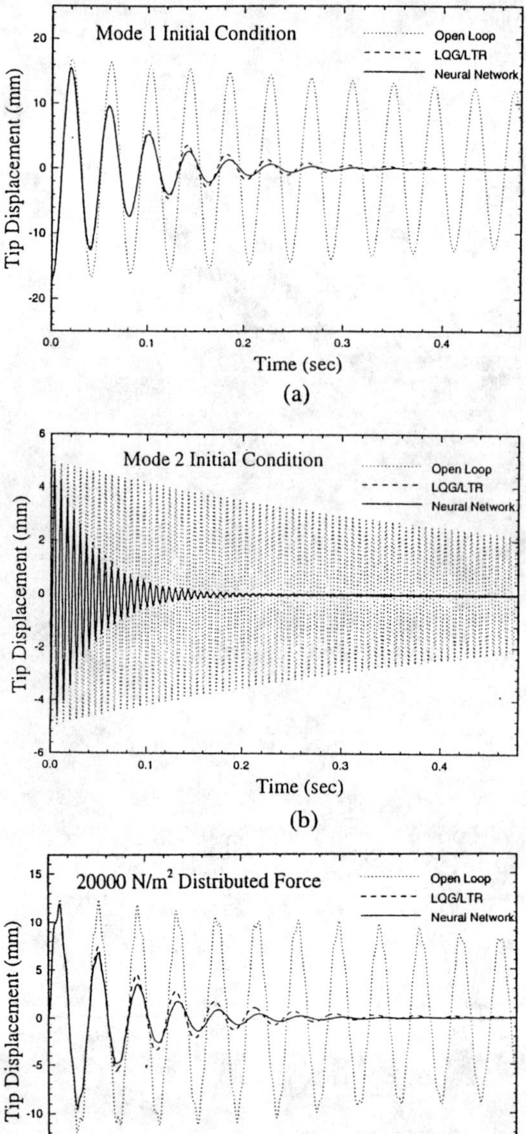


Figure 3: Comparison of performance of LQG/LTR and NN controllers for different excitation

# A NEW CONTROL SCHEME APPLIED TO THE TRAILER-TRUCK BACKING-UP PROBLEM

**R. S. WOODLEY, T. HAN, AND L. ACAR**
*Department of Electrical Engineering and Intelligent Systems Center*
*University of Missouri-Rolla*

**ABSTRACT:**
The trailer-truck backing-up problem has long been accepted as a benchmark to test control algorithms. Its unique blend of nonlinear dynamics coupled with physical limiting conditions makes it a challenging problem. A recent control scheme used on this system took advantage of some path planning techniques. However, this scheme required re-calculation of the control, when the initial conditions changed. The control scheme introduced here will use the data generated by the path planning approach to train a neural network that will be able to create the path data from any initial condition. Since then the path information can be updated in real-time as the trailer-truck traveled, the controller is reduced to tracking the path supplied by the neural network.

The trailer-truck backing-up problem is an interesting control problem. Due to its non-linear dynamics, the control actions must be in the form of either piece-wise linear or general non-linear functions.

In the piece-wise continuous control case, each of the control actions is derived for the linearized dynamics about a specific operating point. However, when the operating point involves a non-linear path in addition to the non-linear dynamics, the control ends up relying on predetermined time intervals or strategically placed zones along the path. In either case, the control is very much dependent on the initial conditions.

The general non-linear control case has enough complications that a comprehensive analytical solution is currently unobtainable. There are some contributions in the area of nonholonomic motion planning, such as the works by Fernandes, Gurvits, and Li (Fernandes, Gurvits, and Li, 1991), by Mukherjee and Anderson (Mukherjee and Anderson, 1993), and many others. Our purpose in this article is to approach to an acceptable non-linear control as one generalization of a piece-wise linear control via neural networks in contrast to approach the problem as a general non-linear control problem.

Neural networks, that have been used to solve the trailer-truck backing-up problem before, were mostly applied on computer simulations of the simplified trailer-truck dynamics. One exception was the hardware setup of Nguyen and Widrow (Nguyen and Widrow, 1989). In their work, Nguyen and Widrow used a form of dynamic programming of backpropagation in order to train a neural network. However, even in this application, the dynamics are simplified by assuming the pivot point and cab rotation point were the same point and by assuming the speed remained constant. The approach also neglected some limiting factors on the cab-trailer angle and a non-realistic steering angle limit. In addition, the controller takes a considerable time to train; since the training is only performed at the end of a run.

There were other intelligent control methods applied to the trailer-truck backing-up problem. Shibata and Fukuda (Shibata and Fukuda, 1993) used a genetic algorithm and a fuzzy critic to perform the control. Kosko (Kosko, 1992) applied a Fuzzy Associative Memory to perform the control. In his work, Kosko used first some rules of thumb to generate the fuzzy rules and then utilized Differential Competitive Learning to automatically train the fuzzy rules in an Adaptive Fuzzy Associative Memory. These works also assume a simplified trailer-truck model and non-realistic steering angles. However, the system is easier to train, since no identification network is needed.

The structures of Widrow, Kosko, and others have another major disadvantage. In the case of changes in the system dynamics, the surroundings, and the goals, the entire controller needs to be re-trained. All the previous training data become invalid, and the training process starts over again.

The primary distinction between the previous controller structures and the structure introduced here lies in the structure. The previous structures required the controller to not only be able to control the trailer-truck but also be aware of the goal of the movement. By breaking the problem into subproblems, our controller brings a modularity to the approach. This modularity enables fast adaptation under various changes without the need for re-training.

The control scheme is made of three parts: the identification, the path planning, and the control networks. For the identification and path planning networks, multi-layer perceptron, hierarchical, and radial basis function networks are considered. These networks are compared in their speed of convergence as well as overall accuracy. The results for the control network will be excluded from this document due to time and space considerations. These results will be reported in the very near future.

# THE PHYSICAL TRUCK SYSTEM

The equations that describe the dynamics of the trailer-truck are more accurate than those in related previous literature. We assume that the system inputs are the speed of the drive tires, $V_b$, and the angular velocity of the steering tires, $\omega_S$. The state variables of the system are the steering angle, $\theta_S$, the angle of the cab with respect to the ground, $\theta_C$, the angle of the trailer with respect to the ground, $\theta_T$, and the two coordinates for the trailer axle with respect to the ground, $X$ and $Y$. The system dynamics based on these variables can be represented as

$$\dot{\theta}_S = \omega_S,$$

$$\dot{\theta}_C = \frac{1}{d_C}\tan(\theta_C)V_b,$$

$$\dot{\theta}_T = \frac{1}{d_T}\sqrt{1+\left(\frac{d_P}{d_C}\tan(\theta_S)\right)^2}\sin\left(\theta_{C\text{-}T} + \tan^{-1}\left(-\frac{d_P}{d_C}\tan(\theta_S)\right)\right)V_b, \quad (1)$$

$$\dot{X} = \sqrt{1+\left(\frac{d_P}{d_C}\tan(\theta_S)\right)^2}\cos\left(\theta_{C\text{-}T} + \tan^{-1}\left(-\frac{d_P}{d_C}\tan(\theta_S)\right)\right)\cos(\theta_T)V_b,$$

and

$$\dot{Y} = \sqrt{1+\left(\frac{d_P}{d_C}\tan(\theta_S)\right)^2}\cos\left(\theta_{C\text{-}T} + \tan^{-1}\left(-\frac{d_P}{d_C}\tan(\theta_S)\right)\right)\sin(\theta_T)V_b.$$

In these equations, $\theta_{C\text{-}T} = \theta_C - \theta_T$, $d_C$ is the length from the steering tires to the drive tires, $d_P$ is the length from the drive tires to the pivot point, and $d_T$ is the length from the pivot point to the axle on the trailer.

The form of the dynamics equations are similar to the ones presented by Divelbiss and Wen (Divelbiss and Wen, 1992b). In fact, the motion planning algorithm derived by Divelbiss and Wen is utilized in obtaining some results of this document. The additional complexity in the dynamics is introduced by the variable $d_P$. For example, without this variable, (1) would reduce to

$$\dot{\theta}_T = \frac{1}{d_T}\sin(\theta_{C\text{-}T})V_b. \quad (2)$$

In the succeeding reports of our trailer-truck controller, we will be able to include the motor dynamics including the opposing forces generated by the tires during steering, and the varying moment of inertia. All of these effects were ignored in the previous work by unrealistically assuming that the speed of the trailer-truck can be kept constant under all circumstances.

# THE CONTROL SCHEME

The control scheme is a modular extension of the path planning approach introduced by Divelbiss and Wen. Once a viable path of the trailer-truck is determined, the controller merely tracks a given trajectory. Overall control scheme is given in Figure 1.

The viable paths are computed for various configurations and a single destination using the motion planning program from the work of Divelbiss and Wen (Divelbiss and Wen, 1992a). These paths form a web about the destination and provide approximate trajectories for the feedback controller. The motion planning program also generates control target data based on the simplified dynamics of the trailer-truck.

Each of the three neural networks in the figure are trained individually. The Identification Neural Network is trained to follow the real system outputs. The Path Planning Neural Network is trained on a set of viable paths. And, the Controller Neural Network is trained using the control target data generated by the path planning program in a manner that diminishes the trajectory tracking error. All the major training processes are done off-line, and very minor training is needed on-line after the trailer-truck starts for its destination.

The Controller Neural Network accepts the incremental path information in terms of a local displacement between the next desired and the current locations. The desired location information is supplied by the Path Planning Neural Network. The Controller Neural Network uses this displacement to compute the speed and the angular velocity that move the trailer-truck to the next desired location.

The modularity of the control scheme is achieved due to this incremental nature of the controller. A trained Controller Neural Network ends up learning a localized function that makes the trailer-truck move in smaller increments, instead of having to learn a global function that moves the trailer-truck to the final destination.

This localization also enables numerous additional elements to be added to perform additional tasks. In the case of changing obstacles, the Controller Neural Network stays the same, and and only portions of the viable paths need to be updated.

# RESULTS

The learning rates of the Identification and the Path Planning Neural Networks were fast. The neural networks learned the desired functions to the desired accuracy.

For comparison, two neural networks were used for identification: a multi-layer perceptron with error backpropagation, and a hierarchical network. The training histories are given in Figure 2. The hierarchical network trained faster; whereas the multi-layer perceptron trained much slower, and it did not perform as well. The backpropagation algorithm used was from the Matlab Neural Network Toolbox.

For path planning, in addition to the same types of neural networks, a radial basis function network from the Matlab Toolbox was also used. The training history of the hierarchical and radial basis function networks are given in Figure 3. Since the size of the training set data for the path Planning network is very large, the multi-layer perceptron and the hierarchical networks learned rather slowly. Once again, the hierarchical network outperformed the multi-layer perceptron. However, the radial basis function network was the best in terms of learning speed. However, the radial basis function network was poor in interpolating between training data.

# CONCLUSIONS

The goal of this work was to approach to an acceptable non-linear control for the trailer-truck backing-up problem as one generalization of a piece-wise linear control via neural networks. The path planning approach offered the information needed to a neural network control scheme that is modular and trainable off-line.

In this scheme, three neural networks are used to identify the system, to generate initial condition independent trajectories, and to control the trailer-truck motion incrementally. As a result, the path planning neural network was designed to learn a collection of paths terminating at the final destination off-line, and the controller neural network was designed to learn how to make the trailer-truck move a small increment.

For this problem and other similar problems, such as in robotics, having to control the motion in small increments is a huge benefit. The problem is still non-trivial, but the task can be reduced to a manageable level. This approach would also provide fast adaptation under various changes in system dynamics, the surroundings, and the goals, without the need for a complete re-training.

## References


Divelbiss, A. W. and Wen, J. T. (1992a). A global approach to nonholonomic motion planning. In *Proceedings of the 31st IEEE Conference on Decision and Control,* 2, Tucson, Arizona, December, 1597–1602. IEEE Control Systems Society.

Divelbiss, A. W. and Wen, J. T. (1992b). A path space approach to nonholonomic motion planning in the presence of obstacles. Technical note, Department of Electrical, Computer, and Systems Engineering, Rensselaer Polytechnic Institute, Troy, New York.

Fernandes, C., Gurvits, L., and Li, Z. X. (1991). A variational approach to optimal nonholonomic motion planning. In *Proceedings of the 1991 IEEE International Conference on Robotics and Automation,* 1, Sacramento, California, April, 680–685. IEEE.

Kosko, B. (1992). *Neural networks and fuzzy systems: A dynamical systems approach to machine intelligence.* Englewood Cliffs, New Jersey: Prentice-Hall Inc.

Mukherjee, R. and Anderson, D. P. (1993). Nonholonomic motion planning using stokes's theorem. In *Proceedings of the 1993 IEEE International Conference on Robotics and Automation,* 3, Atlanta, Georgia, May, 802–809. IEEE.

Nguyen, D. and Widrow, B. (1989). The truck backer-upper: an example of self-learning in neural networks. In *Proceedings of the International Joint Conference on Neural networks,* II, June, 357–363.

Shibata, T. and Fukuda, T. (1993). Intelligent motion planning by genetic algorithm with fuzzy critic. In *Proceedings of the 8th IEEE International Symposium on Intelligent Control,* Chicago, Illinois, August, 565–570. IEEE Control Systems Society.

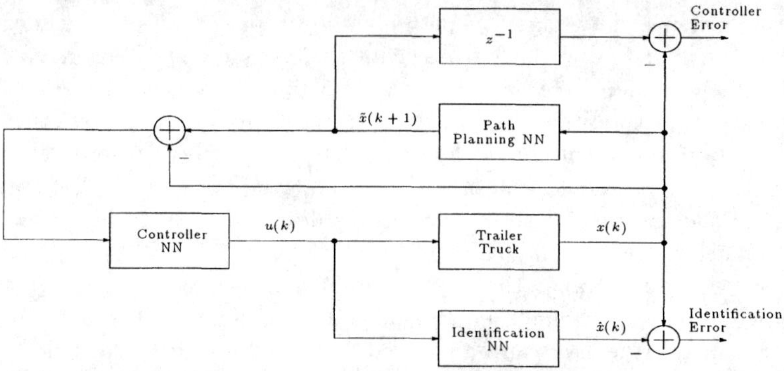


Figure 1: Block diagram for the trailer-truck control.

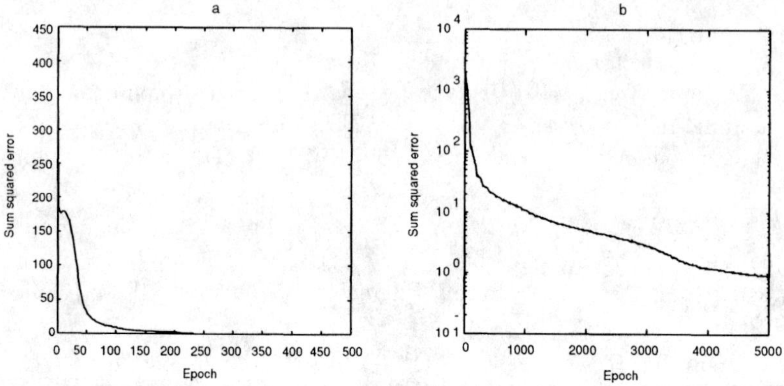


Figure 2: Training history for the identification network, a) hierarchical training algorithm, b) backpropagation training algorithm.

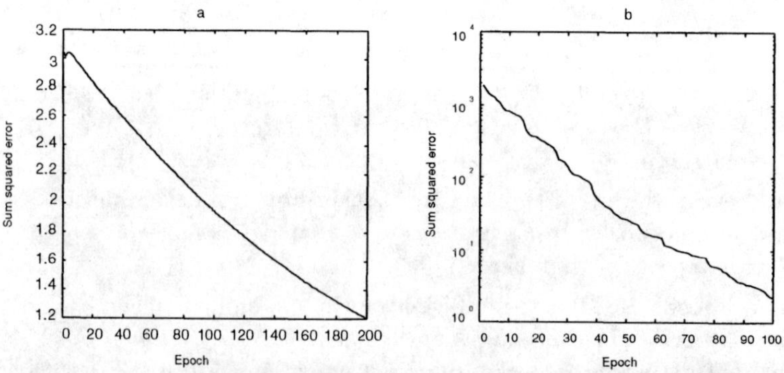


Figure 3: Training history for the path planning network, a) hierarchical training algorithm, b) radial basis function training algorithm.

# Backing Up a Truck and Trailer Using Sets of Three-Neuron Controllers

**John R. Alexander, Jr.**
Computer and Information Sciences Department
Towson State University
Towson, MD 20204
E7C1ALE@TOE.TOWSON.EDU

**Jacob Thomas Cox**
Vitro Corporation
coxj@vitro.com

## ABSTRACT
This work is a continuation of the effort reported in several previous papers presented at ANNIE Conferences (Alexander 1994) (Alexander and Bradley 1995) by the first author. A little-modeled property of biological neurons - the ability of each neuron to fire both above and below its average firing rate - is the feature which our model exploits. The equations developed by the author in his dissertation model neurons possessing an average firing rate. In previous papers we had demonstrated the usefulness of this feature. Thus far, these equations have successfully maintained, at constant setpoint, the height of water in a tank under conditions of randomly changing inflow, and backed a truck to a loading dock. In this paper we demonstrate the ability of another sets of equations which, also employ the average firing rate concept, to successfully back a truck with a trailer attached to a loading dock. Our algorithm for determining the steering angle (for the cab's or truck's wheels) is considerably simpler than the well-known one given by Kosko (Kosko 1992).

## HISTORY
"Only recently has the application of artificial neural networks in control tasks become popular. The word <u>control</u> does not appear in the index of Volume I of Rumelhart and McClelland's (1986) classic PDP book published in 1986. McCulloch and Pitts (1943) modeled an abstract neuron, and proved that nets of such neurons could perform binary logic. However, from that time until the mid 1980's, most authors were concerned with demonstrating or improving methods of recall or association. Thus, the study of artificial neural networks began with studies of memory mechanisms and is now expanding to include studies of control mechanisms. This is in contrast to actual biological evolution. Dr. James Olds (1995) of the National Institutes of Health mentioned to one of the authors (Alexander) that as yet no one has been able to teach a jellyfish any associations whatsoever. However, jellyfish can swim, attack prey, and in general perform all of the tasks necessary of animal life. Control is both necessary and sufficient to life, and memory mechanisms seem to have evolved from control mechanisms." This paragraph began our paper presented at ANNIE95 (Alexander, Bradley 95). These words are also appropriate to introduce this year's paper.

## THEORETICAL BACKGROUND
The first author, in his dissertation, developed a set of artificial neural network equations (Alexander 91) (Alexander 92 - ANNIE92) which incorporate some neuron-like features:

(a) information is transferred only when the sending neurons are firing above or below their average firing rate (Abeles 82) (Alkon 94) and,

(b) the activation of both the sending and receiving neurons are used when calculating the change in activation of the receiving neuron.

This set of ANN's was called the RX equations. It is our intent to study the associative memory capacities and control capabilities of more neuron-like versions of these equations (eventually moving to neural models). To this end, the second version of these equations are the equations for the controller described in this paper. They are called RXV2 (Version 2) and listed and briefly described in Appendix A.

Previously, we employed the RX equations as Three-Neuron Controllers (TNCs) in control of the height of water in a tank subjected to random changes in inflow (Alexander 94) and, to back a truck to a loading dock (Alexander, Bradly 95). In this last paper, we based our model on the truck backer-upper described by Kosko (Kosko 92), substituted our TNC for his fuzzy controller, and compared the performance of these two methods of control.

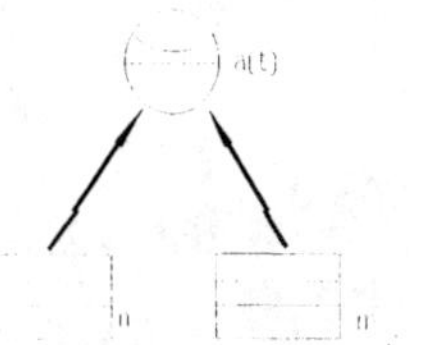

Figure 1. RX two by one net.

A Three-Neuron Controller consists of two input neurons, with one output neuron which integrates the input from the other two. RX nets containing only one output neuron are special cases of general two-layered nets. When used as Three-Neuron Controllers, the RXV2 reduce to the same set of equations the basic RX set did. Figure 1 pictures a TNC. The solid lines in $m_1$, $m_2$, and a(t) represent the resting or average values of these three neurons. The dashed lines represent the current or disturbed values; $m_1$ and $m_2$ being calculated by Equation (1) (or (2) were the current value below average) and a(t) calculated by Equation (3) (or by (4) were the resultant below average). The weights in Equations A4 through A7 of Appendix A are constants, plus or minus one.

The input neurons $m_j$ (in the TNC case j = 1 or 2) possesses some average value $Ave_j$, and their values determined as follows.

$$m_j = \frac{[Observ_j - Ave_j]}{[Max_j - Ave_j]} \qquad If \ Observ_j > Ave \tag{1}$$

$$m_j = \frac{[Observ_j - Ave_j]}{[Ave_j - Min_j]} \qquad Otherwise \tag{2}$$

The difference between the RX and RXV2 equations is small. The denominators, normalizing both the upper and lower case (OUTXY and outxy, where xy represents a capital or small p or an m) "out" terms of the original equations, are omitted in the V2 case. Also, in this case, the $|a_i(t) - \bar{a}_i|$ term in the capital OUT's and the $m_j$ term in the small out's of the original equations have been removed.

The integration of the RX equations as TNCs in our paper in ANNIE95 (Alexander, Bradley 95) contained a typographic error (the coded version was correct, the typo was in the paper). For a TNC the RX and RXV2 equations are the same. The integrated equations of the RXV2, used as the controller for the trailer in our truck and trailer backer-upper in this paper, are listed here correctly.

Although the RX equations are non-linear differential equations, the two-by-one case simplifies them so that they may be integrated in closed form. The integrated solutions are listed below. Note that several cases exist and correspond to both $m_1$ and $m_2$ being positive, both being negative, and both of different sign - one case for the difference between $m_1$ and $m_2$ being positive, and yet another for their difference being negative. The reason for these cases is that different equations describe the flow of activation when a(t) is above its resting value and when it is below this value. Equations (1) through (4) below are (arbitrarily) evaluated at t = 2. This assumes that anytime after a certain minimum time, the receiving neuron has decoded or "understood" the meaning of the input signals.

In Equation (1), $m_p$ represents the sum of $m_1$ and $m_2$ when both are positive.

$$a(t) = 1 - .5 * \exp -2 * [t - 2/m_p (1 - \exp(-m_p * t/2))] \tag{3}$$

In Equation (2), $m_n$ represents the sum of $m_1$ and $m_2$ when both are negative.
For the larger of $m_1$ or $m_2$ ($m_p$) being positive and smaller ($m_n$) negative:

$$a(t) = .5*\exp{-2}*[t-2/m_n(1-\exp(-m_n*t/2))] \quad (4)$$

$$a(t) = 1-.5*\exp-[(1/m_n-1/m_p)-((\exp(-.5*m_n*t))/m_n - (\exp(-.5*m_p*t))/m_p)] \quad (5)$$

and for smaller of $m_1$ or $m_2$ ($m_p$) positive and larger ($m_n$) negative:

$$a(t) = .5*\exp-[(1/m_p-1/m_n)-((\exp(-.5*m_p*t))/m_p - (\exp(-.5*m_n*t))/m_n)] \quad (6)$$

## BACKING THE TRUCK AND TRAILER

The truck and trailer (tractor or cab and trailer) problem is somewhat more difficult from the truck problem alone. Our strategy in controlling the trailer is similar to that used by Kosko (Kosko 1992). Both we and Kosko simply determine what angle the trailer needs to move backwards to arrive at a specific point on the loading dock as if it were a truck alone. We use a different strategy than Kosko, however, in determining the proper steering angle for the cab's front wheels, to back the cab so as to achieve the appropriate movement of the trailer.

The simulation is coded in C++ and written to be run in DOS, using high resolution VGA. We wrote the simulation specifications with reference to that platform and defined the center of the screen to be the origin. These considerations yielded a loading zone for the truck with a range for the x dimension running from -300 to 300 with the negative values to the viewers left and a range for y running from -200 to 200 with negative values on the lower half of the screen. We achieve a one to one correspondence between distance and pixels on the screen using these values. Thus a movement in the x direction of +5 is a motion of 5 five pixels to the right. This placed target point on the loading dock at the coordinates (0, 200).

Concerning direction, we selected 0 degrees for up (towards the loading dock) with 90 degrees to the right and continuing around in clockwise fashion (standard for mapping and navigation). We used this convention for both the orientations of vehicles and the steering angles. Using this absolute measure for all angles seemed to simplify the problem. Kosko and other authors measure steering relative to the body of the vehicle. It is simple to convert the data from our scheme to that required by different controllers. A data conversion function can be placed so that a foreign controller gets the logically equivalent data that it expects. This makes our simulation's standards transparent to foreign controllers. Finally, for comparison of our results with those of Kosko, we accept input in Kosko's geometry (loading dock 100 by 100 and 90 degrees being straight up) but convert it for our internal use.

We use object oriented methodology in coding our simulator. The cab and trailer are both instances of a class "vehicle", designed for this problem. A "vehicle" class contains the data and functions necessary to move a vehicle and to keep track of its position and orientation. The necessary data consists of the following information about itself: Its x and y coordinates of the center of the rear axle, orientation and present steering angle, maximum allowed steering angle, wheelbase (displacement between front and rear axle), length (defined as the displacement from the front axle t to the rear of the vehicle or the hitch to the rear of the vehicle) and width. All of the initial data must be supplied with the instantiation of the vehicle. The class also includes methods to return data the simulation will need (such as the position and orientation of the vehicle), methods to set the steering angle, and a method to tell the vehicle to move a specified distance given its present steering angle.

The class includes two distinct methods to determine the vehicle's motion. The first is used by the cab and takes an input between minus one and positive one, and indicates whether to steer left (negative) or right (positive) and by how much (magnitude). For example, a plus one instructs the vehicle to turn right by the maximum steering angle. The second method is used by the trailer and takes an absolute angle based on the motion of the cab. The class "vehicle" contains all the data necessary to perform the calculation of the motion.

Rather than use Kosko's approximation to determine the motion of the rig as it is backed up, we modified a more accurate simulation, suggested in Neural Networks for Control, Appendix A.5 (Miller, Sutton and Werbos 1992). The geometry of our simulation is pictured in Figure 2.

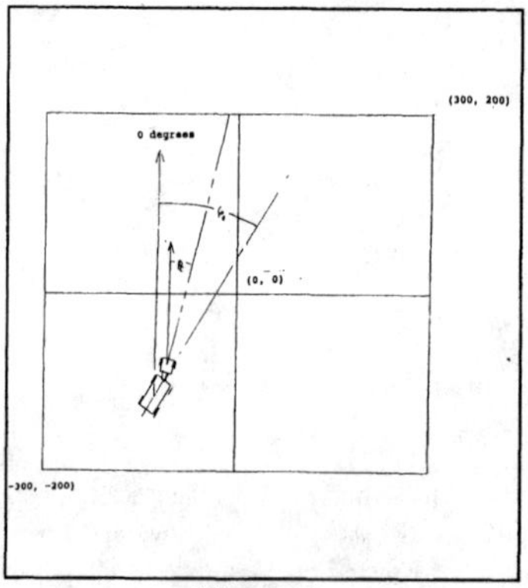


**Figure 2**

The necessary steering angle for the trailer (Kosko's $\beta$) was calculated in two ways: by fuzzy control (primarily to debug the system), with the membership functions being very similar to these presented by Kosko, and by use of a Three-Neuron Controller. This is the same TNC used as a solution to the truck backer-upper problem for ANNIE95 (Alexander, Bradley 95).

Our algorithm for the steering angle for the cab's wheels (Kosko's $\theta$) is simpler than any that either author has thus far encountered. The second author (Cox) stated from personal experience the two following heuristics for backing a cab and trailer:

1) turn the cab the opposite way that you wish the trailer to turn and,
2) do not let the difference in orientation between cab and trailer become too large (in the vernacular, don't jackknife).

Heuristic 1) suggests that the cab had only to attempt to turn onto the negative of the angle requested by the trailer. However, the second heuristic must be also included in any steering algorithm. Control is fairly robust (Alexander 94) and this suggests a linear weighted mixture of the steering angles derived from each of the two heuristics might work. Clearly, as the cab nears a jackknifed position, we must pay less attention to the first heuristic and more to the second. We tried (and it worked):

$$weight_1 = 1 - \left(\frac{(orientationangledifference)}{(maximumangle\ difference\ allowed)}\right)^2 \qquad (7)$$

$$also \qquad weight_2 = 1 - weight_1 \qquad (8)$$

Figure 3 graphs $weight_1$ and $weight_2$ against the angle difference (of the trailer and cab angles).

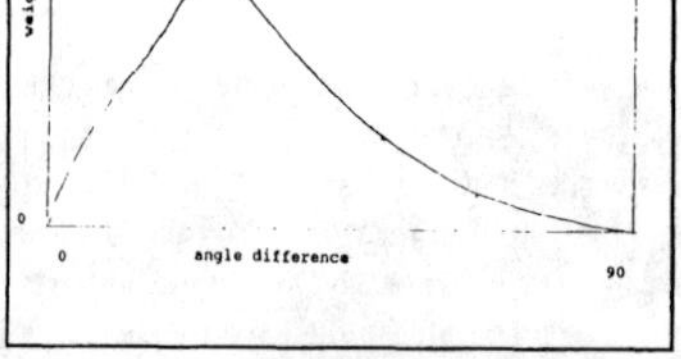


**Figure 3**

The formula we used for the cab steering angle is then:

$$SteeringAngle = weight_1 * (trailerangle) + weight_2 * (FARLEFTORRIGHT) \qquad (9)$$

In Equation 5, "trailer angle" (the term multiplied by $weight_1$), in Kosko's terminology is $-\beta$, and represents the negative of the angle that the trailer would be steered were it a truck alone backing to the dock. The term FARLEFTORRIGHT (which multiplies $weight_2$), represents the maximum correction that the cab can make, i.e. the wheels turned as far one way or the other as they may turn. This corrective term will be in the opposite direction as "trailer angle". That is, it will mitigate the setting of the cabs wheels deemed by our TNC as appropriate to back the rig back to the center of the loading dock.

For comparison, we have backed our rig, using as input the data of the three test cases displayed by Kosko (Kosko 92, page 355). Figures 4, 5, and 6 are from Kosko, and Figures 7, 8, and 9 are the results of Three-Neuron Control.

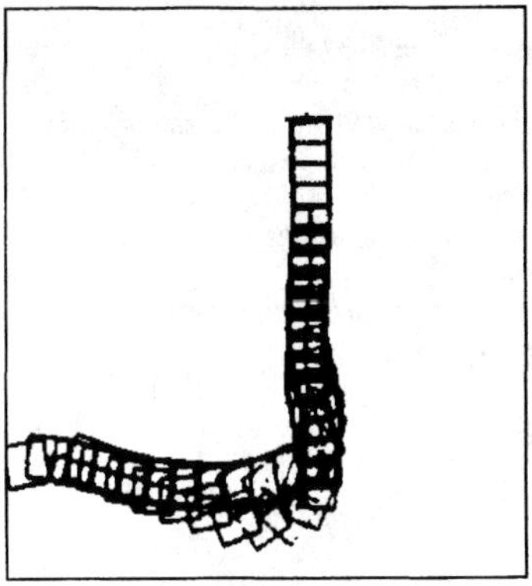

**Figure 4 Kosko's (a)**

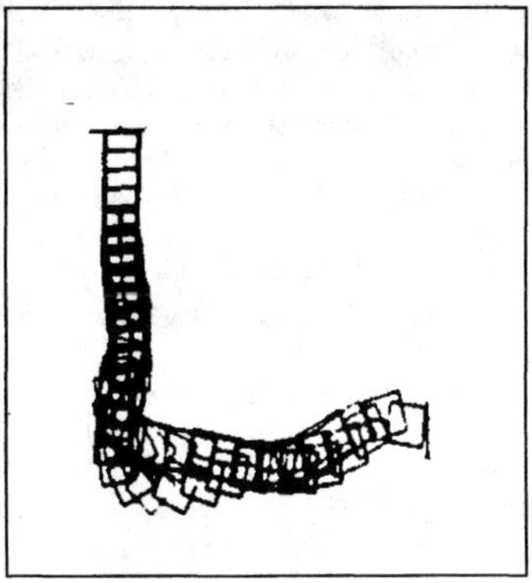

**Figure 5 Kosko's (b)**

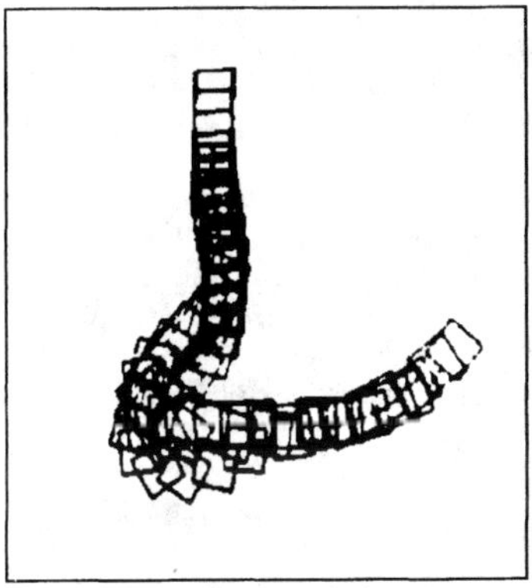

**Figure 6 Kosko's (c)**

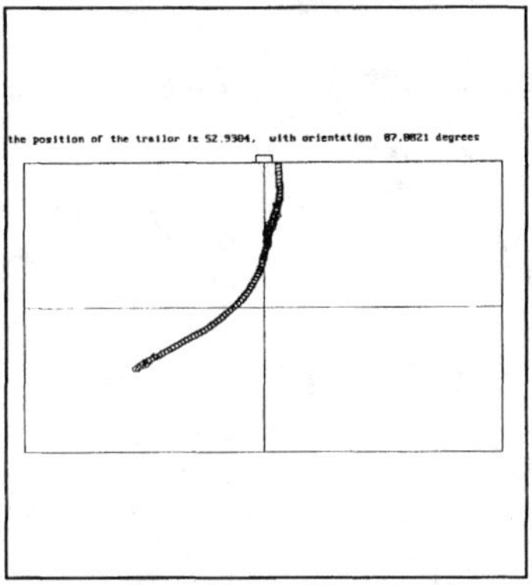


**Figure 7**

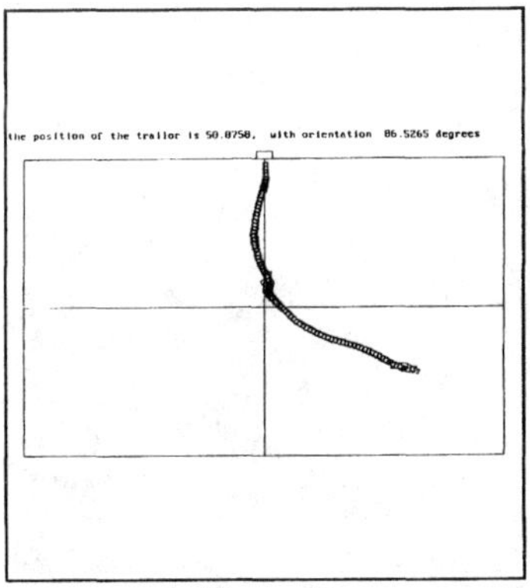


**Figure 8**

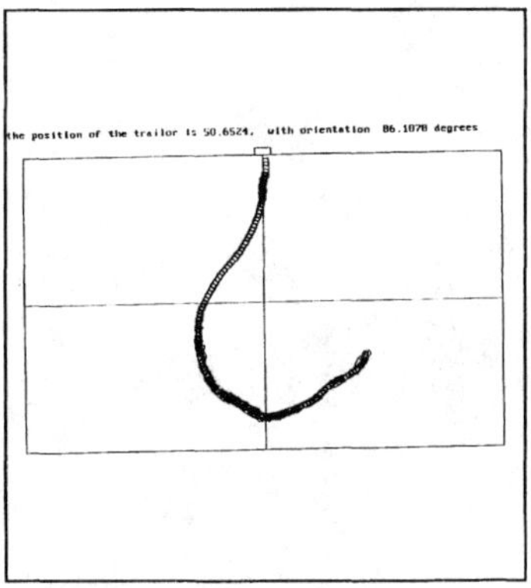


**Figure 9**

## FUTURE WORK AND CONCLUSIONS

At present, we are working on TNC to accomplish the linear mixture needed by the cab controller. Next, we intend to try to back up a "double", that is, a cab drawing two trailers (in an attempt to keep up with the trucking lobby). This seems to be a very do-able task as our coding of the truck and trailer backer-upper will easily allow us to instantiate another "vehicle" and recursively call methods for this new class member.

In the long term, we intend to solve the inverted pendulum via sets of TNCs. This will afford us a set of control problems for our long range goals. Longer term, we intend to keep studying more biologically plausible approximations (dropping back from the RXVx sets) to actual neural behavior.

The calculations accomplished by a fuzzy controller using the two inputs $m_1$ and $m_2$ in its rules to fuzzify, and then a centroid calculation to defuzzify, are apparently similar to the results of the integration as given in Equations (3) through (6) with the input values $m_1$ and $m_2$ as parameters in these formulas (a TNC). Previously, we noted that fuzzy control is extremely robust (Alexander and Bradley 95). It would appear that any set of calculations which produce a control surface which even roughly approximates that produced by a fuzzy controller will offer a reasonable degree of control.

## REFERENCES


Abeles M. (1982). *Local Cortical Circuits An Electrophysiological Study.* Springer-Verlag. New York.

Alexander J.R. (1991). An Analysis of the General Properties and Convergence of a Connectionist Model Employing Probability Data as Weights and Parameters. PhD Dissertation. University of Maryland Baltimore County, Baltimore MD.

Alexander J.R. (1994). Calculating the Centroid of a Fuzzy Logic Controller Using an Artificial Neural Net II, *Intelligent Engineering Systems Through Artificial Neural Networks* **4**, ASME Press, New York, NY.

Alexander J.R. and Bradley, J. B. (1995). The Three Neuron Controller - History, Theoretical Foundations, and Possible Employment in Adaptive Control, *Intelligent Engineering Systems Through Artificial Neural Networks* **5**, ASME Press, New York, NY.

Alkon, D.L. Blackwell, K.T. Barbour, G.S. Werness, S.A. Vogl, T.P. (1994) Biological Plausibility of Synaptic Associative Memory Models. **Neural Networks.** Vol. 7. Nos.6/7

Kosko B. (1992). *Neural Networks and Fuzzy Systems.* Prentice Hall. Englewood Clifts, NJ.

Rumelhart D. & McCelland J. (1986) **Parallel Distributed Processing. Vol 1.** MIT Press. Cambridge, Mass.

McCulloch, W.S. and Pitts, W. (1943) A logical calculus of the ideas immanent in nervous activity. **Bulletin of Mathematical Biophysics.** 5, 114-133.

Miller,W.T. Sutton,R.S. and Werbos, P.J. (1990) **Neural Networks for Control.** The MIT Press Cambridge, Mass.

Olds, J. (1995). Private Communications.

# APPENDIX A

## The RXV2 Equations

Details of the net and its usage The net is two-layered, with the J lower level being the input nodes to the N upper level nodes. The processing units depict a "local" or one-unit-one-concept representation. The net has been used in pattern recognition applications, lower level nodes serving as features and the upper level nodes representing the possible pattern classes. The net's use in control applications is described above. (Note, that since only one upper level node appears in a TNC, the equations for a TNC are much reduced from a general two layered net). The values of the upper level nodes are on [0,1] and called $a_i(t)$. The prior probability of the ith node's existence is called $\overline{a}_i$. Values of $a_i(t)$ greater than $\overline{a}_i$ indicate a higher than average chance of occurrence of the concept presented by node i, and those lower, indicate a less than average chance. The lower level nodes (called $m_j$) are on [-1,1]. Let the range of possible input values to a node be between some $Min_j$ and $Max_j$. Call the average $Ave_j$. Let $Observ_j$ be the observed value. Then define:

$$m_j = \frac{[Observ_j - Ave_j]}{[Max_j - Ave_j]} \qquad If\ Observ_j > Ave_j \tag{A1}$$

$$m_j = \frac{[Observ_j - Ave_j]}{[Ave_j - Min_j]} \qquad Otherwise \tag{A2}$$

When $m_j$ assumes the value 1, then the feature represented by $m_j$ is present, with probability one. When $m_j$ is -1 the feature is absent with probability one. A value of zero indicates that no information exists concerning the absence or presence of the feature ($m_j$.)

Two sets of weights exist. The first is called $w_{ij}$, and in absolute value indicates the probability of occurrence, if positive, and non-occurrence, if negative, of the upper level node (the ith node), given the existence of the lower level node (the jth node). The second weight is called $v_{ij}$, and in absolute value indicates the probability of occurrence, if positive, and non-occurrence, if negative, of the lower level node (the jth node), given the existence of the upper level node (the ith node).

Two auxiliary functions are to be associated with each $a_i(t)$. The function which conveys the excitatory activation is called $Exc_i(t)$, and the one which conveys the inhibitory activation is called $Inh_i(t)$. They are the sums of all the excitatory and inhibitory flow of activation into node $a_i(t)$. The strategy used in defining these functions is to build them as bounded monotone-increasing functions, defined by their derivatives. The monotone-increasing characteristic of the $Exc_i(t)$ term will be achieved by defining it as a product of strictly positive terms in the equation for its derivative. One of these terms is a function of all the $a_i(t)$ and is the variable forcing term. It will for the present be called ForcingFunction and the equations presented shortly. There will be one such function for each lower level (j) node, hence a sum need be taken over all the j nodes connected to any given i node. The bounded characteristic is achieved by including in the equation a factor of the form [N - $Exc_i(t)$]. The choice of N as the bound to which $Exc_i(t)$ approaches is somewhat arbitrary. The differential equation defining $Exc_i(t)$ is then:

$$\dot{Exc}_i(t) = K_1 * \overline{a}_i * [N - Exc_i(t)] * \sum_j ForcingFunction_{ij} \tag{A3}$$

Where $K_1$ is a constant of proportionality. ForcingFunction is a function of all the $a_i(t)$. The $\overline{a}_i$ term as a factor reflects the prior probability of the concept represented by $a_i(t)$ in a meaningful way. $Inh_i(t)$ is defined in a similar fashion.

Since $m_j$ may be above or below average (positive or negative) and the weights $w_{ij}$ may each be either positive or negative, four separate cases may occur. Only four cases can occur, since the signs of the $v_{ij}$ are the same as the $w_{ij}$. These cases are: (1) weight positive with activation positive; (2) weight negative with activation positive; (3) weight positive with activation negative; and (4) both weight and activation negative. As will be shown, all activation transfers (both excitatory and inhibitory) will be calculated as positive, with the inhibitory being subtracted from the excitatory to calculate the total for each case used by the equations in calculating the $a_i(t)$. Each of the four cases will have a term calculated using the $w_{ij}$ weights (called $OUT_{ij}$), and each of these will have a term of opposite sign (offsetting) calculated with the $v_{ij}$ weights (called $out_{ij}$).

Each of the "out" terms will have two letters appended - a (capital or small) p or m. The first of these letters tells if the weight is positive (p) or negative (m) and the second if the activation is positive or negative.

We require two equations per upper level node to describe the change of activation, one when $a_i(t)$ is greater than $\overline{a}_i$, and one when less. Thus for $a_i(t) > \overline{a}_i$:

$$\dot{a}_i(t) = K_3 * c_u * [1 - a_i(t)] * [Exc_i(t) - Inh_i(t)] \tag{A4}$$

When $a_i(t) <= \overline{a}_i$

$$\dot{a}_i(t) = K_3 * c_l * [\quad a_i(t)] * [Exc_i(t) - Inh_i(t)] \tag{A5}$$

Above $K_3$ is a constant of proportionality, and the $c_u$ and $c_l$ are needed to keep the derivative continuous at $\overline{a}_i$. The values of $c_l$ and $c_s$ are:

$$c_u = 1 / (1 - \overline{a}_i),$$
$$c_l = 1 / \overline{a}_i.$$

The complete list of equations to be used with the above is now given.

$$\dot{Exc}_i(t) = K_1 * \overline{a}_i * [N - Exc_i(t)] * \sum_j [k_{11} * OUTPP_{ij}(t) + k_{12} * outpm_{ij}(t) + k_{13} * outmp_{ij}(t) + k_{14} * OUTMM_{ij}(t)] \tag{A6}$$

$$\dot{Inh}_i(t) = K_2 * \overline{a}_i * [N - Inh_i(t)] * \sum_j [k_{21} * outpp_{ij}(t) + k_{22} * OUTPM_{ij}(t) + k_{23} * OUTMP_{ij}(t) + k_{24} * outmm_{ij}(t)] \tag{A7}$$

Here $k_{11}, \ldots, k_{24}$ are included for generality.

$$OUTXY_{ij}(t) = |\, W_{ij} \,| * |\, m_j \,|$$
$$\text{When } W_{ij} > 0;\ m_j > 0;\ XY = PP$$
$$\text{When } W_{ij} > 0;\ m_j < 0;\ XY = PM \tag{A8}$$
$$\text{When } W_{ij} < 0;\ m_j > 0;\ XY = MP$$
$$\text{When } W_{ij} < 0;\ m_j < 0;\ XY = MM$$

$$outxy_{ij}(t) = \sum_{k \neq i} |\, V_{kj} \,| * |a_k(t) - \overline{a}_k| * |a_i(t) - \overline{a}_i|$$
$$\text{When } V_{ij} > 0;\ m_j > 0;\ xy = pp$$
$$\text{When } V_{ij} > 0;\ m_j < 0;\ xy = pm \tag{A9}$$
$$\text{When } V_{ij} < 0;\ m_j > 0;\ xy = mp$$
$$\text{When } V_{ij} < 0;\ m_j < 0;\ xy = mm$$

As written, with absolute value signs included, only positive addends appear in the summation term, which is itself a multiplicand in the equations for $Exc_i(t)$ and $Inh_i(t)$, hence they are both monotone non-decreasing functions.

# TWO METHODS OF ADAPTIVE CONTROLLED CHANNEL RESOURCE ALLOCATION USING REINFORCEMENT LEARNING AND SUPERVISED LEARNING TECHNIQUES

**EDWARD J. WILMES**
*Omnipoint Corporation, Colorado Springs, CO 80907*

**KELVIN T. ERICKSON**
*University of Missouri-Rolla, Rolla, MO 65401*

**ABSTRACT:**
Two methods of dynamic channel allocation for a cellular telephone network using neural networks are investigated. Both methods continuously optimize the mobile network based on changes in calling traffic. The first method uses backpropagation (supervised learning) model predictions to aid the channel allocator. The second method uses the same backpropagation models along with actor-critic (reinforcement learning) models to perform the channel allocation. A comparison shows that both methods significantly outperform fixed channel allocation, even when the call traffic activity deviates from the previously learned models of the call traffic activity.

## INTRODUCTION

The exploding demand for wireless mobile services generates a need for greater cellular network capacity. Capacity can be increased by cell splitting which is limited by the processing power (processing of handoff's) needed to maintain a link as a mobile mitigates through the network of microcells. Also, calling traffic is seldom uniform within a cellular network which increases the complexity of the cell splitting design decisions. However, there is evidence that certain calling density trends exist in a cellular network both spatially and per time of day. For these reasons, it is desirable to dynamically allocate channel resources in an effort to increase cellular network capacity for handling peak traffic conditions. This allows for the cellular network designers to make design decisions for average traffic activity.

In this paper, neural networks are used to uncover trends in network calling traffic, which are used in channel resource allocation decisions. Two approaches are realized using a 41-cell simulated AMPS (Advanced Mobile Phone System) network. The first approach uses backpropagation predictions to identify spatial traffic patterns as a function of the time of day. Future traffic backpropagation predictions are fed into the channel allocator in fixed discrete time increments so that the channel allocator can uncover emerging traffic trends over a short period of time and allocate resources accordingly. This algorithm further contains a direct feedback element to accommodate dramatic traffic deviations away from backpropagation model predictions giving the mobile network real-time adaptation characteristics. The second method builds on the first method by feeding the backpropagation predictions into a real-time learner, which learns a history of traffic activity between two adjacent cells, and makes the allocation decision accordingly. The decision result is immediately reinforced based on the actual traffic results obtained from feedback on lost calls and lack of channel resources. The performance of both methods is described in this paper and is compared to fixed channel techniques. The results demonstrate the applicability of both approaches and illustrate a significant improvement over fixed channel allocation.

## CELLULAR NETWORK

A simulation of a 41-cell network with 7 cells to a cluster is used for the investigations. Figure 1 illustrates the 41-cell network. Each cell is assumed to have an effective propagation radius of one mile. Since the same channels are reused in adjacent cell clusters, the cochannel (same channel) reuse distance, 4.58 miles, must be upheld to prevent lost calls resulting from cochannel interference. Adjacent channel interference was assumed to be negligible.

The call demand is defined as the number of successfully assigned mobiles to valid channels in the cell plus the number of lost hand-offs and the number of lost assignments (new calls) to the cell per time of day. A hand-off is defined as an active mobile being switched from one channel to the next when a mobile leaves a cell and enters a new cell. An assignment is defined as assigning a channel to a new mobile requester.

The traffic model is based on mobile traffic having peak calling times during morning rush hour, lunch time, and during the evening rush hour located along major thoroughfares (shaded area in Figure 1) in a large metropolitan area. The traffic model consists of 200 potential mobiles with calls lasting an average of 3 minutes. The length of the phone calls is generated randomly according to: 45% 2-5 minutes; 30% 2-10 minutes; 15% 2-25 minutes; 10% 2-40 minutes long. The speed of travel for each mobile is 40 miles per hour in one of eight possible directions, North, Northeast, East, etc. The magnitude of calls generated, mobile starting location, and mobile direction of travel all change with the peak and non-peak calling times. During peak times, the traffic is generated at a 3/4 probability of starting in the peak areas with a 9/10 probability of going in the direction of the peak areas. The non-peak mobile traffic is randomly generated over the entire cellular network in random

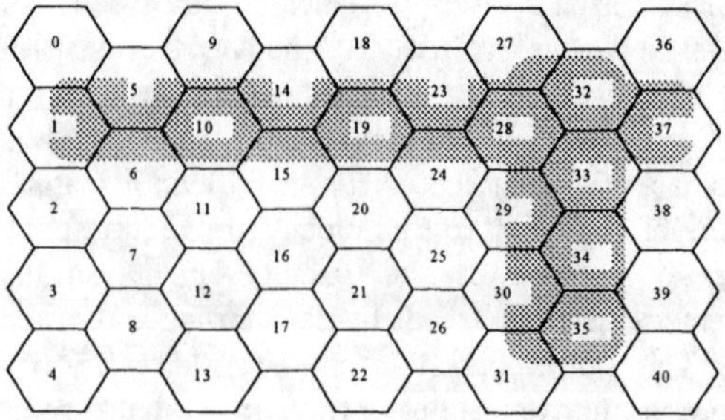


**Figure 1** Forty-one cell network with cell numbers. Shaded area is the area of concentration for peak traffic.

directions. The traffic is generated in the peak areas based on the time of day, as shown in Figure 4. Traffic was generated beyond the network capacity.

## CALL DEMAND LEARNING MODULES

Forty-one two layer backpropagation networks are used to learn the call demand in each of the 41 cells. Each network contains a 31 element input vector representing time, of which 24 components represent each hour of a day, 6 inputs represent 10 minute intervals within the hour, and 1 input to uncover any steady-state phenomena, which is always a value of one (Franklin, 1993). Each network has a 4 element output vector; each element is mutually exclusive. The first neuron output represents the call demand of 0-4 calls requesting channels in the cell; the second neuron output represents 5 calls requesting channels in a cell (full capacity); the third neuron output represents 6-7 calls requesting channels in a cell; the forth neuron output represents 8 or more calls requesting channels in a cell. Each network has one hidden layer consisting of 7 neurons and uses the sigmoid function as the threshold function. The backpropagation network training is stopped when the average root mean squared error is sufficiently low (St. Clair, et al, 1993). Once all of the networks are trained, they are used to quantify the call demand based on the time input vector. When considering all of the outputs of all of the networks, a spatial pattern emerges when looking at the large scale system. The resulting predictions are used as the input vector to the channel allocator and the Adjacent Cell Model Learners.

Once the off-line learning took place, the call demand model learner (backpropagation) outputs were used as predictions for allocating channels 10 minutes into the future (Figure 2). The 41 cells were polled by inputting the time vector at T+10min time into each cell's call demand model learner and storing the predictions for later use. The 41 cells were polled to see if any of the cells were predicted to have low call demand (low traffic), meaning that they would have an excess of channels to lend. Once a "home" cell with available channels was found, the adjacent cells to the "home" cell were then checked to see if there were channels being requested by an adjacent cell. If so, then the cochannel cells to the "home" cell closest to the adjacent requesting cell were checked for channels which could be blocked. The cochannel cells were checked using the polling data and present data, which was a prediction from the call demand model learner 10 minutes into the future. Once it was known that there were open channels in the cochannel cells to

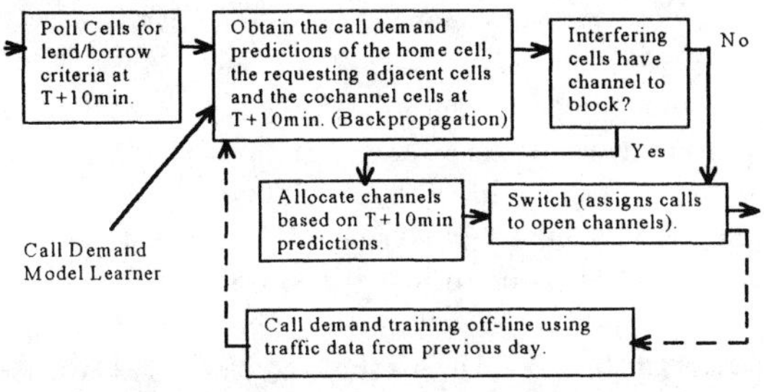


**Figure 2** Mobile Telephone Switching Office with the Call Demand Model Learner.

block, a check was made to see if the channels were the same frequency as the home cell. If the channels were different, hand-offs were made within the same cochannel cells to free up the particular channel targeted for blocking.. The "home" cell then loaned the channel to the requesting adjacent cell and blocked out the corresponding frequencies in the cochannel interfering cells. This pattern was repeated for each cell until all cells were checked and available channel resources were exhausted. This series of events was repeated every 10 minutes. Prior loaned channels were returned to their "home" cells and blocked channels re-enabled when the predictions in the borrowing cells were to "not borrow" and the channels were released from ongoing mobiles.

## ADJACENT CELL MODEL LEARNER

The actor-critic, otherwise known as the adaptive heuristic critic (Sutton, 1988; Sutton, et al, 1991), is a temporal difference method network. Temporal difference method networks learn incrementally and realize the prediction over a series of incremental input vectors. The output prediction is updated as more incremental inputs are added. The actor-critic has an external input (reinforcement) which is used for punishment or reward. The reinforcement affects the actor-critic in this way: if an action given by the critic is satisfactory, reward the critic. This increases the tendency to reproduce that action, given similar inputs to the critic. If an action given by the critic is unsatisfactory, then punish the critic. Punishment suppresses the same action from being produced given similar inputs to the critic.

Actor-Critics were used to determine whether or not a given home cell should lend a channel to a requesting adjacent cell based on the call demand of both cells projected over a 30 minute period into the future. One actor-critic was assigned for each adjacent cell to the home cell. If there were six adjacent cells to the home cell, the home cell contained six actor-critics. The home cell backpropagation output vector and the adjacent cell backpropagation output vector were concatenated to form the input vector for the actor-critic. The ordering of inputs allowed the actor-critic to learn the call demand activity between the home cell and the adjacent cell (Franklin, 1995). Interfering cells were polled for open channels prior to the home cell-adjacent cell lending decision. If, and only if, there was an open channel in the interfering cell(s) to block, then an actor-critic decision was made regarding lending a channel to the adjacent cell. The interfering cells were removed from the actor-critic decision criteria. Channel borrowing, channel blocking, channel lending, and caller activity in interfering cells could not be adequately related to the decision in the home cell-adjacent cell lending criteria. Details about the actor-critic implementation are contained in Wilmes (1995).

The adjacent cell model learner (actor-critic) was consulted after the previously mentioned loaning criteria was satisfied, i.e., a home cell channel was available to lend and the cochannel cells could block out the same channel, see Figure 3. There was one adjacent cell model learner for each of the adjacent cells to the home cell. The actor-critics were trained prior to being incorporated into the system to either lend channels all of the time or to not lend at all. This is similar to the initial training performed by Franklin (1993). All of the actor-critics learned these patterns successfully. The actor-critics were integrated into the same system as mentioned in

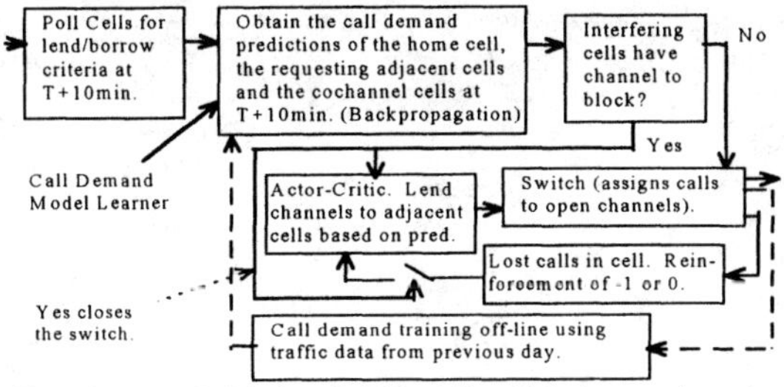


**Figure 3** Mobile Telephone Switching Office with Backpropagation and Actor-Critic.

the Call Demand Learning Modules section. When a home cell was found with available channels to lend and an adjacent cell was requesting channels from the home cell, the cochannel cells were checked as mentioned previously. If these checks all passed, the output prediction from the backpropagation was fed into the actor-critic to determine whether or not to lend the channel. The lending took place as mentioned previously. The performance was compared to that of the fixed channel allocation system, and the results are discussed in the following section.

## RESULTS

The backpropagation networks were trained to the call demand scenario in Figure 4 and were not retrained for the rest of the project for any of the other experiments (including those containing actor-critics). The plot in Figure 5 shows the lost calls per time of day for the Fixed Channel Allocation (FCA) and for both Dynamic Channel Allocation (DCA) methods. Notice that during the peak periods (around 500 minutes and 1100 minutes) the number of lost calls per time of day was reduced by about 20 mobiles per time of day, resulting in an increased network capacity. With the call demand learner model, the number of lost calls for a day was reduced from about 3000 to about 2200. For DCA with the adjacent cell model learners, the number of lost calls was reduced from about 3000 to 2400. The ability of the two DCA methods to perform when the call traffic pattern changes was also determined. The call demand was shifted earlier by one hour and the backpropagation networks were not retrained. As shown in Figure 6, the number of lost calls is higher, but still better than the FCA.

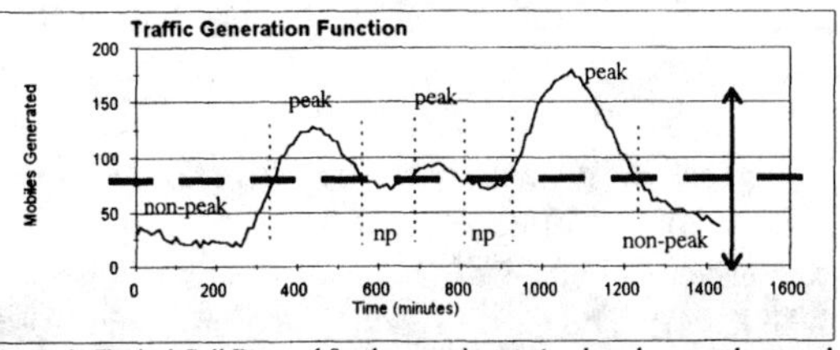


**Figure 4** Typical Call Demand for the experiments (peak and non-peak areas shown).

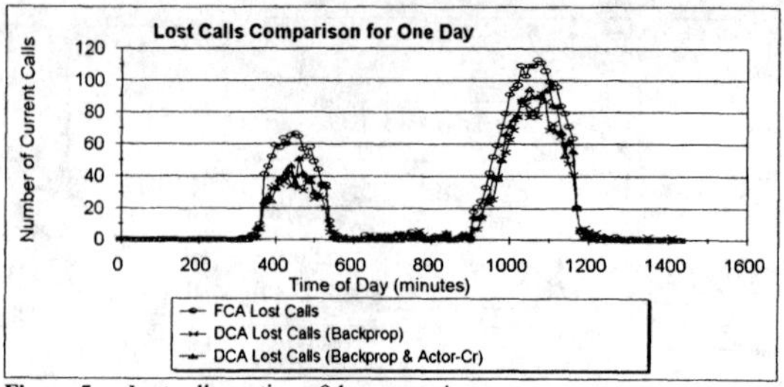


**Figure 5** Lost calls per time of day comparison.

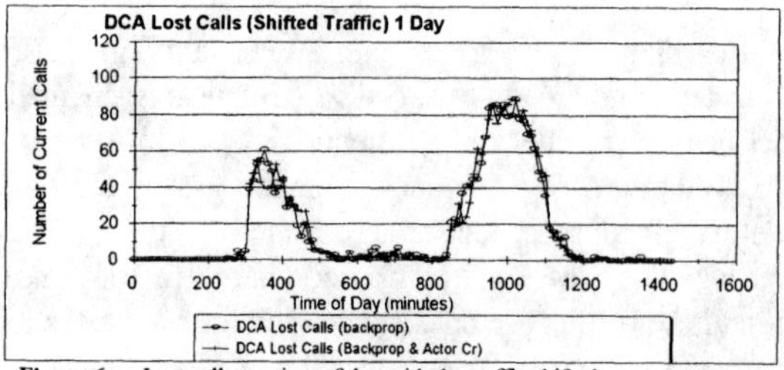


**Figure 6** Lost calls per time of day with the traffic shifted.

# SUMMARY

Both of the methods described in this paper perform better than the fixed channel allocation scheme The results demonstrate the applicability of both approaches and illustrate a significant improvement.

# REFERENCES

Franklin, J.A., " Three Applications of Artificial Neural Networks for Control," GTE Laboratories Incorporated, Waltham, MA 02254, IEEE International Conference on Neural Networks, San Francisco. CA, March-April 1993.

Franklin, J.A., e-mail correspondence, Feb., 17, 1995.

Simpson, P.K., *Artificial Neural Systems Foundations, Paradigms, Applications, and Implementations*, Pergamon Press, Inc., 1990.

St. Clair, D.C., Peterson, G., Aylward, S., and Bond, W., "Design Techniques for the Control of Errors in Backpropagation Neural Networks," *SPIE Conference on Science of Artificial Neural Networks II*, March 1993.

Sutton, R.S., "Learning to Predict by the Methods of Temporal Differences," *Machine Learning*, Vol. 3, p. 9-44, Kluwer Academic Publishers, Boston, MA, 1988.

Sutton, R.S., Barto, A.G., and Williams, R.J., "Reinforcement Learning is Direct Adaptive Optimal Control," *Proceedings of the 1991 IEEE American Control Conference*, Vol. 3, Evanston, IL, June 1991.

Wilmes, E.J., "Neural Network Control of Dynamic Channel Allocation for Mobile Radio Systems," M. S. Thesis, The University of Missouri-Rolla, 1995.

# BACKPROPAGATION FOR SPIKE SORTING IN SYNTHETIC EXTRACELLULAR RECORDINGS

**Ismael Espinosa, Jorge Quiza, Luis Rivera, and Irma Domínguez**
*Laboratory of Cybernetics, Physics Department*
*School of Sciences, México National University (UNAM)*
*México, D.F. 04510*

**Ritaluz Lara**
*ESIME-CULHUACAN, México National Polytechnic Institute (IPN)*
*México, D.F. 04430*

**ABSTRACT:**
The backpropagation learning algorithm is applied for off-line sorting of spikes in simulated extracellular trains of action potentials. A cascade neural net was trained with a collection of spike shapes and it was tested with synthetic arbitrary spike trains formed with such spike shapes. A catalogue of synthetic extracellular spike trains was created where variations in waveform shape, shape mixing, timing and noise were allowed and the neural net sorting capabilities were assessed for different conditions. The results are discussed as well as other alternatives for spike sorting taking into account that the ideal case would be sorting in real time, that is, on-line during the experiment.

## INTRODUCTION

The importance of the ensemble activity of neurons in the brain is strongly recognized nowadays (Eichenbaum, 1993; Deadwyler and Hampson, 1995) and contemporary technology provides all kinds of metal microelectrode arrays with which the natural way of recording the electrical activity of neurons is extracellular. However, experiments where groups of neurons are recorded simultaneously and separably are difficult to perform and if successful produce a huge amount of data that is difficult to analyze. Consequently, simulation of multiple spike activity is useful as an exercise in separating and analyzing the different contribution of every neuron in the ensemble as well as a means for producing a variety of synthetic experiments.

Sorting techniques using a non-neural approach started with the development of electronic instruments for separating spikes in single recordings (Schmidt, 1984a) and then followed computer techniques for separating spikes in multiple recordings where automated systems are required (Wheeler and Heetderks, 1982; Schmidt, 1984b; Yang and Shamma, 1988; Kreiter et al., 1989). Multi-unit recording experiments are becoming of age but they are complex nevertheless. There are few commercial spikes sorters but they are not easy to use in general. The generator of synthetic extracellular recordings proposed here could be used beneficially for training in the use of spike sorters and cross-correlation analyzers.

We are interested in spike sorting because we are studying the functional connectivity among neurons in a group and this requires of the separation of the spike trains in the multiple recording. This is necessary whether one uses cross-correlation and/or gravitational analyses (Gerstein and Aertsen, 1985). In consequence, the generation of the synthetic signal should contain the means to simulate spike trains that have a tailored temporal relationship among them. In the present results signals have been synthesized as having temporal relationships like those found sometimes in real extracellular recordings. In general, the signal can be designed with random firing, isolated groups, couples or trios of bursts, oscillation, synchronization, and so on.

## METHODS

We have developed a computer environment for generating and analyzing multi-recording spike trains. We first worked on the classification of isolated spike shapes in a similar way as it is done with isolated digits, letters or phonemes (Waibel et al., 1989; Bottou et al., 1990) and later on we formed a multi-unit spike train in which the spikes contained in it were detected and classified. We presented in abstract form preliminary accounts of those works (Espinosa and Quiza, 1990, 1991).

Now we have produced a computer program that allows to produce at will multi-unit spike trains. By using a spike database kindly provided by Dr. G.L. Gerstein, spikes are taken from there to form spike trains with different characteristics. Spike detection is performed by an algorithm based on the first derivative of the input signal (Friesen et al., 1990) and the calculation of an adaptive threshold proportional to the mean of the last L samples. A spike is considered present when two consecutive points in the first derivative array exceed threshold. For each detected spike n samples around detection time are stored and this information is used as an input to the three-layer neural network trained with backpropagation. To avoid the poor shift invariance performance of the standard layered feed-forward network under translation in time of the input signal, the architecture was modified in such a way that groups of units in the hidden layer are connected to overlapping windows scanning the input pattern (Waibel et al., 1989). All the units belonging to the same group perform the same task with respect to their inputs.

The results shown were obtained using a net with 128 input units, 13 groups of one unit each in the hidden layer and 8 output units. The inputs to the units in the hidden layer scan 32 consecutive points of the input signal and overlap by 24 points between them. A training set with 240 spikes, 30 samples (including noisy ones) of each class of a total of 8, was employed to train the net.

## RESULTS

We now show representative examples of spike trains generated as well as the spike sorting results for each one.

Fig. 1 shows a multi-unit signal where only the 8 original spike types (without noise) in the training set were chosen to form bursts of 10 spikes each. These spikes are sorted with 100% accuracy and their time of occurrence is exactly as simulated.

This is as expected when the training set is used to verify the neural net. The classification is for an amplitude detection level of 30.

Fig. 2 shows a multi-unit signal that includes the 72 spikes in the database. This is a very unrealistic signal that is used only for the sake of illustration. When this signal is given as input to the classifier the different spikes are clustered as follows according to Fig. 1:

| | | | |
|---|---|---|---|
| Type 1(1) =8 | Type 2(13) =11 | Type 3(24) =4 | Type 4(25) =8 |
| Type 5(42) =20 | Type 6(51) =14 | Type 7(52) =4 | Type 8(64) =3 |

Total number of spikes = 72

The spike type refers to the 8 spikes that were used as the training set as in Fig. 1. The classification is for an amplitude detection level of 15 which is low enough to allow detection of the smallest spikes in the database.

Fig. 3 shows an example of spike similarity in which both spikes were in the training set, that is types 6(51) and 7(52). The spikes are arranged in such a way that they simulate a temporal relationship between the mixed trains. All of the spikes are detected and correctly classified with an amplitude detection level of 30.

Fig. 4 shows another example of spike similarity but in which one spike is in the training set -type 5(42)- and the other one in the test set -spike 43-. In this case the classifier detects all of the 99 spikes in the signal as type 5(42) which is a good guess. The amplitude detection level was of 30, other values produced the same result.

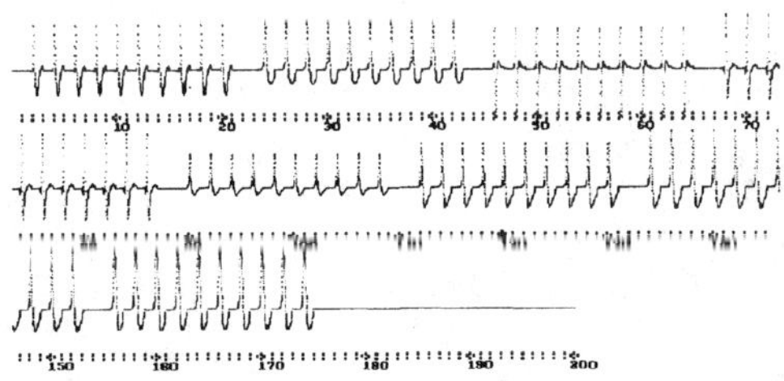

Figure 1: Spike train with eight different bursts.

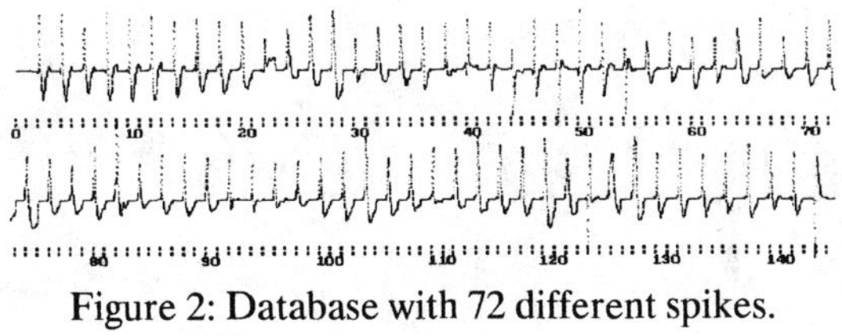

Figure 2: Database with 72 different spikes.

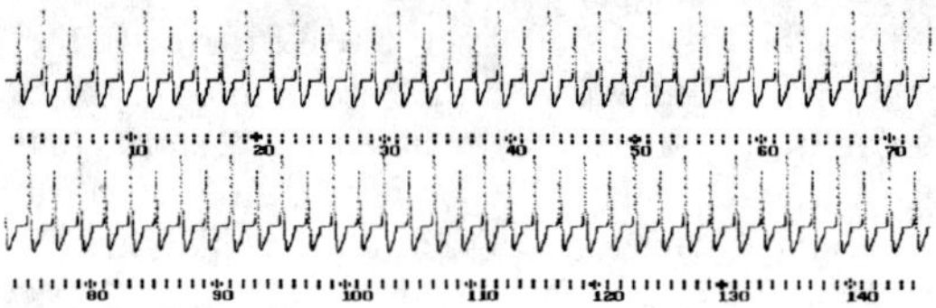


Figure 3: Spike train with two different types of spikes.

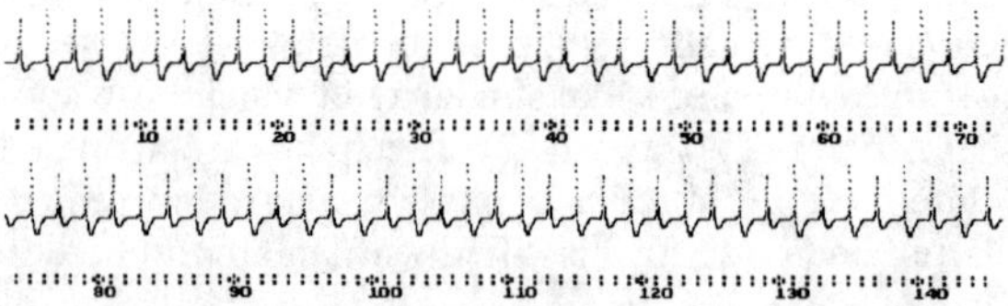


Figure 4: Spike train with two different types of spikes.

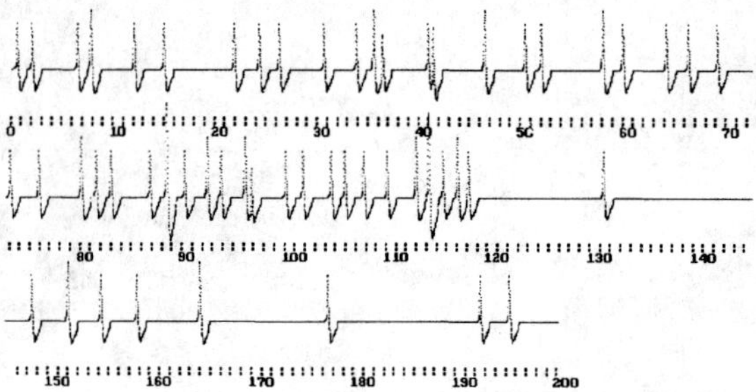


Figure 5: Different spikes ocurring at random times.

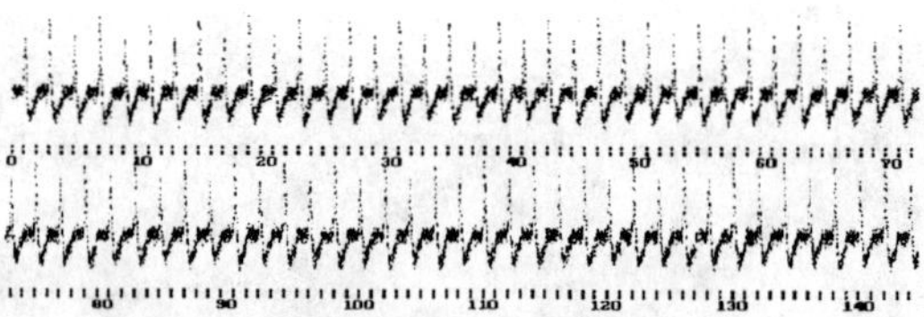


Figure 6: Noisy version of signal in Figure 3.

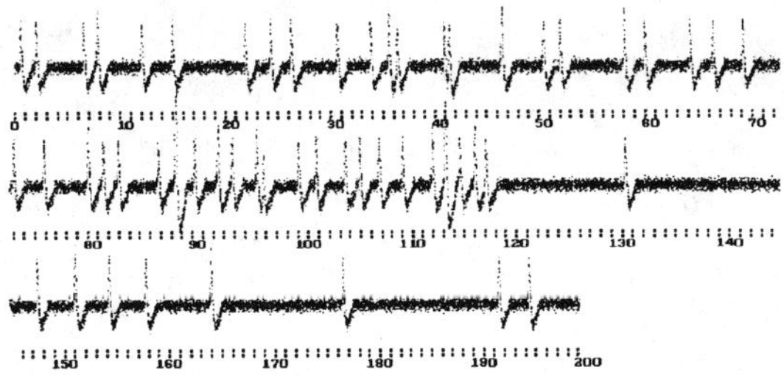

Figure 7: Spike train in Fig. 5 with 20% noise added.

Fig. 5 shows a multi-unit signal where two different spikes appear at random times in the train. The spikes included are 51 and 52, and they appear 43 and 14 times, respectively. The classifier net using an amplitude detection level of 10 sorts the spikes as follows:

39 of type 6(51)    12 of type 7(52)    1 of type 1(1)    1 of type 5(42)

This classification is approximately correct since 4 spikes are missed due to overlap and 2 spikes are misclassified.

Fig. 6 shows the same example as in Fig.3 with 20% of noise added. The classifier gives the following results:

1 of type 2(13)    44 of type 6(51)    54 of type 7(52)

This indicates that 7 spikes are misclassified. When the noise is increased to 40% (not shown) the amount of misclassified spikes goes to 13.

Fig.7 is a noisy version of the signal in Fig. 5. Adjusting the parameters for detection it was possible to detect 52 spikes out of 57 and they were clasiffied as

37 of type 6(51)    14 of type 7(52)    1 of type 1(1)

This shows that classification is as good as in the case of the noiseless signal but the procedure to do it is a little more involved.

## DISCUSSION

The results here described are encouraging because the classification of spikes with a neural net is quite accurate. To classify up to 8 different classes only one hidden layer with 12 to 26 units is usually required. The classifier starts to fail when a certain percentage of noise is added to the signal. However, in real recordings situations where a high amount of electric or movement noise is present are corrected by means of filtering, and dense nervous noise is reduced by adjusting the microelectrode array position to a place where spike separation is improved. In any case, very noisy signals will be avoided or rejected. Thus, simulation of very noisy signals is not useful per se in this synthetic signal generator.

On the other hand, our approach has a big disadvantage: the use of only synthetic spike trains, not real ones, and that the analysis is made off-line. We do not have yet an interface between the neural net and the data recorded during an experiment. This interface should be very fast for accomplishing classification during the experiment with the objective of identifying interacting groups of neurons at that moment. We have made some preliminary attempts for a faster detection stage by using a DSP chip (Espinosa et al., 1991). The next step would be to follow an approach similar to that taken by Kreiter (Kreiter et al., 1989) and another neural network types should be tried like an ARTII type neural net (Grossberg, 1988) to avoid the need of knowing beforehand the spikes' shape since they are different depending on the recording site. Handling overlapping spikes (those with time coincidence in two or more trains) would be very convenient since they have a role in the shape of the cross-correlation histogram around the origin. The present system cannot cope with overlapping spikes. More testing is needed to solve this disadvantage.

## REFERENCES


Bottou, L., Soulie, F., Blanchet, P., and Lienard, J.S. (1990). Speaker-independent, isolated digit recognition: multilayer perceptron vs. dynamic time warping., Neural Networks 3: 453-465.

Deadwyler, S.A. and Hampson, R.E. (1995). Ensemble activity and behavior: What's the code?, Science, Vol. 270, p. 1216.

Eichenbaum, H. (1993). Thinking about brain cell assemblies, Science, Vol. 261, p. 993.

Espinosa, I. and Quiza, J. (1990). Classification of noisy action potentials (APs) by means of a neural network employing back-propagation., Soc. Neurosci. Abstr., Vol.16, Part 2, p. 1092.

Espinosa, I. and Quiza, J. (1991). Off-line sorting of spikes using an artificial neural network., Soc. Neurosci. Abstr., Vol.17, Part 1, p. 124.

Espinosa, I., Quiza, J., and Ayhllon, A. (1991). Detection and classification of neuronal spikes using a DSP chip and a neural network, Proc. 13th Ann. Int. Conf. IEEE Eng. Med. Biol.

Friesen, G.M., Jannett, T.M., Jadallah, M.A., Yates, S.L., Quint, S.R., and Nagle, H.T. (1990). A comparison of the noise sensitivity of nine QRS detection algorithms, IEEE Trans. Biomed. Eng., 37(1),Jan, 85-98.

Gerstein, G.L., Personal Communication.

Gerstein, G.L. and Aertsen, A. (1985). Representation of cooperative firing activity among simultaneously recorded neurons, J. Neurophysiol., 54: 1513-1528.

Grossberg, S. (1988). Neural Networks and Natural Intelligence, MIT Press.

Kreiter, A., Aertsen, A., and Gerstein, G.L. (1989). A low-cost single-board solution for real-time, unsupervised waveform classification of multineuron recordings, J. Neurosci. Meth., 30: 59-69.

Schmidt, E.M. (1984a). Instruments for sorting neuroelectric data: a review., J. Neurosci. Meth. 12: 1-24.

Schmidt, E.M. (1984b). Computer separation of multi-unit neuroelectric data: a review., J. Neurosci. Meth. 12: 95-111.

Waibel, A., Hanazaga, T., Hinton, G., Shikano, K., and Lang, K.J. (1989). Phoneme recognition using time-delayed neural networks., IEEE Trans. Acoustics, Speech, and Signal Processing, 37: 328-339.

Wheeler, B.C. and Heetderks, W.J. (1982). A comparison of techniques for classification of multiple neural signals., IEEE Trans. Biomed. Eng., 29:752-759.

Yang, X. and Shamma, S.A. (1988). A totally automated system for the detection and classification of neural spikes., IEEE Trans. Biomed. Eng., 35: 806-816.

# PART VI : SMART ENGINEERING SYSTEM DESIGN

# Section 5 : Bio-Medical Engineering Systems

# CLASSIFICATION AS UNKNOWN BY RBF NETWORKS: DISCRIMINATING PHYTOPLANKTON TAXA FROM FLOW CYTOMETRY DATA

**COLIN W. MORRIS**
*Department of Computer Studies, University of Glamorgan, Pontypridd, CF37 1DL, UK.*

**LYNNE BODDY**
*School of Pure and Applied Biology, University of Wales, Cardiff, CF1 3TL, U.K.*

***ABSTRACT:***
Identification problems in biology and medicine are often unbounded with the number of possible classes unknown. Often it is more important to reject patterns from classes upon which a network has not been trained than to classify them incorrectly. The ability of radial basis function networks to do this is examined using flow cytometry fingerprints of phytoplankton taxa. Applying the criterion to reject if the hidden layer node with the largest output was less than 0.4, successfully rejected over 95% of novel taxa, while only rejecting incorrectly 6% of test patterns of species upon which the network had been trained. Similar success was achieved by rejecting as novel if the value of the largest output node was less than 0.3.

## INTRODUCTION

Many artificial neural network (ANN) applications are bounded, i.e. all input data must belong to one of the set of n classes upon which the network has been trained. There is, therefore, no requirement to reject as unknown classes upon which an ANN has not been trained. However, many biological and biomedical identification problems are unbounded with the number of possible classes unknown. In such cases it is often essential to be able to reject as 'unknown' examples from classes upon which the network has not been trained rather than to classify them incorrectly (Woolley and Stone, 1987; Bridge, Boddy and Morris, 1994; Morris and Boddy, 1995). Multilayer perceptrons (MLPs; backpropagation networks) and radial basis function (RBF) ANNs have been used increasingly, and with considerable success, to identify biological taxa, including fungal spores (Morris, Boddy and Allman, 1992; Boddy *et al.*, 1994), bees (Echazarreta *et al.*, unpub.), phytoplankton (Frankel *et al.*, 1989; Balfoort *et al.*, 1992; Boddy *et al.*, 1994; Wilkins, Morris and Boddy, 1994; Morris, Boddy and Wilkins, 1994; Wilkins *et al.*, 1996), and bacteria (Goodacre *et al.*, 1994).

Several possible criteria for rejection of novel data can be envisaged. Firstly, for RBF ANNs novel patterns could be rejected if the sum of all the basis functions was less than a threshold value. Secondly, again for RBF ANNs, patterns could be rejected as novel if the output of the closest basis function to the input vector was less than a threshold value. However, a preliminary study showed the second of these to be best as it gave a better proportion of rejected patterns from novel taxa compared with the patterns from known taxa (M.F. Wilkins, L. Boddy and C.W. Morris, unpub.). Thirdly, for both MLP and RBF ANNs, constraints can be applied at the output nodes such that there must be, for example, at least a specified difference between the highest and next highest output node.

This paper examines different criteria for rejection of novel data patterns from RBF ANNs trained to discriminate twenty phytoplankton taxa from flow cytometry data. MLP networks failed to describe the data adequately and were not examined.

## METHODS

### Flow cytometry data

Forty taxa (thirty four different species) of marine and freshwater phytoplankton were grown, under controlled conditions to give pure cultures, and analysed (by R. Jonker) using the EurOPA flow cytometer (Jonker *et al.*, 1995). Eleven light scatter, diffraction and fluorescence parameters were measured for each cell and the data logged to disc. At least 800 cells of each species were measured. Data were gated to eliminate background noise particles and contamination.

### Hardware and software

Neural network simulations were carried out using an IBM compatible PC with an Intel Pentium processor running at 133MHz. The RBF ANN software was custom written (Wilkins, Morris and Boddy, 1994).

### Training MLP and RBF networks

A RBF network with 100 initially randomly placed asymetric kernels was trained on 400 events for each of twenty randomly selected taxa, with the constraint that each taxon was a different species. The Mahalanobis distance metric was employed, and the kernel width parameter was 1.41. A RBF network for all forty taxa was trained similarly. The performance of each trained network was assessed by constructing 'misclassification matrices' from test data containing 400 'unseen' patterns per taxon.

### Rejection of novel data patterns

Networks trained on twenty species were presented with 400 unseen patterns for each of the twenty species upon which the network had been trained and for each of twenty 'unknown' taxa (fourteen different species), and three different criteria were investigated for the rejection of 'novel' patterns: (1) rejection if the HLN with the largest output was less than a threshold value $\theta$. $\theta$ was varied from 0.0 (i.e. no rejection) to 1.0 with intervals of 0.1. (2) Rejection if the value of the winning output node was less than $\alpha$ (where $\alpha$ was 0.3, 0.4 or 0.5). (3) Rejection if the difference between the value of the winning node and the node with the second highest output value was less than $\beta$ (where $\beta$ was 0.1 or 0.15).

The following were then determined: (1) overall proportion of known species that were rejected (incorrectly); (2) overall proportion of novel taxa that were rejected (correctly); (3) confidence of correct classification for each of the known species, estimated from matrices including knowns only and knowns plus novel taxa

TABLE 1: REJECTION OF DATA PATTERNS AS NOVEL EMPLOYING CONSTRAINTS ON OUTPUT NODES. PATTERNS WERE REJECTED IF THE DIFFERENCE BETWEEN THE WINNING NODE AND THE THE SECOND HIGHEST OUTPUT NODE WAS LESS THAN β OR IF THE VALUE OF THE WINNING NODE WAS LESS THAN α.

| Rejection threshold | Knowns | Unknowns | Overall probability of correct classification of knowns (%) | Overall confidence of correct classification of knowns (%) |
|---|---|---|---|---|
| no constraint | - | - | 81 | 80 |
| β = 0.1 | 14 | 86 | 75 | 85 |
| β = 0.15 | 19 | 91 | 72 | 87 |
| α = 0.3 | 5 | 91 | 79 | 82 |
| α = 0.4 | 13 | 97 | 75 | 85 |
| α = 0.5 | 25 | 99 | 68 | 88 |

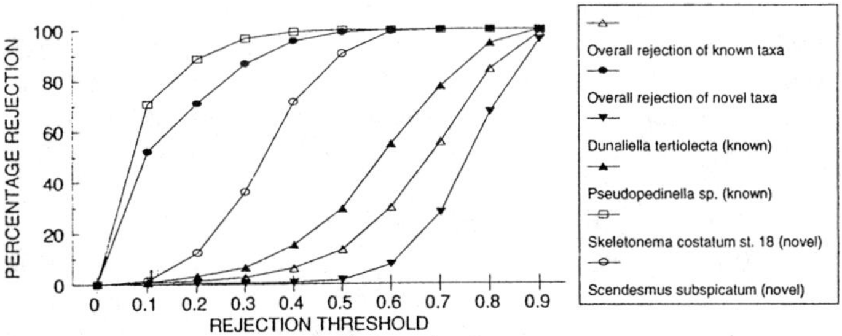


Figure 1: Overall percentage rejection, and upper and lower limits for rejecting taxa as unknown employing a threshold on the largest hidden node output value.

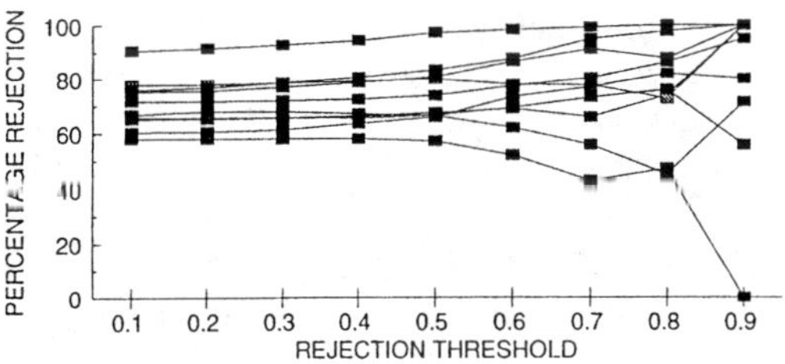


Figure 2: Percentage incorrect rejection of ten representative individual known taxa as unknown employing a threshold on the largest hidden node output value.

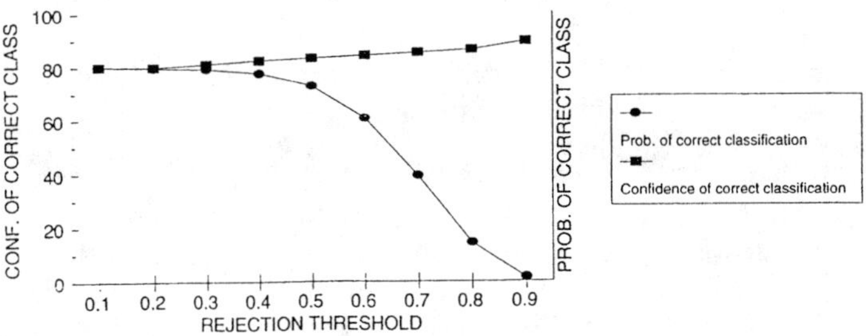


Figure 3: Overall (mean) confidence and probability of correct classification employing a threshold on the largest hidden node output value.

(i.e. for the former the frequency of correct classification as a species divided by the total frequency of that species that was assigned an identification; for the latter the frequency of correct classification divided by the total frequency of that species assigned a species identification or classified as novel); (4) probability of correct classification (the probability that a cell belonging to species x is assigned to class x; (5) for each of the known species the proportion that were rejected (incorrectly); (6) for each of the novel species the proportion that were rejected (correctly).

## RESULTS

The twenty species network discriminated between species with an overall probability of correct classiffication of 80%. Five of the species were identified with less than 70% success. *Gymnodinium simplex* and *Halosphaera russellii* misclassified with several species, but with none of them more than 10% of occasions. *Tetraselmis rubens* was misclassified as *Pyramimonas obovata* 16% of occasions. *Chlorella salina* was misclassified as *Chrysochromulina camella* 25% of occassions, and *Pseudopedinella* sp. was misclassified as *Phaeocystis globosa* and *H. russellii* 24% and 30% respectively.

When the rejection criterion was a constraint on the HLN with the largest output, as the threshold ($\theta$) increased from 0.0 - 0.3 there was a sharp increase in the rejection of novel taxa (Fig. 1). At $\theta = 0.4$, over 95% of novel patterns were correctly rejected, with only 6% of patterns of known species being incorrectly rejected. Above $\theta = 0.5$ the proportion of known species incorrectly rejected increased rapidly. While rejection differed for different individual taxa, 90% of the patterns for the majority of novel taxa were rejected at this threshold, however, less than 75% of *Scendesmus subspicatum* patterns were rejected (Fig. 1). This species was confused slightly, in the forty species matrix though not markedly, with three of the known species (*Dunaliella tertiolecta* (4%), *Ochromonas* sp. (6.5%) and *Prymnesium parvum* (4%)). Similarly, there were slight differences in rejection of individual known species (Fig. 2). Confidence of correct classification estimated excluding those classified as novel was fairly constant at low threshold values, and was $\geq 0.8$ for twelve of the species at a threshold of 0.4, above which for most species it rose slightly (data not presented), following the trend for overall confidence of correct classification (Fig. 3). Confidence of correct classification dropped dramatically for three species (*G. simplex*, *Pseudopedinella* sp., *Pyramimonas obovata*) above a threshold of 0.8. Probability of correct classification declined rapidly at $\theta > 0.5$.

Constraints on the output nodes yielded high rejection of novel patterns (Table 1), and between 5-25% rejection of knowns.

## DISCUSSION

This study has shown that using a threshold value on the maximum HLN output (i.e. rejection of patterns not sufficiently close to any of the basis function centres to cause any of the HLNs to produce sufficiently large output values) is a good criterion for rejection of novel data patterns. An appropriate threshold ($\theta = 0.4$) could be determined which allowed not only high rejection of novel taxa as unknown, but also low rejection of known taxa. Even though there were differences in rejection between species, about 75% of patterns of the least-well-rejected unknown taxa were rejected at $\theta = 0.4$, and no more than 15% for the worst species of knowns rejected as novel. Though the confidence of correct clasification was often improved with increasing $\theta$ (presumably because ambiguous patterns falling between two or more basis functions decreased), it was sufficiently high for all species at $\theta = 0.4$. The dramatic decrease in confidence of correct classification exhibited by a few species at $\theta = 0.9$, though interesting, is of no practical importance, since at this threshold most of the knowns were also rejected as novel. These low confidences of correct classification reflect a disproportionate amount of incorrectly identified species being retained; these species were all confused with others by the trained network.

An appropriate threshold on size of winning output node value ($\alpha = 0.3$) provides an equally good rejection criterion for novel data patterns; rejection of novel patterns was equally high, rejection of known patterns was equally low, and overall confidence of correct classification was the same. A threshold on difference between the winning and the next highest output node value does not, however, form such a good rejection criterion as although high rejection of novel patterns can be achieved, the associated rejection of known patterns is also higher.

Similar constraints on output node values of MLP ANNs may well not be as successful as with RBF ANNs because the two network types represent the data patterns and construct decision boundaries in different ways. With MLPs the decision regions are often infinite in extent, and the network output depends on the position of the pattern to be identified relative to the hyperplanar boundaries represented by the hidden layer nodes. Though in theory MLPs could use surplus hyperplanar boundaries (HLNs) to produce closed, finite decision regions, in practice this would only happen if by so doing the network could reduce its output error, since MLPs learn through gradient decent reduction of error. Moreover, use of excess HLNs may have unpredicatble effects on the form of the decision boundaries, and prevent the network from generalizing. With RBF ANNs, on the other hand, decision regions are finite, the HLN kernel output declining to zero with distance from the kernel centre, reflecting the ability of a threshold on HLN outputs to reject novel taxa. Theoretically, therefore, RBF ANNs should be able to reject novel data patterns, but MLPs are likely to be less good at this. The latter could not be tested here as it did not prove possible to train an MLP network to represent the twenty phytoplankton taxa anywhere near as well as the RBF network; MLP ANNs have previously been trained successfully to discrimate phytoplankton taxa (Boddy *et al.*, 1994; Morris, Boddy and Wilkins, 1994; Wilkins *et al.*, 1996).

The rejection techniques applied here differ from those often used by other workers (e.g. Balfoort *et al.*, 1992) where they have trained an ANN with an extra class into which unknowns are classified. The latter technique does not work particularly well in applications with a high dimensional input space as there is a problem in obtaining training data that is representative of the "unknown class". The ability to reject based on some criterion internal to the network is thus preferred.

## ACKNOWLEDGEMENTS

Thanks to Richard Jonker, University of Amsterdam, who obtained the data during a project funded by the Commission of the European Community, grant # MAS2-CT91-0001 (project # PL910032), and Malcolm F. Wilkins for writing the RBF software, and for valuable discussion on rejecting novel data patterns.

## REFERENCES

Balfoort, H.W., Snoek, J., Smits, J.R.M., Breedveld, L.W., Hofstraat, J.W., Ringelberg, J. (1992). Automatic identification of algae: neural network analysis of flow cytometric data. *J. Plankton Res.*, **14**, 575-589, .

Boddy, L., Morris, C.W., Wilkins, M.F., Tarran, G.A. and Burkill, P.H. (1994). Neural network analysis of flow cytometric data for 40 marine phytoplankton species, *Cytometry*, **15**, 283-293.

Boddy, L., Gimblett, A., Morris, C.W. and Mordue, E. J.M. (1994). Neural network analysis of fungal spore morphometric data for identification of species in the genus Pestalotiopsis. In: Intelligent Engineering Systems Through Artificial Neural Networks, Vol. 4, (ed. C.H. Dagli, B.R. Fernandes, J. Ghosh and R.T.S. Kumara), pp. 605 - 612. A.S.M.E. Press.

Bridge, P.D., Boddy, L. and Morris, C.W. (1994). Information resources for pest identification - an overview of computer-aided approaches. In: *Identification and Characterisation of Pest Organisms* (ed. D.L. Hawksworth), pp. 153-167. CAB International, Wallingford, UK.

Frankel, D.S., Olson, R.J., Frankel, S.L., Chisholm, S.W. (1989). Use of a neural net computer system for analysis of flow cytometric data of phytoplankton populations. *Cytometry*, **10**, 540-550, .

Goodacre, R., Neal, M.R., Kell, D.B., Greenham, L.W., Noble, W.C. and Harvey, R.G. (1994). Rapid identification using pyrolosis mass spectrometry and artificial neural networks of *Propioibacterium acnes* isolated from dogs. *Journal of Applied Bacteriology*, **13**, 157 - 160.

Jonker, R.R., Meulemans, J.T., Dubelaar, G.B.J., Wilkins, M.F. and Ringlelberg, J. (1995). Flow cytometry: a powerful tool in analysis of biomass distributions in phytoplankton. *Water Science and Technology*, **32**, 177 - 182.

Morris, C.W. and Boddy, L. (1995). Artificial neural networks in identification and systematics of eukaryotic microorganisms. *Binary*, **7**, 70 - 76.

Morris, C. W., Boddy, L., and Allman, R. (1992). Identification of basidiomycete spores by neural network analysis of flow cytometry data, *Mycological Research*, **96**, 697-701.

Morris, C. W., Boddy, L., and Wilkins, M.F. (1994). Approaches to applying neural networks to the identification of phytoplankton taxa from flow cytometry data. In: Intelligent Engineering Systems Through Artificial Neural Networks, Vol. 4, (ed. C.H. Dagli, B.R. Fernandes, J. Ghosh and R.T.S. Kumara), pp. 619 - 627. A.S.M.E. Press.

Wilkins, M.F., Morris, C.W. and Boddy, L. (1994). A comparison of radial basis function and backpropagation neural networks for identification of marine phytoplankton from multivariate flow cytometry data, *CABIOS*, **10**, 285-294.

Wilkins, M.F., Boddy, L., Morris, C.W. and Jonker, R. (1996). A comparison of some neural and non-neural methods for identification of phytoplankton from flow cytometry data, *CABIOS*, **12**, 9 - 18.

Woolley, J.B., and Stone, N.D. (1987). Application of artificial intelligence to systematics: SYSTEX - a prototype expert system for species identification, *Systematic Zoology*, **36**, 248-267.

# THE GROWTH OF *ESCHERICHIA COLI O157:H7* - A BACKPROPAGATION NETWORK APPROACH

**Maha N. Hajmeer**
Department of Animal Sciences and Industry
Kansas State University
Manhattan, Kansas 66506.

**Imad A. Basheer & Yacoub M. Najjar**
Department of Civil Engineering
Kansas State University
Manhattan, Kansas 66506.

***ABSTRACT:***
Food associated microbial flora is of major interest to today's modern advances in food engineering. This is due to the rising developments in food formulations, reformulations, technologies, and consumers' changing eating patterns. The control of such flora is essential to provide good quality, safe, and stable food products. Therefore, understanding the nature of microbial growth and its dependence on factors pertaining to ecology or processing modes is crucial. Predictive models that assist in estimating the bacterial growth as function of time are abundant in the literature of microbiology, among which is the most commonly used Gompertz model. The Gompertz model parameters are usually related to environmental conditions via response surface models (RSMs) that are determined by regression technique. The objective of the present study is to demonstrate the capability of backpropagation Neural Networks (NNs) as substitute for the conventional RSMs for determining the microbial growth parameters. The developed NN-based RSMs predicted growth parameters of *Escherichia coli* O157.H7 with high accuracy. Additionally, the models' response to changes in environmental conditions was found to be realistic and in agreement with our intuition about the physical behavior of growth of the studied microorganism.

## INTRODUCTION

Prediction of microbial growth in food engineering systems constitutes a corner stone for issues pertinent to safety and quality. Models for predicting microbial growth have always been the subject of considerable scrutiny in the area of food microbiology. In this paper, NNs were used to develop predictive growth models as an alternative approach to the conventional RSMs. Understanding the effect of various conditions on the growth of microorganisms is essential in evaluating their potential to survive. The USDA Microbial Food Safety Research Unit has encouraged studying the effect of five environmental variables on the proliferation of microorganisms (McMeekin et al., 1993). These variables are temperature, pH, NaCl and $NaNO_2$ concentrations, and the presence of gaseous atmosphere.

Among the kinetic models used for evaluation of the time-dependent microbial growth, the modified Gompertz model has gained most popularity. The Gompertz model parameters are affected by environmental conditions and can be manipulated to determine several kinetic descriptors. The Gompertz model is expressed as $L(t)=A+C \exp\{-\exp[-B (t - M)]\}$, where L is the log count of microorganisms at any time t, A is the initial level of microorganism, C is the microorganisms count as t approaches infinity, and B and M are other two parameters. The Gompertz parameters are determined by fitting an experimentally obtained growth curve for a specific combination of environmental conditions to the Gompertz model equation. RSMs are usually developed to relate the Gompertz parameters (namely B, M) to some environmental conditions such as temperature, pH, and the concentration of both NaCl and $NaNO_2$. Typically, conventional RSMs involve large number of terms and cross terms containing the various environmental variables (Buchanan and Bagi 1994).

In order to investigate the capability of the backpropagation NNs, data on *Escherichia coli* O157:H7 (Buchanan and Bagi 1994) was used. For these data, temperature, pH, and NaCl and $NaNO_2$ concentrations were varied over a wide range of possible environments. The *Escherichia coli* O157:H7 was selected for this study due to the high interest in its characterization and control in the food industry (Doyle 1991). NN-based RSMs for aerobic and anaerobic growth were first developed. Then, the data was combined and used to develop a general model that is sensitive to oxygen availability. The advantage of the general network was two fold: (i) increasing the number of data sets used for training/testing can help the network to achieve better generalization, and (ii) a general network provides easier access for users without the need to consider two separate RSMs.

## DEVELOPMENT OF NN MODELS

In the NN development, only those combinations found (experimentally) to be suitable for growth of *Escherichia coli* O157:H7 were selected. Data sets where no growth was reported were eliminated. A total of 45 aerobic, 46 anaerobic, and 91 aerobic/anaerobic (combined) growth data sets representing different variable combinations were used for developing three NNs. The ranges of the parameters utilized were as follows: temperature (5-42°C), initial pH (4.5-8.5), NaCl (5-50 g/L), and $NaNO_2$ (0-200 $\mu$g/mL). For all data sets used, C= 6.40 and 5.81 for aerobic and anaerobic environments, respectively, and A=3.00. The natural logarithm (ln) of B and M was considered in NNs development in order to compare to RSMs developed by Buchanan and Bagi (1994).

For developing the various neural models, the backpropagation learning algorithm with an adaptive scheme for selecting the optimum network structure was encoded into Fortran program. The collected data were divided into training and testing databases. For the aerobic model, the 45 data case were divided into 37 for training and the remaining 8 data sets were used for validation. Similarly, for the 46 anaerobic data sets, 36 and 10 data sets were used for both training and testing, respectively. For the combined neural model, 73 examples were used for training while the remaining 18 were used for testing the networks accuracy.

## RESULTS AND ANALYSIS

Table 1 summarizes the accuracy indicators of the various developed networks. The three networks were optimized to yield the best prediction for the unseen sets. The aerobic and anaerobic NNs contained 4 inputs in the input layer and two outputs (lnB and lnM). The general NN contained the oxygen availability as an additional input parameter in the input layer. The "oxygen" parameter in the input layer was assigned 0 if anaerobic system prevails and 1 if the growth was aerobic. The aerobic, anaerobic, and general NNs were optimized at 6, 5, and 6 hidden nodes, respectively. The MARE (Mean of the Absolute value of Relative Error of all sets) and $R^2$ (coefficient of determination between predicted and experimental data) for both (lnB) and (lnM) indicated that the general NN gave better predictions as compared to the aerobic and anaerobic NNs separately.

**TABLE 1: ACCURACY INDICATORS OF THE THREE DEVELOPED NETWORKS**

| NN-RSM | Phase | Parameter | No. Points | MARE(%) | $R^2$ |
|---|---|---|---|---|---|
| AEROBIC NEURAL NETWORK | Testing | lnB<br>lnM | 8<br>8 | 11.5<br>6.0 | 0.86<br>0.89 |
| | Training | lnB<br>lnM | 37<br>37 | 8.6<br>4.9 | 0.95<br>0.97 |
| | All Data | lnB<br>lnM | 45<br>45 | 9.2<br>5.1 | 0.94<br>0.96 |
| ANAEROBIC NEURAL NETWORK | Testing | lnB<br>lnM | 10<br>10 | 16.7<br>16.1 | 0.84<br>0.66 |
| | Training | lnB<br>lnM | 36<br>36 | 7.4<br>6.6 | 0.96<br>0.95 |
| | All Data | lnB<br>lnM | 46<br>46 | 9.6<br>8.9 | 0.94<br>0.88 |
| GENERAL NEURAL NETWORK | Testing | lnB<br>lnM | 18<br>18 | 10.8<br>5.0 | 0.91<br>0.95 |
| | Training | lnB<br>lnM | 73<br>73 | 8.1<br>4.7 | 0.96<br>0.97 |
| | All Data | lnB<br>lnM | 91<br>91 | 8.7<br>4.7 | 0.95<br>0.97 |

Combining the aerobic and anaerobic data was found to enhance the capability of the general NN in capturing more of the interrelations among the different variables. Figure 1-(a,b) illustrates the prediction of the unseen data sets by the general network. Additionally, Fig. 1-(c,d) shows the agreement between the predicted and the actual values of (lnB) and (lnM) for the combined data. All the predictions showed that the general NN was able to predict all data with a relatively high accuracy. The testing data was predicted with MARE=10.8% and an $R^2$=0.91 for (lnB), and MARE=5.0% and $R^2$=0.95 for (lnM). For the training and testing data combined, the general NN prediction capability was determined as MARE=8.7% and $R^2$=0.95 for (lnB), and MARE=4.7% and $R^2$=0.97 for (lnM).

Comparison of the aerobic and anaerobic third-order RSMs developed by Buchanan and Bagi (1994) to the NN-based RSMs developed in this study indicated that the later can provide more accurate prediction of growth parameters. Table 2 presents the comparison between the conventional and NN-based RSMs.

**TABLE 2: COMPARISON BETWEEN CONVENTIONAL AND NN-BASED RSMs PREDICTIONS**

| RSM | Aerobic NN | Anaerobic NN | General NN |
|---|---|---|---|
| | lnB / lnM | lnB / lnM | lnB / lnM |
| **Conventional RSMs** | 0.894 / 0.947 | 0.918 / 0.818 | ---------------- |
| **NN-Based RSMs** | 0.940 / 0.960 | 0.940 / 0.880 | 0.950 / 0.970 |

## PARAMETRIC ANALYSIS

The developed general NN was utilized to study the effect of the various environmental conditions on microbial growth. A new growth parameter (called $t_{1000}$) representing collection of all the Gompertz parameters was used for this purpose. The $t_{1000}$ refers to the time required to reach 1000-fold increase in microbial population density from its initial value. An average data set was used to generate 31 data sets with temperature, pH, NaCl, and $NaNO_2$ varying over wide range of values. The general NN-based RSM was run on these sets to calculate lnB, lnM, and $t_{1000}$. The results of these runs for aerobic and anaerobic conditions are displayed in Fig. 2. In Fig. 2-a, it is obvious that increasing the temperature results in decreasing $t_{1000}$; thus implying more favorable growth conditions with increasing temperature. The aerobic conditions are more favorable than anaerobic for growth of *E.coli* O157:H7 especially at temperature below 20 °C. The effect of pH on $t_{1000}$ is portrayed in Fig. 2-b. As can be seen, the microorganism growing under aerobic conditions are more capable to withstand acidic environments than those growing under anaerobic conditions. At pH values above 6.5, the inhibition of growth is almost similar for both the aerobic and anaerobic environments. The addition of NaCl (Fig. 2-c) tends to dramatically increase the time for the microorganisms to multiply; especially at concentrations in excess of 30 g/l. The effect of anaerobic conditions on inhibition of growth by NaCl is well pronounced in a region between 22-60 g/l. The response of the model to $NaNO_2$ concentration is illustrated in Figure 2-d. The addition of $NaNO_2$ increases the inhibition of growth, with stronger retardation for aerobic systems. From microbiological point of view, it is apparent that the above mentioned findings are logical; thus indicating the accuracy of the developed NN-based RSMs to predict the growth dynamics of the studied microorganism.

## REFERENCES


Buchanan, R. L. and Bagi, L. K. (1994),"Expansion of response surface models for the growth of *Escherichia coli* O157:H7 to include sodium nitrite as a variable," Int. J. Food Micro., 23, 317-332.

Doyle, M. P. (1991), "*Escherichia coli* O157:H7 and its significance in foods," Int. J. Food Micro., 12, 289-302.

McMeekin, T. A., Olley, J. N. and Ratkowsky, D. A. (1993), Predictive microbiology: theory and application, Wiley, New York.


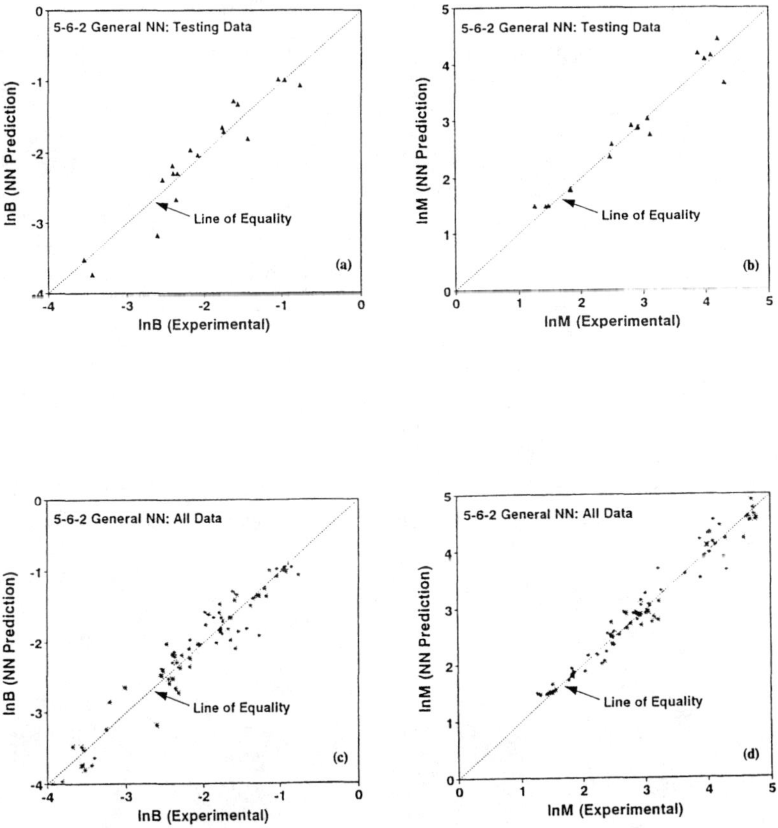


Fig. 1: Prediction by the General Neural Network.

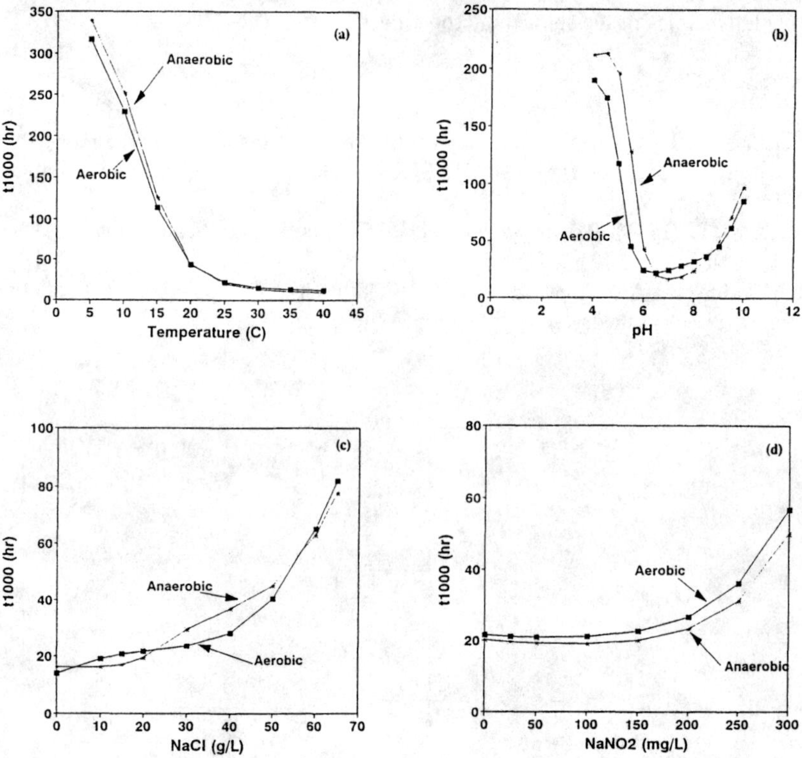


Fig. 2: Results of Parametric Analysis of Aerobic and Anaerobic Growth.

# A Study on Image Analysis of Amoebae From Phase-Contrast Microscope Image

Masao KUBO    Yuuiti SATOU    Hiroshi YOKOI    Yukinori KAKAZU
Email: kubo@complex.hokudai.ac.jp
Tel : 011-706-6445  FAX: 011-700-5060
URL;http://www.complex.hokudai.ac.jp/export/kubo/WWW/index.html
Address:Complex Systems Engineering,
Division of Systems and Information Engineering,,Hokkaido University,JAPAN

## Introduction

*Snake*[3] is one of the most convenient image processing techniques.Usually, a Snake consists of several units and arcs that form a circle(Fig.1). It moves to minimize its enerugy function. To minimize the function ordinary, 2 conditions should be satisfied. 1) the shape of it cirlcle is smooth. 2) the pixels under these arcs are edges on the image. Namely, the region is satisfied with these most correctly when it stop to move. Of couse, if you want to set the enerugy function as you like, you can notice a region that satisfies the condtion.

In this paper, we would like to build a new usage of snake. Normaly, snake is use for detection of target on the image. Therefore, the design of energy functions to detect its target correctly is very important. On the other side, we would like to use as a categorizer of behavior. As you know, which a snake can detect a target on a image, or not, depends on the initial position and the values of pixels around the target, and the target shape. For example, ordinary snake is not into detection of rectangle-like shaped target because of the smoothness condition of energy function. Of couse, if we change the condition of enerugy function, it may be able to detect such shaped target correctly. Namely, we would like to regard success/failure of detection as a categorization of targets based on its enerugy funtion. There is not such standpoint of view of snake, however, there is a lot of requirements.Especially, numerical expression and categorization of behaviors is very important. Fig.2 shows an example for the application of snake. The defference of the ordinary usage of the snake, is only that image where it works, is changing. In Fig.2, the image at T is changing to the image at T+1. When a snake can detect all targets of all of images that are showed to in good order, moreover, the snake can detect other image series, we are able to regard these belong to a same category. This categorization has one good nature. Usually, categorization of behavior means the sum of differenes of images at each time step. Under ordinary categorization, the order of images are not so important. Moreover, important defferences of behavior but small behavior are disappered by trivial behavior but big in sometimes. On the other hand, the new usage of snake will be expected to offer a new categorization methods that consider why and how the target moves to the next positon.

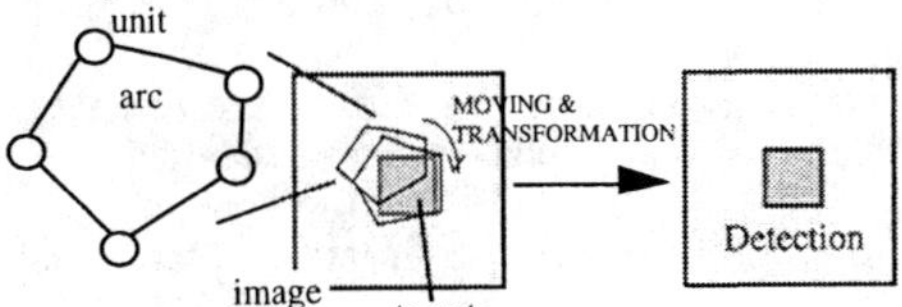


Fig.1 Ordinary Snake Usage

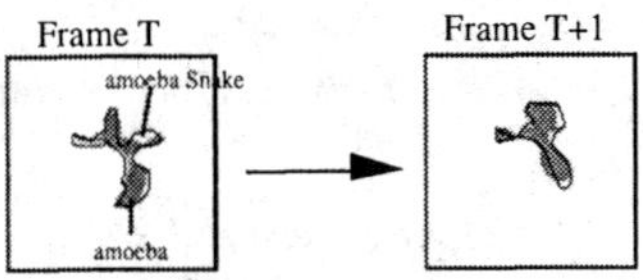


Fig.2 Application Snake to serial images

In the next section, we explain the reason to take up amoeba image as a test.

First, the characters of amoeba images of phase contrast image, are described. In section3, we propoe a new snake for amoeba microscope image. Finally, by simple computer simulations, the expanded model's abilities are shown. However, in this paper, we can not show experiments that deal with real amoeba data because we can not get sufficient video images about them. Then we would like to show results of simple experiment settings to consider the real data. We would like to investigate the detection abilty of snakes results from parameters and moving speed of target.

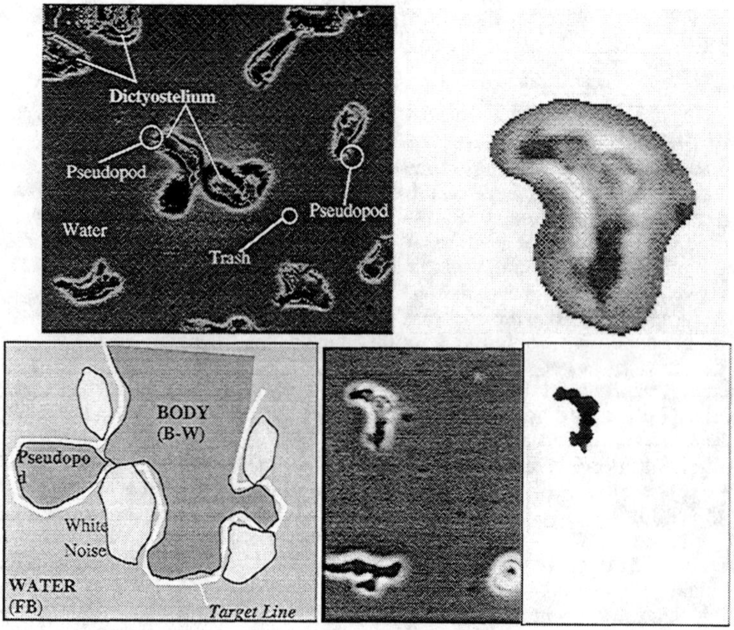


*Fig.3 Image Analysis of Amoebae From Phase-Contrast Microscope Image (above LEFT- a Video Frame,abave Right- Regions of 1 Amoeba,underLeft--parts of an amoeba, Under Right- other amoeba image and hand made region)*

## Character of Amoeba of phase contrast microscope Images

Usually, Amoeba(Dictyostelium discoideum) is living as unicellular organism on water. It moves by transformation of its body. However, a lot of unicellular organisms organize a *pseudoplasmodium* when there is no foods. The *pseudoplasmodium* consist of a lot of amoebae, which amoebae do not occur cellular fusion. After this multicellular state, they make a *fruit body*. Figuratively speaking, *Fruit body* is an office building made by the amoebae. A *fruit body* consists of several parts, for example, *spore* or not.

We take up amoeba image as an example of classification by snake because there are a lot of demands and ordinary image processing methods do not deal with it. For example, the snake may categrize several artificial amoebae on the view of their bahviors. Indeed, when researchers want to present the progress of transformation of an amoeba body, they show the time progress of *Z*-transformation of the bodies that are trimed from caputured each video frame by handwork. In this example, there are 2 problems. First, they can not treat a lot of frames of images because of labor. The second is the reliability and objectivity of time progress of Z-transformation because researcher have each criterion of trimming of amoeba body from a frame. After this section, we show a trimed body as an example, the border between amoeba and water, is very ambiguous. Moreover, we expect that the the snake may be able to chase an amoeba in a large swarm if there are differences of behavior among individuals.

On the other side, we think that image processing methods should tackle several nature of amoeba image when they treat it. However, start points of the snakes on the first video frame, are determined by traditional methods(one of the most simple approach, we think). Then, we would like to think about that. The natures of amoeba image are as follows:
1) assumptions about the shapes, the number of amoeba in a frame, is not able to use because the shape is almost change because of transformation of it body when it moves. 2) In sometimes, an amoeba divides into 2 amoebae. 3) the background(water) and amoeba are not independently each other. The border between each amoeba and water, is vague because wave of water around amoeba is generated by amoeba's trasformation. By the same reason, dusts are forced to move by the transformation. 4)It is too difficult to detect the pseudopod because the expansion and contraction of the pseudopod is very fast and the pseudopod is very ambiguous against water. 5) White noise are occur when several amoebae are situated concentrated.

To the above natures, we approach as follows: noises should be vernished by some fileter from 3). Larger size filter to vernish the white noise of 5), may be needed than the above. From 1),2),4), we can not always set snakes corretly(position and the number of sankes).Then, we must designe that a snake can encircle a target region neverthless its surrounded region does not include the target completely. Moreover, we must detect area on the image where there is no amoebae to delete snakes that are situated on such area. Also,we should introduce a codtion that snakes share same area.

Then, 2 size laplacian filters and differential filters are applied to original video frame.Fig.3-ULEFT shows an image.Fig.3-URIGHT shows a region that researchers detected by hand. Fig.3-DLEFT explain several parts of amoeba on microscope image. Fig.3-DRIGHT shows other image and trimed amoeba by hand. FIG.4 indicates result of application of laplacian and differential filters to Fig.3-ULEFT. As you see, almost areas whereno amoebae are situated, are detecteed(white area).

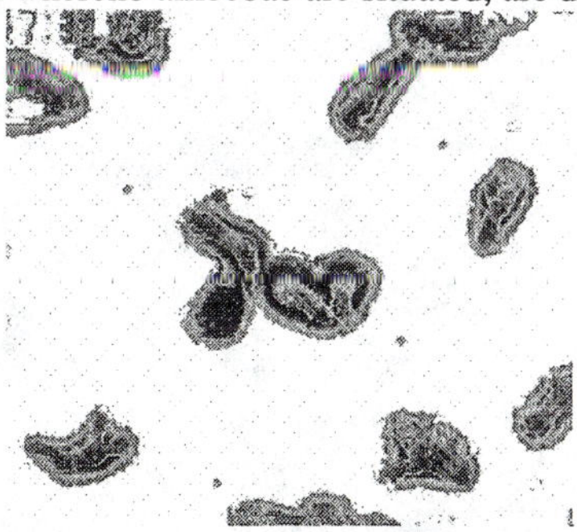

*Fig.4 Result of laplacian Filtering and Edge detection , and segmentation*

## Expansion of Snake to Chase and Enclose Target

In this section, an expansion model of snake to chase target region in amoeba image is proposed. From the last section, the snake can encircle a target region neverthless its surrounded region does not include the target completely. To conquer the condition, we prepare 3 situations about 2 positional relation between a snake and its target.It is the first situation when a snake locates out side of the target region and there no interesection between encircled area and its target. In second situation, a snake is inside of the target. If it is nor situation1,2, we call it situation 3.

When it is in situation 1, the snake should move to the target region. When it is situation 2, the snake should be expand its body to cover all of the target region. In case situation3, the all of region of the snake should move to the inside.

It seems to realize the situation 1 by combination of normal snake and model[5], which is one of derivation of snake. This model can consider some edges that are situated far from a unit of it. However, energy function of the derivation one, is getting decrease according to the expansion of its body size. It will be getting expanding by the model only(see Fig.5 MIDDLE). Then, normal snake's energy function is combined to inhibit the over-expansion. Normal snake size is getting small according to the minimize the energy function(see Fig.5 UPPER).

It seems to materialize by the expansion model, for dealing with the second case. For realization of the third situation, units outside of the target region, should go to the inside. On the other hand, units inside of the area, should cover the area. Thereby, we expect that the required dynamics of the third situation, is realized when the switching operation between situation 1 and 2 works well.

Result of an application of this model to a static image, is shown in Fig.5 BOTTOM. The new model expands its body, passing the region where model[5] did not. However, the new model stop the expansion when the region reach to where a normal snake detects.

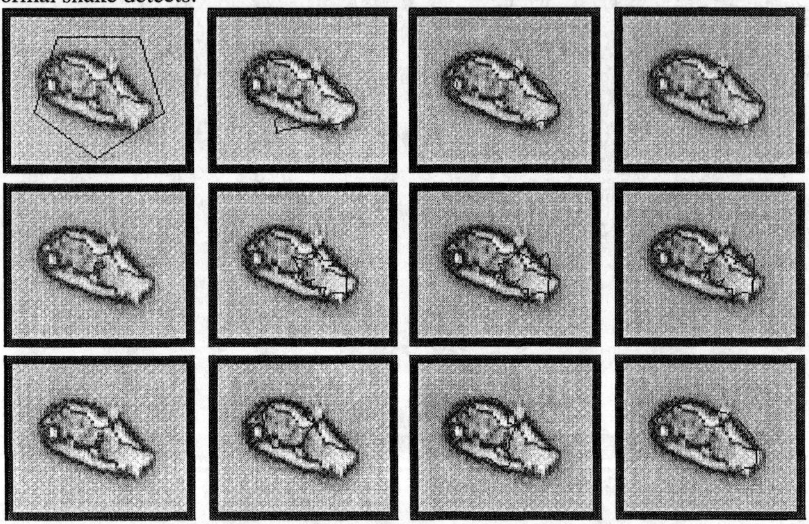

*Fig.5 3 Snake type approaches*
*(upper snake, middle multi-snake, bottom proposed ->->time progress)*

## Computer Simulations

In this section, we confirm our model's chase ability. First, the settings of experiment are described.In this papar, an test serial images are prepared because we would like to evaluate numerically. Especailly, in this paper, we would like to investigate the abilty to chase target under the above 3 situations. In this case, the ability should be evaluated by the rating of covered. If real amoeba images are applied to, we can not evaluate correctly because the region of amoeba is trimed by handwork. Then, a series is prepared that a circle like target is moving on a plane.

The image size is 100*200 pixel. Each pixel value is a number that is selected from 0 to 60 randomly. Next, a circle region (r=10) is corresponded to an amoeba. Each pixel of the circle region, is a number from 60 to 100 randomly too. First, the circle target region is set (25,50), and it moves by vector (5,5) during 10 steps. From the settings, 10 frames are prepared. The circle region is move from (25,50) to(75,100).

Then, the model applied to the frames under 2 different chasing speeds, 30steps per 1 frame and 10steps per 1 frame. The results are shown in Fig.6. The y axis indicates the time step progress. The x axis means the size of the model, the size of intersection of the model and each circle target region, and the ratio*200 of the intersection of total size of the model. As you see, the model can chase and cover under 5PICXEL/30STEP setting. On the other hand, under 5PICXEL/10STEP setting, the model can chase but cannot cover the target. The result of 5PICXEL/20STEP setting is not show in this paper, it is almost same as the result of 5PICXEL/10STEP setting. From these results, the model can not chase if the ratio is smaller than 0.75.

Fig.7 shows the bahaviors. In this snake model, the energy function may be stable when force to expand and force to contract are same. Thereby, in some cases, radiate protuberances like needle, occur. It is very interesting because it just like suggest that the puseudopod of amoeba is occured by same reason. Really, it is one of major dynamics of power of amoeba the dynamics of fluid of internal. Thereby, we suppose that the pseudopod is occure when the balance is disorderd.

However, Fig7 shows that the snake can not chase its target when it covers the target little. Not to lost the chase ability in such situation, we must give some expansiton, and now, we think that learing and predictions are introduced into the units because the energy function must be spicified by itself and noises give influence if there is no predition among units.

## Conclusion

We groped for construction of new usage of *snake*[3] as a categorizor. By simple computer simulations, the proposed snake is able to chase its target. However, In more harsh condition, the snakes could not do it. We think that one of the reason is the character of units that can not predict its environment. In the future, we develop this model to one with learning or some refinement dynamics.

## References


[1]YOKOI,Amoeba Searching Behavior Model for Travelling Salesman Problem,ASME press, Intelligent Engineering Systems Through Artificial Neural Networks Volume5, pp705-710,1995.

[2]SATOU, A basic study of Cellular Automata, graduate school thesis, hokkaido univ.,1996.

[3]KASS,Snakes:active Contour Models,Int.J.Comput.Vision,vol.1,No.4,pp321-331,1988.

[4]UEYAMA,Figure-Ground Separation for Sequential Images with Ginzburg-Landau Equation by use of local clustering, distributed autonomous sypo. proceedings, p41-46,1995.

[5]WADA,Cooperative Distributed Image Segmentation,Transaction of Information Processing Society of Japan,p879-891,1995.

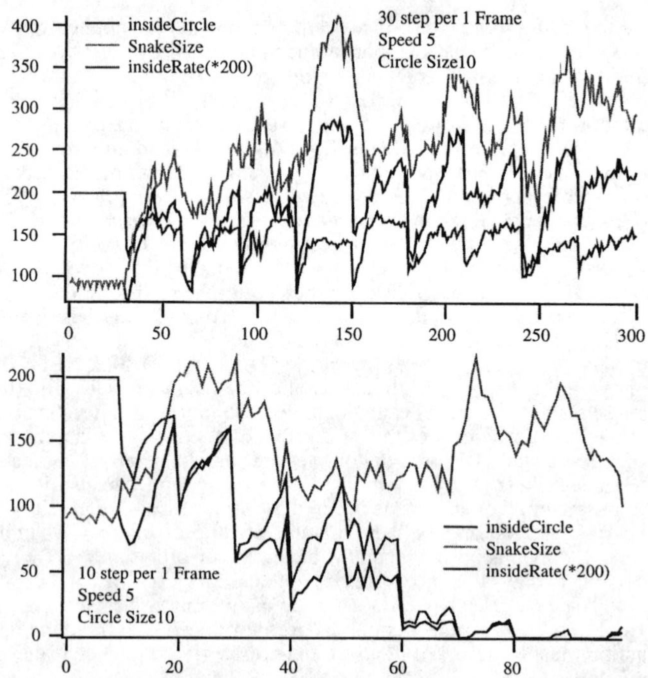


*Fig.6 Time Progress of Chaice of Circle. (Upper -30step/1 Frame, Bottom -10 step/1 Fame)*

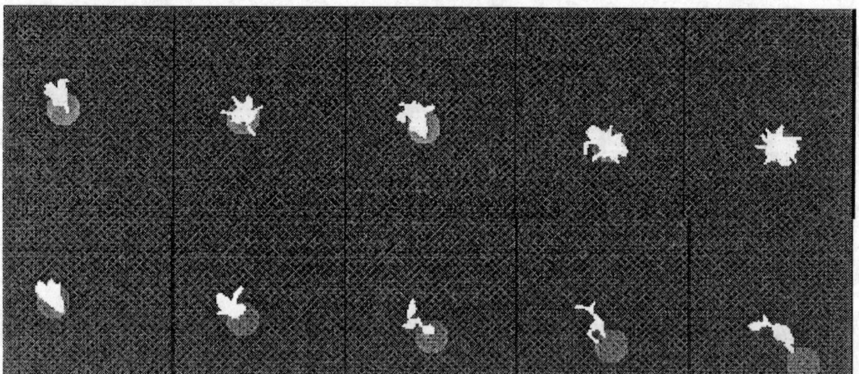

*Fig.7 Time Progress of Chaice of Circle.(Upper -30step/1 Frame, Bottom -10 step/1 Fame)*

# NEURO-FUZZY MODELS FOR MAMMOGRAPHY: DIAGNOSIS OF BREAST CANCER

**KULKARNI A. D., BAKTULA, N. R., AND BOYPATI, S. V.**
**Department of Computer Science**
**The University of Texas at Tyler, Tyler, TX 75701**

*ABSTRACT:*

In this paper, we suggest a computer-aided diagnosis system for mammography that utilizes advanced image processing techniques and a neural-fuzzy inference system to identify regions of interest and to differentiate benign and malignant lesions. The system includes preprocessing, feature extraction, and recognition stages. Spatial filtering and enhancement techniques are used in the preprocessing stage. In the feature extraction stage, a neural network model to extract Fourier transformed domain features is suggested. Features corresponding to the angular and radial bins in the FT domain are extracted. The decision stage utilizes extracted features as input to differentiate benign and malignant lesions. A neural-fuzzy inference system is used for decision making. The training set data was obtained from the database developed and maintained by the Mammographic Image Analysis Society (MIAS). The model was tested with a few sample images using a jack-and-knife method. The results are presented in the paper.

## INTRODUCTION

Here, we suggest a model for a computer vision system for mammography. In mammography the main aim is to detect and characterize regions corresponding to clustered micro-calcification or density masses. Often detection and characterization of an object is a complex task and it is performed in several stages. Stages of a typical image understanding system include: image acquisition, preprocessing, feature extraction, and recognition. These stages essentially correspond to low-level, intermediate-level, and high-level processing. tokens that are obtained in the low-level processing and to extract meaningful entities. One of the well known intermediate-level processing techniques is feature extraction. In our computer vision system, we have developed and implemented a neural network model for extracting FT domain features, and a fuzzy-neural inference system for classification. Computer simulation and results are discussed in the paper.

## FEATURE EXTRACTION

In the feature extraction stage we have used the Fourier transform (FT) domain features. The input to the feature extraction stage is the enhanced mammogram. We have considered angular and radial bins in the FT domain as shown in Figure 1. The radial bins are given by

$$V_{r1r2} \quad = \quad \int\int F^2(u,v)\, dudv \tag{1}$$

where the limits of integration are defined by $r_1^{\,2} \le u^2 + v^2 \le r_2^{\,2}$, and $0 \le u,v \le n\text{-}1$. Radial features are insensitive to rotational differences of input images, and they can be used for extracting scale sensitive features. The angular bins are given by

$$V_{\phi1\phi2} \quad = \quad \int\int F^2(u,v)\, dudv \tag{2}$$

where the limits of integration are defined by $\phi_1 \le tan^{-1}(u/v) \le \phi_2$ and $0 \le u,v \le n\text{-}1$. Angular features can be used for extracting direction sensitive feature. The ANN model for extracting angular and radial features is developed using the Radon transform. The central slice theorem (CST) relates the Radon transform of *f(x,y)* to its Fourier transform. The CST states that the 1-D FT of a projection at an angle $\phi$ is equal to the 2-D FT of *f(x,y)* evaluated along a line in the 2-D space collinear to the unit vector *n* that specifies the projection direction. Thus, the Radon transform along with the central slice theorem is used to evaluate the 2-D FT coefficients from 1-D projections of the function *f(x,y)*(Gindi and Gmitro, 1984). The ANN model for feature extraction using the Radon transform, shown in Figure 2, consists of five layers. Layer $L_1$ is an input layer representing the input image distribution. Layer $L_2$ represents the Radon transform space. Layers $L_{3a}$ and $L_{3b}$ represent real and imaginary parts of the one- dimensional FT coefficients. The weights connecting layers $L_2$ and $L_3$ are represented by the elements of the matrices $\boldsymbol{P_c}$, and $\boldsymbol{P_s}$ and are evaluated as follows. Consider the 1-D discrete Fourier transform (DFT) of the function *f(n)*

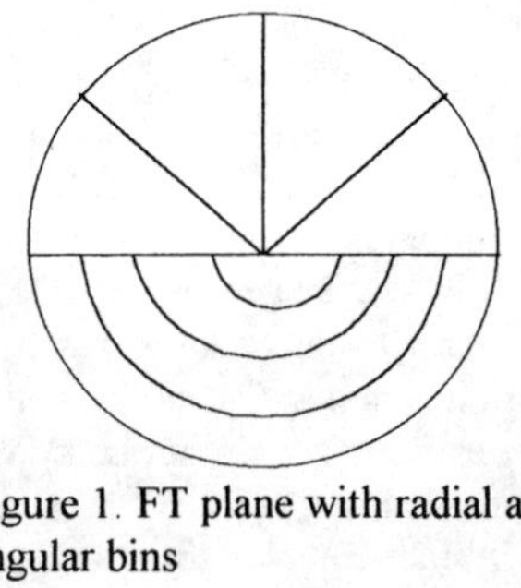

Figure 1. FT plane with radial and angular bins

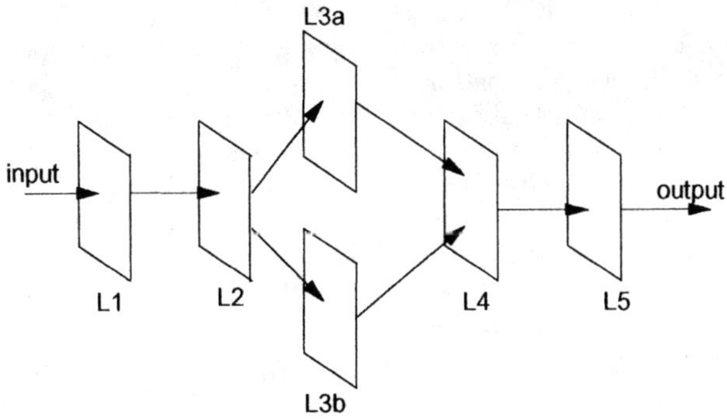


Figure 2. ANN model for feature extraction using Radon transform

$$F(u) = 1/N \sum_{n=0}^{n=N-1} f(n)e^{-2\pi jnu/n} \quad (3)$$

for $u = 0,1,2,...N\text{-}1$. Equation (3) can be represented in a matrix form as

$$f_T = P_c f - j Q_c f \quad (4)$$

where $f$ represents the input vector, $f_T$ represents the Fourier coefficients, and the elements of matrices $P_c$ and $P_s$ are given by $P_c(u,n) = 1/N\, cos(2\pi nu/N)$ and $P_s(u,n) = 1/N\, sin(2\pi nu/N)$ for $u = 1,2, ... ,N\text{-}1$. The units in layer $L_4$ represent amplitudes of the FT coefficients. The rows in layer $L_4$ correspond to FT coefficients obtained from projection of $f(x,y)$ at angle $\phi$. Thus, by adding the outputs of units in each row we get radial features. Similarly by adding outputs of units in each column we get angular features. These features are used as inputs to the recognition stage.

## RECOGNITION

Artificial neural network (ANN) models represent a powerful and reasonable alternative to conventional classifiers (Kulkarni, 1994). A simple three-layer feed-forward network with a backpropagation learning algorithm can be used as supervised classifier. The main disadvantage of a simple backpropagation algorithm is that learning is computationally intensive. Fuzzy logic techniques in the form of approximate reasoning provide decision support and expert systems with powerful reasoning capabilities. The permissiveness of fuzziness in human thought processes suggest that much of the logic behind human reasoning is not traditional two valued logic or even multi-valued logic, but a logic with fuzzy truths, fuzzy connectives, and fuzzy rules of inference (Zadeh, 1973, 1994). During the past decade fuzzy logic has been used in a variety of applications in various fields raging from control to medical diagnosis. There are many neural network based fuzzy logic control/decision systems proposed in literature. Lin and Lee (1991) have suggested a fuzzy logic control/decision network. In their network input/out nodes represent the input states and output control/decision signals, respectively, and in the hidden layers there are nodes functioning as membership functions. The learning algorithm for their network combines unsupervised learning and supervised gradient descent learning procedures. Kulkarni et al. (1994) have developed a neural network fuzzy logic decision system. The model learns the decision rules using a supervised gradient descent procedure. An ANN model for a fuzzy logic decision system is shown in Figure 3. The model consists of six layers. Each layer consists of a number of simple processing units. Layer $L_1$ is the input layer and layers $L_2$, $L_3$ map input feature values to corresponding fuzzy membership values. We have used the triangular and $\pi$-shaped membership functions. These functions can be predetermined or can be adaptive. Layers $L_4$, $L_5$, and $L_6$ represent the backpropagation network. The connection strengths connecting these layers encode decision rules. The algorithm minimizes the mean squared error obtained by comparing the desired output with the actual output. The model works in two phases the training phase and the decision making phase. During the training phase the model is trained using training set data.

## RESULTS AND DISCUSSIONS

In this paper we introduced a computer vision system for mammography: diagnosis of breast cancer. We have considered the task of detection and characterization of clustered micro-calcifications. The test images were obtained from the database developed by the Mammographic Image Analysis Society (MIAS). In the pre-processing stage, we extract features that are used as input to the decision system. Figure 4 shows high-pass images. In the feature extraction stage, we have developed a software package that extracts FT domain features. In our illustrations, we selected a region is of the size 64 scans x 64 pixels. In this case, the FT plane was divided in eight angular and twelve radial bins.

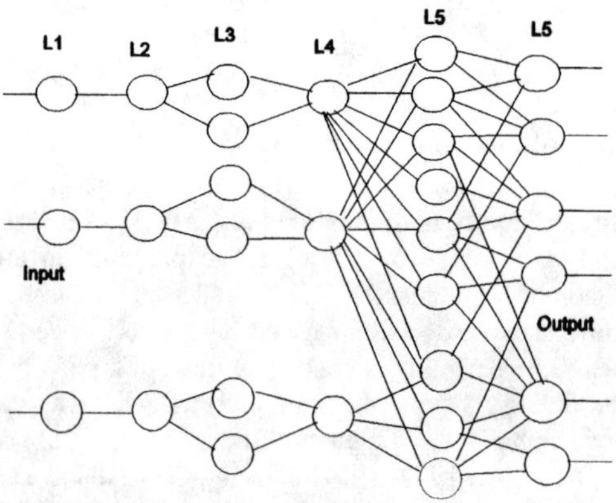


Figure 3. Neural network based fuzzy logic decision system

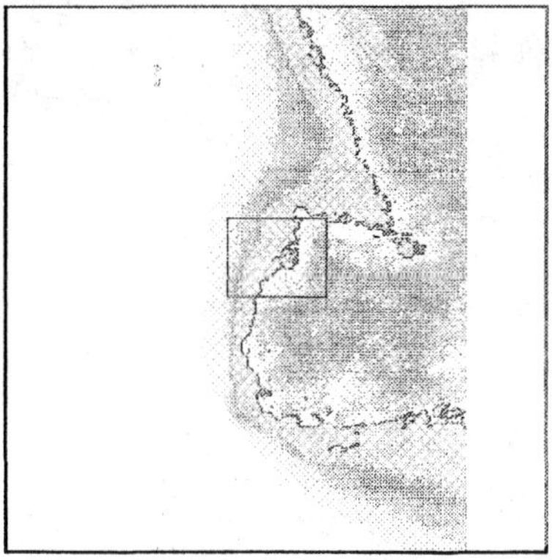

Figure 4a. High-pass image with window on micro-calcification, mdb223.pgm

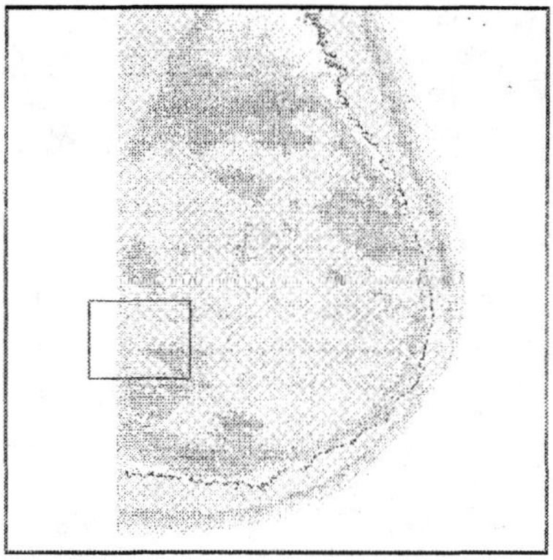

Figure 4b. High-pass image with window on ill-defined masses, mdb030.pgm

| | Sampe No. | Actual Output $Y_1$ | Desired Output $Y_1'$ | Actual Output $Y_2$ | Desired Output $Y_2'$ |
|---|---|---|---|---|---|
| 1 | b218 | 0.996704 | 1.0 | 0.003222 | 0.0 |
| 2 | m211 | 0.059044 | 0.0 | 0.940923 | 1.0 |
| 3 | b223a | 0.881813 | 1.0 | 0.118487 | 0.0 |
| 4 | m231 | 0.091644 | 0.0 | 0.908011 | 1.0 |
| 5 | b226a | 0.996829 | 1.0 | 0.003099 | 0.0 |
| 6 | m239 | 0.062752 | 0.0 | 0.937162 | 1.0 |
| 7 | b226 | 0.948728 | 1.0 | 0.0514221 | 0.0 |
| 8 | m245 | 0.042873 | 0.0 | 0.957018 | 1.0 |

Table 1. Neural-fuzzy inference system output

In the decision stage, we have developed and implemented a neural-fuzzy inference system that combines neural networks with fuzzy logic techniques. The network was trained with training set data shown in Table 1. In our analysis we considered sixteen sub-images each of size 64 scans x 64 pixels. All these sub-images were reviewed by a radiologist and truth data for these images were available. In order to test the system we used a jack-and-knife method. We randomly picked four images from class-1 (benign) and four images from class-2 (malignant). Feature vectors corresponding to these images, were used to train the network. After training all samples in the data set were classified, and results are shown in Table 1. In decision making we considered threshold values of 0.8 and 0.2 that correspond to high and low values, respectively. With these threshold values only one sample out of sixteen samples was classified incorrectly.

We also tried to train a simple three-layer feed-forward network with a backpropagation learning algorithm without fuzzified inputs. The network was unable to converge within 5,000 iterations. The fuzzy-neural system was able to learn from the training set data with in 40 iteration. This is because the fuzzification process is a non-linear mapping which increases dimensions of the feature space; however, it also increases the separability of classes in the feature space. In order to evaluate the effect of fuzzification we used separability measure $J = \| \mathbf{S}_b \| / \| \mathbf{S}_w \|$, where $\mathbf{S}_b$ is the between class scatter matrix, $\mathbf{S}_w$ is the within class scatter matrix, and $\| . \|$ defines the matrix norm. We evaluated the separability of the classes before and after fuzzification and we found increase by 31.31 percent.

## REFERENCES


Gindi, G. R., and Gmitro, A. F. (1984). Optical feature extraction via Radon transform. *Optical Engineering*, vol 23, pp 499-506.

Kulkarni, A. D., (1994). *Artificial neural networks for image understanding*. Van Nostrand Reinhold. New York, NY.

Kulkarni, A. D., Giridhar, G. B., and Coca, P. (1994). Neural network based fuzzy logic decision systems for multispectral image analysis. *International Journal of Neural, Parallel, and Scientific Computing,* vol 3, no 2, pp 205-218.

Lin, Chin-Teng, and Lee George C. S., (1991). Neural network based fuzzy logic control and decision system. *IEEE Transactions on Computers,* vol 40, no. 12, pp 1320-1336.

Zadeh, L. A., (1973). Outline of a new approach to analysis of complex systems and decision processes. *IEEE transactions on Systems, Man, and Cybernetics,* vol 3, pp 28-44.

# WAVELETS FOR SHAPE RECOGNITION WITH APPLICATIONS TO MAMMOGRAPHY

L. M. BRUCE, R. R. ADHAMI, AND J.W. BRUCE
*The University of Alabama in Huntsville*

***ABSTRACT:***
Various methods have been investigated for the automation of lesion detection and classification in mammograms. In this paper, the wavelet transform (WT) is proposed for the use of classifying mammographic lesions. Among the many features that can be measured using the WT, the shape of the lesion was chosen due to the need for both local and global shape measurements. Our results show that the WT can be used to measure the "stellateness" of mammographic lesions.

## INTRODUCTION

Due to limitations on the human visualization of mammograms, researchers are attempting to use modern computers to accomplish automated or computer-aided mammogram analysis. This work includes the enhancement of mammograms (Laine, 1994), the segmentation of lesions (Brzakovic, 1990), and the classification of lesions as benign or malignant (Kilday, 1993; Shen, 1994). The major mammographic signs of carcinoma are related to the tumor mass and include its density, size, shape, borders, and calcification content (Egan, 1988; Feig, 1983). According to Feig and McLelland (Feig, 1983), breast tumors may be categorized as benign or malignant depending on the shape of the lesion boundary, see Figure 1. That is, the more round the lesion, the more likely it is to be categorized as benign. And, the more stellate the lesion, the more likely it is to be categorized as malignant. This paper investigates the use of wavelets to obtain a measurement of the "stellateness" of mammographic lesions.

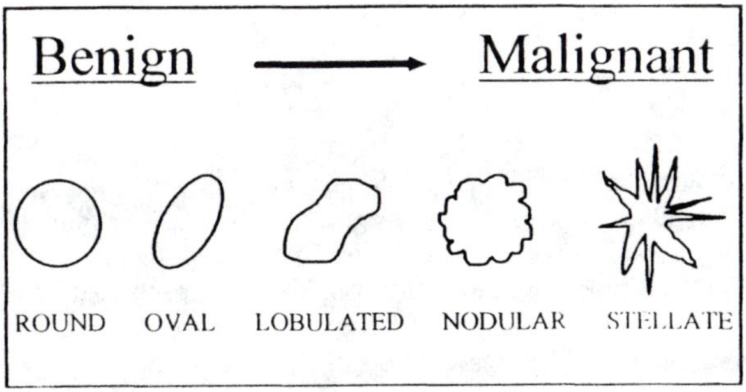


Figure 1 Morphologic spectrum of mammographic lesions

The wavelet transform modulus-maximus (mod-max) representation of a signal can be used to detect and characterize sharp variations (Mallat, 1992). The mod-max method produces numerical descriptors that characterize the magnitude, sharpness, and the Lipshitz order of sharp variations in a given signal. These numerical descriptors can be used as features in a classification system. For the classification of the shape of a mammographic lesion, these numerical descriptors can be used to measure the stellateness.

This paper is organized as follows. Section I. provides a brief introduction. Section II. explains the methodology of using the wavelet transform mod-max representation for detecting and characterizing sharp signal variations. Section III. discusses the use of the mod-max method for shape recognition and its application to mammograms. Section IV. draws conclusions and outlines areas for future research.

# METHODS

In this section, we provide an overview of the use of the mod-max method for shape recognition. It has been shown that the wavelet transform mod-max representation of a signal can be used to detect and characterize sharp variations (Mallat, 1992). The mod-max representation is based on the discrete dyadic wavelet transform (DDWT). The DDWT of a signal *f(x)* is defined as the sequence of discrete signals

$$\left\{S_{2^J} f, \left(W_{2^j} f\right)_{1 \le j \le J}\right\} \quad (1)$$

where the fine signal components $W_{2^j} f(x)$ is the wavelet transform at dyadic scales (powers of 2) and the coarse signal $S_{2^J} f(x)$ provides the signal components below scale $2^J$. The mod-max representation is obtained by detecting the local extrema of the DDWT at each scale $2^j$. These values are then set to zero if the location does not coincide with a local extrema.

The mod-max representation can be used to detect and characterize sharp signal variations. For this use of the mod-max representation, it is important to choose the wavelet function as the first-derivative of a zero-mean Gaussian (Mallat, 1992; Babaud, 1986; Yuille, 1986; Bruce, 1996). Thus the "mother wavelet" is defined as

$$\psi(x) = -\frac{x}{\sigma^2} \cdot \frac{1}{\sigma\sqrt{2\pi}} e^{\frac{-x^2}{2\sigma^2}} . \quad (2)$$

Using the Gaussian wavelet function, the mod-max method can be used to compute numerical descriptors of sharp signal variations. The mod-max method is based on the fact that a sharp variation in the input signal will produce an extremum in each scale of the mod-max representation. That is, each sharp variation in the input signal will produce a sequence of extrema that appears across the scales of the mod-max representation. See Figure 2. b. and Figure 3. b. Using these extrema, the mod-max method obtains numerical descriptors by minimizing the following (Mallat, 1992)

$$\sum_{j=1}^{I}\left(\log_2|a_j|-\log_2(K)-j-\frac{\alpha-1}{2}\log_2((s_1\sigma)^2+2^{2j})\right)^2 \tag{3}$$

where $a_j$ is the local extremum value of the wavelet transform at scale $2^j$. The minimization of this equations produces three numerical descriptors for each signal variation: i) K, the amplitude of the sharp variation; ii) $(s_1\sigma)^2$, the variance or smoothness of the variation; and iii) $\alpha$, the Lipshitz order of the variation. This analysis is valid only for isolated singularities (Mallat, 1992). Bruce (Bruce, 1996) defined "isolated" for adjacent singularities and smooth variations of the input signal. When using the mod-max method to detect and characterize signal variations, the user has a choice of which scales ($j=1, \ldots, I$) to use for computing the numerical descriptors. If the signal variations being analyzed are separated by a minimum distance of $x_o$, the isolation assumption places restrictions on the scales with are appropriate for use. These restrictions are written the form of an isolation definition (Bruce, 1996). For accurate results with the mod-max method, the scales should be in the following range

$$s \leq \sqrt{\left(\frac{x_o}{k+\sigma}\right)^2 - s_1^{\,2}} \tag{4}$$

where $x_o$ is the minimum separation of adjacent variations, $\sigma^2$ and $k$ is the variance and truncation distance of the Gaussian wavelet function respectively, and $(s_1\sigma)^2$ is the variance of the smooth variation (Bruce, 1996). It should be noted that a smooth variation of the input signal are modeled as a singularity convolved with a Gaussian of variance $(s_1\sigma)^2$. If this restriction on utilized scales is not satisfied, the mod-max method can produce inaccurate results.

## SHAPE RECOGNITION

The previous section explained the mod-max method and the limitations on utilized scales. This section explains how the mod-max method can be used for shape recognition. The mod-max shape recognition procedure can be applied toward automatic recognition of mammographic lesions.

In this paper, the mod-max method has been described for one-dimensional (1-D) signals. The method can be extended for two-dimensional (2-D) signals. However, this in not necessary when using the mod-max method for recognition of certain shapes. For the application of mammographic lesion recognition, the 2-D shapes can be transformed into 1-D signals. The mod-max method is then applied to these 1-D signals.

It is assumed that the object to be recognized is already segmented. The edges of the segmented object provide the shape information. The centroid of the object is then computed. Next, the distance from the centroid to the edge is measured. This distance is referred to as the radial distance. The radial distance is measured over a range of angular positions. The number of angular positions determines the resolution of the 1-D representation. See Figure 2. and Figure 3.

The mod-max method is now applied to the 1-D representation of the object shape. Depending on the application, the method may be used in various ways. For the problem of automatic recognition of mammographic lesions, we are interested in the roundness or the stellateness of the lesion. The mod-max method can be used to extract features which numerically describe the roundness or stellateness of the lesions. The method produces the numerical descriptors that specify the magnitude (K), sharpness ($s_1\sigma$), and Lipshitz order ($\alpha$) for each signal variation. In order to classify the shape, the three numerical descriptors need to be analyzed for all major signal variations. The application determines the definition of "major" signal variations.

For our study of the recognition of mammographic lesion shapes, the 2-D shapes were transformed to 1-D radial distance representations at a resolution of five degrees. That is, the radial distance was measured every five degrees. Thus, the 1-D discrete signal consisted of 72 samples. Higher sampling rates were investigated, but they generally caused inconsistencies with the isolation assumption. Next, the mod-max representation of the 1-D signal was computed over the scales $2^1$ to $2^4$. The sampling rate of the 1-D transformation and the choice of appropriate scales would obviously vary depending on the resolution of the digitized mammogram. Further, the choice of appropriate scales depends on the sampling rate of the 1-D transformation. For higher sampling rates, larger scales may be more useful.

Once the mod-max representation was computed, a decision had to be made as to what would be considered a major signal variation. The extrema of only major signal variations would be tracked across scales for the computation of numerical descriptors. First, the system determined which variations could be dependably tracked across all scales. Any variation which could not be tracked was automatically discarded. Next, the numerical descriptors were computed for all remaining variations. The numerical descriptor for magnitude, $K$, was then used for determining major signal variations. A threshold, $K_o$, was set, any variation with magnitude greater than $K_o$ was designated as a major variation. All other variations were discarded.

Once the major signal variations were detected, their numerical descriptors were used as features to estimate the shape classes. The absolute values of the numerical descriptors depend greatly on the digitization resolution, sampling rate, and range of scales used in the mod-max method. Thus, while we provide actual numbers for the observed numerical descriptors, attention should be given to general trends in the features.

It was observed that, in general, "round" lesions consisted of very few major signal variations. These signal variations produced numerical descriptors in the following way: i) $K$ tended to be quite small, near the threshold $K_o$=10; ii) $s_1\sigma$ tended to be relatively large, from 20 to 40; and iii) no conclusion could be readily made about the Lipshitz order $\alpha$. It was observed that "stellate" lesions consisted of many major signal variations. These signal variations produced numerical descriptors in the following way: i) $K$ varied greatly, from 15 to 50; ii) $s_1\sigma$ varied greatly, from 3 to 20; iii) no conclusion could be readily made about the Lipshitz order $\alpha$. Even though the values of $K$ and $s_1\sigma$ varied for the stellate case, they tended to both be large or small simultaneously. Due to the fact that the Lipshitz order did not seem to affect the classification of the lesions, this numerical descriptor was ultimately discarded. However, this descriptor should be investigated more closely.

Through the of lesion shape classes, it was observed that a third class should be included. This is referred to as the "nodular" class. See Figure 1. It was observed that these lesions consisted of many major signal variations. These signal variations produced numerical descriptors in the following way: i) $K$ tended to be small but not distinctively greater than the threshold $K_o$, ranging from 10 to 20; ii) $s_I\sigma$ tended to be consistent but distinctively smaller than those of round lesions, from 10 to 20; and iii) no conclusion could be readily made about the Lipshitz order $\alpha$.

## CONCLUSIONS

The wavelet transform mod-max method can be effectively used to characterize and recognize object shapes. The method can be used to extract features that numerically describe sharp variations in the object shape, such as sharp corners. These numerical descriptors characterize the magnitude, sharpness, and Lipshitz order of the sharp variations. These numerical descriptors can be used to detect specific points in the object shape or used collectively to describe the overall shape.

The wavelet transform mod-max method has been applied to the classification of lesion shapes in mammograms. Breast tumors may be categorized as benign or malignant depending on the shape of the lesion boundary (Feig, 1983). That is, the more round the lesion, the more likely it is to be categorized as benign. And, the more stellate the lesion, the more likely it is to be categorized as malignant. We gave a brief overview of a recent study that investigated the use of the mod-max method to obtain a measurement of the "stellateness" of mammographic lesions (Bruce, 1996). This study concluded that the numerical descriptors produced by the mod-max method can be used as features to form three classes of mammographic lesions. These classes correspond to round, nodular, and stellate lesions.

This paper concludes that the mod-max method is a valid approach to extracting lesion shape features. For a fully automated recognition system, these features could be used in conjunction with other case features, *e.g.* lesion texture, patient age, and calcification information (Egan, 1988; Feig, 1983). Also, the classification portion of the pattern recognition system could take many forms. One obvious choice is a neural network system. The inputs to the neural network components could be 1) the wavelet transform mod-max representation at each scale, 2) the mod-max extrema information for each individual signal variation, or 3) the computed mod-max numerical descriptors. The design of this portion of the recognition system requires much further investigation.

## REFERENCES

Babaud, J., Witkin, A., Baudin, M. and Duda, R., (1986). Uniqueness of the Gaussian kernel for scale-space filtering, *IEEE Trans. Patt. Anal. Machine Intell.*, vol. 8, no. 1, pp. 26-33.

Bruce, L.M., (1996). The use of multiresolutional analysis for automated recognition of non-isolated patterns, Ph.D. dissertation, Dept. of Elec. & Comp. Eng., The University of Alabama in Huntsville.

Brzakovic, D., Luo, X.M., and Brzakovic, P., ( 1990). An approach to automated detection of tumors in mammograms, *IEEE Transactions on Medical Imaging*, vol. 9, no. 3.

Egan, R.L., (1988). *Breast Imaging: Diagnosis and Morphology of Breast Diseases*. Philadelphia: Harcourt Brace.

Feig, S.A. and McLelland, R., (1983). *Breast Carcinoma:Current Diagnosis and Treatment.* New York: Masson Publishing.

Kilday, J., Palmieri, F., and Fox, M.D., (1993). Classifying mammographic lesions using computerized image analysis, *IEEE Transactions on Medical Imaging*, vol. 12, no. 4.

Laine, A., Schuler, S., Fan, J., and Huda, W., (1994). Mammographic feature enhancement by multiscale analysis, *IEEE Transactions on Medical Imaging*, vol. 13, no. 14, pp. 725-740.

Mallat, S. and Zhong, S., (1992). Characterization of signals from multiscale edges, *IEEE Trans. Patt. Anal. Machine Intell.*, vol. 14, no. 7, pp. 710-732.

Shen, L., Rangayyan, R.M., and Leo Desautels, J.E., (1994). Application of shape analysis to mammographic calcifications, *IEEE Transactions on Medical Imaging*, vol. 13, no. 2.

Yuille, A.L. and Poggio, T.A., (1986). Scaling theorems for zero crossings, *IEEE Trans. Patt. Anal. Machine Intell.*, vol. 8, no. 1, pp. 15-25.

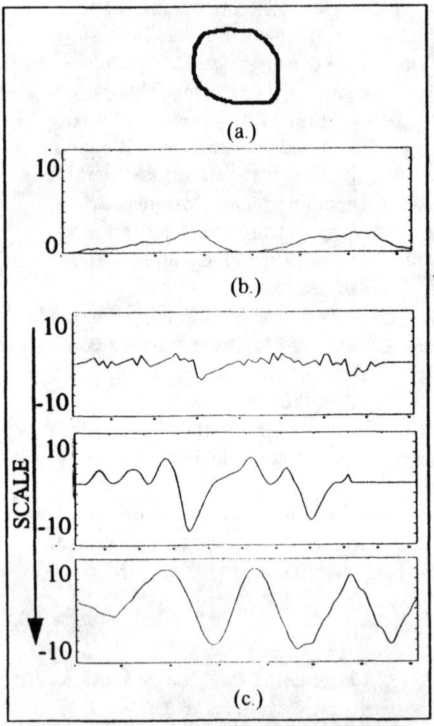


Figure 2. (a.) 2-D representation of "round" lesion, (b.) 1-D radial distance representation, (c.) Wavelet Transform at various scales.

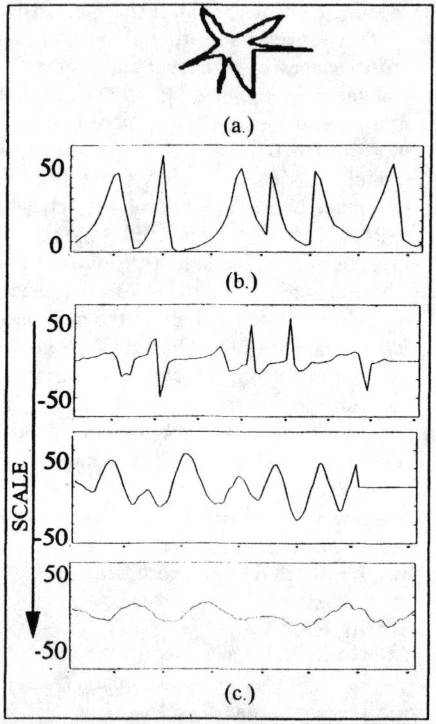


Figure 3. (a.) 2-D representation of "stellate" lesion, (b.) 1-D radial distance representation, (c.) Wavelet Transform at various scales.

# CLASSIFICATION OF THE EFFECTS OF F–ACTIN UNDER TREATMENT OF DRUGS IN ENDOTHELIAL CELLS

**KAMAL J. KHIANI, SAMEH M. YAMANY AND ALY A. FARAG**
Computer Vision and Image Processing Laboratory
University of Louisville, Louisville, Kentucky 40292
email: farag@egypt.spd.louisville.edu

**WILLIAM D. EHRINGER AND FREDERICK N. MILLER**
Center for Applied Microcirculatory Research, School of Medicine
University of Louisville, Louisville, Kentucky 40292

***ABSTRACT:***
This paper deals with developing automatic classification algorithms for identifying images of F–actin distribution in endothelial cells with different treatments of drugs. Statistical and artificial neural network methods are investigated. Among the statistical classification techniques are the Bayes approach and the linear classifier. For the neural network approach, the multilayer feedforward network and the functional link network are investigated.

## INTRODUCTION

Endothelial cells are a single layer of thin flattened cells that lines the vasculature. They have a cytoskeleton that is composed of F–actin. A primary function of vascular endothelium is the mediation and control of water and solute exchange between blood plasma and the interstitial fluid. During inflammatory and immune reactions, the endothelial cells undergo a number of functional and morphological alterations caused by agonists. Many of these agonists are mediators during inflammation and serve to increase permeability in the endothelial cell. This increase in permeability is associated with cytoskeletal F–actin changes in the endothelial cells along with the formation of intercellular gaps. Figure 1 shows the results of the agonists on the endothelial cells.

Developing automatic classification of the effects of agonists on endothelial cell F–actin distribution could be used as a screening process for other agonists. The first–order histogram and second–order texture features are extracted and used to characterize the phenomena of the agonists. The specific agonists used in this study are $\alpha$–thrombin and histamine. This paper investigates different classifiers for automatic classification to determine the most accurate, easy to implement and time efficient classifier for such a complex problem.

---

This work has been partially supported by grants from the Whitaker Foundation and the Center for Applied Microcirculatory Research, University of Louisville.

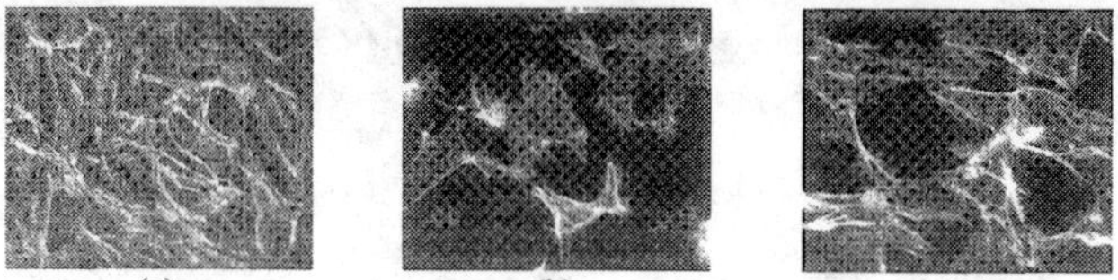

Figure 1: (a) Control endothelial cells, (b) α–Thrombin and (c) Histamine

## FEATURE EXTRACTION

Human umbilical cord vein endothelial cells (HUVEC) were grown to confluence on Transwell membranes. The endothelial cells were plated with α–thrombin and histamine. BODIPY–Phalloidin, which is a fluorescent dye, binds to intracellular F–actin and therefore was used as an indicator of the relative distribution of this cytoskeletal protein (Ehringer, Edwards, et. al., 1996). There were 30 images digitized for each class for a total of 90 images.

First–order tonal and second–order texture features were considered. First–order feature is a type of gray level distribution identifiable with the gray level histogram. There were two features that were considered for the first order descriptors. The first is the mean gray level and the second feature was the tenth percentile which represents the gray level at or below which, lies 10% of the total number of pixels in the image.

The second–order features describe the gray level spatial inter–relationships and measure gray level texture homogeneity (Kadah, Farag, et. al., 1996). These parameters were obtained from the spatial gray level dependence method (SGLDM). The SGLDM characterizes texture by the co–occurrence of its gray levels. The SGLDM is based on the estimation of the second–order joint conditional probability density function, $\Psi(i,j|d,\theta)$. The estimated values can be written in matrix form, the so called co–occurrence matrix which is a function of the image $f$, a displacement vector $d = (d_i, d_j)$ and an angle θ over the domain $D$

$$\Psi_{ij}(f,d) = \#((k_1,l_1),(k_2,l_2)) \quad (k_1,l_1),(k_2,l_2) \in D, \qquad (1)$$
$$f(k_1,l_1) = i, f(k_2,l_2) = j, [k_2,l_2] - [k_1,l_1] = d$$

where # indicates the number of elements in the set. The $d$ used for the paper was 16. The joint probability densities of the co–occurring gray levels are approximated by normalizing the co–occurrence matrices. The angles are quantized to 45° intervals.

Haralick and others (Haralick et. al., 1973) have defined 14 statistical measures of texture which can be extracted from the co–occurrence matrices. From the results of a genetic algorithms based (GA–based) feature selection process (Vafaie and De Jong , 1992), we considered four features for the second order statistics. The first textural feature is the *Angular Second Moment (ASM)*

$$f_1 = \sum_i \sum_j \{P(i,j)\}^2 \qquad (2)$$

where *P(i,j)* is the *(i,j)th* entry in the normalized co–occurrence matrix. This is a measure of homogeneity in the image (Haralick et. al., 1973). The *Contrast* is defined as

$$f_2 = \sum_i \sum_j (i-j)^2 P(i,j) \tag{3}$$

and is a measure of the local variations present in the image. The *Correlation* is

$$f_3 = \frac{\sum_i \sum_j (ij)P(i,j) - \mu_x \mu_y}{\sigma_x \sigma_y} \tag{4}$$

where $\mu_x$, $\mu_y$, $\sigma_x$ and $\sigma_y$ are the means and standard deviation of $p_x(i)$ which is the *ith* entry in the marginal–probability matrix obtained by summing the rows of $P(i,j)$ and $p_y(j)$ is the *jth* entry in the marginal–probability matrix obtained by summing the columns of $P(i,j)$. This feature measures the gray tone linear–dependencies in the image (Haralick, et. al., 1973). The *Entropy* (randomness of image texture) is defined as

$$f_4 = \sum_i \sum_j P(i,j) \log P(i,j) \tag{5}$$

## CLASSIFIER DESIGN

An observation is a feature vector $X = x$ from some space $\chi$, $x \in \chi$, which is in the *sample space*. The link between the sample space and the class $\omega$ to which it belongs is the *classifier*. The total data set is divided into a *training set* (15 images for each class) and *test set* (15 images for each classs).

### Bayes Classifier

The Bayes classifier is based on the assumption that all of the relevant probability values are known (Duda and Hart, 1973). The random variable $x$ is assigned to a class based on a decision rule of probabilities. The *a posteriori* probabilities $P(\omega_i \mid x)$ may be calculated from the *a priori* probabilities $P(\omega_i)$ and the conditional density functions $p(x \mid \omega_i)$ using Bayes' theorem (Duda and Hart, 1973), which is

$$P(\omega_i \mid x) = \frac{p(x \mid \omega_i)P(\omega_i)}{p(x)} \quad \text{where} \tag{6}$$

$$p(x) = \sum_{j=1}^{3} p(x \mid \omega_i)P(\omega_i) \tag{7}$$

The distribution chosen for this classifier was the Gaussian or Normal distribution.

### Linear Classifier Design

The linear classifier has the general form :

$$h_{ij}(x) = V_{ij}^t x + v_{0_{ij}} \tag{8}$$

where $V_{ij}^t$ is the set of coefficients and $v_{0_{ij}}$ is the threshold. We approximated the distribution of $h_{ij}(x)$ to be Gaussian :

$$h_{ij}(x)/\omega_j \sim N(\eta_j, \sigma_j^2) \text{ where} \quad (9)$$

$$\eta_i = V_{ij}^t \mu_i + v_{0_{ij}} \text{ and } \sigma_i^2 = V_{ij}^t \mu_i \Sigma_i V_{ij} \quad (10)$$

and $\mu_i$ and $\Sigma_i$ are the mean and covariance of the samples of class $\omega_i$. Optimum values of $V_{ij}^t$ and $v_{0_{ij}}$ are chosen such that the derivative of a criterion function is equal to 0. Choosing the criterion function $f(\eta_i, \eta_j, \sigma_i^2 \sigma_j^2)$ to be the probability of error, the optimum values become (Fukunaga, 1968)

$$V_{ij} = \left[s\Sigma_i + (1 - s)\Sigma_2\right]^{-1}(\mu_j - \mu_i) \quad (11)$$

$$v_{0_{ij}} = -\frac{s\sigma_{ij}^2 V_{ij}^t \mu_j + (1 - s)\sigma_{ji}^j V_{ji}^t \mu_i}{s\sigma_{ij}^2 + (1 - s)\sigma_{ji}^2} \quad (12)$$

where s is a factor that can take values from 0 to 1. To determine s:

1. Start with s = 0
2. Obtain $v_{ij}$, $\sigma_{ij}^2$ and $v_{0_{ij}}$
3. Calculate the classification error
4. Change s until s = 1
5. Goto 2.

**Neural Network Classifiers**

The multilayer neural network (MLNN) (figure 2a) consists of one input layer, one or more hidden layers and one output layer. The basic element of the network is called the *neuron* (Zurada, 1992). Each neuron consists of a set of weights, one for each input, and an activation function. The neuron computes the weighted sum of the inputs and passes this value to an activation function which generates an output. The weighted sum is referred to as *net*. The activation function used in this paper was the bipolar continuous activation function:

$$f(net) = \frac{2}{1 + \exp(-\lambda net)} - 1 \quad (13)$$

Here $\lambda > 0$ and is proportional to the neuron gain determining the steepness of the

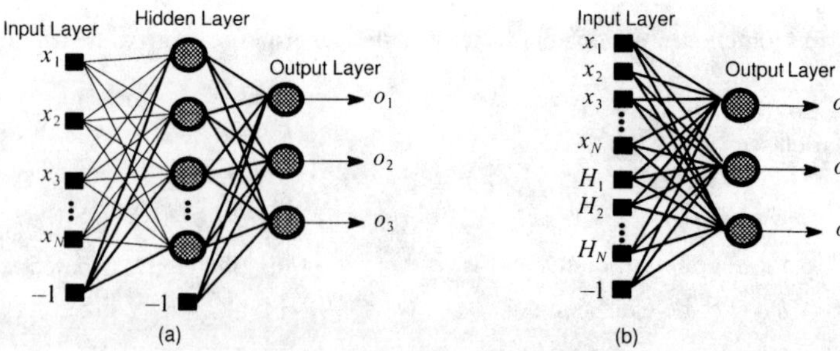


**Figure 2: (a) Multilayer feedforward network and (b) functional link network.**

continuous function. During training, the network weights are adjusted until some maximum error criterion is satisfied. The training is accomplished with the error back–propagation training algorithm (EBPTA) (Zurada, 1992). This is a supervised learning algorithm which performs a gradient descent on a squared error energy surface to arrive at a minimum.

Functional link neural networks (FLNN) (figure 2b) are single–layer networks that are able to handle linearly non–separable classes due to increasing the dimensions of the inputs by using nonlinear combination of the inputs parameters. In the *tensor model,* the additional input terms are obtained for each of the input feature vectors as the product $x_i x_j$ for $\forall_{i,j} \in [1, D]$ where $D$ is the dimension of the input vector. In figure 2b, the $H_1..H_N$ represents the additional input terms. Due to its intrinsic mapping properties, the functional link network in some cases can perform better than the multilayer network and its distinct advantage is the easier training. The simple delta learning rule is used instead of the multilayer general delta learning rule.

## RESULTS

Table 1 shows the results of the classification of both the statistical and neural approaches using the six features. The results show that both methods classify the three class of images well. The best performance of the linear classifier was obtained at s=0.65 which was 93% classification. The linear classifier showed less accuracy as a result of the small mean separability between the classes. Also, linear classifiers work better in cases of equal covariances which is not the case here. In general, the error from the statistical classifiers resulted from some overlap between classes. This is because a few of the images obtained from the well plates had variations due to the non–homogeneous effects of the agonists on the cytoskeleton of the endothelial cells. The multilayer feedforward neural network with a 6–4–3 *(input,hidden,output)* architecture shows a higher classification of 97.7%. The is because the neural network has the power to learn and generalize the recognition of the F–actin distribution. Since the neural method performed better, we investigated the other neural architectures with a reduction in feature dimensionality (using the GA–based feature selection process) to reduce the feature extraction time and complexity. These features were the mean gray level, tenth percentile, entropy and correlation. Different multilayer and functional link topologies were investigated with different training parameters. Table 2 shows the re sults. The functional link neural network using the tensor model was a 10–3 *(input,output)* architecture. Its $\eta$ values were also varied and the highest classification was 95.5%.

## CONCLUSION

Automatic classification of the F–actin distribution in endothelial cells treated with agonists was accomplished using different statistical and neural network classification

**TABLE 1: RESULTS OF THE DIFFERENT CLASSIFIERS**

| Classifier | % Correct |
|---|---|
| Bayes | 95.5% |
| Linear | 93.0% |
| MLNN | 97.7% |

**TABLE 2: RESULTS OF CLASSIFYING WITH NEURAL METHODS (MSE: Mean Square Error)**

| Topology | $\eta$ | Cycles | M.S.E | % Correct |
|---|---|---|---|---|
| MLNN 4–3–3 | 0.4 | 20000 | 0.00118 | 91.1% |
| MLNN 4–3–3 | 0.8 | 10000 | 0.00092 | 91.1% |
| MLNN 4–4–3 | 0.4 | 20000 | 0.00069 | 93.3% |
| MLNN 4–4–4 | 0.8 | 8000 | 0.00024 | 91.1% |
| MLNN 4–5–3 | 0.4 | 40000 | 0.00024 | 93.3% |
| FLNN 10–3 | 6.0 | 20000 | 0.04560 | 95.5% |

techniques. The histogram and texture features characterize the F–actin distribution and other phenomena such as pores and gaps. This enables accurate discrimination of the F–actin distribution on endothelial cells as a result of the agonists. The classification scheme and discrimination from a human observer can complement each other in obtaining overall accuracy. Future work involves other agonists which may have similar F–actin distribution as the agonists used in this paper. This may imply similar mechanisms of permeability and hence can be used as a screening process in reducing the experiments needed to determine its true mechanism. Also a future consideration is to correlate the phenomena of the F–actin distribution with the features. Based on the results obtained, the neural network classifiers were better than the statistical classifiers in terms of speed, accuracy and ease of implementation. In conclusion, it is feasible to discriminate the different F–actin distributions from the types of treatments on endothelial cells.

## REFERENCES

Duda R. and Hart P. (1973). *Pattern Classification and Scene Analysis*, Wiley & Sons, New York, 10–3.

Ehringer, W. D., Edwards, M. J. and Miller, F. N. (1996). "Mechanisms of α–Thrombin, Histamine, and Bradykinin Induced Endothelial Permeability", *Journal of Cellular Physiology.*

Fukunaga, K. (1968). *Introduction to Statistical Pattern Recognition*, 2nd edition, Academic Press, New York.

Haralick, R.M., Shanmugam, K., and Dinstein, I. (1973). "Textural Features for Image Classification", *IEEE Transactions on Systems, Man, and Cybernetics*, Vol. SMC–3 (6), November.

Kadah, Y.M., Farag, A.A., Zurada, J.M. Badawi, A.M. and Youssef, A.M. (1996). "Classification Algorithms for Quantitative Tissue Characterization of Diffuse Liver Disease from Ultrasound Images", *IEEE Transactions on Medical Imaging*, Vol. 15 (4), August.

Vafaie, H., and De Jong, K., (1992). "Genetic Algorithms as a Tool for Feature Selection in Machine Learning", *Proceeding of the 4th International Conference on Tools with Artificial Intelligence*, Arlington, VA, November.

Zurada, Jacek M., (1992). *Introduction to Artificial Neural Systems*, St. Paul, MN: West Publishing Company.

# 3D SEGMENTATION AND LABELING USING UNSUPERVISED CLUSTERING FOR VOLUMETRIC MEASURMENTS ON BRAIN CT IMAGING

**Mohamed N. Ahmed, and Aly A. Farag**
*Computer Vision and Image Processing Lab*
*University of Louisville, Department of Electrical Engineering*
*Louisville, KY 40292*
*E-mail:farag@cairo.spd.louisville.edu, Phone:(502)-852-7510, Fax:(502)852-6807*

**Abstract**

In this paper, we present a new system to segment and label CT Brain slices using differential geometrical invariant features and unsupervised clustering. Our aim is to extract reliable and robust measures from CT images of Traumatic Brain Injury (TBI) patients that can accurately describe the morphological changes in the brain as recovery progresses, and to study the correlation between the morphological changes in progressive CT images after the Day Of Injury (DOI) with cognitive functional changes. Segmentation is performed by assigning a feature pattern to each voxel, consisting of a scaled family of differential geometrical invariant features. The invariant feature pattern is assigned to a certain region using unsupervised clustering. In this paper we used 3 clustering techniques: The ISODATA Algorithm, the Hopfield network, and the Kohonen feature map network. Implementation and performance of these techniques are presented.

**keywords** Pattern Recognition, Biomedical Application, Segmentation, Neural Networks, Clustering, Classification, Hopefield Ntework, Kohonen Feature Map, Unsupervised Learning.

# 1 Introduction

The human brain contains a communicating ventricular system made up of two lateral ventricles, and a third and fourth ventricle. It has been hypothesized that the changes in the ventricles during the course of treatment of a Traumatic Brain Injury (TBI) can help predict prognosis. In order to have volumetric measurements on the brain CT, we must identify the contours of the structures of interest, and develop a method to create the 3D representation of these structures. The first step is to segment the brain CT slices to identify portions of the brain structures of interest. Despite the large variety of methods which have been proposed in the literature, automated segmentation of complex medical images remains at the stage of labotatory research. Several brain image interpretation systems have been reported [1]-[3]. Their quality depends strongly on the efficiency of the low and intermediate level vision algorithms which are applied to extract information from the image data. These techniques may be divided into two broad approaches: region-based and edge-based segmentation.
In this paper, we present a new technique to describe an image volume by assigning a feature pattern to each voxel, consisting of a scaled family of differential geometrical invariant features. The invariant feature pattern is assigned to a specific region using unsupervised clustering. The clustering is performed using a pattern recognition approach (ISODATA algorithm)[4], and a neural network approach (Hopfield and Kohonen networks)[5].
This paper is organized in the following way: Section 2 describes our system overview for volume segmentation. Section 3 is devoted to the feature extraction process, by which a single voxel is converted to a feature pattern. Section 4 details the implementation of three unsupervised clustering techniques to classify the features obtained in section 3 into different regions, corresponding to different anatomical structures. Section 5 contains some illustrative results together with a discussion and directions for future work.

# 2 System Overview

The first step in our system consists of data acquisition and filtering to remove noise. This is followed by a feature extraction process using differential geometrical invariant features. This step is discussed in detail in the following section. 3D brain segmentation is then performed based on voxel classification. A majority filter is then applied to remove holes from regions in the resulting segmented image. This is done by sliding an $(N \times N \times N)$ box over the classified image volume. The new assignment of label for the center voxel is the label associated with the majority class within the box. Multiple iterations can be employed to control the amount of smoothness at the expense of losing some small structure. Finally, to ensure connectivity of the segmented regions, a 3D connected component algorithm is applied to extract the final regions.

# 3 Feature Extraction

Scale plays an important role in image understanding. An image, being a physical observation of a scene, represents the scene on a finite range of scales only. The aim of image understanding is to infer scene strucures, which calls for a multi-scale description of image structure. It has been shown [6] that the only family satisfying the natural front-end vision constraints of linearity, shift-, rotation-, and scale invariance is the Gaussian family and all its partial derivatives. This operator family provides a complete representation of image structure. The kernel for a Three-dimensional Gaussian operator is

$$G_\sigma(x,y,z) = \frac{1}{(\sigma\sqrt{2\pi})^3} e^{-\frac{(x^2+y^2+z^2)}{2\sigma^2}}. \tag{1}$$

The kernel for a three-dimensional Laplacian of Gaussian operator is described as

$$LOG(x,y,z) = \nabla^2(I(x,y,z) * G_\sigma(x,y,z)), \tag{2}$$

where $(*)$ denotes convolution, $I(x,y,z)$ is the 3-D image and $\nabla^2() = \partial^2()/\partial x^2 + \partial^2()/\partial y^2 + \partial^2()/\partial z^2$. Using the rules for derivatives for convolution, we can write

$$LOG(x,y,z) = I(x,y,z) * \nabla^2 G_\sigma(x,y,z). \tag{3}$$

Also we have that

$$\frac{\partial I(x,y,z)}{\partial x} * G_\sigma(x,y,z) = \frac{\partial G_\sigma(x,y,z)}{\partial x} * I(x,y,z) \tag{4}$$

Using the above equations, all image derivatives can be obtained by convolving with the corresponding derivative of the Gaussian. Consequently, a family of scaled differential operators is available, by which image structures may be described completely up to any desired order at various scales.

Thus a voxel can be represented by a feature pattern, $X$, consisting of its gray level value and its value after filtering with different kernels, as shown in Fig.1. In this study, $X$ consists of 4 elements: Original voxel gray level value, and voxel value after applying the LOG operator using $\sigma = 1$, 2, and 2.5.

# 4 Volume Segmentation by Unsupervised Clustering

Once each voxel is represented by a feature vector, $X$, this vector needs to be classified as belonging to a certain region, cluster, using an unsupervised technique to obtain an automatic segmentation. In this section we consider three different cluster-seeking methods in some detail. The methods discussed constitute a cross section of representative approaches to the cluster-seeking problem.

## 4.1 Classification Using the Iterative Self-Organizing Data Analysis (ISODATA) Algorithm

The first method used to classify the feature vector $X$, obtained from the last section, is the ISODATA Algorithm. The ISODATA algorithm is similar in principle to the well-known $K$-means procedure [4] in the sense that cluster centers are iteratively determined by sample means. Unlike the latter algorithm, however, ISODATA represents a fairly comprehensive set of additional heuristic procedures which have been incorporated into a interactive scheme. Before executing the algorithm, it is necessary to specify a set, $N_c$, of initial cluster centers, $Z_1$, $Z_2$,...,$Z_{N_c}$. This set, which need not be equal in number to the desired cluster centers, can be performed by selecting samples from the given data set.
For a set of N samples, which represent the feature vectors $X_1$, $X_2$,...,$X_N$, ISODATA consists of the following principle steps:
**Step 1**. Specify the following process parameters:

$K$=desired number of cluster centers,
$\theta_N$ = minimum number of samples in a cluster,
$\theta_s$=standard deviation parameter,
$\theta_c$=lumping parameter,
$L$=maximum number of pairs of cluster which can be lumped,
$NUM$=number of iterartions.

**Step 2.** Distribute the N samples among the present cluster centers using the relation

$$X \in S_j \qquad if \ ||X - Z_j|| < ||X - Z_i|| \qquad i = 1, 2, ....N_c \ ; \quad i \neq j, \quad \forall X \in N. \tag{5}$$

In this notation, $S_j$ represents the subset of samples assigned to cluster $Z_j$.

**Step 3.** Discard sample subset $S_j$ if $N_j < \theta_n$ and reduce $N_c$ by 1.

**Step 4.** Update cluster center $Z_j$, $j$ = 1,2,...,$N_c$, by setting it equal to the sample mean of its corresponding set $S_j$, that is,

$$Z_j = \frac{1}{N_j} \sum_{X \in S_j} X \qquad j = 1, 2, ..., N_c. \tag{6}$$

where $N_j$ is the number of samples in $S_j$.

**Step 5.** Compute the average distance $D_j$ of samples in cluster domain $S_j$ from their corresponding cluster center, using the relation

$$D_j = \frac{1}{N_j} \sum_{X \in S_j} ||X - Z_j|| \qquad j = 1, 2, ..., N_c. \tag{7}$$

**Step 6.** Compute the overall average distance $D$ of the samples from their respective cluster centers

$$D_j = \frac{1}{N} \sum_{j=1}^{N_c} N_j D_j \tag{8}$$

**Step 7.**

If this is the last iteration, set $\theta_c$=0 and goto Step10.
If $N_c \leq k/2$, go to Step 8.
If $N_c \geq 2K$, go to Step 10.
Otherwise, continue.

**Step 8.** Find the standard deviation vector $\sigma_j = (\sigma_{1j}, \sigma_{2j}, ..., \sigma_{nj})$ for each sample subset as follows

$$\sigma_{ij} = \sqrt{\frac{1}{N_j} \sum_{X \in S_j} (x_{ik} - z_{ij})^2} \qquad i = 1, 2, ..., n; \quad j = 1, 2, ....N_c. \tag{9}$$

where n is the sample dimensionality, $x_{ik}$ is the $i$th component of the $k$th sample in $S_j$, and $z_{ij}$ is the $i$th component of $Z_j$. Each component of $\sigma_j$ represents the standard deviation of the samples in $S_j$ along the principal coordinate axis.

**Step 9.** Find the maximum component $\sigma_{jmax}$ for each $\sigma_j$, $j$ =1,2,...$N_c$. If for any $\sigma_{jmax}$, we have $\sigma_{jmax} > \theta_s$, and

$D_j > D$ and $N_j > 2(\theta_N + 1)$ or,
$N_c \leq K/2$

then split $Z_j$ into 2 new cluster centers $Z_j{}^+$ and $Z_j{}^-$, delete $Z_j$, and increase $N_c$ by 1. Cluster $Z_j{}^+$ is formed by adding a given quantity $\gamma_j$ to the component of $Z_j$ which corresponds to the maximum component of $\sigma_j$; $Z_j{}^-$ is formed by subtracting $\gamma_j$ from the same component of $Z_j$. One way of specifying $\gamma_j$ is to let it equal to some fraction of $\sigma_{jmax}$. If splitting took place in this step, go to Step 2; otherwise continue.

**Step 10.** Compute the pairwise distances $D_{ij}$ between all cluster centers as follows

$$D_{ij} = ||Z_i - Z_j|| \qquad i = 1, 2, ..., N_c - 1; \quad j = i + 1, ..., N_c. \tag{10}$$

Compare the distances $D_{ij}$ against the parameter $\theta_c$. Arrange the $L$ smallest distances which are less than $\theta_c$ in ascending order $[D_{i_1 j_1}, D_{i_2 j_2}, ..., D_{i_L j_L}]$ where L is the maximum number of pairs of cluster centers which can be lumped together. The lumping process is discussed in the next step.

**Step 11.** With each distance $D_{i_l j_l}$, there is associated a pair of cluster centers $Z_{i_l}$ and $Z_{j_l}$. Starting with the smallest of these distances, perform a pairwise lumping operation according to the following rule: For $l$=1,2,...,$L$, if neither $Z_{i_l}$ nor $Z_{j_l}$ has been used in lumping in this iteration, merge these two cluster centers into $Z_l{}^*$ as follows:

$$Z_l{}^* = \frac{1}{N_{i_l} + N_{j_l}} [N_{i_l}(Z_{i_l}) + N_{j_l}(Z_{j_l})] \tag{11}$$

Delete $Z_{i_l}$ and $Z_{j_l}$ and reduce $N_c$ by 1

**Step 12.** If it is the last iteration, the algorithm terminates. Otherwise go to Step 2.

Results for the application of the ISODATA algorithm to segment a brain CT volume are shown in Fig. 2.

## 4.2 Classification Using the Hopfield Network

Hopfield network for the optimization application consists of many interconnected neuron elements [5]. The network minimizes an energy function of the form

$$E = \sum_{k=1}^{M}\sum_{l=1}^{M} T_{kl}V_kV_l - \sum_{k=1}^{M} B_kV_k \tag{12}$$

where $M$ is the number of neurons, $V_k$ is the output of the $k$th neuron, $B_k$ is the bias term, and $T_{kl}$ is the interconnection weight between $k$th and $l$th neuron. The minimization is achived by solving a set of motion equations satisfying

$$\frac{\partial U_i}{\partial t} = -\frac{\partial E}{\partial V_i} \tag{13}$$

where $U_i$ is the input of the $i$th neuron. If the input-output function of the neuron is monotonically increasing, the energy continuously reduces as a function of time and the system converges to a local minimum.

In this section we will follow the ideas proposed in [7]. The neural architecture consists of $M \times K$ neurons with each column representing a class, and each row representing a voxel. Our objective here is to assign $M$ voxels of $P$ features among $K$ classes, such that the assignment of voxels minimizes a criterion function. The criterion function can be based on a metric measure between a voxel to class. The generalized distance measure between the $k$th pixel and the centroid of class $l$ is given by

$$R_{kl} = ||X_k - Z_l||_{A_l^{-1}} \tag{14}$$

where $X_k$ is the P-dimensional feature vector of the $k$th voxel, $A_l$ is a $P \times P$ positive definite weighting matrix, $Z_l$ is the $P$-dimensional centroid of class $l$, and $||X||_A = X^TAX$. If $A_l$ is an idendity matrix, the distance measure is Euclidean and the resulting clusters will be restricted to hyperspherical distributions [4]. If $A_l = \Sigma_l$, the covariance matrix of the class $l$, the distance measure is Mahalanobis, which allows for hyperellipsoidal cluster distributions. The overall energy function $E$ to be minimized is given by

$$E = \frac{1}{2}\sum_{k=1}^{M}\sum_{l=1}^{M} R_{kl}{V_k}^2 \tag{15}$$

By applying the relation in (13) we get a set of equations for the neuronal dynamics given by

$$\frac{\partial U_{kl}}{\partial t} = -R_{kl}V_{kl} \tag{16}$$

where $U_{kl}$ and $V_{kl}$ are the input and output of the $kl$th neuron, respectively. The input-output function for the $k$th row to assign a label n to the (k)th voxel is given by

$$\begin{aligned} V_{kn}(t+1) &= 1 \quad if \;\; U_{kn} = MAX[U_{kl};\ \forall\, l] \\ V_{kn}(t+1) &= 0 \quad otherwise \end{aligned} \tag{17}$$

where t is the iteration step. The algorithm for clustering can be summurized as follows

1. Initialize the input of neurons to random values (which amounts to a random assignment of M voxels to K classes)

2. Apply the input-output relation given in (17) to obtain the new output values for each neuron, establishing the assignment of voxels to classes

3. Compute the centroid (or class mean) $Z_l$ and the covariance matrix $\Sigma_l$ for class l as follows:

$$Z_l = \frac{1}{n_l}\sum_{k=1}^{M} X_kV_{kl} \tag{18}$$

and,

$$\Sigma_l = \frac{1}{(det[\Sigma_l])^{1/P}(n_l-1)}\sum_{k=1}^{M} V_{kl}(X_k - Z_l)(X_k - Z_l)^T \tag{19}$$

where $n_l$ is the number of voxels in class $l$.

4. Solve the set of differential equations in (16) using Euler's approximation to update the inputs of each neuron

$$U_{kl}(t+1) = U_{kl}(t) \; + \; dU_{kl}/dt \tag{20}$$

5. If there is sufficient activity in the input of each neuron, then go to step 2, else terminate.

Results for the application of the Hopfield optimization network to classify feature vectors are shown in Fig. 2.

## 4.3 Classification Using the Kohonen Network

Since regions are not necessarily linearly separable we might not be able to separate them using conventional pattern recognition methods. For this reason we applied an unsupervised classification strategy based on Kohonen network [8]. The Kohonen's self-organizing feature map is adopted to classify the feature pattern $X$ into several categories. In the learning phase, when the vector $X$ is presented to this one-dimensional feature map, as shown in Fig.1., only a single neuron which responds "best" to that input and its neighbors are selected to adjust their weights. This winner $N_k$ is selected based on the distance between the input and the weight of the neuron.

The weight vectors of the winner and all the neurons $N_q$ that lie in a neighborhood $R_c$ of the winner are pulled into the direction of the input pattern. This gives the Kohonen learning rule

$$W_k{}^{t_k+1} = W_k{}^{t_k} + \alpha^{t_k}(X - W_k{}^{t_k}) \tag{21}$$

and

$$W_q{}^{t_q+1} = W_q{}^{t_q} + (\alpha^{t_q})^2(X - W_q{}^{t_q}) \qquad \forall N_q \in R_c \tag{22}$$

where $W_k{}^{t_k}$ is the weight of the winning neuron $N_k$ before updating, and $W_k{}^{t_k+1}$ is the weight after updating, $\alpha^{t_k}$ (a number between 0 and 1) is the learning rate at time $t_k$.
This method was applied to segment a CT volume and the results are shown in Fig. 2.

# 5 Results

In this paper, we have presented a new technique to segment an image volume by assigning a feature pattern to each voxel, consisting of a scaled family of differential geometrical invariant features. The invariant feature pattern is assigned to a specific region using unsupervised clustering. The clustering was performed using a pattern recognition approach (ISODATA algorithm), and a neural network approach (Hopfield and Kohonen networks). Results for the application of these 3 methods are shown in Fig. 2. The classification rate is presented in Table 1. These results illustrate the potential of using neural network to segment the image volume. The integration of the feature extraction step, the neural network classifier, and the connected component algorithm provides an efficient mechanism to learn about the gray level features of the image to the extent where homogeneous regions can be grown with a reasonable amount of smoothing. The overall system proved to be robust, easy to implement, and able to produce accurate regions of interest. Current interests are focused on obtaining a robust 3D surface reconstruction of individual brain structures for accurate volumetric measurements and registration to quantify TBI recovery.

# References

[1] R. Bajcsy, "Multiresolution elastic matching," *Comput. Vision, Graph., and Image Proc.*, vol. 46, no. 1, pp. 1-21 1989.

[2] S. Dellepiane et al., "Model Generartion and model matchingof real images by fuzzy approach," *Pattern Recognition*, vol. 25, no.2, pp. 115-137, 1992.

[3] D. N. Kennedy et al., "Automatic segmentation and volumetric calculations in MRI," *IEEE Trans. Med. Imag.*, vol. 8, no. 1, pp. 1-7 1989.

[4] R. O. Duda and P. O. Hart, "Pattern Classification and scene Analysis," New York: Wiley, 1973.

[5] Jacek M. Zurada, "Introduction to artificial neural systems," West Publishing Company, 1992.

[6] L. Florack and B. M. Romeny, "Scale and the differential structure of images," *Image and Vision Computing*, vol. 10, pp. 376-388, 1992.

[7] S. C. Amartur Y. Takefuji, "Optimization neural networks for the segmentation of MRI images," *IEEE Trans. Med. Imag.*, vol. 11, no. 2, pp. 215-220, 1992.

[8] T. Kohonen, "Self Organization and Associative Memory," Springer-Verlag., 1984.

| *CLASSIFIER* | *% CORRECT* |
|---|---|
| ISODATA | 89.3% |
| Hopfield Network | 93.5% |
| Kohonen Network | 96.5% |

Table 1: Results of the different classifiers to segment the CT volume

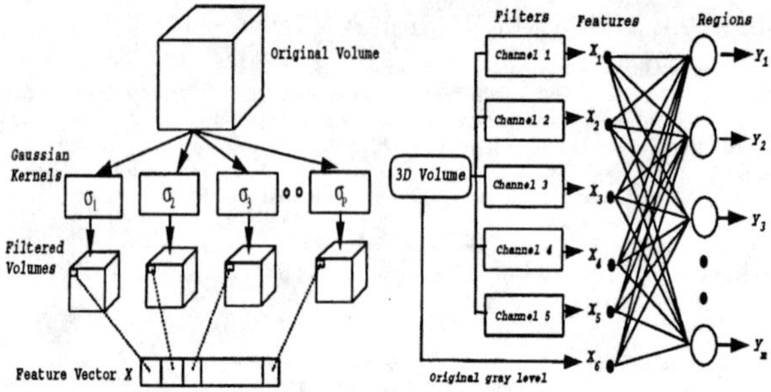


Figure 1: Feature extraction using differential geometrical invariant features, and Clustering using Kohonen feature map.

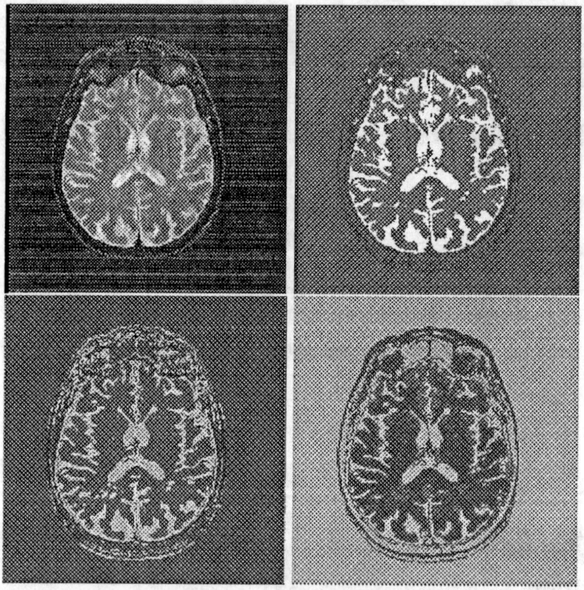

Figure 2: Results for the segmentation of a CT volume using feature extraction method and the ISODATA algorithm, the Hopfield network, and Kohonen feature map. The result shows more region definition in the case of Hopfield and Kohonen methods.

# NEURAL NET CLASSIFICATION OF REM SLEEP BASED ON SPECTRAL MEASURES AS COMPARED TO NONLINEAR MEASURES

**JÜRGEN FELL, JOACHIM RÖSCHKE**

**and MICHAEL GRÖZINGER**

**ABSTRACT:**
Online recognition of REM sleep is of great interest for sleep researchers, since this technique enables to study the effect of selective REM deprivation in depressive illness, recollection of dream contents, etc.. Recently, it was shown that a reliable online REM classification (89 % correctly classified time periods) can be achieved by implementation of a backpropagation network, which processes only one channel of EEG information, but no EOG or EMG data. The input information consists of the RMS values in 6 different frequency bands. For this study, we additionally implemented the nonlinear measures largest Lyapunov-exponent and correlation dimension, and the nonconventional stochastic measures spectral entropy and entropy of amplitudes. We compared the accuracy of neural net classification of REM sleep (epochs of 20,5 sec) of the original network implementing 6 spectral measures, with a network implementing only the 4 additional measures and a combined network using all 10 EEG measures. The main result is, that no improvement of REM classification could be achieved by inclusion of the nonlinear and the nonconventional stochastic EEG measures.

## INTRODUCTION

During recent years neural net classification has become an important tool for the extraction of different EEG characteristics. Artificial neural networks for example were utilized for the recognition of epileptiform EEG patterns (Gabor and Seyal 1992, Jando et al. 1993, Webber et al. 1994), for single trial analysis of event related potentials (Pfurtscheller et al. 1992), for the automatized classification of sleep stages (Mamelak et al. 1991, Roberts and Tarassenko 1992, Schaltenbrand et al. 1993), and for the discrimination between demented patients and healthy controls based on slow wave activity (Anderer et al. 1994). Grözinger and coworkers (1995) demonstrated that a reliable online classification of REM sleep (89 % correctly classified time periods) can be achieved by implementation of a backpropagation network, which processes only one channel of EEG information, but no EOG or EMG data. Hereby, RMS (root mean square) values from 6 different frequency bands served as input information.

Investigations based on so-called surrogate data tests indicate evidence for nonlinearity in EEG signals (see e.g. Rombouts et al. 1995, Pritchard et al. 1995, Fell et al. 1996). Therefore, it appears reasonable that network classification may be improved by inclusion of EEG measures based on nonlinear systems theory. Pritchard and coworkers (1994) investigated the EEG-based neural net classification of Alzheimer's disease versus healthy controls. EEG measures, which served for data preprocessing, were power estimates in the different spectral bands and the nonlinear measures correlation dimension and saturation of the estimated dimension with increasing embedding dimension. It was found that addition of the nonlinear measures enhanced the classification accuracy of Alzheimer's disease versus controls.

In the present study, we addressed the question, whether neural net recognition of REM sleep can be improved by inclusion of nonlinear and nonconventional stochastic measures. We selected the two nonlinear measures largest Lyapunov-exponent L1, correlation dimension D2, and moreover spectral entropy and entropy of amplitudes as additional input variables. Concerning general discrimination of sleep stages (incorporating all stages) these measures previously were shown to be highly efficient as revealed by discriminant analyses (Fell et al. 1996). In short, the correlation dimension D2 yields a lower boundary for the degrees of freedom or in other words the complexity of a system, whereas the largest Lyapunov-exponent L1 quantifies the so-called sensitive dependence on initial conditions or 'chaoticity' (e.g. Grassberger et al. 1991, Röschke 1992). Spectral entropy and entropy of amplitudes estimate the degree of disorder incorporated in the frequency respectively amplitude distribution of a signal. We compared the accuracy of a backpropagation network processing spectral amplitudes from 6 different frequency bands as input, with a network utilizing information from the 4 additional measures and a combined network using information from all 10 EEG measures together.

## MATERIALS AND METHODS

### Sleep EEG data

13 volunteers, who were in self reported good health with regular sleep wake patterns, participated in the investigation. Following an adaptation night polysomnographic recordings were performed for five subjects in two separate nights each (group1, night1 and night2) and for the eight additional subjects (group2) in one night each. Sleep EEG was recorded from 11.00 PM until 7.00 AM next day. Surface electrodes were placed on the skull ($P_Z$, $C_Z$, $C_3$ and $C_4$; 10-20 system) and mastoid to record electroencephalic activity, at the outer canthi on the left and right eye to record eye movements and on the chin to record submental electromyographic activity. The analog EEG data (0.53 Hz high pass filter; 45 Hz low pass filter, 48 dB/octave) from position $C_Z$ were digitized by a 12 Bit analog-digital-converter with a sampling frequency of $f_S$ = 100 Hz. All night EEG was divided into 1440 consecutive segments each consisting of 2048 data points. Visual sleep stage classification of the 1440 EEG epochs was performed according to Rechtschaffen and Kales criteria (1968) by two independent judges.

### Network implementation

The following network architectures were implemented:
network 1 (spectral measures): 6 input neurons, 4 hidden neurons, 1 output neuron; network 2 (additional measures): 4 input neurons, 3 hidden neurons, 1 output

neuron; network 3 (combined measures): 10 input neurons, 6 hidden neurons, 1 output neuron.

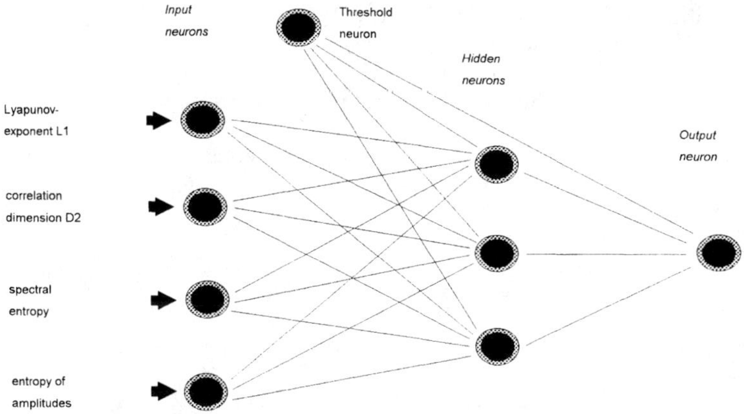


**Figure 1: Illustration of the architecture of network 2 (additional measures).**

Adjacent layers were totally interconnected. Figure 1 as an example illustrates the architecture of network 2. For all networks input neurons were linear, hidden and output neurons had a sigmoid output function and were connected to a threshold neuron. A generalized backpropagation algorithm was used as learning rule for computer simulation. The software program was provided by the Neural Networks esearch Group at the University of Kassel (Klöppel 1990).

Networks were trained using the pooled datasets of group 1 (night 1). Each of the 1440 EEG epochs was assigned $t_i^* = 1$ in case of REM sleep (according to visual scoring) and $t_i^* = 0$ in case of non-REM sleep. The accuracy of neural net classification with respect to visual scoring was quantified by the quadratic mean error:

$$QME = \Sigma_i (t_i - t_i^*)^2, \quad t_i\text{: network output levels}$$

Starting from randomly chosen synaptic weights, training was stopped when the error improved less than $10^{-4}$ within 50 training loops, which occured in a maximum of 2500 loops. Each learning cycle was repeated from at least three different starting points to avoid local minima. After the training phase, networks were applied to the datasets of group 1 (night 2) and group 2.

**Input measures**

Spectral amplitude measures

Each EEG segment was digitally filtered into 6 different frequency bands by FFT, rectangular windowing and backtransforming. The bands were chosen as: 0.5 - 3.5

Hz, 3.5 - 7.5 Hz, 7.5 - 15. Hz, 15. - 35. Hz, 35. - 45. Hz and 0.5 - 45. Hz. Root mean square (RMS) values were calculated for the filtered epochs, revealing the spectral measures for the 6 different frequency bands.

Spectral entropy

The evaluation of the spectral entropy (SEN) was suggested by Inouye and coworkers (1991). The spectral entropy is given by

$$SEN = -\sum S(\omega)*\log S(\omega), \quad \text{with } S(\omega)\text{: spectral power}$$

If the power is concentrated on a single frequency, spectral entropy reaches a minimum, whereas for a broad frequency distribution high values for the spectral entropy are obtained.

Entropy of amplitudes

The entropy of amplitudes (ENA) resulted from the normalized amplitude distribution $P_i$ of the time series.

$$ENA = -\sum P_i * \log P_i$$

For the calculation of amplitude distributions the amplitude domain was partitioned into 100 equidistant boxes covering the interval between maximum and minimum amplitude of the respective EEG segment.

Nonlinear measures

For the estimation of nonlinear measures we applied the reconstruction procedure proposed by Takens (1981) to each EEG segment by embedding the signal into a 20-dimensional (correlation dimension) respectively 10-dimensional (Lyapunov-exponent) phase space. For the time increment $\tau$ we used the first zero crossing of the autocorrelation function. If no zero crossing of the autocorrelation function was found within the first 300 samples, the first minimum of the autocorrelation function is taken for time increment.

The correlation dimension D2 was estimated according to the proposal of Grassberger and Proccacia (1983a).

$$D2 = \lim_{R \to 0} \left( \ln C(R,20) / \ln R \right)$$

$$\text{where } C(R,20) := \sum_{i,j=1;i<j}^{N} \Theta\left(R - \| \mathbf{y}_i - \mathbf{y}_j \|\right),$$

$$\Theta(x) = 1 \text{ for } x > 0, \ \Theta(x) = 0 \text{ otherwise;}$$

$\mathbf{y}_i$ : state vectors in the embedding space;

C(R): correlation sums; R: euklidian distance

50 reference points and a R-axis domain of five decades were implemented. We applied a Theiler-correction of +/- the delay time (Theiler 1986). The value of D2 was automatically extracted from the averaged slopes of the curves (log C(R) vs. log(R)) within the stationarity region $C(R) \in [2.0, C_{max} - 1.0]$ in embedding dimension 20.

The principal Lyapunov-exponent L1 was computed applying a modified version of the Wolf-algorithm (Wolf et al. 1985). Essentially, the Wolf-algorithm computes the vector distance X of two nearby points and evolves its length for a certain propagation time. If the vector length X' between the two points becomes too large, a new reference point is chosen with the properties to minimize the replacement length and the orientation change. Now the two points are evolved again and so on. After m propagation steps the first positive Lyapunov-exponent results from

$$L1 = (1/(t_m - t_0)) \sum_{k=1}^{m} \log_2 ( X'(t_k) / X(t_{k-1}) )$$

We implemented a modified replacement procedure (Frank et al. 1990) and a randomized choice of propagation steps (Fell and Beckmann 1994). For theoretical background and further information on algorithms we refer to Fell and Röschke (1994).

## RESULTS

For all three datasets lowest quadratic mean errors could be reached by network1 (spectral measures) and network3 (combined measures), whereas QMEs for network2 (additional measures) were markedly increased (see figure 2). Two-tailed T-tests revealed no significant differences between quadratic mean errors of network1 and network3 for all three datasets. On the other hand, QMEs for network2 were significantly enhanced compared to network1 and network3 for group1/night1 ($p < 0.002$), group1/night2 ($p < 0.005$ respectively $p < 0.01$) and group2 ($p < 0.005$ respectively $p < 0.01$). Thus, no improvement of neural net based REM classification could be obtained by inclusion of the additional measures spectral entropy, entropy of amplitudes, correlation dimension D2 and largest Lyapunov-exponent L1.

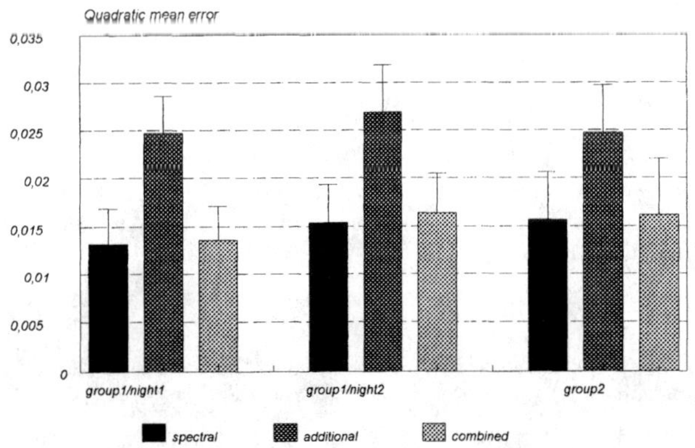


**Figure 2: Quadratic mean errors for the 3 different networks with the datasets of group1/night1, group1/night2 and group2 (average values and standard deviations are plotted).**

## DISCUSSION

Many facts point to the view that brain dynamics are essentially of a nonlinear nature. Nonlinear phenomena were experimentially proven on the neuronal level, and multiple feedback loops are known on each of the hierarchic levels of the CNS (e.g. Lopes da Silva 1991). Schiff and coworkers (1994) successfully applied a controlling chaos technique to pyramidal cells of the hippocampus, which is another strong argument in support of the nonlinear approch to brain dynamics. Moreover, it was shown that EEG signals exhibit nonlinear elements and thus can not be described completely by linear stochastic models (e.g. Rombouts et al. 1995, Pritchard et al. 1995, Fell et al. 1996). In this sense, the application of nonlinear methods to EEG analysis appears reasonable from a theoretical standpoint.

But still the question remains, whether on the practical side the application of nonlinear methods for the experimental situations under study yields information, which can not be obtained by application of conventional EEG measures (e.g. Röschke and Aldenhoff 1991,1992; Röschke et al. 1993,1994). Previous investigations revealed, that nonlinear EEG measures improve neural net classification of Alzheimer's disease (Pritchard et al. 1994), can yield a differentiation between high-IQ and low-IQ groups (Lutzenberger et al. 1992) and improve sleep classification with respect to all stages (Fell et al. 1996). Nevertheless, the present results show that inclusion of nonlinear and nonconventional stochastic EEG measures could not improve neural net based classification of REM sleep. We conclude, that nonlinear and also nonconventional EEG measures do not yield additional information per se. Different kinds of experimental situations respectively classes of EEG data may demand different strategies for analysis. For some kinds of EEG data conventional measures will suffice, whereas in other cases the application of nonlinear and nonconventional stochastic measures may be appropriate.

**ACKNOWLEDGEMENT**

This work was supported by Deutsche Forschungsgemeinschaft (Ro 809/13-1).

## REFERENCES

Anderer P., Saletu B., Klöppel B., Semlitsch H.V., Werner H. (1994). Discrimination between demented patients and normals based on topographic EEG slow wave activity: comparison between z statistics, discriminant analysis and artificial neural network classifiers. *Electroenceph. clin. Neurophysiol.* 91:108-117.

Armitage R., Roffwarg H.P. (1993). Distribution of period-analyzed delta activity during sleep. *Sleep* 16:762-763.

Fell J., Beckmann, P. (1994). Resonance like phenomena in Lyapunov-calculations from data reconstructed by the time-delay-method. *Phys. Lett. A* 190:172-176

Fell J., Röschke J. (1994). Nonlinear dynamical aspects of the human sleep EEG. *Intern. J. Neuroscience*, 76:109-129

Fell J., Röschke J., Mann K., Schäffner C. (1996). Discrimination of sleep stages: a comparison between spectral and nonlinear EEG measures. *Electroenceph. clin. Neurophysiol.* 98:401-410.

Fell J., Röschke J., Schäffner C. (1996). Surrogate data analysis of sleep EEG reveals evidence for nonlinearity. *Biol. Cybern.* 75:85-92.

Frank G.W., Lookman T., Nerenberg M.A.H., Essex C., Lemieux J., Blume W. (1990). Chaotic time series analyses of epileptic seizures. *Physica D* 46:427-438

Gabor A.J., Seyal M. (1992). Automated interictal EEG spike detection using artificial neural networks. *Electroenceph. clin. Neurophysiol.* 83:271-280.

Grassberger P., Procaccia I. (1983a). Measuring the strangeness of strange attractors. *Physica D* 9:189-208.

Grassberger P., Schreiber T., Schaffrath C. (1991). Nonlinear time sequence analysis. *Int. J. Bifurcation and Chaos* 1: 521-547.

Grözinger M., Röschke J., Klöppel B. (1995). Automatic recognition of rapid eye movement (REM) sleep by artificial neural networks. *J. Sleep Res.* 4:86-91.

Gurman G.M. (1994). Assessment of depth of general anesthesia. Observations on processed EEG and spectral edge frequency. *Int. J. Clin. Monit. Comput.* 11:185-189.

Inouye T., Shinosaki K., Sakamoto H., Toi S., Ukai S., Iyama A., Katsuda Y., Hirano M. (1991). Quantification of EEG irregularity by use of the entropy of the power spectrum. *Electroenceph. clin. Neurophysiol.* 79:204-210.

Jando G., Siegel R.M., Horvath Z., Buzsaki G. (1993). Pattern recognition of the electroencephalogram by artificial neural networks. *Electroenceph. clin. Neurophysiol.* 86:100-109.

Klöppel B. (1990). Generalized backpropagation. In: *Proceedings of the Interdisciplinary Symposium on Neural networks*, University of Kassel, pp. 187-212.

Long C.W., Shah N.K., Loughlin C., Spydell J., Bedford R.F. (1989) A comparison of EEG determinants of near-awakening from isoflurane and fentanyl anesthesia. Spectral edge, median power frequency, and delta ratio. *Anesth.-Analg.* 69:169-173.

Lopes da Silva F.H. (1991). Neural meachanisms underlying brain waves: from neural membranes to networks. *Electroenceph. clin. Neurophysiol.* 79:81-93.

Lutzenberger W., Birbaumer N., Flor H., Rockstroh B., Elbert T. (1992). Dimensional analysis of the human EEG and intelligence. *Neuroscience Lett.* 143:10-14.

Mamelak A.N., Quattrochi J.J., Hobson J.A. (1991). Automated staging of sleep in cats using neural networks. *Electroenceph. clin. Neurophysiol.* 79:52-61.

Pfurtscheller G., Flotzinger D., Mohl W., Peltoranta M. (1992). Prediction of the side of hand movements from single-trial multi-channel EEG data using neural networks. *Electroenceph. clin. Neurophysiol.* 82:313-315.

Pijn J.P., Van Neerven J., Noest A., Lopes da Silva F.H. (1991). Chaos or noise in EEG signals; dependence on state and brain site. *Electroenceph. clin. Neurophysiol.* 79:371-381.

Pritchard W.S., Duke D.W., Coburn K.L., Moore N.C., Tucker K.A., Jann M.W., Hostetler R.M. (1994). EEG-based, neural-net predictive classification of Alzheimer's disease versus contrel subjects is augmented by non-linear EEG measures. *Electroenceph. clin. Neurophysiol.* 91:118-130.

Pritchard W.S., Duke D.W., Krieble K.K. (1995). Dimensional analysis of resting human EEG II: Surrogate-data testing indicates nonlinearity but not low-dimensional chaos. *Psychophysiology* 32:486-491.

Rechtschaffen A., Kales A. (1968). A manual of standardized terminology, technics and scoring system for sleep stages of human subjects. Washington, DC: Public Health Service, NIH Publication No.204, US Government Printing Office

Roberts S., Tarassenko L. (1992). A new method of automated sleep quantification. *Med. Biol. Eng. Comput.* 30:509-517.

Rombouts S.A.R.B., Keunen R.W.M., Stam C.J. (1995). Investigation of nonlinear structure in multichannel EEG. *Phys. Lett. A.* 202:352-358.

Röschke J. (1992). Strange attractors, chaotic behaviour and informational aspects of sleep EEG data. *Neuropsychobiology* 25:172-176.

Röschke J., Aldenhoff J.B. (1991). The dimensionality of human's electroencephalogramm during sleep. *Biol. Cybern.* 64:307-313.

Röschke J., Aldenhoff J.B. (1992). A nonlinear approach to brain function: deterministic chaos and sleep EEG. *Sleep* 15:95-101.

Röschke J., Fell J., Beckmann P. (1993). The calculation of the first positive Lyapunov exponent in sleep EEG data. *Electroenceph. Clin. Neurophysiol.* 86:348-352.

Röschke J., Mann K., Fell J. (1994). Nonlinear EEG dynamics during sleep in depression and schizophrenia. *Intern. J. Neuroscience* 75:271-284.

Schaltenbrand N., Lengelle R., Macher J.P. (1993). Neural network model: application to automatic analysis of human sleep. *Comput. Biomed. Res.* 26:157-171.

Schiff S.J., Jerger K., Duong D.H., Chang T., Spano M.L., Ditto W.L. (1994) Controlling chaos in the brain. *Nature* 370:615-620.

Takens F. (1981). Detecting strange attractors in turbulence. *Lect. Notes Math.* 898: 366-381, Springer, Berlin

Theiler J. (1986). Spurious dimension from correlation algorithms applied to limited time-series data. *Phys. Rev. A* 34: 2427-2432

Webber W.R., Litt B., Wilson K., Lesser R.P. (1994). Practical detection of epileptiform discharges (Eds) in the EEG using an artificial neural network: a comparison of raw and parameterized EEG data. *Electroenceph. clin. Neurophysiol.* 91:194-204.

Wolf A., Swift J.B., Swinney H.L., Vastano J.A. (1985). Determining Lyapunov exponents from a time series. *Physica D* 16: 285-317.

# THERAPEUTIC REM SLEEP DEPRIVATION RENDERED POSSIBLE BY APPLICATION OF ARTIFICIAL NEURAL NETWORKS TO ONE CHANNEL EEG DATA.

M.GRÖZINGER, T.UHL, C.SCHÄFFNER AND J.RÖSCHKE
*Department of Psychiatry, University of Mainz,*
*Untere Zahlbacher Straße 8, 55131 Mainz, Germany*

*ABSTRACT*

**In clinical psychiatry sleep deprivation has well known positive impacts on the symptoms of depression. Mainly Rapid Eye Movement (REM) Sleep has been held responsible for the depressiogenic effects of sleep leading to the suggestion of selective REM sleep deprivation. For ANNIE'94 we already described an automatic online algorithm to detect REM sleep from single channel EEG data. In the meantime we were able to improve this system and to collect first experiences with healthy volunteers and depressive patients.**

**In order to be compatible with the manual evaluation the continous EEG recording was analysed in time periods of 20 seconds. For data preprocessing the signal of every time period was split up in segments of 5 respectively 2.5 seconds and digitally filtered in 7 frequency bands (0.5-45, 0.5-3.5, 3.5-7.5, 7.5-11, 11-15 , 15-35 and 35-45 Hz). Afterwards the RMS values were calculated. A resulting vector of 48 real numbers served as input to an artificial neural network. 6 hidden neurons and 1 output neuron together with all possible feedforward connections of adjacent layers completed the architecture.**

**In 2 independent groups of 5 and 8 healthy volunteers this system was able to correctly classify (REM/non-REM) more than 90% of the time intervals not belonging to the training set. As compared to the earlier system the percentage of misclassified time periods could be reduced by about 20%.**

## INTRODUCTION

**Sleep deprivation as applied in many variations in clinical psychiatry has well known positive impacts on the symptoms of depression (Kuhs and Tölle 1991, Giedke et al 1992). Because the physiological pathways of this therapy are not yet identified, different theories offer diverse explanations (Beersma and van der Hoofdakker 1992). Since the seventies Rapid Eye Movement Sleep (REM) was suspected of having depressiogenic effects (Vogel et al. 1975). This hypothesis was leading to the suggestion of selective REM sleep deprivation, which was first studied by Vogel et al. 1980. At that time the EEG had to be evaluated online by humans in order to wake up the subjects at the beginning**

of REM sleep periods. Considering the fact that these events may occur as frequently as 80 times a night under the condition of REM sleep deprivation the task of detecting REM sleep today might rather be taken care of by a computer.

For ANNIE'94 we already described an automatic online algorithm to detect REM sleep using an artificial neural network on the basis of preprocessed single channel EEG data (Grözinger and Röschke 1994). We purposely did not supply EOG and EMG information because we believe this information to be redundant (Röschke and Aldenhoff 1991). This approach showed to be superior to the more classical method of nonparametric discriminant analysis (Grözinger et al. 1994). In healthy subjects REM and non-REM sleep could be identified correctly in close to 90% of the time intervals, each consisting of a 20 second time period of continuous EEG activity (Grözinger et al. 1995a). In the meantime we were able to further improve the system and to collect first online experiences with healthy volunteers and depressive patients.

## MATERIAL AND METHODS

The continous monopolar EEG recording Cz/A1 of each night was analysed in 1440 consecutive time periods, each consisting of 2048 data points sampled with a frequency of fs=100 Hz and digitized with a precision of 12 bit. To avoid distortions at the end-points the resulting periods of about 20 seconds were symmetrically expanded to 30 seconds by inclusion of adjacent signal passages before Fourier-transformation. Digital filtering in 7 frequency bands (0.5-45, 0.5-3.5, 3.5-7.5, 7.5-11, 11-15 , 15-35 and 35-45 Hz) was performed by deletion of the corresponding coefficients in the frequency domain and retransformation. Afterwards the RMS values of the 7 filtered signals were calculated along the 4 consecutive segments of 5.12 seconds (for frequency bands 0.5-45 and 0.5-3.5 Hz ) and the 8 segments of 2.56 seconds (for the remaining frequency bands) of the original time period.

The resulting vector of 48 real numbers served as input to an artificial neural network. 6 hidden neurons and 1 output neuron together with all possible feedforward connections between adjacent layers completed the architecture. The input neurons were linear, the hidden and the output neurons had a sigmoid output function and were connected to a threshold neuron leading to a total of 301 synapses. A generalized backpropagation algorithm was used as the learning rule for computer simulation minimizing the quadratic mean error (QME) between the network output and the desired output during the training process (Klöppel 1990).

Additionally the recordings were scored visually by two independent judges according to Rechtschaffen and Kales using EEG, EOG and EMG information (Rechtschaffen and Kales 1968). This provided the sleep profile and the indicator function of REM sleep, which served as the desired output for the training data and as "gold standard" in respect to the network output during working mode.

The training data set was pooled from first night recordings of 5 healthy subjects, while the corresponding second nights served to schedule the end of the training process. Polysomnographic recordings from another group of 8 healthy subjects were used offline to test the performance of the system. In working mode a time period was assigned "REM" if the network output exceeded a cutoff level, which was set to 0.5 or determined individually for every subject during the adaptation night.

To study selective REM sleep deprivation as a treatment for depressive patients an adaptation night is followed by 10 consecutive nights of REM sleep suppression by automatic alarms of the system as described above. Every time REM sleep is detected for the first time the lights are turned on and music of a basic sound pressure level is played for 30 seconds. The patient is allowed to get back to sleep afterwards. In case this does not cause the patient to get out of stage REM or is directly falling back to REM sleep after a short arousel the loudness of the music is increased in 5 successive steps. After 5 minutes of non-REM sleep there is always a reset of the sound pressure to the basic level. The 10 day procedure is accompanied by different ratings of the psychopathological state.

## RESULTS

In both groups of 5 (second nights) and 8 healthy subjects the offline tests revealed that the system was able to correctly classify more than 90% of the time intervals not belonging to the training set. As compared to the earlier system the percentage of misclassified time periods could be reduced by about 20%.

So far 5 patients suffering from depression (Major Depressive Episode according to DSM III R) were treated with a total of 30 nights of REM deprivation and 2 completed the protocol. They both showed a clear clinical improvement of the depressive symptoms. Unfortunately the rate of patients denying the procedure or dropping out was relatively high.

No REM periods were missed by the system yet, although in some rare cases it took up to 3 minutes to recognize the REM onset. Mostly the delay was about 30 seconds. On the other hand there were definitely some incorrect alarms out of sleep stage I as could be verified by a high EMG level. Under certain circumstances a judgement about the accuracy of the networks decission is not possible on principle because of the EEG changes caused by the alarm. Namely if no typical EOG features are prominent and the EMG level is close to zero stage REM and I are only separable by an ongoing undisturbed EEG (3 minute rule according to Rechtschaffen and Kales).

As an example figure 1 shows the network output and the corresponding network decission indicating REM sleep during the $6^{th}$ night of REM deprivation of a depressive patient. The dotted line shows the cutoff level of the system for turning on the alarm. Figure 2 shows the number of awakenings and the distribution of sound pressure levels in the course of the 10 nights for the same patient. Because of the high number of about 80 alarms during the $6^{th}$ night and the short unevitable delay between each REM onset and the awakening the patient got a total of about 30 minutes of REM sleep. Fortunately this was an extraordinary bad case in terms of the amount of

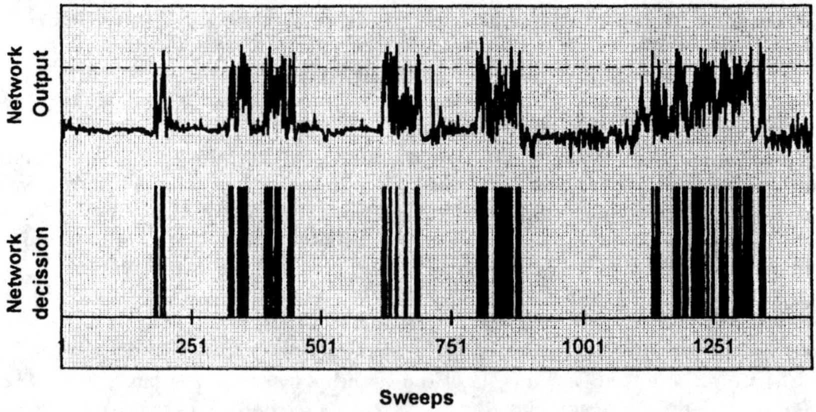


Figure 1: Network output and network decission indicating REM sleep during a night of REM deprivation. The dotted line shows the cutoff level for the waking mechanism.

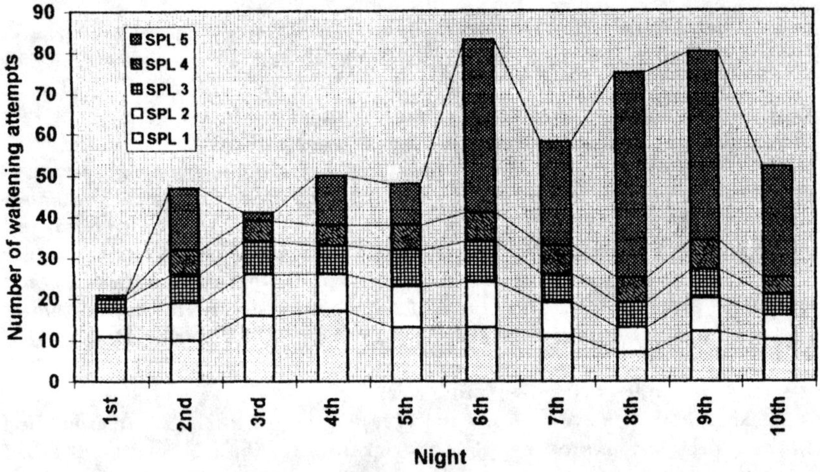


Figure 2: Number of wakening attempts with different sound pressure levels in 10 consecutive nights.

remaining REM sleep. As a preliminary estimate we might be able to take away 70% of the normal REM sleep duration.

An interesting observation to us was the fact that the wakenings took place in blocks meaning that during certain sleep periods the patients got back to REM sleep shortly after the arousel. During other sleep periods there was no effort at all to get into sleep stage REM. This behavior to us seemed to become even more distinct in the course of the 10 day process and might indicate a strengthening of the REM/non-REM oszillator.

## DISCUSSION

As already stated in our contribution for ANNIE'94 we are confident that ANNs in combination with an adequate preprocessing of the EEG data can be used online to recognize REM sleep and are most likely able to detect other sleep stages, too. This is in accordance with other authors (Schaltenbrand et al. 1996). This opens a whole field of clinical applications as for example selective REM sleep deprivation.

As already stated for a very similar system with 56 input neurons (Grözinger et al. 1995b), the accuracy of the network procedure on EEGs of healthy subjects can be improved by considering a smaller time base of about 2.5 seconds instead of 20 seconds. It has still to be examined if this improvement can be reproduced for patients under online conditions

Because some effort has to be put in motivating patients to participate in this procedure, the clinical impact can not be judged before a group of controls will be available receiving the same amount of attention.

Although our experience is still small it already becomes clear, that REM sleep deprivation is physiologically possible meaning that a temporary deficit of REM sleep is not causing insurmountable medical problems. This is confirmed by other studies and is different from slow wave sleep. On the other hand the effort of the physiological system to reach sleep stage REM (REM pressure) is increasing in the course of the procedure, which leads to an increasing number of awakenings and an also increasing frequency of high sound pressure levels of the alarms. It might be suspected that taking away an even higher amount of REM sleep (for example more than 95%) would have more severe medical impacts.

A rather interesting hypothesis of a stabilization of the REM/non-REM oszillator by this kind of therapy was proposed by Vogel et al. 1980. Our few data fit into this picture.

## REFERENCES

Beersma,D. and van der Hoofdakker,R., (1992). Can non-REM sleep be depressogenic? Journal of Affective Disorders 24: 101-108.

Giedke,H., Geilenkirchen,R., Hauser, M., (1992). The timing of partial sleep deprivation in depression. Journal of Affective Disorders 25: 117-128.

Grözinger,M. and Röschke,J., (1994). Artificial Neural Networks in the Online Evaluation of Sleep EEG Data: A Useful Tool for Selective REM Sleep Deprivation in Depressive

Patients. In: Dagli,C. et al. (Eds.): Intelligent Engineering Systems through Artificial Neural Networks, Vol 4: 753-758.

Grözinger,M., Freisleben,B., Röschke,J. (1994). Comparison of a backpropagation network and a nonparametric discriminant analysis in the evaluation of sleep EEG data. In: World Congress on Neural Networks - San Diego: I 462-466, 1994.

Grözinger,M., Röschke,J., Klöppel,B., (1995a). Automatic recognition of Rapid Eye Movement (REM) sleep by artificial neural networks. J.Sleep Res. 4: 86-91.

Grözinger,M., Röschke,J., Frank,C., (1995b). Recognition of REM sleep with artificial neural networks based on preprocessed periods of 2.5 seconds of EEG activity. Soc. Neuroscience Abstr. 21: 1237.

Kuhs,H. and Tölle,R. (1991). Sleep deprivation therapy. Biological Psychiatry 29:1129-1148.

Klöppel, B. (1990). Generalized backpropagating. In: Ewert, J.-P. and Werner, H. (Eds.): Models of Brain Function & Artificial Neuronal Nets: Proceedings of an Interdisciplinary Symposium on Neural Networks, May 28-29, University of Kassel, Germany 187-212.

Rechtschaffen,A. and Kales,A. (1968). A manual of standardized terminology, technics and scoring system for sleep stages of human subjects. Public Health Service. NIH publication 204 , Washington D.C.; US Government, Printing Office.

Röschke, J., Aldenhoff, J.B., (1991). Excitability and susceptibility of the brain's electrical activity during sleep. An analysis of late components of AEP's and VEP's. International Journal of Neuroscience 56: 255-272.

Schaltenbrand,N., Lengelle,R., Toussaint,M. et al. (1996). Sleep stage scoring using the neural net model: Comparison between visual and automatic analysis in normal subjects and patients. Sleep 19(1):26-35.

Vogel,G.W., Thurmond,A.J., Gibbons,P. et al. (1975). REM sleep reduction effects on depression syndroms. Arch. Gen. Psychiatry 32: 765-777.

Vogel,G.W., Vogel,F., McAbee,R.S., Thurmond,A.J. (1980). Improvement of depression by REM sleep deprivation: New findings and theory. Archives of General Psychiatry 37: 247-253.

# Mental Workload Classification Using A Backpropagation Neural Network

**CHRIS A. RUSSELL AND GLENN F. WILSON**
*Fitts Human Engineering Division, Crew Systems Directorate*
*Armstrong Laboratory, Wright-Patterson AFB, OH 45433-7022.*

**CORRINA T. MONETT**
*Logicon Technical Services, Inc.*
*P.O. Box 317258, Dayton, OH 45431-7258.*

***ABSTRACT:***
In this study, mental workload was classified using a backpropagation feedforward neural network with momentum and adaptive learning rates. Multiple channels of eye movement corrected one second FFTs calculated from continuous EEG and peripheral features (heart rate, eye blink and respiration rate) were used as input features to classify two levels of mental workload in a simulated aircraft landing study. Mean classification accuracies of eighty-three percent are reported when differentiating two levels of mental workload during level flight and landing for five pilots.

Electroencephalographic data has been used to classify single task sets, such as a resting baseline and a computer generated math task. These single task classifications have been implemented using numerous classification algorithms, including artificial neural networks. The classification of mental workload in complex environments can be instrumental in the design and modification of new and existing crewstations in order to decrease the crewmember's workload while increasing system efficiency.

## INTRODUCTION

Classification of operator mental workload has numerous applications in the fields of human factors engineering, training, testing and evaluation. For example, knowledge of a pilot's state in an advanced fighter aircraft can increase system efficiency by utilizing this information as feedback to an adaptive control system. The aircraft avionics can assume low priority tasks during periods of operator overload, thereby permitting mental resources to be focused on the highest priority tasks. Accurate and reliable assessment of operator state is the key to successful implementation of adaptive automation. Several approaches have been applied to this problem, none of which have achieved the required accuracy or reliability to permit implementation. Neural networks have several potential advantages that make them attractive for use as classifiers of operator state for adaptive assistance at the workstation. They are adaptive, nonlinear, parallel and they have the ability to generalize.

Several researchers have used neural networks to evaluate electrical brain activity in the study of cognitive workload as well as clinical psychology. Anderson, Devulapalli and Stolz (1995) used neural networks with autoregressive models of the power spectral density of alpha band activity to differentiate between conditions of resting baseline and mental arithmetic with an accuracy of 75%. Klöppel (1994) discriminated between depressive, psychotic and normal individuals using EEG input to a neural network with an overall accuracy of 80%. Most studies of mental workload classification have focused on discriminating between single tasks; however, a more realistic and challenging problem is classifying multitask conditions, such as aircraft flight segments or air traffic control procedures.

Since most real world situations involve simultaneously performing multiple tasks, multiple measures of operator state should be considered (Wilson & Eggemeier, 1991). Individual psychophysiological measures provide relevant information concerning specific aspects of the mental demands placed upon operators. However, combining these measures can provide a more complete picture of the complex mental demands faced by the operator. For example, in a simulated air traffic control study, Brookings, Wilson and Swain (1996) showed differences between mental workload conditions in a multitask situation in both central (EEG) and peripheral (heart rate, eye blinks and respiration) nervous system data. Hence the current study utilizes central and peripheral nervous system measures to classify two segments of simulated flight.

To meet the goal of accurate classification of complex cognitive tasks, a feedforward backpropagation neural network with an adaptive learning rate and momentum is suggested. The adaptive learning rate combined with momentum increases the learning speed and accuracy of the backpropagation neural network (Widrow and Lehr, 1990). The purpose of this study was to utilize this feedforward neural network to classify subject workload in a simulated flight task using both central and peripheral nervous system data as inputs.

## METHODS

### Subjects

Six licensed, general aviation pilots with varying levels of experience served as subjects. Their flying hours ranged from 90 to 1300 (mean 359) and ages ranged from 30 to 48 (mean 35). One subject's data was incomplete and was not used in the analysis.

### Simulator Tasks

In the simulated landing scenario, pilots flew straight and level in clouds for approximately 20 seconds. When they broke out of the clouds, the aircraft was centered at the runway, or positioned either 1,064 or 2,128 feet to the left or right of the runway centerline. The landing segment required approximately 30 seconds to complete. The pilots flew with either a haptically augmented or a normal stick in calm or turbulent conditions. The five runway offsets were flown for each joystick/turbulence combination for a total of twenty trials per session. Each subject completed four sessions in two days. For this study, workload classification was between level flight and the landing segment. The level flight segment was presumed to require a low mental demand while landing was presumed to require a high level of cognitive activity.

### Data Collection and Analysis

Psychophysiological data were collected in a high fidelity simulator consisting of an F-16 cockpit shell with a high resolution, wide field of view (240 by 120 degrees) out-the-window display, six LCD head-down displays, a projected head-up display, and an F-16 Block C throttle and sidestick controller (Brickman et. al., 1996 for a complete description).

Seven channels of EEG data were collected from electrode sites located on the frontal, parietal, central and temporal regions of the head (FP1, FZ, C3, C4, P3, P4, and T6 based on the International 10-20 system) using a Workload Assessment Monitor (Wilson, 1994). One mastoid served as reference. The EEG data were amplified ×60,000 using Coulbourn Instruments bioamplifiers (model S75-01) with filter cutoff settings at 1.0 and 40.0 Hz and sampled at 128 Hz. A 64 point, eyeblink corrected, 1 Hz resolution FFT was calculated every second and saved to disk. The peripheral measures were sampled at 500 Hz and the heart interbeat interval, respiration interbreath interval, minimum and maximum breath amplitudes, and eye-blink intervals, amplitudes and durations were calculated on-line and saved to disk.

The neural networks were implemented on a Pentium P-90 processor with 16 Mbytes of memory. MATLAB™ (version 4.2b) and two MATLAB™ Toolboxes (Signal Processing version 3.0b and Neural Network version 2.0b) were used exclusively to process the data and train the networks.

## PROCEDURE

### Feature Selection

The feature set, including FFTs of EEG activity and peripheral measures was selected based upon observation of the raw data. For example, a display of the heart interbeat interval showed an increase in heart rate during the landing segment; therefore, the mean and slope of the interbeat interval were used as features. The FFTs yielded increases and decreases in variance as well as changes in power between the two flight segments; therefore, the power and variance for each frequency band for each of the seven electrodes were included.

Preprocessing of the FFT consisted of calculating the power spectral density using the direct method (periodogram) over ten second windows with a fifty percent overlap since prior statistical methods of classification (i.e. discriminant analysis and Bayes quadratic classification) indicated that this combination was the most stable. The periodogram was grouped into the following frequency bands: delta (1-3 Hz), theta (4-7 Hz), alpha (8-11 Hz), low alpha (8-9 Hz), high alpha (10-11 Hz), beta I (12-14 Hz), beta II (15-30 Hz), ultrabeta I (31-36 Hz) and ultrabeta II (37-42 Hz). The average power and variance for each band were computed at each electrode to produce 126 features.

Four more physiological features were calculated and added as follows: the mean heart interbeat interval, the mean respiration interbreath interval, the slope of the heart interbeat interval using linear regression, and the number of blinks. Typically, the heart rate increased (interbeat interval decreases) in the landing segments of a session. Past research has demonstrated that heart rate can be used as an indicator of increased physical activity or a high mental workload condition (Itoh et al., 1989). Overall, the number of eyeblinks decreased during the landing segments in all subjects. Wilson and Eggemeier (1991) reviewed literature showing a decrease in blink activity during visually demanding tasks such as aircraft landing. After the addition of these measures, a total of 130 features were input to the neural network.

**Neural Network**

The network used in this study was a feedforward backpropagation neural network with three neuron layers, an adaptive learning rate, and momentum. The input layer consisted of 130 fully connected neurons with a hyperbolic or symmetric sigmoid activation function of the form $f(a) = \tanh(a)$. The hidden layer had 40 fully connected neurons again with the symmetric sigmoid activation function. The output layer consisted of a single neuron with a sigmoid logistic nonlinearity of the form $f(a) = (1 + e^{-a})^{-1}$. The adaptive learning rate allowed the neural network to learn the exemplars as quickly as possible while maintaining the stability of the system. Momentum prevented the neural network from becoming trapped in a local minima which is essential when data contain artifact. The momentum constant was fixed at 0.95 and the change in the weight vector using momentum was based on the Widrow-Hoff learning rule as follow:

$$d\mathbf{W} = mc^{*}d\mathbf{W} + (1 - mc)^{*}lr^{*}\mathbf{d}^{*}\mathbf{p}^{T}$$

$$\mathbf{W}_{k+1} = \mathbf{W}_{k} + d\mathbf{W}$$

where $d\mathbf{W}$ is the weight change, $mc$ is the momentum constant, $lr$ is the learning rate, $\mathbf{d}$ is the derivative of the error vector, and $\mathbf{p}$ is the input vector or exemplar vector. As the learning rate increased, its contribution to the weight change matrix increased.

The neural network was optimized for subject 1 by adjusting the network until the classification accuracy reached asymptote. The parameters manipulated were the number of neurons in the hidden layer, the momentum constant, the training time (maximum number of epochs), and the target sum squared error. After optimization, the neural network algorithm was trained and tested on each condition for each subject using the Leave-One-Out-Method (LOOM) which leaves one exemplar out for testing and trains on the remaining exemplars (Funkunaga, 1972). This process was repeated until all data were tested. The classification accuracy was calculated by determining the ratio between the number of correct classifications and the total number of exemplars.

Two methods were used to prevent overlearning. The training was aborted when one of the two criteria were met. The first method limited the learning time or number of epochs to 1000 and the second method relaxed the training error. The backpropagation neural network trained to a sum square error which prevented the network from overlearning the exemplars.

| TABLE 1: CLASSIFICATION ACCURACY BY SUBJECT | | |
|---|---|---|
| Subject | Accuracy (correct/total) | Percent Accuracy |
| 1 | 472/527 | 89.56% |
| 2 | 423/517 | 81.82% |
| 3 | 390/473 | 82.45% |
| 4 | 376/477 | 78.83% |
| 5 | 370/459 | 80.61% |
| Overall | 2031/2453 | 82.80% |

## RESULTS

The classification process using a feedforward backpropagation neural network with adaptive learning and momentum provided an overall classification accuracy of 83% as shown in Table 1. The data from subject 1 which were used to optimize the neural network showed classification accuracies around 90%. The other four subjects had about ten percent lower classification accuracies.

Reducing the number of nodes in the hidden layer during the optimization process caused the classification accuracy to decrease rapidly after about forty nodes. This reduction also decreased the time required to train the network. A downside to optimizing the network on a single subject is the loss of generalization across subjects. This is evidenced by the reduced classification accuracy for the remaining subjects.

## DISCUSSION

The validity of using neural networks to classify mental workload is demonstrated by the high percentage of correct classifications in this study. The neural network employed, correctly classified mental workload in this complex simulator landing task at rates that are as good or better than those reported in similar studies which use simple, single task classification.

In this case, the reduced accuracy for the four remaining subjects could be the result of individual differences in psychophysiological response patterns between subjects in the multitask environment. Since the neural network was optimized for a single subject, individual differences were not accommodated. However, optimizing the network for each subject, as was done for the first, should increase classification accuracy. The optimized network used here consisted of forty neurons in the hidden layer. This suggests that subject 1's data consisted of about forty salient features out of the total of 130; however, other subjects or conditions may require more than forty features to produce similar results.

In the future, studies should include a saliency measure to determine the features that most contribute to the classification process when using a neural network (Ruck et al., 1990). This would reduce the size of the network, thereby decreasing the learning time and providing information about the processes used in a particular multitask situation. A comparison of the salient features could also help determine which features, if any, are consistent between subjects.

This study resulted in a high level of accuracy in classifying cognitive workload in a complex aircraft simulator landing task. These results suggest that it may be possible to use physiological measures with adaptive interfaces as feedback to a control device to reduce mental workload in overload conditions. Further research in other multitask environments must be conducted to determine if high classification accuracy can be achieved consistently. Depending upon the specific combination of tasks performed, different multitask environments will most likely produce variable results. The features and the interactions between those features can be expected to vary with the immediate task at hand. Therefore, it is imperative that the salient features of any given multitask situation be determined. Given the range of activities that humans undertake, physiological measures in varying multitask environments should be examined further.

## ACKNOWLEDGMENTS

The authors would like to thank Dr. Michael W. Haas and the SIRE (Synthesized Immersion Research Environment) group of Armstrong Laboratory for the physiological data used in this study.

## REFERENCES

Anderson, C. W., Devulapalli, S., Stolz, E. A. (1995). Determining mental state from EEG signals using parallel implementations of neural networks, *Scientific Programming, Special Issue on Applications Analysis*, Vol. 4(3), 171-183.

Brickman, B. J., Hettinger, L. J., Roe, M. M., Lu, L., Repperger, D. W., Haas, M. W. (1996). Haptic specification of environmental events: implications for the design of adaptive, virtual interfaces. *Proceedings of the IEEE Virtual Reality Annual International Symposium*, 147-153.

Funkunaga, K., *Introduction to Statistical Pattern Recognition*, Academic Press, 1972.

Gevins, A., Morgan, N. (1988). Applications of neural network (NN) signal processing in brain research, *IEEE Transactions on Acoustics, Speech, and Signal Processing* , Vol. 36(7), 1152-61.

Itoh, Y., Hayashi, Y., Tskui, I. 7 Saito, S. (1989). Heart rate variability and subjective mental workload in flight task validity of mental workload measurement using HRV method, In Smith, M.J. & Salvendy, G., eds., *Work with Computers: Organizational, Management, Stress and Health Aspects*, Amsterdam: Elsevier, 209-216.

Klöppel, B. (1994). Application of neural networks for EEG analysis: considerations and first results, *Neuropsychobiology* ,Vol. 29, 39-46.

Ruck, D., Rogers, S., Kabrisky, M., (1990). Feature selection using a multilayer perceptron, *Journal of Neural Network Computing*, Vol. 2(2), 40-48.

Widrow, B., Lehr, M. (1990). 30 Years of Adaptive Neural Networks: Perceptron, Madaline and Backpropagation, *Proc. of IEEE*, Vol. 78(9), 1415-1442.

Wilson, G. F. (1994). Workload Assessment Monitor (WAM). *Proceedings of the Human Factors Society* , 944.

Wilson, G. F., Eggemeier, F. T. (1991). Physiological measures of workload in multi-task environments, Damos, D.,ed., *Multiple-task Performance*, London, Taylor & Francis, 329-360.

Wilson, G. F., Swain, C. R., Brookings, J. B. (1995). Workload related changes in eye, cardiac, respiratory and brain activity during simulated air traffic control, Technical Report AL/CF-TR-1995-0156.

# A PRELIMINARY INVESTIGATION OF SELECTION OF EEG AND PSYCHOPHYSIOLOGICAL FEATURES FOR CLASSIFYING PILOT WORKLOAD

**KELLY A. GREENE AND KENNETH W. BAUER**
*Department of Operational Sciences, Air Force Institute of Technology*

**MATTHEW KABRISKY AND STEVEN K. ROGERS**
*Department of Electrical and Computer Engineering, Air Force Institute of Technology*

**CHRISTOPHER A. RUSSELL AND GLENN F. WILSON**
*Performance Assessment Branch, Armstrong Laboratories*

***ABSTRACT:***
Cardiac and respiration measures have shown limited success in classifying pilot workload. In the last few years, electroencephalography (EEG) measures have also been used to classify pilot workload (Sterman et al, 1992, and Wilson et al, 1994). Features of EEG data are currently selected by minimization of pilot workload classification error by a multilayer perceptron (MLP) classifier (Wilson et al, 1994). A set of salient features has not yet been identified for classifying pilot workload.

Saliency metrics provide a way to measure the relative usefulness of a feature. Two types of saliency measures utilized are Ruck's saliency metric (Ruck, 1990, and Ruck et al, 1990) and Tarr's saliency metric (Tarr, 1991). A new saliency metric called the Signal-to-Noise Ratio (SNR) is also utilized (Bauer and Sumrell, 1996). This pilot workload data set is the first non-trivial, real-world application of the SNR saliency metric.

This paper has two objectives. The primary objective is to develop a methodology to identify salient features to classify pilot workload. The second objective is to apply the SNR saliency metric to a non-trivial real-world problem and compare the results of the SNR saliency metric to that of Ruck's saliency and Tarr's saliency measures.

## BACKGROUND

Despite decades of investigation, few measurements exist which are able to provide objective, reliable, and valid estimates of pilot workload. In the last few years, physiological measures have been used to monitor pilot workload and include heart rate, heart rate variability, eye blink rate, and respiration rate (Wilson, 1992, and Wilson, 1993). Measures of the brain's electrical activity, generally on-going activity or evoked potentials (EP), have only been recently added to the arsenal of pilot workload measurements (Sterman et al, 1992, and Wilson et al, 1994). Since the brain is the organ responsible for evaluating sensory information, and making and carrying out decisions, EEG data would seem to hold great potential as a means to measure the mental workload of a pilot.

No one measure has been demonstrated as sufficient and no one measure has been demonstrated as superior. Individual differences in physiological response and in particular mental response should be recognized. As a result, an extremely large number of EEG measures are currently included as input to a neural network that is trained to classify pilot workload as low or high.

Unfortunately, too many input features to a classifier can have negative effects. First of all, insignificant input features to a neural network may reduce classification accuracy (Ruck, 1990). In addition, as the number of features grows, the number of training vectors required grows exponentially (Duda and Hart, 1973). Thus, the number of training vectors required can be reduced which in turn typically reduces the training time if non-salient features are removed.

The Air Force's Flight Psychophysiological Laboratory provided the data set for this preliminary investigation. The data set consists of pre-processed data from a flight simulation. The test subject, an Air Force pilot, flew a simulated aircraft landing scenario. The scenario starts off with the test subject flying his descent in the clouds. While in the clouds, the pilot's workload is classified as low. The test subject then breaks through the clouds on his descent to the airfield. As soon as the pilot breaks through the clouds, his workload is now classified as high. The scenario ends at touchdown.

Throughout the scenario, EEG are collected at six different locations (FP1, FZ, C3, C4, P3, and P4) on the test subject's head. The EEG data are passed through a fast-Fourier transform (FFT) and all frequencies above 64 hertz are removed. "Artifacts" such as head movement and muscle movement which can be caused by speech are also removed. The EEG data are then broken up into nine different bands based upon frequency: $\alpha_1$, $\alpha_2$, $\alpha$, $\beta_1$, $\beta_2$, $\mu\beta_1$, $\mu\beta_2$, $\Delta$, and $\theta$. For each band at each head location, two types of information are collected over a 10-second moving window resulting in a total of 108 features (9 bands x 6 locations x 2 types). The two types of information are:

1. Logarithm (base 10) of the power averaged over the 10-second moving window.
2. Variance of the power over the 10-second moving window.

Four physiological features are also developed for inclusion in the data set. Cardio-respiratory and eye blink data are collected. Cardiopulmonary information is represented by heart rate and heart interbeat interval. Respiratory information is represented by breathing rate. Eye information is represented by the total number of eye blinks.

The multivariate data set contains a total of 112 psychophysiological features (108 EEG features and 4 physiological features). The data set contains only 32 exemplars. Foley's Rule is applicable to this two-class problem (low workload, high workload) which states that the number of training vectors per class should be greater than three times the number of features (Foley, 1972). Since the data set contains 112 features and only 32 exemplars, it is clear that Foley's rule is violated. Two solutions exist:

1. Increase the number of exemplars.
2. Decrease the number of features.

For this investigation, it is desired to decrease the number of features by selecting the most salient features.

## METHODOLOGY

As a first attempt to reduce the number of features, an analysis of the correlation between each feature and pilot workload is conducted. The sample correlation of each feature was calculated and those features with correlation coefficients less than 0.75 were removed and no longer considered as possible salient features.

A MLP was trained using the instantaneous backpropagation method with step size $\eta = 0.3$. 22 exemplars were randomly selected as training vectors and the remaining 10 exemplars were used as test vectors. A "noise" feature was added to the data set for use as a baseline and for use in the SNR saliency. The noise feature exemplars were created by random sampling from a uniform (0,1) distribution. Each of the input features were normalized so that no one feature dominates. There were a total of 19 input features (18 features selected by the correlation screening process and 1 noise feature). One hidden layer had 19 nodes. The output layer had two nodes. All activation functions were sigmoidal. The network was trained until all of the following stabilized: average absolute training error, average absolute test error, weights, and SNRs.

Feature saliency metrics provide a way to measure the relative usefulness of features and a means to rank order the features. Ruck's saliency metric is based upon the sensitivity of an ANN's output to its input and utilizes the derivative of the output with respect to a specific input. This derivative depends upon the inputs and the weights within the trained ANN. In contrast, both the Tarr saliency metric and the SNR saliency metric are weight-based saliency measures that depend only upon the weights within the trained ANN. Three general types of saliency metrics are calculated upon completion of the training: Ruck's saliency, Tarr's saliency, and SNR saliency.

Ruck's Saliency measures feature $i$'s effect on a neural network's output by calculating the sum of the partial derivatives of the outputs to feature $i$. The derivative can be written as a function of the weights and activations as follows:

$$\Lambda_i = \sum_P \sum_I \sum_R \sum_K \left| \frac{\partial z_k^{\,2}}{\partial x_i}(\mathbf{x}_p^{\,i(r)}, \mathbf{w}) \right| \tag{1}$$

where $P$ is the number of training vectors, $I$ is the number of features, $R$ is the number of steps that the range of each feature is uniformly divided into, $K$ is the number of network outputs, $z_k$ is the output of node $k$, $x_i$ is the input of node $i$, $\mathbf{x}_p^{i(r)}$ is the vector with which the partial derivative is calculated using exemplar $p$ with feature $i$ replaced with the value of step $r$, and $\boldsymbol{w}$ is the weight vector. This is typically referred to as Ruck's saliency with pseudo-sampling. The Ruck Saliency metric is used to order the features from most salient to least salient where higher Ruck Saliency metric values indicate higher relative saliency and lower values indicate lower relative saliency.

A variant of the Ruck's saliency is without pseudo-sampling and is computed using:

$$\Lambda_i = \sum_P \sum_I \sum_K \left| \frac{\partial z_k^{\,2}}{\partial x_i}(x_p^{\,i}, w) \right| \tag{2}$$

Tarr's metric determines saliency of feature $i$ by summing the squared values of the weights connecting feature $i$ to the hidden nodes and can be written as follows:

$$\tau_i = \sum_{j=1}^{J} (w_{ij}^{\,1})^2 \tag{3}$$

where $\tau_i$ is the Tarr saliency metric for feature $i$, $J$ is the number of hidden nodes, $w_{ij}^{\,1}$ is the first layer weight between input node $i$ and hidden node $j$. This equation for Tarr's saliency is simply the sum of the squared weights between input node $i$ and hidden node $j$. There are three variants to the Tarr saliency metric. The first variant is the Euclidean norm of the weights of the feature and is typically equated by the square root of the sum of the squared weights emanating from a given input node. The second variant is the "taxi-cab" norm of the weights of the features and is calculated by summing the absolute value of the weights emanating from a given input node. The third variant of the Tarr saliency metric is the infinity norm of the weights of the feature and is simply the largest weight in absolute value of all the weights from a given input node.

The SNR is a new saliency metric. The SNR saliency metric is similar to Tarr's saliency metric in that both rely on the sum of squared first layer weights. This metric is different from both Ruck's and Tarr's though because it directly compares the saliency of a feature to that of an injected noise feature. This novel saliency metric is computed using the following:

$$SNR_i = 10 * \log \left( \frac{\sum_{j=1}^{J} (w_{ij}^{\,1})^2}{\sum_{j=1}^{J} (w_{Nj}^{\,1})^2} \right) \tag{4}$$

where $SNR_i$ is the value of the saliency metric for feature $i$, $J$ is the number of hidden nodes, $w_{ij}^{\,1}$ is the first layer weight from node $i$ to node $j$, and $w_{Nj}^{\,1}$ is the first layer weight from the noise node $N$ to node $j$. The scaled lognormal transformation of the ratio converts the saliency metric to a decibel scale (Bauer and Sumrell, 1996).

The theoretical concept behind the SNR saliency metric is similar to that of the Tarr's saliency metric. If a given feature is not relevant to a neural network's output, the updates of the first layer weights emanating from that feature's input node should be random and simply fluctuate around zero (Tarr, 1991). If, on the other hand, a given feature is relevant to a neural network's output, then the updates of the first layer weights emanating from that feature's input node should be moved in the weight space in a constant direction until the error is minimized (Tarr, 1991). Thus, the SNR saliency metric should be significantly larger than 0.0 for salient features and very close to 0.0 or less than 0.0 for non-salient features.

The SNR saliency metric is appealing because, like both the Ruck saliency metric and the Tarr saliency metric, the SNR saliency metric can be used to rank order the features. The greatest potential of the SNR saliency metric results from its comparison of the saliency of each feature to a baseline noise feature. This allows for the SNR saliency metric to be used "on the fly" while the network is training. Neither the Ruck saliency metric nor the Tarr saliency metric possess this "on the fly" capability.

**RESULTS**

Eighteen of the original 112 features had correlations with pilot workload greater than or equal to 0.75. Seventeen of the screened features are EEG features (seven are EEG variance information and ten are average EEG power information). The only physiological feature that was selected for this screened data set is number of eye blinks.

The MLP was trained in 1200 epochs. The weights were initialized between -0.001 and 0.001 in order to equalize the signal strength for all input features. The test set classification error was 10.0%. Table 1 lists all of the saliency metrics computed based upon the following:

1. Correlation analysis.
2. Ruck's saliency without pseudo-sampling (Ruck1).
3. Ruck's saliency with pseudo-sampling (Ruck2).
4. Tarr's saliency (Tarr1).
5. Euclidean norm of Tarr's saliency (Tarr2).
6. Taxi-Cab norm of Tarr's saliency (Tarr3).
7. Infinity norm of Tarr's saliency (Tarr4).
8. Signal-to-Noise Ratio saliency (SNR).

If we adhere to Foley's rule given that there are only 32 exemplars, at most five features can be used to train the neural network to classify pilot workload for this data set. If five features are selected then 30 exemplars (5 features x 2 classes x 3 Foley's factor) must be randomly selected as training vectors leaving only two test vectors. This would not adequately test the trained net. Thus, it is recommended that only four features are selected to train a MLP using 24 training vectors (4 features x 2 classes x 3 Foley's factor) and 8 test vectors. Table 2 summarizes the top four rankings for Ruck's saliency, Tarr's saliency, and the SNR saliency metrics.

TABLE 1: SALIENCY METRICS

| Feature | Correlation | Ruck1 | Ruck2 | Tarr1 | Tarr2 | Tarr3 | Tarr4 | SNR |
|---|---|---|---|---|---|---|---|---|
| Noise | 0.0187 | 0.0001 | 0.0001 | 0.01 | 0.10 | 0.46 | 0.02 | 00.0000 |
| 2 | 0.7535 | 0.0005 | 0.0008 | 0.78 | 0.88 | 3.94 | 0.20 | 42.7731 |
| 8 | 0.7570 | 0.0004 | 0.0006 | 0.58 | 0.76 | 3.41 | 0.17 | 39.8555 |
| 14 | 0.7684 | 0.0004 | 0.0007 | 0.63 | 0.80 | 3.56 | 0.18 | 40.7045 |
| 16 | 0.8039 | 0.0005 | 0.0008 | 0.88 | 0.94 | 4.20 | 0.22 | 44.0242 |
| 17 | 0.7514 | 0.0005 | 0.0008 | 0.80 | 0.89 | 3.99 | 0.21 | 43.0018 |
| 20 | 0.7801 | 0.0005 | 0.0007 | 0.77 | 0.88 | 3.93 | 0.20 | 42.6814 |
| 22 | 0.7919 | 0.0004 | 0.0007 | 0.66 | 0.81 | 3.63 | 0.19 | 41.1027 |
| 62 | 0.8989 | 0.0006 | 0.0010 | 1.29 | 1.13 | 5.07 | 0.26 | 47.7967 |
| 64 | 0.8620 | 0.0006 | 0.0009 | 1.19 | 1.09 | 4.87 | 0.25 | 46.9922 |
| 70 | 0.8164 | 0.0006 | 0.0009 | 1.14 | 1.07 | 4.78 | 0.24 | 46.6283 |
| 76 | 0.8426 | 0.0006 | 0.0009 | 1.02 | 1.01 | 4.52 | 0.23 | 45.4791 |
| 97 | 0.8607 | 0.0006 | 0.0010 | 1.35 | 1.16 | 5.19 | 0.27 | 48.2794 |
| 98 | 0.8731 | 0.0007 | 0.0010 | 1.45 | 1.20 | 5.38 | 0.27 | 48.9860 |
| 100 | 0.8827 | 0.0006 | 0.0010 | 1.40 | 1.18 | 5.29 | 0.27 | 48.6425 |
| 101 | 0.9702 | 0.0007 | 0.0011 | 1.71 | 1.31 | 5.85 | 0.30 | 50.6575 |
| 102 | 0.8577 | 0.0006 | 0.0010 | 1.28 | 1.13 | 5.06 | 0.26 | 47.7640 |
| 104 | 0.7921 | 0.0006 | 0.0009 | 1.10 | 1.05 | 4.69 | 0.24 | 46.2252 |
| 112 | 0.8965 | 0.0007 | 0.0010 | 1.51 | 1.23 | 5.50 | 0.28 | 49.4052 |
| Constant | NA | 0.0002 | 0.0003 | 0.10 | 0.31 | 1.41 | 0.07 | 22.1355 |

From examination of Table 2, it appears that the traditional saliency metrics of Ruck and Tarr in addition to SNR provide consistent saliency rankings. All of the saliency metrics ranked feature 101 as the most salient feature. Table 3 describes the four most salient features based upon all of the saliency metrics: 101, 112, 98, and 100. It is quite interesting to see that all three of the top salient EEG features were average power of the Δ band. The number of eye blinks was the only physiological feature selected as one of the top salient features.

## CONCLUSIONS

In summary, three saliency metrics were successfully employed to determine the most useful features to classify pilot workload as low or high for this data set. A new saliency metric, SNR, was shown to provide saliency rankings consistent with

TABLE 2: TOP FOUR RANKINGS OF EACH SALIENCY MEASURE

| Ranking | Ruck1 | Ruck2 | Tarr1 | Tarr2 | Tarr3 | Tarr4 | SNR |
|---|---|---|---|---|---|---|---|
| 1 | 101 (tie 1) | 101 | 101 | 101 | 101 | 101 | 101 |
| 2 | 112 (tie 1) | Inconclusive | 112 | 112 | 112 | 112 | 112 |
| 3 | Inconclusive | Inconclusive | 98 | 98 | 98 | 97 | 98 |
| 4 | Inconclusive | Inconclusive | 100 | 100 | 100 | 98 | 100 |

TABLE 3: FOUR MOST SALIENT FEATURES

| Saliency Ranking | Feature Number | Description of Feature |
|---|---|---|
| 1 | 101 | average power of Δ band at location P3 |
| 2 | 112 | number of eye blinks |
| 3 | 98 | average power of Δ band at location FZ |
| 4 | 100 | average power of Δ band at location C4 |

that of Ruck's saliency and Tarr's saliency. Since both Tarr's saliency and the SNR saliency both rely on the sum of the squared weights from the first layer, it makes sense that those two metrics would produce similar results. The SNR saliency has the advantage over Ruck's saliency and Tarr's saliency by providing a way to measure saliency "on the fly" while the ANN is training. A set of salient features can be selected from a data set of EEG and physiological data using several types of saliency metrics.

## REFERENCES


Bauer, K.W. and Sumrell, D.B. (1996). Feature Screening Using Signal-to-Noise Ratios. In preparation.

Duda, R.O. and Hart, P.E. (1973). *Pattern Classification and Scene Analysis.* New York: John Wiley and Sons.

Foley, D.H. (1972). Considerations of Sample and Feature Size, *IEEE Transactions on Information Theory*, IT-18, Sept, 618-626.

Ruck, D.W. (1990). *Characterization of Multilayer Perceptrons and Their Application to Multisensor Automatic Target Detection*, Ph.D. Dissertation, School of Engineering, Air Force Institute of Technology, OH.

Ruck, D.W. et al (1990). Feature Selection Using a Multilayer Perceptron, *Journal of Neural Network Computing,* 2(2), Fall, 40-48.

Sterman, M.B. et al (1992). Quantitative EEG Patterns of Differential In-Flight Workload, *1992 Space Operations Applications and Research Proceedings*, 466-473.

Tarr, G.L. (1991). *Multilayered Feedforward Neural Networks for Image Segmentation*, Ph.D. Dissertation, School of Engineering, Air Force Institute of Technology, OH.

Wilson, G.F. (1992). Applied Use of Cardiac and Respiration Measures: Practical Considerations and Precautions, *Biological Psychology,* Volume 34, 163-178.

Wilson, G.F. (1993). Air-to-Ground Training Missions: A Psychophysiological Workload Analysis, *Ergonomics,* Volume 36, No. 9, 1071-1087.

Wilson, G.F. et al (1994). Evoked Potential, Cardiac, Blink, and Respiration Measures of Pilot Workload in Air-to-Ground Missions, *Aviation, Space, and Environmental Medicine*, Feb, 100-105.

# CLASSIFICATION OF MOVEMENT PATTERNS IN PATIENTS WITH LOW BACK PAIN USING AN ARTIFICIAL NEURAL NETWORK

**JEFFREY B. BISHOP AND MALCOLM H. POPE**
*Iowa Spine Research Center, University of Iowa, Iowa City, IA 52242 USA*

**SANTOSH K. ANANTHRAMAN**
*Neural Applications Corporation, Coralville, Iowa*

**MAREK SZPALSKI**
*Brussels Free University, Brussels, Belgium*

***ABSTRACT:***
Low back pain (LBP) is a problem of enormous medical and economic importance. Lifetime prevalence for low back pain is 75 percent, yet precise diagnosis unknown in 80-90 percent of patients. Accurate evaluation of LBP is very difficult. Impairment ratings of LBP can vary by as much as 70% using current systems which include imaging techniques and functional assessment. An artificial neural network has been developed for analysis of trunk dynamic motion as a diagnostic aid for LBP. A neural network approach was selected because the effect of LBP on back movement is very hard to model based on first principles and can be considered a non-linear multiple input multiple output (MIMO) system. The neural network is trained using motion features extracted from data collected from patients at several clinical sites who have back pain that has been classified by a clinician. Classification results from test data suggest that a neural network approach is promising as a tool for measurement of back pain.

## INTRODUCTION

Low back pain (LBP) is a common source of disability with a lifetime prevalence rate of about 75%. However a precise diagnosis of LBP is unknown in 80 to 90% of patients. It is difficult to quantify LBP, thus qualitative measures are typically used. Accurate measurement of LBP is important for improving diagnosis and treatment of back injuries, for monitoring recovery from injury, and to avoid unnecessary diagnostic tests and surgery. Some traditional measurement methods attempt to identify a pathoanatomic source of LBP. Imaging techniques such as CAT, MRI, and myelography are of this type. Pathoanatomic diagnosis is found in fewer than 15% of patients with LBP and furthermore 25% of asymptomatic subjects show pathoanatomic sources.

Functional assessment is another method practiced to measure back pain and extent of disability. This technique involves the measurement of the functional capacity of the trunk's musculoskeletal system in terms of the amount of force that a patient is willing to generate. However the strength may be limited by pain tolerance level which is highly variable. Another problem is low correlation between strength measures using different apparatus. Testing also may be associated with risk of injury.

A new technique has been proposed by several groups which involves the analysis of unloaded trunk dynamic motion. The hypothesis is that people with LBP move differently than people without LBP. It is well established in the literature that patients with LBP have a reduced maximum range of motion. The work of Marras and Wongsam shows trunk angular velocity is diminished in LBP. An important reason to study unloaded dynamic motion may be that resistance loading may mask motion characteristics that appear in unloaded movement.

## METHODS

Several forms of data have been collected from subjects at clinical sites in the U.S. and one site in Europe. The Quebec Task Force (QTF) questionnaire and Waddell tests were administered to subjects to obtain information about the type and level of pain they experience (Spitzer et al, 1987). A QTF pain score of zero indicates no LBP and a non-zero QTF pain score indicates LBP in present, with higher pain scores generally indicating more serious pain conditions. Back motion was measured in an experimental protocol that included flexion/extension, lateral bending, rotation, and circumduction movements. The BackTracker system from Isotechnologies (Hillsborough, NC) is a triaxial goniometer used to measure the motion of the spine at 100 Hz. Channels for flexion/extension, lateral bending, and rotation were recorded.

Figure 1 shows typical circumduction movement data plotted from clinical data. During circumduction, the subject bends the low back in a pattern such that the subject's head moves in an elliptical pattern when viewed from above. Four repetitions of clockwise and counterclockwise circumduction are recorded in the protocol.

Sixteen subjects (8 with no back pain and 8 with back pain) were chosen from the available data. The shape of each circumduction curve was analyzed using a software system called Image, available from the National Institutes of Health (NIH). An ellipse was fitted (Cramer, 1945) to each circumduction curve, and the following features were recorded: major axis, minor axis, angle of orientation, shape perimeter, and ellipse perimeter.

Since we found in previous work (Pope, 1996) that the range of motion in patients with back pain is generally reduced, it was desirable to avoid classification based merely on the size of the circumduction movement rather than the shape characteristics. Thus the feature vector consisted of the following parameters: ratio of major axis to minor axis, angle of orientation of ellipse, and ratio of shape perimeter to ellipse perimeter. The ratio of the shape perimeter to ellipse perimeter gives a measure of how closely an ellipse was able to fit the circumduction shape.

A neural network with 8 hidden neurons in the hidden layer and 1 neuron in the output layer was trained. Training was done using backpropagation with momentum and adaptive learning. The learning rate was set at 0.01 and a momentum constant of 0.9 was used.

Velocity and shape features were extracted from the flexion/extension, lateral bending, rotational bending, clockwise circumduction, and counterclockwise circumduction tests for 76 subjects, resulting in 380 data points. The features included maximum and average absolute value of velocity in flexion/extension, lateral bending, and rotation during each of the tests. The data was scaled using sigma scaling, and a model was fitted using eleven inputs and one output (the QTF pain evaluation).. A variation of a radial basis function network with 30 clusters was implemented using AEGIS (Neural Applications Corporation, Coralville, IA). The variations used are described in the *AEGIS User Manual*.

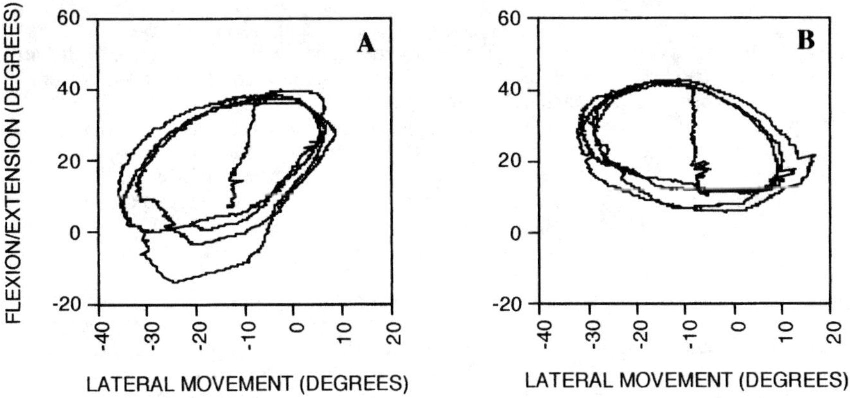


**Figure 1: Circumduction path differs with direction of circumduction a) clockwise or b) counterclockwise.**

## RESULTS

The back propagation network was able to successfully classify all six of the test samples presented based on features extracted from the elliptical fitting of the circumduction shape. The output of the radial basis function network was analyzed under 3 conditions. In Case I, the model was able to correctly distinguish between LBP (non-zero QTF pain score) and no LBP (zero QTF pain score) in 321 of 380 test cases (84% accuracy). In Case II, the predicted and actual Quebec Task Force pain classification scores were divided into 2 pain classes- QTF pain score less than 5 and QTF pain score greater than 5. With these 2 pain classes, the model was correct in 366 of 380 test cases (96% accuracy). In Case III, the predicted and actual Quebec Task Force pain classification scores were divided into 3 classes, with class divisions at QTF pain scores of 3 and 6. With these 3 pain classes, the model was correct in 329 of 380 test cases (87% accuracy).

## DISCUSSION

Why would people with LBP show different patterns of movement? Trunk motion is a product of biomechanical and learned processes. Thus motion could be altered by a mechanical change in the properties of the spine and associated tissues, or by a learned process to minimize pain and energy expenditure during movement. Increased co-activation of muscles could serve as a protective mechanism and would stiffen the trunk and lead to slower velocity and acceleration during movement. Tissue sensitivities could lead to sudden increase or decrease in acceleration. Of particular interest is guarding behavior in which the patient learns to avoid certain painful positions and therefore modifies movement behavior. Particular velocities, such as maximum and average lateral velocity, were found to be the most valuable for classification.

**Feature Extraction**
Many other features besides the shape of the circumduction curves, such as symmetry, consistency, and anticipated shape of movements, are of interest. It had been noted that the shape of the circumduction in some individuals is substantially different depending on whether the movement is clockwise or counterclockwise. Considering the symmetry of the anatomic and control mechanisms that produce the movement, this finding warrants further study. Several approaches to the study of symmetry can be found in the melanoma literature. Digitized images of candidate melanoma are classified based on the color and symmetry of the shape, the same method that dermatologists use. The circumduction data is recorded as position as a function of time which opens up possibilities for velocity analysis and other dynamic methods that are not applicable to still images.

Consistency of movements can be studied over several repetitions. It is expected that healthy subjects would have more consistent movement patterns in general. Trends in movement are also of interest. For example if the size of the circumduction movement grows steadily over several repetitions, this may indicate a 'warm up' effect. If the size decreases steadily, this may indicate fatigue.

Anticipated shape of a circumduction movement may be calculated using data from flexion/extension tests and lateral bending tests. Based on the size and direction of these movements, the shape of the circumduction movement may be anticipated. This allows each patient to be used as his/her own control.

## CONCLUSION

The pattern of movement may contain valuable information of use in measurement and characterization of back pain. A neural network was developed to correctly classify back pain status of patients based on features extracted by motion analysis. Extraction of additional features and training with a larger data set is planned.

## REFERENCES


Cramer (1945). *Mathematical Methods of Statistics*, Princeton Univ. Press, p. 283.

Ferguson, S.A., Marras, W.S., and Waters, T.R. (1992). Quantification of back motion during asymmetric lifting, *Ergonomics* (35), 845-859.

Frymoyer, J.W., Frymoyer, W.W., Wilder, D.G., and Pope, M.H. (1979). The mechanical and kinematic analysis of the lumbar spine in normal living human subjects in vivo, *Journal of Biomechanics*, (12), 165-172.

Frymoyer, J.W., Pope, M.H., Costanza, M.C., Rosen, J.C., Goggin, J.E., and Wilder, D.G. (1980). Epidemologic studies of low-back pain, *Spine* , (5) 419-423.

Frymoyer, J.W., Pope, M.H., Clements, J.H., Wilder, D.G., MacPherson, B., and Ashikaga, T. (1983). Risk factors in low-back pain, *Journal of Bone and Joint Surgery*, (65-A) 213-218.

Gill, K., Krag, M.H., Johnson, G.B., Haugh, L.D. and Pope, M.H. (1988). Repeatability of four clinical methods for assessment of lumbar spinal motion, *Spine* (13) 50-53.

Gomez, T., Beach, G., Cooke, C., Hrudey, W. and Goyert, P. (1991) Normative database for trunk range of motion, strength, velocity, and endurance with the isostation B-200 lumbar dynamometer, *Spine* (16), 15-21.

Gracovetsky, S., Kary, M., Levy, S., Ben Said, R., Pitchen, I., and Helie, J. (1990). Analysis of spinal and muscular activity during flexion/extension and free lifts, *Spine* (15), pp. 1333-1339, 1990.

Harris, S.R., Movement analysis- an aid to early diagnosis of cerebral palsy, *Physical Therapy* (71) 215-221, 1991.

Hertz, J., Krogh, A., and Palmer, R. G. (1991). *Introduction to the therory of neural computation*, Reading, MA: Addison-Wesley Publishing Company.

Marras, W.S. and Wongsam, P.E. (1986). Flexibility and velocity of the normal and impaired lumbar spine, *Arch Phys Med Rehabil*, (67), 213-217.

Marras, W.S. and Mirka, G.A. (1989). Trunk strength during asymmetric trunk motion, *Human Factors* , (31), 667-677.

Marras, W.S. and Mirka, G.A. (1990). Muscle activities during asymmetric trunk angular accelerations, *Journal of Orthopaedic Research*, (8), 824-832.

Marras, W.S. (1992). Toward an understanding of dynamic variables in ergonomics, *Occupational Medicine*, (7), 655-677.

Marras, W.S., Lavender, S.A., Leurgans, S.E., Rajulu, S.L., Allread, W.G., Fathallah, F.A., and Ferguson, S.A. (1993). The role of dynamic three-dimensional trunk motion in occupationally-related low back disorders. The effects of workplace factors, trunk position, and trunk motion characteristics on risk of injury, *Spine*, (18), 617-628.

Marras, W.S., Lavender, S.A., Leurgans, S.E., Fathallah, F.A., Ferguson, S.A., Allread, W.G., and Rajulu, S.L. (1995). Biomechanical risk factors for occupationally related low back disorders, *Ergonomics*, (38), 377-410.

Marras, W.S., Parnianpour, M., Ferguson, S.A., Kim, J.Y., Crowell, R.R., Bose, S., and Simon, S.R. (1995). The classifiacation of anatomic- and symptom- based low back disorders using motion measure models, *Spine*, (20), 2531-2546.

McGregor, A.H., McCarthy, I.D., and Hughes, S.P. (1995). Motion characteristics of the lumbar spine in the normal population, *Spine*, (20), 2421-2428.

Miller, A., Blott, B.H., and Hames, T.K. (1992). Review of neural network applications in medical imaging and signal processing, *Med. & Biol. Eng & Comput.*, (30), 449-454.

Pope, M.H. (1989). Biomechanics of the lumbar spine, *Annals of Medicine*, (21), 347-351.

Pope, M.H., Bishop, J.B., Spratt, K.F., and Szpalski, M, (1996). Characterization of Back Motion in Chronic Low Back Pain, *Proceedings, International Conference on Biomedical Engineering (BME '96)*, pp. 98-99.

Schalkoff, R. (1992). *Pattern Recognition: Statistical, Structural, and Neural Approaches*, New York: John Wiley and Sons, Inc.

Spitzer, W. et al. (1987). Scientific approaches to the assessment and management of activity related spinal disorders, *Spine*, (12) S1-S59.

# NEURAL NETWORK FOR CONTROL OF STIMULATION OF PERIPHERAL MOTOR NEURONS TO ALLOW PATIENT RESPONSIVE AMBULATION

**DANIEL GRAUPE, Ph.D.**
*Dept. of EECS, University of Illinois, Chicago, Il.*

**HUBERT KORDYLEWSKI**
*Dept. of Bioengineering, University of Illinois, Chicago, Il.*

***ABSTRACT:***
The paper discusses an ART-1-based artificial neural network (ANN) adapted to controlling functional electrical stimulation (FES) to facilitate patient-responsive ambulation by paralyzed patients with spinal cord injuries. This network is designed to control FES systems developed by the first author that is presently in use by over 300 patients worldwide (presently without ANN control) and which is the first and the only FES system approved by FDA. The network that is considered discriminates patterns of above-lesion upper-trunk electromyographic (EMG) time series to map patient's posture for activating standing and walking functions under FES and it controls FES stimuli levels using response-EMG signals to overcome muscle fatigue. The network trains itself to adapt to physiological changes of the patient, and it overcomes decision and control errors by simple punishment inputs from a single manual punishment switch. The system thus, is both self adaptive and patient-responsive through a combinations of neural EMG signals and an artificial neural network to achieve patient responsive ambulation using stimulation of the patient's own peripheral motor neurons, namely his own peripheral neural network.

## INTRODUCTION

The ANN presented below is designed for patient responsive control of functional electrical stimulation (FES) for ambulation by spinal cord injured patients via recognition of time series (TS) parameters of above-lesion upper-trunk raw electromyographic (EMG) of these patients time series rather than by manual control. The network also controls stimulation levels in onset of muscles fatigue by analyzing response-EMG signals taken from stimulated sites on these patients extremities. EMG control of FES for ambulation by paraplegics was first proposed by the present senior author [Graupe, 1981] using the time series (TS) parameters of above-lesion raw EMG signals to map patterns of upper-trunk muscle contractions and the related posture to allow complete upper-motor-neuron traumatic paraplegics to ambulate via such posture mapping-based control. The present work is an extension of the above EMG-controlled system when using ANN's for EMG pattern discrimination and subsequent control of FES, and for coordinating patient interface to enhance safety in the face of system errors, to maximize-patient responsive operation, as required for actual patient ambulation. Neural networks applied to the above problem must include two important requirements essential in our implementation: (1) modification of the stored patterns according to changing environments, and (2) the patient's interaction with the network's operation. In the real world EMG parameters may gradually change beyond a given range due to fatigue or other changes in patient's condition. Hence the network must continually modify its weight vectors. However the modification of one pattern should not cause disruption of weights associated with other stored patterns. The proposed implementation adheres to the ART-1 [Grossbert, 1987, 1987] network fundamentals, but because of specific system requirements it differs greatly in details. While the ART-1 network learns new categories during unsupervised training, our network is trained in a supervised mode. In that mode four patterns are created and stored in the network (each pattern is associated with the distinct FES/ambulation function). The same modified ART-1 structure which is used for the above-lesion EMG signal discrimination is also used for controlling FES levels to compensate for patient's fatigue. Easily detectable changes in the response-EMG enable us to predict and to detect onset of fatigue prior to the patient's loosing balance. The new NN incorporates over-ride and retraining functions, which become punishment/reward inputs.

## ABOVE-LESION EMG SIGNAL RECOGNITION AND DISCRIMINATION

### Above-Lesion EMG Signal as a Control Signal for FES

The present ANN is an implementation of the system introduced by Graupe in 1981, [Gaupe] where FES, controlled by pure TS parameter identification of above-lesion EMG signals, was used for non-manual patient-controlled ambulation. (See Figure 1.) This system utilizes a natural skeletal feature that normal above-lesion upper trunk muscles undergo unique patterns of contraction while the patient intends to move a paralyzed lower limb. Because of the near-Gaussian nature of the above-lesion surface EMG signals and their piecewise stationarity [Graupe 1981], it is possible to employ TS parameters of the EMG signals to determine above-lesion posture changes relating to various intended ambulation functions. Having calculated the EMG time series parameters in terms of a pure autoregressive (AR) model, namely:

$$y_k = \sum_i a_i \, y_{k-i} + w_k \qquad (1)$$

where: $y_k$ =recorded EMG signal at time k,. a =AR parameters, w =white noise residual.

such that changes of above-lesion posture are clearly mappable. Thus, analyzing the identified AR parameters ai allows us to determine intended ambulation functions related to posture changes. This system was applied to paraplegics at Michael Reese Hospital and Medical Center in Chicago [Graupe 1983, 1981] since early 1982.

## RECOGNITION AND DISCRIMINATION OF TS PARAMETERS BY ART-1

### Supervised Training of the Neural Network

Most standard NN implementations do not apply the conventional model of the chosen NN system due to the specific and unique problem conditions. To a certain degree this

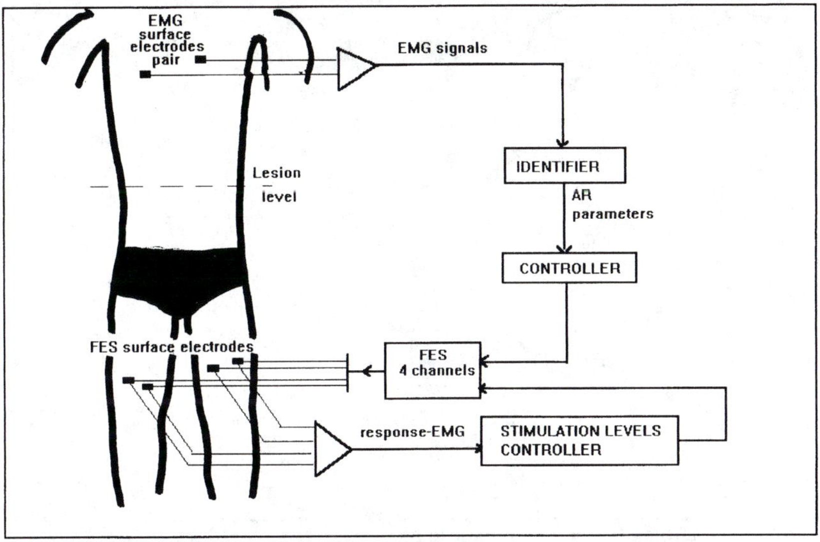


**Figure 1: FES controlled by above-lesion EMG. There are four input groups corresponding to stand up, sit down, step left, and step right.**

is also the case with the standard ART-1 architecture used for the present problem. ART-1 is a network where training is unsupervised and the number of categories into which the input should be classified is unpredictable. In our system the EMG signals need be classified only into one of four groups, each representing a basic ambulation movement to be executed by FES. Instead of initializing recognition and comparison layers' weight vectors and starting unsupervised training, our network is trained in a supervised manner. During training, four categories are created under the patient's manual control.

**Over-ride and Retraining via Over-Ride Input**

In the system presented above, the patient might initiate the intended ambulation function via FES by manual (finger) control. After initial training supervised by the patient's manual inputs, subsequent FES operation is initiated (per each FES/ambulation function) by the patient's upper-trunk posture changes as mapped by the respective above-lesion EMG parameters, using the NN. However, the FES cannot entirely rely on this NN. In cases where the intended ambulation function is not activated or when the wrong FES function is activated, namely when the AR parameters of the EMG pattern are incorrectly recognized, the patient's manual control and retraining of the network are necessary. In our NN, the patient's punishment-input (indicating improper system performance) forces changes in the stored pattern. In the present system there is one situation where the patient manually over-rides the system's operation ( by a single finger touch of one of four input keys (F1,F2,F3,F4) representing distinct FES/ambulation functions as in Fig. 2 and 3), namely, when a wrong FES/ambulation function was activated (Fig. 3). This type of incorrect system interpretation of the EMG signal can be very dangerous for the patient. The execution of an unintended ambulation function

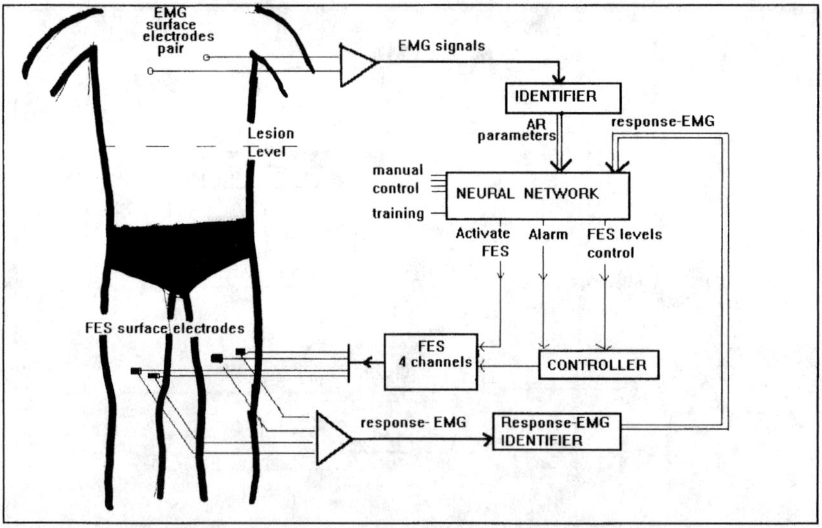


**Figure 2: The FES/ambulation system with the preset neural network.**

might result in loss of balance which would be harmful to the patient. We minimize the probability of this event in two ways: (1) by setting the vigilance parameters to a high value, and (2) by increasing the difference between patterns with different variance parameters. When incorrect activation still occurs, the patient points out (by a manual switch activation) which FES/ambulation function was intended, and network retraining begins. The analyzed input, which had been wrongly classified, is stored as the new pattern for indicated function (Fig 2,3).

**Response-EMG Control of Stimuli Levels**

The system described above and the rules governing its operation are shown in Figures 2 and 3. Before activating the recognition and classification phases, the network is trained in a supervised mode. When the training input T (Fig. 3) is 'on', the patient manually indicates ( via the inputs F1..F4 in Fig. 3) to which ambulation function the just-produced pattern of contraction (represented by the TS parameters of the above-lesion EMG signal) should be assigned. The control signal K activates the pattern updating module, so that the currently identified AR parameter vector is stored in the recognition and comparison layers (L1 and L2 in Fig. 3) as the pattern for the ambulation function indicated by the patient. After the network has stored in the L1 and L2 layers four patterns assigned to the four FES functions (stand up, left and right step,sit down), the training input should be turn off, and normal operation begins. During this phase the network continuously analyzes sets of AR parameters representing the EMG signals, until the input (set of AR parameters) matches any stored pattern within the specified tolerance. When this happens, the network initialize the FES function assigned to the chosen pattern ("ACTIVATE FES" on Fig.2). However, since the paralyzed patient lacks sensation, he does not sense muscle fatigue

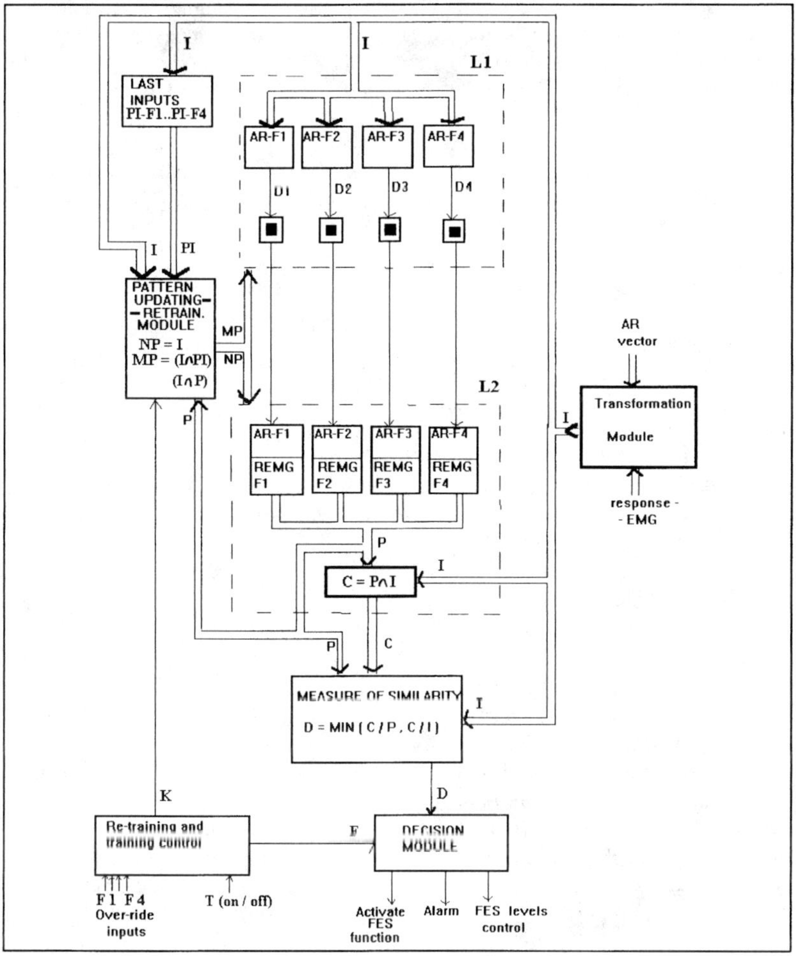


**Figure 3. Block description of the neural network.**

and might not be aware that he must sit down because his muscles no longer support his body weight. Fatigue in paralyzed muscles under FES is thus a major problem. We can overcome this problem by increasing stimulation levels when muscle contraction is insufficient to support body weight. In 1985 Graupe in [Graupe, 1985] introduced a system where progression of fatigue is detected prior to visible knee-angle changes, thus appropriate action can be automatically taken while there is still time. This system utilizes changes in below-lesion EMG patterns produced in response to FES as an indicator of degree of fatigue during ambulation via FES. Changes in this response-EMG signal, as taken at the simulated sites serve to adjust FES stimuli levels. Parts of our NN system (the transformation, the measure of similarity and the comparison layer modules) are thus used for (1) identifying the AR parameters of the above-lesion EMG time series, (2) analyzing response-EMG signals. If no reference pattern of the response-EMG signals was stored in the L2 layer, a new pattern for the response-EMG signal is created (REMG Fn on Fig.3). Otherwise the response-EMG signal is compared with an adequate reference in the L2 layer (REMG F1..REMG F4 in Fig. 3) to determine the progression of fatigue in the stimulated muscles, so that stimuli levels are adjusted accordingly. After activation of the recognized FES function has been executed via above-lesion EMG, the NN updates the chosen pattern and continues the recognition/classification process with the next set of response-EMG parameters.

## PERFORMANCE RESULTS AND CONCLUDING COMMENTS

### System Performance

For simulating FES/ambulation control with the present NN system, we used 35 sets of the AR parameters which represent an actual sequence of above-lesion EMG signals during FES controlled walking by a paralyzed patient. The AR parameters sets were recorded at Michael Reese Hospital and Medical Center in Chicago during ambulation of paralyzed patients via the system described above. The network recognized successfully 95% of the ambulation functions indicated by the changes in patients' EMG signals.

### Concluding Comments

The NN controller described above is shown to allow an integration NN in patient-responsive EMG control of FES-based ambulation by complete upper-motor--neuron (traumatic) paraplegics, using EMG signals taken from the patient These signals are in turn produced by multitudes of neural action potentials of the patient's motor neurons as detected on the skin's surface. The resulting NN controller facilitates such ambulation while allowing close interaction of manual reinforcement commands from manual input pads of the patient to allow effective and safe man-machine interaction. Such interaction is essential for simple system training without external programming. It is applicable to even an untrained user, and allows safe ambulation in the face of system errors, especially noting the non- stationary nature of the EMG parameters that are employed.

## REFERENCES


Graupe,D., 1981. EMG Pattern Analysis for Function Discrimination Purposed, *Proc. NATO Advanced Studies Inst. on Spinal Cord Injuries*, Stake-Mandeville, England, May

Graupe,D., Kralj,A., Kohn,K.H., 1983. Computerized signature discrimination of above-lesion EMG for stimulating peripheral nerves of paraplegics, *Proc. IFAC Symp. on Prosthesis Control*, Ohio, Feb.

Graupe,D., 1989. EMG pattern analysis for patient-responsive control of FES in paraplegics for walker-supported walking, *IEEE Trans. on Biol. Eng.*, vol.36, no.7, July, 711-719

Grossberg, S., 1987. Competitive learning:from interactive activation to adaptive resonance, *Cognitive Science*, vol.11, pp.23-63

Carpenter,G.A., Grossberg,S., 1987. A massively parallel architecture for a self-organizing neural pattern recognition machine, *Comp. Vision, Graphics, and Image Proc.*, vol.37, 54-115

# A LARGE MEMORY STORAGE AND RETRIEVAL NEURAL NETWORK FOR BROWSING AND MEDICAL DIAGNOSIS APPLICATIONS

**DANIEL GRAUPE, Ph.D.**
*Dept. of EECS, University of Illinois, Chicago, Il.*

**HUBERT KORDYLEWSKI**
*Dept. of Bioengineering, University of Illinois, Chicago, Il.*

***ABSTRACT:***
Whereas most conventional artificial-neural-network designs are applicable only to a limited amount of memory, the network described in this paper, termed as LAMSTAR, is concerned with very large memories. The network is based on SOM (Self-Organizing-Map) modules which are modified such that only a limited number of neurons per each module is processed at a given iteration. The LAMSTAR input word is a coded vector X of sub-words (sub-vectors) xi. Each sub-word as above corresponds to a different category (feature, attribute) of the input word and is processed by a different SOM-type module. Three groups of weight vectors that are being controlled by the Task Evaluator, coordinate and prioritize search and retrieval, as follows: Weights Vi determine which sub-word is to be processed first (channeled first to a SOM module), weights Ni,j determine the order in which neurons in SOM module i be evaluated by correlation matching, whereas weights Li,j serve as links between SOM modules. Interpolation/extrapolation is governed by the L-links to fill in unobserved sub-words, or to filter out unobserved/forgotten digits in a given sub-word. A hierarchical structure, feedback, "intelligence" and learning are all shown to be utilized throughout to speed up operation, whereas rigidity is avoided through stochasticity. The network's operation is analyzed through a case study of an application to a simulated medical diagnosis problem - evaluation of patient's condition and long term forecasting after surgical removal of kidney stones.

## MOTIVATION

The neural network discussed in the paper is a concrete neural network for large scale memory storage and retrieval of Information. the LAMSTAR (LArge Memory STorage And Retrieval) network considered attempts to store and retrieve patterns in a computationally efficient manner, using tools of neural networks, especially SOM (Self Organizing Map)-based network modules, combined with statistical decision tools. The LAMSTAR network thus involves:

1. A robust and efficient large scale memory storage and retrieval model, based

on a link-weight vector structure, to allow fast search and retrieval, to consider forgetting and recall, using a probabilistic scheme and incorporating correlation interpolation and interpolation algorithms.

2. An efficient model for representation of inputs through simple ceded vectors.

## AN OUTLINE OF THE LAMSTAR NETWORK

### Basic Structural Elements

The basic processing modules of the LAMSTAR network are somewhat modified Kohonen SOM modules [Kohonen, 1988]. The network is thus organized to find a neural cell (neuron) in a set of neurons of a class (namely, in one SOM module) that best matches (correlates) with the input pattern. The SOM configuration is chosen since it can yield very rapid matching, and it can do so with good error tolerance and because of its generalization capability. We observe that the number of neurons in one SOM module can be huge if there are many patterns of a given class to be considered, Hence, for cases of large scale memory storage/retrieval as considered in the LAMSTAR, only a finite group of neurons (defined by weights Nik) in any SOM module will be searched, and not the whole module. A winning neuron is determined for each input based on the similarity between the input and a weight vector W (stored information). For an input subword Xi, the winning neuron is determined by:

$$||X - Wc|| = \min\{||X - W||\} \qquad (1)$$

where the c refers to the winning unit, W - weights.

The neuron with the smallest error determined by (1) is declared "winner" and its weights are adjusted. The learning process is described by the equation:

$$W_{new} = W_{old} + a * (X - W_{old}) \qquad (2)$$

where: $W_{new(old)}$ = the new (old) value of a weight connecting input X to the winning neuron, a = training rate coefficient.

### Time Correlations, Stochastic Weights and Forgetting Features.

The weights for the links between SOM modules (the correlation links L in Fig. 1, 2) are established on the basis of statistical correlations between the various concepts /words /patterns that are considered (stored or to be stored). Link weights decay over time when not recalled (used) for a length of time, this being a forgetting feature of the network. The correlation links are also changed via a feedback (F1 and F2 in Fig. 1, 2) which employs a punishment/reward of a connection that leads to incorrect recognition through changes in the link weights L and V. The changes are calculated according to eqn. 3:

$$L_{new} = L_{old} - a * (L_{old} - L_{max}) \qquad (3)$$

$L_{new/old}$ = new/old weight links. a = learning coefficient (0.2-0.4).

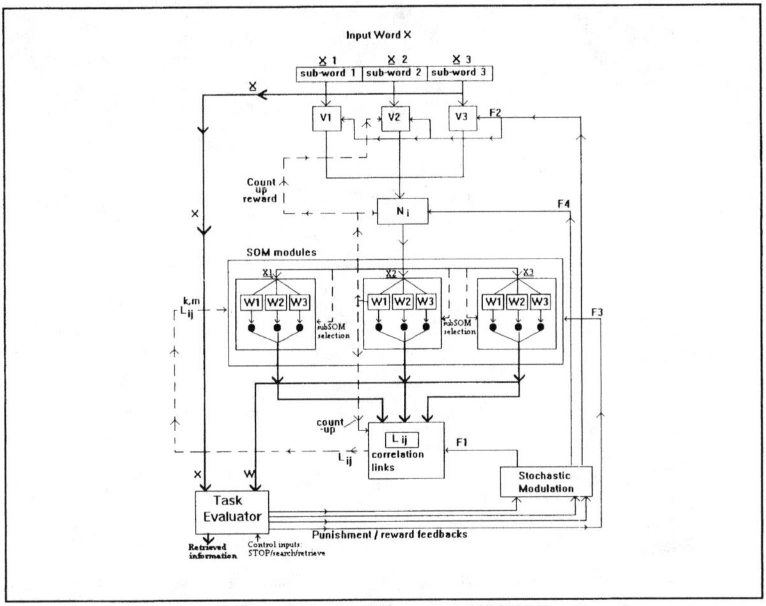


**Figure 1. General block diagram - LAMSTAR network.**

Punishment/reward formula in eqn.3 also applies to weights V.

**Channeling Weights for Fast Retrieval**

The retrieval phase of the LAMSTAR model selects subwords of an input-'word', one at a time and examines correlations with stored subwords of a given class (dimension) of subwords, say, EKG parameters. The channeling of the search that determines the order of exciting the neurons within a given SOM module is first determined by count-up/forgetting feedback from the concerned neurons as in Fig. 1 and 2, subject to priorities determined by the Task Evaluator. These rewards/punishments/priorities determine the mean of a probability distribution function (pdf) applied to pseudo white noise sequence to modulate the deterministic setting which is the mean of that distribution. The latter provides stochastic modulation to finally set the weights Nij, Li, and Vn (see Fig. 1). It is important to observe that in contrast the typical SOM structure, not all neurons of an SOM module are examined, but only a finite set (of p neurons), starting with the neighborhood of top priority (weight) neurons as determined by the weights (Ni) of Fig. 1.

**The SOM Correlations, Links and Wandering Searches.**

The winning output of the SOM module that was reached by the above feedback process and the subsequent stochastic modulation will activate all the nodes associated with it in other SOM modules. Consequently, a single concept subword automatically results in connecting to associated subwords with the highest Lij link-weights. The resulting retrieval is subsequently checked at the SOM level concerning correlations between

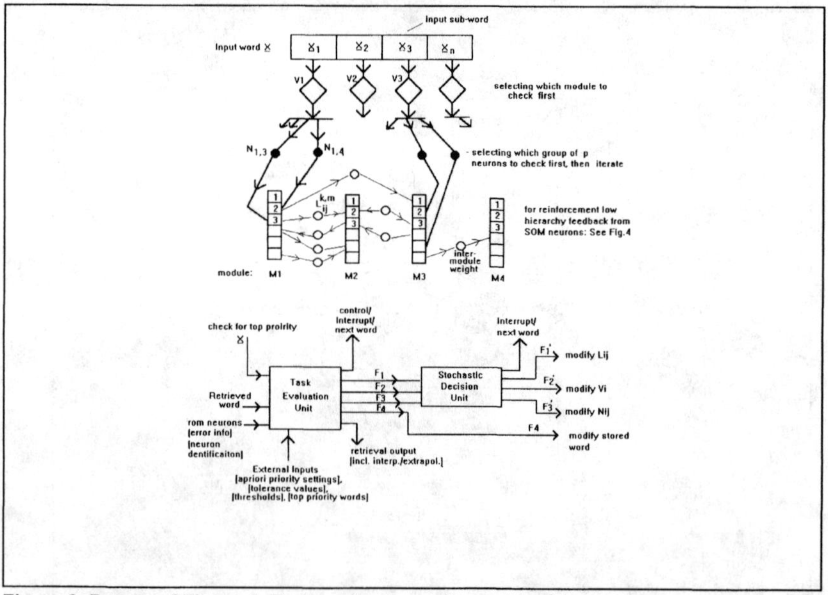


**Figure 2: Details of Figure 1. Task evaluation unit provides highest hierarchy of control - to modify tolerance and thresholds and to output interrupts. Stochastic Decision Unit introduces modulation noise to all settings of weights.**

stored subwords and input subwords. The links are then reinforced in cases of successful retrieval to "strengthen" the weights of such links. Therefore, if for a given SOM module, a winning neuron has been determined (for its corresponding input-subword), then first the highest L-valued links to subwords relating to another dimension of the input word (and which are or may be stored in another SOM module) are being examined by correlations. The final full word retrieval must however await an adequate approval of the Task Evaluator. A similar stochastic adjustment is applied to the weight Vi of Fig. 1 that serve to select which SOM module first receives (processes) a subword of any new input word. Since the decision on which of the input's features should be first examined greatly influences the speed of search, improved skill in that choice of this skill is being 'learnt' by the LAMSTAR network. This learning process (biasing the pseudo-random distribution of selecting the first module) is determined again by feedback from winning neurons. The Task Evaluator serves as an interrupt to stop search, to switch to a new input or to allow wandering to new links (browsing) even after a given search is over.

**Interpolation and Extrapolation of Missing Subwords/Browsing-Capability**

The activation of a subword of an input word propagates to other modules to activate further subwords relating to the same word via Links Li as in Figures 1 and 2. Such activation is interpolative and extrapolative in nature, since it requires the actual subwords of all but the first (the first being determined via Links Nj). Hence, if no subword on a certain category was inputted at all (at the input word considered) then such a subword could and is still retrieved through links Li. Say, if the input word describes a patient, then some link i can be retrieve a certain family history case, though the input contains no such subword. This feature also implies that browsing/wandering takes place.

## LAMSTAR CASE STUDY: AN APPLICATION TO MEDICAL DIAGNOSIS.

### Motivation.

The LAMSTAR network can be applied to problems with patterns and trends too subtle or too complex for direct human observation. While we cannot assimilate more than two or three variables at once, the present network can learn (store) and find (retrieve) correlations among tens of variables. Medical diagnosis is a problem that is based on more than one category of data, symptoms, signs, or measurements, with overlapping diagnosis, recommendation of medication etc. [Baxt, 1995]. Thus, it is almost impossible to create a reliable training set as required for diagnosis by conventional neural networks to handle all the input variables [Wyatt, 1991]. In the LAMSTAR network, the stored knowledge is encoded as a correlation among hundreds of concepts. The proposed network acquires its medical expertise automatically by analyzing and storing correlations among (usually not complete) input data. When the input word with data in only two subwords (e.g. EKG parameters that are already stored, and new patient's physical data) is to be stored, the network creates correlation between the new (physical) data and all data correlated with the already stored subvector (EKG parameters). Its training is self training during normal operation.

### Input Word Coding

The main task of the LAMSTAR network is to find a neuron or a set of neurons that best matches the input word. The input word is coded in terms of a vector X given by:

$$\underline{X} = [\underline{x}_1^T, \underline{x}_2^T, \dots, \underline{x}_N^T]^T \quad (4)$$

where T denotes transposition and where $x_i$ are subvectors .

In the proposed application the network analyzes an input word X that contains patients information, as well as possible diagnosis, recommended medication, and side effects. Each subword is a vector of real numbers. The subword relating to heart/lung contains information on heart rate, EKG, ventilation rate, and partial oxygen pressure of (PO2); the subword relating to physical characteristics contains information on name, profession, gender, age, name, height, etc.. Figure 4 shows the internal organization of three SOM modules related to time, physical characteristic, and heart/lung parameters, as well as, their correlation links, after four different subwords have been stored.

### Results

We tested the network operation by applying it to a problem of evaluation of patients' condition and long term forecasting after surgical removal of kidney stones. The patients' data were obtained from the Urology Department at University of Illinois Medical Center. The input word has 37 variables divided into 17 subwords. The lengths of subwords vary from one to six. The sample subwords are: (1) stone location/volume, (2) calcium, cystine levels, (3) metabolic rate, hypercalciuria, hyperuricousuria. We trained the network using 65 data sets. After the training the network diagnosed successfully 31 (94%) cases out of 33 testing sets. The two cases not recognized correctly are clearly some sort of aberration from the rule. It is important to notice that use of the backpropagation network is not efficient due to error minimization on 37th. dimensional space. The LAMSTAR network doesn't minimize the error but instead makes correlation with reward/punishment feedbacks and the network is fully trained after one presentation of training vectors.

**Conclusion.**

We have developed a network that provides not only very fast memory retrieval and efficient storage, but also interpolation and extrapolation of input data based on stored information. Fast retrieval results partly from analyzing only one sub-word of the input word. Instead of comparing the input with an entire collection of stored words, as in memory addressable storage models, our network compares only one sub-word with a sub-group of stored sub-words in selected SOM network. Thus, if the input word consist of n sub-words, the network analyzes only an n-th part of the memory. The rest of the word is retrieved by analyzing not the input word but correlation links associated the chosen sub-word. Implementation of pseudo-random selection of the sub-group for input analyses prevents the network from getting stuck. If the retrieved information is not accepted, the stochastic selection procedure selects other sub-words or correlation links to create a retrieved word. This retrieval model does not necessarily produce the best possible output, but an output that is acceptable with some tolerance. Also, the data processing is simpler than in most artificial neural network architectures, since an incorrect retrieval does not result in inhibition. Instead of inhibiting incorrectly retrieved data, the network's feedback punishes correlation links, namely, decreases their probability of being chosen. In storage of newly acquired knowledge, the system uses already existing concepts and only creates new association. In the retrieval process, the network not only extrapolates the input word, but also interpolates the input in case of minor variations in data. The extrapolation, which is a primary function of associated memory models, is a result of correlation based retrieval, where only one sub-word is needed to retrieve (extrapolate) all related information. Through SOM modules the network can recognize sub-words that are slightly different from the previously stored patterns (interpolation principle). This class of networks can be very effective in problems where training domain is not well defined, and where it is difficult to create reliable training sets, as in medical diagnosis example.

## REFERENCES

Baxt,W., (1995). Application of artificial neural networks to clinical medicine, *The Lancet*, vol.346, Oct.28,

Graupe,D., Lynn,W.J, (1970). Some Aspects regarding Mechanistic Modelling of Recognition and Memory, *Cybernetica*, No.3, 119-141.

Grossberg,S., (1987). Competitive learning: From interactive activation to adaptive resonance, *Cognitive Science*, 11, 23-63,

Johnston,M., Langton,K., Haynes,B., Mathieu,A., (1994), Effects of Computer-based Clinical Decision Support Systems on Clinician Performance and Patient Outcome, *Ann. Intern. Med*, vol.120, 135-142.

Kohonen,T., (1988). The neural phonetic typewriter, *Computer*, 21(3):11-22.

Wyatt,J., (1991), Computer-based knowledge systems, *The Lancet*, vol.338, Dec.7, 1431-1436

# Section 6 : Prediction

# Periodic Time-Series Analysis Using Neural Networks

*Senshu Ye and C. L. Philip Chen*[♦]

Department of Computer Science and Engineering
Wright State University
Dayton, Ohio 45435
Email: pchen@kiwi.cs.wright.edu
Ph: 513-873-5107 Fax: 513-873-5133

### Abstract

In this paper, we apply artificial neural networks to periodic time series analysis and prediction. Given a time-series *X( t )* with seasonal period *L*, we decompose it into *L* series of *X(i, t)*, each of which consists of data "at time i" of every season. The time-series *X(* t ) of one variable becomes a time-series of *L* variables, which seems to complicate the analysis. However, under certain circumstances, *X(i, t)* becomes easier to analyze for each *i*. With this model, an analytic method for multivariate model can be applied. For purely seasonal autoregressive model, *X(i, t)* is independent of one another for different *i*. For some stationary series, missing data problem can be dealt with by applying a non-time-series method. The proposed approach can estimate the trend function, *TR(t)*, of a traditional model.

**Keywords**: Time-Series Forecasting and Prediction

### Introduction

Given a time series $x(t)$, $t = 1, 2, 3,..., T$, and an integer $L$, define the moving average of lag L, $MA_L(t)$, as follows:

$$(1) \qquad MA_L(t) = \frac{1}{L}\sum_{s=t-K}^{t+K-1} x(s), \quad t = K+1, K+2, \ldots,$$

where

$$K = \begin{cases} L/2, & \text{if } L \text{ is even,} \\ (L-1)/2, & \text{if } L \text{ is odd.} \end{cases}$$

For a periodic time series with period $P$, $MA_P(t)$ is close to a linear function if it has the following form:

$$(2) \qquad MA_P(t) = at + b + e(t)$$

where $e(t)$ is a white noise, if appropriate function transformation (difference, log, for example) is performed before calculating $MA_P(t)$ whenever necessary.

This paper is divided into four parts: (1) introduce a method to find the period $P$ of a time series if there is one; (2) use the period found in part 1 for predicting the series ; (3) introduce and study a new class of time series: Quasi-periodic time series; (4) prediction for general time series.

♦ The support from an AFOSR grant F49620-94-0277 and Air Force contract F33615-94-D-5801 are gratefully acknowledged. Any opinions, findings, and conclusions or recommendations expressed in this article are those of the authors and do not necessarily reflect the views of the funding agencies.

Throughout this paper, we assume certain function transformation has been applied to the time series and so the trend of the series is linear.

**1. The period of a time series**

This section discusses the method of finding the period of a time series. For a given time series $x(t)$, $t = 1, 2, 3, \ldots, T$, let

(3) $$C = \begin{bmatrix} 1\;1\;1\;\ldots & 1 & 1 \\ 1\;2\;3\;\ldots & T-1 & T \end{bmatrix}'$$

and

(4) $$X = \left[x(1)\; x(2)\; x(3)\;\ldots\; x(T-1)\; x(T)\right]'$$

where $A'$ denotes the transpose of $A$. And let

(5) $$R = C(C C)^{-1} C X$$

be the least square linear approximation of $X$[1]. Now for every integer $L$, less than a given bound, we calculate the moving average $MA_L(t)$ of lag $L$, and compare it with $R$. The smallest $L$ that minimizes (locally) $norm(L)$, the norm of ( $MA_L(t)$ -$R$), and satisfies the condition that local minimum point $L'$ of $norm(L)$ is a multiple of $L$ is considered to be the period of $x(t)$. Figure 1 gives the results of two time series. Series 1 (Bowerman and O'Connell [2], Monthly hotel room for 1977-1990, page 312) has period 12, while series 2 [5] has varying 'period', whose $norm(L)$ reaches local minimum at $L = 8, 15, 23, 30, 38, 45$.

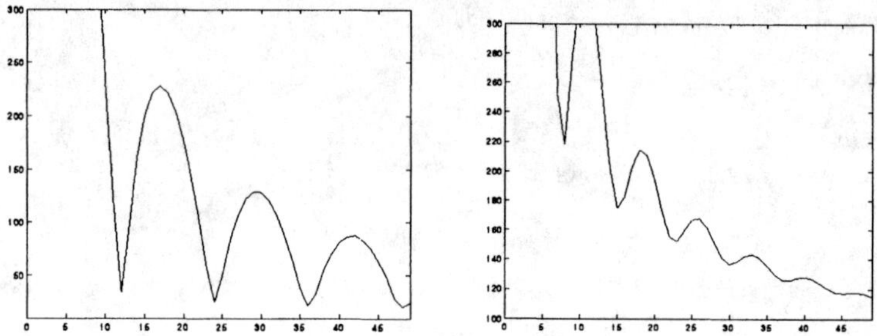

(a) norm(L) for time series 1

(b) norm(L) for time series 2

Figure 1 Periods of two time series

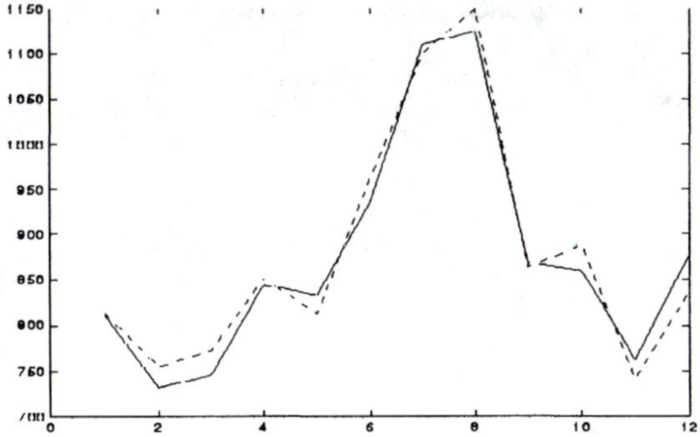


$$\text{Relative error} = \left( \frac{1}{12} \sum \left( \frac{\text{Actual data - prediction}}{\text{Actual data}} \right)^2 \right)^{1/2} = 0.0262$$

$$\text{Absolute average error} = \left( \frac{1}{12} \sum (\text{Actual data - prediction})^2 \right)^{1/2} = 22.3261$$

Figure 2 Prediction for time series 1 using Equation (6) solid line: actual data, dash line: prediction

**2. Prediction of periodic series using moving average.**

Once we have determine the period of a given time series from previous section, prediction is quite simple. Solving equation (1) gives us

$$(6) \qquad x(t) = P\, MA_P\,(t - K + 1) - \sum_{s=t-P+1}^{t-1} x(s)$$

where

$$K = \begin{cases} P/2, & \text{if } P \text{ is even,} \\ (P-1)/2, & \text{if } P \text{ is odd.} \end{cases}$$

As described by Chen [4], the prediction can be done very rapidly. The result of predicting 12 steps ahead is shown in Figure 2. The relative error is less than 3%.

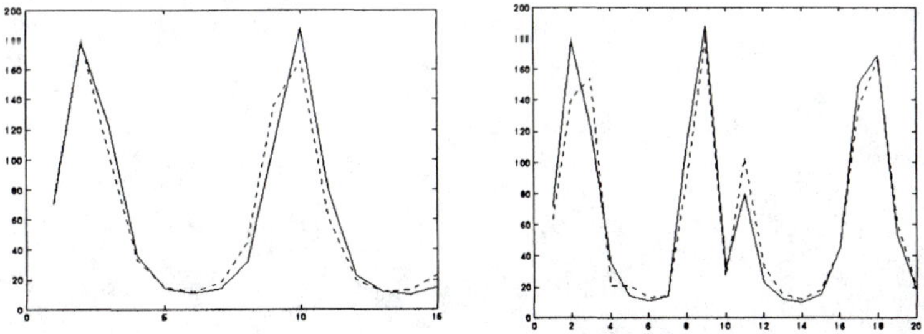


( a ) Period = 15
Relative average error = 0.2323,
Absolute average error = 12.0139.

( a ) Period = 23
Relative error = 0.2360,
Absolute average error = 14.7309.

Figure 3 Prediction for time series 1

## 3. Quasi-periodic time series and prediction using Neural Networks

Time series in real world rarely have a fixed period or they don't have period at all. Time series 2, as shown in Figure 1, has multiple periods (if we still call them period). The possible smallest period is 8. But $norm(L)$ doesn't reach minimum at $8n$ for $n > 1$. Instead, it reaches minimum at $15n$ for $n = 1, 2, 3$. Thus, we choose 15 as the period. Figure 3 is the prediction for series 2 using Equation (6) with $MA_p(t)$ trained using the neural network described in Chen [3].

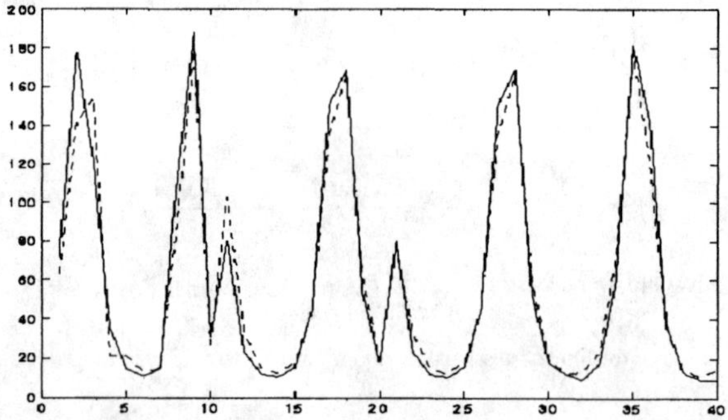


Relative error = 0.2377, Absolute average error = 7.5067.

Figure 4 Prediction of time series 1 using 1010 data in training, period = 23.

It appears that smaller period (period = 15) gives more accurate prediction in the near future, while bigger period (period = 23) gives more accurate prediction in a relative far future. But if we extend $T$ in our training data (from 1000 to 1010 in our example), prediction using period = 23 will be different (Figure 4). Although the relative error doesn't improve, absolute error is half of that before the extension of period. This means that the period varies from time interval to time interval. And we called such time series as *Quasi-period time series.*

## 4. General Prediction

We apply neural networks to AR model for our prediction for general time series. An AR model assumes $x(t)$ is linear function of $x(t-p), x(t-p+1)$, ..., $x(t-1)$. With neural networks,

$$x(t) = F(x(t-p), x(t-p+1), \ldots, x(t-1))$$

where F can be any (continuous) function. Let

$$M = \begin{bmatrix} 1 & x(1) & x(2) & \ldots & x(p) \\ 1 & x(2) & x(3) & \ldots & x(p+1) \\ \ldots & \ldots & \ldots & \ldots & \ldots \\ 1 & x(T-p) & x(T-p+1) & \ldots & x(T-1) \end{bmatrix}, \quad Y = \begin{bmatrix} x(p+1) \\ x(p+2) \\ \ldots \\ x(T) \end{bmatrix}.$$

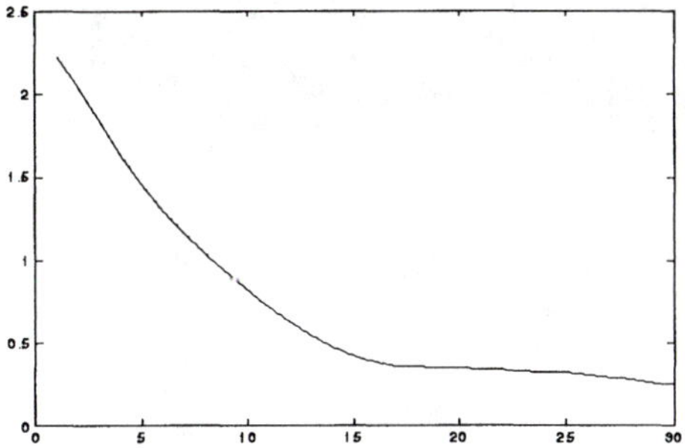


Figure 5 norm(L) for time series 3

Let $W$ be $(p+1)$ $X$ $n$ randomly generated matrix in the interval (-1, 1), where $n$ is a given integer.
And let

$$N = [M \mid \log sig(M\,W)], \text{ and } \quad w = (N'N)^{-1}N'Y$$

where

$$\log sig(x) = \frac{1}{1+e^{-x}}.$$

Now for $m = \left[1 \;\; x(t-p) \;\; x(t-p+1) \;\; \ldots \;\; x(t-1)\right]$, $y(t) = [m \mid \log sig(m\;W)]w$ is our prediction for $x(t)$.

In AR model, to determine the lag $p$ is an exhausting process. Here we suggest that a number between the two smallest periods be considered if it is a quasi-periodic series, and a number around the period be considered if it is a periodic one. In our first time series , $p = 12$ is chosen , in the second time series, any number between 8 and 15 is a candidate for $p$.

What if a time series does not exhibit any period at all? Our time series 3 (ftp ftp.cs.colorado.edu/pub/Time-Series/Santa Fe, A.dat & A.cont) is such one. Figure 5 is obtained after taking difference on the original data. $norm(L)$ decreases as $L$ increases.

From Figure 5 above we observe that the slope of the curve reaches minimum at a point in (3, 4) and another point around 17.(more obvious than the first one). And in practical, $p = 3$ is good for short prediction, and $p = 17$ is better for relative longer prediction. Figure 6 gives a 10 steps ahead prediction for both periods.

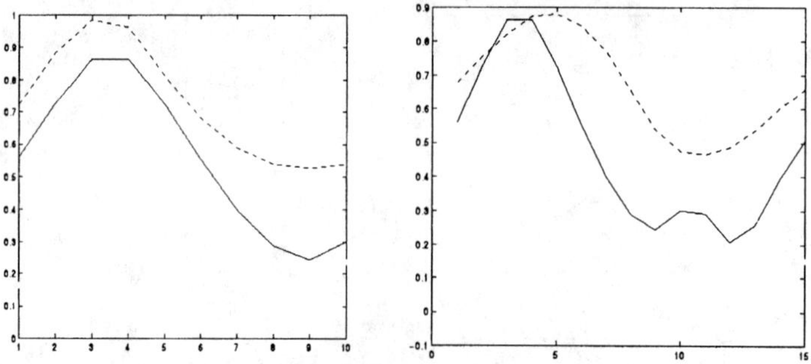

(a) period = 3, prediction up to 10 steps (b) period = 18, prediction up to 15 steps

Figure 6 Prediction of time series 3

## Conclusion

We apply artificial neural networks to periodic time series analysis and prediction. Four topics regarding to periodic time series are discussed. Using this approach, the prediction of a time series can be made easier and more accurate if there is a period in the series. Several examples are given.

## References


[1] George G. Judge et al, *Introduction to the Theory and Practice of Econometrics,* John Wiley & Sons, 2nd edition, 1988.

[2] Bruce L. Bowerman & Richard T. O'Connell, *Forecasting and Time Series, an Applied Approach,* Duxbury Press, 3rd edition, 1993.

[3] C. L. Philip Chen, "A Rapid Supervised Learning Neural Network for Function Interpolation and Approximation," To appear in *IEEE Transaction Neural Networks*, Vol. 7, No. 5, 1996.

[4] C. L. Philip Chen & John Z. Wan, "Forecasting Time Series Using an Artificial Neural Network with an Instant Learning and a Stepwise Updating Algorithm," Proceedings of *Int'l Conference on Artificial Neural Networks in Engineering (ANNIE), 1995, pp. 759-764.*

[5] Andreas S. Weigend & Neil A. Gerbenfeld ed., "Time Series Prediction: Forecasting the and Understanding the Past", *Addison-Wesley Publishing*, 1994

# HYBRID PREDICTION TECHNIQUE BY FUZZY INFERENCING ON THE CHAOTIC NATURE OF TIME SERIES DATA

**SABYASACHI GHOSHRAY**
*School of Computer Science & Department of Electrical Engineering*
*Florida International University, Miami, Florida 33133, USA*
*E-mail : ghoshray@fiu.edu*

***ABSTRACT:***
Predicting foreign exchange rates and stock market indices have been a well researched topic in the field of financial engineering. However, most methods suffer from serious drawbacks due to inherent uncertainty and the data acquisition problems. In this research, we have analyzed the very nature of the time series data from a pure dynamic system point of view and explored the deterministic chaotic characteristic in it. Here, the concept of chaos has been analyzed thoroughly and the relationships among chaos, stability and order have been explained with respect to the concept of time. Finally, a fuzzy reconstruction method based on fuzzy multiple regression analysis method has been used to predict the foreign exchange rates with accuracy.

## INTRODUCTION

An estimation problem of particular importance in the field of financial engineering is the problem of forecasting, or predicting trends in the foreign exchange market or in the stock market. All the various methods that have been applied in predicting the future movements of various financial instruments can be categorized by two broad classes, the fundamental analysis and the technical analysis (Rosenberg, 1996). Fundamental analysis is based on the relationships between the various factors that influences the economy and eventually develops some empirical models, such as, the balance of payments flow model, monetary models of exchange rate determination, currency substitution model, the hybrid monetary/fiscal policy model etc. On the other hand, in the technical analysis, the prediction is done by analyzing a set of past data by various methods, such as; time series analysis, regression analysis, expert systems, etc.

Because exchange rate movements affect so many business, investment and policy decisions, the proper determination of the currency exchange rate has become the primary objective of all market participants. However, the correct determination of exchange rate on a consistent basis has become very difficult. In their continuous effort to come up with the best model, the economists and financial analysts have come up with a wide range of prediction methodologies. Research has indicated that fundamental analysis based models can be used in explaining long-range trend in currency movements, whereas it is inadequate in explaining the short- and -medium range fluctuations in exchange rate data (Rosenberg, 1981). Against this backdrop, many technical-analysis based models have been found to be somewhat successful in predicting currency exchange rates over short-run periods. This relative success of the technical models in currency exchange rate, compared to the lack of success of the fundamental-based models, has made the technical analysis become extremely popular among market participants. However, all these methods suffer from drawbacks in their inability to predict the future values with consistent accuracy (Yasunobu et al, 1993). This occurs due to the lack of reliable knowledge base, as the data bases are not well organized or not coherent enough. Secondly, the movements of both the foreign exchange currency and the stock market follow a chaotic path based on dynamic deterministic systems. Therefore, present methods are not adequate enough to model the dynamic systems that represent the fluctuations in stock market or foreign exchange market. In this research, we have examined several uncertainty modeling techniques, analyzed the behavior of time series data and developed a fuzzy inferencing technique to predict the future values of a time series.

## EXISTING PREDICTION METHODOLOGIES

Where the fundamental-based predictions rely on independent projections of the related economic variables determining exchange rates, the technical models predict exchange rates by analyzing and subsequently extrapolating the set of past data by various methods as mentioned in the previous section. There exists a wide variety of foreign currency exchange rate prediction methods which is based on trend, but, in all fairness, they share a common thread which is the property of extrapolation. Herein comes an inherent drawback, since these models generate forecasts by extrapolating the recent past trend of exchange rates in to the future, the buying or selling can be done only after the currency has already started rising or falling. Therefore, these prediction models can not catch the very top and the bottom of market moves. Instead, they rely on the assumptions that the exchange rates must always move in large swings such that the model can capture enough of a market move in order to be able to earn a sizable profit.

Various studies have been performed to ascertain whether changes in exchange rates can be forecasted solely on the basis of their past behavior (Hsieh, 1989). One trend among the researchers is to seek the existence of stable linear relationships between the current change in the exchange rate (sat at time t), $\Delta E_t$ with the exchange rate in the preceding period $\Delta E_{t-1}$. If there exists a stable positive linear relationship that could confirm the existence of a trend in the currency exchange rates. This subsequently would ascribe validity to the numerous prediction methodologies of today. However, most studies concluded that there exists no stable positive linear relationship between $\Delta E_t$ and $\Delta E_{t-1}$. Series correlation test would determine if there exists a stable linear relationship between $\Delta E_t$ and $\Delta E_{t-1}$. Evidence of positive serial correlation will indicate that a positive change in a currency's value at time t will be followed by another positive change tomorrow.

## CONCEPT OF CHAOTIC MARKET BEHAVIOR

Behind the new tools available for non-traditional research are developments in physical sciences that have a direct impact on financial trading and portfolio management. The new methods have originated from the theories of the physical sciences. The most fundamental characteristic that gives an edge over the traditional methods is the fact that all forms of organizations simultaneously map themselves into two states that apparently are at odds with each other; change and order. Change is characterized by the freedom to move away from the status quo by altering things in a manner reflecting new developments and by taking full account of the forces behind the process of innovation. The representation of this change is complex, and the transition between change and order involves a great number of uncertainties. By a sharp contrast, order is represented by simpler and settled paths to regulation and is often expressed in a structured, hierarchical manner. Both change and order are essential, though they are contradictory to each other. The theory of complexity addresses the problem of transition by interfacing chaos and order. Therefore, in a complex system chaos and order go through transitions. Furthermore, chaotic systems are essentially periodic. What makes them appear random is a continual transition from one periodic orbit to another. To delve deeper into chaos and its applicability in financial systems, it is essential to understand the relationships among chaos, non-equilibrium and instability. A chaotic system could be stable if its particular irregularity persisted in the event of small disturbances. Thus, in spite of having a pattern of irregularity and being locally predictable, chaos may be globally stable. These patterns of disturbances in chaos can be defined as orderly disorder. Therefore, by analyzing the way in which patterns change with respect of time, it is possible to make more sense of chaotic behavior of complex systems, such as financial markets. In this context, the failure of existing models of economy in efficiently predicting future events lies in the fact that these models fail to establish the relationships among chaos, instability and equilibrium. It is to be noted that, there can be no equilibrium condition in the transitions in a dynamic environment if it is steadily characterizing the alterations of change and order. The existing models of economy such as the equilibrium theory and the efficient market hypothesis fail to recognize this aspect. It is therefore, of paramount importance to understand that no dynamic system can be correctly modeled if the analysis either ignores time or treats time as a controllable variable.

### Time Series and Non Linearities

In financial analysis, much of the modeling process aim at the study of time series of values resulting from measurements attributed to specific events. Traditionally, the changes in values of a specific variable has been represented as a function of time. In this regard, mostly linear algorithms are being used to exploit and visualize the relationship that exist between time series values.

Nonlinear approaches to short-term forecasting employ chaotic attractors, making it possible to predict an impending crisis. This is done by dynamically examining time series taken near the catastrophic bifurcation threshold. Chaos theory can be helpful in this context because it represents the possibility that random systems might actually be defined by simple procedures and rules, thereby making them predictable. The objective in chaos theory in the context of financial markets is to develop models that can predict with unprecedented accuracy. Therefore, the aim of this research is to apply prediction capabilities to financial markets and financial instruments so that a high degree of terms of predictability is attained.

The implementing of chaos theory in currency exchange rate prediction essentially revolves around the dynamics of nonlinear solution spaces. The results obtained from chaotic system apparently can look random, which however, is not truly random. In this case, each chunk of values are limited within a given region. This stochastic behavior allows us to predict short range time series data more accurately.

## DYNAMIC BEHAVIOR OF THE TIME SERIES DATA

We begin this section by defining the general concepts of chaotic behavior, stability, and genericity for dynamic systems. Chaos is the mathematical term for the behavior of a system which is inherently unpredictable. A dictionary definition of chaos is "a disordered state of collection; a confused mixture", which is an accurate description of dynamic systems theory. Today a dynamic system is one which changes with time. To understand a dynamic system means to know how it states vary through time - at least to describe their variation, at best to predict it. One of the ingredients of a chaotic system is feedback. Let us explain this with the logistic equation that is introduced to explain the basics of chaos theory: $X_t = K * X_{t-1} * (1 - X_{t-1})$, where *K* is a constant.

In this logistic equation, the results from the previous iteration is fed back repeatedly into the next one. In this context, we can see how trivial the above equation is. It is a simple quadratic formula in the variable $X_{t-1}$. Given $X_{t-1}$ and *K*, we can compute $X_t$ exactly. Our experiments have shown, when *K* is small, the value $X_t$ can be quite predictable. For *K*=0.5, the value $X_t$ tends to 0; whereas for *K*=1.2, 2 and 3, it tends to stabilize or reach a definite limiting value. At above 3, different values of *K* yield significantly different results. This example illustrates in unmistakable terms the consequences of chaos indeterministic systems because of the following observations: First and foremost, it is not possible to accurately predict the result and secondly, the search for a specific solution in a chaotic system is useless. Therefore, in dealing with chaotic systems, we focus on a more global viewpoint and instead of trying to predict the particular behavior of a solution, we seek to identify the totality of all these solutions.

In the search for solutions in a deterministically chaotic dynamic system, an important object is the strange attractors. From a dynamic system point of view, an attractor can be defined as an object toward which all nearby solutions adhere to as time moves on. If we consider a dynamic system, in which the state space *W* can be fined as an open set in some real Banach space *B*, such that *K* is a nonempty compact set having a neighborhood *N* in *W* with the following properties: every trajectory starting in *N* has compact closure in *W* and has all its limit points in *K*; *K* is defined as an attractor. Chaotic dynamic systems are sometimes said to have "strange attractors" or even "strange strange attractors" (Hirch, 1986). In spite of some vagueness associated in this concept, it means, *K* is neither a single stationary point nor a single periodic orbit.

## PREDICTION METHODOLOGY

In this research, the necessary and sufficient condition for the reconstruction of state space of the dynamic system into a 1-dimensional Euclidean space R has been substantiated in accordance to Taken's Theory (Takens, 1981). The first step involves the construction of the n-dimensional time series. We construct the n-dimensional time series from n consecutive quotes of an one-dimensional series. This trajectory in n-dimensional space is similar to the trajectory of the state of the dynamic system. This chaotic dynamic system generates the one-dimensional time series as this state evolves in phase space. In practice, it is not known apriori the dimension of the phase space, henceforth, it is not possible to determine the number n for the reconstruction. In our experiment, we have tested with all the numbers in the set $1 \leq n \leq 10$ and computed the largest Lyapunov exponent as well as the correlation dimension corresponding to each of the reconstructions. According to Takens, the method of determining the correlation dimension of an attractor or *W*-limit set in a physical experiment consists of two parts. First the "reconstruction of the phase space" in $R^n$ and second the determination of the correlation dimension of a bounded subset in $R^n$. In the first part, a sequence , $\{a_0, a_1, a_2, \ldots\}$ obtained as a time sequence of experimental data $(a_i \in R)$, is used to produce a sequence $\{A_0, A_1, A_2 \ldots\}$ in $R^n$ ; this sequence is defined by $A_i = (a_i, a_{i+1}, \ldots a_{i+n-1})$. The experiment involves a finite dimensional deterministic model with phase space X, if the sequence $\{a_i\}$ is obtained by composing the orbit x(i) $\in^x$, describing the evolution during the experiment with the mapping, $y: x \rightarrow R$, such that $a_i = y(x(i))$. Then the sequence $\{A_i\}$ and the set A in $R^n$ are the image of $\{x(i)\}$. It has been proven that if n is sufficiently big, and if y together with the dynamics in x satisfy certain generic conditions, then this transformation form x to $R^n$ is termed as embedding (Takens, 1981).

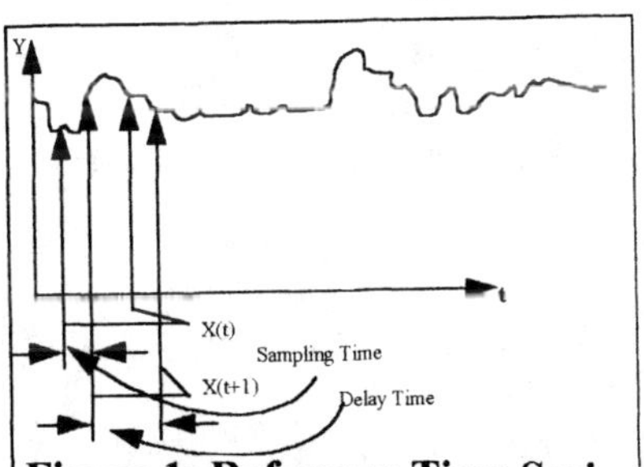


**Figure 1: Reference Time Series**

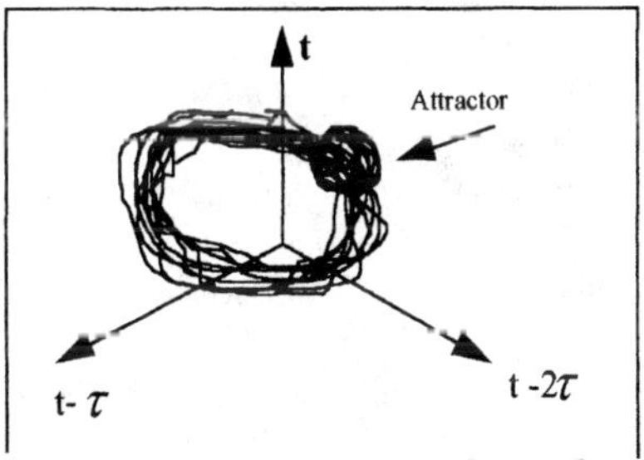


**Figure 2: Reconstructed State Space**

In our experiment, the vector x(t) represents a single point of an n-dimensional reconstructed state space $R^n$ . Figures 1 and 2 above show how the embedding of the original trajectory takes place for a sufficiently large value of the embedding dimension n. Methods of reconstruction a data vector $\psi(T)$ is generated in the state space in which $\Phi(i)$ is the past data, and the state $\Phi$ (i + s) at "S" steps ahead is already known. Analysis of this time series subsequently results in the predicted value $\hat{\Psi}$T +S) of $\Psi$ (T + S). A detailed discussion on fuzzy inferencing is found in our earlier research (Ghoshray, 1996).

**Table 1: Experimental Values For Exchange Rates Between FF/US$**

| Embedding Dimension (n) | Delay Time $(\tau)$ | Data Vector (N) | Correlation Coefficient $(\lambda)$ |
|---|---|---|---|
| 6 | 3 | 6 | 0.982 |
| 5 | 2 | 7 | 0.984 |
| 7 | 3 | 5 | 0.979 |

**Table 2: Experimental Values For Exchange Rates Between SFr/US $**

| Embedding Dimension (n) | Delay Time $(\tau)$ | Data Vector (N) | Correlation Coefficient $(\lambda)$ |
|---|---|---|---|
| 8 | 4 | 5 | 0.979 |
| 6 | 3 | 8 | 0.982 |
| 10 | 2 | 4 | 0.978 |

## CONCLUSION

In this research we have presented a new methodology of predicting foreign exchange rates by fuzzy inferencing method. Time series data of the exchange rates of several European Currencies versus US $ have been tested and the comparison results have been presented in Table 1 and 2. It has been observed that as the dimension of embedding (n) and delay time (τ) increases and the number of neighboring data vector (n) increases, the correlation coefficient between the currencies increases, which verifies the effectiveness of the proposed method.

## REFERENCES


Ghoshray, S., (1995). Foreign Exchange Rate Prediction By Fuzzy Inferencing On Deterministic Chaos, *IEEE/IAFE Computational Intelligence for Financial Engineering*, March 16-18.

Hirch, M., (1986). The Chaos In Dynamical Systems, *Chaos, Fractals, and Dynamics*. pp. 189-196.

Hsieh, P., (1989). Testing for Nonlinear Dependence in Daily Foreign Exchange Rates, *Journal of Business*, 62, No. 3, pp. 339-368.

Rosenberg, M. (1996). Currency Forecasting: A Guide To Fundamental and Technical Models of Exchange Rate Determination, *Irwin Publishing*, pp. 345-377.

Rosenberg, M. (1981). Is Technical Analysis Right For Currency Forecasting?, *Euromoney*, pp. 125-131.

Takens, F., (1981). In Dynamical Systems and Turbuelence, *Springer -Verlag, Berlin*, pp. 366 -381.

Yasunobu, C., Kosaka, M., and Yokomura, K. (1993). A Fuzzy Dealing Support System, *IEEE NO. 32, Denshi Tokyo Section*, pp. 85-88.

# TIME SERIES PREDICTION USING ASSOCIATIVE MEMORY NEURAL NETWORKS

**S. J. SILVER-WARNER AND R. J. GLOVER**
*Neural Networks and Pattern Recognition Research Group,*
*Department of Electrical and Electronic Engineering,*
*Brunel University, Uxbridge, Middlesex, UB8 3PH, England*

***ABSTRACT:***
A normalising preconditioning matrix based on the structure of the activation matrix is presented for Associative Memory Neural Networks (AMNNs), which is shown to improve significantly the convergence time for training on the Mackey-Glass time series. The commonly applied Jacobi preconditioner is found to be unsuitable for this application. A modified Jacobi preconditioner is presented which gives similar performance to the normalising preconditioner while converging faster, but requires a maximum scaling factor to be defined. The generalisation performance of AMNNs can be improved by initialising the basis function weights with values obtained from a global model of the training data. Examples are given for mean and autoregressive global models.

## INTRODUCTION

AMNNs perform non-linear function approximation, but with linear adjustable parameters. AMNNs form local representations of the function being modelled, rather than the global approximation method of the Multi-Layer Perceptron (MLP). One of the simplest AMNN models is the Cerebellar Model Articulation Controller (CMAC), which has seen considerable use in robotics control (Albus, 1981).

An AMNN consists of an input layer, a layer of basis functions with compact support and an output layer which weights and sums the basis function responses. The basis functions occupy locally compact regions of the input space, so only a few are activated by a single input vector. As the output is linear in its adjustable parameters the error surface is quadratic, and training is a linear least mean square optimisation. See (Brown and Harris, 1994) for a comprehensive description.

B-Spline networks are popular AMNNs which consist of piecewise polynomial segments taking the form of basis functions (Lane, Handelman and Gelfrand, 1992). The memory requirements of these networks rise exponentially with the B-Spline order and input dimension, so a content addressable memory is used to store only those basis functions activated by the training set (Silver-Warner, Glover and Stonham, 1996). To improve the network performance when the test set activates basis functions not defined by the training set the response of the existing basis functions are normalised to force a partition of unity (Werntges, 1993).

The Mackey-Glass equation (1), is a chaotic attractor commonly used to evaluate the performance of time-series prediction algorithms. This equation is integrated using a fourth order Runge-Kutta algorithm with an iteration step of 0.025, and equation parameters of $a$=0.2, $b$=0.1 and $T$=17. We are trying to find a prediction equation $f$ such that $y_{t+P} = f(y_t, y_{t-\Delta}, y_{t-2\Delta}, y_{t-3\Delta})$. We use a sampling interval $\Delta$=6, and make predictions at time P=6.

$$\frac{dy_t}{dt} = \frac{a y_{t-T}}{1 + y_{t-T}^{10}} - b y_t \tag{1}$$

The prediction accuracy is measured by the Normalised Mean Square Error (NMSE) which is the root mean square error between the predicted and actual time-series, over the standard deviation of the actual time-series. The B-Spline AMNN models below are trained on 1000 vectors, and tested on the subsequent 1000. Second order B-Splines with basis function widths equal in all dimensions are used.

## ACCELERATING TRAINING CONVERGENCE OF AMNNS

A property of AMNNs used for time series modelling is that the number of training vectors contributing to each basis function is not constant. A preconditioning matrix based on the distribution of training vectors to basis functions can significantly improve the training convergence rate. The output vector $y$ of an AMNN can be characterised by the following equation

$$\mathbf{y} = \mathbf{A}\mathbf{w} \tag{2}$$

$\mathbf{w}$ is a vector of the weights associated with each basis function, $\mathbf{A}$ is a matrix of basis function activations given by,

$$\mathbf{A} = \begin{bmatrix} a_{1,1} & a_{1,2} & \cdots & \cdots & a_{1,m} \\ a_{2,1} & \ddots & & & \\ \vdots & & a_{ij} & & \\ \vdots & & & \ddots & \\ a_{n,1} & & & & a_{nm} \end{bmatrix}$$

where $a_{ij}$ is the activation of the $j$th basis function for input vector $i$, and $m$ is the total number of basis functions, and $n$ is the total number of training vectors.

We are trying to find the least mean square solution between the desired output and the actual output by optimising weight vector $\mathbf{w}$. This is given by the solution to

$$\mathbf{b} = \mathbf{B}\mathbf{w} \tag{3}$$

where $\mathbf{b} = \mathbf{A}\mathbf{y}$, and $\mathbf{B} = \mathbf{A}^T\mathbf{A}$, which can be solved by many methods, including conjugate-gradient (Shewchuk, 1994). In order to reduce the number of cycles needed to converge on the global minimum, equation (3) can be transformed using a preconditioning matrix, $\mathbf{M}$.

$$\mathbf{M}^{-1}\mathbf{b} = \mathbf{M}^{-1}\mathbf{B}\mathbf{w} \tag{4}$$

The gradient vector of the error surface is given by

$$\mathbf{g} = \mathbf{M}^{-1}\mathbf{A}^{T}\mathbf{e} \tag{5}$$

where $\mathbf{e}$ is the residual error between the desired and actual output vectors.

The matrix **M** attempts to make the error surface associated with equation (3) more hyper-spherical, hence a gradient vector will give a better estimate for the direction of the global minimum. The ideal preconditioning matrix is **B**, which will provide a one-step solution, but this of course requires the inversion of **B**.

The Jacobi preconditioner is the simplest preconditioner (Shewchuk, 1994) which consists of the elements on the main diagonal of **B**. It scales the error surface along the co-ordinate axis, and can effectively reduce convergence time on the training set. However, the Jacobi preconditioner often degrades test set performance. The variation in the response makes it difficult to determine the global minimum (see figure 1). This effect is due to the differences between the training set and test set error surfaces. By scaling up the error surface along dimensions which have a low sensitivity to error for the training set the weights associated with those dimensions are adjusted far more in the early stages of training than they would have been without preconditioning. This can result in those weights with low sensitivity to error for the training set fit to noise in the training set, consequently giving a poor test set performance.

Another problem encountered with Jacobi preconditioning is that **M** cannot always be inverted. This situation arises when an input vector falls on the boundary of support of a basis function, activating it, but giving a zero response. If that input vector is the only one to activate a basis function, **B** and hence **M**, will contain a zero on the leading diagonal, making it non-invertable. An extra pre-processing stage could be used to remove these superfluous basis functions.

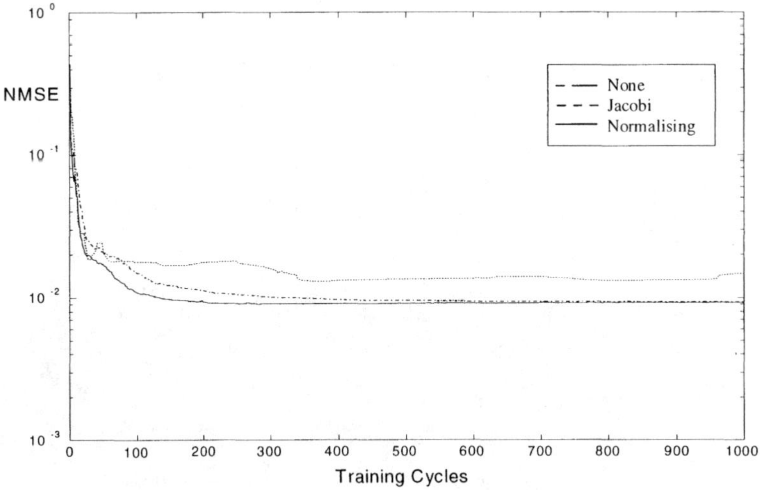


**Figure 1 : Example test set training histories for the Mackey-Glass time series.**

An alternative preconditioning matrix can be formed by considering the structure of the activation matrix **A**. Each row of A has $k$ entries, where $k$ is the number of overlays. The non-uniform distribution of the input vectors results in some columns of **A** having more entries than others. A preconditioning matrix can be formed by calculating the number of non-zero entries in each column and assigning those values to the corresponding elements on the leading diagonal of **M**, with all other entries zeroed. This preconditioner scales the error surface along the co-ordinate axis according to the number of input vectors which activate a given basis function. It does not scale up those dimensions which have a low sensitivity to error for the training set, and is guaranteed to be invertable. Figure 1 shows the network converging more quickly than without the preconditioner. In addition, the minimum error is significantly lower and the curve significantly smoother than with the Jacobi preconditioner.

Figure 2 compares test set performance of the Jacobi preconditioner with the normalising preconditioner for models with a range of basis function widths. Each model is trained for 1000 cycles, and the best result on the test set is given. This strategy removes the dependence on the criteria to stop training, but limits the number of cycles available for training. In general, the normalising preconditioner is performing much better than the Jacobi preconditioner.

The problems with the Jacobi preconditioner are associated with the excessive scaling up of some dimensions in the error surface. Test set performance can be improved by defining a maximum limit to values in $\mathbf{M}^{-1}$. The problem with this approach is in deciding what the limit should be. A high value gives faster training convergence, but if it is too high, the test set performance degrades. The modified Jacobi, with maximum limit of 8, and the normalising preconditioner were compared for a range of basis function widths. The NMSE performances were virtually identical, but modified Jacobi required on average about 30% fewer training cycles.

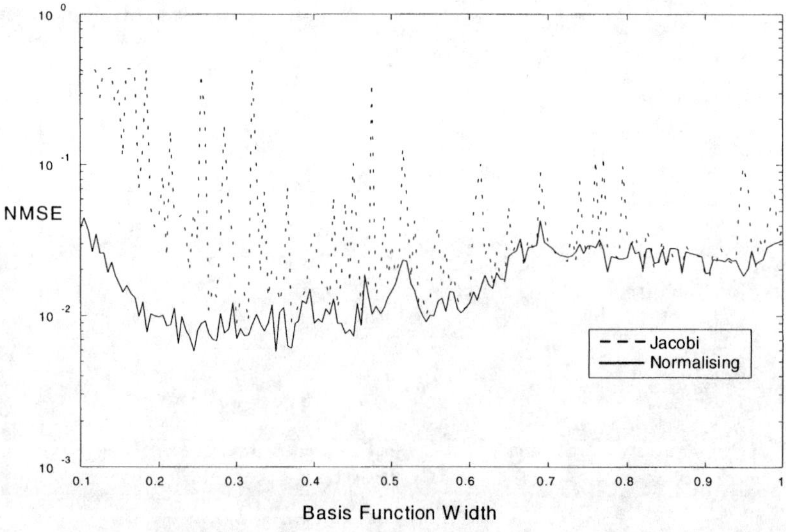


**Figure 2 : Mackey-Glass test set performance comparison between Jacobi and Normalising Preconditioning.**

## TRAINING TERMINATION CRITERIA

Training should be terminated when the test set error after each update reaches its minimum. Problems arise due to noisy traces and false minima, hence the following method is used. The test set is split into interpolation and extrapolation sub-sets. The extrapolation set consists of those input vectors which activate basis functions not created for the training set, the interpolation set consists of the remaining vectors. The performance on the interpolation set is first tested by examining a number of the most recent error measures. If the error in the first test point is less than the errors in the remaining points, a subsequent test is carried out on the overall test set (combining interpolation & extrapolation errors). If the second test also passes, training is terminated. The test window used here spans 20 cycles.

There will not necessarily be an optimal time to terminate training, the test set response can also decay asymptotically, so a limit on the total number of training cycles is imposed. This limit is set to 5000 cycles for the Mackey-Glass time-series.

## IMPROVING GENERALISATION BY PARAMETER INITIALISATION

It has been suggested that an AMNN which has zero initial conditions generalises sensibly (Brown and Harris, 1994). Generalisation can be improved by initialising the weights using a global model of the training data. This enables global relationships to be captured which could otherwise be ignored by a local training algorithm. This is particularly relevant when the lattice resolution is relatively fine, and the model has more degrees of freedom than is used by the training data. The global initialisation pre-defines the degrees of freedom orthogonal to the training data to values more appropriate to the function being modelled. Those basis functions which would fit to noise in the training data if over-trained, are also assigned more suitable weights.

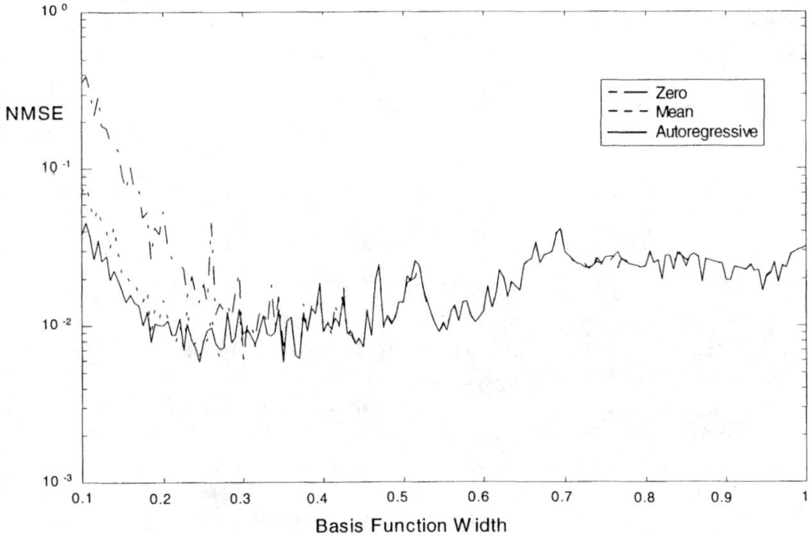


**Figure 3 : Comparing performance using zero, mean and AR initialisation**

The vector defining the centre of each basis function is used as the input to a global model, and the response gives the weight to be assigned to the basis function. For this method to be realistic a global model has to be obtained very rapidly. Two very simple global models are used : a mean model, and an autoregressive model. A mean model is simply a scalar which is the mean of the desired training set responses, and is only useful when the output actually contains an offset. An AMNN which forms a partition of unity can exactly reproduce this model. An autoregressive model is obtained using the conjugate gradient algorithm (Shewchuk, 1994), and provides an AMNN with information about the linear relationships within the training data. A B-Spline AMNN can also exactly reproduce an AR model. Another important aspect of this initialisation procedure is that the improved initial conditions can significantly reduce the number of cycles required to train an AMNN.

Figure 3 shows the generalisation improvement obtained when modelling the Mackey-Glass time series. This graph compares the performance of zero, mean and AR initial conditions for a range of lattice resolutions. The models are trained using a normalising preconditioner. The improvements obtained are more significant for the finer resolution models which increasingly over-parameterise the model. This method of global initialisation is therefore partially overcoming the problem of performance degradation associated with over-parameterisation.

## CONCLUSION

The convergence rate of an AMNN can be improved significantly with the normalising preconditioner. The test set performance of the normalising preconditioner is significantly better than the Jacobi preconditioner, and exhibits less problems with false minima. The modified Jacobi preconditioner provides similar performance to the normalising preconditioner, but with faster convergence. However, further work is required to develop heuristics for determining the maximum scaling factor. By initialising the weights associated which each basis function with values obtained from a global model the generalisation performance can be significantly improved, and the better initial conditions reduce training cycles. Future work will evaluate the performance of the preconditioners and global model initialisation methods on a wider range of data sets.

## REFERENCES


Albus, J.S., (1981). Brains, Behaviour, & Robotics, *Byte Books.*

Brown, M., Harris, C.J., (1994). Neurofuzzy Adaptive Modelling and Control, *Prentice Hall, Hemel Hempstead, UK.*

Lane, S.H., Handelman, D.A., Gelfrand, J.J., (1992). Theory and Development of Higher-Order CMAC Neural Networks, *IEEE Control Systems*, April, pp.23-30.

Shewchuk, J.R., (1994). An Introduction to the Conjugate Gradient Method Without the Agonizing Pain, *Carnegie Mellon University Technical Report Number* : CMU-CS-94-125, March.

Silver-Warner, S.J., Glover, R.J., Stonham, T.J., (1996). Associative Memory Neural Networks for Time Series Prediction, *IEEE Int. Conf. on Electronics, Circuits and Systems,* Rodos, Oct.

Werntges, H.W., (1993). Partitions of Unity Improve Neural Function Approximators, *IEEE Int. Conf. on Neural Networks*, pp. 914-918.

# EVOLVING RECURRENT BILINEAR PERCEPTRONS FOR TIME SERIES PREDICTION

**SATHYANARAYAN S. RAO AND KUMAR CHELLAPILLA**
*Department of Electrical and Computer Engineering*
*Villanova University, Villanova, PA, U.S.A 19085*

***ABSTRACT:***
The development of efficient algorithms, based on optimization techniques, for learning the weights and architecture of recurrent neural networks (RNN), has recently, received much attention. Though many training methods have been successfully used in training RNNs, they suffer from certain fundamental drawbacks. Firstly, it is difficult to devise efficient learning algorithms that guarantee stability of the overall system. Secondly, they rely on some type of approximation for computing the partial derivative, which may be inadequate to train them for tasks involving long-term dependencies. Furthermore, no analytical results concerning the convergence and stability of these schemes have been obtained. In this paper we propose a recurrent bilinear perceptron (RBP) for time series prediction. Evolutionary programming, a multi-agent stochastic search technique, is used to optimally determine the model order and coefficients of the recurrent bilinear perceptrons. Unlike conventional techniques, where the model order is chosen first, and then the parameters of the model are determined, both the model order and the parameters are evolved simultaneously. The minimum description length (MDL) information criterion is used to evaluate each recurrent bilinear perceptron structure as a candidate solution. Experimental results on time series prediction indicate that RBPs are capable of performing better than simple bilinear models in terms of reduced normalized mean squared error (NMSE) and lower evolved model order.

## INTRODUCTION

The development of efficient algorithms, based on optimization techniques, for learning the weights and architecture of recurrent neural networks (RNN), has recently, received much attention. Though many training algorithms have been successfully used in training RNNs, they suffer from certain fundamental drawbacks. Most of these methods rely on the gradient methodology and involve the computation of the partial derivatives with respect to the weights of the network. These algorithms are extensions of backpropagation (BP) for feedforward neural networks. Examples of such extensions are the recurrent back-propagation algorithm (Pinenda 1988), the back-propagation through time algorithms (Werbos 1990), the real time recurrent learning algorithm (Williams and Zipser 1989), and the dynamic backpropagation algorithm (Narendra and Parthasarathy 1991). Because recurrent BP also performs a gradient search in the parameter space, it, as other gradient based algorithms, converges very slowly.

In comparison, it is found that, although recurrent networks generally outperform feedforward networks, they appear more difficult to train optimally. Often their parameters settle in sub-optimal solutions that take into account short term dependencies but not long term dependencies (Bengio 1992), (Mozer 1992). It is shown that BP is not sufficiently powerful to discover contingencies spanning long temporal intervals. Quasi-newton algorithms perform better than BP, but fail more and more as the span of the input-output dependencies is increased, with the increased risk of instability during training. Theoretical and experimental evidence has been presented in (Bengio 1994) that shows that gradient descent of an error criterion may be inadequate to train them for tasks involving long term dependencies. They strongly suggest that when prior knowledge, that is helpful in setting the network connectivity and initial weights, is given, it should be used, since the learning problem itself is so difficult.

One of the reasons behind learning being so difficult is the presence of local minima in the cost function (Bianchini *et al* 1994). Some of the failures reported in finding optimal solutions may not be related to local minima, but to very flat plateaus that may lead to numerical errors. For recurrent networks, this problem is even more serious than in feedforward networks, particularly for long sequences, because of the backpropagation through time of the output error. Obviously, this doesn't mean that it is impossible to train a recurrent neural network on a particular task, but, if the cost function has many local minima, devising an effective algorithm may become very difficult.

In view of the above limitations of gradient based techniques, we need to consider alternative systems and optimization methods that give acceptable results even when the criterion function is not smooth and has long plateaus. In order to overcome these limitations in training recurrent networks, we present a new recurrent bilinear perceptron model. Furthermore, since our aim is to determine both the model order and the parameters simultaneously, we need a technique that is amenable to combinatorial optimization problems. Stochastic search techniques such as simulated annealing, genetic algorithms and evolutionary programming have been found to be robust and effective in solving these types of problems.

Evolutionary programming (EP) has been shown to have both these properties: combinatorial optimization capabilities and the ability to determine model order and parameters for modeling time series. It is particularly suitable for application to problems typified by chaos, chance, temporality and nonlinear interactivity that are not tractable to classical optimization methods. Being a global search technique, EP's performance is robust to the presence of local minima and convergence is guaranteed with a probability of one. The selection scheme can be modified to reject unstable solutions without limiting or constricting the search space. This allows us to guarantee the evolution of a stable recurrent network. The performance of EP is backed by well established analytical results. However, the trade-off is in terms of the convergence rate and computational intensity. Results in the literature indicate that computational requirements of evolutionary techniques are approximately an order of magnitude greater than other linear techniques. These results relate to the implementation on serial computers. On the other hand, it has been shown in (Fogel *et al* 1990) that on parallel machines the backpropagation algorithm and EP are computationally of the same order.

This paper is organized as follows. We first introduce the function optimization technique by evolutionary programming. We then present the bilinear recurrent perceptron and model identification. The technique of evolving the time series prediction models is discussed. Finally, prediction results are reported and conclusions are proposed.

## FUNCTION OPTIMIZATION BY EVOLUTIONARY PROGRAMMING (EP)

A global minimization problem can be formalized as a pair $(V, f)$, where $V \subseteq R^n$ is a bounded set on $R^n$ and $f$: $V \to R$ is an $n$-dimensional real-valued function. The problem is to find a point $x_{min} \in V$ such that $f(x_{min})$ is a global minimum on $V$. More specially, it is required to find an $x_{min} \in V$ such that

$$\forall x \in V : f(x_{min}) \leq f(x) \qquad (1)$$

Here $f$ does not need to be continuous, but it must be bounded.

EP owes its origins to Brooks(Brooks 1958), who introduced a creeping random method and observed its similarity to evolution. Later Fogel *et al.*(Fogel *et al* 1991) proposed a population-based random search strategy termed evolutionary programming, where instead of keeping the single best point in each generation, a population of search points is maintained. Fogel *et al.* (Fogel *et al* 1991) and Bäch *et al.* (Bäch and Schwefel 1993) have indicated that EP with adaptive mutation, meta-EP usually performs better than EP without adaptive mutation. Hence, we shall consider the meta-EP as presented in (Fogel *et al* 1991). The meta-EP was implemented as follows:

1. *Initialization:* Generate the initial population of $\mu$ individuals, and set $k$ = *1*. Each individual is taken as a pair of real-valued vectors, $(x_i, \eta_i)$, $\forall i \in \{1, \ldots, \mu\}$.
2. Evaluate the fitness score for each individual, $(x_i, \eta_i)$, $\forall i \in \{1, \ldots, \mu\}$, of the population based on the objective function, $f(x_i)$.
3. *Mutation:* Each parent $(x_i, \eta_i)$, $\forall i \in 1, \ldots, \mu$, creates a single offspring $(x_i', \eta_i')$ by: for $j = 1, \ldots, n$,

$$x_i'(j) = x_i(j) + \eta_i(j)N(0, \eta_i(j)) \qquad (2)$$

$$\eta_i'(j) = \eta_i(j) + N\left(0, \frac{\eta_i(j)}{c}\right) \qquad (3)$$

   where $x_i(j)$, $x_i'(j)$, $\eta_i(j)$ and $\eta_i'(j)$ denote the *jth* component of the vectors $x_i, x_i', \eta_i$ and $\eta_i'$, respectively. $N(0,\eta)$ denotes a one dimensional random number with a normal distribution centered at the origin and variance equal to 1.
4. *Fitness:* Calculate the fitness of each offspring $(x_i', \eta_i')$, $\forall i \in \{1, \ldots, \mu\}$.
5. Conduct pair wise comparison over the union of parents $(x_i, \eta_i)$ and offspring $(x_i', \eta_i')$, $\forall i \in \{1, \ldots, \mu\}$. For each individual, $q$ opponents are chosen randomly from all the parents and offspring with an equal probability. For each comparison, if the individual's fitness is no greater than the opponent's, it receives a "win".

6. *Selection:* Select the $\mu$ individuals out of $(x_i, \eta_i)$ and $(x_i', \eta_i')$, $\forall i \in \{1, \ldots, \mu\}$, that have the most wins to be parents of the next generation.
7. Stop if the stopping criterion is satisfied; otherwise, $k = k + 1$ and go to step 3.

## THE RECURRENT BILINEAR PERCEPTRON

The recurrent bilinear perceptron is defined as

$$\hat{z}(n) = w_0 + w_1 f\left(\sum_{i=1}^{p} \Phi_i \hat{z}(n-i) + \sum_{i=1}^{k}\sum_{j=1}^{m} \zeta_{ij} a(n-i)\hat{z}(n-j) + a(n) - \sum_{i=1}^{q} \theta_i a(n-i)\right) \tag{4}$$

where $\hat{z}(n)$ is the estimate of the system output and $\{a(n)\}$ is the set of independent zero mean prediction errors. The weights $w_0$ and $w_1$ are added to the model to enable a bias shift and scaling. The function $f$ is the non-linear function. Depending on the data, it is commonly taken as

$$f(x) = \frac{1}{1 + \exp(-x)} \quad \text{or} \quad f(x) = \tanh(x) \tag{5}$$

In all our simulations, the input data was scaled to lie in [-0.8,0.8] and the *tanh* non-linearity function was used.

## MODEL IDENTIFICATION

The identification procedure consists of determination of orders p, q, m and k and for the determined model order, estimation of the parameters $\{\Phi_j\}$, $\{\theta_j\}$, and $\{\zeta_{ij}\}$. The model order is determined as that which minimizes the Minimum Description Length (MDL) criterion which is defined in (Candy 1988) as

$$(N-\gamma)\log(\sigma_e^2) + \left(\frac{1}{2}\right)(\text{number of independent parameters})\log(N-\gamma) \tag{6}$$

where N is the number of observations of the time-series, $\gamma = \max(p, q, m, k)$,

$$\sigma_e^2 = \left(\frac{1}{N-\gamma}\right)\sum_{n=\gamma+1}^{N} (z(n) - \hat{z}(n))^2 \tag{7}$$

and $\hat{z}(n)$ is the predicted output at time $n$ obtained using the model with order $(p, q, m, k)$. Each member in the population is a vector which contains the model order $(p, q, m, k)$ followed by the parameters themselves. So the model order evolves with each generation according to the MDL fitness criterion and the fittest vector after a certain number of generations (after which there is no more improvement in the fitness) contains the desired model order and the model parameters that can be used for time-series prediction.

## EVOLVING TIME SERIES PREDICTION MODELS

The application of EP to the evolution of recurrent bilinear perceptrons for time series modeling is now evident. In all the simulations, the population size was chosen to be 60, that is, 30 parents and 30 offsprings. The parents were initialized to have random model orders between 1 and 9 (that is, each of p, q, m and k were chosen randomly to be integers between 1 and 9). In order to ensure that each member of initial parent population represented a stable recurrent bilinear perceptron, all the feedback and bilinear coefficients, namely the $\zeta_{ij}$'s and the $\theta_i$'s, of the initial parent population were set to zero. The initial variances were chosen uniformly from [0,5] for the model orders, from [0,1] for the feedforward coefficients and from [0,0.1] for the bilinear and recurrent coefficients. The reasons for this choice are as follows. It is well known that the MDL criterion biases the population towards low model orders, so we chose the initial variances for the model order uniformly from [0,5]. The final evolved coefficients mostly lie between -1 and +1, so the initial variances for the feedforward coefficients were chosen uniformly from [-1,1]. In our experiments, we found that in the first few generations, a large fraction of the offspring population contained unstable models, and that this fraction increased with the increase in the initial variance of the bilinear and recurrent coefficients. To avoid a very large fraction of the population in the initial generations from becoming unstable, the variances for the recurrent and bilinear coefficients were chosen uniformly from [0,0.1]. With these initial conditions the EP optimization technique was applied to find the optimum model identification that would minimize the MDL value. During evolution, the fitness of unstable systems was set to a very large value, typically equal to the fitness of the worst (with the largest fitness value) stable system, in the present generation.

The evolutionary optimization technique described above was applied for the modeling of four time series, namely, the sunspot series, the mackey-glass series, the laser generated data series and astrophysical data. The performance of the final evolved models were evaluated using the normalized mean square error(NMSE) which is defined as

$$NMSE = \frac{\sum_k (z(k) - \hat{z}(k))^2}{\sum_k (z(k) - \bar{z})^2} \tag{8}$$

where $z(k), \hat{z}(k), \bar{z}$ correspond to the time series, its prediction and its mean, respectively.

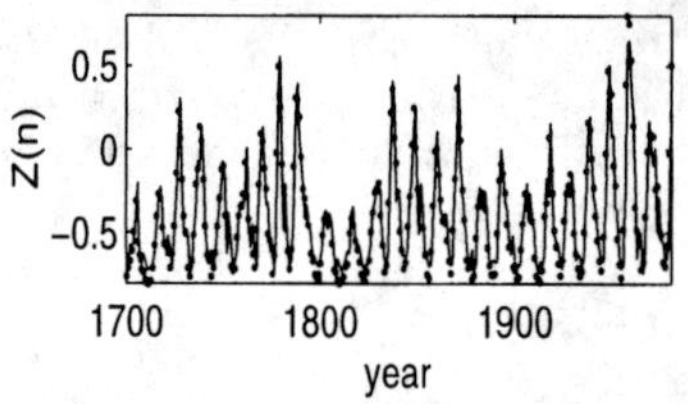


**Figure 1(a): Sunspot series.**
**'...' Desired Output and '-' RBP's output**

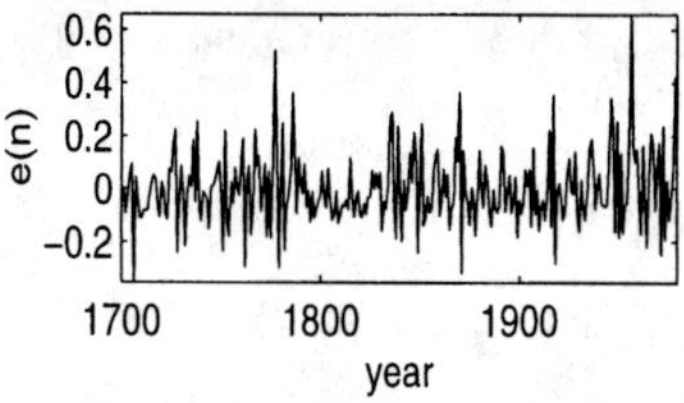


**Figure 1(b): Sunspot series prediction error.**

## PREDICTION RESULTS

### The Sunspot series

The first set of experiments was conducted on Wolf's sunspot series for the years 1700-1988. These numbers are indicative of the average relative number of sunspots observed each day of the year and serve as a standard bench mark for time-series modeling, where the objective is to generate a single - step prediction based on past observations. Consistent with J.R.McDonnel *et al* (McDonnell and Waagen 1994), the data set was partitioned into a training set over years 1700-1920 and test sets for years 1921-1955 and 1956-1979, respectively. The desired and predicted outputs are given in Fig. 1(a) and the prediction error of the evolved model is given in Fig. 1(b).

### The Mackey-Glass equation.

The mackey-glass equation represents a model for white blood cell production in leukemia patients (Mackey and Glass 1977). This model is complicated by the addition of a time delay $\tau$ in the nonlinear differential equation

$$\dot{x}(t) = \frac{ax(t-\tau)}{1 + x^c(t-\tau)} - bx(t) \qquad (9)$$

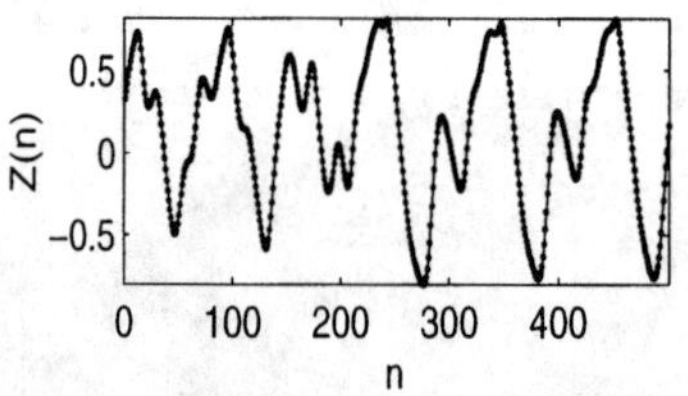


**Figure 2(a): Mackey Glass series.**
**'...' Desired Output and '-' RBP's output**

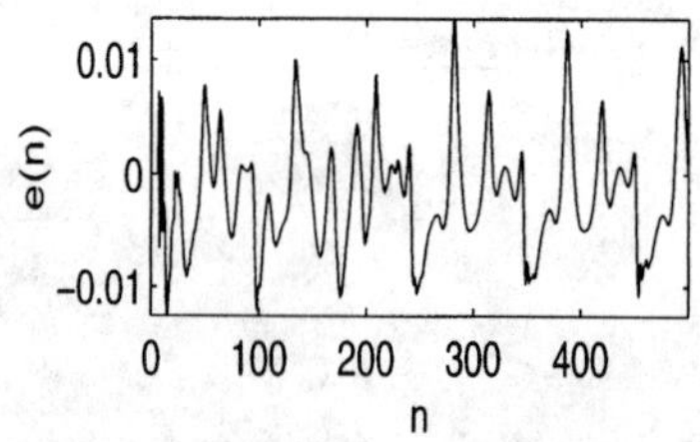


**Figure 2(b): Mackey Glass series prediction error.**

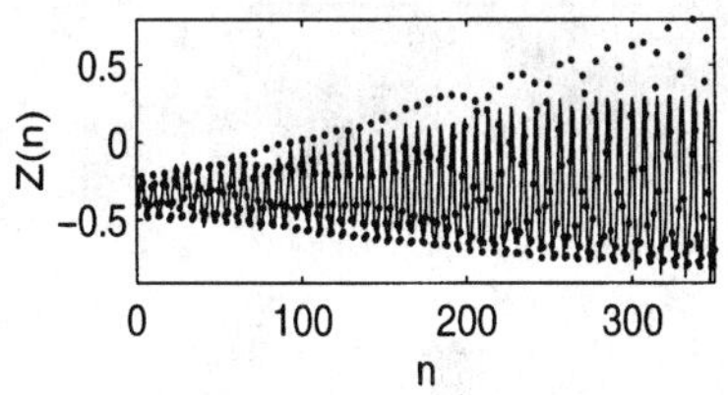


Figure 3(a): Laser data series.
'...' Desired Output and '-' RBP's output

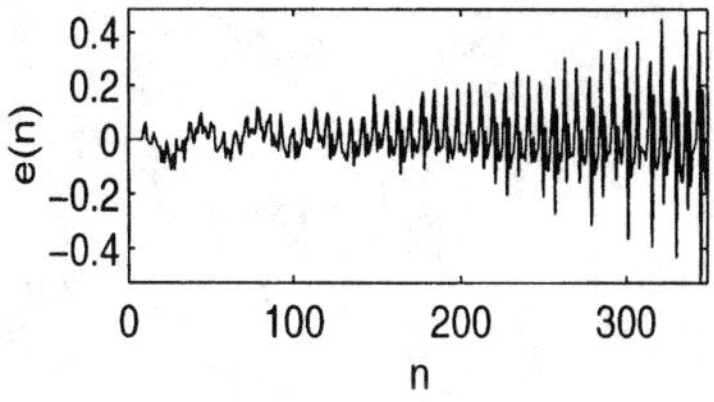


Figure 3(b): Laser data series prediction error.

where the free parameters are selected as a = 0.2, b = 0.1, c = 10, and $\tau$ is set equal to 1 as in (McDonnell and Waagen 1994). The training data consisted of 500 data points, extracted from a longer run so that the initial transients would have no effect. The actual and predicted outputs and prediction error of the evolved model are given in Figs. 2(a) and 2(b), respectively.

**Laser Generated Data**

The laser generated data consists of measurements made on an 81.5 micron 14NH3 cw (FIR) laser, pumped optically by the P(13) line of an N20 laser via the vibrational aQ(8,7) NH3 transition. The intensity data was recorded by a LeCroy oscilloscope. The experimental signal to noise ratio was about 300. The data are a cross-cut through periodic to chaotic intensity pulsations of the laser. The chaotic pulsations follow the theoretical lorenz model of a two level system. The data are a good example of the complicated behavior that can be seen in a clean, stationary, low-dimensional, non-trivial physical system for which the underlying governing equations dynamics are well understood. A training data sequence of 350 data points was chosen. The actual and predicted outputs and prediction error of the evolved model are given in Figs. 3(a) and 3(b), respectively.

**Astrophysical Data**

The Astrophysical data consists of measurements of the light curve (time variation on intensity) of the variable white dwarf star PG1159-035 during march 1989. It was recorded by the Whole Earth Telescope (a coordinated group of telescopes distributed around the earth that permits the continuous observation of an astronomical object). The intensity variations of the star arise from the excited modes, which are spherical harmonics. This data set represents the optical oscillations of a physical system, so, unlike the laser data set, which has very little noise, the measurement process for these data is much noisier. The actual and predicted outputs and prediction error of the evolved model are given in Figs. 4(a) and 4(b), respectively.

The above results are presented in table 1 and are compared with the ones obtained using simple reduced parameter bilinear models presented in (Rao and Chellapilla 1996), given in Table 2. The results in table 2 were generated using fast evolutionary programming (Rao and Chellapilla 1996), which was shown to have better optimization capabilities than meta-EP. From these two tables we observe

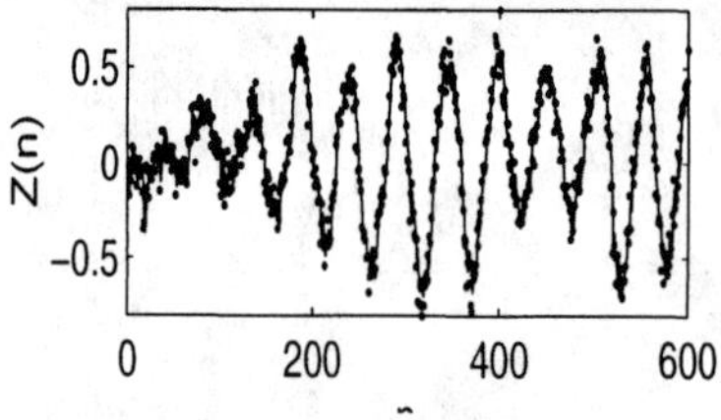


Figure 4(a): Astrophysical data series. '...' Desired Output and '-' RBP's output

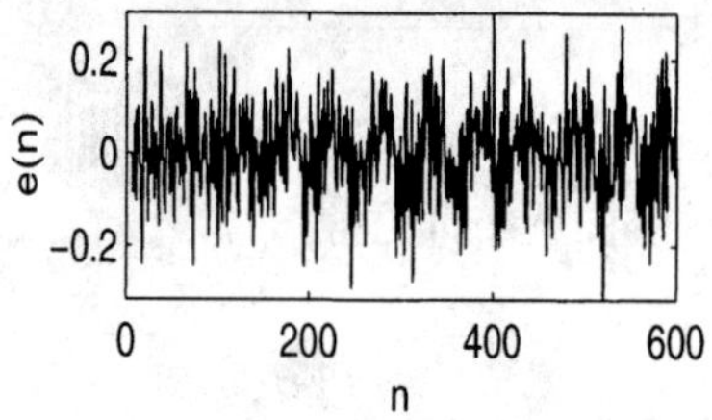


Figure 4(b): Astrophysical data series prediction error

that the recurrent bilinear perceptron is capable of performing better than the simple bilinear model in terms of NMSE and the model order. For comparison, with other techniques please refer to (McDonnell and Waagen 1994).

## CONCLUSION

In this paper, we presented a recurrent bilinear perceptron for time series prediction. The evolutionary programming technique, a multi-agent stochastic search technique, has been successfully applied to optimize the parameters of a recurrent bilinear perceptron and used for prediction of nonlinear chaotic time series. Unlike conventional techniques, where the model order is chosen first, and then the parameters of the model are determined, both the model order and the parameters were evolved simultaneously. The optimum model was determined by minimizing the MDL information criterion. Experimental results on the prediction of the sunspot series, mackey-glass series, laser generated data and astrophysical data indicate that RBPs, that are evolved by EP, are capable of performing better than simple bilinear models in terms of the NMSE and the model order.

**Table 1: Prediction results of the evolved recurrent bilinear perceptrons**

| Series | Recurrent Bilinear perceptrons | | Redeced parameter bilinear models (Results from Rao and Chellapilla 1996) | |
|---|---|---|---|---|
| | Model order | NMSE | Model order | NMSE |
| Sunspot series | (2,0,1,1) | 1700-1920<br>0.0064<br>1921-1955<br>0.0764<br>1956-1979<br>0.0378 | (2,4,9,2) | 1700-1920<br>0.0680<br>1921-1955<br>0.1130<br>1956-1979<br>0.1895 |
| Mackey-Glass series (T = 1) | (1,5,2,1) | 1.3969E-4 | (3,2,3,4) | 2.6312E-4 |
| Laser data series | (2,7,1,1) | 0.0963 | (6,3,2,2) | 0.1026 |
| Astronomical data | (1,4,1,1) | 0.0894 | (2,3,2,2) | 0.0967 |

## REFERENCES


Pinenda F.J., (1988). Generalization of back propagation to recurrent networks, *Phys. Rev. Lett.,* vol. 59, no. 19, pp. 2229-2232, Nov.

Werbos P.J., (1990). Backpropagation through time: what it does and how to do it, *Proc. IEEE,* vol. 78, pp. 1550-1560.

Williams R.J. and Zipser D., (1989). A learning algorithm for continually running fully recurrent neural networks, *Neural Computation,* vol. 1, pp. 270-280.

Narendra K.S and Parthasarathy K., (1991). Gradient methods for the optimization of dynamical systems containing neural networks, *IEEE Trans. Neural Networks,* vol. 2, no. 1, pp. 252-262, Mar.

Bengio Y., De Mori R., Flammia G., and Kompe R., (1992). Global optimization of a neural network - Hidden markov model hybrid, *IEEE Trans. Neural Networks,* vol. 3, no. 2, pp. 252-259.

Mozer M.C., (1992). Introduction of multiscale temporal structure," in *Advances in Neural Information Processing Systems 4,* Moody, Hanson, and Lippman, Eds. San Matteo, CA: Morgan Kaufmann, pp. 275-282.

Bengio Y., Simard P., and Frasconi P., (1994). Learning long-term dependencies with gradient descent is difficult, *IEEE Trans. Neural Networks,* vol. 5, no. 2, pp. 157-166, Mar.

Bianchini M., Gori M., and Maggini M., (1994). On the problem of local minima in recurrent neural networks, *IEEE Trans. Neural Networks,* vol. 5, no. 2, pp. 167-177, Mar.

Fogel D.B., Fogel L.J., and Porto V.W., (1990). Evolving Neural Networks, in *Biol. Cybern.,* vol. 63, pp. 487-493.

Brooks S.H., (1958). A discussion of random methods seeking maxima, *Operations Research.*, vol 6, pp. 244-251.

Fogel D.B., (1991). Fogel, *System Identification through Simulated Evolution: A machine learning approach to modeling*, Needham, MA: Ginn Press.

Fogel D.B., Fogel L.J., Atmar W., (1991). Meta-evolutionary programming, in *Proc. of the 25th Asilomar Conf. on Signals, Systems and Computers*, R. R. Chen, Ed. IEEE computer Society, pp. 540-545.

Bäch T. and Schwefel H.-P., (1993). An overview of evolutionary algorithms for parameter optimization, *Evolutionary Computation,* 1(1):pp. 1-23.

Candy J.V., (1988), *Signal Processing: the model-based approach*, McGraw-Hill Book Company.

McDonnell J.R and Waagen D., (1994). Evolving Recurrent perceptrons for time-series modeling, *IEEE Trans. Neural Networks*, vol. 5, no. 1, Jan.

Mackey M.C and Glass L., (1977). Oscillations and chaos in physiological control systems, *Science,* 197.

Rao S.S. and Chellapilla K., (1996). Evolving reduced parameter bilinear models for time series prediction using fast evolutionary programming. *Proc. of the 1st Annual Conf. on Genetic Programming,* J.R. Koza, D.E. Goldberg, D.B. Fogel and R.L. Riolo. Ed. MIT Press, pp. 528-535.

# MACHINE TOOL FAILURE PREDICTION BY CHAOS AND FUZZY COMPUTING APPROACH

**T. NISHIMURA, D. AKHMETOV AND Y. DOTE**
*Department of Computer Science and Systems Engineering*
*Muroran Institute of Technology, Mizumoto 27-1, Muroran 050, Japan*
*Fax: +81-0143-47-3374; Phone: +81-0143-44-4181*
*E-mail: dote@dote_sun.csse.muroran-it.ac.jp*

*ABSTRACT:*
This paper introduces short-term prediction of the time series having chaotic behavior to find some deterministic regularity in a phenomenon, which was thought to be noisy or to be irregular. First, the observed time series is reconstructed in a multi-dimensional state space according to Takens' theorem of embedding and fuzzy local reconstruction is made including the latest observed time series. This approach is applied to machine tool failure prediction for a boring machine resulting in improved accuracy and processing time in comparison with available neural network based methods.

## INTRODUCTION

Monitoring of machine tool failure has been an active area of research for a long time. A variety of methods have been proposed. Most of them can be implemented without extensive experimentation. As a result very conservative estimates of tool life are used to determine tool changing policies. The predicion or estimation of machine tool failure has been important in machine tool monitoring. Due to the inaccessibility of the cutting tool edge for direct on-line measurement this paper proposes a new chaos and fuzzy prediction method. Then, in order to determine change in the dynamics of the time series data on the ventricle potentials in the normal and abnormal states, the optimal embedding dimension and the optimal time delay which are obtained based upon the above mentioned prediction method are used. Finally, maximum Liapunov exponents and fractal dimension are calculated for machine tool failure detection. This approach is applied to machine tool failure prediction for a boring machine which resulted in improved accuracy and processing time in comparison with available neural network based methods.

## MACHINE TOOL FAILURE PREDICTION BY CHAOS AND FUZZY COMPUTING APPROACH

This method for analyzing nonlinear or chaotic time series is developed in Appendix. It was experimentally applied to a machine tool breakage and wear prediction. A vibration sensor was mounted on a boring machine with a magnet attachment as shown in Fig.1. A vibration measuring system is shown in Fig.2.

The machine tool failure prediction was performed in both states, i.e. when the tool

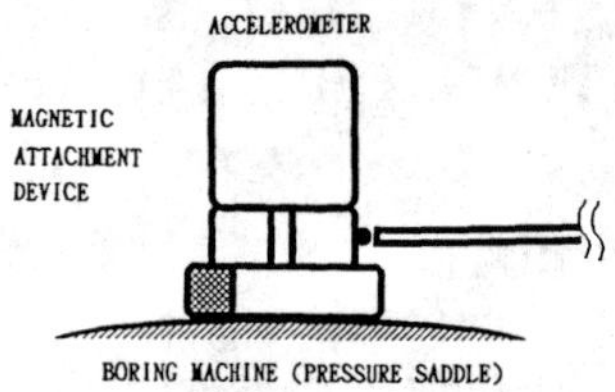


Figure 1: Accelerometer.

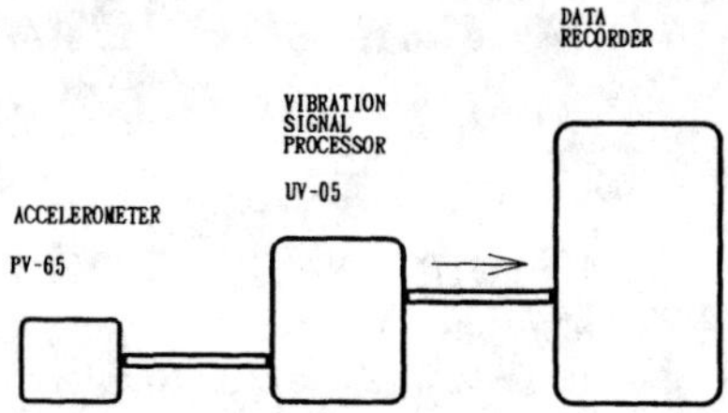


Figure 2: Vibration measuring system scheme.

was in normal state and in the abnormal state. The time series data were acquired at the sampling rate of 1 KHz. The sampled data were tested whether they are chaotic or not. The autocorrelation function is shown in Fig.3. In Fig.4 the autocorrelation function versus step width is almost linear. Because of these characteristics, the time series is said to be chaotic.

Next, the observed time series was reconstructed in a 3-dimensional state space according to the Takens' theorem of embedding with time delay $\tau$. The following vector was obtained.

$$x(t) = (y(t), y(t-\tau), \ldots, y(t-(n-1)\tau)), \quad t = ((n-1)\tau+1)\ldots N \tag{1}$$

where N is the number of data. Local reconstruction is made by using the vectors x(a), x(b), and x(k) neighboring to the latest observed vector z(t).

$$\begin{aligned}
&Z(T)= (y(T), y(T-\tau), y(T-2\tau))\\
&x(a)= (y(a)), y(a-\tau), y(a-2\tau))\\
&x(b)=(y(b), y(b-\tau), y(b-2\tau))\\
&x(c)=(y(c), y(c-\tau), y(c-2\tau))\\
&\\
&Z(T+s)=(y(T+s), y(T+s-\tau), y(T+s-2\tau))\\
&x(a+s)=(y(a+s), y(a+s-\tau), y(a+s-2\tau))\\
&x(b+s)=(y(b+s), y(b+s-\tau), y(b+s-2\tau))\\
&x(c+s)=(y(c+s), y(c+s-\tau), y(c+s-2\tau))
\end{aligned} \tag{2}$$

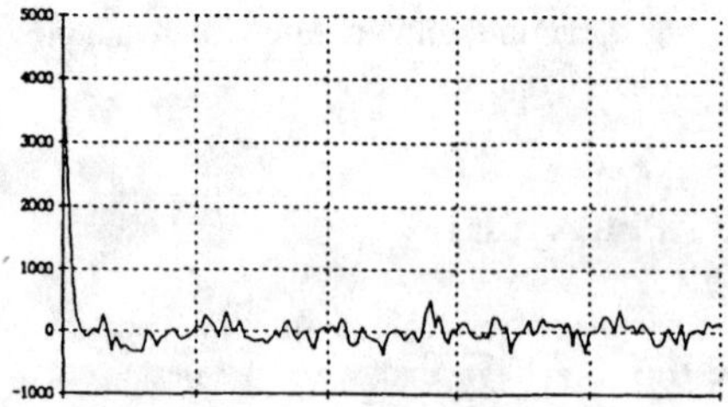


Figure 3: Autocorrelation function.

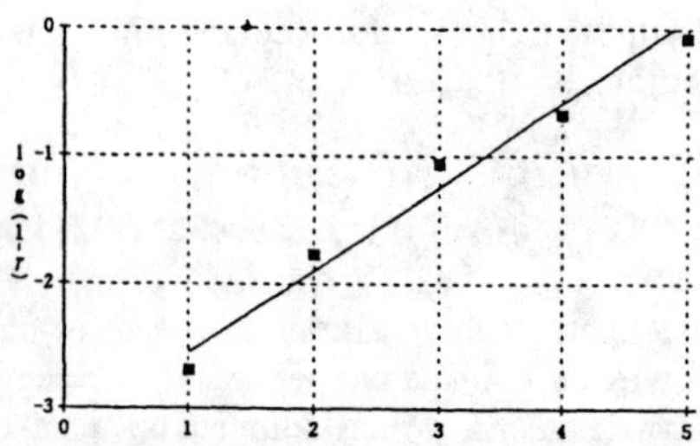


Figure 4: Autocorrelation function (log scale).

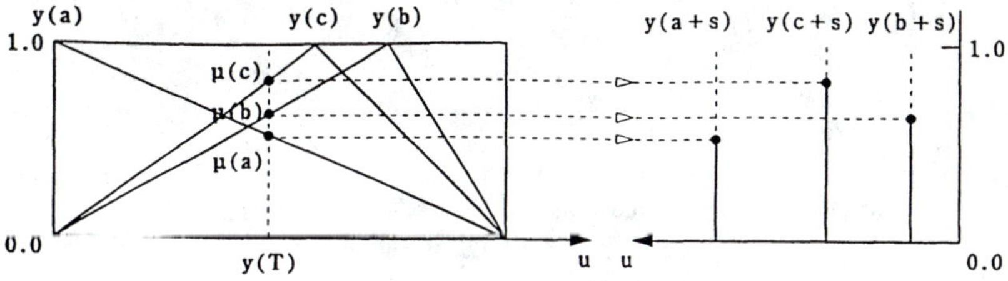


**Figure 5: Local fuzzy reconstruction method.**

The 's' steps ahead vectors Z(T+s) are predicted based upon these vectors in (2). The trajectory from Z(t) to Z(T+s) is influenced by the Euclidean distance from Z(T) to X(i). Therefore, fuzzy rules are introduced as follows:

$$\begin{aligned}&\text{If } Y(T) \text{ is } y(a) \text{ then } Y(T+s) \text{ is } y(a+s)\\&\text{If } Y(T) \text{ is } y(b) \text{ then } Y(T+s) \text{ is } y(b+s)\\&\text{If } Y(T) \text{ is } y(c) \text{ then } Y(T+s) \text{ is } y(c+s)\end{aligned} \tag{3}$$

Calculation method is illustrated in Fig.5. The following fuzzy inference was used:

$$\tilde{y}(T+S) = \frac{\mu(a+s)\, y(a+s) + \mu(b+s) y(b+s) + \mu(c+s) . y(c+s)}{\mu(a+s) + \mu(b+s) + \mu(c+s)} \tag{4}$$

## EXPERIMENTAL RESULTS

The constructed attractor is periodic, then prediction accuracy is high. If the step width is larger, then prediction is more accurate. The optimal data size is 2000. This method gives more accurate prediction than a linear prediction method as shown in Table 1.

**TABLE 1: PREDICTION ACCURACY COMPARISON**

| | Correlation function | PRMSE |
|---|---|---|
| Linear method | 92.449840 | 0.388879 |
| Proposed method | 93.028061 | 0.367077 |

$$\text{RMSE} = \sqrt{\frac{1}{T}\sum_{t=1}^{T}(\hat{x}(t) - x(t))^2} \tag{5}$$

$$\text{PRMSE} = \frac{RMSE}{\sqrt{\frac{1}{T}\sum_{t=1}^{T}(\hat{x}(t) - x_a)^2}} \quad (6)$$

where $x_a$ is the origin.

In order to determine a change in the dynamics of the time series data on the verticle potentials according to the machine tool failure (breakage and wear), the optimal embedding dimension and the optimal time delay obtained based on the above mentioned prediction method are used. Lastly, maximum Liapunov exponents and fractal dimension are calculated for machine tool failure detection.

## CONCLUSION

Machine tool failure prediction by a chaos and fuzzy computing approach is proposed. This method provides improved accuracy and processing in comparison with available neural network based methods.

## REFERENCES

Iokibe, T., and Ohshima, N. (1994) "An Application of Short-Term Prediction by Local Fuzzy Reconstruction Method to Chaotic Time Series", proc. of the IEEE Tokyo Section Electronics Journal, Tokyo. No. 33.

## APPENDIX

**Short-term Prediction on Chaotic Time Series** (Iokibe and Ohshima, 1994)
When the behavior of any time series is chaotic, it can be assumed that the behavior follows a certain deterministic law. Then, if the nonlinear deterministic regularity can be estimated, data accumulated in the near future, until the deterministic causality is lost, can be predicted from the observed data at a certain time point, because chaos has a "sharp dependence on initial value." Some natural phenomena are known to behave chaotically. Also, some of the social phenomena, such as tap water demand, electric power demand and traffic are chaotic. In diverse fields related to those natural and social phenomena, prediction composes an important subject and a variety of techniques have been devised for enhancing prediction accuracy. Particularly, prediction of tap water demand and that of electric power demand exert a serious influence over operation schedules. Therefore, short-term prediction has a great significance, and a chaotic approach, namely, the study on deterministic nonlinear short-term prediction, has been made.

Such a prediction, from the view point of deterministic dynamics theory, is based on the Takens' theory for "reconstructing the state space and the attractor of the original dynamical system from single observed time series data. " The Takens' theory is summarized below.

From observed time series v(t), vector (y(t), y(t−τ), y(t−2τ),..., y(t−(n−1)τ)) is generated where "τ" represents a time delay. The components of this vector are the coordinates of one point of an n-dimensional reconstructed state space $R^n$. Therefore, trajectory can be drawn in the n-dimensional reconstructed state space by changing "$\tau$". Let us assume that the observed time series is obtained through an observation system corresponding to $C^1$ continuous mapping from the state space of the dynamical system to the one-dimensional Euclidean space R.

When the "n" value is sufficiently large, the reconstructed trajectory is an embedding of the original trajectory. Namely, if any attractor is appeared in the original dynamical system, another attractor, which retains the phase structure of the first attractor, will appear in the reconstructed state space. "n" is usually called an "embedding dimension." For such reconstruction to achieve "embedding", it has been proved that the dimension "n" should satisfy the following condition:

$$n \geq 2m + 1 \tag{A1}$$

where "m" represents the state space dimensions for the original dynamical system. However, this is a sufficient condition. Depending on data, embedding can be established, even when "n" is less than 2m+1.

From the observed time series, the state space and attractor are reconstructed according to the Takens theorem of embedding. Further, based on the attractor, such a trajectory, as above is determined and nonlinear dynamics are presumed. This prediction method is termed as deterministic nonlinear prediction. Concretely, the time series y(t) observed at constant sampling time intervals is embedded in an n-dimensional state space with the embedding dimension "n" and time delay "$\tau$" by the above mentioned Takens' embedding theorem as shown in Fig.A1.

This process is called "reconstruction". In consequence, the following vector is obtained.

$$X(t) = (y(t), y(t-\tau), ..., y(t-(n-1)\tau)) \tag{A2}$$

When the reconstruction is repeated on a number of observed data, a smooth manifold consisting of a finite number of data vectors can be composed in the reconstructed

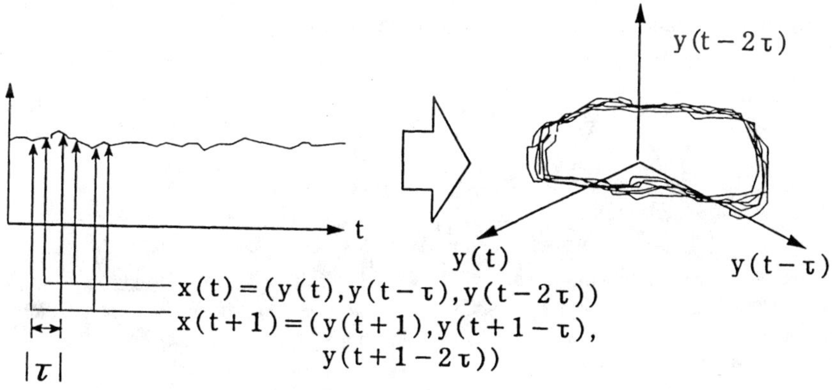


**Figure A1: Embedding of time series into n-dimensional reconstructed state space**

n-dimensional state space. With regard to the state space and attractor trajectory reconstructed by the embedding procedure, the near-future trajectory of the data vector, including the latest time series observed, is presumed by using each trajectory of that data vector and the neighboring one, thereby determining the data vector at "s" steps ahead.

From the data vector Z(T+s) at "s" steps ahead via the local reconstruction, the predicted value y(T+s) of y(T+s) at "s" steps ahead of the original time series is obtained.

**Local Fuzzy Reconstruction Method**
The data vector Z(T) resulting from the latest observation in the n-dimensional reconstructed state space, is plotted and replace the neghboring data vector with X(i). Also, the state of data X(i) at "s" steps ahead is replaced with X(i+s) and similarly the predicted value of data vector Z(T) at s steps ahead with Z(T+s).

Chaotic dynamics can be described in a linguistic form, as shown below.

$$\text{IF } Z(T) \text{ is } X(i) \text{ THEN } Z(T+s) \text{ is } X(i+s) \qquad \text{(A3)}$$

X(i) is the data vector neighboring to Z(T). Therefore, before step "s" loses deterministic causality, it can be assumed that the dynamics of the transition from state Z(T) to state Z(T+s) is approximately equivalent to the one from state X(i) to state X(i+s), because chaos has a "sharp dependency on initial value." When the attractor embedded in the reconstructed n-dimensional state space forms a smooth manifold, as mentioned in the previous section, trajectory from Z(T) to Z(T+s) is influenced by the Euclidean distance from Z(T) to X(i).

When focusing attention on the "j" axis in the n-dimensional reconstructed state space it may be written as follows

$$\text{IF } aj(T) \text{ is } yj(i) \text{ THEN } aj(T+s) \text{ is } yj(i+s), \quad (j = 1\tilde{\ } n) \qquad \text{(A4)}$$

where aj(T): J-axis component of X(i) value neighboring to Z(T) in n-dimensional reconstructed state space; aj(T+s): J-axis component of X(i+s) in n-dimensional reconstructed state space; n : embedding dimension.

Formula (A4) can be expressed by fuzzy function, as follows.

$$\text{IF } aj(T) \text{ is } yj(i) \text{ THEN } aj(T+s) \text{ is } y(i+s), \quad (j=1\tilde{\ } n) \qquad \text{(A5)}$$

Now the following are reiterated.

$$Z(T) = (y(T), y(T-\tau), \dots, y(T-(n-1)\tau))$$

Therefore, the j-axis component of Z(T) in the n-dimensional reconstructed state space becomes equal to yj(T). Accordingly, the j-axis component of the predicted value Z(T+s) for data vector Z(T+s) after "s" steps of Z(T) is obtainable as aj(T+s) by a fuzzy inference with yj(T) substituted into aj(T) of formula (A6). This method is called the "Local Fuzzy Reconstruction method."

# ITERATIVE PREDICTION OF CHAOTIC TIME SERIES USING A RECURRENT NEURAL NETWORK

**MAGDI A. ESSAWY, MOHAMMAD BODRUZZAMAN**
*Department of Electrical and Computer Engineering*
*Tennessee State University, Nashville, TN 37209*

**ABOLGHASEM SHAMSI, STEPHEN NOEL**
*Morgantown Energy Technology Center (METC)*
*Morgantown, WV 26505*

***ABSTRACT:***
Chaotic systems are known for their unpredictability due to their sensitive dependence on initial conditions. When only time series measurements from such systems are available, neural network based models are preferred due to their simplicity, availability, and robustness. However, the type of neural network used should be capable of modeling the highly non-linear behavior and the multi-attractor nature of such systems. In this paper we use a special type of recurrent neural network called the "Dynamic System Imitator (DSI)", that has been proven to be capable of modeling very complex dynamic behaviors. The DSI is a fully recurrent neural network that is specially designed to model a wide variety of dynamic systems. The prediction method presented in this paper is based upon predicting one step ahead in the time series, and using that predicted value to iteratively predict the following steps. This method was applied to chaotic time series generated from the logistic, Henon, and the cubic equations, in addition to experimental pressure drop time series measured from a Fluidized Bed Reactor (FBR), which is known to exhibit chaotic behavior. The time behavior and state space attractor of the actual and network synthetic chaotic time series were analyzed and compared. The correlation dimension and the Kolmogorov entropy for both the original and network synthetic data were computed. They were found to resemble each other, confirming the success of the DSI based chaotic system modeling.

## INTRODUCTION

Chaotic systems are known for their unpredictability, due to their sensitive dependence on initial conditions which is measured by positive Lyapunov exponents. [1] In other words, even when the exact model of a chaotic system is available, it is impossible to predict a chaotic system behavior for a long period of time. [2,3] The reason is that our measurements and calculations are never perfect and are susceptible to errors. Similar errors contribute to the non-exact determination of initial conditions. Any minute error in the initial conditions for a chaotic system will turn, with time, into great differences in the results. However, short term predictions of chaotic systems are still possible. [2,4] How short the time duration is for valid prediction depends on the system average loss of information represented by its Lyapunov exponents and Kolmogorov entropy. Some systems are less predictable than others due to faster loss of information with time, represented by larger positive Lyapunov exponents and larger positive Kolmogorov entropy. Since exact predictions are not possible for such systems, approximate models may produce results as satisfactory as those produced by exact models. This makes neural network based models very good candidates for such applications. Neural network based models are known not to be exact models, but they are easy to implement, robust, fast and data driven. Dynamic neural network models are preferable for such applications, due to their ability to capture time behaviors. [5]

In this work we used a special type of dynamic neural network called the Dynamic System Imitator (DSI). [6] We developed the DSI a few years ago and have used it for several modeling and control applications. [6,7,8] The DSI is biologically motivated and is specially designed to model a wide variety of dynamic systems. It has both short term and long term memory mechanisms that enable the modeling of a system's transient and steady state behavior. In addition, the DSI behavior depends on its initial conditions the same as any differential equation model does, even though no explicit differential equation solving is incorporated in this case. What we know of the DSI characteristics encourages us to recommend it for modeling non-linear systems in general and chaotic systems in particular. More details about the DSI will be discussed below. Since the dynamics of most real systems are accessed via time series measurements, the focus in this paper will be on modeling chaotic time series. The way the chaotic time series model is implemented in this paper is through a one step predictor model. The dynamics of a chaotic time series are modeled through training the DSI to perform a one step prediction. However, at any point of time, the DSI response depends on the initial conditions at time zero, the history of inputs and network state variables, and the current network input. Assuming the network was able to capture the dynamics in the time series, we can start the trained network with any set of initial conditions, use a number of initial data points to put the network on track, and iteratively feed the output of the network back to compute next predicted values. Even though we applied this methodology to several theoretical systems, the current motive is to use it in a strategy to identify certain chaotic behavior modes encountered in a Fluidized Bed Reactor (FBR) system. [9] This identification can be achieved by comparing the actual measurement from the chaotic system with the time series predicted by the DSI iterative predictor model, starting from a short time history of the actual data. In this paper, results from the DSI iterative predictor are discussed for chaotic time series generated using the logistic, Henon and the cubic equations, in addition to one experimental time series measured from the FBR experiment at the Morgantown Energy Technology Center. [9] The DSI network model was evaluated based on comparison made on the time series, phase space trajectories, and chaotic parameters computed from these trajectories. However, in this case, time series similarities are not as important as similar phase space trajectories and similar chaotic parameters. [5] The actual combination of DSI iterative predictor and FBR system is published in other domains. [9]

## THE DYNAMIC SYSTEM IMITATOR (DSI) NEURAL NETWORK

The neural network used for the chaotic time series prediction in this paper is a dynamic neural network called Dynamic System Imitator (DSI). The DSI is a fully recurrent neural network that is specially designed to model a wide variety of dynamic systems. [6,7] As shown in Figure.1, the DSI has a three layer structure: input, hidden, and output layer. Connections have both weights and integrators in parallel to model short term and long term memory mechanisms that handle modeling of time behaviors and time lags in real systems. Every node in the input layer has one input, $x_k(t)$, and two outputs defined by: [6]

$$\begin{aligned} o_{1j}^{i}(t) &= x_j(t)\ , \\ o_{2j}^{i}(t) &= \int_0^t x_j(t)\, dt\ . \end{aligned} \qquad (1)$$

The input layer is fully connected to the hidden and output layers. Every node in the hidden layer is connected to every other node in the hidden and output layers and to itself. The two outputs of every neuron are computed according to the relationship: [6]

$$\begin{aligned} o_{1j}^{h}(t) &= A_j \psi_j (B_j^{-1}[\sum_{k=0}^{m} (w_{1jk}^{hi} o_{1k}^{i}(t) + w_{2jk}^{hi} o_{2k}^{i}(t)) + \sum_{k=0}^{n} (w_{1jk}^{hh} o_{1k}^{h}(t) + w_{2jk}^{hh} o_{2k}^{h}(t))]) \\ o_{2j}^{h}(t) &= C_j \psi_j (D_j^{-1} \int_0^t o_{1j}^{h}(t)\, dt) \end{aligned} \qquad (2)$$

where $\psi_j$ is a nonlinear transformation function, and $A_j$, $B_j$, $C_j$, and $D_j$ are adjustable weights associated with the hidden neuron j, which are used to shape the transfer function for every node. B and D are used to adjust the steepness of the function, while A and C are used to adjust its min-max value. Also m and n are the number of processing nodes in the input and hidden layers respectively; $w_1$ and $w_2$ refer to weights associated with direct and delayed outputs, respectively. The superscript h refers to the hidden layer, i refers to the input layer, hi refers to weights from the input to hidden layer, and hh refers to weights from the hidden to hidden layer. The integrators and feedback connections promote enough asynchrony and interaction in the network to model several system state variables as a function of time. When enough intermediate state variables are generated, the network output can be a function of those state variables. The output layer has only one output per node, which is computed according to the following equation: [6]

$$o_j^{o}(t) = E_j \psi_j (F_j^{-1}[\sum_{k=0}^{m} (w_{1jk}^{oi} o_{1k}^{i}(t) + w_{2jk}^{oi} o_{2k}^{i}(t)) + \sum_{k=0}^{n} (w_{1jk}^{oh} o_{1k}^{h}(t) + w_{2jk}^{oh} o_{2k}^{h}(t))]) \quad (3)$$

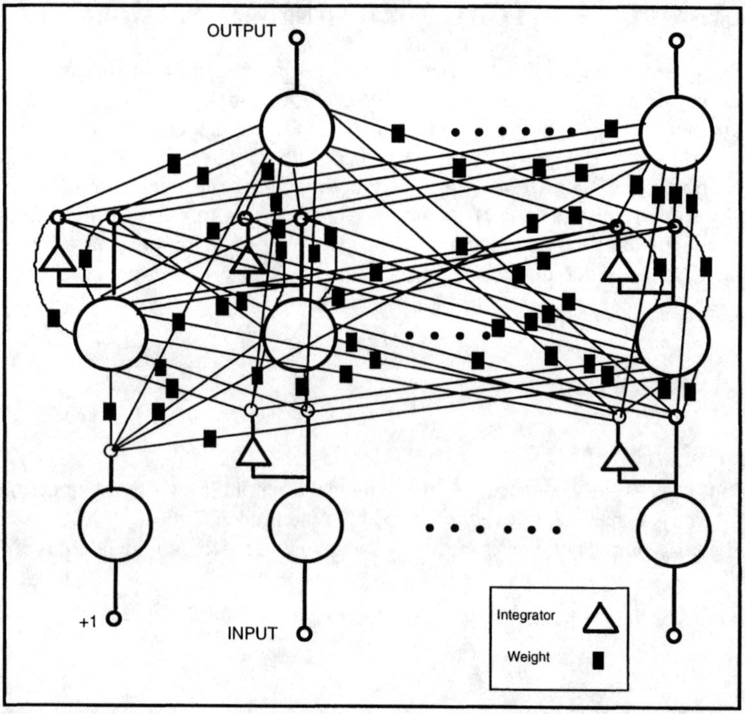

**Figure 1: Schematic diagram of the Dynamic System Imitator (DSI) Network.**

where m and n are the number of nodes in the input and hidden layers respectively, and $E_j$ and $F_j$ are two constants associated with each node in the output layer to shape its own transfer function when needed. The superscript o refers to the output layer, i refers to the input layer, h refers to the hidden layer, oi refers to weights from the input to output layer, and oh refers to weights from the hidden to output layer.

By looking at the complete DSI network design, it is easy to observe that the node interaction, information feedback, and action transfer time lags generate an activity in the network that is similar to the internal activity in real dynamic systems. Even with a simple configuration, the DSI has a complex structure, which makes it very difficult to train. A multi-dimensional optimization technique that adopts the simplex method is used to train the DSI. [6] There are two other difficulties in the training of such a network. One is that a very long time series cannot be introduced to the network at one time and must be divided into reasonably sized sections. The other is that the behavior of the network is dependent on the initial conditions of its state variables, and a certain set of initial conditions has to be found in conjunction with every network design. In other words, whenever the network is updated, a new set of initial conditions must be found. The first problem was overcome by using a moving time window that cascades the introduction of the time series segments to the network during training. The network final conditions at the final training step of every segment is taken as the initial conditions for the next segment, to keep the physical association between the consequent segments of the time series. The second problem was overcome by adding initial conditions search method that runs after every training iterate to find updated initial conditions for every modified version of the network. An arbitrary set of initial conditions can be used at the start of the training process.

## FLUIDIZED BED REACTOR (FBR) PRESSURE DATA COLLECTION

Morgantown Energy Technology Center has built and operated a cold flow model to emulate fluid dynamics in a Fluidized Bed Reactor (FBR). The cold flow verification test facility consists of a ten foot high jetting fluidized bed made of clear acrylic and configured as a half cylinder vessel to facilitate jet observation. A central nozzle, made up of concentric pipes, continuously fed solids at 0 to 8 psig pressures. Separate flow loops controlled the conveyance of solids (inner pipe), the make-up air flow (middle pipe), sparger flow (outer pipe), and six air jets on the sloping conical grid. The half round fluid bed model provided useful information to study fluidization and design issues including jet penetration, chaotic pressure fluctuations, and mass flow rates of particles in various regions of the jetting fluid bed. The fluid bed tests were conducted using cork particles to simulate the relative density of gases to scale for a high pressure coal conversion reactor. As expected, the test generated chaotic pressure fluctuations. The differential pressures were measured at two location with each location consisting of two pressure taps spaced four inches apart. The lower pair of pressure taps were placed at a height just above the nozzle and the upper pair of pressure taps were placed at a height where the jet becomes evenly distributed across the diameter of the reactor. Differential pressure data collected at the higher sensor served as the primary data for the investigation of chaos. It clearly indicated the fluidization regime of the bed supported by visual observations. Data were collected on a data acquisition card at a rate of 50 Hz.

## USING THE DSI FOR ITERATIVE PREDICTION OF CHAOTIC TIME SERIES

A simple configuration of the DSI neural network was used for the iterative prediction of a chaotic time series. This configuration has one node in the input layer, three nodes in the hidden layer, and one node in the output layer. The network was trained to predict one point ahead of the time series, using a set of previous values. These values are not explicitly used for prediction, but are implicitly used by adjusting the state of the network from which the prediction is performed. The prediction method is based upon the idea that once the network is trained to predict one point ahead with good accuracy, this same point can be used as an input to the network to predict the next point. This process can be repeated iteratively to predict many points in the time series. Naturally, the accuracy of prediction will deteriorate over time. During training, a time window of 200 points was used to cascade the time series to the network. The algorithm was applied to three types of simulated chaotic time series generated from the logistic, Henon, and cubic equations, in addition to one experimental time series measurement taken from a Fluidized Bed Reactor (FBR). FBR systems are known for their chaotic behavior, as discussed in several references. [9,10,11,12] The network was able to learn simple one step prediction in a reasonable number of training iterations. After training, the output of the DSI was used iteratively to generate the time series. However, the training initial conditions together with the first actual 25 points of the training time series should be used to start the DSI, in case the time behavior of the training time series must be generated. If not, any network initial conditions and any starting points can be used to generate the state space behavior of the system to which the training time series belongs.

Comparing the predicted time series to the actual time series, we found that the DSI was able to track the training time series time behavior for a short period of time (around 30 points), when started from the training initial conditions , and activated by the first 25 points of the training time series. However, it was able to track the state space attractor to which the training time series belongs, starting from any initial conditions, activated by any arbitrary set of starting points. The only case that fails is zero initial conditions together with zero starting points, which leads to zero solution The actual and predicted time series for all cases are shown in Figures 2-9, while the actual and predicted state space attractors are shown in Figures 10-17. To quantitatively compare these attractors, the correlation dimension and Kolmogorov entropy for the actual and predicted attractors were computed. The results of the correlation dimension and Kolmogorov entropy of the different cases are summarized in Table.1. The correlation dimension is computed, according to the box-counting method, from the slope of the lines representing the correlation integral versus ε (the size of a computing box) on log-log curves for different embedding dimensions. The correlation integral was computed according to the following equation: [13]

$$C(N,\varepsilon) = \frac{2}{N(N-1)} \sum_{j=1}^{N} \sum_{i=j+1}^{N} \theta(\varepsilon - |x_i - x_j|) \tag{4}$$

where $\theta(x)=1$ for $x>0$ and $\theta(x)=0$ for $x<0$. The Kolmogorov entropy was computed according to the equation: [14]

$$K_{m,d} = \frac{1}{\tau m} \ln \frac{C_d(\varepsilon)}{C_{d+m}(\varepsilon)} \tag{5}$$

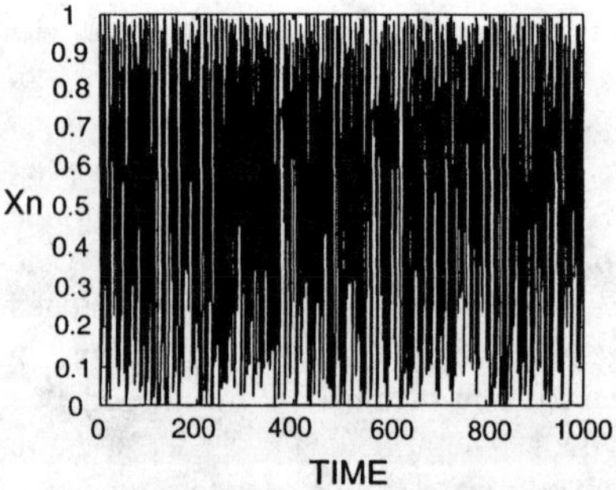

**Figure 2: Actual logistic time series.**

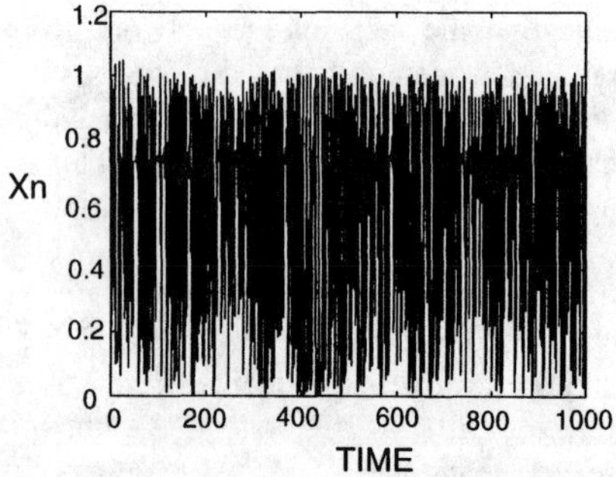

**Figure 3: Synthetic logistic time series.**

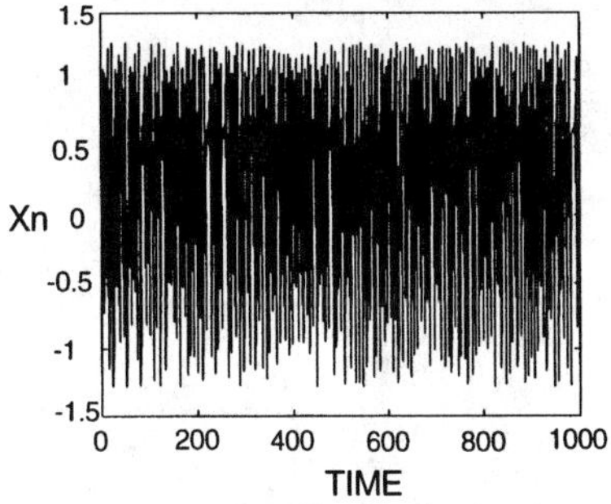


**Figure 4: Actual Henon time series.**

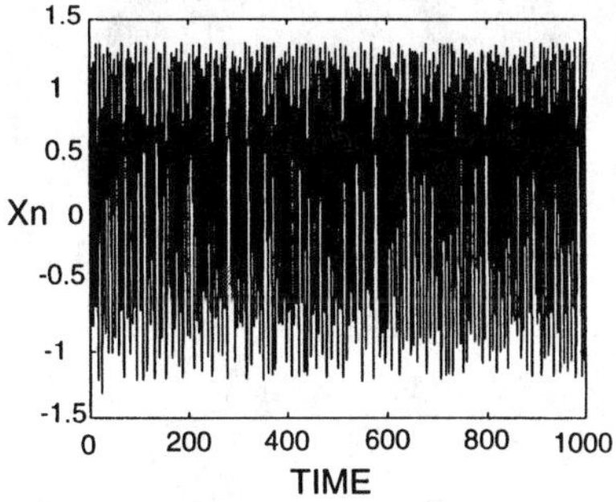


**Figure 5: Synthetic Henon time series.**

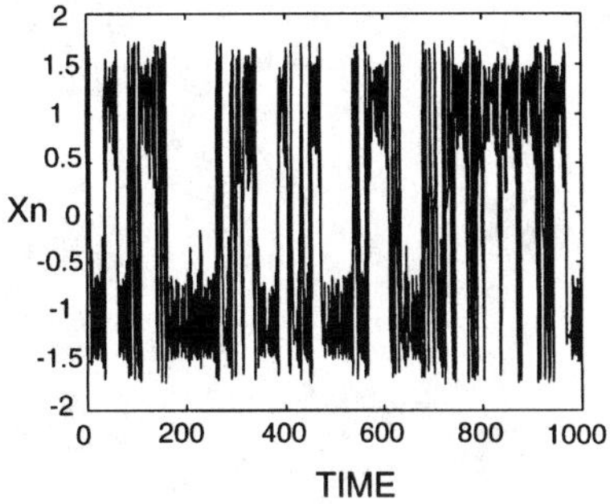


**Figure 6: Actual cubic time series.**

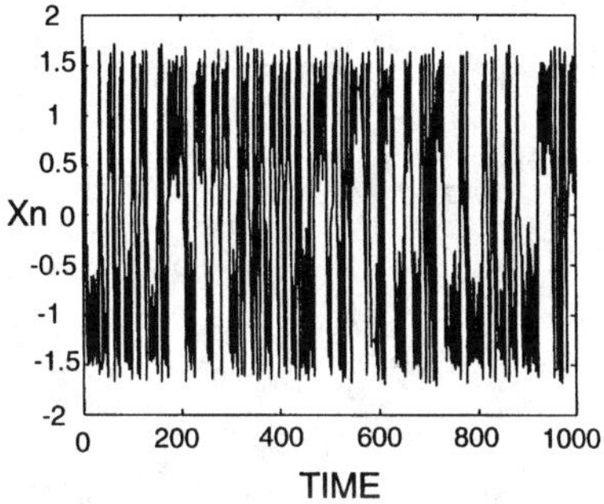


**Figure 7: Synthetic cubic time series.**

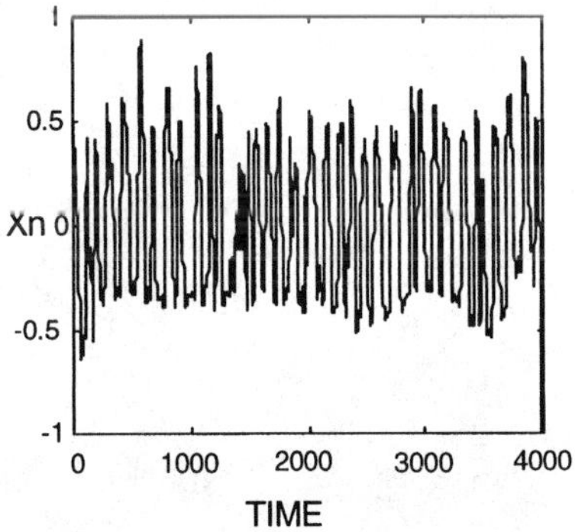


**Figure 8: Actual normal FBC time series.**

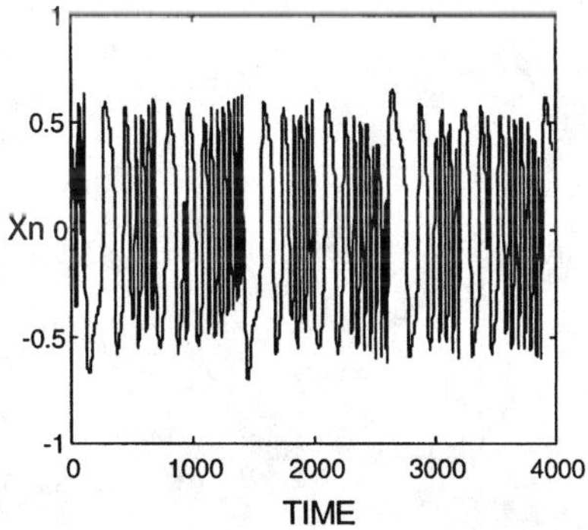


**Figure 9: Synthetic normal FBC time series.**

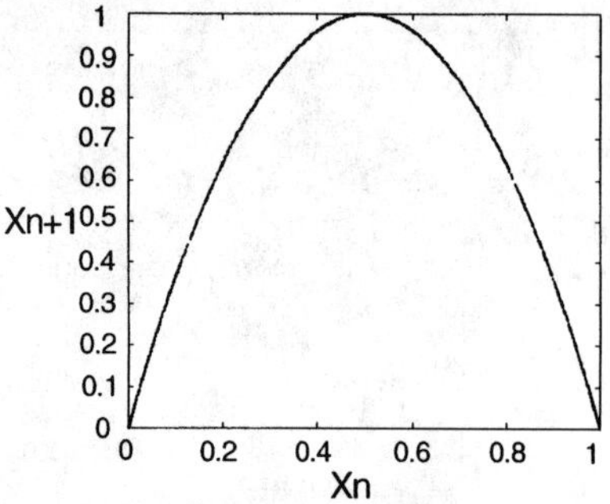


**Figure 10: Actual logistic attractor.**

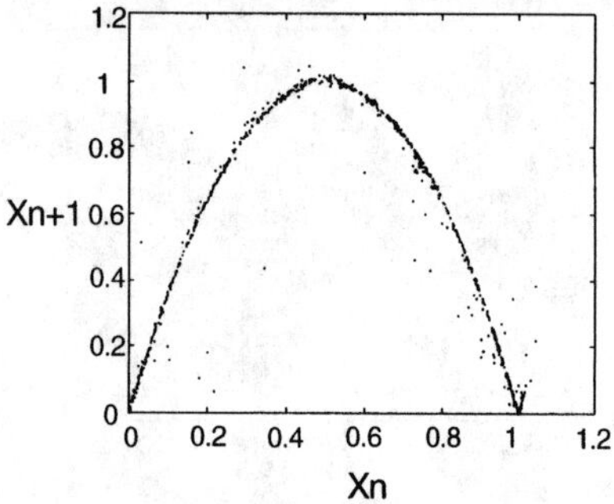


**Figure 11: Synthetic logistic attractor.**

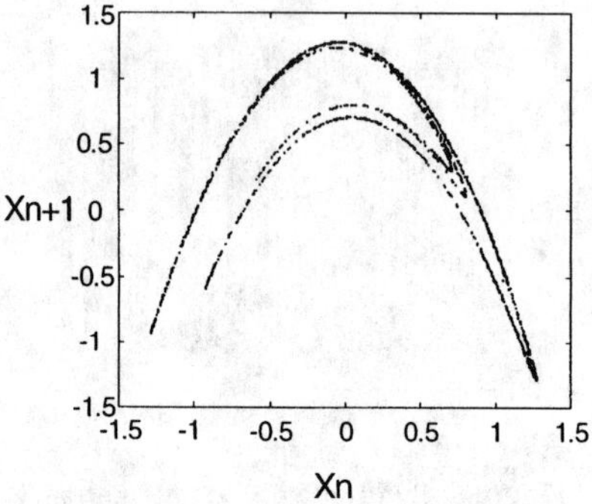


**Figure 12: Actual Henon time attractor.**

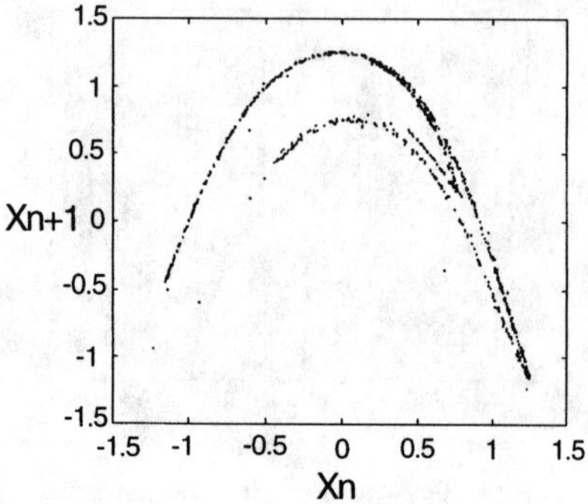


**Figure 13: Synthetic Henon attractor.**

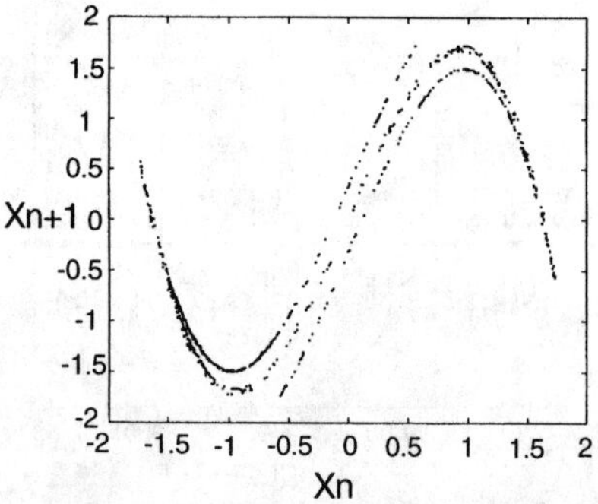


**Figure 14: Actual cubic attractor.**

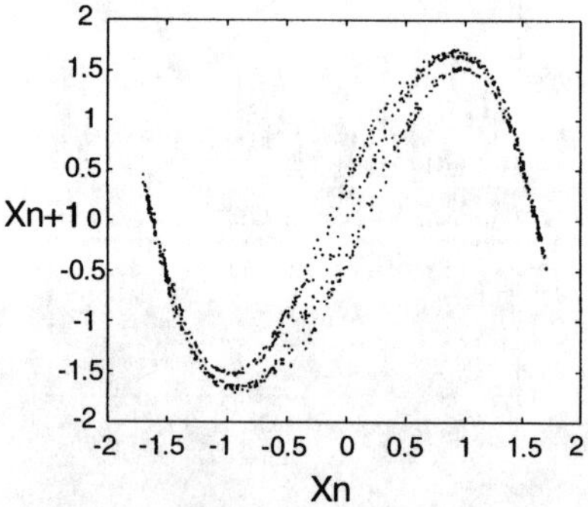


**Figure 15: Synthetic cubic attractor.**

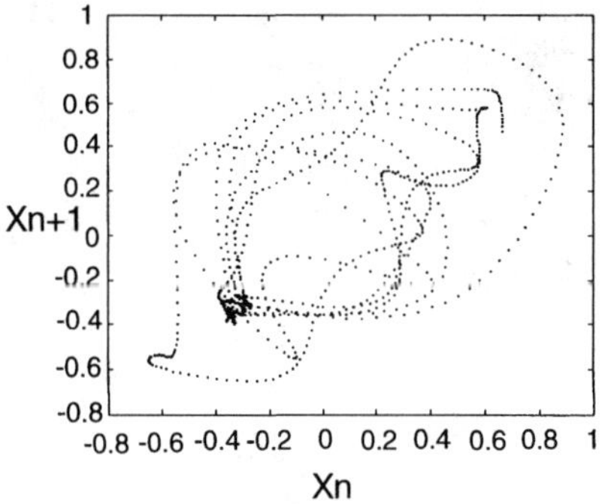


**Figure 16: Actual normal FBC attractor.**

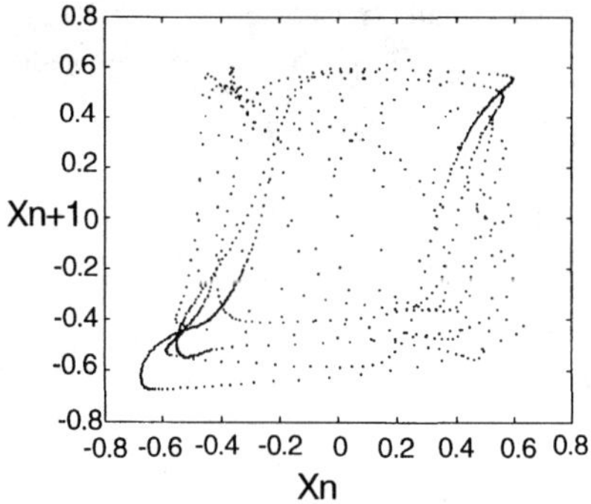


**Figure 17: Synthetic normal FBC attractor.**

**TABLE 1: COMPARISON OF THE ACTUAL AND DSI SYNTHETIC ATTRACTORS PARAMETERS**

| Time Series | Correlation Dimension | Kolmogorov Entropy |
|---|---|---|
| Logistic Map (actual) | 1.0457±0.0057 | 0.6872±0.0198 |
| Logistic map (synthetic) | 1.2316±0.0374 | 0.6013±0.0435 |
| Henon Map (actual) | 1.2607±0.0331 | 0.3267±0.0135 |
| Henon map (synthetic) | 1.3171±0.0725 | 0.2962±0.0239 |
| Cubic Map (actual) | 1.2248±0.0090 | 0.4245±0.0183 |
| Cubic Map (synthetic) | 1.8171±0.01365 | 0.5340±0.0134 |
| FBC normal (actual) | 2.934±.065 | 5.034±.095 |
| FBC normal (synthetic) | 2.12±.07 | 6.3764±1.26 |

## CONCLUSION

In this paper a dynamic neural network based model for chaotic time series has been developed. A one step predictor model was used to iteratively generate chaotic time series. A dynamic neural network called the Dynamic system Imitator (DSI) was utilized. The DSI has distinguishable dynamic features due to its special architecture. The DSI time behavior depends on its initial conditions. After training, the DSI was able to generate the chaotic time series in all test cases. For a short period of time, it was able to generate the same time behavior of the training time series if started with the same initial conditions and the first initial points of the time series. Furthermore, it was able to track the system attractor to which the training time series belongs, for any period of time, for any initial conditions and any initial points, in all test cases. The only case that fails is the zero initial conditions together with zero starting points, which leads to a zero solution. This methodology was applied to three known chaotic models, the logistic, Henon and cubic maps, in addition to one experimental time series taken from differential pressure measurement of a Fluidized Bed Reactor (FBR). The correlation dimension and the Kolmogorov entropy for the actual and DSI network synthetic data were computed and compared. There is a very good match between the actual and synthetic time series parameters in all cases which indicates that the DSI was able to learn the dynamics in those chaotic time series to a very good extent.

## ACKNOWLEDGMENT

This work has been supported by the DOE Grant No. DE-FG22-94MT94015. We wish to thank the DOE for their support of this research.

## REFERENCES

1. E. Ott, "Controlling chaos," Physics Today, pp 34-40, May 1995.
2. J.D. Farmer, J. Sidorowich, "Predicting Chaotic Time Series, " Phys. Rev. Lett. 59, pp 845, 1987.
3. M. Casdagli, "Non-linear Prediction of Chaotic Time Series," Physica D 35, pp 335, 1989.
4. G. Sugihara, R. M. May, "Non-linear Forecasting as a way of Distinguishing Chaos from Measurement Error in Time Series," Nature 344, pp 734, 1990.
5. J. C.Principe, J-M Kuo, "Dynamic Modeling of Chaotic time Series with Neural Networks," Advances in Neural Information Processing Systems 7, Ed. Tesauro, Touretzky, Leen, pp 311-318, 1995.
6. Magdi A. Essawy, "Artificial Neural Networks for System Modeling Monitoring and Control," Ph.D. Dissertation, University of Tennessee, Knoxville, July 1995.
7. Magdi A. Essawy, Robert E. Uhrig, "The Dynamic System Imitator: A New Approach for a Dynamic Neural Network Architecture," the 9th Power Plant Dynamics, Control & Testing Symposium, Knoxville, Tennessee, May 24-26 1995.
8. Magdi A. Essawy, Mohammad Bodruzzaman, "Using A Recurrent Neural Network for Chaotic Behavior Control," to appear in the World Congress on Neural Networks '96, Sep. 15-20 1996.
9. Magdi A. Essawy, Mohammad Bodruzzaman, "Recurrent Neural Network-Based Monitoring and Control of Chaotic Fluidized Bed Combustion Systems," the 4th Annual HBCU Private Sector Energy Research and Development Technology Transfer Symposium, Greensboro, NC, April 2-3 1996.
10. Fuller, "Interpretation of Pilot Scale, Fluidized Bed Behavior Using Chaotic Time Series Analysis," 12th international FBC Conference, LA Jolla, CA, May, 1993.
11. S. Daw, "Characterization of Voidage and Pressure Signals from Fluidized Beds, Using Deterministic Chaos Theory," Fluidized Bed Combustion, ASME Conference 1991, pp 777-785.
12. S. Halow, "Characterizing Fluidized Bed Behavior, By Decomposition of Chaotic Phase Space Trajectories," 1993 Annual AIChE Meeting, St. Louis, MO, November, 1993.
13. Ding, C. Grebogi, E. Ott, T. Sauer, J. A. Yorke, "Plateau Onset for Correlation Dimension: When Does It Occur ?", Physics Review Letters 70, 3872, 1993.
14. Peter Grassberger, "Estimation of the Kolmogorov Entropy from a Chaotic Signal", Physical Review A, Vol. 28, No. 4, pp 2591-2593, October, 1983.

# A CONFIDENCE MEASURE FOR GRNN-BASED SYSTEM MODELING AND PREDICTION

**G. DAN HAMMACK** hammack@ew.esy.com
*Raytheon E-Systems, Garland Div., 1200 Jupiter Rd, Garland, Texas 75042*

**ALIREZA KHOTANZAD** kha@seas.smu.edu
*Electrical Engr. Dept., Southern Methodist University, Dallas, Texas 75275*

**ABSTRACT:**

The General Regression Neural Network (GRNN) provides a system model using Parzen estimates of probability density functions formed from the training set and produces an estimate based on the conditional mean of the output given an input. This formulation lends itself to a direct network implementation. In engineering applications, a confidence measure is useful to determine the accuracy of a given estimate. For the GRNN the conditional variance of the output given the input can be used as a confidence measure. It is shown that this measure can be implemented as an addition to the GRNN network structure. In this paper, we examine the performance of the network on time series data and compare results using Gaussian and Cauchy kernels. It is found that a Cauchy kernel may be better suited for sparse data sets. The network is also used for prediction of the daily low temperature for Dallas, Texas using multi-variate climatic inputs from various geographic regions.

## INTRODUCTION

This paper focuses on the problem of system modeling where we are given a set of input/output pairs and try to find a network that produces similar outputs. Artificial neural networks have been successful in many signal processing applications, including system modeling [1]. The multi-layer perceptron (MLP) has become very popular due to its well documented back-propagation learning algorithm. The MLP provides a complex non-linear network that is used to model the training set and has been found to provide excellent results in many areas. Because the MLP is non-linear in its weights, the weights of the system generally cannot be solved for in closed form. Instead the weights are randomly guessed and a gradient decent approach is used to optimize the weights in a least mean square error sense which may require extensive iteration over the training set.

As an alternative to the MLP, Specht [4] introduced the General Regression Neural Network (GRNN) which provides a network structure whose weights are determined directly from the training set based on probability theory. The GRNN uses as its estimate the conditional mean of the output, $y$, given an input, $x$, $E[y|x]$. To compute this mean, the joint probability density function of $y$ and $x$ is required. This density function is approximated using a Parzen estimator [3] that is created using Parzen windows centered at the points in the training set. In the network, the training points correspond to hidden nodes that provide a weighting function in relation to the distance from the given input. To limit the size of the network, training points that are close to one another can be clustered together to form a single node. Once the nodes are established the only parameter that must be calculated is the Parzen window spreading parameter, sigma. This parameter can be optimized by running the network with a representative test set and determining which value yields the smallest mean square error over the entire test set.

Like other types of neural networks, the GRNN produces an estimate for any given input. Unfortunately, no additional information is supplied as to the accuracy of this estimate. It would be advantageous if some measure of confidence would be given for each estimate based on information obtained from the training set. In [2] the cross-validation error was used in a Radial Basis Function Network to provide confidence limits.

Since the GRNN bases its estimate on the conditional mean of the output given an input, it is also possible to compute the conditional variance of the output given an input in a similar manner. The variance should be low in regions where training set points have similar values and increase in areas where the training points vary greatly. Since the estimate is based on these values from the training set, the confidence on the accuracy of the estimate will vary according to the variances of the points used to create the estimate. In this paper we evaluate the effectiveness of using the conditional variance as a confidence measure and compare the performance of the network using the typical Gaussian kernel versus a Cauchy kernel.

## CONDITIONAL VARIANCE AS CONFIDENCE MEASURE

To produce a confidence measure we can extend the formulation of the GRNN estimate to include the conditional variance. This requires the joint probability density function of $x$ and $y$ which we will estimate using a Parzen density estimator. The Parzen estimate uses kernel functions that place "lumps" of probability at each point in the training set which are summed together to obtain the estimate of the density function. To insure that our final form lends itself to a logical network implementation we have three requirements on the function used to estimate the joint probability density function. First, the density must be separable in $x$ and $y$. Second, the density of $y$ must have a finite mean and thirdly, it must have a finite mean square. Given the first requirement we can write the joint probability density function as:

$$f(x,y) = \frac{1}{N}\sum_{i=1}^{N} f_{X_iY_i}(x,y) = \frac{1}{N}\sum_{i=1}^{N} f_{X_i}(x) f_{Y_i}(y) \tag{1}$$

The conditional probability density function is given as:

$$f(y|x) = \frac{f(x,y)}{f(x)} = \frac{f(x,y)}{\int_{-\infty}^{\infty} f(x,y)\partial y} \tag{2}$$

For the estimate we compute the conditional mean of the output given an input.

$$E[y|x] = \int_{-\infty}^{\infty} y f(y|x)\partial y = \frac{\int_{-\infty}^{\infty} y f(y,x)\partial y}{f(x)} \tag{3}$$

$$= \frac{\int_{-\infty}^{\infty} y \frac{1}{N} \sum_{i=1}^{N} f_{X_i}(x) f_{Y_i}(y) \partial y}{f(x)} = \frac{\frac{1}{N} \sum_{i=1}^{N} f_{X_i}(x) \int_{-\infty}^{\infty} y f_{Y_i}(y) \partial y}{f(x)} \quad (4)$$

The final integral in Equation 4 is the mean of the probability density of $y$, defined by our second stipulation.

$$E[y|x] = \frac{\frac{1}{N} \sum_{i=1}^{N} f_{X_i}(x) E_{Y_i}[y]}{\frac{1}{N} \sum_{i=1}^{N} f_{X_i}(x)} = \frac{\sum_{i=1}^{N} f_{X_i}(x) E_{Y_i}[y]}{\sum_{i=1}^{N} f_{X_i}(x)} \quad (5)$$

The conditional variance of $y$ given $x$ is defined as:

$$var[y|x] = E[y^2|x] - (E[y|x])^2 \quad (6)$$

To compute the variance we need to find the conditional mean of $y$ squared given $x$.

$$E[y^2|x] = \int_{-\infty}^{\infty} y^2 f(y|x) \partial y = \frac{\int_{-\infty}^{\infty} y^2 f(y,x) \partial y}{f(x)} \quad (7)$$

$$= \frac{\int_{-\infty}^{\infty} y^2 \frac{1}{N} \sum_{i=1}^{N} f_{X_i}(x) f_{Y_i}(y) \partial y}{f(x)} = \frac{\frac{1}{N} \sum_{i=1}^{N} f_{X_i}(x) \int_{-\infty}^{\infty} y^2 f_{Y_i}(y) \partial y}{f(x)} \quad (8)$$

The final integral in Equation 8 is the mean square of the probability density of $y$, defined by our third stipulation.

$$E[y^2|x] = \frac{\frac{1}{N} \sum_{i=1}^{N} f_{X_i}(x) E_{Y_i}[y^2]}{\frac{1}{N} \sum_{i=1}^{N} f_{X_i}(x)} = \frac{\sum_{i=1}^{N} f_{X_i}(x) E_{Y_i}[y^2]}{\sum_{i=1}^{N} f_{X_i}(x)} \quad (9)$$

Now we substitute the value from Equation 9 back into our conditional variance equation.

$$var[y|x] = E[y^2|x] - (E[y|x])^2 = \frac{\sum_{i=1}^{N} f_{X_i}(x) E_{Y_i}[y^2]}{\sum_{i=1}^{N} f_{X_i}(x)} - \left[ \frac{\sum_{i=1}^{N} f_{X_i}(x) E_{Y_i}[y]}{\sum_{i=1}^{N} f_{X_i}(x)} \right]^2 \quad (10)$$

This general form allows us to use any probability density function for $f_{X_i}(x)$ as our distance weighting function and any density with defined mean and mean square for $f_{Y_i}(y)$. For example a Gaussian kernel would have a mean of $Y_i$ and mean square of $\sigma^2 + Y_i{}^2$. Where $Y_i$ is the $i$th training pair output. Using the results from Equation 10, implementation of the confidence measure in the original GRNN network becomes possible as shown in Figure 1. Only a single additional weight that corresponds to $E_{Y_i}[y^2]$ for each hidden node needs to be added to the network. This implementation also supports the network modifications used when clustering training set pairs to reduce network size.

The Gaussian function is often used for the Parzen windows in GRNN. Although this function provides good estimates in many cases, it has the disadvantage that its high rate of decay causes the nearest point to highly bias the estimate. This may be advantageous in dense portions of the training set but in sparse areas it forces the estimate to the nearest point when an averaged result may be more appropriate. A weighting functions that does not decay as rapidly as the Gaussian function is the Cauchy kernel:

$$f_{X_i}(x) = \frac{1}{\pi\sigma}\frac{1}{1+(\frac{x-X_i}{\sigma})^2} \tag{11}$$

This kernel is used with a sparse data set and the results compared to those using a Gaussian kernel.

## RESULTS

The propossed GRNN network was tested with a training set similar to [2] using both Gaussian and Cauchy kernels. Twenty-five training points were created using $y(x) = \frac{1}{2}\sin(1.5\pi x + \frac{\pi}{2}) + 2 + v$, where $v$ is a zero mean Gaussian noise source with variance 0.003, and $x$ is sampled from a nonuniform distribution with probability density function $f(x) = 0.5 + 0.45x$ in the interval $[-1, +1]$. To optimize sigma, the network was executed using a test set that included 100 evenly spaced samples over the interval $[-1, +1]$. Figure 2 shows the resulting output using a Gaussian kernel with an optimal sigma of 0.085. The graph displays the original 25 training samples, the GRNN estimate of the test set, and the confidence limits created by adding and subtracting one standard deviation (i.e. $\pm\sqrt{var[y|x]}$) from the estimate. At the bottom of the graph the standard deviation of the training and test set points is displayed.

At zero on the graph, the variance is low, implying the confidence is high, where the points are evenly distributed. On the left side of zero there is a gap between points and the variance increases as the function transitions between Parzen windows. On the right side of zero the variance increases slowly and then flattens out. At the extremes of the graph there is a problem with the confidence measure. With the Gaussian kernel as we move beyond the extent of the training set the estimate remains close to the last point encountered. This provides a misleading confidence of the estimate. We would prefer the confidence to decrease outside the training set.

In Figure 3, the results from the network are displayed when the Gaussian kernel is replaced with a Cauchy kernel. This causes a drastic change in the confidence measure. Between training samples, the variance increases as the distance between the samples increases. A major advantage of this kernel function can be seen as points fall outside the training set. As the input moves away from the last training point the estimate starts to approach the sample mean of the training set and in a similar manner, the variance approaches the sample variance of the training set. This provides a result desirable in a confidence measure whereas the confidence is higher in regions where points are more dense and lessens in sparse regions. Additionally, in tests using this training set with and without clustering the Cauchy kernel produced a lower mean square error than the Gaussian kernel.

In the next set of experiments the GRNN was used to predict the next day low temperature for Dallas, Texas, by taking into account the previous days temperature profile of adjacent regions. The data set used contained daily high

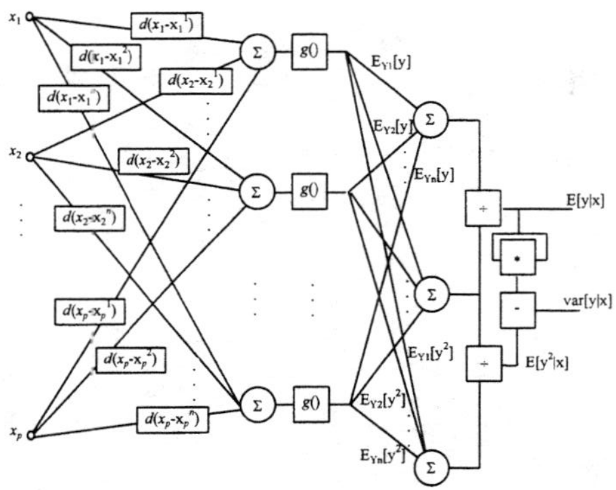


Figure 1. GRNN signal graph with confidence measure. The functions d() and g() combine to create $f_{Xi}()$.

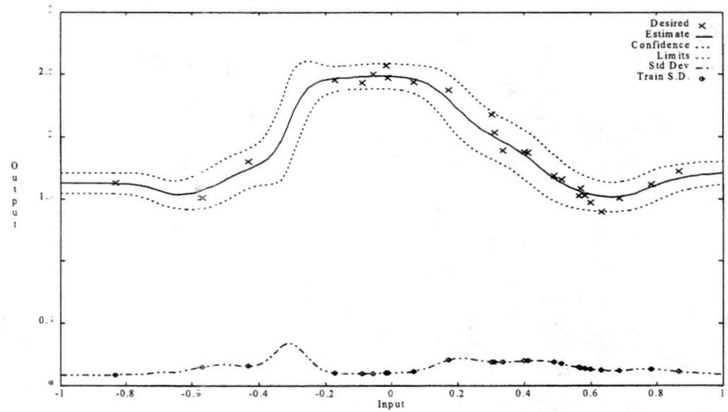


Figure 2. Function estimation results of GRNN with confidence measure using Gaussian kernel.

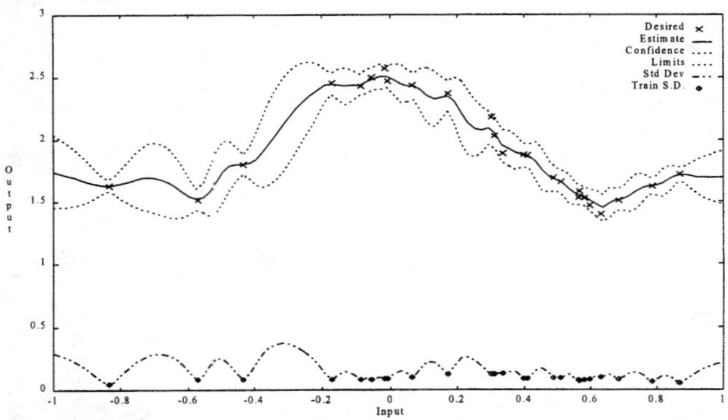


Figure 3.Function estimation results of GRNN with confidence measure using Cauchy kernel.

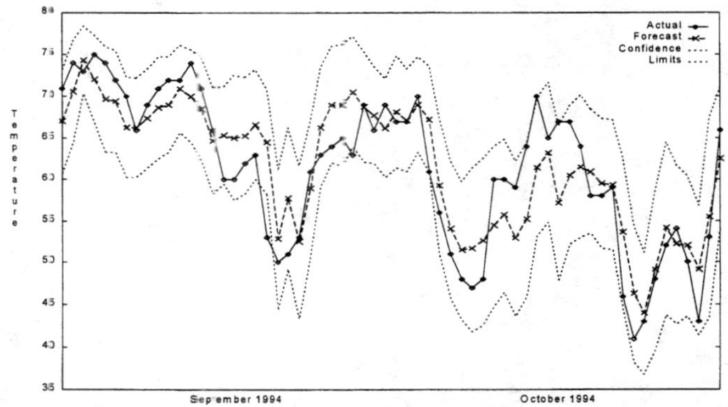


Figure 4. Prediction of daily low temp. for Dallas, TX using GRNN with confidence measure.

and low temperatures from June, 1992, through May, 1995, for Dallas and three cities: Amarillo (TX), Oklahoma City (OK), and Little Rock (AR). Note: this produced a multi-variate input of dimension eight. The network was trained using the first two years of data and tested over the last year of data.

Table 1 shows the Mean Absolute Error (MAE) of the predicted temperature in degrees and the corresponding confidence measure for each month in the test set. The confidence measure is in agreement with the MAE, increasing as the error gets larger. This information could be used to provide error bounds with each estimate. Representative temperature prediction results and confidence limits are displayed in Figure 4.

| Year | 1994 | | | | | | | 1995 | | | | |
|---|---|---|---|---|---|---|---|---|---|---|---|---|
| Month | Jun | Jul | Aug | Sep | Oct | Nov | Dec | Jan | Feb | Mar | Apr | May |
| MAE | 1.95 | 1.79 | 2.14 | 3.61 | 3.60 | 5.13 | 4.39 | 5.52 | 5.01 | 4.37 | 3.48 | 4.18 |
| Confidence | 4.48 | 4.69 | 4.82 | 6.61 | 8.19 | 8.36 | 7.64 | 7.55 | 8.26 | 8.32 | 7.54 | 8.24 |

Table 1: GRNN Temperature Prediction Results and Confidence Measure

## SUMMARY

The GRNN provides a network that is very useful in many engineering applications due to its fast training and limited number of unknown parameters. The system weights are determined directly from the training set and the only additional parameter left to determine is the Parzen window spreading factor, sigma. A modification to the network provides a confidence measure related to the accuracy of the network based on the conditional variance of the output given the input. In certain applications, such as those with sparse data sets, it may be more advantageous to use probability density functions other than the Gaussian function commonly used in GRNN networks.

## REFERENCES


[1] A. Lapedes, R. Faber "Nonlinear Signal Processing Using Neural Networks: Prediction and System Modeling," *LA-UR-87-2662*, 1987.

[2] J.A. Leonard, M. A. Kramer, and L. H. Ungar "Using Radial Basis Functions to Approximate a Function and Its Error Bounds," *IEEE Transactions on Neural Networks*, vol. 3, no. 4, pp. 624-627, 1992.

[3] E. Parzen, "On estimation of a probability density function and mode," *Annals Mathematic Statistics*, vol. 33, pp. 1065-1076, 1962.

[4] D. F. Specht, "A General Regression Neural Network," *IEEE Transactions on Neural Networks*, vol. 2, no. 6, pp. 568-576, 1991.

# Experiments in Predicting Financial Time Series Using Modular Neural Network Architectures

**KARSTEN SCHIERHOLT AND CIHAN H. DAGLI**
*Smart Engineering Systems Laboratory*
*Engineering Management Department*
*University of Missouri - Rolla, Rolla, MO 65409-0370*
*E-Mail: karsten@umr.edu, dagli@umr.edu*
*Tel.: (573) 341-4374, Fax.: (573) 341-6567*

***ABSTRACT:***
Forecasting of stock returns and stock market behavior has received special attention as application of smart engineering systems. In previous experiments, this problem has been most commonly attacked using multi-layer perceptrons or Probabilistic Neural Networks. An adaptive modular network, as developed in this paper, weighs the decisions of several architectures for a final output depending on the input pattern and can thus combine the advantages of both architectures.
Test runs show that the proposed architecture is clearly able to outperform the index, although results are not as good as in previous experiments. The network achieves an average final portfolio value about 2% higher than the index.

## I. INTRODUCTION

With growing computational power available to the research community, the interest in complex problems such as the analysis and prediction of economical time series has grown rapidly. While much energy is spent on understanding and predicting bonds and currencies, the forecasting of stock returns and stock market behavior has received special attention [CMMR92]. In addition to statistical and econometric analysis methods, much research in recent years was conducted to develop applications of chaos theory and to improve to neural network architectures. These attempts were encouraged by various evidence that economic markets do not behave randomly, but rather perform in a chaotic manner [Mal94]. An investment in stocks, that combines stocks in the same ratio as the components of a stock market's index, changes its value identical to the index' changes. It is therefore sufficient to model an index in order to describe the behavior of a market.
Several models can be used to predict the next day's value of a time series, the upcoming change, or just to give a binary rise-fall output. In this paper the latter approach is favored. A single lost chance may hurt more than several negligible missed opportunities. If a model manages to predict only the most crucial 20% of all decisions correctly and all other decisions wrongly, the model would still perform better than the index. The main focus is therefore directed upon maximizing the performance of the portfolio rather than maximizing the percentage of correct decisions.
In the following sections, the environment for the presented network, the used data and the classification model, is described. The modular network is explained in detail before results of previous and current research are compared.

## II. DATA

The input data for the proposed network consists of market data collected between February 1994 and September 1995. Best results are obtained by using recent S&P500 Index values as well as a currency index, which is composed of the exchange rates of the three currencies Japanese Yen, British Pound and German Mark. Only the closing values are used. For the prediction of the next day's change, today's value, the change with respect to the five previous days, and the change from two weeks previous are investigated as possible inputs to the network. All inputs are normalized. Due to the nature of stock markets, most

changes are only of very small amplitude compared to the maximum changes. In order to receive a more equal distribution over the input space, a sigmoidal normalization function is used.
The full data set consists of more than 400 patterns. 100 patterns of these are used for initial training. After being used for testing and evaluation of the network, more patterns are added to the training set. Further details of this process are described in section IV.

# III. CLASSIFICATION MODEL

The network is designed to give advice about whether to buy or to sell stocks. In order to put special emphasis on training and prediction of large-scale changes in the time series, the output should be differentiated into more than two output classes. Those classes representing only a minor change do not need to be trained as intensively as those who have a major influence on the network's overall performance.
The raw structure of the network in its prediction phase is sketched in Figure 1. The Classification Network uses the input pattern to determine a membership degree for each output class. In the second stage, the Declassification Module transforms this interim output into a recommendation for behavior in the market, the trading advice.
The network's performance is measured by a portfolio, a collection of stocks, that has an initial value of 100 and changes by the same percentage as the index every time a "buy" advice is given. The portfolio keeps its value upon a "sell" advice until the network suggests again to enter the market. This portfolio's performance, i.e. its value at the end of the prediction period, is compared to the performance of the index.

# IV. PROPOSED NETWORK ARCHITECTURE

## IV.a. Overview

Combining multi-layer perceptrons (MLP) and Probabilistic Neural Networks (PNN) in one network might merge advantages of both architectures to improved results. In this introductory section, a general picture of the network developed in this paper and its training process is drawn. Figure 2 shows a crude data flowchart. More details about the hierarchical neural network (HNN), the multilayer perceptron,

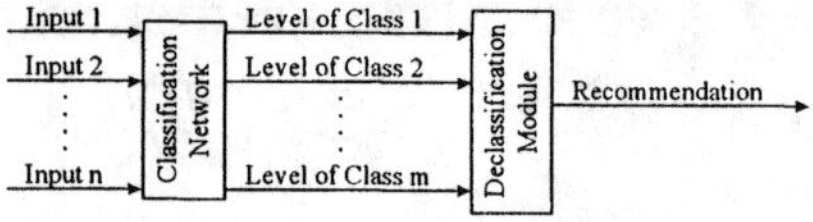


Figure 1: Structure during prediction phase

and the probabilistic neural network is given in the following subsections. Subsection IV.e. will further describe the Confidence Estimation Module (CEM).
The kernel of the network is a hierarchical neural network. An initial number of patterns is only used for training. The learning process of the HNN is controlled by the Confidence Estimation Module. After every training pass, the CEM uses a set of estimation patterns to determine the HNN's present prediction quality, the present confidence. The estimation patterns are made up of a moving window of the most recent (in the time series) unlearned and data patterns that were already used for prediction purposes. Training continues until the confidence level has saturated or a maximum number or training passes was performed.
Now the one input pattern that follows the latest estimation pattern is used to predict the respective output. The oldest estimation pattern is then included into the training set and the estimation patterns are filled up by the pattern previously used for prediction. Training resumes again until the CEM signals to predict the next pattern. This process continues as long as prediction data is available.

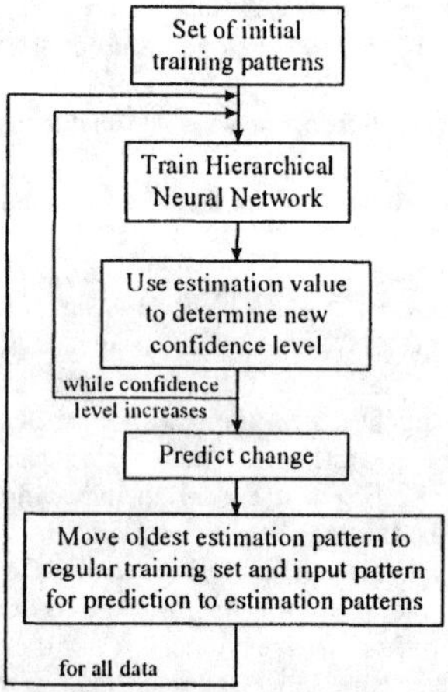


Figure 2: Data flow through the network

## IV.b. Hierarchical Neural Network

A hierarchical neural network consists of several specialized networks (Expert Networks), that are trained to perform well for only part of the input space. These Expert Networks are mediated by an integrating unit, the Gate Network. The Gate Network decides on how the outputs of the experts should be combined to form the final output of the system as well as on the degree to which every expert learns the given training pattern [JJNH91, JJ91]. The combined output of experts on one level, calculated by a summation unit, can perform as the input to the next higher hierarchy level. A detailed description of the used learning algorithm can be found in [Hay94].
For this research, a network with three levels was used. The topmost Gate Network mediates among the outputs of two summation units on the second level, each of which combines two experts, one MLP and one PNN (Figure 3). Every network can be configured with its own set of parameters.

## IV.c. Multilayer Perceptron Expert

MLP experts function like stand alone MLP networks. During the feed-forward phase, an input is processed and each dimension of the output reflects the determined membership degree for the respective output class. The learning takes place in form of changing weights to minimize the difference between actual and desired output [RM86]. The error in a HNN is passed down from the topmost level and adjusted by the gate networks such that the weights in a MLP change only accordingly to their importance during the feed-forward phase.
In order to adjust faster to changes in the environment, recent training patterns are trained more often than older ones. It can therefore happen that during a training pass one pattern is trained twice or more often while others are not trained at all. Since every training pattern belonged to the group of recent patterns at some point, there is no danger for patterns to be totally ignored.
As mentioned in the introduction, the patterns that have a large effect on the performance of the network as a whole should receive special attention. This is taken into account by using a learning rate that increases with the importance of the pattern to be trained.
So far, only MLP experts with one hidden layer are used.

## IV.d. Probabilistic Neural Network Expert

IV.d.i. Introduction

Probabilistic Neural Networks, as proposed by Specht [Spe90], perform a statistical classification task by calculating the probability density functions for each class over the input space from previously learned examples and then deciding to which of these classes an unknown input vector most likely belongs to. Still, in the original form, they are not very well applicable as experts in an HNN structure. The major disadvantage of the PNN develops from the fact that it requires all input vectors to be stored [Spe92], which negatively influences the computation time. Data compression in form of clustering close patterns can help. Other modifications focus on the network acting in a changing environment. Older

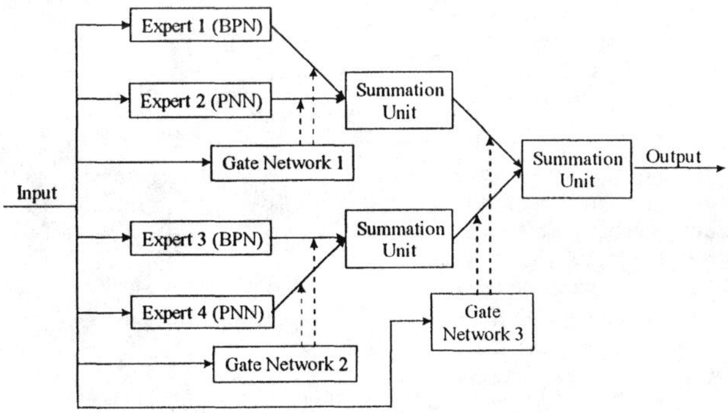


Figure 3: Structure of the Hierarchical Neural Network

patterns should not receive the same high attention as recent ones. Also, PNN parameters should be adapted to achieve the best possible performance of the expert. These subjects are addressed in the following sections.

IV.d.ii. Modification of Parameters

For each category, a probability density function is calculated from the training patterns. In this paper, a Gaussian function is used as the Parzen Weighting function, such that the probability of an input $X$ belonging to a class $A$ can be written as

$$f_A(X) = \frac{1}{(2\pi)^{\frac{p}{2}} \prod_{i=1}^{p} \sigma_i} \cdot \frac{\sum_{i=1}^{m} w_i n_i e^{-\sum_{j=1}^{p} \frac{(X_j - X_{Aij})^2}{2\sigma_j^2}}}{\sum_{i=1}^{m} W_i n_i}$$

where $p$ denotes the dimensions of the input, $m$ the number of clusters in class $A$, $w_i$ the weight of cluster $i$, $n_i$ the number of patterns stored in cluster $i$, $X_j$ the $j^{th}$ component of input vector $X$, and $X_{Aij}$ the $j^{th}$ component of the $i^{th}$ cluster in class $A$.

This method is identical to the one given by Specht, except for the consideration of clusters and the use of separate smoothing parameters $\sigma_i$ for each input category. It is suggested in [Spe92] that adapting separate $\sigma$'s for each dimension improves the generalization ability. With every new training of the PNN, one of the parameters ($\sigma_i$ or the cluster variable) is slightly changed. If the present confidence increases after the next evaluation, this perturbation is manifested. Otherwise the changes are rejected.

The PNN does not, like the MLP, use an incremental learning method but rather builds the network with given parameters in one pass. Every time, when one of the parameters is changed, that is after each confidence evaluation or after every HNN training pass, the network is reconstructed.

IV.d.iii. Weights and Clustering of Patterns

Every pattern (or rather cluster) is assigned a weight of initially one when integrated into the network. With every new pattern learned by the network, the weight decreases by a small percentage. The importance of a cluster therefore decays exponentially with time. This process emphasizes on the more recently learned patterns in the evaluation process.

As mentioned earlier, the basic PNN stores all input patterns in its structure and therefore uses large amounts of space and time during the training and classification process. Several ways, such as in [Tra91] or [Spe92], have been suggested to combine patterns to clusters. The technique used in this approach is slightly different due to the highly separated input space. Patterns or clusters belonging to an identical class are combined only if they are not further apart than a maximum distance. The position of a cluster after including a new pattern depends not only on the former positions of the components but also on the cluster's weight and the number of patterns already represented through the cluster.

In order to keep the clusters from being filled up with old patterns, and therefore through their inertia from adjusting the centers quickly to changing environments, the number of patterns in a cluster should be decreased. Every pattern has its age, which increases every time, when a new pattern is learned by the net. Patterns that reach a certain maximum age, are removed from the cluster.

The learning process of the modified PNN is shown in Figure 4.

## IV.e. Confidence Estimation Module

The CEM uses a number of untrained patterns to determine the network's prediction ability and thus its ability of generalizing the trained data to predict unknown patterns. Not only the correctness of the final decision (buy or sell) is

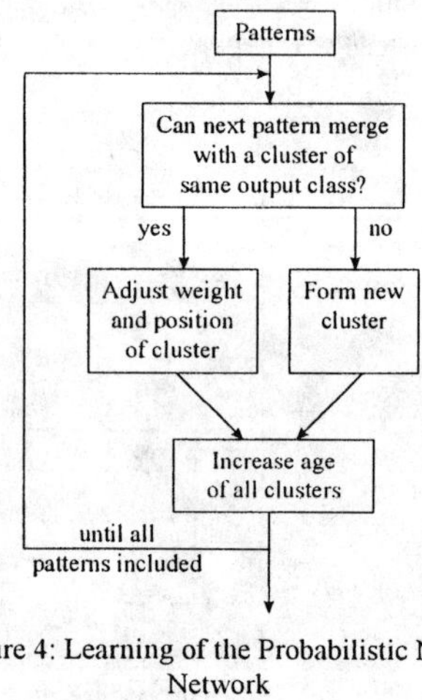


Figure 4: Learning of the Probabilistic Neural Network

considered. Attention is also paid to the difference between the exact desired and predicted output. Furthermore, the prediction over the whole period is included into the calculation of the present confidence.
The confidence level is estimated newly after every training pass. At the beginning of each training period (after adding a new pattern to the training set) the initial confidence level is set back to 0 and then slowly adjusted to the calculated present confidence using exponential smoothing. This gives the network some time to adjust to the new situation and prevents overreactions upon a temporarily lower present confidence.

## V. RESULTS OF PREVIOUS RESEARCH

Previous attempts to model the behavior of the S&P500 Index have used MLPs with a back-propagation learning algorithm as well as PNNs. While MLPs are the most commonly used neural networks for predicting economic time series, promising results have been received with PNNs [TPW95, Che94].
Both architectures performed extremely well while the index is decreasing. This is an inherent characteristic of the given problem structure: in a declining market, the portfolio can only perform better than or equal to the index. Beating the index during an inclining market requires not only the increasing steps to be predicted correctly, to keep track with the index, but also small declines must be detected to gain advantage over the index.
MLP and PNN were able to outperform the index using the above described performance measure. Most often, the PNN performed slightly better than the MLP. MLPs have a very good generalization ability by trying to learn a functional relationship. Nevertheless, they underlie a high danger of overfitting if trained too much. The PNN, due to its exclusivity [Wer92], seems to be suited better to perform the classification task over a highly separated input space. Most important, while MLPs usually work in static environments with separate training and testing phases, the stock market is a fast changing environment, that calls for an adaptive approach for forecasting change. Due to its fast and easy learning, the PNN seems well suited for this purpose.
A more detailed description of the experiments and their results can be found in [SD96].

## VI. EXPERIMENTAL RESULTS AND CONCLUDING REMARKS

Test runs show that the proposed architecture is able to outperform the index. In the time frame used for prediction, the index rose about 1%. The results of ten tests averaged to final portfolio values of 103.2 with a variance of 2.6. With more different network structures being examined, the performance can probably be improved to a level similar to the results presented in [SD96].
One representative test run is shown in Figure 5. Many of the important decisions are made correctly. Improved results might be received by introducing other kinds of expert networks and thus enlarge the options for construction the hierarchical network.

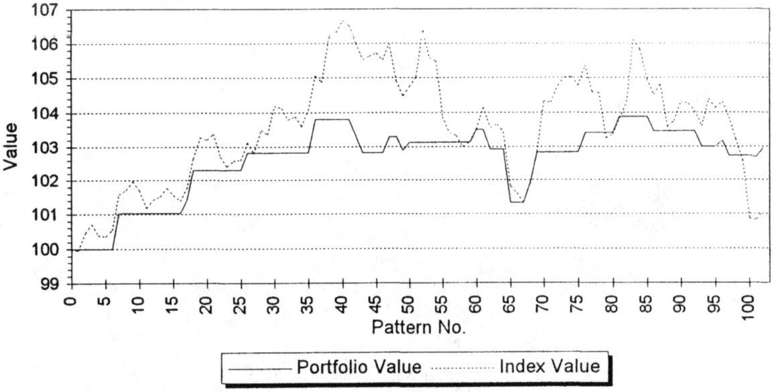


Figure 5: Performance during test run

## VII. REFERENCES


[CMMR92] Chakraborty, Kanad; Kishan Mehrotra, Chilukuri K. Mohan, and Sanjay Ranka (1992). Forecasting the Behavior of Multivariate Time Series Using Neural Networks. *Neural Networks*, **5**, 961-970.

[Che94] Chen, C.H. (1994). Neural Networks for Financial Market Prediction. *Proc. 1994 IEEE International Conference on Neural Networks*, 1199-1202.

[Hay94] Haykin, Simon (1994). *Neural Networks.* Englewoods Cliffs, NJ, Macmillan College Publishing Company, Inc.

[JJNH91] Jacobs, R.A.; M.I. Jordan, S.J. Nowlan, and G.E. Hinton (1991). Adaptive Mixtures of Local Experts. *Neural Computation*, **3**, 79-87.

[JJ91] Jacobs, R.A. and M.I. Jordan (1991). A Competitive Modular Connectionist Architecture. In: *Advances in Neural Information Processing Systems 3* (R.P. Lipmann, J.E. Moody, and D.J. Touretzky, eds.). San Mateo, CA, Morgan Kaufmann.

[Mal94] Malliaris, Mary E. (1994). Modeling the Behavior of the S&P500 Index: A Neural Network Approach. *Proc. 10th IEEE Conference on AI for Applications*, 86-90.

[RM86] Rummelhart, David E., and James L. McClelland (1986). *Parallel Distributed Processing*, **1**, 318-362. Cambridge, MA, MIT Press.

[SD96] Schierholt, Karsten, and Cihan H. Dagli (1996). Stock Market Prediction Using Different Neural Network Classification Architectures. *Proc. IEEE/IAFE 2nd Conference on Computational Intelligence in Financial Engineering*, 72-78.

[Spe90] Specht, Donald F. (1990). Probabilistic Neural Networks. *Neural Networks*, **1**, 109-118.

[Spe92] Specht, Donald F. (1992). Enhancements to Probabilistic Neural Networks. *Proc. IEEE/INNS International Joint Conference on Neural Networks*, **I** 761-768.

[Tra91] Traven, H.C.G. (1991). A Neural Network Approach to Statistical Pattern Classification by 'Semiparametric' Estimation of Probability Density Functions. *IEEE Trans. on Neural Networks*, **2**, 366-377.

[TPW95] Tan, Hong; Danil V. Prokhorov, and Donald C. Wunsch II (1995). Probabilistic and Time-Delay Neural Network Techniques for Conservative Short-Term Stock Trend Prediction. *Proc. World Congress on Neural Networks*, **II**, 44-47.

[Wer92] Werbos, Paul (1992). Neurocontrol and Supervised Learning: An Overview and Evaluation. In: White, David A., and Donald A. Sofge, ed.: *Handbook of Intelligent Control.* Van Nostrand Reinhold, 65-89.

# AN EFFICIENT NEURAL NETWORK MODEL TO PREDICT STOCK MARKET BY USING MODIFIED CONDITIONAL DENSITY ESTIMATION

**SABYASACHI GHOSHRAY**
*School of Computer Science & Department of Electrical Engineering*
*Florida International University, Miami, Florida 33133 USA*
*Email: ghoshray@fiu.edu*

***ABSTRACT:***
In this research, we have analyzed the performance of several density estimating neural networks, in predicting the return and volatility of stock market indices. Presence of outliers and heteroscheduostic errors present drawbacks in some of the conditional density estimating neural networks. In this regard, we use the conditional density estimating network, in which the performance is indicated in terms of the mean squared error and the correlation between the target and the prediction on training and test set. Our method uses a parameter precision network with two outputs, one for the conditional expectation and the other for the conditional variance. We have compared our proposed model with some of the existing neural network models to provide a volatility estimate for the Dow Jones and NASDAQ stock indices and the results obtained so far indicate superior performance in the proposed method.

## INTRODUCTION

One of the heavily researched problem in the scientific community is predicting the future of temporal sequences that ranges from forecasting the weather to estimating stock market indices. The desire to know the future is what shapes the laws in science. The ability to forecast the behavior of a system hinges on two types of knowledge. The first and the most powerful one is the knowledge of the laws controlling a given phenomenon. This knowledge when expressed in the form of equations, helps us in predicting the future outcome of an experiment. The second type of knowledge relies on the discovery of empirical regularities without taking resort to knowledge of the underlying mechanisms. The key problem in this context is to determine which aspects of the data are irrelevant and which are the true indicators of the inherent characteristics. This issue has surfaced quite frequently in the application of neural networks for prediction in financial environment. The complexity and the seriousness of this issue lies in the fact that here we are dealing with real world data, which are limited in precision and sample size.

Neural network based regression models have been used quite extensively in financial forecasting (Refenes, 1994, Hutchinson et al, 1994). These models use sum-squared error measures and they provide the asymptotically best estimators under some distribution assumptions. But, this assumptions can not be maintained under practical situations, which therefore reduces their significance in terms of predictability. In this regard, density estimating neural networks (DEN) can be used to provide a framework for developing further predictability in financial markets. The reason is simple; probability density estimating neural networks are more adequate in describing probabilistic relations than common feedforward network (White, 1992, Ormoneit, 1993, Ormoneit, 1994). The additional information that is inherent in the conditional density can be exploited to represent multi-valued mappings in the form of multi-modal densities to approach inverse problems (Bishop, 1994). In this research, the performance of different density estimating neural networks, in predicting the return and volatility of stock market index, have been analyzed and a more efficient prediction methodology have been presented.

## CONDITIONAL DENSITY ESTIMATION BY INDIRECT METHODS

In order to predict a future event in the areas of stochastic control or financial forecasting, the data distribution can not be obtained via statistical estimation techniques, such as simple regression analysis. In this context, we need to estimate the complete probability density of the given data. One of the most efficient of the estimators is the one that can estimate the conditional densities p(y|x) given empirical data points (y, x) that are obtained by a time series. The conditional density p(y|x) can be defined as the probability for a future value *y* provided the value *x* has been observed. We begin a generalized discussion of conditional density estimation by defining a function p(w) as a linear combination of a set of mulit-dimensional Gaussians p(w|i), such that:

$$\mathrm{p(w)} = \sum_{i=1}^{n} p(i)p(w|i)$$

where p(i)s are the weighting factors that satisfy the following conditions:

(i) $\mathrm{p(i)} \geq 0$

(ii) $\sum_{i} p(i) = 1$

Here, p(i) is a prior probability for the ith component of the probability distribution and p(w|i) is the probability of *w* given that i has occurred, or it can also be interpreted as the measure of belief that the data point *w* was produced by component i. The joint probability density for *x* and y is written as p(x,y) and the conditional probability density of *y*, given *x* has occurred is written as p(y|x). The conditional density calculation therefore uses the relation

$$\mathrm{p(y|x)} = \frac{p(y,x)}{p(x)}$$

There are several conditional density estimation networks (CDEN) that have been used.

**Probabilistic Neural Network**

This network was first invented by Parzen in 1962 and was later modified by Specht in 1990 (Parzen, 1962, Specht, 1990). In this network, the data points define the centers of the Gaussian component densities. Here all the apriori probabilities are assumed to be equal; thus, the variances of the Gaussians are the only parameters that are adjusted via exhaustive search. The drawback of this probabilistic Neural Network (PNN) is that the quality of the probability density estimate is extremely sensitive to the appropriate choice of the variance.

### Gaussian Mixture Network

The working principles of this network is based on the estimation of the joint density and the subsequent computation of the conditional density using the relationship described earlier. In this Gaussian Mixture Network (GMN), the hidden units represent component densities p(w|i) and the activation values of the hidden units are summed up to a linear output neuron (Ormoneit et al). The weightings for the hidden-to-output neurons represent the p(i)s in the linear combination described above. Figure 1 shows a schematic of the type of network being described here.

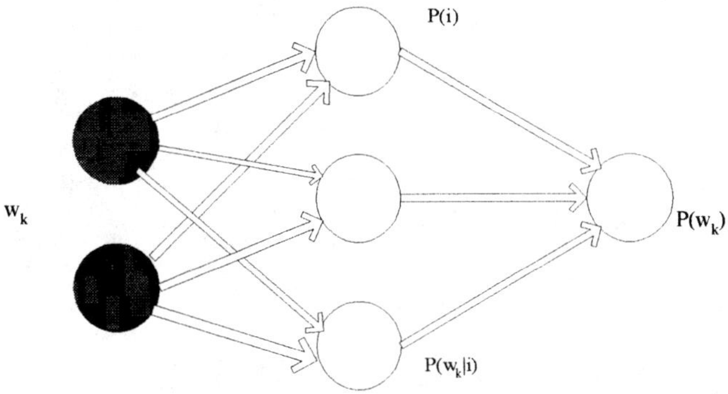


Figure 1: Gaussian Mixture Network

Here, the GMN gets the set of training patterns that are given by $w_k = (y_k, x_k)$ such that the joint likelihood $L = \Pi_k p(w_k)$ is maximized. This maximizing technique is in synchronization with the back propagation algorithm in the sense that the maximization function is equivalent to the minimization of the logarithm of the inverse joint likelihood. This produces the error function E as follows:

$$E = -\log L = -\sum_k \log p(w_k)$$

### Conditional Density Estimation Network

The above two methods, PNN and GMN constitute the indirect approaches to the conditional density estimation. The requirement of large data set and inferior performance of these two networks propel us towards investigating in a different direction in which the conditional density can be estimated directly. In this Conditional Density Estimation Network (CDEN), given a set of data points $w_k = (y_k, x_k)$, the conditional likelihood $L^c = \Pi_k p(y_k|x_k)$ is maximized. This is the same as minimizing :

$$E = \sum E_k^c = -\sum_k logp(y_k|x_k)$$

$$= \frac{\sum_k logp(y_k, x_k)}{\int p(y, x_k)dy}$$

In this network realization, the network outputs are used as density parameters during the forward propagation. On the other hand, the derivatives of the error function with respect to the parameters are used to initialize the external error during the error back proportion as seen in figure 2.

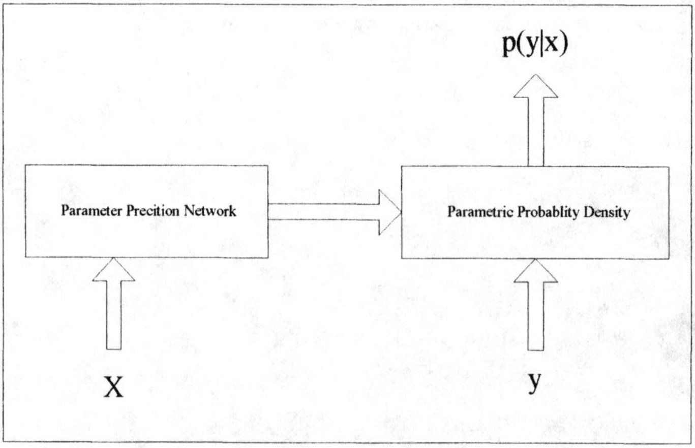


Figure 2: Conditional Density Estimation Network

**Outliers and Heteroscedasticity**

Outliers frequently present drastic drawbacks in the prediction models of time series. Therefore, the probability model which perceives the presence of outliers leads to robust estimators. Outliers can be proven to be an exception by an appropriate choice of the mixture weightings that can be regarded as prior probabilities for each Gaussian distribution. In our method, we used a parameter precision network with two outputs, one for the conditional expectation and another for the conditional variance. This helped in minimizing the heteroschedastic error. Therefore, training the CDEN corresponded to deriving a generalized least square estimator (GLSE).

## IMPLEMENTATION OF THE PROPOSED NETWORK

We use the conditional Density Estimation Network (CDEN) to predict the performance of stock market indices in the future. We have tested our methods in both Dow Jones and NASDAQ returns. Our method focused on outliers and heteroscedasticity. Outliers in financial times series are caused by *information shocks* such as the announcements of dividends paid by companies that are very different from market expectations. In the trajectory of an affected asset, these shocks appear as discontinuities. In the stock market, the variance of a stock's return is commonly related to companies debt-to-equity (D/A) ratio. In this context, heteroscedasticity arises due to changes in the risk associated with an investment. Difficulties in financial times series arises due to heteroscedasticity or situations where the variance of $\varepsilon_t$ changes over time. We have considered the case, where the variance is input-dependent: $\sigma_k^2 = \sigma^2(x_k)$.

The conditional density p(y/x) may be fundamental as $p_{\psi(x)}(y)$, where $p_{\psi}(+)$ is a parametric density family. The conditionality on x is expressed by realizing the parameter vector $\psi$ as a nonlinear function $\psi(x)$. Both $\psi(x)$ and $p_{\psi}$ (+) may be implemented as neural networks, with the output of the first determining the weights of the latter. Denoting the weights contained in the parameter prediction network as *w*, we can express

$p(y/x,w)=p_{\psi(x,w)}$ (y). To yield an estimate of the weights in the parameter prediction network, denoted as *w*, we minimize the error function:

$E(z^*, w) = - \sum_{k=1}^{m} \log\ p(y_k|x_k,\ w)$ by performing gradient descent. In order to approach specific problems with this model, therefore, we determine an appropriate density family $p_{\psi}$ (.). A Gaussian mixture such as:

$$p_{\psi}(y) = \sum_{i=1}^{n} k_i p\ (y|i,\ \mu_i,\ \sigma_i),\ k_i \geq 0,\ \sum_i k_i = 1$$

is a good choice in which the normal densities $p(y|i,\ \mu_i\ \sigma_i)$ is defined as

$$p(y|i,\ \mu_i\ \sigma_I) = \left(\sqrt{2\Pi\sigma_i}\right)^{-1} \exp\left\{\frac{(y - \mu_i)^2}{2\sigma_i^2}\right\}$$

**Forecasting the Volatility of the Stock Indices**

Volatility estimates have gained major attention in financial analysis as a basis for directional market forecasts to evaluate and hedge derivative market instruments. Since the speed of the market changes frequently, and observations are disturbed by large moves caused by information shocks or discontinuities in time, it is relatively difficult to predict intraday stock market data. Here we have taken the value of the NASDAQ stock exchange observed in minute basis October 1995 as the basic time series. We calculated the volatility by averaging the last absolute differences of the index. Our job is to predict the value of the volatility 60 minutes ahead, based on values at t=0, t=30, t=50, t=90, t=150. We used the first 3000 data points for training and the next 2000 for testing. The performance is indicated in terms of the mean squared error (MSE) and the correlation between target and prediction on training and test set. Following comparisons in Table 1 reveals the efficiency of the proposed CDEN method with respect to other methods described earlier.

**Table 1 :Comparison Among Different Neural Networks**

| NETWORK | MSE | | CORRELATION | |
|---|---|---|---|---|
| | TRAIN | TEST | TRAIN | TEST |
| PNN | 0.8591 | 1.410 | | |
| GMN | 0.7402 | 1.321 | 0.521 | 0.292 |
| PROPOSED CDEN | 0.6869 | 1.239 | 0.471 | 0.386 |

## CONCLUSION

In this research, we have presented an efficient methodology to predict the future values of stock market index based on Neural Network modeling of time series data. We have analyzed several existing neural network algorithms that attempts to predict the time series values based on probability density estimation. Due to space constraint, we are able to publish only partial results of the comparisons among several neural networks including the proposed one. However, our analysis indicates the superior performance of CDEN with respect to other neural network models. Our future research includes investigating towards developing a more efficient time series prediction model by integrating fuzzy engineering and neural networks.

## ACKNOWLEDGMENT

The author wishes to thank Ms. Jennifer Schulke of the Decision Science Department at Florida International University, for the creation of all the graphs, tables and the editing for the papers presented at the 1996 Annie Conference.

## REFERENCES

Bishop, C.M.,(1994). Mixture Density Networks, *Technical Report* Aston University.

D. Ormoneit.,(1993). Estimation of Probability Densities Using Neural Networks, *Master's Thesis*, University of Munchen.

Hutchinson, J.M., Lo, A., & Poggio, T., (1994). A Nonparametric Approach to Pricing and Hedging Derivative Securities Via Networks, *Technical Report MIT Artificial Intelligence Laboratory*.

Neuneier, R., Hergert, F., Finnoff, W., & Ormoneit, D., (1994). Estimation of Conditional Densities: A Comparison of Neural Network Approaches. *International Conference on Artificial Neural Networks*, I:689-692

Parzen, E., (1962). On Estimation of a Probability Density Function and Mode, *Annals of Mathematical Statistics* 33.

Refenes, A.N, & Azema-Barac, M., (1994). Neural Network Applications in Financial Asset Management. *Neural Computing and Applications*, 2:13-39.

Spect, D.F., (1990). Probabilistic Neural Networks, *Neural Network* 3.

White, H.,(1992) Parametrical Statistical Estimation with Artificial Neural Networks, *Technical Report University of California*.

# ARTIFICIAL NEURAL NETWORK IN TIME SERIES ANALYSIS: AN OCEAN ENGINEERING EXPERIENCE

**K. RAMANITHARAN , C.W. LI**
*Department of Civil and Structural Engineering, The Hong Kong Polytechnic University, Hong Kong.*

***ABSTRACT:***
The applicability of an Artificial Neural Network (ANN) to the ocean wave system as a forecasting tool is analyzed. ANNs without hidden layers and with linear sigmoid functions are used. Single step ahead forecasting of significant wave period is considered with minimum Root Mean Square (RMS) error as the basis of the stopping criterion. Different networks are tried with different pre-processing of real time data before inputting at input neurons. For comparison, appropriate Auto Regressive Integrated Moving Average (ARIMA) models for the same data set are developed. Results from these models are checked for their statistical correctness. The ANN approach is found more flexible and superior as compared to the ARIMA modeling approach in terms of accuracy and efficiency.

## INTRODUCTION

In spite of the domination of deterministic numerical models developed from the radiative transfer equation (e.g., Li, 1992) in ocean wave forecasting, the ocean wave is mostly stochastic because of the dominant energy input of wind to waves. Hence, the prediction of the wave parameters can be based on its time history.

ARIMA models are stochastic models for which the exponentially weighted moving average forecast yields the minimum square error. ANN is a kind of computing system that, rather than computing model parameters explicitly, is "trained" to give acceptable results from the past in order to yield its best predictions for the real time values. In the field of ocean wave modeling, ARIMA model has been utilized a lot (Mandal et al., 1992). The ANN approach was employed for significant wave height and significant wave period predictions using past wave data and wind data directly without preprocessing (Ramanitharan and Li, 1996).

The aim of this paper is to compare the conventional ARIMA model approach and the less used ANN approach in the ocean wave time series forecasting.

## CASE STUDY

Hourly significant wave period data together with corresponding wind speed and direction data measured at Waglan Island, Hong Kong from 28$^{th}$ July 1994 15:00 hrs to 8$^{th}$ September 1994 14:00 hrs are used for this case study.
Training set: 28 July 1994 (15:00) - 31 August 1994 (06:00) - 808 Data
Validation Set: 31 August 1994 (15:00) - 04 September 1994 (14:00) - 96 Data
Test set: 04 September 1994 (15:00) - 08 September 1994 (14:00) - 96 Data

This time series is found to be non-Gaussian and non-stationary. For the comparison with the ARIMA models, all the neural networks used (except one, net 15 in Table 1) are single layer perceptron NNs with linear transfer function. The net 15 employs wind data additionally as input and uses asymmetric sigmoid function instead of linear transfer function.

The pre-processing of input data varies in the type of transformation used, order of difference involved and the number of input data per pattern used. Except networks 11 and 13, all the networks use the minimum RMS error of the validation set as the stopping criterion for the determination of the optimal weights. In networks 11 and 13, the stopping criterion, the minimum RMS error of the training set, is used for studying the analogy between the ARIMA (in which the maximum likelihood estimates are mostly obtained by least squares) and the ANN models.

Linear correlation coefficients (lr crcf), maximum and minimum values of data, number of correctly / incorrectly (ct/in) identified turning points, ac and pac functions of data and forecasting sets, an error measurement based on RMS error, number of iterations involved, number of errors above certain error percentages and statistical properties ( of both validation and test sets) are used for choosing the best network.

## RESULTS

The following notations are used for the results given in Table 1 and Figures 1 and 2.

1. Neural Networks (except No 15): I J K S M
2. Neural Network No 15: I $H_1$ $H_2$ O W R $T_H$ $T_O$
3. ARIMA Models: ARIMA(p,d,q)

The results of predictions of one hour ahead significant wave period is given in Table 1. Apart from the comparison among themselves, the results are compared with the values of "one hour shift forecasting." Since comparisons during the modeling are made with networks with inputs differenced by 24 hours to check the seasonal effects (not shown), all the final analysis are made and given for the data except the first 26 in the validation and testing tests, though calculations include the whole 96 data of each test. Figures 1 and 2 picturize two important sets of comparisons.

ARIMA(9,1,0) model is selected based on FPE, BIC, AIC and Chi square test values the residuals. Also another ARIMA model, ARIMA(10,1,0) which is less efficient than ARIMA(9,1,0) according to the above tests, is included for comparison purpose.

1. ARIMA(9,1,0) model

$$w_t=-.463w_{t-1}-.252w_{t-2}-.148w_{t-3}-.144w_{t-4}-.158w_{t-5} -.043w_{t-6}-.061w_{t-7}-.01w_{t-8}-.133w_{t-9}+a_t \quad (1)$$

2. ARIMA(10,1,0) model

$$w_t=-.479w_{t-1}-.253w_{t-2}-.155w_{t-3}-.15w_{t-4}-.177w_{t-5} -.061w_{t-6}-.079w_{t-7}-.041w_{t-8}-.189w_{t-9}-.121w_{t-10}+a_t \quad (2)$$

TABLE : FORECASTING RESULTS FOR SIGNIFICANT WAVE PERIOD

| | Actual | 1 Hr shift | 2,0,0,NT | 2,0,0,T | 2,1,0,NT | 2,1,0,T | 2,1,1,T | 3,0,0,NT | 3,0,0,T | ARI(9,1,0) | 10,0,0,T | 10,0,0,T,M | 9,1,0,T | 9,1,0,T,M | ARI(10,1,0) | 986121NN |
|---|---|---|---|---|---|---|---|---|---|---|---|---|---|---|---|---|
| Model No | | 1 | 2 | 3 | 4 | 5 | 6 | 7 | 8 | 9 | 10 | 11 | 12 | 13 | 14 | 15 |
| Input nodes | | | 2 | 2 | 2 | 2 | 3 | 3 | 3 | | 10 | 10 | 9 | 9 | | 9 |
| Valdn err | | 0.1004 | 0.0849 | 0.0843 | 0.0822 | 0.0815 | 0.0824 | 0.0828 | 0.0814 | 0.0815 | 0.0816 | 0.082 | 0.0809 | 0.0842 | 0.084 | 0.0837 |
| Test err | | 0.0533 | 0.0442 | 0.0445 | 0.0421 | 0.0431 | 0.0427 | 0.0434 | 0.0432 | 0.0439 | 0.0436 | 0.0443 | 0.0449 | 0.439 | 0.0445 | 0.0428 |
| Iterations | | | 200 | 255 | 116 | 172 | 28 | 153 | 277 | | 359 | 883 | 148 | 753 | | 519 |
| Val (ct\|in) | | | 12 58 | 12 58 | 21 49 | 21 49 | 15 55 | 14 56 | 14 56 | 16 54 | 16 54 | 16 54 | 17 53 | 16 54 | 16 54 | 16 54 |
| Tst (ct\|in) | | | 13 57 | 14 56 | 17 53 | 17 53 | 12 58 | 17 53 | 15 55 | 15 55 | 15 55 | 15 55 | 19 51 | 16 54 | 15 55 | 22 48 |
| Val(lr crcf) | | 0.6848 | 0.7307 | 0.7351 | 0.752 | 0.7656 | 0.76 | 0.7515 | 0.7542 | 0.7421 | 0.7543 | 0.7536 | 0.7697 | 0.7421 | 0.74 | 0.7406 |
| Tst(lr crcf) | | 0.5696 | 0.6495 | 0.6597 | 0.6862 | 0.6835 | 0.6895 | 0.6587 | 0.6682 | 0.6716 | 0.6689 | 0.6691 | 0.6627 | 0.3721 | 0.6597 | 0.6913 |
| Val (max) | 14.22 | 14.22 | 12.43 | 11.94 | 12.5 | 12.43 | 12.51 | 11.95 | 11.84 | 12.33 | 12.01 | 11.87 | 13.32 | 12.31 | 12.15 | 12.24 |
| Val (min) | 4.13 | 4.13 | 4.98 | 4.76 | 4.38 | 4.4 | 4.4 | 5.38 | 4.84 | 4.6 | 4.71 | 4.71 | 4.35 | 4.6 | 4.78 | 5 |
| Tst (max) | 10.67 | 10.67 | 10 | 9.82 | 9.35 | 9.73 | 9.86 | 9.81 | 9.75 | 10.15 | 9.92 | 9.84 | 10.17 | 10.14 | 10.21 | 9.77 |
| Tst (min) | 6.4 | 6.4 | 7.11 | 7.02 | 6.94 | 7 | 6.99 | 7.37 | 7.16 | 6.96 | 7.04 | 6.98 | 6.98 | 6.96 | 7 | 6.89 |
| Val(mean) | 8.3 | 8.38 | 8.35 | 8.21 | 8.33 | 8.32 | 8.31 | 8.48 | 8.29 | 8.29 | 8.24 | 8.17 | 8.33 | 8.28 | 8.27 | 8.28 |
| Tst(mean) | 8.58 | 8.57 | 8.51 | 8.41 | 8.53 | 8.56 | 8.55 | 8.63 | 8.5 | 8.56 | 8.48 | 8.41 | 8.59 | 8.55 | 8.56 | 8.51 |
| Val(var) | 1.99 | 2.05 | 1.54 | 1.49 | 1.79 | 1.78 | 1.78 | 1.41 | 1.52 | 1.68 | 1.55 | 1.53 | 1.79 | 1.68 | 1.62 | 1.59 |
| Tst(var) | 0.92 | 0.92 | 0.67 | 0.65 | 0.79 | 0.79 | 0.78 | 0.62 | 0.67 | 0.78 | 0.7 | 0.69 | 0.81 | 0.78 | 0.77 | 0.78 |
| Val(skew) | 0.54 | 0.53 | 0.46 | 0.27 | 0.37 | 0.36 | 0.37 | 0.42 | 0.25 | 0.37 | 0.31 | 0.31 | 0.42 | 0.37 | 0.3 | 0.53 |
| Tst(skew) | 0.18 | 0.16 | 0.3 | 0.34 | 0.41 | 0.41 | 0.41 | 0.36 | 0.39 | 0.3 | 0.31 | 0.3 | 0.29 | 0.3 | 0.27 | 0.33 |
| Val(kur) | 3.96 | 3.7 | 3.53 | 3.51 | 3.47 | 3.47 | 3.45 | 3.43 | 3.43 | 3.28 | 3.37 | 3.36 | 3.5 | 3.27 | 3.25 | 3.36 |
| Tst(kur) | 2.27 | 2.29 | 2.28 | 2.29 | 1.91 | 1.92 | 1.97 | 2.01 | 2.05 | 2 | 2.07 | 2.07 | 2.06 | 2 | 2.02 | 1.98 |
| 5% (v \| t) | | 51 50 | 46 40 | 40 40 | 39 36 | 40 36 | 46 38 | 44 39 | 41 37 | 49 36 | 45 36 | 46 38 | 45 36 | 49 37 | 46 39 | 46 39 |
| 10%(v \| t) | | 28 22 | 26 17 | 30 15 | 24 13 | 22 18 | 25 16 | 25 18 | 26 17 | 28 18 | 27 17 | 28 17 | 27 17 | 28 18 | 28 16 | 25 14 |
| 15%(v \| t) | | 23 8 | 18 5 | 18 5 | 15 4 | 16 4 | 18 5 | 16 5 | 17 4 | 18 5 | 17 4 | 17 4 | 17 8 | 18 5 | 21 5 | 14 4 |
| 20%(v \| t) | | 14 4 | 11 1 | 10 1 | 12 [illegible] | 12 1 | 11 1 | 12 1 | 11 1 | 13 1 | 14 1 | 14 1 | 12 1 | 13 1 | 14 1 | 13 0 |
| 25%(v \| t) | | 10 1 | 8 0 | 7 0 | 7 [illegible] | 6 0 | 6 0 | 7 1 | 5 0 | 7 0 | 5 0 | 5 0 | 6 0 | 7 0 | 9 0 | 8 0 |
| 30%(v \| t) | | 8 0 | 4 0 | 3 0 | 6 [illegible] | 6 0 | 5 0 | 6 0 | 4 0 | 3 0 | 5 0 | 5 0 | 3 0 | 3 0 | 5 0 | 5 0 |
| 35%(v \| t) | | 5 0 | 3 0 | 3 0 | 5 [illegible] | 5 0 | 4 0 | 5 0 | 4 0 | 3 0 | 2 0 | 2 0 | 3 0 | 3 0 | 2 0 | 4 0 |
| 40%(v \| t) | | 4 0 | 2 0 | 2 0 | 5 [illegible] | 4 0 | 3 0 | 3 0 | 2 0 | 2 0 | 2 0 | 2 0 | 2 0 | 2 0 | 2 0 | 3 0 |
| 45%(v \| t) | | 3 0 | 2 0 | 1 0 | 3 [illegible] | 2 0 | 2 0 | 3 0 | 2 0 | 2 0 | 2 0 | 1 0 | 2 0 | 2 0 | 2 0 | 3 0 |
| 50%(v \| t) | | 3 0 | 2 0 | 1 0 | 3 [illegible] | 1 0 | 1 0 | 3 0 | 1 0 | 1 0 | 1 0 | 1 0 | 1 0 | 1 0 | 1 0 | 2 0 |

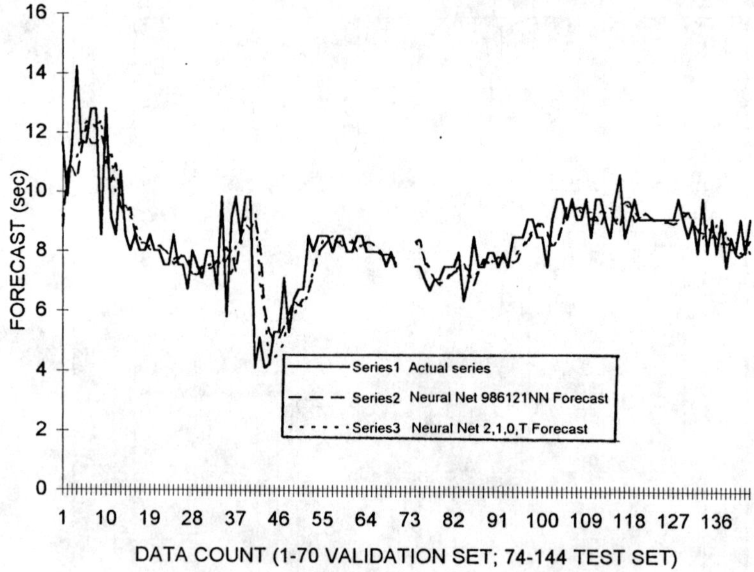


**FIGURE 1: SIGNIFICANT WAVE PERIOD FORECASTING.**

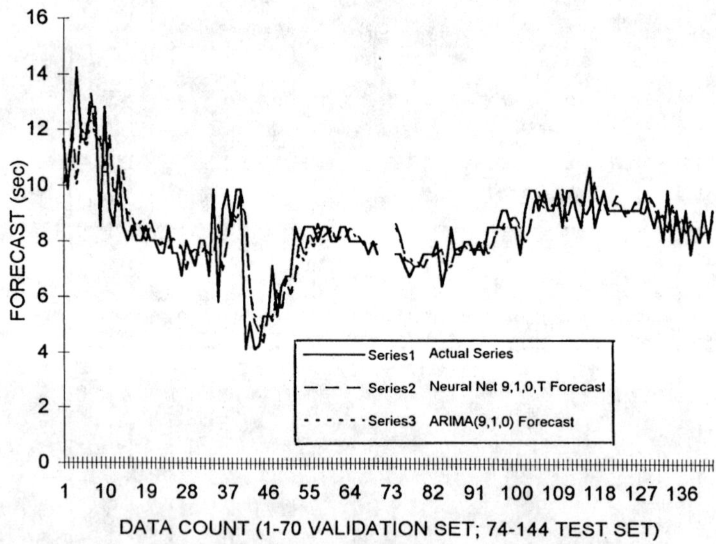


**FIGURE 2: SIGNIFICANT WAVE PERIOD FORECASTING**

## CONCLUSIONS AND DISCUSSION

### Conclusions

The network 5 (2,1,0,T) seems to be the best model in the consideration of minimum deviation in the outputs from the targets, number of correctly identified turning points, comparatively strong correlation and model size. Also, even the worst network yields RMS error which is about 15% - 20% less and linear correlation coefficient which is about 10% - 15% better than those obtained by the "one hour shift forecasting."

### Discussion

From Table 1, it can be observed that when the raw data (i.e., non-gaussian) are used in networks the number of occurrences of predicted values with high percentage errors are more and the number of occurrences of predicted values with low percentage errors are less than those obtained in the case that the transformed (normalized) data are used regardless the type of model and the order of difference involved. These show that those networks with raw input data are less adaptive to the sudden changes in the data as compared to the networks with normalized input data. However, as regard to the accuracy of predication in terms of other statistical properties, both approaches give more or less the same results.

Using the differenced data as input (the data series becomes stationary after differencing) produces better results in terms of error reduction, capability of identifying turning points correctly and linear correlation coefficients.

The conventional ARIMA model, ARIMA(9,1,0) is not so effective as compared to the 2,1,0,T neural net. Further, it yields almost the same results as what the neural net 9,1,0,T gives. And, the net 9,1,0,T,M that uses the weight matrix and threshold values of those of minimum RMS error of training set (which is literally equivalent to the least square error of the ARIMA model) has the weighting coefficients which are close to those obtained in ARIMA(9,1,0). Hence, it can be considered that a single layer perceptron neural network with linear transfer function acts as an ARMA model in which the parameters can be expressed as a linear function of the corresponding weights and threshold values. Further, using cross-correlation criterion to get the best neural net parameters may lead to more accurate results as compared to the ARIMA model using the least squares stopping criterion in many cases (unless another better method is used to achieve the exact maximum likelihood estimate of the parameters for the ARIMA model).

Forecasting using ANNs with second order differenced data as input, together with detrending of seasonal effect(24 hrs seasonality is assumed) has also been carried out. However the results are not quite satisfactory.

The lengthy ARIMA models may be replaced by reduced parsimonious ARIMA models either by System Identification process (Mandal, 1992) or by accepting non-Gaussian Gamma or Log-Normal residuals (Jeyawardena and Lai, 1991).

For the improvement on the accuracy of prediction using ANNs, the followings can be considered. At the preprocessing stage, instead of using traditional methods, we may employ a neural net to give an output data for the real time data input, which may subsequently be used as the best input data for the current forecasting network.

At the post-processing stage, reduction of the error due to the measurement noise may be achieved by adding a filter (either another network which updates the network parameters for every input pattern or a Kalman filter) after the existing filter.

The last improvement is the reduction of errors due to the sudden level changes in the data. These changes may be detected and treated with a Bayesian approach included Kalman filtering (Jun (1985)). Hence, for the combined adaptive neural network, this Bayesian approach may be introduced to overcome the shortcoming of the network at sudden level changing points.

## ACKNOWLEDGMENTS

The authors wish to thank the Hong Kong Polytechnic University for its support for this research work.

## NOMENCLATURE

$a_{t-j}$ Value of the deviation (random residual) series at time t-j
d Order of differences involved to bring the stationarity in the series
$H_i$ Number of neurons in hidden layer I
I Number of input neurons
J Number of times the data differenced
k A constant
K Number of previous residual involved
L A symbol for showing that no cross correlation involved
M Stopping Criterion ( M is L or if cross correlation involves, no notification)
N Asymmetric Sigmoid function, $f(x)=1/(1+e^{-kx})$ (Here, $T_H$ and $T_O$ are N)
NT A symbol for showing that no transformation involved
O Number of output neurons
p Number of past differenced data considered
q Number of past residuals considered
R Number of previous neural net output recurrently used
S Type of transformation (S is T or NT)
T Log-Normal
$T_H$ Type of transfer function in hidden layers
$T_O$ Type of transfer function used in the output layer
W Number of wind data used
$w_t$ Value of the differenced series at time t ( $=\nabla \ln z_t$)
$z_t$ Value of the original real time data at time t
$\nabla$ Differencing operator

## REFERENCES

Jayawardena, A.W., Lai, F., (1991). Non- gaussian ARMA modelling and forecasting of water quality data in Hong Kong: In Lee & Cheung (eds.), *Environmental Hydraulics:* 1147-1153.

Jun, D.B. (1985). ARMA Forecasts with Random Level Changes.

Li, C.W. (1992). A split operator scheme for ocean wave simulation. *International Journal for Numerical Methods in Fluids, Vol. 15:* 579-593.

Mandal, S., Witz. J.A., and Lyons, G.J. (1992). Reduced order ARMA spectral estimation of ocean waves. *Applied Ocean Research (14):* 303-312.

Ramanitharan, K., Li, C.W. (1996). Forecasting Ocean waves using neural Networks: Accepted for the HYDROINFORMATICS 96 Conference, Zurich.

# A Seasonal and Modular Network Approach to Power Load Forecasting

**YASSER AL-RASHID,** *Dept of Electrical Engineering*
**JANET M. TWOMEY,** *Dept. of Industerial and manufacturing Engineering.*
*Wichita State University, Wichita, Kansas*

**ABSTRACT:**
Short-term forecasting (daily peak load, hourly load, average daily load) allows an electric utility company to run a more reliable and efficient operation, and in turn create a major economic advantage. In this paper, the electric load characteristics for the Wichita, Kansas area are presented and the factors which effect the load discussed. As an alternative a series of modular networks is proposed and some preliminary work is presented. Results are compared to a single network approach.

Electric load forecasting seeks to predict load, given past load and weather information and current and forecasted weather information. Short-term forecasting (daily peak load, hourly load, average daily load) allows an electric utility company to run a more reliable and efficient operation, and in turn creates a major economic advantage for the company. In the past, various forecasting techniques have been utilized to provide accurate load forecasts. One popular statistical model is Autoregressive Moving Average (ARMA) model (Paarmann and Najar, 1994). ARMA models suffer from the inability to adapt to sudden changes. As an alternative, systems of combined approaches from areas of computational intelligence and knowledge based systems have significantly improved load forecasts. Artificial Neural Networks (ANN) are a major component (Ho and Yang, 1992) The ANN forecasting model is typically a single network.

A single ANN model built and trained for the Wichita, Kansas area did not adequately forecast power load within the requirements given by the local utilities (± 50MW). The Wichita area poses a particularly difficult task for load forecasting due to its highly changeable weather conditions and large industrial base. As an alternative to the single network approach, we propose a series of seasonal network models composed of seven daily modules, where each daily module contains 24 hourly network modules. In this paper, the electric load characteristics for the Wichita area are presented and the factors which effect the load discussed. Some preliminary results are given on the proposed series of modular networks. These results are compared to a single network model.

## ELECTRIC LOAD CHARACTERISTICS FOR THE WICHITA, KS AREA

The power load during the year followed the same patterns, with some cyclical and random changes. Figure 1 depicts the contribution of temperature on daily electric load peak. The data comes from Kansas Gas and Electric company (KG&E) of KS.

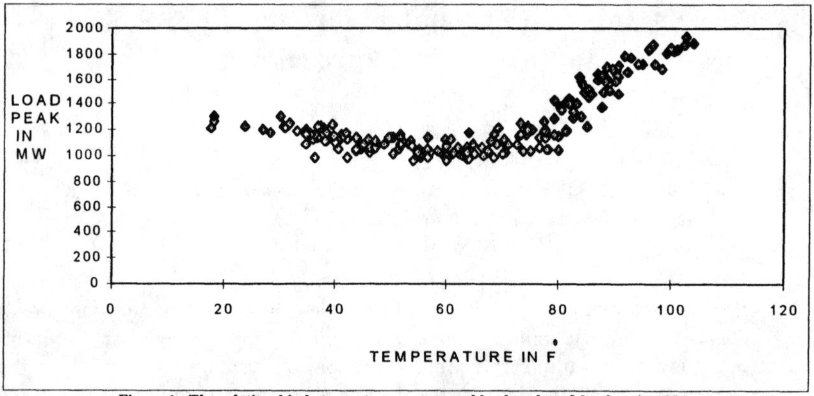


**Figure 1 : The relationship between temperature and load peak on Mondays in 1991.**

As expected, during the summer months temperature increases result in a greater demand for electricity. During the winter months, the reverse relationship exists because of the need for space heating; i.e. as temperature decreases, load demand increases. Summer temperatures are highly correlated with load peaks. Temperature is the main reason people feel hot in the summer; however relative humidity determines how hot it feels. The Heat Index measures how hot the air feels. Winter weather has a less dramatic effect and an inverse relationship to the load peak. Again, the temperature drop has a major effect on increasing demand, but wind chill temperature, which is analogous to the Heat Index, represents how cold it actually feels.

Industrial operations affects electric load most notably when considering: a) weekday (high) vs. weekend use (low), b) holiday vs. non-holiday use, c) increased or decreased production in terms of monthly and yearly variation, and d) unexpected activities, such as major labor strikes or overtime operation.

## MODULAR NETWORKS

According to Haykin, a neural network is said to be modular if the input-output relationship is mapped by decomposing the problem into several sub-problems and training more than one module to solve each part of the overall problem (Haykin, 1994). The performance of a modular network is superior to a single network in solving complex problems for the following reasons: a) Speed of learning. Complex problems are decomposed into several simpler subproblems, thus each module can be trained faster. b) Accurate mapping. Each module addresses a simple part of the overall complex problem; whereas a single network addresses all parts at once. c) Flexibility in modeling. By using a modular network, different architecture can be used for each module in order to get the best overall result from the network

Power load forecasting presents a very complex nonlinear problem. The added flexibility of such an approach is expected to offer significant improvement over the more typical single network approach to load forecasts.

## PROPOSED SYSTEM OF MODULAR NETWORKS

To date, no information is readily available regarding industrial power use. Thus the proposed system of modular networks will improve load forecasts by deconstructing the diurnal trend in the data, and by specifically addressing the demand for power as a function of weather variables, and past load demand.

### Seasonal networks

The overall system is composed of a series of four seasonal networks built to forecast power loads during the spring, summer, fall, and winter months. By separating the individual seasons, the complexity of forecasting will be reduced, insuring that an 80 degree day during the winter months will not produce the effects of an 80 degree day during the summer months. Each seasonal network consists of seven daily modules (one for each day of the week), and each daily module contains 24 individual hourly networks. Figure 2 depicts a single seasonal network composed of hourly modules.

Parameter descriptions:

* **X** is the raw data vector containing the hourly raw past load, temperature, and relative humidity.
* $\mathbf{I}_i$ is the pre-processed data vector containing a combination of the past load, past temperature, and forecasted future temperature for the daily module.
* $\mathbf{L}_{Mon.}$(t) is the forecasted load for Mondays.
* *Data pre-processing unit:* Historical raw data, consisting of hourly weather parameters and the corresponding load, are preprocessed and input to the daily modular networks. Preprocessing of the data includes:
  1) calculation of Heat Index for the Summer forecasting case
  2) averaging of temperature and Heat Index
  3) sorting the processed data into final vector form.
* *The daily modules.* Each daily module consists of 24 hourly networks.

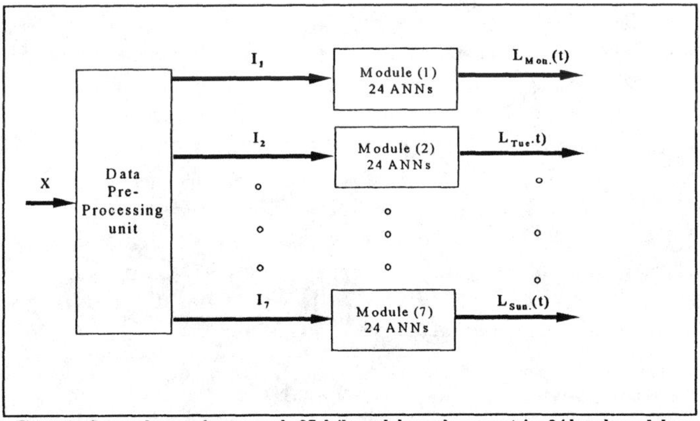


**Figure 2: Seasonal network composed of 7 daily modules each one contains 24 hourly modules.**

**Hourly network modules**
An hourly network modular is a multi-layer fully-connected network, trained by backpropagation. The input layer has eight input neurons. For summer months, forecasting temperature and relative humidity are combined to form a single Heat Index input. The number of hidden layers and units are chosen according to the problem. A single node outputs the load forecast for the indicated hour. All hourly networks are of similar construction. Table 1 provides a summary of inputs and output of an hourly network module.

TABLE 1 SUMMARY OF INPUTS AND OUTPUT OF HOURLY FORECAST MODULE

| *The input/output parameters for each hourly network:* |
|---|
| * $x_1$ is yesterday's actual load at this hour. |
| * $x_2$ is the actual load of the same day the previous week at this hour. |
| * $x_3$ is the max. temperature/Heat Index of the forecasted day. |
| * $x_4$ is the average temperature/Heat Index for the past three hours from this hour. |
| * $x_5$ is yesterday's max. temperature/Heat Index at this hour. |
| * $x_6$ is the max. temperature/Heat Index of the same day the previous week at this hour. |
| * $x_7$ is yesterday's average temperature/Heat Index for the past three hours from this hour. |
| * $x_8$ is the average temperature/Heat Index for the past three hours of the same day the previous week at this hour. |
| * Y is forecasted load at this hour. |

## SOME PRELIMINARY WORK

To examine the effects of the modular network approach to load forecasting, three training cases from years 1990 and 1991 were selected for the Wichita area. As a preliminary analysis, data was restricted to hourly load for Wednesdays during the winter, spring and summer months. Testing consisted of 24 hour ahead forecasts for Wednesdays in January, April and August of 1992. Data was preprocessed and hourly modules were constructed as above. Each hourly module was constructed of a single hidden layer of 10 nodes. Note: the forecasted data was not available for the years 1990, 1991 and 1992. Therefore, in testing, the *actual* weather information for the forecasted day was used instead of the weather bureau forecasted data.

**Modular networks results**
The results for the system for the month of August of modular networks are depicted in Figure 3. The figure indicates relatively low forecast error for one day ahead. Forecasts. The network captures diurnal variations completely. Figure 4 represents the forecast error distribution for the year 1992. Ideally KG&E prefers forecasts to be within $(\pm 50)MW$ of actual demand. The proposed modular models produced forecasts within $(\pm 50)MW$ of the actual load 75% of the time.

**Modular networks models vs. a single ANN model**
To examine the approximate improvement due to a module network approach over the single network approach, two ANN models were built and trained using the same data as above. The entire set of data for years 1990 and 1991 was subdivided for

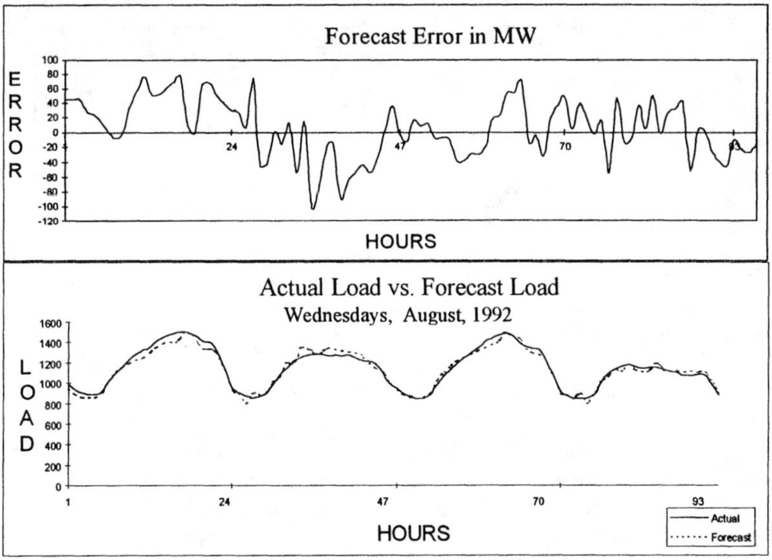


**Figure 3 Modular network results (August)**

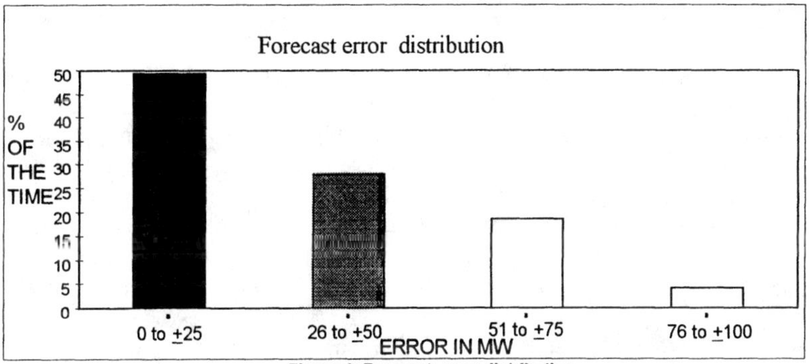


**Figure 4: Forecast error distribution**

training a single network corresponding to winter and fall months, and a separate network for spring and summer months. Both single ANN models consisted of an input layer of 12 neurons, 2 hidden layers of 10 and 6 neurons, and one output neuron (Al-Rashid, 1995). The module is fully connected and trained by backpropagation. The output neuron is used to forecast the next 24 hour load by using the previous day and week load, temperature, and relative humidity, in addition to the future forecasted temperature and relative humidity.

Tables 2 and 3 contain summary information forecast errors for Wednesday 1992 obtained from the modular network models and the single ANN models respectively. Included is a commonly used error metric for power load forecasting: *Mean Absolute Percentage Error* (MAPE) for the peak load. MAPE is the mean of the "%$err_t$" over some given range of the time, t.

TABLE 2 THE MODULAR NETWORK 'S FORECAST ERROR FOR WED. JAN., APRIL, AND AUG. 1992

| | Mean (MW) | Max. Error (MW) | Min. Error (MW) | MAPE |
|---|---|---|---|---|
| January | 19.23 | 77.95 | -89.32 | 2.16 |
| April | 29.5 | 76.34 | -96.87 | 2.09 |
| August | 30.28 | 78.95 | -98.93 | 2.21 |

TABLE 3 THE SINGLE NETWORKS FORECAST ERROR FOR WED. OF JAN., APR., AND AUG. 1992

| | Mean (MW) | Max. Error (MW) | Min. Error (MW) | MAPE |
|---|---|---|---|---|
| January | 22.35 | 102.0 | -104.5 | 2.35 |
| April | 31.07 | 111.29 | -90.3 | 2.37 |
| August | 33.0 | 122.52 | -112.06 | 3.15 |

Although not significant, the tables indicate in all cases improved performance by the modular networks. Mean absolute error is reduced and well as MAPE. The major improvement is in the reduction of minimum and maximum errors (e.g. 76.34 vs. 101.0 for April) . The single networks produced forecasts within $(\pm 50)MW$ of the actual load 70% of the time. Thus there is a 5% increase in the number of forecasts outside the required limits using a single network approach.

## CONCLUSION

The series of seasonal networks composed of hourly modular network models for short-term power forecasting are presented in this paper. The modules are used to forecast load 24 hours ahead by using the previous day and week load, temperature, and relative humidity, in addition to the future forecasted temperature and relative humidity. The performance of the above models is quite favorable when compared with the single network model Future work toward increasing the accuracy of the proposed system will include: 1) Error Feedback - will enhance the performance of each of the individual modules and force the modules to adapt to changes in the load characteristics.; 2) More Training Data - will reduce the forecast error. Successful methods for load forecasting using the ANN model have used at least three years of data for training. Introduction of industrial use variables should be used to account for unexpected variations not due to weather.

## REFERENCES


Al-Rashid, Yasser. 1995. *Short-Term Electric Load Forecasting Using Neural Network Models.* M.S. Thesis, Dept. of Electrical Engineering. Wichita, KS: Wichita State University.

Haykin, Simon, *Neural Networks: A Comprehensive Foundation*, New York: Macmillan College Publishing Company, 1994.

Ho, K.,Hus, Y.and Yang,. C., "*Short Term Load Forecasting Using A Multi- Layer Neural Network with an Adaptive Learning Algorithm.*" IEEE Trans. PWRS, Vol. 7, No. 1, Feb. 1992.

Paarmann, L. D. and Najar, M., " *Short-term Power Load Forecasting Based on Autocorrelation Function Optimization*," 37th Midwest Symp. Cir. and Sys., Lafayette, LA, August, 1994.

# AN ARTIFICIAL NEURAL NETWORK-BASED MEDIUM-TERM ELECTRICAL LOAD AND ENERGY FORECAST FOR A RAPIDLY GROWING UTILITY

**SALEH M. AL-ALAWI AND SYED M. ISLAM**
*Dept. of Electrical & Electronics, Sultan Qaboos University, Oman*
*Dept. of Electrical & Computer Engineering, Univ. of Newcastle, Australia*

**KHALED A. ELLITHY**
*Dept. of Electrical Engineering, University of Cairo, Cairo, Egypt*

**CIHAN H. DAGLI**
*Engineering Management Dept., UMR, Missouri, USA*

***ABSTRACT:***
In this paper, two ANN-based models have been developed to forecast electrical load and energy for 24 months ahead. These models have been applied to historical energy, load, and weather data available for the Muscat Power System from 1986-1990. Forecast results, when compared to actual data for 1991-1992, show that monthly peak load and electrical energy can be predicted within a mean absolute percentage error of 2.86% and 4.77%, respectively.

## INTRODUCTION

Load forecasting methodologies may be classified into two broad categories: autonomous models and conditional models. Autonomous models attempt to relate future growth of electricity demand on a system based to its past growth and the conditional models attempt to relate electricity demand growth to other variables, mainly economic indicators. Forecasting approaches like the statistical techniques, socio-economic models, etc. belong to the autonomous class, and approaches like econometric and microanalysis methods belong to the conditional class. Statistical techniques based on Box-Jenkins models (Box and Jenkins, 1976), exponential smoothing (Christiaanse, 1971), multiple regression (Thompson, 1976), adaptive and weather models (Gupta and Yamada, 1972), and Kalman filtering (Frissari et al., 1982) have been proposed and applied with reasonable success for short-term load forecasting. The general problem with the statistical methods based on time-series analysis has been their relative inaccuracy and numerical instability (Park et al, 1991). The reason for inaccuracy may be traced to the fact that these models ignore the high correlation between weather variables and electricity demand. Statistical methods based on adaptive and weather models rely on a stationary linear functional relationship between electricity demand and weather. In practice, however, the weather-load relationship is neither stationary nor linear, whilst multiple regression approaches either do not have the ability or are too complex to address the issue. Statistical methods will furthermore fail to provide accuracy when applied to longer-term forecasting.

To alleviate these problems, some authors have developed a more specialized model such as the socio-economic model (Barakat and Al-Rashad, 1993). This model produced a longer-term forecast under conditions of high growth and was reported to have achieved better results than the Box-Jenkins and Gompertz models. Others used Artificial Intelligence (AI) techniques such as Artificial Neural Networks (ANNs) and Expert Systems to develop electrical demand forecasting models. ANN has been successfully applied to electric load forecasting by many authors reporting higher accuracy compared to other methods. However, applications have been limited to 1 hour to 24 hour ahead short-term load forecasts. Although short-term load forecasting is very useful in deciding system economic dispatch, spinning reserve and unit commitment, it does not provide any information on seasonal load variation and monthly energy use. These are of extreme importance from a utility point of view to decide on monthly fuel procurement, scheduling maintenance on units, diversity interchanges between neighboring utilities and system expansion planning.

In this paper, novel and comprehensive ANN models are developed to forecast monthly electrical peak load and energy consumption for months ahead. The models are based on historical load and energy data for the capital area's (Muscat) power system of the Sultanate of Oman.

## WEATHER-LOAD AND WEATHER-ENERGY RELATIONSHIP

The electrical load and energy consumption pattern of a power system depends heavily on weather. On the other hand, the growth in load and energy demand depends largely on the number of consumers connected to the system. Some recent papers (Low et al., 1990), (Park et al., 1991), and (Al-Alawi and Islam, 1995) have exploited the weather-load relationship in short-term demand forecasting using ANN. In these applications, however, the weather parameters used were limited to temperature and humidity. Another paper (Rahman and Hazim, 1993) based on expert systems points to the importance of including other weather variables such as wind speed, accumulated heat, wind-chill index, and temperature-humidity index.

### Selection of Weather Variables

In this paper, fifteen different weather variables such as Absolute Maximum Temperature ($T_{max}$), Average Maximum Temperature ($T_{avmax}$), Average Maximum Relative Humidity ($RH_{max}$), Average Relative Humidity ($RH_{av}$), Wind Speed (W), Duration of Bright Sunshine (S), Global Radiation (R), Vapor Pressure (VP), and Precipitation (PR) have been tested for their correlation and contribution to energy, as shown in Table 1. Any weather variable which leads to less than ten percent correlation together with less than four percent contribution to the energy forecast has been eliminated (Al-Alawi and Islam, 1996). As a result of this test, the variables PR and VP were eliminated from the energy model. Similarly, weather variables such as Temperature ($T_{pkld}$) and Relative Humidity ($RH_{pkld}$) at historical peak loads, $T_{max}$, $RH_{max}$, and W have been tested for their correlation and contribution to electrical load. Although some variables like W and $RH_{max}$ had strong correlation to monthly electrical energy, they had very little correlation to monthly peak load and, therefore, were eliminated from the load model.

**TABLE 1: CORRELATIONS AND CONTRIBUTIONS OF INPUT VARIABLES TO ENERGY**

| Input variable | Correlation factor | Contribution percent |
|---|---|---|
| $T_{av}$ | 0.90137 | 7 |
| $RH_{max}$ | (0.14021) | 4 |
| W | 0.16604 | 9 |
| S | 0.63579 | 6 |
| R | 0.59722 | 7 |
| $RH_{av}$ | 0.12575 | 10 |
| DD | 0.92201 | 27 |
| CI | 0.92607 | 9 |
| CON | 0.32741 | 21 |

## Energy Model

While the monthly energy consumption pattern is dependent on the weather, energy growth is largely dependent on the number of electrical connections. Therefore, the proposed energy model contains both weather and electrical connection components. Since the weather-energy relationship is never a linear one, the following nonlinear model is developed:

$$E_i = f_i\,(T_{avmaxi},\ RH_{maxi},\ RH_{avi},\ W_i,\ S_i,\ R_i,\ DD_i,\ CI_i,\ CON_i) \tag{1}$$

for $i = 1, 2,$ ---- $n$; $n$ being the number of historical months for which the ANN model is to be trained; where, $E_i$ = energy consumption for the i-th month in MW-hrs. and $CON_i$ = number of electrical connections at the end of i-th month. DD (degree days, °C-days) and CI (comfort index) are two new variables introduced to include the effects of HVAC requirement and level of discomfort and are defined as [12]

$$DD_i = \left|\frac{T_{max}+T_{min}}{2} - 20^\circ\right| * \textit{days per month 'i'} \tag{2}$$

where $T_{max}$ and $T_{min}$ are on a 24 hourly basis, and $CI_i = (70\text{-}TH_i)$; where,

$$TH_i = \left(\frac{9}{5}T_{avmaxi} + 32\right) - 0.55\left(1 - \frac{RH_{maxi}}{100}\right)\left(\frac{9}{5}T_{avmaxi} - 26\right) \tag{3}$$

### Load Model

Since the peak load in a month occurs at one particular instant, it may have no bearing on the variables like $T_{avmax}$, $T_{av}$, $RH_{av}$, S, R, DD, and CI. Therefore, these variables have been eliminated from the load model by intuition. Wind speed and $RH_{max}$ were also dropped from the proposed load model after their correlation was found to be minimal. Instead, two important variables from the historical database were added, i.e., $T_{pkld}$ and $RH_{pkld}$. $T_{maxi}$ shows a strong correlation to load and, therefore, has been retained in the load model for a possible temperature-load lag relationship. The final load model is then obtained as

$$L_i = g_i\,(T_{pkldi},\ RH_{pkldi},\ CON_i,\ T_{maxi}) \qquad (4)$$

for i = 1,2, --- m; m being the number of months for which the ANN model is to be trained.

## ARTIFICIAL NEURAL NETWORKS

Artificial neural networks (ANNs) are massively parallel information processing architectures composed of many simple, usually adaptive, processing elements interconnected to achieve a certain collective computational structure. ANNs possess features which are particularly attractive for data analysis. These features include graceful degradation, robust recall with fragmented and noisy data, speed inherent to parallel distributed architectures, generalization, and the capability of modeling nonlinear systems. In general, ANN consists of three basic elements: an organized architecture of interconnected processing elements, a method for encoding information during training, and a method for recalling information during testing. Simpson (1990) provides a coherent description of these elements and presents comparative analyses, applications, and implementation of 27 different ANN paradigms.

### ANN-Based Energy Model Development

As a result of using the correlation and contribution techniques, the following variables were qualified to be the candidate input to the energy model: $T_{avmax}$, $RH_{max}$, $RH_{av}$, W, S, R, DD, CI, and CON. With these input variables, the energy model given in equation 1 was developed using the NeuroShell ANN simulator based on the backpropagation paradigm (Rumelhart and McClelland, 1988). The network for the energy model consisted of three layers, an input layer with 9 input neurons, a hidden layer with 15 neurons, and an output layer with 1 output neuron. Historical data for the above variables, in addition to the monthly energy available from 1986 to 1990, was used in the training of the ANN model. After a series of network modifications during the training process, the network converged. The forecast produced by this network was then tested against the 1991 and 1992 actual data. The best forecast accuracy obtained was with a mean absolute percentage error of 4.77%. The energy model developed above provided a test $R^2$ of 0.943. $R^2$ is the most common indicator of the BP network generalization capability.

**ANN-Based Load Model Development**

As shown by Figure 1, the input variables selected for the load model were $T_{pkld}$, $RH_{pkld}$, $T_{max}$, and CON. With these input variables, the load model given in equation 4 was developed using the same ANN simulator. The network was also trained using the historical data available from 1986 to 1990 for the above input variables and the monthly load. Using the 1991 and 1992 data for testing, the best forecast accuracy obtained using the load model was with a maximum mean absolute percentage error of 2.863%. This result was obtained with a testing set $R^2$ value of 0.9832.

## RESULTS AND DISCUSSION

The historical data used to develop the artificial neural network (ANN) model was also used to develop the socio-economic (S-E) model (Barakat and Al-Rashad, 1993), the Box & Jenkins (B&J) model (Box and Jenkins, 1976), and the generalized parametric estimation model (PEM) (Soderstrom and Stoica, 1989). To determine the forecasting accuracy of these models, actual historical monthly consumer connections and weather data for the years 1991 and 1992 were then used to produce a forecast of the monthly peak load and energy consumption for this period. The forecast produced by each of these models was then compared with the actual load and energy values for the same period. Figure 2 shows the actual and forecasted energy for the 1991 and 1992 years. Statistical error comparisons for these models were then performed, and the results obtained are illustrated in Table 2 for the load model. Figure 2 shows that the ANN energy forecast was in closer agreement with the actual values when compared to the socio-economic model and the statistical models' results. Similar results were also obtained for the load models. Table 2 revealed that the proposed ANN load model produces lower mean absolute error (MAE), mean squared error (MSE), mean absolute percentage error

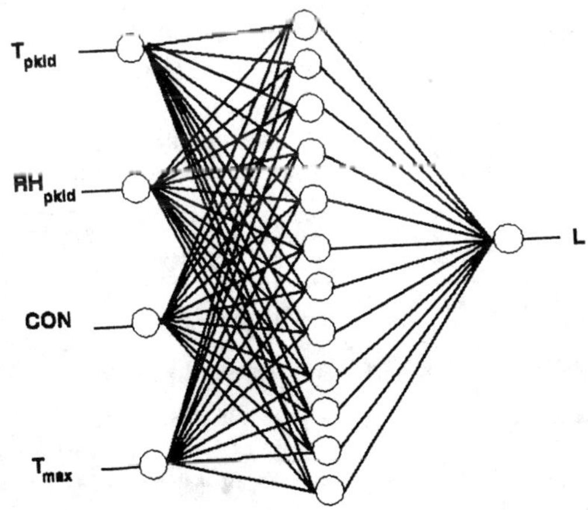


Figure 1. The ANN-based load model

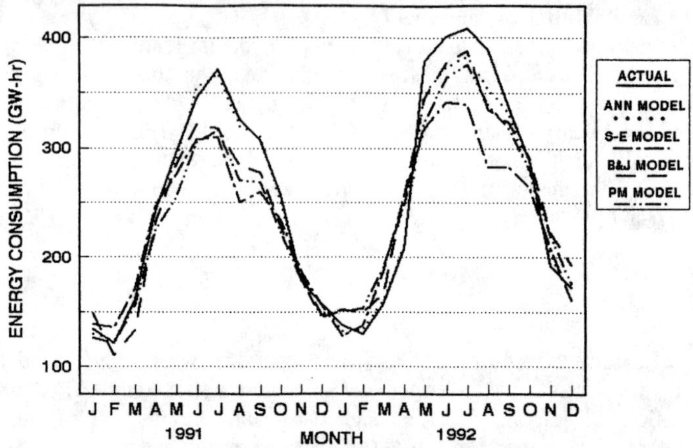


**Figure 2. Monthly actual & forecasted energy for '91-'92**

(MAPE), standard deviation of error (SDE), and higher $R^2$ when compared to the other models. This clearly demonstrates the capability of the proposed ANN model in outperforming the other statistical models.

## CONCLUSION

ANN-based medium-term electrical load and energy forecast models have been developed in this paper. ANN input variables have been chosen as a result of correlation and contribution to decision analysis. The model is shown to be valid for 24 months. This is very useful in planning fuel procurement, scheduling unit maintenance, and diversity interchange. The developed model is also inherently capable of taking care of the effect of special lunar and other cyclic festive events. The proposed model is capable of forecasting monthly peak load and energy within a mean absolute percentage error of 2.86% and 4.77%, respectively.

**Table 2. Statistical Error Comparison for 1991-1992 Load Forecast**

| Error Type | ANN Model | S-E Model | B&J Model | PEM Model |
|---|---|---|---|---|
| Mean Error (ME) | -4.21917 | 10.2329 | -60.797 | -7.7083 |
| Mean Absolute Error (MAE) | 16.0425 | 56.001 | 67.208 | 47.2917 |
| Mean Squared Error (MSE) | 516.879 | 6370.855 | 6402.542 | 3698.375 |
| Mean Absolute Percentage Error (MAPE) | 2.8625 | 11.6084 | 13.75267 | 10.40744 |
| Standard Deviation of Error (SDE) | 11.3438 | 39.597 | 47.52347 | 33.4403 |
| Coefficient of Determination $(R)^2$ | 0.9832 | 0.7798 | 0.8967 | 0.8748 |

## ACKNOWLEDGMENT

The authors would like to gratefully acknowledge the help of Mr. Redha Hassan Ali, Director-General, MEW, and Mr. Said A. Al-Adhaly, Chief, Department of Climate and Meteorology, MOC, Sultanate of Oman, for providing valuable data.

## REFERENCES

Al-Alawi, S.M. and Islam, S.M. (1995). Short-term load forecasting using artificial neural networks, Second IEEE International Conference on Electronics, Circuits and Systems, ICECS'95, 381-384.

Al-Alawi, S.M. and Islam, S.M. (1996). Principles of electricity demand forecasting, part 1 methodologies, IEE Power Engineering Journal, vol. 10, no. 3, 139-143.

Barakat, E.H. and Al-Rashad, S.A. (1993). Social environmental and economic constraints affecting power and energy requirements in fast developing areas, IEE Power Engineering Journal, vol. 7, no. 4, 177-184.

Box, G.E. and Jenkins, G.M. (1976). Time series analysis-forecasting and control, Holden-Day, San Francisco.

Bunn, D.W. and Farmer, E.D. (1985). Comparative models for electrical load forecasting, John Wiley & Sons.

Christiaanse, W. (1971). Short-term load forecasting using exponential smoothing, IEEE Trans. PAS, vol. PAS-90, 900-910.

Frissan, G.D., Widergren, S.E., and Yehsakul, P.D. (1982). On-line load forecasting for energy control center application, IEEE Trans. PAS, vol. 101, 71-78.

Gupta, P. and Yamada, K. (1972). Adaptive short-term forecasting of hourly loads using weather information, IEEE Trans. PAS, vol. 91, 2085-2094.

Low, K.L., Hsu, Y. , Liang, C., and Lai, T.(1990). Short-term load forecasting of Taiwan Power System using a knowledge-based expert system, IEEE Trans. PAS, vol. 5, no. 4, 1214-1221.

Park, D.C., El-Sharkawi, M., and Marks II, R. J. (1991). Electric load forecasting using an ANN, IEEE Trans. PAS, vol. 6, no. 2, 442-449.

Rahman, S. and Hazim, O. (1993). A generalized knowledge-based short-term load forecasting technique, IEEE Trans. PAS, vol. 8, no. 2, 508-514.

Rumelhart, D. and McClelland, J. (1988). Parallel distributed processing, explorations in the microstructure of cognition, vol. 11: Foundations, Cambridge, MA, MIT Press/Bradford Books.

Simpson, P.K. (1990). Artificial neural systems, foundations, paradigms, applications, and implementations, Pergamon Press.

Soderstrom, T. and Stoica, P. (1989). System identification, Prentice Hall, Inc.

Thompson, R. P. (1976). Weather sensitive demand and energy analysis on a large geographically diverse power system application to short-term hourly electric demand forecasting, IEEE Trans. PAS, vol. 95, 384-393.

# HYBRID BOX-JENKINS AND NEURAL NETWORK FORECASTING OF POTABLE WATER DEMAND

**QIN WANG**
*NeuroCorp, Inc., 4151 SW Frwy., Ste. 606, Houston, Texas 77027*

**MIRIAM HELLER**
*Department of Industrial Engineering, University of Houston, Houston, Texas 77204-4812*

***ABSTRACT:***
A "hybrid" Box-Jenkins neural network (HBNN) model developed by Lachtermacher and Fuller (1995) is extended to seasonal non-stationary time series. A HBNN employs Box-Jenkins time series analysis to identify "lagged" inputs to a backpropagation predictor. Univariate and multivariate HBNN models of real water demand time series data prove superior to individual Box-Jenkins and all neural network models of the same data. The results suggest that identifying the critical univariate inputs to a network plays a more important role in performance than does additional information from multivariate inputs. Moreover, the HBNN models train more efficiently than their pure backpropagation analogs.

Neural networks have been studied and successfully applied to predict time series of water demand (Heller and Thind, 1994; Cubero, 1991; Canu et al., 1990), electricity demand (Chen 1991), and scores of other areas (Weigend, 1993). Despite success in these applications and, particularly in capturing non-linear time series, much room remains to improve neural network forecasting accuracy and training efficiency. The greatest opportunities to improve neural networks derive from their data-driven nature: training and forecasting depend directly on available data. Several researchers observed that neural network have historically been viewed as black boxes with topologies, including inputs, determined *a priori* with little or no exploratory data analysis (Gordon, 1994; Lachtermacher and Fuller, 1995). Consequently, the models often required large quantities of data and converged slowly during training.

One complementary approach to time-series forecasting is the Box-Jenkins method: a statistical, model-driven approach in which relationships among variables are identified first and model parameters are estimated afterwards. One drawback of the Box-Jenkins technique, however, is that the underlying model is strictly linear. In practice, this assumption prohibits accurate analysis and forecasting of many real time series which frequently exhibit non-linearity.

Lachtermacher and Fuller (1995) recently developed a hybrid Box-Jenkins and neural network (HBNN) method and applied it to small non-seasonal non-stationary time series with promising results. This hybrid approach uses the Box-Jenkins procedure to identify the 'lag components' of the data which define the input variables of the neural network. The result is a small but adequate network which requires less data as well as training. The preliminary phase of input identification renders the hybrid method a model-driven approach. Finally, the neural network with appropriately selected input helps to improve the forecasting performance of real time series when significant non-linearity is present in the system being modeled.

In this research, the HBNN is extended to address more complicated time series, specifically a large seasonal non-stationary data set consisting of real daily water demand. By experimenting with these data, which exhibit both trend and seasonality, the generality of the hybrid method is established. Hybrid Box-Jenkins neural networks using multivariate inputs are also explored to further enhance the model.

## METHODOLOGY

The development of a hybrid Box-Jenkins and neural network model (HBNN) consists of two phase: the exploratory phase and the network training phase. In the first phase, the Box-Jenkins procedure is performed (Box and Jenkins, 1976). The time series is plotted to identify the trend and seasonality. A differencing procedure was executed to transform the data into a stationary time series. The autocorrelation and partial autocorrelation functions of the transformed time series were calculated to formulate an autoregressive integrated moving average (ARIMA) model of the data. The general ARIMA model for water demand on day t, $Y_t$, is given by

$$\Phi_p(B)\Phi_P(B^L)(1-B^L)^D(1-B)^d(Y_t-\mu)=\Theta_q(B)\Theta_Q(B^L)\epsilon_t \tag{1}$$

where

B is the backshift operator such that $B^kY_t = Y_{t-k}$,

$\Phi_p(B) = 1-\varphi_1B-...-\varphi_pB^p$, the non-seasonal autoregressive operator of order p,

$\Phi_P(B^L) = 1-\varphi_{1,L}B^L-...-\varphi_{P,L}B^{PL}$, the seasonal autoregressive operator of order P,

$\Theta_q(B) = 1-\theta_1B-...-\theta_qB^q$, the non-seasonal moving average operator of order q,

$\Theta_Q(B^L) = 1-\theta_{1,L}B^L-...-\theta_{Q,L}B^{QL}$,the seasonal moving average operator of order Q,

d is the order of differencing at the non-seasonal level,

D is the order of differencing at the seasonal level,

$\varphi_i$, $\varphi_{i,k}$, $\theta_j$, and $\theta_{j,l}$, $\forall$i,j,k,and l, are linear parameters estimated from sample data,

$\mu$ is the mean of the time series, and

$\epsilon_t$ is random error at t, which, $\forall$t, is i.i.d. with mean $\mu$ and variance $\sigma$.

Using inversion, the ARIMA model was transformed into a strictly autoregressive (AR) model. After further simplification, the remaining autoregressive terms in the approximate model defined the "lagged" inputs to the neural network.

In the second phase, the backpropagation algorithm was used to train the neural network for the HBNN (Rumelhart et al., 1986). The neural network was a fully connected feedforward network with an input layer defined by the lagged variables identified in the exploratory phase, one hidden layer, and a single nonlinear unit output layer. The HBNN was tested by forecasting 1986 summer water demand.

One-step ahead forecasts of daily water demand from the HBNN were compared with those from feedforward networks using input schemes corresponding to those reported by Heller and Thind (1994). The univariate backpropagation model used seven lagged inputs. The multivariate backpropagation also introduced three lagged variable each for normalized maximum temperature and daily precipitation. A binary input indicated holidays and some adjacent weekends to capture invention effects. This selection inputs was consistent with those used by Canu et al. (1990) and Cubero (1991) for neural predictors of water demand time series and of Smith (1988) for a semi-parametric water demand predictor. With month and day of the week indicators, the total number of inputs to the multivariate BP was twenty-one.

The multivariate HBNN like the univariate HBNN used the lagged daily water variables determined by Box-Jenkins analysis. Multivariate inputs to the HBNN were similar to those used for the multivariate backpropagation, but the precipitation and temperature variables were determined using correlation calculations.

Daily potable water demand data for the Washington, DC Metropolitan area were obtained from the Washington Suburban Sanitation Commision. Summer data, defined as the period from May 1 to September 30 in each year, is graphed in Figure 1. Annual data exhibit a high degree of seasonality and a positive trend according to Thind (1994). All models were developed using 1983-1985 summer data and tested on 1986 data. All neural networks used learning rates of 0.01 and slopes of 0.5 for the hyperbolic tangent transfer function.

Performance based on the residual forecasting errors was measured as relative root mean squared, absolute of maximal, average errors, and bias. The number of errors exceeding 10% of actual demand was deemed relevant by experts. Visual inspection of plotted residuals also provided valuable information.

The backpropgation algorithm was simulated using code written in ANSI C under DEC OSF/1 on a DEC ALPHASTATION 200 4/166. Box-Jenkins time series analyses were carried out using SAS™ under CMS on an IBM ES 9000 and Minitab™ under Windows NT on a Gateway 2000 PC.

## RESULTS AND DISCUSSION

The Box-Jenkins analysis yielded the first experimental results; the "lag components" were identified. The autocorrelation (ACF) and partial autocorrelation (PACF) functions of the 1983-1985 summer data time series, transformed using first regular and first seasonal differencing were examined. Since the ACF cut off both at non-seasonal and seasonal levels, the transformed time series was considered stationary. Since the PACF died down both at non-seasonal and seasonal levels, the tentative Box-Jenkins model for the daily water demand was taken as

$$(1-\varphi_{1,153}B^{153})(1-B^{153})(1-B)(Y_t-\mu) = (1\ \theta_{1,1}B)\epsilon_t. \qquad (2)$$

Model residuals were tested to prove statistical significance of the estimated parameters. The estimation and a diagnostic check showed that the tentave model was not adequate since the ACF of the residuals at lag 5 is significantly large. By iteratively including moving average terms at lags 5 and 6 and an autoregressive term at lag 1, the modified model assumed the form

$$(1-\varphi_{1,1}B)(1-\varphi_{1,153}B^{153})(1-B^{153})(1-B)(Y_t-\mu) = (1-\theta_{1,1}B-\theta_{1,5}B^{7}-\theta_{1,6}B^{6})\epsilon_t \qquad (3)$$

Parameter estimation and the subsequent diagnostic check verified (3) as an appropriate model for 1983-1985 summer data. The residuals passed a normality test and showed no significantly large spikes in their ACF and PACF. The Ljung-Box statistic was calculated at lags 6, 12, 18, 24, 30, 36, and 42. The smallest prob-value at 0.018 indicated no residual serial correlation with over 98% confidence.

Substituting the estimated parameters and applying the backshift operators into

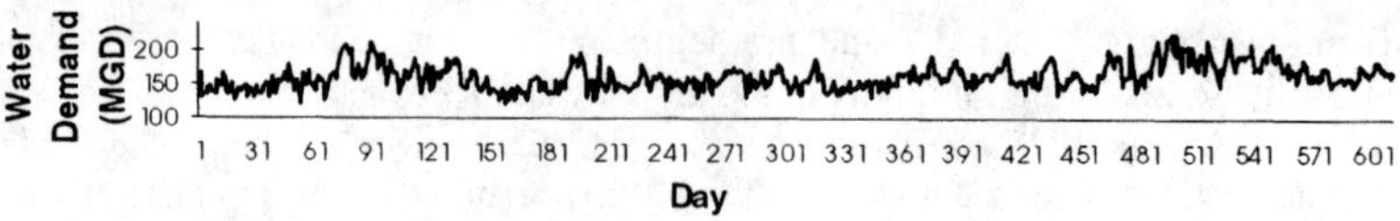

Figure 1 - Daily water demand during the summers from 1983-1986 in MGD

equation (3), the model was estimated as

$$Y_t - 1.306Y_{t-1} + 0.306Y_{t-2} - 0.229Y_{t-153} + 0.229Y_{t-154} + 0.142Y_{t-155} - 0.465Y_{t-306} - 0.607Y_{t-307} - 0.142Y_{t-308} = 0.002 + \epsilon_t - 0.797\epsilon_{t-1} - 0.254\epsilon_{t-5} + 0.050\epsilon_{t-6} \quad (4)$$

The moving average terms on the right-hand side of (4) approximately cancel. Autoregressive terms with lags greater than 153 on left-hand side can be dropped since they exceed the seasonal lag and can be correlated recursively to lags 1, 2 and 153. So the inversion of the model is defined as

$$Y_t = \gamma_0 + \gamma_1 Y_{t-1} + \gamma_2 Y_{t-2} + \gamma_3 Y_{t-153}. \quad (5)$$

This AR model identifies the first, second and seasonal lags of the daily water demand time series as inputs to the univariate HBNN. The most significant difference between these inputs and those for the pure neural networks is the inclusion of an input representing the seasonal lag. While the addition of a seasonal lag seems intuitively obvious, given the seasonal nature of this type of data, it was omitted from the neural models for water demand prediction reported in the literature.

Correlation analysis was used to identify precipitation and temperature time series inputs to the multivariate HBNN. The results are given in Table 1. Weak autocorrelation in the rainfall time series and the equally weak cross-correlation between the rainfall and the demand time series resulted in the inclusion of a single rainfall input. This can be thought of as an intervention factor. The temperature time series showed strong autocorrelation and cross-correlation with demand data at lags one, two, three and four was also high. Testing revealed that the cross-correlation could be represented well by a single one lagged temperature variable in the multivariate HBNN. The multivariate HBNN involved three lags identified from the Box-Jenkins model, one temperature variable and one precipitation variable both at lag one. With indicators for day of week, month and holiday plus a bias term, the HBNN used fourteen inputs. Thus the multivariate HBNN achieves a simpler and more efficient input structure than the multivariate BP with twenty-one input nodes.

Forecasting results obtained from the Box-Jenkins (BJ), univariate and multivariate backpropagation BP and HBNN models are summarized in Table 2.

**TABLE 1: CROSS- AND AUTOCORRELATION FUNCTIONS OF MULTIPLE INPUT VARIABLES**

| Lag | 1 | 2 | 3 | 4 |
|---|---|---|---|---|
| Autocorrelation (Rainfall) | 0.0211 | -0.0362 | 0.0225 | 0.0143 |
| Cross-correlation (Rainfall,Demand) | -0.0381 | 0.0412 | 0.0675 | 0.1193 |
| Autocorrelation (Temperature) | 0.6957 | 0.4129 | 0.2639 | 0.2262 |
| Cross-correlation (Temperature, Demand) | 0.5323 | 0.3721 | 0.2912 | 0.2513 |

**TABLE 2: FORECASTING RESULTS BY MODEL TEST ON SUMMER 1986 DATA**

| Model | Relative RMS | Maximum Error | Average Error | Bias | 10% Error |
|---|---|---|---|---|---|
| Box-Jenkins | 0.0942 | 46.66 | 7.33 | 0.0012 | 44 |
| Univariate BP | 0.0948 | 49.71 | 12.18 | 0.0332 | 38 |
| Multivariate BP | 0.0893 | 47.60 | 11.83 | 0.0300 | 37 |
| Univariate HBNN | 0.0586 | 29.95 | 8.07 | -0.0028 | 13 |
| Multivariate HBNN | 0.0383 | 26.25 | 4.82 | -0.0045 | 5 |

Neural-based model performance measures represent averages over five simulations, except maximum error, which is the maximum absolute error over all simulations.

The BJ and BP models performed comparably, except the BP models were an order of magnitude more biased toward unestimation. The univariate HBNN outperforms both the BJ and BP in nearly all measures except average error. Visual inspection of Figures 2, 3, and 5 show that the HBNN residuals are randomly distributed with no apparent bias; residuals associated with the univariate HBNN are generally smaller than those associated with either BJ or BP models; the maximum error dropped over 35% in the HBNN model compared with all BJ and BP models. The number of days with over 10% error dropped by over 60% as well. Since HBNN is based on BJ, a linear method, these results can be interpreted as the improvement associated with capturing the non-linearity in the data by the neural network.

The superiority of the HBNN over the BP confirms the importance of the input structure to a neural network. Although the data-driven univariate BP has more inputs to the network, the critical seasonal lag is not present. The hybrid univariate HBNN has a simpler input structure yet the model-driven component has included the seasonal component in its topology and results in more accurate forecasts.

The importance of the input structure is further confirmed by comparing the performance between the univariate HBNN and the multivariate BP in Table 2 and Figures 4 and 6. The multivariate BP produces larger errors and the residuals show obvious underestimation of demand. Better forecasts are generated by the univariate HBNN than the multivariate BP, demonstrating that the inappropriate input structure of the univariate BP can not be remedied simply by including multivariate inputs. For this problem, adding information from multiple variables cannot render an inadequate univariate model competitive with an adequate univariate model.

The univariate HBNN results, though, can be enhanced through the inclusion of multivariate inputs. The multivariate HBNN exhibits improved performance over the univariate HBNN as shown in Table 2 and Figures 5 and 6. The magnitude of the forecasting error is reduced significantly.

Simulation times during training for the univariate BP, the multivariate BP, the univariate HBNN and the multivariate HBNN are approximately 2, 4, 1.5 and 6 minutes of elapsed time, respectively. Simplified HBNN input structures compared with the BP models explains the training efficiency and reduced simulation time.

## CONCLUSIONS

In this research the hybrid Box-Jenkins and neural network method (HBNN) developed by Lachtermacher and Fuller (1995) was extended to seasonal nonstationary time series. In the exploratory phase, differencing was used to detrend

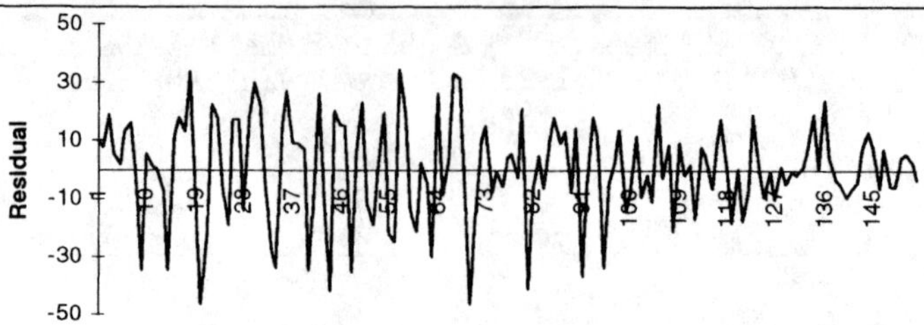


Figure 2. Box-Jenkins model forecast residual for summer 1986 in MGD

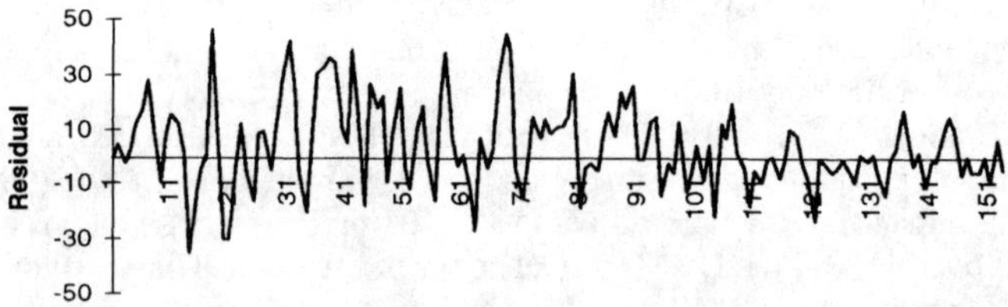


Figure 3. Univariate Backpropagation model forecast residuals for summer 1986 in MGD

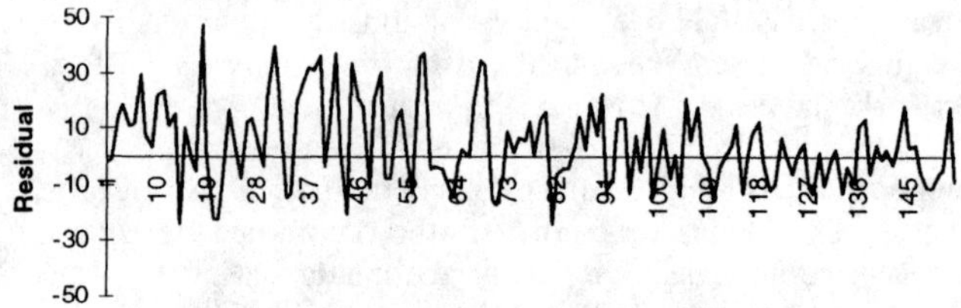


Figure 4. Multivariate Backpropagation model forecast residuals for summer 1986 in MGD

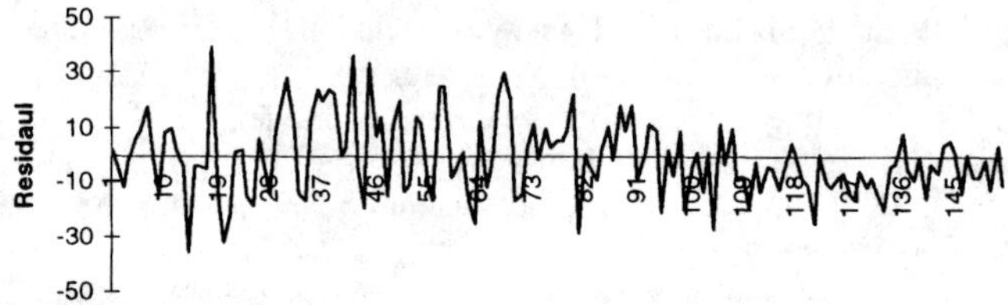


Figure 5. Univariate hybrid Box-Jenkins neural network model forecast residuals for summer 1986 in MGD

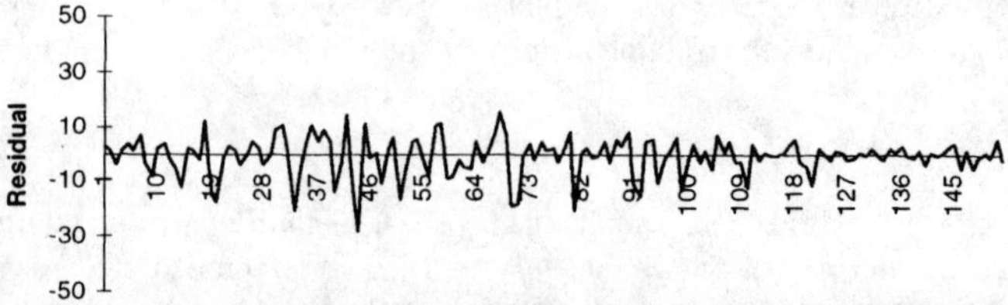


Figure 6. Multivariate hybrid Box-Jenkins neural network model forecast residuals for summer 1986 in MG

and deseasonalize daily water demand data. HBNN forecasts were compared with those from individual Box-Jenkins (BJ) and univariate backpropagation (BP) models.

The univariate BP and the BJ performed similarly, except BP models were biased. The univariate HBNN outperformed the BJ and BP models on most measures. The superiority of the univariate HBNN to the BP models suggests that inputs selection controls the predictor's bias. Conversely, even though the univariate HBNN model used the same input variables as the linear BJ model, the HBNN's superior performance confirms the advantages of the ability of neural network to learn nonlinearity. The improved performance of the HBNN can be attributed to the identification of the seasonal lag, which, although intuitive, has been excluded in all neural predictors of water demand reported in the literature.

The value of selecting appropriate inputs to the HBNN is further demonstrated by the fact that introducing multiple input variables does not significantly improve BP performance.The multivariate BP can not compete with even the univariate HBNN. The univariate HBNN, though, was improved by the introduction of multivariate inputs. The multivariate HBNN offers an alternative to multivariate Box-Jenkins modeling that is easy to develop and more appropriate for non-linear time-series.

The univariate and multivariate HBNN prove superior not only in terms of effectiveness but also in terms of efficiency. Neural network training time using the hybrid method is reduced significantly.

## REFERENCES


Box, G. and Jenkins, G. (1976). Time Series Analysis: Forecasting and Control, Holden Day: CA.

Canu, S., Sobral, R. and Lengelle, R. (1990). Formal Neural Networks as an Adaptive Model for Water Demand, *Proceedings of the Int'l Neural Netowrk Conference,* Paris, France, 131-135.

Chen, S. (1991). Weather Sensitive Short-Term Load Forecasting Using Non-Fully Connected Artificial Neural networks, Ph. D. Thesis, University of Wisconsin-Milwaukee.

Cubero, R.G. (1991). Neural Networks for Water Demand Time Series Forecasting, In *International Workshop on Artificial Neural Networks IWANN '91,* Prieto, A. (ed.), Springer-Verlag, 453-460.

French, M. N., Krajewski, W. F. and Cuykendall, R. R. (1992). Rainfall Forecasting in Space and Time using a neural Network, *Journal of Hydrology*, 137:1-31.

Heller, M. and Thind, H. (1994). Forecasting with Cascade Correlation: An Application to Potable Water Demand, *Intelligent Engineering Sys. Through Artificial Neural Networks*, 4: 1155-1160.

Lachtermacher, G. and Fuller, J. D. (1995). Back Propagation in Time Series Forecasting, *Journal of Forecasting*, 14(4):381-393.

Rumelhart, D. E., McClelland, J. L., and The PDP Research Group (1986). Parallel Distributed Processing: Exploration in the Microstructure of Cognition, Vol. 1, MIT Press: Cambridge, MA.

Smith, J.A. (1988). A Model of Daily Municipal Water Use for Short-Term Forecasting, *Water Resources Research,* 24:210-206.

Thind, H. S. (1994). Forecasting Nonlinear Time Series Using Artificial Neural Networks: An Application to Daily Municipal Water Demand Forecasting, Master's Thesis, Department of Industrial Engineering, University of Houston.

Weigend, A. S. And Gershenfeld, N.A. (Eds.). (1993). Time Series Prediction: Forecasting the Future and Undrstanding the Past, Addison-Wesley: Redwood, CA.

# PREDICTION OF SHEAR STRESS-STRAIN BEHAVIOR OF SOIL WITH RECURRENT NEURAL NETWORK

**Jian-Hua Zhu, Musharraf M. Zaman**
*School of Civil Engineering & Environmental Science*
*The University of Oklahoma, Norman, Oklahoma*

**Theodore B. Trafalis**
*School of Industry Engineering*
*The University of Oklahoma, Norman, Oklahoma*

***ABSTRACT:***
Shear stress-strain behavior of soil is greatly influenced by such factors as stress history, stress path and void ratio. Prediction of stress-strain behavior of soil by traditional constitutive models is a challenging task because of variabilities of the above-mentioned factors. A recurrent neural network (RNN) provides an efficient and novel approach to the prediction of shear stress-strain behavior of soil, since it possesses learning and self-organizing capabilities. Two kinds of test data with different shear stress paths are used for training and predicting soil behavior using the designed RNN. Both dilative and contractive shearing behavior are predicted by the network. A good agreement between the measured data and predicted results demonstrates that the RNN is able to capture significant variabilities inherent in the soil samples, and has promising applications in many soil mechanics and geotechnical engineering problems.

Understanding of shear behavior of soil is very important in dealing with many problems related to soil mechanics and geotechnical engineering. Many engineering parameters including shear strength and elastic modulus can be obtained from a series of relevant shear stress-strain curves. The desired shear stress-strain curves can also be used for the design and analysis of foundation engineering, such as evaluation of settlement of buildings, cracking potential and overall safety of structures. The stress-strain behavior of a soil is greatly influenced by such factors as stress history, stress path and void ratio. Traditional constitutive models for soil behavior often yield poor predicted response because a majority of these models have difficulties in capturing some significant attributes of the soil behavior, such as strain softening/hardening, dilatancy, anisotropy and rate dependency. In order to develop a constitutive model various assumptions are made without giving sufficient consideration of the specific behavior of the soil in question. These assumptions result in natural deficiency in a constitutive model in terms of satisfying the conservation law of mechanics and the invariance requirements. Therefore, although constitutive models developed during the last several decades have received significant attention, the application of such models to practical engineering problems has been rather restricted and the results of numerical calculations are sometimes difficult to convince the engineers and clients.

A recurrent neural network (RNN) provides an efficient, novel approach to the prediction of stress-strain behavior of soils, because it has strong capabilities of learning, self-organizing and self-adjustment. One of its distinctive properties is that it is based on experimental data rather than on weakly-founded assumptions. The RNN learns from experimental data and forms neural connection stimuli from the learning process, functioning analogous to a human brain. Because of its unique learning, training and predicting characteristics, the RNN model has great potential application in geotechnical engineering.

This paper presents a framework of RNN modeling of soil behavior. Two different kinds of test data, namely isotropic consolidation undrained (ICU), and anisotropic consolidation undrained (ACU) are utilized to train the designed RNN. The predictions made by using the trained RNN showed that RNN is able to model complex soil behavior in a fairly consistent and accurate manner. The significant variations inherent in the soil behavior can be successfully smoothened by using an appropriate algorithm function and design parameters of neural networks.

## RECURRENT NEURAL NETWORK

It is now recognized that recurrent neural networks (RNN) are superior to other kinds of networks in that they have both feedforward and feedbackward connections between inputs and outputs (Elman, 1990). The typical architecture of the RNN is shown in Figure 1. The initial input **X** is propagated to the hidden layer **Z** through the weights V and bias term $V_0$. The results from the hidden layer are then transmitted to the output layer **Y** through weights W and bias $W_0$, and feedbackward to the input layer as external data input to the hidden layer again. The action of the RNN shown in Figure 1 can be represented by the following equations:

$$\mathbf{Z}_j = F_1(\Sigma V_{ij} * X_{ij} + V_0 + \Sigma V_{zi\text{-}1j} * Z_{i\text{-}1}) \quad (1)$$

$$\begin{aligned}\mathbf{Y}_k &= F_2(\Sigma W_{jk} * Z_j + W_0)\\ &= F_2(\Sigma W_{jk} * F_1(\Sigma V_{ij} * X_{ij} + V_0 + \Sigma V_{zi\text{-}1j} * Z_{i\text{-}1}) + W_0)\end{aligned} \quad (2)$$

where $F_1$ and $F_2$ are the activation functions from the input layer to the hidden layer and from the hidden layer to the output layer, respectively; $V_{ij}$ refers to the weight connecting node j in the hidden layer with the node i in the input layer; $W_{jk}$ is the weight connecting node k in the output layer with the node j in the hidden layer.

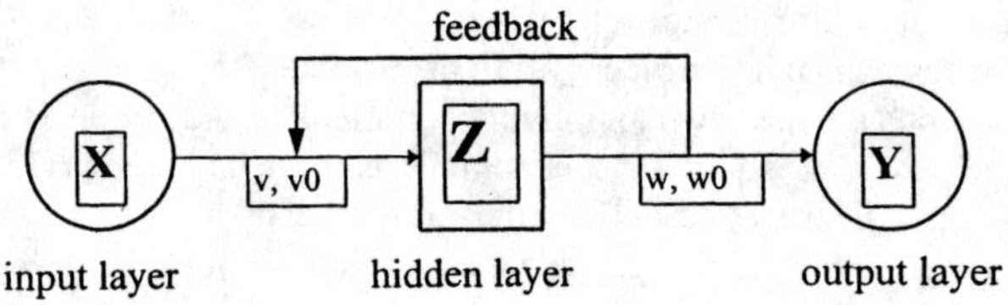


**Figure 1 Typical Architechture of Recurrent Neural Network**

The updates of the weights are evaluated by minimizing the sum squared error between the target values and output value, using delta rules (Rumelhart et al., 1986; Vogl et al., 1988; Widrow and Lehr, 1990 ).

## DATA BASE FOR RNN MODELING

In order to test the ability of RNN modeling, laboratory data from a series of triaxial shearing tests performed on a fine-grained residual soil are used. The soil contains about 10% clay, 15%-25% silt and 65%-75% sand particles. The average liquid limit of the soil is 78% and the average plastic index is 19 (Zhu, 1995). Two kinds of stress path tests were conducted for the simulation of RNN modeling. The first set is isotropic consolidation undrained (ICU) tests, while the other set is anisotropic consolidation undrained (ACU) tests. In a ICU test the soil sample was first consolidated under an isotropic stress condition, i.e. vertical stress is the same as the horizontal stress. During the consolidation process a drainage line was kept open so that the excess pore water could be drained out from the sample. After consolidation, the drainage line was closed and the vertical stress was gradually increased until the sample failed. While in the ACU test, the sample was first consolidated under an anisotropic stress condition. In this case, a vertical consolidation stress was about 2.5 times of the horizontal stress. This stress condition resulted in a non-zero initial shear stress condition for the shearing stage. For some slope stability problems involving a steep slope, ACU consolidation stresses are more realistic than ICU stress conditions. The soil behaved differently with respect to the shearing behavior due to the significant variabilities of the soil structure and constituents (Zhu, 1995). Some specimens dilated during the shearing, while others underwent compaction. Ten sets of original experimental data were used in the study, with each five sets of data as training and testing data, respectively.

## RNN ARCHITECTURE FOR PREDICTING SOIL BEHAVIOR

Generally, shear stress-strain relationship of the soil can be expressed as:

$$\{\Delta\sigma\} = [D]\{\Delta\varepsilon\} \tag{3}$$

where $\{\Delta\sigma\}$ is the incremental stress vector; [D] is the coefficient matrix usually called elastic modulus matrix or elastic-plastic modulus matrix; $\{\Delta\varepsilon\}$ is the incremental strain vector that are known in advance.

The major objective in the RNN modeling is to search for something like the coefficient matrix [D] from which $\{\Delta\sigma\}$ can be found. To optimize modeling, a series of choices required for setting up a RNN architechture were made based on knowledge of the problem, as listed in the following:

(1) Input variables and output results: In the specified shear tests, a strain rate is given and the strain value at a specific time is fixed, so the axial strain $\varepsilon_1$ and increment of the strain $\Delta\varepsilon_1$ should be the input variables. The output, from the problem itself, should be shear stress $\Delta\sigma$. According to our understanding and experience, the soil stress-strain behavior is also greatly influenced by such important factors as stress history, consolidation condition and void ratio. Therefore, the effective confining

pressure $\sigma_3'$, previous shear stress $\sigma_i$, pore water pressure $u_i$, and void ratio e are considered as input variables.

(2) Number of hidden layer and nodes : One hidden layer with respect to both input and output is used in the study. It is recognized now that one hidden layer can yield satisfactory results in most modeling works, although sometimes two hidden layers may efficiently deal with some complex problems. The number of nodes in the hidden layer is determined by a trial and error method. In the preliminary study, a RNN model with different nodes in the hidden layer was trained for the same number of epochs, and it is found that the sum of squared errors achieve their minimum value when there are 20 nodes in the hidden layer.

A hyperbolic tangent function ( $F_1=(e^x-e^{-x})/(e^x+e^{-x})$) is used as an activation function from the input layer to the hidden layer, while in the connection between the hidden layer and the output layer a linear transform function is adopted.

(3) Training process: The training process is initialized by inputting a set of "comprehensive" data to the network. The so called "comprehensive" data imply that the data are most representative and contain all the necessary information for the given problem. A network trained with a set of comprehensive data can have strong predictability. The flexibility of the network itself ensures that the network can be trained ceaselessly until the comprehensive data are included in the training data base.

(4) Testing of networks : With the weights and biases obtained in the training process, the network was tested with both trained data and untrained data to test its simulation capability and predictability.

## PREDICTION RESULTS

Figures 2 and 3 show a comparison between the predicted results from the RNN model with the test data. Although the RNN model did not learn anything from these test data, the network can predict the soil behavior very well based on the information learned from the training data.

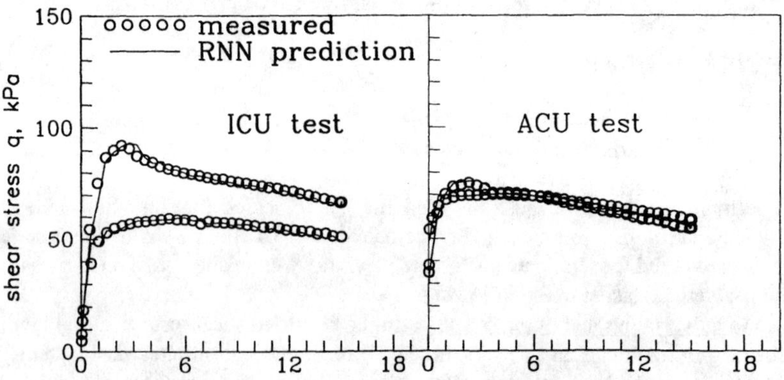


**Fig. 2 Prediction of shear stress–strain behavior of soil from ICU and ACU tests**

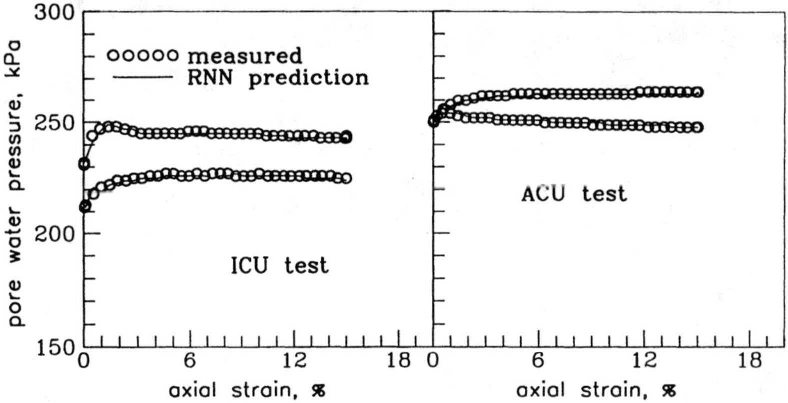


**Fig. 3 Prediction of pore water pressure of soil from ICU and ACU tests**

Figure 2 shows the relationships of shear stress versus axial strain of the soil for both ICU and ACU tests. From this figure one can see that there is a peak shear stress beyond that the shear stress decreases with the axial strain (soften response). When using a conventional mathematical model, it is very difficult to predict a softening response. It should also be pointed out that the ACU shearing tests are initialized at a non-zero shear stress that is also difficult to be modeled by most traditional constitutive models (Duncan & Chang, 1970; Desai, 1990) It may be noted that the pore water pressure generated during the undrained test is simultaneously simulated by the RNN model, as shown in Figure 3. This feature is very important in evaluation of safety of earth structures and this is beyond the capability of many constitutive models. The facts that both the zero and non-zero initial shear stress conditions can be captured by the RNN modeling, and that different shear behavior including dilation and contraction can be present within one RNN model demonstrate a strong capability of the RNN modeling.

From the study stated above, it is evident that with the RNN modeling one does not need to find a series of parameters from which mathematical equations are defined. This is one of the important advantages of the RNN modeling over traditional constitutive modeling approach.

## CONCLUSIONS

Application of the neural network approach to the modeling of soil behavior is a novel research area. The successful modeling of the shear behavior of a residual soil given in this study demonstrates the capability of the RNN model. Although the soil exhibited both dilative and contractive behavior during shearing, the RNN model captured these features efficiently. The architecture design of the RNN model varies with the characteristics of the problem. In this study one hidden layer with twenty nodes is found adquate for the modeling of both strain-controlled undrained tests and stress-controlled drained tests.

## ACKNOWLEDGMENT

The data used here were taken from a research work supported by National Science Foundation Grant No. BCS92-11927, supervised by Dr. S. Anderson at University of Hawaii at Manoa. The financial support by School of Civil Engineering and Environmental Science at the University of Oklahoma is also gratefully acknowledged.

## REFERENCES

Anderson, S. A. and Sitar, N. (1995). Analysis of rainfall-induced debris flows, *J. Geotech.* Engr., Vol. 121(7), ASCE,. 544-552.

Anderson, S. A. and Zhu, J. H. (1996). Assessing the stability of a tropical residual soil slope, *Proc. of the 7th Int. Symp. On Landslides*, Norway, in press.

Desai, C. S. (1990). Modeling and testing : Implementation of numerical models and their application in practice, *Numerical Method and Constitutive Modeling in Geomechanics*. Edited by Desai and Gioda, Springer-Verlag, Wien-New York, 1-168.

Duncan, J. M. and Chang, C. Y. (1970). Nonlinear Analysis of stress and strain in soils, *J. Soil Mechanics and Foundations Division*, ASCE, Vol. 96, No. SM5,1629-1653.

Elman, J. L. (1990). Finding structure in time, *Cognitive Science,* Vol. 14, 179-211.

Ghaboussi, J. (1992). Neural-Biological Computational Models with Learning Capabilities and Their Applications in Geomechanical Modeling, *Proc. of Recent Accomplishments and Future Trends in Geomechanics in the 21st Century,* U.S. Canada Workshop. Edited by Zaman, et.al. Norman, Oklahoma, 131-134.

Rumelhart, D., Hinton, G. E. and Williams, R. J. (1986). Learning internal representations by error propagation, *Parallel Data Processing*, Rumelhart and McClelland eds., M.I.T. Press, Cambridge, 318-362.

Vogl, T. P., Mangis, J.K., Rigler, A. K., Zink, W.T. and Alkon, D. L. (1988). Accelerating the convergence of the back propagation method, *Biological Cybernetics*, Vol. 59, 257-263.

Widrow, B. and Lehr, M. A. (1990). 30 Years of Adaptive Neural Networks: Perceptron Madaline and Backpropagartion, *Proc. IEEE.*, 78.(9), 1415-1442.

Zhu, J. H. (1995). Shear strength and soil behavior on a residual soil slope, Civil Engineering Master's thesis, University of Hawaii at Manoa, Honolulu, Hawaii.

# PREDICT BUILDING CONSTRUCTION PRODUCTIVITY USING NEURAL NETWORKS

**Jingsheng Shi**
*Department of Building and Construction, City University of Hong Kong*
*83 Tat Chee Ave, Kowloon, Hong Kong*

*ABSTRACT:*
This paper identifies the key factors which affect building construction productivity. Then two neural network models are studied to investigate the relationships between these factors and the productivity value. The advantages and limitations of these neural network architectures are analyzed for this purpose. Computer simulation technique has been applied to generate training data sets, and has shown an effective alternative for obtaining training data when actual collected data are not enough or available.

## INTRODUCTION

The prevailing method for obtaining productivity value in the industry is historical values kept by the company or industry wide averages published in many commercial manuals (e.g. R.S. Means). In most cases the estimator applies considerable personal judgment in selecting a productivity value that best represents the specific conditions of the project being estimated. Although the historical records of completed projects can provide valuable information for estimating a new project, the characteristics of uniqueness in construction project prevents an estimator from directly using the productivity values of the previously completed projects. The similarity and differences between the current project and the previous projects have to be analyzed before these records can be used, for instance, working conditions, site conditions, equipment, crew, and construction methods. These complicated relationships make a realistic estimate be very difficult to achieve for a project. To cover the risks associated with the uncertainties of a project, contingency is subjectively added to an estimated bid according the levels of risks, which may lead to lose a bid; otherwise, a contractor may lose money for its low bid without considering all factors and uncertainties involved in the project. The industry wide averages don't reflect the specific characteristics of a specific company, for instance, the skills of its labor, equipment, management. It is very difficult to come up with a realistic estimate for a project based on that. Therefore, determining the productivity of a project is always a challenge for estimators who must use excessive amount of subjective judgment in selecting productivity values for use in preparing a new bid.

Extensive research has been attempted in measuring, predicting, and improving productivity at various institutions. The work sampling technique has been used to measure and predict the unit rate productivity for power plant construction (Liou and Borcherding 1986). The work sampling technique and learning curves have been employed to measure labor utilization during the construction process of framing operations (Handa and Abdalla 1989). A factor model has been studied for forecasting labor productivity for masonry (Thomas and Yiakoumis 1987, Thomas et. al 1989, and Thomas and Sakarcan 1994). A study of the factors affecting construction durations in Hong Kong has been reported by Chan and Kumaraswamy (1995). Recent years, neural networks have been attempted in forecasting construction productivity (Moselhi and Hegazy 1991, and Chao and Skibniewsky 1994). The neural network based approach is still under study at institutions. Moreover, there is no a ready-to-use computer system being developed to facilitate the estimating process for the user.

## NEURAL NETWORKS FOR PREDICTING

Supervised networks which learn from correct patterns (e.g. past experience) have gained widely applications for predicting purpose in engineering. The typical supervised networks employed for predicting are backpropagation and GRNN (General Regression Neural Networks).

### Backpropagation

Backpropagation has a multilayered and feedforward topology (Chester 1993). It has an input layer, an output layer and hidden layer(s). The number of neurons in the input layer and output layers corresponds to the number of input and output of the real world problem respectively. It can have more than one hidden layer and as many hidden neurons as necessary in each hidden layer although the most widely employed has one hidden layer. All the neurons on two close layers are forward fully connected through synapses. A backpropagation network starts outs with random weightings on its synapses. Then a training set of input patterns is presented to the network as inputs, and the outputs will be calculated. The differences between the calculated outputs and the correct outputs will be evaluated and will be used to adjust the network's weights to reduce the differences. As the training proceeds, the network's weights are continuously adjusted until it can accurately respond to the inputs. Backpropagation is the most widely implemented and employed among all neural network paradigms. Its advantages are its flexibility and ability to store great number of patterns. However, it usually requires vary large training sets; it may converge to local minimal; and it usually takes long time to train a complex problem.

### General Regression Neural Network (GRNN)

GRNN is an another type of supervised network. It is capable to train quickly on sparse data sets. GRNN applications (Ward Systems Group, Inc. 1995) are able to produce continuous valued outputs. They sometimes responds much better than backpropagation to some types of problems. It is especially useful for continuous function approximation. Because GRNN networks evaluate each output independently of the other outputs, they may be more accurate than backpropagation networks when there are multiple outputs.

## PREDICTING PRODUCTIVITY OF AN EARTHMOVING OPERATION

Earthmoving is a common construction operation typical to foundation work of building projects, dam construction, airport construction, road construction, and strip-mining. It is an equipment intensive construction system. Studying an earthmoving operation mainly focuses on selecting the models of equipment and the number of each type of equipment to maximize the system's production rate, minimize the unit cost, or maximize the profit based on the project's objective.

**Factors affecting the productivity of an earthmoving operation**

According to our observations, the factors which affect an earthmoving operation can be classified into three groups associating with loading equipment, hauling trucks, and weather conditions respectively.

Loading equipment affects productivity mainly on the loading time required to fill a truck, which is jointly determined by the capacity of the loading equipment, the capacity of the truck, and the working conditions. Because the loading time can vary from one truck to another, the mean loading time (LT) and standard deviation (LT_S) are used to describe the loading process.

Hauling trucks affect the productivity mainly on the capacity of trucks (TC), number of trucks (TN), and time required to complete a hauling cycle (associating with the hauling distance and road conditions). Like loading time, the hauling time varies from cycle to cycle. The mean hauling time (HT) and standard deviation (HT_S) are used to describe the random nature of the hauling process.

Weather conditions (W) can affect the productivity of an earthmoving operation by influencing the loading and hauling time. In this paper, the weather conditions are simplified into two discrete states (1 or 2). State 2 is the normal weather condition. State 1 is the raining state at which the productivity will be reduced accordingly. More states can be defined for more complicated weather conditions.

**Productivity factors**

A project manager's basic interest is the mean productivity value (P) that can be achieved in a project. However, because of the random nature involved in construction operations, the mean value does not explain the whole story. The standard deviation (P_S) of the productivity value is also very important to let the project manager understand the possible uncertainties or risks involved.

**Training data**

The training data sets should be representative and consistent. For instance, if the training data sets only include the operating information for hauling distance between 2 and 5 km, the trained network may not work for hauling distance beyond this range. On the other hand, construction operations are random in nature. A small sample will not be able to provide a global picture of an operation. The author observed an earthmoving operation on site for a week, and found out that the daily production of the same fleet achieved could vary up to 20% or 30%. This random nature brings two problems to applying neural networks: 1) the training process will be very long and unstable; 2) accuracy of the predicted value. To avoid these, a reasonable size of sample for each situation should be processed into one training pattern using statistics.

**Using simulation to generate training data sets**

There is a huge demanding for training data sets. It may not be possible to collect all required data to cover a wide range of situations such as different hauling distance and different type of equipment. Computer simulation can repeat a real operation on a computer, and might be a very effective alternative approach to obtain enough training data sets.

An earthmoving contractor mainly uses either CAT-777 or Titan-2200 trucks for hauling, and either a Hitachi EX-1800 or EX-1100 backhoe for loading in his earthmoving projects. Number of trucks can vary from 2 to 12. A simulation model was constructed for this sample application by using SLAMSYSTEM (Pritsker 1995). Experimenting with the model under various situations, the results show that the daily production can vary from 1000 to 12000 $m^3$. Total 49 training patterns have been generated.

**Backpropogation network**

Various backpropogation networks have been experimented with above training data by using NeuroShell 2 (Ward Systems Group, Inc. 1995). The Ward backpropogation network with two hidden slabs can give the best fit. This network has four slabs: one input, one output and two hidden slabs.

The net-perfect function of NeuroShell was used to train the network with the 49 patterns, the actual daily production (obtained from simulation) versus calculated production for the training patterns is plotted in Figure 1. The contribution factors of the seven input variables illustrates that the capacity of trucks, number of trucks, and loading time are the most important factors to the productivity and standard deviation. However, contribution factors can not always give an accurate picture for their real influence, especially when there is more then one output variable, and the influence of the same input variable has different impact to different output variables.

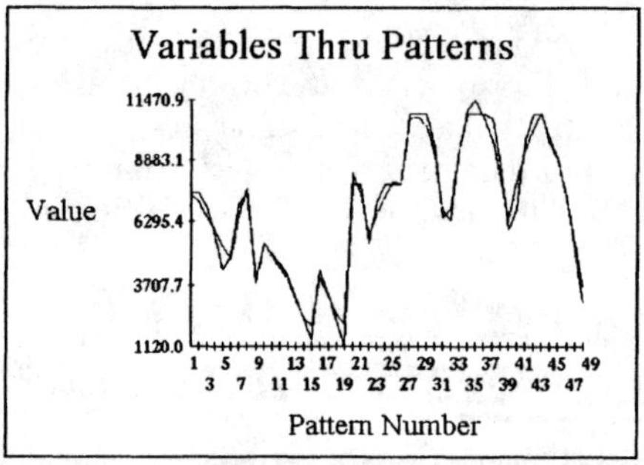


**Figure 1 Actual production versus predicted production**

The two curves in Figure 1 show certain difference. The reasons could be: 1) backpropogation usually works better for only one output; and 2) complicated relationships between input factors and predicting factors. It is recommended to use one specific network model for predicting one output when using a backpropogation network. In this case, two network models should be adopted to predict productivity and standard deviation separately. Regarding to the complicated relationships, a detailed study on the network architecture and training algorithm based on these special relationships will be necessary. For instance, productivity will linearly increase with the number of trucks within a certain range, then increase very little (non-linear) beyond that range until no any increase can be achieved after a top ceiling is reached.

**Training with GRNN**

A GRNN network with 49 hidden neurons has also been experimented. The actual daily production versus calculated production is plotted in Figure 2. The results show that a better fit for the training data sets can be achieved than from backpropogation.

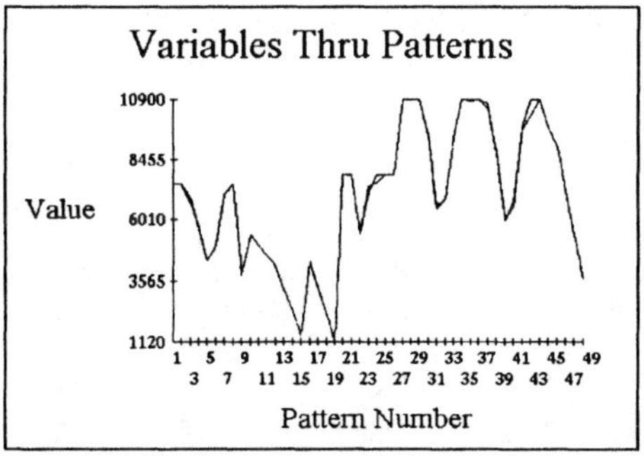


**Figure 2 Actually production versus calculated production from GRNN**

**Test patterns**

Six sets of data have been extracted from an actual operation, and the results are listed in Table 1.

**TABLE 1 RESULTS FOR TEST PATTERNS**

| Input information | | | | | | | Actual | | GRNN | | Backprop. | |
|---|---|---|---|---|---|---|---|---|---|---|---|---|
| TC | TN | LT | LT_S | HT | HT_S | W | $P_A$ | $P\_S_A$ | P | P_S | P | P_S |
| 50 | 5 | 4.36 | 0.47 | 18.2 | 1.37 | 2 | 6920 | 56.50 | 6713 | 56.33 | 6457 | 54.66 |
| 50 | 3 | 4.38 | 0.47 | 18.2 | 1.38 | 2 | 4350 | 38.20 | 4372 | 49.84 | 5301 | 43.57 |
| 50 | 7 | 4.2 | 0 | 14.4 | 0 | 1 | 7850 | 0.00 | 7850 | 5.00 | 8328 | 17.61 |
| 90 | 2 | 5.47 | 0.46 | 12.0 | 1.22 | 2 | 6800 | 65.80 | 6524 | 73.29 | 7997 | 71.16 |
| 90 | 4 | 5.47 | 0.46 | 12.0 | 1.21 | 2 | 10900 | 86.80 | 10270 | 98.27 | 10294 | 87.55 |
| 90 | 6 | 6 | 0 | 60.0 | 0 | 2 | 5400 | 0.00 | 5400 | 5.00 | 5066 | 13.00 |

**Table 2 A comparison**

| $\lvert P_A - P_{GRNN} \rvert$ | $\lvert P\_S_A - P\_S_{GRNN} \rvert$ | $\lvert P_A - P_{GRNN} \rvert$ | $\lvert P\text{-}S_A - P\_S_{GRNN} \rvert$ |
|---|---|---|---|
| 463 | 1.84 | 207 | 0.17 |
| 951 | 5.37 | 22 | 11.64 |
| 478 | 17.61 | 0 | 5.00 |
| 1197 | 5.36 | 276 | 7.49 |
| 606 | 0.75 | 630 | 11.47 |
| 334 | 15.0 | 0 | 5.00 |
| Mean = 671.5<br>S = 332.7 | Mean = 7.65<br>S = 7.01 | Mean = 189.2<br>S = 245.6 | Mean = 6.80<br>S = 4.39 |

The results in Table 2 show that there is a significant difference in productivity value predicted from backpropogation and from GRNN. Statistic methods can be applied to support this conclusion. GRNN can provide a better prediction in this case. There is no significant different in the standard deviation predicted from both networks.

## CONCLUSION AND LIMITATIONS

This paper attempted to apply neural networks for predicting construction productivity. This study has shown the promise. As an alternative approach for obtaining extensive training and test patterns, this paper applied simulation technique to generate these data sets when there are not enough actual data sets available. The author also found some limitations in the current studies. For instance, there are some special characteristics for predicting productivity. The general purpose network architectures and algorithms may not be able to satisfy these specific requirements. A detailed study is under way to provide an easy-to-use and accurate computer system specifically for predicting productivity purpose.

## REFERENCES


Chan, D. W. M. And M. Kumaraswamy (1995). "A study of the factors affecting construction durations in Hong Kong." Construction Management and Economics, 13, 319-333.

Chao, L., and M.J. Skibiewski (1994). "Estimating Construction Productivity: Neural Network based Approach." Journal of Computing in Civil Engineering, ASCE, 8(2), 234-251.

Chester, M. (1993). Neural Network: A tutorial. Prentice-Hall, Inc., Englewood Cliffs, NJ 07632.

Everett, J.G. and S. Farghal (1994). "Learning curve Predictors for Construction Field Operations." Journal of Construction Engineering and Management, ASCE, 120(3), 603-617.

Handa, V.K. and O. Abdalla (1989). "Forecasting Productivity by Working Sampling." Construction Management and Economics, 7, 19-28.

Herbsman, Z. and R. Ellis (1990). "Research of Factors Influencing Construction Productivity." Construction Management and Economics, 8, 49-61.

Liou, F. and J.D. Borcherding (1986). "Work Sampling Can Predict Unit Rate Productivity." Journal of Construction Engineering and Management, ASCE, 112(1), 90-103.

Moselhi, O. T. Hegazy, and P. Fazio (1991). "Neural Networks as Tools in Construction." Journal of Construction Engineering and Management, ASCE, 117(4), 606-625.

Pritsker, A. A. B. (1995). Introduction to Simulation and SLAM II. Fourth edition, John Wiley & Sons, NY.

Rayburn, L.M. (1989). "Productivity Database and Job Cost Control Using Microcomputers." Journal of Construction Engineering and Management, ASCE, 115(4), 585-601.

Tam, C.M., A.T. So and A.P.C. Chan (1995). "Application of Neural Network in Predicting Contractor Performance." Proceeding of ICEC/AIQS Sumposium, May 1995, 94-102.

Thomas, R. and A. Sakarcan (1994). "Forecasting Labour Productivity Using Factor Model." Journal of Construction Engineering and Management, ASCE, 120(1), 220-239.

Thomas, R., C.T. Mathews and J.G. Ward (1986). "Learning Curve Models of Construction Productivity." Journal of Construction Engineering and Management, ASCE 112(2), 245-258.

Ward Systems Group, Inc. (1995). NeuroShell 2. Third edition. Ward Systems Group Inc., Frederick, MD 21702.

# Section 7 : Machining and Robotics

# A STUDY ON AUTONOMOUS ROBOT WITH EMPIRICAL MEASURE

**JUN HAKURA, HIROSHI YOKOI**
*Autonomous Systems Engineering, Complex Systems Engineering, Hokkaido University, kita-13, nishi-8, kita-ku, sapporo, 060 Japan.*

**YUKINORI KAKAZU**
*Autonomous Systems Engineering, Complex Systems Engineering, Hokkaido University, kita-13, nishi-8, kita-ku, sapporo, 060 Japan.*

***ABSTRACT:***
This paper proposes a new design concept to develop an autonomous robot to have its own empirical measure of the environment. Empirical measure is expected to make the robot into an autonomous and intelligent one without coming across the Frame problems. This paper considers "Invariant" in Affordance theory as an empirical measure. A perception mechanism that would be indispensable to the realization of an empirical measure, is proposed. The mechanism introduces sonar sensing system, perceptual model, information processing system, and learning mechanisms using several ANNs. Computational results show that proposed mechanism has possibilities to satisfy the four conditions.

## INTRODUCTION

How to realize am autonomous is the target of this paper. For this purpose, we need methodologies to obtain robot's own measurement, we call the measure as "Empirical Measure". The Empirical Measure in this paper corresponds to the functional property of the environment related to the behavior of the robot by means of the fixed uninterpreted sensors. Namely, the functional properties of the environment enable the robot to make certain behaviors. Functional property of the environment is considered as the Affordance proposed by the cognitive psychologist Gibson (Gibson 1979), as Reed pointed out in (Reed 1988). Affordance originally requires no information processing but the extraction of the functional properties of the environment. Namely, it assumes no physical and objective properties to be perceived in the environment. Instead of that, it assumes functional and subjective properties those should be perceived in relation with the behaviors as discussed in Section 2.

The autonomous robot is expected to acquire the measure through its own experience. The SSA ( e.g., Brooks 1991) has been succeeded in the behavior-based system. It realizes the direct perception by coupling the sensory inputs with the modules corresponding to the behavior. However, the coupling is given by the designer. Therefore, the functional property in the environment should detected by the designer and the robot should behave as it designed. With a perception mechanism introduced in this paper, the intervention by the designer is expected to be reduced. Therefore, the autonomous robot with the perception mechanism is expect to adapt to the environment through its own experiences.

In the perception mechanism described in Section 3, we are trying to represent the perception of the robot as "grounded symbols" ( Harnad 1990, and Pfeifer 1996) in the Inner Model. The use of this representation is mainly on the reduction of the enormous amount of the information in the real world and on the preservation of the similarity between situation in the environment. Moreover, the symbolic representation of the perception and the behavior would enable the robot to utilize the intelligent process proposed in the field of the symbolic AI. The further discussion on the "grounded symbol" is made in Section 2. We introduce an information processing system to achieve the Empirical Measure together with some learning mechanisms described in Section 4.

The paper is consisted of five sections: the requirement to solve the two problems to the realization of the autonomous exploring robot are discussed in section 2. Then, the proposed perceptual mechanism is described in section 3. Learning mechanism to achieve an empirical measure of the environment is appeared in section 4. Finally, computational experiments show the environment perception by means of proposed mechanism have possibilities to achieve the inner model that satisfies the four conditions described in Section 2, and the possibilities of the realization of empirical measure of the environment within this framework.

## PROBLEMS: DESIRED AUTONOMOUS ROBOTS

Figure 1 shows the typical example of what this paper aims. A robot in front of walls with an opening between the walls, is considering what to do. The robot has sensory system through that it can observe the environment. If the robot had a cognition mechanism based on some conventional AI method, the robot is expected to check every properties of the environment as forced by the designer. What if the robot is able to know the possible behaviors as soon as it observes the environment according to what it perceived with the environment? The robot will never be required to infer which objects should take into account and which are not (Frame Problem: e.g., McCarthy & Hayes 1969 for reference). Moreover, it will be able to have a room to make deliberately think, i.e., reasoning and planning on its behavior.

To achieve this story in the real autonomous robot, mainly two problems are in front of us: empirical measure acquisition problem, and the grounding problem.

### Empirical Measure Acquisition Problem

The intelligent robot based on the conventional AI is often separated from the environment and the environment is recognized by the robot by means of the objective

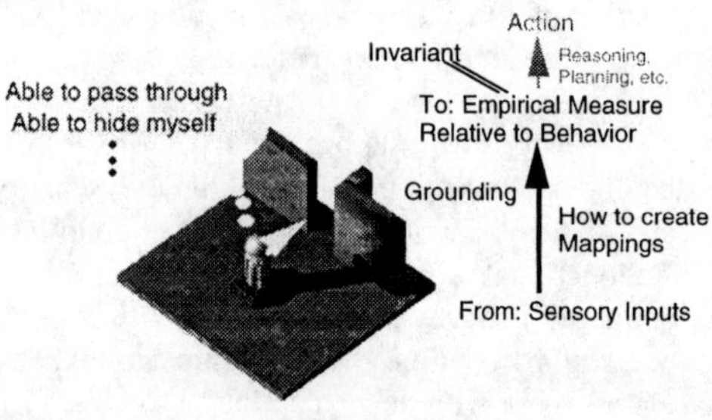


**Figure 1. Invariant treated as a self-organized measure of environment.**

and absolute measure that is previously given by the designer. The robot, therefore, should extract objects from the environment, reason the relationships between the objects, infer the meanings of the relations, and plan the action altogether. Each of these processes contains difficulties such that it would stand as a problem.

We have inspired by the concept Invariant in Affordance to find solution to the problem above. Moreover we regard the Invariant as the measurement to the environment acquired by the robot itself through experience, i.e., Empirical Measure in this paper. The empirical measure is very local representation of the information from the environment coupled with its behaviors. Therefore, the quite different measure will be emerged among the robots according to the size, use, and the experience of the robot, etc. Those differences are ought to be observed in each robot just the same as our human perception is somehow different by the person.

Achieving the empirical measure requires a grounded information on the environment that should be particular to each robot. In this paper, we consider this grounded information as perceptual model, i.e., inner reconstruction of the environment. Namely, a kind of grounding problem should be taking into account to achieve the aim.

### Grounding Problem

"How is symbol meaning to be grounded in something other than just more meaningless symbols?" ( Harnad 1990). To acquire the empirical measure of the environment, some intermediate field that would realize the robot's own representation of the environment, is indispensable. Namely, grounding the sensory inputs onto some symbols or concepts is considered to be required the inner representation of the environment. This paper considers the inner model as the intermediate field and set the inner reconstruction method to be learned.

What if a robot would reconstruct its own perceptual world according to some brief rules and criteria and learn the reconstruction method of its perception? The robot will be much more free from the designer's intention, because the designer would not be able to forecast the perception reconstructed by the robot. Moreover, robots are expected to grow personality according to what they have expected or according to the rules they intrinsically have.

Within the perception reconstructed by the robot, an empirical measure ought to be emerged. We settle the following four conditions for perception to be satisfied to achieve the empirical measure:

1. Distinction ability of the difference among the different environment.
2. Successiveness in accordance with action.
3. Dependence on the environment.
4. Successiveness on the experience.

A Perception mechanism that satisfies these four conditions is proposed in the next section.

## PERCEPTION MECHANISM

As discussed above, a mechanism that would make an autonomous robot to have an Affordance-like Cognition, would be a breakthrough in robotics. A mechanism that

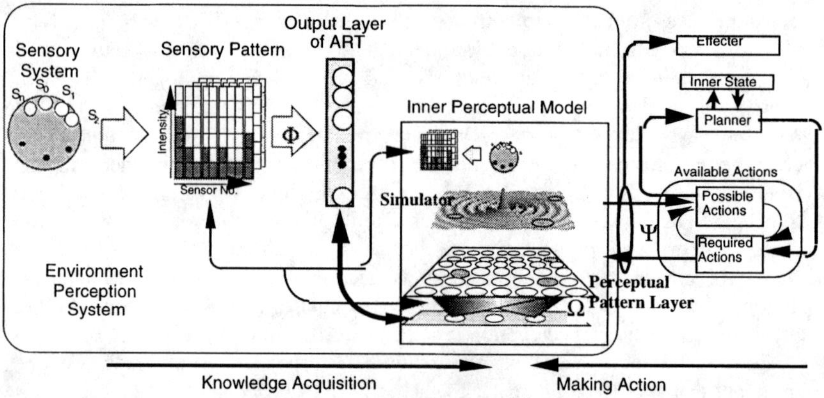


**Figure 2. Outline of Perception Mechanism**

would be a clue for this purpose is proposed in this section. The mechanism consists of roughly three parts as illustrated in Figure 2.

### Active observation system of the environment

The robot generates the pulse wave and senses the reflection wave by the sensors whose output $s_i$ is defined as equation (1). The reflection wave is the result of the interactions among the objects in the environment so that the reflection wave contains the information on the environment. Inputs to the sensors are the series of strength of the wave integrated through a certain time.

$$s_i = \int_{T}^{T+OT} o_i(t)\, dt \qquad (1)$$

In equation (1), $o_i(t)$ is a wave strength of the i-th sensor at time t, and OT corresponds to the refractory period of the sensor.

### Inner Model

The inner model introduced in this paper corresponds to the reconstructed perception of the environment that will be indispensable to the grounding the sensory inputs. The model consists of a field, simulator of the active observation system and modules.

Field is an inner environment that will be a medium of the observation wave generated by the active observation system simulator. Active observation system simulator observes the reconstructed inner perception quite in the same way as the real observation system. This simulator is used in reconstruction process to observe the state of the perception. A module is a reflection body with a certain size and shape. In this paper the reflection rate of the module is fixed as 100%. Module is placed in the inner field by the robot itself.

The reconstructed perception is quite particular to the robot and no forecast is available for the designer. This perception will be a grounded information of the robot. With this perception, the robot would be able to plan its behavior with respect to the recollected action by the perception.

### Information Processing Mechanism

As depicted in Figure 2, the information processing system consists of ART (Carpenter & Grossberg 1987), and several simple artificial neural networks (ANN). Those ANN work as not only a device for the long term memory but an accelerator. The information processing is needed to achieve the Empirical Measure autonomously by the robot. While, once the measure is acquired, the processing is not essential to the robot unless it is facing with some new situation or some change is required to its perception by the inconvenience suffered by the perception. For further description on the information processing in the proposed mechanism ( Hakura et al. 1996) should be referenced.

The patterns reconstructed in the inner model would be modified according to the relations between the patterns and the behaviors. The modified pattern is considered to be corresponding to the Invariant in the Affordance Theory.

## LEARNING MECHANISMS

The mappings shown in Figure 2 is generated by means of inner model. These mappings are acquired through mainly trial and error in the beginning of the activity. As learning proceeds, the robot is able to reconstruct the inner model quickly and the available action/behavior set recalled in accordance with the perceptual pattern in the inner model, would be more correct one. While We settle three mappings to be realized, namely, Φ, Ω, and Ψ in the figure, we describe only on W for the sake of the paper restriction.

### Acquisition of Mapping Ω

Mapping Ω is realized by the weighted links between the output units of the ART and the units on the perceptual pattern layer in the inner model. Because the robot selects the single unit in the perceptual pattern layer according to the output value of the unit to put a module, each place corresponding to the unit is evaluated by means of the simulator. The evaluation is carried out under a simple rule: left the module, if the sensory pattern from the simulator $S^{in}$ is included in the sensory pattern $S^{re}$ observed in the real world. Otherwise, remove the module from the layer. With this procedure, the learning process is carried out in parallel according to the following rule:

$$w_{kij} = w_{kij} + \alpha\Diamond(o_{ij} - 0.5)x_k \tag{2}$$

$$o_{ij} = \begin{cases} 1.0, & \text{iff } S^{re} \supseteq S^{in} \\ 0.0, & \text{otherwise} \end{cases} \tag{3}$$

where, $\alpha$ is a learning coefficient, $w_{kij}$ is the weight in the ART. $x_k$ is the input to the ART. The same procedure is applied when the accepted module is turned out to be a rejected one by the movement of the robot.

## COMPUTER SIMULATIONS

The results of one of the computer simulations are illustrated to show the proposed perception mechanism has an ability to realize four conditions described in Section 2. For the others, please refer ( Hakura et al. 1996). The simulation space is set up as 2D for simplification.

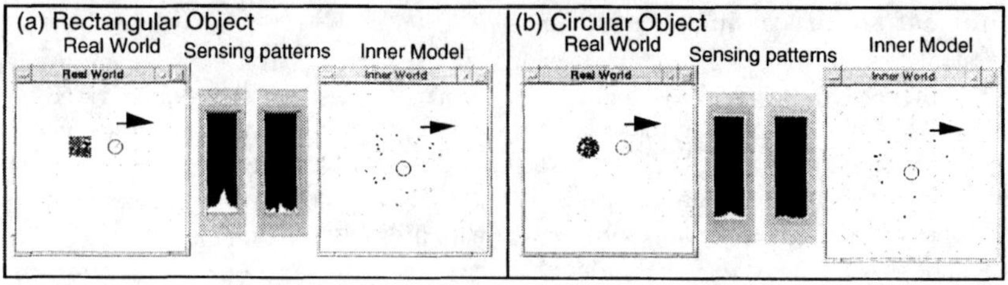


**Figure 3. The result of the computer simulation for condition 1.**

## Ability to Distinct Different Environments

This simulation attempts to illustrate the ability of the proposed mechanism to distinguish the different environments. This ability is indispensable for the perception of the robot not only with the Empirical Measure, but those robots somehow reducing the amounts of information on the environment. In this simulation, the number of objects is restricted to one, the distance between the robot and the object is fixed for the sake of simplicity. Then two cases, i.e., the shapes of the objects are different, are compared. As shown in Figure 3, the reconstructed perceptual pattern reflects the difference of the objects, even though the two environments are different only in their shapes.

## CONCLUSIONS

A robot with the Empirical Measure is aimed at in this paper. For the sake of the realization of the robot, importance of an empirical measure, that enables the robot to catch the environment as its own style and tightly coupled with action, is discussed. Then, the perception reconstructed by the robot itself is required to achieve the empirical measure. This paper, therefore, introduces four conditions to be satisfied by the reconstructed perception. The computer simulation shows that the conditions are satisfied by the proposed mechanism. With this perception mechanism, an empirical measure for the realization of autonomous robot will be acquired in the near future.

## REFERENCES

Brooks, R. A., (1991). Intelligence without Representation, *Artificial Intelligence*, 47, pp. 139-160.

Carpenter, G. A. and Grossberg, S., (1987). ART 2: Self-organization of stable category recognition codes for analog input patterns, *Applied Optics*, 26, pp. 4919-4030.

Gibson, J. J., (1979). The Ecological Approach to Visual Perception, *Houghton Mifflin*.

Hakura, J., Yokoi, H., Kakazu, Y., (1996) A Study on Autonomous Robot with Empirical Measure, *Intelligent Engineering Systems through Artificial Neural Networks*, vol. 6, *to be published*..

Harnad, S., (1990). The Symbol Grounding Problem, *Physica D*, 42, 335-346.

McCarthy, J. & Hayes, P. J., (1969). Some Philosophical Problems from the Standpoint of Artificial Intelligence, *Machine Intelligence*, 4, pp. 463-502.

Pfeifer, R., (1996). Symbols, patterns, and behavior: beyond the information-processing metaphor, *Encyclopedia of Microcomputers*, vol. 17, pp. 253-275.

Reed, E., S., (1988). James J. Gibson and the Psychology of Perception, *Yale University Press*.

# ON THE APPLICATION OF NEURAL NETWORKS TO A PETRI NET-BASED INTELLIGENT WORKSTATION CONTROLLER FOR MANUFACTURING

**YI-HUI MA AND NICHOLAS G. ODREY**
*Department of Industrial & Manufacturing Systems Engineering, Lehigh University*

***ABSTRACT*:**
The objective of this paper is to investigate an intelligent manufacturing workstation controller which integrates Petri net-based models and neural network techniques for preliminary diagnostics. A timed, colored Petri net (TCPN) model is used to provide an effective modeling scheme for workstation operations and serve as the basis for constructing neural networks. A multilayer feedforward neural network architecture is proposed to facilitate real-time preliminary diagnostic processes and to provide necessary information for further diagnostics. The proposed approach provides one possible way to relate high level operational information with low level sensor feedback and has the advantages of being able to learn and make decisions with minimum human intervention.

## INTRODUCTION

In a hierarchical control context, manufacturing workstation controllers have two major functions : 1) generating detailed processing steps in order to accomplish the operation request by a higher cell level, and 2) regulating the execution of manufacturing activities within a workstation such that adaptive and responsive reactions can be issued in the event of error occurrence. A general framework for constructing such an intelligent workstation controller has been presented previously by Odrey & Ma [1]. In their proposed framework, Petri nets were identified as the basic representation for modeling of processing steps and providing necessary information for the control purpose. The control of the workstation consisted of three primary subfunctions : monitoring, diagnostics and recovery plan generation. Neural networks and expert systems were suggested as two potential techniques for diagnosing errors and generating a recovery plan. Their approach gives the possibility of integrating control subfunctions by a hybridization of a Petri net representation with neural networks and expert systems.

As an extension of the previous work, this paper presents the detailed implementation along with an example of utilizing Petri nets and neural networks for the purpose of preliminary diagnostics. Preliminary diagnostics is the first subtask of the diagnostic subfunction and is used to facilitate the diagnostic process. The approach taken here contains three different neural networks as shown in Fig. 1. NN1 generates the expected system status by converting a Petri net representation into a neural network structure for real-time control. NN2 implements a sensor fusion and/or logical sensors concept [2] to provide NN3 with the actual system status such that a sensory-based control system can be realized. NN3 is a multilayer feedforward neural network for classifying data obtained from NN1 and NN2 into different categories for preliminary diagnostics.

The focus of this paper includes detailing a formal methodology for constructing NN1 and presenting an example to evaluate the performance of neural networks. The

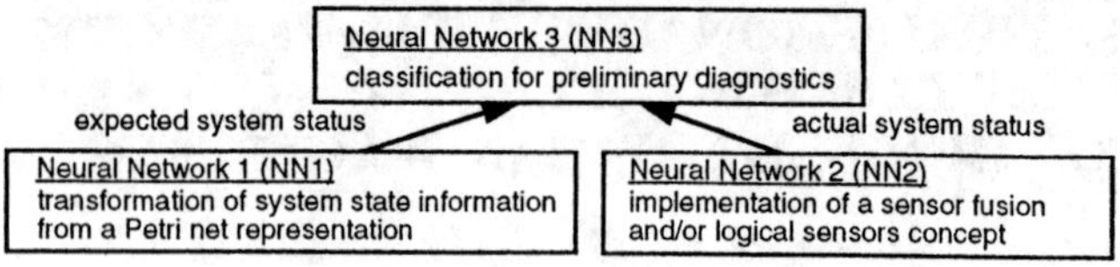


**Figure 1. Neural networks architecture for preliminary diagnostics**

detailed structure of NN2 involves different techniques for processing sensor data and is the subject of a separate paper.

## PETRI NET MODELING OF MANUFACTURING WORKSTATION OPERATIONS

A Petri net (PN) is a graphical and mathematical tool for modeling, analysis, and performance evaluation of discrete event dynamic systems such as workstation operations [3]. The basic definition and related information about Petri nets can be found in [4]. To demonstrate the modeling process using Petri net representation, a workstation example similar to the one used by Zhou, et al. [5] is considered.

As shown in Fig. 2, a top-down Petri net decomposition approach is performed in our research to construct a hierarchical PN model for the given workstation example. The hierarchical PN model differs from Zhou's work in the sense that high level Petri nets such as TCPN and TPN are included to enhance the modeling capability and a hierarchical concept is introduced to provide necessary task decomposition. The first (highest) sublevel is a timed-colored Petri net (TCPN) which is a general PN with two additional parameters : 1) a time factor to represent the operation time for each operational place, and 2) color tokens to distinguish between parts. At the second sublevel, a timed Petri net (TPN) is constructed where color tokens are not required because different parts (color tokens) are modeled separately. The third sublevel of the model further decomposes the operations at the assembly table into detailed processing steps such as "pick up", "transport", and "place".

## NN1 CONSTRUCTION

Several attempts toward combining PNs with neural networks have been reported ([6],[7]). Their applications mainly focused on embedding neural network structures in a Petri net model and are called Neural Petri Nets (NPN). The approach taken here differs in that we embed a Petri net model in a neural network structure termed Petri Neural Nets (PNN). The purpose of a PNN is to facilitate the process of obtaining state evolution information (the expected system status) by taking advantage of the parallel computational structure provided by neural networks and utilizing the T-gate threshold logic concept proposed by Ramamoorthy, et al. ([8]).

The state evolution of a system modeled by Petri nets can be expressed using the following matrix equation ([3] [9]) :

$$M(K+1) = M(K) + U^T(K)A, \quad K = 1, 2, .... \qquad (1)$$

M(K) is a (1xm) row vector representing the system marking at the Kth stage. U(K) is a (nx1) column vector containing exactly one nonzero entry "1" in the position corresponding to the transition to be fired at the Kth firing. The matrix A is a (nxm)

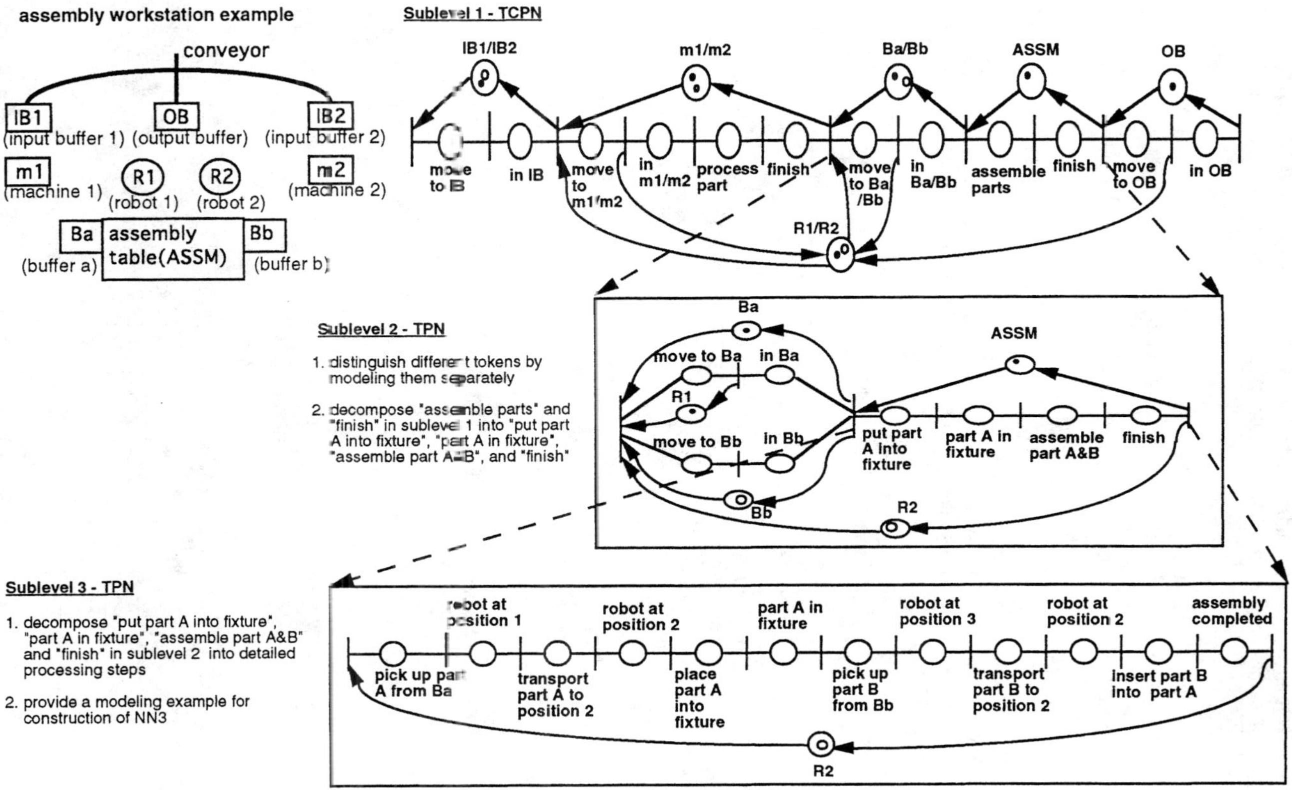


Figure 2. Hierarchical Petri net model of an assembly workstation.

transition-to-place incidence matrix. Based on the state equation (1), a three-layered PNN with an embedded T-gate threshold logic which simulates the state evolution of a general PN from M(K) to M(K+1) is built as follows (see Fig. 3):

1. The input vector $\mathbf{I_k} = [I_1, I_2, \ldots, I_m]$ (where m = number of places) is set equal to M(K). The expected output vector $\mathbf{O_i} = [O_1, O_2, \ldots, O_m]$ is M(K+1).
2. The second layer of the PNN contains three vectors : (i) $\mathbf{V_j}$ (j=1, 2, ... , m) representing M(K), (ii) $\mathbf{G_r}$ (r=1, 2, ... , n where n = number of transitions) representing $U^T(K)$ which is determined by the execution rules of Petri nets. For a decision-free PN [10], the execution rules can be implemented using AND T-gate threshold logic. The T-gate threshold logic is a neural network with hard weights (fixed) and can be used to implement a rule-based expert system for time-critical applications [8]. (iii) $\mathbf{H_h}$ (h=1, 2, ..., m) representing $U^T(K)A$ .

3. The weights in the PNN are hard weights and are assigned according to the following rules :

$$W_{kj} = \begin{cases} 1 & \text{if } k=j \\ 0 & \text{otherwise} \end{cases} \qquad W_{kr} = \begin{cases} 1 & \text{if } p_k \text{ is an input place of } t_r \\ 0 & \text{otherwise} \end{cases}$$

$$W_{ji} = \begin{cases} 1 & \text{if } j=i \\ 0 & \text{otherwise} \end{cases} \qquad W_{hi} = \begin{cases} 1 & \text{if } h=i \\ 0 & \text{otherwise} \end{cases}$$

$$W_{rh} = a_{rh} \qquad \text{where } a_{rh} = \text{element of } A$$

4. The output function for each layer is defined as follows :

$$V_j = \sum_{k=1}^{m} W_{kj} I_k \quad j=1,\ldots,m \qquad O_i = \sum_{j=1}^{m} W_{ji} V_j + \sum_{h=1}^{m} W_{hi} H_h \quad i=1,\ldots,m$$

$$H_h = \sum_{r=1}^{n} W_{rh} G_r \quad h=1,\ldots,m \qquad G_r = \begin{cases} 1 & \text{otherwise} \\ 0 & \text{if } \sum_{k=1}^{m} W_{kr} I_k \le T-1 \end{cases} \quad r=1,\ldots,n$$

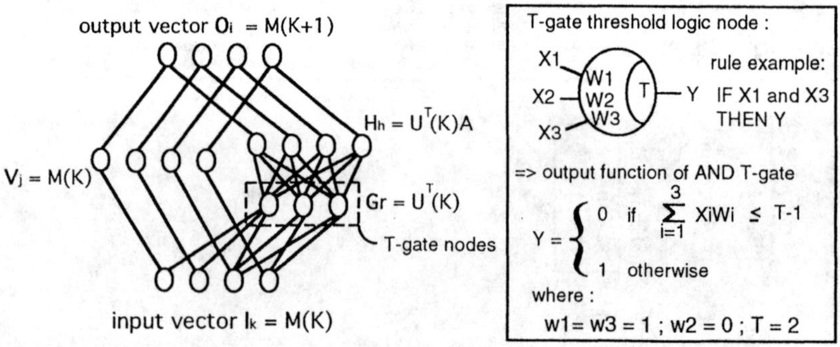


**Figure 3. NN1 (Petri Neural Network) architecture**

## PRELIMINARY DIAGNOSTICS USING NN3

The purpose of preliminary diagnostics is to classify operation conditions occurring in the workstation into several categories, each one associated with a preliminary action. The classification into a preliminary action is similar to the initial stage of diagnosis in the Integrated Diagnostic Model (IDM) developed by Fink, et al. [11]. The preliminary actions can be considered analogous to the entry points in the IDM system and are used to provide an instruction for further analysis.

### Development of Network Architecture

The development of NN3 for preliminary diagnostics is based on the Petri net model of the operations and the available sensor information. For example, the assembly operations modeled by the sublevel 3 - TPN model (see Fig. 2) has a basic architecture described by the following :

1. The input vector of NN3 contains 19 nodes. The first 13 nodes $[p_1, p_2, p_3, \ldots, p_{13}]$ (as shown in the sublevel 3 - TPN model of Fig. 2) represent the expected system status and are obtained from the output of NN1 (i.e. M(K+1) of the corresponding sublevel 3-TPN model of Fig. 3). The rest of the nodes $[s_1, s_2, \ldots, s_6]$ represent six assumed categories of sensor information (see Table 1) which can be obtained from NN2. One possible realization of NN2 is the work of Zurada & Graham [12] who presented a neural network-based robot safety system using two sensors to determine whether or not an emergency stop decision should be generated.
2. The output vector $[o_1, o_2, o_3, o_4]$ of NN3 represents the four preliminary actions : shutdown ($o_1$), call operator ($o_2$), continue operation ($o_3$), and invoke further diagnostics ($o_4$). The value of these output are either "0"-representing not activated, or "1"-representing activated.
3. Training and testing data are obtained using diagnostic rules based on common knowledge about the system. Six representative fault scenarios are listed in Table 2.

### Input/Output Representations

In general, the actual operation status of a system at any instant is the set of readings of all the sensor outputs. However, the actual system status information given by the sensor outputs is not sufficient for determining preliminary actions. Both the actual system status and the expected system status are required. The determination of a preliminary action in the assembly operations can thus be stated as follows :

IF *"the expected system status"* = $[p_1, p_2, p_3, \ldots, p_{13}]$
AND *"the actual system status"* = $[s_1, s_2, \ldots, s_6]$
THEN *"preliminary action"* = $o_i$ $(i = 1, 2, 3, 4)$

**TABLE 1: SENSOR DESCRIPTION AND STATUS**

| Sensor | Description | Status |
|---|---|---|
| 1 | robot status | operative (1), non-operative (2) |
| 2 | gripper status | empty(1), with part (2), collision occurred (3) |
| 3 | part status at Ba | no part (1), part in (2), wrong part (3) |
| 4 | part status at Bb | no part (1), part in (2), wrong part (3) |
| 5 | ASSM status | no part (1), part A in (2), part B in (3) |
| 6 | safety check | no intruder (1), intruder (2) |

TABLE 2: FAULT SCENARIOS AND THE CORRESPONDING PRELIMINARY ACTIONS

| Fault | Description | Corresponding Sensor Status | Desired Preliminary Action |
|---|---|---|---|
| 1 | robot malfunction | S1 = 2 | call operator |
| 2 | intruder entered | S6 = 2 | shutdown |
| 3 | part error | S3/S4 = 3 | further diagnostics |
| 4 | collision | S2 = 3 | further diagnostics |
| 5 | gripper slip | S2 = 1 | call operator |
| 6 | insufficient supplies | S3/S4 = 1 | call operator |

Based on the sublevel 3-TPN model, NN1 generates 13 different outputs, corresponding to 13 possible expected system status M(K), K=1, 2, ..., 13. Because each of the six representative fault scenarios in table 2 can happen at a different system state, specific fault scenarios are generated for each expected system state (see Table 3). These specific fault scenarios are the bases for simulating the actual system status and generating diagnostic rules.

In this research, one diagnostic rule was generated for each specific fault scenario listed in Table 3, resulting in 42 diagnostic rules. In addition, 13 normal operation scenarios, resulting in a "continue operation" action, were simulated for each expected system status. Each rule represented one pattern for the training/testing data set. The training/testing data set is a set of patterns. Each pattern consists of two parts : an input pattern and the corresponding output pattern. A typical pattern generated from a diagnostic rule would be $[p_1, p_2, \dots, p_{13}, s_1, s_2, \dots, s_6]$ and $[o_1, o_2, o_3, o_4]$.

Among the 55 patterns generated, 13 of them were used for testing the trained neural network and the rest were used as training data set. The 13 testing data patterns were randomly selected from each of the system states such that each state had one testing pattern.

## Training and Testing Phase

The neural network model (NN3) simulated in this research is a 3-layer perceptron network with backpropagation. As noted by Mercharvi & Kumara [13], this architecture is best suited for pattern matching/classification tasks, and the backpropagation algorithm has powerful convergence behavior as well as great potential for learning arbitrary patterns.

TABLE 3: DETAILED FAULT SCENARIOS FOR DIFFERENT SYSTEM STATES

| System state K | Description | Fault scenarios considered (referring to table 2) |
|---|---|---|
| 1 | pick up part A from Ba | (1) (2) (3) (6) |
| 2 | robot at position 1 | (1) (3) (4) (5) |
| 3 | transport part A to position 2 | (1) (2) (5) |
| 4 | robot at position 2 | (1) (4) (5) |
| 5 | place part A into fixture | (1) (2) (5) |
| 6 | part A in fixture | (1) (4) (5) |
| 7 | pick up part B from Bb | (1) (2) (3) (6) |
| 8 | robot at position 3 | (1) (3) (4) (5) |
| 9 | transport part B to position 2 | (1) (2) (5) |
| 10 | robot at position 2 | (1) (4) (5) |
| 11 | insert part B into Part A | (1) (2) (5) |
| 12 | assembly completed | (1) (4) (5) |
| 13 | robot available | (1) (2) |

In this research, the number of hidden nodes was varied in order to obtain a best suited neural network for NN3. The basic setting of the neural networks were selected as follows : a hyperbolic tangent function and a delta-bar-delta rule were used in the hidden and output nodes; the learning coefficient and the momentum term of the backpropagation algorithm were chosen as 0.5 and 0.4, respectively. NeuralWorks Explorer software by Neuralware Inc. was used to simulate the neural network. A 486-33 PC was used in this research. Four different 3-layer (input-hidden-output) perceptron networks (19-5-4, 19-10-4, 19-15-4, and 19-20-4) were simulated from 5000 cycle to 10,000 cycle in 1000 cycle increments. In addition, a 19-15-4 perceptron network with a sigmoid function was also tried but it took over 20,000 cycle with 220 sec. of training time to converge to a RMS error close to 0.15. The training results of these networks are summarized in Table 4 using representative training cycles of 5000, 7000, and 9000. All the networks exhibited convergence behavior during the training cycle and were able to classify the training data correctly (one exception being the 19-20-4 network with a 5000 training cycle having a classification rate of 0.92). Training time ranged from 40 sec. to 110 sec. with about 5 to 10 sec. increments for each five additional hidden nodes added to the network.

After the neural networks were trained, the testing data patterns were presented to test the trained networks. The actual output values generated from the trained networks were converted to 1 if the value was greater than or equal to 0.5 and 0 otherwise. The testing results for the four different networks are summarized in Table 4 (column 5). The test results showed that 19-15-4 perceptron outperformed other networks in the sense that only one misclassification out of 13 patterns (i.e. 7% of misclassification) resulted. One interesting result is that by increasing the hidden nodes from 10 to 15, the ability of the network to classifying new patterns improved. As the number of hidden nodes increased from 15 to 20, the ability of generalization got worse. This result indicated that too many hidden nodes may leave too much freedom for the weights to adapt during the training [14].

TABLE 4. SUMMARY OF FOUR NEURAL NETWORKS SIMULATION RESULTS

| Perceptron network * | Number of training cycle | Actual training time required (sec) | RMS error | % of misclassification from testing data |
|---|---|---|---|---|
| 19 -5 -4 | 5000 | 40 | 0.07 | 15 |
| | 7000 | 60 | 0.10 | 15 |
| | 9000 | 75 | 0.03 | 15 |
| 19 -10 -4 | 5000 | 45 | 0.15 | 23 |
| | 7000 | 65 | 0.08 | 23 |
| | 9000 | 85 | 0.05 | 15 |
| 19 -15 -4 | 5000 | 50 | 0.14 | 7 |
| | 7000 | 75 | 0.12 | 7 |
| | 9000 | 95 | 0.06 | 7 |
| 19 -20 -4 | 5000 | 60 | 0.14 | 31 |
| | 7000 | 85 | 0.09 | 23 |
| | 9000 | 100 | 0.05 | 23 |

* basic setting of neural networks : hyperbolic tangent function and delta-bar-delta rule with 0.5 of learning coefficient and 0.4 momentum

## CONCLUSION

A neural network-based approach for preliminary diagnostics in a manufacturing workstation has been presented. The proposed approach consists of three different neural networks. For NN1, a 3-layer PNN with an embedded T-gate threshold logic was constructed based on a hierarchical PN model. For NN3, 3-layer neural networks with different hidden nodes were investigated. The 19-15-4 perceptron network gave the lowest percent misclassification. The approach taken in this research represents preliminary but quite promising research. Future work will consider the following :

1. extend the general Petri net model in PNN to a higher level Petri net
2. investigate proper sensor fusion techniques to be used in a workstation and construct the NN2 for implementation
3. collect real processes data instead of using simulation data to test the proposed neural network for NN3
4. investigate different neural network architectures to gain more knowledge about the possibility of improving the performance of the network studied in this paper

## REFERENCE


1. Odrey, N. G. and Ma, Y. H., "Intelligent Workstation Control : An Approach to Error Recovery in Manufacturing Operations," Proceedings of the 5th International FAIM Conference, pp. 124 - 141, 1995.
2. Henderson, T., Shilcrat, E., "Logical Sensor Systems," J. of Robotic Systems, Vol. 1, No. 2, pp. 169-193, 1984.
3. Zhou, M. C., DiCesare, F., "Petri Net Synthesis for Discrete Event Control of Manufacturing Systems," Kluwer Academic Publishers, 1993.
4. Murata, T., "Petri Nets : Properties, Analysis, and Approaches," Proceedings of IEEE, Vol. 77, no. 4, April, 1989.
5. Zhou, M.C., DiCesare, F., Desrochers, A. A., "A Hybrid Methodology for Synthesis of Petri Net Models for Manufacturing Systems," IEEE Trans. on Robotics and Automation, Vol. 8, no. 3,pp. 350 - 361, 1992.
6. Hadjinicolaou, M. G., Abdelrazik, M. B. E., Musgrave, G., "Structured Analysis for Neural Networks Using Petri Nets," Proceeding of Midwest System on Circuit Control , pp. 770 -773, 1991.
7. Ahson, S. I., "Petri Net Models of Fuzzy Neural Networks," IEEE Trans. on System, Man, and Cybn., Vol. 25, no. 6 pp. 926 -932, 1995.
8. Ramamoorthy, P. A., Huang, S., "Implementation of Rule-Based Expert Systems for Time-Critical Application Using Neural Networks," Proc. IEEE Software Engineering, pp. 147-150, 1989.
9. Liu, C. S., "Real-time Control of Flexible Manufacturing Cells with Alternative Routings," Ph.D. Dissertation, Department of IMS&E, Lehigh University, 1992.
10. Krogh, B. H., Sreenivas, R. S., "Essentially Decision Free Petri Nets for Real-Time Resource Allocation," Proceedings of IEEE on Robotics and Automation, pp. 1005 -1011, 1987.
11. Fink, P. K., Lusth, J. C., and Duran, J. W., "A General Expert System Design for Diagnostic Problem Solving," IEEE Trans. on Pattern Analysis and Machine Intelligence, Vol. PAMI-7, No. 5, pp. 553-560, 1985.
12. Zurada, J., Graham, J. H., "Sensory Integration in a Neural Network-based Robot Safety System," the International J. of Human Factors in Manufacturing, Vol. 5(3), pp. 325 - 340, 1995.
13. Merchawi, N. S., Kumara, R. T., "Neural Networks in Continuous Process Diagnostics," ANN for Intelligent Manufacturing, Edited by Dagli, C. H., Chapman & Hall, London, pp. 435 - 461, 1994.
14. Sorsa, T., Koivo, H. N., Koivisto, H., "Neural Networks in Process Fault Diagnostics," IEEE Trans. on Systems, Man, and Cybn., Vol. 21, no. 4, pp. 815 - 825, 1991.

# NEURO-FUZZY CONTROL OF A ROBOTIC ARM

**WALLACE E. KELLY III**
*Department of Electrical Engineering*
*Texas A&M University, College Station, Texas 77843-3128*

**RAJAB CHALLOO, ROBERT MCLAUCHLAN, S. IQBAL OMAR**
*Intelligent Control Systems Laboratory*
*Texas A&M University – Kingsville, Kingsville, Texas 78363*

***ABSTRACT:***
This paper first presents a discussion of the reasoning and method for combining neural networks and fuzzy logic. The problem of moving a robotic arm in the presence of an obstacle is discussed. Several neuro-fuzzy controllers are trained using sample data obtained from a human's control of a robotic arm. Their performance is quantified and compared. It is shown that the definition of the fuzzy membership functions plays a significant role in the ability of the neuro-fuzzy controller to learn and generalize. Possible directions for future work are suggested.

## NEURO-FUZZY SYSTEMS

Recently, the combination of neural networks and fuzzy logic has received attention. Neural networks bring into this union the ability to learn, but also require an excessive number of iterations for training of complex systems. Fuzzy logic offers a system model based on membership functions and a rule base, but require an explicit stating of the IF/THEN rules.

Several methods for combining neural networks and fuzzy logic have been studied (Khan, 1993) (Lin and Song, 1994) (Nauck, et. al., 1993). In this paper, the authors implement the inference stage of a fuzzy system using a neural network (Challoo, et. al., 1994) (Keller, et. al., 1992). Figure 1 illustrates the system architecture for the described combination of neural networks and fuzzy logic. By replacing the rule base of a fuzzy system with a trainable neural network, complex input-output relationships can be achieved which can not be easily specified by IF/THEN rules. With fuzzification and defuzzification stages augmenting a neural

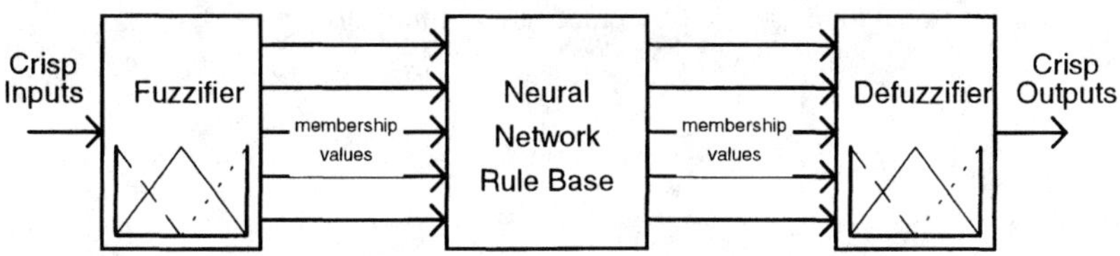


**Figure 1. A fuzzy system with neural network rule base**

network, significant improvements in the training time, in the ability to generalize, and in the ability to find minimizing weights can be realized. Furthermore, the fuzzy membership functions give the designer more control over the neural network inputs and outputs.

## THE ROBOTIC ARM PROBLEM

For robots to become effectively used in a wide range of applications, they must gain the ability to work in unpredictable environments. This paper addresses the problem of planning the trajectory of a three-link robotic arm in the presence of an obstacle.

The arm operates in two dimensions in an environment containing a randomly placed obstacle and goal. The starting position of the arm is arbitrary as well. For the purpose of designing the control system, the positions of the obstacle and goal and the joint angles of the arm are assumed to be available from position feedback sensors in the arm.

The arm is modeled as a three-link planar manipulator, as shown in Figure 2. The model is strictly geometric. That is, the dynamics of moving a finite mass arm are not considered for the purpose of this study. The controller will determine a series of joint angles, $\Theta(t)$, that move the end effector from a given starting position $(x_s, y_s)$ to a desired final position $(x_g, y_g)$ without colliding with the obstacle at $(x_o, y_o)$.

Previous approaches to this problem using fuzzy logic have focused on choosing the joint angles of redundant manipulators given a desired path (Kim and Lee, 1993) (Xu, et. al., 1993) (Wang, et. al., 1994) or specifying criteria for choosing those joint angles using fuzzy rules (Palm, 1992). Unlike these approaches, this work uses a fuzzy controller that *learns the strategy* for moving the arm to the goal position without touching the obstacle. That strategy is learned by observing a human's control of the arm.

The goal must be assumed to lie within the possible reach of the end-effector and the obstacle must be assumed to lie outside the path of the first link. The problem can be summarized as follows:

Given a robotic arm with:

- the current joint angles, $\Theta(0)$,
- an obstacle position, $(x_o, y_o)$, and
- a desired position, or goal, $(x_g, y_g)$;

Find a trajectory, $\Theta(t)$, such that:

- the end-effector reaches the goal,
- the arm does not touch the obstacle, and
- the calculations can be performed in real-time with current hardware.

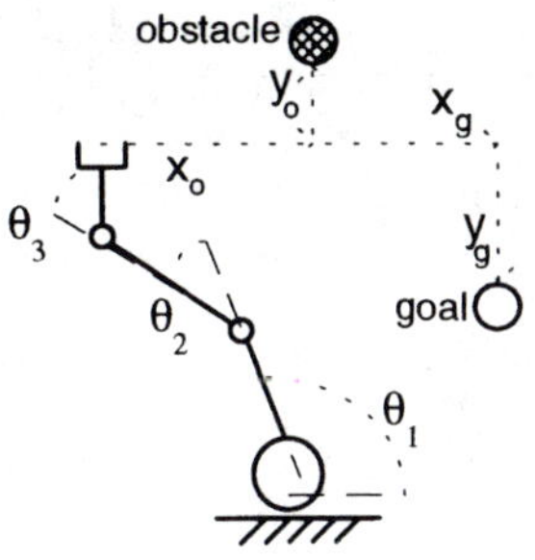


**Figure 2. The state representation for the goal-obstacle problem.**

## STATE REPRESENTATION

One of the primary considerations in the design of the neuro-fuzzy controller is the representation of the state of the system. The ability of the final neuro-fuzzy controller to generalize a solution from training data depends largely on the data representation scheme. The controller must be provided with the joint angles and the locations of the obstacle and goal.

There are actually several ways to provide information about the locations of the goal and obstacle. Providing Cartesian or polar coordinates of the two objects would result in seven, independent inputs. This scheme, however, would require that the controller repeatedly learn the same strategy for similar configurations of the end-effector, goal and obstacle if that configuration happened to occur in a different Cartesian region. If, on the other hand, the location of the goal and obstacle are represented by their relative distance to the end-effector, then strategies can be generalized for similar configurations regardless of where they occur in Cartesian space. Table 1 and Table 2 list the inputs and outputs that were chosen to represent the system.

## NEURO-FUZZY SOLUTION

As with most systems involving neural networks, training samples were required which demonstrated the desired input-output relationship. The samples were

TABLE 1: INPUTS TO THE NEURO-FUZZY CONTROLLERS

| Input | Description |
|---|---|
| $\theta_1$ | joint angle of link1 to the base |
| $\theta_2$ | joint angle of link2 from link1's axis |
| $\theta_3$ | joint angle of link3 from link2's axis |
| $x_o$ | horizontal distance to the obstacle |
| $y_o$ | vertical distance to the obstacle |
| $x_g$ | horizontal distance to the goal |
| $y_g$ | vertical distance to the goal |

TABLE 2: OUTPUTS OF THE NEURO-FUZZY CONTROLLERS

| Outputs | Description |
|---|---|
| $\Delta\theta_1$ | a change in the angle between link1 and the base |
| $\Delta\theta_2$ | a small change in the angle between link2 and the link1 axis |
| $\Delta\theta_3$ | a small change in the angle between link3 and the link2 axis |

obtained by cycling through a series of goal and obstacle positions on a simulation of the robotic arm. The simulated arm was manually maneuvered past the obstacle and to the goal. Goal and obstacle positions were taken from the entire reachable region.

As the training samples were recorded, each movement created a new input to the neuro-fuzzy system. Therefore, selecting starting positions of the arm at the two extremes of possible movements generated training samples that extended over the entire state space. Obtaining training samples over an entire range of possible state space is important for neuro-fuzzy design.

The recorded samples were used to train seven neuro-fuzzy controllers to emulate the strategies demonstrated in the human's control of the arm. Testing the controllers consisted of placing the goal and obstacle objects at random locations and assigning a random starting position for the arm. The controller simulation quantified the performance of each neuro-fuzzy controller by determining the percentage error in reaching the goal, detecting collisions, performing all the necessary calculations so that a judgment can be made about the possibility of real-time application.

## EXPERIMENTAL RESULTS

The training samples were used to train seven different neuro-fuzzy controllers, named FNN1, FNN2, ... FNN7 for reference. The controllers varied in the definition of their fuzzy membership functions and in the number of hidden nodes.

Figure 3 through Figure 9 show the seven FNN fuzzy membership functions. Notice the change in granularity and also the change in the concentration of the fuzzy values. With the increased granularity in the fuzzy sets, more hidden layer nodes were required. For example, FNN1 partitioned each of the seven inputs into three fuzzy sets yielding 21 input nodes. In this case, 35 hidden layer nodes were used. Figure 10 shows the final RMS training error which indicates which fuzzy membership function definitions were best able to emulate the training samples.

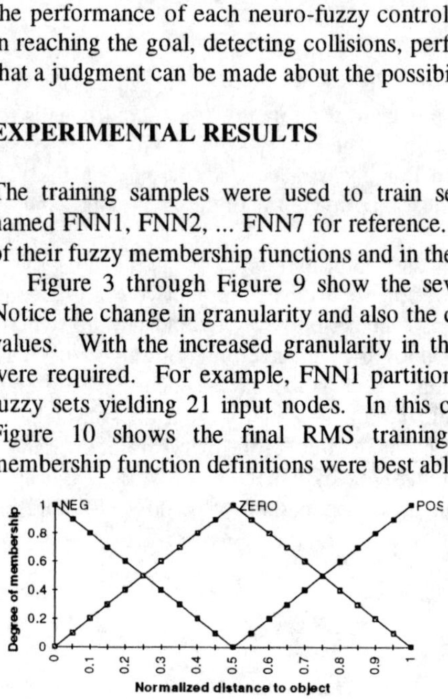


**Figure 3. FNN1 - 3 fuzzy sets per input and 35 hidden layer nodes.**

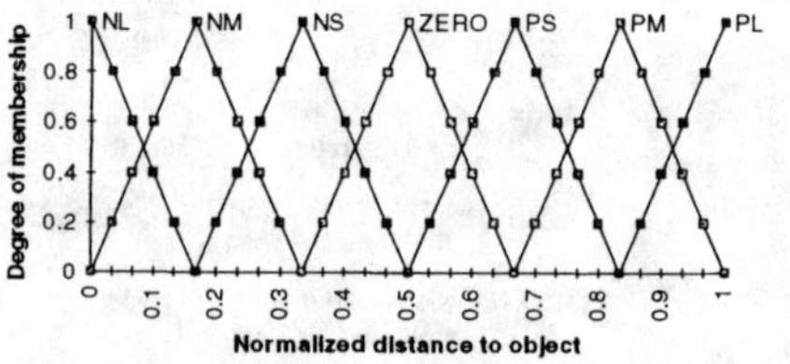


**Figure 4. FNN2 - 7 fuzzy sets per input and 65 hidden layer nodes.**

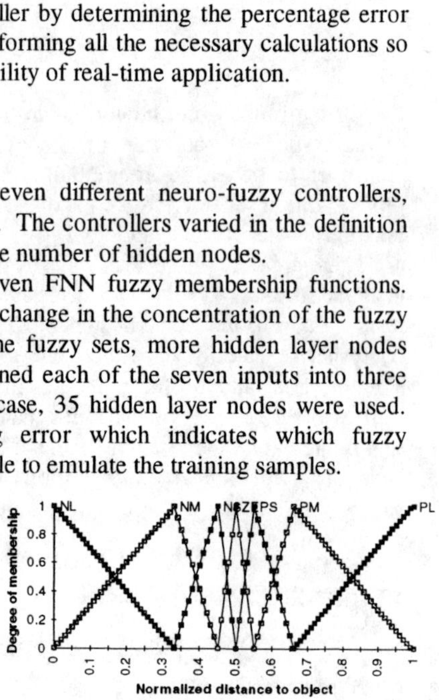


**Figure 5. FNN3 - 7 fuzzy sets per input and 65 hidden layer nodes.**

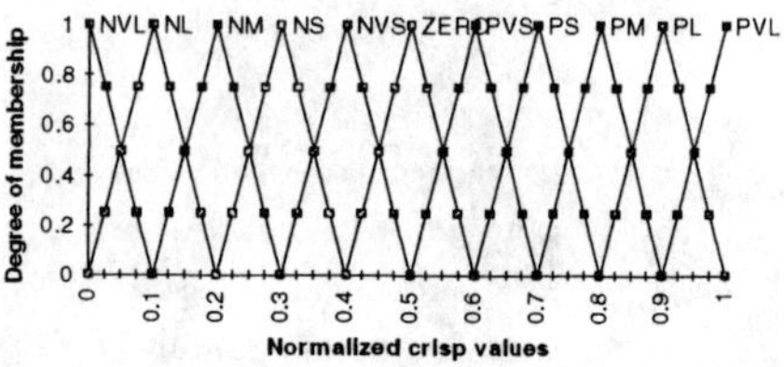


**Figure 6. FNN4 - 11 fuzzy sets per input and 85 hidden layer nodes.**

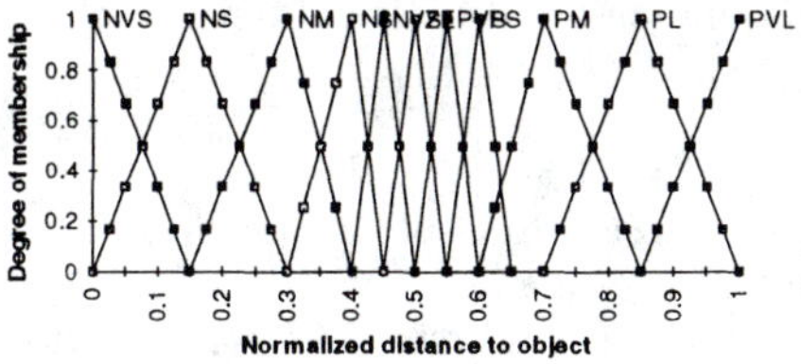


**Figure 7. FNN5 - 11 fuzzy sets per input and 85 hidden layer nodes.**

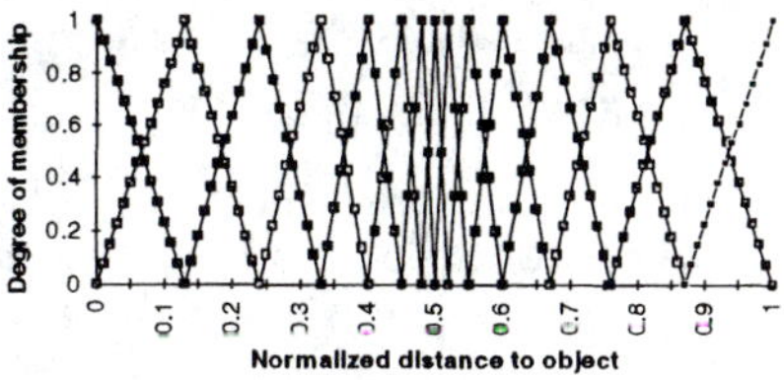


**Figure 9. FNN7 - 15 fuzzy sets per input and 100 hidden layer nodes.**

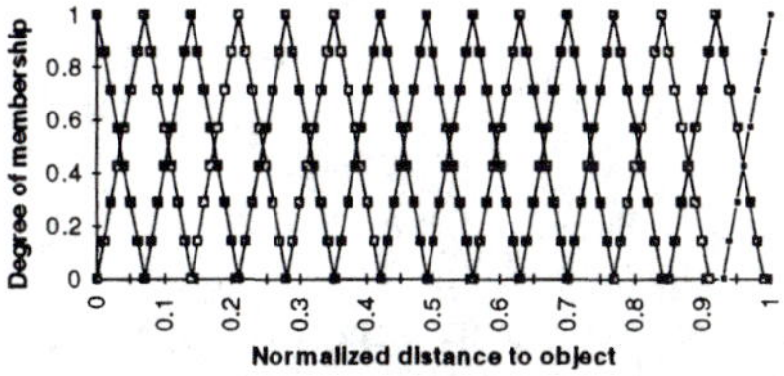


**Figure 8. FNN6 - 15 fuzzy sets per input and 100 hidden layer nodes.**

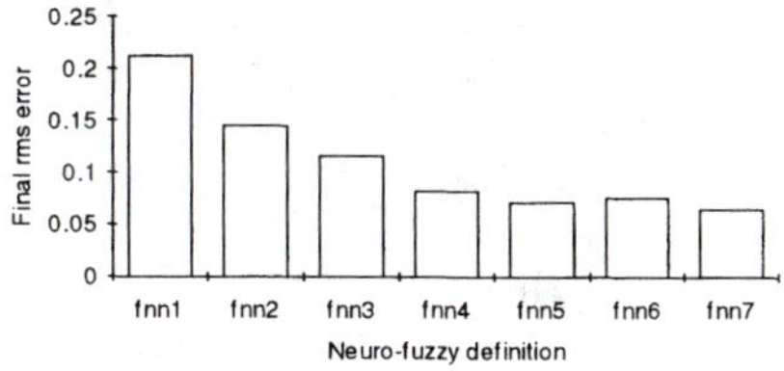


**Figure 10. A comparison of the final RMS training errors.**

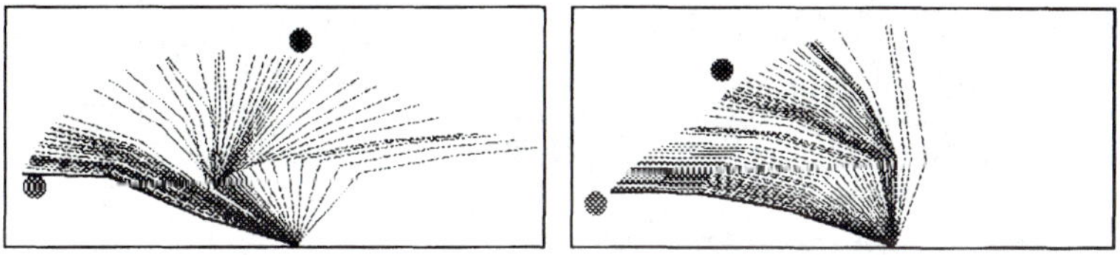

**Figure 11. Examples of successful control by the neuro-fuzzy controllers.**

Figure 11 shows two screen shots of successful control by the trained controller. Figure 12 compares the collision and success rate for the seven controllers. The controller was successful if the end-effector touched the goal and the arm never touched the obstacle. A collision occurred if any part of the arm touched the obstacle. The percentage of runs that resulted in the arm attempting to move past a physical constraint, and the percentage of runs that the arm did not reach the goal are not shown. Those two cases account for the remaining percentages.

Finally, many attempts were made to train a back propagation neural network to control the simulated arm. The neural networks without the fuzzification layer were

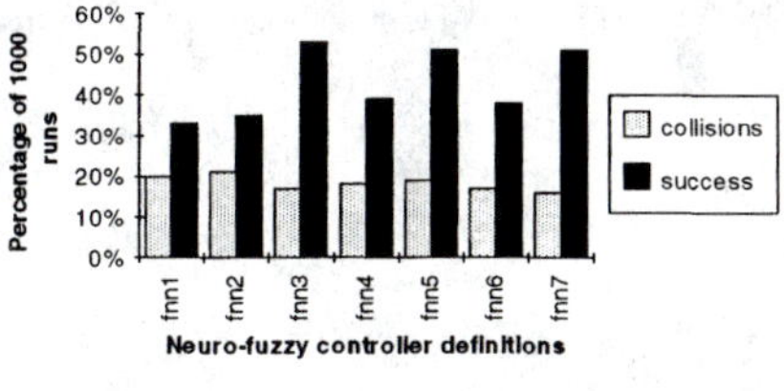


**Figure 12. A comparison of the performance of all the controllers.**

not able to generalize strategies from the sample data.

## CONCLUSIONS

Two conclusions can be drawn from the experimental results. First, the membership functions are an important part of the neuro-fuzzy system. Figure 10 shows that the final training error was reduced 50% by increasing the number of fuzzy values in a fuzzy set. Figure 12 shows that membership functions, defined with an important aspect of the problem in mind, consistently improved performance by more than 25%. Second, the fuzzification stage allowed the neural networks to learn more complex functions than a neural network without fuzzification.

The performance of the neuro-fuzzy controllers is less than perfect. Even the best controller had a collision rate of 17%. A large percentage of the failures resulted from the controller attempting to move past the physical joint constraints. Another situation that decreased performance occurred when the arm started oscillating between two points. A few heuristics in the control program could decrease these problems. The seven FNN controllers of this project were trained to 2150 training samples. Future work will determine the adequacy of the training samples. A study of the relationship of the training samples and the fuzzy membership functions would be particularly helpful.

The use of neuro-fuzzy systems for control has been examined. Fuzzification of a neural network's inputs and outputs will become a standard procedure in neural network applications.

## REFERENCES


Challoo, R., Clark, D., McLauchlin, R., and Omar, S. (1994). "A Fuzzy Neural Hybrid System," *Proceedings of the 1994 IEEE International Conference on Neural Networks*, pp. 1654-1657.

Keller, J., Yager, R., and Tahani, H., (1992). "Neural Network Implemention of Fuzzy Logic," *Fuzzy Sets and Systems*, Vol. 45, pp. 1-12.

Khan, E., (1993). "An Elegant Combination of Fuzzy Logic & Neural Nets", *Proceedings of Fuzzy Logic '93*, pp. A223-1 - A223-7.

Kim, S., and Lee, J., (1993). "Inverse Kinematics Solution Based on Fuzzy Logic for Redundant Manipulators", *Proceedings of the 1993 IEEE/RSJ International Conference on Intelligent Robots and Systems,* pp. 904-910.

Lin, J., and Song, S., (1994). "A Novel Fuzzy Neural Network for the Control of Complex Systems," *Proceedings of the 1994 IEEE International Conference on Neural Networks,* pp. 1668-1673.

Nauck, D., Klawonn, F., and Kruse, R., (1993). "Combining Neural Networks and Fuzzy Controllers" *Fuzzy Logic in Artificial Intelligence (FLAI93),* ed. Klement, Erich Peter and Slany, Wolfgang, pp. 35-46.

Palm, R., (1992). "Control of a Redundant Manipulator Using Fuzzy Rules," *Fuzzy Sets and Systems,* Vol. 45, pp. 279-298.

Wang, D., and Gu, M., (1994). "Fuzzy Logic Joint Path Generation for Kinematic Redundant Manipulators with Multiple Criteria," *Proceedings of the 1994 IEEE/RSJ/GI International Conference on Intelligent Robots and Systems,* pp. 649-656.

Xu, Y., and Nechyba, M., (1993). "Fuzzy Inverse Kinematic Mapping: Rule Generation, Efficiency, and Implemention," *Proceedings of the 1993 IEEE/RSJ International Conference on Intelligent Robots and Systems*, pp. 911-918.

# A MULTI-NEURAL NETWORK INTELLIGENT PATH- PLANNER FOR A ROBOT ARM

**JAMES P. JOHNSON, RAJAB CHALLOO, ROBERT A. MCLAUCHLAN, AND S. IQBAL OMAR**
*EE/CS Department, Box 192*
*Texas A&M University - Kingsville, Kingsville, Texas 78363*
*Tel. (512)595-2628 Fax (512) 595-2110*
*E-mail: R. Challoo@TAMUK.EDU*

***Abstract:***

A novel multi-neural network based control program written in C, has been applied to a robot arm. The control algorithm uses two Adaptive Resonance Theory II (ART2) neural network classifiers and a Bi-directional Associative Memory System (BAM System). The control enables obstacle avoidance, and on-line mapping of a robot's workspace. The controlling algorithm is considered a back-up system which could be used with other navigational sensory and control subsystems in a variety of applications

## INTRODUCTION

The practicality of multiple subsystems available in a complex assemblage, such as a robot, is the protection provided by a back-up system in case of primary control system failure, or the inability of a sensor to detect obstacles in various situations.

Many papers have been written on various methods of robotic path-planning, navigation, path optimization, and intelligent control. Recent path-planning methods have included those which tend to minimize a cost function in order to obtain an optimal path. The use of artificial potential fields is an area where a more complex sensing system must be included in the control to provide information about the locations of objects in the robot's work area (Ratering, 1993) (Chuang, 1993) (Abdalla, 1992). Configuration space mappings (delPobil, 1992), neural network control of robots (Sethi, 1990) (Henaff, 1994), fuzzy controllers and navigators (Cela, 1994) (Liu, 1994), and the many other types of robot path-planning systems for robots usually require the use of expensive sensing devices, or are solely simulations. Most papers written in these areas require vision or ultrasonic sensing for their systems. The sensing in the back-up system presented here is entirely in software. Although the idea of actual physical contact of the robot to its surroundings may not be a popular solution, as a secondary sensing system, in case of primary system failure, this less expensive sensing alternative may be more attractive than an additional clone back-up system.

Many papers have based their work entirely on simulation. Simulations can provide a means of working out problems in system design, but the even greater obstacles lie in realization. Thus the motivation for the design and implementation of this intelligent robotic control system and robot arm includes not only to develope a back-up system, but also to create an educational intelligent robotics test-bed where intelligent control experience can be gained through actual realization, not achievable in simulation analysis / design alone, or within safety limits required in the use of larger industrial robots.

Another consideration is processing time. In most of the previous applications cited, processing times were rather long in determining paths, or the necessity of great numbers of parallel processors were required. This sub-system has the potential, with a uniprocessor, to encode and recall path data in fractions of a second.

## BACKGROUND

Two ANNS, which are used in this intelligent control system, are the Bi-directional Associative Memory (BAM) and the BAM System, and the Adaptive Resonance Theory 2 (ART2) (Simpson, 1990) (Fausett, 1994) (Blum, 1992). For the sake of brevity, background information on these ANNS should be sought in the cited references.

## PATH-PLANNING SUBSYSTEM

In the subsystem presented, a vigilance parameter of the ART2s was chosen as a value which would allow approximately a 10% noise variance on each element of the input analog vector presented to the ART2's inputs before the ART2 would attempt to create a new class, for a previously known input. The input pattern presented to the ART2 consisted of the present and next positions the arm would

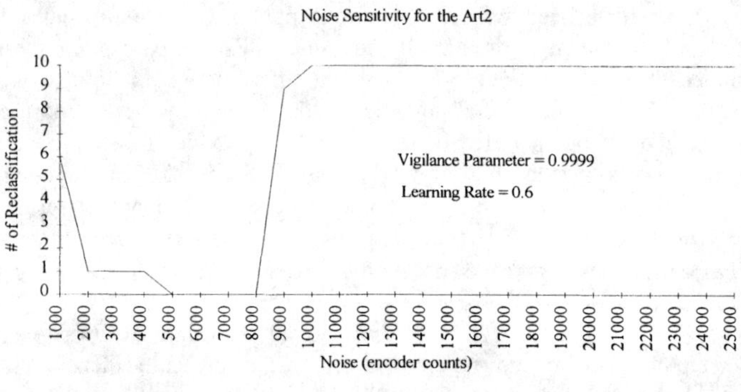


**Figure 1. Noise Sensitivity for Vigilance Parameter = 0.9999**

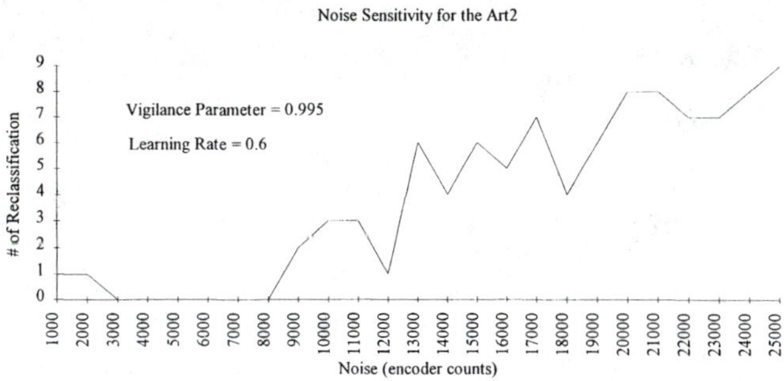


**Figure 2. Noise Sensitivity for Vigilance Parameter of 0.995**

assume, read from an input "move" file. If the elements of a vector presented to an ART2 were not within 10% of a previously known pattern, the ART2s used would not create new classes, but merely return a “no” to the calling process, determining that the upcoming move was never before known to cross an obstacle’s position. The workspace was divided before any moves were made, allowing movement in one direction to be considered forward, and the other reverse. A position pair is entered from a file which contains a list of object positions (which are to be moved) and final positions for each object (where the object is to be moved). The move is calculated considering the arm's current position and the object's position. If any software sensors of the arm are triggered, a test procedure insues, allowing the arm to proceed to its final destination and grasp (or release) the object to be retrieved or delivered. If an obstacle was encountered, the positions of the arm as it breached the obstacle are remembered, and the ANNS in the system updated at the end of the movement. An ART2 classifier receives the input file data and determines whether the move could be made by a conventional move algorithm or whether, in a previous move there had been an obstacle recorded in a path, similar to this move's path, and a sequence of BAM patterns must be followed to intelligently avoid the obstacle. If the classifiers determined that the BAMs must be used to make the move, knowing the direction, the initial position of the input data is fed into either the forward or reverse BAMs. The arm moves toward the goal, and does not stop its move unless an obstacle is present in the path. If an obstacle is present, as expected, the position of the arm at contact with the obstacle and the first recall of the BAM System are compared. If the two positions are within a reasonable distance, the BAMs assume control with the help of an interfacing algorithm which keeps the output of the BAM going into the input and being used in the position control provided to the robot arm. If the positions are not close enough it is assumed that the arm had encountered a different obstacle and the testing procedure must resume control. Because the outputs of the Bam System are directly applied to the inputs of the Bam System when it is in control, the recalled positions are exactly those encoded, unless an input or output pattern which has been encoded is not unique. If new obstacles are encountered at anytime during the move, testing procedures assume control until the new obstacle is breached, the positions are remembered, the ANNS are updated, and the process continues.

## IMPLEMENTATION

The system's success was based on the ability of the arm to learn the location of an obstacle in its workspace and then move around the obstacle in future moves. The controlling program was written in C. The motors of the robot were controlled via two 8-channel motor controller boards in a VME bus. The master of the VME was a MVME-147 with a 68030 processor. Power amplifiers and interface cards were built for the motors allowing pulse-width-modulated control and optical encoder feedback. Only two degrees of freedom were used in the system's analysis for simplicity, using only the shoulder's horizontal and vertical movements. The move file read by the robot in testing for results was a four position move file. The arm was to begin at the origin of a right-hand coordinate system, move 120 degrees horizontally and raise 17 degrees, and remaining at this vertical height move back to the origin horizontally, back to 120 degrees horizontally, and again back to the origin horizontally.

Without obstacles in the workspace, the arm moved freely between its points, and did not update the weights of the ANNS between moves.

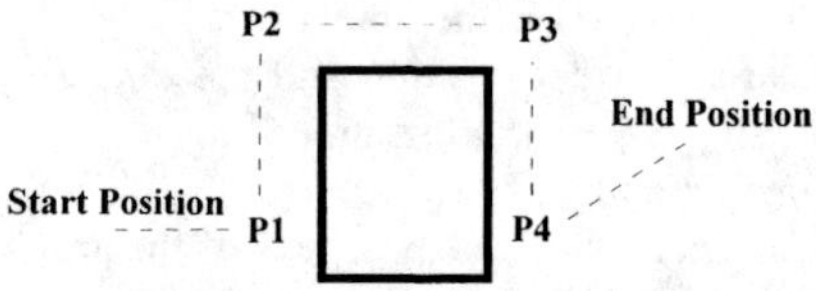


**Figure 3. Generalized Obstacle Avoidance Path**

A general idea of how the path was determined around the obstacle can be extracted from Figure 3. It was found that for the arm to successfully move around the box after the first movement in which the testing procedure was called (as a collision was made with the box), the points-of-success had to be carefully located by the testing procedure. The location of the points-of-success was the most crucial factor in this method. The desired locations of points-of-success found in the up-and-over testing algorithm used are shown in the generalized diagram of Figure 3 as P1 through P4. These encoded positions are the positions recalled during the BAM System recall procedure.

The success, at this time, is the ability of the arm to map a single obstacle and move around this obstacle in future movements, if the paths are recognized by the ART2 to be those in which a collision will be made with an obstacle.

The portion of the code which caused the greatest delay was the BAM encoding function. The delays for all other portions of the code were generally less than a second or two on the 68030. The times required to encode 5 pattern pairs in the BAM System on various machines is shown in the Table 1.

| TABLE 1: PROCESSOR SPEED | |
|---|---|
| Machine Type | Time to BAM Encode 5 pattern pairs |
| Motorola 68030 25 MHz | 16 seconds * |
| Sun SPARCStation 2 | 7 seconds * |
| Intel 80486 33MHz | 3 seconds * |
| Sun SPARCStation 10 | 1 second * |
| Sun SPARCStation 20 | 236 milliseconds (processor time) |

*real time ≈ processor time

## CONCLUSIONS

In conclusion, it was determined that this method of path-planning is a feasible approach. Feasibility measures include the speed of the robot's recall of the C-Space points to which the robot should move in order to avoid the obstacle and complete its desired move. Speed of the robot's recall was based on a few factors. The major factors considered here included the controlling program and the speed of the processor used. The speed considerations of the controlling program included the absolutely necessary calculations needed to implement the system, the language(s) used to write the program, and the knowledge of what was and what was not efficient code. The other main consideration, the computer system speed, had a much more significant effect on the runtime. The difference between using a 80386 and a 80486, or a 68030 and 68040 were substantial. In this system, a 68030 was used. With a 68040, the run-time of the controlling program would be reduced to less than one half of the time required on the 68030.

## FUTURE WORK

Future work which could be done with this subsystem includes 1. A more detailed and systematic testing procedure, 2. The program could be altered in such a way that when it is determined that a previously known obstacle is no longer present, the ART2 and the BAM System could remove the encoded classifications, 3. The testing procedure could be altered so that the points-of-success would be remembered for an object in such a way that they lie in positions which could be connected by a smoother trajectory using B-Splines.

## REFERENCES


Abdalla, M., *Fuzzy Logic Controller Navigator Design and Simulation.*1992 Texas A&I University Thesis.

Blum, A., *ANNS in C++, An Object-Oriented Framework for Building Connectionist Systems.* John Wiley and Sons, Inc., New York,1992.

Cela, A., et al., *Robust Fuzzy Logic Controller for Trajectory Tracking of Robotic Systems.* 1994 . IEEE International Conference on Systems, Man, and Cybernetics, (pg. 459.), The IEEE, Inc., 1994.

Challoo, R., Johnson, J., et al., *Intelligent Control of a Stanford JPL Hand Attached to a 4 Degree of Freedom Robot Arm.* 1994 IEEE International Conference on Systems, Man, and Cybernetics, The IEEE, Inc., 1994.

Chuang, J.H., *Potential-Based Modeling of Three Dimensional Workspace for Obstacle Avoidance.* IEEE International Conference on Robotics and Automation, (pg. 19.), IEEE Computer Society Press, Los Alamitos, California, 1993.

del Pobil, A. P., et al., *A New Representation for Collision Avoidance and Detection.*IEEE International Conference on Robotics and Automation, (pg. 246.), IEEE Computer Society Press, Los Alamitos, California, 1992.

Fausett, L., *Fundamentals of Neural Networks.* Prentice Hall, Englewood Cliffs, New Jersey, 1994.

Henaff, P., et al., *Adaptive Nerual Control in Mobile Robotics: Experimentation for a Wheeled Cart.* 1994 IEEE International Conference on Systems, Man,and Cybernetics, (pg. 1139.), The IEEE, Inc., 1994.

Liu, K., and Lewis, F.L., *Fuzzy Logic-Based Navigation Controller for an Autonomous Mobile Robot.* 1994 IEEE International Conference on Systems, Man, and Cybernetics, (pg. 1782.), The IEEE, Inc., 1994.

Ratering, S., and Gini, M., *Robot Navigation in a Known Environment with Unknown Moving Obstacles.* IEEE International Conference on Robotics and Automation, (pg. 25.), IEEE Computer Society Press, Los Alamitos, California, 1993.

Sethi, I.K., and Yu, G., *A Neural Network Approach to Robot Localization Using Ultrasonic Sensors.*Proceedings 5th IEEE International Symposium on Intelligent Control 1990, (pg. 513.), IEEE Computer Society Press, Los Alamitos, California, 1990.

Simpson, P. K., *Artificial Neural Systems.* Pergamon Press, Inc., New York, 1990.

# CMAC HARDWARE DESIGN WITH VERILOG SIMULATION

**C.-Y. WU and C.-S. LIN**
*Department of Electrical Engineering*
*University of Missouri-Columbia*
*Columbia, MO 65211*
*ecelin@mizzou1.missouri.edu*

## ABSTRACT

This paper presents a pipelined CMAC hardware architecture. The hardware consists of three major portions. In the first stage, state values are mapped into sets of *block numbers*. Each set of block numbers together with a *layer number* are used as an address for retrieving distributively stored data from a memory in the second stage. Memory contents retrieved are added up in the last stage. Software written in Verilog was developed to simulate the operation of the design.

## INTRODUCTION

Although there have been a lot of studies on CMAC [1-5], very few considered the problem of hardware realization. Most were on the fundamental study, the improvement of CMAC structures and applications. Most studies were done through software simulation or with the use of a general-purpose computer. Drawbacks by using a computer include the low speed, low efficiency, and the need of a CMAC expert in designing a CMAC-based system. These drawbacks can be eliminated if a special-purpose unit is available. Being motivated by these issues, we have been working on the development of a CMAC hardware model. In this paper, we will first briefly introduce the CMAC technique and then present the CMAC hardware architecture. The design has been verified through Verilog [6] simulation.

## INTRODUCTION TO CMAC

In the CMAC technique, each state variable is quantized and the problem space is divided into discrete states. A vector of quantized input values specifies a discrete state and is used to generate addresses for retrievmng output information from memory for this state. Information for a state is distributively stored in the memory. Figure 1 illustrates the scheme. This example has two input variables with each quantized into five discrete regions, called *blocks*.

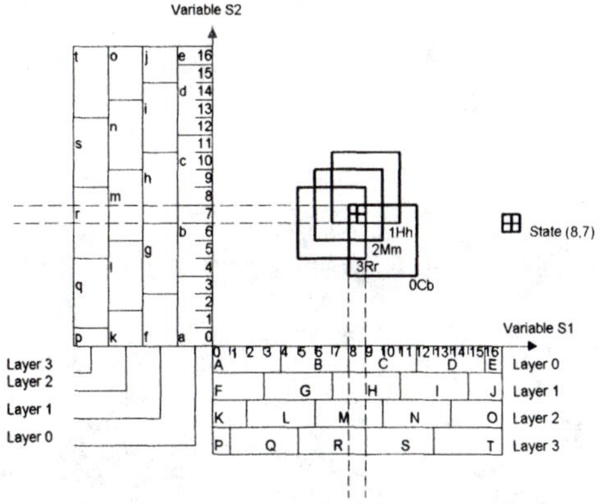


Figure 1. Block Division of CMAC for a Two-Variable Case

For instance, *s1* can be divided into A, B, C, D and E and *s2* divided into a, b, c, d, and e. Areas formed by quantized regions are called hypercubes; for instance, Aa, Ab, Ce, etc. are *hypercubes*. By shifting each variable a small interval (called an *element*), different hypercubes can be obtained. For example, F, G, H, I, J in the *2nd* row for *s1* and f, g, h, i, j in the *2nd* column for *s2* are possible shifted regions. Fh, Fi, Hg, *etc*. are new hypercubes from the shifted regions. We restrict that hypercubes are formed by the corresponding quantizations, *e.g*. row *p* with column *p*. With this kind of composition, one can imagine that there are $N_e$ layers of hypercubes, where $N_e$ is the number of elements in a complete block. The *p*th layer is divided into hypercubes using the *p*th row division and the *p*th column division. Each state is covered by $N_e$ different hypercubes, one from each layer. The CMAC associates each hypercube to a physical memory element. Information for a discrete state is the sum of contents of memory elements associated with the hypercubes that cover this state. Information can be created through learning. The target value must be provided and compared to the retrieved one. The difference is divided by $N_e$ and added back to each of the $N_e$ involved memory elements. The target values must be provided externally by a teacher or a learning unit that decides necessary updating.

## A CMAC HARDWARE ARCHITECTURE

In the section, we present the proposed hardware architecture. We will first introduce the overall diagram and then the diagram for each submodule.

*Basic Functions and the Block Diagram*

The proposed architecture consists of three major portions: (1) block numbers and physical memory address generation, (2) memory contents retrieval, (3) hypercube data summation. The considered module in this study has four 6-bit inputs, one 20-bit output and uses eight layers (*i.e.*, $N_e$=8). Each input variable is quantized into 58 elements with eight elements in each complete block. Figure 2 shows a block diagram for the module. At first, input values X1[5:0], X2[5:0], X3[5:0] and X4[5:0] (one for each input variable) are used to create block numbers C1[2:0], C2[2:0], C3[2:0] and C4[2:0], respectively, in the first stage (block numbers and memory address generation). The block numbers, one from each input variable, are put together with the layer number L[2:0] to form a physical memory address. The address is used to extract a datum from the memory for the hypercube in layer L. Memory contents for different layers are retrieved sequentially in the pipeline structure. Finally, the last part adds up all data in the pipeline and sends the sum to the output. In this early stage of study, we emphasize on data retrieval and assume that training is performed off-line.

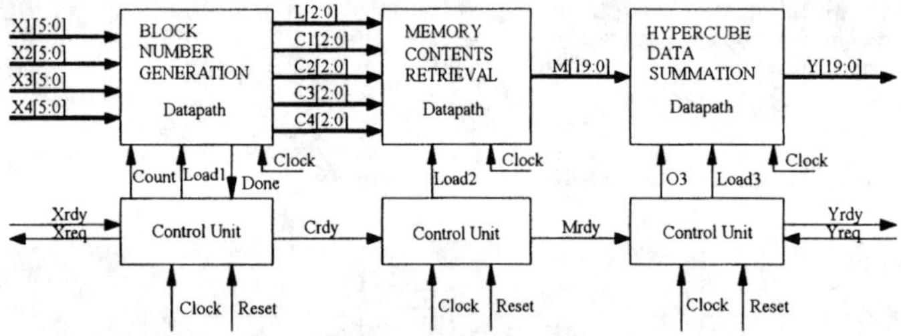


Figure 2. System Diagram of the Pipelined CMAC

*Block Numbers and Memory Address Generation*

X1[5:0] to X4[5:0] in the first stage as shown in Figure 3 denote four 6-bit inputs. The four block numbers, of which each is for one input (also see Figure 1), are generated for one layer at each clock click. These block numbers are expressed in 3 bits as C1[2:0] to C4[2:0]. L[2:0] from the counter denotes the layer number. C1 through C4 and L specify a hypercube. Xrdy and Xreq for the control unit (see Figure 2) are handshaking signals for synchronization with external devices (probably sensors). Xreq is one-bit signal that requests data from the outside system. Xrdy from the external system indicates that data are ready at the input buses. Crdy is similar to Xrdy except that it is the output data ready signal for the next stage. The four quantized input variables are loaded into registers (regBN1, regBN2, regBN3, regBN4) at the clock click when Xreq is 1 and the counter output is 0. On the other hand, the 3-bit counter (lgen) generates a layer number, which is between 0 and 7. This layer number is also sent to adders (codeBN1, codeBN2, codeBN3, codeBN4). Adders implement the following operation and calculation, and decode the block numbers for the next stage:

$$B_n = \textit{Integer part of}\ \frac{X + L_n}{N_e} \tag{1}$$

where $B_n$ : the block number for $L_n$ ($L_n = 0, \ldots 7$)
$X$ : a quantized input (0~57)
$L_n$ : the layer number (0~7)
$N_e$ : the number of elements for each block (8 in this design)

The division in the above equation is implemented by getting the three most significant bits from the output of the 6-bit adder.

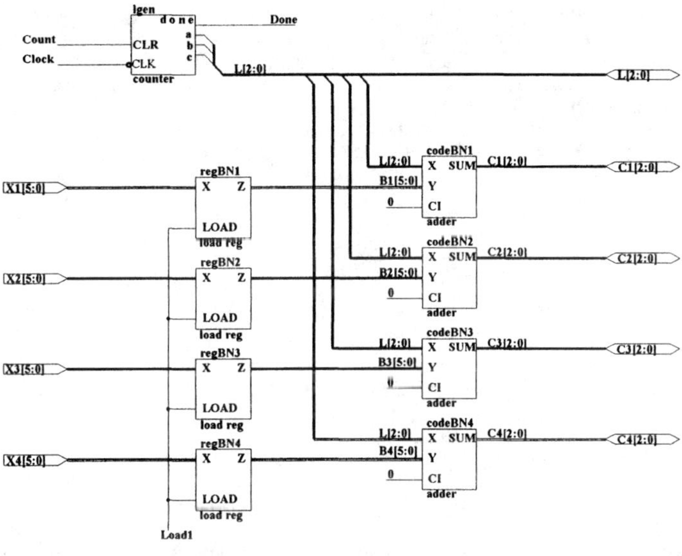


Figure 3. Stage 1- Block Numbers and Memory Address Generation

*Memory Contents Retrieval*

The diagram for the data path for this stage is shown in Figure 4. The five input buses (L[2:0], C1[2:0], C2[2:0], C3[2:0], C4[2:0]) are connected to load registers (regL, regBN1, regBN2, regBN3, regBN4), which will store the layer number and the block numbers generated from stage 1. The outputs of the load registers (Layer[2:0], E1[2:0], E2[2:0], E3[2:0], E4[2:0]) are used as an address for retrieving a stored hypercube datum from a ROM or an EPROM. The output is sent to M[19:0]. The size of the memory is 32k. The weights should be generated by a CMAC learning program.

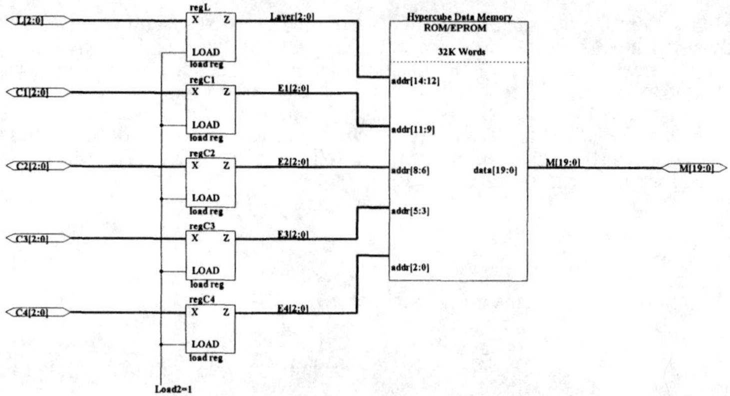


Figure 4. Stage 2 - Memory Contents Retrieval

*Summation of Distributed Data*

All the extracted memory contents must be added up. It can be done in a sequential order. Figure 5 shows the design. The partial sum is latched and fed back to the adder. Additions can be completed in 8 clock cycles. Comparing to the use of general purpose computer (4 or more cycles for each instruction for typical Intel microprocessors), the speed will be much faster. The 3-stage CMAC hardware is capable of performing necessary mapping if needed information has been learned/created and loaded into the memory. However, if the on-line learning capability is desired on the CMAC module, the structure must be modified.

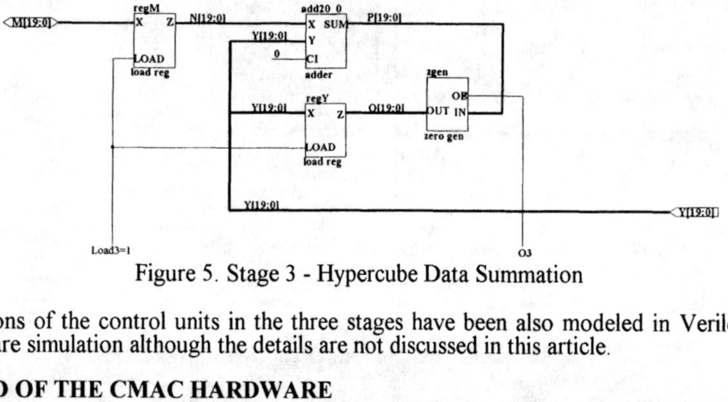


Figure 5. Stage 3 - Hypercube Data Summation

Functions of the control units in the three stages have been also modeled in Verilog for hardware simulation although the details are not discussed in this article.

## SPEED OF THE CMAC HARDWARE

The speed of the CMAC hardware is determined by the propagation delay in basic components. Table 1 shows the evaluated delay times for the three stages. These times are evaluated based on the assumed delay times for basic gates NOT, NAND, and NOR (~5ns in our simulation models). While the longest one among the three stages is 106 ns, it is reasonable to set the clock period to 110 ns. The first CMAC output will come out in a period of 11 clocks for the pipeline structure. Then the new output will be generated in every 9 clocks. Simulations have been done using Verilog to verify the expected operation at the suggested clock rate. However, the detail simulation results and the complete program are too long to be listed in this paper. We select the Verilog module for stage 1 and list it in Appendix A to provide additional information for the readers who are interested in more details regarding the simulation.

Table 1. Propagation Delays of the Three Stages

| Stage | Propagation Delay (the worst case) |
|---|---|
| Stage 1-Block Number Generation | 101.5 ns |
| Stage 2-Data Memory Retrieval | 106.0 ns |
| Stage 3-Hypercube Data Summation | 98.5 ns |

## OUTPUT ACCURACY OF THE HARDWARE CMAC

The CMAC module uses the fixed-point calculation. This will limit the precision in approximating a function or modeling a system. Simulation has also been performed to evaluate the output accuracy. The accuracy of the CMAC hardware is compared to that generated from software learning. The following nonlinear equation has been used in evaluation of the accuracy:

$$Y = \sin(tx1) + \cos(tx2) + \sin(tx3) + \cos(tx4)$$

$$tx_i = x_i * \pi \qquad \text{with } x_i \in [-1,+1] \tag{2}$$

$$x_i = (2 * \frac{srand(seed)}{RAND\_MAX} - 1) / 4$$

To generate random patterns for CMAC training and testing, RAND_MAX and srand(), two functions of the C language, have been used for computing $x_i$. The srand() function generates the random numbers and RAND_MAX is the maximum value that can be returned by srand(). In the simulation, the weights were generated by training 1000 patterns for 20 epochs. Table 2 shows the results for 4 random test patterns. The table includes the target value from (2), the output value from software learning ($Y^+$) and the hardware output ($Y^*$). The difference between the output from the software learning and the hardware output is around 5%. Such error is acceptable for applications in most learning control problems.

Table 2 Output Accuracy

| Patter No. | X1 | X 2 | X 3 | X 4 | Target | Software | Hardware |
|---|---|---|---|---|---|---|---|
| | | | | | $Y$ | $Y^+$ | $Y^*$ |
| 0 | 25 | 32 | 29 | 31 | 1.626 | 1.656 | 1.531 |
| 1 | 35 | 29 | 34 | 30 | 3.279 | 3.277 | 3.156 |
| 2 | 33 | 21 | 35 | 31 | 2.908 | 2.935 | 2.813 |
| 3 | 22 | 24 | 29 | 24 | 1.348 | 1.260 | 1.219 |

## DISCUSSION AND CONCLUSION

Most previous studies on CMAC were done by simulation or with the use of a computer. The proposed special-purpose architecture makes the design of CMAC-based system easier and increases the processing speed. Additional merits of the architecture include expandability and flexibility. If there are more than four inputs, additional Block Number Generators must be used. One consequence is that the number of address line for the next stage will increase and hashing circuits may be required. If there is less inputs, the unused memory in the block number generation portion can be simply set to zero. If multiple outputs are required from the CMAC, multiple memory modules and adders can be used. The address generation portions do not need to be changed. Using a smaller number of quantized blocks reduces the memory size but provides less output precision. Sometimes such quantization is desired. The structure can be modified for a smaller number of quantized blocks. With a smaller number of elements, the number of layers is reduced. Not all 8 clock periods are necessary. This can be done by comparing the counter content in Figure 5 with the number of elements (hard wired). If the counter content is the same as the number of elements, the counter is reset at the beginning of the next clock cycle.

## REFERENCES


1. Albus, J. S., "A New Approach to Manipulator Control: The Cerebellar Model Articulation Controller (CMAC)," Journal of Dynamic Systems, Measurement, and Control, Transactions of ASME, p.220-227, September 1975.
2. Albus, J. S., "Data Storage in the Cerebellar Model Articulation Controller (CMAC)," Journal of Dynamic Systems, Measurement, and Control, Transactions of ASME, p.228-233, September 1975.
3. Miller, T.W., Hewes, R.P., Glanz, F.J., and Kraft, L.G., "Real-Time Dynamic Control of an Industrial Manipulator Using a Neural-Network-Based Learning Controller," IEEE Transactions on Robotics and Automation, vol. 6, no. 1, p.1-9, Feb, 1990.
4. Wong, Y. and Sideris, A. "Learning Convergence in the Cerebellar Model Articulation Controller," IEEE Transactions on Neural Networks, vol. 3, no. 1, p. 115-121, Jan 1, 1992.
5. Lin, C. S. and Chiang, C. T., " Theoretical Foundation for CMAC Technique," Proceedings of ANNIE '93, St. Louis, MO, p. 11-19, November 14-17, 1993.
6. Thomas, D.E. and Moorby, P. R., *The Verilog Hardware Description Language*, Kluwer Academic Publishers, 1991.


## APPENDIX A. Verilog Module for Stage 1

The Verilog module for stage 1 is shown to provide some insight into the simulation software. NAND, NOR and NOT are the three basic gates used to build up all components. Verilog codes for components regBN's, codeBN's, etc. are not listed here.

```
//STAGE 1 – Block Number Generation
module stage1 (Xreq,Crdy,C1,C2,C3,C4,L,Xrdy,X1,X2,X3,X4,Clock,Reset);
//Parameter Declaration
    reg      [1:0]  Q, NQ;                   //Q is the present state and NQ is the next state
    input           Xrdy, Clock, Reset;      //Control input lines
    input    [5:0]  X1, X2, X3, X4;          //Input buses
    output          Crdy, Xreq;              //Handshaking lines
    output   [2:0]  C1, C2, C3, C4, L;       //Output buses
    wire     [5:0]  B1, B2, B3, B4;          //Buses for connection
    wire            Kout1, Kout2, Kout3, Kout4, Done, Load1, Count; //Wires for connection
//Data Path
eldreg_6  #(6,6,6,6,6,6,6,6,6,6,6) regBN1 (B1,, X1, Load1, Clock), regBN2 (B2,, X2, Load1,
Clock),
                                      regBN3(B3,, X3, Load1, Clock), regBN4 (B4,, X4,
Load1, Clock);          code_6t3 #(2,2,2,2,2,2,0) codeBN1 (C1, Kout1, B1, L, 1'b0), codeBN2
(C2, Kout2, B2, L, 1'b0),
                              codeBN3 (C3, Kout3, B3, L, 1'b0), codeBN4  (C4, Kout4, B4, L,
1'b0);
lgen      #(26,26,26,26,26,26) L3_7 (L, Done, Clock, Count);
//Behavioral Description of the Control Unit
always
  begin
    @(negedge clock)
        begin
          if(Reset==1) NQ=0;
          else if(Q==0) if (Xrdy==1) NQ=1;
                             else NQ=0;
        else if(Q==1) if (Done==1) NQ=0;
                             else NQ=1;
        end
    #50 Q = NQ;
  end
  assign      //default value is zero
    Load1 = (Q==0),     Count = (Q==0),
    Crdy  = (Q==1),     Xreq  = (Q==0);
endmodule
```

# A NEURAL NETWORK PREDICTOR FOR REENGINEERING OF RESIN MANUFACTURING

**YU-TO CHEN, PRATAP KHEDKAR**
Information Technology Lab, GE R&D, Schenectady, NY 12301

**MAARTEN P. TER WEEME**
GE Plastics, Mt. Vernon, IN 47620

**ABSTRACT:**
We describe the design and testing of a neural network for chemical process prediction and its use for process optimization. Our objective is to design an improved, robust, accurate, and adaptive model for Resin Melt Index prediction, leading to a better prediction of addition of chain stopper during polymerization. We synthesize a neural network to predict the melt index and optimize the chain stopper, while providing learning capabilities to account for process mean drifts as well as accommodating future changes in the lab analysis. We show that this approach results in a predictor that is superior to the existing nonlinear regression approach, as measured by a number of performance indices.

## INTRODUCTION

Our objective is to design an improved, robust, accurate, and adaptive model for predicting the Melt Index of a resin, leading to a better prediction of addition of chain stopper during polymerization. This is expected to lead to a higher plant rate and better product quality. For a typical plant, even a 1% increase in annual throughput could amount to millions of pounds per year [1].

A nonlinear regression predictor is currently in use for this purpose. However, it is incapable of learning automatically, and needs to be redesigned manually. The predictor uses the analysis of the incoming raw material, a solution of monomer and solvent, to determine the anticipated melt index, which is used iteratively by an optimizer to determine the amounts of chainstopper, monomer and co-monomer, that should be added to achieve the targeted melt index.

This paper is organized as follows. We describe the overall system architecture for the reengineered melt index prediction. We then briefly describe data acquisition, followed by the neural network training issues, including the network structure, the fuzzy accelerator, and the performance indices. After that we present the preliminary results and improvements of the neural network prediction based on the principle of *divide-and-conquer*. Finally, concluding remarks and avenues for further development are outlined.

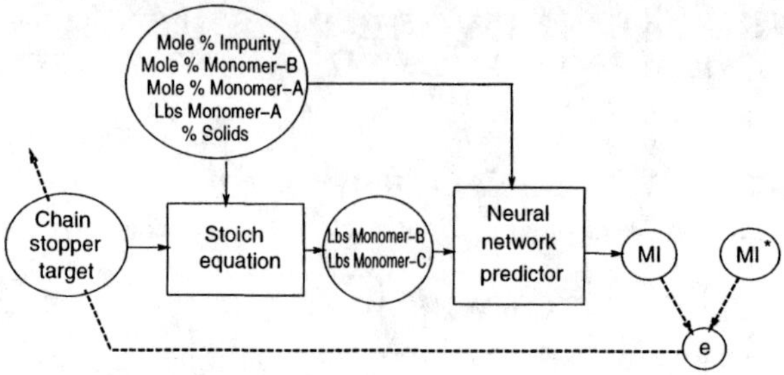


Figure 1: Schematic for the prediction.

## SYSTEM ARCHITECTURE

The overall scheme for the Resin melt index prediction is shown in Figure 1. It consists of two principal components: stoichiometry equations and neural network predictor. The stoichiometry equations take six inputs and gives *lbs of Monomer-B added* and *lbs of Monomer-C added* as output. Its six inputs are: Mole % total Impurity, Mole % total Monomer-C, Mole % total Monomer-A, Lbs of Monomer-A, % Solids, and $CST$ (Chain Stopper Target.)

The neural network predictor uses seven inputs and outputs *MI*, the predicted melt index. The seven inputs are: Mole % total Impurity, Mole % total Monomer-C, Mole % total Monomer-A, Lbs of Monomer-A, % Solids, Lbs of Monomer-B added, and Lbs of Monomer-C added. The learning techniques and NN architecture is described in detail later.

Given an initial guess of $CST$ and a targeted $MI^*$, the system iteratively updates $CST$ so as to minimize the absolute error between *MI* and $MI^*$. After *MI* and $MI^*$ agree to within a tolerance, say $10^{-6}$, the system gives $Monomer - B^*$ and $Monomer - C^*$ as final outputs.

The above process can be formulated as an optimization problem :

$$minimize \;\; F = |\, y - y^* \,| \;\; s.t. \;\; y = G(H(x)) \;\; and \;\; x = [a, b]$$

where $y$ and $y^*$ are the actual and targeted *melt index*, respectively, and $x$ is the *chain stopper target* which is bounded by two real numbers, $a$ and $b$, and $F$, $H$ and $G$ are the objective function, the stoichiometry equation, and the neural network predictor, respectively. In essence, we try to minimize the difference between $y$ and $y^*$. The latter is the target melt index and the former is the melt index derived from our system. Note that $a$ and $b$ are set to 2.0 and 4.5 in our present application. Also note that we assume that the values of *mole % Impurity, mole % Monomer-C, mole % Monomer-A, lbs of Monomer-A* and *% solids* are constants.

Note that the above is equivalent to inverting the function $G(H(.))$. However, this function is **not** monotone, and hence does not have a unique inverse. Even the $H()$ function, which is modeled by a neural net, is not invertible uniquely, so its inverse cannot be directly learnt by a supervised learning process. That is why the optimization uses a root-finding approach and an initial guess, since a sufficiently different initial guess can conceivably lead to a different optimal solution.

Golden section search is employed to solve the above minimization problem [2]. This is similar to the bisection method for root-finding, and it iteratively brackets a minimum by a triplet of points until the distance between the two outer points of the triplet is tolerably small, say $10^{-6}$.

## DATA ACQUISITION

The data we used here were obtained from actual Resin batch formulations. There are 328 patterns in the data set, each of which has eight fields: Mole % Impurity, Mole % Monomer-A, % Monomer-C, %Solids, Lbs of Monomer-A added, Lbs of Monomer-B added, Lbs of Monomer-C added, and Melt Index (MI). In each pattern, the first seven are inputs, and the last is the predicted variable.

The data set was divided into two parts: a training set consisting of the first 200 patterns in the time series, and a testing set consisting of the remaining 128 patterns. The reason we did not randomly shuffle the data set is twofold. On one hand, we expect the neural network to be able to cope with the short term fluctuations or gradual drifting of input/output variables. On the other hand, this division implies there is a periodic training of the neural network for every 200–300 batches.

## TRAINING ISSUES

In this section we describe the proposed neural network architecture, the training method used, and the performance indices for the prediction.

A three-layer feed-forward neural network is constructed for melt index prediction. The network architecture is $7 \times 7 \times 1$: seven input nodes, seven hidden nodes, and one output node. The activation function for both hidden and output nodes is a sigmoid.

The training method employed in this paper is error backpropagation [3]. The weight update equation is as follows:

$$W_{ij}^{s+1}(l) = W_{ij}^{s}(l) - \eta \frac{\partial E}{\partial W_{ij}^{s}(l)} + \alpha \left[ W_{ij}^{s}(l) - W_{ij}^{s-1}(l) \right]$$

where $W_{ij}(l)$ is the weight between the $i$th neuron at the $l$th layer and the $j$th neuron at the $(l-1)$th layer, $E$ is the sum of squared error between target and actual output, $s$ is the iteration step, $\eta$ is the learning rate and $\alpha$ is the momentum coefficient. In addition, the choice of activation function, $f(x)$ could also influence the learning process: $f(x) = 1/(1 + \exp(-2\beta x))$, where $\beta$ is the steepness parameter of the activation function.

The Fuzzy Accelerator [4] used here is essentially a fuzzy inference system which adaptively tunes $\eta$, $\alpha$, and $\beta$ simultaneously after each weight updating iteration. This has the effect of creating a "smart" way to update the learning rate, momentum term, and the sensitivity of the neural network to small perturbations around the nominal input. This speeds up the learning times substantially, and has been used in several applications.

Three performance indices were constructed for comparison of the original regression analysis and the proposed neural network. They are $MSE$ (Mean Squared Error), $APE$ (Average Percentage Error), and $R^2$ (R Squares). $MSE = \frac{1}{p} \times \sum_{i=1}^{p}(T_i - O_i)^2$ where $p$ is the number of patterns in training/testing, and $T_i$ and $O_i$ are $i$th targeted output and calculated output, respectively. The smaller the MSE, the closer the calculated output is to the targeted output. $APE = \frac{1}{p} \times \sum_{i=1}^{p} \frac{|T_i - O_i|}{|T_i|} \times 100\%$ In essence, APE reveals on the average how far the calculated output is from the targeted output in percentage. $R^2 = 1.0 - \left(\frac{\sum_{i=1}^{p}(T_i - O_i)^2}{\sum_{i=1}^{p}(T_i - \bar{T})^2}\right)$ The closer the $R^2$ value is to 1, the better the performance.

## RESULTS OF NN PREDICTION

We used the Fuzzy Accelerator to speed up Backpropagation training on a 7×7×1 neural network. One training cycle consisted of 20,000 iterations, which took less than 15 minutes on a Sun Sparc 10. The following results were averaged over ten independent simulation runs, as each run used a different random seed to initialize the weight matrices of the neural net.

The result is shown in the following tabular form.

| *Training* | | *Testing* | |
|---|---|---|---|
| Regression | NN | Regression | NN |
| $MSE = 0.0338$ | $MSE = 0.0176$ | $MSE = 0.7131$ | $MSE = 0.1763$ |
| $APE = 11.52\%$ | $APE = 7.36\%$ | $APE = 24.78\%$ | $APE = 20.73\%$ |
| $R^2 = 0.9134$ | $R^2 = 0.9549$ | $R^2 < 0$ | $R^2 = 0.5679$ |

It is clear that the proposed neural network outperforms the regression in all performance categories for the training data as shown in the first two columns. For instance, the average percentage error (APE) for the regression is 11.52%, while it is 7.36% for the neural net. This means that the predicted output from the regression deviated from the correct output by approximately 11.5% on the average, while the neural net only deviated by 7.4%. In addition, the above result confirms that the neural network is indeed a much better predictor than the regression for the testing data as shown in the last two columns. For instance, the mean squared error (MSE) for the regression is 0.7131, while it is 0.1763 for the neural net. Note that $R^2$ of the regression analysis is less than 0 for the testing data. It results from the fact that the unexplained variance, $\sum_{i=1}^{p}(O_i - \bar{T})$, is greater than the total variance, $\sum_{i=1}^{p}(T_i - \bar{T})$.

## NN DECOMPOSITION

As shown in Figure 2, it is clear that the running measurement of the melt index clusters into two natural clusters according to magnitude. One cluster of the scaled MI is centered around *0.35*, while the other is centered around *0.15*. As a result, it is intuitive to use two neural networks to predict the two

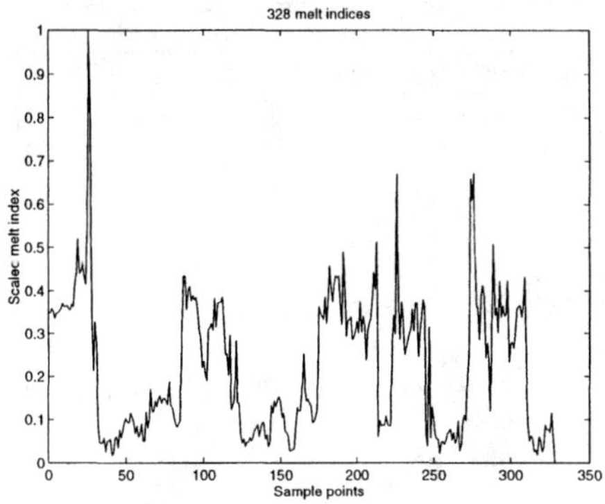


Figure 2: Running measurement of the melt index.

clusters of MI separately, as motivated by the principle of "*divide-and-conquer.*" In other words, we would like to substitute two smaller and less complicated neural networks for a larger one so as to improve accuracy of prediction and to speed up its computation, as well as to reduce the required memory for such computation.

Rather than any complicated clustering algorithms, our first attempt was to use the mean value of MI as a threshold to divide the data set. That is, the data set is divided into two sets: one using its MI value greater than or equal to the original mean, the other with its MI value smaller than the original mean. Due to the fact that the targeted $MI^*$ is known in advance, the system will be able to select the *appropriate* neural network for its prediction by comparing $MI^*$ against the pre-defined threshold.

## RESULTS OF DECOMPOSITION

Two three-layer feedforward neural networks are formulated via the decomposition described earlier. The architectures for both networks is $7 \times 5 \times 1$. Note that the original network is $7 \times 7 \times 1$.

We terminated the training process after 2,000 iterations, which is ten times less than that of the original training. This took less than 2 minutes on a Sun Sparc 10. Similarly, the following results were averaged over ten independent simulation runs, as each run used a different random seed to initialize the weight matrices of the neural net.

The result is shown in the following tabular form along with the previous experiments.

| *Training* | | *Testing* | |
|---|---|---|---|
| Original | Decomposition | Original | Decomposition |
| $MSE = 0.0176$ | $MSE = 0.0147$ | $MSE = 0.1763$ | $MSE = 0.1487$ |
| $APE = 7.36\%$ | $APE = 6.94\%$ | $APE = 20.73\%$ | $APE = 17.20\%$ |
| $R^2 = 0.9549$ | $R^2 = 0.9625$ | $R^2 = 0.5679$ | $R^2 = 0.6356$ |

From the above table, it is clear that the decomposition of neural networks outperforms the original network in all performance categories for both the training and testing data. Note that the above performance improvements of the decomposition of neural networks were obtained with simultaneous complexity and training time reduction. Thus it makes this approach even more appealing to practical applications that usually deal with huge training patterns and complex network architectures.

## SUMMARY AND CONCLUSIONS

We have described the design and testing of a neural network for chemical process prediction and its use for process optimization. We designed a neural network trained with a fuzzy accelerator for this purpose, which improved prediction accuracy while providing adaptive capabilities to account for process mean drifts and future changes in the measurement process. This approach resulted in a predictor that was superior to the existing nonlinear regression based approach. Cross-validation was used to avoid overfitting. Further improvements of the neural network prediction can be gained from the idea of "divide-and-conquer."

The neural network predictor was implemented in C on a Sun Sparc 10, then ported to a PC running MS-DOS, while the nonlinear regression predictor currently used on the factory floor is running MS-Windows using Excel functions. A simultaneous, side by side field comparison in a plant trial will be conducted in the near future. Future work will focus on trending analysis of the melt index to determine the optimal time window for periodic re-training of the neural net predictor.

### References

[1] P. Khedkar and Y. T. Chen. Resin melt index prediction. Technical report, GE Corporate Research and Development, Schenectady, NY, 1995.

[2] W. H. Press, S. A. Teukolsky, W. T. Vetterling, and B. P. Flannery. *Numerical Recipes in C.* Cambridge University Press, New York, NY, 1992.

[3] D. E. Rumelhart, J. L. McClelland, and G. E. Hinton. *Parallel Distributed Processing, Vol. 1 and 2.* The MIT Press, Cambridge, MA, 1986.

[4] Y. T. Chen and S. Kumara. Design of grinding process via inversion of neural nets. In C. H. Dagli, I. L. Burke, B. R. Fernandex, and J. Ghosh, editors, *Intelligent Engineering Systems through Artificial Neural Networks*, pages 715–720, New York, NY, 1993. ANNIE, ASME.

# PARAMETRIC DETERMINATION OF W-M FUNCTION FOR MACHINED SURFACE CHARACTERISATION

***Guang Yang and Caroline Tite***
*Southampton Institute, UK*

***Graham Smith and Tony Hope***
*Southampton Institute, UK*

***ABSTRACT*:**
In contrast to the existing surface analytical method, the technique based on the Weierstrass-Mandelbrot ( W-M ) function can recognise the pattern of a surface profile spectrum with fractal features and introduces an iterative rather than analytical approach for characterisation of the surface profile of hardened steel turned components. The method generates an artificial spectrum with fractal features on a computer screen to simulate the real surface profile obtained from experimentation. If the simulated spectrum is statistically similar to that of experiment, it can represent the real surface profile and the parameters determined from the W-M function could be used to characterise the surface profile. The work has also investigated the estimation of fractal parameters and how they affect the spectrum. An interactive software package has been developed to implement the simulation which could provide both industry and academics with a means of more accurate assessment of surface measurement.

Key word: Fractals, Iterative Function Systems, Surface Characterisation.

## INTRODUCTION

The methodology of analysing surface topography has been developed for more than fifty years. Over thirty parameters are used to quantify topography in the National and ISO Standards of surface roughness of a machined component. It has been realised that much of the difficulty in using standard measures of surface roughness occur due to the non-stationary and multi-scale nature of surface topography. Therefore a fractal approach was introduced to describe this surface characterisation (Majumdar and Bhushan, 1990 ). Moreover, it is the fractal features that are independent of measurement scales which are thought to be intrinsic parameters of surface topography and show great potential in such fields as tribology.

The most common use of functionality for surface profile characterisation is the W-M function. The function and its transformation have found application to spectral analysis, such as signal processing and machined surface characterisation ( Everts and Mandelbrot, 1992, Agterberg, 1994 ). The W-M function can also be used to simulate surfaces that have different dimensions for differing spectral regions ( Lopez, et al, 1995 ). The advantage of this method is that the characterisation parameters are determined to fulfil the requirements of a complete simulation by the W-M function which satisfies the mathematical properties of real surfaces.

Surface simulation using the W-M function may be divided into two stages: parameter determination and surface spectral generation. Successful simulation depends largely on the parameter determination and most previous research work has concentrated on this. Majumdar and Tien ( 1990 ) used power spectra to determine parameters whereas Zhou, et. al. ( 1995 ) used a modified structure function to determine the parameters from bearing area distribution. The research was conducted using mathematics, hence some derived equations are complicated and further assumptions were introduced. Although these researchers contributed to a good understanding of a theoretical solution for characterising the parameters, it is difficult to determine them by a few purely analytical performances.

A few researchers used algorithmic methods to experimentally characterise the parameters. Lim and Shi ( 1993 ) selected different fractal dimensions, D, and the ratios of spectral density, R, to create a fractal surface based on a fractal model of light scattering. Kristakova and KupKova ( 1994 ) obtained fractal dimensions by studying profiles of sections through the sample fracture surface. The current work introduced a fractal method which combines the two stages into one iterative process to determine the fractal parameters in terms of statistical similar. That is the appropriate set of parameters are obtained by experimentally computing the data from an artificial spectra and compared with that of experimental spectra. The effects of these parameters on spectra and their interaction were also investigated to obtain the knowledge of influence. An initial software package has been developed to implement the computation and aid its subsequent presentation. The computation and comparison processes are in the background and only the final presentation is shown in the foreground screen.

## SURFACE SIMULATOR AND EFFECTS OF PARAMETERS

The spectra generated by the W-M function is called the fractal surface spectra. It is not necessarily visually similar to that of a real surface since it is well recognised that a machined surface profile is a non-stationary random process ( Sayles et al, 1978 ), but the statistical data which describes its characteristics, such as average roughness, Ra, the largest roughness, Rp, and the smallest roughness, Rv, should be close to those of experiment. A successful simulation relies on the correct determination of fractal parameters of the W-M function, scaling constant, G, ratio of spectral density, R, and fractal dimension, D. The current research work considered all of these parameters in the same priority and introduced an iterative method to determine them. The procedure is shown in Fig. 1. In the iterative process, initial parameters were set up to generate an artificial spectrum in computer memory; then the statistical data, Ra, Rp and Rv are calculated and compared with that of experiment. If the comparison results were outside a pre-defined tolerance margin, the iterative process would continue until values are within tolerance. The aim of the work is to algorithmically determine fractal parameters of the W-M function while conserving its analytical capability. To more efficiently simulate the spectra, it is essential to investigate the effects of these parameters on the simulated spectrum, especially on its amplitude.
The W-M function is:

$$z(x) = G^{(D-1)} \sum_{n=n_1}^{\infty} \frac{\cos 2\pi R^n x}{R^{(2-D)n}} \quad ........(1)$$

As $|\cos 2\pi R^n x| \le 1$, we have:

$$\frac{\cos 2\pi R^n x}{R^{(2-D)n}} \le \frac{1}{R^{(2-D)n}} \quad .......... (2)$$

When $R^{(2-D)} > 1$, We have:

$$\sum_{n=0}^{\infty} \frac{1}{R^{(2-D)n}} = \frac{R^{(2-D)}}{R^{(2-D)} - 1} \quad .......... (3)$$

where: G is a scaling constant. R can be estimated from ( 3 ) for given G and D, It usually has a value between 1 and 2. Previous researchers treated R = 1.5 as a constant when the simulation was carried out (Gagnepam and Roques-CAmes, 1986, Zhou et al, 1995, Majumdar and Tien, 1990 ). D is the fractal dimension, in the case of a spectrum; $1 < D < 2$.

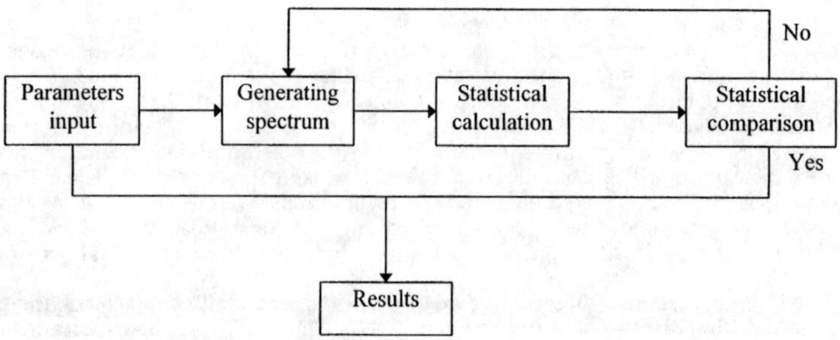


Fig. 1 Iterated process of parametric determination

Equation (3) is named as the " largest possible form" ( LPF ) of the W-M function. It was used for the analysis and the interactive effects of fractal parameters throughout this study.

The x-axis was divided into 450 points, named x1, x2, x3, ......x450. The number of iterative items in this present study was 50, due to the fact that the value of the sum of the remaining items was too small to contribute to the graphical display on a screen. The initial values of the parameters of the W-M function were set up as: G = 0.9, R = 1.5, D = 1.5, n1 = 0. For every run of data, one variable is given and the rest are regarded as temporary constants substituted by initial values

In the W - M function, G is defined as scale constant. The LPF with G as a variable and the other variables substituted with default values is:

$$z(\tfrac{n}{R^n}) = 5.443 \times G^{0.5} \quad ..........(4)$$

The value of G does not have an effect on the shapes of the spectra, it only affects the amplitude. Spectra with different scaling constants, G, appear the same shape, but change in amplitude. The trend of change in amplitude responding to different values of G is shown in Fig 2. Although the variance of the amplitude for G is quadratic, the appearance of the curve is approximately linear so that G is used as an adjustable factor for trimming the spectrum.

The value of R is defined as the ratio of spectral density, where R is a factor which causes geometrical divergence which may be regarded as a non-stationary random process. The LPF with R as a variable gives:

$$z(\tfrac{n}{R^n}) = \frac{0.948R^{0.5}}{R^{0.5}-1} \quad ..........(5)$$

The value of R significantly changes the shape, amplitude and peak number of a spectrum. With a decrease in R's value, the peaks become sharp, the amplitude increases and the number of peaks increases. When R increases, the amplitude and number of peaks decrease and widen, which is regarded as the merging of two adjacent peaks. The effects on a spectrum are non-linear and are complicated. When R has a small value, the amplitude significantly reduces as R decreases. When R > 1.5, then the ratio of change in amplitude becomes more flattened.

The letter D is used to define the fractal dimension in the W - M function and is the most important parameter of a fractal feature. Most previous research work has been performed on the characterisation of D whilst other factors were usually fixed. Again, the possible largest form with default values and D variables is shown by:

$$z(\tfrac{n}{R^n}) = \frac{0.9^D 1.5^{(2-D)}}{1.5^{(2-D)}-1} \quad ..........(6)$$

It should be noted that D affects spectra in the same manner as R, but in the opposite direction, and has a slight difference in amplitude properties.

The diagram shown in Fig. 3. illustrates that the variance of amplitude is similar to that of R, but with the opposite effect. In contrast, the variance is flatter when D is small, whereas it becomes significantly more sharp when D is > 1.5. The contribution of R and D to the amplitude are the same where the R and D values are equal to 1.5. This might be the reason that several of the previous researchers have chosen R = 1.5 as a fixed parameter to perform fractal simulation for surface characterisation. In fact, it seems their action compensates each other, but there is a difference in amplitudes between R and D. To accurately simulate the spectral generation the values of R should not be regarded as a constant.

## EXPERIMENTAL VALIDATION

Important components of machines which are working under loading and contact conditions are usually made up of hardened steel. The surface profile of the component will influence the contact and lubrication of two jointed parts and to some extent will influence the machine performance. The hardened steel turning process, which is replacing the grinding process in the automobile industry prior to the polishing procedure, may produce the simplest surface profile among all machining processes. The turning process is a continuous single point cutting process so that the feed marks on the surface are mainly produced by both feed rate and nose radius of the tooling. The feed marks are influenced by

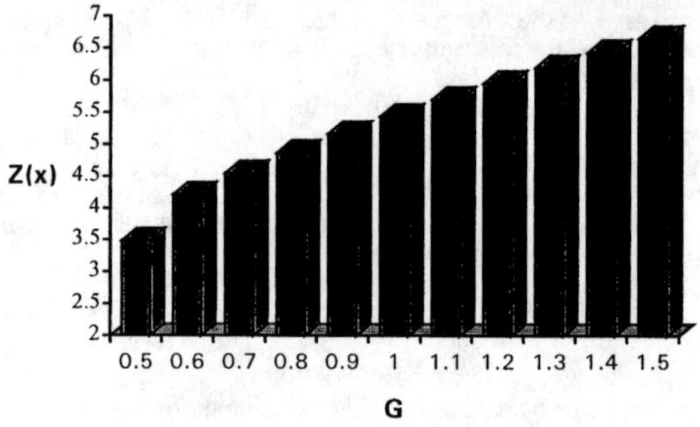


Fig. 2 Amplitude z (micron) versus scaling constant G

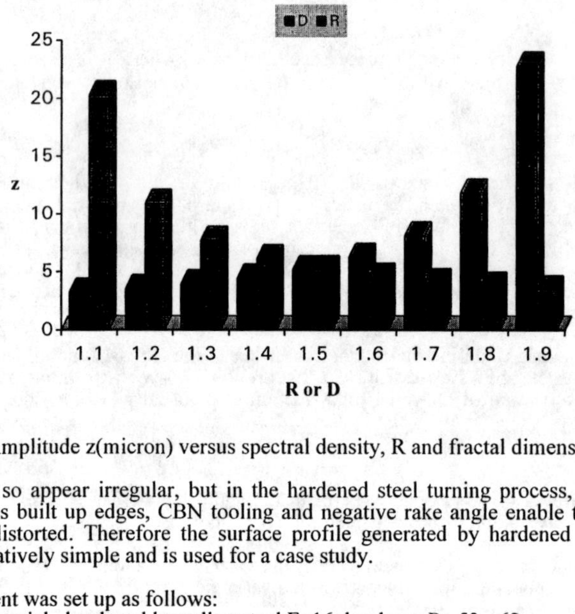


Fig. 3. Amplitude z(micron) versus spectral density, R and fractal dimension, D

other factors so appear irregular, but in the hardened steel turning process, less material side flow, less built up edges, CBN tooling and negative rake angle enable the surface to appear less distorted. Therefore the surface profile generated by hardened steel turning process is relatively simple and is used for a case study.

The experiment was set up as follows:
workpiece material   hardened low alloy steel En16, hardness Rc 53 - 62;
machine tool   CINTURN200-15 CNC Turning Centre
tooling   DBC50 (CBN), with -7 rake angle, nose radius 0.8 mm
cutting parameters: depth of cut 0.3 mm, surface speed 156 m/min, feed rate 0.05-0.25. The experiment is a typical finish cutting process.
surface measurement equipment:   Taylor-surf6 with graphic device.

In this study, the statistical data, average roughness Ra, longest value Rp and smallest value Rv taken from experimental measurement are used as criteria to compare the Ra', Rp' and Rv' taken from a simulated spectrum. The spatial values, Rsm and Sm are used to select spatial fractal parameters R and S for specific instruments.

The parameters to be determined are G, R, and D, and the appropriate parameters are achieved by means of fractal surface simulation. In additional to these parameters, a x-axis factor S was introduced. S is related to computer screen resolution in pixels and the width of a line expressed in pixels. S can give R to some extent freedom from device dependency. Once S has been determined, R is associated with the spatial characters of a spectrum. The tolerance margin was set as 10%. The first iteration is to find the spectra whose Ra' is close to Ra and refine the D, G, R to make Rp' and Rv' close to Rp and Rv respectively. When $R > 1.5$, $D > 1.8$, $\Delta D/\Delta R = 4$, and $R > 1.5$, $1.5 < D < 1.8$, $\Delta D/\Delta R = 2, 3$, the variance of Ra' is not significant. When $R < 1.5$, $D < 1.5$, the tendency is opposite. This phenomenon is consistent with Fig. 3.

The comparison of experimental and simulated data is shown in table 1.

Table 1

| Ra | Rp | Rv | Ra' | Rp' | Rv' |
|---|---|---|---|---|---|
| 0.46 | 1.87 | 1.6 | 0.464 | 1.878 | 1.649 |
| 0.51 | 2 | 1.87 | 0.510 | 2.087 | 1.874 |
| 0.6 | 2.38 | 2.29 | 0.608 | 2.418 | 2.341 |
| 0.65 | 2.41 | 2.32 | 0.655 | 2.49 | 2.396 |
| 0.92 | 3.6 | 2.87 | 0.914 | 3.792 | 3.074 |
| 1.11 | 4.1 | 3.4 | 1.087 | 4.759 | 4.039 |

From the table it is found that the simulated spectra agree well with fine hardened steel cutting, when Ra < 1 μm. With an increase in Ra, the difference between experimental and simulated data becomes significant. It indicates that the surface pattern in rough cutting, even in turning, is more complicated, perhaps requiring a multi-fractal model, and needs further research. In this study, only fine cutting is considered due to the fact that the process is used to some extent to replace the grinding process in manufacture. It also makes a significant contribution to the surface finish of a machined component. Fig. 5 and Fig. 6 are fractal spectra with $D = 1.4$, and $D = 1.5$ respectively, other values, $G = 1.5$, $R = 1.98$ are the same for both figures. They reflect the roughness increase with D increase. The fractal spectra can represent those of experiment as patterns or symbols and fractal parameters can be used as parameters to characterise the surface profile.

**Conclusions**

This investigation has introduced an iterative method based upon the W-M function to simulate a spectrum which can be used to characterise a machined surface. The correct combination of these parameters has to be selected for successful spectral simulation. Thus it is essential to look at the effects of these parameters on a spectrum individually and in some detail. To quantitatively describe the effects of fractal parameters, the largest possible form derived from the W-M function was adopted with the aim of reducing the simulation time. An initial software package that integrated the W-M function computation and presentation on a screen, has been developed to implement the spectral simulation corresponding to fractal parameters.

The effects of such fractal parameters appear as a non-linear relationship. The scaling constant G does not change the shape of a spectrum, it only enlarges and shrinks the spectrum. The ratio of spectral density R affects both the amplitude and spatial properties of the spectrum. To some extent, the effect of fractal dimension D - which is regarded as the most important fractal factor - is similar to that of R but in the opposite direction.

A simulated spectrum could be employed to represent that from experiment and their fractal parameters could be regarded as additional to the data set of surface measurement for surface profile characterisation. The simulated spectra are agree well with experimental results for fine cutting of hardened steel. The spectra can be regarded as a spectral fractal pattern of the real surface profile. Further analysis, such as correlation and synthesis, could be conducted on the pattern.

**REFERENCES**:

Encarnacao, J. L., Peitgen, H. O., Sakas, G., and Englert, G., 1992, Fractal Geometry and Computer Graphics, Springer-Verlag.

Gagnepam, J. J. and Roques-Cames, C., 1986, "Fractal Approach to two-dimensional and three-dimensional Surface Roughness, " Wear, vol. 109. pp. 119-126

Iim, S.P., and Shi, W.M., 1994, "Fractal Surface and its Measurement by Computer Simulation", Interfrerometry VI: Application, V.2004, CH.38, pp. 182-186.

Kristakova, Z., and Kupkova, M., 1994, "Correlation of Fractal Surface Description Parameters with Fracture Toughness", International Conference on Fractals and Dynamics Systems in Geoscinece, Ch.31, pp 77-85
Lin, T. Y., Lu, S. H., and Stout, K. J., 1995, "Model-Based Topography Characterisation of Machined Surfaces in Three Dimensions", International Journal of Machining Tools Manufacture, Vol. 35, No. 2, pp.239-245.
Lopez, J., Hansali, G., Bosse, J. C. L., and Mathia, T., 1995, "3D Fractal-based Characterisation for Engineered Surface Topography", International Journal of Machining Tools Manufacture, Vol. 35, No. 2, pp.211-217.
Majumdar, A., and Bhushan, B., 1990, "Role of Fractal Geometry in Roughness Characterisation and Contact Mechanics of Surface", ASME, Journal of Tribology, Vol. 112, pp.205-216.
Majumdar, A., and Tien, C. L., 1990, "Fractal Characterisation and Simulation of Rough Surfaces", Wear, Vol. 136, pp.313-327.
Mandelbrot, B., Passoja, D. E., and Paullay, A. J., 1984, " Fractal Character of Fracture surface of Metals, " Nature, vol. 308, pp. 1571-1572.
Mandelbrot, B . B. 1982, The fractal Geometry of Nature, W. H. Freeman, New York.
Sayles, R.S., and Thomas, T.R., 1978, "Surface Topography as a Nonstationary Random Process, Nature, Vol. 271. pp. 431-434
Thomas, A., and Thomas, T. R., 1988, "Digital Analysis of very Small Scale Surface Roughness", Journal of Wave-Material Interaction, Vol. 3, No. 4., pp. 341-350.
Voss, R. F., 1985, "Random Fractal Forgeries," in Fundamental Algorithms for Computer Graphics, ed. R. A. Eamshaw, pp. 805-835.
Zhou, G., Leu, M., and Blackmore, 1995, "Fractal Geometry Modelling with Application in Surface Characterisation and Wear Prediction", International Journal of Machining Tools Manufacture, Vol. 35, No. 2, pp. 203-209.

# *AN INVESTIGATION OF THE SENSITIVITY OF A SINGLE-LAYER PERCEPTRON NEURAL NETWORK TO TOOL WEAR INCEPTION DURING A TURNING PROCESS*

*DIMLA E DIMLA, JNR., PAUL M LISTER, NIGEL J LEIGHTON*
Engineering Research Group, SCET, University of Wolverhampton,
Wolverhampton, West Midlands, WV1 1SB, UK

*ABSTRACT:*
Artificial intelligence (AI) has considerably assisted engineers and manufacturers make moderate optimisations and improve turn-over in the shop environment. Implementation of neural networks to integrate sensor signals in the cutting tool condition monitoring (TCM) problem has been widely pursued, but the methods used have either been complicated or required detailed sensor signal pre-processing. In this paper, the authors present a simple sensor integration method via a Single-Layer Perceptron neural network to the TCM problem. Three triaxial sensor signals, namely the static cutting force, dynamic cutting force and the vibration signature were used, and incorporated into the input vector pattern were the cutting condition parameters. It was found that the perceptron learnt to correctly detect the ensuing tool state with over 90% accuracy.

## *INTRODUCTION*

With the increasing tendency to automate in most modern production systems, the role of "the experienced machinist" is increasingly being replaced by smart sensors [1]. An alternative way to determine the ensuing extent of tool wear or be forewarned of imminent tool failure is needed [2]. One of the major obstacles to metal cutting automation is that of tool state prediction where tool failure is an important factor in productivity and manufacturing efficiency. Tool monitoring and diagnostics have therefore become a necessary precondition to automation [3]. Untended operations normally involve the integration of sensors in production machines, and the intelligent processing of these signals with the goal of increasing productivity, reliability and assures quality, as in Flexible Manufacturing Systems (FMS) and Computer Integrated Manufacture (CIM). Implementation of intelligent machining systems that can perform specified machining operations in rapidly changing and unpredictable environments consequently becomes paramount [4].

A drawback to the implementation of the aforementioned system has been the fact that the development of a sensor to replicate the human operator exactly is impossible. Instead, numerous researchers have carried out extensive work on the various parameters related to tool wear in cutting operation. Sensor signal processing technologies (emerging and established) and advanced sensor design permits information gathering of the cutting process condition thus enabling adequate measurement or counter measure to be taken in order to optimise the process.

Some researchers in TCM have proposed direct measuring methods [5], whereas the majority have advocated indirect monitoring methods [1-16] [18]. Direct methods as the name suggests, relies on sensing techniques that measure the wear of the tool in-situ. These methods have included touch trigger probes, optical, radioactive, proximity sensors and electrical resistance measurement techniques [5]. Direct measurement of tool wear on-line, usually on a very small wear zone, and in a continuously moving system (environment) to an extent is practically unachievable. Indirect methods on the other-hand, measure other factors that are the cause of the tool wear. Indirect methods that have been pursued include the cutting forces [5-9], acoustic emission [6] [7] [11], vibration [12] [13], spindle motor current [6] and the cutting conditions [1] [7] [8] [10] [11]. These factors, reflect far more than tool wear alone, and parameters associated with tool wear must therefore be extracted from them and correlated to give a measure or extent of tool wear.

The use of these indirect methods requires accurate predictive models that link the un-measurable parameter of interest (tool wear), with the indirectly detectable variables [16]. Effectively, the combined process of tapping useful signals or process parameters, directly or indirectly, from a cutting process constitutes what Burke [8] symbolically terms, the detection level. This level in any TCM System is primarily concerned with choosing the relevant sensory signal.

The choice of the suitable sensor type as well as the point of application (sensor location) are inextricably linked. Thus a suitable location for any such sensor is where the specified signal has the highest concentration and best reproducibility. Certain limitations, however, do exist due to the structure of the machine. Therefore, information from a variety of different sensors has to be collected, and these signals of varying reliability integrated (fused). Sensor fusion serves to enhance the richness of the underlying information (wear level) contained in each signal [6], thereby increasing the reliability of the monitoring process. The lack of an accurate model for wear prediction, has led to researchers resorting to other methods of sensor integration, which do not rely on any theoretical or empirical process.

The expertise of integrating sensor signals is viewed as constituting the identification of wear, thus the characteristic identification level [8]. The utilisation of single sensor signals offers incomplete sensory information as to the wear state (condition) of an engaged cutting tool. This shortfall has been obviated by pursuance of a sensor fusion strategy by many researchers [9]. Predictively, this method has proven highly successful as more reliable information on the tool state was extractable. Dornfeld [6], initially used linear discriminant functions to implement TCM, a laborious pattern recognition method, which he reports had a high sensitivity to the cutting conditions and therefore inefficient. In addition to pattern recognition techniques; hiperplane decision surfaces, mapping techniques, neural networks, clustering techniques and perceptron techniques are reported by Noori-Khajavi and Komanduri [14] in their literature survey of other techniques. Methods primarily based on neural networks have proven particularly popular because tool wear intrinsically is a highly non-linear process; and its representation via neural networks a viable, reliable and an attractive alternative to previously employed methods [15].

In this paper, the authors report preliminary results in an on-going more elaborate research programme investigating the applications of neural networks to TCM. Herein, efforts have largely concentrated on investigating the sensitivity of one of the most basic neural networks, a single layer perceptron, to cutting tool wear. To the best of the authors' knowledge, no in depth attention to date has been paid to SLP types of neural network on their own as a possible device worth investigating in the TCM problem. Efforts have often largely focused on complex and hierarchical systems. Readers seeking further information on perceptrons are advised to consult neural network text books.

## *EXPERIMENT*

The instrumentation set-up presented in Figure 1 essentially consisted of a Lang Swing J6 centre lathe onto which were attached a Kistler dynamometer platform type 9263A and a Kistler triaxial accelerometer type 8730A. A Kistler charge amplifier type 5006 and a power supply/coupler (Kistler type 5134) were used to amplify and decouple the cutting forces and acceleration signals respectively. The decoupled and amplified voltages were then sampled using a PC in which was an Amplicon PC-30 range data acquisition card. Recording 1024 samples per channel at 20 kHz, 6 data samples were collected per cut and saved on the computer HDD for further analysis. Key features of the tests are summarised in Table 1.

The instrumentation set-up was such that the dynamic forces could not be sampled simultaneously with the static forces, hence were calculated as the oscillatory component in the static forces resulting in a total of 9 data points (triaxial static and dynamic forces, and triaxial acceleration signals) per test cut. The dynamic signals (dynamic forces and accelerations) were transformed using the forward FFT and the accumulative sum total of energy in the spectra calculated for the first 512 samples in each test. This was because after the transformation, only the power or energy distribution in the first half of the frequency band was of interest. It was therefore thought not necessary to obtain the complete energy distribution in the entire frequency range, thus the total power in the FFT spectrum from $f_k = 0$ to $f_k = N/2$ was required and obtained.

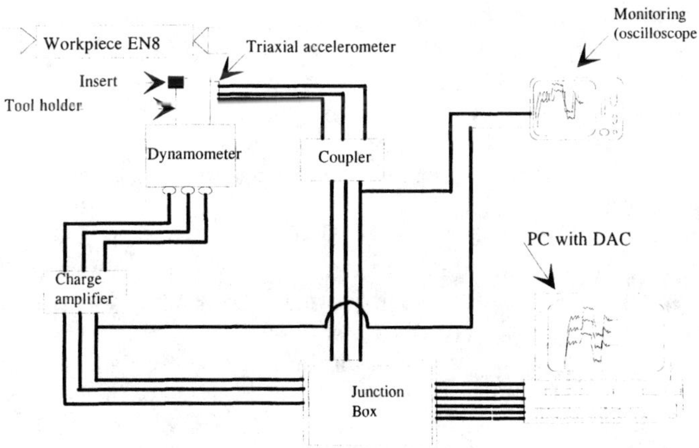


**Figure 1: The Experimental set-up**

**Table 1: Test Conditions, material type and tool details**

| | | |
|---|---|---|
| Machine Tool | | Lang Swing J6 |
| Workpiece Material | | |
| | Work Material | EN8 BS 970 |
| | Hardness | Brinell 170 |
| | Composition | 0.35% C, 0.05% Si, 0.6% Mg |
| Tooling Material | | |
| | Tool Holder | Kennametal KSBPR 2525 M12 |
| | Tool Type | SPUN 12 03 08 |
| | Tool Material | Sandvik Coromant S1P P10 2361 |
| | Overhang /mm | 45 |
| Cutting Conditions | | |
| | Cutting Speed Range m/min | 80 - 300 |
| | Feed Rate Range mm/rev | 0.081 - 0.163 |
| | Depth of Cut Range /mm | 0.5 - 3 |
| Cutting fluid used | None | |

Feature extraction was neither required nor performed. Test results were normalised and divided into two groups : a training set and a test set. Normalisation involved dividing all data samples in a particular data set (e.g. the feed-force component of the static cutting force) by a normalising factor, the magnitude of which would ensure that all data points in that set were subsequently reasonably well distributed between 0 and 1. In total, 83 cutting tests were performed using a range of cutting conditions (see Table 1). Data from 41 of these tests was used to train the network, with the remaining 42 tests then being used to test network performance.

Using the MATLAB Neural Network Toolbox, a single-layer perceptron with a 12-vector input (3 components of static force, 3 components of dynamic force, 3 components of acceleration, and the depth of cut, feed-rate and cutting speed information) and one binary output depicting tool state was designed. An output of '0' was used to denote a sharp tool, whilst a '1' represented a worn tool. Clearly this type of output represents two extremes in the range of possibilities of tool wear, and does not distinguish between the type of wear (e.g. flank or crater wear). Cutting tests involving 'sharp' tools used either a new cutting edge at the start of cut or were nominally sharp (i.e. wear features evident, but of minor proportions). Those involving worn tools involved cutting with a tool exhibiting at least 0.3 mm of flank wear (mean) and 0.02 mm cater depth measured using a tool maker's microscope and a dent tool indenter respectively.

The data set used for training, comprising sharp and worn states, was designated P, and contained 41 x 12 normalised values. P was targeted to a 41 element column vector of binary values of tool status (1 for worn and 0 for sharp). Such a configuration did not permit successful classification, i.e. it was not possible for the perceptron training algorithm to converge on finite values of weights and a bias. For successful classification using a perceptron, data must be linearly separable. Visualisation of a 2-vector input perceptron is straightforward, with successful classification reliant upon separating output states with a straight line. Similarly, a 3-vector input perceptron involves separating output states with a plane surface. Thereafter, visualisation becomes difficult, but clearly it is natural to assume that some data points of the 12-input vector data sets are encroaching into what may be considered to be the wrong sector.

By trial and error, the first 31 sets of data from the initial training set of 41 were found to converge after less than 1000 epochs (training cycles). It was therefore evident that 'undesirable' patterns occurred between the 32nd and 41st data set. These undesirable patterns were labelled as those that encroached into the wrong class and manually removed from the training set. Omission of the spoiled patterns for training but inclusion for cross-validation could be expected to result in a reduction in the maximum classification accuracy as fewer than initial patterns were used to train the network. Therefore, instead of a 100% success rate, prediction may be limited to a maximum success rate of 75.61% (i.e. 31/41). Alternatively, if the values beyond the 31st pattern data were used neither for training nor cross-validation, then a 100% success classification rate would again be achievable.

After training the perceptron, its weights and bias were noted, and used to test its performance off-line.

## *CROSS-VALIDATION OF THE TRAINED PERCEPTRON*

Testing involved presenting a perceptron with a hard limit transfer function on its output and based upon values of weights and bias found in training, with data which had not been used for training. The output results were then compared to the actual values of tool wear state. It was found that 11 of the 42 data sets were classified wrongly, giving a success rate of 73.81%. On further examination, it was evident that 9 of these 11 were associated with data sets that were broadly similar to data sets omitted in the training process, and as such it might be argued that the network had no knowledge of this type of data set. By omitting the 11 data sets from the training and testing, classification improved significantly, with only 2 of the 31 data sets used for testing failing to classify correctly (i.e. a 93.56% success rate).

## *DISCUSSION*

The training technique employed for the perceptron was the Perceptron Learning Rule, details of which are described in full in Ref. [17]. The SLP is simple in operation and design, thus is uniquely suitable for this type of pattern classification problem.

A comparison of the results obtained from the final classification showed a decrease in the predicted and expected classification rate of 1.8% (i.e. from 75.61% to 73.81%) when all 41 patterns were considered. But the overall classification success rate increased when only the 31 patterns 'actually' employed in training were considered (i.e. from 73.81% to 93.56%).

The simplest form of the perceptron network such as that in use by the present authors has just one layer and this does place restrictions on its computational ability. Moreover, the hard limit restricts the perceptron output to 0s and 1s, which would be the output only if the inputs were linearly separable, thus a case of mere linear interpolation. Another difficulty arises when some vectors form outliers i.e. some inputs might be much larger or smaller than others, and this would slow down the convergence process. The existence of outliers in the validation data set could have an adverse effect in that predicted class could in fact be wrong as the SLP cannot handle its descriptive data.

## *CONCLUSIONS*

A TCM system based on a multi-sensor integration of signals through application of a SLP has been investigated and presented. This study showed that a SLP could satisfactorily classify worn and sharp tool data from a turning process by partitioning the input data into two subspaces using a hyperplane. Furthermore, the study has shown a SLP as not being 'ideally' suited for TCM since it cannot cope with the non-linearity of metal cutting process. However, if one is concerned in the state of the cutting tool only (worn or sharp defined by some criteria), then it suffices, as its classification performance is comparable to that of hybrid neural networks. For a higher degree of confidence in the obtained results, networks that can handle the non-linearity of the cutting data should be investigated. This line of investigation is currently being pursued by the present authors.

## *ACKNOWLEDGEMENT*

The authors would like to thank the Research Support Unit at the University of Wolverhampton for its support and financial assistance provided.

## *REFERENCES*

1. Rangwala S. S., "Machining Process Characterisation and Intelligent Tool Condition Monitoring Using Acoustic Emission Signal Analysis", PhD Dissertation, Mechanical Engineering Department, University of California at Berkeley, 1988.

2. Lee L. C., Lee K. S. and Gan C. S., " On the correlation between dynamic cutting force and tool wear", Int. J. Mach. Tools Manufact., Vol. 29, No. 3, pp. 295-303, 1989.

3. Ravinder H. V., Srinivasa Y. G. and Krishnamurthy R., "Modelling of tool wear based on cutting forces in turning", Wear, Vol. 169, pp 25-32, 1993.

4. Choi G. S., Wang Z. X., Dornfeld D. A. and Tsujino K., " Development of an Intelligent On-line tool wear monitoring system for turning operations", Proc. USA-Japan Symposium on Flexible Automation, A Pacific Rim Conf. ISCIE Kyoto Japan, pp 683-690, 1990.

5. Tansel I. N., Rodriguez O. and Mekdeci C.; "Detection of tool breakage in micro-drilling operation with RCE Neural networks", ASME, PED-Vol. 47-1, Eng. Sys. Design and Analysis, Vol. 1, pp. 83-88, 1992.

6. Dornfeld D. A., "Neural network sensor fusion for tool condition monitoring" Annals of the CIRP, Vol. 39/1, pp. 101-105, 1990.

7. Rangwala S. and Dornfeld D., "Integration of sensors via neural networks for detection of tool wear states", ASME PED v 25. pp 109-120, 1987.

8. Burke L. I., "Automated Identification of Tool Wear States in Machining Processes : An Application of Self-organising Neural Networks", PhD Dissertation, Department of Industrial Engineering and Operations Research, University of California at Berkeley, USA, 1989.

9. Liu T. I. and Ko E. J., "On-line recognition of drill wear via artificial neural networks", Winter Annual Meeting of the ASME, Monitoring and Control for Manufacturing Processes, PED-Vol. 44, Dallas USA, pp 101-110, 1990.

10. Purushothaman S. and Srinivasa Y. G., "A backpropagation algorithm applied to tool wear monitoring"; Int. J. of Mach. Tools and Manufact. Vol. 34 No. 5 pp. 625-631, 1994.

11. Emel E.; "Tool wear detection by neural network based acoustic emission sensing", ASME DSC-Vol. 28/PED-Vol. 52, Control of Manufacturing Processes, pp. 79-85, 1991.

12. Tansel I. N., Wagiman A. and Tziranis A.; "Recognition of chatter with neural networks", Int. J. Mach. Tools Manuf., Vol. 31, No. 4, pp. 539-552, 1991.

13. Kamarthi S. V., Sankar G. S., Cohen P. H. and Kumara S. R. T.; "On-line Tool Wear Monitoring Using a Kohonen Feature Map", Proc. Artificial Neural Networks in Engineering (ANNIE), St. Louis, Missouri, USA, pp 639-644, 1991.

14. Noori-Khajavi A. and Komanduri R., "Frequency and time domain analyses of sensor signals in drilling-II. Investigation on some problems associated with sensor integration", Int. J. Mach. Tool Manufact., Vol. 35, No. 6, pp. 795-815, 1995.

15. Elanayar S. V. T., Shin Y. C. and Kumara S.; "Machining condition monitoring for automation using neural networks", ASME's Winter Annual Meeting, Monitoring and Control of Manufacturing Processes, PED Vol. 44, pp. 85-100, 1990.

16. Oraby S. E. and Hayhurst D. R., "Development of models for tool wear force relationships in metal cutting", Int. J. Mach. Tools Manufact., Vol. 33, No. 2, pp. 125-138, 1991.

17. MathWorks Inc., "User guide, Neural network Toolbox for use with MATLAB", 1994.

18. Kamarthi S. V. K., "On-line tool wear estimation through sensor data fusion and neural networks", PhD Dissertation, Department of Industrial and Manufacturing Engineering, The Pennsylvania State University, USA, 1994.

# APPLICATION OF NEURAL NETWORK IN THE ANALYSIS OF ROLLING PROCESSES

Zhengjie Jia and Jay S. Gunasekera
Mechanical Engineering Department
Ohio University

Luis C. Rabelo
Industrial Manufacturing Systems Engineering Department
Ohio University

Mal Gunasekera
Computer Science Department
Ohio University

key word - I: Manufacturing Application

key words - II: Nonlinear Function Approximation, Neural Network Application, Mathematical Modeling, Manufacturing Application,

## ABSTRACT

This paper addresses a new method for the analysis of rolling processes which applies a neural network technique to predict the rolling force. In the development of the system for the optimization of an entire rolling process sequence, a quick and accurate prediction of the forming load in rolling process is needed to meet the requirement of optimization procedure which iterates several thousand times per minute. Since the results from finite element model is accurate but it is time consuming, a neural network predictor based on a back-propagation algorithm has been developed to meet this goal. A non-linear mathematical model was developed to guide and supervise the learning procedures. A near-optimal neural network structure determined by using a development process that emphasized second derivative order information can give prediction very quickly and meet the requirement of the system for the optimization of an entire forging sequence. The application of neural networks has made it feasible the quick and accurate prediction of forming load during high speed iteration for optimization.

## INTRODUCTION

A neural network is an information-processing system that has certain performance characteristics in common with biological neural networks. Neural networks have been widely used in the areas of pattern recognition, signal processing, vision, speech processing, forecasting and modeling, and robotics. One successful application of neural networks has been the modeling of nonlinear systems. The ability of neural networks to serve as generic functional approximators makes them a useful tool for synthesizing such mechanisms. In that capacity, neural networks can realize non-linear models with high accuracy levels.

All neural networks are special cases of vector mapping [1]. The mapping relationship is encoded in its structure. The nature of the mapping relationship between input and output vectors is defined by the weights within the network. Neural networks provide a novel approach for the interpretation of results from numerical models. Neural networks can solve problems which are difficult to be simulated using the logical, analytical techniques, and traditional approaches, where rules are not known or there are incomplete or noisy data. As an alternative of traditional techniques, neural networks can offer the advantage of execution speed, once the network has been trained.

Most of the work in neural networks in manufacturing has centered in manufacturing processes. However, little investigation has addressed the use of neural techniques to the flat rolling process. In this paper, the modeling of the flat rolling process is investigated using a backpropagation neural network. A non-linear mathematical model was used to train and test the neural structure. An algorithm based on the second derivative order information was tailored to our specific case. This method, called Optimal Brain Damage (OBD) developed by Cun, Denker, and Solla [2], was modified from the original implementation of backpropagation with shared weights to our particular case. Results of this neural network model show that the performance of the developed neural network in the learning errors, prediction errors, and training times has been significantly improved by the second-derivative order information. A 40% reduction of the network size has been reached without the error increase.

## BACK-PROPAGATION PARADIGM

Neural networks are information processing systems motivated by the goals of reproducing the cognitive processes and organizational models of neurobiological systems. The specific characteristics of a neural network depend on the paradigm utilized. The paradigm is determined by the architecture and the neurodynamics employed. The architecture defines the arrangement of the neurons and their interconnections (see Figure 1). The neurodynamics specifics how the inputs to the neurons are going to be combined together (i.e., short-term memory), what type of function or relationship is going to be used to develop the output, and how the weights (i.e., long-term memory) are going to be modified.

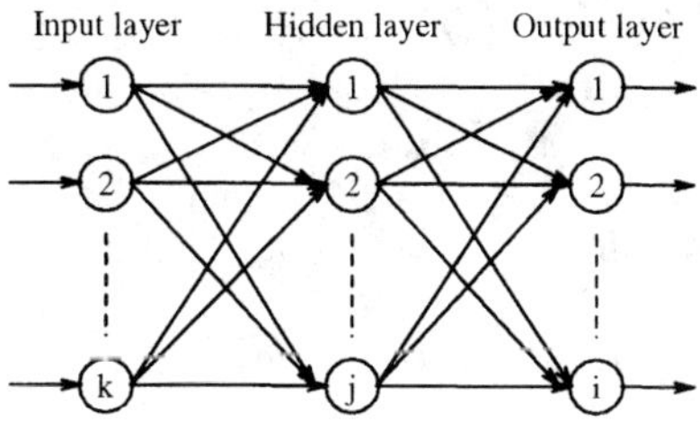


Figure 1. A two layer back-propagation network archtitecture

The learning mechanism which handles modifications to the weights can be classified under supervised, unsupervised, and reinforcement learning. The spectrum of different paradigms is quite extensive. For example, the network architectures range from simplistic perceptrons to the hierarchical neocognitron. In addition, there is a large number of algorithms for the modification of the weights. The various paradigms developed have their limitations and strengths; hence, one must identify the suitable application areas for which they lend themselves.

**Back-Propagation Algorithm**

Back-propagation is a specific technique for implementing gradient descent in weight space for a multilayer network. When applied to multilayer networks, the back-propagation technique adjusts the weights in the direction opposite the instantaneous error gradient. In its simplest form, backpropagation training begins by presenting an input pattern vector $X$ to the network, sweeping forward through the system to generate an response vector $O$, and computing the errors at each output. The next step involves sweeping the effects of the errors backward through the network to associate a "square error derivative" $\delta$ with each connection, computing a gradient from each $\delta$, and finally updating the weights of each connection based upon the corresponding gradient. A new pattern is then presented and the process is repeated. The initial weight values are normally set to small random numbers. A two layer back-propagation network architecture is shown in Figure 1.

The training of a network by backpropagation involves three stages: the feedforward of the input training pattern, the calculation and backpropagation of the associated error, and the adjustment of the weights. After training, application of the net involves only the computations of the feedforward phase. Even if training is slow, a trained net can produce its output very rapidly.

The total error on the outputs when one training example is presented is calculated, that is, the error (a function of the weights) to be minimized is

$$E = \frac{1}{2}\sum_i (t_i - O_i)^2$$

The back-propagation algorithm consists of carrying out a gradient descent minimization process on $E$. In general, each connection weight is modified following each presentation of example $t$, using following weight updating rules. For the output layer:

$$\Delta w_{ij} = -\eta \frac{\partial E}{\partial w_{ij}} = \eta(t_i - O_i) f'(Net_i) O_j$$

and for the hidden layer:

$$\Delta w_{jk} = -\eta \frac{\partial E}{\partial w_{jk}} = \eta[\sum_i (t_i - O_i) f'(Net_i) w_{ij}] f'(O_j) O_k$$

**Second Derivative Order Information**

Backpropagation neural networks have successfully been applied in manufacturing. But the training process is a very complicated problem. The structure of a network can be modified by adjusting the number of neurons as well as the number of connections per layer. An algorithm based on the second derivative order information can be used to conduct this procedure. This algorithm is derived from the Optimal Brain Damage and given below.

The Optimal Brain Damage (OBD) was proposed by Cun, Denker, and Solla in 1990 [2]. They use information-theoretic ideas to derive a class of practical and nearly optimal schemes for adapting the size of a neural network. By removing unimportant weights from a network, several improvements can be expected: better generalization, fewer training examples required, and improved speed of learning and/or classification. The basic idea is to use second-derivative information to make a tradeoff between network complexity and training set error.

OBD reduces the size of a learning network by selecting deleting weights and suggests better architecture. The basic idea of OBD is that it is possible to take a perfectly reasonable network, delete half (or more) of the weights and wind up with a network that works just as well, or better. It can be applied in situations where a complicated problem must be solved and the system must make optimal use of a limited amount of training data. For a fixed amount of training data, networks with too many weights do not generalize well. On the other hand, networks with too few weights will not have enough power to represent the data accurately. The best generalization is obtained by trading off the training error and the network complexity.

A simple strategy consists in deleting connections with small "saliency", i.e. those whose deletion will have the least effect on the training error. Other things being equal, small-magnitude connections will have the least saliency, so a reasonable initial strategy is to train the network and delete small-magnitude connections in order. After deletion, the network should be retrained. Of course, this procedure can be iterated. Such methods include the implicit hypothesis that the appropriate measure of network complexity is the number of connections (or sometimes the number of neurons) in the network. This methods uses the second derivative of the objective function with respect to the connections to compute the saliencies. The objective function can be expressed in a Taylor series. A perturbation $\delta U$ of the connection vector will change the objective function by [2]

$$\delta E = \sum_i g_i \delta u_i + \frac{1}{2}\sum_i h_{ii}\delta u_i^2 + \frac{1}{2}\sum_{i \neq j} h_{ij}\delta u_i \delta u_j + O(\|\delta U\|)^3$$

where, the $\delta u$'s are the components of $\delta U$, the $g_i$'s are the components of the gradient $G$ of $E$ with respect to $U$, and the $h_{ij}$'s are the elements of the Hessian matrix $H$ of $E$ with respect to $U$:

$$g_i = \frac{\partial E}{\partial u} \qquad and \qquad h_{ij} = \frac{\partial^2 E}{\partial u_i \partial u_j}$$

The goal is to find a set of connections whose deletion will cause the least increase of $E$. The equation can be simplified as follows:

$$\delta E = \frac{1}{2}\sum_i h_{ii}\delta u_i^2$$

In order to find an efficient way of computing the diagonal second derivatives $h_{ij}$, the network state is computed using the standard formulae:

$$O_i = f(Net_i) \qquad and \qquad Net_i = \sum_j w_{ij} O_j$$

where $O_i$ the output of neuron $i$, $Net_i$ its total input (weighted sum), $f$ is activation function and $w_{ij}$ is the weight going from neuron $j$ to neuron $i$. By the chain rule, the diagonal terms of H are given by

$$h_{ij} = \sum_{(i,j)\in v_k} \frac{\partial^2 E}{\partial w_{ij}^2}$$

Since

$E = (t_i - O_i)^2$ and $O_i = f(Net_i) = \dfrac{1}{1+e^{-Net_i}}$ Thus, the second derivatives back-propagated from last-layer to hidden layer is given by

$$h_{ij} = \frac{\partial^2 E}{\partial W_{ij}^2} = \frac{\partial^2 E}{\partial Net_i^2} O_j^2 = [2O_i^2(1-O_i)^2 -$$
$$2(t_i - O_i)O_i(2O_i^2 - 3O_i + 1)]O_j^2$$

and the second derivatives back-propagated from hidden layer to input layer is given by

$$h_{jk} = \frac{\partial^2 E}{\partial W_{jk}^2} = \frac{\partial^2 E}{\partial Net_j^2} O_k^2 = [O_j^2(1-O_j)^2 \sum_i W_{ij}^2 \frac{\partial^2 E}{\partial Net_i^2}$$
$$-O_j(2O_j^2 - 3O_j + 1)\sum_i (-(t_i - O_i)O_i(1-O_i)W_{ij})]O_k^2$$

**Saliencies for each connection:**

$$S_{ab} = \frac{(\sum_{i=1}^{n} \frac{h_{ab}^{i}}{n}) w_{ab}^{2}}{2}$$

where $w_{ab}$ is the weight between neuron a and neuron b, and $i$ is the $i$th pattern. Deleting a connection is defined as setting it to 0 and freezing it there.

# NEURAL NETWORK APPROACH

Manufacturing practices are generally established by trial and error. These processes are not only cumbersome but also expensive. Neural networks process information dynamically in response to external inputs and are thus classified as self adapting systems. The development of a neural network for the flat rolling process involves: the development of a mathematical model of the process which is used to generate the training and testing data for the neural network; the identification of process parameters; data collection; training; optimization; and testing of the neural network.

## Analytical Modeling of The Flat Rolling Process

Rolling is a major metal forming process. During rolling, the desired shape of metal is obtained by plastic deformation taking place between two rolls. The mathematical modeling of rolling process has been long recognized to be a desirable approach to aid in investigating rolling operating practice and the design of mill equipment to improve productivity and quality. However, many factors such as strain hardening and strain rate hardening, friction condition, flattening of rolls, deflection of rolling mill, temperature, as well as their interactions make the theoretical analysis of rolling process very complex and time-consuming.

Since the earlier part of this century, many theories of rolling have been proposed, beginning with the pioneering works of Siebel and von Karman in 1924 and 1925, as discussed in the review by Ford [3]. The most comprehensive work was done by Orowan [4]. His attempt to mathematically describe rolling process started a trend toward greater complexity in rolling theory. After that, Bland and Ford [5], Sims [6], Sabatini [7], and Stone [8] made major contributions to the investigations of rolling. During the last decade, the finite element method has been widely employed by many researchers in the analysis of rolling process.

An analytical model for the flat rolling process is given here based on assumptions that deformation is under plane strain conditions, plane sections remain plane, material remains isotropic and homogeneous, and the deformed arc of contact remains circular. This analytical model is used to generate the data for the training and testing of the developed neural networks.

## Equilibrium Differential Equations

Considering the equilibrium of forces in x and y directions:

$$\Sigma x = 0: \quad 2\sigma_x yB - 2(\sigma_x + d\sigma_x)(y+dy)B + 2(p\sin\theta R' d\theta \mp t\cos\theta R' d\theta)B = 0$$

$$\Sigma y = 0: \quad \sigma_y B dx + (pB\cos\theta R' d\theta \pm tB\sin\theta R' d\theta) = 0$$

where $\sigma_x$, $\sigma_y$ are the local normal stress on slab in x and y direction, respectively shown in Figure2, *dx* is the width of slab, *dy* is the half reduction at the considered slab, *p* is the local, pressure normal to roll surface, $\theta$ is the angle to the line of centers from the considered slab section, *B* is the width of billet at the considered slab, *R'* is the radius of deformed roll, $y = h_x/2$, $h_x$ is the thickness of slab, *t* is the friction stress between roll and slab, The lower of the ± signs are applicable in the backward slip region and the upper signs are applicable in the forward slip region.

Assuming cold flat rolling to be carried out with a forward tension $q_h$ and a backward tension $q_H$, and the yield shear stress k varies from the entry to the exit. The Coulomb's friction law is used in the model: $t = \mu p$. The max bite angle $\alpha_{max}$ is given by: $\mu \geq \tan\alpha_{max}$. The constitutive equation is given by [9]

$$Y = Y_0(1 + B\bar{\varepsilon})^n \times (1 + D\dot{\bar{\varepsilon}})^m$$

Where Y is equivalent flow stress associated with $\bar{\varepsilon}$ and $\dot{\bar{\varepsilon}}$, $\bar{\varepsilon}$ is equivalent strain, $\dot{\bar{\varepsilon}}$ is equivalent strain rate, $Y_0$ is initial yield stress, B, D, n, m are material constants.

In cold rolling, the rolling pressure causes significant flattening of rolls and thus increases the length of contact arc as shown in Figure 3. The projection of the arc of contact $A_2B_2C$ is given by

$$l = A_2D + B_{12}C = \sqrt{R\Delta h + x_2^2} + x_2$$

where $x_2 = \frac{8p_{av}R}{\pi}(\frac{1-\nu_1^2}{E_1} + \frac{1-\nu_2^2}{E_2})$, $p_{av}$ is average roll pressure, $E_1$, $E_2$, and $\nu_1$, $\nu_2$ are elastic moduli and Poisson's ratios for the material of rolls and billet, respectively. During rolling, the originally cylindrical rolls deform in a very complex manner. The radius of deformed roll R' is given by [10]

$$R' = R(1 + \frac{2CF}{\Delta h})$$

where $C = \frac{8}{\pi}(\frac{1-\nu_1^2}{E_1} + \frac{1-\nu_2^2}{E_2})$, R is the original roll radius, R' is the radius of the deformed roll.

### Solution of the Differential Equation

The differential equation of flat rolling process can be solved by the introduction of the above boundary conditions. The solution of the equilibrium equation for the backward and forward slip regiones are

$$p = \frac{2k}{\delta}[(\xi_H\delta - 1)(\frac{H}{h_x})^\delta + 1]$$

$$p = \frac{2k}{\delta}[(\xi_h\delta + 1)(\frac{h_x}{h})^{\delta} - 1]$$

where $\xi_H = (1 - q_H/2k - t\ tan\alpha\ /2k)$ and $\xi_h = (1 - q_h/2k\ )$, $\delta = 2l\mu/\ h$, The above equations describe the roll pressure distribution along the arc of contact. The final formula for the roll separating force per unit width in cold rolling process can be obtained by integrating these equations:

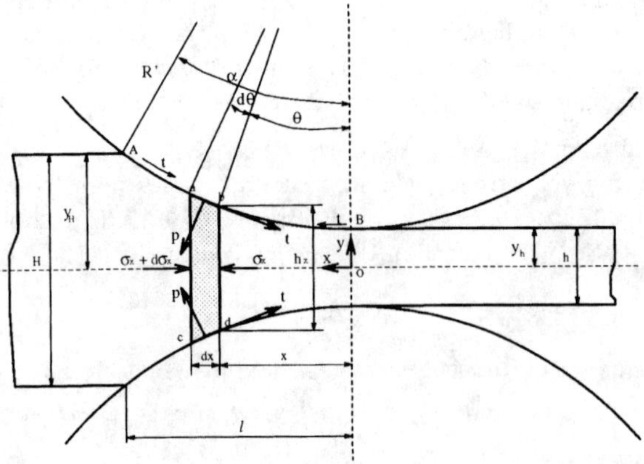


Figure 2. Forces acting in the deformation zone

$$F = \int_{o}^{l} p dx$$

## Data Collection

Knowledge representation is the key to the success of any neural networks. Information about the given problem domain should be collected accurately and should be represented in the correct format. The data should be selected in an optimal fashion by conducting only a minimum number of experiments which would yield the maximum information. The data from the analytical model needs to be comprehensive to cover the spectrum of the process conditions. The critical process parameters for flat rolling process such as billet thickness, roll speed, material properties, forward/backward tensions, etc., were identified as inputs to develop the neural network. Based on the preceding analytical model of flat rolling, a program was used to generate the training and testing data. This program can automatically remove non-practical combinations of the selected parameters, and check the entry condition.

## Training the Neural Network

The neural network was trained and tested with the data from the preceding analytical model. NeuralWare II/Plus was used as the neural network software paradigm for the system. The neural network was trained with 4832 patterns containing 8 inputs, and 1 output. The first step is to determine the number of hidden neurons. The second step is to determine the threshold, learning rate, and momentum. The threshold 0.0065 and momentum 0.2 were set. The learning rates of 0.25 and 0.35 were set for the first 500,000 epochs of the output and the input layers, respectively.

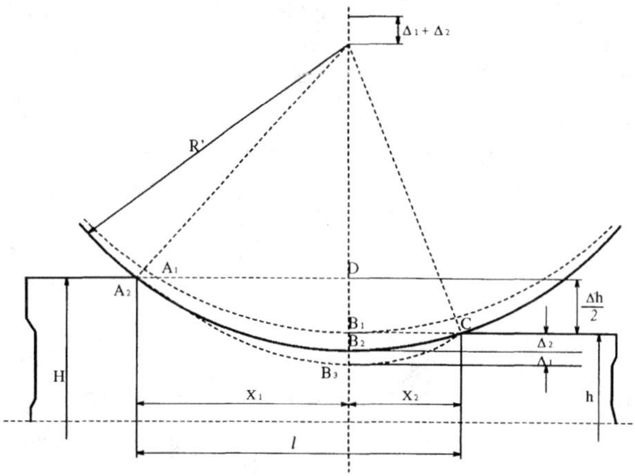


Figure 3. The arc of the contact of deformed roll

### Optimizing the Neural Network Using Second Derivative Order Information

The second derivative order information was used to optimize the network structure by selecting deleting connections with very small saliencies ($S_{ab} \leq 0.5$) each time. After the deletion of the connections with very small saliencies, the weights were reinitialized with small random numbers, and the network was retained again, and after retraining, the connections with very small saliencies were removed again, and weights were reinitialized and the network was retrained again. This procedure was repeated several times till the learning speed was no longer improved. A C program coding the second derivative order information algorithm was used to conduct these procedures.

### Testing Neural Network

The trained neural network was tested with the testing data. The results from the neural network were compared with the output from the analytical model. The testing results show very good co-relationship each other. All prediction error was almost within 5 percentage.

## RESULTS AND DISCUSSIONS

### Prediction Error for Training and Testing Data

The four initial neural networks were generated with the hidden neurons 10, 20, 30, and 40. Four networks run for 450,000 iterations. Upto 450,000 iterations, the networks with the hidden neurons 10 and 40 have larger errors, whereas, the networks with the hidden neurons 20 and 30 have smaller errors. Thus, the networks with hidden neurons 20, 25, 27, 28, 29, 30 were selected. Each network run 600,000 iterations. The results show that the neuron networks with hidden neuron 28 and 29 have lower errors. The small difference between PE=28 and PE=29 is that after 450,000 iterations, the learning of the network with PE=29 becomes very slow, whereas, the network with PE=28 is a little faster.

Therefore, the neural network with hidden neuron PE=28 was selected for further study using the second derivative order information. The selected neural network has 28 hidden neurons, 8 inputs, 1 output, 1 bias, and total connections 281. The threshold was set 0.0065. The average prediction error was calculated at each reduction of the connections using the second derivative order information. The average prediction error is defined by

$$e_{average} = \frac{\sum_{i=1}^{n} \left| \frac{t_i - O_i}{t_i} \right|}{n}$$

where $t_i$ is target value, $O_i$ is predicted value, $n$ is the number of patterns.

The average prediction errors (simply prediction errors) in percentage for training data are shown in Figure 4. There are total 7 times reduction of connections using the second derivative order information. In Figure 4 for the first time using the second derivative order information, a very large reduction of the prediction error was achieved after re-training, but in subsequent time using the second derivative order information, the reduction of the prediction continuously gradually decreased. When the reduction of connections reached at the maximum 37 - 40 % reduction of the original connections, the network can well predict the rolling force even without the need of re-training the network.

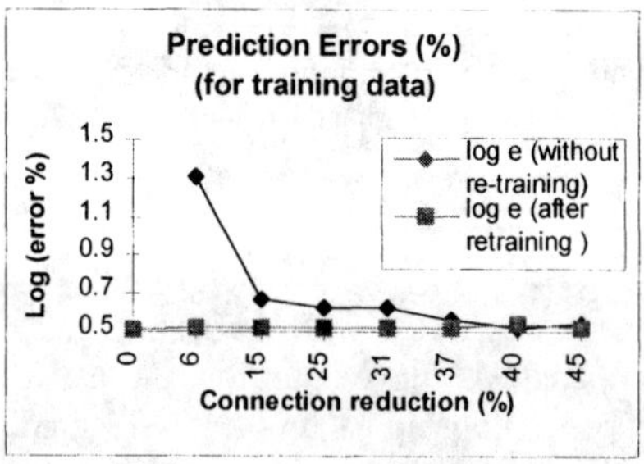


Figure 4. Prediction errors (%) (for training data)

The average prediction errors for testing data are given in Figure 5. There are also total 7 times reduction of connections using the second derivative order information. Figure 5 shows similar situation in Figure 4, except the prediction error is about 5% after retraining in Figure 5, whereas, it is about 3% in Figure 4, which indicates that there is a little increase of the prediction error with testing data.

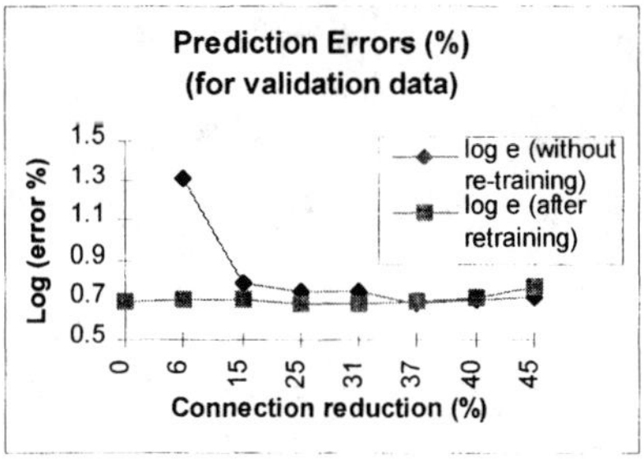


Figure 5. Prediction errors (%) (for validation data)

Figure 6 shows the learning speed in the number of epochs of the selected neural network at the threshold 0.0065. The curve very clearly shows that the number of epochs has two times quick reductions at 6% and 37% connection reduction. The maximum connection reduction corresponding to maximum reduction of epochs is between 30-40%. Therefore, the training speed of the selected neural network can be expected to speed up approximate 40%.

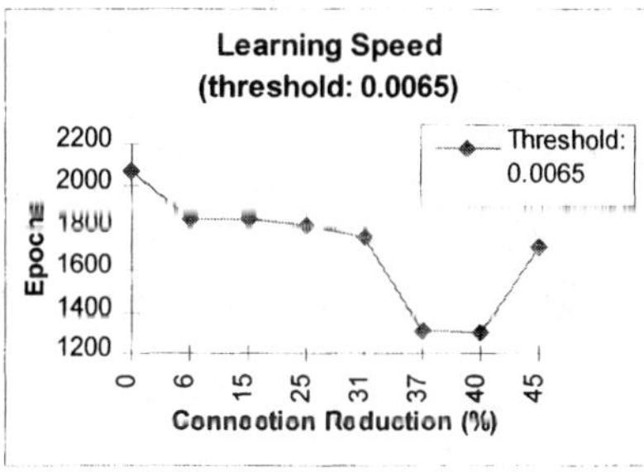


Figure 6. Learning Speed

Comparing Figure 4, 5, and 6, it can be seen that between 30-40% connection reduction using the second derivative order information, the selected neural network reached the maximum reduction of the neural network size without the prediction error increase for both training data and testing data.

**Comparison Of Results From Neural Network Approach And Analytical Model**

The trained neural network used to predict rolling force has hidden neurons 28 and 177 connections. A new set of testing data was used for the comparison of target value and predicted value. In Figure 7, the billet thickness $H_1$ was assumed to vary from 0.192 to to 0.242 inches. Figure 7 shows that the prediction value matches the target value very well. The unit of rolling force is tons per unit width.

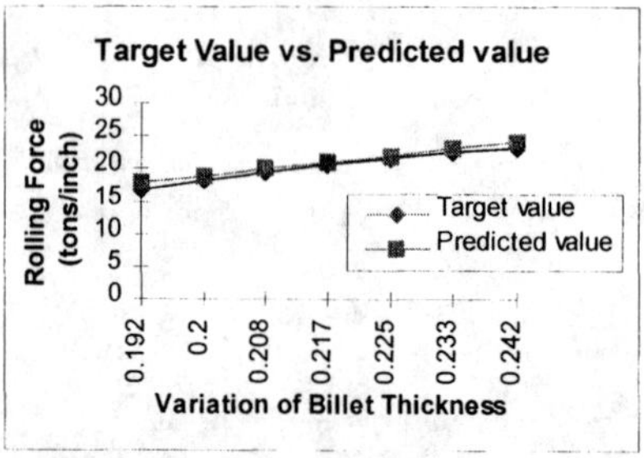


Figure 7. Target Value vs. Predicted Value (I)

In Figure 8, the friction coefficient was assumed to vary from 0.07 to 0.13. The prediction value of rolling force (tons per unit width) is little less than the target value at lower friction coefficient and little greater than the target value at higher friction coefficient as shown in Figure 8.

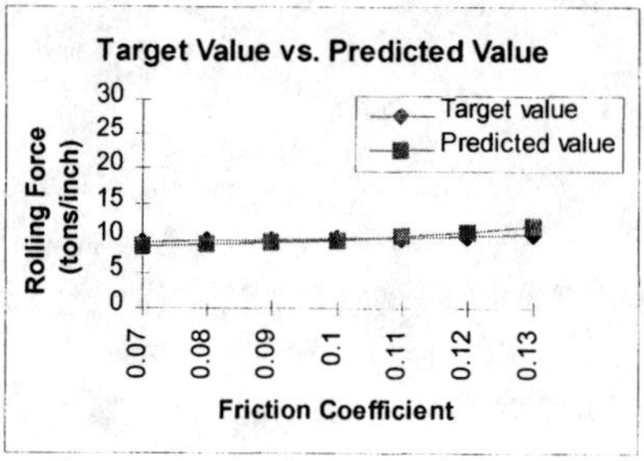


Figure 8. Target Value vs. Predicted Value (II)

## CONCLUSIONS AND FUTURE WORK

A Neural Network Approach based on a back-propagation paradigm has been developed to predict the flat rolling force. The performance of the developed neural network in the learning errors, prediction errors, and training times has been significantly improved by the second derivative order information. As the advantages of neural networks are explored in this research, the neural network approach can become a promising modeling tool for use in analysis of flat rolling processes to meet the requirements of both quick and accurate prediction of the forming load during high speed iterations for the system optimization.

## REFERENCES


[1] Philip D. Wasserman, Advanced Methods in Neural Computing, Van Nostrand Reinhold, 1993

[2] Yann Le Cun, John S. Denker, and Sara A. Solla, "*Optimal Brain Damage*", Proc. AIIE 1990 , pp598-605

[3] D. Ford, *The Theory of Rolling*, Metallurgical Review, Vol. 2, 1957, pp1

[4] E. Orowan, *"The Calculation of Roll Pressure in Hot and Cold Flat Rolling"*, Proc. I. Mech. E., Vol. 150, 1943, pp140

[5] D. R. Bland and H. Ford, *"Cold Rolling with Strip Tension, Part II, Comparison of Calculated and Experimental Results"*, J.I.S.I., Vol. 171, 1952, pp245

[6] R. B. Sims, *"The Calculation of Roll Force and torque in Hot Rolling Mills"*, Proc. I. Mech. E., Vol. 168, 1954, pp191

[7] Sabatini B and K. A. Yeomans, *"An Algebra of Strip Shape and Its Application to Mill Scheduling"*, J.I.S.I., Dec. 1968

[8] M. D. Stone, et al, *"Theory and Practical Aspects of Crown Control"*, AISE Yearly Proceedings, 1965, pp657

[9] J. M. Alexander, R. C. Brewer, and G. W. Rowe, Manufacturing Technology, Ellis Horwood Limited, UK, 1987

[10] Wang Kezhi, Theory of Rolling, Beijing University of Science and Technology, Beijing, P.R. China, 1986

# ARTIFICIAL NEURAL NETWORK APPROACH TO THE CONTROL OF A WAVE SOLDERING PROCESS

**Janet M. Twomey**
*Department of Industrial and Manufacturing Engineering*
*Wichita State University, Wichita, KS 67260*

**Alice E. Smith**
*Department of Industrial Engineering*
*University of Pittsburgh, Pittsburgh, PA 15261*

***ABSTRACT:***
This research proposes a novel artificial neural network based approach for the control of wave soldering processes of electronic circuit boards. The network approach is committee networks by cross-validation. This method maximizes the use of the entire data set for model building and evaluation. The approach is compared to the two most common methods of building and evaluating neural networks.

Wave soldering takes into account (1) preheating, (2) fluxing, (3) soldering using a wave of solder, (4) cleaning, and (5) quality control. The goal is to optimize quality via predetermined processing parameters, where optimal quality is defined in terms of minimal solder defects. The controllable process parameters involve preheat temperatures and line speed. In the past, the manufacturer has relied on trial and error, and simple linear models to control the solder process [1]. An artificial neural network (ANN) model was shown to outperform linear regression models [2]. The results presented in this paper extend the research reported in [2,3]. Coit, et al used a sample of 204 observations to build and validate their ANN model. Here we are interested in building and evaluating ANN models in cases where data is scarce. While not a problem for [2], scarcity of observations in real world applications of ANN is a common occurrence. This paper proposes an ANN model building and evaluation approach that maximizes the use of the entire number of observations available to the ANN modeler. The approach combines the committee network (CN) method of ANN model building with the cross-validation (CV) or leave-one-out method of model validation.

## TYPICAL METHODS OF ANN MODEL BUILDING

The most commonly used method of ANN model building is to subdivide the total available sample of n observations into a training set of size $n_1$ and a test set of size $n_2$; where $n = n_1 + n_2$. A single network is constructed of $n_1$ observations and is tested for generalization capabilities on $n_2$ observations. This method is referred to as the Train-and-Test (T/T) method.

The error estimate according to the train-and-test (T/T) method is given by:

$$\hat{Err}_{T/T} = \frac{1}{n_2}\sum_{j=1}^{n2}\left|y_j - \hat{f}[T_{(n1)}, \mathbf{x}_j]\right| \quad (1)$$

where $\hat{f}[T_{(n1)}, \mathbf{x}_j]$ is the network constructed on $n_1$. ANN [4] like other nonparametric models require large amounts of data to achieve the best performing model. In cases where samples are small, the ANN model built on a reduced number of observations, $n_1$, will not perform as well as an ANN built on n observations. In addition, the error estimate is often inaccurate and highly variable [5]. A second method builds and evaluates the ANN model on the total number of observations available to the modeler n. While making the best use of n for model building, this approach has the obvious drawback of underestimating the error of the ANN model, since the same data that is used to train the ANN model is also used to test the ANN model. This method is referred to as the resubstitution (Resub) method. The error estimate is given by:

$$\hat{Err}_{Resub.} = \frac{1}{n}\sum_{j=1}^{n}\left|y_j - \hat{f}[T_{(n)}, \mathbf{x}_j]\right| \quad (2)$$

where $\hat{f}[T_{(n)}, \mathbf{x}_j]$ is the network constructed on n. Both methods produce good estimates of error when sample sizes are large. Since building and evaluating ANN is more problematic in cases where data is scarce, this research examines a new approach for building and evaluating ANN models under conditions of limited data.

## COMMITTEE NETWORK (CN) APPROACH TO ANN MODEL BUILDING

The committee network approach to ANN building has also been referred to as "majority fusers" or "democratic networks" [6,7,8]. It is compared to the more commonly used single network approach. There are numerous variations on the committee network theme, but the general idea is as follows: a number of networks, k, are built and trained to solve the same problem, where the problem is either classification or function approximation. An input vector $\mathbf{x_j}$, is provided to each trained network:

$$\textit{trained net}_i(\mathbf{x}_j) \rightarrow y_{ij};\ \text{for } i = 1,2,\ldots,k. \quad (3)$$

where i is the ith network, and j is the jth input vector. The output $y_{ij}$ of each network are combined using some rule or fuser, *f*. The output, $Y_j$, that is the result of the rule is then used as the final prediction:

$$f(y_{ij}) \rightarrow Y_j;\ i = 1,2,\ldots,k. \quad (4)$$

In order to evaluate the performance of $Y_j$, $Y_j$ is compared to target $t_j$. The performance of the committee network is evaluated following the usual T/T methodology. The theory behind committee networks is that a composite system of combined agents, in this case ANN agents, will be better than the best of the agents that went into the composite system. Methods to combine the system of agents for classification purposes have been proposed and studied [6,7,8]. The focus of this research is on CN used as function approximators.

**Cross-validation (leave-one-out) Method of Network Validation**

The cross-validation [9] method (CV) of estimating the error is a popular method for estimating the error statistical model, $\hat{Err}_{CV}$. CV uses data resampling to determine the estimate of a model's error that has been constructed on all n observations; i.e. the estimate of the model's error is determined by constructing n models on sub-sets of all data points leaving out one (size n-1), then testing on the one omitted point (size 1) in all n possible ways:

$$\hat{Err}_{CV} = \frac{1}{n}\sum_{j=1}^{n}\left|y_j - \hat{f}[T_{(j)}, \mathbf{x}_j]\right| \qquad (5)$$

where j is the excluded point and $T_{(j)}=\{(\mathbf{x}_1, y_1), (\mathbf{x}_2, y_2), \ldots, (\mathbf{x}_{j-1}, y_{j-1}), (\mathbf{x}_{j+1}, y_{j+1}), \ldots, (\mathbf{x}_n, y_n)\}$ is the data set minus point j used to construct the network model $\hat{f}[T_{(j)}, \mathbf{x}_j]$.

**Combined approaches of the Committee Networks and Cross-validation**

The combined committee network approach with the CV method constructs the network that is ultimately used to predict the wave solder temperature, as a committee network formed by fusing the outputs of the networks constructed according to the CV method of resampling. The committee network is then evaluated by $\hat{Err}_{CV}$. This method is referred to as committee networks by cross validation (CN by CV).

## EXPERIMENTS

A number of experiments were performed to investigate the proposed ANN training methodology that enables the formulation of the best prediction model while also providing an accurate and reliable measure of performance of the network, in cases where the data available to the modeler is scarce. The committee network performance is compared to the performance of single networks built according to the T/T and Resub. methods The estimates of the network's error as determined by the corresponding methodologies is also compared according to the accuracy and variability of the estimate.

**Experimental Design**

For this research the ANN models were constructed using four input variables and one output variable shown in Table 1. This is a subset of the sixteen original input variables used in the study conducted by Coit, et al [2]. The output variable, mean card surface temperature at the wave of solder, was found to be a good indicator of solder quality. The total amount of data available for experimental purposes was limited to 204 observations, collected and supplied by the manufacturer.

For this study, 100 of the 204 observations, were selected at random and set aside to evaluate the three methodologies (T/T, Resub, and CN by CV) in terms of the *performance* of the networks used for prediction. The remaining 104 observations were used to construct samples of data, n, available to the modeler for both model construction and evaluation. Two levels of scarce data were examined, n = 12 and 40, which were chosen randomly from the 104 observations. The three methods were applied using the same sample of n = 12 and 40. To examine the variability associated with the individual approaches to network construction and corresponding methods of error estimation, 100 replications of each experiment were conducted. A single experiment is replicated by taking independent samples from the 104 observations, thereby creating 100 samples of size 12, and 100 samples of size 40. Each sample was unique in makeup; i.e. no two samples were alike. A single experiment is described here:

Train-and-Test: Historically, the percentage of n used to train a network has varied. Thus 3 levels of percentage of n used to train a network were examined: 75%, 50%, and 25%; i.e. n = 12 (9/3), (6/6) and (3/9), and for n = 40, (30/10), (20/20), and (10/30). Networks were trained and tested at each level of T/T.

Resubstitution: One network is built on n observations and tested on the same n observations.

Committee network by cross-validation: From the n observations, n networks were trained on the resampled data sets of n-1 observations. The CN fuser chosen for this research was the mean over the predictions of output Y given by the n networks. The CV error of the CN was determined according to eq (5).

Network Performance: The T/T networks, the Resub network, and the CN by CV were evaluated on the same 100 observations not included for training.

Table 1 Network Model: Inputs and Output

| | |
|---|---|
| Input 1 | location 1: pre-heat temperature prior to wave solder |
| Input 2 | location 2: pre-heat temperature prior to wave solder |
| Input 3 | loaded mass of circuit board |
| Input 4 | conveyor speed |
| Output 1 | Temperature: mean thermal condition at the wave |

## Network Training

In order to compare methodologies networks architecture, training algorithm and training parameters were held constant for all networks. Networks were constructed according Table 3, with 2-hidden layers of 3 nodes each. The network training algorithm was the original version of the backpropagation algorithm with momentum term. All network weights were randomly initialized to the same starting point between -0.5 and 0.5. All training parameters were held constant with a sigmoidal transfer function, a training rate of 0.10, and momentum term of 0.90. Networks continued to train until the network no longer continued learn: i.e. as long as the weights continued to change within a given tolerance.

## RESULTS

The 95% confidence intervals for the average performance, measured in terms of the mean absolute error (MAE), of the networks obtained from the 100 independent

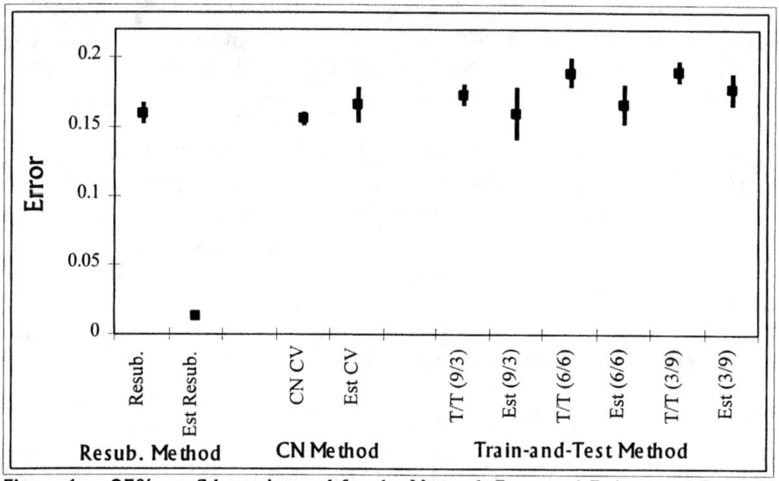


Figure 1. 95% confidence interval for the Network Error and Estimate of Error (Est) for three methods of constructing and evaluating networks: n = 12.

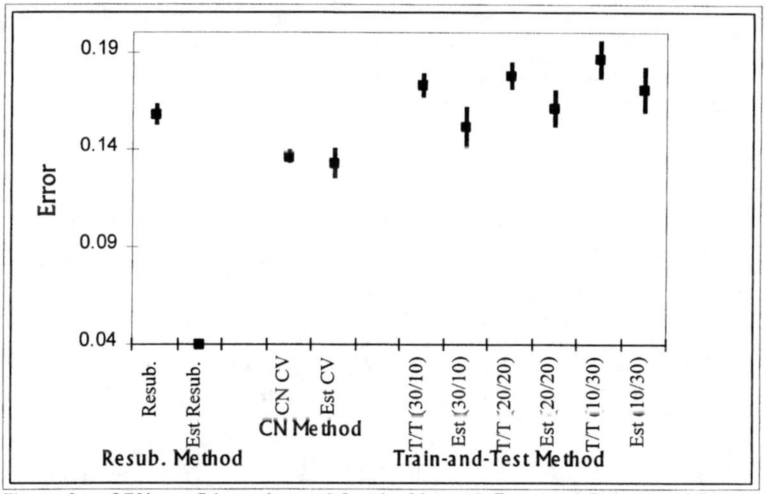


Figure 2. 95% confidence interval for the Network Error and Estimate of Error (Est) for three methods of constructing and evaluating networks: n = 40.

observations are depicted in Figures 1 and 2 for n = 12 and 40, respectively. Also depicted are the error estimates (Est.) as determined by the corresponding method. Network performance is considered first. As expected, performance improved for all methods of network construction as n increased from 12 to 40. CN by CV on average provides the best performing networks as exhibited by low average error and associated low variability. Although the difference in average error when compared to the Resub method at n = 12 is not great, the smaller confidence width indicates that the performance of the CN by CV is more stable over a variety of samples, while the Resub and T/T methods are sensitive to the changes in the composition of the sample used to construct the network. As expected, due to the decrease in training set size, the performance of the T/T networks were poorest. What is not expected is the large associated variability.

The goodness of the error estimates is considered next. The most striking result is the error estimate according to the Resub method. As anticipated, the Resub method underestimates the error of the network to a very large degree. The variability of the Resub estimates are minimal. This result reflects the consistency in training; i.e. all networks were trained equally well. The T/T methods also tend to underestimate the error of the corresponding networks, but with greater variability. The CV estimates tend to overestimate the error of the CN by CV, while providing the most accurate and least variable estimates.

## CONCLUSIONS

The results have indicated that on average the committee networks by cross-validation produce more accurate and less variable performing networks with the best performing estimates of the networks error, when compared to the most common approaches to ANN construction and evaluation, under conditions of limited data. This was demonstrated using historical data for the control of the wave soldering process of circuit boards. Future work will include investigations into the bias/variance and theoretical underpinnings.

## REFERENCES


1. Brinkley, Paul A. (1993). Northern Telecom achieves improved quality by combining DOE and SPC, *Industrial Engineering*, **25**(5), 63-65.

2 Coit, D.W., Billa, J. Leonard, D.,Smith, A.E., Clark, W., and El-Jaroudi, A. (1994). Wave solder process control neural modeling using a neural network approach. *Intelligent Engineering Systems Through Artificial Networks: Volume 4, ASME Press, 999-1004.*

3 Twomey, J. M. and Smith, A. E. (1995), "Committee Networks by resampling", *Intelligent Neural Engineering Systems Through Artificial Networks, Volume 4,* ASME Press.

4 Geman, S., Bienenstock, E., and Doursat, R. (1992). Neural networks and the bias/variance dilemma. *Neural Computation*, **4**, 1-58.

5 Twomey, J.M. and Smith, A.E. "Bias and variance of validation methods for function approximation neural networks under conditions of sparse data" Submitted to *IEEE Transactions on Systems, Man, and Cybernetics.*

6 Rao, Nageswara S.V. and Oblow, E.M. (1994) Majority and location_based fusers for systems of PAC concept learners. *IEEE Transactions on Systems, Man, and Cybernetics*, **24**, 715-727.

7 Battiti, R. and Colla, A.M. (1994). Democracy in neural nets: voting schemes for classification. *Neural Networks*, **7**, 691-707.

8. Hasham, S. and Schmeiser, B. (1992) Improving model accuracy using optimal linear combination of trained neural networks. *Technical Report SMS92-16,* School of Industrial Engineering, Purdue University.

9. Lachenbruch, P. and Mickey, M. (1968). Estimation of error rates in discriminant analysis. *Technometrics*, **10,** 1-11.

# Section 8 : Process Monitoring Control and Automated Inspection

# PERFORMANCE RELIABILITY PREDICTION: ESTIMATING CONFIDENCE INTERVALS FOR FFNS USING SELF-ORGANIZING FEATURE MAPS

**RATNA BABU CHINNAM**
*Department of Industrial Engineering and Management,*
*North Dakota State University, Fargo, ND 58105*

***ABSTRACT:***
This paper presents an approach that uses a temporal processing network and a self organizing feature map to predict the real-time conditional performance reliability measure of risk relative to tool failure. The specific application considered is in-process monitoring of the condition of the drill bit in a drilling process utilizing the torque and thrust signals, and the results are very promising.

## INTRODUCTION

Feed forward neural networks (FFNs) have proven to be very effective in function approximation and time-series forecasting [Sharda and Patil, 1990]. When used for time series forecasting, they focus primarily on accurate prediction of future values. However, they normally do not provide adequate information about the confidence with which they make these predictions. Knowing the "Confidence Interval" is of paramount importance in many applications because it indicates the extent to which one can rely on predictions for making future decisions. In contrast to globally generalizing networks such as MLP, locally generalizing networks such as RBF networks and Cerebellar Model Arithmetic Computer (CMAC) networks have a naturally well defined concept of local neighborhood.

The principal objective of this paper is to introduce a unique approach to compute confidence intervals/quantiles for any FFN by combining the network with a Self-Organizing Feature Map (SOFM). A secondary objective is to introduce an approach that utilizes the above unique concept to monitor physical systems, and determine their performance reliability as they degrade with time. In particular, the application considered in this paper is a machining process, and the intent is to predict the performance reliability of a drill-bit as it degrades with usage.

## AN APPROACH TO COMPUTE CONFIDENCE INTERVALS USING FFNS

This section introduces an approach that integrates SOFMs with FFNs to estimate the confidence intervals/quantiles associated with network predictions. The intent is to utilize the SOFM to introduce the concept of a "local neighborhood" even with globally approximating FFNs, critical for determination of confidence intervals.

Let, $X$ denote a spatially *continuous input space*, the topology of which is defined by the metric relationship of the vectors $\mathbf{x} \in X$. Let $A$ denote a *spatially discrete output space*, the topology of which is endowed by arranging a set of neurons ($N$) as the computation nodes of a lattice. Let $\Phi$ denote a non-linear transformation called a *feature map*, which maps the input space $X$ onto the output space $A$, as shown by $\Phi: X \rightarrow A$. This may be viewed as an abstraction that defines the location of a winning neuron $i(\mathbf{x})$ developed in response to an input vector $\mathbf{x}$. Given an input vector $\mathbf{x}$, the SOFM algorithm proceeds by first identifying a best-matching or winning neuron $i(\mathbf{x})$ in the output space A, in accordance with the feature map $\Phi$. The synaptic weight vector $\mathbf{w}_i$ of neuron $i(\mathbf{x})$ may then be viewed as a *pointer* for that neuron into the input space $X$.

**Introduction of a Definition of Local Neighborhood for FFNs Using SOFMs**

Let $M$ represent the total number of training patterns spanning the entire input space $X$. Let $M_j$ the "membership" of neuron $j$ in the discrete output space $A$ represent the subset of training patterns from input space $X$ that activate it. This is shown by:

$$i(\mathbf{x}) = j \text{ for all } \mathbf{x} \in M_j, \quad j = 1, 2, ..., N \tag{1}$$

It is also true that the sum of the memberships of the neurons in the lattice output space must equal the total number of training patterns for the SOFM, as shown by:

$$\sum_{j=1}^{N} M_j = M \tag{2}$$

The properties of: Approximation of the Input Space, Topological Ordering, and Density Matching (see Haykin [1994]) exhibited by SOFMs provide the motivation to utilize the SOFM to break the input space $X$ into $N$ distinct regions (denoted by $X_j$) that are mutually exclusive, and hence satisfy the following relationship:

$$\sum_{j=1}^{N} X_j = X \tag{3}$$

All the signal patterns from any given distinct region $X_j$, when provided as input to the feature map $\Phi$, will activate the same output neuron $j$. This is shown by:

$$i(\mathbf{x}) = j \text{ for all } \mathbf{x} \in X_j, \ j = 1, 2, ..., N \tag{4}$$

Thus, using SOFMs, one can introduce the concept of a "local neighborhood," the resolution depending on the number of neurons ($N$) in the discrete output space.

**Construction of Confidence-Intervals/Quantiles for FFNs**

From the definition of local neighborhood introduced earlier, input signal patterns can be associated unambiguously with one of the distinct regions $X_j$. Assuming that a FFN is being used for function approximation or time-series forecasting, an estimate of the covariance matrix for the FFN model residuals within the domain of region $X_j$ is given by:

$$Cov_j = \begin{bmatrix} S_{11} & S_{12} & \cdots & S_{1O} \\ S_{21} & S_{22} & \cdots & S_{2O} \\ \vdots & \vdots & \ddots & \vdots \\ S_{O1} & S_{O2} & \cdots & S_{OO} \end{bmatrix} \tag{5}$$

where:

$S_{pq} = \frac{1}{(M_j - 1)} \sum_{k=1}^{M_j} (E_{kp})(E_{kq})$, and denotes the covariance between output variables $p$ and $q$

$E_{kp}$ denotes the FFN model residual for output variable $p$ for pattern $k$
$O$ denotes the number of output variables predicted by the FFN

Assuming that the residuals are independent and Gaussian distributed with a constant covariance matrix over the domain of any region but varying from domain to domain, the (1-α) quantile is given by the point **x** satisfying the following condition:

$$(\mathbf{x} - \mu_j)^T Cov_j^{-1} (\mathbf{x} - \mu_j) \leq \chi_O^2(\alpha) \tag{6}$$

where:

$\chi_O^2(\alpha)$ denotes the (1-α) quantile of the Chi-Square distribution with $O$ degrees of freedom.
$\mu_j$ denotes the mean residual vector for domain $X_j$, and
$Cov_j^{-1}$ is the inverse of the matrix $Cov_j$.

If the FFN has adequate representational capacity, the fit should not be significantly biased, and the mean residual vector can be a null vector.

## PERFORMANCE RELIABILITY PREDICTION USING FFNS AND SOFMS

Traditional approach to reliability analysis was based on life tests that record only time-to-failure [Lu and Meeker, 1993]. Such analyses provide information regarding the reliability of an "average" system, subsystem, or component. If there are system, subsystem, or component characteristics whose degradation over time can be related to reliability, then collecting "degradation data" can provide information about the system, subsystem, or component reliability as it degrades with time. This section introduces an approach that utilizes the unique concepts introduced earlier to monitor physical systems, and determine their performance reliability as they degrade with time.

Given an appropriate definition of failure in terms of the performance characteristic, *performance reliability* can be defined as "the conditional probability that the performance measure $Y$ is less (or greater) than a performance critical limit $Y^{pcl}$, given operating environmental conditions, for a specified period of time or cycles $t$." This definition applies to all three basic possibilities that exist: 1) lower-is-better performance measure with an upper critical limit, 2) higher-is-better performance measure with a lower critical limit, and 3) nominal-value-is-best performance measure with two-sided critical limits.

Let F($t$) denote the probability that failure takes place at a time or usage less than or equal to $t$ (i.e. P($\mathbf{t} \leq t$)), where the random variable **t** denotes the time to failure. From the definition of conditional probability, the conditional reliability that the system, subsystem, or component will fail at some time or usage $\mathbf{t} > t + \Delta t$, given that it has not yet failed at time $\mathbf{t} = t$ will be:

$$\begin{aligned} R((t+\Delta t)|t) &= P(\mathbf{t} > t + \Delta t | \mathbf{t} > t) \\ &= [1 - P(\mathbf{t} \leq t + \Delta t)]/P(\mathbf{t} > t) \end{aligned} \tag{7}$$

Let $\mathbf{Y} = \left[Y_1 Y_2 .. Y_O\right]^T$ denote the vector of $O$ physical performance characteristics being monitored from the system, subsystem, or component being evaluated. Let $\mathbf{Y}^{pcl} = \left[Y_1^{pcl} Y_2^{pcl} .. Y_O^{pcl}\right]^T$ denote the vector of critical performance limits for the $O$ physical performance measures. The above definition of performance reliability can be easily extended to incorporate this multiple performance measure case. Let us make the assumption here that all the $O$ performance measures are of lower-is-better type characteristics.

Since a FFN can provide us with a predicted performance measure into the future (at time $t_f$) denoted by $\hat{\mathbf{Y}}(t_f)$, under the assumption that the change in the performance measures from the current time point ($t_c$) to the predicted time point ($t_f$) is either monotonically decreasing or increasing, the reliability that the system, subsystem, or component will operate without failure until $t_f$ is given by:

$$R\left(\left(\mathbf{t} \geq t_f\right) \middle| t_c\right) = \int_{y_1=-\infty}^{y_1=Y_1^{pcl}} \int_{y_2=-\infty}^{y_2=Y_2^{pcl}} .. \int_{y_O=-\infty}^{y_O=Y_O^{pcl}} f\left(\mathbf{Y}\left(t_f\right)\right) dy_1 dy_2 .. dy_O \tag{8}$$

where:

$f\left(\mathbf{Y}\left(t_f\right)\right)$ denotes the probability density function of the predicted performance measure into the future at time $t_f$.

Assuming that the residuals are independent and Gaussian distributed, with a constant covariance matrix over the domain of any region (SOFM partitioned input space) but varying from domain to domain, $f\left(\mathbf{Y}\left(t_f\right)\right)$ in the above reliability equation can be replaced by the following:

$$f\left(\mathbf{Y}\left(t_f\right)\right) = \frac{1}{(2\pi)\left|Cov_j\right|^{1/2}} \exp\left[-\frac{1}{2}\left(\mathbf{Y}\left(t_f\right) - \hat{\mathbf{Y}}\left(t_f\right) - \mu_j\right)^T Cov_j^{-1}\left(\mathbf{Y}\left(t_f\right) - \hat{\mathbf{Y}}\left(t_f\right) - \mu_j\right)\right] \tag{9}$$

where:

$\mathbf{Y}\left(t_f\right)$ denotes the predicted performance measure into the future at time $t_f$,

$\mu_j$ denotes the mean residual vector for domain $X_j$ of the SOFM,

$Cov_j$ is the covariance matrix for the FFN model residuals within the domain of region $X_j$ of the SOFM, and

$\left|Cov_j\right|$ denotes determinant of the covariance matrix for the FFN model residuals within the domain of region $X_j$ of the SOFM.

**Drilling Process Example**

To demonstrate the concepts developed above, a series of drilling tests were conducted on 0.25 inch hot rolled steel bars using 16 quarter-inch HSS twist drill-bits of M-1 grade on a HAAS VF-1 Machining Center, at the following conditions: speed 2000 rpm, feed 20 ipm, using no coolant. The resulting data were used to train a Temporal Processing network for time-series forecasting and a SOFM for introducing the definition of local neighborhood.

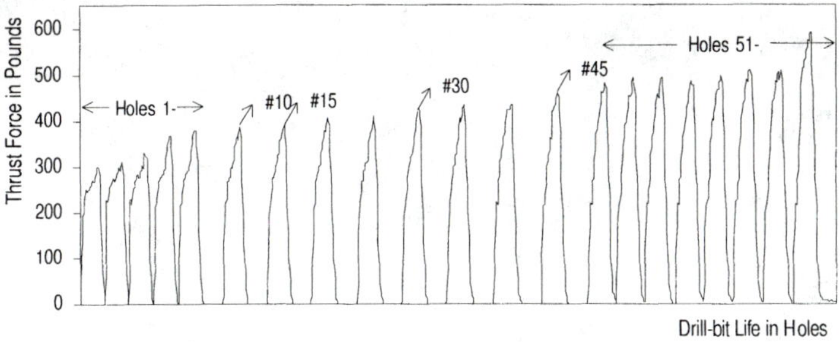


**Figure 1. Drill-bit #1 performance in terms of thrust force in pounds.**

The 16 drill-bits varied in life any where between 33 holes and 75 holes. The data collection sampling rate was approximately 150 data points per second (hertz). We decided to work with the *thrust force* and *torque* acting on the drill-bit to monitor the performance reliability of the drill-bit. The thrust signature collected from drill-bit #1 is shown in Figure 1. This drill-bit lasted 58 holes, and not all the information is used for training purposes. In order to reduce the network training time, it was decided to use the information collected during the first 5 holes, last 8 holes, and every 5th hole in between.

**Feed Forward Neural Network Design and Training Issues**

A temporal processing network [Wan, 1990] with the following configuration appeared to be good at maintaining generalization with respect to predicting the thrust force and torque levels into the future.[1]

Input layer:

Number of neurons[2]: 2; Number of taps per synaptic filter: 20

Hidden layer:

Number of neurons: 50; Number of taps per synaptic filter: 0

Output layer:

Number of neurons[3]: 6

A SOFM with a two-dimensional lattice of neurons (8x8) was used in dividing the 42 dimensional continuous input space into 64 distinct regions in an adaptive, topologically ordered fashion.[4] The adaptive training scheme and parameter selection process discussed by Haykin [1994] was utilized in training the network.

The conditional performance reliabilities are calculated based on the assumption that the critical plane for the drill-bit with respect to thrust is 700 pounds and critical plane for the

[1] Of the information collected from 16 drill-bits, information from 12 randomly picked drill-bits was used for training purposes, and the information from the remaining 4 drill-bits was used for testing purposes. The network was designed to reduce the mean-square-error associated with testing patterns.

[2] One for each of the two (thrust force and torque) performance measures.

[3] Three neurons for each of the two performance measures. First neuron is used for predicting 1 step into the future, second neuron for predicting 3 steps into the future, and the third neuron is used for predicting 6 steps into the future.

[4] (20+1) dimensional thrust force input vector and (20+1) dimensional torque input vector equals 42 dimensions.

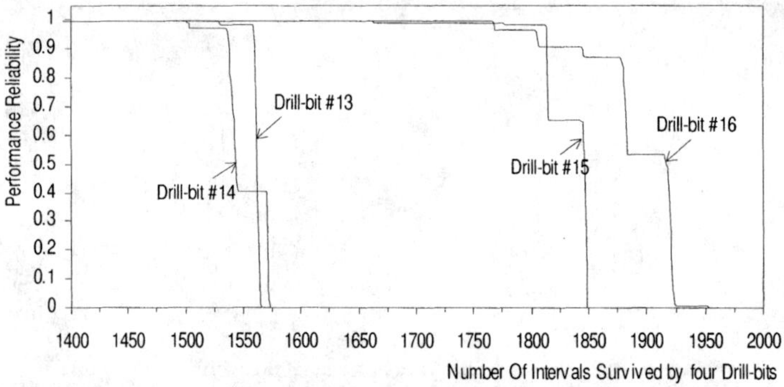


**Figure 2. Performance reliability exhibited by drill-bits #13, #14, #15, and #16.**

drill-bit with respect to torque is 42.5 inch-pounds. Figure 2 depicts the changes in the performance reliabilities (unconditional) for the 4 drill-bits used for testing. Since the temporal processing network has 21 taps in the input layer, these calculations are based on the assumption that each drill-bit will survive the first 21 intervals (0.14 seconds) with certainty.

## CONCLUSION

In contrast to locally generalizing networks such as RBF networks and CMAC networks that have a naturally well defined concept of local neighborhood, globally generalizing networks FFNs, such as MLP networks and Temporal Processing networks, are not inherently capable of providing confidence intervals for their predictions. A unique approach to compute confidence intervals/quantiles for any FFN by combining the network with a SOFM was introduced. In addition, an approach that utilizes the above unique concept to monitor physical systems, and determine their performance reliability as they degrade with time is introduced.

## REFERENCES

Haykin, S., 1994, *Neural Networks: A Comprehensive Foundation*, Macmillan College Publishing Company, New York, NY.

Lu, C. J. and Meeker, W. Q., 1993, "Using Degradation Measures to Estimate a Time-to-Failure Distribution," *Technometrics* 35, pp. 161-174.

Sharda, R. and Patil, R. B., 1990, "Neural Networks as Forecasting Experts: An Empirical Test," *Proceedings of the IJCNN Meeting*, pp. 491-494, Washington DC.

Wan, E. A., 1990, "Temporal Backpropagation for FIR Neural Networks," *Proceedings of the IEEE International Joint Conference on Neural Networks*, Vol. 1, pp. 575-580, San Diego, CA.

# RAPID IDENTIFICATION OF NUCLEAR POWER PLANT MALFUNCTIONS WITH ARTIFICIAL NEURAL NETWORKS VIA FOURIER TRANSFORMED SIGNALS

**SANDOR BENEDEK**
*Institute for Electric Power Research, 1368 Budapest, POB 233, Hungary*

**MARK J. EMBRECHTS** (embrem@rpi.edu)
*Rensselaer Polytechnic Institute, Troy, NY 12180-3590, USA*

***ABSTRACT:***
Rapid identification of malfunctions is of premier importance for the safe operation of nuclear power plants. In order to provide sufficient lead time malfunctions have to be identified within 60 seconds. A novel method that relies on the Fourier transform of power plant transient signals shows the promise to satisfy this goal well within this time frame. A feedforward neural network trained with the backpropagation algorithm was developed to model simulated nuclear power plant malfunctions for a pressurized water reactor (PWR) and this model was then successfully applied to identify real malfunctions of the Hungarian Paks Nuclear Power Plant.

## 1. INTRODUCTION

The proper and timely recognition of nuclear power plant malfunctions is of premier importance to guarantee the safe and reliable exploitation of nuclear power plants. Previous methods to identify nuclear power plant transients were based on time series data of various transient signals. They relied mainly on rule based procedures to pre-select proper aggregate networks of malfunctions. Various neural networks were then trained and used to select the proper malfunction from the appropriate aggregate.

The methodology discussed in this paper is novel in the sense that selected Fourier components of transient signals are retained to train one single artificial neural network (ANN) in order to learn to characterize various nuclear power plant malfunctions -- rather than time series of various transient signals. The key advantages of this method are the avoidance of heuristically developed rules and the potential for a faster and more reliable identification of the transient in question. Training neural networks based on the original time series data in a single neural network consistently led to significantly inferior performances when compared to networks based on Fourier transformed data.

The artificial neural networks employed in this paper are of the basic feedforward type trained with the traditional backpropagation algorithm (Haykin, 1994). The results illustrated in this paper are mainly for 80 seconds of transient signals, however the FFT method is applicable to successfully recognize reactor malfunctions within a shorter time frame (i.e., 48 seconds).

The terminology related to nuclear power plant malfunctions and the fundamental ingredients to develop ANNs for nuclear power plant malfunctions are discussed by Parlos et al. (Parlos, Atiya, Chong and Tsai, 1992). Different improved ANN methods related to nuclear power plant malfunction recognition with the aid of artificial neural networks have been described in the literature: see e.g., (Bartlett and Uhrig, 1992), (Cheon and Chang, 1993). A real-time ANN learning methodology with an adaptive real-time monitoring capability is described for plant wide data from an operating nuclear power plant by Nabeshima et al. (Nabeshima, Turkan, and Cifcioglu, 1994). The employed network type in this reference is auto-associative (i.e., inputs and outputs are the same).

This paper introduces the use of Fast Fourier Transformed signal identifiers for the rapid recognition of nuclear power plant malfunctions with the aid of artificial neural networks. One of the earlier applications of the FFT for signal recognition is discussed by Embrechts et al. (Embrechts, Lahey, and Yapo, 1992). Different on-line recognition methods for PWR fault diagnosis are compared by Palancz et al. (Palancz, Benedek, and Magyarné). In that paper five aggregate networks of malfunctions developed with a rule-based system - based on time series data of various transients - are further refined with the aid of ANNs. The FFT methodology discussed in this paper uses the same basic transient data as the paper by Palancz, et al. However, in the FFT-based ANN all signals (rather than heuristic rule-based subsets) are shown to one single neural network (rather than 5 different aggregate networks). The FFT based methodology results in fewer inputs overall and a better performance when compared to a single network and rule based subsets based on time series data.

## 2. NEURAL MODELLING

### 2.1 Nuclear Power Plant Malfunctions

The nuclear power plant malfunctions were generated with the Hungarian full scale training simulator of Paks NPP. Seventeen transient signals (Table I) were monitored during 80 seconds with time intervals of one second. Seventeen distinct basic cases of malfunctions were used for training (Table II) and twenty two malfunction patterns were used for testing. The last five of these testing malfunctions were of a type not included in the seventeen distinct base cases.

### Table I. Plant variables used as input signals

| VARIABLE | DESCRIPTION |
|---|---|
| 1 | Relative power of reactor |
| 2 | Water level in pressurizer (cm) |
| 3 | Pressure in pressurizer (bar) |
| 4 | Pressure drop in reactor core (bar) |
| 5 | Pressure in main steam collector nr. 1 (bar) |
| 6 | Pressure in main steam collector nr. 2 (bar) |
| 7 | Position of control rods (cm) |
| 8 | Containment pressure (mbar) |
| 9 | Cold-leg coolant temperature in primary loop nr. 1 (C) |
| 10 | Cold-leg coolant temperature in primary loop nr. 2 (C) |
| 11 | Electrical power of generator nr. 1 (MW) |
| 12 | Electrical power of generator nr. 2 (MW) |
| 13 | Pressure in bubble condensator after safety valve of pressurizer |
| 14 | Rotation of steam turbine nr. 1 (rotations/min) |
| 15 | Rotation of steam turbine nr. 2 (rotations/min) |
| 16 | Pressure connecting broken steam line in steam generator (nr. 3) |
| 17 | Difference between steam and feedwater flow rates (kg/s) |

**Table II. MALFUNCTION PATTERNS FOR NETWORK TRAINING**

| TRAINING CASE | MALFUNCTION |
|---|---|
| 1 | 100 percent rupture of main steam collector (nr. 1) |
| 2 | 50 percent rupture of main steam collector (nr. 2) |
| 3 | Rupture of steam line near to a steam generator (nr. 3) |
| 4 | Failure of steam turbine due to electrical problem (speed valve does not close) |
| 5 | Failure of steam turbine due to mechanical problem (speed valve closes) |
| 6 | Failure of steam turbines due to electrical problem (house service set works) |
| 7 | Drop of electrical load of steam turbines (house service set does not work, speed valve does not close) |
| 8 | Plant station black-out |
| 9 | Small break in primary pipeline |
| 10 | Steam generator tube leak |
| 11 | Failure opening of safety valve in pressurizer, but it closes after 30 secs |
| 12 | Failure opening of safety valve in pressurizer (remains open) |
| 13 | Break-down three main pumps out of six pumps (reactor power reduced to 50 %) |
| 14 | Same as case 13 (but power reduction fault, reduction to 83 percent) |
| 15 | Same as case 13 (but power reduction fault, reduction to 67 percent) |
| 16 | Same as case 14 (and turbine power reduced by operator to 50 percent) |
| 17 | Same as case 15 (and turbine power reduced by operator to 50 percent) |

### 2.2 Fourier Preprocessing

The transient signals (i.e., deviation from the steady state values) are Fourier transformed. The Fourier transforms were generated by padding the relevant time signal with zeros to 128 inputs(one input each second) for each of the seventeen indicative signals for each transient. The moduli of the Fourier components are calculated with the FFT and the DC components are discarded. The retained FFT components are logarithmically scaled between zero and unity for each pattern. The length of the meaningful time steps for the signals and the number of retained low-frequency FFT components varied in this study in order to determine the optimal choice for the FFT components and the minimal time frame for which malfunction recognition is feasible.

The transient record data started seven seconds before the time point when the deviation of one of the variables (numbers. 1, 2, 3, 11, 12, or 16 in Table I) exceeds a threshold value. If the eleven low-frequency components were retained each malfunction would consist of 11x17, or 187 input data. The eleven FFT components are multiplied here by seventeen, because seventeen distinct signals are monitored (see Table I).

### 2.3 Neural Network Training

The neural network is trained for 17 distinct malfunctions and tested on 22 malfunctions. The neural net is of the feedforward type, fully connected, and trained with a standard backpropagation algorithm. The data are scaled between zero and unity and a standard sigmoid function was used as transfer function. All studies were carried out with the MetaNeural™ code (version 5.1) which is an in-house neural network code for personal computers and workstations developed by one of the authors (MJE) at Rensselaer Polytechnic Institute. The neural network outputs for these malfunctions are categorized with seventeen distinct output values: depending on which transient a different output value will be set to unity, all other

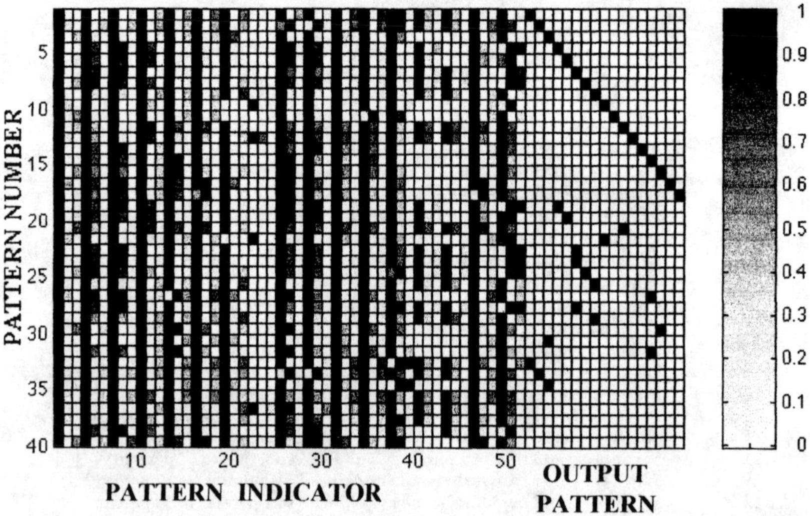


**Figure 1. Gray-scale plot for the network training (top 17 rows) and testing patterns (next 22 rows) for 3 Fourier components. There are 51 inputs (3 FFTs x 17 signals). The last 17 columns code up the identification of the 17 possible reactor malfunction types.**

outputs are zero. Various network configurations were tried. The most successful networks had two hidden layers and a relatively large number of neurons in the hidden layer. All cases were trained with 35 and 20 neurons in the first and second hidden layer respectively. Training was stopped as soon as 100 percent of the malfunctions were recognized during the training phase and when testing yielded the best recognition (i.e., minimal test error). When 11 components for the FFT were retained this would result in a 187x35x20x17 network. Figure 1 shows a gray-scale plot of 39 transients (17 training and 22 testing) for the case where the 3 lowest-frequency Fourier components (after discarding the DC component) were retained. The pattern identification coding can be seen in the last 17 columns of the figure). It is actually possible from this figure to match up some patterns from the test set to corresponding patterns of the training set with the naked eye (without a neural network). This would be impossible with the raw time series signals and illustrates the appropriateness of using Fourier transforms for this pattern classification problem.

## 3. TESTING OF ANNs WITH SIMULATED TRANSIENTS

Twenty two test cases generated with the PWR simulator were retained for testing the neural network. Note that the last five test cases were transients of a type not seen before, and these "unlabeled transients" were therefore coded up as seventeen zero's in the output. The number of retained Fourier components varied between 3 and 15, and the number of meaningful time steps varied between 32 and 80. The

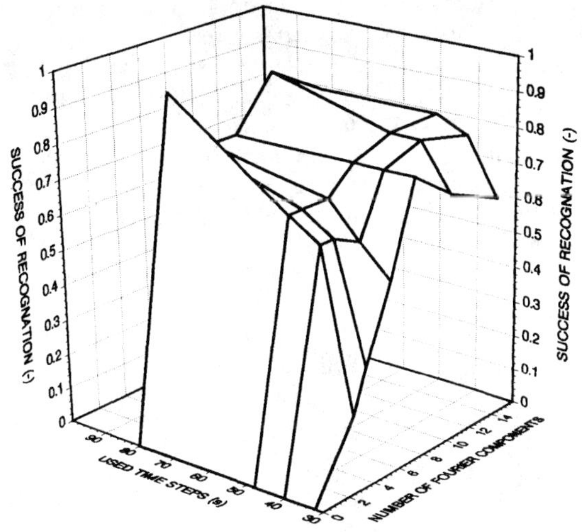


**Figure 2. Success of recognition for testing in function of the number of retained Fourier components and the number of time steps used.**

recognition for the test cases is summarized in figure 2. The results were not very sensitive to the number of neurons in the hidden layers, as long as a large number of neurons was utilized. Note that the largest recognition rates were 91 and 95 percent. The best recognition is achieved for the largest meaningful signal time frame (80 seconds). Note that retaining a very small number of FFT components (3) or a relatively large number of FFT components (11) resulted in the best test cases.

## 4. RESULTS AND DISCUSSION

An analysis of various neural network trials leads to the following observations:

(1) Excellent recognition (better than 90 percent) can be achieved when data representing 80 seconds of transient signals are utilized.

(2) None of the tried networks resulted in a 100 percent recognition. More simulator cases will probably lead to networks with a better performance.

(3) The FFT technique leads to only one network with a good malfunction recognition (81 percent) for relatively small time scales (32-48 seconds of relevant data) provided that a relatively large number of FFT components is retained. Note that none of the cases cited in the literature -- which relied on a rule-based pre-selection scheme with 5 subnets -- resulted in good recognition rates at shorter time frames. If the full time series was used for all 17 transient signals for 48 time steps in a single network (816x60x35x17) the best recognition rate was only 77 percent. The advantage of the FFT preprocessing versus full signals (cfr. Palancz et al., 1994) is that the number of inputs

presented to the network can be reduced and that the FFT based networks have a better recognition of the unlabeled transients.

(4) The best networks were relatively large in size (2 hidden layers with 35 and 20 neurons respectively). This would suggest that the network might acts as a memorizer rather than a generalizer. However, the fact that only these large networks could often successfully identify the unlabeled malfunctions shows that they properly generalize, rather than memorize.

(5) A single network trained with time series data rather than Fourier transformed data gives worse recognition rates (77 percent recognition at best).

## ACKNOWLEDGEMENTS

This work was sponsored partially by the US.-Hungarian Joint Science and Technology Fund (nr. 413) and by the Hungarian National Foundation for Scientific Research (Grant T016073).

## REFERENCES

Bartlett, E.B., Uhrig, R.E., (1992). Nuclear power plant status diagnostics using an artificial neural network, *Nuclear Technology*, Vol. **97**, 272 - 280.

Cheon, S.W., Chang, S.H., (1993). Application of neural network to a connectionist expert system for transient identification in nuclear power plants, *Nuclear Technology*, Vol. **102**, 177 - 191.

Embrechts, M.J., Lahey, Jr., R.T., and Yapo, T.C., (1992). The application of neural networks to two-phase flow regime identification, *Intelligent Engineering Systems through Artificial Neural Networks*, Vol. **2**, Cihan H. Dagli, Soundar R. T. Kumara, and Yung C. Shin, Eds., ASME Press, New York, 865 - 870.

Simon Haykin, (1994). Neural Networks: A Comprehensive Foundation, MacMillan, New York.

Nabeshima, K., Türkcan, E., and Cifcioglu, Ö., (1994). Nuclear power plant monitoring using real-time learning neural networks, *Proceedings of the Special Meeting on Applications of Artificial Intelligence and Robotics to Nuclear Power Plants*, May 30, Tokai, Japan, 313 - 322.

Palancz, B., Benedek, S., and Magyarné, M.J., (1994). Fault diagnostics for PWR using different on-line recognition methods, *Proc. 1994 Fall Meeting of the Atomic Energy Society of Japan*, Sapporo, 161.

Parlos, A.G., Atiya, A.F., Chong, K.L., and Tsai, W.K., (1992). Nonlinear identification of process dynamics using neural networks, *Nuclear Technology*, Vol. **97**, 79 - 96.

# Detecting Cavitation In Hydraulic Pumps Using Artificial Neural Networks

**John P. H. Steele, Michelle Archuleta, and Tyler Hooley**
*Laboratory for Intelligent Automated Systems*
*Colorado School of Mines*
*Golden, Colorado 80401*

***Abstract:***

One of the key components of an automated condition monitoring system is the detection techniques. For pumps, particularly hydraulic pumps, the occurrence of cavitation, if left unchecked, represents a sufficient condition for system failure. Thus detection of pump cavitation is a very important condition to monitor. In this paper, we present a comparison of detection approaches, one using multivariable linear analysis and another using artificial neural networks. The data, which consisted of pressure, temperature and flow rate data, were collected from a wide variety of system operating conditions for a large capacity swashblock displacement pump. The comparison shows that while the behavior is largely linear, there are non linear coupling effects. This is supported by the somewhat better detection results provided by the multi-layered neural network.

## INTRODUCTION

The strenuous demands placed upon equipment and people who are tasked with environmental restoration and waste management within the DOE complex, bring with them special opportunities to develop new technology, to enhance productivity and to increase worker safety. System Health Monitoring and Prediction, is such a technology.

The particular approach to the problem of on-line monitoring in real-time that has been developed at Colorado School of Mines is called SHARP (System Health Assessment and Real-time Prediction)[1]. This technology provides continuous active monitoring of machine condition in real-time. In addition, trending information is developed by recording and comparing historic parameter values against current values.

SHARP technology is a valuable tool for those tasked with maintaining high cost equipment and machinery in a variety of industries. The goal of this technology is to provide on-line monitoring of system behavior (machine condition) and assessment of the operational condition of the equipment to the operator and management in real-time, i.e., assessment based upon current measurements as well as historic trending

[1] This work has been supported by the Department of Energy through CSM contract 4-41360; technical direction is provided by Sandia National Laboratories and the Cross Cutting and Advanced Technologies Program.

information. In addition, the long term goal is to provide management with predictions of the remaining useful life of the machine before maintenance, repair, and/or replacement activities are required (predictive maintenance). This will allow management to plan and prepare for the removal of this expensive and critical equipment from service rather than being forced to react to an unscheduled outage, or to pull fully functional equipment from service for scheduled maintenance when it is not really necessary.

To obtain these goals, sensor based measurements technologies are melded to knowledge based systems to record, characterize, and interpret data in an automated fashion; resulting in information which can support continuous and automatic health assessments using intelligent software. This will allow the human operator to focus on the task at hand.

Previous hydraulic system diagnostics techniques employ the use of expert systems to perform the health assessment (Mitchell 1991). These types of systems apply hydraulic system parameters to expert system rules to infer the system health. Artificial neural networks have also been applied to vibration analysis for monitoring and diagnosis of various types of machinery (Alguindigue et al 1993) (Kuo and Chou 1993). Our approach combines these two ideas and applies artificial neural networks to the tasks of monitoring and diagnosis of hydraulic systems, specifically, the identification of cavitation.

## CAVITATION

Cavitation in pumps is the implosion of void bubbles within a liquid. The bubbles are first formed by the liquid (in this case hydraulic fluid) being subjected to pressure conditions which are below the vapor pressure of the liquid. When the bubble collapses, shock waves, which impinge on nearby materials, cause erosion and vibration in the material. Erosion of the pump parts will result in particulate generation which will accelerate wear in the system and can cause failure of other subsystems, e.g., jamming of the servo control valves. In addition, cavitation can lead to vibration of the pump which, if left unchecked, can result in catastrophic failure of the pump even after a few hours of operation.

Typically, the cavitation is detected by a human operator or mechanic who notes a change in the audible operating signature of the pumping system. When these changes denote cavitation, the machine is halted, and remediation action is commenced. However, in the case of remote operation, cavitation is likely to go unnoticed and therefore, unchecked, leading to failure of the pump and loss of operation. Thus, for remote or automated operation, an on-line detection system is very desirable. Such a system must rely on available sensor technologies and the correct interpretation of the corresponding sensor information.

### Detection of Distinct Data

In previous work done in the LIAS (Laboratory for Intelligent Automated Systems) lab at CSM, (Hooley 94) demonstrated that neural nets could be used to classify sensor patterns as either cavitation or non-cavitation. This work was based upon data provided by Idaho National Engineering Laboratories(Austad 94). The data provided was developed from testing gear pumps which were systematically cycled between "normal" operation and full cavitation. The results of training a multi-layer net with this data resulted in convergence and identification of the two states, cavitation and non-cavitation. However, Hooley noted that the data was linearly separable, i.e., the hidden layer was not needed. This lead to an analysis using a multi-variable statistical approach, called the Fisher Linear Discriminant (FLD). This approach was successful for the INEL data. While this work demonstrated the viability of the approach, it used very segregated data and gave no indication as to the effectiveness of this approach for a more typical operation of the pump.

Further work in LIAS for INEL, lead to the development of *on-line* monitoring capabilities as described previously, i.e., SHARP (Steele 95). This installation allowed on-line testing of pump behavior over a wide range of conditions. A series of tests were undertaken to understand the conditions and associated parameter values for cavitation. A matrix of response values for the three independent variables: inlet pressure, outlet pressure, and motor speed were developed.

### FLD Results

One can explain the process of developing the FLD as the process of seeking to find a plane in five dimensional hyperspace (note the importance of each variable is individually weighted) which allows the "best" clustering of the cavitating and non-cavitating data examples. Once this plane has been found, all of the data is then collapsed onto this plane, and the resulting clusters are displayed. A line drawn on the plane represents the best line of separation between the cavitating and non-cavitating clusters, see Figure 1. The correlation for this discrimination was 0.71. Using the "mean" criterion, i.e., finding the half way point between the two centroids of the two groups (non cavitation [o], and cavitation [1], the probability of committing a Type I error is 7% (predicting no cavitation, when in fact there is cavitation). The probability of a Type II error (saying it is cavitation when is not) is 3%.

## NEURAL NETWORKS

While the results from the FLD for this new data set were quite useable, we wanted to see if there were, in fact, *nonlinear aspects* to this problem. In order to investigate the nonlinear characteristics of the relationship between the pressure/temperature/flow parameters, and the resulting state of cavitation or not-cavitation, both multi-layer and single layer neural networks were developed. To explore these issues, a whole series of neural networks were developed and tested.

### Approach

The neural networks used for this work were created, trained, and simulated using the Neural Networks Toolbox in Matlab. Standard backpropagation was chosen as the network training algorithm. Network parameters that were varied were the learning rate and the number of hidden neurons. The data used for training and testing consisted of 499 data sets. Each set contained inlet and outlet temperatures and pressures, the flow rate, and a classification of 1 for cavitation and 0 for not-cavitation. Training was performed using a randomly chosen subset of 450 of the data sets, and then the trained network was tested with the remaining 49 data sets. Each time the network trained, a different 450 data sets were chosen at random for training with the remaining data sets

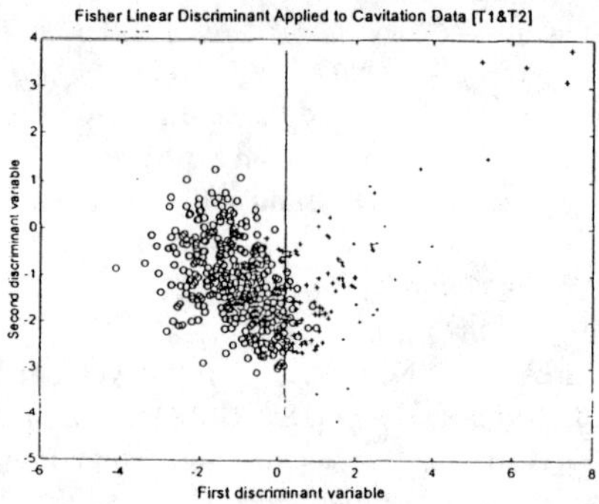


**Figure 1: Linear Discriminant Results, Classification of Cavitation.**

used for testing. The temperatures, pressures, and flow rate were the inputs to the network, and the desired output was a classification of the system state as cavitation or not cavitation. To aid in training, the inputs to the net were normalized between 0 and 1, and the desired, or target, outputs were set to 0.1 for not cavitation and 0.9 for cavitation.

After the network trained, the test data were presented to the network which produced an output for each set of data. Different thresholds were applied to these outputs to establish the classification. Three different thresholds were applied. The first classified any values greater than or equal to 0.7 as cavitating and less than or equal to 0.3 as not-cavitating. Any of the values between 0.3 and 0.7 were left as undetermined. The second was the same as the first except 0.7 and 0.3 were replaced with 0.6 and 0.4, respectively. The third threshold classified anything greater than or equal to 0.5 as cavitating and everything less than 0.5 as not-cavitating. The results using these thresholds were then compared to the desired results, or classifications. The performance of the network was measured by what percent of the test data the network classified correctly. A final variable in the training sequence was to vary the learning rate of the network to see if there was an optimal value. In addition, the performance of the network was also compared to the output of the Fisher Linear Discriminant for the same test data.

## Results

Figure 2 shows a plot of the average percent correct as a function of the number of nodes. As can be seen from this plot, the best results for the test data came when we used three or four nodes on the hidden layer. The next three graphs (3, 4, & 5) show the results for various threshold values, along with comparisons with corresponding single layer nets. As would be expected, the threshold of 0.5 performs the best however, the results for the 0.6/0.4 were almost as good. Note also that the performance of the net is learning-rate dependent. The next graph shows similar results, but highlights the fact that the nets have been trained with fewer number of epochs, with the multi-layer net, but give better results, generally. Again, the most generous threshold combined with the lowest learning rate produced the best results.

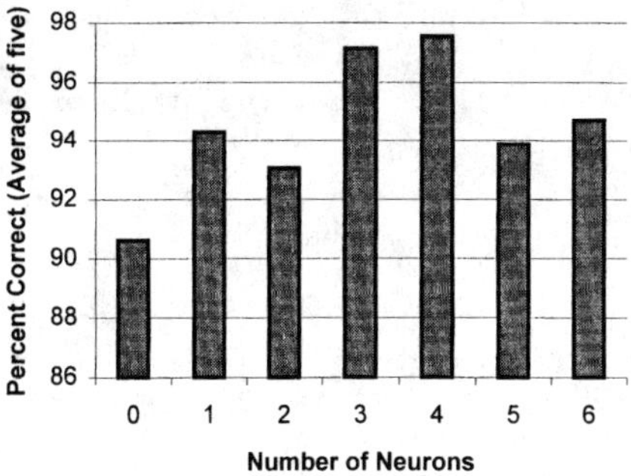


Figure 2: Test Results as a function of the number of hidden nodes.

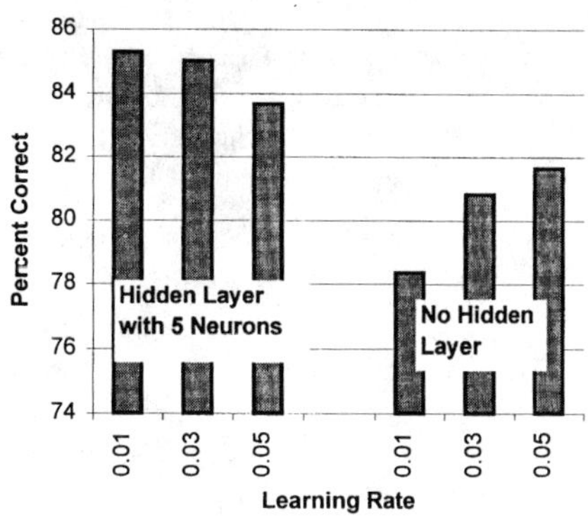


Figure 3: Results for Thresholds of 0.7/0.3.

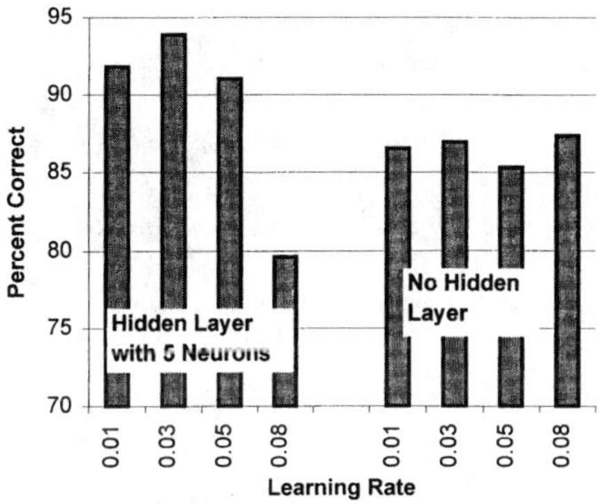


Figure 4: Results for Thresholds of 0.6/0.4.

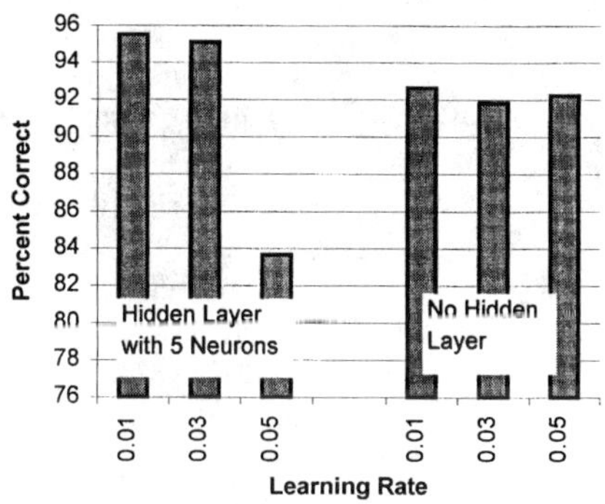


Figure 5: Results for Threshold of 0.5.

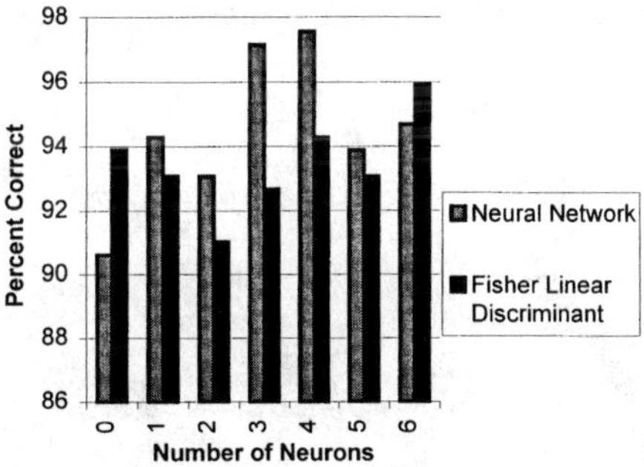


Figure 6: Comparison of Testing Results for ANN and Fisher Linear Discriminant for Various Numbers of Hidden Nodes.

Finally, in Figure 6, the results for the multi-layered net are compared against the Fisher Linear Discriminant for various numbers of hidden nodes. This plot clearly demonstrates that the neural network provides better results, although the difference in performance on the test data is relatively small.

## CONCLUSIONS

Cavitation can be detected using state sensors, i.e., pressure, temperature and flow rate. The correlation between the system state and these parameters is approximately linear as demonstrated by the Fisher Linear Discriminant analysis. However, there are non-linear aspects to this relationship, since the neural network is able to consistently do better at classifying cavitation.

The ability to fuse data from several difference sensors is demonstrated by these results. The best generalization is provided when four nodes are used in the hidden layer, for this five input problem. Lower learning rates result in better test results. The use of 0.5 as the separator between classification gave the best results.

The strongest discriminator correlation for cavitation is with supply pressure followed by outlet temperature and then flow rate. Supply pressure alone is not sufficient to predict cavitation, additional information is needed, i.e., the outlet temperature.

These results provide the analysis needed to implement an on-line cavitation monitoring system. As can be seen, this system could be either a linear discriminant, or if somewhat better results are desired, an artificial neural network could be used. In addition, the neural network approach could be used to classify between different types of faults, whereas, the linear discriminant approach can only detect cavitation. Detection of multiple faults remains for future work.

## REFERENCES

Alguindigye, Israel E., Anna Loskiewicz-Buczak, and Robert E. Uhrig, 1993, "Monitoring and Diagnosis of Rolling element Bearings Using Artificial Neural Networks," *IEE Transactions on Industrial Electronics*, v 40, n3.

Austad, Wayne, 1994, personal communications, WR Robotics Development Group, Idaho National Engineering Laboratories.

Hooley, Tyler. L., 1995, "An Automatic Health Assessment System Prototype", *Thesis*, Engineering Division, Colorado School of Mines, T-4705.

Kuo, R.J. and C.L.Chou, 1993, "Multi-Sensor Fusion Based Diagnosis for Turbine Blade Faults via Artificial Nerual Networks," *Intelligent Engineering Systems through Artifial Nerual Networks, Vol.3*, New York, NY: ASME Press.

Mitchell, Julie, 1991, "Diagnostic Expert System Techniques for Improving Hydraulic Maintenance of a Continuous Mining Machine," *Mining Engineering*, April 1991.

Steele, John P. H., 1995, "System Health Monitoring and Prediction Program, Technology Assessment Report", Laboratory for Intelligent Automated Systems, CSM, technical report *LIAS-TR-951001*.

# THE APPLICATION OF NEURAL NETWORKS TO A CONDITION MONITORING SYSTEM FOR THE OPTICAL FIBRE DRAWING PROCESS

**HONG SHI AND CAROLINE TITE**
*Systems Engineering Faculty, Southampton Institute, Southampton, U.K.*

**TONY HOPE AND SIAMAK NOROOZI**
*Systems Engineering Faculty, Southampton Institute, Southampton, U.K.*

**LOUISE POYNTZ-WRIGHT AND ANDY MARSHALL**
*Pirelli Cables Limited, Bishopstoke, Hampshire, U.K.*

***ABSTRACT:***
It is well known that optical fibre quality can be affected by the conditions present during fibre drawing, particularly in the drawing furnace. Early furnace decay can lead to problems with fibre production and thus can be detrimental in terms of fibre properties. Currently, a furnace rebuild is carried out regularly, based on a scheduled maintenance plan. A monitoring system which could recognise the need for a rebuild would be beneficial since (i) furnace life could be extended which could reduce changing time and increase machine utilisation or (ii) reduction in fibre yield could be avoided through recognition of early furnace decay. The drawing tension and speed were found to be good indicators of furnace decay and these have been reported previously (Shi, 1996). However, drawing tension is also affected by some other drawing parameters such as furnace power, speed and preform diameter. In this paper, neural networks are used to integrate information from different drawing parameters in order to recognise the furnace state during the optical fibre drawing process and thus decide on the necessity of the furnace rebuild. Different structures of neural networks for application in a furnace condition monitoring system were investigated and the results are presented.

## INTRODUCTION

Optical fibres have achieved tremendous success in long distance telecommunication applications. Silica-based optical fibres are manufactured in two steps: a preform of glass containing the core-cladding structure is made first, followed by fibre drawing and coating. In the fibre drawing process, the preform is lowered into a high temperature furnace operating at 2100-2300 $^{o}$C, where the end of the preform softens enough to be drawn into fibre. A capstan provides the pulling tension, and the preform feed rate and capstan speed determine the drawn-down ratio. Protective coatings are applied to the pristine fibre for protection before it is touched by the drawing mechanism (Menachem, 1993). The quality of optical fibre is highly dependent upon the conditions in the drawing furnace. A graphite resistance furnace is often used since it is capable of operating at high and accurately controlled temperatures with relatively low cost. However, the furnace has a relatively short heating element lifetime (Payne, 1976, Vasil'ev, 1988). Furnace decay can result in the formation of dust particles of the materials of the heating element which affect the temperature distribution within the heating zone and thus are detrimental to the optical and mechanical properties of optical fibres (Ainslie, 1981,1982, Paul, 1983, Schonhorn, 1978). A furnace rebuild is carried out regularly, based on a scheduled maintenance plan. It would obviously be beneficial if furnace decay could be monitored and predicted since the furnace life could be extended which could reduce changing time and increase machine utilisation. Also reduction in fibre yield could be avoided through recognition of early furnace decay.

## PARAMETERS CONCERNED WITH FURNACE CONDITIONS

The drawing tension and speed were found to be good indicators of furnace decay and these have been reported previously (Shi, 1996). The drawing tension in the bare fibre increases during the furnace lifetime and the drawing speed fluctuates when furnace decay is serious. So the drawing tension and the standard deviation of the drawing speed can be used to indicate the furnace decay. However the drawing tension also varies with the furnace power setting and drawing speed. Preform diameter and variations in furnace power also affect the drawing speed and hence the standard deviation. Therefore, using only these two parameters to predict furnace condition cannot be expected to give a high accuracy.

To achieve a good result for classifying furnace conditions, the drawing tension should be modified by the furnace power and drawing speed thus enabling comparison of results. Figure 1 shows the changes of the normalised bare fibre tension during the furnace life time. It is clear that the normalised tension increases during the furnace life time. Figure 2 shows the variation in the standard deviation of the drawing speed and it can be seen that the standard deviation increases rapidly towards the end of the furnace lifetime. However this can also occur at the beginning or middle of the furnace lifetime in some cases, which may be due to the changes of preform diameter or furnace power. The characteristics of variations in the drawing speed caused by changes of the preform diameter and furnace decay are quite different. The speed fluctuation caused by preform diameter appears to be a long period change while the speed fluctuation due to the furnace decay appears to be a short period change. Therefore, the frequency of speed fluctuation can be used to indicate if the preform diameter is the cause of the speed changes. Figure 3 shows the fluctuation frequency of the drawing speed during the furnace life time. It can be seen that the fluctuating frequency of the drawing speed is high when the furnace runs near the end of its life time. However, in some cases this frequency can be high at the beginning or middle of the furnace lifetime and it was found that this is due to the fluctuation of furnace power. Therefore the standard deviation of furnace power is another parameter used to indicate the drawing furnace condition. Figure 4 shows the interrelation of various variables. Since the evaluation of furnace conditions is very complicated and affected by so many parameters , an artificial neural network is required to recognise furnace conditions.

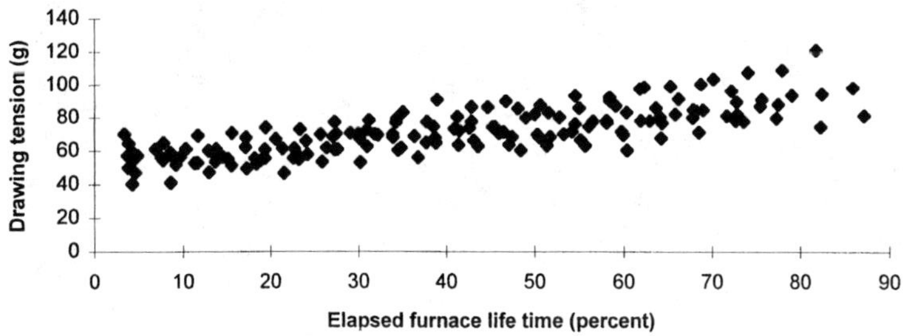


**Figure 1 Changes of drawing tension normalised to line speed and furnace power during furnace lifetime**

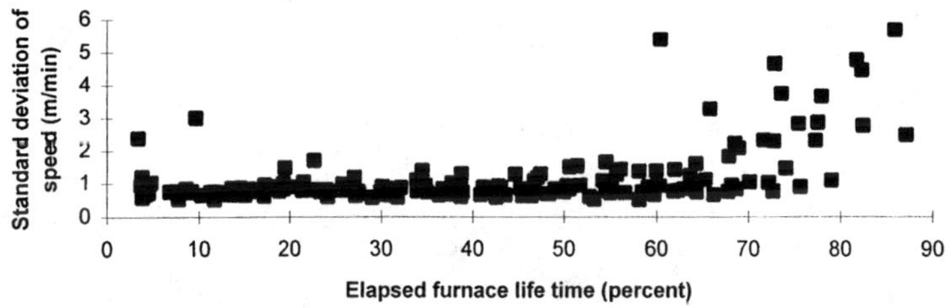


**Figure 2 Changes of standard deviation of drawing speed during furnace lifetime**

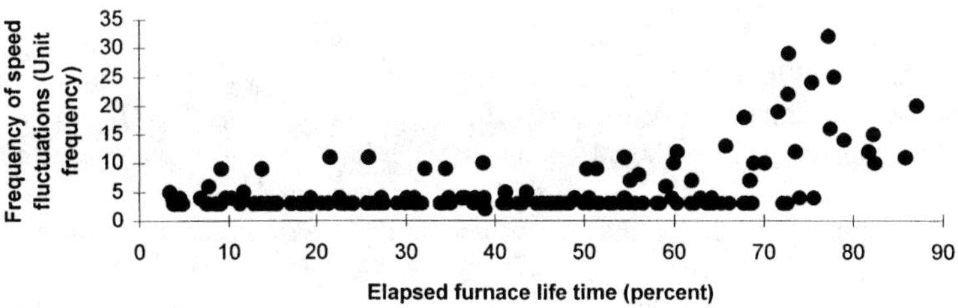


**Figure 3 Variation of drawing speed fluctuation frequency during furnace lifetime**

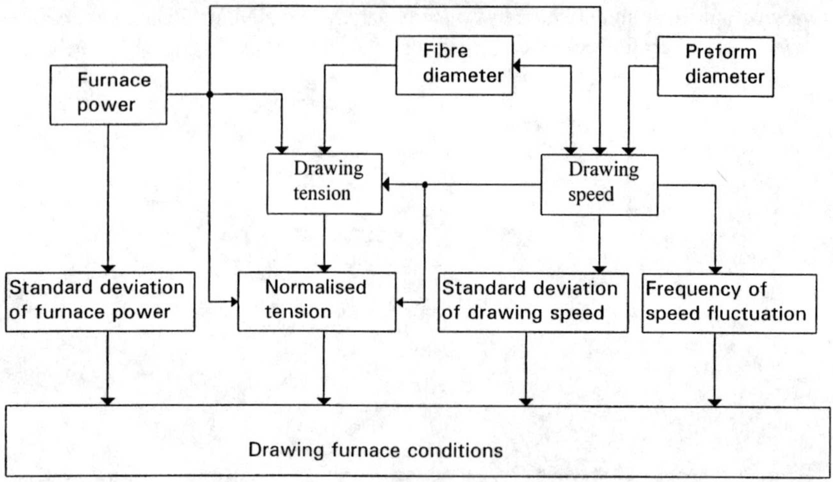


**Figure 4 The interrelation of various variables**

## THE ARTIFICIAL NEURAL NETWORKS

In this paper, a multilayer backpropagation neural network is used for furnace condition monitoring of the optical fibre drawing process. The backpropagation network is a very popular neural network model. It does not have feedback connections, but errors are backpropagated during training (Michael, 1993). A software implementation of this network on a serial computer is used as the learning and decision-making component. There are three kinds of processing units in such networks: input layer nodes which accept information from the external world, output layer nodes which generate outputs to the external world and hidden nodes which do not directly interact with the external world. The role of the hidden nodes is to perform feature extraction on the information presented at the input layer (Rangwala, 1990).

The input signals to each of these neurones are weighted and then summed together with a bias term. The neurone output is then obtained by passing the summed, weighted inputs through a non-linear activation function. This function is typically sigmoidal and threshold or hyperbolic tangent (Demuth, 1994). The actual output is compared with the target output for each input vector, and the root mean square error is calculated. This error is then propagated backwards through the neural connections, and the process is repeated until the error is within an acceptable threshold.

Once the network architecture has been selected and is trained on actual process data, it can be used to associate patterns presented at the input layer with appropriate patterns at the output layer. For the furnace condition monitoring system, the input pattern can represent the features of the drawing parameters whereas the output layer nodes can be used to indicate the states of the furnace.

## FURNACE CONDITION PREDICTION

Although useful information may be contained in the different drawing parameters, from practical use, only a few of these parameters can be utilised for training and pattern association purposes. In the present work, the normalised tension, standard deviation of drawing speed, fluctuating frequencies of the drawing speed and the standard deviation of the furnace power, which show high sensitivity to the furnace condition, are chosen as the input parameters. Three nodes with binary outputs are used in the output layer, which can present eight different combinations in which seven conditions are used to indicate the different furnace conditions: new furnace, normal, high tension, power fluctuation, speed fluctuation, failure warning and total failure. Several different networks with one or two hidden layers and different numbers of nodes were trained in order to find the best one for this application. Sigmoidal nonlinearities were used in each node, the learning rate constant was set between 0.1 to 0.3 and the momentum constant was set between 0.5 to 0.9. Stochastic Back Propagation and Standard Back Propagation are used as a training algorithm referred to as "supervised training". Stochastic Back Propagation differs from Standard Back Propagation by the point at which the weight changes are made. With Standard Back Propagation, the rate of change of error with respect to weight for a particular weight are summed for every pattern in an epoch and then the weight changes are made based on this sum. With Stochastic Back Propagation, the weight changes are made after each pattern presentation. Because the order of the patterns may influence learning, the patterns are presented in random order. Development work is being conducted using Nerframe 1.1, a neural network development package. 245 training data sets were constructed from real data collected from the fibre drawing process with these seven furnace conditions, where the target output could be generated as a result of known furnace conditions due to inspection of the drawing parameters, drawing performance, fibre quality and furnace rebuilt by scheduled maintenance. The prediction accuracy of these neural networks was then tested on 112 data sets collected later in the fibre drawing process with seven furnace conditions. Table 1 shows the results of applying the different sized networks.

**TABLE 1 PERFORMANCE OF NEURAL NETWORKS**

| Number of hidden layers | Number of nodes in the first hidden layer | Number of nodes in the second hidden layer | Algorithms | Success rate |
|---|---|---|---|---|
| 1 | 7 | 0 | stochastic | 81.5 % |
| 2 | 5 | 4 | stochastic | 86.7 % |
| 2 | 7 | 5 | standard | 88.9 % |
| 2 | 8 | 6 | standard | 90.0 % |
| 2 | 7 | 5 | stochastic | 96.7 % |

The best results were obtained using a 4-7-5-3 neural network with Stochastic Back Propagation, in which the success rate for recognising furnace conditions in the fibre drawing process reached a maximum of 96.7%

## CONCLUSION

This paper has described the feature parameters extracted from the parameters of the optical fibre drawing process which are sensitive to the decay of drawing furnace and a novel condition monitoring method which was devised for the drawing furnace. The monitoring system utilised the feature parameters and a neural network as the learning and decision-making component. This allows for greater reliability in making decisions on the necessity of the furnace rebuild and enables a high success rate for recognising furnace conditions in optical fibre drawing process. The results are very encouraging and indicate possibilities of using neural networks to recognise furnace conditions.

## REFERENCES


Ainslie B.J., Beales K., Day C R., and Rush J. D., (1981), Interplay of design parameters and fabrication conditions on the performance of monomode fibres made by MCVD. IEEE Journal of quantum electronics, Vol.une pp 854-857.

Ainslie B.J., Beales K.J., Cooper C.R., Day C.R., and Rush J.D., (1982), Drawing-dependent transmission loss in Germania-doped silica optical fibres. Journal of Non-Crystalline Solids, Vol.47, No. 2 pp 243-246.

Demuth, H, Beale, M, (1994), Neural Network Toolbox for MATLAB, Math Works.

Menachem Lewin and Jack Preston, (1993), High Technology Fibers, Part C, Marcel Dekker,Inc., New York.

Michael Chester, (1993), Neural Networks, PTR Prentice-Hall, Inc. , New Jersey.

Paul S.OH, John J. Mcalarney, and Dilip K. Nath, (May 1983),Effect of Fibre Drawing Tension on Optical and Mechanical Properties of Optical Fibre Waveguides. Communications of the American Ceramic Society, pp C-84-85.

Payne, D.N. and Gambling,W.A., (1976), A resistance-heated high temperature furnace for drawing silica-based fibers for optical communications, Am. Ceram. Soc. Bull., Vol. 55, No.2, pp195-197.

Rangwala S, Dornfeld D, (1990), Sensor Integration Using Neural Networks for Intelligent Tool Condition Monitoring, ASME, Journal of Engineering for Industry, Vol. 112, pp 219-228.

Schonhorn,H.;Vazirani,H.N.; Frisch,H.L.; (1978), Relationship between fiber tension and drawing velocity and their influence on the ultimate strength of laser-drawn silica fibers. Journal of Appl. Phys. Vol. 49 (7), July , p 3703-3706.

Shi H., Tite C., Hope A. D., Noroozi S., Poyntz-Wright L. and Marshall A., (1996) Using Condition Monitoring to Predict Furnace Decay during the Optical Fibre Drawing process" INSIGHT, The Journal of The British Institute of Non-Destructive Testing. Vol. 38, No. 1, pp 57-59.

Vasil'ev,V.N.;Dul,nev,G.N.; Naumchik,V.D., (1988), Thermal mode of a resistance furnace for drawing optical fiber. I. Generalized mathematical model of the heating unit., Journal of Engineering Physics, Vol. 54, No. 2, Aug, p 176-182.

Vasil'ev,V.N.;Dul,nev,G.N.; Naumchik,V.D., (1988), Thermal mode of a resistive furnace for drawing optical fiber. II., Results of a mathematical experiment. Journal of Engineering Physics, Vol. 54, No. 3, pp 320-324.

# CLASSIFYING WIRE BONDS: NEURAL NETWORKS VERSUS DISCRIMINANT ANALYSIS

**XIAOYUN SUN**
*BehavHeuristics, Inc.*
*7240 Parkway Drive*
*Suite 170*
*Hanover, Maryland 21076*

**BRUCE GOLDEN**
*College of Business and Management*
*University of Maryland*
*College Park, Maryland 20742*

**EDWARD WASIL**
*Kogod College of Business Administration*
*American University*
*Washington, DC 20016*

***ABSTRACT:***
A wire bond found in a microcircuit can be screened for quality and classified as either good or bad and a bad wire bond can then be further classified as exhibiting one of five different defects. In this paper, we use ultrasonic data gathered during the production of nearly 3,100 wire bonds for the Westinghouse Electric Corporation to construct neural network models that classify the quality of a wire bond. We construct two types of neural network models using our own computer code and compare the results of the neural network models to three different discriminant analysis models. The neural network models are better at classifying the wire bonds than the discriminant analysis models with the best neural network model correctly classifying on average over 97% of the wire bonds.

## INTRODUCTION

Electronics manufacturers connect two different surfaces in a microcircuit with a very thin gold wire. A ball that is formed at one end of the gold wire is bonded onto one surface called a pad. The gold wire is then looped and ultrasonically welded to a second pad to form a crescent-shaped stitch bond. The ball bond, gold wire loop, and stitch bond make up what we call a *wire bond* in the remainder of this paper. The three-part process of manufacturing a wire bond takes about one second on a wire bonder machine.

The quality of a wire bond can be determined through a visual inspection or a pull test. In a visual inspection, the manufacturer can determine, for example, if a gold wire is lifted from either pad. In a pull test, a gold wire is pulled with a small amount of force to see if it breaks or to see if either bond breaks contact with its pad. Wire bonds that fail the visual inspection or the pull test are defective, while those that pass are regarded as nondefective (we call these good wire bonds). The screening methods can be costly to apply, so that it is common for manufacturers to screen a small, random sample of wire bonds. In very high-quality applications, manufacturers might need to screen all wire bonds that are produced.

Westinghouse Electric Corporation manufactures wire bonds and determines wire bond quality by using both screening methods. In a project for Westinghouse reported in Wang et al. [1993] and Sun et al. [1994], we demonstrated that a neural network model is an accurate, low-cost way to screen a wire bond. We showed that a neural network model can learn unknown relationships between ultrasonic measurements and wire bond quality from empirical data in order to predict a wire bond's pull-strength value (that is, the amount of force that is required to cause a wire bond to fail). Our best neural network model identified nearly 97% of the failed wire bonds in a test set.

Given the success of our neural network model for prediction, Westinghouse was interested in a related classification problem: Can a neural network model successfully classify a wire bond as good or bad? In particular, Westinghouse wanted to classify a wire bond into one of the six categories shown in Table 1. A good wire bond is placed in Category 1 while a defective wire bond is placed in Categories 2 through 6. The five categories of defectives are the ones most frequently encountered by Westinghouse during the production of wire bonds. We point out that sometimes the wire bonder machine does not produce a gold wire that connects two surfaces. This type of defective wire bond is placed in Category 6.

In the remainder of this paper, we describe our effort to construct neural network models that successfully classify a wire bond into one of the six categories. We compare the performance of our neural network models to three types of discriminant analysis models and evaluate the cost of screening policies based upon the results of neural network models and random sampling methods.

## DATA GATHERING AND VARIABLE GENERATION

A total of 3,072 wire bonds were produced by the Advanced Microwave Technology Center of Westinghouse on a Hughes 2460-III automatic wire bonder. There were about 1,300 ultrasonic waveform measurements recorded for each wire bond. These data gave rise to three types of curves: (1) current vs. time, (2) current vs. frequency, and (3) impedance vs. frequency. Westinghouse subjected each wire bond to a visual inspection and then classified each wire bond into one of the six categories shown in Table 1.

**TABLE 1: CLASSIFICATION CATEGORIES**

| Category | Description |
|---|---|
| 1 | Good wire bond |
| 2 | Ball bond off pad |
| 3 | Stitch bond off pad |
| 4 | Epoxy on ball bond |
| 5 | Epoxy on stitch bond |
| 6 | No gold wire between pads |

Westinghouse provided us with the measurements for 2,304 wire bonds (we would build our models on this data set) and held back the measurements for 768 wire bonds (Westinghouse would validate our final models on this data set). Five wire bonds were removed (we discovered that they had erroneous or missing measurements), so that we were left with ultrasonic measurements for 2,299 wire bonds. For the 2,299 wire bonds in our modeling data set, there were 379 wire bonds in Category 5 and 384 wire bonds in each of the other five categories.

Using our insight into the characteristics of the ultrasonic signatures combined with statistical analyses and feedback from discussions with wire bonding experts, we generated 54 independent variables that could be used to classify a wire bond into one of the six categories. For example, one of the variables that we tracked was the maximum current in the frequency domain when a ball bond was produced (see Sun [1995] for more details about the other independent variables and analysis techniques such as visual data plots that we developed and used). The key dependent variables were the six classification categories.

## EXPERIMENTAL DESIGN

Using the 2,299 wire bonds in the modeling data set, we set out to construct discriminant analysis models and neural network models that were capable of classifying a wire bond into the six categories. We summarize our model-building effort in Table 2. We point out that details on the computer code that we used to construct our models (developed at the University of Maryland and known as the Maryland Network Code), as well as additional details on network pruning (links with unimportant weights are removed during training), the skip-layer network (there are connections from the inputs directly to the outputs in a network), and the hierarchical neural network are provided by Sun et al. [1994] and Sun [1995].

First, we randomly split the modeling data set into two subsets -- 1,149 wire bonds in a training set and 1,150 wire bonds in a test set. Second, we performed a stepwise discriminant analysis on the training set using all 54 independent variables and ran the same analysis on the test set. Based upon both discriminant analysis results, we identified 22 key independent variables that could be used to classify a wire bond. Third, we performed three modeling experiments (see Table 2). In Experiment 1, we built three types of discriminant analysis models (LDA, QDA, and NPDA) and two types of neural network models (NN and HNN) on the training set of 1,149 wire bonds and then tested these models on the test set of 1,150 wire bonds. To gain some confidence about the applicability of our models to this classification problem, we flipped the training set and the test set and rebuilt the five models in Experiment 2 (see Ramani and Patrick [1992] for details about a similar experimental design). In Experiment 3, we built five models (LDA, QDA, NPDA, NN, and HNN) using all 2,229 wire bonds and tested the performance of each model on the 768 wire bonds that Westinghouse held back.

## RESULTS

In conducting Experiments 1 and 2, we learned that of the five models we built (that is, LDA, QDA, NPDA, NN, and HNN), HNN was the most accurate classifier. On average, HNN correctly classified 96.52% of the wire bonds in Experiment 1 and

**TABLE 2: SUMMARY OF MODEL-BUILDING EFFORT**

**Data Sets**

| Experiment | Number of Wire Bonds: Training Set | Test Set |
|---|---|---|
| 1 | 1149 | 1150 |
| 2 | 1150 | 1149 |
| 3 | 2299 | 768 |

**Variables**

- ◆ 22 independent variables
- ◆ 6 dependent variables

**Discriminant Analysis Models**

- ◆ Linear Discriminant Analysis (LDA), Quadratic Discriminant Analysis (QDA), and Nonparametric Discriminant Analysis (NPDA)
- ◆ Used SAS v6.03 run on an IBM 370 mainframe

**Neural Network Model(NN)**

- ◆ Three layer or skip-layer feedforward system
  22 input nodes, 6, 8, or 10 hidden nodes, 6 output nodes
- ◆ Sigmoid activation functions with different slopes
- ◆ Backpropagation learning rule with decreasing learning rate and network pruning
- ◆ Used Maryland Network Code run on a Sun Sparc-10 workstation

**Hierarchical Neural Network Model(HNN)**

- ◆ Three layer feedforward system
  24 input nodes, 6 or 10 hidden nodes, 6 output nodes
- ◆ Targets are the six classification categories and the inputs are the predictions from NN, LDA, QDA, and NPDA
- ◆ Backpropagation learning rule with decreasing learning rate and network pruning
- ◆ Used Maryland Network Code run on a Sun Sparc-10 workstation

97.56% of the wire bonds in Experiment 2. NN was close behind with 95.50% and 95.90% in the two experiments, respectively. We point out that the neural network models required about 600 to 800 iterations for training. This took from 17 to 37 minutes on a SUN Sparc-10 workstation and the training time depended on whether or not a skip-layer structure was used and whether or not network pruning was used.

Among the three discriminant analysis models, QDA exhibited better performance than NPDA which, in turn, exhibited better performance than LDA. The complete set of results for the first two experiments is provided by Sun [1995].

The results of the first two experiments revealed that our statistical models and neural network models were reasonably accurate classifiers. Next, we placed all 2,299 wire bonds in a training set and then set out to construct a final set of models in Experiment 3. We built five models and then tested the classification accuracy of each model on the set of 768 wire bonds that Westinghouse had held back. The accuracy results are shown in Table 3.

In Table 3, the numbers under the columns marked 1 to 6 give the percent of wire bonds in the held-back test set that were correctly classified by a model. For example, LDA correctly classified 89.06% of the good wires (Category 1). On average, LDA

TABLE 3: CLASSIFICATION ACCURACY OF FINAL MODELS

| | Percent Correctly Classified | | | | | | | | |
|---|---|---|---|---|---|---|---|---|---|
| Model | 1 | 2 | 3 | 4 | 5 | 6 | Average | Misc* | Other |
| LDA | 89.06 | 99.22 | 96.09 | 92.97 | 92.19 | 96.09 | 94.27 | 1 | |
| QDA | 99.22 | 90.63 | 94.53 | 96.09 | 96.88 | 95.31 | 95.44 | 12 | |
| NPDA | 92.97 | 97.66 | 96.09 | 96.09 | 92.27 | 100.00 | 95.06 | 1 | 8** |
| NN | 97.66 | 96.88 | 93.75 | 97.66 | 96.88 | 94.53 | 96.22 | 4 | |
| HNN | 96.88 | 95.31 | 96.09 | 98.44 | 98.44 | 97.66 | 97.14 | 6 | |

* Number of defective wire bonds misclassified as good wire bonds

** Number of wire bonds that NPDA could not classify into the six categories

correctly classified 94.27% of the wire bonds and misclassified only one bad wire bond as a good wire bond.

From Table 3, we see that HNN is the best classifier. On average, HNN correctly classified 97.14% of the wire bonds in the held-back test set. NN was close behind with 96.22%. QDA was the best discriminant analysis model. On average, QDA correctly classified 95.44% of the wire bonds in the held-back test set. Westinghouse was very pleased with these results, although managers were a bit concerned that even a very small number of defective wire bonds were misclassified as good wire bonds.

## COST ANALYSIS

To evaluate the usefulness of our final results, we now compare the cost of an inspection policy based upon the NN model to the cost of a policy based upon random sampling. We use the results in Table 3.

Recall that there are 768 wire bonds in the held-back test set. There are 128 wire bonds in each of the six categories; therefore, there are 128 good wire bonds and 640 defective wire bonds. In Table 3, we see that our final NN model correctly classified 97.66% (that is, 125/128) of the good wire bonds in Category 1 and misclassified 2.34% (that is, 3/128) of the good wire bonds as bad wire bonds. NN correctly classified 99.38% (that is, 636/640) of the defective wire bonds and misclassified 0.62% (that is, 4/640) of the defective wire bonds as good wire bonds.

From information provided by Westinghouse engineers, suppose that wire bond quality has the following distribution: Prob(wire bond is good) = 0.99745 and Prob (wire bond is defective) = 0.00255. Then we have

$$\begin{aligned}\text{Prob (NN rejects a wire bond)} &= \text{Prob(NN rejects | wire bond is good)} \times \text{Prob (wire bond is good)}\\ &\quad + \text{Prob(NN rejects | wire bond is defective)} \times \text{Prob (wire bond is defective)}\\ &= (0.0234)(0.99745) + (0.9938)(0.00255) = 0.026.\end{aligned}$$

Suppose we consider a typical production run of 166,725 wire bonds of which 0.00255(166,725) = 425 wire bonds are defective. The NN model would flag 0.026(166,725) = 4,335 wire bonds as being defective. These wire bonds would then be subjected to further screening such as a pull test. Using an inspection cost of \$0.583 per wire bond, performing a pull test on all 4,335 wire bonds would cost \$2,527. Furthermore, the NN model would not catch 0.0062(425) = 3 defective wire bonds. NN would classify these three defective wire bonds as good wire bonds. The overall yield for the NN model would be 0.99998 = 166,722/166,725.

In contrast, a 10% random inspection policy would flag 0.10(166,725) = 16,673 wire bonds for a pull test. The cost of performing this test would be \$9,721. This policy would catch only 43 of the 425 defective wire bonds. The overall yield would be 0.99771. To achieve the same overall yield as the NN model, Westinghouse would have to use a random sampling policy in which 99.3% of the wire bonds are screened by a pull test. This policy would cost \$0.583 (165,558) = \$96,520.

## CONCLUSIONS

In this paper, we developed three types of discriminant analysis models and two types of neural network models for classifying the quality of wire bonds produced by Westinghouse. The neural network models were the most accurate classifiers on the held-back test set of data. Furthermore, we demonstrated that results from neural network models could form the basis of a high-reliability, low-cost wire bond quality assurance system for Westinghouse.

## ACKNOWLEDGMENT

We thank Hau Cheong and Ken Hines of Westinghouse for their help in producing and analyzing the wire bond data.

## REFERENCES

Ramani, N. and Patrick, P. (1992). Fish detection and identification using neural networks -- some laboratory results, *IEEE Journal of Oceanic Engineering*,*17*, 364-368.

Sun, X. (1995). Neural network models for the wire bonding process, Ph.D. dissertation, College of Business and Management, University of Maryland, College Park, Maryland.

Sun X., Golden, B., Kwon, O., and Wasil, E. (1994). Enhancing the performance of neural network models for the wire bonding process, in *Intelligent Engineering Systems through Artificial Neural Networks* (C. Dagli, B. Fernandez, J. Ghosh, and R. Kumara, editors), ASME Press, New York, 1041-1048.

Wang, Q., Sun X., Golden, B., DeSilets, L., Wasil, E., Luco, S., and Peck, A. (1993). A neural network model for the wire bonding process, *Computers & Operations Research*, *20*, 879-888.

# ARTIFICIAL NEURAL NETWORK AND THE TAGUCHI METHOD APPLICATION FOR OPTIMUM ULTRASONIC WELDING PROCESS DESIGN

**JUNGEUI HONG**
*Department of Industrial Engineering*
*Chung Ju National University*
*Chung Ju City, Chung-Chong Buk Do S. Korea*

**GERALD W. GREENWAY**
*Department of Engineering Management*
*University of Missouri-Rolla*

**KENNETH M. RAGSDELL**
*Director, Design Engineering Center*
*University of Missouri-Rolla*

*ABSTRACT:*
In quality engineering, and in most experimental design cases, understanding the relationship between design factors and product or process performance is essential for improving quality. Because most manufacturing conditions are so complicated, a large number of experiments are generally required. Taguchi's approach can systematically reduce the number of experiments required. But in certain case, where interaction exist between design factors, the level average analysis will not select optimum condition. To overcome this weakness after orthogonal array experiments are performed, an artificial neural network is used to select the optimum conditions instead of traditional level average analysis. Actual application, ultrasonic plastic welding process design, proved that this approach can select real optimum point both within and between factor levels.

## INTRODUCTION

Ultrasonic welding is a process for joining similar and dissimilar materials by introducing high-frequency, typically 20 to 40 kHz, vibrations into the overlapping materials in the area to be joined. The pulses are transmitted from a generator to the part by a resonant vibrating tool called a horn, causing two plastic materials to vibrate against each other. The part are heated and fused together(Crawford, 1955).

Ultrasonic welding for the thermoplastic materials requires more attention because each of these has special characteristics. The need to join these with parent materials has created a new sensitivity to using ultrasonic welding as an assembly technique. Therefore, the requirement of this forming method should be considered before their part hit the assembly line. Weld strength is related to the amount of interface between the two parts and to the depth of the welding. For conservative calculations, however, assume that the joint strength is approximately one-half of the base resin strength. The essential first step in getting reliable welds is designing a suitable part and joint.

Although design mistakes can be corrected, they are usually costly and time consuming . Key points to consider include the material's welding characteristic, how parts will be supported during welding, how they will be aligned, and the contacted areas for the ultrasonic horn.

Supportive fixtures are critical when designing parts for ultrasonic welding. Welding material should be firmly supported to prevent movement during the welding process. Also, the fixture must not interface with the ultrasonic horn. Alignment is done by providing a step to the weld interface. In the past, a 30 to 45o angle was recommended for the base of step. However, research shows that a 90o step provides a stronger weld in a shorter time(Kenny, 1992). If a step configuration is neither possible nor disadvantageous, some other method of aligning parts should be substituted. Another important design factor is horn location. The horn is a custom designed tool that transmits vibrations to the weld surface. It should bear directly over the joint area, and must be at a uniform distance from the joint interface.

## PARAMETER DESIGN

**Selecting Control Factors** The Branson 400 model ultrasonic welding machine was used for this experiment. This welding machine has three major control factors, which are weld time controller, hold time controller, and pressure controller. The other control factors are the gap thickness between the ultrasonic horn and the support fixture and the type of support fixture. The welding time is the application time of the high frequency to the materials. The length of welding time is affected greatly by low-density polyethylene film welding strength. Hold time, the pressing time after finishing the ultrasonic frequency to the materials, is another important factor and affects the welding strength. Pressure is a critical factor. Too much causes the parts to vibrate as an integral structure with no heating. Too little pressure does not provide enough contact friction or heating. Gap thickness divided by three levels like 0.012mm, 0.016mm and 0.02mm. This factor depends on the type of material and the physical properties of that material. The last factor is the type of supportive fixture. Because, a critical type of fixture was used, the radius of the fixture directly related to the contact area of the ultrasonic horn and welding materials.

**Conducting the Experiment** The $L_{27}$ orthogonal array was used for this experiment. Five samples are collected each row of the $L_{27}$ orthogonal array. The order of experimental runs can be randomly selected. In this experiment, replacing the supportive fixture was not simple, and that reason, the run order of the first nine rows are randomized for die type I. Table I shows the controllable factor levels for this experiment.

Another way to randomize is by replicating the experiment. Each row of the $L_{27}$ orthogonal array is replicated five times. All 135 runs are randomized. the maximum tensile strength(kg) is measured from the strength measurement instrument. Each orthogonal array row has five measurements of data. Because the higher tensile strength is preferred and this design characteristic is continuous and nonnegative, as a results, this design can be defined as having a larger-the-better quality characteristic.

**TABLE I: CONTROLLABLE FACTOR LEVELS**

| *Control factor* | *Level I* | *Level II* | *Level III* |
|---|---|---|---|
| Die Type (Radius mm) | 397 | 276 | 173 |
| Gap Thickness | 0.012 | 0.016 | 0.02 |
| Hold Time | 1.5 | 2.0 | 2.5 |
| Weld Time | 2.0 | 3.0 | 4.0 |
| Pressure(psi) | 60 | 70 | 80 |

**Calculating Signal-to-Noise ratio** The purpose of computation and the use of the signal-to-noise ration is to represent the effect of environmental condition, internal deterioration, and variation from unit to unit of those factors that can be controlled. In the specific application of the signal-to-noise ratio concept to the larger-the-better characteristic, those combinations of control factor settings that result in a consistently high value are preferred. One extremely high value under ideal environmental conditions will not do if the results are unsatisfactory the remainder of the time.

The formula for the larger-the-better signal-to-noise ratio quality characteristic is also a logarithmic function and can be written very similar to the smaller-the-better signal-to-noise formula. The signal-to-noise ration can be calculated from:

$$\eta = -10\log_{10}\left(\frac{1}{n}\sum_{i=1}^{n}\frac{1}{y_i^2}\right) \tag{1}$$

where n is the number of replications for each row, and $y_i$ is the measured maximum tensile strength(kg). For the larger-the-better quality characteristic, the signal-to-noise ratio increase as the average results increase. Improved consistency or reduced variability between unit to unit will again raise the signal-to-noise ratio.

## ANALYSIS OF EXPERIMENT DATA

**Analysis of Variance(ANOVA)** Different control factors affect the tensile strength to different degrees. The relative magnitude of the factor effects could be judged from the average signal-to-noise ratio for each factor level. More sophisticated techniques using analysis of variance (ANOVA) can provide more quantifiable analysis.

Table II shows the ANOVA table for this experiment. In the ANOVA table, die type and gap thickness show high $F$ values and small $p$-values, which means these affect mainly the signal-to-noise of the ultrasonic welding experiment. In most experimental design case, the relationship between design factor and performance can be approximated adequately by the additive model. The equation of additive model can be written

$$\eta(D_i, G_j, H_k, W_l, P_n) = \mu + d_i + g_j + h_k + w_l + p_n + \varepsilon \tag{2}$$

Where D, G, H, W and P are the effect of die type, gap thickness, hold time, weld time, and pressure, respectively. The $\varepsilon$ implies that the error of additive approximation plus the error of repeatability of measuring the signal-to-noise ratio for a given experiment.

**TABLE II: ANOVA TABLE**

| *Source* | *DF* | *Sum of Square* | *Mean of Square* | *F Value* | *Pr>F* |
|---|---|---|---|---|---|
| Die | 2 | 60.98 | 30.49 | 8.77 | 0.0027 |
| Gap | 2 | 45.72 | 22.86 | 6.57 | 0.0082 |
| Hold | 2 | 10.30 | 5.15 | 1.48 | 0.2567 |
| weld | 2 | 27.48 | 13.74 | 3.95 | 0.0403 |
| Pressure | 2 | 0.14 | 0.07 | 0.02 | 0.9797 |
| Error | 16 | 55.64 | 3.48 | | |
| Total | 26 | 200.27 | | | |

Table II shows the ANOVA table of this model. From table III, the model fitness can be tested using the f-test. The $p$-value of this model is:

$$P - value = p[F_{10,16} \geq 4.16] < 0.0057 \tag{3}$$

This means that this model fits this experiment within the 0.57% level Because the p-value is small enough, the additive model adequately represents this experiment.

**TABLE III: ANOVA TABLE FOR ADDITIVE MODEL**

| *Source* | *Df* | *Sum of Square* | *Mean Square* | *F value* | *pR>F* |
|---|---|---|---|---|---|
| Model | 10 | 144.63 | 14.46 | 4.16 | 0.0057 |
| Error | 16 | 55.64 | 3.48 | | |
| Total | 26 | | | | |

**Selecting Optimum Factor Level** The primarily goal in conducting the orthogonal array experiment is to optimize the product or process design. In this experiment, a higher signal-to-noise ratio means a higher average welding strength.

Table IV shows the average signal-to-noise ratio for each factor level. The highlighted values are the optimum condition for this experimental design. The level I of die type and gap thickness, level II of hold time and pressure, and level III of weld time are selected as an optimum condition. The signal-to-noise ratio of this optimum condition can be predicted from equation(4). Since the sum of square error due to factor hold time and pressure is small enough, that these terms are included as error.

$$\eta_{predicted} = \eta_m + (\eta_{D_1} - \eta_m) + (\eta_{G_1} - \eta_m) + (\eta_{W_1} - \eta_m) \quad (4)$$
$$= 22.56\text{dB}$$

Where $\eta_m$ is the average of signal-to-noise ratio and $\eta_{D_1}, \eta_{G_1}$ and $\eta_{W_1}$ are the averages of the factors at the given levels.

**TABLE IV: AVERAGE S/N RATIO FOR EACH FACTO LEVEL**

| *Factor* | *Level I* | *Level II* | *Level III* |
|---|---|---|---|
| Die type | 20.41 | 19.48 | 19.86 |
| Gap Thickness | 20.10 | 19.56 | 17.11 |
| Hold Time | 18.06 | 19.47 | 19.91 |
| Weld Time | 17.54 | 19.31 | 19.91 |
| Pressure | 18.97 | 19.21 | 18.58 |

**Confirmation Test** After determining the optimum condition and predicting the corresponding response, a confirmation experiment with the optimum parameter setting is conducted, and the observed value of η is compared with the predicted value. Table V shows the confirmation test results and corresponding signal-to-noise ratios. The observed signal-to-noise are close to each other, and it is concluded that this model is adequate for describing the dependence of the signal-to-noise ration on the various parameters.

**TABLE V: CONFIRMATION TEST RESULTS**

| *Test No* | *exp1* | *exp2* | *exp3* | *exp4* | *exp5* | *S/N Ratio* |
|---|---|---|---|---|---|---|
| 1 | 11.24 | 10.08 | 11.46 | 11.72 | 9,76 | 20.88 |
| 2 | 10.46 | 11.34 | 11.44 | 11.04 | 9.84 | 20.87 |

## NEURAL NETWORK APPLICATION

Five controllable factors exist, and each of these factors has three levels. The full factorial combination of the experiment is 35=243. Because the L27 orthogonal array is used, a total of 27 data points exist for neural network training. The input data for the neural networks are the controllable factor levels and the output data of the neural network is the signal-to-noise ratios of a corresponding combination.

The total number of weights of the neural networks are usually less than the total data point. In The fully connected neural network shows a very good performance compared to the number of hidden nodes. Using the fully connected backpropagation neural network, the input and output data are normalized. The fully connected neural network was trained with two hidden nodes and 0.1 learning rate.

**TABLE IV: CONFIRMATION TEST RESULTS SELECTED BY ANN**

| *No* | *Die type* | *Gap thickness* | *Hold Time* | *Weld Time* | *Pressure* | *S N Ratio* |
|---|---|---|---|---|---|---|
| 1 | I | 0.012 | 1.5 | 2 | 80 | 20.65 |
| 2 | I | 0.012 | 1.5 | 2.5 | 80 | 21.83 |
| 3 | I | 0.012 | 2.0 | 2 | 80 | 21.4 |
| 4 | I | 0.012 | 1.5 | 3 | 80 | 21.24 |
| 5 | I | 0.012 | 2.0 | 2 | 70 | 21.05 |

After training the neural network, the five highest combinations were selected and confirmation tests conducted. The test results are in table VI. Because four of the control factor have quantitative data, the optimum condition between factor levels can be selected by response surface analysis. From figure I, the highest estimated signal-to-noise ratio is near the 2 to 2.5 sec weld time. Confirmation test number 2 in table VII is a combination selected from weld time between level I and II.

## CONCLUSIONS

The best ultrasonic welding parameters are selected by this new approach. In ultrasonic welding, the controllable parameters are welding time, holding time, welding pressure, gap thickness between ultrasonic horn and die, and die type. Because these factors are interrelated with each other, the optimum welding condition is difficult to select.

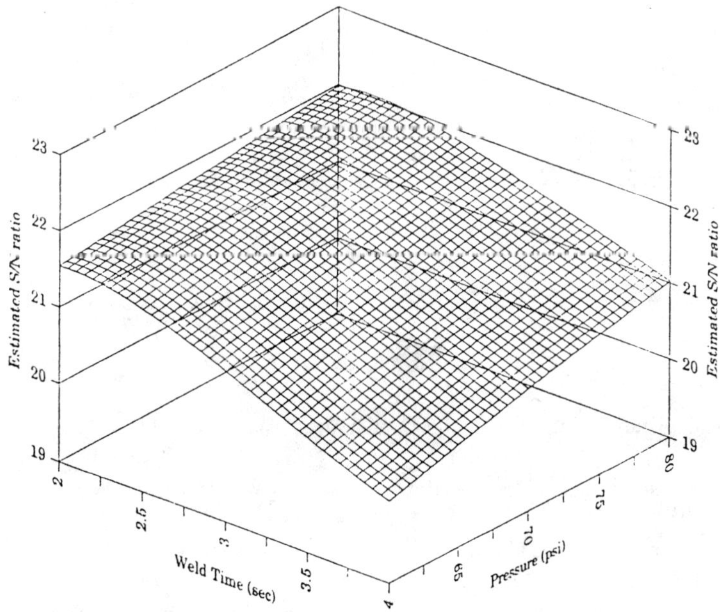


**Figure 1: Response surface with constant level I of die type, gap thickness and hold time**

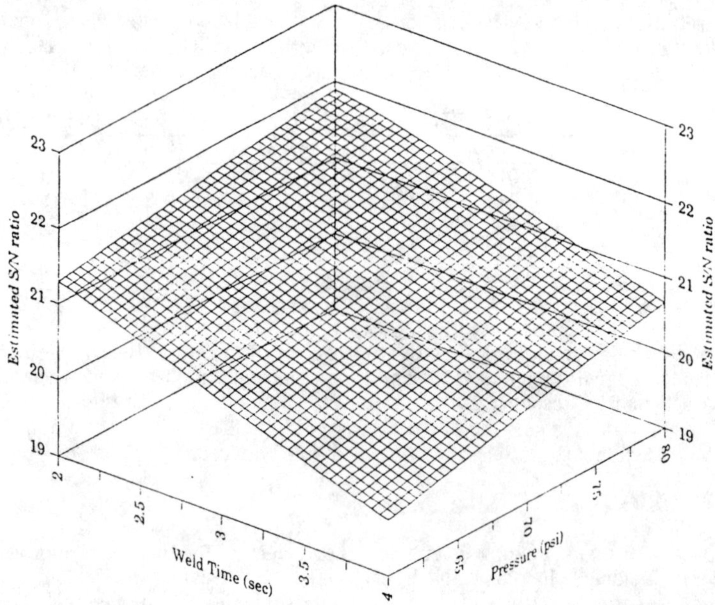


**Figure 2: Response surface with constant level I of die type, gap thickness and level II of hold time**

Parameter design result show the optimum welding condition is a die radius of 397mm, a gap thickness of 0.012mm, a hold time of 2 seconds, a weld time of 4 seconds, and pressure of 70 psi. In confirmation test, the signal-t- noise ratio of that condition is 20.875dB. With the orthogonal experiment data, the fully connected neural network trained with 18 weight and a 0.2 learning rate., Figure (1), (2) show the response surface resulting from the data

The test results show that artificial neural network result are very successful for selecting optimum conditions of ultrasonic welding process not only within the control factor level, but between the control factor levels.

## REFERENCES

Crawford, Alan E., Ultrasonic Welding: with Particular Reference to High Power Application. *Academic Press*, New York, 1955

Kenny, Warren E., Designing Plastic Parts for Ultrasonic Assembly, *Machine Design*, 1992, pp 65-68

Taguchi, Genishi., Taguchi Method-Case studies from U. S. and Europe. *ASI press*, 1989

Taguchi Genishi., System of Experimental Design Vol. I, II. *Kraus International Publications*, 1987.

Taguchi, Genishi., Taguchi Methods - Signal to Noise Ratio for Quality Evaluation. *ASI Press*, 1989

Werbos, Paul, Beyond Regression: New tools for prediction and analysis in the behavioral science, *Ph.D Dissertation*, Harvard University, 1974

# PERFORMANCE MONITORING SYSTEM OF VALVES OPERATION

**M H Wu**
*School of Engineering, University of Derby, U.K*

***Abstract*:**

On-line computer aided performance monitoring systems become more and more important in Industrial Automation and Modernisation. This paper introduces a novel on-line monitoring system of valve's performance in fuel injection system. The operating parameters used in the machine control are collected as an essential data for monitor system to analyse the valves working status. The monitoring systemis based on Fuzzy Sets Theory. As a result, the methodology of employing the fuzzy sets theory to create the monitoring model is mentioned details in the paper. It includes the method of blurring the input parameters into fuzzy input, method of input/output fuzzy sets definition, fuzzy decision model, etc.
The paper outlines the fuel injection at the beginning for reader to understand the whole research easily. The result of the test of the system shows the system is successful and reliable under the real working environment.

## INTRODUCTION

The correct functioning of a fuel injection system is dependent on that of its Pneumatic Hydraulic System and electric circuits (Fig.1). If a defect occurs in either of these, it could lead to the faulty operation of one or more of the valves, namely the Air Valves, Va, Gas Valve, Vg, Inlet/Exhaust Valve, Vi/Ve. Obviously this would cause the fuel injection system to malfunction.

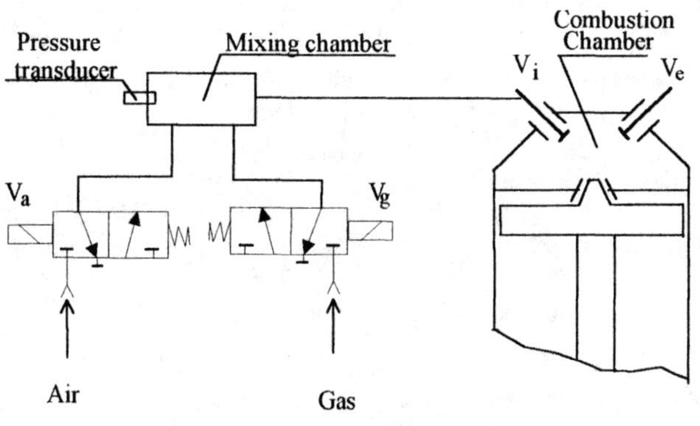


Figure 1: The fuel injection system

The valves' normal states have been shown in Table 1. The paper will discuss how to monitor the valve performance in two cases: one or more valve(s) break down and the leak of the valve(s).

| Valve | Air | Gas | Inlet | Exhaust |
|---|---|---|---|---|
| Normal states | closed | closed | closed | open |

Table 1. The valves' states

## FUZZY DIAGNOSIS SYSTEM FOR THE BROKEN VALVES DETECTION

Suppose it is required to determine whether an element is a member of a certain fuzzy set. The first step is to define the conditions of membership of the sets using Fuzzy Sets theory and then to assess if the element in question satisfies that condition.

In the case of the system, the variation of pressure in the Mixing chamber can be employed to describe the operational states of the valves. It can be assumed that the elements of the universe are pressure values, and the Fuzzy Sets are $\underset{\sim}{G}$ (the pressures associated with a faulty gas valve), $\underset{\sim}{A}$ (the pressures associated with a faulty air valve), $\underset{\sim}{I}$ (the pressures associated with a faulty inlet valve) and $\underset{\sim}{E}$ (the pressures associated with a faulty exhaust valve), etc.. In an ideal valve the operating conditions would be constant, and the components of the electric and hydraulic circuits would respond instantly and repeatable, and the pressures Pa, Pg would have fixed values. However during actual operation the pressure will vary about these values. Fuzzy Sets Theory can be applied to determine which fuzzy set the variable P (pressure) belongs to and the defective valve can be traced without delay.

### Relationship between defective valves and pressure value

The valves in the fuel injection system can be divided into two groups depending on their location relative to the compressed air/gas sources and the pressure transducer. From fig.1 it can be seen that Va and Vg are directly linked to the compressed air/gas sources respectively, so that in their working states they will cause variation of pressure in the mixing chamber, regardless of the states of Vin and Vex. Vin and Vex will make the pressure in the mixing chamber vary when and only when Va and Vg is switched on correctly. In this case the elements of the fuzzy sets $\underset{\sim}{G}, \underset{\sim}{A}, \underset{\sim}{I}$ and $\underset{\sim}{E}$ which have been defined above are related to values of Pa and Pg. Fig.2 shows the relationship between the fuzzy sets (faulty valves) and the pressure values.

### Function of Grade of Membership

The key problem of determining which fuzzy set an element belongs to is to obtain the functions of grade of membership for each fuzzy set.

Figure 3 shows the four kinds of relationship between the grade of membership and pressure Pa or Pg. These relationships were observed experimentally. In Fig.3 a and P represent respectively Psa (or Psg) and Pa (or Pg).

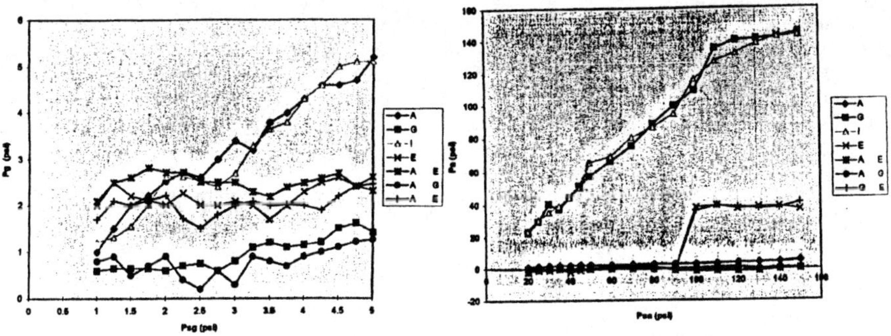

Figure 2: Relationship between the fuzzy sets (faulty valves) and the pressure values

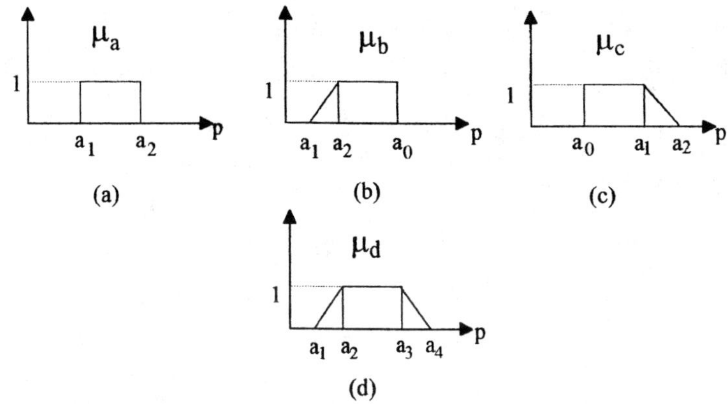


Figure 3: Relationship between the grade of membership and pressure Pa or Pg

From the figure, the functions of grades of membership are:

$$\mu_a = \begin{cases} 0; & x < a_1 \text{ or } \quad x > a_2 \\ 1; & a_1 \leq x \leq a_2 \end{cases}$$

$$\mu_b = \begin{cases} 0; & x < a_1 \text{ or } \quad x > a_0 \\ \frac{x-a_1}{a_2-a_1}; & a_1 \leq x \leq a_2 \\ 1; & a_2 < x \leq a_0 \end{cases}$$

$$\mu_c = \begin{cases} 0; & x < a_0 \quad \text{ or } x > a_2 \\ 1; & a_0 \leq x \leq a_1 \\ \frac{a_2-x}{a_2-a_1}; & a_1 \leq x \leq a_2 \end{cases}$$

$$\mu_d = \begin{cases} 0; & x < a_1 \quad or \ x > a_4 \\ \frac{x-a_1}{a_2-a_1}; & a_1 \le x < a_2 \\ 1; & a_2 \le x < a_3 \\ \frac{a_4-x}{a_4-a_3} & a_3 \le x \le a_4 \end{cases}$$

For the different sets the functions of the grade of membership are:

Set $\underset{\sim}{A}$:

$\mu_{\underset{\sim}{A},a}(p_a) = \mu_a(x); \quad a_1 = 0, a_2 = 2$

$\mu_{\underset{\sim}{A},g}(p_g) = \mu_b(x); a_0 = Psg + 2.5, a_1 = 0.8Psg, a_2 = Psg$

$\mu_{\underset{\sim}{A}}(p_a, p_g) = 0.4\mu_{\underset{\sim}{A},a} + 0.6\mu_{\underset{\sim}{A},g}$

Set $\underset{\sim}{G}$:

$\mu_{\underset{\sim}{G},a}(p_a) = \mu_b(x); \qquad a_0 = 200, a_1 = 0.8Psa, a_2 = Psa$

$\mu_{\underset{\sim}{G},g}(p_g) = \mu_c(x);$ when $0.6Psg < 3, a_0 = 0, a_1 = 0.6Psg, a_2 = 3$

$= 0;$ when $0.6Psg > 3$ and $0.6Psg \le Pg$

$\mu_{\underset{\sim}{G}}(p_a, p_g) = 0.5\mu_{\underset{\sim}{G},a} + 0.5\mu_{\underset{\sim}{G},g}$

Set $\underset{\sim}{I}$ :

$\mu_{\underset{\sim}{I},a}(p_a) = \mu_c(x);$ when $Psa \le 100, a_0 = Psa, a_1 = 1.2Psa, a_2 = 1.4Psa$

$= \mu_a(x);$ when $Psa > 100, a_1 = 1.1Psa, a_2 = 1.6Psa$

$\mu_{\underset{\sim}{I},g}(p_g) = \mu_b(x);$

when $Psg \le 2.1, a_0 = Psg + 2, a_1 = Psg - 0.6, a_2 = Psg - 0.4$

when $Psg > 2.1, a_0 = Psg + 1.5, a_1 = Psg - 0.8, a_2 = Psg - 0.6$

$\mu_{\underset{\sim}{I}}(p_a, p_g) = 0.6\mu_{\underset{\sim}{I},a} + 0.4\mu_{\underset{\sim}{I},g}$

Set $\underset{\sim}{E}$:

$\mu_{\underset{\sim}{E},a}(p_a) = \mu_a(x);$ when $Psa \le 110, a_1 = 0, a_2 = 2$

when $Psa > 110, a_1 = 20, a_2 = 60$

$\mu_{\underset{\sim}{E},g}(p_g) = \mu_{\underset{\sim}{A} \cap \underset{\sim}{E},g}(p_g);$

$\mu_{\underset{\sim}{E}}(p_a, p_g) = 0.5\mu_{\underset{\sim}{E},a} + 0.5\mu_{\underset{\sim}{E},g}$

Set $\underset{\sim}{A} \cap \underset{\sim}{E}$:

$\mu_{\underset{\sim}{A} \cap \underset{\sim}{E},a}(p_a) = \mu_a(x); \qquad a_1 = 0, a_2 = 2$

$\mu_{\underset{\sim}{A} \cap \underset{\sim}{E},g}(p_g) = \mu_d(x); \qquad a_1 = 0.8, a_2 = 1.2, a_3 = 3, a_4 = 3.5$

$\mu_{\underset{\sim}{A} \cap \underset{\sim}{E}}(p_a, p_g) = 0.5\mu_{\underset{\sim}{A} \cap \underset{\sim}{E},a} + 0.5\mu_{\underset{\sim}{A} \cap \underset{\sim}{E},}$

Set $\underset{\sim}{A} \cap \underset{\sim}{G}$:

$\mu_{\underset{\sim}{A} \cap \underset{\sim}{G},a}(p_a) = \mu_a(x); \quad a_1 = 0, a_2 = 10$

$\mu_{\underset{\sim}{A} \cap \underset{\sim}{G},g}(p_g) = \mu_{\underset{\sim}{G},g}(p_g);$

$\mu_{\underset{\sim}{A} \cap \underset{\sim}{G}}(p_a, p_g) = 0.5\mu_{\underset{\sim}{A} \cap \underset{\sim}{G},a} + 0.5\mu_{\underset{\sim}{A} \cap \underset{\sim}{G},g}$

Set $\underset{\sim}{G} \cap \underset{\sim}{E}$:

$\mu_{\underset{\sim}{G} \cap \underset{\sim}{E},a}(p_a) = \mu_{\underset{\sim}{E},a}(p_a);$

$\mu_{\underset{\sim}{G} \cap \underset{\sim}{E},g}(p_g) = \mu_{\underset{\sim}{G},g}(p_g);$

$\mu_{\underset{\sim}{G} \cap \underset{\sim}{E}}(p_a, p_g) = 0.5\mu_{\underset{\sim}{G} \cap \underset{\sim}{E},a} + 0.5\mu_{\underset{\sim}{G} \cap \underset{\sim}{E},g}$

where:

Psa : Set air pressure; Psg : Set gas pressure.

As can be seen from the above equations, in the functions of grade of membership, the points a0, al, a2, a3 and a4 depend on the variables Psa and Psg. Fig. 2 shows the variation of Pa and Pg in each set for different Psa and Psg.

**Determination of defective valves**

According to the principle of largest subordination mentioned before, the following comparison will be made:

$$\mu(p_{g0}, p_{a0}) = \underset{*=1}{\overset{7}{\forall}} \mu_*(p_{g0}, p_{a0})$$

where:

* means $\underset{\sim}{A}, \underset{\sim}{G}, \underset{\sim}{I}, \underset{\sim}{E}, \underset{\sim}{G} \cap \underset{\sim}{E}, \underset{\sim}{A} \cap \underset{\sim}{G}$ and $\underset{\sim}{A} \cap \underset{\sim}{E}$;

The value of (Pg, Pa) will be equal to the largest of all seven functions of grade of membership. The results of the equations above will determine which valves are defective. It is obvious that appropriate auxiliary checks are required to verify the correctness of the judgement and resolve the ambiguity in the low pressure area (Fig. 3). As a result, the computer will arrange to actuate automatically each of the valves suspected to be defective and give the final diagnosis according to whether changes in the measured pressure are observed. The result will be printed to show the valve or electric element to be replaced.

The above discussion relates to the case when the valves are break down and cannot reach their states of operation.

## MONITORING THE LEAKAGE OF THE VALVES

Valve leakage normally happens if the valve is worn out. It is a long team action, so that the on-line monitoring is not necessary. The leakage of the valves are tested automatically each day when the system is switched on.

The valve is identified as faulty and recommended for replacement if the measured value as follow:

Gas valve is faulty, if the measured pressure is about 90% Psg;

Air valve is faulty, if the measured pressure is about 70% Psa;

Inlet valve is faulty, if the charged pressure, Psa, is reduced 20% within 200ms;
Exhaust valve is faulty, if the charged pressure, Psa, is reduces 20% within 100ms.

## CONCLUSION

The special electronic circuit is designed to switch the Solenoid-valves, Va, Vg, Vin and Vex, for test purpose.. The valves has been switch off (equivalent of the valve broken) individually or in group randomly. The faulty valve(s) has been pointed out correctly..

## REFERENCES

Zadeh, L., (1965). Fuzzy sets. J. Information Control, 1965, 8,338-353
Zimmermann, H.J., (1991). Fuzzy sets theory and its applications, (Kluwer, Boston).
Wang, P..Z., (1983). Fuzzy sets theory and its applications, (Shanghai Science and Technology Press, Shanghai, PRC).
Lee, C. and Tobias, S.A., (1985) Micro-computer control of petro-forge machine. Int.J.Mach.Tool , 25(3),245-267

# EVALUATION OF MODEL DISCRIMINATION TECHNIQUES IN ARTIFICIAL NEURAL NETWORKS WITH APPLICATION TO GRAIN DRYING

**FIRST, SECOND, AND THIRD AUTHORS**
*ANGELITA D.N. GARTH, DERRICK K. ROLLINS, AND JUN ZHU*
*DEPARTMENTS OF STATISTICS AND CHEMICAL ENGINEERING*
*IOWA STATE UNIVERSITY*
*AMES, IOWA 50011*

**FOURTH AUTHOR**
*VICTORIA C.P. CHEN*
*SCHOOL OF INDUSTRIAL AND SYSTEMS ENGINEERING*
*GEORGIA INSTITUTE OF TECHNOLOGY*
*ATLANTA, GEORGIA 30332*

## INTRODUCTION

Artificial neural network (ANN) models have had considerable growth as a modeling technique for physical phenomena in chemical engineering in recent years. Their popularity as an empirical technique has caused researchers to overlook weaknesses. In particular, ANNs have a tendency to overfit data. This occurs when the ANN has an excessive number of parameters and is modeling the random error in the data as well as the underlying structure. In an attempt to overcome this drawback, researchers employ various existing statistical model discrimination (MD) techniques to determine the correct artificial neural network model. Unfortunately, because one can seldom be confident of what the true model is in real world applications, there is no way of knowing whether or not the MD technique is choosing the optimal model. As a consequence, there has been little testing of statistical MD techniques in ANN applications. This paper evaluates six common techniques for ANNs: the *F*-test, Mean Squared Error (MSE), Akaike's Information Criterion (AIC), Schwarz's Bayesian Criterion (SBC), Mallow's $C_p$, and Data Splitting.

For the first phase of testing, data with added random error were generated from ten three-layer ANN models consisting of between three to six inputs, one to four hidden nodes, and one output. A $3^n$ factorial experimental design was used, where $n$ is the number of inputs. Under varying combinations of error variances and replications, the techniques were evaluated as to how often they chose the true model. For the second phase of testing, five of the MD techniques (not including Data Splitting) were applied to a mathematical grain drying model provided by researchers in the Agriculture Engineering Department at Iowa State University.

MSE and the *F*-test are methods from classical statistics. AIC, SBC, and $C_p$ are commonly used in model discrimination for linear regression. AIC and SBC are similar in that they both contain a term for lack-of-fit and a penalty term for increasing the complexity of the model (i.e., increasing the number of nodes). Mallow's $C_p$ computes an estimate of the total mean squared error scaled by an unbiased estimate of the error variance. For all three, the best model is that which minimizes the MD measure.

Data Splitting partitions the data into a training set and a test set. First, we fit the ANN model to the training set. Then using the test set, we compute two lack-of-fit measures, Sum of Squared Prediction Errors (SSP) and Mean Squared Prediction Error (MSP). Again, the smallest model which minimizes these quantities is best.

The MD techniques are described in more detail in the next section. Simulations and results for cases when the true model is an ANN are described under the subheading of "When the True Model Is a Neural Network." Finally, the application to grain drying is presented in the section titled "Application to Grain Drying."

## MODEL DISCRIMINATION TECHNIQUES

The purpose of this study was to evaluate existing MD techniques to see how often, when given competing ANN models, each would select the correct model. In this section, the six techniques that were evaluated are described for feedforward ANNs with one hidden layer and one output. For a single output $y$, $n$ inputs $x_1, \ldots, x_n$, $k$ hidden nodes, and noise $\epsilon$, our ANN model equation using a logistic activation function is:

$$y = \left[1 + e^{-\sum_{j=1}^{k} w_{j1} z_j}\right]^{-1} + \epsilon \tag{1}$$

where for each hidden node $j$:

$$z_j = \left[1 + e^{-\sum_{i=1}^{n} v_{ij} x_i}\right]^{-1}$$

where $k$ can vary from 1 to $n$ - 1. We assume $\epsilon$ is normally distributed with mean zero and variance $\sigma^2$. For a data set with $N$ observations, denote the $N$ x 1 vector of observed output values $\mathbf{y}$ and the corresponding predicted values using ANN model with $k$ hidden nodes by $\hat{\mathbf{y}}_k$. Then the Nx1 vector of residuals for the ANN model with $k$ hidden nodes is:

$$\hat{\boldsymbol{\epsilon}} = \mathbf{y} - \hat{\mathbf{y}}_k \tag{2}$$

### Mean Squared Error

The Mean Squared Error (MSE), or residual mean square, is a measure of the amount of variability in the data which cannot be explained by the fitted model. For our ANN model in Eq.1, MSE is computed as:

$$MSE = \frac{\sum_{i=1}^{N} (y_i - \hat{y}_i)^2}{N - k(n + 1)}$$

where $y_i$ is the i-th observed value for the single output variable, $\hat{y}_i$ is the i-th fitted value. The denominator, $N - k(n + 1)$, is the degrees of freedom for error and is computed as the number of observations minus the number of parameters in the model.

Starting with one node in the hidden layer, MSE should decrease as more nodes (and therefore more parameters) are added to the model. Close to the "best" fit to the data,

MSE should begin to level off. The model with the smallest MSE is selected as the best model based on this criterion with the smaller (fewer parameters) model chosen in the case of a tie.

**The *F*-Test**

The *F*-test is a classical statistical test derived for linear models in which the observations of the response are normally distributed (or, equivalently, the random error is normally distributed). The purpose of this test is to compare two nested linear models, a larger (full) model and a smaller (reduced) model.

For ANN models, we compare the (full) model with $k + 1$ nodes in the hidden layer to the (reduced) one with $k$ nodes. From Eq. 1, our null and alternative hypotheses are, respectively:

$$H_0 : Y = \left[1 + e^{-(w_{11}Z_1 + ... + w_{k1}Z_k)}\right]^{-1} + \epsilon \qquad \textit{(reduced model)}$$

$$H_a : Y = \left[1 + e^{-(w_{11}Z_1 + ... + w_{k1}Z_k + w_{k+1}Z_{k+1})}\right]^{-1} + \epsilon \qquad \textit{(full model)}$$

The sum of squares for the full model is computed as $\hat{\epsilon}_{k+1}^T \hat{\epsilon}_{k+1}$ and for the reduced model as $\hat{\epsilon}_k^T \hat{\epsilon}_k$ where $\hat{\epsilon}_{k+1}$ and $\hat{\epsilon}_k$ are the residuals as in Eq. 2. If these models were to have the properties of a linear regression model, since the residuls are normally distributed, the ratio $\hat{\epsilon}_{k+1}^T \hat{\epsilon}_{k+1}/\sigma^2$ would have a chi-squared distribution with $N - (k + 1)(n + 1)$ degrees of freedom, and $(\hat{\epsilon}_k^T \hat{\epsilon}_k - \hat{\epsilon}_{k+1}^T \hat{\epsilon}_{k+1})/\sigma^2$ would have a chi-squared distribution with $n + 1$ degrees of freedom. Then the *F*-statistic:

$$F_{k,k+1} = \frac{(\hat{\epsilon}_k^T \hat{\epsilon}_k - \hat{\epsilon}_{k+1}^T \hat{\epsilon}_{k+1})/(n + 1)}{\hat{\epsilon}_{k+1}^T \hat{\epsilon}_{k+1}/(N - (k + 1)\ (n + 1))} \qquad \textbf{(3)}$$

would have an *F*-distribution with $n + 1$ numerator and $N - (k + 1)(n + 1)$ denominator degrees of freedom.

Since ANN models are nonlinear, even when $\epsilon$ is normally distributed with mean zero and variance $\sigma^2$, $F_{k,k+1}$ is not exactly *F*-distributed, especially when $N$ is small. As we compare the *F*-statistic in Eq. 3 to *F*-values with the corresponding degrees of freedom and a particular $\alpha$-level of significance, the actual significance level is not exactly $\alpha$.

In spite of this discrepancy, Eq. 3 will be compared to *F*-values with $n + 1$ numerator and $N - (k + 1)(n + 1)$ denominator degrees of freedom. If the calculated *F*-statistic is greater than the corresponding *F*-value with a particular $\alpha$, then the reduced model with $k$ nodes is rejected. The model with $k + 1$ nodes is then compared to a model with $k + 2$ nodes. At any time, if the null hypothesis fails to reject, the procedure is stopped and the current null hypothesis model is selected as the best model. Otherwise, the process continues until the null hypothesis is no longer rejected. We studied *F*-tests with significance levels of $\alpha = 0.01$, 0.05, and 0.10, but to conserve space, we only present the results for $\alpha = 0.01$. All three *F*-tests had similar characteristics with the 0.01-level *F*-test performing best.

**Akaike's Information Criterion**

Akaike's Information Criterion (Akaike,1974) is typically used in regression model fitting to give a measure of how well a proposed regression model fits the observed data. It has also been used in ANN model discrimination. AIC is calculated in the following manner for our ANN model with $k$ hidden nodes:

$$AIC = N \ln (MLE\ of\ \sigma^2) + 2k(n+1)$$

$$= N \ln \left( \frac{\hat{\epsilon}_k^T \hat{\epsilon}_k}{N} \right) + 2k(n+1)$$

where MLE represents the maximum likelihood estimator of $\sigma^2$. Notice that the first term in this sum is a monotone decreasing function of the residual sum of squares, and the second term in the sum is a penalty paid for increasing the number of parameters. The model that has the smallest AIC is the one selected by this technique.

**Mallow's $C_p$ Statistic**

Mallows' $C_p$ (Mallows, 1973) was created to find a proper compromise between the effects of underfitting and overfitting a model. An underfitted model may have biased coefficients and biased predictions ($\hat{y}$), whereas an overfitted model may have large variances for both the coefficients and the predictions. An adequate model should be a compromise between these two effects. Although we will not go into its derivation, the resulting statistic for our ANN model with $k$ hidden nodes is:

$$C_p(k) = \frac{SSE_k}{\hat{\sigma}^2} - (N - 2k(n+1))$$

$$= \frac{\hat{\epsilon}_k^T \hat{\epsilon}_k}{\hat{\sigma}^2} - (N - 2k(n+1))$$

where $\hat{\sigma}^2$ is a properly chosen estimate of $\sigma^2$, the unknown variance of the true model.

A drawback of this method is the subjective choice of $\hat{\sigma}^2$. In this study, it will be estimated by the mean squared error of the largest model being compared. The model with the minimum value of $C_p$ is selected as the value corresponding to the best model for this criterion.

**Schwarz's Bayesian Criterion**

Schwarz's Bayesian Criterion (Schwarz, 1978) treats the problem of selecting the correct model by finding its Bayes solution and evaluating the leading terms of its asymptotic expansion. It is quite similar to AIC yet differs in that the dimension is multiplied by ½ in $N$, as shown below for our ANN model with $k$ hidden nodes:

$$SBC = \ln \left( \frac{\hat{\epsilon}_k^T \hat{\epsilon}_k}{N} \right) + k(n+1) \frac{\ln N}{N}$$

or equivalently,

$$SBC = N \ln \left( \frac{\hat{\epsilon}_k^T \hat{\epsilon}_k}{N} \right) + k(n+1) \ln N$$

This difference causes SBC to lean more toward lower-dimensional models than AIC, if there are more than eight observations. However, mathematically, both procedures follow the principle of parsimony (smaller is better) in model building. In spite of their similarities, the two procedures are said to give quite different results for large samples (Schwarz, 1978). Like AIC, the model corresponding to the smallest SBC value is chosen as the best model.

**Data Splitting**

Another method of model discrimination is called Data Splitting. This method involves the partitioning of the data into two sets: a training set and a test set. This concept has been around for several years in the statistical literature (See Snee, 1977 and Stone, 1974), but is also now widely used in ANN MD. It appears to be one of the most popular methods of MD used by ANN researchers at this time.

The Data Splitting method consists of training, or fitting, an ANN model to a training data set. This model is then used to estimate the output responses in a test set. The prediction errors (difference between estimated response and response in the test set) are used to generate statistics. For our ANN models, we computed the sum of squared prediction errors (SSP) and the mean squared prediction error (MSP) as follows:

$$SSP = \sum_{i \in test\ set} (y_i - \hat{y}_i)^2 \quad and \quad MSP = \sum_{i \in test\ set} \frac{(y_i - \hat{y}_i)^2}{N_{test} - k(n+1)}$$

where $y_i$ is i-th observation in the test set, $\hat{y}_i$ is the estimated response for the i-th observation in the test set using the fitted model from the training set, and $N_{test}$ is the number of observations in the test set.

These statistics should decrease as more parameters are added to the model, up to the true model, but once too many parameters are in the model, these statistics should not decrease further, and may even increase. Thus, the minimum value of the statistic is the criterion used to select the best model. In the case of a tie, the smallest model is chosen. In the statistical literature, it is suggested that the test and training sets be approximately equal in size (Snee, 1977). We split our data according to the number of replications being used.

## WHEN THE TRUE MODEL IS A NEURAL NETWORK

Ten true ANN models with the functional form in Eq. 1 were chosen to test the seven MD techniques described under "Model Discrimination Techniques." These models are shown with their true parameters in Table 1.The parameters were chosen so that the output responses ranged between 0.25 and 0.75. Since we found the training time to be significantly slower when the number of nodes was greater than or equal to the number of inputs, we only fit models with fewer nodes than inputs. We chose the true ANN models, so that we could fit at least one competing model with more nodes than the true model. Thus, for the true ANN model with *n* inputs, we fit competing *n*-input ANN models with 1 to *n* - 1 hidden nodes. For more details on this study, see Garth (1995).

## Simulation Setup

For the true models with $n$ inputs, we determined the input vectors using a $3^n$ full factorial design. This experimental design sets each input at one of three levels and includes all possible combinations of these three levels for $n$ inputs. For each true model, we created data sets for testing MD techniques, by computing the true model's output responses for each of the $3^n$ settings of the input values and adding random error and replications. The number of replications was two or three.

**Table 1: TEN TRUE MODELS AND THEIR PARAMETERS.**

| Model | | Parameters | | | | | | |
|---|---|---|---|---|---|---|---|---|
| Inputs | Node $j$ | $v_{1j}$ | $v_{2j}$ | $v_{3j}$ | $v_{4j}$ | $v_{5j}$ | $v_{6j}$ | $w_{j1}$ |
| 3 | 1 | 1.6 | -0.09 | -2.25 | | | | 1.3 |
| 4 | 1 | 1.2 | -1.8 | -3 | 1.5 | | | 1.1 |
| 4 | 1 | 0.7 | -1.9 | -2.1 | 0.65 | | | 2.1 |
| | 2 | -2.1 | -0.9 | 1 | 0.75 | | | -1.2 |
| 5 | 1 | 1.2 | -1.6 | -3.5 | 2.4 | 0.25 | | 1.1 |
| 5 | 1 | 0.25 | -1.9 | -2.1 | 0.65 | 0.25 | | 2.9 |
| | 2 | -2.4 | -0.9 | 0.2 | 0.6 | 0.8 | | -1.2 |
| 5 | 1 | -2 | -1.3 | 1.1 | -0.9 | 0.7 | | 2.5 |
| | 2 | -1.7 | -2.5 | 3.4 | 0.2 | 0.5 | | -1.3 |
| | 3 | 1.1 | 0.1 | -0.6 | 3 | -2 | | 0.9 |
| 6 | 1 | 1.2 | -1.6 | -3.5 | 2.4 | 0.25 | -0.65 | 1.1 |
| 6 | 1 | -2.5 | 2.5 | -0.5 | 3.5 | -2.5 | 1.3 | 0.8 |
| | 2 | -2.5 | 0.1 | -1.9 | 2.5 | -0.1 | 2.5 | -1.1 |
| 6 | 1 | -2.5 | 0.5 | -1.5 | -0.5 | 4.5 | -0.5 | 0.9 |
| | 2 | -0.5 | 3.5 | -0.5 | -1.5 | -0.5 | -1.5 | -1 |
| | 3 | 1.5 | 0.5 | -3.5 | 2.5 | 0.5 | 1.5 | -0.1 |
| 6 | 1 | 1.5 | -0.5 | -1.5 | 0.5 | -2.5 | -0.5 | -2.9 |
| | 2 | 3.5 | -0.5 | -2.5 | 0.5 | -1.5 | -1.5 | 0.7 |
| | 3 | -1.5 | 1.5 | 0.5 | -2.5 | 0.5 | 3.5 | -1.1 |
| | 4 | -0.5 | 2.5 | 0.5 | -1.5 | 0.5 | 3.5 | 0.5 |

**TABLE 2: SAMPLE SIZES AND REPLICATIONS.**

| Inputs | Repl | Total No. of Objects |
|---|---|---|
| 3 | 2 | 54 |
| | 3 | 81 |
| 4 | 2 | 162 |
| | 3 | 243 |
| 5 | 2 | 486 |
| | 3 | 729 |
| 6 | 2 | 1458 |
| | 3 | 2187 |

Table 2 shows the total number of output response observations in a data set for *n* inputs. The random error was generated from normal distributions with mean zero and variance $\sigma^2$. For the error levels we chose $\sigma = 0.01$ (4% of the smallest output response value) and $\sigma = 0.05$ (20% of the smallest response), so that the output observations would fall between zero and one.

For the Data Splitting technique, we needed to create a training set and a test set. When using only two replications, the data was split in half, with one replication making up the training set, and the other replication making up the test set. Four training data sets (two levels of replication and two levels of error) and four "total" data sets (for the other) were associated with each of the ten true models.

To fit the competing ANN models we employed the nonlinear least squares parameter estimation algorithm GREG (for General REGression) developed by Stewart et al. (1992). It addresses estimation problems with upper and/or lower constraints on the parameters and uses a gradient optimization method developed by Caracotosis (1986). GREG has been successfully applied to some very difficult nonlinear estimation problems (Caracotios, 1986; Rollins, 1987).

Before implementing GREG, we needed to choose starting parameter values. For a true model with *n* inputs, there were *n* - 1 competing ANN models (with 1 to *n* - 1 nodes) for which we generated starting values. The parameters were bounded between -10 and 10. Using small random weights from a Uniform (-1,1) distribution as starting values, we trained each competing model on a $3^n$ data set of true model output responses with one replication and no random error. This training process was repeated fifty times with different weights, and the parameter values corresponding to the smallest residual sum of squares and not on the boundary were chosen as starting values for GREG. The choice of "optimal" parameter starting values significantly decreased the computational time for our simulations.

Given the starting values, one simulation case for a specific true model with *n* inputs consisted of generating the four total and four training data sets, fitting the competing ANN models with 1 to *n* - 1 nodes, and computing the MD measures to determine their selections for the "best" model. For each true model, we ran simulations until we acquired 100 convergent cases. A case was only considered convergent if all the competing models converged. Convergence was indicated by GREG if the relative change in each parameter between two consecutive iterations was 0.00001 or less. The maximum number of iterations to convergence was set at 200.The computational time to acquire 100 convergent cases varied approximately between 30 minutes and 12 hours on a DEC 5000/25 workstation or Alpha workstation.

**Results**

Table 3 shows how often the true model was chosen by each of the seven MD techniques for the different replications and error levels. The technique using MSE performed worst, followed by the Data Splitting criterion SSP. All seven of them performed best on the simplest true model (3-input 1-node) and had difficulty with two complex true models, the 5-input 3-node and the 6-input 4-node models. Tables 4 through 7 take a more detailed look at the results for the 5-input 3-node and 6-input 4-node models. Tables 4 and 5 present the number of times the techniques chose each competing model. To determine the stability of our simulation study, we conducted an

TABLE 3: SUMMARY OF MD TECHNIQUE SELECTIONS BASED ON 100 SIMULATIONS.

| Model | | | | % Correct Selections of the True Model | | | | | | |
|---|---|---|---|---|---|---|---|---|---|---|
| Inputs | Nodes | σ | Repl | MSE | $F_{.01}$ | AIC | $C_p$ | SBC | SSP | MSP |
| 3 | 1 | 0.01 | 2 | 76 | 100 | 95 | 95 | 100 | 70 | 97 |
| 3 | 1 | 0.01 | 3 | 84 | 100 | 95 | 95 | 100 | 68 | 92 |
| 3 | 1 | 0.05 | 2 | 72 | 99 | 94 | 95 | 99 | 75 | 98 |
| 3 | 1 | 0.05 | 3 | 78 | 100 | 97 | 98 | 100 | 71 | 91 |
| 4 | 1 | 0.01 | 2 | 39 | 99 | 91 | 92 | 100 | 69 | 95 |
| 4 | 1 | 0.01 | 3 | 39 | 100 | 92 | 92 | 100 | 66 | 93 |
| 4 | 1 | 0.05 | 2 | 39 | 100 | 83 | 87 | 100 | 75 | 95 |
| 4 | 1 | 0.05 | 3 | 42 | 100 | 90 | 92 | 100 | 69 | 93 |
| 4 | 2 | 0.01 | 2 | 55 | 100 | 79 | 82 | 100 | 84 | 95 |
| 4 | 2 | 0.01 | 3 | 57 | 100 | 88 | 90 | 100 | 68 | 90 |
| 4 | 2 | 0.05 | 2 | 48 | 97 | 80 | 83 | 100 | 79 | 95 |
| 4 | 2 | 0.05 | 3 | 56 | 96 | 84 | 87 | 100 | 78 | 93 |
| 5 | 1 | 0.01 | 2 | 13 | 97 | 88 | 88 | 100 | 73 | 96 |
| 5 | 1 | 0.01 | 3 | 18 | 99 | 83 | 84 | 100 | 68 | 91 |
| 5 | 1 | 0.05 | 2 | 19 | 98 | 83 | 83 | 100 | 84 | 99 |
| 5 | 1 | 0.05 | 3 | 14 | 97 | 83 | 85 | 100 | 74 | 97 |
| 5 | 2 | 0.01 | 2 | 39 | 100 | 97 | 98 | 100 | 69 | 86 |
| 5 | 2 | 0.01 | 3 | 37 | 100 | 92 | 93 | 100 | 63 | 89 |
| 5 | 2 | 0.05 | 2 | 33 | 100 | 83 | 86 | 100 | 75 | 95 |
| 5 | 2 | 0.05 | 3 | 40 | 99 | 90 | 91 | 100 | 69 | 90 |
| 5 | 3 | 0.01 | 2 | 26 | 73 | 64 | 66 | 9 | 62 | 57 |
| 5 | 3 | 0.01 | 3 | 20 | 65 | 41 | 44 | 35 | 64 | 68 |
| 5 | 3 | 0.05 | 2 | 23 | 6 | 15 | 15 | 0 | 14 | 4 |
| 5 | 3 | 0.05 | 3 | 32 | 10 | 23 | 23 | 0 | 23 | 13 |
| 6 | 1 | 0.01 | 2 | 11 | 98 | 85 | 86 | 100 | 79 | 94 |
| 6 | 1 | 0.01 | 3 | 9 | 95 | 81 | 81 | 100 | 65 | 89 |
| 6 | 1 | 0.05 | 2 | 5 | 99 | 92 | 92 | 100 | 82 | 98 |
| 6 | 1 | 0.05 | 3 | 7 | 98 | 82 | 83 | 100 | 62 | 90 |
| 6 | 2 | 0.01 | 2 | 10 | 90 | 67 | 67 | 100 | 75 | 92 |
| 6 | 2 | 0.01 | 3 | 10 | 85 | 66 | 66 | 100 | 66 | 86 |
| 6 | 2 | 0.05 | 2 | 12 | 98 | 83 | 84 | 100 | 89 | 98 |
| 6 | 2 | 0.05 | 3 | 15 | 97 | 81 | 81 | 100 | 78 | 95 |
| 6 | 3 | 0.01 | 2 | 38 | 97 | 85 | 86 | 0 | 80 | 99 |
| 6 | 3 | 0.01 | 3 | 45 | 97 | 92 | 93 | 0 | 75 | 93 |
| 6 | 3 | 0.05 | 2 | 17 | 96 | 78 | 81 | 61 | 80 | 87 |
| 6 | 3 | 0.05 | 3 | 17 | 96 | 79 | 79 | 94 | 67 | 86 |
| 6 | 4 | 0.01 | 2 | 0 | 7 | 0 | 0 | 40 | 23 | 35 |
| 6 | 4 | 0.01 | 3 | 0 | 0 | 0 | 0 | 40 | 13 | 19 |
| 6 | 4 | 0.05 | 2 | 36 | 5 | 18 | 16 | 0 | 14 | 2 |
| 6 | 4 | 0.05 | 3 | 23 | 5 | 14 | 13 | 0 | 24 | 12 |

**TABLE 4: THE PERCENTAGE S FOR THE NUMBER OF NODES CHOSEN BY EACH MODEL DISCRIMINATION TECHNIQUE FOR THE 5-INPUT 3-NODE TRUE MODEL.**

| | | $k$ | % Selections of $k$ Nodes | | | | | | |
|---|---|---|---|---|---|---|---|---|---|
| σ | Repl | Nodes | MSE | $F_{0.01}$ | AIC | $C_p$ | SBC | SSP | MSP |
| 0.01 | 2 | 1 | 0 | 0 | 0 | 0 | 0 | 0 | 0 |
| 0.01 | 2 | 2 | 0 | 18 | 4 | 4 | 91 | 25 | 38 |
| 0.01 | 2 | 3 | 26 | 73 | 64 | 66 | 9 | 62 | 57 |
| 0.01 | 2 | 4 | 74 | 9 | 32 | 30 | 0 | 13 | 5 |
| 0.01 | 3 | 1 | 0 | 0 | 0 | 0 | 0 | 0 | 0 |
| 0.01 | 3 | 2 | 0 | 11 | 3 | 3 | 65 | 13 | 19 |
| 0.01 | 3 | 3 | 20 | 65 | 41 | 44 | 35 | 64 | 68 |
| 0.01 | 3 | 4 | 80 | 24 | 56 | 53 | 0 | 23 | 13 |
| 0.05 | 2 | 1 | 0 | 0 | 0 | 0 | 0 | 0 | 0 |
| 0.05 | 2 | 2 | 16 | 94 | 71 | 72 | 100 | 82 | 96 |
| 0.05 | 2 | 3 | 23 | 6 | 15 | 15 | 0 | 14 | 4 |
| 0.05 | 2 | 4 | 61 | 0 | 14 | 13 | 0 | 4 | 0 |
| 0.05 | 3 | 1 | 0 | 0 | 0 | 0 | 0 | 0 | 0 |
| 0.05 | 3 | 2 | 14 | 90 | 64 | 66 | 100 | 72 | 86 |
| 0.05 | 3 | 3 | 32 | 10 | 23 | 23 | 0 | 23 | 13 |
| 0.05 | 3 | 4 | 54 | 0 | 13 | 11 | 0 | 5 | 1 |

**TABLE 5: THE PERCENTAGES FOR THE NUMBER OF NODES CHOSEN BY EACH MODEL DISCRIMINATION TECHNIQUE FOR THE 6-INPUT 4-NODE TRUE MODEL.**

| | | $k$ | % Selections of $k$ Nodes | | | | | | |
|---|---|---|---|---|---|---|---|---|---|
| σ | Repl | Nodes | MSE | $F_{0.01}$ | AIC | $C_p$ | SBC | SSP | MSP |
| 0.01 | 2 | 1 | 0 | 0 | 0 | 0 | 0 | 0 | 0 |
| 0.01 | 2 | 2 | 0 | 0 | 0 | 0 | 0 | 0 | 0 |
| 0.01 | 2 | 3 | 0 | 4 | 0 | 0 | 50 | 4 | 13 |
| 0.01 | 2 | 4 | 0 | 7 | 0 | 0 | 40 | 23 | 35 |
| 0.01 | 2 | 5 | 100 | 89 | 100 | 100 | 0 | 73 | 52 |
| 0.01 | 3 | 1 | 0 | 0 | 0 | 0 | 0 | 0 | 0 |
| 0.01 | 3 | 2 | 0 | 0 | 0 | 0 | 0 | 0 | 0 |
| 0.01 | 3 | 3 | 0 | 0 | 0 | 0 | 10 | 2 | 3 |
| 0.01 | 3 | 4 | 0 | 0 | 0 | 0 | 40 | 13 | 19 |
| 0.01 | 3 | 5 | 100 | 100 | 100 | 100 | 50 | 85 | 78 |
| 0.05 | 2 | 1 | 0 | 0 | 0 | 0 | 0 | 0 | 0 |
| 0.05 | 2 | 2 | 0 | 0 | 0 | 0 | 0 | 0 | 0 |
| 0.05 | 2 | 3 | 12 | 94 | 68 | 71 | 100 | 83 | 98 |
| 0.05 | 2 | 4 | 36 | 5 | 18 | 16 | 0 | 14 | 2 |
| 0.05 | 2 | 5 | 52 | 1 | 14 | 13 | 0 | 3 | 0 |
| 0.05 | 3 | 1 | 0 | 0 | 0 | 0 | 0 | 0 | 0 |
| 0.05 | 3 | 2 | 0 | 0 | 0 | 0 | 0 | 0 | 0 |
| 0.05 | 3 | 3 | 7 | 95 | 69 | 71 | 100 | 67 | 86 |
| 0.05 | 3 | 4 | 23 | 5 | 14 | 13 | 0 | 24 | 12 |
| 0.05 | 3 | 5 | 70 | 0 | 17 | 16 | 0 | 9 | 2 |

**TABLE 6: RESULTS FROM ELEVEN SIMULATION STUDIES (INCLUDING THE STUDY PRESENTED IN TABLE 3) FOR THE 5-INPUT 3-NODE TRUE MODEL. THE HIGHEST AND LOWEST PERCENTAGES OUT OF THE 11 STUDIES ARE GIVEN FOR NUMBER OF NODES CHOSEN BY EACH MODEL DISCRIMINATION TECHNIQUE.**

| | | *k* | % Correct Selections of the True Model | | | | | | |
|---|---|---|---|---|---|---|---|---|---|
| σ | Repl | Nodes | MSE | $F_{0.01}$ | AIC | $C_p$ | SBC | SSP | MSP |
| 0.01 | 2 | Low | 16 | 59 | 43 | 47 | 9 | 60 | 51 |
| 0.01 | 2 | High | 27 | 73 | 64 | 66 | 23 | 68 | 62 |
| 0.01 | 3 | Low | 13 | 65 | 41 | 44 | 28 | 55 | 59 |
| 0.01 | 3 | High | 25 | 82 | 59 | 60 | 42 | 68 | 73 |
| 0.05 | 2 | Low | 21 | 0 | 15 | 14 | 0 | 11 | 4 |
| 0.05 | 2 | High | 40 | 8 | 27 | 26 | 0 | 22 | 13 |
| 0.05 | 3 | Low | 22 | 0 | 15 | 14 | 0 | 14 | 4 |
| 0.05 | 3 | High | 34 | 10 | 23 | 23 | 0 | 23 | 13 |

**TABLE 7: RESULTS FROM ELEVEN SIMULATION STUDIES (INCLUDING THE STUDY PRESENTED IN TABLE 3) FOR THE 6-INPUT 4-NODE TRUE MODEL. THE HIGHEST AND LOWEST PERCENTAGES OUT OF THE 11 STUDIES ARE GIVEN FOR NUMBER OF NODES CHOSEN BY EACH MODEL DISCRIMINATION TECHNIQUE.**

| | | | % Correct Selections of the True Model | | | | | | |
|---|---|---|---|---|---|---|---|---|---|
| σ | Repl | | MSE | $F_{0.01}$ | AIC | $C_p$ | SBC | SSP | MSP |
| 0.01 | 2 | Low | 0 | 3 | 0 | 0 | 29 | 16 | 22 |
| 0.01 | 2 | High | 5 | 12 | 11 | 11 | 49 | 38 | 42 |
| 0.01 | 3 | Low | 0 | 0 | 0 | 0 | 40 | 8 | 12 |
| 0.01 | 3 | High | 1 | 3 | 2 | 2 | 61 | 19 | 28 |
| 0.05 | 2 | Low | 18 | 2 | 9 | 9 | 0 | 6 | 0 |
| 0.05 | 2 | High | 36 | 10 | 27 | 25 | 0 | 19 | 7 |
| 0.05 | 3 | Low | 21 | 5 | 14 | 13 | 0 | 15 | 6 |
| 0.05 | 3 | High | 31 | 14 | 26 | 25 | 0 | 24 | 12 |

additional ten simulation studies for these two difficult two models. The results, for which high and low values are shown in Tables 6 and 7, indicate that our first simulation was fairly stable. The range of high and low percentages also helps us determine whether or not the effects of increasing replications or increasing variances are significant.

No effect due to error level (σ) was apparent for MSE. For the 5-input 3-node, 6-input 2-node, and 6-input 4-node true models, the 0.01-level *F*-test, AIC, and $C_p$ tended to select models with fewer nodes as the error level increased. This effect was also seen by SBC with the exception of the 6-input 2-node true model, for which it performed perfectly. For the 6-input 3-node true model, SBC, and to a much lesser extent AIC and $C_p$, tended to choose models with more nodes as error level increased. The Data Splitting criteria, SSP and MSP, tended to select models with fewer nodes as error level increased.

No effect due to replication was apparent for MSE, the 0.01-level $F$-test, or $C_p$. The Data Splitting criteria tended to select models with more nodes as replications increased, and AIC had mixed effects. Some of the effects due to replications and error level may be seen in Tables 4 and 5. Overall, having a higher number of replications and/or a smaller error level did not appear to significantly improve the MD techniques' ability to discriminate.

Excluding the 5-input 3-node and 6-input 4-node true models from the following discussion, SBC discriminated almost perfectly, except on the 6-input 3-node true model. The $F$-test with $\alpha = 0.01$ was nearly as perfect as SBC *and* performed well on the 6-input 3-node true model. The Data Splitting criterion MSP performed well, but was definitely one step below the 0.01-level $F$-test. Finally, AIC and $C_p$, which performed similarly to each other, discriminated adequately, except on the 6-input 2-node true model.

Of the six true models with more than one node, all the techniques except SBC tended to select too many nodes when they were incorrect. On the three true models for which SBC had difficulty, the technique tended to choose the model with one less than the true number of nodes. For the 4-input 2-node true model, only the Data Splitting criteria selected the one-node model, and both did so exactly once. For the 5-input 2-node and 6-input 3-node true model, only SBC (as mentioned above) and the Data Splitting criteria selected models with too few nodes. Finally, for the 5-input 3-node and 6-input 4-node true models, as seen in Tables 4 and 5, the MD techniques made incorrect selections of too few and too many nodes.

## APPLICATION TO GRAIN DRYING

The MD techniques MSE, the 0.01-level $F$-test, AIC, $C_p$, and SBC were computed for ANN models fit to data generated with random error from a theoretical counter-flow model for drying shelled corn by Thompson (1968). The experimental design for the data was a central composite design for seven factors consisting of 83 points. Two data sets were created, one with an error level of 0.01 and one with an error level of 0.05. For the competing ANN models, the number of hidden nodes varied from 1 to 10 (the highest possible for an 83-point data set).

Since the true grain drying model is *not* an ANN, we could not predetermine the "correct" number of nodes in the hidden layer. For the error level of 0.01, AIC and SBC selected 10 nodes (the maximum number nodes), MSE selected 9, $C_p$ chose 8, and the $F$-tests with $\alpha = 0.01$, 0.05, and 0.10 all selected 5 nodes. For the error level of 0.05, AIC selected 9 nodes, MSE, the 0.01 level $F$-test selected only 2 nodes. We believe that the selections of 8, 9, and 10 nodes are excessive. The lack of agreement among the MD techniques points towards an overall inability to discriminate well when the true model is not an ANN.

## CONCLUSIONS

When the true model was an ANN, none of the models performed well in all cases although SBC and the 0.01-level $F$-test were outstanding discriminators for the less complex true models. Incorrect selections by SBC and the 0.01-level $F$-test were only off by one node. A proper understanding of the statistical properties of these MD techniques must be developed for ANN models.

Empirical modeling of the grain drying process is ongoing research. Currently, linear regression models as well as ANN models are being considered for various size experimental designs. The application of the MD techniques on grain drying data sets yielded inconclusive results which indicate that MD criteria specifically for ANNs may be needed. We are currently seeking to develop a nonlinear MD technique with α-level performance and high power for ANN models.

For further research, it may be informative to investigate a number of different true parameter combinations for each model structure. Since we only tested one set of parameters for each model structure, it is difficult to judge whether a test performed well on a particular model structure in general, or just for that particular set of parameter values. However, for the 5-input 3-node true model, we repeated the simulations with different parameter values and obtained essentially the same results as the ones reported in this paper.

Other important simulation studies include testing ANN models for which the number of nodes could be greater than the number of inputs, testing the MD techniques on data sets with more than two varying sizes and more than two varying error levels, and testing other MD techniques.

## LITERATURE CITED


Akaike, H. "A New Look at the Statistical Identification Model." *IEEE Trans. Auto. Control,* 19, 716 (1974).

Caracotsios, M. *Model Parametric Sensitivity Analysis and Nonlinear Parameter Estimation, Theory and Applications.* Ph.D. Dissertation, University of Wisconsin-Madison (1986).

Garth, A.D.N. *Evaluation of Existing Model Discrimination Techniques in Artificial Neural Networks.* M.S. Thesis, Iowa State University (1995).

Mallows, C.L. "Some Comments on $C_p$." *Technometrics*, 15, 661 (1973).

Rollins, D.K. *Model Assessment and Multiresponse Estimation for a Fractionation Model with Coupled Ordinary Differential Equations.* M.S. Thesis, Ohio State University-Columbus (1987).

Schwarz, G. "Estimating the Dimension of a Model." *Annals of Statistics*, 6, 461 (1978).

Snee, R.D. "Validation of Regression Models: Methods and Examples." *Technometrics*, 19, 415 (1977).

Stewart, W.E., M. Caracotsios, and J.P. Sorensen. "Parameter Estimation from Multiresponse Data." *AIChE Journal*, 38, 641 (1992).

Stone, M. "Cross-Validatory Choice and Assessment of Statistical Predictions (with discussion)" *J. Royal Statistical Society*, 36, 111 (1974).

Thompson, T.L., R.M. Peart, and G.H. Foster. "Mathematical Simulation of Corn Drying- A New Model." *Trans. Of ASAE.*, 11, 582 (1968).


## Acknowledgements


We would like to acknowledge partial support for this project by the National Science Foundation under Grant No. 9453534 and NASA under Grant No. NAG 2-909. We are deeply grateful to Tae-sung Shin for his contribution to the simulation work.

# FUZZY CONTROLLER FOR ANTI-COLLISION BRAKING SYSTEM

**Naveen K. Ramagiri and Peng-Yung Woo**
Electrical Engineering Department, Northern Illinois University, Dekalb, IL 60115

**ABSTRACT:**

The purpose of this article is to design a fuzzy controller for a anti-collision braking system and to study the stability of the controller. The designed controller works for a majority of situations.

## Introduction

The rapidly growing need for mobility has brought with it a major challenge for improvement in the operation and utilization of automobile systems. The economical, environmental and safety constraints imposed by the increase in the number of the road vehicles and subsequent governmental policies also require substantial product development through application of information technology. This involves the enhancement of vehicle informatics and telematics with additional sensors and systems. Within the automobile industry, it is known that quiet technological revolution is progressively taking place in the area of vehicle system performance and control. This revolution is fueled by the fact that at present automotive component and vehicle manufacturers are confirmed with ever increasing demands from their customers and government legislation for high quality, reliable, safe and environmentally less damaging products. Currently there is a drive towards the development of the so-called 'smart' or intelligent vehicles [1]. If the vehicle of the future is to be made more intelligent with some elements of autonomous control, integrated sensory systems will form the most important and central aspects of the vehicle.

## The Automobile Braking System and Its Design

The braking system is one of the important systems in an automobile. This is the system on which heavy research is being done to improve the safety of the passengers and the automobile itself. The Anti-lock Braking System (ABS) and the Anti-skid Systems are some to support the statement.

This article is involved in improving the safety of the automobile as well as its passengers by automating the braking system itself. To control the system we need two input signals, the relative distance and the relative velocity of the vehicle with respect to its obstruction. The obstruction can be any moving object in front of it. This is illustrated in Figure 1.

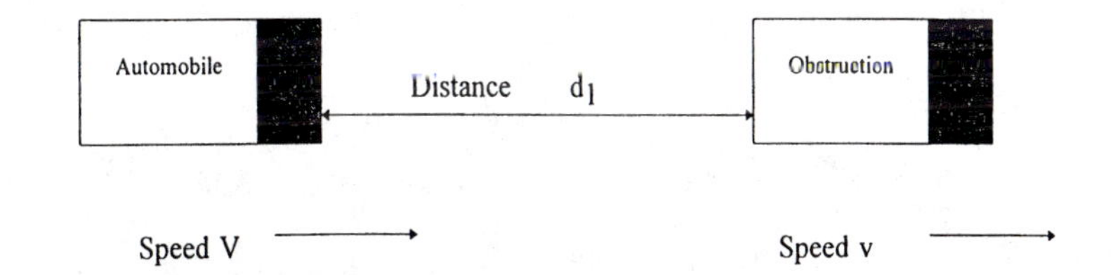


Fig.1 Automobile and its obstruction

Let $a_a$ and $a_b$ be the accelerations, and $j_a$ and $j_b$ be the jerks of the vehicle and the obstruction respectively. $K_a$ and $K_b$ are constants. The change in relative distance in time $\Delta t$ is given by

$$\Delta d = (V - v)*\Delta t - 1/2(a_a - a_b)*(\Delta t)^2 + (K_a*j_a - K_b*j_b)*(\Delta t)^3 + .... \quad (1)$$

Since the equation of motion involves higher orders of distance, the rules have to be written such that the deceleration of the vehicle has to be dependent on the relative velocity, the relative distance and the deceleration of the obstruction.

These signals can be obtained by installing sensors in the vehicle that will detect the distance of the obstruction in front of the vehicle and the velocity of the obstruction, say every 0.05 seconds.

Figure 2 shows the control system diagram.

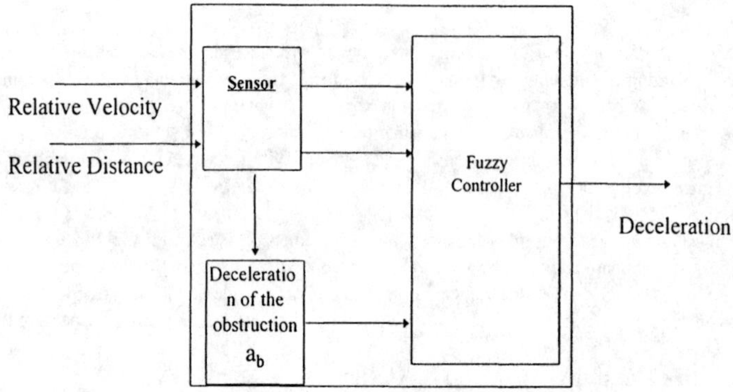


Fig.2 Anti-collision braking system

There is a single control loop in this system. The control strategy for the braking system can be summarized as follows:

1) *The relative velocity of the vehicle with respect to the obstruction should be zero before the relative distance goes to zero, in order to avoid any collision.*

2) *When the relative velocity is negative, it means that the obstruction of the vehicle is moving faster than the vehicle itself and therefore there is no need of any braking.*

3) *When the relative velocity is more than the speed of the vehicle itself, which means that the obstruction is traveling in the opposite direction, then there is no need of any deceleration.*

4) *When the relative velocity is equal to the velocity of the vehicle which implies that the obstruction is stationary, then there is no need of any deceleration.*

5) *The deceleration of the vehicle must be proportional to the deceleration of the obstruction.*

## The Controller Design for the Braking System

### Label Overlap

In fuzzy control systems, the degree of overlap between membership functions impacts the system performance. A fuzzy system without overlapping membership functions is reduced to a system based on Boolean logic. Given a set of crisp inputs, multiple rules are fired due to the overlap of membership functions defining each input variable. Inappropriate overlap of membership functions may lead to haphazard behavior of the resultant fuzzy control system. While defining the overlapping of the labels the following overlap indices should be satisfied quantitatively.

1) Overlap ratio = (overlap scope)/(adjacent MF scope)
2) Overlap Robustness = (Area of summed overlap)/( Maximum area of summed overlap)

As a rule of thumb the overlap ratio must be kept between .2 and .6. The value of overlap robustness is usually greater than overlap ratio; it should be kept between .3 and .7 [4][5][6].

For the braking system, the names of the membership functions are chosen as follows for each input: For the relative distance (Rd), the membership functions are named as RDZ, RDXXN, RDXN, RDVN, RDN, RDM, RDF, RDVF and RDXF. For the relative velocity (Rv), the membership functions are named as RVVL, RVL, RVM, RVH and RVVH and for the deceleration of the obstruction (Do), the membership functions are named as DoH, DoM and DoL. The membership functions for the output, deceleration of the automobile (D), are named as DA, DB, DC, DD, DE, DF, DG, DH, DI, DJ, DK, DL and DM.

**Rules for the Braking System**

The anti-collision braking system has to operate when it senses that the driver is not alert. This can be done by knowing the following two facts:

1) When the input signals to the controller show that the relative distance is very low,
2) When the relative speed of the vehicle with respect to its obstruction is very high.

Both of the above conditions indicate the non-alertness of the driver. The rules therefore have to be written such that when the relative distance falls into the input labels 'RDF' or 'RDVF' or 'RDXF', the deceleration of the vehicle should be less and it should be increased only for lower values of relative distance. Similarly the deceleration should be increased for higher values of relative speed and for higher values of deceleration of the obstruction [7].

Since this system has three inputs and one output, the relationship between the inputs and the output is a little complex to be illustrated on a three dimensional coordinate axis. Figure 3 illustrates the relationship between the relative velocity (Rv), the relative distance (Rd) and the deceleration (D) of the vehicle when the deceleration of the obstruction (Do) is DoH. Now, the relationship between the output and the three inputs can be considered as a group of surfaces like that shown in Figure 3 placed together, one above the other.

Keeping Figure 3 in view, taking the cross sectional view of the curve parallel to the Z-Y plane, we get the curve of Figure 4. There are some critical points on the curve, which are indicated, and can be used for controller design. Between those points the curve is relatively linear and we can use those points to give dimensions for the membership functions of the input and the output variables. Rules can thus be obtained from the control curve.

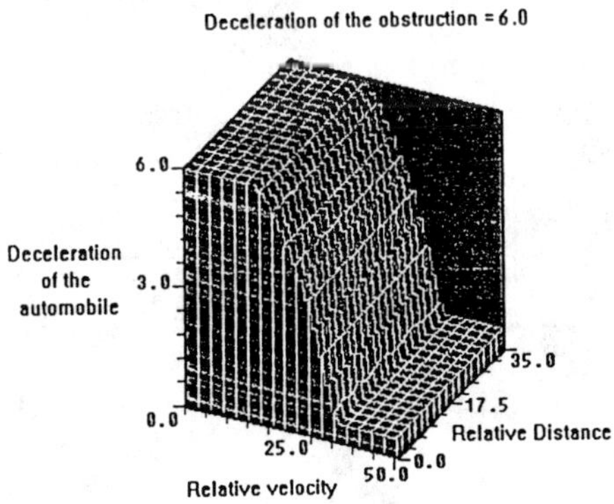


Figure 3

Surface view of the relationship between Rv, Rd & D for DoH

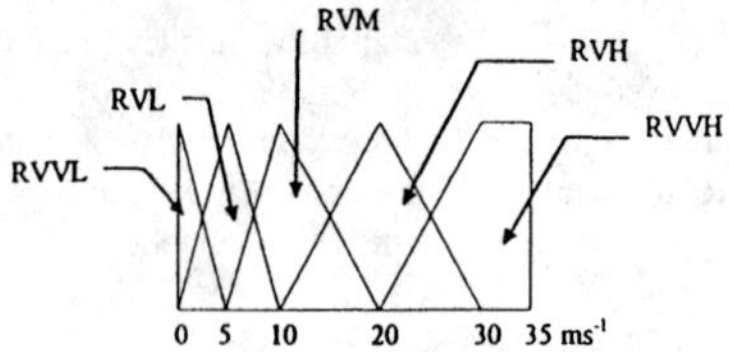


Membership functions for the relative velocity

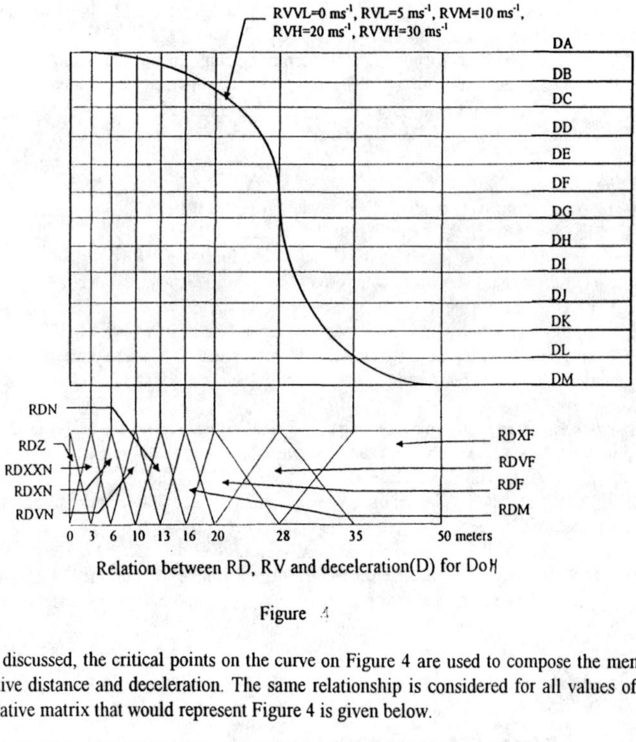


Relation between RD, RV and deceleration(D) for DoH

Figure 4

As discussed, the critical points on the curve on Figure 4 are used to compose the membership functions for the relative distance and deceleration. The same relationship is considered for all values of Rv for DoH. The fuzzy associative matrix that would represent Figure 4 is given below.

| | **RDZ** | **RDXXN** | **RDXN** | **RDVN** | **RDN** | **RDM** | **RDF** | **RDVF** | **RDXF** |
|---|---|---|---|---|---|---|---|---|---|
| **RVVH** | DA | DA | DA | DA | DA | DA | DB | DF | DL |
| **RVH** | DA | DA | DA | DA | DA | DA | DB | DF | DL |
| **RVM** | DA | DA | DA | DA | DA | DA | DB | DF | DL |
| **RVL** | DA | DA | DA | DA | DA | DA | DB | DF | DL |
| **RVVL** | DA | DA | DA | DA | DA | DA | DB | DF | DL |

Fig.5 Fuzzy Associative Memory Matrix for Anti-collision System (DoH)

With the same approach, we obtain Figure 6 and Figure 7.

| | RDZ | RDXXN | RDXN | RDVN | RDN | RDM | RDF | RDVF | RDXF |
|---|---|---|---|---|---|---|---|---|---|
| **RVVH** | DA | DA | DA | DA | DA | DA | DB | DF | DL |
| **RVH** | DA | DA | DA | DA | DA | DA | DB | DF | DL |
| **RVM** | DA | DA | DD | DE | DF | DG | DG | DI | DM |
| **RVL** | DA | DD | DF | DG | DG | DG | DI | DM | DM |
| **RVVL** | DD | DG | DG | DG | DG | DI | DK | DM | DM |

Fig.6 Fuzzy Associative Memory Matrix for Anti-collision System (DoM)

| | RDZ | RDXXN | RDXN | RDVN | RDN | RDM | RDF | RDVF | RDXF |
|---|---|---|---|---|---|---|---|---|---|
| **RVVH** | DA | DA | DA | DA | DA | DA | DB | DF | DL |
| **RVH** | DA | DD | DG | DH | DJ | DK | DL | DM | DM |
| **RVM** | DA | DG | DJ | DK | DL | DM | DM | DM | DM |
| **RVL** | DA | DJ | DL | DM | DM | DM | DM | DM | DM |
| **RVVL** | DG | DM | DM | DM | DM | DM | DM | DM | DM |

Fig.7 Fuzzy Associative Memory Matrix for Anti-collision System (DoL)

## Simulation Results and Conclusions

### Simulation Results

For all the simulations, the vehicle weight, the gravity, the friction between the tire and the road, the air density, the frontal area of the vehicle and the drag coefficient are considered constant vary from vehicle to vehicle and from road to road [2][3]. Also for all the simulations it is assumed that the driver has a reaction time of two seconds to respond to any obstruction.

Simulation of the anti-collision braking system is done for a variety of inputs. Two cases are taken into consideration which are most commonly experienced every day.

*Situation I:* The velocity of the vehicle is taken as 67 mph, i.e. 30 $ms^{-1}$ and that of the obstruction is 50mph, i.e. 22$ms^{-1}$. The distance between the two is 30m. A wind velocity of about 5$ms^{-1}$ is assumed. This situation is encountered most of the times on the highways, where the difference between the speeds of the vehicle and the obstruction can be more.

*Situation II:* The velocities of the vehicle and the obstruction are assumed to be 40mph, i.e. 18$ms^{-1}$ and 30mph, i.e. 13$ms^{-1}$ respectively. An initial distance of 15m is also assumed. This situation represents city driving. No wind velocity is assumed for this case.

For each of the above situations four kinds of deceleration of the obstruction are considered. The equations of the deceleration with respect to the time are taken as

1) $D_o = 3$ [constant] (2)
2) $D_o = 1.5*t - 0.04$ [first order] (3)
3) $D_o = t^2 - 2*t + 2$ [second order] (4)
4) $D_o = t^3 - 4*t^2 + 4*t + 2$ [third order] (5)

where t is the time in seconds.

Also for both of the situations, it is assumed that the driver has a reaction time of two seconds. Until that time, the vehicle without the anti-collision braking system is assumed to be going at a constant speed. After those two seconds, the vehicle with the braking system and the vehicle without the braking system is assumed to be slowing down at a deceleration of 4.5$ms^{-2}$.

The simulation results are summarized in Table 1.

## Discussion

Table 1 gives the values of the relative velocity, the relative distance and the time of collision (if any) for all the simulations. The velocities are given in miles per hour, the distance in meters and the time in seconds. A '--' in the table indicates that the value is inappropriate. Looking at the simulations' table (Table 1), we can see a drastic difference between the relative velocities and the relative distance for the automobiles with or without the anti-collision braking system, after three seconds. The simulations are shown for various orders of the deceleration to illustrate that the braking system will operate for any kind of situation that it might experience. Although it is virtually impossible to avoid the accidents, the vehicle with the anti-collision braking system will reduce the speed of the vehicle considerably and therefore gives the driver a lot more time to avoid an accident, compared to the vehicle without the braking system. Also as the braking system is running all the time, the vehicle with the controller will not encounter most of the situations that are shown here.

## Contributions

The main contributions of this article are:

1) The development of the concept of the anti- collision braking system.
2) The design of a three-input and one-output fuzzy controller.

| | **Simulation 1** Figures 14, 15 & 16 | | | | **Simulation 2** Figures 17, 18 & 19 | | | | **Simulation 3** Figures 20, 21 & 22 | | | | **Simulation 4** Figures 23, 24 & 25 | | | |
|---|---|---|---|---|---|---|---|---|---|---|---|---|---|---|---|---|
| Deceleration of the obstruction→ | $D_o = 3$ | | | | $D_o = 1.5*t - 0.04$ | | | | $D_o = t^2 - 2*t + 2$ | | | | $D_o = t^3 - 4*t^2 + 4*t + 2$ | | | |
| Situation → | Situation 1 | | Situation 2 | | Situation 1 | | Situation 2 | | Situation 1 | | Situation 2 | | Situation 1 | | Situation 2 | |
| Controller is→ | on | off | on | off | on | off | on | off | on | off | on | off | on | off | on | off |
| Relative velocity after 3 seconds (mph)→ | 12.0 | -- | 5.5 | -- | 9.0 | 18.0 | 6.1 | -- | 6.3 | 13.3 | 3.4 | -- | 7.3 | 17.7 | 3.0 | -- |
| Relative distance after 3 seconds (meters) → | 7.1 | -- | 2.5 | -- | 11.2 | 2.9 | 2.4 | -- | 12.4 | 6.1 | 3.7 | -- | 10.6 | 1.0 | 3.3 | -- |
| Relative velocity at the time of collision (mph) | -- | 27.0 | -- | 24.0 | -- | -- | -- | 14.0 | -- | -- | -- | 8.0 | -- | -- | -- | 17.0 |
| Time of collision→ (seconds) | -- | 2.65 | -- | 1.95 | -- | -- | -- | 2.5 | -- | -- | -- | 2.95 | -- | -- | -- | 2.2 |

Table 1 Simulations summary

## REFERENCES


[1] K. M. Passino, "Intelligent Control for Autonomous Systems", IEEE Spectrum, June 1995

[2] C.O. Nwagboso, *Automotive Sensory System,* Chapman & Hall, New York, 1993

[3] G. Sovran, T. Morel and W.T. Mason Jr., *Aerodynamic Drag Mechanisms of Bluff Bodies and Road Vehicles,* Plenum Press, New York, 1978

[4] *Fuzzy Logic Version 1,* Motorola Inc. Products.

[5] *FIDE Version 1*, Aptronix Inc. Products.

[6] *Application Notes,* Aptronix Inc. Products.

[7] D. Driankov, H. Hellendoorn and M. Reinfrank, *An Introduction to Fuzzy Control,* Springer-Verlag, New York, 1993

# Section 9 : General Engineering Applications

# BLIND CHANNEL EQUALIZATION BY NEURAL NETWORKS WITH LONG- AND SHORT-TERM MEMORIES

James Ting-Ho Lo
Department of Mathematics and Statistics
University of Maryland Baltimore County
Baltimore, MD 21228

**Abstract**

Artificial neural networks (ANNs) are known to outperform adaptive linear filters for channel equalization. However, the adaptation of an ordinary ANN involves excessive computation, undesirable local extrema, and long and poor transient performances.

To remedy these shortcomings, this paper proposes to use the ANNs with long- and short-term memories (LASTMs) as the channel equalizers. Two blind equalization situations, that with known signal moments and that for white signal sequence, are discussed to illustrate the advantages.

## INTRODUCTION

It was reported (e.g., [2, 4, 5]) that properly trained artificial neural networks (ANNs) outperform adaptive linear filters in simulations for both trained and blind equalizations of both linear and nonlinear communication channels. However, the ANNs used as channel equalizers have three shortcomings: a relatively large amount of on-line computation, many undesirable local extrema on the adaptation performance surface, and relearning the entire operating environment after a change in an unknown channel parameter.

This paper proposes using ANNs with LASMs (long- and short-term memories), which are described in [6], for blind channel equalization to alleviate these shortcomings. The weights (and/or parameters) of these ANNs are divided into LASMs. The long-term memory is determined in an a priori off-line training, leaving only the short-term memory to be adapted to the unknown channel parameter on-line.

A large number of research papers have been published on blind channel equalization in recent years. Due to the page limit, only the blind equalization with known signal moments (BEKSM) [3, 4] and the blind equalization for white signal sequence (BEWSS) [8, 1] are discussed here to illustrate our approach.

The following notations will be used: The nonlinear and linear weights are denoted by $u$ and $v$ respectively. Superscripts and subscripts of $u$ and $v$ are used in exactly the same manner as are those of $w$. For instance, if $w_{ij}^{l}$ is linear, it is denoted by $v_{ij}^{l}$. Similarly, if the delay weight $w_{ij}^{rl}$ (i.e. weight on a connection with a time delay) is nonlinear, it is denoted by $v_{ij}^{rl}$. The activation level of neuron $j$ in layer $l$ at time $t$ is denoted by $\beta_j^l(t)$.

## EQUALIZATION WITH KNOWN SIGNAL MOMENTS

A general multi-dimensional (1-D is norm) communication channel to be equalized is described by the vector equation: For $t = 0, 1, \ldots,$

$$y(t+1) = f(y(t), \ldots, y(t-p+1), x(t), \ldots, x(t-q+1), \theta(t), \epsilon(t)), \quad (1)$$

where the vector-valued function, $f$, and the integers, $p$ and $q$, are given; $x(t)$ is the signal vector at time $t$; $\epsilon$ is a random vector sequence with given joint probability distributions; $\theta(t)$ denotes a vector-valued unknown environmental parameter at time $t$; and the initial state, $(y_0, \ldots, y_{-p+1}, x_{-1}, \ldots, x_{-q+1})$, is a random vector with given probability distribution.

Assume that the unknown parameter $\theta(t)$ is a constant $\theta$ or varies so slowly that it can be viewed as constant in a reasonably short period of time. ANNs of many different paradigms can be used. For illustrating the idea, a multilayer perceptron (MLP) with tapped delay lines (or alternatively an MLPWIN, i.e. MLP with at least one hidden layer of interconnected neurons) is considered. The tapped delay lines are used to receive $y(t)$ at time $t$ and provide $(y(t), \ldots, y(t-B+1))$, where $B$ is a positive integer such that $x(t-\Delta)$ for $\Delta \geq 1$ is uniquely determined by $(y(t), \ldots, y(t-B+1))$ for $t = \Delta+1, \Delta+2, \ldots,$ as the input vector to the MLP at the same time $t$, then the MLP with all its weights denoted by $w$ outputs an estimate $\hat{x}(t-\Delta, w)$ of $x(t-\Delta)$. (Alternatively, if an MLPWIN with an identity activation function in each output neuron is used, the MLPWIN with all its weights denoted by $w$ inputs $y(t)$ at time $t$ and outputs an estimate $\hat{x}(t-\Delta, w)$ of $x(t-\Delta)$ at the same time $t$.)

Assuming that the signal moments $E\left[x_i^k(t)\right], k = 1, \ldots, m$, for all components $i$ and all $t$, a blind channel equalization problem, which is basically the same as that in [4], is to determine $w$ to minimize, at time $t$, the unsupervised adaptation criterion,

$$Q_1(w,t) := \sum_{k=1}^{m} \alpha_k \sum_{i=1}^{n} \left(E_t[\hat{x}_i^k(t-\Delta)] - E[x_i^k(t-\Delta)]\right)^2, \quad (2)$$

where $n$ is the dimension of $x(t)$; $E_t$ denotes the estimated mean value using the first $t$ outputs of the ANN; and $\alpha_k$ are positive constants. Modifying a formula in [4], the estimated mean $E_t$ can be computed by the recursive formula,

$$E_t[\hat{x}_i^k(t-\Delta)] = \frac{1-\rho}{1-\rho^t}\left(\frac{1-\rho}{\rho-\rho^t} E_{t-1}[\hat{x}_i^k(t-1-\Delta)] + \hat{x}_i^k(t-\Delta, w)\right), \quad (3)$$

where $\rho$ is a forgeting factor. Notice that $w$ enters $Q_1(w,t)$ through this equation.

### A Priori Training for BEKSM

The equation (1) are simulated to generate nonadaptive training data as follows: Select a set $\Theta$ of examplary values, $\theta_1, \ldots, \theta_N$, of $\theta$. Using a pseudo-random number generator to simulate the initial state and the stochastic sequences $\epsilon$ and $\xi$, generate a collection $S_{\theta_n}$ of exemplary realizations $\omega$ of the input/output sequences, $\omega = \{(x(\tau, \theta_n, \omega), y(\tau.\theta_n, \omega)), \tau = \Delta - B + 2, \Delta - B + 3, \ldots, T\}$, $\omega \in S_{\theta_n}$, for each selected exemplary value of $\theta_n$. The positive integer $T$, the set $\Theta$ and the collections $S_\theta$ should be selected and generated such that the union $S := \cup_{n=1}^{N} S_{\theta_n}$ reflects the operating environments sufficiently. Then $u$ is determined by minimizing following nonadaptive training criterion by the variation of both the nonlinear weights $u$ and the variables $V(\theta) := (v(\theta_1), \ldots, v(\theta_N))$:

$$P_1(u, V(\theta)) = \sum_{n=1}^{N} \sum_{\omega \in S_{\theta_n}} \sum_{\tau=\Delta+1}^{T} \frac{\lambda^{T-\tau}}{(T-\Delta)\#S} \|x(\tau-\Delta, \theta_n, \omega) - \hat{x}(\tau-\Delta, \theta_n, \omega, u, v(\theta_n))\|^2, \quad (4)$$

where $\#S$ is the number of elements in the set $S$; $\lambda$ is a positive real number less than or equal to 1; and the MLP's (or alternatively the MLPWIN's) outputs $\hat{x}(\tau-\Delta, \omega, \theta_n, u, v(\theta_n))$ are obtained at the output neurons of the MLP (or alternatively MLPWIN) with its adaptively adjustable weights $v$ equal to the variables $v(\theta_n)$, right after feeding $y(s, \theta_n, \omega)$, $s = \Delta - B + 2, \Delta - B + 3, \ldots, \tau$ consecutively one at a time, at the terminals of the MLP's tapped delay lines (or alternatively the MLPWIN's input neurons). The values of the variables, $u, v(\theta_1), \ldots, v(\theta_N)$, that result from the nonadaptive training, are denoted by $u^*, v^*(\theta_1), \ldots, v^*(\theta_N)$ and called the best-fit values of these variables.

### On-Line Adaptation for BEKSM

Since the optimal value $u^*$ of $u$ has been obtained for all possible values of $\theta$ in an a priori training, the only weights of the ANN that need be adapted are $v$. Therefore, the unsupervised adaptation criterion $Q_1$ in (2) becomes

$$J_1(v, t) := Q_1(u^*, v, t) \quad (5)$$

If $v$ are the ANN's linear weights, which are the weights, $w^L$, of its last layer, layer $L$, the term $\hat{x}_i^k(t-\Delta, w)$ in (3) is replaced by

$$\hat{x}(t-\Delta, u^*, v) = \left[\sum_{j=0}^{n_{L-1}} v_{1j}^L \beta_j^{L-1}(t, u^*), \ldots, \sum_{j=0}^{n_{L-1}} v_{n_L j}^L \beta_j^{L-1}(t, u^*)\right]^T \quad (6)$$

where $\hat{x}(t-\Delta, u^*, v)$ denotes the output of MLP (or alternatively MLPWIN) with weights $(u^*, v)$ at time $t$ after $y(s)$, $s = \Delta - B + 2, \Delta - B + 1, \ldots, t$, have been fed to the input terminals of the MLP's tapped delay lines (or alternatively MLPWIN's input neurons). Note that $n_L = n$. Hence the gradient of $J_1(v, t)$ with respect to $v$ can be easily calculated by the formula,

$$\frac{\partial J_2(v,t)}{\partial v_{ij}^L} = \frac{1-\rho}{1-\rho^t} \beta_j^{L-1}(t, u^*) \sum_{k=1}^{m} \alpha_k \sum_{i=1}^{n} \left(E_\tau[\hat{x}_i^k(t-\Delta)] - E[x_i^k(t-\Delta)]\right).$$

Neither backpropagation nor RTRL is required.

## EQUALIZATION FOR A WHITE SIGNAL SEQUENCE

A widely publicized approach to blind equalization is that based on the concept of maximizing the magnitudes of the measurements' normalized cumulant estimates [8, 1]. This cumulant approach is valid only under the restrictive assumption that the signal process $x(t)$ is a white (noise) sequence and the channel is a linear FIR system. The approach involves a large amount of on-line computation.

Here we consider a zero-mean white signal sequence communicated through the following nonlinear channel: For $t = 0, 1, \ldots,$

$$\begin{aligned} y(t+1) &= cx(t) + \zeta_t, \\ \zeta_t &:= f(y(t), \ldots, y(t-p+1), x(t-1), \ldots, x(t-q+1), \theta(t), \varepsilon(t)), \end{aligned} \tag{7}$$

where $c$ is an unknown constant; and $f, \theta(t), \varepsilon(t)$ are as defined for 1.

Because the signal sequence is white, it is easy to see that

$$y(t+1) - E\left[y(t+1)\,|y^t, \theta(t)\right] = cx + \zeta_t - E\left[\zeta_t|y^t, \theta(t)\right] =: c\hat{x}(t). \tag{8}$$

If $\hat{x}(t)$ is used as the output of the equalizer, $\zeta_t - E[\zeta_t|y^t, \theta(t)]$ is the equalization error. This observation allows us to convert a blind equalization problem of estimating $x(t)$ into a nonlinear filtering problem of predicting $y(t+1)$ given $y^t := \{y(s), s = 1, \ldots, t\}$. If $f$ and $\varepsilon(t)$ are such that $\zeta_t - E[\zeta_t|y^t, \theta(t)]$ is acceptably small for an application, then we say that the channel (7) is equalizable for the application. Since $y(t+1)$ becomes available after a unit time delay, an ANN as a predictor for $y(t+1)$ can be adapted by minimizing the supervised adaptation criterion:

$$Q_2(w,t) := \frac{1-\lambda^t}{1-\lambda} \sum_{\tau=1}^{t} \lambda^{t-\tau} \left\| y(\tau) - \hat{y}(\tau, w) \right\|^2, \tag{9}$$

where $\hat{y}(\tau, w)$ denotes the predicted value of $y(\tau)$ given $y^{\tau-1}$, that the ANN produces. If (7) or a sufficient set of realizations of it is available, ANN with LASMs can be used.

### A Priori Training for BEWSS

The observation (8) shows that if an optimal estimate $\hat{y}(t+1)$ given $y^t$ can be obtained, an optimal estimate $c\hat{x}(t)$ given $y^t$ is easily obtained by $c\hat{x}(t) = y(t+1) - \hat{y}(t+1)$. Therefore, an ANN with LASTMs is used to input $y(t-1)$ and produce $\hat{y}(t)$ at time $t$.

The equation (7) is simulated to generate a priori training data as follows: Select a set $\Theta$ of examplary values, $\theta_1, \ldots, \theta_N$, of $\theta$. Using a pseudo-random number generator to simulate the initial state and the stochastic sequences $\epsilon$, we generate a collection $S_{\theta_n}$ of exemplary realizations $\omega$ of the output sequence, $\omega = \{y(\tau, \theta_n, \omega),\ \tau = -B+2,\ -B+3,\ \ldots,\ T\}$, $\omega \in S_{\theta_n}$, for each selected exemplary value of $\theta_n$. The positive integer $T$, the set $\Theta$ and the collections $S_\theta$ should be selected and generated such that the union $S := \cup_{n=1}^{N} S_{\theta_n}$ reflects the operating environments sufficiently. Then the long-term memory $u$ is determined by minimizing following a priori training criterion by the variation of both $u$ and the variables $V(t) := (v(\theta_1), \ldots, v(\theta_N))$:

$$P_2(u, V(t)) = \sum_{n=1}^{N} \sum_{\omega \in S_{\theta_n}} \sum_{\tau=0}^{T-1} \frac{\lambda^{T-\tau}}{T(\#S)} \|y(\tau+1, \theta_n, \omega) - \hat{y}(\tau+1, \theta_n, \omega, u, v(\theta_n))\|^2, \quad (10)$$

where $\#S$ is the number of elements in the set $S$; $\lambda$ is a positive real number less than or equal to 1; and the MLP's (or alternatively the MLPWIN's) outputs $\hat{y}(\tau+1, \omega, \theta_n, u, v(\theta_n))$ are obtained at the output neurons of the MLP (or alternatively MLPWIN) with its short-term memory $v$ equal to the variables $v(\theta_n)$, right after feeding $y(s, \theta_n, \omega)$, $s = -B+2, -B+3, \ldots, \tau$ consecutively one at a time, at the terminals of the MLP's tapped delay lines (or alternatively the MLPWIN's input neurons). The values of the variables, $u, v(\theta_1), \ldots, v(\theta_N)$, that result from the nonadaptive training, are denoted by $u^*, v^*(\theta_1), \ldots, v^*(\theta_N)$ and called the best-fit values of these variables.

### On-Line Adaptation for BEWSS

Recall that an ANN with LASTMs is a priori trained to input $y(t-1)$ and produce a predicted value of $y(t)$. The best-fit value of the long-term memory is obtained by minimizing 10 and denoted by $u^*$. Since $y(t)$ is available on-line during the operation of the ANN with LASTMs, the short-term memory $v$ of this ANN can be adapted on-line by minimizing at time $t$ the following supervised adaptation criterion:

$$J_2(v, t) := \frac{1-\lambda^t}{1-\lambda} \sum_{\tau=1}^{t} \lambda^{t-\tau} \|y(\tau) - \hat{y}(\tau, u^*, v)\|^2, \quad (11)$$

where $\hat{y}(\tau, u^*, v)$ denotes the output of ANN (an MLP with tapped delay lines or an MLPWIN) with weights $(u^*, v)$ at time $\tau$ after $y(s)$, $s = -B+2, -B+1, \ldots, \tau-1$, have been fed to the input terminals of the ANN (the MLP's tapped delay lines or alternatively MLPWIN's input neurons). If $v$ are the ANN's linear weights, which are the weights, $w^L$, of its last layer, layer $L$, it follows that

$$\hat{y}(\tau, u^*, v) := \left[ \sum_{j=0}^{n_{L-1}} v_{1j}^L \beta_j^{L-1}(\tau, u^*), \ldots, \sum_{j=0}^{n_{L-1}} v_{n_L j}^L \beta_j^{L-1}(\tau, u^*) \right]' \quad (12)$$

where $\beta_0^{L-1}(\tau, u^*) := 1$ and for $j = 1, \ldots, n_{L-1}$, $\beta_j^{L-1}(\tau, u^*)$ is the activation level of neuron $j$ in layer $L-1$. $J_2$ is obviously quadratic in $v$, whose $(i \times j)$-th component is $v_{ij}^L$. Thus, a wide range of fast algorithms for adaptive linear filters can be applied [7].

## CONCLUSION

The purpose of the nonadaptive training is to shift the dependency of the ANN on the unknown channel parameter $\theta$ to the linear weights. If a mathematical model of the channel is not available and the cost of collecting real data is high, we can just collect a small number of the I/O realizations of some typical channel(s) and modify them to generate more for achieving the same pupose. This modification requires the mathematical modelling of the thermal and impulse noises, modulation/demodulation process, crosstalk interference, effects of amplifiers and converters, and differences between channels, which cause the various "values of the unknown channel parameter $\theta$."

In this paper, we assume that $y(t)$ is available at $t$ to simplify our discussion, which can be easily generalized if only a noisy measurement $z(t)$ of $y(t)$ is available at time $t$.

## REFERENCES


[1] J. A. Cadzow. Blind deconvolution via cumulant extrema. *IEEE Signal Processing Magzine*, 13:24–42, 1996.

[2] G. J. Gibson, S. Siu, and C. F. N. Cowan. The application of nonlinear structures to the reconstruction of binary signals. *IEEE Transactions on Signal Processing*, 39:1877–1884, 1991.

[3] S. Haykin. *Adaptive Filter Theory, 2nd ed.* Prentice Hall, Englewood Cliffs, New Jersey, 1991.

[4] G. Kechriotis, E. Zervas, and E. S. Manolakos. Using recurrent neural networks for adaptive communication channel equalization. *IEEE Transaction on Neural Networks*, 5:267–278, 1994.

[5] N. Li, T. Adali, and X. Liu. Finite alphabets least relative entropy algorithm for channel equalization. *Proc. 1995 World Congress on Neural Networks*, 3:III76–79, 1995.

[6] J. T. Lo. Adaptive system identification by nonadaptively trained neural networks. In *Proceedings of the 1994 International Conference on Neural Networks*, Washington, D. C., June, 1996.

[7] A. H. Sayed and T. Kailath. A state-space approach to adaptive RLS filtering. *IEEE Signal Processing Magazine*, II:18–60, 1994.

[8] O. Shalvi and E. Weinstein. Universal methods for blind deconvolution. In S. Haykin, editor, *Blind Deconvolution*. Prentice Hall, Englewood Cliffs, New Jersey, 1994.

# A NEURAL NETWORK SOLUTION TO THE MULTICAST PACKET RADIO TRANSMISSION PROBLEM

**THOMAS L. HEMMINGER**
*School of Engineering and Engineering Technology*
*Penn State University, Behrend College*
*Erie, PA 16563*
*e-mail: hemm@eetsg33.bd.psu.edu*

**CARLOS A. POMALAZA-RAEZ**
*Indiana University-Purdue University*
*Fort Wayne, IN 46805*
*e-mail: raez@sunstorm.ipfw.indiana.edu*

***ABSTRACT:***
This work describes a practical technique for the computation of optimum or near optimum paths from a single source to multiple destinations in a packet radio network (PRN) environment. The problem is common and is usually solved by making copies of the packet, addressing each copy appropriately, then forwarding each copy to its final destination. Although this solution requires a greater communications channel bandwidth it is frequently tolerated because determination of an optimal solution yielding a minimal number of transmissions is NP-complete. This paper proposes a resolution to the problem by employing a Hopfield neural network. Results are compared against optimal solutions derived through exhaustive search.

## INTRODUCTION

The primary function of a packet radio network (PRN) is the efficient transfer of information between source and destination nodes. For example, one algorithm may be favored over others if it handles the greatest traffic with a minimal amount of routing and error correction codes. Problems of this nature have been addressed in several papers (Tobagi, 1987; Silvester, 1987; Hou and Li, 1984) but these are mainly concerned with algorithms that transfer data from a single source to a single destination. This paper deals with the problem of multicasting, which is the simultaneous transmission of data to multiple destinations.

Source and destination nodes are rarely within direct radio contact in multihop PRNs, thus requiring retransmission of information packets. In order to forward the packets to their final destination each node must have sufficient knowledge of the network topology. Depending on the degree of connectivity among the nodes, a packet may travel through several hops before reaching its destination. The problem of efficient multicast transmission, when applied to PRNs, becomes one of minimizing the number of transmissions when routing a packet from its source to multiple destinations. This goal is of paramount importance considering the increasing volume of radio traffic experienced today. The least cost tree and the tree that minimizes the number of transmissions in a multicast communication is a Steiner tree (Hakimi, 1971). The problem of finding a Steiner tree is known to be NP-complete (Karp, 1972).

## PROBLEM DESCRIPTION

Consider the radio network whose connectivity is represented by the graph shown in Fig. 1(a). Assume that a multicast packet is sent from $a$ to $h$, $l$, and $m$. If no effort is made to minimize costs or the number of transmissions a copy of the packet can be addressed with the optimum path to each intended destination. These paths $\{a,d,h\}$, $\{a,c,f,k,m\}$, and $\{a,b,e,i,l\}$ are shown as a tree in Fig. 1(b). With an appropriate link layer protocol the first transmission can deliver the packet simultaneously to intermediate destinations $b$, $d$, and $c$ (assuming here a broadcast channel). For each path from the second hop onwards the routes are independent and each copy of the packet requires a transmission to complete each hop. A total of eight transmissions are required to deliver the packet to all three destinations, assuming no retransmissions. An alternative solution is the one shown in Fig. 1(c). The paths are now $\{a,d,h\}$, $\{a,d,g,j,l\}$, and $\{a,d,g,j,m\}$. The total number of transmissions in this scenario is four, a 50% reduction when compared to the first solution. For our case, a solution that minimizes the number of transmissions is a Steiner tree but for a given network configuration not all Steiner trees minimize the number of transmissions.

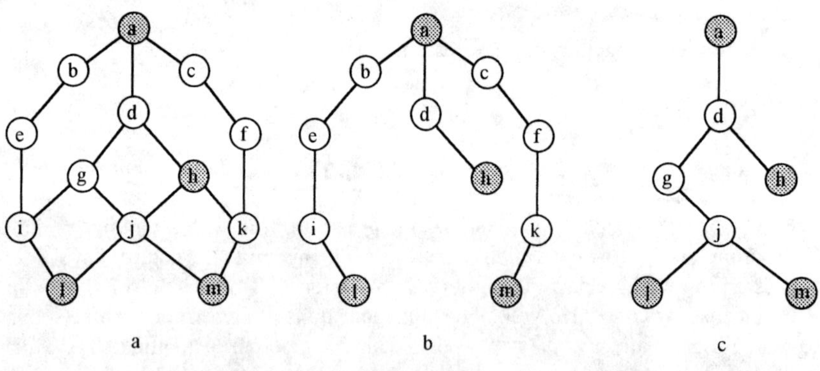


**Figure 1. (a) Multihop packet radio network with destinations *h, l,* and *m*. (b) Simple Solution. (c) Optimal solution requires 50% less transmissions.**

## DESCRIPTION OF ALGORITHM

To formulate a solution it is necessary to develop a scheme which can represent the connections of the graph as neurons of the Hopfield net (Hopfield, and Tank, 1985; Lawler, Lenstra, Rinnooy, and Shmoys, 1985; Ali, and Kamoun, 1993). The first step in configuring the network is to re-label the nodes of the graph with Arabic numerals as in Fig. 2(a). The links can be modeled with an *NxN* matrix with each element representing a neuron addressed by indices $x,i$. Diagonal elements are eliminated since nodes in the graph are not self-connected. As shown in Fig. 2(b), the neurons represent links in the graph, e.g. neuron 6,3 corresponds to the link from node 6 to node 3. The number of elements in the matrix is a maximum of $n(n-1)/2$, although in practice the number is frequently much smaller. Connections from lower rows upward are read row-wise, e.g., node 9 is connected to nodes 6 and 7. A solution is illustrated by the tree shown in Fig. 2(b). For this discussion we define an *upper link* as a link coming from the direction of the source and a *lower link* as a link outward toward targets further from the source.

There are very few numerical constraints for a problem of this nature, considering solutions are based primarily on structure. For this reason it becomes necessary to observe several characteristics of successful routing solutions and attempt to incorporate these more or less quantitatively. We have noted the following characteristics of successful solution trees: (1) When developing minimal length trees it is useful to combine as many lower links as possible into a single node thereby increasing the fan-out from ancestor nodes. This is the purpose of the $A$ term in motion equations (1) and (2). This term totals and subtracts all active lower links of a PRN node from the maximum possible for that node, yielding a lower energy state as the fan-out increases. (2) There is rarely any reason to have two upper links entering a node. The $C$ term forces there to be, at most, only one upper link per node. (3) The $D$ term ensures that if an upper link enters a non-target node there should be at least one lower link for that node.

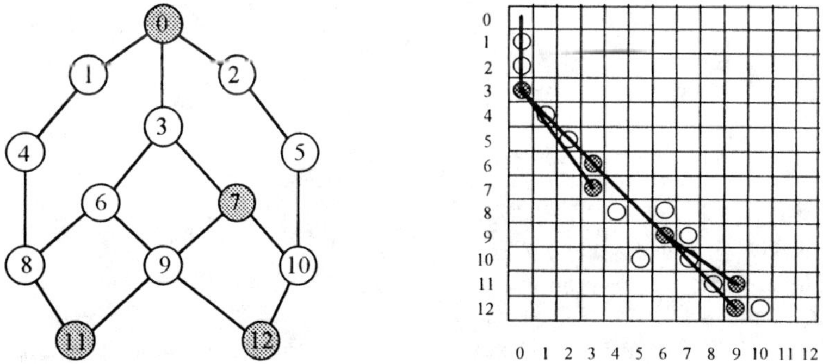


(a) (b)

**Figure 2. (a) Example from Figure 1 with Arabic labeling. Solution indicated with bold edges. (b) Neural network representing PRN links. A neuron is not required for source node 0. Solutions are illustrated with filled circles for paths 12→9→6→3→0, 11→9→6→3→0, and 7→3→0.**

The $B$ term forces all neurons to values of either 0 or 1 (stable states). All of these constraints have been included in equation (1) for non-target nodes:

$$\frac{dU_{xi}}{dt} = -A\sum_{\substack{y=1\\y\neq x}}^{N}\left(-V_{yi}\right) - B\left(1-2V_{xi}\right) - C\sum_{\substack{j=1\\j\neq i}}^{N} V_{xj} - D\sum_{y=1}^{N}\left(1-V_{yx}\right) \tag{1}$$

Equation (2) for target nodes differs from equation (1) with the omission of the C and D terms. An additional E term ensures that target nodes have *at least* one upper link, yet will permit a greater number if the coefficient is not too great in magnitude.

$$\frac{dU_{xi}}{dt} = -A\sum_{\substack{y=1\\y\neq x}}^{N}\left(-V_{yi}\right) - B\left(1-2V_{xi}\right) - E\left[\sum_{j=1}^{N} V_{xj} - 1\right] \tag{2}$$

In solving this routing problem we have determined the coefficients by first employing some results stated in (Aiyer, Niranjan, and Fallside, 1990) then refining the parameters by trial and error, thus yielding the following set of parameters:

$$A = 4;\ \ B = 1;\ \ C = 12;\ \ D = 10;\ \ E = 2$$

The activation function used in this work is described by equation (3).

$$V_i = f(U_i) = \begin{cases} 0, & U_i \leq -0.1 \\ U_i + 0.5, & -0.1 < U_i < +0.1 \\ 1, & U_i \geq +0.1 \end{cases} \tag{3}$$

In all trials the neural inputs $U_{xi}$ are initialized to random numbers on the order of $1/N^2$ to ensure all outputs $V_{xi} = 0$. The time constant $\tau$ is set to 1 and the time interval $\Delta t$ is set by experiment to 0.020. The simulation is halted when: $(\alpha \text{ and } \beta)$ or $\gamma = \text{true}$ where:

$\alpha$) All $U_{xi}$ and $dU_{xi}$ have the same sign indicating approach to a stable state.

$\beta$) There only exists at most a single path into a node unless each of the nodes above needs to transmit anyway.

$\gamma$) Time out is reached at 2500 iterations.

## RESULTS

To investigate the performance of the proposed technique we tested network configurations with as few as 15 nodes to as many as 30 nodes with varying levels of complexity. Each time the number of nodes was increased we generated 50 topologies, recorded the performance of the proposed technique, and averaged the results. For most graphs the neural network normally required from 100 to 2000

| TABLE 1 COMPARISON BETWEEN EXACT SOLUTION AND NEURAL NETWORK *Relative time units are used* | | | | | | |
|---|---|---|---|---|---|---|
| Network Complexity | | | Exhaustive Search | | Neural Network | |
| Network nodes | Targets | Possible Paths | trans. | time | trans. | time |
| 15 | 5.05 | 808 | 4.93 | 3 | 5.22 | 75 |
| 16 | 5.31 | 369 | 5.21 | 3 | 5.61 | 91 |
| 17 | 5.48 | 1272 | 5.24 | 7 | 5.57 | 121 |
| 18 | 6.11 | 37657 | 6.18 | 209 | 6.41 | 152 |
| 19 | 6.28 | 29678 | 6.48 | 182 | 6.77 | 214 |
| 20 | 6.71 | 25451 | 6.75 | 158 | 7.21 | 225 |
| 21 | 6.92 | 98643 | 6.98 | 751 | 7.38 | 282 |
| 22 | 7.31 | 166558 | 7.47 | 1189 | 7.86 | 330 |
| 23 | 7.75 | 2147658 | 7.91 | 17732 | 8.04 | 471 |
| 24 | 8.08 | - | - | - | 8.56 | 494 |
| 25 | 8.35 | - | - | - | 9.12 | 554 |
| 26 | 8.62 | - | - | - | 9.53 | 651 |
| 27 | 9.13 | - | - | - | 9.71 | 672 |
| 28 | 9.27 | - | - | - | 9.62 | 739 |
| 29 | 9.74 | - | - | - | 9.91 | 841 |
| 30 | 9.92 | - | - | - | 10.62 | 943 |

iterations to converge; a maximum of a few seconds on a workstation. The results from an exhaustive search have been included to provide a performance benchmark. In Table 1 we see that the time required for the exhaustive search surpasses that required by the neural network very early in the trials. From these results we infer that the Hopfield approach would be the preferred choice even taking into consideration the slightly larger number of transmissions yielded from this technique. The convergence rate of the neural network remains nearly constant in terms of iterations while the size of the PRN is increased, and the elapsed time increases minimally with the complexity of the system. A plot of transmissions versus graph complexity is illustrated in Fig. 3(a) for both methods. On average the neural network yielded results within 4% of the optimum number of transmissions, which we consider to be very reasonable. The plot in Fig. 3(b) dramatically illustrates the need for more processing power if all paths are to be examined, yet it also demonstrates the efficiency of the proposed method in terms of CPU time. For the neural network the time requirements increase mainly due to the larger number of neurons which need servicing, since the number of iterations remains roughly constant throughout the experiment.

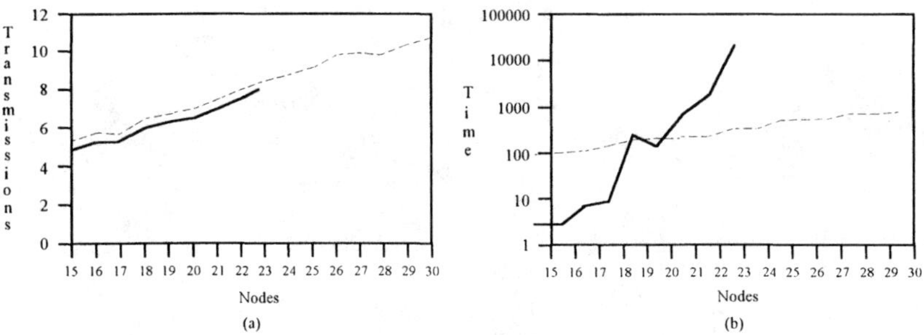


**Figure 3. (a) Number of transmissions determined by exhaustive search (solid line), and neural network (dashed line). (b) Log plot of time required by each method.**

## CONCLUSION

A neural network-based solution to the multicast PRN problem has been introduced. Extensive tests illustrate the efficacy of this approach and its simplicity. The algorithm is extremely easy to implement and converged to a valid solution in over 98% of our simulations. Through extensive simulation with different network sizes and topologies we conclude that this model can be effective in solving the multicast problem both efficiently and rapidly.

## REFERENCES


Aiyer, S.V.B., Niranjan, M., and Fallside, F., (1990). A Theoretical Investigation into the Performance of the Hopfield Model, *IEEE Transactions on Neural Networks*, Vol. 1, No. 2, 204-215.

Ali, M.K.M., and Kamoun, J., (1993). Neural Networks for Shortest Path Computation and Routing in Computer Networks, *IEEE Transactions on Neural Networks*, Vol. 4, No. 6, 941-954.

Hakimi, S.L., (1971). Steiner's problem in graphs and its implications, *Networks*, vol. 1, June, 113-133.

Hopfield, J.J., and Tank, D.W., (1985). Neural Computation of Decisions in Optimization Problems, *Biological Cybernetics*, Vol. 52, 141-152.

Hou, T., and Li, V.O., (1984). Performance analysis of routing strategies in multihop packet radio networks, *Proceedings of GLOBECOM*, November.

Karp, R.M., (1972). Reducibility among combinatorial problems, *Complexity of Computer Computations,* Miller, R.E. and Thatcher, J.W., Eds. New York, NY: Plenum, 85-103.

Lawler, E.L., Lenstra, J.K., Rinnooy, K., and Shmoys, P.B., (1985). *The Traveling Salesman Problem*, Wiley, New York .

Silvester, J.A., (1987). Spatial reuse in multihop packet radio networks, *Proceedings of the IEEE*, vol. 75, Jan, 156-172.

Tobagi, F.A., (1987). Modeling and performance analysis of multihop packet radio networks, *Proceedings of the IEEE*, vol. 75, Jan, 135-155.

# OPTIMISING INFORMATION TRANSMISSION FOR DATA COMPRESSION USING FEEDFORWARD NEURAL NETWORKS

**C. WHITROW, K.A. OSMAN, A.M. HIGGINSON**
***Faculty of Engineering and Computer Technology***
***University of Central England, Birmingham, B24 2SU, England***

*ABSTRACT*
This paper considers the problem of reducing the dimensionality of a data set that is to be the input to a neural network, for pattern classification or regression. Starting from Linsker's principle of maximum information transmission, it is proposed that the entropy of the reduced representation should be maximised. A method for approximating this entropy by using a kernel function estimate of the associated probability density function is presented and gradient ascent is used to optimise the entropy. It is then shown how to use a neural network with unsupervised learning to perform dimensionality reduction according to the foregoing principle. The performance of the resulting algorithm has been tested on multidimensional asymmetric gaussian data and a comparison is made with principal components analysis and the use of autoassociative networks.

## INTRODUCTION

The aim of this work is to find a method of maximising the amount of information preserved between two layers of a neural network (cf. Plumbley and Fallside, 1988). This follows the *principle of maximum information preservation* as stated by Linsker (Linsker, 1987, 1988), which can be used as the basis for reducing the dimensionality of a data set, hitherto achieved using autoassociative networks (Cottrell et al., 1987). The autoassociative technique will be used as a comparison for the novel scheme proposed here. Information preservation offers an alternative approach, comparable, but not equivalent, to using autoassociative networks (Bell and Sejnowski, 1995). The analysis presented here is less rigorous than that of Bell and Sejnowski, in that reasonable approximations have been used. These approximations, however, allow the application of this technique to more powerful network architectures.

### Autoassociative Networks

Autoassociative networks are trained to reproduce a set of input patterns at the output layer. However, as the hidden layer contains fewer units than the input layer, the activities of the hidden units can be interpreted as a reduced encoding of any vector presented at the input units. The network, after training, may be analysed into two parts with complementary functions. The first part, including the hidden units, performs a compression of the original data. The second part then reconstructs the data from the compressed representation on the hidden units (Cottrell et al., 1987).

This work is concerned only with the first half of the network architecture, the encoder. Error signals will be presented directly to the hidden units, using the principle of maximum information preservation, in order to train the first layer of weights. Reconstruction of the original data is not the primary purpose, so training of the second half of the network is unnecessary. The network with a single hidden layer is limited in its capabilities and, with logistic or sigmoidal activation functions, can only perform linear principal components analysis (Bourlard and Kamp, 1988). This limitation can be overcome by using more than one layer of non-linear hidden units (Bishop, 1995, pp. 313-317; Kramer, 1991; DeMers and Cottrell, 1993). The results presented in this paper will not include general non-linear dimensionality reduction. However, the technique which will be developed can easily be applied to these more powerful architectures. Again, it would only be necessary to consider the encoding part of the network.

**Maximum Information Preservation**

Linsker (1993) applied the 'infomax' principle to a particular class of 'weakly non-linear' models with noise on both inputs and outputs. In this work it is assumed that there is no noise on the inputs and the noise on the outputs is allowed to become small. The aim is to maximise the mutual information (Shannon, 1948), $I(\mathbf{x};\mathbf{y})$, between the activations of the units on the input ($\mathbf{x}$) and output ($\mathbf{y}$) layers of an MLP. This is equivalent to maximising the entropy, $h(\mathbf{y})$, of the distribution of $\mathbf{y}$ with respect to the network parameters. The density $p(\mathbf{y})$ is approximated by a linear combination of gaussian kernels centred on the patterns in the training set. This is also known as a *parzen window estimator* (Parzen, 1962). Gradient ascent is then used to maximise $h(\mathbf{y})$.

With two non-linear output units whose activations are bounded between 0 and 1, the output 'space' may be considered as a square with sides of length 1. The input patterns are transformed by the network into this output space. The ideal situation is one in which the output patterns 'fill' the output space uniformly. In this case, the entropy of the density of the output is maximised and as much information as possible is preserved about the input patterns. Success depends both upon the structure of the input data and the appropriateness of the output unit transfer functions for representing the input. Linear units would be inappropriate since the output space must be bounded in order for h($\mathbf{y}$) to have a finite maximum.

**Maximising the Entropy of a Gaussian Mixture**

The probability density in n-dimensional space can be described by:

$$p(\mathbf{x}) = \frac{1}{N}\sum_{i=1}^{N} g_i(\mathbf{x}) \qquad (1)$$

where $\mathbf{x}$ is any random variable (not the input to a neural net). There are $N$ kernels, each denoted $g_i(\mathbf{x})$, in this mixture. The centre (or mean) of the ith kernel is $\mathbf{m}_i$ and the jth component of this centre is $m_{ij}$. Note that the kernels are identical and spherically symmetric, with width $\sigma$. More general models lead to similar results.

If $H$ is the entropy of p($\mathbf{x}$), the derivative of $H$ with respect to $m_{ij}$ can be found by:

$$-\frac{\partial H}{\partial m_{ij}} = \frac{\partial}{\partial m_{ij}} \int_{V_\infty} p(\mathbf{x}) \log p(\mathbf{x}) \, d\mathbf{x} \tag{2}$$

After some manipulation , this leads to :

$$N\sigma^2 \nabla H(\mathbf{m}_i) = \sum_{r=1}^{N} \int_{V_\infty} \frac{g_i(\mathbf{x})\,(\mathbf{x}-\mathbf{m}_r) g_r(\mathbf{x})}{\sum_{k=1}^{N} g_k(\mathbf{x})} \, d\mathbf{x} \tag{3}$$

As it stands, (3) is analytically intractable. Fortunately, further simplification is possible if interest is restricted to the *direction* rather than the magnitude of $\nabla H(\mathbf{m}_i)$. If the width, $\sigma$, of the ith kernel is allowed to tend to zero whilst leaving the others unchanged, $g_i(\mathbf{x})$ becomes a dirac delta function centred on $\mathbf{m}_i$ and the integral in (3) disappears, leaving a sum. Although the denominator in the integrand diverges to infinity, as it is a scalar it has no influence on the direction of the resulting (infinitesimal) vector. So :

$$N\sigma^2 \nabla H(\mathbf{m}_i) \approx \sum_{r=1}^{N} \frac{(\mathbf{m}_i - \mathbf{m}_r) g_r(\mathbf{m}_i)}{\sum_k g_k(\mathbf{m}_i)} \tag{4}$$

In equation (4), the denominator is infinite because of the term where k=i but it is a scalar constant for all r, so it may be ignored as only the direction of the vector derivative is of interest, which is determined solely by the numerator. (4) can be considered an approximation to (3) in the case where $g_i(\mathbf{x})$ becomes highly localised. Then the denominator does not actually become infinite and may be disregarded. So, if the change in $\mathbf{m}_i$ is denoted by $\Delta\mathbf{m}_i$, it is required that :

$$\Delta\mathbf{m}_i = \varepsilon \sum_{r=1}^{N} (\mathbf{m}_i - \mathbf{m}_r) g_r(\mathbf{m}_i) \tag{5}$$

Where $\varepsilon$ is a small arbitrary constant (the learning rate). What (5) says is that $\mathbf{m}_i$ must move directly away from every other kernel centre and $g_r(\mathbf{m}_i)$ will act as a weighting factor, depending solely upon the distance between $\mathbf{m}_i$ and $\mathbf{m}_r$. The learning rule is analogous to electrostatic repulsion. Each pattern (kernel centre) is akin to a point charge both repelling and being repelled by the other identical charges represented by the other patterns. The difference is that a gaussian law instead of an inverse square law of force is used. Intuitively, (5) is what might be expected when attempting to maximise entropy.

**Application To Neural Networks**

Having calculated $\Delta\mathbf{m}_i$ at the output, it is easy to calculate the necessary changes in the network weights by backpropagating this error signal. However, heuristics must be employed to control $\sigma$ during training. Figure 1 shows the $\Delta\mathbf{m}_i$ values (given by straight lines, not to scale) attaching to each pattern (represented by crosses) in the case where $\sigma$ is large. This is the kind of situation needed in order to force the output patterns to fill the entire output space. Equally, when training is well advanced and the patterns are approaching the edges of the output space, it may be desirable to reduce $\sigma$ so that the model can stabilise. Figure 2 shows how the patterns are changing in this case.

## RESULTS

### Data, Network Architecture and Training

The input data comprised 50 real-valued points generated from a 5-dimensional gaussian. The network had 5 input units, one layer of weights and 2 logistic sigmoidal output units, corresponding to the 'encoder' part of the architecture. This configuration enabled the output patterns to be visualised readily in 2 dimensions. The weights and biases were initialised to small random values to ensure that the output patterns began in a small region near the centre of the output space. It was decided that $\sigma$ should be adapted during training, to be equal to the root mean squared distance between nearest neighbours in the output space. Training was terminated when the output patterns began to approach the boundaries of the output space.

### Observations

As expected, the output patterns eventually filled the output space fairly evenly after 40 epochs, and the weights tended to settle into the subspace spanned by the 2 principal components of the data. Figure 3 shows how the angle ($\theta_{maj}$) between the subspace spanned by the 2 weight vectors and the subspace spanned by the 2 principal components changed during training. The final value of this angle was 0.48 radians. It can be seen that most of the training is actually complete within 18 epochs, when $\theta_{maj}$ reached 0.39 radians. Using this measure, performance degraded as the output patterns reached the boundary of the output space.

A variety of 2-layer MLPs were also trained to autoassociate the data using backpropagation. They all had 5 input, 2 hidden and 5 output units. After 4000 epochs, the sum-squared reconstruction error on one trial was 1412, compared to the theoretical optimum of 1200.

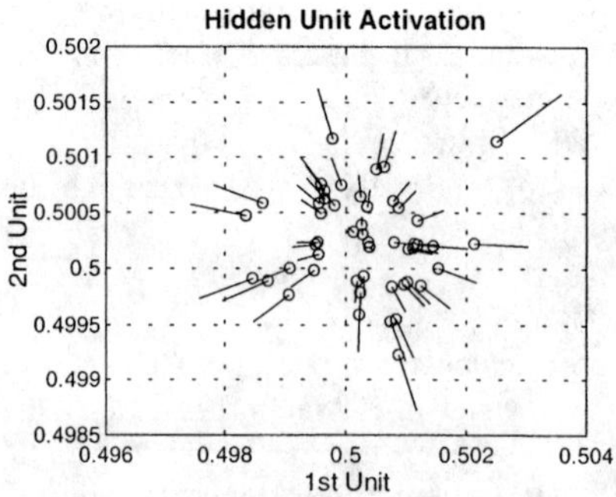


Figure 1: Large sigma

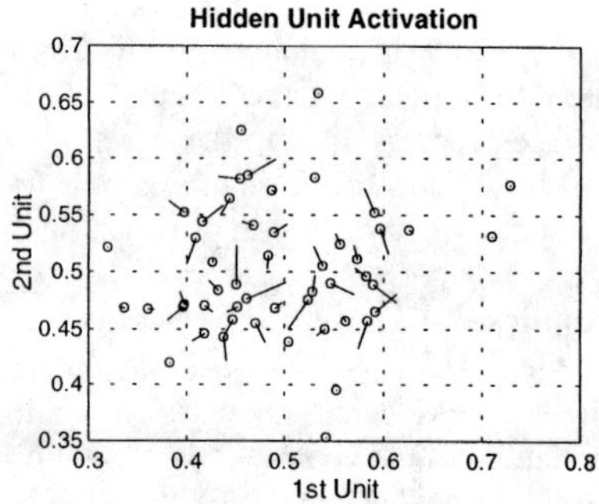


Figure 2: Small sigma

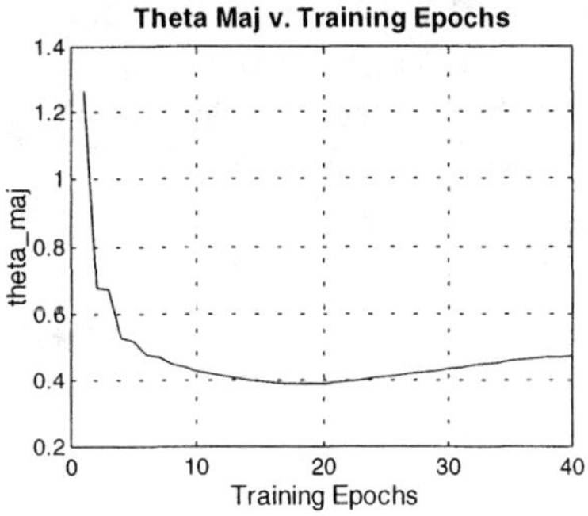


Figure 3: Infomax network

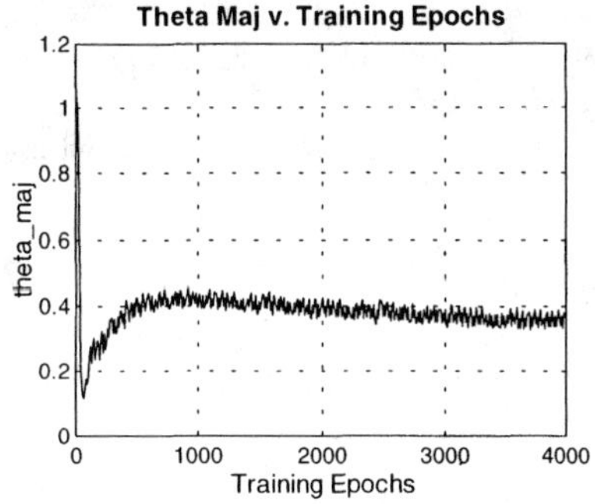


Figure 4: Autoassociative network

The final value of $\theta_{maj}$ (defined as above) was 0.37 although it reached its minimum value of 0.08 after only 47 epochs. At this point though, the sum-squared error (SSE) was about 6000; well above the theoretical minimum. Figure 4 shows how $\theta_{maj}$ varied during training. Again, the training patterns eventually filled the output space.

**Discussion**

It seems that dimensionality reduction may be achieved by directly maximising information transmission using feedforward neural networks. Such a strategy does not require reconstruction of the data and so reduces the number of free parameters in the network, usually by half. It also allows the use of simple criteria for terminating training. This is not always the case for autoassociative networks, where the theoretical minimum SSE for a given compression ratio is normally unknown. Furthermore, a large SSE could be due to poor reconstruction rather than poor encoding at the hidden units. Also, a small SSE does not necessarily imply a 'good' representation of the data by the hidden units. Here, there may be scope for using 'infomax' principles to modify a least-squares algorithm in order to optimise its performance.

The 'infomax' algorithm should be augmented with the use of a 'regulariser' to incorporate prior information about model complexity (see Bishop, 1995, pp. 385-390). By incorporating weight decay (Hinton, 1987), for example, it is more likely that the model will reject small components of the data, if that is what is required.

The next step should be to test the method using more complicated data together with a network architecture capable of extracting non-linear features. It will then be necessary to backpropagate the $\Delta\mathbf{m}_i$ values through another layer. In general, the 'infomax' principle should not be equivalent to least-squares autoassociative learning. There are major differences in the implementation of the two methods and the choice between them will doubtless depend upon what is to be achieved. If the aim is to minimise the reconstruction error, the direct approach with a least-squares algorithm may be preferred. However, even in this case, 'infomax' principles should be borne in mind and could form part of a hybrid algorithm, possibly capable of faster convergence than either method. If the aim is to explore the structure of a data set, the 'infomax' method offers an alternative which may be capable of revealing structures different from those found using a least-squares autoassociative network. By removing the need to reconstruct the data, the problem of having to estimate the optimum reconstruction error can be avoided.

**Acknowledgements**

Many thanks to Luxfer Gas Cylinders and to the UK Engineering and Physical Sciences Research Council (EPSRC) for funding the research programme under which this work was carried out.

## REFERENCES

Bell, A.J., Sejnowski, T.J., (1995). An information maximisation approach to blind separation and blind deconvolution, Neural Computation **7** (6), 1129-1159.

Bishop, C.M., (1995). Neural Networks for Pattern Recognition, Clarendon Press, ISBN 0-19-853849-9.

Bourlard, H.A., Kamp, Y., (1988). Auto-association by multilayer perceptrons and singular value decomposition, Biological cybernetics **59**, 291-294.

Cottrell, G.W., Munro, P., Zipser, D., (1987). Learning internal representations from grey-scale images: an example of extensional programming, Proc. 9th annual conference of the Cognitive Science Society, 461-473.

DeMers, D., Cottrell, G., (1993). Non-linear dimensionality reduction, in Advances in Neural Information Processing Systems **5,** eds. S.J. Hanson, J.D. Cowan, C.L. Giles, pub. Morgan Kaufmann, 580-587.

Hinton, G.E., (1987). Learning translation invariant recognition in massively parallel networks. Proc. Parle Conference on Parallel Architectures and Languages Europe, eds. J.W. de Bakker, A.J. Nijman, P.C. Treleaven pub. Springer-Verlag, 1-13.

Kramer, M.A., (1991). Non-linear principal components analysis using autoassociative neural networks. AICHE Journal **37** (2), 233-243.

Linsker, R., (1987). Towards an organising principle for perception: Hebbian synapses and the principle of optimal neural encoding. IBM research report RCL2820, IBM Research, Yorktown Heights.

Linsker, R. (1988). Self-organisation in a perceptual network, Computer **21**, 105-117.

Linsker, R. (1993). Deriving receptive fields using an optimal encoding criterion, in Advances in Neural Information Processing Systems **5,** eds. S.J. Hanson, J.D. Cowan, C.L. Giles, pub. Morgan Kaufmann, 953-960.

Parzen, E., (1962). On estimation of a probability density function and mode, Annals of Mathematical Statistics **33**, 1065-1076.

Plumbley, M.D., Fallside, F., (1988). An information-theoretic approach to unsupervised connectionist models, Proc. 1988 Connectionist Models Summer School, San Mateo, CA., eds. D. Touretzky, G. Hinton, T. Sejnowski, pub. Morgan-Kaufmann, 239-245.

Shannon, C.E., (1948). A mathematical theory of communication, The Bell System Technical Journal 27 (3).

# AN IMPROVED HOPFIELD NETWORK APPROACH TO CHANNEL ASSIGNMENT IN A CELLULAR MOBILE COMMUNICATIONS NETWORK

KATE SMITH
*Department of Business Systems, Monash University, Clayton, Victoria, 3168, Australia.*

M. PALANISWAMI
*Department of Electrical and Electronic Engineering, The University of Melbourne, Parkville, Victoria 3052, Australia.*

**ABSTRACT:**
In this paper, an improved Hopfield network is presented which guarantees feasibility of the solutions, and permits escape from local minima of the objective function to improve solution quality. The network is applied to the problem of assigning channels to calls in a cellular mobile communications network, where the Hopfield network results are compared to those obtained using traditional heuristic techniques.

## INTRODUCTION

Over recent years, we have seen a steady increase in the popularity of cellular mobile communication systems. The fact that the electromagnetic spectrum available for this purpose is a limited resource places severe limitations on the size and performance of such systems. Careful design of a network is necessary to ensure efficient use of the available spectrum. A competitive communications industry also forces the network design to accommodate certain performance criteria. Operators of such networks need to assign a caller from a certain region to a channel within the available frequency spectrum in such a way that the probability of the signal-to-interference (S/I) ratio being below some predefined limit is low. Potential interferences to this call come from three possible sources: another caller within some range using the same channel (a *cochannel* interference); another caller in an adjacent region using an adjacent channel in the frequency domain (an *adjacent channel* interference); and another caller within the same region using another channel within some range (a *co-site* interference). The combinations of regions and channels which cause interferences are determined by the radio frequency (RF) propagation (obtained from the regional topography and morphostructure) and the spatial density of the expected traffic, which are also used to predict the demand for channels.

The channel assignment problem is then to assign the required number of channels to each region in such a way that interference is precluded and the frequency spectrum is used efficiently. Hopfield neural networks have been used by several researchers to solve the channel assignment problem (Funabiki and Takefuji, 1992; Kunz, 1991a; Kunz, 1991b), although the types of energy functions chosen have been of the Hopfield-Tank kind (Hopfield and Tank, 1985) which involve many terms for each of the constraints and objectives. As has been shown in the literature (Wilson and Pawley, 1988), such choices of energy function are likely to lead to infeasible solutions. The strict descent nature of the Hopfield network dynamics also means that even if the final solution is feasible, the network will have converged to the first local minima it encounters, and the solution quality is therefore likely to be sub-optimal. These were the conclusions reached by Kunz (1991b) when testing the standard Hopfield network on the channel assignment problem.

In this paper, we reformulate the channel assignment problem as a generalised quadratic assignment problem, treating the non-interference constraints as soft constraints in the objective function and demand satisfaction as a hard constraint. The advantage of this approach is that a solution which minimises the severity of any interferences will always be found, which is particulary useful if the demands and constraints are such that no interference-free solution exists within the number of channels which are available for the network. An improved Hopfield network approach is then described in which feasibility of the solutions are guaranteed through an appropriate choice of energy function representation, and local minima of the objective function are escaped via the incorporation of a hill-climbing mechanism. Results of this Hopfield network are presented for a 21-cell regular hexagonal network, and are compared to solutions obtained using simulated annealing and steepest descent heuristics, as well as a commercial optimisation package.

## MATHEMATICAL FORMULATION

Suppose there are $N$ *cells* (or regions) in the network, and the number of channels available is given by $M$. The channel requirements (or expected traffic) for cell $j$ are given by $D_j$. The constraints specify the minimum distance in the frequency domain by which two channels must be separated in order that an acceptably low S/I ratio can be guaranteed within the regions to which these channels have been assigned. These minimum distances are stored in a symmetric *compatability matrix* **C** which is of dimension $N \times N$. Let us define a set of binary variables:

$$X_{j,k} = \begin{cases} 1 & \text{if cell } j \text{ is assigned to channel } k \\ 0 & \text{otherwise} \end{cases}$$

for $j = 1, \ldots, N$ and $k = 1, \ldots, M$. The channel assignment problem can then be formulated as:

$$\text{minimise} \quad F(\mathbf{X}) = \sum_{j=1}^{N}\sum_{k=1}^{M} X_{j,k} \sum_{i=1}^{N}\sum_{l=1}^{M} P_{j,i,(|k-l|+1)} X_{i,l} \tag{1}$$

$$\text{subject to} \quad \sum_{k=1}^{M} X_{j,k} = D_j, \quad \forall j = 1, \ldots, N, \tag{2}$$

$$X_{j,k} \in \{0,1\} \ \forall j = 1, \ldots, N, \text{ and } k = 1, \ldots, M. \tag{3}$$

Here, $\mathbf{P}$ is a *proximity factor* tensor (which assigns a penalty to each assignment), generated according to the recursive relation

$$P_{j,i,m+1} = max(0, P_{j,i,m} - 1), \qquad P_{j,i,1} = C_{ji}, \qquad P_{j,j,1} = 0$$

for $m = 1, \ldots, M-1$ and for all $j, i \neq j$. We refer the interested reader to Smith (1996) for further details of this derivation. The formulation can readily be converted to the form

$$\begin{aligned} \text{minimise} \quad f(\mathbf{x}) &= \mathbf{x}^T \mathbf{Q} \mathbf{x} & (4) \\ \text{subject to} \quad & \mathbf{A}\mathbf{x} = \mathbf{b} & (5) \\ & x_i \in \{0,1\} \qquad \forall i = 1, \ldots, n, & (6) \end{aligned}$$

where $n = N \times M$, and $\mathbf{x}$ is the n-dimensional vector obtained by concatenating the rows of $\mathbf{X}$. This standard quadratic form will enable solution by our improved Hopfield network.

## AN IMPROVED HOPFIELD NETWORK APPROACH

Problems of infeasibility and poor solution quality in the standard Hopfield network can be essentially eliminated by an appropriate form of energy function and modification of the internal dynamics of the Hopfield network (Smith, 1996). By expressing all constraints of the problem in a single term, the overall number of terms and parameters in the energy function can be reduced. Consider the general energy function:

$$\mathcal{E}(\mathbf{x}) = f(\mathbf{x}) + \frac{\gamma}{2} \|\mathbf{x} - (\mathcal{P}\mathbf{x} + \mathbf{s})\|^2, \qquad (7)$$

where $\mathcal{P} = \mathbf{I} - \mathbf{A}^T(\mathbf{A}\mathbf{A}^T)^{-1}\mathbf{A}$ and $\mathbf{s} = \mathbf{A}^T(\mathbf{A}\mathbf{A}^T)^{-1}\mathbf{b}$. The first term of the energy function is the objective function, while the second term measures the deviation of the vector $\mathbf{x}$ from its projection on the constraint plane given by $\mathbf{A}\mathbf{x} = \mathbf{b}$. The advantage of this energy function is that only one penalty parameter, $\gamma$, needs to be selected. If $\gamma$ is large enough, then validity of the solution is ensured, since the constraint term will be forced to vanish.

We now modify the internal dynamics of the Hopfield network to permit temporary increases in this energy function (in particular the cost function $f(\mathbf{x})$) in order to allow escape from local minima. The HCHN, like the standard Hopfield network (Hopfield and Tank, 1985), consists of a fully inter-connected system of $n$ neurons. Neuron $i$ has internal state $u_i$ and output level $x_i$ (bounded by 0 and 1). The internal state $u_i$ incorporates a bias current (or negative threshold) denoted by $i_i$, and the weighted sums of outputs from all other neurons. The weights, which determine the strength of the connections from neuron $j$ to $i$, are given by $W_{ij}$. The relationship between the internal state of a neuron and its output level is determined by an activation function $g_i(u_i)$, which is bounded below by 0 and above by 1. The HCHN differs from the standard Hopfield network in its internal dynamics which permit hill-climbing. The modified differential equation is given by:

$$\frac{du_i}{dt} = \alpha(t)(\sum_j W_{ij} x_j + i_i), \qquad \text{and} \qquad u_i = g^{-1}(x_i). \qquad (8)$$

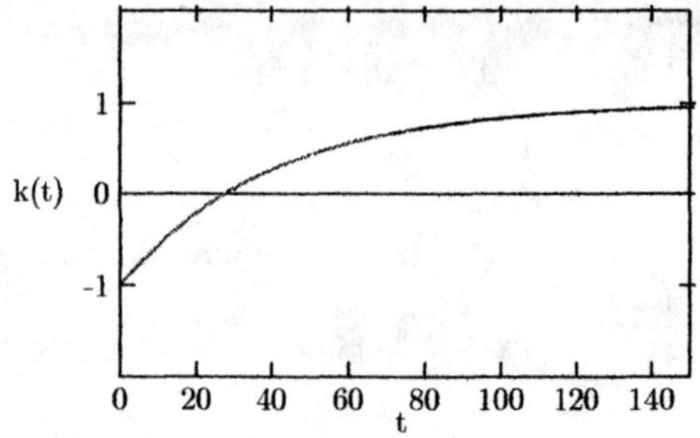


Figure 1: Graph of $k(t) = 1 - 2e^{-t/\tau}$ with $\tau = 40$.

Furthermore, the usual sigmoidal activation function is replaced by the piecewise linear function

$$x_i = g(u_i) = \begin{cases} 0 & u_i \leq 0 \\ u_i & 0 < u_i < 1 \\ 1 & u_i \geq 1 \end{cases} \tag{9}$$

so that $u_i = x_i$ within the unit hypercube. The modified differential equation means that the direction of the change in any neuron is now controlled by a new parameter $\alpha(t)$, where

$$\alpha(t) = random[k(t), 1] \text{ and } k(t) = 1 - 2e^{-t/\tau}.$$

Figure 1 shows how the value of $k$ changes with time for $\tau = 40$. If $k(t) = 1$ always, then HCHN is simply the original Hopfield network and the states of the network evolve to minimise the standard energy function. Thus, steepest descent and ascent of the objective function on the constraint plane are achieved when $\alpha(t) = \pm 1$ respectively. Initially, $k(0) = -1$, and so $\alpha(0)$ is randomly selected from the range [-1,1]. Consequently, the energy value (which is equivalent to the objective cost provided $\mathbf{x}$ lies on the constraint plane) will often increase initially. As $t \to \infty$ however, $k(t) \to 1$, and so $\alpha(t)$ will also approach unity which is needed for strict Liapunov descent. Provided the value of the parameter $\gamma$ is large enough, the convergence trace will be forced to lie on the constraint plane, resulting in a feasible solution.

## RESULTS FOR CHANNEL ASSIGNMENT

The class of test problems is based upon the 21 cell regular hexagonal network used by Sivarajan and McEliece (1989), as shown in Figure 2. Two sets of demands for the 21 cells are used and are shown below:

$$\mathbf{D}_1^T = (2,6,2,2,2,4,4,13,19,7,4,4,7,4,9,14,7,2,2,4,2),$$

$$\mathbf{D}_2^T = (1,1,1,2,3,6,7,6,10,10,11,5,7,6,4,4,7,5,5,5,6).$$

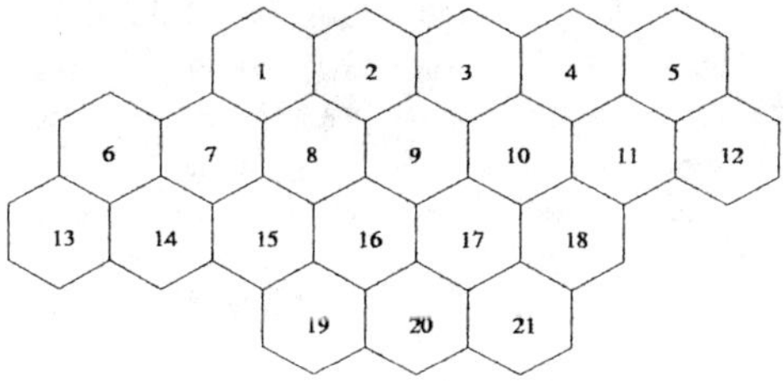

Figure 2: 21 cell hexagonal network

| TABLE 1: PROBLEM DESCRIPTIONS | | | | | | |
|---|---|---|---|---|---|---|
| Problem | $N$ | $M$ | $\mathbf{D}$ | cochannel | adjacent | $C_{ii}$ |
| HEX1 | 21 | 37 | $\mathbf{D}_1$ | yes | no | 2 |
| HEX2 | 21 | 91 | $\mathbf{D}_1$ | yes | yes | 3 |
| HEX3 | 21 | 21 | $\mathbf{D}_2$ | yes | no | 2 |
| HEX4 | 21 | 56 | $\mathbf{D}_2$ | yes | yes | 3 |

The compatability matrix is generated by considering the first two rings of cells around a particular cell as interferers. Adjacent channel constraints and co-site constraints are varied, as is the size of the network. The details of these hexagonal test problems (denoted by HEX1 through to HEX4) are shown in Table 1, where the number of available channels $M$ is the lower bound obtained using the lemmata of Gamst (1986).

We solve these test problems using a variety of techniques. The first of these is the commercial optimisation package GAMS with the nonlinear solver MINOS-5 (Brooke *et al.*, 1990). Next we consider two heuristics: steepest descent (SD) and simulated annealing (SA) which both operate on the solution matrix $\mathbf{X}$ by starting with a feasible solution (satisfying demands) and swapping columns to improve the solution quality. Finally, we solve the test problems using the improved Hopfield network (HCHN) and the same network with $k(t) = 1$ for all $t$ (which is a strict descent Hopfield network using the energy function (7) and is labelled HN). For the HCHN, the length of the Markov Chain is given by $10MN$ and $\tau = 40$. These results are shown in Table 2. Each of the techniques (except GAMS/MINOS-5) is run from 10 different random starts. 'Min' represents the minimum objective function found during these 10 runs, while 'Av.' is the average objective value.

The results in Table 2 demonstrate quite clearly that the Hopfield network

| TABLE 2: RESULTS OF TEST PROBLEMS | | | | | | | | | | |
|---|---|---|---|---|---|---|---|---|---|---|
| | GAMS | SD | | SA | | HN | | HCHN | | |
| Problem | Min | Av. | Min | Av. | Min | Av. | Min | Av. | Min | |
| HEX1 | 54 | 56.8 | 55 | 50.7 | 49 | 49.0 | 48 | 48.7 | 48 | |
| HEX2 | 27 | 28.9 | 25 | 20.4 | 19 | 21.2 | 19 | 19.8 | 19 | |
| HEX3 | 89 | 88.6 | 84 | 82.9 | 79 | 81.6 | 79 | 80.3 | 78 | |
| HEX4 | 31 | 28.2 | 26 | 21.0 | 17 | 21.6 | 20 | 18.9 | 17 | |

solutions perform well when compared to more traditional techniques. The strict descent version of the Hopfield network (HN) outperforms the steepest descent heuristic (SD), and matches the performance of the simulated annealing heuristic (SA). The hill-climbing Hopfield network (HCHN) manages to locate better quality solutions more often by escaping from local minima of the objective function, and performs the best of all the techniques. Both the Hopfield network approaches (like the heuristics) yield feasible solutions in 100% of the trials.

## CONCLUSIONS

In this paper, we have presented an improved Hopfield network approach which guarantees the feasibility of the final solution and permits escape from local minima in order to improve solution quality. We have tested this approach using the channel assignment problem, which has been reformulated to allow for a soft-objective approach. The results demonstrate that the improved Hopfield network approach is able to outperform the more traditional heuristic methods of steepest descent and simulated annealing, as well as a commercial optimisation package. It is clear that since the traditional problems associated with the Hopfield network (infeasibility and poor solution quality) have been overcome, the Hopfield network is potentially a very powerful optimisation technique.

## REFERENCES

Brooke, A., Kendrick, D., and Meeraus, A., (1990). *GAMS – A User's Guide*, Scientific Press, California.

Funabiki, N., and Takefuji, Y., (1992). "A Neural Network Parallel Algorithm for Channel Assignment Problems in Cellular Radio Networks", *IEEE Transactions on Vehicular Technology*, 41(4), 430-437.

Gamst, A. (1986). "Some Lower Bounds for a Class of Frequency Assignment Problems", *IEEE Transactions on Vehicular Technology*, 35(1), 8-14.

Hopfield, J. J., and Tank, D. W. (1985). " "Neural" Computation of Decisions in Optimization Problems", *Biological Cybernetics*, 52, 141-152.

Kunz, D. (1991a). "Channel Assignment for Cellular Radio Using Neural Networks", *IEEE Transactions on Vehicular Technology*, 40(1), 188-193.

Kunz, D. (1991b). "Suboptimal solutions obtained by the Hopfield-Tank neural network algorithm", *Biological Cybernetics*, 65, 129-133.

Sivarajan, K. N., and McEliece, R. J. (1989) "Channel Assignment in Cellular Radio", *Proceedings IEEE Vehicular Technology Conference*, 846-850.

Smith, K. (1996). "Solving Combinatorial Optimisation Problems Using Neural Networks", Ph.D. thesis, University of Melbourne.

Wilson, G. V., and Pawley, G. S. (1988). "On the stability of the TSP algorithm of Hopfield and Tank", *Biological Cybernetics*, 58, 63-70.

# A NEURAL NETWORK-BASED DISTRESS MODEL FOR KANSAS JPCP LONGITUDINAL JOINTS

**Imad A. Basheer & Yacoub M. Najjar**
Department of Civil Engineering
Kansas State University
Manhattan, KS 66506

***ABSTRACT:***

In Jointed Portland Cement Pavements (JPCP), longitudinal joints constitute essential elements for the overall functionality and performance of the pavement system. Data relevant to spalling of the longitudinal joints (LJs) were collected from 31 highway sections in Kansas City along with other data pertaining to design parameters for both the LJs and the pavement. The spalling extent of the sampled LJs was studied using a newly defined nondimensional parameter which represents combination of the spalling areal size and frequency of occurrence. Artificial neural networks (ANNs) were utilized in this study for modeling the spalling given several parameters pertinent to pavement age, accumulated traffic volumes, and other pavement design elements. The neural models were able to predict spalling with relatively high accuracy. Moreover, the developed models were successfully employed to address issues related to quality of construction and improvement of design procedures.

## PURPOSE OF STUDY

One of the reasons behind the increased popularity of concrete pavements is related to their strength to withstand high traffic volumes and the associated low annual maintenance cost. Unfortunately, the functionality of concrete pavements is influenced by cracking of slabs and damage of the transverse and longitudinal joints (LJs). The observed LJs failure of some interstate highways in the Kansas City area was found to exist in both the newly constructed and old pavements. Spalling was noticed to be the predominant mode of failure of the LJs. Spalling is defined as the cracking, breaking, or chipping of the slab edge within 2 ft (0.60 m) of the LJ (Huang 1993). Generally, it may occur due to excessive traffic load, accumulation of incompressible materials in joints, and other factors pertaining to poor construction. Additional factors include concrete disintegration by freezing/thawing, weak concrete at joints caused by overworking, and poorly designed and constructed load transfer devices (Darter et al. 1985). The overall impact of spalling is to increase annual maintenance costs, and to adversely affecting the life of the associated pavements, riding safety and quality, as well as aesthetic appearance of the highways.

Preliminary field survey on several sections of interstates I-70, I-35, and I-435 in the Kansas City metropolitan area was conducted to investigate the modes of the LJs deterioration. The study also aimed at modeling the mechanism of deterioration and studying the effect of various design parameters and traffic load conditions on the extent and frequency of LJ deterioration (Najjar and Basheer 1996).

## EXPRESSIONS OF SPALLING

The 31 sections surveyed totalled 21,883 ft (6625 m) in pavement length. Spalling was measured by determining the number of spalled spots and the total spalling surface area within a prespecified length of the observed distressed pavement. These spalling quantities were converted to (i) Average Spalling Area per unit length of the LJ (ASA) expressed in units of $in^2/ft$ and (ii) Average Spalling Frequency (ASF) interpreted as the average spacing between successive spalled spots. Because both indicators are important, a new spalling extent measure was suggested which combines ASA and ASF. The new spalling extent measure is nondimensional and referred to as the Nondimensional Spalling Index (NDS). This index is defined as the product of the total spalling area and number of spall occurring within a stretch of the distressed LJ. Using the previously defined measures of spalling, NDS=ASA/ASF. The NDS parameter is a useful indicator to reflect the LJ damage level. For the sections studied, NDS varied roughly between $1x10^{-10}$ and $1x10^{-2}$. LJs with NDS$<1x10^{-5}$ were denoted undamaged and those with NDS$>1x10^{2}$ were assumed to be severely damaged. Other intermediate levels reflect several degrees of deterioration of the LJs and equally implies different levels of action to be taken for maintenance and repair of these LJs.

## MODELING

Few studies dealing with distress prediction of concrete pavement joints can be cited in the literature. COPES study (Darter et al. 1985) investigated effect of several pavement parameters on distress of transverse joints and slab cracking. The RIPPER study by Smith et al. (1990) indicated that most of the tested models were inadequate to predict distress, serviceability, and roughness of the pavements. The PEARDARP prediction models (van Wijk 1985) presented empirical equations for estimating pumping, faulting and spalling of transverse joints, cracking of slabs, and slabs roughness. From these studies one can conclude (i) in all of these studies, large discrepancy between model predictions and measured values was observed, (ii) no research on modeling of the LJs deterioration was given any attention in these studies.

### ANN Modeling

The spalling extent expressed in NDS was modeled using the backpropagation algorithm with momentum. For the two ANNs developed in this study, a single output node (NDS) constituted the entire output layer. In the first ANN, the input layer contained 11 parameters; namely the age of both the pavement and sealant, accumulated total and heavy traffic counts, and an additional set of seven design-related parameters (i.e. slab thickness, slab length, concrete compressive strength, width of LJ, and reinforcement diameter, length, and spacing. In the second ANN, the contractor type (name) who constructed the pavement was included in the input layer to implicitly reflect construction quality/practice. The ranges of the various input parameters used in developing the models are: slab thickness (8-12 in.), slab age (0.1-20 yrs.), compressive strength (3000-7000 ksi), slab length (10-70 ft), LJ width (0.1-1.2 in.), sealant age (0.1-20 yrs.), tie-bar diameter (0-0.8 in.), tie-bar length (0-40 in.), tie-bar spacing (10-100 in.), accumulated total traffic (0-400 million), and accumulated heavy traffic (0-25 million). Since only a small number of data sets (31 data sets) existed, all the data were used in the training phase. The proper generalization and absence of overfitting (memorization) was assured from the size of the hidden layer of the designed ANNs as will be shown later.

**Uniform-construction quality model**

In this network, the contractor name was not included in the input layer. This is equivalent to the case where all the pavements are constructed under similar construction quality/practice. Optimization of the first ANN was found at : one hidden neuron, 5000 training cycles, learning rate ($\alpha$)=0.50, and momentum coefficient ($\mu$)=0.80. The prediction accuracy was estimated with an $R^2$=0.942 between predicted and measured values. The single hidden unit needed and the relatively small number of learning cycles required may insure that the network did not memorize the data. Figure 1 shows the uniform-construction quality ANN with input nodes 1 through 11. Comparison between predicted and measured values of NDS for the developed network (denoted by 11-1-1) is shown in Fig.2-a. The sum of squared differences (SSD) for all sets was calculated as $2.49\times10^6$. Because NDS varied over a wide range, Fig. 2-b is the same agreement plot redrawn on logarithm scales. It is obvious from Fig. 2-b that the model predicts NDS more closely for pavements with high spalling. Fortunately, since high-NDS pavements are those demanding maintenance and design/construction reconsideration, the developed model is an efficient tool. The high deviations from the observed values occurred mostly for pavements with relatively low spalling. In order to use the model, Table 1 summarizes all the connection weights and biases needed to calculate NDS.

**Contractor-based model**

This ANN was designed to address effect of construction quality/practice on the spalling extent. In addition to the 11 input variables, the contractor type (name) was included in the input vector. Seven different contractors were involved in constructing the 31 sections surveyed for this study. The contractor name was encoded in the input layer using seven-digit string such that the digit representing one contractor is activated. With $\alpha$=0.50 and $\mu$=0.80, a network with one hidden node trained for 5000 learning cycles was found to produce the best mapping of the data ($R^2$=0.975). This ANN is denoted by 18-1-1. Figures 3-a (normal scales) and 3-b (log- scales) illustrate the agreement between predicted and observed NDS. It is obvious that this model predicts spalling with higher accuracy for pavements with appreciable amount

**TABLE 1: CONNECTION WEIGHTS AND BIASES FOR THE 11-1-1 ANN**

| **Input Node to Hidden Node** | Con.Weight | **Input Node to Hidden Node** | Con.Weight |
|---|---|---|---|
| Slab Thickness | 2.4358 | Tie-bar Length | 2.1681 |
| Pavement Age | -7.1734 | Tie-bar Spacing | -4.2768 |
| Compressive Strength | 5.0815 | Acc. Total Traffic | -6.8473 |
| Slab Length | -3.4777 | Acc. Heavy Traffic | -15.008 |
| Width of LJ | -2.0657 | **Hidden Node to Output Node** | 6.3893 |
| Sealant Age | 1.2382 | **Bias at the Hidden Node** | 3.9032 |
| Tie-bar Diameter | 1.7663 | **Bias at the Output Node** | 5.9143 |

of LJ damage. Points which seem to deviate considerably from the line of agreement belong to pavements with minimally damaged LJs. The SSD was calculated as $1.04 \times 10^{-6}$, thus implying a significant improvement over the uniform-construction quality ANN. This improvement may indicate the importance of including the contractor type (or alternatively the construction quality) in the model. Additionally, this observation reflects that the involved contractors have performed paving at various degrees of construction quality. Figure 1 is also a schematic representation of the developed 18-1-1 ANN with input nodes 1 through 18.

## RANKING OF CONSTRUCTION QUALITY/PRACTICE

With the large number of variables, it is extremely difficult to use the raw data for ranking the contractors based on their construction quality. Alternatively, the 18-1-1 ANN was used to compute average NDS by allowing each of the 7 contractors to construct the 31 sections once. The smaller the average NDS the better is the construction quality. This analysis showed that contractor C was the best in performing the paving job with average NDS=$5.3 \times 10^{-5}$. Using t-statistics at 95% confidence level, the contractors were ranked as C first (best), E second, F, R, and T third, H fourth, and K fifth (worst).

## SENSITIVITY ANALYSIS

The purpose of sensitivity) analysis is to examine the accuracy and soundness of the developed distress model and to address influential parameters affecting LJ deterioration extent. The 11-1-1 ANN was run on a large number of cases that vary in values of the different parameters. Since quantitative conclusions might only be representative for the example cases studied, the following conclusions are only qualitative in nature: (i) spalling decreases significantly with increasing slab thickness, (ii) spalling reduces by increasing slab length, (iii) concrete compressive strength beyond 3500 psi (24 MPa) has only slight effect on spalling extent, (iv) tie-bars diameter, length, and spacing slightly influence spalling, and (v) spalling increases with increasing both accumulated total traffic and accumulated heavy traffic.

## REFERENCES


Darter, M.I., Becker, J.M., Snyder, M.B., and Smith, R.E. (1985), Portland cement concrete pavement evaluation system (COPES), NCHRP Report, No. 277, Transportation Research Board.

Huang, Y.H. (1993),Pavement analysis and design, Prentice Hall, Englewood Cliffs, NJ.

Najjar, Y.M. and Basheer, I.A. (1996), Modeling deterioration of longitudinal joints of jointed plain concrete pavements, Final Report, Kansas Department of Transportation.

Smith, K.D., Peshkin, D.G., Darter, M.I., Mueller, A.L., and Carpenter, S.H. (1990), Performance of jointed concrete pavements, Vol. II. Evaluation and modification of concrete pavement design and analysis models, Report No. FHWA-A-RD-89-137, Federal Highway Administration, Final Report.

van Wijk, A.J. (1985), Purdue economic analysis of rehabilitation and design alternatives for rigid pavements: a user's manual for PEARDARP, Federal Highway Administration, Final Report.


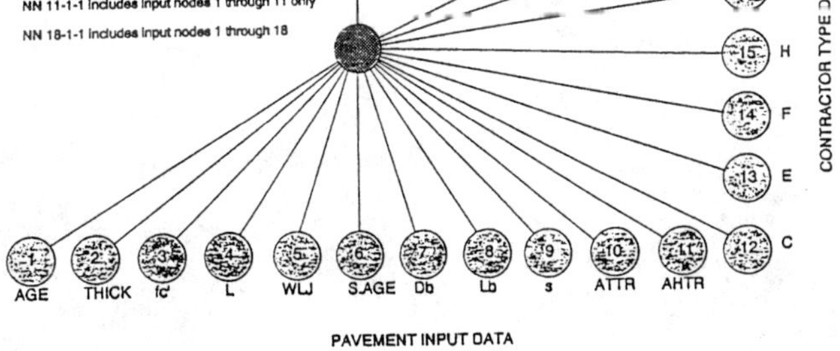


Fig. 1. Schematic of the Two ANNs. Input nodes 1-11 Correspond to Uniform-construction Quality ANN and 1-18 Correspond to Variable Constriction Quality ANN.

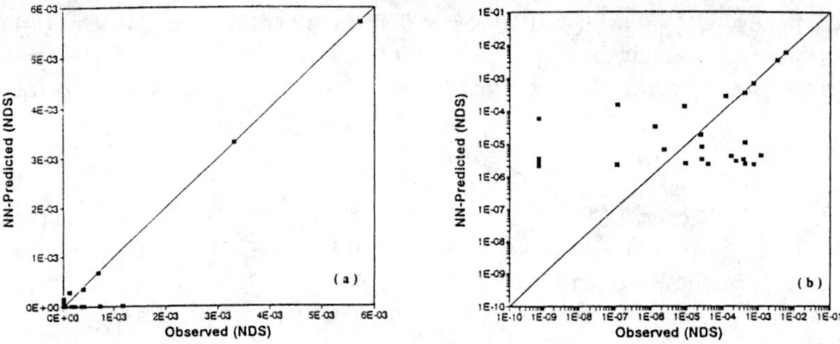


Fig. 2. Agreement Between Observed and ANN-Predicted NDS Using the Uniform Construction Quality ANN. (a) normal scales, (b) logarithmic scales.

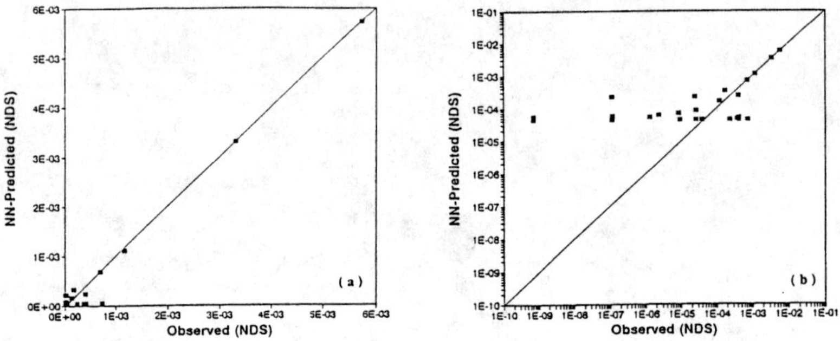


Fig. 3. Agreement Between Observed and ANN-Predicted NDS Using the Variable Construction Quality ANN (a) Normal Scales, (b) Logarithmic Scales.

# SEPARATION OF HASHISH SIGNATURES WITH NEURAL NETWORKS

**FIRST, SECOND, AND THIRD AUTHOR**
*Andreas Tenhagen (*tenhaga@math.uni-muenster.de*),*
*Thomas Feuring (*feuring@math.uni-muenster.de*),*
*Wolfram-Manfred Lippe (*lippe@math.uni-muenster.de*),*
*Institut für Informatik, Universität Münster,*
*Einsteinstr. 62, 48149 Münster, Germany.*

**FOURTH AND FIFTH AUTHOR**
*H. Lahl, G. Henke,*
*Institut für Pharmazeutische Chemie, Universität Münster,*
*Hittorfstr. 58-62, 48149 Münster, Germany.*

**ABSTRACT:**
The exact analysis of a hashish sample is impossible, so that it is difficult to tell wether two different analyses result from the same sample. For criminalistic and forensic reasons it is desirable to be able to associate a certain hashish sample with other samples that all stem from the same source.
We use neural networks to solve this classification task. We train and test these nets with data that was aquired using two different methods of analysis. Thereby we want to learn which method yields the "better" data.

## INTRODUCTION

Classifying hashish samples with regard to concentration of certain elements is the usual method for identifying samples as belonging to the same bigger sample. It is a common problem in criminology to be able to prove the origin of certain hashish samples. This proof can be used to trace the way of a hashish sample back to the original sample it was taken from. The classification can be done by statistically analyzing a lot of sample data. This takes a long time and has to be done separately for each sample. The results of the statistical analysis then have to be interpreted correctly.
So there is no classification-tool which simply can be fed with element analysis data and then tells the user if two or more samples match or stem from different sources. Our aim was to train neural networks so that they can serve as easy-to-use classification tool.

Neural networks are adaptive systems which can be trained to solve classification problems. Furthermore they lead to good classification results even if the input data are noisy. In a first step we tried to train a feedward network with a small training set to solve an easier task: the network had to classify 20 different hashish signatures.

Table 1: Statistics for twelve times analysed sample (ER)

| | Sc/Fe | Sc/Ce | Fe/Ce | Fe/Cr | Pa/Co |
|---|---|---|---|---|---|
| mean | 3.537 | 5.989 | 1.694 | 2.289 | 1.928 |
| std.dev. | 0.018 | 0.089 | 0.028 | 0.047 | 0.043 |
| | Fe/Sc | Fe/Ce | Fe/Pa | Fe/Cr | Fe/Co |
| mean | 0.283 | 1.694 | 2.350 | 2.289 | 4.528 |
| std.dev. | 0.0015 | 0.028 | 0.037 | 0.047 | 0.050 |

As (Casale et al., 1993) have shown for a similar problem, neural networks are able to solve such tasks.

In this paper we first present the different methods of data aquisition. In the next section the classification methods we used are described before we go on to present the neural network we applied. Results are given in the last section.

## DATA AQUISITION

Twenty different hashish samples, which where confiscated in Irak, Iran, Bulgaria, Switzerland , Great Britain and Germany were analysed with different methods. These methods detect the concentration of certain elements in a given hashish sample. This element-analysis data is then used in two ways:

### Element Relationships (ER)

The concentration of Fe, Sc, Ce, Pa, Cr, Co in each sample was measured. The element relationships

1. Sc/Fe, Sc/Ce, Fe/Ce, Fe/Cr, Pa/Co (ER1)

2. Fe/Sc, Fe/Ce, Fe/Pa, Fe/Cr, Fe/Co (ER2)

where chosen to represent the sample, because statistical analysis showed them to be least correlated. So this method yields two different data sets — five values each — for every sample. The mean and standard deviation over all twenty samples was also computed.

One sample was analysed twelve times, which produced the statistical data in Table 1.

### Multi-Element Analysis (MEA)

By means of the ICP-AES multiple-element analysis the concentration of Cu, Cr, Zn, Sr, Mn, Ti, Mg, Fe, Al and Ca was detected. Again, the mean and standard deviation over all twenty samples was calculated. Also, one sample was analysed multiple — sixteen — times, which provided us with the data in Table 2.

So data is aquired in three ways (ER1, ER2, MEA) and the question arises, if one of these methods yields data that is easier to classify than the data from the others.

Table 2: Statistics for sixteen times analysed sample (excerpt) (MEA)

| | Zn | Sr | Mn | Ti | Mg | Fe | Al | Ca |
|---|---|---|---|---|---|---|---|---|
| mean | 0.0549 | 0.0558 | 0.2703 | 1.2404 | 4.0008 | 9.4721 | 10.37 | 21.9757 |
| std.dev. | 0.0049 | 0.0024 | 0.0151 | 0.0882 | 0.2188 | 0.5749 | 0.85 | 1.1185 |

### Artificial Data

With this small amount of real world data we obviously had to generate artificial data. This is no difficult task, since we know the typical error in measurement for each data aquisition method (we interpret the standard deviation in Tables 1 and 2 as error in measurement). We then varied the values of each sample within the "error-bound" and with respect to the mean and standard deviation over all twenty samples. Thereby we can generated an arbitrary amount of "reasonable" training and testing samples.

## CLASSIFICATION WITH NEURAL NETWORKS

We have twenty different classes that we want to be identified. There are basically two ways of defining the desired output:

### One-dimensional (1D)

The output is a real number. We chose the desired outputs to be the numbers from 1 to 20.

For this, networks with only one output node are needed — the advantage being the faster training. Additionally, there is no need to adapt the network architecture, when the number of classes changes. The disadvantage of this approach lies in the output for not correctly classified samples: If the output is not very near one of the twenty desired outputs, it does not tell the user anything about the most probable classification.

### Multi-dimensional, Fuzzy (MD)

The output is a twenty-dimensional fuzzy vector (i.e. a vector whose components are $\in [0, 1]$). The desired outputs are the twenty vectors with all components zero except one which is $= 1$.

This leads to large networks and slower training, the architecture depends on the number of classes. But if the network does not give a clear answer for some input it is easier to interpret its output, that is, if the generalization capability of the network is good.

## NEURAL NETWORK

### Multilayer Perceptron (MLP)

The first network is a standard multilayer perceptron which is trained with a modified backpropagation algorithm.

The method of steepest descent, which is implemented by the classic backpropagation learning law has some weak points, that take effect on typically shaped error surfaces. (Jacobs, 1988) examined the behaviour of steepest descent on backpropagation error surfaces and suggested the delta-bar-delta learning law. This law uses individual (one for each weight) and variable learnrates. Our modified learning law works like this (it is presented in detail in (Tenhagen et al., 1994)):

- The weights are updated using individual learnrates and the well known momentum-version of backpropagation training.
- The learnrates are changed according to the delta-bar-delta rule.
- Some parameters are varied by a fuzzy controller.

The algorithm is notably faster than standard backpropagation and tunes the most important parameters by itself.

**Fuzzy Neural Network (FNN)**

Fuzzy neural networks are fully fuzzified multilayer feedforward networks. They consist of one input-layer, one output-layer and at least one hidden-layer. The activation and the output function of a formal neuron in a fuzzy neural network must be computed on fuzzy numbers because all weights of the network are fuzzy numbers, too. This network can be trained with crisp and fuzzy numbers and it computes crisp or fuzzy numbers as output data. So these networks naturally generalize crisp neural networks. In order to minimize the computational expense, we use *triangular* fuzzy numbers for weights, input and output data.

Because of the small set of training data, the standard deviation is used as fuzziness value of the fuzzy numbers.

In (Feuring, 1995) has shown that the output behaviour of a trained fuzzy neural network can be predicted for arbitrary input data. This leads to good generalization capability of fuzzy neural networks. To train a fuzzy network a relative small training set is required which is another advange of these networks.

Only a few learning algorithms for fuzzy neural networks exist (see (Feuring, 1996)). The main difficulty is the adjustment of the fuzzy weights while training the network. Fuzzy neural networks with different sizes are trained with fuzzy data. We get best results with a 2-layered net with 5 input neurons 25 units in the hidden layer and 20 output neurons. This network reaches a minimum error of 0.032629 after 360,602 pattern presentations. Because this error is still too large, we are looking for better learning algorithms to make full use of the theoretical benefits of fuzzy neural networks.

## RESULTS

We trained the **MLP** with data that was aquired using the three different methods ER1,ER2 and MEA. We employed the classification methods 1D and MD. All combinations of these were used.

Tables 3 and 4 show the average number of trainings epochs needed for test success over 10 different training-runs with different starting weights and slighlty different learning parameters (as pointed out above, our learning law adjusts the

Table 3: Average number of training-epochs needed for test success (1D)

| | ER1/1D | ER2/1D | MEA/1D |
|---|---|---|---|
| MLP (5/10–35–1) | 27400 | 30700 | 31900 |
| MLP (5/10–40–1) | 18200 | 14500 | 15000 |
| MLP (5/10–45–1) | 19800 | 17300 | 14200 |
| MLP (5/10–20–20–1) | 24300 | 23500 | 32200 |
| MLP (5/10–30–30–1) | — | 42300 | — |

most important parameters itself). 2000 training samples were used. "Test success" means, the trained net classified all of the test-data (2000 samples) correctly.

In all cases we were able to find net-architectures that solve the classification problem. But the time and number of training steps needed vary:

The 1D classification approach yields faster convergence than the MD approach, as could be expected: The MD net is a lot larger than the 1D net and consequently more training steps are needed to adjust the high number of weights. This is only true of course, when the 1D net is provided with enough hidden neurons. 35 hidden neurons are not sufficient, as can be seen in the first row of Table 3 — the training has to go on nearly twice as long as when using 40 hidden neurons. Nevertheless, training succeeded with 35 hidden neurons, nets with less hidden neurons did not achieve this.
Also, when the 1D net contains too many hidden neurons (row 5 in Table 3), training may not succeed. The 'moving target problem' can not be solved when that many weights have to be adjusted: There is no convergence.
For the 1D/ER approach we found nets with a $5-40-1$ structure to be the fastest in training.
The 1D/MEA approach worked best with a $10-45-1$ net. This is not suprising, since one can expect that more input data needs more hidden neurons to be more easily classified.

The MD classification approach uses 20 neurons in the output layer. Our experiments show, that at least 20 hidden neurons have to be used in these nets. As a result the MD nets have more than twice the number of weights 1D nets use. This leads to considerably longer training *time*. As can be seen in Table 4 the number of training steps needed for success is also much larger than for 1D nets. Altogether training a MD net takes about four times longer than training a 1D net.
Again - as with the 1D approach - the ER data can be learned fastest by a 'small' $(5-20-20)$ net. MEA data needs a $10-25-20$ net for fastest convergence.

As expected the MD classification method takes a lot more epochs to be successful. But taking into account that the output of these nets (fuzzy vector) can be more easily interpreted one would probably still want to use this method.

Comparison of the best results for ER and MEA data shows that there is no significant difference in the number of epochs needed. However the ER2 method yields better results than ER1 data which is an interesting thing to notice.

Table 4: Average number of training-epochs needed for test success (MD)

| | ER1 | ER2 | MEA |
|---|---|---|---|
| MLP (5/10–20–20) | 25100 | 23000 | 27900 |
| MLP (5/10–25–20) | 29400 | 26700 | 24300 |
| MLP (5/10–30–20) | 33700 | 35900 | 28600 |
| MLP (5/10–10–10–20) | 46300 | 49200 | — |
| MLP (5/10–10–15–20) | 43200 | — | — |

In Tables 3 and 4 "MLP ($5/10$–$n_1$–$n_2$–$m$)" stands for a Multilayer Perceptron with 5 (for ER data) or 10 (for MEA data) input nodes, $n_1$ nodes in the first hidden layer, $n_2$ nodes in the second hidden layer (if existent) and $m$ output nodes.
If "—" is given instead of a number this means that test success could not be accomplished after training for 60000 epochs.

## CONCLUSION AND FURTHER WORKS

We learn, that neural nets can solve the described classification tasks, independent from the kind of analysis data and classification type we use. When employing the MLP as a hashish classification tool, ER2 or MEA data and the MD classification method should be used. The training takes longer than for the 1D method but the output of a MD net is easier to work with.

We are currently applying the **FNN** to the classification problem. We do not need artificial data for training the FNN, so the data from the original samples is used. Triangular fuzzy numbers are created that represent these samples and completely cover the input space.

Also self-organizing neural networks will be trained to act as classifiers.

Hashish samples that where grown in the same area contain similar concentrations of certain elements. In future — depending on available data — we can possibly extend our work to classifying samples according to cultivation areas.

## REFERENCES


Casale, J.F., Watterson, J.W., (1993). A Computerized Neural Network Method for Pattern Recognition of Cocaine Signatures, *J. of Forensic Sciences* **38***(2) 292–301*

Jacobs, R.A., (1988). Increased Rates of Convergence Through Learning Rate Adaption, *Neural Networks, 1, 295-307.*

Lippe, W.-M., Tenhagen, A., Feuring, Th., (1994). Fuzzy descent in feedforward networks, *Intelligent Engineering Systems through Artificial Neural Networks, Proceedings of ANNIE '94, ASME Press, New York, 955-960*

Feuring, Th., Lippe, W.-M., (1995). A Goodness Function for trained fuzzy neural networks, *Proc. of IEEE Int. Conf. on Neural Nets, Perth, 2264-2269*

Feuring, Th., (1996). Learning in fuzzy neural networks, *Proceedings of IEEE International Conference on Neural Networks, Washington, 1061-1066*

# NEURAL NETWORKS FOR ACCELERATING THE TRANSISTOR DESIGN PROCESS

**Ryan Ferguson and David J. Roulston**
*Silicon Devices and Integrated Circuits Laboratory*
*University of Waterloo, Waterloo, Ontario, Canada*

**ABSTRACT**
We have developed a neural network based tool which accelerates the transistor design process. Our tool, REED (Rapid Engineering of Electronic Devices), replaces the conventional, iterative scheme of designing transistors with a single step scheme. In this paper we present a brief overview of the conventional transistor design process and describe how REED may be applied to accelerate the design of a bipolar transistor.

## INTRODUCTION

In the semiconductor industry, a short time to market is critical for corporate survival. Market dominance and obsolescence both occur so quickly that following a rival to market often results in the failure of an otherwise promising product. In this paper, we discuss how neural networks can be used to accelerate the design of transistors, the fundamental building block of all micro-electronic products.

The process of transistor design involves the optimization of the physical structure of a transistor to meet stated design goals such as switching speed, current gain and power consumption. When designing a new transistor, device designers typically start with an existing transistor and make educated guesses at the structural changes necessary to meet the new performance specifications. Device simulators are then employed to ascertain the performance of the proposed structure. If necessary, the engineers then propose changes to the structure to improve performance. The rounds of 'modify and verify' continue until the engineers are satisfied with the performance of their design.

In the next section, we will describe how neural networks can eliminate the 'trial and error' design process by *learning* how to approximate the functions which map the electrical performance of a transistor to its physical structure. In this way, given the electrical requirements of the desired transistor, the neural network can describe the optimal physical structure.

## CHARACTERIZATION OF BIPOLAR TRANSISTORS

The functions that map a transistor description in the electrical domain to a description in the structural domain are non-linear and very complex [1]. When device designers propose a new transistor structure, they apply their knowledge and intuition to approximate these functions, but due to the complexity of these functions their proposal is often sub-optimal.

We have developed a neural network tool, REED (Rapid Engineering of Electronic Devices), which can learn the functions that map the electrical performance of a bipolar transistor to its structural characteristics. This mapping can be made arbitrarily accurate by adjusting the number of hidden units of the neural network [2], [3]. These mapping functions allow REED to map from the electrical domain to the structural domain in just one step.

To investigate the performance of REED thoroughly, it is necessary to understand how transistors are described both structurally and electrically. The lateral dimensions of a transistor are determined by a series of masks, each of which defines a particular fabrication step. For example, a bipolar transistor is characterized by separate masks for the buried layer, the base region, the emitter region, the contact openings and the metal interconnect. The sizes of the structures on each of these masks will impact the electrical performance of the transistor. In addition, the dopants (impurities added to the silicon to modify the conductivity) vary in the vertical direction. The shape of this impurity profile has a profound impact on transistor performance. Figure 1 shows two views of a bipolar transistor: a cross-section revealing the different functional regions, and a projection of the dopant density versus the depth below the transistor surface.

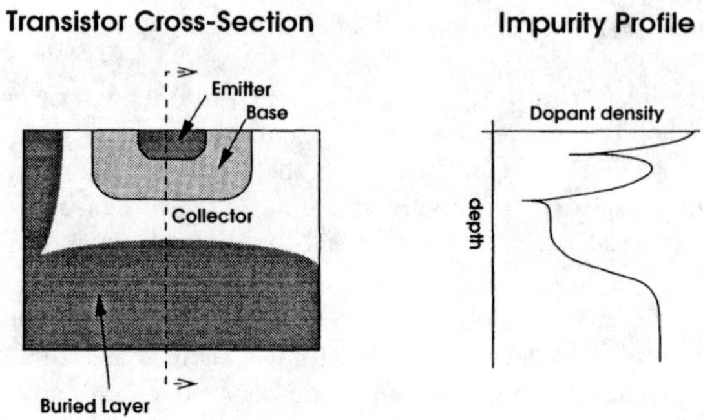


Fig. 1. Two projections of a transistor. The vertical profile of a transistor highlighting the different functional regions is shown on the left. On the right is a plot of doping density versus depth. The emitter and buried layer regions are the most heavily doped.

Since most structures defined by masks are regular geometric shapes, the masks can be described analytically with a small number of parameters. The impurity profile cannot be described so simply. The mechanisms governing the diffusion and implantation of impurities into silicon are not fully understood [4]. However, reasonable results are obtained by approximating the dopant density, $n(x)$, with equations such as Equation (1) which describes the emitter diffusion.

$$n(x) = \mathbf{ne} \times e^{-\left(\frac{x}{\mathbf{xe}}\right)^{\mathbf{nxe}}} \tag{1}$$

**ne** is the peak dopant density, **xe** is the characteristic length of the impurity profile, and **nxe** is related to the mechanics of the diffusion [5]. The independent variable '$x$' is the distance below the upper surface of the transistor. By using these equations, the entire impurity profile can also be represented by a small number of parameters. In total, 15 parameters are often sufficient to describe the physical structure of a bipolar transistor.

REED uses SPICE[1] model parameters to characterize the electrical performance of transistors. SPICE models conveniently describe the electrical operation of a transistor with a small number of parameters and are well known in the microelectronics community because of the prevalence of SPICE–based circuit simulators. Important design goals such as current gain and maximum switching speed are directly correlated to SPICE model parameters.

## IMPLEMENTATION

### Neural Network Structure

At the core of REED is a multi-layer feed-forward neural network trained by a modified back-propagation of errors method. The number of units in the input layer corresponds to the number of SPICE model parameters. In the case of the bipolar transistor, 34 parameters are used. The number of output units equals the number of free parameters used to describe the structure of the transistor. As previously noted, 15 parameters are often sufficient to describe the physical structure of a transistor. One hidden layer with a user–specified number of units is used. For most applications we have found 15 hidden units to be sufficient. Additional work has to be done to relate the number of units in the hidden layer to the performance of the neural network. The network topology of REED is shown in Figure 2.

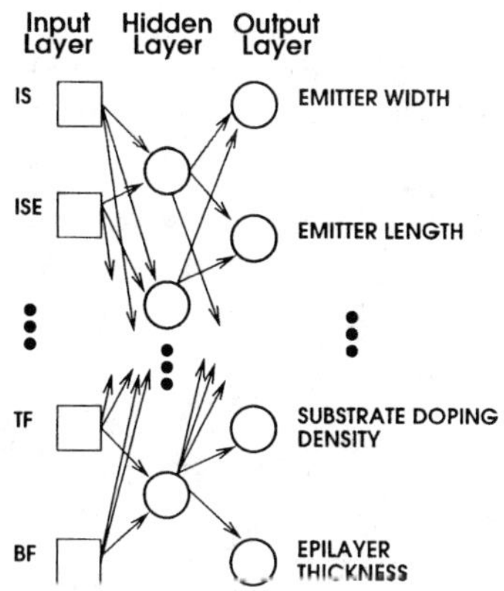


Fig. 2. The topology of the REED neural network. The SPICE parameters, applied to the input layer, give rise to outputs corresponding to the physical description of a transistor.

### Training Set

The most important consideration when using REED is the selection of an appropriate training set. The training set contains the structural parameters and the associated electrical parameters for a randomly generated collection of transistors. A seemingly logical choice for the training set would be prototype transistors that had been fabricated and have had their SPICE parameters extracted through electrical measurements. Unfortunately, this training set is infeasible for two reasons:

1. Fabricating enough transistors for a full training set would be prohibitively expensive and time-consuming.
2. Neural networks are tools for interpolation. Designing the next generation of bipolar transistors is necessarily an extrapolation from the current technology. The transistors of the next generation will have features smaller than any of the transistors in the present generation. Training with the present

[1]SPICE stands for Simulation Program with Integrated Circuit Emphasis. SPICE is a public domain tool for simulating circuit operation that was developed at U.C. Berkeley in the 1970s. Many commercial variants exist.

generation of transistors will not allow REED to accurately learn the mapping rules for transistors with smaller feature sizes.

Instead, a device simulator is used to generate the training patterns. Randomly generated transistor structural descriptions, representative of transistors which are feasible with the available manufacturing technology, have their SPICE parameters extracted by a device simulator. The extracted SPICE parameters and the associated transistor structural description then form a training pattern. For rapid extraction of the SPICE model parameters of a bipolar transistor, REED uses the BIPOLE3 bipolar device simulator [6], a fast simulator with proven accuracy for a wide range of modern bipolar transistors. Using a relatively coarse approximation, BIPOLE3 can extract the SPICE model parameters of a transistor structure in less than 10 seconds on a SparcStation20 workstation. Thus, thousands of randomly generated transistors and their associated SPICE models can be created overnight.

## TRAINING

The process of training the REED neural network consists of several stages:

a) the generation of random transistor structural descriptions and the extraction of their associated SPICE model parameters,
b) the initialization of the neural network,
c) the application of the back-propagation algorithm to update the weights of the network, and
d) the decision to terminate training.

Each of these stages is elaborated upon in this section.

### Generation of Training Patterns

When REED is invoked, the input file is read into memory and is used as a template to generate the training data. The neural network parameter **epoch_size** indicates how many randomly generated transistors are to be created for the training set. Each transistor in the training set follows the rules laid out in the input file, and is matched to a set of SPICE model parameters extracted by BIPOLE3.

### Neural Network Initialization

The neural network is then initialized with the physical dimensions specified in the input file. **input_layer_size**, **hidden_layer_size**, and **output_layer_size** correspond to the number of units in each layer, while **W1_initrange** and **W2_initrange** set the bounds on the randomly generated initial weight values.

### Updating the Weights Using Back-Propagation

The training of the REED neural network proceeds on a pattern by pattern basis. Training patterns are selected in random order to ensure that the gradient descent in weight space is stochastic [7].

REED uses a modified back-propagation scheme to speed up convergence and improve generalization. A momentum term has been added to the standard back-propagation algorithm. With momentum, the change in a weight is not only dependent on the error, but also on the amount of change in the weight during the previous update phase [8]. Weight decay is also employed in our modified back-propagation algorithm. Weight decay improves the generalization of networks by penalizing large weights that do little to improve the mean square error of the network [9]. REED employs weight decay by decreasing each weight by a fixed percentage after the presentation of each training pattern. With these modifications, the weight update rule becomes [10]:

$$\Delta W_{kj}(t+1) = -\eta \frac{\partial Error}{\partial W_{kj}} + \alpha \Delta W_{kj}(t) - \beta W_{kj}(t) \tag{2}$$

The first term on the right hand side is the standard back-propagation term. The second term on the right hand side implements momentum and the third term implements weight decay. The parameters $\eta$, $\alpha$ and $\beta$ are all specified in the REED input file.

**Termination of Training**

Presentation of training patterns continues until all of the patterns in the set have been exhausted. After an epoch is completed, REED evaluates whether its neural network is fully trained.

By coupling REED with BIPOLE3, we have developed a simple method to test the convergence of the REED neural network. We apply the SPICE model parameters representing the electrical specification of the transistor being designed to the input layer of the neural network. The values propagated to the output layer then represent an approximation to the structural parameters of this transistor. The quality of the approximation can be evaluated by BIPOLE3. BIPOLE extracts the SPICE model parameters from the approximated structure; if these parameters are closely matched to the SPICE model initially applied to the input layer, REED is correctly mapping SPICE parameters to their appropriate structure. Training is then terminated.

If REED determines the approximation to be poor, then the training patterns will be presented for another epoch of training. This 'train and test' scheme continues until a suitable description of the transistor is found or a user-defined upper limit of epochs is reached.

## RESULTS

We have successfully employed REED to directly determine the structural parameters of bipolar transistors. To demonstrate the effectiveness of REED at approximating the structural parameters of a transistor, we have selected a target transistor and presented its SPICE model parameters to REED. Figure 3 shows the impurity profile of the target transistor as well as the profile of the approximation generated by REED. The approximation is excellent. Both profiles were plotted with the BIPOLE3 bipolar device simulator.

The SPICE model parameters of the target transistor were presented to REED after its network was trained with 9 epochs of 3000 training transistors. REED optimized 2 mask parameters and 7 impurity profile parameters. Two other parameters related to the manufacturing process were also optimized. Generating the 3000 transistor structural descriptions and their associated SPICE model parameters took 5 hours on a SparcStation20 workstation. Total time for training was about 30 minutes; an average of less than 4 minutes per epoch.

Further evidence of the success of REED at approximating the structural parameters of the target transistor is given in Figure 4 where the transistor unity current gain frequency is plotted versus the collector current for both the target transistor and the approximation. The unity current gain frequency is particularly sensitive to variations in the impurity profile.

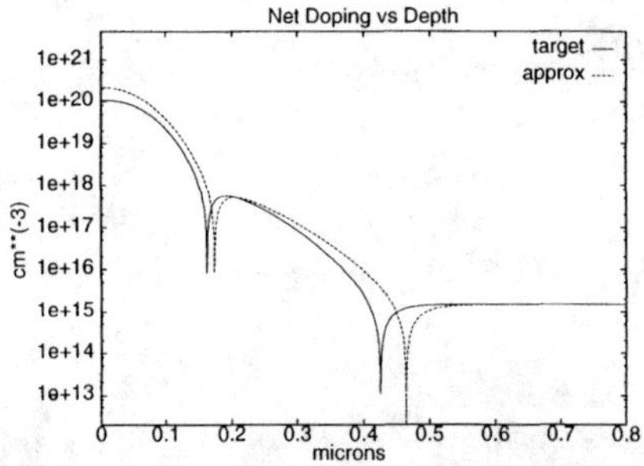


Fig. 3. The impurity profiles for the target transistor and REED's approximation.

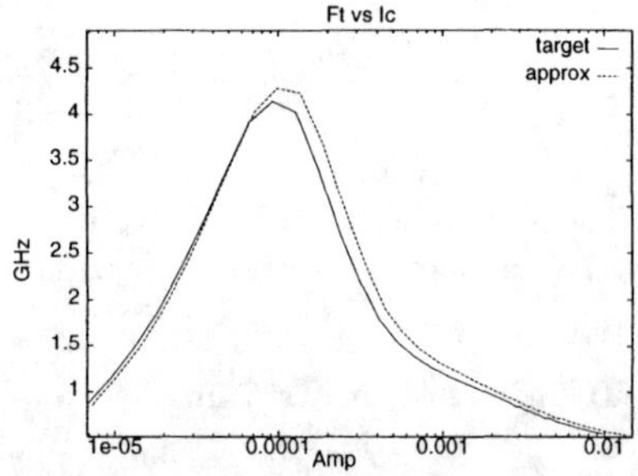


Fig. 4. BIPOLE3 simulation of unity current gain frequency for both transistors.

## ACKNOWLEDGMENTS


The authors would like to acknowledge the assistance of Marian Spiteri.


## REFERENCES


[1] I. E. Getreu, *Modeling the Bipolar Transistor*, Elsevier Scientific Publishing, 1978.

[2] G. Cybenko, "Approximation by superposition of a sigmoidal function", *Mathematics of Control, Signals and Systems*, vol. 2, no. 4, pp. 303–314, 1989.

[3] L. P. J. Veelenturf, *Analysis and Applications of Artificial Neural Networks*, Prentice-Hall, Inc., 1995.

[4] W. R. Runyan and K. E. Bean, *Semiconductor Integrated Circuit Processing Technology*, Addison-Wesley, 1990.

[5] D. J. Roulston, *Bipolar Semiconductor Devices*, McGraw-Hill, Inc., 1990.

[6] D. J. Roulston, *BIPOLE3 User's Manual*, BIPSIM Inc., March 1995.

[7] S. S. Haykin, *Neural Networks : A Comprehensive Foundation*, Macmillan, 1994.

[8] D. Plaut, S. Nowlan, and G. Hinton, "Experiments on learning by back propagation", Tech. Rep. CMU-CS-86-126, Department of Computer Science, Carnegie Mellon University, 1986.

[9] A. S. Pandya and R. B. Macy, *Pattern Recognition with Neural Networks in C++*, CRC Press, Inc., 1996.

[10] J. Hertz, A. Krogh, and R. G. Palmer, *Introduction to the Theory of Neural Computation*, Addison-Wesley, 1991.

# CODING A CONCEPTUAL MODEL INTO A NEURAL NETWORK IN MODELING ICE-CORRECTION [1]

**MARKUS HUTTUNEN[2] AND BERTEL VEHVILÄINEN**
*Finnish Environment Institute, Finland*

**ESKO UKKONEN**
*University of Helsinki, Department of Computer Science*

***ABSTRACT:***
In this work three models, a neural network, a conceptual model and a combination of these two, a hybrid model are applied to model the backwater effect of ice in a river. The neural network is a black-box model, based on data. The conceptual model is based on a physical description of the system. The data is used in optimizing the free parameters of the description. In the hybrid model, the neural network is modified so that the physical description of the conceptual model can be coded into the structure of the network. In the beginning of fitting, the hybrid network already performs as well as the conceptual model. During fitting also the physical description is optimized, not only the parameters of the description.

## THE PROBLEM

River ice causes backwater effects on the water level. The discharge of a river is observed by measuring the water level. The water level is transformed to discharge using stage-discharge relation, which defines discharge for each water level. During the ice period, however, the relation may vary and discharge observations have to be corrected. Inputs of the models are daily values of short wave radiation, temperature, precipitation and observed uncorrected discharge (Figure 4). Output is an estimate of the ice-corrected discharge, with the backwater effect eliminated.

In this work three different models are applied. One is a black-box model, a neural network (Herz et al., 1991). Second is a conceptual model, where a simple physical description of ice-cover is created based on expert knowledge and then the free parameters in the description are fitted to data. Third model is a hybrid one, a combination of these two; the physical description is coded into a neural network like network, which is fitted to data using neural network fitting algorithms. During fitting the form of the physical description is also optimized, not only the parameters of the description.

The benefit of the hybrid model is that it utilizes the existing expert knowledge, not only data as the neural network. Another benefit is that the hybrid model is not restricted to the fixed description of the conceptual model, but the equations of the description are modified automatically on the basis of the data. Many similar methods to code knowledge into and extract it from a neural network have been presented in

[1]This work is supported by the IHP Committee of the Academy of Finland and by the Imatran Voiman Säätiö (IVO) Foundation.

[2]e-mail: Markus.Huttunen@Vyh.Fi, http://www.vyh.fi/syke/people/Markus_Huttunen.html, postal: Finnish Environment Institute, P.O.Box 140, FIN-00251 Helsinki, Finland.

the literature. In (Towell, 1991) is described a method for inserting and extracting symbolic knowledge and in (Giles et al., 1992) there is a method for processing finite state automata. The most similar application is in (Cozzio–Büeler, 1995) where differential equations are coded into a neural network. However, the knowledge in our conceptual ice-correction model does differ from these. The knowledge consists of extremely simple difference equations. After fitting the model it is necessary to get the result back into similiar equations to be able to verify consistency of the model. This is why here is presented a new hybrid model, which has the speciliaty that computation in the network is very similar to the conceptual model. It is not necessary to explicitly extract rules from the network after fitting because the network itself is understandable.

## CONNECTIONIST BLACK-BOX MODEL

The neural network model for ice-correction is a feedforward neural network with a few backward connections. The network consists of seven layers and the backward connections are between two fixed layers (Figure 1). The model is fitted to data and to a set of boundary conditions (see Section Results). Ice thickness and ice-correction are modeled using separate networks. Parameters of the network are the number of nodes, and the existence and weights of the connections.

Parameters are optimized using an evolutionary optimization algorithm (Fogel and Fogel, 1994), which optimizes simultaneously a set of networks. New network is created by combining two networks in the set. The two networks are chosen randomly, networks are cutted into two pieces by picking randomly a layer and a node where to cut. The new network is a combination of the first part of the one network and the second part of the other network. The new network is modified randomly by adding and deleting nodes and connections and changing weights of the connections. After this the weights are optimized using a direct search optimization algorithm (Hooke and Jeeves, 1961). The resulting network is added to the set and the worst network in the set is removed. The direct search optimization is based on comparing the values of the error function when the values of the parameters are changed by small steps. The gradient of the error function is not needed and hence the method can be used with boundary conditions. When used whith evolutionary optimization, direct search optimization makes convergence faster in the beginning of optimization.

An evolutionary algorithm is robust; it can avoid local minima of the error function, at least to some extent. Unfortunately the algorithm is also time demanding. We used set size 20, each network had 30-50 nodes and 1000-5000 new networks were created during the optimization. Computation time can be weeks in Alpha VAX.

## CONCEPTUAL MODEL

Our conceptual ice-correction model is based on simple description of formation, melt and backwater effect of ice-cover. In the equations of the model there is a number of free parameters. Initial values for the parameters are set by an expert that knows the ice process and the model is fitted to data by optimizing the free parameters.

Equilibrium temperature $T_{eq}$ (Equation 1) defines the point when ice is not melting nor new ice is formating, this idea is from (Bengtsson, 1976). Degree-day temperature sum is defined in (2) and relative ice-cover area ($0 \leq Open \leq 1$) in (4). Indexes *t-1* and *t* mean consecutive days. Ice formation and melt is defined in (7). Values of

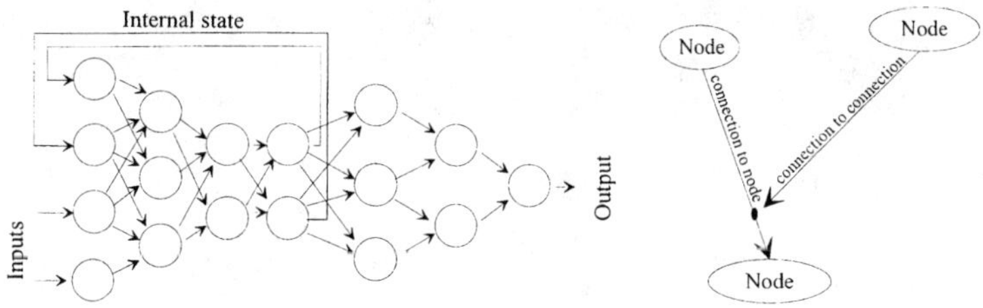


**Figure 1: (Left) An example of a neural network. Value of each node is updated once in each timestep. Backward connections feed the state of the system to the next timestep. (Right) A connection of the hybrid network can end to another connection. In this case the value of the connection is coefficient to the target connection.**

parameters in (5) and (6) depend on if $T_{eq}$ is greater than or less than zero. Finally the ice-correction depends on observed discharge and ice thickness as in (8).

$$T_{eq} = T - (ShortWaveRadiation - parameter_1)/parameter_2 - parameter_3 \quad (1)$$

$$\begin{aligned}
TemperatureSum_t &= TemperatureSum_{t-1} + T_{eq} && (2)\\
parameter_4 &\leq TemperatureSum_t \leq parameter_5 && (3)\\
Open_t &= Open_{t-1} + parameter_6 * TemperatureSum_t && (4)\\
COEFF_{ice} &= 1/\left(1 + Icethickness * param_{(7,8)}\right) && (5)\\
COEFF_{snow} &= 1/\left(1 + PrecipitationOnIce * param_{(9,10)}\right) && (6)\\
Icethickness_t &= Icethickness_{t-1} + param_{11} * T_{eq} * Open + \\
&\quad param_{12} * T_{eq} * (1\text{-}Open) * COEFF_{ice} * COEFF_{snow} && (7)\\
CorrectedDischarge &= ObservedDischarge - param_{13} * Icethickness^{param_{14}} * \\
&\quad \max\left(0, ObservedDischarge - param_{15}\right)^{param_{16}} && (8)
\end{aligned}$$

## HYBRID MODEL

A black-box model does not utilize already existing knowledge. A conceptual model does utilize knowledge but only the parameters of the equations are fitted to data. Our hybrid model offers a method that is able on the one hand to utilize the knowledge as in the conceptual model and on the other hand is able to change the coded knowledge by optimizing the structure of the model using an evolutionary algorithm.

The hybrid model is a modification of neural network. In the hybrid network, the value of a connection is piecewise linear function of the value of the start node (Figure 2). A connection can end to another connection instead of a node, in this case the value of the connection is a multiplicative coefficient to the value of the target connection (Figure 1). The nodes have a linear activation function. This hybrid model can be made probabilistic by defining distribution for the values of the connection function (Figure 2). Here the distribution is uniform and distribution parameters are optimized to get confidence limits as narrow as possible (Figure 4).

Let's define that the variables of a conceptual model consist of the inputs and the outputs of the model and of all the intermediate results defined by the equations. Hence the variables are the same as the left hand sides of the equations (see

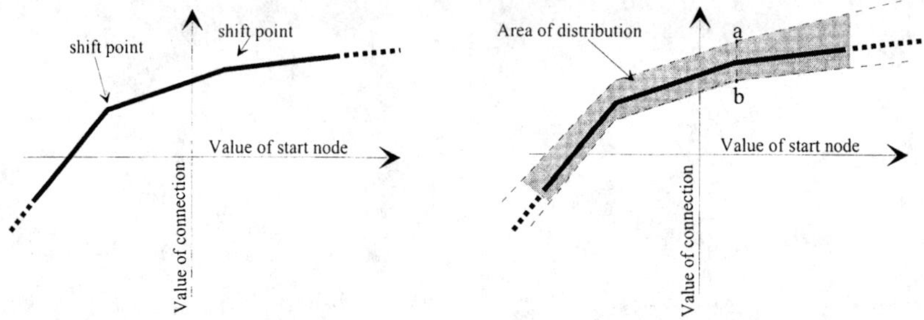


**Figure 2: (Left)A connection in the hybrid network is a piecewise linear function. (Right) The model is made probabilistic by adding probability distribution to the values of the connection function. The width of this uniform distribution ((a) and (b)) is defined piecewise linearly at each shift point of the function.**

Section Conceptual Model). Into the hybrid network it is allowed to code all functions of one variable (like $a = f(b)$). With two or more variables only the basic arithmetic operations, i.e. addition, subtraction, multiplication and division are available ($a = f(b,c) = b + c$). This means that equations of a conceptual model have to be converted so that the right hand side of the equations contains either only one variable (and the function to be applied on it) or only basic arithmetic operations. The hybrid network has one node for each variable of the conceptual model. The nodes in Figure 3 are the variables of the conceptual ice-correction model. An exception is that complicated equations may have to be coded using additional nodes for intermediate results. Another exception is that for some variables one needs only a connection, for example Equations (5) and (6).

Equations that contain function of one variable are represented by one connection, which is simply a piecewise linear approximation of the equation. If the equation contains sum of two or more variables as in Equation (1), then one needs as many connections as there are variables in the sum. Equation (1) is coded by (a) and (b) in Figure 3. Connections ending to another connection are needed when the equation contains a product of variables. One of such variables is coded as a connection ending to the target node and the other parts of the product are coded as connections ending to that connection. For example, in the coding of Equation (8) in Figure 3 there are connections (c) and (d) for the sum and connection (e) for the product. Before starting the evolutionary algorithm, that modifies the structure of the network the hybrid model is equivalent to the conceptual model, except for the error that comes from approximating equations as piecewise linear functions.

It is a known fact that a piecewise linear function can approximate any function with an infinitesimal error. However, in coding the equations there will arise approximation errors. Then an expert is needed who knows the conceptual model and is able to define amount of acceptable approximation error for each equation. The number and placement of shift points in the piecewise linear connection functions are initialy set on the basis of original equations and this definition of acceptable error.

The parameters of the hybrid model consist of the number of nodes, the existence of connections between the nodes and the number and placement of shift points in the piecewise linear functions of the connections. If the model is probabilistic then there is also parameters for the width of distributions. Width is defined by two

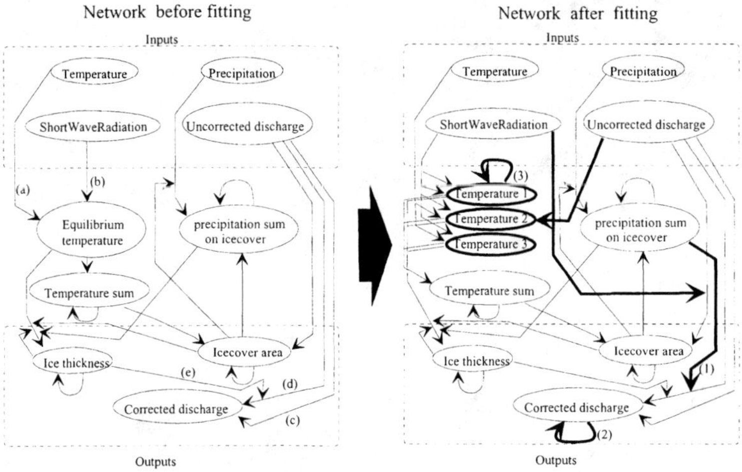


**Figure 3: Nodes and connections in the conceptual model (network before fitting) and hybrid model (network after fitting). Thick line marks nodes and connections that were added by the fitting algorithm.**

parameters at each shift point (Figure 2). Parameters are fitted using the same evolutionary optimization algorithm as in fitting the neural network ice-correction model.

An example of a network after fitting is in Figure 3. The evolutionary algorithm has added some meaningful nodes and connections to the network; connection (1) in Figure 3 means that amount of ice-correction depends on precipitation sum on ice-cover. Connection (2) means that corrected discharge depends on corrected discharge at previous day and in (3) the new node and connection means that equlibrium temperature depends on equlibrium temperature at previous day.

## RESULTS

We have applied our models so far to one observation station, Tornionjoki Karunki in the Northern Finland. The models are fitted to two datasets (Figure 4), ice thickness and manual ice-correction (Leppäjärvi and Vehviläinen 1994). The optimized error function (9) used in fitting of the models is weighted sum of errors from the data and from the set of boundary conditions. The purpose of the boundary conditions is to reject models which fit well on the data but in other respects do not correspond to the existing knowledge of the system. There are coefficients $c_i$ for errors in each dataset and in each boundary condition. New boundary conditions may have to be created during fitting and values of the coefficients may have to be modified manually in order to guarantee that the model corresponds to expert's knowledge of the system.

$$Error = \sum_{i=1}^{2} c_i * DataSetError_i + \sum_{i=1}^{n} c_{i+2} * BoundaryConditionError_i \qquad (9)$$

Mean effect of the ice at the station between 1 October and 30 June is 100.36 $m^3/s$. Mean error of the hybrid model is 13.10 $m^3/s$, the conceptual model 13.33 $m^3/s$ and the neural network 15.52 $m^3/s$. Also in ice thickness modeling the best model is

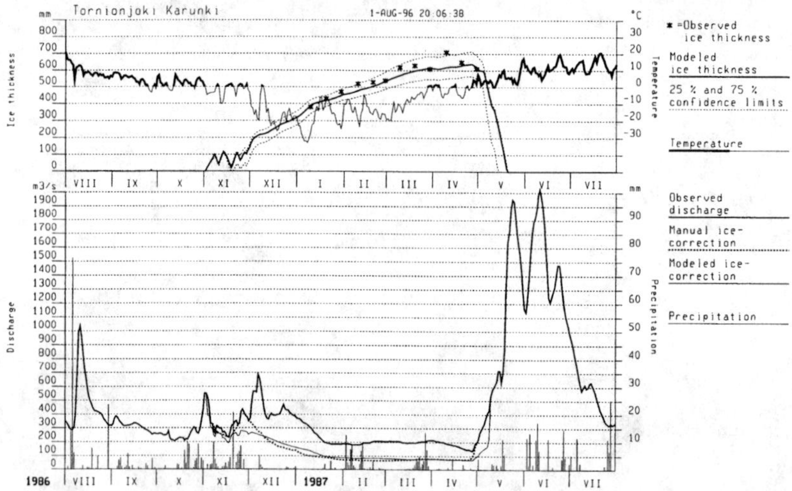


**Figure 4: Observed and modeled (hybrid model) values for ice thickness and ice-correction at the observation station Tornionjoki Karunki for the winter 1986-87. The ice thickness model is probabilistic, with confidence limits within which at least 50 % of the observations lay. Observation data used in training the models are from the years 1961-93.**

the hybrid model. Based on the data, the fitting algorithm of the hybrid model has also modified the initialy coded knowledge (Figure 3), these modifications are meaningful and useful when the conceptual description is developed further.

## REFERENCES


Bengtsson, L., (1976). Snowmelt Estimated from Energy Budget Studies, *Nordic Hydrology*, Vol. 7, 1976, pp. 3-18.

Cozzio–Büeler, Rico A., (1995). The Desing of Neural Networks using a priori Knowledge, *A dissertation, Swiss Federal Institute of Technology*, 1995.

Fogel, D.B., Fogel, L.J., Eds., (1994). *IEEE Transactions on Neural Networks:* Special Issue on Evolutionary Programming, Vol. 5, No. 1, January 1994.

Giles, C.L., et al., (1992). Learning and Extracting Finite State Automata with Second-Order Recurrent Neural Networks, *Neural Computation*, No. 4, pp. 393-405, 1992.

Herz, J., Krogh, A., Palmer, R.G., (1991). Introduction to the Theory of Neural Computation, *Santa Fe Institute studies in the sciences of complexity*, Addison Wesley 1991.

Hooke, R., Jeeves, T.A., (1961). "Direct Search" Solution of Numerical and Statistical Problems, *Journal of the Assosiation for Computing Machinery*, Vol. 8, 1961.

Leppäjärvi, R., Vehviläinen, B., (1994). Ice reduction of winter discharge by watershed models, *Publications of the Water and Environment Research Institute, Finland*, ISBN 951-47-9749-3.

Towell, G.G., (1991). Symbolic Knowledge and Neural Networks: Insertion, Refinement and Extraction, *P.D.Thesis, University of Wisconsin - Madison*, 1991.

# UTILIZING NEURAL NETWORKS TO PREDICT WING LEADING EDGE SLATS AIRLOADS

**ELIAS BOUNAJEM AND ALAN VAN DER VEEN**
Cessna Aircraft Company
Wichita, KS 67277

***ABSTRACT:***

This paper describes a method to predict pressure on the leading edge slats of a business jet using Artificial Neural Networks. Neural Networks are used to avoid the problem with unbiased estimators that stems from multicollinearity in the data. Pressure prediction is achieved through mathematical modeling of flight test data collected during the aircraft certification process. Mathematical model parameter selection and pressure port groupings are achieved through a combination of statistical analysis and engineering understanding of the problem. Use is made of Kohonen's Self-Organizing Map as a means of detecting redundant and conflicting data from the flight tests, and in selection of the training and test sets for the predictor networks. The backpropagation paradigm is then used to carry out the prediction process. Prediction errors were statistically examined to detect and correct any bias in the modeling process.

## INTRODUCTION

Data gathering for aircraft, whether it is for the purpose of type certification, operational stability determination, testbed and research, or simulator building, is a costly and time consuming process. The most common methods of data acquisition are flight testing, wind tunnel testing, predictive modeling and empirical techniques [Kohlman, 1989].

Flight testing is the only environment where all the aircraft and atmospheric variables come together, giving a unique opportunity for gathering data pertaining to the characteristics of the aircraft and its subsystems in real situations. However, testing in the real environment creates some special problems. For instance, finding sensors and transducers that will survive the extremes of temperature, pressure and vibration, and yet have the required accuracy and frequency response, is not a trivial task. Transducers must be properly calibrated. Pre- and post- flight checks must be performed to verify that the instrumentation is operating properly. Data must be reviewed on a timely basis. Last, but not least, flight testing is a very expensive operation. It is evident that increases in the size of the required dataset are linked to increases in cost.

A more controlled environment for data gathering is wind tunnel testing. Wind tunnel testing is most useful during the early stages of the aircraft design process, before a prototype has been built. As in flight testing, identifying reliable and adequately sensitive transducers, conducting and applying transducer calibration, timely data reviewing and careful planning are equally necessary to make data collected in the tunnel usable. There is an additional set of technical issues: capability of calculating balances and allowing for drifts, ability to measure model attitude, unsteadiness and turbulence in tunnel flow, geometric fidelity of the model, wall interference correction, transition location and aeroelastic distortion of the model. Wind tunnel time is similarly expensive. The cost to build the model, run the tunnel and man the test itself is very high, and increases with the amount of data to be collected.

Additional sources of data are predictive and empirical methods. However, the data needs to be validated once wind tunnel or flight test data become available

Flight and wind tunnel testing remain the most trustworthy sources of data in spite of the challenging technical issues. During the testing phase, extensive quantities of data are collected and analyzed. If these amounts of data can be reduced and then supplemented by other means without jeopardizing the integrity and airworthiness of the design, the result would be a shorter design and certification cycle and attendant reduction in development cost, coupled with increased profit by early introduction of the product into the market.

The proposed approach consists of collecting the reduced set of data, either from wind tunnel or flight test, at strategically selected fight conditions. Neural networks are then used to predict the desired data after being trained and tested with the reduced set.

In this paper, Neural Networks are used to predict pressure distributions on the leading edge slats of a business jet. Pressure on the slats is modeled as a function of relevant parameters. Training and test sets were extracted from the available flight test data. Once the network has been trained, the pressure distribution on the slats can then be generated.

Within the framework of the proposed approach, the available flight test data can be considered a reduced data set. Data for additional flight conditions can be generated from the neural network.

## DESCRIPTION OF DATA SET

The data used in this investigation was collected during the certification phase of a business jet model. Data was collected over the span of one flight. Pressure transducers were installed on the leading edge slat upper and lower surfaces and were grouped at four wing stations. Parameters that were available from that flight included angle of attack, angles of sideslip, aircraft attitudes and rate of change of the attitude, longitudinal, lateral and vertical accelerations, control surface deflections, flap setting, airspeed, altitude, MACH number, dynamic pressure and outside air temperature. The data set was in engineering units. Types of maneuvers that were flown ranged from acceleration and deceleration at different load factors to roller coaster type maneuvers, to ensure a good coverage of angle of attack in the negative range. Data was recorded with slats in the extended and retracted positions. In selecting training and test sets, no restrictions on rates, accelerations or any other parameter were applied in order to insure unbiased sets.

## METHOD OF ANALYSIS

### Variable Selection

In general, pressure distribution on a given body in the presence of a flowfield is a function of the body geometry, upstream flow velocity and angle of attack. In this study, slat geometry was not included explicitly as an input for the neural network, since pressure predictions were performed on a port by port or a group of ports basis.

Input variables selected for the analysis included flap deflection, angle of attack and dynamic pressure that, normally, dictate the pressure magnitude and pattern around the body. Figures 1 shows the pressures at a pressure port, located on the slats upper surface, as a function of angle of attack.

Once pressure has been normalized with dynamic pressure and plotted against angle of attack (see Figure 2), the data appears to follow a well-defined trend. However, at a given angle of attack there appears to be some variation in pressure which, upon further investigation, could be attributed to the load factor $n_Z$. Longitudinal acceleration was also included to account for any transient pressure response.

## Dataset Selection

Selection of test and training data sets are one of the most critical aspects of neural network implementations. The most frequently utilized approach entails the use of some random number generator to parse the data into training/testing/validation sets. This implementation uses Kohonen's SOM as the mechanism for dataset selection.

Since the available data was derived from a single test flight, a finite set of data vectors were available. The set was constant, that is, this was not an adaptive environment. The boundary regions of the data were also non-regular, since there were certain flight conditions that could not be met, such as 35 degrees of flap at mach 0.8. In addition, the data was subject to degradation of the sensors used during the flight test. However, the extent to which this degradation might have occurred is indeterminable for slow sensor degradation.

Use was made of Kohonen's SOM for test and training set selection, where the m-dimensional input vector is compressed to a 2-dimensional output vector representing a locational coordinate pair. The activation level for each Kohonen layer neuron is determined by a measure of the distance between the input vector and the vector of weights leading from the input neuron to the Kohonen layer neuron. The distance metric used was the Euclidean measure *d(k)*,

$$d(k) = \min_j \left[ \sum_{n=1}^{N} \left( x_n(k) - w_{n,j}(k) \right)^2 \right] \quad (1)$$

where $x_n(k)$ is the $k^{th}$ vector presented to input neuron n, and $w_{n,j}(k)$ is the connection weight between input neuron n and Kohonen neuron j for input vector k, and N is the dimensionality of the input vector. The number of neurons within the Kohonen layer is set to the desired number of vectors in the training set for the predictive network, with the constraint that the Kohonen layer be rectangular in shape, and preferably square. The implementation software (Neuralware's Professional II/Plus) allowed for a modified winning neuron strategy in the Kohonen layer, where the three neurons with the highest activation experience a weight adjustment that is a function of the magnitude of their normalized activation's. Network convergence has occurred when the three winning neurons are adjacent to each other for a unique presentation of every input vector in the training set. The network in then tested with the full set of test and training vectors, and the output coordinate pairs used to form clusters of data vectors. A single vector can be selected randomly from each cluster and assigned to the training set, with the balance of the

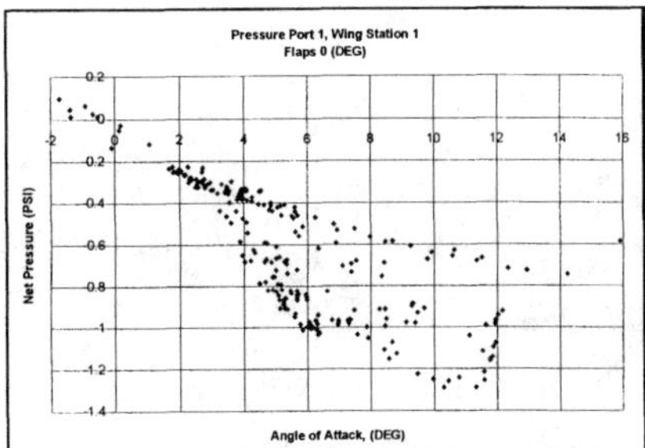


**Figure 1: Port pressure versus Angle of Attack. Variations in pressure at each aoa can be attributed to load factor $n_z$**

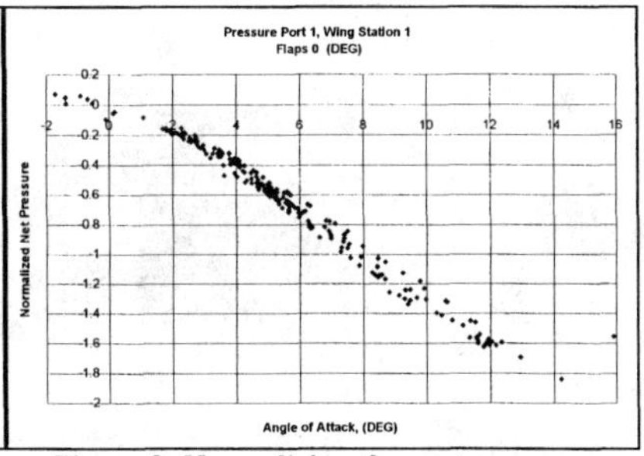


**Figure 2: Normalizing the port pressure with respect to dynamic pressure linearizes the relationship between aoa and port.**

vectors within each cluster being assigned to the test set. A common problem encountered in industry is that of conflicting and redundant data. Common causes for this are an incomplete set of input variables that explicitly model the output, or the lack of unique solutions due to linear dependencies between input vectors. Examination of the clusters formed by the Kohonen network clearly reveals redundant data, which can be deleted. The presence of conflicting data is also almost invariably revealed, with the frequency of occurrence serving to indicate the lack of a suitable set of descriptive input variables for the predictive network.

### Predictive Network Implementations

All predictive parts of this implementation were accomplished using the standard backpropagation network, which seeks to reduce the network output error $e_j[s]$.

$$e_j^{[s]} = f'(I_j^{[s]}) \cdot \sum_k (e_k^{[s+1]} \cdot w_{kj}^{[s+1]}) \qquad (2)$$

where $f'(z)$ refers to the derivative of the transfer function used in layer *s,* and $w_{kj}$ refers to the weighted connections from layer *[s+1]* to layer *[s]*. Initial explorations of appropriate backpropagation network topologies were performed using the Cascade-Correlation Learning paradigm [Fahlman, 1990], which led to a network topology with four inputs, a single hidden layer with ten processing elements, and one or more output neurons. This topology performed satisfactorily whether predicting for a single pressure port or a group of pressure ports, not surprising since adjacent ports had pressure distributions that were similar in shape, but shifted due to their location on the curved leading edge surface. The Extended-Delta-Bar-Delta learning rule was used [Minai, 1990], with a hyperbolic tangent transfer function

$$f(x) = \frac{e^x - e^{-x}}{e^x + e^{-x}} \qquad (3)$$

The network was considered to have converged when the decrease in root mean square error of the test set fell below 2.5 percent of what it was in the prior learning iteration.

The first prediction attempt focused on trying to train the network to recognize pressure recordings for each transducer across all flap settings, without much success. The pressure data for each transducer was subsequently broken up into quartiles, and cross tabulations run on flap setting versus pressure quartile for each transducer. A typical mosaic plot [SAS JMP] is shown in figure 3. It was evident that there was a marked contrast between quartiles

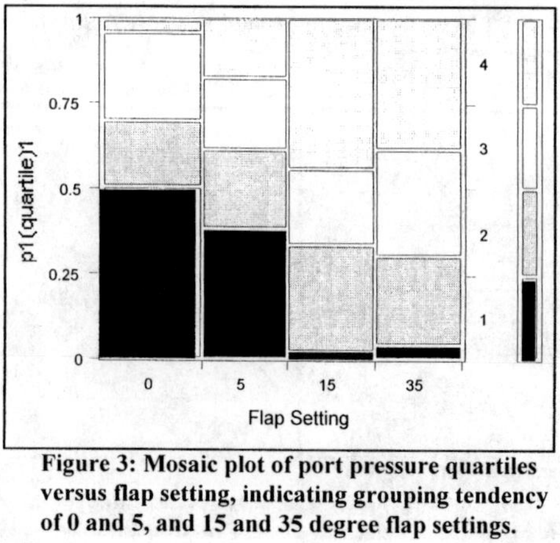


**Figure 3: Mosaic plot of port pressure quartiles versus flap setting, indicating grouping tendency of 0 and 5, and 15 and 35 degree flap settings.**

for flap settings 0 and 5 degrees, versus those for 15 and 35 degrees. The chi-square values bore this out. A less significant difference was noted between quartile proportions for 0 degrees versus 5 degrees flap.

The results suggested partitioning the data for each transducer into at least two sets, one for 0 and 5 degree flap settings, and the other for 15 and 35 degree flaps. The Kohonen SOM for dataset selection was used within each set to develop training and testing data for the predictive networks.

There was significant correlation evident between some of the potential input variables, where the term "significant" is taken to mean greater than 60 percent correlation [Johnson, 1996]. The significant correlation between $n_X$ and $n_Z$ is explained in the following way: as $n_X$ changes, airspeed and dynamic pressure will change, since dynamic pressure is directly proportional to the airspeed squared. Lift is directly proportional to dynamic pressure, and $n_Z$ is a function of lift. Therefore, a change in dynamic pressure will result in a change in $n_Z$. This was part of the problem with using unbiased estimators for pressure in a least squares solution, when the physics of the problem dictated the use of the linearly dependent variables. The solution also decayed substantially when either $n_X$ or $n_Z$ was deleted as an independent variable.

## RESULTS AND DISCUSSION

### Predicting for a Single Port

Initial efforts focused on training a backpropagation network to predict port pressures from a single port, with data partitioned into either 0 and 5 degrees, or 15 and 35 degree flap settings.

Figure 4 shows the output from a network predicting net pressure of one port at the inboard section of the slats. One or two residuals were always very large relative to the others, and were traced back to where the rate of change of aircraft attitudes in any one of the three major axes reached very large values.

### Predicting Simultaneously for Multiple Ports

The second focus was to train a backpropagation network to simultaneously estimate pressures for more than one port, where the grouped ports were selected based on a high positive correlation between them. The training set usually encompassed 75 percent of the available data vectors. The number of iterations to convergence of the network's typically centered around 200,000.

Figure 5 shows the residuals versus actual port pressure for Ports 1 through 3, wing station 1. In this case, for a flap setting of 0 and 5 degrees, very few residuals approached 0.1 psi, which was typical for ports in this location relative to the leading edge slats. These high residuals were determined to exist in cases where the rate of change of aircraft attitudes rate of change in any one of the three major axes was very high with respect to the norm in the dataset, similar to what was experienced when a network was trained to predict a single port's pressure. In addition, residuals for all ports for that input vector showed marked increases over those generally experienced for those ports.

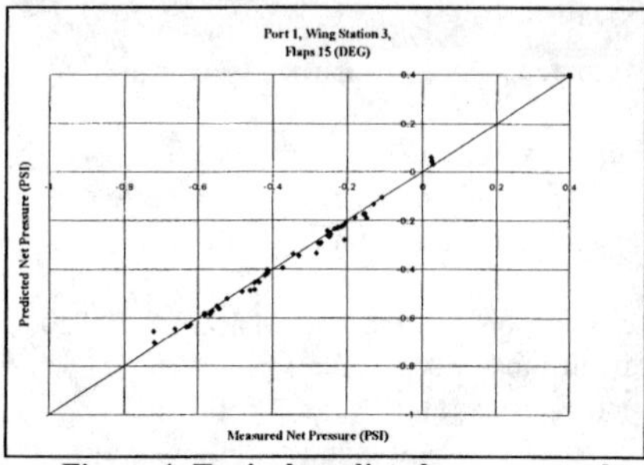


**Figure 4: Typical predicted versus actual port pressures for a single port.**

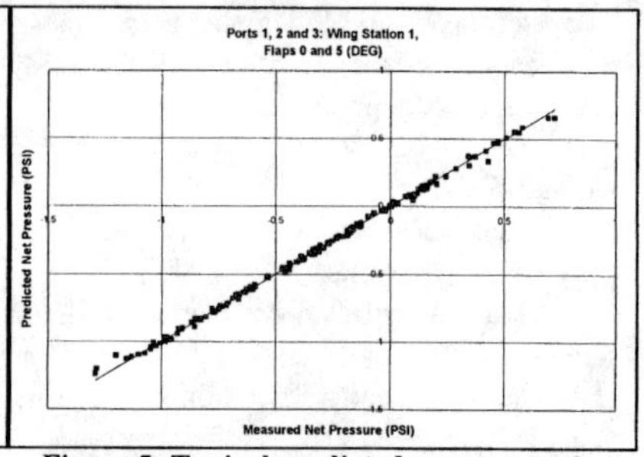


**Figure 5: Typical predicted versus actual port pressures for simultaneous prediction of multiple port pressures.**

## CONCLUSIONS

Neural networks have proven to be successful when used to predict pressure distributions on the leading edge slats of a business jet. It has also been demonstrated that, with careful choice of training data, input variables and network topology, multiple pressure distributions can be predicted with a single network without any appreciable degradation in output quality or precision. This becomes more significant the larger the quantity of dependent variables to be predicted. In this implementation, the number of potential networks was reduced from one hundred and sixty, that is, forty transducers with four flap settings per transducer, to just over thirty by taking advantage of similarities between dependent variable distributions.

Project deadlines did not allow the exploration of recurrent networks for this particular application. Preliminary investigation indicates that time history might be of significant benefit where high rates of pitch, roll or yaw are encountered in the available flight profile data.

## REFERENCES

1. Fahlman, S.E. and Lebiere, C. *The Cascade Correlation Learning Architecture.* Adv. Ner. Inf. Proc. Syst, 2. 1990. Los Altos, CA: pp 524-532

2. Johnson, Dallas. *Applied Multivariate Methods.* Seminar notes, April 1996: Washington D.C.

3. Kohlman, D.L. and Schweikhard, W.G. *Flight Test Principles and Practices.* Seminar notes, University of Kansas, 1989.

4. Minai, A.A. and Williams, R.D. *Acceleration of Back-Propagation through Learning Rate and Momentum Adaptation.* International Joint Conference on Neural Networks. Volume 1. January: pp 676-679.

5. NeuralWare, Inc. *Neural Computing,* 1993.

6. SAS JMP Software, Version 3.1.6. The SAS Institute Inc. Cary, N.C.

# OPTIMISED CONTROL OF MULTIPLE POWER SOURCES USING GENETIC ALGORITHMS, NEURAL NETWORKS AND FUZZY LOGIC

**K. A. OSMAN, A. M. HIGGINSON**
*Faculty Of Engineering And Computer Technology*
*University Of Central England In Birmingham*
*Birmingham B42 2SU, UK*

*ABSTRACT:*

This preliminary study has investigated methods to achieve efficient utilisation of multiple power source systems. A range of methods have been evaluated to solve the optimisation problem and the results have been encapsulate into a form usable in a control strategy. These ideas are applicable to many systems, including control of power generation plant, compressor stations and automobile powertrain.

## INTRODUCTION

The problem of controlling multiple power sources is presented here in a generic way, where a given output power demand must be satisfied in the most efficient method, by using one or more power sources in combination. This is a constrained optimisation problem, where the desired output power (the constraint) must be attained while minimising the total input power. The optimisation was approached using a sophisticated constrained optimisation algorithm based on SQP and compared with results obtained from real valued genetic algorithms.

A subsequent implementation phase considered how the results could be incorporated into a conventional control strategy using a variety of methods including simple feedforward neural networks and fuzzy logic using a Sugeno type FIS, [Sugeno, 1985, Jang, 1993]. The results from this aspect of the work will not be reported here for reasons of brevity but will form the basis for a separate paper.

### Problem Definition

A generic problem was formulated, initially using two power sources, but subsequently extended to five, so the results could be more readily displayed and interpreted. A non-linear transfer function exhibiting a minimum activation energy and high level saturation was defined in the form of a log-sigmoid, with two constants, the maximum power, $P_{max}$ and power offset, $P_{offset}$:

$$P=P_{max}./(1+exp(-(P_{in}+P_{offset}))) \qquad [1]$$

The responses of the power sources can be seen in Figure 1 and displayed as a surface in Figure 2. It is clear that $P_1$ alone should be used for a power demand less than 15, and, $P_2$ alone represents the best solution. between 15 and 22. However, for demands exceeding 22, both sources must be used and the outputs combined to

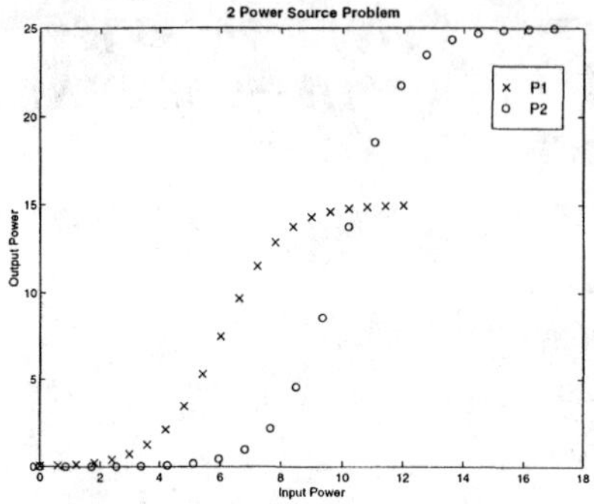


**Figure 1: 2 Power Source Responses**

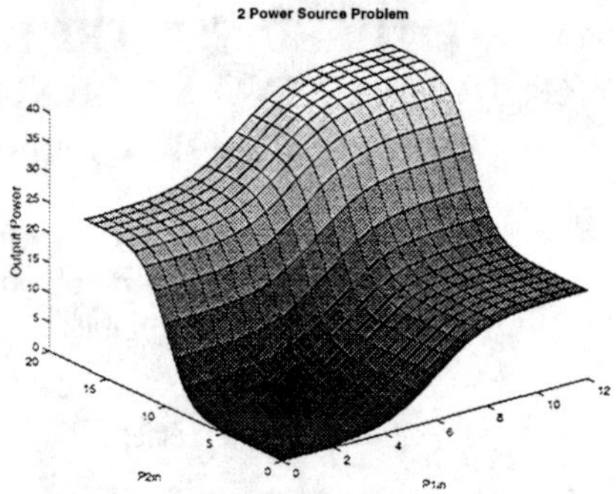


**Figure 2: Power Source Response Surface**

match the demand, while minimising the total input power.

## THE OPTIMISING PROCESS

This identified the optimum combination of output powers to satisfy a given output power demand, while minimising total input power. The problem was solved at a series of discrete points that spanned the total output range. The input power was minimised while achieving a given output demand, as this was more appropriate to the end use in a control system, where the demanded output power should be achieved in the most efficient way. No attempt has yet been made to account for the dynamic responses of real devices and the transfer characteristics were assumed to be independent and time invariant.

### Constrained Optimisation

The constrained optimisation routine, CONSTR, contained in the MATLAB Optimisation Toolbox was used for this work. This finds the minimum of constrained multivariable functions using an SQP method, where a quadratic programming (QP) sub-problem is solved at each iteration [Schittowski, 1985]. An initial starting point must be provided and upper and lower bounds can be imposed on the solutions [Grace, 1995].

### Genetic Algorithm

The genetic algorithm (GA) is a stochastic search method that mimics some of the selection processes that occur in biological evolution, and adopts a global search strategy using a population of potential solutions, rather than a line or gradient based search. Although the first GAs used binary string coding [Goldberg, 1989], real-valued coding can offer a number of advantages [Wright, 1991, Michaelwicz, 1992]. The fitness of each solution is assessed by an objective function, and highly fit individuals have a higher probability of selection for breeding, crossover is used to exchange genetic information to produce, on average, fitter individuals. Point mutation is used at a very low level of background operation to ensure that the probability of searching a particular subspace of the problem is never zero. The GA Toolbox for MATLAB from the University of Sheffield was used for this work [Chipperfield, 1995].

## SET-UP OF THE OPTIMISATION PROBLEM

A vector of 30-50 uniformly spaced output power points was defined to span the entire output power range. The chromosome coding for the GA was set-up in the same way as the solution vector for CONSTR. In the two power source problem the three element potential solution vector appeared as [$P_{1out}$ $P_{2out}$ $P_{demand}$]. Constrained optimisation problems generally take the form *minimise f(x) subject to g(x)<=0,* where g represents the matrix of constraints and f(x) represents a scalar function of x. In this case, f(x) was the value of total input power for the combination of power source outputs specified in x, and g(x) was the power output error, $P_{demand}$-$\Sigma P_{out}$. $P_{demand}$ was constant at each point in the output power vector, and $P_{1out}$ and $P_{2out}$ were varied by the GA or CONSTR, and were bounded for each output demand to improve the search efficiency.

### GA Set-Up

The GA fitness function combined the total input power and output demand error into a single merit figure, defined in the following way:

$$\text{Fitness} = 100 - [\, \Sigma P_{in} + \text{ErrorWeight} \times \text{abs}\,(P_{demand} - \Sigma P_{out})] \qquad [2]$$

After initial experimentation, a value of 3 was used for the ErrorWeight term. A standardised GA set-up was used, with an initial population of 100 chromosomes, divided into 5 equal sub-populations. 50 generations only were allowed, and a migration rate of 0.2 was allowed every five generations. Individuals from each sub-population were selected using stochastic universal sampling. Although 50 generations is a small number, it allowed the overall effectiveness of the GA method to be investigated.

### CONSTR Set-Up

The termination of the optimisation was defined by a maximum number of iterations or by attainment of limits on precision of the independent variables, x, the precision required in the objective function, f and the worst case acceptable constraint violation of g. The maximum number of iterations was set to be 100 times the number of independent variables, and in most cases termination was due to satisfying one of the other three constraints, except where problems with singularity or poor scaling were encountered. The limits on f, x and g were standardised to 0.01, providing a compromise between accuracy and excessive iteration. Four different sets of initial conditions were used for CONSTR:

1. Initialise all outputs to their minima
2. Initialise all outputs to their maxima
3. Set the largest power-source output to the demand or 0.9 of its maximum, then use the next largest etc.
4. Set the smallest power-source output to equal the demand or 0.9 of its maximum, then use the next largest power source

The situation to be avoided is for the start point to provide an initial solution i.e. where $\Sigma P_{out} = P_{demand}$ and then try to minimise.

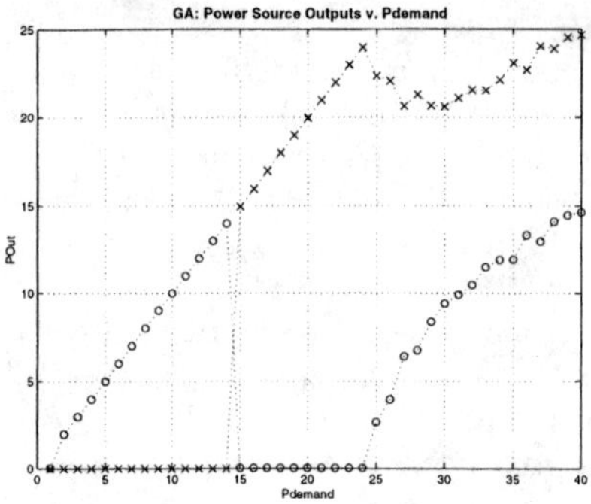


Figure 3: GA Results 2 Sources

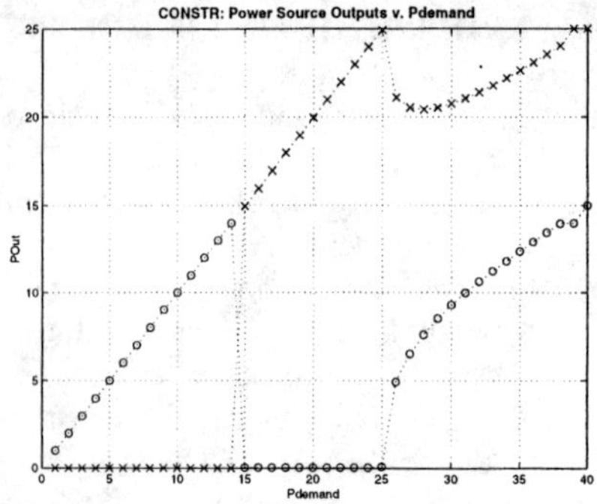


Figure 4: CONSTR Start Condition 4

## RESULTS

### Two Power Sources

The results from the GA are shown in Figure 3, where the output power from each source is plotted against total output demand. As expected, $P_1$ is used until close to its saturation region when $P_2$ is used until close to its saturation, after which both are "mixed" up to a maximum demanded power of 40. The output error is very close to zero for the entire range of output powers. The best result from CONSTR is shown in Figure 4 for start condition four. Changing the start conditions had a dramatic effect on the solutions obtained, and the GA performance was equalled by CONSTR only when type four initial conditions were used, i.e. an understanding of how the problem should be solved was needed during initialisation.

### Three Power Sources

An additional, smaller power-source was added to the problem. This could be used at intervals in the output range to improve efficiency. The results from the GA can be seen in Figure 5, which shows a performance similar to that expected. $P_1$ is used first, then $P_2$, $P_3$ and then combinations of $P_1$, $P_2$ and $P_3$ to achieve the maximum output power. The results from CONSTR were more problematic, with increased sensitivity to start conditions and problems with poorly conditioned or singular matrices, resulting in some unsolved values. These points in the output power vector have been omitted from the figures. The best results, again from start condition 4 are shown in Figure 6, and are close to the GA results. As further sources were added, the GA still performed well, but the constrained optimisation encountered problems at many more points in the output space and was again very dependent on initial starting conditions.

### Rule Based Scheduling

A simple rule based control was also implemented, defining a maximum percentage allowable output for each power-source to avoid operating in the inefficient saturation region. A comparison of the total input power required for the two rule based methods and the GA is shown in Figure 7 and in enlarged form in Figure 8, where the superiority of the GA, especially at demands above the maximum of $P_2$ is clearly apparent. In the two power-source problem, the best CONSTR solution

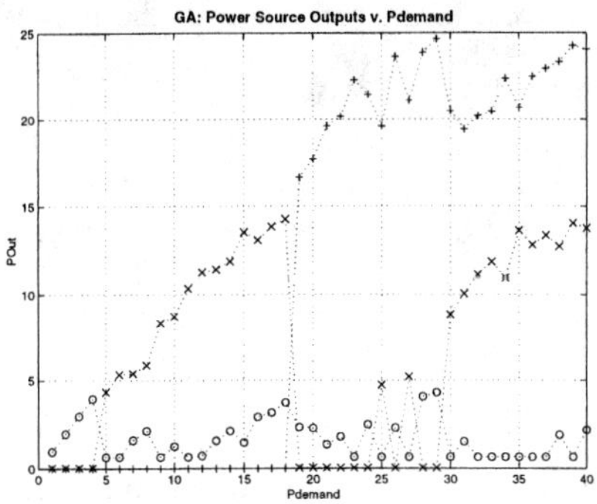


**Figure 5: GA 3 Power Sources**

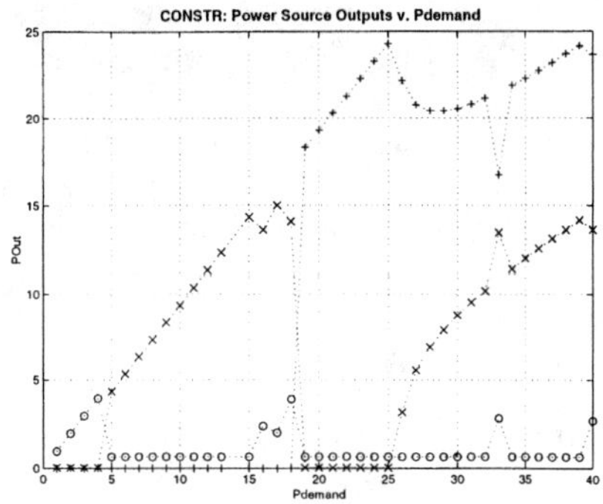


**Figure 6: CONSTR Start Condition 4**

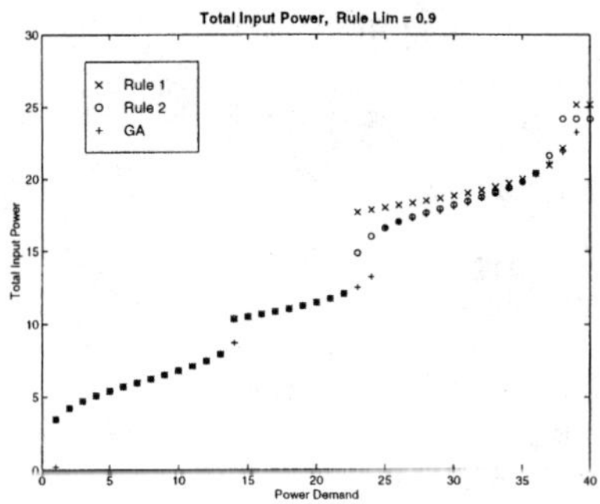


**Figure 7: GA v. Rule Base**

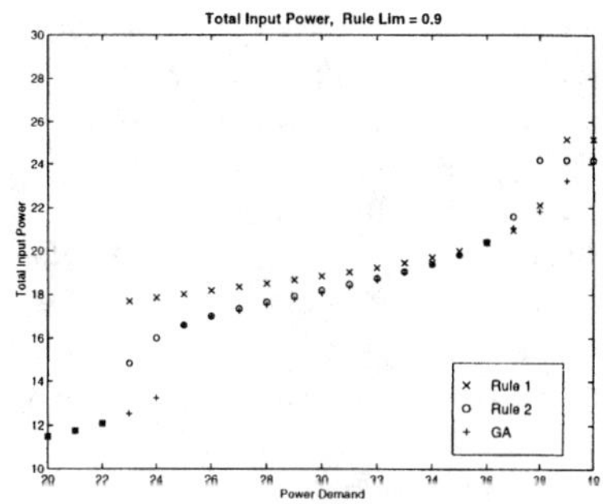


**Figure 8: Enlarged View of Figure 7**

matched the GA solution, so this has not been plotted to avoid confusion.

As the number of power-sources increases and the transfer functions become more complex constructing the rule base becomes increasingly difficult, and it is believed that a rule-based controller is impracticable except for very simple cases of this form of problem.

## DISCUSSION

The GA was able to find solutions that were intuitively correct and solved the optimal mixing of the power sources effectively. Both GA and CONSTR produced a similar solution for the two power-source problem. The fittest solution only was selected from the GA solutions returned, and the fitness function did not include "non-smoothness" of response. It is possible that one of the almost as fit solutions may provide a smoother output response, which may also improve with an increased population size and more generations, at the expense of increased computation. The GA should be able to equal the performance of the constrained optimisation, provided it is given sufficient time and that the fitness function is defined appropriately. The GA was rather slow, typically 20 seconds per output power point compared with 0.5- 2.0 seconds for CONSTR. However, in some cases, especially with more complex problems, CONSTR failed to find either a good solution or any solution due to ill-conditioned or singular matrices, and great care was needed with the selection of the start point, due it is believed, to the saturation of the power-source responses.

The responses used were artificial and smooth, but in reality may be empirical data. While the GA would have little trouble with this, the constrained optimisation may find more problems. It is therefore likely that the GA will provide a more robust solution where real device data are used. One possibility is to combine both approaches by using the GA to identify likely optimal local search regions for CONSTR, which would then carry out the more localised searches.

## IMPLEMENTATION

Having solved the optimisation problem, the results in a variety of forms suitable for a control system, using Neural Networks, especially MLP and RBF, Fuzzy Rule bases especially ANFIS, and a simple linear interpolation method. A comparison of the individual system performance was simulated using the MATLAB SIMULINK toolbox, into which the trained ANN blocks, ANFIS and rule-based systems were readily incorporated. These results will not be reported here for reasons of brevity.

## CONCLUSIONS

This work has shown that real-valued GAs are capable of solving a wide range of quite complex optimisation problems and that the use of sub-populations with migration improves search efficiency. The mathematically based method was less robust and required considerable care with initial starting conditions, and was affected by badly scaled matrices, singularities, etc. The GA is immune to such problems as it makes no use of properties of the underling function, such as partial derivatives. Therefore the GA may provide the better tool for complex optimisation when real device performance is modelled from empirical data, at the expense of increased computation time.

## REFERENCES


Chipperfield, A., Fleming P., Pohlheim H., Fonseca, C., "Genetic Algorithm Toolbox Users Guide", Dept. Automatic Control and Systems Engineering, University of Sheffield, UK

Goldberg, D.E., "Genetic Algorithms In Search, Optimisation And Machine Learning", Addison Wesley, January 1989.

Grace, A., "Optimisation Toolbox Users Guide", The Mathworks Inc., November 1995

Jang, J.-S.R., "ANFIS: Adaptive-Network Based Fuzzy Inference System", IEEE Transactions On Systems, Man And Cybernetics, Vol. 23, No. 3, Pp. 665-685, May 1993.

Michaelwicz, Z., "Genetic Algorithms + Data Structures = Evolution Programs", Springer Verlag, 1992

Sugeno, M., "Industrial Applications Of Fuzzy Control", Elsevier Science, 1985

Schittowski, K., "NLQPL A Fortran Subroutine Solving Constrained Non-Linear Programming Problems", Operations Research, Vol. 5, 485-500, 1985

Wright, A.H., "Genetic Algorithms For Real Parameter Optimisation", Foundations Of Genetic Algorithms, Morgan Kauffman, 205-218, 1991

# GENETIC GENERALIZED BAR-DELTA NEURAL NETWORK FOR ASSESSING POWER SYSTEM TRANSIENT STABILITY

**M. Moechtar, T.C. Cheng, and X. Jin**
*Department of Electrical Engineering*
*University of Southern California, Los Angeles, CA 90089-0271*

**A.S. Farag**
*Department of Electrical Engineering*
*King Fahd Univ. of Petroleum & Minerals, Dhahran, Saudi Arabia*

*ABSTRACT:*

The goal of this paper is to develop a method that uses neural network and genetic algorithm to predict power system transient stability. To justify the use of neural network, we prove that it is capable of approximating any nonlinear functions. Similarly for genetic algorithm, we prove its selection property converges to a near optimal solution. We also develop a generalized bar-delta (GBD) rule that excels the backpropagation (BP) for training of neural network. Finally, we establish a genetic and generalized bar-delta (GGBD) neural network as an intelligent predictor.

## INTRODUCTION

Most of the power system transient stability analytical tools assess, particularly in the post-fault condition, the critical clearance time (CCT). This parameter constitutes a very complex functional relationship between the pre-fault condition, type and location of fault, and the clearance sequence. In [1], we reviewed the available methods for evaluating that parameter. One of the methods uses the multilayer perceptron (MLP) neural network with sigmoidal nonlinearity and backpropagation (BP) learning algorithm. Two main attractive features of this approach are its functional approximation capability and learning ability. However, this network often poorly performs the generalization and is slow in learning.

To overcome those problems, we propose a modification on BP learning algorithm, called generalized bar-delta (GBD) rule. It excels the BP algorithm in both generalization and training time, while maintaining the approximation to arbitrary accuracy. Then, we incorporate genetic algorithm (GA) to optimize the set of input variables for training the GBD network. In doing so, this combine network, called genetic and generalized bar-delta (GGBD) neural network, can be expected to satisfy the criteria for fewer input variables, faster training and more accurate recall.

To verify our proposed approach, we apply it to predict the CCT values of a four-machine power system. More details on this power system, and how to select the features for input data set were presented in [2]. Then, we validate our propose approach by comparing the GBD net to BP net, and to GGBD net as well. Either GBD or GGBD nets estimated the CCT values within an arbitrary accuracy.

## APPROXIMATION CAPABILITY OF NEURAL NETWORK

The approximation capability of a MLP neural network had been investigated by many researchers [3-5]. Basically, any continuous function $f(x)$: $U \subset \mathrm{R}^n \rightarrow \mathrm{R}^m$ can be uniformly approximated to within an arbitrary accuracy by an $\hat{f}(x)$ on $U$, and formulated as follows:

$$\left| f(x) - \hat{f}(x) \right| < \varepsilon \tag{1}$$

and the approximated function is defined as

$$\hat{f}(x) = \sum_{i=1}^{n} w_i \sigma\left(\sum_{j=1}^{m} a_{ij} x_j + b_i\right) \tag{2}$$

where $x_j$ is the input data, and $x_j \in U$; $w_i$, $a_{ij}$ , and $b_i$ are adjustable parameters, and $\sigma(\ )$ is the activation function. In [2] we extended this results as follows:

$$\left| f(u) - \sum_{i=1}^{n} w_i \sigma\left( \sum_{j=0}^{m} a_{ij} u(x_j) + b_i \right) \right| < \varepsilon \tag{3}$$

where $U$ is a compact set in $C[p, q]$, a space of functions; and for any $\varepsilon > 0$, there exists $m + 1$ points $p = x_0 < x_1 < \ldots < x_m = q$ such that $u(x_j\ )$ can be defined as$(u(x_0),\ldots,u(x_m))$; $u \in U$.

The significance of our extension is the fact that it discusses approximation to continuous functionals defined on some compact set in $C[p, q]$ (a space of infinite dimensions). While most of the papers published [3-5] discuss the problem of approximating a continuous function defined on some compact set in $\mathrm{R}^n$ (a space of finite dimensions).Therefore, it is able to deal with a more difficult and complicated problem. It is also not only solve the representation capability by multi-layer neural nets, but also give an explicit form of the approximant. These cannot be obtained by the Stone-Weierstrass theorem, which forms the basis of other papers [6].

## GENERALIZED BAR-DELTA LEARNING ALGORITHM

In this section, we describe the learning algorithm and our proposed modification. Learning in the sense of neural networks is the adjustment of the connection weights between nodes in the adjacent layers. Thus, during learning those connection weights are adjusted such that the error between the desired output and the approximated output can be minimized.

There are many learning algorithms had been proposed for a MLP neural network. Among the most elegant heuristic techniques is the one proposed by

Rumelhart et al. [7], i.e., momentum heuristic, which modifies the learning equation to be

$$\Delta w_{kj}(n) = -\eta \frac{\partial E}{\partial w_{kj}} + \mu \Delta w_{kj}(n-1) \tag{4}$$

where $\mu$ is a momentum factor. However, this algorithm uses a uniform learning rate for all weight, and step size in each direction is proportional to the slope along that direction. Descent along that direction can vary greatly in terms of the actual step size, that may cause local minima across the steep slopes to be jumped over. It also leads to very slow descent along shallow slopes. To overcome this problem, Jacobs [8] proposed the Delta-Bar-Delta (DBD) algorithm. Although the DBD algorithm does provide significant speed-up, it suffers from certain drawbacks. One of them is that using momentum along with DBD can sometimes cause the search to diverge drastically. To improve the performance of this algorithm, we proposed a more general algorithm, called generalized bar-delta (GBD) rule [2]. The followings are the descriptions of our proposed algorithm:

- the weight changes

$$\Delta w_{kj}(n) = -\eta_{kj}(n) \frac{\partial E(n)}{\partial w_{kj}} + \mu_{kj}(n) \Delta w_{kj}(n-1) \tag{5}$$

- the learning rate:

$$\eta_{kj}(n) = \min[\eta_{max}, \eta_{kj}(n-1) + \Delta\eta_{kj}(n-1)] \tag{6}$$

with its change is defined as

$$\Delta\eta_{kj}(n) = \begin{cases} \kappa_\eta e^{(-\beta_\eta |\bar{\delta}_{kj}(n)|)} & \text{if } \bar{\delta}_{kj}(n-1)\delta_{kj}(n) > 0 \\ -\phi_\eta \eta_{kj}(n) & \text{if } \bar{\delta}_{kj}(n-1)\delta_{kj}(n) < 0 \\ 0 & \text{else} \end{cases} \tag{7}$$

- the momentum:

$$\mu_{kj}(n) = \min[\mu_{max}, \mu_{kj}(n-1) + \Delta\mu_{kj}(n-1)] \tag{8}$$

with its change is defined as

$$\Delta\mu_{kj}(n) = \begin{cases} \kappa_\mu e^{(-\beta_\mu |\bar{\delta}_{kj}(n)|)} & \text{if } \bar{\delta}_{kj}(n-1)\delta_{kj}(n) > 0 \\ -\phi_\mu \mu_{kj}(n) & \text{if } \bar{\delta}_{kj}(n-1)\delta_{kj}(n) < 0 \\ 0 & \text{else} \end{cases} \tag{9}$$

where $\delta_{kj}(n)$ is the gradient descent of error, and $\bar{\delta}_{kj}(n)$ is basically an exponential decaying trace of gradient values.

## GENETIC SELECTION CAPABILITIES

Since Holland's seminal work [9], genetic algorithm (GA) had been investigated recently for solving optimization of a large and complex space in an adaptive way. The adaptation was guided by the equivalent biological evolution mechanisms of selection (differential reproduction), crossover, and mutation. The effects of each should be treated separately, then combined to produce a single equation, a well-known Fundamental Theorem of Genetic Algorithms:

$$n(S, t+1) \geq [n(S, t) f(S) / f(P)]\{1-P_c[\delta(S) / (l - 1)] - o(S)P_m\} \qquad (10)$$

where $n(S, t)$ = number of representative of schema $S$ at time $t$; $f(S)$ = average fitness function for schema $S$; $f(P)$ = average objective function over the population; $P_c$ = probability of a crossover occurring on a mating; $\delta(S)$ = defining length of schema $S$; $l$ = string length; $o(S)$ = order of $S$ (number of 1 or 0 positions); $P_m$ = probability of a mutation at a given bit position.

Having establishing the fundamental theorem, next we establish the fundamental concept and theoretical aspects of one of the most powerful genetic process, i.e., genetic feature selection. This capability is very relevant to solve a problem with no *a priori* knowledge of its multitude of available features (variables). There are three potential benefits in applying this genetic feature selection for input to other algorithm, or network. First, the performance of the network can be improved by reducing the possible inputs to a set of relevant uncorrelated variables. Second, the use of smaller set of features not only reduces the network complexity, but also reduces the processing time needed to produce the features, thus speeding up the response time of the system. Third, reduction of feature set means that smaller off-line training sets can be used with all the ensuing benefits in terms of data collection and training times.

In [2], we proved that increasing average fitness function yields the convergence to global maximum value. In proving our proposed theorem, we defined a population by its probability density function (pdf) where its sequences

$$f_{k+1}(S) = \frac{[f_k(S).f(S)] * f_{w_k}(S)}{\int_F f_k(S).f(S).dS} \qquad k = 0, 1, \ldots.. \qquad (11)$$

yields to increasing average fitness converging to a global maximum value. Namely, $E[f_{k+1}] \geq E[f_k]$ and $\lim_{k\to\infty} E[f_k] = f^*$. In (11), we defined the fitness function $f(S) \geq 0$ for all $S \in F$; and the global maximum value equal to $f^*$ (at possible many locations); then there exists a sequence of $m$-dimensional mutation densities $\{f_{w_k}\}_{k=0}^{\infty}$ if the initial probability mass is nonzero in at least one of such neighborhoods ($f^*$). It is obvious that the fitness function has an essential effect on the performance of genetic operators. In our study we used the negative exponential fitness function

$$f(S) \underline{\underline{\Delta}} \exp[-\tfrac{1}{2}(S - S^*)^{\mathrm{T}} Q(S - S^*)] \qquad (12)$$

where $S, S^* \in F$, and $Q = Q^{\mathrm{T}} > 0$ is a covariance matrix.

## SIMULATION RESULTS

To show the performance of our proposed approach, we used the four-machine power systems as in [2]. Prior to simulate the system in order to generate the training and testing data sets for the neural network, we have to perform a feature selection.

From the immense number of variables that may characterize the behavior of a power system, we require that the chosen set should be able to represent system states, yet be small enough to avoid a large number of neural network parameters.

In this study, the features are defined by direct measurements of power system variables, such as: i) active power of all generators at the instant of the fault; ii) power of the prime movers; iii) electrical angle of all generators at the instant of the fault; and iv) mechanical inertia. A total of 90 simulated cases were performed for this power system. We divided the simulation cases into 63 of them for training, and the rest for the test set.

From a series of runs, we found that the neural network with GBD rule required only 4200 cycles to obtain a better prediction than the BP algorithm that required 20,000 cycles. This is as shown in Figure 1. For both networks, we used the hidden nodes, $r = 8$; learning rates for hidden and output nodes, $\eta_h = 0.05$ and $\eta_o = 0.15$, respectively; and momentum term, $\mu = 0.2$.

After 22 generations with fitness factor of 0.003845, genetic algorithm selected six out of twelve features. We used the uniform crossover for parent selection with probability, $P_c = 0.7$, and probability of mutation, $P_m = 0.05$. The comparison results are as shown in Figure 2.

## CONCLUSIONS

In this paper, a neural genetic intelligent assessor was used to predict the CCT values of a generator in a power system. This assessor, called a "Genetic Generalized Bar-Delta" (GGBD) neural network, was proven to be capable of approximating nonlinear real continuous function on a compact set to arbitrary accuracy.

The GBD rule was also proven to outperform the BP algorithm, in matching the desired multi input-single output pairs. Prior to inputting the data set to train the neural network, it was preprocessed by a genetic variable selector. This variable selector uses a set of genetic operators to guide the set of input variables in the search for the fittest one.

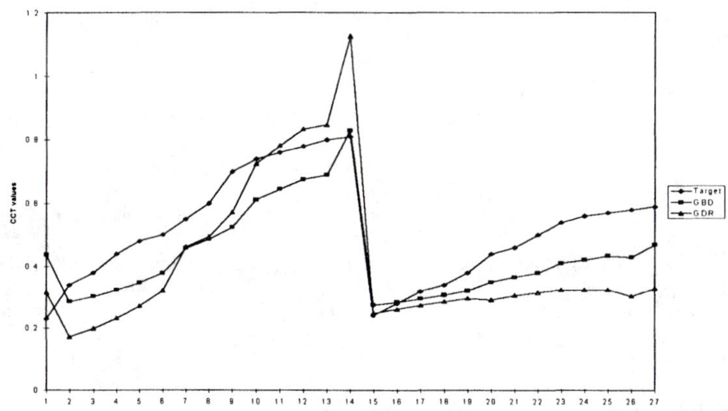


Figure 2: Comparison of BP and GBD rules on network performance in predicting CCT values.

We also observed that: 1) GBD rule outperformed BP algorithm in the sense of generalization and faster training; 2) BP algorithm was proven to be easily trapped to local minima; and 3) GGBD based assessor achieved similar or better performance than the GBD net, despite fewer variables were used to train it. Its only drawback is iterative training, which is impractical for real-time application. However, its quick recall and arbitrary accuracy in approximating a function, may be utilized in other power system applications, or other prediction problem in general.

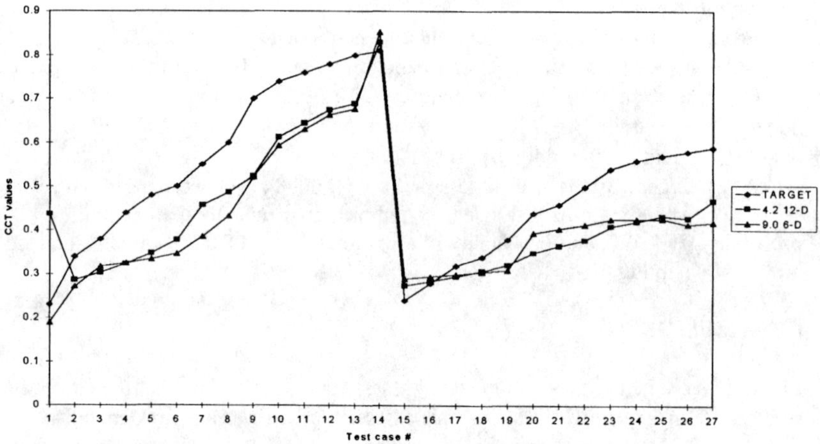


Figure 2: Performance comparison of GBDNN (4200 cycles, $r = 8$) and GGBDNN (9000 cycles, $r = 5$).

## REFERENCES


1. Moechtar, M., Cheng, T.C., and L. Hu, (1995). Transient stability of power system - a survey, *1995 WESCON Conference Record (WESCON / 95)*, Nov., 166-171.
2. Moechtar, M., (1996). Design and analysis of neural genetic predictors for transient stability of power systems, *Ph.D. Dissertation*, Department of Electrical Engineering, University of Southern California.
3. Hornik, K., Stinchcombe, M., and White, H., (1989). Multilayer feedforward networks are universal approximators, *Neural Networks*, v. 2, 359-366.
4. Broomhead, D.S. and Lowe, D., (1988). Multivariable functional interpolation and adaptive networks, *Complex Systems*, v. 2, 321-355.
5. Park, J. and Sandberg,I.W., (1991). Universal approximation using radial-basis-function networks, *Neural Computation*, v. 3, 246-257.
6. Cotter, N.E., (1990). The Stone-Weierstrass theorem and its application to neural network, *IEEE T-NN*, v. 1, n. 4, 290-295.
7. Rumelhart, D.E., Hinton, G., and Williams, G., (1986). Learning internal representations by error propagation, *Parallel Distributed Processing: Explorations in the Microstructure of Cognition, Vol. 1*, D.E. Rumelhart, J.L. McClelland, and the PDP Research Group (Eds.), MIT Press.
8. Jacobs, R.A., (1988). Increased rate of convergence through learning rate adaptation, *Neural Networks*, v. 1, 295-307.
9. Holland, J.H., (1992). *Adaptation in Natural and Artificial Systems*, MIT Press.

# Section 10 : Decision Analysis

# AN AFFINE SCALING SCATTER SEARCH APPROACH FOR CONTINUOUS GLOBAL OPTIMIZATION PROBLEMS

**THEODORE B. TRAFALIS AND SUAT KASAP**

*University of Oklahoma, School of Industrial Engineering*
*202 West Boyd, Room 124, Norman, OK, USA 73019*
*EMail: trafalis@mailhost.ecn.uoknor.edu*

***ABSTRACT:***
The objective of this study is to develop an algorithm to search for the global optimum of a continuous optimization problem. The Affine Scaling Scatter Search Method (ASSSM) is proposed to solve this problem. This method is a combination of an affine scaling method and a scatter search method and it can be considered as a hybrid of an evolutionary programming and an interior point method. The ASSSM will be used for neural network training in the next phase of this study. The development of the algorithm, parameter settings, experimentation, and efficiency of the algorithm are discussed. Preliminary computational results are also reported.

## INTRODUCTION

The objective of this study is to develop an algorithm to search for the global optimum of a continuous optimization problem. The development of the algorithm, parameter settings, experimentation, and efficiency of the algorithm will be discussed.

The problem to be considered in this paper is the search a global minimum of a general nonlinear function, *f(x)*, which is defined within a convex set $D \subseteq \Re^n$. This problem can also be stated as follows:

$$\text{find a point } x^* \in D \text{ such that } f(x^*) < f(x), \ \forall x \in D \qquad (1)$$

The solution methodologies to solve this problem are classified into two categories as deterministic and stochastic methods. Deterministic methods attempt to create sequences of successive approximations that converge to points that satisfy the local optimality criteria. They are efficient when the starting point belongs to a close neighborhood of a global minimum. Stochastic methods attempt to cover efficiently the entire space so that all the local optima will be identified, and hopefully one of them will be a global optimum. They try to identify either a collection of points or convergent sequences of points whose limits satisfy some property characterizing an optimum solution. The main difference between deterministic and stochastic methods is that the stochastic methods do not strictly improve the objective function.

Deterministic and stochastic methods have been developed and used by researchers to solve the problem above. Pardalos and Rosen (1987) summarize work on constrained global optimization deterministic solution methods with special emphasis on concave minimization problem and their applications. Recent advances in global optimization and a set of test problems are collected by Floudas and Pardalos (1990, 1992) in another study. Horst et al. (1995) give an introductory comprehensive study to cover the fundamentals of deterministic global optimization techniques. Interior point methods for global optimization are reviewed by Pardalos and Resende (1996). Some extensive review studies of the heuristic techniques for stochastic methods are described by Kan and Timmer (1989), and Törn and Zilinkas (1989) and edited by Reeves (1993).

In this paper the Affine Scaling Scatter Search Method (ASSSM) is proposed to solve the problem above. This method is a hybrid of an interior point method and evolutionary approach. Specifically it extends an affine scaling method in an evolutionary programming setting by combining benefits of an efficient deterministic interior point and a stochastic method approach. The ASSSM will be used for neural network training in the next phase of this study.

## THE AFFINE SCALING SCATTER SEARCH METHOD (ASSSM)

In this paper, constrained and unconstrained cases of global optimization problem will be considered. The unconstrained problem deals with the optimization of a given function without any constraints. The constrained problem deals with the optimization of a given function subject to a given set of constraints. In order to solve these problems, a new method, ASSSM, is proposed. This method is a combination of an affine scaling method (Dikin, 1967) and scatter search (Glover, 1994).

### Affine Scaling Method

The basic affine scaling method was first presented by Dikin. After Karmarkar's (1984) revolutionary interior point approach to solve Linear Programming (LP) problems, the affine scaling approaches have attracted researchers' attention. Later, it was independently rediscovered by Barnes (1986) and Vanderbei et al. (1986). An extensive explanation and basic principles of the affine scaling method are expressed by Fang and Puthenpura (1993) as follows. The method begins with an initial solution, and transforms the feasible space so that the current solution is at the center of the transformed feasible space. Then, it moves in the direction of steepest descent to find a better solution. Thereafter, an inverse transformation is applied to map the improved solution back to the original space. This process is repeated until an optimum solution is obtained within the desired tolerance.

### Scatter Search

Scatter search was introduced by Glover (1994). Recently scatter search was used to solve continuous optimization problems by Fleurent et al. (1995), Trafalis and Al-Harkan (1995, 1996). The basic principles of scatter search algorithm are explained by Trafalis and Al-Harkan (1995) as follows. The algorithm starts with a random solution and uses the search directions to generate a random collection of solutions. Then, it computes the weighted center of gravity by using these solutions and assigns a weight to each solution. Next, a new set of search directions is generated by using the old search directions. The weighted center of gravity is the new solution. The process is repeated until an optimum solution is obtained within the desired tolerance.

**ASSSM for Global Optimization Problems**

The ASSSM begins with an initial '*base point*', and uses the random search directions to generate a collection of '*test points*'. Then, it transforms the base and test points so that the base point is at the center of the transformed feasible space. Then, the new base point is computed in the transformed space and the search directions are updated. Thereafter, an inverse transformation is applied to map the new base point and updated search directions back to the original space. The process is repeated until an optimum solution is obtained within the desired tolerance. The steps of the ASSSM can be described as follows:

- **Step 1:** Generate a base point, $X_0$, and name it as '*base point*'.
- **Step 2:** Generate M test points, $x_i$, and name them as '*test points*'.

$$x_i = X_0 + \mu_i\, D_i, \quad \text{where} \quad i = 1\ldots M \tag{2}$$

  Where $\mu_i$ is step length and $D_i$'s are generated and normalized from a random initial set of M search directions $d_i$'s.

- **Step 3:** Use the following affine scaling transformation to transform base point, $X_0$, into the vector $e$ and test points, $x_i$, into $y_i$.

$$Y_0 = T^{-1} X_0 = e = (1,1,\ldots,1)$$
$$y_i = T^{-1} x_i, \text{ where } i = 1,..,M \text{ and } T = diag(x_0) \tag{3}$$

  Where T is the scaling matrix for the affine scaling transformation.

- **Step 4:** Evaluate the objective function for each of the test points, $f(y_i)$.
- **Step 5:** Assign a weight, $w_i$, to each test point according to its $f(y_i)$.
- **Step 6:** Compute a new point as a *weighted center of gravity*, $Y_{1_1}$, by using linear combination of points, $y_i$, and weights, $w_i$.

$$Y_1 = \sum_{i=1}^{M} w_i\, y_i, \quad \text{where} \sum_{i=1}^{M} w_i = 1 \tag{4}$$

- **Step 7:** Sort the search directions, $D_i$, of each $y_i$ by using $w_i$.
- **Step 8:** Select sorted directions in pairs to create new search directions by performing a convex combinations of the old pairs.

$$D_K^{k+1} = v\, D_K^k + (1-v)\mathrm{D}_\mathrm{L}^\mathrm{k} \quad \text{and} \quad D_L^{k+1} = v\, D_L^k + (1-v) D_K^k, \tag{5}$$

  where $0 \leq v \leq 1$.. Then normalize the new directions.

- **Step 9:** Use inverse affine scaling transformation to transform $Y_1$ and new $D_i$ to the original space.

$$X_1 = T\, Y_1 \text{ and } D_i^* = T\, D_i, \text{ where } i = 1,..,M \tag{6}$$

- **Step 10:**. Check for optimality. If it is satisfied, the algorithm stops with the best solution $X_1$. Otherwise, it sets $X_0 = X_1$ and goes to Step 2.

**Parameter Selection and Implementation**

The initial base point, $X_0$, is uniformly distributed between 0 and 1. Initial search directions, $d_i$, are normally distributed with zero mean and unit variance. The number of directions, M, needed to attempt to cover the whole solution space, are expressed in terms of the dimensionality of the problem.

$$M = (n-1)\,2^3 \qquad (7)$$

where $n$ is the dimension of the problem. To determine the step size, $\mu_i$, line search is used. Weights, $w_i$, are calculated by using the following formula.

$$w_i = \frac{f(x_i)}{\sum_{k=1}^{M} f(x_k)}, \quad \text{where } f(x_i) \geq 0, \;\; i = 1,..,M \qquad (8)$$

When $f(x_i) \leq 0$, problem could be solved by translating all the test points so that a strictly positive objective value exists. To perform the convex combination of the old search directions, three different ν values are selected as 0.75, 0.5, and 0.25. During the test runs, it is observed that for smaller ν values, results are better.

During implementation, selection sort is used, and ν is selected to be 0.25. The following two stopping criteria are used at the same time.

$$\left|x^{k+1} - x^k\right| \leq \varepsilon \;\text{ and }\; \left|\mathrm{f}\left(x^{k+1}\right) - \mathrm{f}\left(x^k\right)\right| \leq \varepsilon \qquad (9)$$

**ASSSM for Artificial Neural Network Training**

Artificial Neural Network (ANN) training can be considered as a nonlinear programming problem. Specifically, minimizing a nonlinear error function between the calculated output and the given desired output with respect to parameter weights. To find the optimum solution to this problem the ASSSM will be used in the next phase of this study which is to develop a fast training algorithm.

**Computational Results**

To test ASSSM, it is coded in PASCAL for IBM compatible PC by using BORLAND TURBO PASCAL 6.0 compiler. Then a series of computational tests are performed by using the following 2-D test problems. The optimum and the experimental solutions of these problems are shown in the table 1 and table 2 respectively.

*Problem 1:*

$$f(x) = 1.41x_1^4 - 12.76x_1^3 + 39.91x_1^2 - 51.93x_1 + 24.37 + (x_2 - 3.9) \qquad (10)$$

*Problem 2:* Goldstein and Price's function (Floudas and Pardalos, 1990)
*Problem 3:* N-Dimensional test function, N=2 (Koon and Sebald, 1995)
*Problem 4:* 2-D Rastrigin function (Koon and Sebald, 1995)
*Problem 5:* Constrained problem, Example 9.4 (Reklaitis et al., 1983)

| TABLE 1: SOLUTIONS OF TEST PROBLEMS | | | |
|---|---|---|---|
| Pr. # | $x_1$ | $x_2$ | f(x) |
| 1 | 3.483 | 3.9 | -3.9872 |
| 2 | 0 | -1 | 3 |
| 3 | -2.90354 | -2.90451 | -78.3323 |
| 4 | 0 | 0 | -2 |
| 5 | 1.2 | 1.6 | 0.2 |

| TABLE 2: EXPERIMENTAL SOLUTIONS OF TEST PROBLEMS | | | | | |
|---|---|---|---|---|---|
| Pr. # | $x_1$ | $x_2$ | f(x) | iter. # | % |
| 1 | 3.48267 | 3.89994 | 3.98717 | 23.7 | 100 |
| 2 | -0.00005 | -1.00003 | 3.00004 | 31.8 | 90 |
| 3 | -2.90361 | -2.90451 | -78.3339 | 16.5 | 100 |
| 4 | -0.00001 | 0.00012 | -1.999987 | 19.1 | 100 |
| 5 | 1.1634 | 1.5817 | 0.2022 | 12.7 | 100 |

During experiments we observe that straddled and cycling around the optimum becomes a problem. The number of iterations for each run are written down without these problems by observing the results of the each iteration.

On table 2, average iteration numbers are computed from the best solution's total iteration number of each cycle. The last column gives the percentage of finding a global minimum. The percentage of obtaining the global minimum using the ASSSM is measured by running the algorithm for ten independent cycles for each problem. Each cycle contains five independent runs and the best of each run is considered the best solution of this cycle for the problems above. For each cycle, if the best solution is around the global solution, we assumed that the global optimum is found.

## CONCLUSIONS AND RESULTS

According to the computational results the ASSSM algorithm is quite effective in identifying the global optimal solutions for the test problems above. To be able to conclude that, the ASSSM is an effective algorithm in identifying the global optimal solutions, we need to do more testing by using higher dimensional problems.

As we mentioned before, straddled and cycling are frequently encountered problems around local minima. By improving stopping criteria, these problems can be solved. Since this is a preliminary study, we did not consider the efficiency of the implementation. But, it is possible to improve the implementation by using efficient searching and sorting techniques.

Comparative studies with other methods should be done to get more realistic ideas about the ASSSM. For future studies, other heuristic techniques, like simulated annealing, genetic algorithms and tabu search should be compared with the ASSSM.

In the next phase of this study which is to develop a fast training algorithm for ANN training, the ASSSM will be used.

## REFERENCES


Barnes, E.R., (1986). A variation of Karmarkar's algorithm for solving linear programming problems, *Mathematical Programming*, 36, 174-182.

Dikin, I.I., (1967). Iterative solution of problems of linear and quadratic programming, *Doklady Akademiia Nauk USSR* 174, 747-748. Translated in English in *Soviet Mathematics Doklady* 8, 674-675.

Fang, S., and Puthenpura, S., (1993). *Linear optimization and extensions: theory and algorithms.* New Jersey, Prentice-Hall Inc.

Fleurent, C. et al., (1995). A Scatter Search Approach for Unconstrained Continuous Optimization, Working Paper, University of Colorado, Boulder.

Floudas, C.A., and Pardalos, P.M., (1990). *A Collection of Test Problems for Constrained Global Optimization Algorithms.* Germany, Springer-Verlag.

Floudas, C.A., and Pardalos, P.M., (1992). *Recent advances in global optimization.* New Jersey, Princeton University Press.

Glover, F., (1994). Tabu search for nonlinear and parametric optimization (with links to genetic algorithms), *Discrete Applied Mathematics*, 49, 231-255.

Horst, R., Pardalos, P.M., and Thoai, N.V., (1995). *Introduction to Global Optimization.* Dordrecht, Kluwer Academic Publishers.

Kan, A.H.R., and Timmer, G.T. (1989). Global Optimization, G. L. Nemhauser et al. Eds., in *Handbook in OR & MS*, 631-662. North-Holland, Elsevier Science Publisher.

Karmarkar, N. (1984). A new Polynomial time algorithm for linear programming, *Combinatorica*, 4, 373-395.

Koon, G.H., and Sebald, A.V., (1995). Some interesting test functions for evaluating evolutionary programming strategies, in McDonnell, J.R., Reynolds, R.G., and Fogel, D.B., Eds., *Proceedings of the fourth annual conference on evolutionary programming*, 479-499.

Pardalos, P.M., and Rosen, J.B., (1987). *Constrained global optimization: algorithms and applications.* Germany, Springer-Verlag.

Pardalos, P.M., and Resende, M.G., (1996). Interior Point Methods for Global Optimization, to appear as a chapter in *Interior Point Methods in Mathematical Programming*, T. Terlaky eds., Kluwer Academic Publishers.

Reeves, C.R., (1993). *Modern heuristic techniques for combinatorial problems.* New York, John Wiley & Sons.

Reklaitis, G.V., Ravindran, A., and Ragsdell, K.M., (1983). *Engineering Optimization Methods and Applications.* New York, John Wiley & Sons.

Törn, A., and Zilinskas, A., (1989). *Global Optimization.* Lecture Notes in Computer Science, Heidelberg, Springer-Verlag.

Trafalis, T.B., and Al-Harkan, I.M., (1995). A continuous scatter search approach for global optimization, Extended Abstract in: *Conference in Applied Mathematical Programming and Modelling (APMOD'95)*, London, UK.

Trafalis, T.B., and Al-Harkan, I.M., (1996). A hybrid scatter genetic tabu approach for continuous global optimization problems, Technical Report, University of Oklahoma, Oklahoma.

Vanderbei, R.J., Meketon, M.S., and Freedman, B.A., (1986). A modification of Karmarkar's linear programming algorithm, *Algorithmica*, 1, 395-407.

# DATA MODELING USING CONSTRAINED CATEGORICAL REGRESSION[†]

**Harvey L. Bodine, Steven S. Henley, and Robert L. Dawes**
*Martingale Research Corporation, Allen, TX 75002 mrc@metronet.com*

**Richard M. Golden**
*School of Human Development, University of Texas at Dallas, Richardson, TX 75083-0688 golden@utdallas.edu*

**T. Michael Kashner**
*UT Southwestern Medical Center at Dallas, 8267 Elmbrook, Suite 250, Dallas, TX 75247-9141*

***ABSTRACT:***
We apply a sparsely-connected neural network to the problem of recognizing statistical regularities and patterns in a National Labor and Alcohol Survey database. The network architecture, called Constrained Categorical Regression (CCR), is designed to identify valid statistical inferences even in the presence of a mis-specified model and offers fast training with guaranteed convergence. Each weight within the network can be tested for statistical significance and the overall network is interpretable for meaning and validity.

## PROJECT OVERVIEW

**A sparsely-connected network architecture that utilizes recent techniques from the field of econometrics for comparative evaluation of data models has been applied to the problem of extracting statistically valid information from longitudinal survey data.** This network architecture is known as Constrained Categorical Regression (CCR), where the name represents the derivation of this network paradigm from classical statistical analysis. The technique for applying the network is known as CCR modeling, which refers to the structured development of a network using a iterative testing-and weight pruning process. The CCR modeling approach allows the development of causal and/or explanatory models that may or may not fit the data, but which can still be tested for statistical validity even when the model is mis-specified.

## NETWORK ARCHITECTURE

**The Constrained Categorical Regression network provides probabilistic estimates for the network's categorical outcomes.** Let S = [S(1), S(2), ..., S(d)] be

---

[†] The authors gratefully acknowledge the support of the National Institute on Alcohol Abuse and Alcoholism. This work is based on data supplied by NIAAA and technology developed for NIAAA under Contract N43AA42008. Any opinions, findings, conclusions or recommendations expressed in this publication are those of the authors and do not necessarily reflect the views of the NIAAA.

a d-dimensional vector representing the values of a particular set of d binary-valued state variables. A heuristic rule is represented by an ordered set [**W**(i),$\phi_1$(**S**)]. The scalar-valued Boolean function $\phi_1$(**S**) =1 if the logical constraints on a given set of state variables are satisfied and $\phi_1$(**S**) = 0 otherwise. Suppose the $j^{th}$ element of the m-dimensional column vector W(i) is equal to 1 and the remaining elements of the vector **W**(i) are zeros. Then this heuristic rule is interpreted as:

IF $\phi_1$(**S**) =1, THEN outcome j is implied.

Let there be R heuristic rules and let $\phi$(**S**) be a $R$ - dimensional column vector whose $i^{th}$ element is $\phi_1$(**S**). Let $\boldsymbol{\beta}$ be a diagonal $R$-dimensional matrix whose $i^{th}$ on-diagonal element is β(i). Let **W** be a $m$ by $R$ dimensional matrix whose $i^{th}$ column is **W**(i). Let the $j^{th}$ element of a vector **h**, h(j), indicate the amount of evidence supporting the presence of the $j^{th}$ outcome. The probability, p(k), of classifying the set of state variables, **S**, as belonging to the $k^{th}$ outcome is computed as follows:

$$p(k) = \frac{\exp[\mathbf{h}(k)]}{\sum_{j=1}^{m} \exp[\mathbf{h}(j)]} \quad \text{where } \mathbf{h} = \mathbf{W}\beta\phi(\mathbf{S})$$

The R free parameters of the CCR model are β(1), β(2), β(3),..., β($R$) where β(i) is interpreted as the influence of the $i^{th}$ heuristic rule in the heuristic rule database with respect to explaining the observed data. All remaining variables in the model are known constants.

The CCR model may be interpreted as a special type of feed-forward neural network model. [Ref. 2, Ref. 8] Let S(v) be the activity level or state of the $v^{th}$ assertion state variable. Let S =[S(1) ... S(d)] be an activation pattern corresponding to a particular pattern of state variables presented to the input of the neural network model. This activation pattern is then mapped via a nonlinear logical mapping using the functions $\phi_1$(**S**), $\phi_2$(**S**),..., $\phi_K$(**S**) into the first layer of hidden units. Each hidden unit in the second layer is "linear" and simply computes a weighted sum of incoming connection strengths and assertion unit activity levels. The connection strength from the $k^{th}$ hidden unit in the first layer of hidden units to the $j^{th}$ unit in the second layer of hidden units is given by the formula: **β**(i)w(j,k) where w(j,k) is the $jk^{th}$ element of **W**. The activation of the $j^{th}$ hidden unit, h(j), in the second layer of hidden units can be interpreted as the evidence supporting the $j^{th}$ value of the outcome variable. The mapping from the second layer of hidden units to the model's output units is a fixed nonlinear mapping which transforms the evidence for the $j^{th}$ outcome variable's value into the probability of the $j^{th}$ outcome. The hidden unit to output unit transformation involves "forward lateral inhibition" which has the effect of decreasing the probability for the $j^{th}$ outcome if evidence supporting alternative outcomes is present. That is, the hidden to output unit transformation implements a necessary and appropriate nonlinear response competition mechanism, as described by Bridle. [Ref. 8]

The maximum likelihood estimates of the parameters of the CCR probability model are unique given a sufficiently large "non-degenerate" data set. [Ref. 3]. Explicit formulas for the maximum likelihood estimates, however, do not necessarily exist so these estimates must be computed by maximizing the likelihood of the data set with respect to the free parameters β(1), β(2), ... β(i) using nonlinear optimization methods. The CCR model algorithms incorporate an explicit formula for the log-likelihood function whose unique global maximum is the desired set of parameter estimates.

The process for achieving the optimum set of parameter estimates is an iterative process, alternating between construction of heuristic rules and evaluation of the Hessian, condition number, prediction rates, and betas produced during parameter estimation and model evaluation. During this iterative process, the connections are pruned from the candidate model, using a within-groups statistical test that defines which asserted rules $\Phi(x)$ can be removed due to low significance. [Ref. 7]

## PROBLEM ADDRESSED

**We utilized the CCR neural network architecture to extract information about causal relationships from a national survey database**. This database contained information relating to alcoholism among the U.S. population. A number of such databases are currently available and an important use of such databases is to identify <u>longitudinal patterns</u> of disease-related symptoms and administrative strategies, as well as medical and psychiatric conditions which effectively predict patient outcomes.

Our project utilized the results of the National Longitudinal Survey - Youth (NLSY) study, which was commissioned by the Department of Labor, with support from the National Institute on Alcohol Abuse and Alcoholism (NIAAA). The NLSY database contains the result of a study of 12,868 young men and women aged 14 to 21 as of January 1, 1979. Since 1979, over 90% of the respondents participated in an annual personal interview approximately one hour in length. Their responses were to a wide variety of questions about their work experience, military experience, training, schooling, family background, marital status, income and assets, health, fertility, child car, drug and alcohol use. For five of the survey years (1983, 1984, 1985, 1988, 1989), specific questions were asked which allowed the Diagnostics Standard Method - Version III (DSM-III) and Version IV (DSM IV) diagnoses to be used.

Survey or epidemiological data is subject to the same exploitation limitations as are found in pattern recognition applications. The interpretation of a database containing disease-related information is limited by the quality and the quantity of the data. However, if appropriate prior knowledge and expectations are introduced directly into the statistical model used to analyze the database, improved statistical inferences can be obtained from the same data set. This rather non-traditional perspective upon statistical inference is now possible given recent advances in the field of econometrics. [Refs. 4, 6. 7]

## ANALYSIS RESULTS

**The final models developed for this project examine the effects of various demographic and economic factors on both male and female sub-populations.** These models are known as FM16 (female subpopulation) and ML14 (male subpopulation). These rulebases were both derived using rule (beta) pruning from the initial 96 rule model. The end result of the model development procedure was an 11-rule model for FM16 and a 6-rule model for ML14. Each model was trained against subset databases (male/female), then tested against the original database. The confusion matrices are shown below in Figure 2. As expected, given the weakly predictive nature of the input stimulus variables, the classification (prediction) performance was relatively low (40.3% - 40.09%). An important note however, is that the CCR model training procedure was *successful at discovering structural* *relationships* between the stimulus variables and the target outcomes, even though the classification was poor.

**FM16 Female Submodel**

| Tgt/Pdct | Dep | Abuse | None |
|---|---|---|---|
| Dep | 14 | 41 | 2 |
| Abuse | 14 | 64 | 20 |
| None | 125 | 313 | 270 |

**ML14 Male Submodel**

| Tgt/Pdct | Dep | Abuse | None |
|---|---|---|---|
| Dep | 1 | 15 | 41 |
| Abuse | 3 | 30 | 65 |
| None | 24 | 369 | 315 |

**Figure 2. Confusion matrices for male and female models for full datasets.**

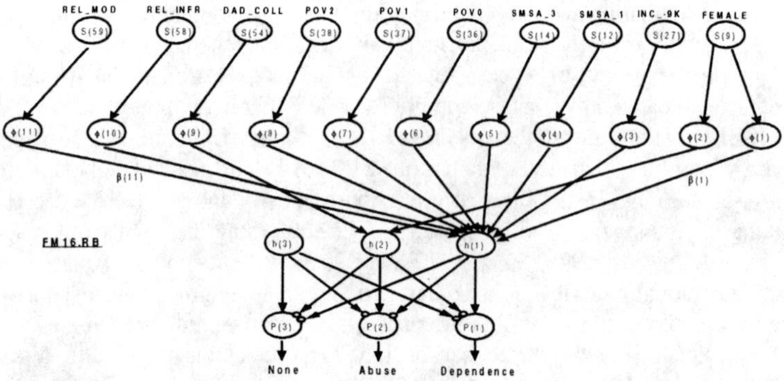


**Figure 3. Female Subpopulation Model (FM16).**

Valid statistical comparisons were made using the CCR modeling software even though the input stimulus variables within the data are not strongly predictive. We were able to compare the relative performance of the models, the effects of the different inputs on the target outcomes, and the effects of a single variable on the likelihoods of the target outcomes. For reference, Figure 3 shows the network diagram of the FM16 alcohol model and Figure 4 below shows the network diagram of the ML14 alcohol model. There are 11 rules in the FM16 model, represented by the $\phi$ units. These units are shown grouped by target output unit in order to simplify the network pictorial. The ML14 model has 6 rules, represented by the $\phi$ units.

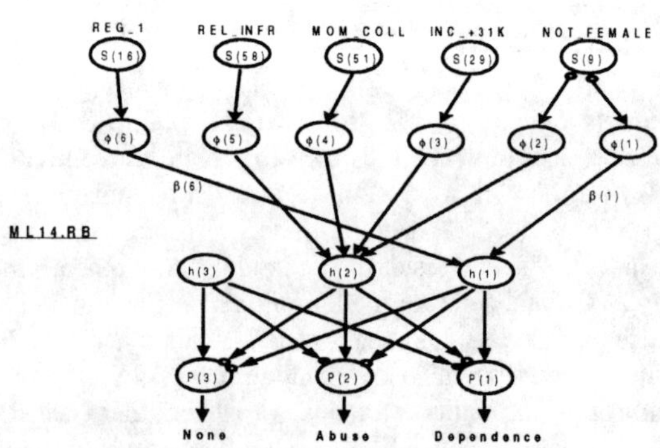


**Figure 4. Male Subpopulation Model (ML14)**

The beta strength and significance level associated with each rule in the rulebase for FM16 is shown in Table I. Note that the significance level of a rule is simply the probability of incorrectly rejecting the null hypothesis that the population beta strength of the rule is equal to zero. Also note that the rules in the table have been rank-ordered according to their significance levels. Rules 1 and 2 are "bias" rules which adjust for the inherent relative probabilities of the three possible outcomes in the lack of other factors. The *increase* or *decrease* in the relative outcome probability is determined by the sign of the beta strength coefficient.

**Table I. FM16 Beta strength and significance levels**

| Beta ID Number | Beta Strength | Sig. Level | Rule Interpretation Using Beta Strength Coefficient Sign |
|---|---|---|---|
| 2 | -2.283 | p = 1E-13 | IF Female RELATES TO OUTCOME = Abuse (Bias Rule) |
| 7 | 1.979 | p = 9E-12 | IF Family Lived in Poverty for at least 1 Year RELATES TO OUTCOME = Dependence |
| 6 | 1.657 | p = 6E-9 | IF Family Lived in Poverty for at 0-1 Years RELATES TO OUTCOME = Dependence |
| 8 | 2.083 | p = 1E-5 | IF Family Lived in Poverty for at least 2 Years RELATES TO OUTCOME = Dependence |
| 3 | 0.870 | p = 0.002 | IF Family Annual Income Decreased (0 to -$9K) RELATES TO OUTCOME = Dependence |
| 4 | 0.817 | p = 0.009 | IF SMSA Location #1 RELATES TO OUTCOME = Dependence |
| 9 | 0.462 | p = 0.014 | IF Father had college education RELATES TO OUTCOME = Abuse |
| 10 | 1.180 | p = 0.025 | IF Infrequent religious attendance at Age 18 RELATES TO OUTCOME = Dependence |
| 5 | 0.810 | p=0.030 | IF SMSA Location #3 RELATES TO OUTCOME = Dependence |
| 11 | 1.103 | p=0.038 | IF Moderate Religious Attendance at Age 18 RELATES TO OUTCOME = Dependence |
| 1 | 0.375 | p=0.554 | IF Female RELATES TO OUTCOME = Dependence (Bias Rule) |

The beta strength and significance level for each rule in the rulebase for ML14 is shown in Table II. For this model there were 6 rules, and as before, rules 1 and 2 are "bias" rules. The beta strength coefficients for all rules in the female rulebase model FM8 were statistically significant at the 0.05 level with the exception of rule 1. All beta strength coefficients in the derived male rulebase model ML14 were statistically significant at the 0.05 level.

**Table II. ML14 Beta strength and significance levels**

| Beta ID Number | Beta Strength | Sig. Level | Rule Interpretation Using Beta Strength Coefficient Sign |
|---|---|---|---|
| 1 | -1.807 | p = 4E-22 | IF Male RELATES TO OUTCOME = Dependence (Bias Rule) |
| 2 | -2.681 | p = 8E-18 | IF Male RELATES TO OUTCOME = Dependence (Bias Rule) |
| 3 | -1.255 | p = 8E-5 | IF Family Annual Income Increased ($31K - $57K) RELATES TO OUTCOME = Abuse |

| Beta ID Number | Beta Strength | Sig. Level | Rule Interpretation Using Beta Strength Coefficient Sign |
|---|---|---|---|
| 5 | 0.361 | p = 0.009 | IF Infrequent religious at Age 18 RELATES TO OUTCOME = Abuse |
| 4 | 0.352 | p = 0.021 | IF Mother had college education RELATES TO OUTCOME = Abuse |
| 6 | 0.410 | p= 0.027 | IF Region of residence #1 RELATES TO OUTCOME = Dependence |

## CONCLUSIONS

**Using the Constrained Categorical Regression network, combined with a principled approach for pruning connections from candidate data models, results in statistically useful data analysis models.** When compared to a 'bias-only' model, Wilk's Generalized Likelihood Ratio test showed the FM16 model to be significantly better at the 0.01 level and the ML14 model to be significantly better at the 0.05 level. We also found using the prior expert knowledge of the user to define the models is very important in avoiding extra effort during the model refinement phase. If all available stimulus inputs are simply thrown "into the pot" without forethought, then problems such as collinearity, redundant data, and insufficient data precision will exist. As with other neural network paradigms, there is no substitute for careful and thorough data preparation to achieve optimum data representation.

In summary, we have shown that the CCR neural network architecture is effective at analyzing ill-conditioned data sets, and can provide valid statistical inferences even when the data model is mis-specified. The resulting network is directly interpretable in terms of input-output relationships (rules); and the model's rules can be rigorously evaluated using standard statistical testing procedures.

## REFERENCES


[1] Sullins, John. R., (1992). Rule Induction and Correction in Expert Networks. *Intelligent Engineering Systems Through Artificial Neural Networks*, Vol. 2, 957-962

[2] Rumelhart, D. E., Hinton, G. E., and Williams, R. (1986). Learning internal representations by error propagation. *Parallel Distributed Processing: Explorations in the Microstructures of Cognition*, Vol. 1, 318-362.

[3] Golden, R. M. (1995). Making correct statistical inferences using a wrong probability model. *Journal of Mathematical Psychology*, 38, 3-20.

[4] Vuong, Q. H. (1989). Likelihood ratio tests for model selection and non-nested hypotheses. *Econometrica*, 57: 307-333.

[5] Golden, R. M. (1994). Analysis of categorical time-series text recall data using a connectionist model. *Journal of Biological Systems*, 2, 283-305.

[6] White, H. (1982). Maximum likelihood estimation of misspecified models. *Econometrica*, 50: 1-25.

[7] White, H. (1989). Learning in artificial neural networks: A statistical perspective. *Neural Computation*, 1: 425-464.

[8] Bridle, J. S. (1990). Probabilistic interpretation of feedforward classification network outputs, with relationships to statistical pattern recognition. In F. Fougelman-Soulie & J. Herault (Eds.), *Neuro-computing: Algorithms, Architectures, and Applications* (pp. 227-236). New York: Springer-Verlag.

# DISCOVERING RELATIONSHIPS IN RELATIONAL DATABASES THROUGH DATA MINING

**DAVID J. HARRIER**
*Monsanto Agricultural Group*

**DANIEL C. ST. CLAIR**
*University of Missouri-Rolla*
*Engineering Center, St. Louis*
*stclair@umr.edu*

***ABSTRACT:***
Mining data in a relational database environment differs from mining data in less structured environments, since the design of the relational database itself implies that certain key dependencies exist. This paper presents a methodology for applying symbolic machine learning data mining tools to extract previously unknown relationships from within a relational database environment.

## INTRODUCTION

With the increase in the use of information warehouses, the quality of data within application databases is of primary concern of both computer professionals and customers. Current research has focused on defining key attributes of data quality: accuracy, completeness, relevance, timeliness, and interpretability (Kon et al 1993). The introduction of third and fourth generation languages, entity-relationship data modeling, data normalization, computer aided software engineering, and various programming methodologies helped produce noticeable improvements in data quality. The development of the relational model (Codd 1990) and the ensuing database management systems (DBMS) supported the definition of robust integrity constraints: domain, column, entity, referential, and user-defined within the DBMS.

The advent of data mining provides new and effective tools for improving the quality of data within a database. These tools consist of algorithms such as those found in the symbolic machine learning and neural network communities. The relationships discovered by these tools take the form of logic statements or of mathematical models such as networks or statistical models. The relationships discovered provide information which is not only of value outside the database, it provides information which can be used to assess and maintain data quality. Knowing that attributes and/or derived attributes are related in certain, unanticipated ways, helps in the recognition of questionable entries in the database.

This paper presents a methodology for applying symbolic machine learning data mining tools to extract previously unknown relationships from within a relational database environment. The approach differs from the current approaches used in data mining and knowledge discovery (Frawley & Piatetsky-Shapiro 1991) in that some relationships such as functional and multi-valued dependencies are already a part of the database architecture and need not be "rediscovered."

Symbolic machine learning algorithms (Quinlan 1986, Clark et al 1989) are tools for discovering relationships. These algorithms learn by analyzing the relationships between attribute db-values within the database. They build a decision tree or a set of rules which can be used to classify inputs. Each path in a decision tree, illustrated in Figure 1, from the root node to a leaf node, represents a learned relationship. A classification rule can be expressed as a function that returns a set of possible classifications, $f(a_{1i}, ..., a_{nq}) = O_j$, where $O_j$ is a set of classifications, and $a_{1i}, ..., a_{nq}$ are the db-values for the attributes $A_1, ..., A_n$ respectfully. This differs from the standard machine learning approach in that $|O_j| \geq 1$.

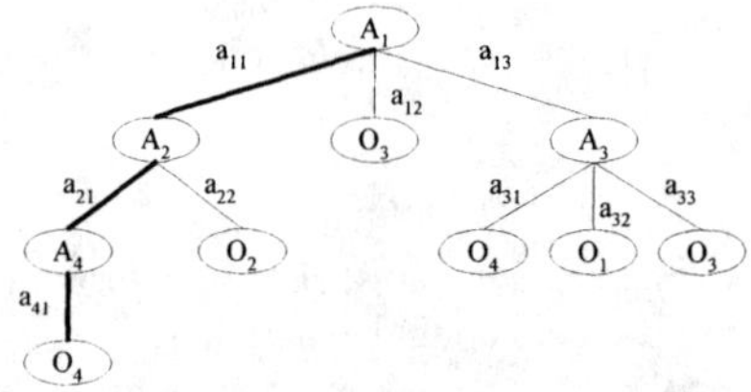


**Figure 1. Decision Tree**

## THE RELATIONAL MODEL AND SYMBOLIC LEARNING ALGORITHMS

A relational database models a known *universe* U, where U is described by a finite set of *attributes* $\{A_1, A_2, ..., A_m\}$. The *domain* $D_k$ is the finite set of db-values which can be assumed by attribute $A_k$. The db-value $a_{im} \in D_i$ is the m-th db-value for $A_i$. The *value function* $V(A_i)$ returns the db-value, $a_{ij}$, for the attribute $A_i$ of the current tuple under consideration. Attributes are grouped in $R^n$, where $R^n$ is a *relation* of degree $n \geq 1$, which contains a subset of n attributes from U. The *tuple function* $T(j,R^n)$ returns the j-th tuple of relation $R^n$. No ordering is implied on $R^n$ by this function. When $j = 0$, then $T(0,R^n)$ is the null tuple, otherwise $T(j,R^n) = <a_{1p}, a_{2q}, ... a_{nz}>$ for $1 \leq j \leq |R^n|$, where $|R^n|$ is the number of tuples in $R^n$.

Symbolic classification learning algorithms are designed to acquire knowledge about a goal attribute, $A_g$, by analyzing its relationships to other attributes $A_i \in U$, where $i \neq g$. Let $Q = \{<a_{1j}, a_{2k}, ... a_{mz}, a_{gx}> \mid <a_{1j}, a_{2k}, ... a_{mz}, a_{gx}> = T(j,P^m)$ for some $1 \leq j \leq |P^m|\}$
be the collection of *candidate training tuples* used as the first step in constructing a training set. The relation $P^m$ is the result of one or more relational operators applied to one or more relations that include $A_g$, where $a_{gx} \in D_g$ and $A_g \in Q$.

A *training set* $S \subset Q$ is used by the induction task. The $V(A_g)$ is known to be correct for each tuple selected. Each *training tuple* $T(i,S)$ is of the form $<a_{1j}, a_{2k}, ..., a_{g-1,p}, a_{g+1,q}, ..., a_{mw}, a_{gx}>$, where $V(A_g) = a_{gx}$ is known to be the correct classification when $V(A_1) = a_{1j}$, $V(A_2) = a_{2k}$, ..., and $V(A_m) = a_{mw}$.

The set of learned *relationships*, $F = \{f_1, f_2, ..., f_n\}$, is the result of the induction task. Each relationship, $f_i$, will have associated with it a set $O_p$, which contains classification db-values from domain $D_g$.

## THE ID3 SYMBOLIC CLASSIFICATION LEARNING ALGORITHM

Quinlan's ID3 will be used to take a training set S and produce a set of learned relationships F. The algorithm represents its knowledge in the form of a decision tree where each path represents a rule. The development of the tree is accomplished in a recursive manner by first selecting the attribute which is "most important" in properly classifying the goal attribute, $A_g$. This attribute becomes the root of the tree. Branches represent db-values for the root attribute. The ID3 algorithm is then applied recursively to each of the branches.

The ID3 algorithm has three key characteristics. First, it converges on set F rather quickly, even for large training sets. Second, decision trees generated by ID3 are usually wide and shallow, thus producing rules which generalize well. Third, the set F is easy to use and apply on unseen data.

In a decision tree F, each path, beginning at the root node and ending at a leaf node, is a classification rule $f_i$, while set $O_p$ is a leaf node where $O_p = \{a_{gi}, a_{gj}, ..., a_{gw}\}$, and $|O_p| \geq 1$. The bolded path in Figure 1 represents the relationship $f_1$.

$$f_1 : (V(A_1) = a_{11}) \wedge (V(A_2) = a_{21}) \wedge (V(A_4) = a_{41}) \Rightarrow O_4$$

Each function in set F will classify the goal attribute's db-value as successful or unsuccessful. A successful classification occurs when the goal attribute's db-value is found in the function's classification set, while an unsuccessful classification occurs when the goal attribute's db-value is not found in the function's classification set.

## METHODOLOGY FOR DISCOVERING RELATIONSHIPS

The methodology for guiding the discovery of relationships in a relational database uses the principles of data mining and knowledge discovery (Frawley & Piatetsky-Shapiro 1991), and uses the ID3 machine learning algorithm as the discovery tool. The methodology consists of six steps: selecting the goal attribute, $A_g$; building the training set S; building the decision tree or rule set, set F; analysis of set F; testing the performance of set F on unseen data; and applying set F to the database.

### Selecting the Goal Attribute

Choosing which attribute, $A_i$, to use as the goal attribute $A_g$ is straight forward. Usually, $A_g$ is of special interest, for many reasons. For example, attribute $A_g$ may be of high importance to the database, or erroneous db-values may have been found in $A_g$. Before selecting the goal attribute, the following guidelines should be considered.

The $V(A_g)$ should be predictable. If the domain for $A_g$ consists of a large set of db-values, the ability to build a representative training set S may be prohibitive. It may be acceptable to verify that the $V(A_g)$ is in a group. This is achieved by partitioning the db-values of $A_g$ into range groups prior to building the candidate training set Q. This is similar to grouping non-goal attributes (Gallion et al 1993).

Learning relationships, set F, for $A_g$ is not necessary when $A_g$ is the primary key for a relation since relationships on this attribute are already defined. The same may be true for certain components of a composite primary key.

### Building the Training Set

Building the training set S requires two steps: building set Q, which gathers candidate training tuples using the answer set of a relational database query; and building set S by selecting tuples from Q that are known to be correct.

The first step in building S involves collecting candidate training tuples, $T(j,P^m)$, into set Q from set U. The process begins with the construction of a relational database query command, called the derived relation $P^m$, based on the domain knowledge of the data administrator and application specialists. Relation $P^m$ includes $A_i$'s $\in$ U that are believed useful in discovering relationships.

Domain knowledge can be used to significantly reduce the number of attributes to be analyzed. Extreme care must be given in not being too selective of which attributes are chosen, since discarding the wrong attributes may result in restricting the search and deliberately ruling out valuable discovery.

To assist in determining which $A_i$'s to select, both inter-attribute and intra-attribute relationships must be analyzed. Inter-attribute relationships are between those attributes within the same relation. Therefore, analysis of functional dependencies are useful in finding inter-attribute relationships. Intra-attribute relationships are between attributes from many relations. Therefore, foreign keys are useful in finding intra-attribute relationships.

The creation of derived attributes should also be considered. This can include a variety of possibilities such as; aggregation functions, boolean functions that indicate if an attribute's db-value is marked as missing or present, and the partitioning of an attribute's db-values into range groups.

**Building and Analyzing Decision Tree Structures**

The induction algorithm uses set S to generate a decision tree, F (see Figure 1). Before set F is applied to the subject database, it must be analyzed to insure the elements are useful relationships. The analysis will focus on the attributes selected by the induction task, the structure of the decision tree, and the relationships learned.

Each attribute used in set F will have an influence on the tree's ability to correctly classify unseen tuples. Therefore, each attribute should be carefully reviewed with the data administrator and application specialists to insure their continued use. Flachsbart et al (1994) provide additional discussion as to how to use ID3 to determine what attributes should be removed.

If the decision tree contains a root node and leaf nodes, each having a single conclusion, it is likely that the induction task discovered a functional dependency relationship. For example, let attribute $A_i$ be the primary key for relation $R^n$, and attribute $A_j \in R^n$, then $A_j$ is functionally dependent on attribute $A_i$. Assume $A_j$ was chosen as the goal attribute, and $A_i \in S$, then the generated relationships found in set F will be of the form $f_k : A_i(a_{im}) \Rightarrow O_p$, where $O_p = \{a_{jn}\}$, and $|O_p| = 1$. To prevent the induction task from learning known functional dependencies, remove the primary key attribute, $A_i$, for the goal attribute, $A_j$, from the candidate training set Q. This may also apply to all components of a composite primary key.

A choice must be made to accept, or not accept, a set F that contains leaf nodes with multiple conclusions, $|O_p| \geq 1$. In the context of relationship learning, it will be common to accept a set F that contains leaf nodes with multiple conclusions. Multiple conclusion leaf nodes could be caused by either too few attributes in the candidate training set Q, or the training set S contains noise. Analysis of all $V(A_i)$, for each tuple that mapped to the same multiple conclusion leaf node, will be required to find the erroneous data.

Set F should be reviewed when the learned relationships contain many conjunctions. These relationships are usually too specific and are prone to classification failures for unseen data. Pruning attribute(s) from set S produces more generalized rules which often perform better on unseen data.

**Analyzing Performance on Unseen Data**

It is important to determine how well set F performs on unseen tuples that are known to be correct. The V-fold cross validation technique is useful in verifying the accuracy of set F (Breiman et al 1984).

The V-fold technique randomly partitions a training set S into N subsets of equal cardinality, where $S = W_1 \cup W_2 \cup ... \cup W_n$, $W_i \cap W_j = \phi$ for $i \neq j$, and $|W_1| = |W_2| = ... = |W_N|$. Classification accuracy tests are performed by selecting each $W_i$ as a test set, with its associated training set $X_i = W_1 \cup W_2 \cup ... \cup W_{i-1} \cup W_{i+1} \cup ... \cup W_n$. A classification accuracy percentage for each test set $W_i$ is calculated by the number of correct classifications from $W_i$ using the decision tree built with $X_i$. The mean and standard deviation of these percentages is calculated for the N tests.

Another statistical measure of interest is the failure percentages of both path and leaf failures. A path failure occurs when none of the branches from the attribute node being tested are labeled with a db-value that matches the corresponding attribute's db-value of the test tuple. A high percentage of path failures may indicate that the |S| is too small. Path failures for attributes which are based on ordered domains can be eliminated without increasing the |S| by the use of a technique called "nearest neighbor" (St. Clair et al 1992).

A leaf failure can occur either when the goal attribute's db-value for the test tuple does not match any db-value in the reached leaf node, or the test tuple reaches a multiple conclusion leaf node when multiple conclusions are not acceptable.

Leaf failures due to no db-value match in the leaf node can be reduced by increasing the size of the training set S. Failures due to an incorrect match in a multiple conclusion leaf node can either be accepted as an actual failure classification, or can be reduced by providing additional attributes to set S.

Separate path and leaf failure percentages for each test set $W_i$ should be computed in a similar manner as the classification accuracy percentage. The mean and standard deviation percentages are calculated.

**Applying the Decision Tree to the Database**

The final step of the methodology involves applying the learned relationships in set F against the goal attribute $A_g$, in order to find possible data quality problems. This step can be performed using a variety of techniques.

The first technique suggested, and probably the easiest to perform, would be to "sweep" all tuples in database U against set F that contain the goal attribute $A_g$. This technique simply tests each tuple against the decision tree. This technique is passive, since it does not require set F to become an integral part of the subject database.

An integrated technique is to transform set F's relationships, one for one, into DBMS user-defined constraints and then store them in the subject database U. With the learned relationships stored in U, any future problems for the goal attribute $A_g$ could be prevented. This technique is useful when it is believed that set F has a long lasting quality. Two key long lasting qualities for set F are; generality and stability.

Harrier (1995) illustrated the methodology's discovery approach using a data set from a large customer sales information warehouse database that represents real domains and includes noise. Table 1 summarize these results.

**TABLE 1: SALES DATA SET - AVERAGE RULE SET ACCURACY**

| | Successfully Classified | Percent Successfully Classified | Leaf Errors | Percent Leaf Errors | Path Errors | Percent Path Errors |
|---|---|---|---|---|---|---|
| Average | 3,738 | 95.9 | 151 | 3.9 | 8 | 0.20 |
| (Std. Dev.) | (21.329) | (0.5) | (13.301) | (0.5) | (5.447) | (0.1) |

## CONCLUSIONS

The use of data mining has been presented as a valuable tool to improve data quality. This paper describes a methodology that guides both the discovery of relationships within a relational database and the implementation of these learned relationships as an integral part of a database's constraints.

## REFERENCES


Breiman, L., Freidman, J. H., Olshen, R. A., and Stone, C. J., (1984). *Classification and Regression Trees*, Wadsworth International Corp.

Clark, P., and Niblett, T., (1989). The CN2 Induction Algorithm, *Machine Learning*, Vol. 3, pp 261-283.

Codd, E. F., (1990). *The Relational Model For Database Management: Version 2*, Addison-Wesley Publishing Co., Inc.

Flachsbart, B., St. Clair, D.C., Holland, J., and Bond, W.E., (1994). Using the ID3 Symbolic Classification Algorithm To Reduce Data Density, *Proceedings of the 1994 ACM Symposium on Applied Computing*, pp. 292-296.

Frawley, W. J., and Piatetsky-Shapiro, G., (1991). *Knowledge Discovery in Database*, AAAI Press / MIT Press.

Gallion, R., St. Clair, D. C., Sabharwal, C., Bond, W. E., (1993). Dynamic ID3: A Symbolic Learning Algorithm For Many-Valued Attribute Domains, *Proceedings of the 1993 Symposium on Applied Computing*.

Harrier, D. J., (1995). *Discovery of Integrity Relationships in Relational Databases*, Masters Thesis, Department of Computer Science, University of Missouri - Rolla.

Kon, H. B., Lee, J., and Wang, R. Y., (1993). *A Process View of Data Quality*, TDQM-93-01, Sloan School of Management, MIT.

Quinlan, J. R., (1986). Induction of Decision Trees, *Machine Learning*, Vol. 1, pp 81-106.

St. Clair, D. C., Bond, W. E., Rigler, A. K., and Aylward, S., (1992). An Evaluation of Learning Performance in Backpropagation Neural Networks and Decision-Tree Classifier Systems, *1992 ACM/SIGAPP Symposium on Applied Computing*, pp. 636-642.

# A NEW CHAOS BASED APPROACH TO DATA CONCEPTUALIZATION

**RICHARD T. GORDON AND RICHARD J. CANTOR**
Computational Intelligence Laboratory
Information Technology Business Innovation Group, Chubb Group of Insurance Companies,
477 Martinsville Road, Liberty Corner, NJ 07938-0803
Email: rtg@aip.org & rickdba@haven.ios.com

**MARK J. WOYSHVILLE**
Case Western Reserve University
Mood Disorders Program
University Hospitals of Cleveland
11400 Euclid Avenue, Cleveland, OH 44106-5000
Email: mjw14@aol.com

**ABSTRACT:**
Data conceptualization is the process of clustering data according to attributes so that grouping relationships can be inferred. This concept is the cornerstone of data mining. Utilization of neural networks for this task sees an unsupervised architecture, such as the self-organizing map, applied to the data to perform the grouping. While generally effective, the application of this technique to nonlinear time-series data presents many challenges. In this paper we present a technique where the original time-series data is reduced to only the measure of its correlation dimension, with the correlation dimension then utilized as the input to a self-organizing map for the data conceptualization. We present an example of this technique applied to time-series of human mood data in a blind study. We show that this technique allows the essential behavior of the dynamics of the time-series to be retained while reducing the complexity of the problem in both dimensional and computational terms. We further show in the mood data example that meaningful groupings can be found utilizing this technique.

## INTRODUCTION

In recent years data mining has become increasingly important to both business and research. A cornerstone concept in data mining is data conceptualization, the process of clustering data according to attributes so that grouping relationships can be inferred. Unsupervised artificial neural network models, such as the self-organizing map (Kohonen, 1984) and adaptive resonance theory (Carpenter and Grossberg, 1986), are widely used to perform data conceptualization. While generally effective, the application of these techniques to nonlinear time-series data presents many challenges. Problems can arise from time-series having differing lengths or exhibiting chaotic behavior.

In this paper we present a technique for dealing with these problems where the original time series is reduced to only the measure of its correlation dimension (Grassberger and Procaccia, 1983a&b, Takens, 1981, Takens, 1984). The correlation dimension for each series is then utilized as the only input to a self-organizing map for the data conceptualization. Utilizing an example of this technique applied to time-series of human mood data in a blind study, we show that meaningful groupings can be found even though two abstractions removed from the original data. We further show that this technique allows the essential behavior of the dynamics of the time-series to be retained while reducing the complexity of the problem in both dimensional and computational terms.

**CORRELATION DIMENSION**

An essential element in the analysis of a chaotic time-series is the calculation of the correlation dimension. Of interest here is the determination of whether the dynamics of the time-series arise from a deterministic chaotic system or from a system exhibiting a significant random component. The embedding theorem (Takens, 1981, Takens, 1984) holds that if $\{X_n\}$ is a discrete-time scalar cross-section of a continuous-time multidimensional system which has an attractor within a manifold of dimension I, there will generally exist an embedding dimension $J \leq 2I+1$ such that J vectors $X_{n-1} \ldots X_{n-J}$ comprise a set corresponding to the same dimension as the underlying attractor. This then frames the problem of the diagnosis of chaos in a system as one of determining whether a set of vector observations lie within a subset of fractal dimension.

The correlation dimension is an estimation of the fractal dimension of the time-series and as such captures the nature of the nonlinear behavior. The correlation dimension is calculated based on nearest-neighbor distances (Grassberger and Procaccia, 1983a&b) by taking a count of the data points within hyperspheres of various radii which are centered on each data point in a phase space reconstruction which is dimensionally embedded. The embedding dimension utilized should be greater than the anticipated dimension of the time-series data.

**CHAOS BASED GROUPING**

It has been shown (Gordon, 1994, Weigend and Gershenfeld, 1994) that exploratory data analysis is essential in neural network modeling. As a representation of the fractal dimension of a time series, the correlation dimension is a fundamental metric of the nonlinear behavior of a time-series. As such, the correlation dimension should provide a good means for grouping data and also high information content as input to a data conceptualization engine such as the self-organizing map. In this research, we utilize two separate techniques for performing the groupings. First, we construct surface plot visualizations to see if any groupings are easily identifiable. Next, we develop a self-organizing map for the data conceptualization.

**APPLICATION TO HUMAN MOOD DATA**

In this research, we test the efficacy of the chaos based grouping by application to time-series of human mood data. There is importance in the application of techniques from complex systems theory to psychiatric/psychological data. Phenomenological data, the bulk of which exhausts linear paradigms and routine descriptive statistics, is largely ignored due to analytic intractability. Therefore, it is essential that techniques from the study of complex adaptive systems be extended to the analysis of such data. In addition to the technique presented in this paper, we applied spectral analysis, correlation functions, recurrence time reconstruction, calculation of largest Lyapunov exponent, and other nonlinear techniques to the analysis of the mood data.

In this research, we perform a blind analysis on time-series of mood ratings for 51 subjects taking part in a mood disorders program. All that is known of the subjects is that control subjects as well as subjects with diagnosed affective (mood) disorders are represented. Our data is comprised of daily mood ratings on a 100mm analog scale anchored at 0mm by the lifetime lowest mood, 100mm by the lifetime most elevated mood, and 50mm by mood which is neither low nor elevated. An average of 90 days worth of mood ratings was generated by each subject. Further, the data is comprised of three time-series per subject representing the means and extremes of daily mood: the prevailing mood of the day is marked on the analog scale, flanked by the extremes of low and elevated mood for that day, however briefly experienced. The analysis was approached from the assumption that each of the three series represented one dimension of the overall mood system. Therefore, the correlation dimension for each series represents the fractal dimension and hence a description of the nature of the chaotic dynamics for each dimension of the mood system for each subject.

The three correlation dimensions for each subject showed a wide range of behavior from low to high order chaos, ranging from 1.80 to 7.26. Three subjects also exhibited time-series which converged to a fixed point and hence had a zero correlation dimension. Two of these subjects showed the convergence to a fixed point in two of the three series, while the third subject showed the fixed point convergence in one series.

As a first attempt at grouping, surface plots were constructed from the correlation dimensions for all 51 subjects. Three separate surface plots were constructed utilizing an ascending order of the correlation dimensions by subject. In the first plot the ordering was done based on the correlation dimensions of the low mood of the day series and thus plotted on the x-axis, with the prevailing mood correlation dimensions plotted on the y-axis and the high mood of the day plotted on the z-axis. The second surface plot was ordered based on the correlation dimensions for the prevailing mood of the day plotted on the x-axis, with the high mood on the y-axis and the low mood on the z-axis. The third plot was ordered on the correlation dimensions for the high mood which was plotted on the x-axis, with the low mood on the y-axis and the prevailing mood on the z-axis.

A first visual examination of the surface plots clearly showed two main groupings, an outer smooth area surrounding a central spiky area. Closer examination of the three plots revealed the central area as two different spiky areas consisting of an outer spiky shell separated by a thin smooth transition from a central spiky area. The smooth areas corresponded to subjects who had similar correlation dimensions for either all three series or at least two of the three series. The spiky areas corresponded to subjects whose correlation dimensions varied widely between series. In addition to providing an initial grouping of the subjects, the surface plot visualizations would also be utilized as a cross validation for the self-organizing map.

Traditional information theory would suggest that any signal in data would be lost when analysis is performed two abstractions removed from the original data. This idea is based in a traditional reductionist view of modeling as characterized by regression analysis. An application of linear regressions to the time-series in this research yielded no significant results. In our attempting to perform the data conceptualization based only on the correlation dimensions, we hoped to advance a technique based on the theories of complex adaptive systems, a decidedly none reductionist view. In looking to the correlation dimensions, it was hoped that the fractal dimension would represent an emergent property of the overall dynamics of the system. This then constituted the premise we tested through the self-organizing map conceptualization.

The self-organizing map consisted of three inputs corresponding to the correlation dimensions for each subject. Two different maps were developed with the first dividing the subjects into two groups and the second dividing the subjects into three groups. Here again we made an assumption about the subjects. Even though blind, we knew that we had control subjects as well as subjects diagnosed with affective disorder, hence the initial two grouping map. Our assumption was that we would find a third group corresponding to subjects that were not clearly identifiable as members of either group, hence the three grouping map. Training was performed on a 90MHz Pentium PC and was run for 1,000 iterations. Training time was approximately 1.5 minutes.

**RESULTS**

Both self-organizing map models grouped the subjects in agreement with the groupings found in the surface plot visualizations. This, in theory, supports the efficacy of using correlation dimensions rather than the original time-series in doing data conceptualization. A more stringent test is whether the groupings found by the self-organizing maps have significance in a clinical sense.

First, in removing the blind from the subjects for the purpose of the analysis of results, it was learned that the subjects represented three different groups. The first was a control group of 18 subjects with no psychiatric disorders. The second was a coenrolled group of 6 subjects with psychiatric disorders, but not necessarily affective disorders. The third group consisted of 27 patients diagnosed with affective disorders, but all had in common unstable, rapidly fluctuating affect as a unifying phenomenological theme.

For the purpose of this discussion, we will label the groupings found by the self-organizing maps as spiky and smooth for the two group map and spiky, smooth and transitional for the three group map. The two group map identified 14 of the control subjects as belonging to the spiky group with 4 categorized in the smooth group. This map categorized 3 of the coenrolled subjects in each of the two groups. The patient subjects were categorized with 16 falling into the spiky group and 11 falling into the smooth group. More telling results were obtained from the three group self-organizing map. The three group map categorized 9 of the controls into spiky group, 7 into the transitional group and 2 into the smooth group. The coenrolled subjects were grouped with 0 in the spiky group, 1 in the smooth group and 5 in the transitional group. The patient subjects were classed with 8 falling into the spiky group, 10 into the smooth group and 9 into the transitional group.

The significance of these findings lies in that the patient subjects were fairly evenly sorted into three groups, whereas the control subjects were more likely to occupy the spiky and transitional groups at the expense of occupancy of the smooth group. The coenrolled subjects clearly showed a preference for occupancy of the transitional group. This strongly suggests a different dynamic domain than that associated with "normality" or affective disorders. These findings further suggest that there is a dynamic area of "normality" which is routinely occupied by control subjects and to a significant extent by patient subjects, but that there is an additional domain occupied predominantly by patient subjects. It is important to note that with most psychiatric illness of the sort represented by the patient subjects in this study, their psychopathology represents an extreme of normal experience. Therefore, at least metaphorically, there seems to be some consistency between the clinical model of the illness and the sorting of the dynamic variables characteristic of the time-series across the groups.

**CONCLUSIONS**

The use of only the correlation dimensions as input to a self-organizing map holds much promise for overcoming limitations placed on data conceptualization by the quasi-random nature of chaotic time-series. Further, this technique reduces the computational resources necessary to perform attribute based grouping. That it yielded useful results on the mood data application offers encouragement that this technique can be generalized to other applications in medicine, business and research. Future work will focus on extending the technique to encompass other measures of the overall dynamic behavior of chaotic time series.

**REFERENCES**


Carpenter, G., and Grossberg, S, (1986). Adaptive Resonance Theory: Stable Self-Organization of Neural Recognition Codes in Response to Arbitrary Lists of Input Patterns, *Eighth Annual Conference of the Cognitive Science Society*, Lawrence Earlbaum Associates, Hillsdale, NJ, 45.

Gordon, R., (1994). Improving Neural Network Modeling by Understanding the Dynamics of Time Series Data, *Intelligent Engineering Systems Through Artificial Neural Networks*, **4**, Dagli et al Eds., ASME Press, New York, 1129.

Grassberger, P., and Procaccia, I., (1983a). Characterization of Strange Attractors, *Physical Review Letters*, **50**, 346.

Grassberger, P., and Procaccia, I., (1983b). Measuring the Strangeness of Strange Attractors, *Physica D*, **9**, 189.

Kohonen, T., (1984). *Self-Organization and Associative Memory*, Springer-Verlag, Berlin.

Takens, F., (1981). On the Numerical Determination of the Dimension of an Attractor, *Lecture Notes in Mathematics*, **898**, Rand and Young Eds., Springer-Verlag, New York, 366.

Takens, F., (1984). On the Numerical Determination of the Dimension of an Attractor, *Lecture Notes in Mathematics*, **1125**, Springer-Verlag, New York, 99.

Weigend, A.S., and Gershenfeld, N.A., (1994). *Time Series Prediction*, Addison-Wesley, Reading, MA.

# AN ALGORITHM FOR CLUSTERING NON-DETERMINISTIC DATA

Charles Wittmaier
AAA Auto Club of Missouri
12901 North Outer Forty
St. Louis, MO 63141

Chaman L. Sabharwal
University of Missouri - Rolla
8001 Natural Bridge Road
St. Louis, MO 63121
E-mail: chaman@umr.edu, URL: http://www.umr.edu/~chaman

## ABSTRACT

This paper presents an algorithm for clustering non-deterministic discrete data for use in classification applications. The clustering algorithm analyzes input data and ranks the relative predictive strength of each value of each attribute. There are two levels of clustering: intra-attribute and inter-attribute. The values of each attribute are clustered to a level that provides the relatively deterministic dataset. The clustering technique is guided by the number of binary input units for the Artificial Neural Network (ANN) and training time requirements. Experimental results, on predicting the lapse intentions of Automobile Club of Missouri members, show that the networks trained with clustered data classify as well or better than networks trained with data clustered by other methods. The clustering algorithm improved the accuracy of the ANN algorithm over prediction accuracy achieved on raw data.

## INTRODUCTION

Non-deterministic data is typified by human behavior. Two individuals with identical stimuli may respond differently. There is no mathematical function governing the reaction to stimuli. Non-deterministic data defies common training methods used with feed forward ANNs [Hinton et al. 1986, Masters 1993] The works by Lebowitz [1985] and Gallion et al. [1993] address the problem of grouping data. These techniques are similar to frequency distribution methods. Hitt [Hitt 1994, Ruck et al. 1990] described a method for preprocessing input data for an ANN without clustering where only values with high correlation to the output values are selected, all values with low correlation are effectively ignored, this method is called 'Key Value' distribution in this paper. Since these methods are basically suitable for deterministic data, they have proved to be ineffective on non-deterministic data.

This paper describes an algorithm for clustering discrete non-deterministic data to improve the classification accuracy. The clustered data is classified by an ANN algorithm to show the improvement in prediction accuracy. The new clustering method is applied to a commercial classification problem: prediction of membership lapse for the Automobile Club of Missouri. Experimental results using a feed forward ANN show improvement in classification accuracy over the other methods.

## PREVIOUS RELATED WORK

Two methods of clustering values are based on correlation to a particular outcome. Though each method has its drawbacks, they provide a foundation for the new method presented in this paper. Each of these is described briefly as follows: *Method 1 - Frequency Based Method.* The strength of a value of an attribute is measured by the *frequency* of occurrence of the outcome associated with it. If an attribute has k values, they are first ordered by frequency of an outcome. These values are grouped into n clusters with n < k. This clustering is called *intra-attribute* clustering. Each of the n clusters represents the same percentage of the total population. This method is commonly used to produce quick learning and effective neural networks [Masters 1993, Stubbs 1990], but it does not always work well for several reasons. *Method 2 - Correlation Based Method.* The values of attributes that are most highly correlated with a particular conclusion are selected in preference to those with weaker *correlation.* This approach is also used in industry [HITT 1994]. The n values of chosen attributes makeup the input units to an ANN; data clustering is not performed.

## DESCRIPTION OF APPLICATION DATA

The new algorithm is developed in the context of an application to illustrate the concepts. This application concerns the Automobile Club of Missouri for membership lapse prediction where the algorithm is in production. Auto Club of Missouri member data consists of about 48,000 memberships. The attribute values directly represent product and service use, and demographic information. Nine attributes pertaining to the membership are selected for use in the prediction of membership lapse. The nine attributes constitute both symbolic and ordinal data. These attributes are described as: 1. *Member Type* - indicates level of membership, regular or AAA Plus. 2. *Solicitation Code* - promotional offer and rates in effect at time member initially joined. 3. *Member Count* - total number of members within the membership. 4. *Branch Code* - code (based on zipcode) for branch office to which the member is assigned. This code is geographic in nature. 5. *Demographic Code* - code assigned by nine digit zipcode. This code implies certain spending habits and education level of the individual. 6. *Marital Status* - marital status reported by member on membership application. 7. *Auto Insurance Count* - number of current auto insurance policies associated with all members on the membership. 8. *Current Year Emergency Road Service (ERS) Usage* - number of times within current membership term that ERS was used. 9. *Domestic Travel Usage* - number of times in current membership term that Triptiks were requested. The universe of data for this experiment was chosen to proportionally represent the monthly population of the membership. Members of three or fewer years duration have been identified by the marketing department as prime targets for membership retention tactics.

## ANALYSIS -- MAXIMUM PREDICTABILITY

The sample data set is divided into the two classification clusters: active (renewers) and lapse (non-renewers). We first calculate the relative frequencies of the values of attributes. A measure, the lapse-to-active ratio (L/A), is used to rank the values of an attribute in order of propensity to lapse. It is defined as follows:

$$L/A = \frac{\%\ of\ population\ that\ lapsed}{\%\ of\ population\ that\ renewed}$$

The new method extends the concept of correlation and the ratio L/A becomes the cornerstone of the new method. First , *Maximum Predictability* (Max Pred) algorithm is developed to predict network performance based on a given training set. A precise definition of this method follows. A given dataset is composed of N feature vectors, e.g., each vector corresponds to one Auto Club member. A feature vector consists of n+1 elements. The first n elements are attribute values , e.g., characteristics of a member. The (n+1)st element is a binary classification variable, e.g., the renewal status of the member. Any two vectors with first n identical elements are said to be equivalent. The set of all equivalent vectors forms an equivalence class. If there are K equivalence classes and an equivalence class consists of k vectors then

$$1 \leq K \leq N, \qquad 1 \leq k \leq N$$

The number of vectors in an equivalence class with one conclusion are denoted by a, and with the other conclusion are denoted by p, where

$$0 \le a \le k, \qquad 0 \le p \le k, \qquad a+p = k$$

There are at most two possible prediction accuracy measures associated with any equivalence class. They are:

$$\frac{\max(a,p)}{k} * 100 \qquad \text{and} \qquad \frac{\min(a,p)}{k} * 100$$

Max Pred, the highest attainable prediction accuracy for the entire dataset is defined by the equation

$$\frac{\sum_{i=1}^{K} k_i \frac{\max(a_i,p_i)}{k_i}}{\sum_{i=1}^{K} k_i} * 100 = \frac{\sum_{i=1}^{K} \max(a_i,p_i)}{N} * 100$$

where $a_i$ and $p_i$ refer to the number of active and lapse elements of the i_th equivalence class corresponding to complimentary conclusions. The highest attainable prediction accuracy is determined by the Clustering Interval Length (CIL) value(s) yielding the highest Max Pred rating for all different CIL values. The clustering with maximum predictability are used to create training datasets for the ANN.

The Figure 1 shows the predictive strength associated with various clustering levels. The hybrid clustering algorithm is given in the next section. It is clear that the highest Max Pred rating occurs when the CIL is set to 0.010 for 78 nodes, 0.055 for 36 nodes, and 0.075 for 22 nodes. Note that the Max Pred is used to determine the best CIL, not necessarily the number of input nodes.

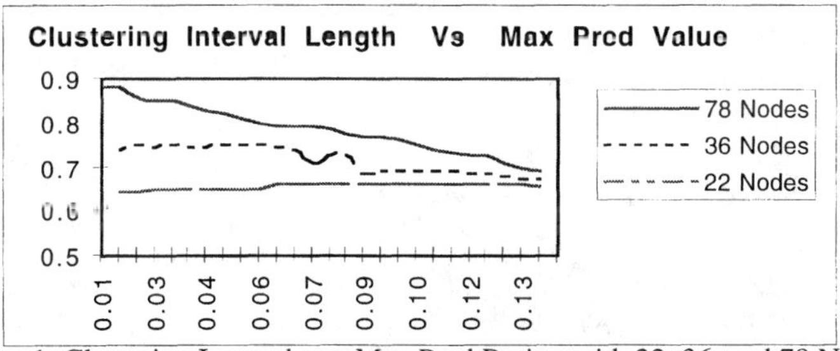


Figure 1. Clustering Intervals vs. Max Pred Rating with 22, 36, and 78 Nodes

## AUTOMATED CLUSTERING METHOD

### ANN Configuration

The UMR Intelligent Systems Center project [UMR-ISC 1993] showed that an ANN with integer-valued inputs could not be reliably trained on Auto Club data using Neuroshell software [1993]. Using the ANN software package [Masters 1993], feedforward networks are trained with unclustered data in equilateral binary form. Each training vector for the ANN consists of a binary string of 141 digits (input nodes), representing the nine attributes used. The 141 units were translated into 22, 36, and 72 unit by the *inter-attribute* and *intra-attribute* clustering algorithm given below. Networks with 22, 36, and 78 input units were trained. The number of input nodes was chosen arbitrarily to test the use of relatively few to many input nodes. Experience shows that varying the number of nodes in the hidden layer from three to six does not significantly affect network performance. The performance goes down for internal nodes less than three or more than six. The final experiments were performed on 22-3-1, 36-3-1, and 78-3-1 networks.

## The Hybrid Method

This new method is a hybrid of the Flat Distribution and Key Value methods. This clustering scheme is a two-phase iterative approach that examines many levels of *intra-attribute* and *inter-attribute* clustering. The first part of algorithm is used to determine of the optimum CIL. The second part uses the optimum CIL to create the binary training datasets for the ANN algorithm. In this method, the ratio L/A is used to generate three new quantitative measures: B1 - unbiased strength measure, B2 - directed strength measure, and B3 - decision measure.

The *unbiased strength measure* B1 is defined in terms of a ratio between lapse and active percentages for a particular value. A value of B1 close to 0 indicates weak intentions to lapse or renew, a value of B1 close to 1 indicates a strong intention to lapse or renew. It is defined as follows:

B1 = 1 - (Active percent / Lapse percent) if Lapse percent > Active percent,
B1 = 1 - (Lapse percent / Active percent) if Lapse percent < Active percent,
B1 = 0 if Lapse percent = Active percent.

The *directed strength measure* B2 represents the likelihood of this attribute to be associated with a particular outcome of lapse or renew. Like B1, B2 is defined in terms of a ratio between lapse and active percentages for a particular value. A value of B2 close to 0 indicates a strong intention to renew, a value of B2 is close to .5 indicates weak intention to lapse or renew, a value of B2 close to 1 indicates a strong intention to lapse. It is defined as follows:

B2 = 1 - (Active percent / Lapse percent)/2 if Lapse percent > Active percent,
B2 = (Lapse percent / Active percent)/2 if Lapse percent < Active percent,
B2 = 0.5 if Lapse percent = Active percent.

The *decision measure* B3 is used to select candidates for unique input to the neural network. It is defined as B1 * PP (percentage of population represented by a value or cluster of values). After *intra-attribute* clustering has been done, the decision measure B3 values are used in global competition; the values of attributes with the n largest values of B3 are winners to represent individual values and all of the remaining values are combined to represent one clustered value. This represents *inter-attribute* clustering. For example, suppose an arbitrary attribute has values which result in 5 discrete values after clustering. Let n = 3 in this case. The other two values represent the insignificant values. The binary representation of this attribute is: 1 0 0; 0 1 0; 0 0 1; corresponding to the first, the second, and the third of the 3 values, and 0 0 0 - pertaining any value other than the 3 selected values. These three binary digits represent three input nodes, but correspond to all five discrete values.

## Clustering Algorithm Description - Learning Phase

The clustering algorithm is divided into six major steps. Step 1. *Clustering Measures* - For each attribute value, calculate the unbiased strength measure B1, the directed strength measure B2, Population Percentage PP. Step 2. *Initialize clustering parameters* - Initialize the clustering interval length *CIL*, CIL *Increment*, number of *iterations* and the number of binary *input nodes*. Step 3. *Perform clustering* - Sort in ascending order of B2 values within each attribute, *intra-attribute sorting*. The use of B2 accurately portrays the relative strength of a tendency (toward lapse or active) in relation to neutral intentions. Step 4. *Select values for input nodes* - Calculate B3 = B1 * PP. Select the n attribute values for which B3 is maximum. Step 5. *Encode the data for Max Pred algorithm* - Each attribute is represented by a number of binary values. If the value of an attribute belongs to one of the n clusters, its input value is one, else a zero is used in the encoded vector. Step 6. *Apply Max Pred algorithm* - Evaluate the encoded dataset by the Max Pred algorithm resulting in the CIL that yields the highest Max Pred score.

Now we select the clustered dataset, the corresponding CIL and the encoded dataset which leads to optimal Max Pred value obtained in the learning phase. The dataset for this CIL is used for training. The test data is encoded by the same criteria.

## TESTING

The predictive performance of a neural network is defined as follows:

$$\text{Lapse Prediction \%} = \frac{\textit{Lapse Cases Predicted Correctly}}{\textit{Lapse Cases in the Test Set}} * 100$$

$$\text{Active Prediction \%} = \frac{\textit{Active Cases Predicted Correctly}}{\textit{Active Cases in the Test Set}} * 100$$

To minimize the occurrence of false positive and false negative lapse predictions, the overall score for rating the effectiveness of the network is calculated by multiplying the lapse prediction percentage by the active prediction percentage. Therefore,

Overall Score = Lapses Prediction Percentage * Active Prediction Percentage

On *unclustered data*, a fifty percent prediction accuracy level leads to an overall score of .2500, slightly less than the average overall score using 3, 4, 5, and 6 hidden units. For testing and comparison purposes the data is clustered so that it yields the highest Max Pred value (*optimal clustering*) and at levels that yield less than the highest Max Pred value (*suboptimal clustering*). For each network variation (22-3-1, 36-3-1, or 78-3-1 nodes) tests were performed using these ANNs. As expected the ANNs trained with optimally clustered data yield higher prediction accuracy than those trained with suboptimally clustered data. At each clustering level, the training data was divided into five segments of approximately the same size. For each of the five experiments, the network was trained on four of the segments and tested on the fifth. This testing method is referred to as *V-Fold testing* [Breiman 1984].

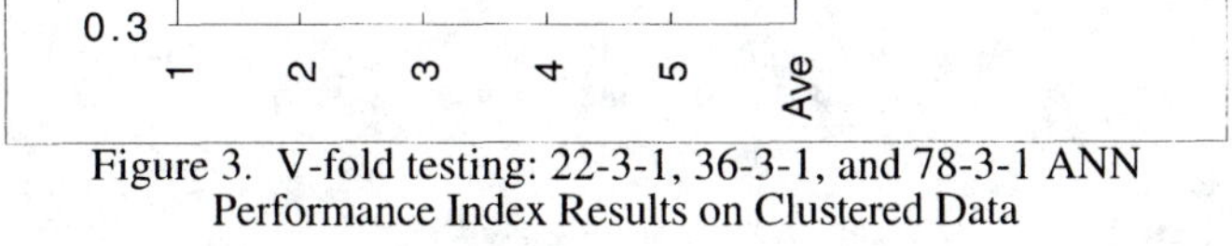


Figure 3. V-fold testing: 22-3-1, 36-3-1, and 78-3-1 ANN Performance Index Results on Clustered Data

With all three methods the highest score was achieved using a ANN with only 22 inputs. This can be attributed to a higher level of recognition of the available data. From Figure 1, it is apparent that as the number of input nodes go up the so does the Max Pred rating for the Optimal CIL. Thus the Max Pred is used to determine the best CIL. It has been shown that the highest accuracy for a configuration is achieved by the ANN trained with the optimal CIL. Figure 4 shows that clustering outperforms the previous techniques with each of the 22-3-1, 36-3-1, and 78-3-1 ANNs.

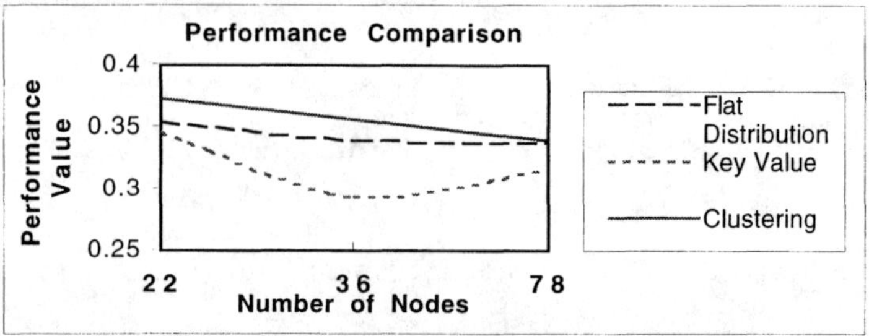

Figure 4. Comparison of three methods with 22-3-1, 36-3-1, and 78-3-1 ANN

## CONCLUSION

A new algorithm is developed for clustering non-deterministic data. This algorithm is applied to membership lapse prediction for the American Automobile Association. Use of multi-level clustering with B2-directed strength measure and B3-decision measure, in conjunction with the Max Pred algorithm provides improvement over previous methods used for clustering non-deterministic data for the purpose of lapse prediction. Experimental results show that tremendous savings are realized while maintaining a prediction accuracy comparable to, or better than, the previous methods.

Future opportunities for research include mathematical analysis of the coupling function(s) which link B1-unbiased strength measure, and PP-percentage population, to achieve higher predictive accuracy. In this paper, the clustering parameter CIL was constant with respect to all of the attributes in any particular experiment, varying the CIL from attribute to attribute may lead to a better and efficient algorithm. This work may be extended from binary to multiple conclusion data by partitioning the data into multiple sets instead of two sets, active dataset and lapse dataset.

## REFERENCES


Gallion, Roger., Sabharwal, C. L., Bond, W. E., St. Clair, D.C. "DYNAMIC ID3: A Symbolic Learning Algorithm For Many-Valued Attribute Domains," Association For Computing Machinery Applied Computing: States Of The Art And Practice - 1993.

Hitt, Benjamin. Personal interview. March 29 1994.

Lebowitz, Michael. "Categorizing Numeric Information For Generalization," COGNITIVE SCIENCE, VOL 9, NUMBER 3, 285-308.

Masters, Timothy. Practical Neural Network Recipes in C++. San Diego: Academic Press, Inc., 1993.

Quinlan, J. R. "Discovering rules by induction from large numbers of examples: a case study," Expert systems in the micro-electronic age, Edinburgh University Press, 1979.

Ruck, Dennis W., Rogers, Steven K., and Kabrisky, Matthew. "Feature Selection Using a Multilayer Perceptron," Journal of Neural Network Computing, VOL 2, NUMBER 2, 40-48., 1989.

Stubbs, Derek F. "Six Ways to Improve Back-Propagation Results," Journal of Neural Network Computing, VOL 1, NUMBER 4, 64-67., 1990

UMR-ISC, "Supplementary Documentation of Membership Lapse Prediction System Development for Auto Club of Southern California," Intelligent Systems Center University of Missouri-Rolla, May 1993.

# A DRAFT STANDARD FOR THE CERTIFICATION OF NEURAL NETWORKS USED IN SAFETY CRITICAL SYSTEMS

**D. F. BEDFORD, G. MORGAN, J. AUSTIN**
*Advanced Computer Architecture Group,*
*Department of Computer Science,*
*University of York.*

**ACKNOWLEDGMENT:**
We thank John P. Oxley and Michael J. Turner.

**ABSTRACT:**
This paper constitutes a draft standard for the certification of neural networks in safety critical systems. We take a combined process and product certification approach, introducing a methodology for development and testing. We consider network construction in three main sections: network design, network training, and network-system interfacing. We divide testing into: development, verification and validation testing. Standards already exist for the development, verification and validation, and ultimate certification of conventionally engineered high-integrity software. The acquisition and treatment of the training data is vitally important in neural computing. With this in mind, we determine how to specify, verify and validate, and finally certify the neural network component of a safety critical system. In developing this draft standard, we indicate how such standards should be adapted, to allow for the certification of neural network modules within safety critical systems.

## INTRODUCTION

The use of artificial neural networks (ANNs) as embedded modules within larger systems is growing. However, before ANNs can be used in safety critical (SC) systems, a certification process must be developed, and later incorporated into the certification process for the whole SC system. A system is safety critical if a failure in its operation may lead to death, injury, or damage to material or the environment (Poucet *et al.*, 1993). Examples include: aircraft, nuclear power plants and monitoring systems for the very seriously ill.

Our intention in writing this document is to draft a general standard, which can then be adapted to the particular needs of a specific class of applications, or to the style of an existing certification standard. We think it unlikely that one document could adequately cover the needs of all neural networks in all safety critical systems, because of the variation in style, emphasis and content in existing standards, for examples see (Jesty *et al.*, 1990; Eurofighter, 1989; Systems Integration Task Group,

1994; MoD, 1991; MoD, 1993). Rather, this document should be viewed as a concrete starting point for adaptation or criticism and improvement. We anticipate that most ANNs will continue to be simulated in software on digital computers; we will not discuss neural hardware implementations. This standard is not appropriate for the certification of neural networks which learn on-line, because the verification and validation (V&V) process cannot be applied to such networks.

The motivation for enabling the certification of neural networks is two-fold. Firstly, there exists a class of problems for which it is prohibitively difficult or even impossible to write an exact specification for the intended operation of the system, even though examples of correct behavior are available. These are the types of problems for which neural networks may provide a suitable solution. A certification process would enable their use in the SC domain for the first time. Secondly, even in problems where a specification could be written, there may be situations in which it is safer to use a certified neural network. If the conventional specification is long and complicated it may be very expensive or impossible to certify it, whether by formal methods or rigorous argument, as described in (MoD, 1991; MoD, 1993). Testing a conventional software implementation can never dispel the possibility of an error in an untested state (Parnas *et al.*, 1990; Poucet *et al.*, 1990; Knight and Levenson, 1986), and even for simple tasks there are too many states in a computer to search the exhaustively. Conversely, it may be possible to use formal methods to certify the software which simulates neural networks because such kernels would be comparatively small. Also, there are better grounds for interpolating between outcomes when testing a neural network because of the summate and threshold operation of each unit (Partridge, 1995).

The following sections will be written as if this draft standard were a document of mandatory practice. Although this is not the case, we feel that this style will most usefully serve the purpose of making this document a suitable template for subsequent documents, which will eventually carry legislative weight. The first section deals with the construction of the network: its design, the collection of its training data and the relation of the network to the rest of the system. The second section covers testing of the network at different states of its development.

## CONSTRUCTING THE NETWORK

In general, the developer's aim in documenting the construction of the neural network should be to ensure that the construction process is **replicable** and **unambiguous**. A second general point is that system development is a cyclical, not linear process, so that the developer will go through early stages of network development and of the certification process many times.

### Designing the Network

*The high-level goals for the ANN module must be established and documented.* This must be done with care, otherwise, all subsequent certification of intermediate steps to fulfill these high-level goals will be in vain. Validation of the ANN module will depend on its ability to fulfill these goals across a wide range of situations.

*The developer must unambiguously record what type of network was used.* The type of network includes the architecture (the number of layers, the connectivity) and the learning rule used. The design record must include details of every parameter used to characterize the network as well as details of how it was initialized. For example, if the network is initialized with random numbers, the developers must record which random number generator was used and with what number it was seeded, and in which order the weights were initialized.

*The ANN module must generate a measure of confidence or accuracy in its primary output.* It is not enough that the network generate the output function or classification for which it was built; there must also be available to other parts of the system in which the network is embedded an indication of how well versed it is in the particular part of the input space in which it is operating. There are various ways of doing this, such as training a network to detect novelty (Roberts and Tarassenko, 1994; Bishop, 1994), generating a confidence measure directly (Roberts *et al.*, 1994), and other methods (Corrieu, 1994).

**Training the Network**

*The training data must adequately reflect the high-level goals.* This means having enough examples (Maass, 1995), and sufficient examples from each of the desired operations of the network (DTI, 1994).

*The developer must properly document the collection of the data.* As the performance of the trained network ultimately depends on the quality of the data with which it was trained, great care must be taken in obtaining the data. The documentation must include details of the units in which quantities were measured; how, when and by whom the equipment for data collection were calibrated; and also collateral data (DTI, 1994), to ensure that no systematic errors were introduced in collection. For example, the statistics of the data should have no correlation with the identity of the person gathering the data.

*The developer must document the pre- and post-processing of the data and outputs of the network.* Any transforms, for example, scaling or normalization of network inputs or outputs must be recorded.

*The developer must investigate the coverage of input space achieved by the training data.* Care should be taken that the training set has good coverage of each input. For instance, if data are clumped towards one end of the input space, a logarithmic transform may be appropriate. Conspicuous gaps in the input space should be identified and investigated. Visualization tools may help the developer to do this.

*The developer must record the performance function used in training the network.* The performance function us used to calculate how well the network has been trained, given the set of input-output states. This function must be recorded along with the values of any parameters used in the function.

**Interfacing the Network**

*The developer must specify how the network was attached to the rest of the system.* This requirement is closely related to the issue of pre- and post-processing, but also includes such things as determining whether and how the network's operation will be synchronized in time with the other components in the system.

**Process Certification Requirements**

Because of the difficulty in certifying the conventional software product, certification of the software engineering process is used (Buckley *et al.*, 1990). Some of the techniques can be adapted to neural computing.

*The manager of the development team must be able to demonstrate an understanding of neural computing.* Qualification requirements are a recognized element of process certification (BSI, 1992; Buckley *et al.*, 1990). However, even in the fields considerably more mature than neural computing, defining which qualifications are acceptable and which are not is difficult.

*Individuals on the development team must be able to demonstrate an understanding of their particular part of the development process.* For example, an agent gathering training data must understand that the quality of the operation of the trained network ill directly depend upon his or her care in gathering the training data.

*Wherever possible, formal methods such as Z or VDM, should be used to develop the software simulations which implement the neural network. If this is not possible, a rigorous argument, as defined in (MoD, 1991; MoD, 1993), that the simulations are correct should be developed.* We anticipate the development of small libraries of certified neural network kernels which can be assembled into networks. The overall certification of the network can never be of greater integrity than the integrity of the certification of the software with which it is simulated. Therefore, it is important to attempt to engineer these software libraries with the highest possible integrity.

*Within the development team, assignment of responsibilities must reflect the underlying structure of neural network development.* Separate sub-tasks can be identified within network development: network design, network implementation, and data collection and treatment. One person should be responsible for ensuring dependability in a specific area of development.

*Quality assurance methods must be used in gathering the training data for the network.* Because of the direct impact the quality of the training data has on the quality of the network's performance, assurance methods such as (BSI, 1987; MoD, 1989) must be used in gathering the data, and the actions of the data collection team should be carefully documented.

## TESTING THE NETWORK

As mentioned previously, neural network development and the associated certification process is cyclical, rather than linear (design, build test). However, particular parts of the testing cycle, such as verification and validation tests, should not be attempted until the network design has matured. Otherwise there would exist a great temptation to tweak the network until it passed the V&V tests, but not discover the underlying cause of failure.

*The V&V team are to be managerially and technical independent of the development team.* This requirement is no different from conventional V&V. Such independence means that data used to verify and validate the network by the V&V team will never be available to the developer.

*The V&V team are to verify the network has learned the underlying principles embedded in the training data by using test of the network's ability to generalize.* As a neural network learns from its training set, there comes a point when its error rate *when tested on unseen data* reaches a minimum, even though the network's error rate on its training set will continue to decrease. This is the point when the network has learned as about the underlying principles embodied in the training set as it is going to learn. As this training set reflects the high-level goals which were set for the network, verification is the process of testing to see if the network has learned what it was supposed to learn i.e. can it perform well on unseen data. (Conventionally, verification means testing to see if a system's operation matches its specification.)

*The V&V team are to validate the network's operation by investigating the operation of the whole system in which the network is incorporated when the network is exposed to the whole range of inputs which it might operationally receive.* According to requirements to be placed on the developers in the next section of this draft standard, the developers are to obtain information about the possible failure modes of components which act as inputs to the neural network. The V&V team must evaluate the impact of such failures when "seen" by the network on the whole system in which the network module is embedded.

**SUMMARY**

We have presented a draft standard for the certification of neural networks in safety critical systems. It is clear that there remains much to do in this area before certification with legislative weight becomes a reality (for instance, deeper investigation of the possibilities of rule extraction and the impact of this technology on ANN certification). However, while this standard is unlikely to be the last word in neural network certification, it does constitute a concrete starting point.

As a final comment, our society, while tolerant of occasional mistakes in the SC domain, when made by biological neural networks, remains intolerant of mistakes made by machines, neural or otherwise. The black-box nature of neural networks may well exacerbate this difficulty. However, we believe that such an issue, while of undoubted importance, would be better addressed in a psychological or sociological study, rather than in a draft standard such as this.

**REFERENCES**

Andrews R., Diedrich, J., Tickle, A.B., (1995) A survey and critique of techniques for extracting rules from trained artificial neural networks. *Technical Report QUTNRC-95-01-02,* Queensland University of Technology Neurocomputing Research Centre.

Bishop, C.M., (1994). Novelty detection and neural network validation. *IEE Proc.-Vis. Image Signal Process,* 141(4):217-222.

British Standards Institute, (1987). BS 5750 parts 1-6, *Quality systems, ISO9000 series equivalent,* Technical Report, BSI.

British Standards Institute, (1992). Draft-functional safety of electrical/electronic/programmable electronic systems, generic aspects, *part 1: General Requirements (IEC 65a(secretariat)123), Technical Report,* British Standards Institute.

Buckley, T.F., Jesty, P.H., Hobley, K.M., West, M., (1990). Review of current practices for the certification of safety critical software. *Technical Report, University of Leeds, School of Computer Studies*, Technical Report Series.

Corrieu, P., (1994). Three algorithms for investigating the domain of validity feedforward neural networks, *Neural Networks,* 7:169-174

Eurofighter, (1989). Eurofighter system safety assessment procedure QAP-J-O-E-0048, *Technical Report,* NATO.

Ford Motor Company's Engineering and Manufacturing Staff, (1988). Potential failure modes and effects analysis (FMEA), *Technical Report,* Ford Motor Company.

Jesty, T.F., Buckley, K.M., Hobley, K.M., West, M., (1990). Review of current standards for safety critical software. *Technical Report, University of Leeds, School of Computer Studies*, Technical Report Series.

Knight, J.C., Leveson, N.G., (1986). An experimental evaluation of the assumption of independence in multiversion programming. *IEEE Transactions on Software Engineering,* SE-12(1).

Laprie, J.C., (Ed.), (1994). Dependability: Basic concepts and terminology, *Technical Report, International Federation for Information Processing.*

Maass, M., (1995). Perspectives of current research about the complexity of learning on neural nets. *Technical Report, Institute for Theoretical Computer Science,* Technische Universiaet Graz, Graz, Austria.

MoD, (1989). Practices and procedures for reliability and maintainability, *Defense Standard 00-41*, Ministry of Defense, UK.

MoD, (1993). Requirements for defense systems containing programmable electronics, *Defense Standard 00-56,* part1: requirements and part2: guidance. Ministry of Defence, UK.

Parnas, D.L., van Schouwen, J., Kwan, S.P., (1990). Evaluation of safety-critical software, *Communications of the ACM.*

Partridge D., (1995). Dependable software and inductive programming. *Proceedings Neural Networks Producing Dependable Systems,* National Motorcycle Museum, Solihull, West Midlands, UK.

Poucet, A., Wand, I.C., Wilikens, M.A., (1993). The licensing of safety critical systems containing software. *Joint Research Centre; Commission of European Communities,* Institute for Systems Engineering and Informatics.

Roberts S., Tarassenko, L., (1994). A probabilistic resource allocating network for novelty detection. *Neural Computation,* 6:270-284.

Roberts, S., Tarassenko, L., Pardey, J., Siegwart, D., (1994). A confidence measure for artificial neural networks, *Proceedings of the International Conference on Neural Networks and Expert Systems in Medicine and Healthcare,* pages 23-30.

Systems Integration Requirements Task Group, (1994). Certification considerations for highly-integrated or complex aircraft systems ARP4753, final draft edit 411. *Technical Report, SIRTG,* AS-1C, ASD.

UK Department of Trade and Industry, (1994). Best practice guidelines for developing neural computing applications.

# USING ARTIFICIAL NEURAL NETWORKS TO SOLVE GENERALIZED ORIENTEERING PROBLEMS


**QIWEN WANG**
*Guanghua School of Management*
*Peking University, Beijing, China 100871*

**XIAOYUN SUN**
*BehavHeuristics, Inc.*
*7240 Parkway Drive, Suite 170*
*Hanover, MD 21076*

**BRUCE L. GOLDEN**
*College of Business and Management*
*University of Maryland, College Park, MD 20742*



***ABSTRACT:***

In the orienteering problem (OP), we are given a transportation network in which a start point and an end point are specified. Other points have associated scores. Given a fixed amount of time, the goal is to determine a path from start to end through a subset of locations in order to maximize the total path score. This problem has received a considerable amount of attention in the last ten or so years. In previous work, we applied a modified, continuous Hopfield neural network to attack this NP-hard optimization problem. The approach was shown to perform quite well. In this paper, we attack a generalization of the OP in which each point has a score with respect to a number of attributes (e.g., natural beauty, historical significance, cultural and educational attractions, and business opportunities) and the overall objective function is nonlinear. This generalized orienteering problem (GOP) is, therefore, more difficult than the OP. We present a modified, continuous Hopfield neural network solution procedure and apply it to a specific problem involving 27 cities in China. The paper indicates that neural networks can, indeed, be used to solve certain combinatorial optimization problems effectively.


## INTRODUCTION

Consider the following hypothetical problem. A traveler comes to China and has time to visit several cities. He has several goals in mind (e.g., enjoy the natural beauty of the country, visit sites of historical interest, attend some major cultural events, and identify promising business opportunities). Since travel between Chinese cities is expensive and our traveler must adhere to a budget restriction, we impose a limit on the overall distance his journey can consume. Given this distance constraint, he seeks to do well (in terms of score) with respect to each of his goals. We refer to this problem as the GOP. The GOP is a generalization of the OP.

The OP has been studied extensively in the literature since the early 1980s. In recent papers, Chao et al. [1996] apply a metaheuristic and Wang et al. [1995] apply neural networks to obtain high-quality solutions in a reasonable amount of time.

The main difference between the GOP and the OP lies in the objective function. In the OP, the total score is obtained by summing the scores associated with each point along the path from start to end. In the GOP, the total score is a more complicated function of the points visited.

The motivation for the objective function is based on the following observation: The Tai mountain is quite beautiful and its score with respect to natural beauty might be 8. The Huang mountain, however, is the most beautiful in all of China; its score might be 10. If one visits the Tai mountain alone, a score of 8 is achieved. On the other hand, visiting both mountains generates a score of 10 or perhaps a little more. (A Chinese proverb says that having seen Huang mountain, no other mountains are worth seeing.) In other words, the utility or total score for a path is determined, mainly, by the largest score with respect to each goal. The GOP is, therefore, a combinatorial optimization problem with multiple goals and a nonlinear (mathematical programming) objective function. In this paper, we develop a modified, continuous Hopfield neural network to solve the GOP. An example problem involving 27 Chinese cities is solved using the approach.

## FORMAL DEFINITION OF THE GOP

Let V be the set of N points and E the set of edges between points in V. G = {V,E} is a complete graph. Each edge in E has a symmetric, non-negative cost d(i,j) which is the distance or travel time between point i and point j. Assume the starting location is point 1 and the end location is point N. Each point i in V has a non-negative score vector $S(i) = (S_1(i), S_2(i), ..., S_m(i))^T$, where m is the number of independent goals, and $S_g(i)$ is the score of point i with respect to goal g.

Now we discuss the objective function. Given the dominant role played by the point with the largest score with respect to each goal, we define the utility or total score of a path P, which starts at point 1 and ends at point N, as

$$Z = \sum_{g=1}^{m} W_g \{\max_{i \in P} (S_g(i))\} \tag{1}$$

where $W_g$ is the weight that the traveler attaches to goal g. While notationally simple, the above objective function is difficult to work with since it is not continuously differentiable.

We, therefore, approximate Z using Z′, where

$$Z' = \sum_{g=1}^{m} W_g \, [\{\sum_{i \in p} [S_g(i)]^k\}^{1/k}]. \tag{2}$$

It can be shown that as k grows very large, the term within the outer brackets in (2) approaches the term within the braces in (1) and Z′ approaches Z. The OP is clearly a special case of the GOP, e.g., let k=1 and m=1. Since the OP is an NP-hard problem, the GOP is at least as hard.

## THE NEURAL NETWORK MODEL

For the traveling salesman problem and the OP, many heuristic techniques have been applied, including neural network algorithms (from Hopfield and Tank [1985] to Wang et al. [1995]). In general, traditional heuristic techniques have clearly outperformed neural network algorithms. On the other hand, given the nonlinear objective of the GOP, a neural network algorithm may be well-suited to solve this problem.

We now present a modified, continuous Hopfield neural network for the GOP. The neural network is represented by an N by N matrix of neurons or nodes. Each row corresponds to a particular point, while each column corresponds to a particular position in the path from origin to destination. A sigmoid function is associated with each neuron, and the output value V(i,j) is the state or activation level of the node located in row i and column j. At the completion of our computations, each V(i,j) is either 0 or 1. If V(i,j) = 1, then point i is the $j^{th}$ point visited in the path.

The (fourth order) energy function, denoted by E, consists of five terms and is shown below. We seek to minimize E.

$$E = a/2 \sum_{j=1}^{N} \sum_{i=1}^{N} \sum_{h \neq i}^{N} V(i,j)\, V(h,j) + b/2 \left[ \sum_{i=1}^{N} \sum_{j=1}^{N} V(i,j) - N \right]^2$$
$$+ c[2 - V(1,1) - V(N,N)] + d/2\; H\left( \sum_{i=1}^{N} \sum_{h=1}^{N} \sum_{j=1}^{N-1} d(i,h) V(i,j) V(h,j+1) - D \right)$$
$$- e \sum_{g=1}^{m} W_g \left\{ \sum_{i=1}^{N} Q\left( \sum_{j=1}^{N} V(i,j) \right) [S_g(i)]^k \right\}^{1/k} \qquad (3)$$

where $H(x) = x^2 \text{ if } x \geq 0$, $H(x) = 0 \text{ if } x < 0$, $Q(x) = x \text{ if } x \leq 1$, and $Q(x) = 1 \text{ if } x > 1$.

With respect to the "a" term, the goal is to penalize any column with more than one activated node, since only one point can be the $j^{th}$ point in a path. The "b" term seeks to ensure that the number of activated nodes will be equal to N, one for each column. Note that if fewer than N nodes are in the path, then some rows will have a string of consecutive activated nodes. This follows from the minimization of E. The path is encouraged to start at point 1 and end at point N by the "c" term. The "d" term is designed to prevent the total path length from exceeding D units, which is the upper bound. Finally, the "e" term is used to maximize the objective function.

Because the origin and destination are fixed (i.e., V(1,1) = V(N,N) = 1), we do not need to update these two states or the corresponding rows and columns. In other words, the "c" term is forced to equal zero and it, therefore, vanishes. For 1 < i < N and 1 < j < N, the partial derivative of the energy function with respect to state V(i,j) can now be computed. We do not show it here due to space limitations.

We use discrete-time dynamics to update this continuous-valued neural network. Let U(i,j) be the input value of the sigmoid function at neuron (i,j), i.e., $V(i,j) = [1 + e^{-U(i,j)}]^{-1}$. Using the gradient descent method, we update the state at neuron (i,j) via the formula

$$\Delta U(i,j) = -\partial E/\partial U(i,j) * \Delta t = -\partial E/\partial V(i,j) * \partial V(i,j)/\partial U(i,j) * \Delta t$$
$$= -\partial E/\partial V(i,j) * V(i,j) * [1 - V(i,j)] * \Delta t \quad \textbf{(4)}$$

where $\Delta t$ is a pre-specified value. From the chain rule, we can show that $\partial E/\partial t < 0$. Thus, the energy decreases over time, when the step $\Delta t$ is small enough.

To speed up convergence, we select at most two neurons in one row or column at a time, such that $\Delta U(i,j)$ for one neuron is the largest positive value and $\Delta U(i,j)$ for the other neuron is the most negative value within the row or column. In addition, the absolute value of these $\Delta U(i,j)$ should be greater than a threshold, denoted $\delta$. Two other numbers are specified in order to ensure a proper variety of solutions from which to select the best one. N1 represents the number of replications. At the beginning of each replication, the state matrix is randomly initialized. N2 is the number of iterations per replication. Within a replication, the state matrix is gently perturbed at the end of each iteration. The state updating procedure is as follows:

Step 0. Set parameter values for N1, N2, a, b, d, e, $\Delta t$, and $\delta$. Let r = 0.
Step 1. Randomly initialize the state matrix to very small positive values. Let r = r + 1 and i = 1.
Step 2. Randomly choose one row or column and calculate $\Delta U(i,j)$ for each neuron in the row or column. Choose the neuron with the largest positive value and the neuron with the most negative value of $\Delta U(i,j)$. If they exist and exceed $\delta$ in absolute value, then update their states. Repeat this process until the state matrix reaches stability, i.e., until all $|\Delta U(i,j)| < \delta$.
Step 3. Select the largest state in each column. Set this state to 1 and set all other states in the column to 0. Construct the path corresponding to the state matrix. Apply the 2-opt procedure (see Wang et al. [1995]) to improve the path.
Step 4. If the length of the path does not exceed D, compare it with the best solution obtained so far. Store the best solution. If i = N2, go to Step 6.
Step 5. If the length of the path exceeds D, the path is infeasible. Increase d and decrease e. If the path is feasible, increase e and decrease d. Perturb the state matrix slightly. Let i = i + 1. Go to Step 2.
Step 6. If r ≤ N1, go to Step 1. Otherwise, stop and display the best result.

Note that since we never alter the first row, last row, first column, or last column, the neural network could be represented by (N-1)*(N-1) neurons.

## AN EXAMPLE

To illustrate the GOP, we generated an example involving 27 Chinese cities. In Table 1, the longitude, latitude, and a score vector are presented for each city. $S_1$, $S_2$, $S_3$, and $S_4$ are the scores with respect to natural beauty, historical significance, cultural and educational attractions, and business opportunities. Note that these entries in Table 1 are approximate, rather than definitive. They are scaled from 1 to 10.

The distance matrix is calculated using trigonometric formulas on spherical surfaces. The average radius of the earth is taken as 6371 kilometers.

In order to specify a robust set of parameter values, we sought to standardize the raw data. For example, in the "d" term of the energy function, each distance is divided by the largest one, so that each one is contained in (0,1]. The distance limit is also divided by the same number. If k=5 in the "e" term, then the following parameter settings seem to work well: a=10, b=1, d=300, e=1, N1=10, N2=10, $\Delta t=1$, $\delta=0.1$. Using the above parameters and algorithm steps, we wrote a computer program to solve the GOP. The program runs on a PC 486/50 and requires approximately 3 minutes to solve the 27 Chinese cities problem.

In Table 2, we present results for six cases. Each case corresponds to a combination of weight vector and distance limit. The total path distance, the resulting score, and the path itself are provided. In all cases, Beijing is both the origin and destination. The five weight vectors used include W0 = (0.25, 0.25, 0.25, 0.25), W1 = (1, 0, 0, 0), W2 = (0, 1, 0, 0), W3 = (0, 0, 1, 0), and W4 = (0, 0, 0, 1). The first of these gives equal weight to each of the four goals. The four other weight vectors each stress one goal and ignore the other three. The two distance limits are 5000 km. and 11000 km.

In comparing the paths from Table 2, the impact of weight vector and distance limit on the results is clear. In case 2, where the traveler is interested in natural beauty only, the most beautiful city, Guilin (point 19), is visited. In case 5, where the traveler is pursuing business opportunities only, the resulting path clings to the east coast of China. In other words, by varying the weight vector, we can generate different paths for travelers with different purposes. We point out that in case 6, where the distance limit is large enough, all cities are visited, as with the traveling salesman problem.

## CONCLUSION AND FUTURE WORK

This paper builds upon the earlier, successful application of neural networks to the OP. The GOP is a more general and more complex combinatorial optimization problem than the OP. Unlike the OP, the GOP has a nonlinear (mathematical programming) objective function. Nonetheless, the modified, continuous Hopfield neural network approach seems to work well, even when applied to the GOP. An interesting question arises: How would more traditional heuristics perform on the GOP? The answer is not clear. Perhaps future work will address this question.

Another direction for future research would be to use a modification of the proposed GOP algorithm to model the score as time-dependent. In other words, the score at point i with respect to goal g would depend upon the amount of time spent at point i pursuing goal g. The total amount of time spent at all points plus the total travel time would be limited by D. To begin, we might focus on the time-dependent OP, i.e., a time-dependent GOP with a single goal.

## REFERENCES


Chao, I-M., Golden, B., and Wasil, E., (1996). "A Fast and Effective Heuristic for the Orienteering Problem," European Journal of Operational Research, (88), 475-489.

Hopfield, J., and Tank, D., (1985). "Neural Computation of Decisions in Optimization Problems," Biological Cybernetics, (52), 141-152.

Wang, Q., Sun, X., Golden, B., and Jia, J. (1995). "Using Artificial Neural Networks to Solve the Orienteering Problem," Annals of Operations Research, (61), 111-120.

TABLE 1: LOCATIONS AND SCORES OF 27 CITIES IN CHINA

| | City | Longitude | Latitude | $S_1$ | $S_2$ | $S_3$ | $S_4$ | | | City | Longitude | Latitude | $S_1$ | $S_2$ | $S_3$ | $S_4$ |
|---|---|---|---|---|---|---|---|---|---|---|---|---|---|---|---|---|
| 1 | Beijing | 116.40 | 39.91 | 8 | 10 | 10 | 7 | | 15 | Wuhan | 114.30 | 30.55 | 6 | 6 | 8 | 6 |
| 2 | Tianjin | 117.18 | 39.16 | 6 | 5 | 8 | 8 | | 16 | Changsha | 113.00 | 28.20 | 6 | 6 | 6 | 5 |
| 3 | Jinan | 117.00 | 36.67 | 7 | 7 | 5 | 6 | | 17 | Guangzhou | 113.15 | 23.15 | 6 | 6 | 5 | 10 |
| 4 | Qingdao | 120.33 | 36.06 | 7 | 4 | 5 | 7 | | 18 | Haikou | 110.35 | 20.02 | 7 | 3 | 4 | 8 |
| 5 | Shijiazhuang | 114.50 | 38.05 | 5 | 4 | 5 | 5 | | 19 | Guilin | 110.29 | 25.28 | 10 | 4 | 4 | 4 |
| 6 | Taiyuan | 112.58 | 37.87 | 5 | 6 | 5 | 5 | | 20 | Xi'an | 108.92 | 34.28 | 5 | 9 | 8 | 6 |
| 7 | Huhehaote | 111.70 | 40.87 | 6 | 6 | 5 | 5 | | 21 | Yinchuan | 106.27 | 38.48 | 5 | 7 | 5 | 5 |
| 8 | Zhengzhou | 113.60 | 34.75 | 5 | 6 | 5 | 5 | | 22 | Lanzhou | 103.80 | 36.03 | 7 | 6 | 5 | 6 |
| 9 | Huangshan | 118.29 | 29.73 | 9 | 3 | 2 | 2 | | 23 | Chengdu | 104.07 | 30.66 | 6 | 7 | 6 | 5 |
| 10 | Nanjing | 118.75 | 32.04 | 7 | 8 | 8 | 6 | | 24 | Guiyang | 106.70 | 26.59 | 8 | 5 | 4 | 5 |
| 11 | Shanghai | 121.45 | 31.22 | 5 | 4 | 9 | 9 | | 25 | Kunming | 102.80 | 25.05 | 9 | 7 | 7 | 6 |
| 12 | Hangzhou | 120.15 | 30.25 | 9 | 8 | 7 | 6 | | 26 | Shenyang | 123.40 | 41.80 | 5 | 8 | 5 | 6 |
| 13 | Nanchang | 115.88 | 28.35 | 7 | 6 | 5 | 5 | | 27 | Dalian | 121.60 | 38.92 | 7 | 5 | 6 | 7 |
| 14 | Fuzhou | 119.30 | 26.10 | 6 | 5 | 5 | 7 | | | | | | | | | |

TABLE 2: RESULTS OF SIX CASES

| Case | Weight Vector | Distance Limit | Total Distance | Score | Path |
|---|---|---|---|---|---|
| 1 | W0 | 5000 | 4993.4 | 12.38 | 1-2-3-10-11-12-9-13-17-19-16-20-6-5-1 |
| 2 | W1 | 5000 | 4987.7 | 13.05 | 1-2-3-4-10-11-12-9-13-16-19-24-20-6-5-1 |
| 3 | W2 | 5000 | 4875.1 | 12.51 | 1-2-26-27-3-10-11-12-9-13-15-20-6-5-1 |
| 4 | W3 | 5000 | 4987.5 | 12.78 | 1-2-3-5-6-20-8-15-16-13-9-12-11-10-4-27-1 |
| 5 | W4 | 5000 | 4989.8 | 12.36 | 1-2-3-10-9-13-16-17-14-12-11-4-27-1 |
| 6 | W0 | 11000 | 10428 | 13.35 | 1-7-6-5-3-8-20-21-22-23-25-24-19-18-17-16-15-13-14-9-12-11-10-4-27-26-2-1 |

# A SYSTEMATIC APPROACH FOR DESIGNING NEURAL NETWORKS WITH APPLICATION TO CLASSIFICATION, DECISION, AND CONTROL

**NADER M. BOUSTANY**
*General Motors Energy Center*

***ABSTRACT:***
This paper describes a systematic approach for designing artificial neural networks for solving a wide class of classification and decision-making problems. This approach allows the determination of the number of layers, the number of neurons in each layer, and provides a natural interpretation of the function of the hidden layers. It relies on a hyperplane realization of classification in n-space as well as neural realizations of arbitrary Boolean Functions. Furthermore, the approach also shows how, in principle, the synaptic weights can be directly calculated for the class of problems being considered.

**Key Words:** Neural Net Architecture, Decision Making, Classification, Neuro Control.

## 1.0 Definition of the Classification Problem:

This paper deals only with neural networks of the feedforward multilayer variety (see Fig. 1) consisting only of neurons having a {0, 1} threshhold function (see Fig. 2). That is, neurons only capable of having an output of 0 or 1.

Classification in n-space with neural nets will be looked at as the approximation of the classes such that the characteristic functions differ over a set that can be made to have arbitrarily small measure.

We shall further restrict our attention to sets **U** that can be approximated arbitrarily closely by sets bounded by a finite number of hyperplanes in $\mathbf{R}^n$. In other words we will not be considering classification dealing with sets like {all the rationals on a segment of a real line} or subsets of $\mathbf{R}^n$ of fractal dimension. The ability of neurons to represent linear inequalities on $\mathbf{R}^n$ is well documented (see for example Nilsson[1] and Jain , Mao and Mohiuddin [2]). Furthermore, in [1] and [2] various interpretations of how the hidden layers realize the entire characteristic function are provided. The major contribution of this paper is to provide an alternative interpretation of how the various layers contribute to the realization of the characteristic function. The first layer will essentially consist of n neurons, each connected to all the inputs $\{x_i\}_{i=1}^{n}$. The synaptic weight from input i to neuron j is

$w_{ij}$. Each neuron j has a bias $d_j$ (see Fig. 2). We will call this layer the hyperplane layer.

Now, consider the outputs of the hyperplane layer as logical variables (i.e. 1=True and 0=False). Then if we connect these outputs to the appropriate Boolean function (e.g. one consisting of AND-gates, OR-gates, and NOT-gate), we will essentially get the realization we are looking for.

If the neural net in question mechanizes a classifier, then the Boolean function maps the {0, 1} outputs of the hyperplane layer into a binary vector (i.e. one whose elements are in {0, 1}) which encodes the various classes. If we are talking about a decision-maker, then the outputs of the Boolean function encode the decisions. If analog outputs are desired, a bank of D/A converters could be used to generate the analog outputs. A D/A converter is essentially a specialized neuron that does not have any threshholding and whose synaptic weights (in case of binary word coding) are powers of 2 (multiplied by a scaling constant).

## 2.0 Neural Net Realizations of Boolean Functions:

In this section of the paper we will show how to realize AND-gates OR-gates and NOT-gate each with a single neuron. Any Boolean function can be realized with a combination of these gates.

### 2.1 Realization of an AND-Gate:

An n-input AND-gate is defined as follows:

Output=0 if $x_i$=0 for at least one i

Output=1 if $x_i$=1 for all i

A symmetry argument tells us that all the inputs have to have the same synaptic weight w. Therefore, the above two conditions imply the following relationship between the number of inputs n, the synaptic bias d, and the synaptic weight w.

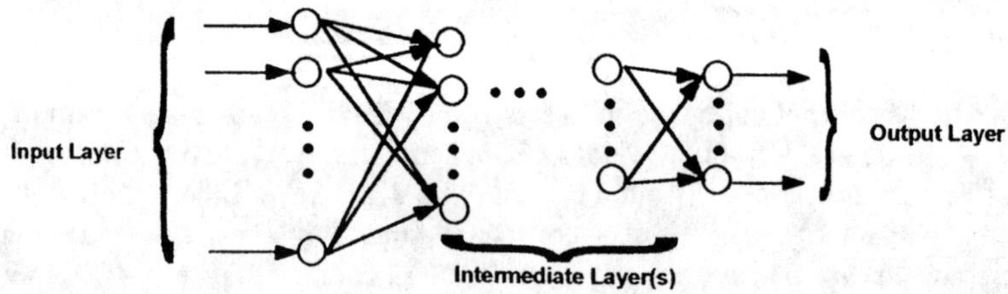


Fig. 1 Feedforward Network

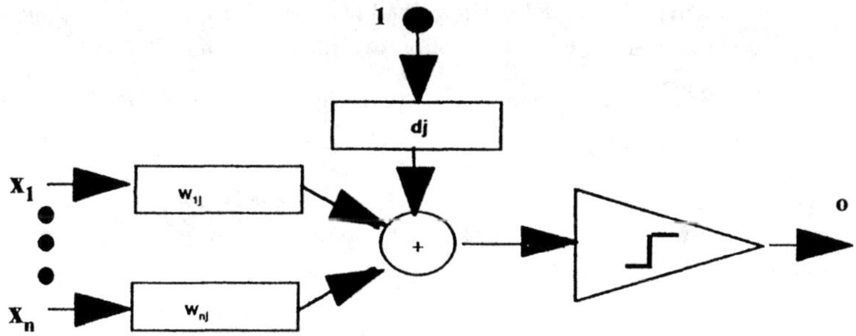


$\sum$ (x.w)+d >0 o=1

otherwise o=0

Fig. 2 General Model for Neuron (in this case Neuron j in the Hyperplane Layer)

$$(n-1)w+d<0$$

These two inequalities reduce to the following condition:

$$-d/n \leq w < -d/(n+1) \text{ and } d<0$$

## 2.2 Realization of the OR-Gate:

An n-input OR-gate is defined as follows:

Output=1 if $x_i$=1 for at least one i

Output=0 if $x_i$ =0 for all i.

This definition implies the following:

$$w + d > 0 \text{ and } d<0$$

## 2.3 Realization of the NOT-Gate:

In this case we need ( as the reader can readily verify):

$$w+d<0, \text{ and } d<0$$

## 2.4 A Comment on Synaptic Weights:

For the calculation of the synaptic weights we will distinguish between two cases. The first case is when the classification regions are known from the underlying principles. In this case the regions are to be approximated with hyperplanes, which in turn determine the synaptic weights and bias terms of the first layer. The second case is when the classification is to be determined from statistical samples. In this case discriminant analyisis can be used to determine the hyperplanes (see for example Reference [3]).

## 3.0 An Illustrative Example:

In this section a simple example will be worked out in some detail to illustrate how to compute the synaptic weights (at least in principle) and to provide some insight on the function of the hidden layers.

Refer to Fig. 3 and consider Regions I, II, III, and IV in $\mathbf{R}^2$ that are defined by the parabola

$$g_1(\mathbf{x})=x_1^2-x_2=0$$

and the horizontal axis,

$$g_2(\mathbf{x})=x_2=0.$$

These regions will be approximated (in the sense defined in Section 1) with the three hyperplanes L1, L2, and L3 defined in Fig. 3. The labels (with no risk at confusing the reader) L1, L2, and L3 will now be used for the following logical propositions:

$$\text{L1:}\ \ -x_2>0$$

$$\text{L2:}\ \ x_2+1.667x_1+0.5>0$$

$$\text{L3:}\ \ x_2-1.667x_1+0.5>0$$

From Section 1 we know that the hyperplane layer will consist of three neurons each having $x_1$ and $x_2$ as inputs and whose outputs are logical variables L1, L2, and L3 corresponding to the propositions above. Moreover, for each neuron the synaptic weights are the coefficients of $x_1$ and $x_2$ in the appropriate proposition. The constant term in the proposition will be the bias weighting in each case. The outputs of the hyperplane layer are inputs to a Boolean function.

We desire to classify the points in $\mathbf{R}^2$ as belonging to either Region I, II, III, or IV by building a neural net with two inputs $x_1$ and $x_2$ and two outputs O1 and O2 defined in Table 1. A standard text on switching theory (e.g. Kohavi [4]) will provide a methodology for deriving the Boolean functions for our neural net. These turn out to be:

$$\text{O1} = \text{L1} \wedge \text{L2}$$

$$\text{O2} = (\text{L1} \wedge \text{L2}^{\text{c}}) \vee (\text{L1} \wedge \text{L3}^{\text{c}})$$

| $\{x_1, x_2\}$ in Region | O1 | O2 |
|---|---|---|
| I | 0 | 0 |
| II | 0 | 1 |
| III | 1 | 0 |
| IV | 1 | 1 |

Table 1 Binary Coding for the Outputs of the Parabola Classifier

## 4.0 Concluding Remarks:

The resulting parabola classifier will turn out not to be in the form illustrated by Fig. 1. By defining two additional neurons it can be transformed to that form at the cost of adding neurons. The benefit, however, is that we will have a standard six layer general purpose neural net that is amenable to automated design. Its synaptic weights and bias terms can in principle be calculated. Furthermore the number of neurons in each layer can be determined exactly. The alloted space , however, does not permit a detailed discussion of these aspects of the design.

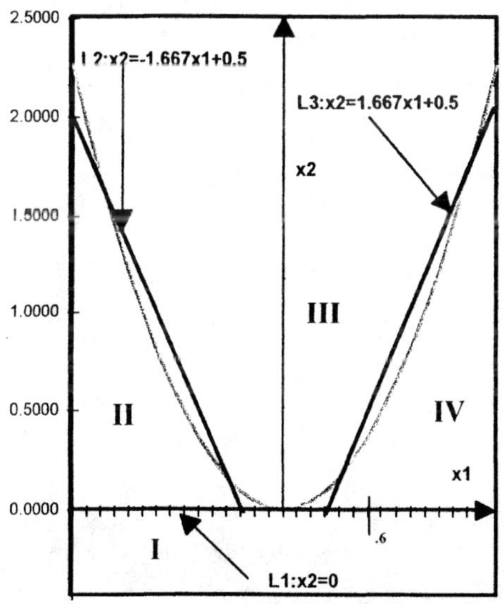


Fig. 3 Parabola Classifier

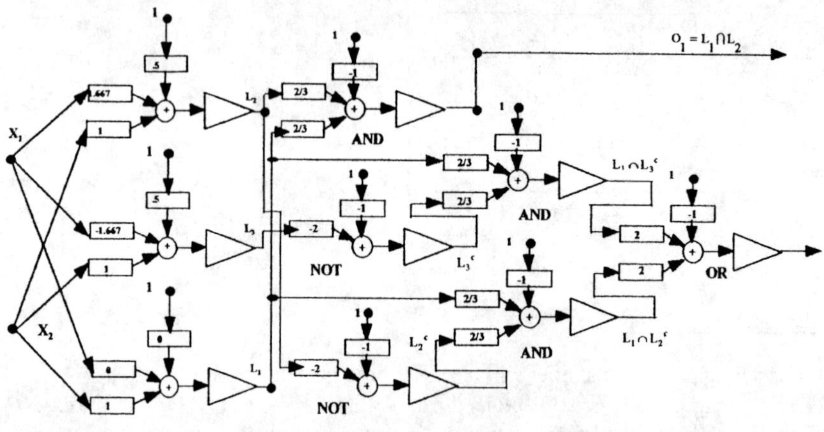


Fig 4 The Neural Net Realization of the Parabola Classifier

## Bibliography


[1] Nilsson, N. J., The Mathematical Foundations of Learning Machines, Morgan Kaufman, San Mateo, 1990.

[2] Jain, A. K.; Mao, J.; Mohiuddin, K. M., Artificial Neural Networks: A Tutorial, Computer (IEEE), March 1996.

[3] Johnson, R. A.; Wichern, D.W., Applied Multivariate Statistical Analysis, Prentice Hall, Englewood Cliffs, 1982.

[4] Kohavi, Z. , Switching and Finite Automata Theory, Mc Graw Hill, New York, 1970.

# FUZZY LOGIC APPROACH FOR SERVICEABILITY EVALUATION

**JIANGPING WANG AND VENKAT ALLADA, Ph.D.**
*Computer Integrated Manufacturing Laboratory*
*Department of Engineering Management*
*205 Engineering Management Building*
*University of Missouri-Rolla*
*Rolla, MO 65401*

***ABSTRACT:***
Serviceability evaluation during early design stages is a process of evaluating the product design in terms of ease of service or maintenance during a product usage life cycle. The methodological focus of this study is on evaluating serviceability index using a multi-attribute, fuzzy logic approach. The aim of this study is to offer an evaluation tool that assesses various design alternatives in terms of serviceability.

## INTRODUCTION

Life-cycle engineering design addresses not only the functional requirements of a product but also manufacturability, serviceability, reliability, and recyclability (Ishii *et al.*, 1993). All of these life-cycle elements interact with each other. The ultimate goal of green design is to reduce the overall environmental impact of producing goods and services.

There are various approaches for the development of environmentally oriented products, all of which can obtain good results. However, their success depends on the situation in which they are applied (Van der Horst *et al.*, 1993). Green design evaluation in the product design stage is a process to evaluate the product design in terms of environmental compatibility or greenness. Quantifying approaches are needed to assist the evaluation. The most widely used approach for quantifying evaluation of product designs is index (or indicator) based approach which combines the diversity of design considerations into a unique index. While serviceability index is examined in this study, the proposed approach can easily be expanded to develop a comprehensive product evaluation tool comprising wider spectrum of considerations.

## SERVICEABILITY INDEX CONSIDERATIONS

Serviceability is a measure of the ease of performing all service-related operations on a product. The ability to effectively and economically service products provides the opportunity to extend product life and contribute to its sustainable development. There are many factors or considerations that affect serviceability. Due to the high variability in product types and configurations, serviceability considerations can vary from one product to another. However, there are some general serviceability considerations which can be defined in the areas of disassembly, reassembly, handling and service cost. These serviceability considerations are further subdivided as shown in Figure 1. Each of these serviceability considerations is discussed in this section.

Disassembly is a measure of ease of disassembling components from a given product. It is affected by the basic attributes as the ones listed below.

- Accessibility: Easy access to product areas requiring service is desirable.
- Fastener number: Fewer number of fasteners in a given product means fewer disassembly operations.
- Fastener variability: Fewer variety of fastener types in a given product means simplified tooling requirements for disassembly purposes.
- Fastener type: Some types of fasteners can be easily disassembled than others.
- Direction variability: Fewer disassembly directions are preferable for disassembly purposes.
- Common tooling: Commonly available tools are preferable over specialty tools for disassembly purposes.

A large number of analogies and one-to-one correspondence often exist between disassembly and reassembly during product service. However, although the basic attributes are similar, their values may vary. This means that easier disassembly procedures do not necessarily result in easier reassembly procedures. Scheuring *et al.* (1994) provide three categories of design for disassembly influence on assembly: positive effect, negative effect and relatively little effect. For instance, reduction in the number of fasteners has positive effect on assembly, while two way snap fits or break points on snap fits have negative effect on assembly.

Some kinds of handling and identification aids make it easier to protect from repair damage and to provide easier service. These include,

- No-rigid parts: Parts with bending behavior are difficult to manipulate for both disassembly and reassembly.
- Grasping: Grasping "features," such as handles and hooks, on a product facilitate serviceability.
- Marking/coding: In-mold coding, bar coding and color coding provide easier identification of components and subassemblies.
- Modularity: Modular design generally facilitates easier testing, replacement of the individual modular components of a product.

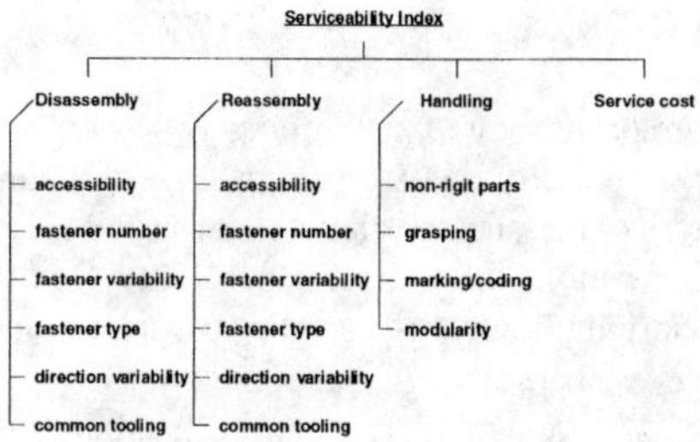


**Figure 1**: Serviceability index.

The service cost is an aggregate of the labor cost, material cost, transport cost and other incidental costs, which is affected by factors such as frequency for the service or mean time between failure (MTBF), time or steps involved in the service, unit service rate and necessary tooling factor (Ishii, 1995, Gershenson *et al.,* 1993).

## FUZZY LOGIC APPROACH

The fuzzy logic approach uses the fuzzy system architecture which consists of four principal components: fuzzification interface, fuzzy rule base, fuzzy inference machine, and defuzzification interface (Haykin, 1994) as shown in Figure 2.

Fuzzification interface performs a scale mapping that changes the range of values of input variables into corresponding universe of discourse, and fuzzification that converts non-fuzzy (crisp) input data into suitable linguistic values, which may be viewed as labels of fuzzy sets. Fuzzy rule base consists of a set of linguistic fuzzy associative memory (FAM) rules written in the IF-THEN form. Fuzzy inference machine is a decision making logic that employs rules from the fuzzy rule base to infer fuzzy output in response to fuzzified inputs. Defuzzification interface performs a scale mapping that converts the range of values of output variables into corresponding universe of discourse, and defuzzification that yields a non-fuzzy (crisp) output value from an inferred fuzzy output.

### Fuzzy Variables

The variables of basic level attributes may be expressed with fuzzy values to characterize their uncertainty. Interval variable membership functions are used to evaluate some attributes in terms of most likely and largest likely as shown in Figure 3(a). The fuzzy values of variables can be expressed by a trapezoidal function. As shown in Figure 3(a), values *b* and *c* stand for lower- and upper-bound values of most likely interval, and values *a* and *d* stand for lower- and upper-bound values of largest likely interval. An example of interval membership functions is service cost which is estimated as most likely between $20 and $30 and largest likely between $15 and $40.

Linguistic variable membership functions are used to express some attribute with linguistic terms. As shown in Figure 3(b), the membership function for this study is defined with linguistic variables LOW (L), MEDIUM LOW (ML), MEDIUM (M), MEDIUM HIGH (MH), and HIGH (H).

Examples of linguistic variables are: accessibility is medium, marking or coding measure is high and modularity is medium high.

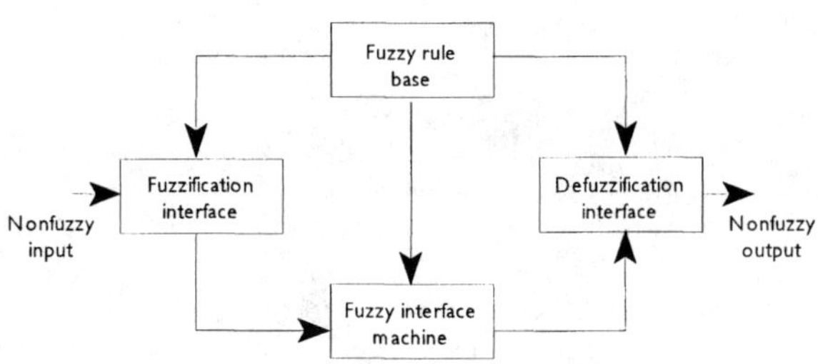


**Figure 2**: System architecture.

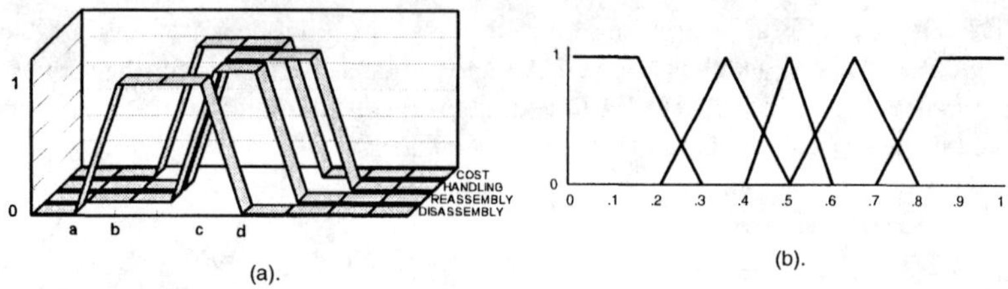


**Figure 3**: Membership function.

## Fuzzy Value Transformation

Because the units and the range of raw values for basic level attributes are different and thus difficult to compare them directly, the raw value of each basic variable should be transformed into an index that is bounded in the uniform range 0 to 1. Using the best value and worst value for the basic attributes, raw values can be transformed into an index value of their own. This process normalizes the fuzzy values in relation to the best and worst values for a particular criterion.

The transformation could be linear or nonlinear as shown in Figure 4. On most occasions, the membership functions are assumed to be linear for values between the most likely and largest likely values, thus the transformation is based on best values (BV) and worst values (WV). For example, if the attribute is cost, then lower values are better, thus the best value is less than the worst value. If the attribute is percentage of modularity, then higher values are better, thus the best value is greater than the worst value.

There are also occasions when the transformation is nonlinear in nature. An example of such transformation is the index for service cost (Ishii, 1995, Eubanks *et al.*, 1993).The index is based on a user-specified maximum reasonable cost (mrc) and calculated using an exponential decay function as follows:

$$Index = e^{\left(\frac{cost}{mrc}\right)\times\ln(0.5)}$$

The index ranges from 1 to 0, with 1 representing a good design accumulating zero service cost, 0.5 representing a design accumulating the maximum reasonable cost, and 0 representing a design accumulating infinite service cost.

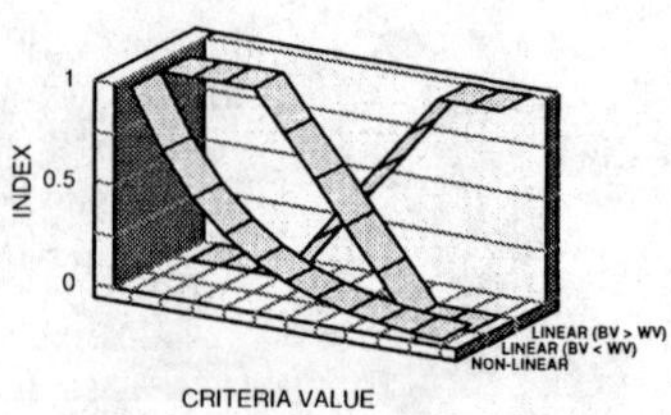


**Figure 4**: Value transformation

| TABLE 1: PAIRED-ATTRIBUTE MATRIX | | | | |
|---|---|---|---|---|
| | Disassembly | Reassembly | Handling | Cost |
| Disassembly | 1 | 1 | 2 | 0.5 |
| Reassembly | 1 | 1 | 2 | .5 |
| Handling | .5 | .5 | 1 | .33 |
| Cost | 2 | 2 | 3 | 1 |

## Higher Level Indices

The higher level index values, such as indices for disassembly and reassembly, are calculated by using weighting factors, *w*, as following:

$$Index = \sum w * S$$

The weighting factors, which are used to indicate the relative importance of each criterion, are determined in a method of a paired attribute matrix. A paired attribute matrix is prepared by using the relative importance of each attribute. The eigenvectors of the maximum eigenvalues are then calculated (Saaty, 1988). The weighting factors are obtained by normalizing the eigenvectors such that all the weighting factors for each comparison sum to a value of 1. The example of the paired matrix for weighting factors is provided in Table 1 and the corresponding eigenvector is (0.23, 0.23, 0.12, 0.42).

FAMs can be used for transformations and fuzzy sets to fuzzy sets (Kosko, 1992). The FAM rules have the IF-THEN form. For the serviceability considerations here, the transformations can be expressed with FAM rules, such as (MH, MH; MH) and (H, H; H) for the mapping between accessibility, common tooling and disassembly. We can represent the bank of FAM rules as linguistic matrix as shown in Table 2.

## Ranking Fuzzy Indices

To defuzzify and rank the alternative design evaluations, the maximizing and minimizing set concepts of fuzzy logic are used (Paek *et al.,* 1992). This ranking method is illustrated in Figure 6. The maximizing set $\mu_M$ intersects each trapezoidal fuzzy set in two places. The value for the right utility $U_R$ for each fuzzy set is the largest of these two intersection values. Similarly, the value for the left utility value $U_L$ is the maximum of the two intersection values of the minimizing set $\mu_G$ and the fuzzy set. The total utility or ordering value *U* for each alternative is found by the following equation:

$$U = (U_R + 1 - U_L)/2$$

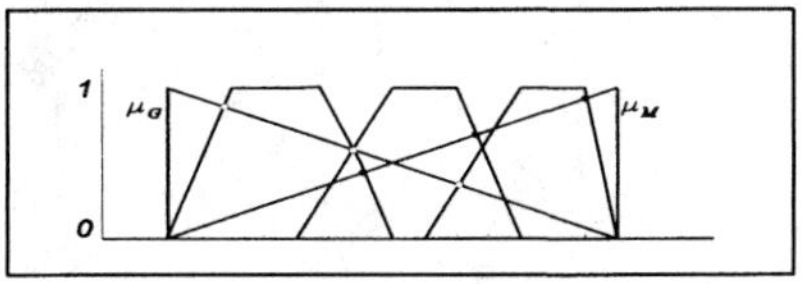


**Figure 5**: Ranking fuzzy indices.

**TABLE 2: FAM RULES BANK**

| Disassembly | | Common Tooling | | | | |
|---|---|---|---|---|---|---|
| | | L | ML | M | MH | H |
| | L | L | L | ML | L | L |
| | ML | ML | ML | ML | M | L |
| Accessibility | M | ML | M | M | M | MH |
| | MH | L | M | MH | MH | MH |
| | H | L | L | MH | H | H |

## CONCLUSIONS

This paper presents a fuzzy logic approach for serviceability evaluation. The uncertainty of the basic attributes for the evaluation can be expressed with fuzzy values. Fuzzy analysis accommodates uncertainty about the meaning of words, in this case, of attributes, their levels, and their relative importance. Further investigation in the use of fuzzy neural networks for serviceability evaluation is currently being pursued by the authors.

## REFERENCES


Eubanks, C. F., and Ishii, K., (1993). AI methods for life-cycle serviceability design of mechanical systems, *Artificial Intelligence in Engineering*, Vol. 8, pp. 127-140.

Gershenson, J., and Ishii, K., (1993). Life-cycle serviceability design, *Concurrent engineering: automation, tools, and techniques,* edited by Andrew Kusiak, John Wiley & Sons, Inc., NY.

Haykin, S., (1994). Neural networks, a comprehensive foundation. Macmillan College Publishing Company, NY.

Ishii, K., (1995). Life-cycle engineering design, *Design for Manufacturability, ASME 1995,* DE-Vol. 81, pp. 39-45.

Ishii, K., Eubanks, C. F., and Marks, M., (1993). Evaluation methodology for post-manufacturing issues in life-cycle design, *Concurrent engineering: research and applications,* Vol. 1, pp. 61-68.

Kosko, B., (1992). Neural networks and fuzzy system, a dynamical systems approach to machine intelligence. Prentice-Hall, NJ.

Paek, J. H., Lee, Y. W., and Napier, T. R., (1992). Selection of design/build proposal using fuzzy-logic system, *Journal of construction engineering and management,* Vol. 118, No. 2, pp. 303-317.

Saaty, T. L., (1988). Multicriteria decision making: the analytic hierarchy process. University of Pittsburgh, Pittsburgh, PA.

Scheuring, J. F., Bras, B., and Lee, K., (1994). Significance of design for disassembly in integrated disassembly and assembly processes, *International journal of environmentally conscious design and manufacturing,* Vol. 3, No. 2, pp. 21-33.

Van der Horst, T. J. J., and Zweers, A., (1993). Environmentally oriented product development: various approaches to success, *Proceedings of ICED 93*, pp. 739-746.

# INJECTION MOLD COMPLEXITY EVALUATION MODEL USING A BACKPROPAGATION NETWORK IMPLEMENTED ON A PARALLEL COMPUTER

**RAWIN RAVIWONGSE AND VENKAT ALLADA, Ph.D.**
*Computer Integrated Manufacturing Laboratory,*
*Department of Engineering Management University of Missouri-Rolla,*
*Rolla, MO 65401*

***ABSTRACT:***
Mold design is one of the most important activities in the injection molding process. It is a complex task which affects several downstream processes including mold construction, quality of part produced, and cost of mold manufacture. Various factors such as part dimensions, number of undercuts, parting line, cavity detail, tolerances, and number of cavities per mold have been found influencing the complexity of a mold design. This paper demonstrates the application of a backpropagation neural network, running on a parallel computer, to evaluate the complexity level of a mold. The outputs from the network are classified into three levels: easy, moderate, and difficult. Ten part samples have been used to determine the ability of the network in classifying the levels of mold complexity.

## INTRODUCTION

Injection molding process is a major processing technique for converting plastic resins into different types of products. It accounts for approximately one-third (by weight) of the plastic parts produced (Rosato, 1986). The advantages of injection molding process include the ability to produce at high production rates, little or no finishing requirements, process flexibility, reusability of scrap, etc. The injection molding process basically operates in sequence starting from i) raising the temperature of the plastic resin to a point where it will flow under pressure, ii) allowing the plastic to solidify in the mold during mold closure cycle, and iii) opening the mold to eject the plastic part (Rubin, 1972).

The injection mold is a mechanism where the hot plasticized material is injected and maintained under pressure while it cools into a desired shape. A mold consists of ten major components: mold base, guide pins, a sprue, runners, gates, cavities and forces, water channels, sides, vents, and an ejector mechanism. Figure 1 illustrates the mold components.

A backpropagation network (BPN) is currently the most general purpose and commonly used neural network paradigm (Skapura 1996). BPN is a multilayer network

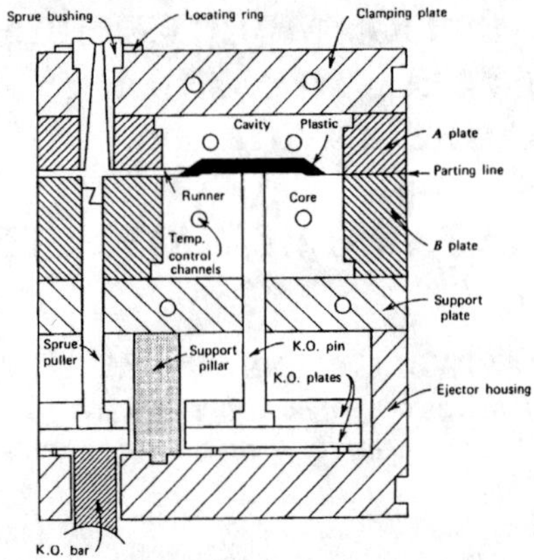


Figure 1. Mold components (Rubin, 1972)

which compares the target results with the actual results, calculates the errors and modifies the network weight matrix to minimize the errors. It is useful for tasks that require classification, pattern recognition, or mapping continuous input values to continuous output values.

The CNAPS (Connected Network of Adaptive ProcessorS) system, used in this study, is a complete parallel system based on a custom digital chip containing up to 64 processors (Adaptive Solution, 1993). The CNAPS software consists of the following modules: i) the BuildNet, which is used for conducting experiments and developing applications using prewritten algorithms such as Backpropagation (BP), Linear Vector Quantization (LVQ2), and Self Organizing Map (SOM), ii) the CNAPS-C, which is a specialized C compiler, and iii) the CodeNet, which is the assembly language programming environment. In this project, the standard backpropagation software of the BuildNet module is applied to the problem of mold complexity evaluation.

## INPUT ATTRIBUTES

This section provides details regarding the network's input attributes, which are adapted from a handbook (Dixon and Poli, 1995). Each input attribute consists of two or more values. The level of each attribute are shown in the bracket following the attribute type.

**Part Type** [Flat, Box]

Parts which are used for evaluation are classified as either *flat* or *box* shaped. It is determined by the ratio of the sides of the *basic envelope* for the part. The basic envelope is simply the smallest rectangular prism which completely encloses the part.

The length of the sides of the basic envelope are represented by L, B, and H; where $L \geq B \geq H$. A part is considered flat if L/H is greater than 4.00; otherwise it is considered as box shaped. Flat parts typically yield the lower complexity for mold design and construction.

**Part Size** [Small, Medium, Large]
Parts are also classified according to their size. The longest side of the basic envelope of a part (L) determines the size of a part. If 'L' is greater than 480 mm, a part is considered as a 'large' part. If 'L' falls between 250 to 480 mm, a part is considered 'medium', and if 'L' is less than 250 mm, a part is considered as a 'small' part. The complexity level increases if the size of the part increases.

**Number of Internal Undercuts** [No Internal Undercut, One, More than One]
Generally, undercuts are combinations of part features whose directions are not parallel to the mold closure direction. Internal undercuts are features, either protrusion or depression, on the inner surface of a part which prevent mold cores from being withdrawn along the line of closure. Internal undercuts result in greater mold complexity because special devices such as cores or pins have to be included in the design of a mold.

**Number of External Undercuts** [No Internal Undercut, One, Two, More than Two]
External undercuts are usually depression features on the external surface of a part which are not parallel to the direction of mold closure. Protrusion features which are placed on the external surface of a part in such a way that a single dividing surface, either planar or non planar, cannot pass through all of them are also considered external undercuts. External undercuts result in greater mold complexity because a special device such as a side cavity or a side core must be added to the mold design.

**Parting Line** [Planar, Non-Planar]
Parting line is refer to the imaginary line which is related to the primary opening of the mold. The selection of parting line is influenced by the shape of the part, method of fabrication, mold type, etc. Parting line maybe either planar or irregular-shaped. Planar parting line has only one plane of parting surface and is usually preferable because it is less complex for mold construction.

**Cavity Detail** [Low, Moderate, High, Very High]
Cavity detail refers to the complexity of the features present on a mold. Features such as ribs, bosses, holes, lettering, etc., aligned with the mold closure direction contribute to cavity detail. The mold complexity increases as the cavity detail becomes high.

**Tolerances** [Commercial, Tight]
Tolerance is the total permissible variation of size, form, or location. The strictness of required tolerances on the part also contribute the complexity to the mold design. Tighter tolerances always increases the complexity level of a mold.

**Number of Cavities per Mold** [One, Two, Three,..., etc.]
A multicavity mold is more complicated to design because the designer has to consider other factors such as number and location of runners, gates, and venting and cooling systems.

## METHODOLOGY

In this study, a three-layer backpropagation network is built with one input layer, one hidden layer, and one output layer. The input layer consists of eight nodes. The hidden layer and the output layer each has three nodes. Table 1 summarizes the network structure and its parameters.

The training and testing data have been adapted from the handbook (Dixon and Poli, 1995) and a 10% random noise is manually added to the training data. The initial weight matrix, ranging from -0.3 to 0.3, is generated randomly using a built-in BuildNet software command. A sigmoid transfer function was employed.

The network was trained with several learning rates in order to compare its performance; and a 0.1 learning rate was found to give the best results. During training, the network reads a training input files and adjusts its weight matrix. After the network has been trained, another data set with the same size of the training data set is applied to the network for testing purpose. In the testing process, the network reads a testing input file and calculates error levels without changing its weight. The epoch which yields the best network performance is then selected and the weight matrix from that particular epoch is saved. The network is now ready to perform the evaluation of the designs.

**TABLE 1: NETWORK STRUCTURE AND TRAINING PARAMETERS**

| Type | Parameters | Value |
|---|---|---|
| Network Structure | Number of Hidden Layer | 1 |
| | Number of Input Nodes | 8 |
| | Number of Hidden Nodes | 3 |
| | Number of Output Nodes | 3 |
| | Number of Training Vectors | 500 |
| | Number of Testing Vectors | 500 |
| | Number of Test Parts | 10 |
| Training Parameters | Number of Epochs (train) | 1000 |
| | Best Epochs (test) | 215 |
| | Best Learning Rate | 0.1 |
| | Error Limit | 0.02 |

**TABLE 2: CHARACTERISTICS OF TEST PARTS**

| No. | Part Type | Part Size | Int. U/C | Ext. U/C | Parting Line | Cavity Detail | Tol. | Cavities per Mold |
|---|---|---|---|---|---|---|---|---|
| 1 | box | small | 0 | 2 | P | low | std. | 1 |
| 2 | box | small | 0 | 2 | P | low | std. | 4 |
| 3 | box | small | 0 | 0 | NP | low | tight | 1 |
| 4 | box | small | 0 | 0 | NP | low | tight | 2 |
| 5 | box | small | 0 | 2 | NP | low | std. | 1 |
| 6 | box | small | 0 | 2 | NP | low | std. | 2 |
| 7 | box | small | 1 | 1 | NP | low | std. | 1 |
| 8 | box | small | 1 | 1 | NP | low | std. | 2 |
| 9 | flat | small | 0 | 0 | P | high | std. | 1 |
| 10 | flat | small | 0 | 0 | P | high | std. | 4 |

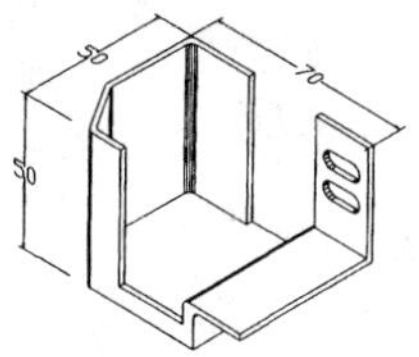


(a) Test part #7 for a single-cavity mold
Test part #8 for a multi-cavity mold

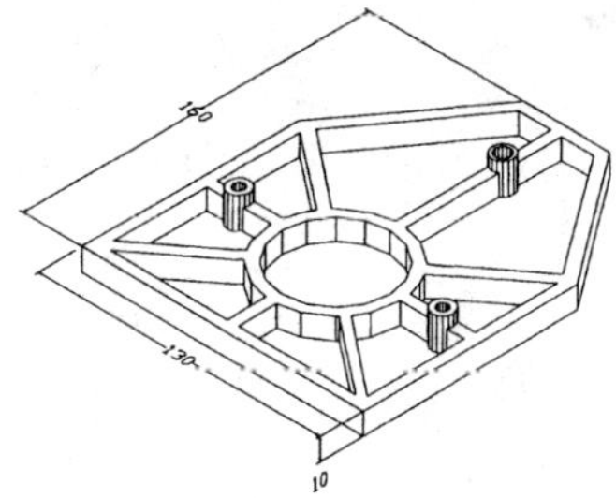


(b) Test part # 9 for a sungle-cavity mold
Test part # 10 for a multi-cavity mold

Figure 2. Examples of test parts (Dixon and Poli, 1995)

**TABLE 3: SUMMARY OF THE EVALUATION**

| No. | Easy (%) | Moderate (%) | Difficult (%) | Evaluation Result |
|---|---|---|---|---|
| 1 | 88 | 11 | 1 | Easy |
| 2 | 22 | 68 | 10 | Moderate |
| 3 | 94 | 6 | 0 | Easy |
| 4 | 77 | 22 | 1 | Easy |
| 5 | 33 | 66 | 1 | Moderate |
| 6 | 0 | 2 | 98 | Difficult |
| 7 | 1 | 44 | 55 | Difficult |
| 8 | 0 | 4 | 96 | Difficult |
| 9 | 42 | 58 | 0 | Moderate |
| 10 | 0 | 2 | 98 | Difficult |

Ten part designs were used to determine the ability of the network in classifying the levels of complexity. Some of the testing samples are shown in Figure 2 and Table 2 summarizes the characteristics of the test parts.

## CONCLUSION

This study demonstrates the ability of the backpropagation network in classifying the level of mold complexity. The system described in this paper serves as a preliminary evaluation model for designers to determine whether mold design and construction would be complex for a given part design. Further investigation is being carried out by the authors to incorporate a more comprehensive neural network model for mold complexity evaluation by considering factors such as mold material type and plating requirements.

## REFERENCES


Adaptive Solutions, Inc. "CNAPS Back-Propagation Guide", *Instruction Manual*, Adaptive Solutions, Inc. Volume 2, Oregon, 1993.

Dixon, J.R., and, Poli, C., "Engineering Design and Design for Manufacturing: A Structured Approach", Field Stone Publisher, Massachusetts, 1995.

Rosato, Donald V. and Rosato, Dominick V., "Injection Molding Handbook", Van Nostrand Reinhold Publishing, New York, 1986.

Rosato, D.V., Di Mattia, D.P., and Rosato, Dominick V., "Designing with Plastics and composites; A Handbook", Van Nostrand Reinhold Publishing, New York, 1991.

Rubin, I.I., "Injection Molding Theory and Practice", *Society of Plastics Engineers*, John Wiley & Sons, New York, 1972.

Skapura, D.M., "Building Neural Networks", Addison-Wesley Publishing, New York, 1996.

Smith, A. "Characterizing Backpropagation Neural Network Architecture Response to Domain Entropy: Implemented with a Fuzzy Logic Based Expert System", *Ph.D. Dissertation*, Department of Engineering Management, University of Missouri-Rolla, 1991.

# SENSITIVITY ANALYSIS VALIDATION FOR EXPERT NETWORKS

**RICHARD T. GORDON, MICHAEL C. JASMAN AND RICHARD J. CANTOR**
Computational Intelligence Laboratory
Information Technology Business Innovation Group, Chubb Group of Insurance Companies,
477 Martinsville Road, Liberty Corner, NJ 07938, Email: rtg@aip.org

**ABSTRACT:**
One drawback to the acceptance of expert networks in a corporate environment is a low comfort level with the black box aspects of these models. This is especially true where the decision space is mainly a grey area with much subjectivity utilized in the human task. Standard cross-validation techniques, while offering support as to the efficacy of expert networks, do little to address the black box nature of these models. In this paper, the technique of sensitivity analysis is presented as a means to validate the learned knowledge of expert networks. We discuss the technique and present an example utilizing a complex decision task. In the example, an expert network is trained to underwrite oil/gas lease and pipeline operations for general liability insurance coverage. We show that sensitivity analysis can be used to tangibly demonstrate what an expert network has learned and how the network compares to human utilization of the same information.

**INTRODUCTION**
Expert networks are artificial neural networks which are utilized to perform human decision tasks. Expert networks require no explicit knowledge of the rules or steps followed in a decision task. Thus, the use of expert networks to perform human decision tasks has grown dramatically in recent years. One drawback to the acceptance of expert networks in a corporate environment is a low comfort level with the black box aspects of these models. This is especially true where the decision space is mainly a grey area with much subjectivity utilized in the human task. Two surveys of the neural network community illustrate this point. The first study, done by NASA (Lin, 1993), was of neural network researchers in NASA, universities and industry. The respondents make the point that it is hard to convince people to adopt a neural network solution even when it performs better than the technique currently in use. The study also reports the respondents feeling that there is insufficient funding and

resources. The second study was done by a group at the College of Business at the University of Maryland (Golden, 1995). This study reported that only 45% of the respondents believed that neural network modeling efforts had strong corporate support. The study also brought out that many respondents cited management resistance as their greatest obstacle.

Cross validation is the standard paradigm for assessing neural network performance and has become ubiquitous in neural network development. Techniques such as leave-one-out, k-fold and jackknife cross validation (Weiss and Kulikowski, 1991) allow a measure of the network error rate free of sample generated bias. However, when developing systems based on inductive inference (supervised neural networks), work has shown that it is not possible, using only a priori reasoning, to justify a correlation between reproduction of a training set and generalization error off the training set (Wolpert, 1992). Particularly, one can over minimize cross validation error thereby increasing generalization error. The above issues become increasingly problematic when the human decision being emulated by an expert network is not clearly right or wrong. In this paper, the technique of sensitivity analysis is presented as a means to validate the learned knowledge of expert networks, thereby addressing the black box question. We discuss the technique of sensitivity analysis and present an example utilizing a complex decision task. In the example, an expert network is trained to underwrite oil/gas lease and pipeline operations for general liability insurance coverage. The human performance of this task is characterized by a high number of variables being integrated into the decision. We show that sensitivity analysis can be used to tangibly demonstrate what an expert network has learned and how the network compares to human utilization of the same information.

**SENSITIVITY ANALYSIS**

An ancient Arabian quatrain states that: *He who knows and knows that he knows is wise - follow him*. This ancient wisdom is at the heart of the issue of the black box aspect of expert networks. For expert networks to become fully integrated into critical decision tasks, how the network utilizes information and what forms the basis of network decisions must be known. Published research (Garson, 1991) suggested that rank ordering based on the weights of backpropogation networks could be used to interpret network learning. However, attempts to replicate the study utilizing the equations and network presented in (Garson, 1991) failed. Another study (Garavaglia, 1993) has shown that output values from counter-propagation networks can be utilized to construct rank orderings which are valuable in interpreting the network learning.

In this research, sensitivity analysis is presented as a means for discerning how the expert network "knows what it knows." While this technique has been known for some time, it has seen only sporadic use by neural network developers. The sparse literature discusses applications such as pruning of inputs (Peterson and Ladage, 1992) or neurons (Mozer and Smolensky, 1989), sensor fusion (Lin et al, 1993) and corporate bond rating ( Utans and Moody, 1991). Sensitivity analysis is the process of systematically testing the effect of individual input variables on the output of trained networks. In this technique (Belsley et al, 1980), one input variable is varied in both the positive and negative directions within a given range while the remaining input variables are frozen, with the corresponding change in output values being recorded. This procedure is then repeated for all input variables over all training patterns. The change in the output values is recorded and utilized to construct tables showing the amount of change each variable induced in the output. There are many ways to represent the change, such as absolute minimum, absolute maximum, mean and standard deviation. The goal here is to construct rank orderings of all input variables based on the change induced in each output variable as well as a rank ordering based on the overall change induced for all output variables. Although sensitivity analysis is computationally expensive, there are many advantages to performing this procedure. In addition to providing a tangible method for determining how the network utilizes input information in comparison to the human decision process, it allows for "what-if" scenarios to be performed. Sensitivity analysis is also a valuable tool in the development of expert systems, in that it can be utilized to both create rules and to develop weightings for the application of rules. The application of sensitivity analysis to expert system development reduces the labor intensive nature of the process.

**EXPERT NETWORK EXAMPLE**

As stated in the introduction, for expert networks to be accepted in human decision tasks, it is of paramount importance to quantify how the network arrives at a particular solution and also how it utilizes input information in an overall sense. In this research we apply sensitivity analysis to a complex decision task, underwriting oil/gas lease and pipeline operations for general liability insurance coverage. General liability insurance provides companies with coverage for damages due to bodily injury or property damage caused to others. It also affords legal defense for suits stemming from tort liability. Oil/gas lease and pipeline operations represent an extremely difficult insurance decision due to the inherent risk of the operations.

The underwriting decision is based on an application which collects 125 separate pieces of information detailing operations, financial condition, experience, and specific exposures. While there are certain conditions which clearly make a risk unacceptable, the vast majority of decisions are based on the overall impression that an underwriter forms on a particular company after weighing all of the information. In a large number of cases, not all application information is available. The decision is also subject to underwriter experience, regional differences and market conditions. In many cases there is no clear right or wrong, and as such, this constitutes the epitome of the grey area decision. Underwriter expertise is built only after much experience is gained. This expertise is easily lost through employee turnover and promotion to management positions. Expert systems developed for this task are limited to the most simplistic case of perfectly acceptable or perfectly unacceptable outcomes by the greyness of the decision and also by the frequently incomplete information. In performing knowledge engineering with expert underwriters, it became clear that although rules could only be formulated covering simple cases, these underwriters were able to communicate which variables had the most impact on their decisions. As a function of training others, the expert underwriters were able to communicate, in a qualitative sense, what variables and combinations were of importance in the decision. It was also clear that if the neural network could be interpreted in these terms, then validation could easily be performed by the experts, thus creating a comfort level with the machine decision.

**NETWORK TRAINING AND TESTING**

A feedforward backpropagation architecture was chosen for the oil/gas expert network. This architecture was chosen for the qualities of having well understood training characteristics. The input layer consisted of 125 neurons corresponding to the separate pieces of information gathered on the underwriting application. All underwriting information was transformed to numerical values for training. Hidden layer size was varied between a maximum of 125 neurons and a minimum of 75 neurons. The network had two output neurons corresponding to the accept-decline decision. The learning rate and momentum were held constant during training at .30 and .90 respectively. The total number of training/testing patterns was 100, with both a 4-fold and a 10-fold cross validation procedure employed. Training was performed on a 90 MHz Pentium PC and took five minutes on average to complete. A tolerance procedure varying between .49 and .10 was utilized to determine when training should be terminated. The same tolerance procedure was also employed for network testing.

Training/testing patterns consisted of actual underwritten applications with the accept-decline decision used for network training/testing being the decision taken by the human underwriter on the account, as opposed to some idealized criterion. The training set represented decisions from 6 different regional locations and roughly 10 different individual underwriters. Given the grey nature of this underwriting decision, it was anticipated that there would be poor decisions as well as some conflicting decisions represented in the training/testing set. The nature of the training/testing patterns provided a good experimental environment in which to test sensitivity analysis as a validation procedure. The testing performance of the networks showed little variance across hidden layer size or tolerance level. The networks did, however, show considerable variance based on cross validation group. In the 4-fold cross validation, the networks testing performance varied between 40% and 80% correct. The 10-fold cross validation, saw the expert networks testing performance vary between 50% and 90% agreement with the human decisions. This result is not unexpected, given the subjective nature of the decision. Further these type of results are at the root of the inadequacy of standard cross validation procedures when evaluating expert network performance on a complex human decision task.

**SENSITIVITY ANALYSIS VALIDATION**

Sensitivity analysis was applied to the oil/gas expert networks by varying the value of each input variable ±10% across all training patterns. In this regime, a single input was varied at a time while all others were frozen. The computations took, on average, ten minutes on a 90 MHz Pentium PC. Results were compiled for the amount of change induced in the outputs corresponding to the absolute minimum, absolute maximum, mean and absolute mean. The results were then utilized to construct tables showing the ranking, in descending order based on the amount of change, ordered on the mean value. Tables were constructed for each of the two output variables as well as a table corresponding to the absolute mean change for both outputs. The rank order tables were next compared to how human underwriters were taught to utilize the information which comprised the input values to the network. Based on discussions with high level underwriters and rules from an expert system developed for underwriter training, the relative importance was established for each input variable. The rank ordering showed that the expert network had indeed learned the underwriting task in a way analogous to what would be expected of an expert underwriter. Variables corresponding to subcontractor controls, the presence of poison gas, years of experience, drilling operations, offshore wells and others, which represent the most complex and hazardous operations, showed on the rank ordering to have the most effect on the network output. Variables which are considered to be necessary file information, but the have little effect on the underwriting decision showed at the bottom of the rank ordering.

**CONCLUSIONS**

Convincing a company's management to utilize a computer system to automate a human task has always been difficult. This has been especially true with neural networks. Given the results obtained in this research, sensitivity analysis offers encouragement as a method for addressing the black box nature of expert networks. Sensitivity analysis provides a means to tangibly document the learned knowledge of expert networks. Additionally it provides a context specific framework in which to assess a network's generalization behavior. Finally, when used in conjunction with traditional cross validation, sensitivity analysis provides a strong regime for expert network development.

**REFERENCES**


Belsley, D.A., Kuh, E., and Welsch, R.E., (1980) *Regression Diagnostics: Identifying Influential Data and Sources of Colinearity*, Wiley & Sons, New York.

Garavaglia, S., (1993). *Interpreting Neural Network Output*, PCAI Magazine, May/June, 52.

Garson, G.D., (1991). *Interpreting Neural Network Connection Weights*, AI Expert, April, 47.

Golden, B.L., Wasil, E.A., and Coy, S.P., (1995). *Neural Networks in Practice*, Technical Report, College of Business and Management, University of Maryland.

Lin, C.S., (1993). *An Overview on Development of Neural Network Technology*, Technical Report, Space Technology Initiatives Office, NASA Langley Research Center.

Lin, C.S., Meyer, C.M., and Guo, T.H., (1993). *Neurocomputing Techniques for Selecting Measurements for Sensor Validation*, Intelligent Engineering Systems Through Artificial Neural Networks, **3**, Dagli et al Eds., ASME Press, New York.

Mozer, M.C., and Smolensky, P., (1989). *A technique for Trimming the fat from a Neural Network via Relevence Assessment*, Advances in Neural Information processing Systems, **1**, D.S. Touretzky Ed., Morgan Kaufmann, San Mateo, CA.

Peterson, G.E., and Ladage, R.N., (1992). *On Using Sensitivity Analysis to Prune the Inputs to a Neural Network*, Intelligent Engineering Systems Through Artificial Neural Networks, **2**, Dagli et al Eds., ASME Press, New York.

Utans, J., and Moody, J., (1991). *Selecting Neural Network Architectures via the Prediction Risk: Application to Corporate Bond Rating Prediction*, Proceedings of the First International Conference on Artificial Intelligence Apllications on Wall Street, IEEE Computer Society press, Los Alamitos, CA.

Weiss, S.M., and Kulikowski, C.A., (1991). *Computer Systems That Learn*, Morgan Kaufmann, San Mateo, CA.

Wolpert, D.H., (1992). *On the Connection Between In-Sample Testing and Generalization Error*, Santa Fe Institute Technical Report, 92-01-006.

# USING GROUPING AND UNCERTAIN REASONING DURING ID3 DECISION TREE CONSTRUCTION AND TESTING

**CARMELA R. C. SANTOS**
*Electronic Data Systems Corp.*

**DANIEL C. ST. CLAIR**
*University of Missouri-Rolla*
*Engineering Education Center*
*stclair@umr.edu*

**PETER E. MAHER**
*Open Computing Institute, Inc.*

***ABSTRACT:***
Quinlan's ID3 algorithm produces classification trees that perform poorly when used with noisy or continuous-valued data. The UR-ID3 algorithm combines uncertain reasoning with the ID3 rule set to handle approximate test data. This paper presents a variation, called UR-ID3e, that assumes that training data is also uncertain. Experimental results are presented that compare the performance of UR-ID3e with other algorithms.

## INTRODUCTION

Quinlan's ID3 algorithm is a decision tree-based classification system [Quinlan 1986] that produces a clear set of rules but tends to perform poorly when classifying noisy data or continuous-valued data. In comparison, neural network-based systems produce highly accurate results on such data but do not produce rule sets that clearly show how each classification was reached.

A variation of ID3, called UR-ID3, achieves high accuracy with uncertain test data by using uncertain reasoning to compare test samples with an ID3 tree's nodes in order to determine the most likely classification [Maher and St.Clair 1993]. An enhanced version, UR-ID3e, goes further by pre-processing the training data. During training, ID3 creates enough tree nodes from data points to correctly classify a sample. An unnecessarily complicated tree results when continuous data is used for training. Furthermore, the tree will mirror all the inconsistencies inherent in noisy data. To solve these problems, the UR-ID3e algorithm analyzes the training samples and groups the data points into clusters. The weighted midpoints of the clusters are used to produce a summarized ID3 tree. Information gathered during pre-processing is used with UR-ID3 to find the most likely classification of each test sample.

## THE ID3 ALGORITHM

Quinlan's ID3 uses a collection of training examples to produce a classification tree. Each training instance contains a set of attribute values and an associated classification. An arbitrary collection of objects that is either empty or contains only objects of one class produces a decision tree that has only one leaf labeled with the class name. Otherwise, some test T is performed on each object generating possible outcomes $O_1, O_2, \ldots, O_w$. The test thus partitions the collection into w subsets, with each subset containing those objects having the same outcome. For ID3, the test determines which attribute becomes the root of the tree. This process is repeated for each subset until there are only single-object subsets left that satisfy the one-class requirement for a leaf. In the end, a decision tree is produced that correctly classifies each object in the collection from which it was trained. The rules produced from ID3 decision trees require exact matches to arrive at a classification, so that the presence of noise in the data results in multiple or erroneous conclusions.

## THE UR-ID3 ALGORITHM

The UR-ID3 algorithm [Maher and St. Clair 1993] permits the use of uncertain or approximate attribute values within the ID3 framework through uncertain reasoning using fuzzy logic [Zadeh 1965]. This approach addresses uncertainty in the comparison problem as one traverses the tree.

Suppose there are two real numbers N and M that need to be compared. Strict comparison is useless for cases where some error in measurement introduces noise in the data, making a number slightly different from another. Fuzzy logic allows comparison of such approximate numbers.

A "fuzzy number," denoted by ~N, is a set that contains values that approximate a number N. The degree of membership in this set can be defined by a membership function, mem(x), whose values fall within the range [0,1]. If mem(x) is 1, then x is the same as N. If mem(x) is 0, then x does not even approximate N. The example in Figure 1 depicts mem(x) as a triangular function whose base is determined by $\alpha$, which falls within the range [0,1). The function returns 0 whenever x falls outside the triangle's base. Other ways can be used to determine the function.

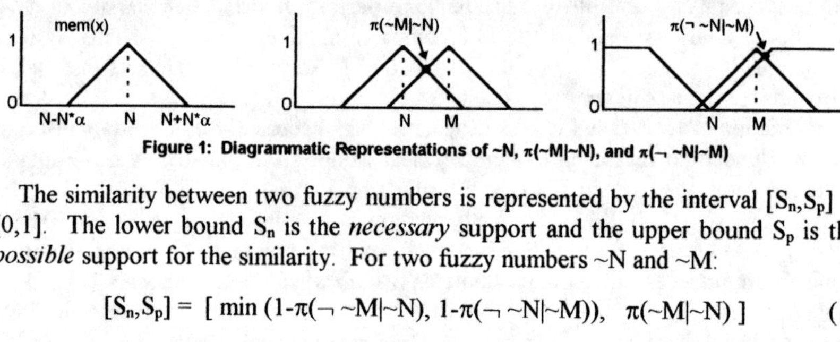


**Figure 1: Diagrammatic Representations of ~N, π(~M|~N), and π(¬ ~N|~M)**

The similarity between two fuzzy numbers is represented by the interval $[S_n,S_p] \subseteq [0,1]$. The lower bound $S_n$ is the *necessary* support and the upper bound $S_p$ is the *possible* support for the similarity. For two fuzzy numbers ~N and ~M:

$$[S_n,S_p] = [\ \min(1-\pi(\neg \sim M|\sim N),\ 1-\pi(\neg \sim N|\sim M)),\ \ \pi(\sim M|\sim N)\ ] \qquad (1)$$

where $\pi(\sim M|\sim N)$ denotes the maximum membership value at the points of intersection of the membership functions for ~M and ~N. The notation "¬ ~N" denotes the inverse of the membership function for ~N. Figure 1 shows the diagrammatic representations of $\pi(\sim M|\sim N)$ and $\pi(\neg \sim N|\sim M)$. The interval [1,1] implies full support *for* a classification, while [0,0] implies full support *against* a classification. The UR-ID3 algorithm produces a support interval for each possible classification of a test sample.

### Step 1: Construct an ID3 tree

The UR-ID3 algorithm constructs an ID3 decision tree from training values that are considered as certain. After the tree is built, the values associated with each branch are considered as approximate, or "fuzzy," with triangular membership functions used to calculate support intervals. An uncertainty factor $\alpha$ is used to compute the width of the triangle base, as in Figure 1.

Test data is also considered approximate. For a given sample in the test dataset, a set of support intervals for each possible classification can be produced by traversing the tree as described in Steps 2 and 3.

**Step 2: Determine the weighted support interval $\theta'_j$ at each leaf $L_j$**

For a given test sample, use equation (1) to compute the support intervals along the path from the root to the leaf $L_j$ by comparing each attribute value from the test sample with its corresponding branch value until a leaf is reached. This results in the set of support intervals $I_j$ being associated with the nodes along the path leading to each leaf node $L_j$. Combine the support intervals in $I_j$ into a single interval $\theta_j$ at each leaf node $L_j$ using the AND operator, which is defined on two intervals as: $[a_1,b_1]$ AND $[a_2,b_2]$ = $[a_1*a_2,\ b_1*b_2]$.

To get the final support interval at each leaf, apply a weighting factor to the interval $\theta_j$ based on the number of instances $T_j$ at the leaf node occurring during the training period. At each leaf node $L_j$, compute the interval $\theta'_j = \theta_j * (T_j / \Sigma T_t)$, where $\Sigma T_t$ is the total number of training samples, and the "*" multiplies the endpoints of the intervals by the weight. That is, if $\theta_j = [a,b]$, then $\theta_j * (T_j / \Sigma T_t) = [a * (T_j / \Sigma T_t), b * (T_j / \Sigma T_t)]$.

**Step 3: Determine a single support interval for each possible classification**

Look at all leaf nodes that produce the same classification $C_q$. These nodes provide support for $C_q$. Combine the support intervals associated with these leaf nodes using the logical OR operator, which is defined on two intervals as: $[a_1,b_1]$ OR $[a_2,b_2]$ = $[a_1+a_2 - a_1*a_2,\ b_1+b_2 - b_1*b_2]$. The combined intervals produce a single interval $\Phi_{Cq}$ that gives the likelihood that $C_q$ is the correct classification.

The highest lower bound, $S_n$, of the support intervals for the possible classifications is used to pick the winning classification for the test sample. The upper bound, $S_p$, is used as a tie-breaker.

## THE UR-ID3E ALGORITHM: AN ENHANCED UR-ID3

The ID3 algorithm creates a node for each value in the training data, producing a complicated tree when given continuously-valued samples. Since these values are usually mapped into a relatively much smaller set of distinct classes, pre-processing the training data to reduce the number of samples should simplify the tree. The enhanced UR-ID3 algorithm, UR-ID3e, provides a pre-processor that partitions training data into a small number of intervals. These intervals are then used to create a summarized dataset that is presented to the ID3 algorithm to produce a decision tree. When the tree is tested, the interval information will determine the membership functions used by UR-ID3 for its classification support intervals.

**Step 1: Determine Intervals**

The training data for each attribute is partitioned into L intervals using Lebowitz's algorithm [Lebowitz 1985], which finds the L-1 biggest relative gaps in the data and uses the midpoints of those gaps as the endpoints of the intervals.

The Lebowitz intervals are then used to compute the points that define the triangular membership functions. Each interval has a corresponding membership function made up of three points: the minimum endpoint, the maximum endpoint, and the midpoint ($Xmid_i$, $Ymid_i$). To make some allowance for test data that may fall beyond the first and last data points, adjustments are made to Lebowitz's algorithm to double the width of the first and last intervals.

The x-coordinate of the midpoint is computed using the weighted mean of the data points that fall within the interval. This positions the highest value of the membership function above the part of the interval with the highest concentration of data points. For interval i, the x-coordinate of the midpoint is:

$$Xmid_i = \left[ \sum_{k=1}^{V_i} (Value_k * ValueFrequency_k) \right] \div IntervalTotal_i \qquad (2)$$

where $V_i$ is the number of distinct data points within interval i; $Value_k$ is the value of the $k^{th}$ data point in interval i; $ValueFrequency_k$ is the frequency of the $k^{th}$ data point in interval i; and $IntervalTotal_i$ is the total number of data points in interval i.

Three possibilities were implemented for the y-coordinate of the midpoint $Ymid_i$:

1. Set $Ymid_i = 1$. This approach, based on examples used in Maher and St. Clair's UR-ID3 [1993], results in uniform function heights.
2. Set $Ymid_i$ as a function of the frequency of data points in the interval. Intervals that have a higher frequency of data points will produce membership functions that are taller, thus they will more likely get better support intervals when compared with the test data.

$$Ymid_i = 1 - [\, 1 \div (IntervalTotal_i + 1)\, ] \qquad (3)$$

3. Set $Ymid_i$ as a function of the density of data points in the interval. The function heights will reflect the distribution of the training data, based on the assumption that the test data will closely follow the training data.

$$Ymid_i = \frac{Density_i}{\max_i(Density_i)} \; ; \; Density_i = \frac{IntervalTotal_i}{LebowitzMax_i - LebowitzMin_i} \qquad (4)$$

The minimum and maximum endpoints of the membership functions are determined by making an adjustment to the Lebowitz intervals, widening them by either a fixed value or by a percentage of the width of the interval. This produces overlapping membership functions that allow for the possible ambiguity around the edges of each interval.

**Step 2: Summarize the Training Set**

Examine the training set's attribute values and find the Lebowitz interval in which each particular value belongs. Then replace each value with the midpoint of the corresponding membership function for that interval. This process will reduce the number of points in the dataset, effectively summarizing it.

**Step 3: Generate an ID3 Tree and Perform Tests**

Use the summarized training data to generate an ID3 tree. The midpoints, from Step 2, can be used to associate a membership function with each branch. Use these membership functions with UR-ID3 to compare fuzzy numbers from test data with the attribute values at each branch of the tree.

## EXPERIMENTAL EVALUATION

Experiments were conducted to compare the accuracy of the UR-ID3e algorithm against ID3, nearest-neighbor ID3 (ID3nn), UR-ID3, and backpropagation neural networks (BP). The ID3nn algorithm is like ID3 except that it picks the closest path for test sample attribute values that are not found in the tree [Gallion et al 1993].

Four different datasets were used to test the algorithms. These are available at the University of California-Irvine's site on the World Wide Web at the following URL address: http://www.ics.uci.edu/AI/ML/MLDBRepository.html. A summary of each dataset is given in Table 1.

**TABLE 1: DATASET SUMMARY**

| Dataset | #Attrib | #Classes | #Samples | # Unique Values per Attribute |
|---|---|---|---|---|
| Iris | 4 | 3 | 150 | 35, 23, 43, 22 |
| Glass | 9 | 7 | 214 | 178, 142, 94, 118, 133, 65, 143, 34, 32 |
| Breiman | 21 | 3 | 399 | 3, 269, 271, 269, 270, 399, 399, 399, 269, 399, 269, 399, 269, 399, 399, 399, 269, 270, 270, 269, 4 |
| Pima | 8 | 2 | 768 | 17, 136, 47, 51, 186, 248, 517, 52 |

Experiments were performed on the datasets using four-fold cross validation, where each dataset was randomly subdivided into four roughly equal subsets [Breiman et al 1984]. The means and standard deviations were calculated for each set of tests.

Table 2 shows the parameters used when the UR-ID3e algorithm was tested. All possible combinations of the chosen parameter values were used, and the best results were noted for each of the dataset partitions.

**TABLE 2: PARAMETERS USED FOR UR-ID3E**

| Parameter | Description | Values Used |
|---|---|---|
| reduction | Indicates the %reduction done to the number of intervals produced from the training set | 0, 10%, 25%, 40%, 60% (also 75% and 85% for the Breiman dataset) |
| overlap | Determines how each membership function's minimum and maximum endpoints are computed from the original Lebowitz intervals | 0, 5%, 20% of the original interval width |
| uncertainty | Determines the width of the base of the triangular membership functions for each value of each attribute in the test set | 0, 5%, 10% |
| midpoint | Determines the height of the triangular membership functions in the training set | (1) as a function of interval data frequency, or (2) as a function of interval data density |

**TABLE 3: PERCENTAGE OF CORRECTLY-CLASSIFIED TEST SAMPLES**

| Algorithm | Glass | | Iris | | Breiman | | Pima | |
|---|---|---|---|---|---|---|---|---|
| | average | std dev | average | std dev | average | std dev | average | std dev |
| ID3 | 15.85 | 5.49 | 76.08 | 6.09 | 33.35 | 4.32 | 23.70 | 2.04 |
| ID3nn | 44.88 | 7.81 | 93.98 | 1.20 | 68.43 | 4.52 | 55.20 | **1.61** |
| UR-ID3 | 46.27 | 8.94 | 94.65 | 3.27 | 69.19 | 4.30 | 58.85 | 3.01 |
| UR-ID3e | **54.65** | 6.02 | **97.99** | **1.16** | **86.72** | **1.76** | **67.32** | 1.74 |
| BP | 49.05 | **2.71** | 87.95 | 4.15 | 76.18 | 6.27 | 65.50 | 2.38 |

Table 3 contains a summary of the results for UR-ID3e, UR-ID3, ID3, ID3nn, and BP. The results for BP represent the percentage of test patterns that had an error less than or equal to a target error of 0.001. During training, the networks were constantly tested with the test sets to get the percentage of good patterns. The best results are given in Table 3. A complete description of the experiments is contained in Santos et al [1996].

Results from the experiments show that the performance of UR-ID3e is consistently better than that of the other ID3 algorithm variations tested. The UR-ID3e algorithm always produced the best average accuracies and its standard deviations were usually better than those of the other algorithms. Better results may be produced for BP, UR-ID3, and UR-ID3e with additional fine-tuning of the parameters. In most cases, increasing the value for the reduction parameter improved the accuracy of UR-ID3e. It is thus useful to try a wide range of values for the reduction parameter in order to find the best results for a given dataset. The effects of changing the other parameters were not as straightforward as that of the reduction parameter. The midpoint type used sometimes affected the results for the better, sometimes for the worse, but only slightly. These parameters are best regarded as means towards fine-tuning the accuracy of the algorithm.

## CONCLUSIONS

The UR-ID3e algorithm consistently performed better than ID3, ID3nn, and UR-ID3 in these experiments. Results for many-valued datasets like Breiman are much improved and are comparable to results from backpropagation neural networks. Thus UR-ID3e is a well-performing decision tree-type algorithm that provides an alternative to backpropagation neural networks. The UR-ID3e algorithm is preferable over neural networks in cases where easily-interpreted rule sets are important.

## REFERENCES


Breiman, L., Friedman, J. H., Olshen, R. A., and Stone, C. J. (1984). Classification and Regression Trees, Belmont, California: Wadsworth & Brooks/Cole.

Fisher, R. A. (1936). "The Use of Multiple Measurements in Taxonomic Problems." In *Annual Eugenics*, 7, Part II: 179-188.

Gallion, R., Sabharwal, C. L., St. Clair, D. C., and Bond, W. E. (1993). "Dynamic ID3: A Symbolic Learning Algorithm for Many-Valued Attribute Domains." In *Proceedings of the 1993 Symposium on Applied Computing*. Indianapolis, Indiana.

Lebowitz, M.(1985). "Categorizing Numeric Information for Generalization." In *Cognitive Science*, Vol. 9: 285-308.

Maher, P. E., and St. Clair, D. C. (1993). "Uncertain Reasoning in an ID3 Machine Learning Framework." In the *Second IEEE International Conference on Fuzzy Systems*.

Quinlan, J.R. (1986). "Induction of Decision Trees." In *Machine Learning*, Vol. 1: 81-106.

Santos, C. C., St. Clair, D. C., and Maher, P. E. (1996). "Using Grouping and Uncertain Reasoning During ID3 Decision Tree Construction and Testing." M.S. thesis, University of Missouri-Rolla.

Smith, J. W., Everhart, J. E., Dickson, W. C., Knowler, W. C., and Johannes, R. S. (1988). "Using the ADAP Learning Algorithm to Forecast the Onset of Diabetes Mellitus." In *Proceedings of the Symposium on Computer Applications and Medical Care*, pp. 261-265.

Zadeh, L.A. (1965). "Fuzzy Sets." In *Information and Control*, 8: 338-313.

# A TABLE–BASED APPROACH TO EXPERT FAULT DIAGNOSIS NETWORKS

**Alan P. Levis, Kristin L. Adair and Susan I. Hruska**
Department of Computer Science
Florida State University
Tallahassee, FL 32306-4019

## ABSTRACT

In an implementation of expert networks for diagnosing machine and sample faults in automated gas chromatography, we present a technology for constructing a network based on a knowledge table rather than the traditional rule based expert system. This system uses linguistic qualifiers to relate symptoms observed in the chromatogram to faults in the chromatograph or sample preparation and diagnoses the possible cause of those faults. Automated tools for constructing the expert network from the table and refining both the structure and the inferencing of the system are presented. Results of the use of this expert network on a functioning chromatograph reveal the efficacy of this new technology.

Keywords: Diagnosis, Artificial Neural Networks, Neural Network Applications, Diagnosis, Diagnostic Systems, Expert Systems, Fault Identification, Feedforward Neural Networks, Hybrid Artificial Neural Network Systems, Knowledge Extraction, Knowledge Representation, Machine Monitoring, Multilayer Feed Forward NN, Neural Network Applications, Neural Network Architectures, Neural Network Systems

## KNOWLEDGE ENGINEERING

*Knowledge engineering* encompasses the tasks of acquiring expert knowledge, determining an appropriate representation of that knowledge and refining that knowledge to facilitate its use by automated reasoning processes. Typically, knowledge acquisition involves a process of querying domain experts to determine the relationships between inputs and outputs of the system under development. Further, the experts are often asked to provide some form of quantification of these relationships. Given that people utilize their own individual reasoning processes, it is often difficult for the experts to provide either the relationships or their quantification in terms which are useful to a computer–based reasoning process.

Traditional expert systems typically use a set of *if–then* rules to represent symbolic knowledge and traditional logic as an inferencing mechanism. Knowledge encoded in these systems can generally be understood by a person and their sequential nature of inferencing easily supports an explanation facility. However, these systems tend to be brittle [ Giarratano and Riley (1994) ].

Alternatively, artificial neural network (ANN) methodologies have used *directed graphs* to encode knowledge while the firing functions of the network define the inferencing. These systems operate on a sub–symbolic level and therefore the reasoning behind a given conclusion can be difficult to extract. However, the ability to learn from example data and the property of graceful degradation are extremely attractive in a dynamic environment [ Gallant (1993) ].

*Expert network* (EN) methodologies are well known [ Kuncicky et al. (1992) ; Gallant (1993) ; Fu (1993) ]. This hybrid approach seeks to maintain the strengths of both traditional expert systems and ANNs while minimizing the drawbacks of each. This approach makes use of expert knowledge to define the network architecture and encodes the inferencing mechanism in the firing functions of the network. Thus, ENs incorporate existing knowledge while exploiting the ability to refine inferencing through data based training.

## TEST-BED APPLICATION

Gas chromatography (GC) is a process in which a sample is vaporized and its absorption is measured across some spectrum of electro-magnetic radiation. One such trial is called a *run*. From the resulting graph, called a *chromatogram*, specific chemical constituents can be identified and the constituents of the original sample inferred. However, before a determination of the composition of the original sample can be made, the chromatogram must be analyzed by highly trained technicians for indications of *faults*, either errors in sample preparation or machine failure. If a fault is detected, the chromatogram may not be suitable for constituent analysis and the run may need to be discarded at significant cost in time and materials [ Elling et al. (1994) ] . Chromatographers have stated that, in an automated environment, approximately 5 – 20% of runs may show some degree of fault.

As the experts perform it, fault diagnosis for gas chromatography is a process of evidence accumulation. Salient features, *symptoms*, of a chromatogram are identified and the combination of symptoms in conjunction with their relative strengths form the basis of the experts' conclusions. In the testbed system, symptom identification and scaling are performed by a preprocessing system designed by members of the research team at Los Alamos National Laboratory [ Lahiri et al. (1995) ] . A single fault is assumed to be the cause of failure for any given chromatogram.

## TABLE-BASED KNOWLEDGE ACQUISITION

The knowledge base for GC fault diagnosis is derived from three sources of domain expertise: an existing table of GC diagnosis knowledge, GC equipment manufacturers' documentation and interviews with experts in the field. Previous work of Lahiri and Stillman produced a two-way lookup table which relates symptoms to faults for GC data using a flame ionization detector [ Lahiri and Stillman (1992) ] . Additionally, a trouble shooting guide from Varian Chromatography Systems, a GC equipment manufacturer, was used as a source of expert knowledge. Both of these knowledge sources provide crisp *True/False* relationships between symptoms and faults. In addition, the Lahiri/Stillman table was developed using a Flame Ionization Detector while the target GC system was equipped with an Electron Capture Detector.

Human experts were consulted to determine if the existing knowledge was accurate and applicable to the testbed system. During the course of these interviews it became clear that the experts were not comfortable with crisp relationship values. Instead they utilized the terms *Always*, *Usually*, *Sometimes*, *Infrequently*, and *Never* to describe the relationships of symptoms to faults. For example, experts might characterize the relationship between symptoms and the fault *Makeup Gas Loss* in the following manner, "When you are losing make up gas, you would always see a change in sensitivity and sometimes see tailing peaks in the chromatogram." The use of linguistic qualifiers greatly improved the productivity of the expert interviews and thus aided the knowledge acquisition process. As a result, the experts agreed upon a set of 26 symptoms and 12 faults related by one of five linguistic qualifiers.

## KNOWLEDGE REPRESENTATION

Knowledge representation in the Table-Based Expert Network environment makes use of two equivalent formats, the *knowledge table* and the *table-based expert network*. The former is a two-way table relating symptoms to faults. This format is easily understood by humans and is the format in which the knowledge is presented to the experts for review and revision. The latter format, the table-based expert network, is derived from the knowledge table and is suitable for automated reasoning and inferencing refinement.

#### The Knowledge Table

The knowledge table is a table with symptoms appearing as rows and faults as columns. The intersection of each row and column contains an indicator for the linguistic qualifier used by the experts to relate the symptom to the fault or a blank to indicate that no relationship exists between symptom and fault. It should be noted that the exact meanings of the linguistic qualifiers are not indicated, nor even known. In this format, the domain experts are free to classify the relationships according to their own semantics.

#### Table–Based Expert Network

The expert network is generated by the NetMaker tool [ Adair et al. (1996) ]. This tool accepts an ASCII file containing the expert knowledge in table form. It then translates the linguistic qualifiers to default inferencing parameters and produces a network specification file which describes a four–layer, feed forward network comprised of four distinct node types: *Input*, *Filter*, *Combination* and *Output*. See Figure 1 for an example network.

**Input Nodes** Input values to the network are accepted at the Input nodes. These values, which correspond to the degree of severity of the symptoms observed in the chromatogram, are determined by a signal processing package [ Lahiri et al. (1995) ] which designates values in the range 0.0 to 1.0 to indicate degree of severity, or −2.0 to indicate that a particular symptom was not evaluated. The value of each symptom node is set equal to the corresponding symptom value. Input nodes do not process their input values in any way, but simply pass those values to subsequent layers.

**Filter Nodes** Filter nodes act as evidence discriminators by comparing their inputs against a threshold which is intended to eliminate noise from the data. If the input value surpasses the threshold, the value is considered positive evidence. Values not exceeding the threshold are considered negative evidence and −2.0 values are considered neither positive nor negative. Positive evidence is then weighted by a reinforcement factor, while negative evidence is weighted by a penalty factor. The weighting factor values are determined by the Filter node type which in turn corresponds to the linguistic qualifier used by the experts to describe the relationship of symptom to fault. The Filter node then takes on the value of the weighted evidence. This value is then passed to subsequent layers of the network along a connection where the weight is initially set heuristically based on the Filter node type.

**Combination Nodes** Combination nodes, as their name implies, combine positive and negative evidence to form a conclusion based on a unique set of symptoms. An EMYCIN–like [ Shortliffe and Buchanan (1984) ] computation is used to perform the combination of evidence. The output of each Combination node represents the strength of the conclusion for a fault based on the Symptom–Filter chains which are connected to it, called a *symptom set*. Multiple symptom sets for a single fault offer multiple methods of reaching a common conclusion, *i.e.*, multiple paths of reasoning [ Adair (1996) ].

**Output Nodes** Each Output node selects the most likely conclusion from the symptom sets connected to it. Each fault corresponds to a single Output node. In the example shown in Figure 1, two symptom sets are related to the fault *Leaking Syringe*. The first set consists of *Ghost Peaks* (Usually), *Sensitivity Change* (Always) and *Replicate Precision* (Usually). The second set consists of *Sensitivity Change* (Always), *Replicate Precision* (Usually) and *No Peaks* (Infrequently).

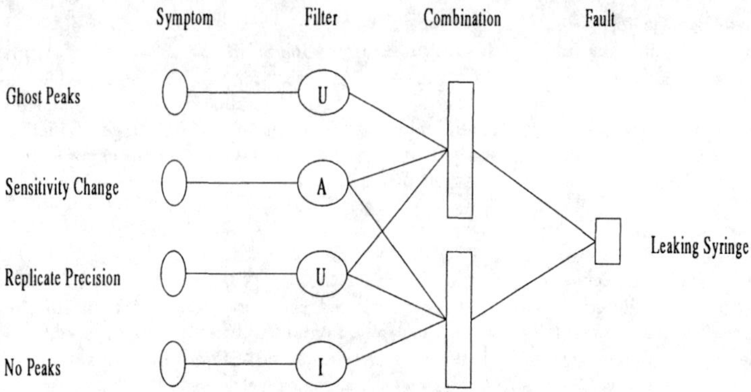


**Figure 1. Network with multiple paths of reasoning.**

The result of propagating symptom values forward through the network is a ranked list of possible faults.

## TABLE–BASED KNOWLEDGE REFINEMENT

Since the intelligent system is based on expert knowledge, the diagnostic capabilities are directly related to the quality of that knowledge. Automated tools are needed to refine existing knowledge, and thus improve diagnostic performance. Additionally, given the ambiguity associated with the use of linguistic qualifiers, a method of refining inferencing under a static knowledge base is required.

### Table Refinement

The NetMedic [ Adair et al. (1996) ] tool is used to refine expert knowledge based on a collection of example data. It first processes all data examples through the inferencing mechanism and then performs a statistical analysis of the data and inferencing results. The results of this analysis are then used to suggest changes to the knowledge table. Possible NetMedic suggestions include adding symptom to fault relationships that do not exist in the table and changing the linguistic qualifier which labels an existing relationship.

In previous trials, NetMedic has been shown to discover new knowledge as well as refine the existing knowledge [ Timpany (1995) ] .

### Inferencing Refinement

The training mechanism used to refine table–based expert network inferencing is a batch mode, supervised method based on backpropagation of error [ Dayhoff (1990) ] . Data labeled with known faults is presented to the network and adjustments are made to reduce error at the Output nodes. In the table–based expert network this error is defined in terms of correct fault identification and maximizing separation between the the output values of the two highest Output nodes. Weights on the connections between Filter and Combination nodes are adjusted to minimize error.

The training algorithm offers several options to the user. Among these is $\Delta w$ *averaging.* This technique averages the weight changes contributed by each individual fault before applying the weight change. This technique improves training performance on data sets with unequal distributions of fault examples. $\Delta w$ averaging decreases the magnitude of the effective weight changes and thus requires a significant increase in the training rate [ Levis (1995) ] as compared with traditional expert network training.

Another training option is to allow negative weight values. This option seems counter-intuitive since EMYCIN evidence combination assumes positive inputs indicate supporting evidence and negative inputs indicate contrary evidence. Allowing negative weights circumvents this convention and presents a positive output of the Filter node (assumed to be positive evidence) as negative evidence at the Combination node. It is believed that negative weight values in a trained network indicate that the knowledge encoded by that connection is not supported by the examples in the data set. That portion of the knowledge may then be reconsidered by the experts. If it is deemed that the knowledge is correct, then allowing negative weight values permits, to some extent, the network to assimilate anomalous data.

## RESULTS

The data set consists of symptom values for 667 chromatograms representing twelve faults. Each example is labeled with the correct fault. Initial evaluation of the intelligent system showed correct fault identifications were made only 44.4% of the time. After employing the knowledge refinement techniques described in this paper and techniques, based on unsupervised clustering of the data, to identify multiple paths of reasoning [ Adair (1996) ], the diagnostic capabilities improved to 71.2% correct identifications. Training then refined the inferencing parameters to produce a system which correctly identifies 95.7% of the example chromatograms.

To evaluate the ability of the system to identify previously unseen data, the data set was randomly partitioned into training and testing sets in each of three trials. In each trial the training set contained 162 examples and the testing set contained 505 examples. The network was then trained using only the training set. On average, this produced a trained network which

**Table 1. Average performance of trained networks on new data.**

| Fault name | Number of examples | Average number correct | Average percent correct |
|---|---|---|---|
| Leaking Syringe | 50 | 46.0 | 92.0% |
| Column Bleed | 24 | 23.7 | 98.8% |
| Column Degradation | 71 | 67.3 | 94.8% |
| Makeup Gas Loss | 79 | 71.3 | 90.2% |
| Bad Liner | 17 | 11.0 | 64.7% |
| Leaking Septum | 33 | 22.0 | 66.7% |
| Detector Saturated | 5 | 5.0 | 100.0% |
| Sample Too Concentrated | 33 | 32.7 | 99.1% |
| Low Carrier Gas | 39 | 37.7 | 96.7% |
| No Carrier Gas | 36 | 36.0 | 100.0% |
| Contaminated Sample | 24 | 23.0 | 95.8% |
| No Fault | 94 | 91.7 | 98.64% |
| All Faults | 505 | 467.3 | 92.5% |

correctly identified 97.3% of the training examples. The testing examples were then presented to the trained network. Table 1 shows the averaged abilities of the trained networks to identify faults in new data.

## CONCLUSIONS

Use of the Table–Based Expert Network approach to fault diagnosis offers improvements over traditional expert system and neural network methods in each phase of the knowledge engineering process. The use of linguistic qualifiers to develop a tabular knowledge representation simplifies the knowledge acquisition phase of knowledge engineering by allowing experts to express their knowledge in a more natural manner. Automated tools such as NetMedic may then be used in an iterative approach to refine the knowledge acquired. The NetMaker tool may then be used to translate the tabular knowledge representation to an expert network representation. Once in a network representation, connectionist training techniques may be applied to refine inferencing to improve diagnostic accuracy and discriminatory separation.

## REFERENCES


Adair, K. L., 1996, "Data Driven Optimization of Expert Networks," Master's thesis, Florida State University, Tallahassee, Florida.

Adair, K. L., Levis, A. P., Hruska, S. I., 1996, "An Expert Network Development Environment for Automating Machine Fault Diagnosis," In Rogers, S. K., Ruck, D. W., editors, *Applications and Science of Artificial Neural Network II*, Vol. 2760 of *SPIE Proceedings*, pp. 506–522.

Dayhoff, J., 1990, *Neural Network Architectures: An Introduction* Van Nostrand Reinhold.

Elling, J. W., Klatt, L. N., Unruh, W. P., 1994, "Automated Data Interpretation in an Automated Environmental Laboratory," *Laboratory Robotics and Automation*, Vol. 6, pp. 73–78.

Fu, L. M., 1993, "Knowledge–Based Connectionism for Revising Domain Theories," *IEEE Tansactions on Systems, Man, and Cybernetics*, Vol. 23, pp. 173–182.

Gallant, S. I., 1993, *Neural Network Learning and Expert Systems* The MIT Press.

Giarratano, J. C., Riley, G., 1994, *Expert systems : principles and programming* The PWS series in computer science. PWS Pub. Co., 2nd edition.

Kuncicky, D. C., Hruska, S. I., Lacher, R. C., 1992, "Hybrid Systems: The Equivalence of Rule-Based Expert System and Artificial Neural Network Inference," *International Journal of Expert Systems*, Vol. 4, pp. 281–297.

Lahiri, S., Roberts, R., Elling, J., 1995, "Feature Extraction Algorithms for the Qualitative Assessment of Gas Chromatograms," Los Alamos National Laboratory Accepted for publication: *Journal of Chromatographic Science.*

Lahiri, S., Stillman, M. J., 1992, "Expert Systems: Diagnosing the Cause of Problem AAS Data," *Analytical Chemistry*, Vol. 64, pp. 283–291.

Levis, A. P., 1995, "Inferencing and Training in a Development Environment for Table–Based Expert Networks," Master's thesis, Florida State University, Tallahassee, Florida.

Shortliffe, E. H., Buchanan, B. G., 1984, *Rule-Based Expert Systems: The MYCIN Experiments of the Stanford Heuristic Programming Project*, chapter 11, pp. 233–262 Addison-Wesley Publishing Company, Reading, Massachusetts.

Timpany, R. G., 1995, "Data Assisted Knowledge Acquisition for Alteration of Table–Based Expert Networks," Master's thesis, Florida State University, Tallahassee, Florida.

# INTELLIGENT COMPUTING BUDGET ALLOCATION FOR STOCHASTIC SIMULATION[1]

**CHUN-HUNG CHEN**
*Dept. of Systems Engineering, University of Pennsylvania, Philadelphia, PA 19104-6315, chchen@eniac.seas.upenn.edu*

**LIYI DAI**
*Dept. of Systems Science and Mathematics, Washington University, St. Louis, MO 63130, liyi@zach.wustl.edu*

***ABSTRACT:***
Simulation plays a vital role in analyzing many stochastic systems. Unfortunately, using simulation to solve such problems is very time consuming. In this paper we present an effective approach which can intelligently determine the best simulation lengths for all simulation experiments, significantly reducing the total computation cost.

In order to efficiently manage and operate large man-made stochastic systems such as communication networks, traffic systems, and automated manufacturing plants, it is often necessary to apply extensive simulation to study their performance since no closed-form analytical solutions exist for such problems. Unfortunately, using simulation to solve such problems can be both expensive and time consuming due to their massive search space and their evolution in time according to complex man-made rules and the influence of random occurrences.

Suppose we want to compare $n$ different stochastic systems (designs or alternatives), we do $T$ simulation replications for all $n$ designs. Totally, we need $nT$ simulation replications. The simulation results become more accurate when $T$ increases. If the accuracy requirement is not low ($T$ is not small), and if the total number of designs in a decision problem is not small ($n$ is large), then $nT$ can be very large, which may easily make total simulation cost extremely high and preclude the feasibility of simulation approach. To reduce total simulation time, one can either develop more efficient simulation technology or use faster computers to reduce the simulation time of each simulation experiment. In this paper, we present another approach to improve the overall simulation efficiency. We reduce the total cost from $nT$ to a much smaller number by intelligently determining the number of simulation replications for each design.

To intelligently allocate computing budget, first of all, one must have an efficient way to estimate the confidence level based on the results of the completed simulation. Further, one must have easy ways to anticipate how the confidence level will change if some computing budget is allocated and additional simulation replications are completed. Goldsman and Nelson (1994) provide an excellent survey on current approaches to estimating simulation confidence level for problems not having large number of designs (2 ~ 20 designs). For real-life stochastic problems, the number of

---

[1] This work was partially supported by National Science Foundation under grant No. ECS-9624279 and the University of Pennsylvania Research Foundation.

designs can grow up to numerous orders of magnitude easily. Chen (1996) presents a feasible way to quantify confidence level for *large* DES simulation. Further, when the approach in Chen (1996) is applied the sensitivity information of the confidence level with respect to simulation replication numbers can be easily obtained, which will provide the basis to determine how to allocate computing budget among designs in this paper. Next section describes the notation used in this paper and a brief overview of Chen (1996)'s approach to quantify confidence level for large problems.

## A BAYESIAN APPROACH TO CONFIDENCE LEVEL QUANTIFICATION

Denote

$n$: the total number of designs,

$T$: the number of replication, or the total number of samples,

$\hat{J}_j(t)$: the $t$-th sample of the performance measure of design $j$,

$\bar{J}_j(T)$: the sample mean of design $j$, namely, $\bar{J}_j(T) = \frac{1}{T}\sum_{t=1}^{T}\hat{J}_j(t)$, and

$J_j$: the performance measure, or more specifically, the mean performance measure of design $j$, i.e., the mean of $\hat{J}_j(t)$.

Assume that

i) $\hat{J}_j(t)$ is i.i.d. for all $t$,

ii) the simulations for designs $i$ and $j$, $i \neq j$, are independent. Thus, $\hat{J}_i(t)$ and $\hat{J}_j(t)$ are independent.

As the strong law of large numbers, with probability 1, $\bar{J}_j(T) \rightarrow J_j$, as $T \rightarrow \infty$. Without infinitely long simulations or infinite number of simulations replication, the sample mean $\bar{J}_j(T)$ is an approximation to $J_j$. With the approximation results, the best we can do is to pick the observed best design. Usually, the observed best is not necessarily the true best design. Then what is the confidence level?

Chen (1996) adopts the Bayesian model to analyze such confidence probability. Under the Bayesian model, $J_j$ is treated as a random variable and has a prior distribution which describes the knowledge or the subjective belief about $J_j$ before any sampling. The posterior distribution is updated after we observe the samples { $\hat{J}_j(t)$, $t$=1,..,$T$}. With some normal assumptions, the posterior

$$\tilde{J}_j \equiv J_j \mid \{\hat{J}_j(t), t = 1,2,\cdot\cdot,T\} \sim N(\frac{1}{T}\sum_{t=1}^{T}\hat{J}_j(t), \frac{\sigma_j^2}{T})$$

Chen (1996) also shows that the Confidence Probability (for maximization problems) CP $\equiv$ Pr{the observed best design ($o$) is actually the true best design}

$$\geq \text{Approximated Confidence Probability ACP} \equiv \prod_{j=2, j\neq o}^{n} \Pr\{\tilde{J}_o > \tilde{J}_j\},$$

ACP can be computed very easily. This Bayesian model is ideally suitable for developing our intelligent computing budget allocation approach because the sensitivity information of the confidence level with respect to simulation replication numbers can be easily obtained.

## INTUITIVE EXPLANATIONS AND PROBLEM FORMULATION

The intuitive explanations are as follows. Some bad designs can be discarded before completing all of the $T$ replications. We don't have to waste efforts on simulating bad designs and so reduce overall simulation time. Then the question is how to systematically do this? When? And which designs? Ideally, we want to optimally choose the number of simulation replications for all designs to minimize the total simulation cost, while obtaining the desired confidence level. In fact, this question is equivalent to optimally decide which designs will receive computing budget for continuing simulation. Figure 1 illustrates the ideas by comparing a typical solution to this problem with the conventional approach using equal simulation lengths. Chen (1995) formulates this question and obtained promising preliminary results using very simple heuristics. In this paper, we present some further results and the ideas of the application of an expert system.

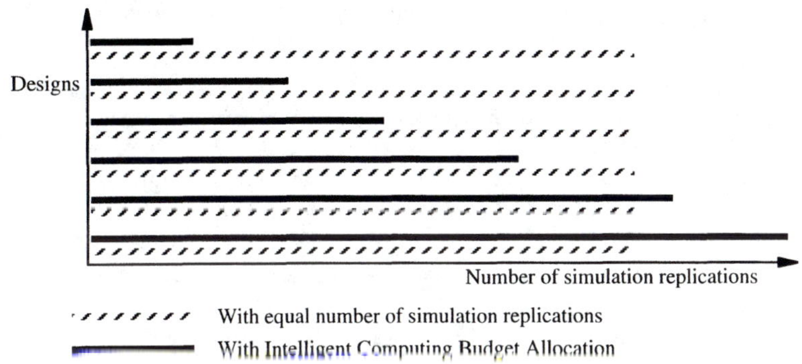


Figure 1. Comparison of simulation budget allocations for obtaining the same Confidence Level.

Consider simulating $n$ systems and let $T_j$ be the number of simulation replications of design $j$. If simulations are performed on a sequential computer, the computation cost can be approximated by $T_1 + T_2 + \cdots + T_n$. Mathematically our goal is to solve

$$\textbf{(P)} \quad \min_{T_1, \cdots, T_n} (T_1 + T_2 + \cdots + T_n), \text{ s.t. ACP} \geq \text{(a satisfactory level).}$$

There are two major difficulties in solving (P):

1. ACP($T_1, T_2, \cdots, T_n$) can be computed only after doing simulations until $T_1, T_2, \cdots, T_n$. Before performing simulations until $T_1, T_2, \cdots, T_n$, how can we predict or estimate the ACP at $T_1, T_2, \cdots, T_n$?

2. $T_1, T_2, \cdots, T_n$ are integers. Even if we have techniques to estimate ACP at $T_1, T_2, \cdots, T_n$, an extremely large combinatorial space must be searched to find a solution to (P), especially when n is large.

Note that the purpose of solving (P) is to further reduce computation cost for obtaining a desired confidence level. We should not exert too much effort solving (P) during simulation. Otherwise, the additional cost of solving (P) will cancel the benefits of computing budget allocation. Efficiency is more crucial than optimality in this application.

## A SEQUENTIAL APPROACH

This section presents a sequential procedure to overcome the difficulties in solving (P). To Optimally allocate computing resource, it is equivalent to determine which designs we should do more simulation. We will sequentially decide it, although this is usually not an optimal solution any more.

Before doing simulation there is neither knowledge about ACP nor a basis for choosing $T_j$. First, all designs are simulated until length $t_0$ to obtain statistical information about sample means and sample variances. Then we try to determine how to further allocate computing budget using available statistical information. When simulation is stopped at $t_0$, the posterior distribution of design $j$ is

$$\tilde{J}_j \equiv \tilde{J}_j(t_0) \equiv J_j \mid \{\hat{J}_j(t), t = 1,2,\cdot\cdot,T\} \sim N(\frac{1}{t_0}\sum_{t=1}^{t_0} \hat{J}_j(t), \frac{\sigma_j^2}{t_0})$$

At this moment, we have some ideas about each design and then can decide which designs are worthy of being allocated more computing budget. To determine how to further allocate computing budget, we have to be able to know how the ACP will change if some computing budget is allocated to some designs (Difficulty 1). More specifically, based on statistical information at $t_0$, we want to anticipate the posterior distribution at $t_0 + \Delta$ ,where $\Delta$ is a positive integer. To do this, we assume the sample mean and variance at $t_0$ are near those at $t_0 + \Delta$, and approximate the posterior distribution at $t_0 + \Delta$ using the **estimated** posterior distribution

$$N(\frac{1}{t_0}\sum_{t=1}^{t_0} \hat{J}_j(t), \frac{\sigma_j^2}{t_0 + \Delta})$$

Note that the denominator of variance portion is $t_0 + \Delta$ rather than $t_0$. Using the estimated posterior distributions, we can estimate the ACP at $t_0 + \Delta$ using the statistical information at $t_0$, and call it the "Estimated ACP" or EACP.

Similarly, when simulation proceeds until $T_1, T_2, \cdots, T_{i-1}, T_i, T_{i+1}, \cdots, T_n$, we can also use the available information to estimate how ACP will change if design i is given additional budget $\Delta$ using the estimated posterior distribution

Since ACP will become larger as simulation proceeds, we sequentially add computing budget by $b$ each time until ACP reached some satisfactory level (say $P_{sat}$). In order to minimize to the total computation cost, at each step, this budget $b$ is allocated among some designs such that the EACP is maximized. Thus, at step $k$, $k$=1,2,3,.., we have

**(P-*k*)** $\max_{\tau_1^k, \cdots, \tau_n^k} \text{EACP}(T_1^k + \tau_1^k, T_2^k + \tau_2^k, \cdots, T_n^k + \tau_n^k)$, s.t. $\tau_1^k + \tau_2^k + .. + \tau_n^k = b$

A sequential *Intelligent Computing Budget Allocation* (**ICBA**) algorithm is

**Step 0.** PERFORM SIMULATION until length $t_0$ for all designs, $k \leftarrow 0$. $T_1^k = T_2^k = \cdots = T_n^k = t_0$.

**Step 1.** If ACP( $T_1^k, T_2^k, \cdots, T_n^k$ ) $\geq P_{sat}$, stop; otherwise, go to Step 2.

**Step 2.** Solve (P-$k$). $T_i^{k+1} = T_i^k + \tau_i^k$, for $i = 1, .., n$. $k \leftarrow k+1$,

**Step 3.** PERFORM SIMULATION until ( $T_1^k, T_2^k, \cdots, T_n^k$ ). Go to Step 1.

## SOLVING PROBLEM (P-*k*)

The next question is how to **efficiently** solve (P-$k$). While solving (P-$k$) is easier than solving (P), again, efficiency is much more important than optimality here. Our approaches are as follows:

### A Simple Greedy Method

In the approach, $m$ designs are chosen and then the computing budget is equally distributed to them (each design has $b/m$ ). For each design, we calculate the anticipated increment of ACP if computing budget $b/m$ is allocated to it. Then those designs are chosen if their anticipated ACP increments are among top-$m$.

### A Steepest-Descent Method

Chen, Chen and Dai (1996) presents a simple way to estimate the gradient of ACP with respect to $T_i$. A steepest-descent method (Luenberger 1984) can be applied to solve (P-$k$) by utilizing the gradient information.

### An Expert System

This is one of our ongoing research subjects. A few factors will be emphasized in developing our expert system: i) the differences of sample means among designs; ii) the sample variances; iii) problem characteristics.

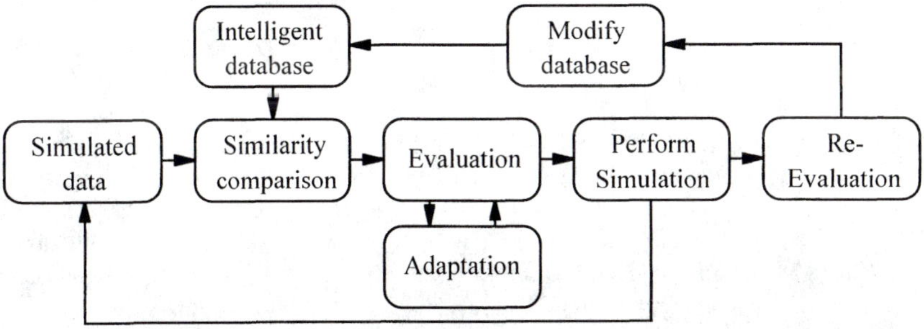


Figure 2. An Intelligent Computing-Budget-Allocation Expert System.

The expert system which we are developing for intelligently allocating computing budget is illustrated in Figure 2. Given simulated data, some knowledge is acquired from database module to compare the similarity between knowledge and simulated data. A decision of computing budget allocation is made based on similarity comparison and knowledge from the database. Evaluation of the decision can be conducted using the previous simple approaches. Then we adapt the budget allocation to obtain higher confidence level improvement. After graduate adaptation and evaluation, the decision is finalized and real computer simulation is performed as the budget allocation.

## NUMERICAL TESTING

We test our approaches on a simple single-node queue. The interarrival time is Uniform[0.1, 1.9]. We consider 10 designs with different service times, which are Uniform[0.1, 1.85-$i$*0.05] for design $i$, $i$=1,..,10. Suppose we are interested in the average system time of the customers served between time 0 and time 10. We set $b$ = 12 and $t_0$ = 5. 10,000 independent experiments are done to estimate the average cost for using different approaches. Table 1 shows the average total numbers of simulation replications for obtaining the confidence level $P_{sat}$.

From Table 1, we see that smartly allocating computing budget can reduce computation cost significantly. We also observe that different parameter choices may obtain different performances. How to choose these parameters and the development of an expert system are two of our ongoing research topics.

Table 1. Comparison of Average Total Number of Simulation Replications ($T_0$=5).

| | Simple Greedy Method | | | | Steepest-Descent Method | | | None |
|---|---|---|---|---|---|---|---|---|
| $P_{sat}$ | $m$ = 4 | $m$ = 3 | $m$ = 2 | $m$ = 1 | 1 itrn | 2 itrns | 4 itrns | |
| **60%** | 127.6 | 127.6 | 129.9 | 139.9 | 130.1 | 122.3 | 125.9 | 305.2 |
| **70%** | 186.3 | 183.6 | 182.4 | 192.3 | 185.9 | 175.3 | 185.3 | 475.5 |
| **80%** | 305.6 | 287.0 | 275.4 | 283.6 | 290.4 | 273.1 | 285.5 | 790.0 |
| **90%** | 576.9 | 507.1 | 470.0 | 470.1 | 496.2 | 467.3 | 492.1 | 1462.0 |

## REFERENCES


Chen, C. H. (1995). "An Effective Approach to Smartly Allocate Computing Budget for Discrete Event Simulation," *Proceedings of the 34th IEEE Conference on Decision and Control*, 2598-2605.

Chen, C. H., Chen, H. C., and Dai, L. (1996) "A Gradient Approach of Smartly Allocating Computing Budget for Discrete Event Simulation," To appear in *Proceedings of the 1996 Winter Simulation Conference,* Dec.

Chen, C. H., and Ho, Y. C. (1995). "An Approximation Approach of the Standard Clock Method for General Discrete Event Simulation," *IEEE Transactions on Control Systems Technology*, Vol. 3., No. 3., pp. 309-317, Sept.

Chen, C. H. (1996). "A Lower Bound for the Correct Subset-Selection Probability and Its Application to Discrete Event System Simulations," To appear in *IEEE Transactions on Automatic Control*, August.

Goldsman, G., and Nelson, B. L. (1994). "Ranking, Selection, and Multiple Comparison in Computer Simulation," Proceedings of the 1994 Winter Simulation Conference, pp. 192-199, Dec.

Luenberger, D. G., (1984). *Linear and Nonlinear Programming*, Addison-Wesley.

# FINANCIAL PLANNING WITH FUZZY LINEAR PROGRAMMING

**ZÜLAL GÜNGÖR**
*Dept. of Indutrial Engineering*
*Gazi University of Engineering and Architecture*
*Ankara, TURKEY*

*ABSTRACT:*

When solving financial planning problems with multiple goals by means of multiobjective programming, the presence of fuzziness with ranking or weighting of the goals leads to technical difficulties. Most of the existing methods generate a set of nondominated solutions or construct a single compromise solution for the linear multiobjective programming problem. In this paper Fuzzy Linear programming approach for solving such financial planning model with multiple goal variables is introduced. An example of financial planning model is also given.

**Key words: Financial Planning, Fuzzy Linear Programming, Multiple-Criteria, Decision making.**

## INTRODUCTION

In this paper, we define financial planning in a firm as a structured process of identification and selection of present and future capital investment projects. The theory of financial management needs to consider together with the all investment and financial decision together with dividend option facing up the firm. Therefore financing is difficult and planner must consider financial decision together with objectives and aspects of the firm. Financial plans may consider to evaluate the objectives by financial theory such as; The maximization of the share- holders ' wealth (the market valueof the firm), the stability of earnings over time, employment level in the company. This shows that finacial planning essentially has complex and multiple conflicting goals. Therefore we are faced with multictriteria decision problem. In the literature several multicriteria decision methods are applied to the financing planning problem such as goal programming and other interactive methods (Zeleny 1982; White, 1990).

At this point it may be helpful to state the main features of our approach in very general terms. Especially planners attempt to supply financial options in proper sequence, In general these objectives. Further, other enviromental factors often present additional conflicting, "unclear" objectives. This unclearness suggests "Fuzzy" linear programming (FLP) may be applicable. Fuzzy linear programming is easy to use and relatively simple to apply to many problems of the other area of the economics and production.

## FUZZY LINEAR PROGRAMMING MODEL

Fuzzy linear programming occurs when fuzzy set theory is applied to linear multicriteria decision making problems. In fuzzy set theory, an element X has a degree of membership in a set A, denoted by a membership function $\mu_A(X)$. The range of the membership function is [0, 1]. Degree of the membership function for each objective represents its satisfaction level. If the membership function of the objective lies in (0, 1) then the objective is partially achieved. Zimmermann first applied the fuzzy set theory concept with some suitable membership functions to solve multiobjective linear programming problems. The solution of the fuzzy linear programming are always efficient (Zimmermann, 1987).

The linear multiobjective financial planning problem is a special type of vector minimum or maximum problem when the objectives are conflicting in nature. Here a fuzzy programming technique is presented to solve such a multiobjective financial planning problem. One can use the proposed fuzzy linear programming model for multiobjective financial planning problems when importance of each objective is unkonuwn. The formulation of the general FLP models is dicussed in Ignizio (1976) and is also formulated similarly to the LP model is initially solved. The multiobjective financial planning problem as single objective problem using each time only one objective (all others are ignored). Then corresponding values for every objective at each solution are derived. An upper bound and lower bound for each objective function may be found. Let $L_K$ and $U_K$ denote the lower and upper bounds, respectively, for the k-th objective ($Z_k$). For the multiobjective financial planning problem, the membership function $\mu_k(X)$ (k= 1,2,... k) for the k-th objective function is defined as:

$$\mu_k(X) = \begin{cases} 0 & \text{if } Z_k \le L_k \\ 1 - \left( \dfrac{U_k - Z_k}{U_k - L_k} \right) & L_k \langle Z_k \le U_k \\ 1 & Z_k \rangle U_k \end{cases} \qquad (1)$$

Provided $U_k \neq L_k$ for all k;
$\mu_k(X) = 1$ if $U_k = L_k$ for all k.

The equivalent FLP model of finacial planning model is formulated by adding the membership function together as

Max $\lambda$

Subject to problem contraints together with

$$\lambda \le \frac{Z_k - L_k}{U_k - L_k}$$

We are now faced with a problem in which 2 rows representing our fuzzified objectives.

$\lambda$ is called the fuzzy achievement (or maximum degree of overall satisfaction) function or fuzzy decision function. This is a single objective linear programming problem. The problem can be solved by simplex method, since the linear membership function fits the multiobjective linear programming problem, the solution is nondominated and occurs at extreme point, which is nearer to the best compromise solution.

## DESCRIPTION THE MODEL

Assume that there are 10 investment projects, and decision maker wants to choose from these projects which satisfy the company's objectives. The variables of the program are investment project $(X_i, 0 \leq X_i \leq 1, \text{for } i = 1..\ 10)X_0$ is existing assets, amount of debt and lending is $D_t$, and cash held is $L_t$ in each period. All debts have a maturity of one year, that means borrowed at the beginning and repaid at the end of the same period t. The planning takes 10 periods, tax rate T is 40 %, debt interest rate is 10 %. The characterictics of projects consist of the unlevered present values $A_i$, the after-tax cash flows in each period $C_t^i$, the contribution to the periodic earnings before interest and taxes $E_t^i$, and the total employment level in the company's $W^i$. The first goal of the company is maximize the company's total market value. Total market value is defined as the sum of the unlevered present value of its investment project plus the present value of its financing decisions (the value of tax saving produced by debt financing). In this case the unlevered present value of investment projects is assumed to be estimated directly by management. First goal is defined as follows, as we stated earlier this gives us the maximization of the total market value (Myers, 1974).

$$\text{Max } Z_1 = \sum_{i=1}^{10} A_i X_i + \sum_{t=0}^{9} \frac{r_d . T . D_t}{(1+r_d)^t} - D_0 - \sum_{t=0}^{9} \frac{r_f . L_t}{(1+r_f)^t} + L_0 \qquad (3)$$

$r_f$ =risk free interest rate ($r_f \leq r_d$)
$r_d$ = debt interest rate
$D_t$ =Debt in t perod
$L_t$ = Cash held in period t
T = Tax rate

$$\sum_{t=0}^{9} (r_d . T . D_t)/(1+r_d)^t = \text{Present value of debt}$$

The numerator represents the tax saving caused by a shift from all- equity financing to partial debt financing (Goedhart 1995).

$$-\sum_{t=0}^{9}\frac{(r_f . L_t)}{(1+r_f)^t} = \text{Present value of having an annual cash level of Lt.}$$

It is negative, because no interest is gained. If liquid assest are equivalent to negative debt, then liquid assests create an additional tax rather than a tax shield. Result of this fact is the negative value of $A_i$ (Myers, 1974).

The second goal is to keep growth rate constant, providing the stabilization of the financial earning over planning period. In the present model second goal is assumed as;

$$\text{Min } Z_2 \geq e_t^- \qquad t = (1.........9) \qquad (4)$$

with $$\sum_{i=1}^{9} E_t^i . X_i - e_t^+ + e_t^- = E_t^*, \quad t = (1....9)$$

$E_t^*$ = is the target for earning in period t.
Et = $E_{t-1}$ $(1+g_E)$
$g_E$ = 5 % is the target growth rate (defined by decision maker).

The third goal is taken from organization pattern of the firm, the maximization of the emloyment level is given as;

$$\text{Max } Z_3 = \sum_{i=1}^{10} w^i X_i$$

$w^i$ = Incremental number of jobs per investment project. Model is taken as deterministic and all uncertainities are accounted as exogenous factors (Myers, 1974).

As stated this plan is solved by fuzzy linear programming, fuzzy upper and lower bound value and acceptable membership function is presented in Table 1.

**TABLE 1. RESULTS OF LINEAR PROGRAMMING AND FUZZY LOWER AND UPPER BOUNDS**

| | $L_k$ | $U_k$ | $d_k$ |
|---|---|---|---|
| Max $Z_1$ | -821 | 8260 | 9081 |
| Max $Z_3$ | 24 | 125 | 101 |
| $Z_1$ $(X^1)$= 8260 | $Z_2$ $(X^2)$= -821 | $Z_1$ $(X^3)$= 7755 | |
| $Z_3$ $(X^1)$= 93 | $Z_3$ $(X^2)$= 24 | $Z_3$ $(X^3)$=125 | |

Finacial plan models are solved seperately for every objective function. The results of this solutions are given in table 1. Then the following approach is used. First result of the linear programming is assumed as aspiration level fuzzy linear programming. Then membership function $\mu_1$ (X) and $\mu_2$ (X) of the fuzzy sets characterizing the objective functions rise linearly from 0 to 1 at the highets achievable value.

$Z_1$ $(X^1)$=8260, $Z_3$ $(X^3)$= 125 respectively. Fuzzy membership functions are given as;

$$\mu_1(x)=\begin{cases}0 & Z_1(X)\le -821\\ \dfrac{Z_1(X)+821}{9081} & -821\,\langle Z_1(X))\le 8260\\ 1 & 8260\langle Z_1(X)\end{cases}\qquad (6)$$

$$\mu_2(x)=\begin{cases}0 & Z_2(X)\le 24\\ \dfrac{Z_3(X)-24}{101} & 24\,\langle Z_2(X)\le 125\\ 1 & 125\langle Z_2(X)\end{cases}\qquad (7)$$

## SELECTION THE FINANCIAL PLAN

$X^1$ is the optimum results with respect to objective function $Z_1(X)$ (improvement of total market value) and $X^3$ is the optimum result with respect to objective function $Z_3(X)$ (total employment level in the company). The optimum values are $Z_1$ (X)= 8260 and $Z_3$ (X)= 125 respectively, and other result $X^2$ is the result of objective function $Z_2$ (X). $Z_1(X^2)$= -821, $Z_3$ $(X^2)$= 24 are the lowest justifiable value of the objective functions, that means a further decrease of the value of one objective function can not be balanced on even counteracted by an increase in the value of the other objective function. To solve this problem we first assume aspiration levels for the objective functions, then we would have to construct two fuzzy sets; one of them is the solution acceptable with respect to objective function 1 and the other solution is acceptable with respect to objective function 3.

The membership function $\mu_1$ (X) and $\mu_2$ (X) of the fuzzy sets characterizing the objective function rise linearly from 0 to 1 at the highest achievable values $Z_1$ (X) = 8260 and $Z_3$ (X) = 125 respectively.

It means that we assume that the level of total market value rises from 0 for -821 loss or more to 1 for market value 8260 and more; the satisfaction level rises with respect to employment level in company from 0 to if employment level 24 or less to 1 if employment level is 125 or more.

## CONCLUSION

We have obtained the solution process for our financial planning problem. the value of Goals are presented (see table 2) and the goal variables are obtained by optimizing each of goal variables seperately without any restrictions on the values of the other goals. At the first goal which is maximixing the overall wealth is obtained as 8260. The value of third goal is 125 jobs level. According to the plan level of cashs are always large and the level of debts is close to the restriction level in the second goal. In the final plan (Fuzzy linear programming) two investment projects are fully implemented. In the first goal the level of debt is close to the restriction level of 600 during the first years and almost zero during the remainder of planning period when required funds for investment are much lower. The earning in the last planning period are 295 below the % 5 target growth path. The earning underattainment is mainly due to low earnings in the last period of the planning.

TABLE 2. OPTIMAL VALUE FOR INSTRUMENT VARIABLE

| | Objective1 | Objective2 | Objective3 | Fuzzy Linear Programming |
|---|---|---|---|---|
| | $Z_1$ (X)= 8260 | $Z_2$ (X)= -295 | $Z_3$ (X)=125 | = 0.2712 |
| $X_0$ | 1 | 0.326 | 1 | 0 |
| $X_1$ | 1 | 0 | 0 | 0 |
| $X_2$ | 1 | 0 | 0 | 0 |
| $X_3$ | 1 | 0 | 0 | 0 |
| $X_4$ | 1 | 1 | 1 | 0 |
| $X_5$ | 1 | 1 | 1 | 0 |
| $X_6$ | 0 | 0 | 1 | 1 |
| $X_7$ | 1 | 1 | 0 | 0 |
| X8 | 0 | 0 | 0 | 1 |
| $X_9$ | 1 | 1 | 1 | 0 |
| $X_{10}$ | 1 | 0 | 0 | 0 |
| $D_0$ | 600 | 600 | 600 | 600 |
| $D_1$ | 176 | 300 | 300 | 300 |
| $D_2$ | | 300 | 0 | 8.3 |
| $D_3$ | | 300 | 0 | 300 |
| $D_4$ | | 300 | 300 | 0 |
| $D_5$ | | 300 | 300 | 300 |
| $D_6$ | | 300 | 300 | 0 |
| $D_7$ | | 300 | 300 | 0 |
| $D_8$ | | 300 | 0 | 0 |
| $D_9$ | | 300 | 0 | 0 |
| $D_{10}$ | | 0 | 0 | 0 |
| $L_0$ | 10 | 10 | 10 | 10 |
| $L_1$ | 0 | | 674 | 0 |
| $L_2$ | 888 | 291 | 1306 | 0 |
| $L_3$ | 1963 | 583 | 2256 | 300 |
| $L_4$ | 2888 | 775 | 3406 | 592 |
| $L_5$ | 3921 | 1084 | 4356 | 1202 |
| $L_6$ | 4704 | 1178 | 5056 | 997 |
| $L_7$ | 4837 | 1260 | 5106 | 975 |
| $L_8$ | 5020 | 1243 | 5006 | 1102 |
| $L_9$ | 5308 | 1376 | 5564 | 1270 |
| $L_{10}$ | 5576 | 1208 | 5504 | 1438 |

## REFERENCES


Goedhart, M.H., Spronk, J., (1995). Financial planning with fractional goals, *European Journal of Operational Research* (82), 111-124.

Ignizio, James P., (1976). goal Programming and Extensions. *Lexington, Mass, Lexington Books*.

Myers, S.C., and Pogue, G.A., (1974). A programming approach to corparate financial management, *Journal of Finance* (29), 579- 599.

Myers, S.C., (1974). Interactions of Corporate Financing and Investment Decision-Implications for Capital Budgeting, *Journal of Finance*, March.

White, D.J., (1990). A bibliography on the applications of matematical programming multiobjective methods, *Journal of the Operational Research Society* (41), 669-691.

Zeleniy, M., (1982). Multiple Criteria Decision Making, *McGraw Hill New York*.

Zimmermann, H.J., (1987). Fuzzy set theory and its applications, 2nd revised edition, *Kluwer- Nijhoff, Dordrecht- Boston*.

# THE THEORY OF DISCRETE OPTIMIZATION AND ITS RELATIONSHIP WITH FACILITIES LAYOUT

**ORHAN TÜRKBEY**
*Gazi University, Engineering and Architecture Faculty, Industrial Engineering Department, Maltepe, 06570, Ankara, Turkey,*

**Abstract :**
Discrete Optimization is a branch of Combinatorical Mathematics. Certainly Discrete Optimization does not encompass all of the Combination Theory which is a very vast field. The study of plant layout constitutes a vital aspect of combinatorics.This stance is based on the analysis and selecting of the elements which have a feasible solution space among mutually exclusive alternatives.In our study,using the Discrete Optimizing Theory,an algorithm is developed to obtain a near - optimal plant layout for the Net Present Value of material handling cost (the transportation cost),and the Net Present Value of penalty cost for breaking up departments into sections,and the Net Present Value of calculation cost of computer (the cost of CPU time) the man - machine - material variables in any plant. Furthermore,The Discrete Optimizing Theory is summarized.
**Key words: Discrete Optimization; Combinatorial Analysis; Material Handling; Integer Programming; Plant Layout**

## INTRODUCTION

Facilities Layout problem lies within the plant design which is one of the sub-items of plant planning. It is the determination or arrangement of the locations of production or sales units in a plant, offices or departments in administrative buildings, basic productive units in a fabrication site or the machine sections/spares/equipment within these sites taking into consideration various criteria. The number and location of the plants, and the quantitative values of their interactions with the available plants are identified as decision variables. Decision criteria in plant layout are the p of the handling distance, minimization of materials handling cost, maximization of closeness to the desired interaction degree or the best possible flow patterns.

The plant is an integrated system within its own scope. The plant system is established targeting the lowest cost and optimum productivity. Not only the industrial plants but also agricultural and commercial plants, hotels, restaurants, military plants and hospitals are faced with Plant Layout problems. When the scope is extended to cover all types of plants, then the use of the term "Plant/workplace Layout" would be more convenient (Türkbey,1983).

In this study, the cost of the intermittent materials handling between the fields or machines from the point of view of handling materials such as fork trucks, hand trucks, cranes etc. will be considered as a variable (Pegels,1966). As stated by a manager, "the cost of the material handling from one production site to the other is a complete loss, since it has an important effect on the cost of the product". For this reason, the mentioned cost should be taken into consideration during the production phase since it has to be eliminated to achieve the minimum cost (Cameron,1952).

The objective of this study is to present the method for the achievement of optimum or near-optimum Plant Layout which minimizes the handling distance depending on the number of the loads handled between the departments. This method will be more useful in a job-shop where sections manufacturing is employed or in a manufacturing process of job-shop qualifications.

## REVIEW OF LITERATURE AND ADVENCES ON THE SUBJECT

In the past,dicrete optimization was dealt with to a large extent. The examples are net-flows by Ford and Fulkerson (Ford and Fulkerson,1962), the first study on Cutting Planes developed by Gomory (Gomory,1963), and studies on optimum matching by Edmonds (Edmonds,1965). During the last 10-15 years, the Complexity Theory which still has an important place in literature and developed by Cook (Cook,1971) and then Karp (Karp,1975) searched the classification methods of the problems by mathematical hierarchy with the help of digital computer algorithms. During this period , the four common subjects were transformed into the Status Theory which appeared to be a Complete Theory of Cutting Planes. Lagrangean technics were adapted into discrete optimisation and ellipsoidal and related methods which provide a polynomial solution plan for linear programs (Grötschel,Lovasz and Schrijver,1986) (Schrader,1982).

However during the recent years, noteworthy progress on theoretical base have not been provided, but application-oriented development has been observed. In fact Knapsack problems can be effectively solved at emprical level (Balas,1975) (Balas,1980) (Martello and toth,1979). In addition to certain scheduling problems, even the Travelling Salesman problem can be solved to a certain degree (Parker and Rardin,1983,1984). Moreover, the problems of big dimension which existed 10 years ago, can now be solved in routine.

## THE APPLICATION OF DISCRETE OPTIMIZATION THEORY TO PLANT LAYOUT

Travelling Salesman problem, Assignment problem and Knapsack problem constitute the sample family within discrete optimisation. DOP has been designed to solve problems like Travelling Salesman problem, except complete enumeration (Parker and Rardin,1988). Layout problem is a travelling salesman problem which is very difficult to be solved by complete enumeration. As the number of the variables in the problem increases, complete enumeration will not be possible. For example, in a travelling salesman problem where he has to travel n=11 cities, the number of the total routes will be 11 or approximately $50.10^{6.}$ For this reason, the solution time of the big dimensioned problems with increased number of variables is a burden on cost. A solution derived from a complete enumeration where the calculation cost is not important will not be optimal. In DOP developed by Reiter and Sherman (Reiter and Sherman,1965) statistical sampling is used to avoid complete enumeration and the calculation cost is considered in order to decide when the sampling is to be terminated.

**The Formulation of the Application**

As known, our purpose here is to find the layout f(p) which minimizes the present value of the total cost function formed by the variables such as material handling, breaking up or split up into departments and CPU time, by using discrete optimisation theory. Let $V_1$= handling cost,$V_2$= Cost of breaking up into departments, $V_3$ = CPU cost of the new layouts obtained by combinatorial calculation.

In order to simplify the problem, from a theoretical point of view, the shape of the plant area is considered to be rectangular grid. Length of the plant (L) (index cell length l = 1,2,...,L) width W (index cell width w = 1,2,...,W) and the plant area is LW. If suitable units are used to express L and W, LW will be an integer and the number of L-W combinations will be factorial.

Each x department can be divided into sections of specified sizes where x = 1,2,...,X. For example, a department consisting of 100 cells can be divided into 4 sections with sizes 40,20,20,20. The size of the sections is limited with the manufacturing process in each department section. Vector b, defines the number of the cells covered by each department section, that is, department section j occupies $b_j$ cells. Then,

$$\sum_{j=1}^{n} b_j = LW \tag{1}$$

The number of the loads to be handled between the department sections can be defined by the symmetrical A matrix, the elements of which are real numbers.

$$A = \|a_{jk}\| ;\ j,k = 1,2,\ldots,n \ ;\ a_{jk} = a_{kj} \ ;\ \ j = k \ \text{ için } \ a_{jk} = 0 \tag{2}$$

The distances between the department sections can be defined by a symmetrical D matrix with real numbers.

$$D = \|d_{jk}\| \ ;\ j,k = 1,2,\ldots,n \ ;\ j = k \ \text{ ve } \ d_{jk} = d_{kj} \ \text{ için } \ d_{jk} = 0 \tag{3}$$

The distances between the department sections can be defined in terms of Euclidean unit as below (Francis,Jr.McGinnis and White,1992):

$$d_{jk} = \left[ \left(\bar{\ell}_j - \bar{\ell}_k\right)^2 + \left(\bar{w}_j - \bar{w}_k\right)^2 \right]^{1/2} \tag{4}$$

Here $l_j$ , $l_k$ , $w_j$ , $w_k$ , are the average values of l and w for the department sections j and k. Figure 2 shows the grid illustration of a department.

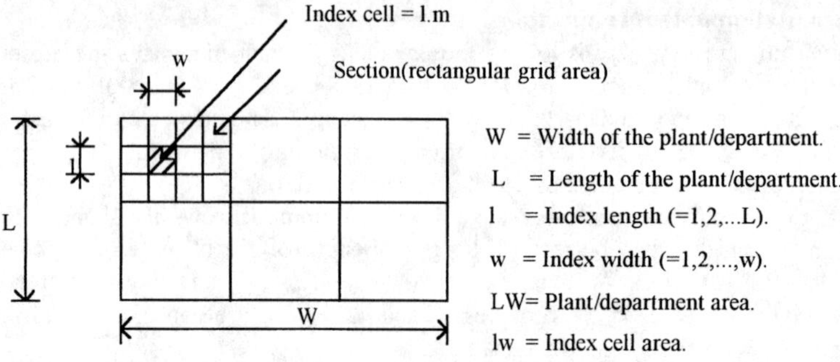


Figure 2: Illustration of a Plant/Department as Sections in Grid Form.

The present value of the material handling cost between the department sections can be defined as follows:

$$V_1 = \frac{v_1(1-e^{-rT})}{r} \sum_{j=1}^{n} \sum_{k=j+1}^{n} a_{jk} d_{jk} \tag{5}$$

Here $v_1$ is the fixed cost of each load distance and $\left(\frac{1-e^{-rT}}{r}\right)$ is the present value factor (Davidson,Smith and Wiley,1962)(Fabrycky and Thuesen,1974). If the capital cost r or the interest is known, the present value factor (within an equivalent present cost) can change the flow cost of the future T length within a time period.Here e = 2,7183.

As an example, let's assume that the capital cost is annual 9%; the operational life of the planned production plant is 120 months and fixed cost of each load distance is $v_1 = 0{,}10$ \$ .The present value factor will be:

$$\frac{1-2{,}7183^{-0{,}90}}{0{,}0075} = 79{,}124$$

If the number of loads which defines the number of loads to be handled within a distance during a production planning period is $\sum_{j=1}^{n} \sum_{k=j+1}^{n} a_{jk} d_{jk} = 100.000$ units, the present value of the handling cost will be: $V_1 = (79{,}124)(0{,}10)(100.000) = 791.240$ \$ /period .The present value of the penalty cost of dividing (splitting up) the departments into different sections will be :

$$V_2 = \frac{v_2(1-e^{-rT})}{r} \sum_{x=1}^{X} (N_x - 1) \tag{6}$$

Here, N = The number of the sections (x = 1,2,...,X) of x department; and $v_2$ = Penalty cost for each division. Let's assume that x department was divided into 3 sections (Nx = 3), $v_2$ = 50 \$/month and the monthly penalty cost of dividing all departments into sections is 1000 $ .In this case, the cost of dividing x department into 3 sections will be $v_2 (N_x - 1) = (50)(2) = 100$ $. And the present value of the total penalty cost of dividing all departments into sections will be $V_2 = (79{,}124)(1.000) = 79.124$ $.

For a randomly selected p sample layout M(p) from the all possible layouts set (P), let the CPU cost be $v_3$ and process time be t. In order to define P possible layouts set (M(1),M(2),...,M(P) which was calculated and formed in combinatory form, the present value of CPU cost which covers the calculations of handling and dividing of each different layout M(p) will be:

$$V_3 = t \cdot v_3 \frac{(1 - e^{-rT})}{r} \qquad (7)$$

Let's assume that CPU time for obtaining M(1) layout form in the P possible layouts set is $t = 5{,}3$ sec and the unit cost of CPU time is $v_3 = 10$ $/sec. The present value of M(1) layout cost will be $V_3 = (5{,}3)(10)(79{,}124)/4.193{,}572$ $. Here t process time will be different for each layout form.

If Total Plant Layout Cost is TC(p) = f(p) , TC(p) = f(p) = Net Present Value of the Handling Cost + Net Present Value of the Penalty Cost for Dividing (split-up) the Departments into Sections + Net Present Value of the Computer Application in the Layout = $V_1 + V_2 + V_3$

$$f(p) = \frac{(1 - e^{-rT})}{r} \left[ v_1 \sum_{j=1}^{n} \sum_{k=j+1}^{n} a_{jk} d_{jk} + v_2 \sum_{x=1}^{X} (N_x - 1) + t.v_3 \right] \qquad (8)$$

In this case, if P is considered as the all possible layout set for the department sections, the total cost of each $p \in P$ will be: f(p) = 791.240 + 79.124 + 4.193,57 = 874.557,57 $.Within a p = 1,2,...,P layout set ($p \in P$), a p* optimal layout can be the layout obtained when $f(p^*) \leq f(p)$.

The algorithm presented in this study is only one of the many algorithms within the basic DOT. The selection of an algorithm itself is an optimisation problem because each small neighbourhood/proximity with low calculation cost provides the possibility of finding a smaller absolute optimum from the bigger proximities with higher calculation costs. There is a real need for the selection of such an algorithm with respect to obtaining data on the size of the proximities to be used and related to the decision making process.

**Algorithm**

Step 1. Define the measurement unit for L and W which defines the size of the cells within the rectangular grid. The size of the cells within the grid will be defined by the unit of the measurement in this algorithm and the hypotethical sample. The measurement unit, although can be a complex problem to a certain extent, it is possible to use either smaller or bigger cells than the measurement unit. If L and W are measured in meters, the area of the cell will be in square meters.

Step 2. Select at random L which should be an integer in the range $0 < L < LW$ ( =the area of plant). Then W will be an integer and $W = LW/L$.

Step 3. Define at random, the number of the sections in each department to be divided. The number of the sections of the x department will be Nx in order that the limits are defined by Nx. In this case the total number of the sections of the plant will be :

$$n = \sum_{x=1}^{X} N_x \tag{9}$$

As known, the departments are divided into sections. The number of (b) vectors will be the same with the number of the sections obtained by dividing the departments. (b) vector defines the number of the cells in each section.

Step 4.Define the $a_{jk} = 0$ matrix for the real, symmetrical matrix $A=\|a_{jk}\|$; j,k= 1,2...,n; for $a_{jk}=a_{kj}$ and j=k, $a_{jk}=0$ which shows the (b) vector and the number of the loads to be handled between the departments.

Step 5. Select a (l,w) from the grid divided into LW sections (from the plant area with integer properties). Select the biggest j section from the b vector (The biggest section with the smallest index should be selected in another place where the sections are equal in a different way). Enter in a suitable order, the $b_j$ cells of the j section which is on the cellular area divided into LW sections. For example enter in the LW grid which continues as l,w; l,w-1; l,w-2 untill the $b_j$ cells. Select again at random, an l,w and repeat the above procedure for the subsequent largest section untill all the department sections take place on LW grid. Define this first layout as $p_1$.

Step 6. For the real, symmetrical matrix $D=\|d_{jk}\|$; j,k = 1,2,...,n ; j = k which expresses the distance between the department section, the elements of $d_{jk} = 0$ and $d_{jk} = d_{kj}$ matrix can be calculated at this point with Euclidean distance ;

$$d_{jk} = \left[\left(\bar{\ell}_j - \bar{\ell}_k\right)^2 + \left(\overline{w}_j - \overline{w}_k\right)^2\right]^{1/2} \tag{10}$$

Step 7. Calculate the value of $f(p_1)$ according to $f(p)=V_1+V_2+V_3$.

Step 8. Change at intervals the proximal sections or numerically the sections which are equal in size, in terms of area units at the given order. The order of interval change for a four-department plant will be 1-2, 1-3, 1-4, 2-3, 2-4, 3-4. If none of the interval changes is possible, that change will be ignored. [D] matrix is defined and f(p) is calculated for each interval change. The layout with the lowest f(p) is kept and compared to each successful layout. In case of the layout with minimum cost or equality, the first layout with minimum cost is defined as $p_2$. If none of the layouts can provide a cost lower than $p_1$, then the iteration is stopped. $p_1$ will be the local minimum layout of its proximity.

Step 9. If $f(p_2) < f(p_1)$ the step 8 is repeated starting with $p_2$ layout. If a layout of lower cost is found after all possible interval changes are completed, that layout will be defined as $p_3$. This procedure is continued until no improvement in the repetance of step 8 is observed. The layout of lowest cost found at that time will be the local minimum layout of the analysed proximity.

**An Example**

In order to explain the algorithm better, a simple example is given below. Table 1, shows the process-route schema of three products (A,B,C) in a hypotethical plant and Table 2 shows the production and load requirements. The required area (LW) has 6 units. The above given algorithm is changed a little bit in order the solve manually a managable problem. Instead of a repeated calculation, a layout change is tried to be made by repeating the steps 1,2,3 and 4 only once. L=3 and W=2 units. Table 3 shows the possible divisions of 3 sections within the departments. Department 1 can be divided into 3 or 2 sections (small sections) and Department 2 can be divided into 2 sections. The purpose of the problem is to make the layout of the three given departments in such a way to minimize the handling distance determined by the total number of the loads.

**TABLE 1: PROCESS - ROUTE SCHEMA**

| Departments | Sections | Dept. no | Products (or parts) A | B | C |
|---|---|---|---|---|---|
| Machine | 3 | 1 | 2 | 2 | 3 |
| Assembling | 2 | 2 | 3 | 3 | - |
| Control and Loading | 1 | 3 | - | - | - |

**TABLE 2:TABLE OF PRODUCTION AND LOAD REQUIREMENTS**

| | Parts A | B | C |
|---|---|---|---|
| Monthly Production Requirements | 15 | 20 | 25 |
| Number of each load section | 3 | 5 | 5 |
| Monthly Number of loads | 5 | 4 | 5 |

**TABLE 3: MINIMUM SIZE OF THE DEPARTMENT SECTIONS**

| Depts. | Size of the Sections | | | Department Area |
|---|---|---|---|---|
| 1 | 1 | 1 | 1 | 3 |
| 2 | 1 | 1 | 0 | 2 |
| 3 | 1 | 0 | 0 | 1 |

**TABLE 4: THE NEW, DEFINED SIZE OF THE DEPTS. SECTIONS**

| Depts. | Size of the Sections | | Department Area |
|---|---|---|---|
| 1 | 2 | 1 | 3 |
| 2 | 2 | 0 | 2 |
| 3 | 1 | 0 | 1 |

At random selection $N_1=2$ and $N_2=1$ is defined. The new sizes of the department sections are given at Table 4. In order to simplify the analysis, the sections are renumbered as section 11 to be 1, section 12 to be 2, section 21 to be 3 and section 31 to be 4. Vector will be $b = (b_1,b_2,b_3,b_4) = (2,1,2,1)$ and Matrix will be $A = \|a_{jk}\|$; j,k=1,2,3,4.

$$\text{Let}\quad A=\begin{bmatrix}0&5&7&4\\5&0&2&1\\7&2&0&9\\4&1&9&0\end{bmatrix}\quad \text{and}\quad v_1=r/(1-e^{-rT})\qquad v_2=v_3=0$$

$$f(p)=\sum_{j=1}^{4}\sum_{k=2}^{4} a_{jk}d_{jk} \tag{11}$$

is calculated for each layout.Table 5 shows the layout forms from 1 to 11 and their layout costs f(p).

**TABLE 5: SUITABLE PLANT LAYOUTS**

| | p=1 | p=2 | p=3 | p=4 | p=5 | p=6 |
|---|---|---|---|---|---|---|
| Row 1 | 1 1 | 2 2 | 2 2 | 2 2 | 3 4 | 3 4 |
| Row 2 | 2 4 | 3 4 | 1 3 | 1 4 | 2 2 | 1 1 |
| Row 3 | 2 3 | 1 1 | 1 4 | 1 3 | 1 1 | 2 2 |
| | f(p)=39 | f(p)=35 | f(p)=34 | f(p)=34 | f(p)=40 | f(p)=33 |

| | p=7 | p=8 | p=9 | p=10 | p=11 |
|---|---|---|---|---|---|
| Row 1 | 3 4 | 3 4 | 1 4 | 2 4 | 2 4 |
| Row 2 | 1 2 | 2 1 | 1 2 | 2 1 | 2 3 |
| Row 3 | 1 2 | 2 1 | 3 2 | 3 1 | 1 1 |
| | f(p)=35 | f(p)=36 | f(p)=47 | f(p)=45 | f(p)=36 |

Here, the first department is divided into two sections , 1 and 2, both with two cells. The second and third departments are also divided into two sections, 3 and 4, both having one cell. This is shown in Table 4. Therefore, 3 departments have been divided into 4 sections having 6 cells. LW area will then have 6 grids. Since the vector is $b = (b_1,b_2,b_3,b_4) = (2,1,2,1)$, and $p \in P$, different layout forms of p will be obtained in number P = 4! = 24. As seen in Table 5, 11 forms of suitable layouts are considered and the other layouts (13) are neglected. The point here is that the split-up of the departments into grids as cellular area can be made either in accordance with the plant properties or optional as in this example. In the layouts of combinatorial form there are layouts of equal value.

For example;

$$\begin{matrix}11\\22\\34\end{matrix}\;\rightarrow\;\begin{matrix}11\\22\\43\end{matrix}\;\rightarrow\;\begin{matrix}34\\22\\11\end{matrix}\;\rightarrow\;\begin{matrix}43\\22\\11\end{matrix}$$

The above layouts are either equivalent or identical. However this equivalence is controversial.Applying the 5.,6.,7.,8.,and 9th steps of the algorithm, the costs of the different layouts in Table 5 are percepted to be the lowest and highest points taking into consideration the upper and lower values used in branch algorithms; and the problem was handled in tree form. The vertices of this tree are shown as layout forms considering the layout costs, and lower and upper limit values. The layouts where the layout costs are taken as upper and lower limits, tree sets (Z) which are not mimimum for any tree, and minimum elements are formed. For the start-up vertex (layout) of the tree, local minimum layout will be formed. This is shown in Figure 3. It is seen that there are 4 minimum layouts with the same cost value. We can say that each minimum layout has a local minimum for one or more proximity.

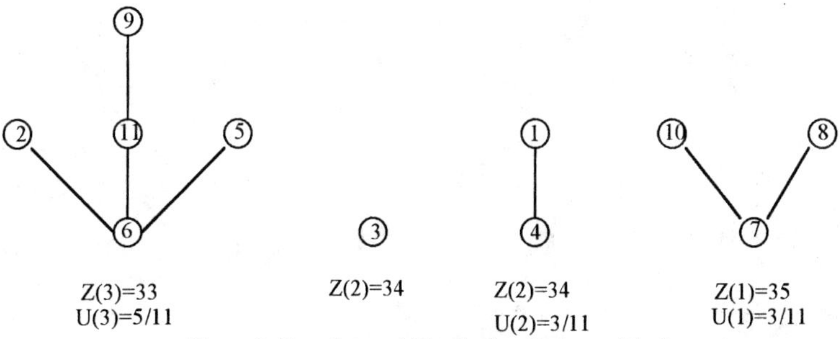


Figure 3: Tree Sets and Distribution Options of the Layouts.

As in Figure 3, assuming that the layout costs are distributed uniformly, 34 -35 layouts of absolute minimum costs can be made for 33 layout forms (Z(3) tree set) of absolute minimum cost with the probability of 5/11 and each 3/11.That is, within a LW area of 3 departments, the layouts providing absolute minimum are designed in tree form, in a combinatorial layout set consisting of 4 sections and 6 cells formed by b vector which expresses the number of the cells occupied by each department section.

## CONCLUSIONS

An iterative logarithm is presented which aims to minimize the net present value of each expenditure item for "handling, split-up of departments into sections, and calculation time" related to the production activity of man-machine- material complex within a plant by use of Discrete Optimization Theory. It is important that the algorithm takes as a basis and integrates this 3 groups of expenditure items in a Plant Layout Analysis and makes a combinatorial evaluation of multi-variables/ purpose for each layout form. Moreover, the algorithm considers the flow or change of money within a time period, and plays a defining role in financial decision-making.

The resultant layout and its cost can not be said to be optimal due to following reasons: A number of subjective prescients (such as the division of the departments into sections) have been used to form the cost variables used in the algorithm; A combinatorial calculation was required at algorithmic level; and Simulation Algorithm was used for interval change of place of the proximities.

Since that the main purpose of our study is to present the algorithm, only the first (handling) of the above three expenditure items was taken into consideration and the other two (Splitting-up the departments into sections and calculation time) were neglected in order to provide simplicity of scope in the example. However in real problems, these two expenditure items can be easily included in the application. The mentioned algorithm appears to be a strong tool not only for small-sized problems but also for real problems of large size.

## REFERENCES


Balas, E., (1975). "Facets of the Knapsack Polytope", *Mathematical programming*, (8),146-164.

Balas, E.,.Zemel,E., (1980). "An Algorithm for Large Zero-One Knapsack problems", *Oper.Res.*, (28),1130-1154.

Block, T.E., (1979). "On the Complexity of Facilities Layout Problems", *Man.Sci.*, (25),3,280-285.

Burkard, R.E., (1984). "Locations with Spatial Interaction-Quadratic Assignment Problem", in: Francis,R.L., Mirchandani,P.B., (eds.), *Discrete Location Theory*, Academic press, N.Y.

Cameron, D.C., (1952). "Travel Charts: A Tool for Analyzing Material Movement Problems", *Modern Materials Handling*,(8),1.

Chandrasekaran, R., Kabadi,S.N.,and Murty,K.G., (1981). "Some NP-Complete Problems in Linear Programming", *Oper.Res.Letters*, (1),101-104.

Cook, S.A., (1971). "The Complexity of Theorem-Proving Procedures", Proceedings 3rd Annual ACM Symposium on Theory of Computing, *Association for Computing Machinery*, N.Y.

Davidson, R.K., Smith,W.L.,and Wiley, J.W.,(1962). Economics: An Analytical Approach, *Homewood*, Illinois, Irwin.

Donald, R.P., McMillan, C.Jr., (1971). Discrete Optimization: Integer Programming and Network Analysis for Management, *Prentice-Hall, Inc.*, Englewood Cliffs, N.J.

Edmonds, J., (1965). "Maximum Matching and a Polyhedron with 0-1 Vertices", *J. of Res. of the National Bureau of Standards (B),(*69),125-130.

Fabrycky,W.J., Thuesen,G.J., (1974). Economic Decision Analysis, *Prentice-Hall, Inc.,* Englewood Cliffs, N.J.

Ford, L.R.Jr., Fulkerson, D.R.,(1962). Flows in Networks, *Princeton University Press*, Princeton, N.J.

Francis, R.L.,McGinnis,L.F.Jr.,and White,J.A., (1992). Facility Layout and Location, *Prentice-Hall, Inc.*, Englewood Clifss, N.J.

Gomory, R.E., (1963). "An Algorithm for Integer Solutions to Linear Programs", in: Recent Advances in Mathematical Programming (Graves,R.L. and Wolfe,P., eds.), *McGraw-Hill*, N.Y.

Grötschel, M., Lovász,L., and Schrijver,A., (1986). The Ellipsoid Method and Combinatorial Optimization, *Springer*, Heidelberg.

Hax, A.C., Candea, D.,(1984). Production and Inventory Management, *Prentice-Hall, Inc.*, Englewood Cliffs, N.J.

Herrera, F., Verdegay, J.L., (1995).,"Theory and Methodology : Three Models of Fuzzy Integer Linear Programming", *European J. of Oper.Res.*,(83),581-593.

Karp, R.M., (1975). "On the Computational Complexity of Combinatorial Problems", *Networks*,(5),45-68.

Kusiak, A., Heragu,S.S.,(1987). "The Facility Layout Problem", *European J. of Oper.Res.*, (29),229-251.

Martello, S., Toth,P., (1979). "The 0-1 Knapsack Problem", in: Combinatorial Optimization (Christofides,N., Mingozzi,A.,Toth,P.,Sandi, C.,and eds.), *Wiley*, Chichester,237-279.

Parker, R.G., Rardin,R.L.,(1982). "An Overview of Complexity Theory in Discrete Optimization: Part I. Concepts", *IIE Transactions*,(14),3-10.

Parker, R.G., Rardin,R.L., (1982). "An Overwiew of Complexity Theory in Discrete Optimization: Part II. Results and Implications", *IIE Transactions*, (14),83-89.

Parker, R.G., Rardin,R.L., (1983). "The Travelling Salesman Problem: An Update of Research", *Naval Res. Logistics Quarterly*, (30),69-96.

Parker, R.G., Rardin,R.L., (1984). "Guaranteed Performance Heuristics for the Bottleneck Travelling Salesman Problem", *Oper.Res.Letters*, (2),269-272.

Parker, R.G., Rardin,R.L., (1988). Discrete Optimization, *Academic Press, Inc.*, Boston.

Pegels, C.C., (1966). "Plant Layout and Discrete Optimizing", *The Int. J. of Pro.Res.*, (5),81-92.

Reiter, S., Sherman, G.R., (1965). "Discrete Optimizing, *SIAM Journal*, (13).

Sahni, S., Gonzales, T.,(1976). "P-Complete Approximation Problem", *J.of Associated Comput. Machi.*, (23),3,555-565.

Schrader, R., (1982). "Ellipsoid Methods", in: Modern Applied Mathematics-Optimization and O perations Research (Korte,B., eds.), *North Holland*, Amsterdam, 265-311.

Taha, A.H., (1975). Integer Programming Theory, Applications and Computations, *Academic Press*, N.Y.

Türkbey, O., (1983). An Application of a Heuristic Method with Computer Aided for Plant Layout in Industrial Business (Ph.D.Thesis), Dept.of Industrial Eng., Faculty of Eng.and Arch., Gazi University,Ankara,Turkey.

# AN APPLICATION OF GRAPH THEORY TO COMMUNICATION SYSTEMS: A CASE STUDY

**ORHAN TÜRKBEY**
*Gazi University, Engineering and Architecture Faculty, Industrial Engineering Department, Maltepe, 06570 Ankara, Turkey*

***Abstract:***
In telephone networks, every subscriber at any distance must be provided by an individual line from telephone exchange. The area box system, which is one of the elements of the distribution system of the rural telephone networks, is widely used for this purpose. Because the system has a simple structure, troubleshooting is easy and it requires minimum labor input. For these reasons, the system is commonly used in Turkey. The main problem is the minimization of network consisting of local and principal cables, that is, determination of the location of the operator an boxes. To solve this problem, computer aided Prim's Algorithm is used and Minimum Spanning Trees (Local networks) are built for each area. Since the local networks should be connected to the central station through principal cables, we have produced special algorithm. Using this, the local network is connected to the principal network and then to the central station to form principal network. In our application, two different computer programs have been used integrally. Finally, a reduction of 22,30 % and 49,32 % have been obtained for the local and principal networks, respectively, which form the telephone network system.
**Key words: Facilities , location , graph theory , network programming , routing.**

## INTRODUCTION

In practice, some problems are modelled in graph form, using arcs or edges and variables such as weight and cost. In a communication network, the edges (arcs) express the cables. The natural weights to be used should be the length or cost of the cables. For example, in a work assignment chart, the weights to be used may be the time or cost of the realized work. In practice, we are faced with these weights in the problems requiring the minimization of cost. In such situations, the problem is either to find the vertices on the path of minimum cost between two vertices. The first consists of the problems related to electrical circuits, telecommunications, and sewerage networks; and the suitable form of graph to apply these is "the Minimum Spanning Tree (MST)". The second consists of air maps, route determination, project management etc. which are on "the Shortest Route/Path Problem (SRP)" graph form (Hartsfield and Ringel,1990).

A graph can be defined as a set of vertices and edges joining the vertices. A system on a structure can be expressed by a graph where the lines indicate the relationship between the vertices, that is, the elements of the system. The reader should keep in mind that the graph theory is a theory of relations, not definitions (Hu,1982). Vertices are also called nodes, and edges are also called arcs, branches, or links. We usually assume that there are n vertices in the graph G and at most one edge joining any two vertices and there is no edge joining a node to itself.

Many traditional graph algorithms have been developed for the problems related to MST and SRP. One of the algorithms to be used in our study is the Prim's algorithm, and the other one is developed by us. The traditional methods provide a general approach to solutions. The algorithms developed by us provide short and efficient results in terms of both data structures and performance. While examining the relations between certain quantities or concepts for the solution of theoretical problems or for different areas of application, it is inevitable or most rational to convert interconnections into graph form. Because of the nature of this idea, Leonhard Euler (1707-1782) who investigated the problems related to the bridges in Königsberg city, reduced the matter to the main connections of some vertices and lines and formed the basis for graph theory (Hartsfield and Ringel,1990), (Lawler,1976). During many years, only mathematicians have studied the graph theory and no area of application have been available other than puzzles and games. However, with its specific definitions and concepts, the graph has developed into a separate discipline in the midst of this century. When the history of graph is reviewed, the contributions of other researchers, like Hamilton, Heawood, Whitney, Tuttle, Ore, Erdös, Katona, Polya, Seshu, have been valuable (Harary,1972).

Graph theory having a countless number of application areas, is becoming an area of interest in Turkey in the recent years. One of application areas is the communication networks which is operated within PTT (Post, Telegram and Telephone). In this study, distribution boxes, local cabinets and centrals form the vertices of mentioned networks and the distance between the vertices are the edges of graph.

The purpose of our study is to establish such communication network in a definite region. The system to be established is desired to be the most economical way to meet the communication needs of the region. The system required each subscriber to be connected by different a line to central. Location routes among some distribution boxes, local cabinets and central are combinational and the distances are different. In this case, how will a minimum cost project be developed to link the subscribers in our region?.

To solve such a communication problem a tree, the sum of its edges of which is minimum, should be found. For this purpose, the minimum sum of a tree weighted edges and without internal circles have to be developed. In graph theory, such trees are called the MST. Prim's, Kruskal, Sollin etc. algorithms were developed to obtain a tree of such features.

In our application, an algorithm being a mixture of Prim's (Hu,1982) and Sollin (Kaufmann,1968) algorithms were used to solve the problem mentioned above. In order to indicate the importance and necessity of graph theory a completely new project system of communication network has been developed and presented here, being different in many ways from the methods applied by the organization which established the communication system in Turkey.

## THE MINIMUM SPANNING TREE (MST)

Given an undirected network G, we can select a subset of edges which form a tree T in which there is a path between every pair of vertices in the network. This kind of tree is called a spanning tree of the network. If there are costs $d_{ij}$ associated with the edges, the cost of a spanning tree is the sum of $d_{ij}$ for edges in the tree. A spanning tree with minimum cost among all spanning trees is called a MST (Hu,1982). In general, the MST is different from the tree of shortest routes.

In an electrical network, there are n terminals, in fixed positions, which must all be electrically connected by wires. Which pairs of terminals should be joined, if the total length of wire used is to be as small as possible?

Such a problem can be presented in graph-theoretic terms, as follows. Let G=(V,E) be a simple graph in which every edge e is assigned a real number $\ell(e)$, called its length (Carré,1979). It is required to find a connected partial graph H = (V,F) of G whose total length.

$$\ell(H)= \sum_{e\in F} \ell(e) \tag{1}$$

is as small possible. Here, H must be a spanning tree of G and such a graph is defined as a MST.

### The Construction of any MST

A MST can be grown from an arbitrarily chosen vertex $v(0)$, one edge at a time, by constructing a sequence of graphs

$$H^{(k)}=\left(V^{(k)},F^{(k)}\right) \ , \ (k=0,1,2,...,n-1), \tag{2}$$

as follows.

Step 1. Let $v(0)$ be any vertex of G, and let $V^{(0)}=\{v^{(0)}\}$, $F^{(0)}=\varnothing$.

Step 2. For k = 1,2,...,n-1, form V(k) and F(k) as follows: let C(k) be the cut set of edges which have exactly one endpoint in V(k-1), let e(k) be a minimum edge of C(k), and let v(k) be the endpoint of e(k) which does not belong to V(k-1)

Then set,
$$V^{(k)}=V^{(k-1)}\cup\{v^{(k)}\} \ , \ F^{(k)}=F^{(k-1)}\cup\{e^{(k)}\}. \tag{3}$$

On termination the graph $H^{(n-1)}=\left(V^{(n-1)} \ , \ F^{(n-1)}\right)$ is a MST of G.

## INTRODUCING THE APPLICATION SYSTEM

The telephone, telex, data telex, faximile and similar communication systems use a telephone network. In general, telephone networks are extremely different from other networks (electric, gas etc.). In denoted network, every subscriber must be fed by an individual line beginning from telephone central no matter how far it is. A telephone network consists of a distribution box, local cable, local cabinet, principal cable, junction cable and repertitor.

The place of application for our study is the area of Balgat neighbourhood central, in Ankara. PTT which is the responsible state enterprise for the subject estimates that 16 local cabinets and about 400 distribution boxes will be required for the telephone network to be built in the mentioned area. The concerned public enterprise has not been using a scientific algorithm to determine the places of local cabinets, distribution boxes or central places. In our study, we have tried to find the MST of the problem first by using our algorithm and then Prim's algorithm, and determined the places of main instruments of telephone system as central, local cabinet and distribution boxes. With the help of the obtained MST, we have tried to minimize the length of the principal cable coming from the central and feeding local cabinets, and also the length of the local cables coming from local cabinets and feeding the distribution boxes. In this way, we have reduced the cost of cable as well as the costs of underground excavations, workmanship, time etc. Our study is based upon the formation of MST network and selection of route in order to determine the best places for the central, local cabinets and distribution boxes.

## APPLICATION

As mentioned above, in our application in a defined area with definite coordinates, the places (vertices) of central, local cabinets and distribution boxes used in the telephone communication system, have been determined through the use of the algorithms of MST. Later in order to minimize the lengths of principal and local cables to be located between the distances connecting the main instruments, or to find MST, Prim's algorithm for the local cable and a special algorithm developed by us for the principal cable, were applied. Both algorithms were written in PASCAL language.

Network plans defining the graph, data tables, print outs of computer programs, MST diagrams of local and principal cables were not included here because of the scale of the application. But the mentioned information is given in the appendices at the end of the manuscript. First, how the data to be used in the two mentioned algorithms have been prepared, will be explained. Later the software of the algorithms will be presented according to the nature of the problem.

### Developing the Model for the Definition of the Application Graph

The data system required for our algorithms is as follows :

1. The application of central area is divided into 16 local cabinet areas according to demand size. In the application, the area of local cabinets and coding system of the concerned public foundation were used.
2. The tree network of each area is obtained in chart form.
3. The distribution boxes in the area of each cabinet are determined. The coordinates of each box were found according to its own project and the original main project of the central area formed by joining six projects.
4. The intervertices distance matrix was built by using the rectilinear statement of $U=|x_2-x_1|+|y_2-y_1|$ (Francis,McGinnis and White,1992). The elements of this matrix were calculated by a computer program which was written by us.

5. The matrices of assignment intervertices and the development steps of MST (minimum spanning local trees) were calculated by the application of PASCAL computer program using Prim's algorithm, to the intervertices distance matrix data.
6. After the local cable trees were developed, the principal tree program which will connect the local networks, was solved by a PASCAL programming language using the special algorithm which was developed by us and similar to the Sollin's Algorithm.
7. The lengths of local and principal network which should be used in the system, were obtained comparatively.

**The MST Algorithm of Local Network**

The basis of the algorithm presented below, is the Prim's algorithm and it was adapted to the nature of our problem.

Step 1. Calculate the intervertices distance matrix.

Step 2. Develop two sets using the original vertices.

Step 3. Select two the vertices with the shortest distance and connect them to eachother.

Step 4. Transfer the mentioned vertex to the set of N and write the distance between the two connected vertices.

Step 5. Find the nearest vertex connected to the set of N and go to Step 4.

N : The set of connected vertices.

n : The set of non-connected vertices. Initially, all the vertices is in the set of n.

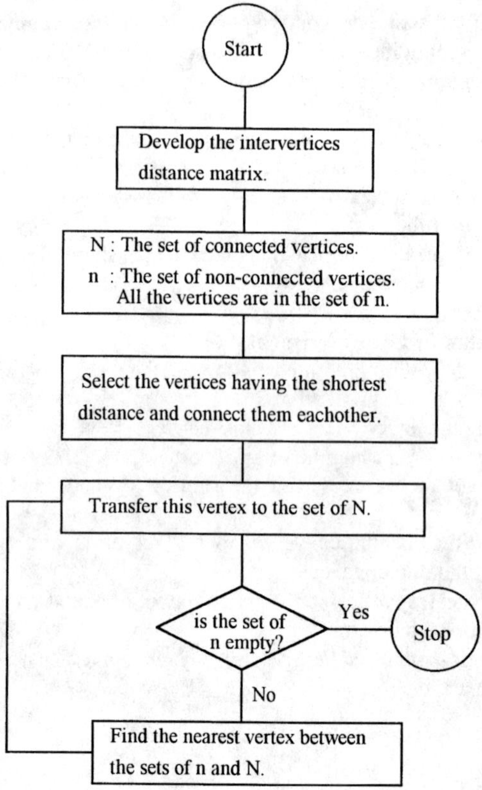


Figure 1: The flow chart Prim's MST algorithm adapted to the local network.

**The MST Algorithm of Principal Network (Special Algorithm Developed for This Project)**

Step 1. Develop the interareas distance matrix with the nearest vertices which connect all areas.

Step 2. Develop two sets using the original vertices.

Step 3. Connect two areas which have the shortest distance in the set of n.

Step 4. Transfer this vertex to the set of N. Fix the connected vertices and recalculate the interareas distance matrix, omitting the other vertices in denoted area.

Step 5. Find the nearest vertex which is connected to the set of N and go back to step 4.

Step 6. Stop, when the vertices are completed.

N : The set of connected vertices.

n : The set of non-connected vertices. Initially, all the vertices are in the set of n.

The difference of the developed algorithm from Kruskal's and Prim's algoritms, is the way of forming the branches.

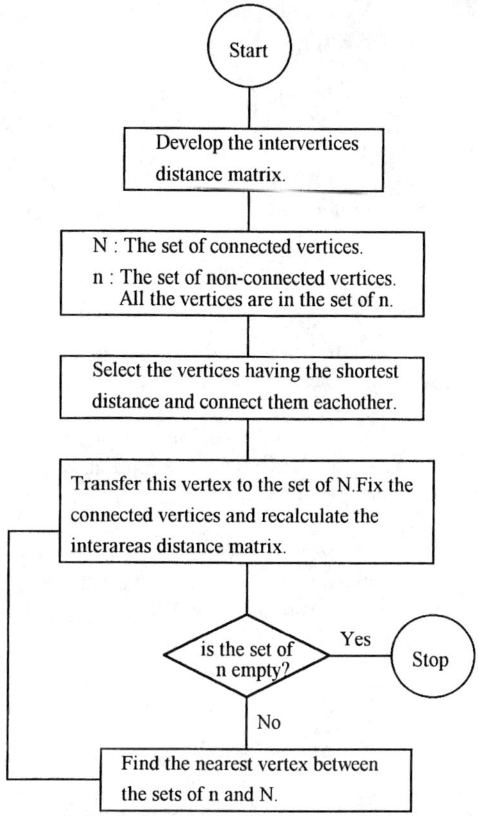


Figure 2: The flow chart of special MST algorithm developed by us according to the principal network.

**TABLE 1: THE ALGORITHMIC COMPARISON OF CABLE LENGTHS WHICH SHOULD BE USED IN THE LOCAL AND PRINCIPAL NETWORK ON THE BASIS OF PERFORMANCE**

| The kinds of cable used in network | The traditional method of public foundation | Prim and developed special algorithm | Performance percentage |
|---|---|---|---|
| Local cable length (meter) | 19.048 | 14.800* | 22.30 |
| Principal cable length (meter) | 9.264 | 4.695* | 49.32 |

(*) The Prim's algorithm is used in the calculation of local network/cable length. Due to the characteristics of Ericsson local cabinets, a special algorithm which is a developed form of Prim's algorithm is used in the calculation of principal network length.

## CONCLUSIONS

In Turkey, PTT (Post, Telegram and Telephone) Enterprise is responsible from the development and operation of the communication systems. The telephone central, cable network and the subscriber post form 35.50 and 15 percent, respectively of total investment cost of telephone network system.

As seen, the cable network constitutes 50 percent of the total telephone network investments. Due to the importance of the issue, our study was carried on the cable network of telephone network system to be built by the concerned enterprise in a defined region. In order to minimize investment cost and time for workmanship, it is important to reduce the principal and local network system and to select the shortest route (to find MST) for reduced excavation work. However, due to economic, technical and political reasons, the mentioned public enterprise have not been able to benefit from the related scientific algorithms, in the network projects carried out until now.

Our study, includes the adaptation of telephone system in the mentioned region to the graph theory which is widely used. The assumptions in our study are that the central area is within the municipal plan, and telephone demands are known from the beginning.

All the demand points (distribution boxes) within the central area are the vertices of graph, and the distances between the points are the edges of graph. There are 16 cabinet areas and a total of 500 distribution boxes, with 10-80 boxes in each area of the central region. The purpose is to form a connex graph connecting the vertices in each cabinet area and to minimize the local network.

For each area the MST (local networks) are produced by using the Prim's algorithm in the computer environment. As of the network system requirement, the local networks should be connected to central with the principal cables. So, the local network is connected to the principal network using a special algorithm developed by us and these are connected to central to obtain the principal network.

In order to form the communication graph and find MST by use of Prim's algorithm, the intervertices distance matrix should be found. To obtain that, a matrix should be produces with the dimensions (500x500) and the 250.000 measuring process should be made on the intervertices. Since that each measuring process takes 3 minutes, the total measuring time will be 750 thousand minutes, 12 thousand hours, 1562 days or 4.28 years.

Since that the time needed for measurements is too long, a computer programme using "rectilinear distance" term has been prepared for the calculation of intervertices distance. The matrix elements have been approximately calculated and the matrix is formed.

In table I, the traditional method which is not based upon any scientific algorithm used by PTT enterprise is compared to our method which is based upon two algorithms. One is already known and the other prepared by us in the table, the minimum cable lengths that should be used in local and principal networks and performance percentages are given comparatively.

As can be seen from the table, both the Prim's algorithm and the special algorithm developed to solve the MST problem, give better results than the traditional method of the mentioned enterprise. Consequently, 22,30 and 49,32 percent savings have been provided at the local and principal cable distances, respectively.

As seen from the application in Balgat Central area in Ankara, by the use of scientific algorithms in practice, the investment projects may be realized in shorter time and at less cost. But, such an application requires to process large scale of data sets, to use developed computers and employ skilled personnel in the organizations. The interested data and documents used in the application, are given in appendices.

## REFERENCES


Alavi,Y.,Chertrend,G.,Dellermann,O.R., and Schwenk,A.J.,(1991).Graph Theory Combinatorics and Applications ($1^{st}$ ed. Vol.2), *Western Michigan University,*

Berge, *Graphs ($2^{nd}$ ed.),* North-Holland, Amsterdam, 1989,26.

Bondy,J.A., Murty,U.S.R.,(1979). Graph Theory with Applications, *North-Holland,* Amsterdam, 218.

Carré,B.,(1979). Graphs and Networks, *Clarendon press,* Oxford,26, 64, 158, 160.

Francis,R.L.,McGinnis.Jr.L.F., and White,J.A.,(1992).Facility Layout and Location: An Analytical Approach, *Prentice-Hall, Inc.*, N.J.,164-169.

Gould,R.,(1988). Graph Theory, *The Benjamin/Cummings Publishing Co., Inc.*, California, 4, 68, 70-72.

Harary,F.,(1972). Graph Theory, *Addison-Wesley Publishing Co.*, California, 2.

Hartsfield,N., and Ringel,G.,(1990). Pearls in Graph Theory: A Comprehensive Introduction, *Academic Press, Inc.*, Boston, 49, 122-126.

Hillier,F.C.,and Lieberman,G.J.,(1989). Introduction to Operations Research *($4^{th}$* ed.), *McGraw-Hill Book Co.,* N.Y., 300.

Hu,T.C.,(1982). Combinatorial Algorithms, *Addison-Wesley Publishing Co.*, California, 1, 27-30.

Kaufmann,A.,(1968). Méthodes Et Modéles (Premiére partie), *Dunod,* Paris, 34-36.

Lawler,E.L.,(1976). Combinatorial Optimization: Networks and Matroids, *Rinehart and Winston,* N.Y., 36.

Tucker,A.,(1894). Applied Combinatorics ($2^{nd}$ ed.), *John Wiley and Sons, Inc.*, N.Y., 27.

# SUBJECT INDEX

## A



## B

**C**



**D**

**E**



**F**

## G



## H



## I

S



T

**U**



**V**



**W**

# AUTHOR INDEX

**W**



**Y**



**Z**